W9-CHQ-776
$49·50

Essentials of
Human Anatomy & Physiology

Essentials of Human & Anatomy Physiology

Fourth Edition

John W. Hole, Jr.

Wm. C. Brown Publishers

Book Team

Editor *Colin H. Wheatley*
Developmental Editor *Jane DeShaw*
Production Editor *Gloria G. Schiesl*
Designer *Elise A. Burckhardt*
Art Editor *Carla M. Goldhammer*
Photo Editor *Laura Fuller*
Permissions Editor *Vicki Krug*
Visuals Processor *Amy L. Saffran*

Wm. C. Brown Publishers

President *G. Franklin Lewis*
Vice President, Publisher *George Wm. Bergquist*
Vice President, Operations and Production *Beverly Kolz*
National Sales Manager *Virginia S. Moffat*
Group Sales Manager *Vincent R. Di Blasi*
Vice President, Editor in Chief *Edward G. Jaffe*
Marketing Manager *John W. Calhoun*
Advertising Manager *Amy Schmitz*
Managing Editor, Production *Colleen A. Yonda*
Manager of Visuals and Design *Faye M. Schilling*
Production Editorial Manager *Julie A. Kennedy*
Production Editorial Manager *Ann Fuerste*
Publishing Services Manager *Karen J. Slaght*

WCB Group

President and Chief Executive Officer *Mark C. Falb*
Chairman of the Board *Wm. C. Brown*

Cover photos: © Stephen Marks/Stockphotos, Inc.; Insets: © Edwin A. Reschke

Copyeditor: Barbara R. Day

The credits section for this book begins on page 549, and is considered an extension of the copyright page.

Library of Congress Catalog Card Number: 91–76062

ISBN 0–697–09703–X (Cloth)
0–697–16790–9 (Paper)

Printed in the United States of America by Wm. C. Brown Publishers, 2460 Kerper Boulevard, Dubuque, IA 52001

10 9 8 7 6 5 4 3 2 1

For my former students,
who taught me how to teach

Brief Contents

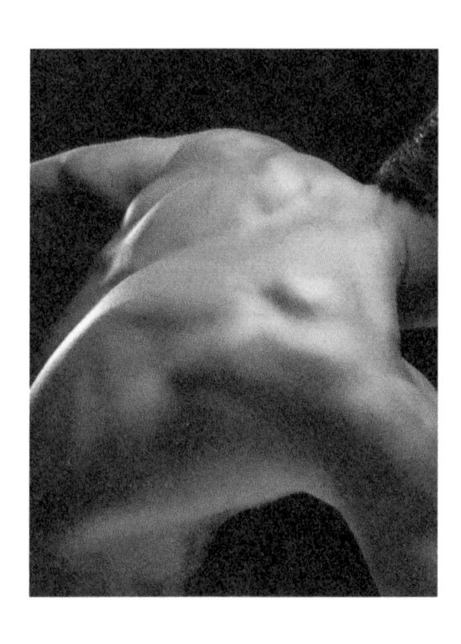

Unit 1 Levels of Organization

Unit 2 Support and Movement

Unit 3 Integration and Coordination

Unit 4 Processing and Transporting

Unit 5 Reproduction

Expanded Contents

Unit 1
Levels of Organization

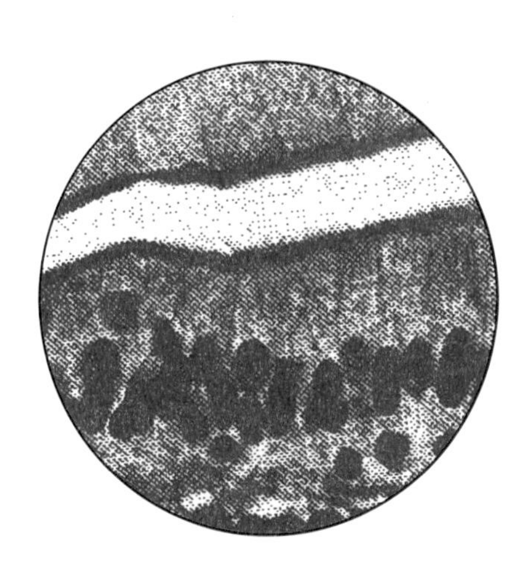

1 Introduction to Human Anatomy and Physiology 4

2 Chemical Basis of Life 34

3 Cells 50

4 Cellular Metabolism 74

5 Tissues 94

6 Skin and Integumentary System 114

Unit 2
Support and Movement

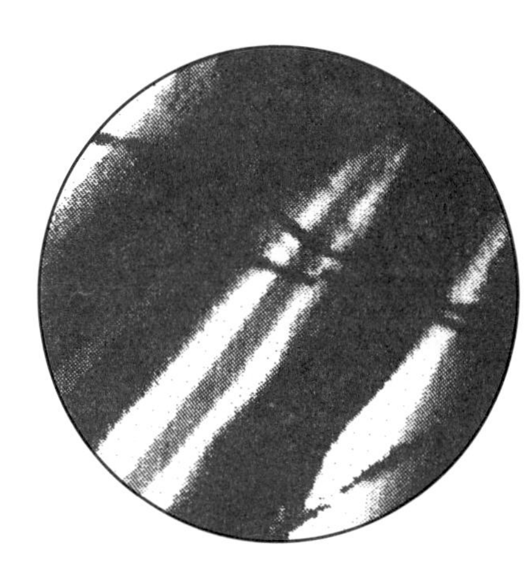

7 Skeletal System 128

8 Muscular System 168

Unit 3
Integration and Coordination

Unit 4
Processing and Transporting

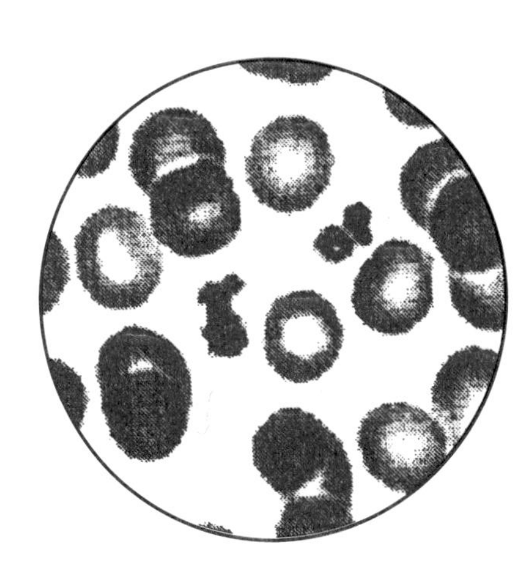

Unit 5
Reproduction

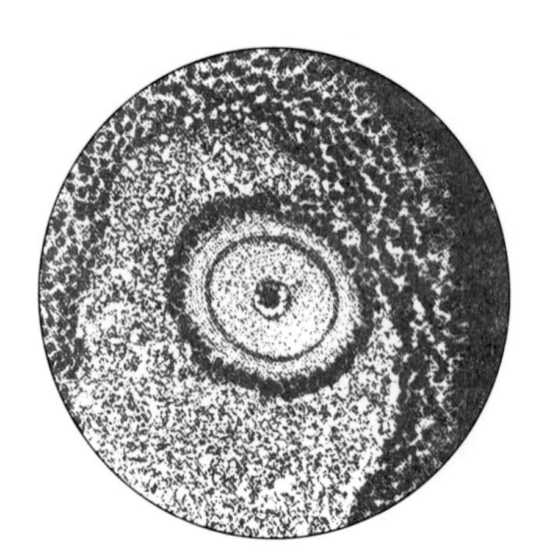

20 Pregnancy, Growth, and Development 512

Preface

Essentials of Human Anatomy and Physiology, Fourth Edition, is designed to provide accurate information about the structure and function of the human body in an interesting and readable manner. It is especially planned for students of one-semester courses in anatomy and physiology who are pursuing careers in allied health fields and who have minimal backgrounds in physical and biological sciences.

Organization

The text is organized in units or groups of related chapters. These chapters are arranged traditionally beginning with a discussion of the physical basis of life and proceeding through levels of increasing complexity.

Unit 1 is concerned with the structure and function of cells and tissues; it introduces membranes as organs and the integumentary system as an organ system.

Unit 2 deals with the skeletal and muscular systems, which support and protect body parts and make movements possible.

Unit 3 concerns the nervous and endocrine systems, which integrate and coordinate body functions.

Unit 4 discusses the digestive, respiratory, circulatory, lymphatic, and urinary systems. These systems obtain nutrients and oxygen from the external environment; transport these substances internally; utilize them as energy sources, structural materials, and essential components in metabolic reactions; and excrete the resulting wastes.

Unit 5 describes the male and female reproductive systems, their functions in producing offspring, and the early growth and development of this offspring.

Biological Themes

In addition to being organized according to levels of increasing complexity, the narrative emphasizes the following: the complementary nature of structure and function, homeostasis and homeostatic-regulating mechanisms, interaction between humans and their environments, and metabolic processes.

Changes in the Fourth Edition

The fourth edition of *Essentials of Human Anatomy and Physiology* has been reviewed throughout, and special attention has been given to suggestions received from users of the book. Major changes in the new edition include the following:

1. Many figures have been revised or replaced with new art, and many new micrographs and photographs have been included. See chapters 8 and 10 for examples.
2. Sections of the narrative that are new concern the cavities of the head, atomic weights, reticular fibers, homeostasis of bone, repair of fractures, ratchet theory of muscle contraction, pain nerve pathways, and the pineal gland.
3. Nineteen longer asides called "Current Topics" have been added. These include radioactive isotopes, cancer, mutations and mutagenic factors, elevated body temperature, use and disuse of muscles, factors affecting synaptic transmission, diabetes, dental caries, hepatitis, emphysema and lung cancer, leukemia, hypertension, atherosclerosis, autoimmunity, water balance disorders, sodium and potassium imbalances, acid-base imbalances, birth control, and teratogens.
4. Fourteen new boxed asides have been included covering the topics of anabolic steriods, hip fracture, pain and muscle contraction, corneal transplantation, cellular turnover, Valsalva maneuver, anorexia nervosa, lipid requirements, hyperoxia, atrial natriuretic factor, endothelium, cyclosporine, testicular cancer, and in vitro fertilization.
5. Certain sections of the narrative have been revised or expanded. Among these are the composition of cell membranes, cranial nerves, pineal gland, thyroxine, tonsils, lymphocytes, immune responses, and AIDS.

6. Seven new summary charts have been included. These concern protein synthesis, muscle contraction, nerve impulses, release of neurotransmitters, impulses to hearing receptors, gas exchanges, and birth control.
7. Some new items have been added to the clinical application of knowledge questions at the end of each chapter.

Pedagogical Devices

The text includes an unusually large number of pedagogical devices intended to involve students in the learning process, to stimulate their interests in the subject matter, and to help them relate their classroom knowledge to their future vocational experiences. These include the following:

Unit Introductions Each unit opens with a brief description of the general content of the unit and a list of the chapters within the unit (see p. 3 for an example). This introduction provides an overview of the chapters that make up a unit and tells how the unit relates to other aspects of human anatomy and physiology.

Chapter Introductions Each chapter introduction previews the chapter's contents and relates that chapter to others within the unit (see p. 6).

Chapter Objectives Chapter objectives at the beginning of each chapter indicate what the reader should be able to do after mastering the information within the narrative (see p. 5). Review activities at the end of each chapter (see p. 22) are phrased like detailed objectives, and it is helpful for the reader to review them before beginning a study of the chapter. Both sets of objectives are guides that indicate important sections of the narrative.

Key Terms A list of terms and their phonetic pronunciations is given at the beginning of each chapter to help the reader build a scientific vocabulary. The words included in each list are used within the chapter and are likely to be found in subsequent chapters as well (see p. 5). There is an explanation of phonetic pronunciation on pages 5 and 535.

Aids to Understanding Words Aids to understanding words, at the beginning of each chapter, should help to build the reader's scientific vocabulary. This section presents a list of word roots, stems, prefixes, and suffixes to help the reader discover and remember word meanings. Each root, stem, prefix, or suffix and an accompanying example in which it occurs are defined.

Review Questions within the Narrative Review questions occur at the ends of major sections within each chapter to test the reader's understanding of previous discussions (see p. 6). If the reader has difficulty answering a set of these questions, the previous section should be reread.

Illustrations and Tables In each chapter, numerous illustrations and tables are placed near their related textual discussion. They are designed to help the reader visualize structures and processes, clarify complex ideas, summarize sections of the narrative, or present pertinent data.

Sometimes the figure legends contain questions that will help the reader apply newly acquired knowledge to the object or process the figure illustrates (see fig. 1.9). The ability to apply information to new situations is of prime importance, and such questions concerning the illustrations will provide the practice needed to develop this skill.

Boxed Information Short paragraphs set off in colored boxes also occur throughout each chapter (see p. 8). These asides often contain information that will help the reader apply the ideas presented in the narrative to clinical situations. In some of the boxes, there are descriptions of changes that occur in body structure and function as a result of disease processes.

Current Topics Longer asides are entitled Current Topics. They discuss pathological disorders, physiological responses to environmental factors, and other topics of general interest (see page 37).

Clinical Terms At the end of certain chapters is a list of related terms, with their phonetic pronunciations and definitions, that are sometimes used in clinical situations (see p. 19). Although these lists are often brief, they will be a useful addition to the reader's understanding of medical terminology.

Chapter Summaries A summary in outline form occurs at the end of each chapter to help the reader review the major ideas presented in the narrative (see p. 20).

Clinical Application of Knowledge At the end of each chapter are questions (see p. 22) that help the reader gain experience in applying information to clinical situations. They may also serve as topics for discussion with other students or with the instructor. Suggested answers to these questions appear in the Instructor's Manual that accompanies the textbook.

Review Activities Review activities at the end of each chapter (see p. 22) will serve as a check on the reader's understanding of the major ideas presented in the narrative.

Appendix The appendix consists of lists of clinical laboratory tests commonly performed on human blood and urine. These lists give the tests' names, normal adult values, and brief descriptions of their clinical significance (see p. 531).

Readability

Readability is an important asset of this text. The writing style is intentionally informal and easy to read. Technical vocabulary has been minimized, and illustrations, summary tables, and flow diagrams are carefully positioned near the discussions they complement. Other features provided to increase readability and aid student understanding include the following:

Bold and Italic Type identifying words and ideas of particular importance.

Glossary of Terms with phonetic pronunciations, supplying easy access to definitions.

Index, which is complete and comprehensive.

Supplementary Materials

The following supplementary materials, designed to help the instructor plan class work and presentations and to aid students in their learning activities, are also available:

Instructor's Resource Manual and Test Item File by John W. Hole, Jr., and Karen A. Koos, which contains chapter overviews, instructional techniques, a suggested class schedule, discussions of chapter elements, lists of related films, and directories of suppliers of audiovisual and laboratory materials. It also provides test items for each chapter of the text, designed to help evaluate student understanding.

WCB TestPak is a computerized testing service offered free upon request to adopters of this textbook. It provides a call-in/mail-in test preparation service. A complete test item file is also available on computer diskette for use with IBM compatible, Apple IIe or IIc, or Macintosh computers.

A Student Study Guide to Accompany Essentials of Human Anatomy and Physiology, Fourth Edition, by Nancy Sickles Corbett, which contains overviews, chapter objectives, focus questions, mastery tests, study activities, and answer keys corresponding to the chapters of the text.

A Set of 100 acetate transparencies, designed to complement lectures or to be used for short quizzes.

Visuals Testbank is a set of 50 transparency masters that are available for use by instructors. These feature line art from the text with labels deleted for student quizzing or for student practice.

Laboratory Manual to accompany *Essentials of Human Anatomy and Physiology,* Fourth Edition, by John W. Hole, Jr., contains 45 exercises planned to stimulate interest in the subject matter, to involve students in the learning process, and to guide them through a variety of laboratory activities.

Also available from WCB. . . .

- Study Cards for Human Anatomy and Physiology by Van De Graaff/Rhees/Creek
- *The Coloring Review Guide to Human Anatomy* by McMurtrie/Rikel
- *Atlas of the Skeletal Muscles* by Robert and Judith Stone
- The WCB Anatomy and Physiology Video Series
- *Anatomy and Physiology of the Heart* videodisc
- *Computer Review of Human Anatomy and Physiology* software by Davis/Zimmerman/Van De Graaff
- *Knowledge Map of Human Anatomy Systems* software (Macintosh) by Craig Gundy of Weber State College

Acknowledgments

Once again, I want to express my sincere gratitude to those who used or reviewed previous editions of *Essentials of Human Anatomy and Physiology* and provided suggestions for its improvement. I am especially indebted to those who read all or portions of the Fourth Edition manuscript while it was being prepared. They include:

Dr. Sheila Cagle
Eli Lilly and Company

Redding I. Corbett, III
Midlands Technical College

Anne Funkhouser
University of the Pacific

James D. Houston
Westark Community College

Eunice R. Knouse
Spartanburg Methodist College

Constance B. Looke
Southern Maine Technical College

Without their valuable assistance, the task of revising this book would have been immeasurably more difficult.

Essentials of
Human Anatomy & Physiology

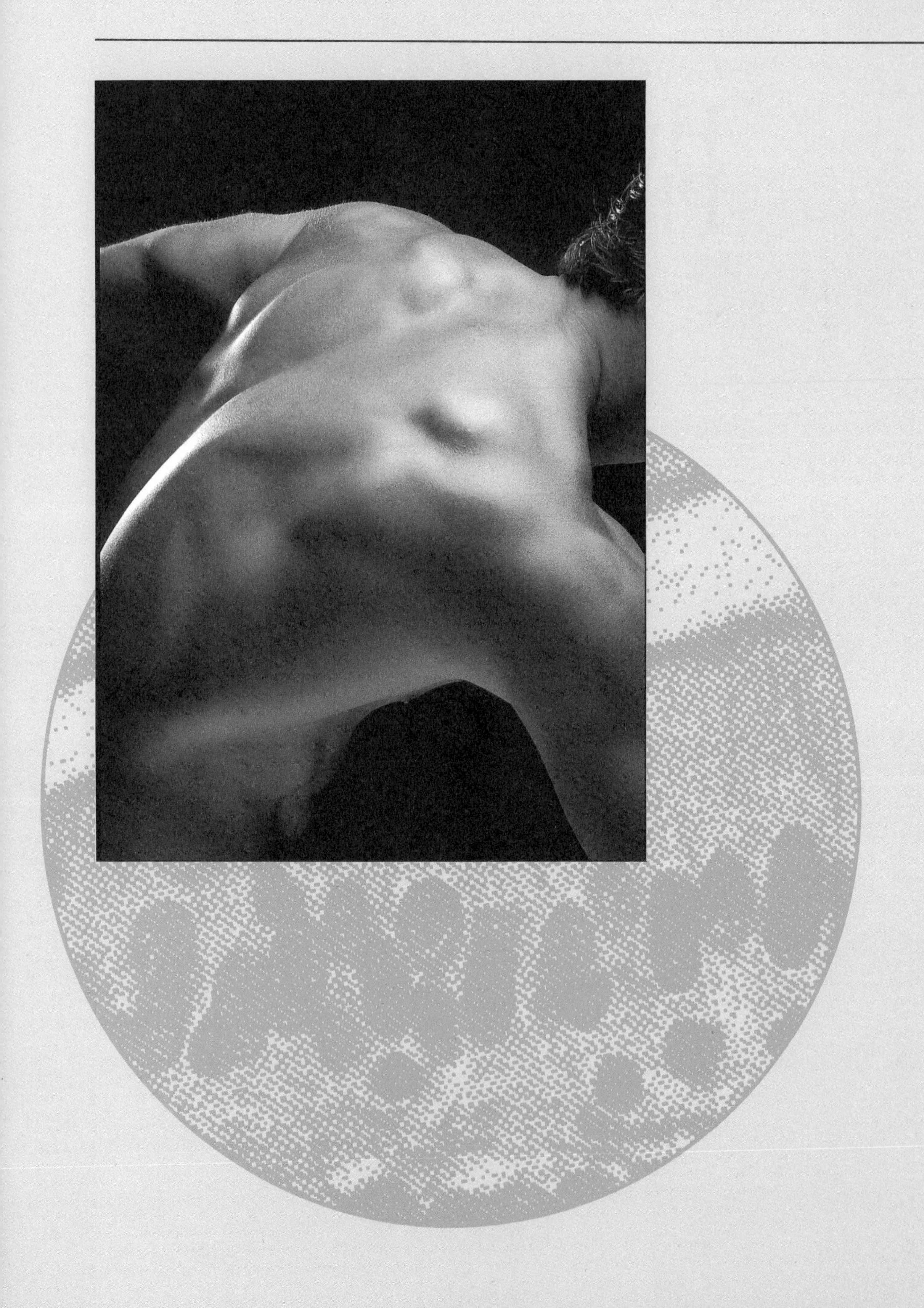

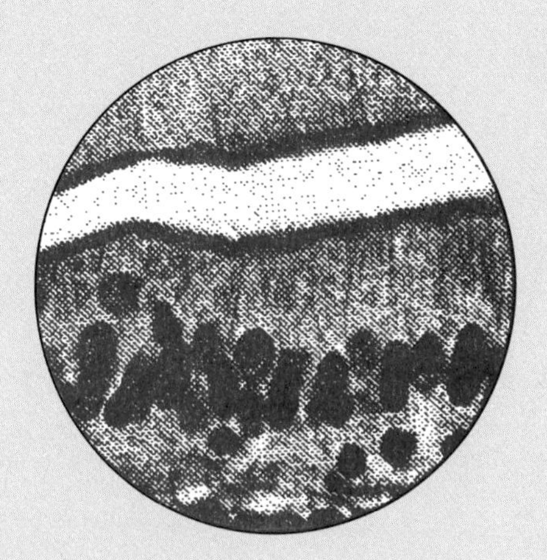

Unit 1
Levels of Organization

The chapters of unit 1 introduce the study of human anatomy and physiology. They present the similarities between humans and other living organisms, the interdependent relationship between the structure and function of human body parts, and the way survival of the human organism is dependent upon interactions between its parts—chemicals, cells, tissues, organs, and organ systems. This unit includes:

These circular images are a computer-manipulated version of figure 5.2b.

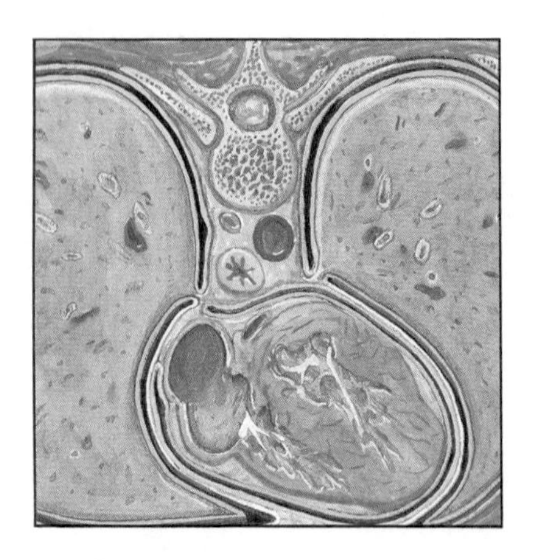

1 Introduction to Human Anatomy and Physiology

Although a human represents a particular kind of living organism, it shares traits with other organisms. For example, it carries on life processes and thus demonstrates the characteristics of life. It has needs that must be met if it is to survive, and its life depends upon maintaining a stable internal environment.

A discussion of the traits that humans have in common with other organisms and of the way the complex human body is organized provides a beginning for the study of anatomy and physiology in chapter 1. This chapter also introduces a set of special terms used to describe body parts.

Chapter Objectives

After you have studied this chapter, you should be able to

1. Define *anatomy* and *physiology*, and explain how they are related.
2. List and describe the major characteristics of life.
3. List and describe the major needs of organisms.
4. Define *homeostasis*, and explain its importance to survival.
5. Describe a homeostatic mechanism.
6. Explain what is meant by levels of organization.
7. Describe the location of the major body cavities.
8. List the organs located in each major body cavity.
9. Name the membranes associated with the thoracic and abdominopelvic cavities.
10. Name the major organ systems, and list the organs associated with each.
11. Describe the general functions of each organ system.
12. Properly use the terms that describe relative positions, body sections, and body regions.
13. Complete the review activities at the end of this chapter. Note that the items are worded in the form of specific learning objectives. You may want to refer to them before reading the chapter.

Key Terms

absorption (ab-sorp′shun)

anatomy (ah-nat′o-me)

appendicular (ap″en-dik′u-lar)

assimilation (ah-sim″ĭ-la′shun)

axial (ak′se-al)

circulation (ser-ku-la′shun)

digestion (di-jest′yun)

excretion (ek-skre′shun)

homeostasis (ho″me-ō-sta′sis)

metabolism (mĕ-tab′o-lizm)

organelle (or″gan-el′)

organism (or′gah-nizm)

parietal (pah-ri′ĕ-tal)

pericardial (per″ĭ-kar′de-al)

peritoneal (per″ĭ-to-ne′al)

physiology (fiz″e-ol′o-je)

pleural (ploo′ral)

reproduction (re″pro-duk′shun)

respiration (res″pĭ-ra′shun)

thoracic (tho-ras′ik)

visceral (vis′er-al)

The accent marks used in the pronunciation guides are derived from a simplified system of phonetics standard in medical usage. The single accent (′) denotes the major stress, meaning that emphasis is placed on the most heavily pronounced syllable in the word. The double accent (″) indicates the secondary stress. A syllable marked with a double accent receives less emphasis than the syllable that carries the main stress, but more emphasis than neighboring unstressed syllables.

Aids to Understanding Words

append-, to hang something: *append*icular—pertaining to the arms and legs.

cardi-, heart: peri*cardi*um—a membrane that surrounds the heart.

cran-, helmet: *cran*ial—pertaining to the portion of the skull that surrounds the brain.

dors-, back: *dors*al—a position toward the back of the body.

homeo-, same: *homeo*stasis—the maintenance of a stable internal environment.

-logy, study of: physio*logy*—the study of body functions.

meta-, change: *meta*bolism—the chemical changes that occur within the body.

pariet-, wall: *pariet*al membrane—a membrane that lines the wall of a cavity.

pelv-, basin: *pelv*ic cavity—a basin-shaped cavity enclosed by the pelvic bones.

peri-, around: *peri*cardial membrane—a membrane that surrounds the heart.

pleur-, rib: *pleur*al membrane—a membrane that encloses the lungs within the rib cage.

-stasis, standing still: homeo*stasis*—the maintenance of a relatively stable internal environment.

-tomy, cutting: ana*tomy*—the study of structure, which often involves cutting or removing body parts.

Introduction

THE STUDY OF the human body probably began with our earliest ancestors, who must have been curious about their body parts and functions just as we are today. At first their interests most likely concerned injuries and illnesses, because healthy bodies demand little attention from their owners. Certainly primitive people injured themselves, suffered from aches and pains, bled, broke bones, and developed diseases. When they were sick or felt pain, these people probably sought relief by visiting shamans who relied heavily on superstitions and notions about magic. As they tried to help the sick, these early medical workers began to discover useful ways of examining and treating the human body. They observed the effects of injuries, noticed how wounds healed, and attempted to determine the causes of death by examining dead bodies. They also began to learn how certain herbs and potions affected body functions and could sometimes be used to treat coughs, headaches, and other common problems.

In ancient times, however, it was generally believed that natural processes were controlled by spirits and supernatural forces that humans could not understand. Then, about twenty-five hundred years ago attitudes began to change, and the belief that humans could understand natural processes grew in popularity.

This new idea stimulated people to look more closely at the world around them. They began asking more questions and seeking answers. In this way the stage was set for the development of modern science. As techniques for making accurate observations and performing careful experiments evolved, knowledge of the human body expanded rapidly (fig. 1.1). At the same time, new terms were being coined to name body parts, to describe their locations, and to explain their functions. These terms, most of which originated from Greek and Latin words, formed the basis for the language of anatomy and physiology.

A list of some modern medical and applied sciences appears on page 19.

1. What factors probably stimulated an early interest in the human body?
2. What idea sparked the beginning of modern science?
3. What kinds of activities helped promote the development of modern science?

Anatomy and Physiology

Anatomy is the branch of science that deals with the structure (morphology) of body parts—their forms and arrangements. **Physiology,** on the other hand, is concerned with the functions of body parts—what they do and how they do it.

Figure 1.1
The study of the human body has a long history, as indicated by this illustration from the second book of *De humani corporis fabrica* by Andreas Vesalius, issued in 1543.

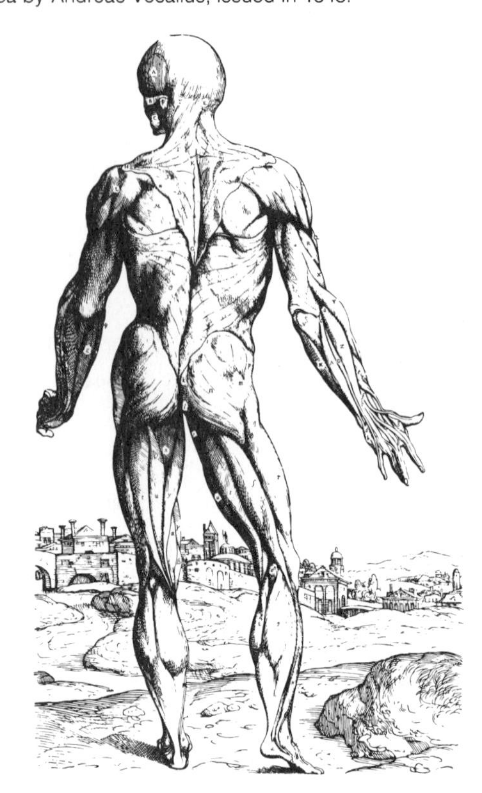

Actually, it is difficult to separate the topics of anatomy and physiology because the structures of body parts are so closely associated with their functions. These parts are arranged to form a well-organized unit—the **human organism**—and each part plays a role in the operation of the unit. This role, which is the part's function, depends upon the way the part is constructed—that is, the way its subparts are organized. For example, the arrangement of parts in the human hand with its long, jointed fingers is related to the function of grasping objects; the hollow chambers of the heart are adapted to pump blood through tubular blood vessels; and the shape of the mouth is related to the function of receiving food; and teeth are designed to break solid foods into small pieces (fig. 1.2).

1. Why is it difficult to separate the topics of anatomy and physiology?
2. List several examples to illustrate the idea that the structure of a body part is closely related to its function.

Figure 1.2
The structures of body parts are closely related to their functions. (*a*) The hand is adapted for grasping; (*b*) the heart for pumping blood; and (*c*) the mouth for receiving food.

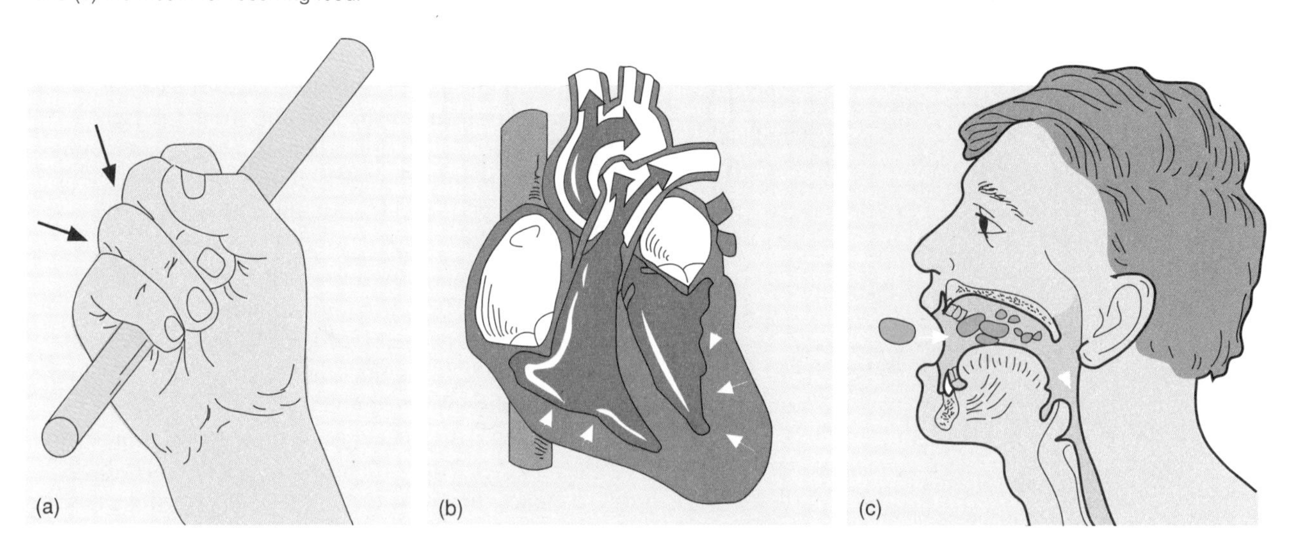

Characteristics of Life

Before beginning a more detailed study of anatomy and physiology, it is helpful to consider some of the traits humans share with other organisms. These traits, which are called *characteristics of life,* include the following:

1. **Movement** often refers to a self-initiated change in an organism's position or to its traveling from one place to another. However, the term also applies to the motion of internal parts, such as a beating heart.
2. **Responsiveness** is the ability of an organism to sense changes taking place inside or outside its body and to react to these changes. Seeking drinking water to quench thirst is a response to a loss of water from the body tissues; moving away from a hot fire is another example of responsiveness.
3. **Growth** refers to an increase in body size, usually without any important change in shape. It occurs whenever an organism produces new body materials faster than old ones are worn out or used up.
4. **Reproduction** is the process of making a new individual, as when parents produce an offspring. It also indicates the process by which microscopic cells produce others like themselves, as they do when body parts are repaired or replaced following an injury.
5. **Respiration** is the process of obtaining oxygen, using oxygen in the release of energy from food substances, and removing the resultant gaseous wastes.
6. **Digestion** is the process by which various food substances are chemically changed into simpler forms that can be taken in and used by body parts.
7. **Absorption** refers to the passage of digestive products through the membranes that line digestive organs and into the body fluids.
8. **Circulation** is the movement of substances from place to place within the body by means of the body fluids.
9. **Assimilation** is the changing of absorbed substances into forms that are chemically different from those that entered the body fluids.
10. **Excretion** is the removal of wastes that are produced by body parts as a result of their activities.

Each of these characteristics of life—in fact, everything an organism does—depends upon physical and chemical changes that occur within body parts. Taken together, these changes are referred to as **metabolism.**

Vital signs are among the more common observations made by physicians and nurses working with patients. Assessment of vital signs includes measuring body temperature and blood pressure, and monitoring rates and types of pulse and breathing movements. There is a close relationship between these signs and the characteristics of life, since vital signs are the results of *metabolic activities.* In fact, death is recognized by the absence of such signs. More specifically, a person who has died displays no spontaneous muscular movements (including those of the breathing muscles and beating heart), does not respond to stimuli (even the most painful that can be ethically applied), exhibits no reflexes (such as the knee jerk reflex and pupillary reflexes of the eye), and generates no brain waves (which reflects an absence of metabolic activity in the brain).

1. What are the characteristics of life?
2. How are the characteristics of life related to metabolism?

Maintenance of Life

With the exception of an organism's reproductive structures, which function to ensure that its particular form of life will continue into the future, the structures and functions of all body parts are directed toward achieving one goal—maintaining the life of the organism.

Needs of Organisms

Life is fragile, and it depends upon the presence of certain environmental factors for its existence. These factors include the following:

1. **Water** is the most abundant substance in the body. It is required for a variety of metabolic processes and provides the environment in which most of them take place. Water also transports substances from place to place within the organism and is important in the process of regulating body temperature.
2. **Food** refers to substances that provide the body with necessary chemicals in addition to water. Some of these chemicals are used as energy sources, others supply raw materials for building new living matter, and still others help regulate vital chemical reactions.
3. **Oxygen** is a gas that makes up about one-fifth of ordinary air. It is used in the process of releasing energy from food substances. The energy, in turn, is needed to drive metabolic processes.
4. **Heat** is a form of energy. It is a product of metabolic reactions, and the rate at which these reactions occur is partly governed by the amount of heat present. Generally, the greater the amount of heat, the more rapidly chemical reactions take place. (Temperature is a measurement of the amount of heat present.)
5. **Pressure** is an application of force to something. For example, the force acting on the outside of the body due to the weight of air above it is called *atmospheric pressure.* In humans, this pressure plays an important role in breathing. Similarly, organisms living under water are subjected to *hydrostatic pressure*—a pressure exerted by a liquid—due to the weight of water above them. In complex organisms, such as humans, the heart action creates blood pressure (another form of hydrostatic pressure), which forces the blood through certain blood vessels.

Although organisms need water, food, oxygen, heat, and pressure, the presence of these factors alone is not enough to ensure their survival. The quantities and the qualities of such factors are also important. For example, the amount of water entering and leaving an organism must be regulated, as must the concentration of oxygen in the body fluids. Similarly, survival depends on the quality as well as the quantity of food available—that is, food must supply the correct chemicals in adequate amounts.

Homeostasis

As an organism moves from place to place or as the climate around it changes, factors in its external environment also change. If the organism is to survive, however, the conditions within the fluids surrounding its body cells must remain relatively stable. In other words, body parts function efficiently only when the concentrations of water, food substances, and oxygen, and the conditions of heat and pressure remain within certain narrow limits. Thus, it is not surprising that the metabolic activities of an organism are directed toward maintaining such steady conditions. This tendency to maintain a stable internal environment is called **homeostasis.**

To better understand this idea of maintaining a stable environment, imagine a room equipped with a furnace and an air conditioner (fig. 1.3). Suppose the room temperature is to remain near 20° C (68° F), so the thermostat is adjusted to a set point of 20° C. Because a thermostat is sensitive to temperature changes, it will signal the furnace to start whenever the room temperature drops below the set point. If the temperature rises above the set point, the thermostat will cause

Figure 1.3
A thermostat that can signal an air conditioner and a furnace to turn on or off maintains a relatively stable room temperature. This system provides an example of a homeostatic mechanism.

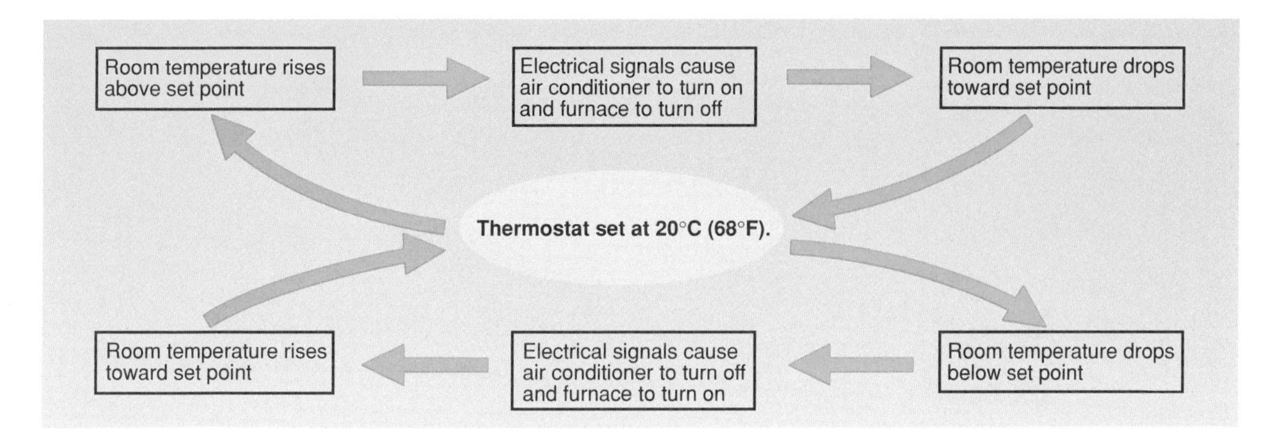

Figure 1.4
The homeostatic mechanism involved in body temperature regulation.

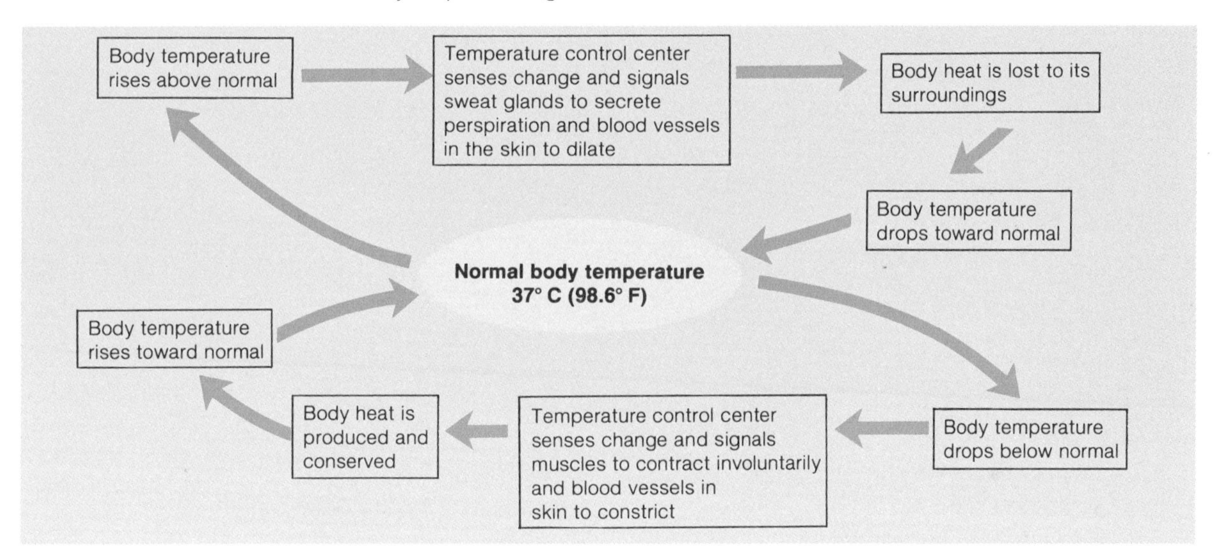

the furnace to stop and the air conditioner to start. As a result, a relatively constant temperature will be maintained in the room.

A similar *homeostatic mechanism* regulates body temperature in humans. The "thermostat" is a temperature-sensitive region in a temperature control center of the brain. In healthy persons, the set point of the brain's thermostat is at or near 37° C (98.6° F).

If a person is exposed to a cold environment and the body temperature begins to drop, the brain senses this change and triggers heat-generating and heat-conserving activities. For example, small groups of muscles may be stimulated to contract involuntarily, an action called *shivering*. Such muscular contractions produce heat, which helps to warm the body. At the same time, blood vessels in the skin may be signaled to constrict so that less warm blood flows through them, and heat that might otherwise be lost is held in deeper tissues.

If a person is becoming overheated, the temperature control center of the brain may trigger a series of changes that promote the increased loss of body heat. For example, it may stimulate sweat glands in the skin to secrete perspiration. This water evaporates from the surface, some heat is carried away and the skin is cooled. At the same time, the brain center causes blood vessels in the skin to dilate. This allows the blood carrying heat from deeper tissues to reach the surface where some heat is lost to the outside. The brain also stimulates an increase in heart rate, causing a greater volume of blood to move into surface vessels, and it stimulates an increase in breathing rate, allowing more heat-carrying air to be expelled from the lungs (fig. 1.4).

Figure 1.5
A human body is composed of parts within parts, which vary in complexity.

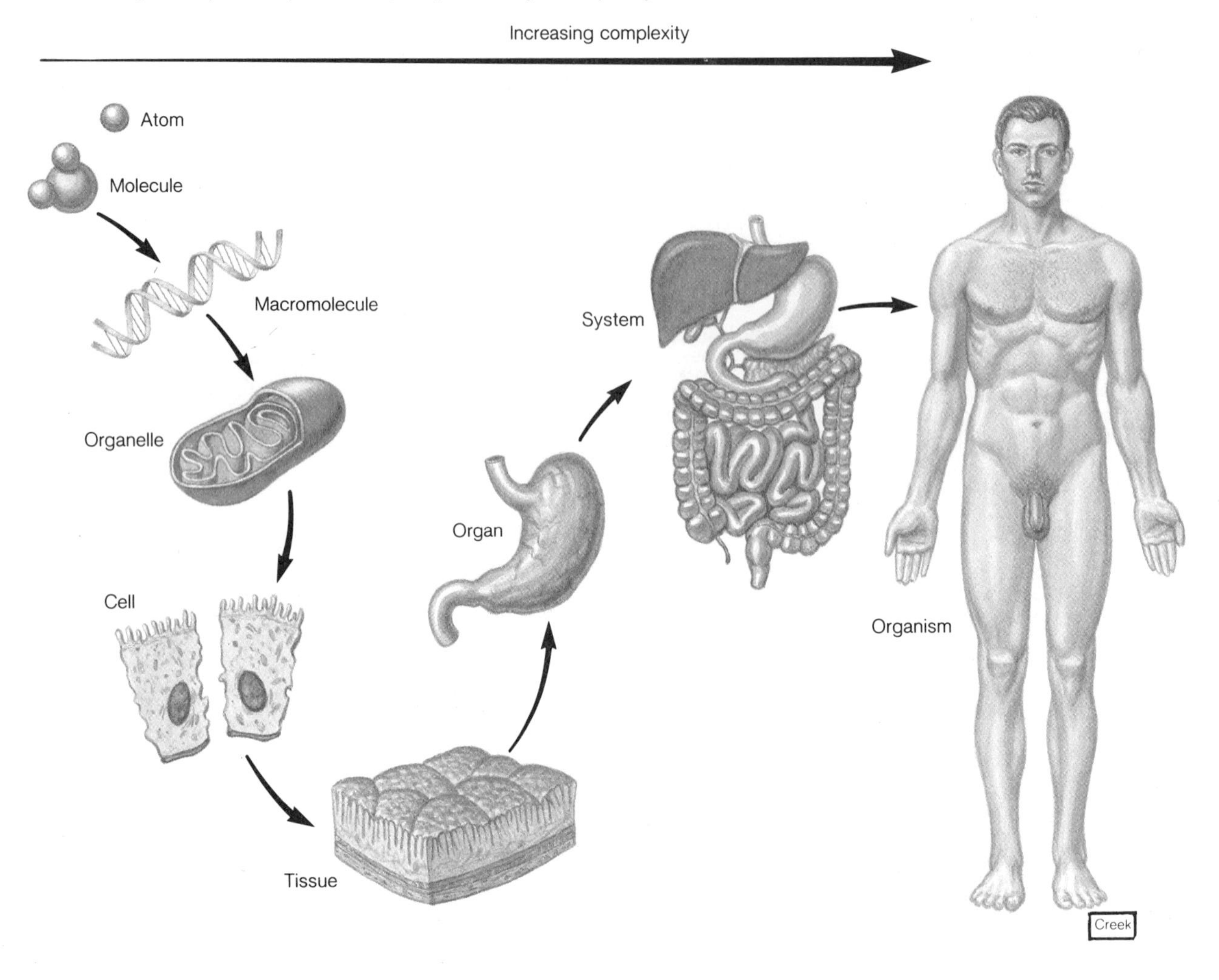

Another homeostatic mechanism functions to regulate the blood pressure in the vessels leading away from the heart (arteries). In this instance, pressure-sensitive sensors in the walls of these vessels are stimulated when the blood pressure increases above normal. When this happens, the sensors signal a pressure control center of the brain, which then signals the heart, causing its chambers to contract more slowly and with less force. Because of this decreased heart action, less blood enters the blood vessels, and the pressure inside the vessels decreases. If the blood pressure is dropping below normal, the brain center signals the heart to contract more rapidly and with greater force so that the pressure in the vessels increases.

1. What needs of organisms are provided from the external environment?
2. Why is homeostasis important to survival?
3. Describe two homeostatic mechanisms.

Levels of Organization

Early investigators focused their attention on larger body structures, for they were limited in their ability to observe small parts. Studies of small parts had to wait for the invention of magnifying lenses and microscopes, which came into use about four hundred years ago. Once these tools were available, it was discovered that larger body structures were made up of smaller parts, which, in turn, were composed of even smaller ones.

Today scientists recognize that all materials, including those that comprise the human body, are composed of chemicals (fig. 1.5). These substances are made up of tiny, invisible particles, called **atoms,** which are commonly bound together to form larger particles, called **molecules;** small molecules may be combined in complex ways to form larger molecules called **macromolecules.**

Within the human organism, the basic unit of structure and function is a microscopic part called a **cell.** Although individual cells vary in size, shape, and specialized functions, all have certain traits in common. For instance, all cells contain tiny parts called **organelles,** which carry on specific activities. These organelles are composed of aggregates of large molecules, including those of such substances as proteins, carbohydrates, lipids, and nucleic acids.

Cells are organized into layers or masses that have common functions. Such a group of cells forms a **tissue.** Groups of different tissues form **organs**—complex structures with specialized functions—and groups of organs are arranged into **organ systems.** Organ systems make up an **organism.**

Thus, various body parts can be thought of as occupying different *levels of organization,* such as the *atomic level, molecular level,* or *cellular level.* Furthermore, body parts vary in *complexity* from one level to the next. That is, atoms are less complex than molecules, molecules are less complex than organelles, and so forth.

Chapters 2 through 6 discuss these levels of organization in more detail. Chapter 2, for example, describes the atomic and molecular levels. Chapter 3 deals with organelles and cellular structures and functions, and chapter 5 describes tissues. Chapter 6 presents membranes as examples of organs, and the skin and its accessory organs as an example of an organ system.

Beginning with chapter 6, the structures and functions of each of the organ systems are described in detail.

1. How can the human body be used to illustrate the idea of levels of organization?
2. What is an organism?
3. How do body parts that occupy different levels of organization vary in complexity?

Organization of the Human Body

The human organism is a complex structure composed of many parts. Its major features include several body cavities, layers of membranes within these cavities, and a variety of organ systems.

Body Cavities

The human organism can be divided into an **axial portion,** which includes the head, neck, and trunk, and an **appendicular portion,** which includes the arms and legs. Within the axial portion there are two major cavities—a **dorsal cavity** and a larger **ventral cavity** (fig. 1.6*a*). The organs within such a cavity are called *visceral organs.* The dorsal cavity can be subdivided into two parts—the **cranial cavity** within the skull, which houses the brain; and the **spinal cavity,** which contains the spinal cord and is surrounded by sections of the backbone (vertebrae). The ventral cavity consists of a **thoracic cavity** and an **abdominopelvic cavity.**

The thoracic cavity is separated from the lower abdominopelvic cavity by a broad, thin muscle, called the *diaphragm.* The wall of the thoracic cavity is composed of skin, skeletal muscles, and various bones. The viscera within it include the lungs and a region called the *mediastinum* (fig. 1.6*b*). The mediastinum separates the thorax into two compartments, which contain the right and left lungs. The remaining thoracic viscera—heart, esophagus, trachea, and thymus gland—are located within the mediastinum.

The abdominopelvic cavity, which includes an upper abdominal portion and a lower pelvic portion, extends from the diaphragm to the floor of the pelvis. Its wall consists primarily of skin, skeletal muscles, and bones. The visceral organs within the *abdominal cavity* include the stomach, liver, spleen, gallbladder, and most of the small and large intestines.

The *pelvic cavity* is the portion of the abdominopelvic cavity enclosed by the pelvic bones and below the pelvic brim (chapter 7). It contains the terminal portion of the large intestine, the urinary bladder, and the internal reproductive organs.

In addition to the relatively large dorsal and ventral body cavities, there are several smaller cavities within the head. They include the following:

1. *Oral cavity,* containing the teeth and tongue.
2. *Nasal cavity,* located within the nose and divided into right and left portions by a nasal septum. Several air-filled *sinuses* are connected to the nasal cavity (see chapter 7).
3. *Orbital cavities,* containing the eyes and associated skeletal muscles and nerves.
4. *Middle ear cavities,* containing the middle ear bones (fig. 1.7).

Thoracic and Abdominopelvic Membranes

The walls of the right and left thoracic compartments, which contain the lungs, are lined with a membrane, called the *parietal pleura* (fig. 1.8). The lungs themselves are covered by a similar membrane, called *visceral pleura.* (Note: *parietal* refers to the membrane that is attached to the wall and forms the lining of a cavity; *visceral* refers to the membrane that is deeper—toward the interior—and covers the organs contained within a cavity.)

Figure 1.6
Major body cavities; as viewed (*a*) from the side and, (*b*) from the front.

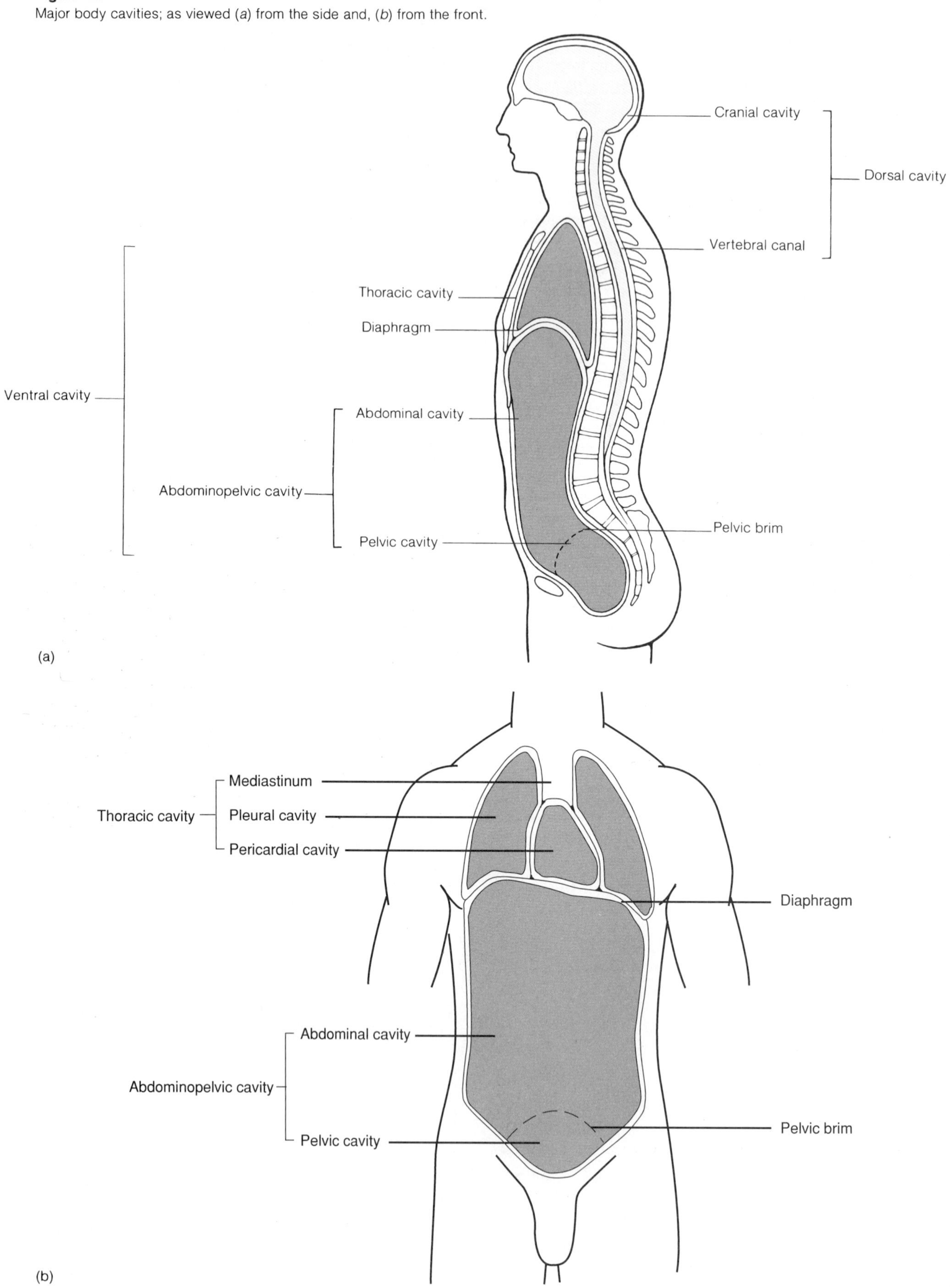

Figure 1.7
The cavities within the head include the oral, nasal, orbital, and middle ear cavities, as well as several sinuses.

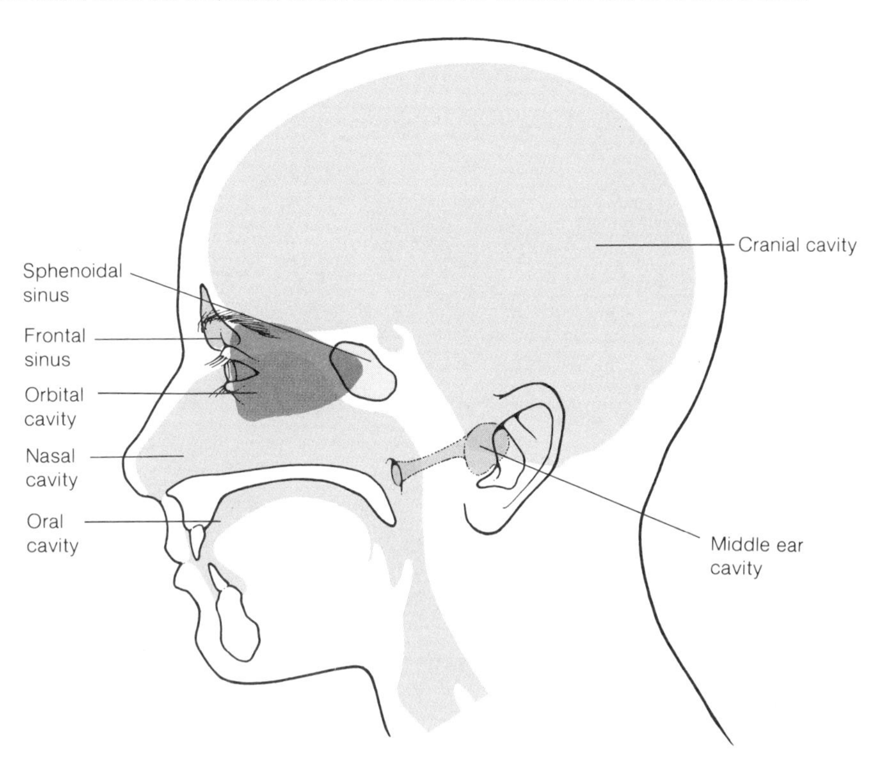

Figure 1.8
A transverse section through the thorax reveals the serous membranes associated with the heart and lungs (*superior view*).

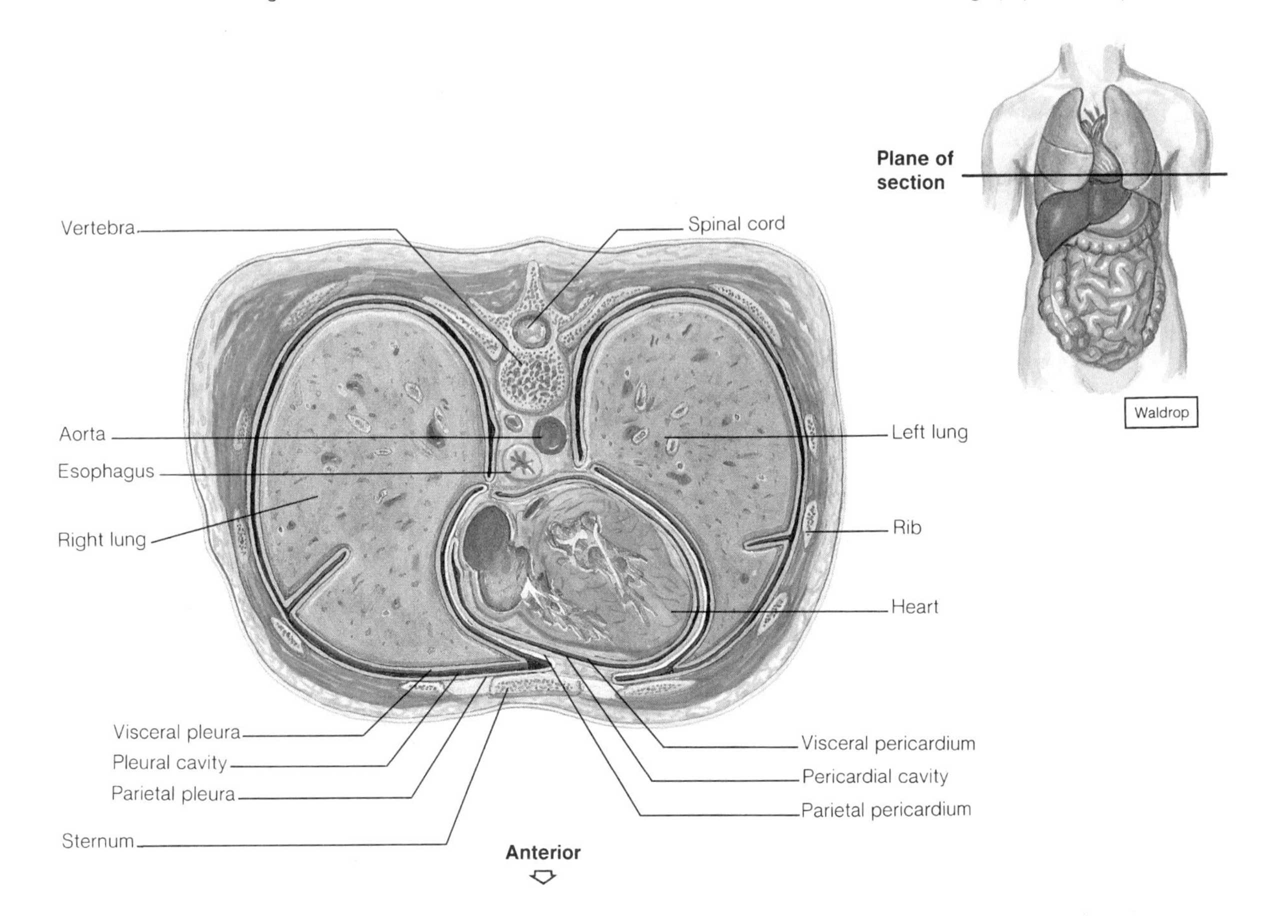

Figure 1.9
A transverse section through the abdomen (*superior view*). How would you distinguish between the parietal and visceral peritoneum?

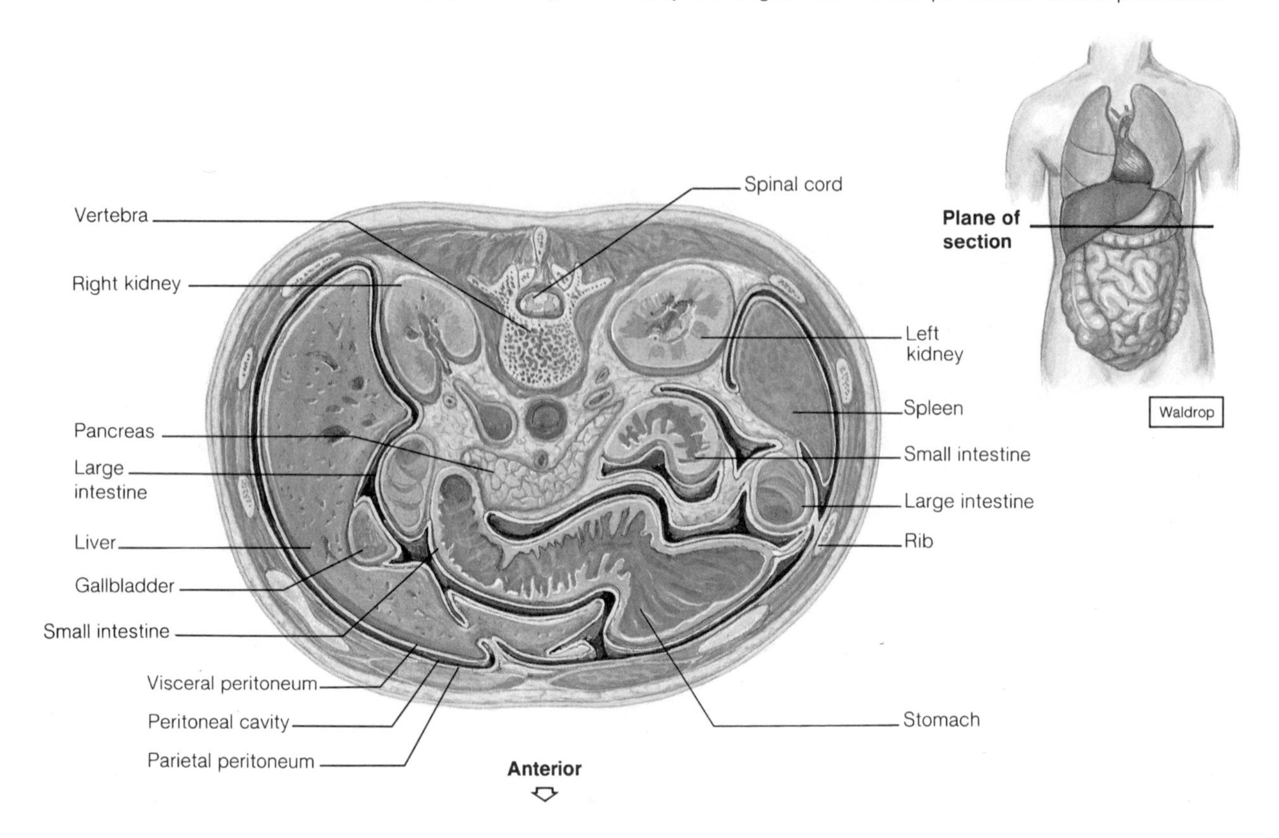

The parietal and visceral **pleural membranes** are separated by a thin film of watery fluid (serous fluid) that they secrete. Although there is normally no actual space between these membranes, the potential space between them is called the *pleural cavity.*

The heart, which is located in the broadest portion of the mediastinum, is surrounded by **pericardial membranes.** A thin *visceral pericardium* covers the heart's surface and is separated from a much thicker, fibrous *parietal pericardium* by a small amount of fluid. The potential space between these membranes is called the *pericardial cavity.*

In the abdominopelvic cavity, the lining membranes are called **peritoneal membranes** (fig. 1.9). A *parietal peritoneum* lines the wall, and a *visceral peritoneum* covers each organ in the abdominal cavity. The potential space between these membranes is called the *peritoneal cavity.*

1. What is meant by the term visceral organs?
2. What organs occupy the dorsal cavity? The ventral cavity?
3. Name the cavities of the head.
4. Describe the membranes associated with the thoracic and abdominopelvic cavities.

Organ Systems

The human organism consists of several organ systems. Each system includes a set of interrelated organs that work together to provide specialized functions. As you read about each system, you may want to consult the illustrations of the human torso and locate some of the organs listed in the description (see Reference Plates, pages 25–32).

Body Covering

The organs of the **integumentary system** (chapter 6) include the skin and various accessory organs such as the hair, nails, sweat glands, and sebaceous glands. These parts protect underlying tissues, help regulate body temperature, house a variety of sensory receptors, and synthesize certain products.

Support and Movement

The organs of the skeletal and muscular systems (chapters 7 and 8) function to support and move body parts.

The **skeletal system** consists of bones as well as ligaments and cartilages, which bind bones together. These parts provide frameworks and protective shields for softer tissues, serve as attachments for muscles, and act together with muscles when body parts move. Tissues within bones also produce blood cells and store inorganic salts.

Muscles are the organs of the **muscular system.** By contracting and pulling their ends closer together, they provide the forces that cause body movements. They also maintain posture and are the main source of body heat.

Integration and Coordination

For the body to act as a unit, its parts must be integrated and coordinated. That is, their activities must be controlled and adjusted so that homeostasis is maintained. This is the general function of the nervous and endocrine systems.

The **nervous system** (chapter 9) consists of the brain, spinal cord, nerves, and sense organs. The nerve cells within these organs use electrochemical signals called *nerve impulses* to communicate with one another and with muscles and glands. Each impulse produces a relatively short-term effect on the parts it influences.

Some nerve cells are specialized to detect changes that occur inside and outside the body. Others receive the impulses transmitted from these sensory units and interpret and act on this information. Still others carry impulses from the brain or spinal cord to muscles or glands and stimulate these parts to contract or to secrete various products.

The **endocrine system** (chapter 11) includes all the glands that secrete chemical messengers called *hormones*. The hormones, in turn, travel away from the gland in some body fluid such as blood or tissue fluid. Usually a particular hormone only affects a particular group of cells, which is called its *target tissue*. The effect of a hormone is to alter the metabolism of the target tissue. Compared to nerve impulses, hormonal effects occur over a relatively long period of time.

The organs of the endocrine system include the pituitary, thyroid, parathyroid, and adrenal glands, as well as the pancreas, ovaries, testes, pineal gland, and thymus gland.

Processing and Transporting

The organs of several systems are involved with processing and transporting nutrients, oxygen, and various wastes. The organs of the **digestive system** (chapter 12), for example, receive foods from the outside. Then, they convert various food molecules into simpler ones that can pass through cell membranes and thus be absorbed. Materials that are not absorbed are eliminated by being transported back to the outside. Certain digestive organs also produce hormones and thus function as parts of the endocrine system.

The digestive system includes the mouth, tongue, teeth, salivary glands, pharynx, esophagus, stomach, liver, gallbladder, pancreas, small intestine, and large intestine.

The organs of the **respiratory system** (chapter 13) provide for the intake and output of air and for the exchange of gases between blood and air. The nasal cavity, pharynx, larynx, trachea, bronchi, and lungs are parts of this system.

The **circulatory system** (chapters 14 and 15) includes the heart, arteries, veins, capillaries, and blood. The heart functions as a muscular pump that helps force the blood through the blood vessels. The blood serves as a fluid for transporting gases, nutrients, hormones, and wastes. It carries oxygen from the lungs and nutrients from the digestive organs to all body cells, where these substances are used in metabolic processes. The blood also transports hormones from various endocrine glands to their target tissues and carries wastes from body cells to excretory organs, where wastes are removed from the blood and released to the outside.

The **lymphatic system** (chapter 16) is sometimes considered to be part of the circulatory system. It is composed of the lymphatic vessels, lymph fluid, lymph nodes, thymus gland, and spleen. This system transports fluid (tissue fluid) from the tissues back to the bloodstream and carries certain fatty substances away from the digestive organs. Lymphatic organs also aid in defending the body against infections by removing particles such as microorganisms from the tissue fluid. In addition, it supports the activities of certain cells (lymphocytes) that produce immunity by reacting against specific disease-causing agents.

The **urinary system** (chapter 17) consists of the kidneys, ureters, urinary bladder, and urethra. The kidneys remove various wastes from the blood and assist in maintaining the body's water and electrolyte balances. The product of these activities is urine. Other portions of the urinary system are concerned with storing urine and transporting it to the outside.

Reproduction

Reproduction is the process of producing offspring (progeny). Cells reproduce when they divide and give rise to new cells. The **reproductive system** of an organism, however, is involved with the production of whole new organisms like itself (chapter 19).

The male reproductive system includes the scrotum, testes, epididymides, vasa deferentia, seminal vesicles, prostate gland, bulbourethral glands, penis, and urethra. These parts are concerned with producing and maintaining male sex cells, or sperm cells (spermatozoa). They also function to transfer these cells from their site of origin into the female reproductive tract.

The female reproductive system consists of the ovaries, fallopian tubes, uterus, vagina, clitoris, and vulva. These organs produce and maintain female sex cells, or egg cells (ova), receive the male cells, and transport the male and female cells within the female system. The female reproductive system also provides for the support and development of embryos and functions in the birth process.

1. Name each of the major organ systems, and list the organs of each system.
2. Describe the general functions of each organ system.

Anatomical Terminology

To communicate effectively with one another, investigators over the ages have developed a set of terms with precise meanings. Some of these terms concern the relative positions of body parts, others relate to imaginary planes along which cuts may be made, and still others are used to describe various body regions.

When such terms are used, it is assumed that the body is in the **anatomical position,** that is, it is standing erect, the face is forward, and the arms are at the sides, with the palms forward.

Relative Position

Terms of relative position are used to describe the location of one body part with respect to another. They include the following:

1. **Superior** means a part is above another part or closer to the head. (The thoracic cavity is superior to the abdominopelvic cavity.)
2. **Inferior** means situated below another part or toward the feet. (The neck is inferior to the head.)
3. **Anterior** (or *ventral*) means toward the front. (The eyes are anterior to the brain.)
4. **Posterior** (or *dorsal*) is the opposite of anterior; it means toward the back. (The pharynx is posterior to the oral cavity.)
5. **Medial** relates to an imaginary midline dividing the body into equal right and left halves. A part is medial if it is closer to this line than another part. (The nose is medial to the eyes.)
6. **Lateral** means toward the side with respect to the imaginary midline. (The ears are lateral to the eyes.)
7. **Proximal** is used to describe a part that is closer to a point of attachment or closer to the trunk of the body than another part. (The elbow is proximal to the wrist.)
8. **Distal** is the opposite of proximal. It means that a particular body part is farther from the point of attachment or farther from the trunk than another part. (The fingers are distal to the wrist.)
9. **Superficial** means situated near the surface. (The epidermis is the superficial layer of the skin.) **Peripheral** also means outward or near the surface. It is used to describe the location of certain blood vessels and nerves. (The nerves that branch from the brain and spinal cord are peripheral nerves.)
10. **Deep** is used to describe parts that are more internal. (The dermis is the deep layer of the skin.)

Body Sections

To observe the relative locations and arrangements of the internal parts, it is necessary to cut or section the body along various planes (fig. 1.10). The following terms are used to describe such planes and sections:

1. **Sagittal** refers to a lengthwise cut that divides the body into right and left portions. If a sagittal section passes along the midline and divides the body into equal parts, it is called *median* (midsagittal).
2. **Transverse** (or *horizontal*) refers to a cut that divides the body into superior and inferior portions.
3. **Frontal** (or *coronal*) refers to a section that divides the body into anterior and posterior portions.

Sometimes a cylindrical organ such as a blood vessel is sectioned. In this case, a cut across the structure is called a *cross section,* an angular cut is an *oblique section,* and a lengthwise cut is a *longitudinal section* (fig. 1.11).

Sections of internal organs can be visualized by using a procedure called *computerized tomography* or CT scanning (fig. 1.12). In this procedure, an X-ray-emitting part is moved around the body region being examined. At the same time, an X-ray detector is moved in the opposite direction on the other side. As the parts move, an X-ray beam is passed through the body from hundreds of different angles. Since tissues and organs of varying composition within the body absorb X rays differently, the amount of X ray reaching the detector varies from position to position.

The measurements made by the X-ray detector are recorded in the memory of a computer, which later combines them mathematically and creates a sectional image of the internal body parts on a viewing screen.

Figure 1.10
To observe internal parts, the body may be sectioned along various planes.

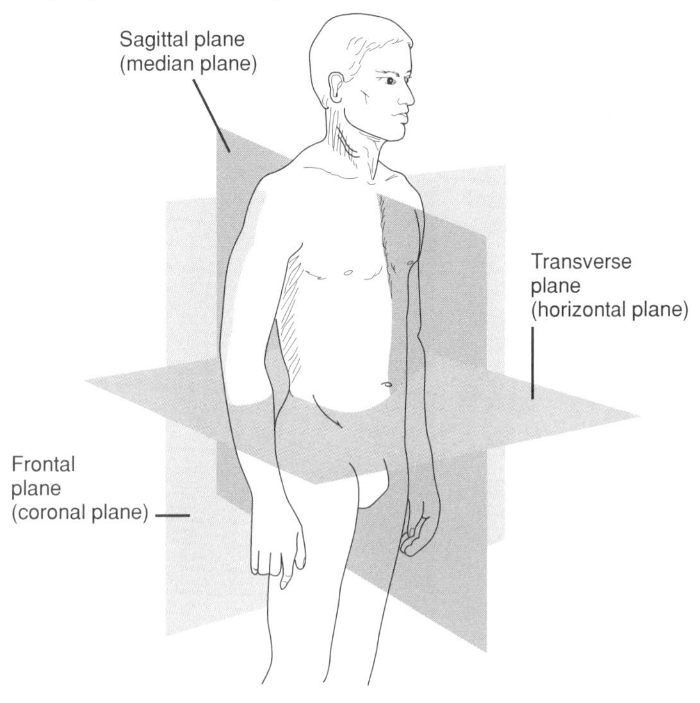

Figure 1.11
Cylindrical parts may be cut in (*a*) cross section, (*b*) oblique section, or (*c*) longitudinal section.

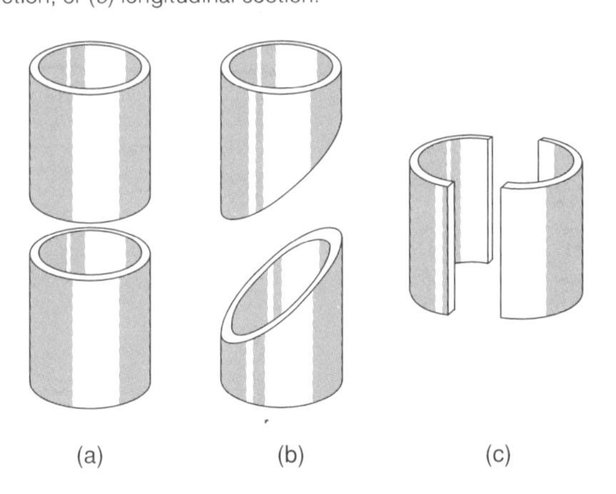

Figure 1.12
CT scans of the (*a*) head and (*b*) abdomen. What features can you identify?

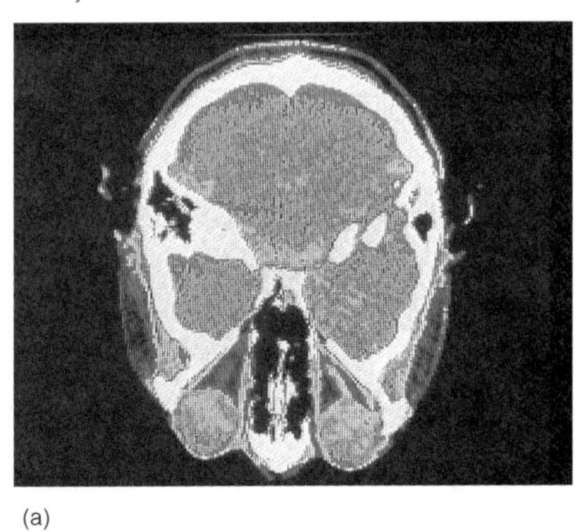

(a)

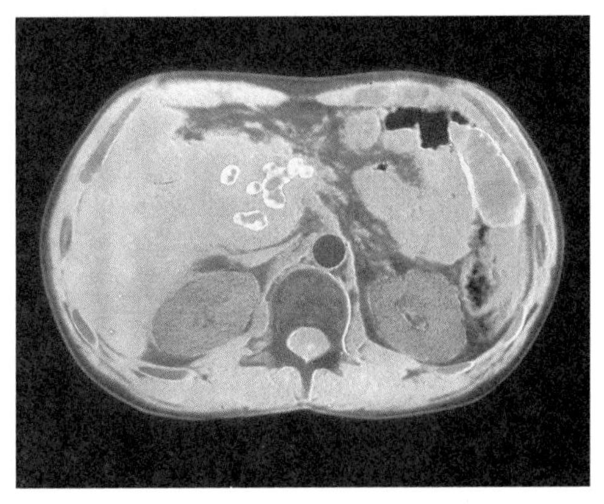

(b)

Figure 1.13
The abdominal area is subdivided into nine regions. How do the names of these regions describe their locations?

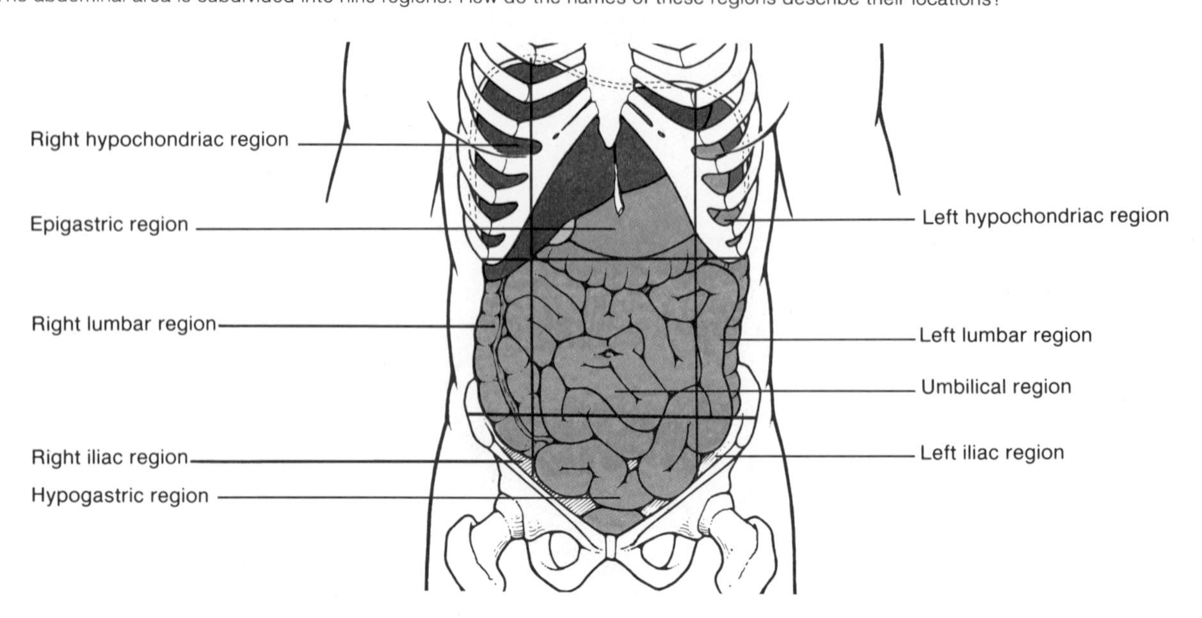

Body Regions

A number of terms are used to designate various body regions. The abdominal area, for example, is subdivided into the following nine regions, as shown in figure 1.13:

1. **Epigastric region**—the upper middle portion.
2. **Left** and **right hypochondriac regions**—on each side of the epigastric region.
3. **Umbilical region**—the middle portion.
4. **Left** and **right lumbar regions**—on each side of the umbilical region.
5. **Hypogastric region**—the lower middle portion.
6. **Left** and **right iliac regions**—on each side of the hypogastric region.

The following terms are commonly used when referring to various body regions (figure 1.14 illustrates some of these regions):

abdominal (ab-dom'ĭ-nal)—the region between the thorax and pelvis.
acromial (ah-kro'me-al)—the point of the shoulder.
antebrachial (an"te-bra'ke-al)—the forearm.
antecubital (an"te-ku'bĭ-tal)—the space in front of the elbow.
axillary (ak'sĭ-ler"e)—the armpit.
brachial (bra'ke-al)—the upper arm.
buccal (buk'al)—the cheek.
carpal (kar'pal)—the wrist.
celiac (se'le-ak)—the abdomen.
cephalic (sĕ-fal'ik)—the head.
cervical (ser'vĭ-kal)—the neck.
costal (kos'tal)—the ribs.
coxal (kok'sal)—the hip.
crural (kro͞or'al)—the leg.
cubital (ku'bĭ-tal)—the elbow.
digital (dij'ĭ-tal)—the finger.
dorsal (dor'sal)—the back.
femoral (fem'or-al)—the thigh.
frontal (frun'tal)—the forehead.
genital (jen'ĭ-tal)—reproductive organs.
gluteal (gloo'te-al)—the buttocks.
inguinal (ing'gwĭ-nal)—the depressed area of the abdominal wall near the thigh (groin).
lumbar (lum'bar)—the region of the lower back between the ribs and the pelvis (loin).
mammary (mam'er-e)—the breast.
mental (men'tal)—the chin.
nasal (na'zal)—the nose.
occipital (ok-sip'ĭ-tal)—the lower back region of the head.
oral (o'ral)—the mouth.
orbital (or'bi-tal)—the eye cavity.
otic (o'tik)—the ear.
palmar (pahl'mar)—the palm of the hand.
pectoral (pek'tor-al)—the chest.
pedal (ped'al)—the foot.
pelvic (pel'vik)—the pelvis.
perineal (per"ĭ-ne'al)—the region between the anus and the external reproductive organs (perineum).
plantar (plan'tar)—the sole of the foot.
popliteal (pop"lĭ-te'al)—the area behind the knee.
sacral (sa'kral)—the posterior region between the hipbones.
sternal (ster'nal)—the middle of the thorax, anteriorly.
tarsal (tahr'sal)—the instep of the foot.
umbilical (um-bil'ĭ-kal)—the navel.
vertebral (ver'te-bral)—the spinal column.

Figure 1.14
Some terms used to describe body regions: (*a*) anterior regions; (*b*) posterior regions.

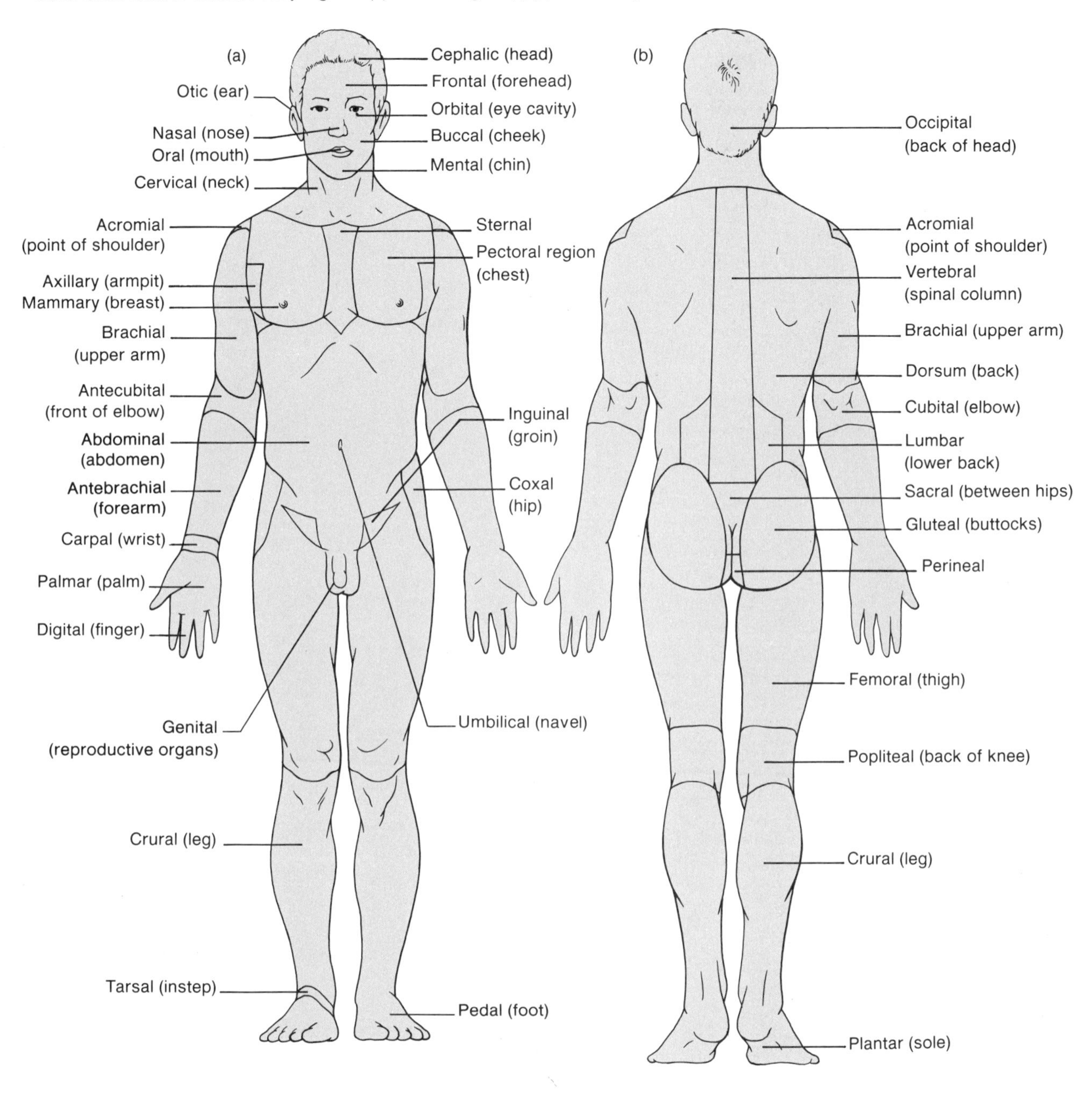

1. Describe the anatomical position.
2. Using the appropriate terms, describe the relative positions of several body parts.
3. Describe three types of body sections.
4. Name the nine regions of the abdomen.

Some Medical and Applied Sciences

cardiology (kar″de-ol′o-je)—a branch of medical science dealing with the heart and heart diseases.

cytology (si-tol′o-je)—the study of the structure, function, and diseases of cells.

dermatology (der″mah-tol′o-je)—the study of the skin and its diseases.

endocrinology (en″do-krĭ-nol′o-je)—the study of hormones, hormone-secreting glands, and the diseases involving them.

epidemiology (ep″ĭ-de″me-ol′o-je)—the study of the factors determining the distribution and frequency of the occurrence of various health related conditions within a defined human population.

gastroenterology (gas″tro-en″ter-ol′o-je)—the study of the stomach and intestines, and diseases involving these organs.

geriatrics (jer″e-at′riks)—a branch of medicine dealing with elderly persons and their medical problems.

gerontology (jer″on-tol′o-je)—the study of the process of aging.
gynecology (gi″nĕ-kol′o-je)—the study of the female reproductive system and its diseases.
hematology (hem″ah-tol′o-je)—the study of the blood and blood diseases.
histology (his-tol′o-je)—the study of the structure and function of tissues.
immunology (im″u-nol′o-je)—the study of the body's resistance to disease.
neonatology (ne″o-na-tol′o-je)—the study and treatment of disorders of newborn infants.
nephrology (nĕ-frol′o-je)—the study of the structure, function, and diseases of the kidneys.
neurology (nu-rol′o-je)—the study of the nervous system in health and disease.
obstetrics (ob-stet′riks)—a branch of medicine dealing with pregnancy and childbirth.
oncology (ong-kol′o-je)—the study of tumors.
ophthalmology (of″thal-mol′o-je)—the study of the eye and eye diseases.
orthopedics (or″tho-pe′diks)—a branch of medicine dealing with the muscular and skeletal systems and the problems of these systems.
otolaryngology (o″to-lar″in-gol′o-je)—the study of the ear, throat, larynx, and diseases of these parts.
pathology (pah-thol′o-je)—the study of structural and functional changes produced by diseases.
pediatrics (pe″de-at′riks)—a branch of medicine dealing with children and their diseases.
pharmacology (fahr″mah-kol′o-je)—the study of drugs and their uses in the treatment of diseases.
podiatry (po-di′ah-tre)—the study of the care and treatment of feet.
psychiatry (si-ki′ah-tre)—a branch of medicine dealing with the mind and its disorders.
radiology (ra″de-ol′o-je)—the study of X rays and radioactive substances and their uses in the diagnosis and treatment of diseases.
toxicology (tok″sĭ-kol′o-je)—the study of poisonous substances and their effects upon body parts.
urology (u-rol′o-je)—a branch of medicine dealing with the urinary system, male reproductive system, and diseases of these systems.

Chapter Summary

Introduction (page 6)

1. An early interest in the human body probably concerned injuries and illnesses.
2. Primitive doctors began to learn how certain herbs and potions affected body functions.
3. In ancient times, it was believed that natural processes were caused by spirits and supernatural forces.
4. The belief that humans could understand forces that caused natural events led to the development of modern science.
5. A set of terms originating from Greek and Latin words forms the basis for the language of anatomy and physiology.

Anatomy and Physiology (page 6)

1. Anatomy deals with the form and arrangement of body parts.
2. Physiology deals with the functions of these parts.
3. The function of a part depends upon the way it is constructed.

Characteristics of Life (page 7)

Characteristics of life are traits shared by all organisms.

1. These characteristics include the following:
   a. Movement—changing body position or the moving of internal parts.
   b. Responsiveness—sensing and reacting to internal or external changes.
   c. Growth—increasing size without changing shape.
   d. Reproduction—producing offspring.
   e. Respiration—obtaining oxygen, using oxygen to release energy from foods, and removing gaseous wastes.
   f. Digestion—changing food substances into forms that can be absorbed.
   g. Absorption—moving digestive products into body fluids.
   h. Circulation—moving of substances through the body in body fluids.
   i. Assimilation—changing substances into chemically different forms.
   j. Excretion—removing body wastes.
2. Together these activities constitute metabolism.

Maintenance of Life (page 8)

The structures and functions of body parts are directed toward maintaining the life of the organism.

1. Needs of organisms
   a. Water is used in a variety of metabolic processes, provides the environment for metabolic reactions, and transports substances.
   b. Food is needed to supply energy, to provide raw materials for building new living matter, and to supply chemicals necessary in vital reactions.
   c. Oxygen is used in releasing energy from food materials; this energy drives metabolic reactions.
   d. Heat is a product of metabolic reactions and helps govern the rates of these reactions.
   e. Pressure is an application of force to something; in humans, atmospheric and hydrostatic pressures help breathing and blood movements, respectively.
   f. The survival of an organism depends upon the quantities and qualities of these factors.

2. Homeostasis
   a. If an organism is to survive, the conditions within its body fluids must remain relatively stable.
   b. Metabolic activities are largely directed toward maintaining stable internal conditions.
   c. The tendency to maintain a stable internal environment is called homeostasis.
   d. Homeostatic mechanisms help regulate body temperature and blood pressure.

Levels of Organization (page 10)

The body is composed of parts that occupy different levels of organization.

1. Material substances are composed of atoms.
2. Atoms join together to form molecules.
3. Organelles contain groups of large molecules.
4. Cells, which are composed of organelles, are the basic units of structure and function within the body.
5. Cells are organized into tissues.
6. Tissues are organized into organs.
7. Organs are arranged into organ systems.
8. Organ systems constitute the organism.
9. These parts vary in complexity from one level to the next.

Organization of the Human Body (page 11)

1. Body cavities
   a. The axial portion of the body contains the dorsal and ventral cavities.
      (1) The dorsal cavity includes the cranial and spinal cavities.
      (2) The ventral cavity includes the thoracic and abdominopelvic cavities.
   b. The organs within a body cavity are called visceral organs.
   c. Other body cavities include the oral, nasal, orbital, and middle ear cavities.
2. Thoracic and abdominopelvic membranes
   a. Thoracic membranes
      (1) Pleural membranes line the thoracic cavity and cover the lungs.
      (2) Pericardial membranes surround the heart and cover its surface.
      (3) The pleural and pericardial cavities are potential spaces between these membranes.
   b. Abdominopelvic membranes
      (1) Peritoneal membranes line the abdominopelvic cavity and cover the organs inside.
      (2) The peritoneal cavity is a potential space between these membranes.
3. Organ systems
   The human organism consists of several organ systems. Each system includes a set of interrelated organs.
   a. Body covering
      (1) The integumentary system includes the skin, hair, nails, sweat glands, and sebaceous glands.
      (2) It functions to protect underlying tissues, regulate body temperature, house sensory receptors, and synthesize various substances.
   b. Support and movement
      (1) Skeletal system
         (a) The skeletal system is composed of bones, cartilages, and the ligaments that bind bones together.
         (b) It provides a framework, protective shields, and attachments for muscles; it also produces blood cells and stores inorganic salts.
      (2) Muscular system
         (a) The muscular system includes the muscles of the body.
         (b) It is responsible for body movements, the maintenance of posture, and production of body heat.
   c. Integration and coordination
      (1) Nervous system
         (a) The nervous system consists of the brain, spinal cord, nerves, and sense organs.
         (b) It functions to receive impulses from sensory parts, interpret these impulses, and act on them by causing muscles or glands to respond.
      (2) Endocrine system
         (a) The endocrine system consists of glands that secrete hormones.
         (b) Hormones help regulate metabolism.
         (c) This system includes the pituitary, thyroid, parathyroid, and adrenal glands; the pancreas, ovaries, testes, pineal gland, and thymus gland.
   d. Processing and transporting
      (1) Digestive system
         (a) The digestive system receives foods, converts food molecules into forms that can pass through cell membranes, and eliminates materials that are not absorbed.
         (b) It includes the mouth, tongue, teeth, salivary glands, pharynx, esophagus, stomach, liver, gallbladder, pancreas, small intestine, and large intestine.
         (c) Some digestive organs produce hormones.
      (2) Respiratory system
         (a) The respiratory system provides for the intake and output of air and for the exchange of gases between the air and the blood.
         (b) It includes the nasal cavity, pharynx, larynx, trachea, bronchi, and lungs.
      (3) Circulatory system
         (a) The circulatory system includes the heart, which pumps the blood, and the blood vessels, which carry the blood to and from the body parts.

(b) The blood transports oxygen, nutrients, hormones, and wastes.

(4) Lymphatic system

(a) The lymphatic system is composed of lymphatic vessels, lymph nodes, thymus, and spleen.

(b) It transports lymph from the tissues to the bloodstream, carries certain fatty substances away from the digestive organs, and aids in defending the body against disease-causing agents.

(5) Urinary system

(a) The urinary system includes the kidneys, ureters, urinary bladder, and urethra.

(b) It filters wastes from the blood and helps maintain water and electrolyte balance.

e. Reproduction

(1) The reproductive systems are concerned with the production of new organisms.

(2) The male reproductive system includes the scrotum, testes, epididymides, vasa deferentia, seminal vesicles, prostate gland, bulbourethral glands, urethra, and penis, which produce, maintain, and transport male sex cells.

(3) The female reproductive system includes the ovaries, fallopian tubes, uterus, vagina, clitoris, and vulva, which produce, maintain, and transport female sex cells.

Anatomical Terminology (page 16)

Terms with precise meanings are used to help investigators communicate effectively.

1. Relative position
   These terms are used to describe the location of one part with respect to another part.
2. Body sections
   Body sections are planes along which the body may be cut to observe the relative locations and arrangements of internal parts.
3. Body regions
   Various body regions are designated by special terms.

Clinical Application of Knowledge

1. In health, the body parts function together effectively to maintain homeostasis. In illness, the maintenance of homeostasis may be threatened and various treatments may be needed to reduce this threat. What treatments might be used to help control (*a*) a patient's body temperature, (*b*) blood oxygen concentration, and (*c*) water content?
2. Suppose two individuals are afflicted with benign (noncancerous) tumors that produce symptoms because they occupy space and crowd adjacent organs. If one of these persons has the tumor in the ventral cavity and the other has a tumor in the dorsal cavity, which would be likely to develop symptoms first? Why?
3. If a patient complained of a "stomachache" and pointed to the umbilical region as the site of the discomfort, what organs located in this region might be the source of the pain?
4. How could the basic needs of a human be provided for a patient who is unconscious?

Review Activities

Part A

1. Briefly describe the early development of knowledge about the human body.
2. Distinguish between anatomy and physiology.
3. Explain the relationship between the form and function of body parts.
4. List and describe ten characteristics of life.
5. Define *metabolism.*
6. List and describe five needs of organisms.
7. Describe two types of pressure that may act on the outside of an organism.
8. Define *homeostasis* and explain its importance.
9. Explain how body temperature is controlled.
10. Describe a homeostatic mechanism that helps to regulate blood pressure.
11. List the levels of organization within the human body.
12. Distinguish between the axial and appendicular portions of the body.
13. Distinguish between the dorsal and ventral body cavities, and name the smaller cavities that occur within each.
14. List the cavities of the head and contents of each cavity.
15. Explain what is meant by *visceral organ.*
16. Describe the mediastinum and its contents.
17. Distinguish between a parietal and a visceral membrane.
18. Name the major organ systems, and describe the general functions of each.
19. List the major organs that comprise each organ system.

Part B

1. Name the body cavity in which each of the following organs is located:
   a. Stomach
   b. Heart
   c. Brain
   d. Liver
   e. Trachea
   f. Rectum
   g. Spinal cord
   h. Esophagus
   i. Spleen
   j. Urinary bladder

2. Write complete sentences using each of the following terms correctly:
   a. Superior
   b. Inferior
   c. Anterior
   d. Posterior
   e. Medial
   f. Lateral
   g. Proximal
   h. Distal
   i. Superficial
   j. Peripheral
   k. Deep
3. Prepare a sketch of a human body, and use lines to indicate each of the following sections:
   a. Median
   b. Transverse
   c. Frontal
4. Prepare a sketch of the abdominal area, and indicate the location of each of the following regions:
   a. Epigastric
   b. Umbilical
   c. Hypogastric
   d. Hypochondriac
   e. Lumbar
   f. Iliac
5. Provide the common name for the region to which each of the following terms refers:
   a. Acromial
   b. Antebrachial
   c. Axillary
   d. Buccal
   e. Celiac
   f. Coxal
   g. Crural
   h. Femoral
   i. Genital
   j. Gluteal
   k. Inguinal
   l. Mental
   m. Occipital
   n. Orbital
   o. Otic
   p. Palmar
   q. Pectoral
   r. Pedal
   s. Perineal
   t. Plantar
   u. Popliteal
   v. Sacral
   w. Sternal
   x. Tarsal
   y. Umbilical
   z. Vertebral

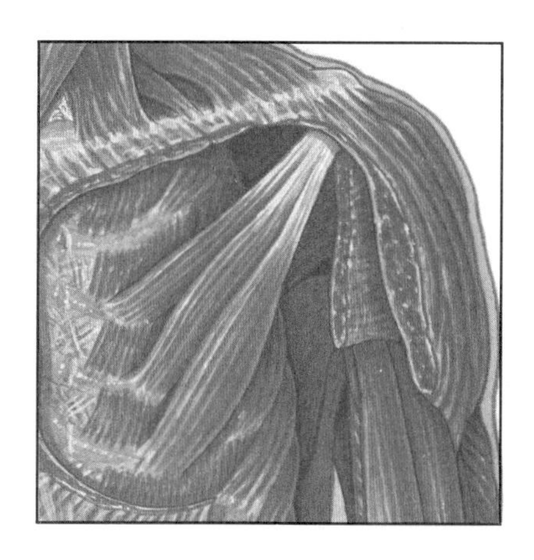

Reference Plates
The Human Organism

The following series of illustrations shows the major organs of the human torso. The first plate illustrates the anterior surface and reveals the superficial muscles on one side. Each subsequent plate exposes deeper organs, including those in the thoracic, abdominal, and pelvic cavities.

The purpose of chapters 6–19 of this textbook is to describe the organ systems of the human organism in detail. As you read them, you may want to refer to these plates to help yourself visualize the locations of various organs and the three-dimensional relationships that exist among them.

Plate 1

A human torso showing the anterior surface on one side and the superficial muscles exposed on the other side. (*m.* stands for *muscle,* and *v.* stands for *vein.*)

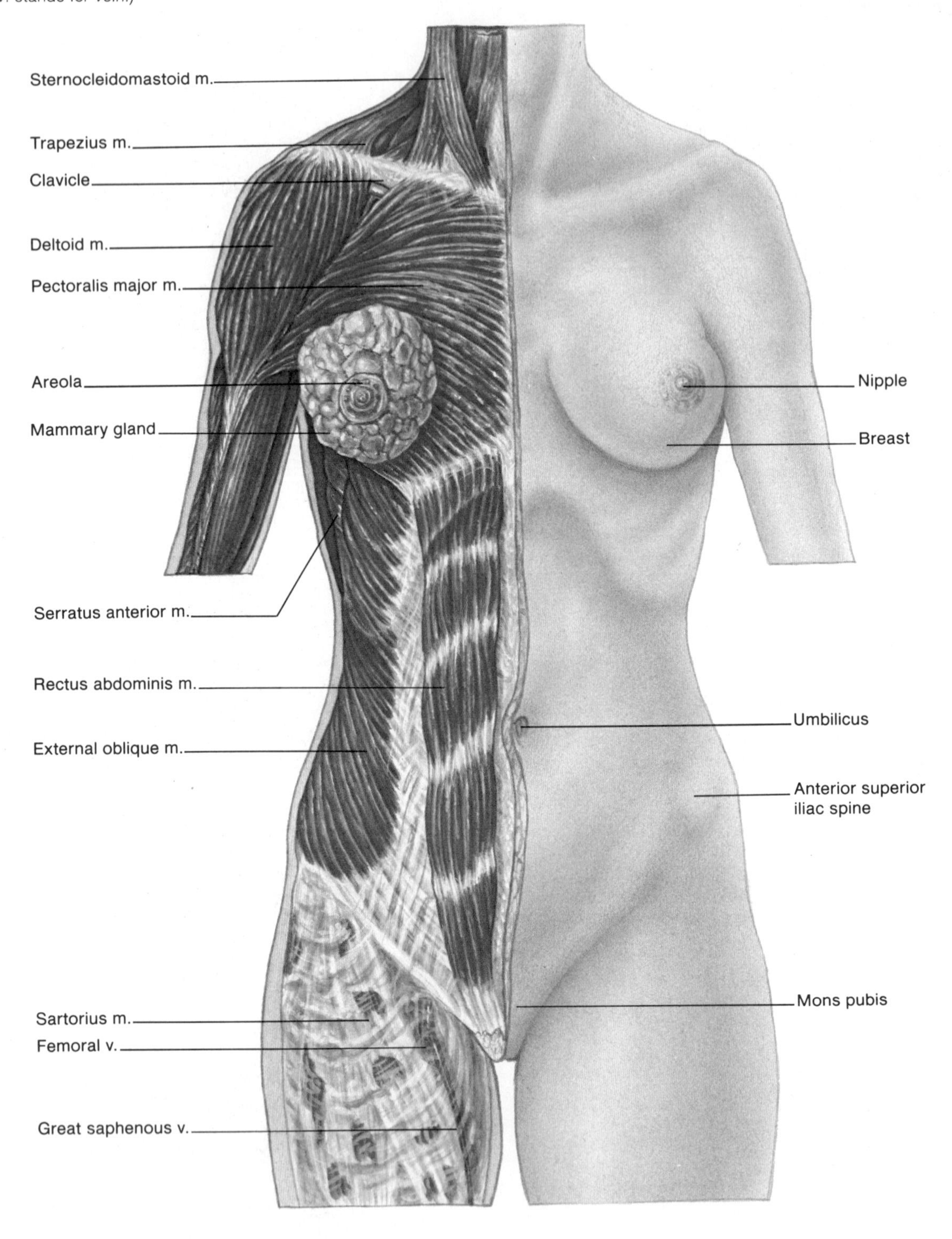

Plate 2

A human torso with the deeper muscles exposed. (*n.* stands for *nerve.*)

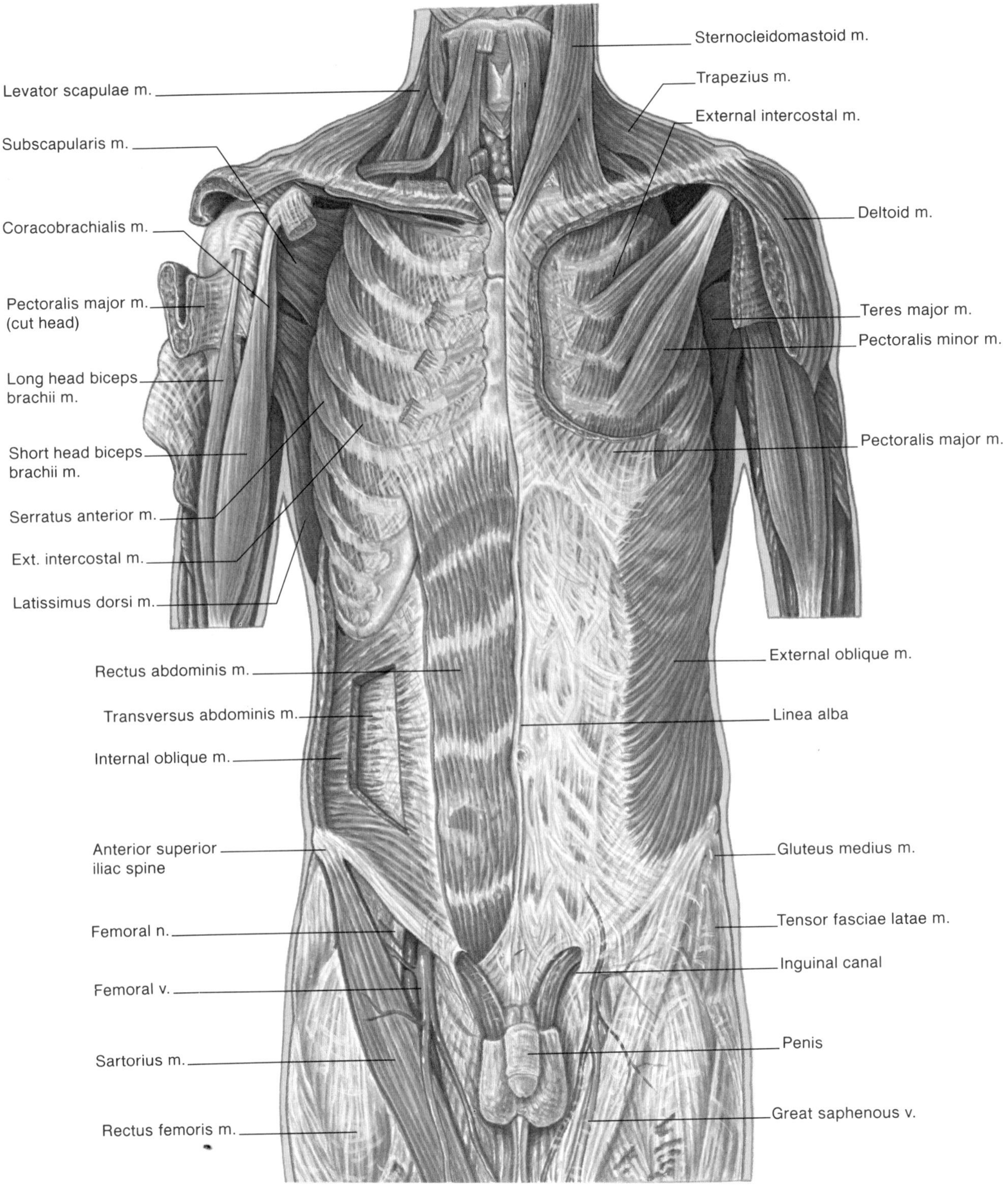

Levator scapulae m.
Subscapularis m.
Coracobrachialis m.
Pectoralis major m. (cut head)
Long head biceps brachii m.
Short head biceps brachii m.
Serratus anterior m.
Ext. intercostal m.
Latissimus dorsi m.
Rectus abdominis m.
Transversus abdominis m.
Internal oblique m.
Anterior superior iliac spine
Femoral n.
Femoral v.
Sartorius m.
Rectus femoris m.

Sternocleidomastoid m.
Trapezius m.
External intercostal m.
Deltoid m.
Teres major m.
Pectoralis minor m.
Pectoralis major m.
External oblique m.
Linea alba
Gluteus medius m.
Tensor fasciae latae m.
Inguinal canal
Penis
Great saphenous v.

Plate 3

A human torso with the deep muscles removed and the abdominal viscera exposed. (*a.* stands for *artery.*)

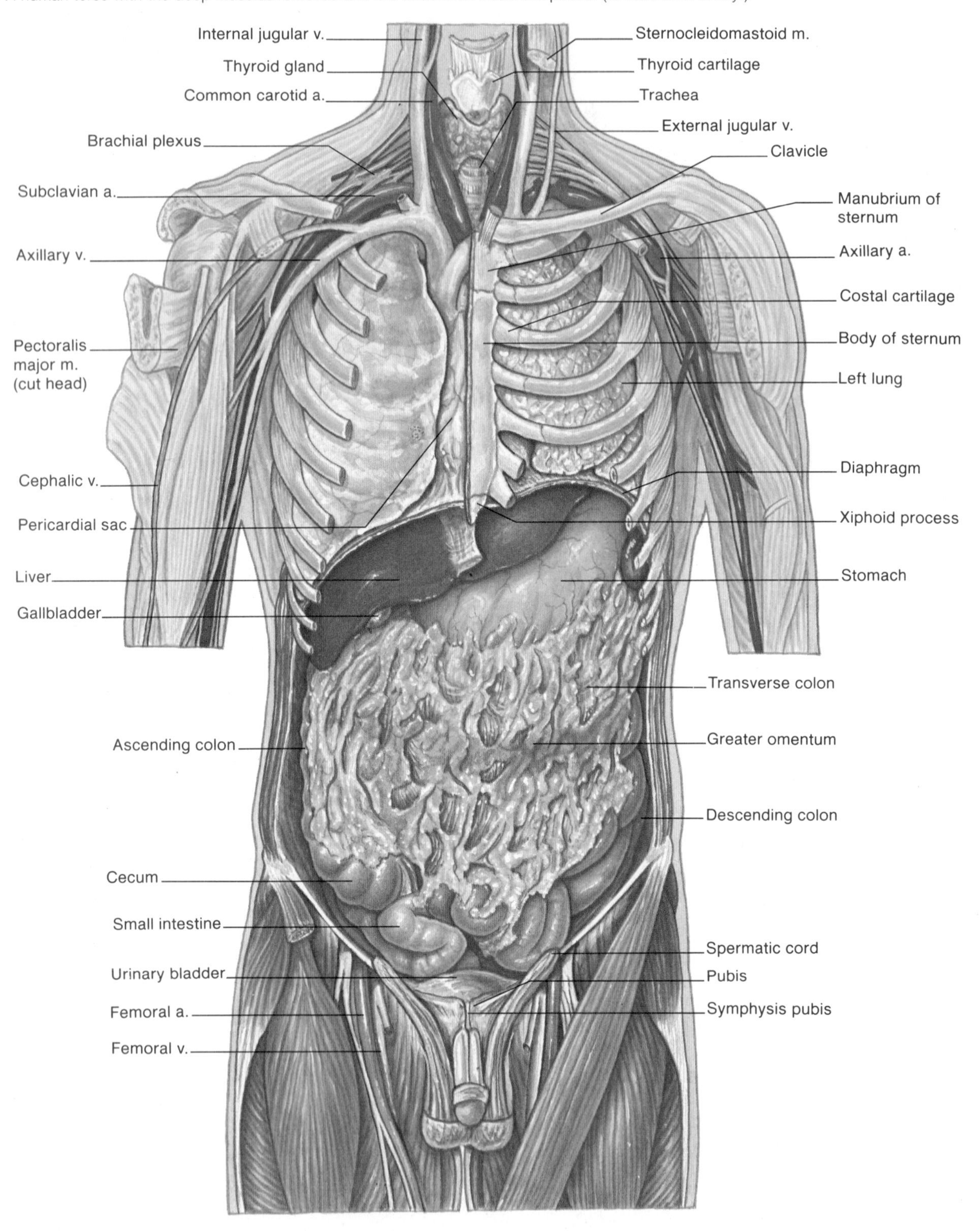

Plate 4
A human torso with the thoracic viscera exposed.

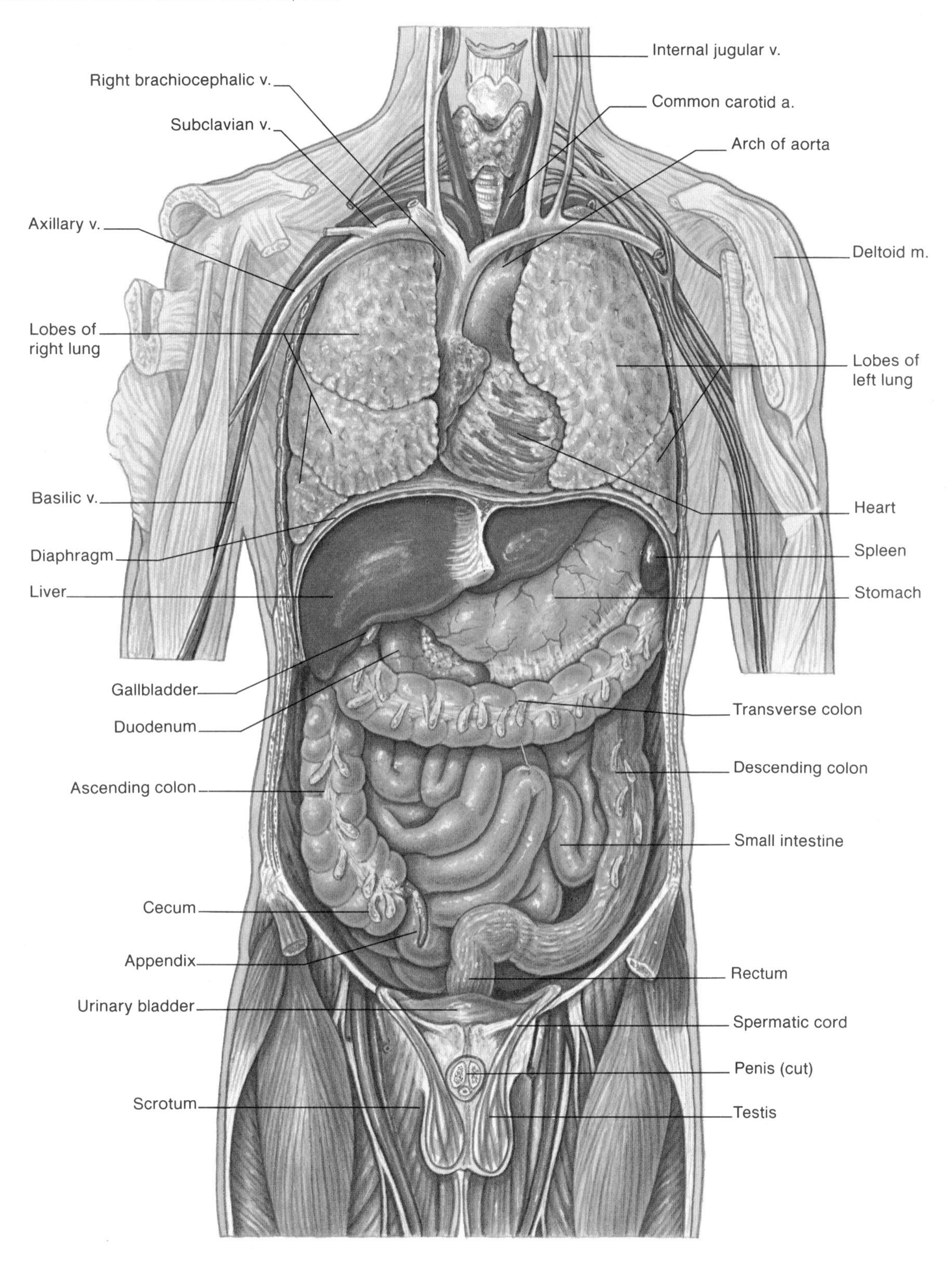

Plate 5
A human torso with the lungs, heart, and small intestine sectioned.

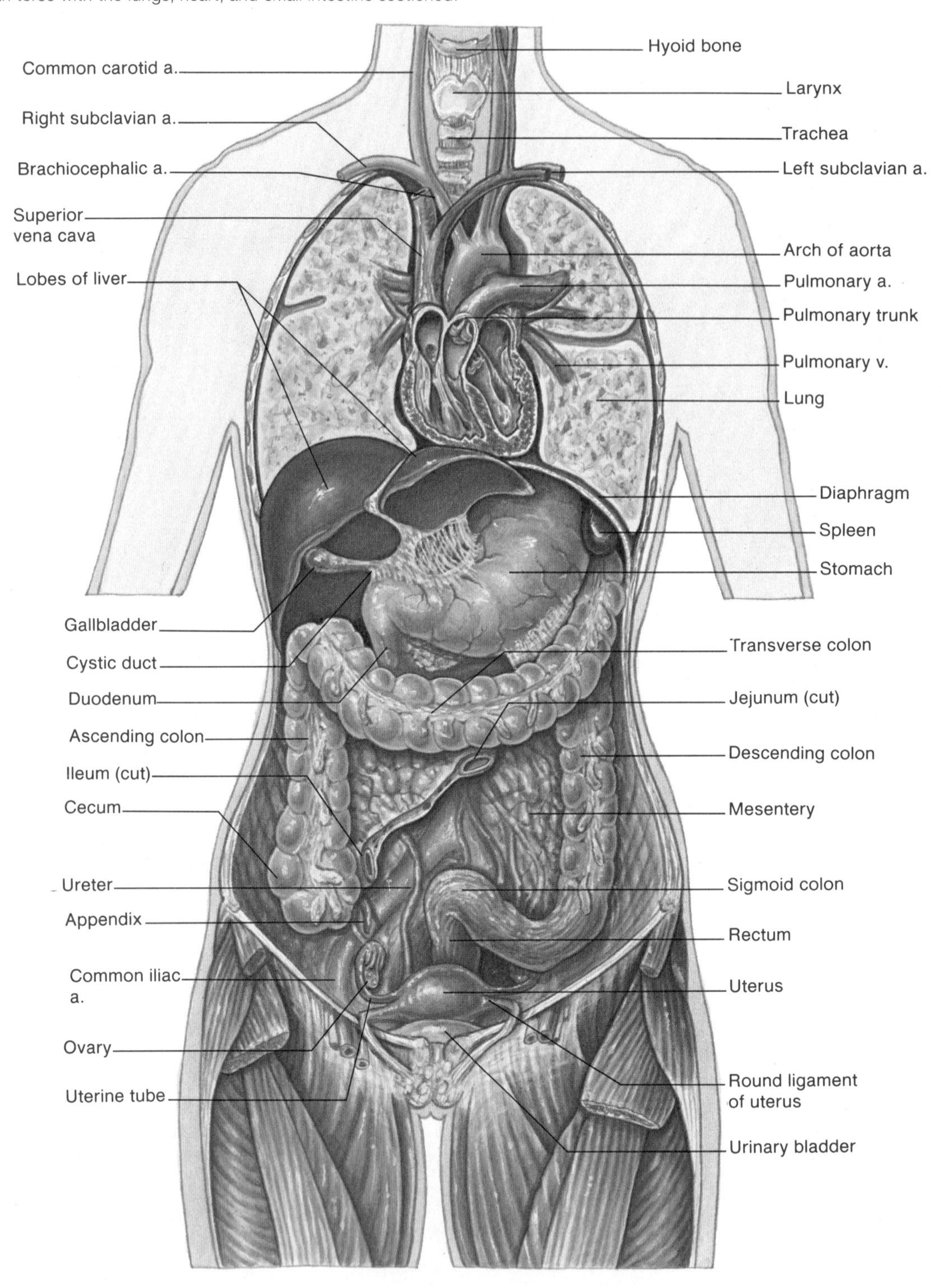

Plate 6

A human torso with the heart, stomach, and parts of the intestines and lungs removed.

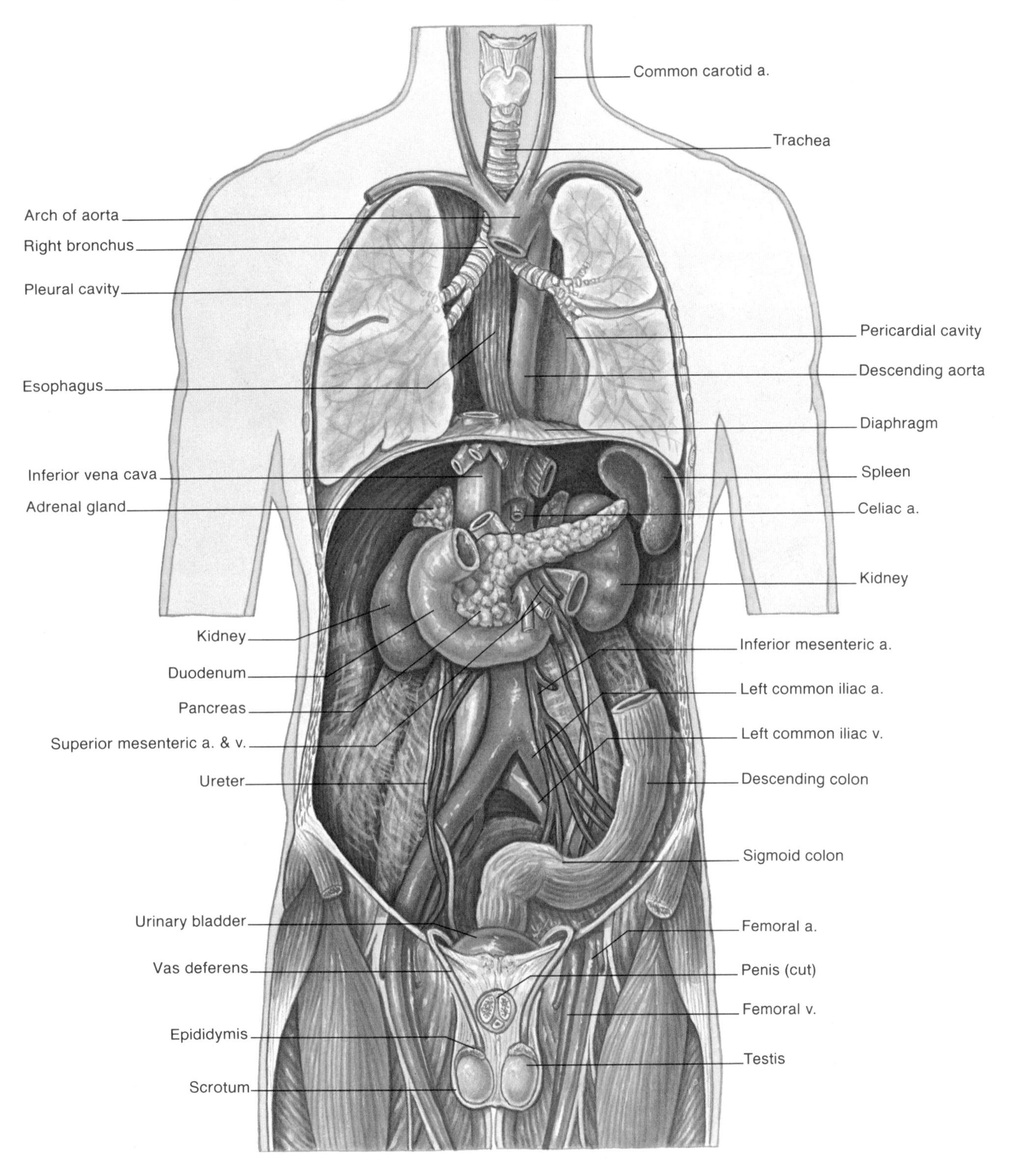

Plate 7
A human torso with the thoracic, abdominal, and pelvic visceral organs removed.

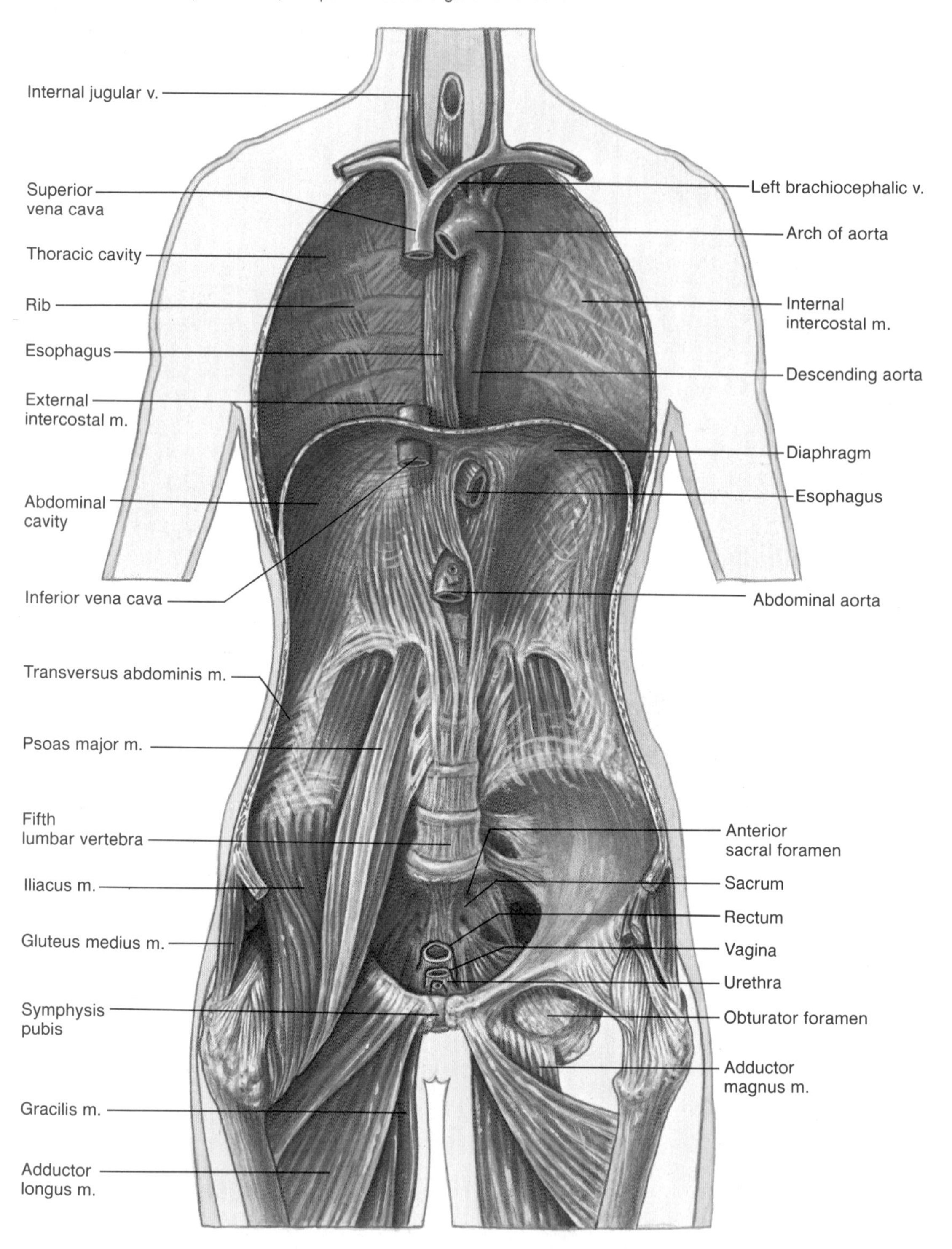

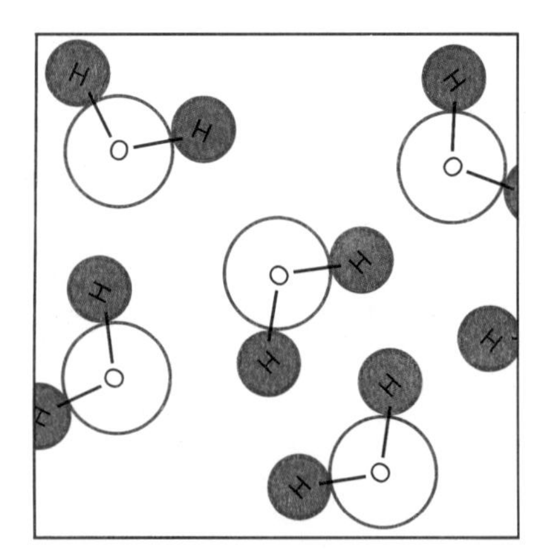

2 Chemical Basis of Life

As chapter 1 explained, an organism is composed of parts that vary in levels of organization. These parts include organ systems, organs, tissues, cells, and cellular organelles.

Cellular organelles, in turn, are composed of chemical substances made up of atoms and groups of atoms called molecules. Similarly, atoms are composed of even smaller units.

A study of living material at the more fundamental levels of organization must be concerned with the chemical substances that form the structural basis of all matter and that interact in the metabolic processes of organisms. This is the theme of chapter 2.

Chapter Objectives

After you have studied this chapter, you should be able to

1. Explain how the study of living material is dependent on the study of chemistry.
2. Describe the relationships between matter, atoms, and molecules.
3. Discuss how atomic structure is related to the ways in which atoms interact.
4. Explain how molecular and structural formulas are used to symbolize the composition of compounds.
5. Describe two types of chemical reactions.
6. Discuss the concept of pH.
7. List the major groups of inorganic substances that are common in cells.
8. Describe the general roles played in cells by various types of organic substances.
9. Complete the review activities at the end of this chapter. Note that the items are worded in the form of specific learning objectives. You may want to refer to them before reading the chapter.

Key Terms

atom (at′om)

carbohydrate (kar″bo-hi′drāt)

decomposition (de″kom-po-zish′un)

electrolyte (e-lek′tro-līt)

formula (for′mu-lah)

inorganic (in″or-gan′ik)

ion (i′on)

lipid (lip′id)

molecule (mol′ĕ-kūl)

nucleic acid (nu-kle′ik as′id)

organic (or-gan′ik)

protein (pro′te-in)

synthesis (sin′thĕ-sis)

Aids to Understanding Words

di-, two: *di*saccharide—a compound whose molecules are composed of two saccharide units bound together.

glyc-, sweet: *glyc*ogen—a complex carbohydrate composed of sugar molecules bound together.

lip-, fat: *lip*ids—a group of organic compounds that includes fats.

-lyt, dissolvable: electro*lyte*—a substance that dissolves in water and releases ions.

mono-, one: *mono*saccharide—a compound whose molecules consist of a single saccharide unit.

poly-, many: *poly*unsaturated—a molecule that has many double bonds between its carbon atoms.

sacchar-, sugar: mono*sacchar*ide—a sugar molecule composed of a single saccharide unit.

syn-, together: *syn*thesis—a process by which substances are united to form new types of substances.

Introduction

CHEMISTRY IS THE branch of science dealing with the composition of substances and the changes that take place in their composition. Although it is possible to study anatomy without much reference to this subject, a knowledge of chemistry is essential for understanding physiology, because body functions involve chemical changes that occur within cells.

Structure of Matter

Matter is anything that has weight and takes up space. This includes all of the solids, liquids, and gases in our surroundings as well as inside our bodies.

Elements and Atoms

Studies of matter have revealed that all things are composed of basic substances called **elements.** At present, 108 such elements are known, although naturally occurring matter on earth includes only 91 of them. Among these elements are such common materials as iron, copper, silver, gold, aluminum, carbon, hydrogen, and oxygen. Although some elements exist in a pure form, they occur more frequently in mixtures or united in chemical combinations.

About 20 elements are needed by living things, and of these oxygen, carbon, hydrogen, and nitrogen make up more than 95% (by weight) of the human body. A list of the major elements in the body is shown in chart 2.1. Notice that each element is represented by a one- or two-letter *symbol.*

Elements are composed of tiny, invisible particles called **atoms.** Although the atoms that make up an element are similar to each other, they differ from the atoms that make up other elements. Atoms vary in size, weight, and the ways they interact with each other. Some, for instance, are capable of combining with atoms like themselves or of combining with other kinds of atoms, while others lack these abilities.

Chart 2.1 Major elements in the human body

Major elements	Symbol	Approximate percentage of the human body (by weight)
Oxygen	O	65.0%
Carbon	C	18.5
Hydrogen	H	9.5
Nitrogen	N	3.2
Calcium	Ca	1.5
Phosphorus	P	1.0
Potassium	K	0.4
Sulfur	S	0.3
Chlorine	Cl	0.2
Sodium	Na	0.2
Magnesium	Mg	0.1
	Total	99.9%
Trace elements		
Cobalt	Co	Together less than 0.1%
Copper	Cu	
Fluorine	F	
Iodine	I	
Iron	Fe	
Manganese	Mn	
Zinc	Zn	

Atomic Structure

An atom consists of a central portion, called the **nucleus,** and one or more **electrons** that are in constant motion around the nucleus. The nucleus contains some relatively large particles, called **protons** and **neutrons** (fig. 2.1).

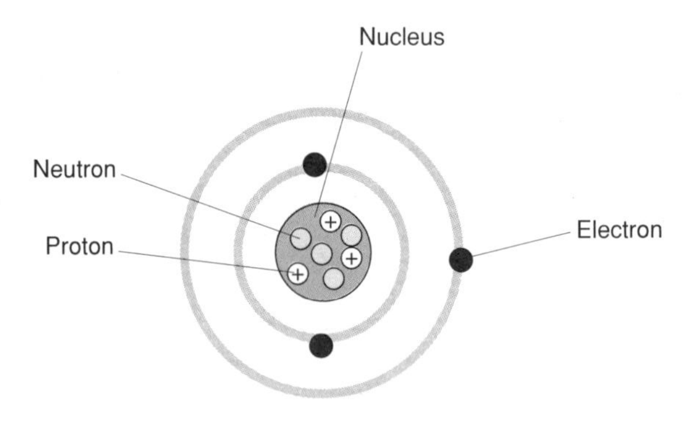

Figure 2.1
An atom of lithium includes 3 electrons in motion around a nucleus, which contains 3 protons and 4 neutrons.

Electrons, which are extremely small, carry a single, negative electrical charge (e^-), but protons each carry a single, positive electrical charge (p^+). Neutrons are uncharged and thus are electrically neutral (n^0).

Since the nucleus contains the protons, this part of an atom is always positively charged. However, the number of electrons outside the nucleus is equal to the number of protons, so a complete atom is electrically uncharged, or neutral.

The atoms of different elements contain different numbers of protons. The number of protons in the atoms of a particular element is called its *atomic number.* Hydrogen, for example, whose atoms contain 1 proton, has the atomic number 1; and carbon, whose atoms have 6 protons, has the atomic number 6.

A Current Topic

Radioactive Isotopes

All the atoms of a particular element have the same atomic number because they have the same number of protons and electrons. However, the atoms of an element vary in the number of neutrons in their nuclei; thus, they vary in atomic weight. For example, all oxygen atoms have eight protons in their nuclei. Some, however, have eight neutrons (atomic weight 16), others have nine neutrons (atomic weight 17), and still others have ten neutrons (atomic weight 18). Atoms that have the same atomic numbers but different atomic weights are called **isotopes** of an element.

The ways atoms interact with one another are due largely to the number of electrons they possess. Because the number of electrons in an atom is equal to its number of protons, all the isotopes of a particular element have the same number of electrons and react chemically in the same manner. Therefore, any of the isotopes of oxygen can play the same role in the metabolic reactions of an organism.

Although some of the isotopes of an element may be stable, others may have unstable atomic nuclei that decompose, releasing energy or pieces of themselves. Such unstable isotopes are called *radioactive,* and the energy or atomic fragments they give off are called *atomic radiations.*

Atomic radiations include three common forms called alpha (α), beta (β), and gamma (γ). Alpha radiation consists of particles from atomic nuclei that travel relatively slowly and have weak abilities to penetrate matter. Beta radiation consists of much smaller particles (electrons) that travel more rapidly and penetrate matter more deeply. Gamma radiation is similar to X-ray radiation and is the most penetrating of these forms.

Each kind of radioactive isotope produces one or more forms of radiation, and each tends to lose its radioactivity at a particular rate. The time required for an isotope to lose one-half of its radioactivity is called its *half-life.* Thus, the isotope of iodine called iodine-131, which emits one-half of its radiation in 8.1 days, has a half-life of 8.1 days. Similarly, the half-life of iron-59 is 45.1 days; that of phosphorus-32 is 14.3 days; that of cobalt-60 is 5.26 years; and that of radium-226 is 1,620 years.

Because it is possible to detect the presence of atomic radiation by using special equipment, such as a scintillation counter, radioactive substances are useful in studying life processes. A radioactive isotope, for example, can be introduced into an organism and then traced as it enters into metabolic activities. Since the human thyroid gland makes special use of iodine in its metabolism, radioactive iodine-131 can be used to study its functions and to evaluate patients with thyroid disease. Likewise, thallium-201 that has a half-life of 73.5 hours is commonly used to assess heart conditions, and gallium-67 with a half-life of 78 hours is used to detect and monitor the progress of certain cancers and inflammatory diseases.

Atomic radiations also can cause changes in the structures of various chemical substances, and, in this way, they can alter vital cellular processes. This is the reason radioactive isotopes, such as cobalt-60, are sometimes used to treat cancers. The radiation coming from the cobalt can affect the chemicals within cancerous cells and cause their deaths. Because cancer cells are more susceptible to such damage than normal cells, the cancer cells are destroyed more rapidly than the normal ones.

Chart 2.2 Atomic structure of elements 1 through 12

Element	Symbol	Atomic number	Atomic weight	Protons	Neutrons	Electrons in shells		
						First	Second	Third
Hydrogen	H	1	1	1	0	1		
Helium	He	2	4	2	2	2	(inert)	
Lithium	Li	3	7	3	4	2	1	
Beryllium	Be	4	9	4	5	2	2	
Boron	B	5	11	5	6	2	3	
Carbon	C	6	12	6	6	2	4	
Nitrogen	N	7	14	7	7	2	5	
Oxygen	O	8	16	8	8	2	6	
Fluorine	F	9	19	9	10	2	7	
Neon	Ne	10	20	10	10	2	8	(inert)
Sodium	Na	11	23	11	12	2	8	1
Magnesium	Mg	12	24	12	12	2	8	2

The weight of an atom of an element is due primarily to the protons and neutrons in its nucleus, because the electrons have so little weight. For this reason, an atom of carbon with six protons and six neutrons weighs about twelve times more than an atom of hydrogen, which has only one proton and no neutrons.

The number of protons plus the number of neutrons in each of its atoms is approximately equal to the *atomic weight* of an element. Thus, the atomic weight of hydrogen is 1, and the atomic weight of carbon is 12 (chart 2.2).

Figure 2.2
The single electron of a hydrogen atom is located in its first shell; the 2 electrons of a helium atom fill its first shell; and the 3 electrons of a lithium atom are arranged with 2 in the first shell and 1 in the second shell.

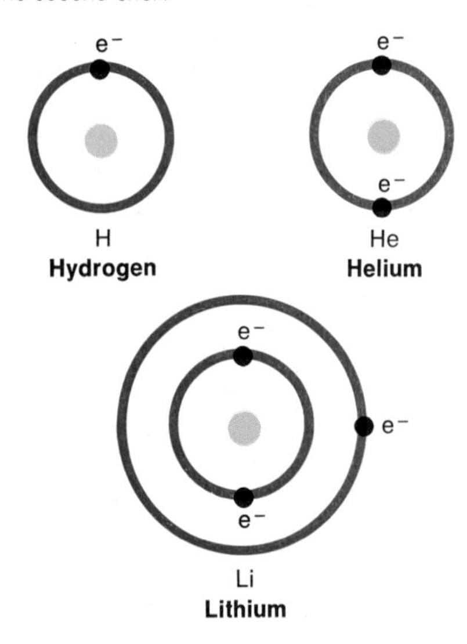

Figure 2.3
A diagram of a sodium atom.

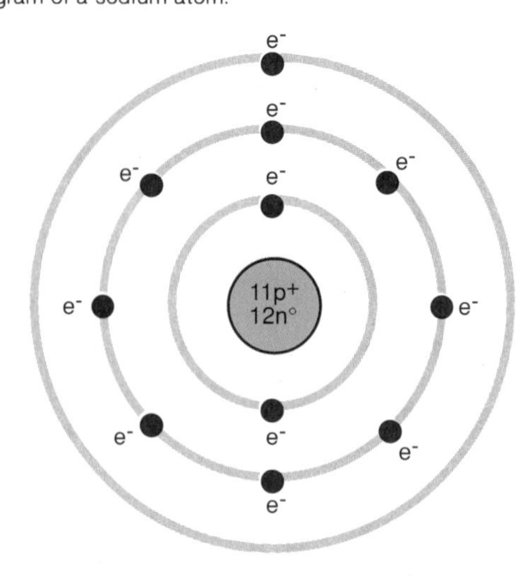

Na atom contains
11 electrons (e⁻)
11 protons (p⁺)
12 neutrons (n°)

Atomic number = 11

1. What is the relationship between matter and the elements?
2. What elements are most common in the human body?
3. How are electrons, protons, and neutrons positioned within an atom?
4. What is the difference between atomic number and atomic weight?

Bonding of Atoms

When atoms combine with other atoms, they either gain or lose electrons, or share electrons with other atoms.

The electrons of an atom are arranged in one or more *shells* around the nucleus (chart 2.2). The maximum number of electrons that each of the first three inner shells can hold is as follows:

First shell (closest to the nucleus)	2 electrons
Second shell	8 electrons
Third shell	8 electrons

(The third shell can hold 8 electrons for elements up to atomic number 20. More complex atoms may have as many as 18 electrons in the third shell.)

The arrangements of electrons within the shells of atoms can be represented by simplified diagrams such as those in figure 2.2.

Atoms, such as helium, whose outermost electron shells are filled have stable structures and are chemically inactive (inert). Atoms with incompletely filled outer shells, such as those of hydrogen or lithium, tend to gain, lose, or share electrons in ways that empty or fill their outer shells. In this way they achieve stable structures.

An atom of sodium, for example, has 11 electrons (fig. 2.3). This atom tends to lose the single electron in its outer shell, which leaves the second shell filled and the form stable.

A chlorine atom has 17 electrons arranged with 2 in the first shell, 8 in the second shell, and 7 in the third shell. An atom of this type will tend to accept a single electron, thus filling its outer shell and achieving a stable form (fig. 2.4).

Since each sodium atom tends to lose a single electron and each chlorine atom tends to accept a single electron, sodium and chlorine atoms will react together. During this reaction, a sodium atom loses an electron and is left with 11 protons (11+) in its nucleus and only 10 electrons (10−). As a result, the atom develops a net electrical charge of 1+ and is symbolized Na^+. At the same time, a chlorine atom gains an electron, which leaves it with 17 protons (17+) in its nucleus and 18 electrons (18−). Thus it develops a net electrical charge of 1− and is symbolized Cl^-.

Atoms that have become electrically charged by gaining or losing electrons are called **ions,** and two ions with opposite electrical charges are attracted to one another electrically. When this happens, a chemical bond, called an *electrovalent bond* (ionic bond), is created between them, as shown in figure 2.4. When sodium ions (Na^+) and chloride ions (Cl^-) unite in this manner, the compound sodium chloride (NaCl, common salt) is formed.

Figure 2.4
If a sodium atom loses an electron to a chlorine atom, the sodium atom becomes a sodium ion and the chlorine atom becomes a chloride ion. These oppositely charged particles are attracted electrically to one another and become united by an electrovalent bond.

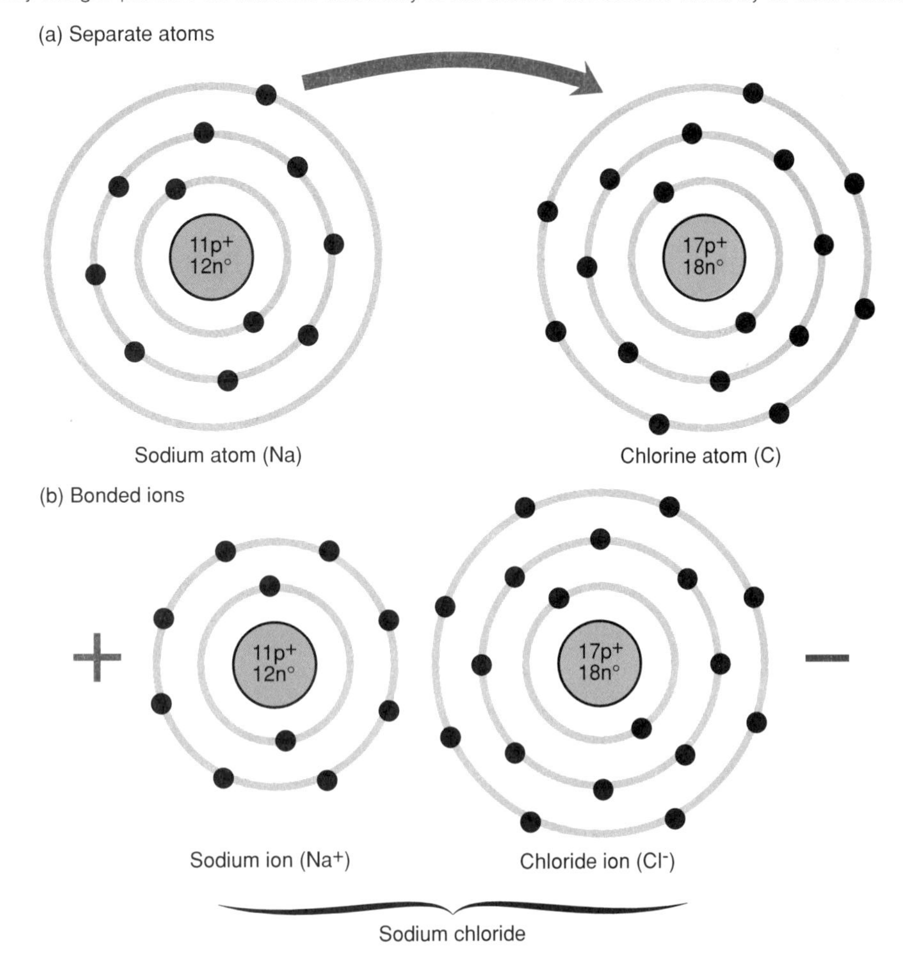

Figure 2.5
A hydrogen molecule is formed when 2 hydrogen atoms share a pair of electrons and are united by a covalent bond.

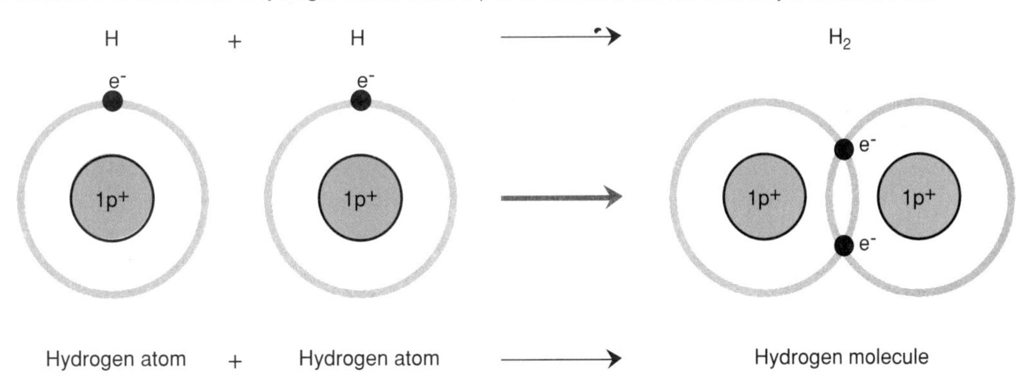

Atoms may also bond together by sharing electrons rather than by exchanging them. A hydrogen atom, for example, has 1 electron in its first shell, but needs 2 to achieve a stable structure (fig. 2.5). It may fill this shell by combining with another hydrogen atom in such a way that the 2 atoms share a pair of electrons. The 2 electrons then encircle the nuclei of both atoms so that each achieves a stable form. In this case, the chemical bond between the atoms is called a *covalent bond.*

Carbon atoms, with 2 electrons in their first shells and 4 electrons in their second shells, form covalent

Figure 2.6
Hydrogen molecules combine with oxygen molecules to form water molecules.

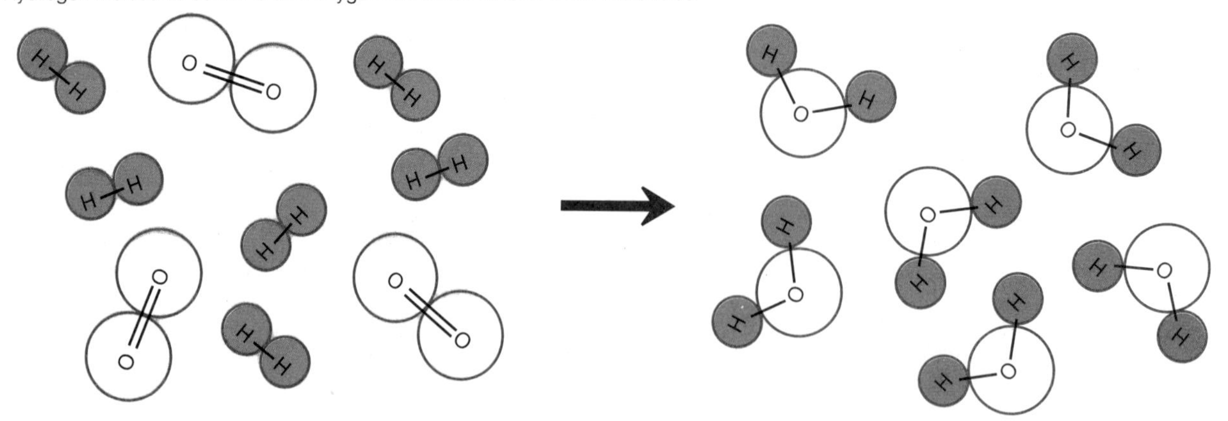

bonds when they unite with other atoms. In fact, carbon atoms may bond to other carbon atoms in such a way that 2 atoms share one or more pairs of electrons. If one pair of electrons is shared, the resulting bond is called a *single covalent bond;* if two pairs of electrons are shared, the bond is called a *double covalent bond.*

Molecules and Compounds

When two or more atoms bond together, they form a new kind of particle, called a **molecule.** If atoms of the same element combine, they produce molecules of that element. The gases of hydrogen, oxygen, and nitrogen contain such molecules (fig. 2.5).

When atoms of different elements combine, molecules of substances called **compounds** are formed. Two atoms of hydrogen, for example, can combine with one atom of oxygen to produce a molecule of the compound water (fig. 2.6). Sugar, baking soda, natural gas, alcohol, and most drugs are examples of compounds.

A molecule of a compound always contains definite kinds and numbers of atoms. A molecule of water, for instance, always contains 2 hydrogen atoms and 1 oxygen atom. If 2 hydrogen atoms combine with 2 oxygen atoms, the compound formed is not water, but hydrogen peroxide.

Chart 2.3 lists some particles of matter and their characteristics.

Chart 2.3 Some particles of matter

Name	Characteristic
Atom	Smallest particle of an element that has the properties of that element
Electron (e^-)	Extremely small particle; carries a negative electrical charge and is in constant motion around a nucleus
Proton (p^+)	Relatively large particle; carries a positive electrical charge and is found within a nucleus
Neutron (n^0)	Relatively large particle; uncharged and thus electrically neutral; found within a nucleus
Ion	Atom that is electrically charged because it has gained or lost 1 or more electrons
Molecule	Particle formed by the chemical union of 2 or more atoms

1. What is an ion?
2. Describe two ways that atoms may combine with other atoms.
3. Distinguish between a molecule and a compound.

Formulas

The numbers and kinds of atoms in a molecule can be represented by a *molecular formula.* Such a formula consists of the symbols of the elements in the molecule together with numbers to indicate how many atoms of each element are present. For example, the molecular formula for water is H_2O, which means there are 2 atoms of hydrogen and 1 atom of oxygen in each molecule (fig. 2.7). The molecular formula for a sugar called glucose is $C_6H_{12}O_6$, which means there are 6 atoms of carbon, 12 atoms of hydrogen, and 6 atoms of oxygen in a molecule.

Usually the atoms of each element will form a specific number of bonds—hydrogen atoms form single bonds, oxygen atoms form 2 bonds, nitrogen atoms form 3 bonds, and carbon atoms form 4 bonds. The bonding capacity of these atoms can be represented by using symbols and lines as follows:

$$-\mathrm{H} \quad -\mathrm{O}- \quad \overset{\diagdown\;\diagup}{\underset{|}{\mathrm{N}}} \quad -\overset{|}{\underset{|}{\mathrm{C}}}-$$

Figure 2.7
Structural formulas for the molecules of hydrogen, oxygen, water, and carbon dioxide.

H—H	O=O	H–O–H	O=C=O
H_2	O_2	H_2O	CO_2

These representations can be used to show how atoms are joined and arranged in various molecules. Illustrations of this type are called *structural formulas* (fig. 2.7). Three-dimensional models that reflect the formulas can be built of colored parts representing different kinds of atoms (fig. 2.8).

Chemical Reactions

When chemical reactions occur, bonds between atoms, ions, or molecules are formed or broken. As a result, new chemical combinations are created. For example, when two or more atoms (reactants) bond together to form a more complex structure (end product), as when atoms of hydrogen and oxygen bond together to form molecules of water, the reaction is called **synthesis.** Such a reaction is symbolized in this way:

$$A + B \rightarrow AB$$

If the bonds within a reactant molecule break so that simpler molecules, atoms, or ions form, the reaction is called **decomposition.** Thus, molecules of water can decompose to yield the products hydrogen and oxygen. Decomposition is symbolized as follows:

$$AB \rightarrow A + B$$

Synthetic reactions are particularly important in the growth of body parts and the repair of worn or damaged tissues, which involve the buildup of larger molecules from smaller ones. When food substances are digested or energy is released from food substances, the processes are largely decomposition reactions.

Many chemical reactions are *reversible.* This means the end product (or products) of the reaction can change back to the reactant (or reactants) that originally underwent the reaction. A **reversible reaction** can be symbolized using a double arrow, as follows:

$$A + B \rightleftharpoons AB$$

Whether a reversible reaction proceeds in one direction or another depends on such factors as the relative proportions of the reactant (or reactants) and product (or products), the amount of energy available to the reaction, and the presence or absence of *catalysts*—particular atoms or molecules that can change the rate of a reaction without themselves being consumed by the reaction.

Figure 2.8
(*a*) A water molecule (H_2O) can be represented by a three-dimensional model, in which the white parts represent hydrogen atoms and the red part represents oxygen; (*b*) in this model of a glucose molecule ($C_6H_{12}O_6$), the black parts represent carbon atoms.

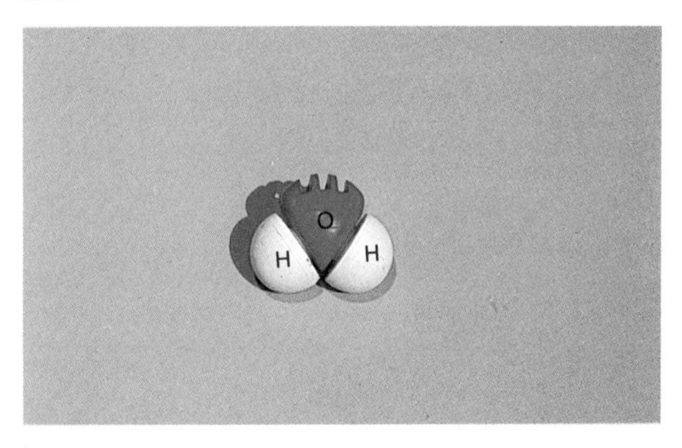

(a)

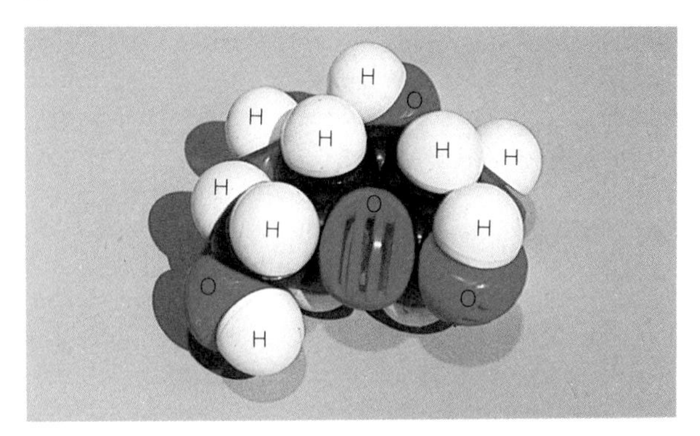

(b)

Acids and Bases

Some compounds release ions when they dissolve in water or react with water molecules. Sodium chloride (NaCl), for example, releases sodium ions (Na^+) and chloride ions (Cl^-) when it dissolves, as is represented by the following:

$$NaCl \rightarrow Na^+ + Cl^-$$

Since the resulting solution contains electrically charged particles (ions), it will conduct an electric current. Substances that release ions in water are, therefore, known as **electrolytes.** Electrolytes that release hydrogen ions (H^+) in water are called **acids.** For example, in water the compound hydrochloric acid (HCl) releases hydrogen ions and chloride ions.

$$HCl \rightarrow H^+ + Cl^-$$

Electrolytes that release ions which combine with hydrogen ions are called **bases.** The compound sodium hydroxide (NaOH), for example, releases hydroxyl ions (OH^-) when placed in water.

$$NaOH \rightarrow Na^+ + OH^-$$

Figure 2.9

As the concentration of hydrogen ions (H^+) increases, a solution becomes more acidic, and the pH value decreases. As the concentration of ions that combine with hydrogen ions (such as hydroxyl ions) increases, a solution becomes more basic, and the pH value increases.

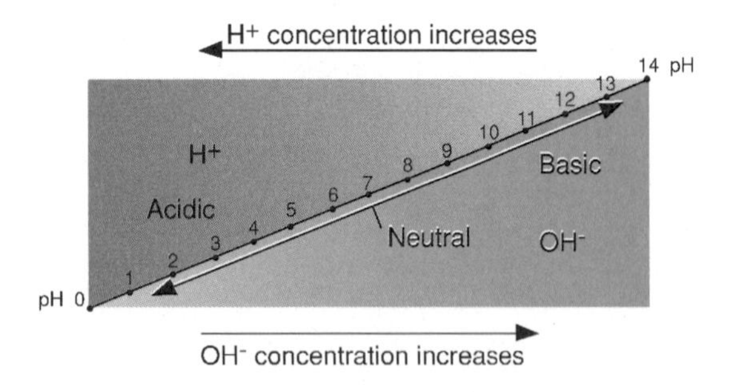

The hydroxyl ions, in turn, can combine with hydrogen ions to form water; thus, sodium hydroxide is a base.

The chemical reactions involved with many life processes, such as those functioning to regulate blood pressure and breathing rate, are affected by the presence of hydrogen and hydroxyl ions; therefore the concentrations of these ions in body fluids are of special importance. Such concentrations are measured in units of **pH.**

The pH scale includes values from 0 to 14 (fig. 2.9). A solution with a pH of 7.0, the midpoint of the scale, contains equal numbers of hydrogen and hydroxyl ions and is said to be *neutral.* A solution that contains more hydrogen than hydroxyl ions has a pH less than 7.0 and is *acidic;* whereas, one with fewer hydrogen than hydroxyl ions has a pH above 7.0 and is *basic* (alkaline). Figure 2.10 indicates the pH values of some common substances.

There is a tenfold difference in the hydrogen ion concentration between each whole number on the pH scale. For example, a solution of pH 4.0 contains 0.0001 grams (gm) of hydrogen ions per liter; a solution of pH 3.0 contains 0.001 grams of hydrogen ions per liter, which is ten times as much as 0.0001 gm/l ($0.0001 \times 10 = 0.001$).

Note in figure 2.10 that the pH of human blood is normally 7.4. If this pH value drops below 7.4, the person is said to have *acidosis;* if it rises above 7.4, the condition is called *alkalosis.* Without medical intervention a person usually cannot survive if the pH drops to 6.8 or rises to 8.0 for more than a few hours.

The regulation of the hydrogen ion concentration in body fluids is discussed in chapter 18.

Figure 2.10

Approximate pH values of some common substances.

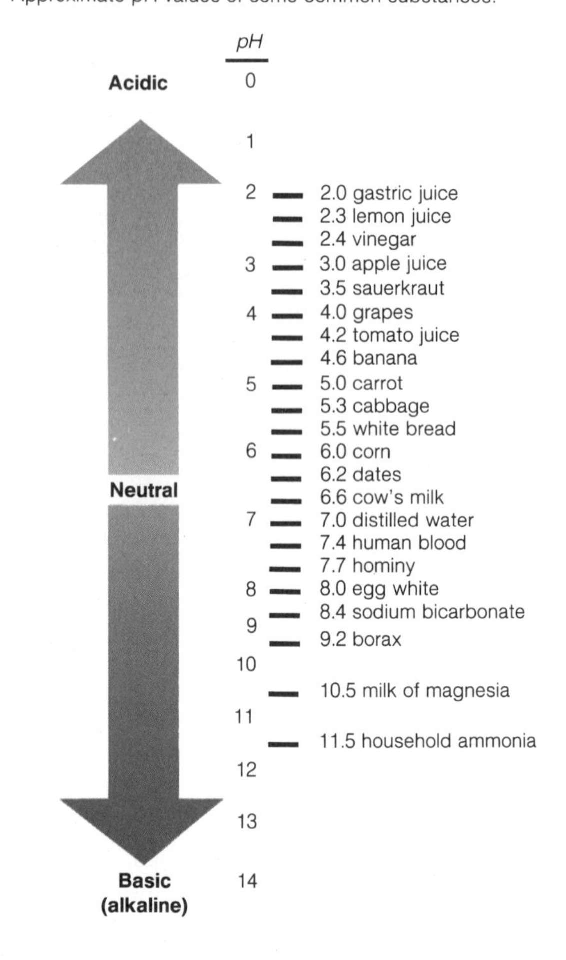

1. What is meant by a molecular formula? A structural formula?
2. Describe two kinds of chemical reactions.
3. Compare the characteristics of an acid with those of a base.
4. What is meant by pH?

Chemical Constituents of Cells

Chemicals that enter into metabolic reactions or are produced by them can be divided into two large groups. Except for a few simple molecules, those that lack carbon atoms are called **inorganic.** Those molecules that contain both carbon and hydrogen atoms are called **organic.**

Generally, inorganic substances will dissolve in water or react with water to release ions; thus they are *electrolytes.* Some organic compounds will dissolve in water also, but they are more likely to dissolve in organic liquids like ether or alcohol. Those that will dissolve in water usually do not release ions and are, therefore, called *nonelectrolytes.*

Inorganic Substances

Among the inorganic substances common in cells are water, oxygen, carbon dioxide, and a group of compounds called inorganic salts.

Water

Water (H_2O) is the most abundant compound in living material and is responsible for about two-thirds of the weight of an adult human. It is the major ingredient of the blood and other body fluids, including those within cells.

When substances dissolve in water, relatively large pieces of the materials break into smaller ones, and eventually molecular-sized particles or ions result. These tiny particles are much more likely to react with one another than were the original large pieces. Consequently, most metabolic reactions occur in water.

Water also plays an important role in the transportation of chemicals within the body. The aqueous portion of the blood, for example, carries many vital substances, such as oxygen, sugars, salts, and vitamins, from the organs of digestion and respiration to the body cells.

In addition, water can absorb and transport heat. Thus, the heat released from muscle cells during exercise can be carried by the blood from deeper parts to the surface, where it may be lost to the outside.

Oxygen

Molecules of oxygen (O_2) enter the body through the respiratory organs and are transported throughout the body by the blood and specialized blood cells (red blood cells). The oxygen is used by cellular organelles in the process of releasing energy from glucose and certain other molecules. The released energy is used to drive the cell's metabolic activities.

Carbon Dioxide

Carbon dioxide (CO_2) is a simple carbon-containing compound of the inorganic group. It is produced as a waste product when energy is released by certain metabolic processes and is given off into the air within the lungs.

Inorganic Salts

Inorganic salts are abundant in body parts and fluids. They are the sources of many necessary ions, including ions of sodium (Na^+), chlorine (Cl^-), potassium (K^+), calcium (Ca^{+2}), magnesium (Mg^{+2}), phosphate (PO_4^{-3}), carbonate (CO_3^{-2}), bicarbonate (HCO_3^-), and sulfate (SO_4^{-2}). These ions play important roles in metabolic processes, including the transport of substances into and out of cells, the contraction of muscles, and the conduction of nerve impulses.

Figure 2.11
(*a*) Some glucose molecules ($C_6H_{12}O_6$) have a straight chain of carbon atoms; (*b*) more commonly, glucose molecules form a ring structure, which can be symbolized by this shape:

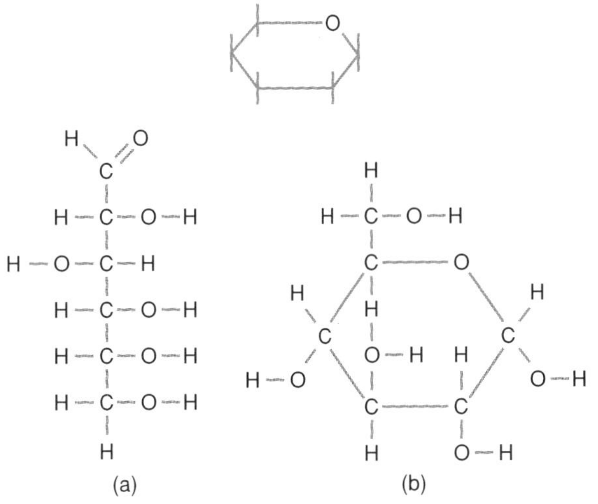

(a) (b)

1. What is the difference between an inorganic molecule and an organic molecule?
2. What is the difference between an electrolyte and a nonelectrolyte?
3. Name the inorganic substances common in body fluids.

Organic Substances

Important groups of organic substances found in cells include carbohydrates, lipids, proteins, and nucleic acids.

Carbohydrates

Carbohydrates provide much of the energy that cells require. They also supply materials to build certain cell structures and often are stored as reserve energy supplies.

Carbohydrate molecules contain atoms of carbon, hydrogen, and oxygen. These molecules usually have twice as many hydrogen as oxygen atoms, the same ratio of hydrogen to oxygen that occurs in water molecules (H_2O). This ratio is easy to see in the molecular formulas of the carbohydrates glucose ($C_6H_{12}O_6$) and sucrose ($C_{12}H_{22}O_{11}$).

The carbon atoms of carbohydrate molecules are joined in chains whose lengths vary with the kind of carbohydrate. For example, those with shorter chains are called **sugars.**

Sugars with 6-carbon atoms are known as *simple sugars,* or **monosaccharides,** and they are the building blocks of more complex carbohydrate molecules. The simple sugars include glucose, fructose, and galactose. Figure 2.11 illustrates the structural formula of glucose.

Figure 2.12

(*a*) A monosaccharide molecule consists of one 6-carbon atom building block; (*b*) a disaccharide molecule consists of two building blocks; (*c*) a polysaccharide molecule consists of many building blocks.

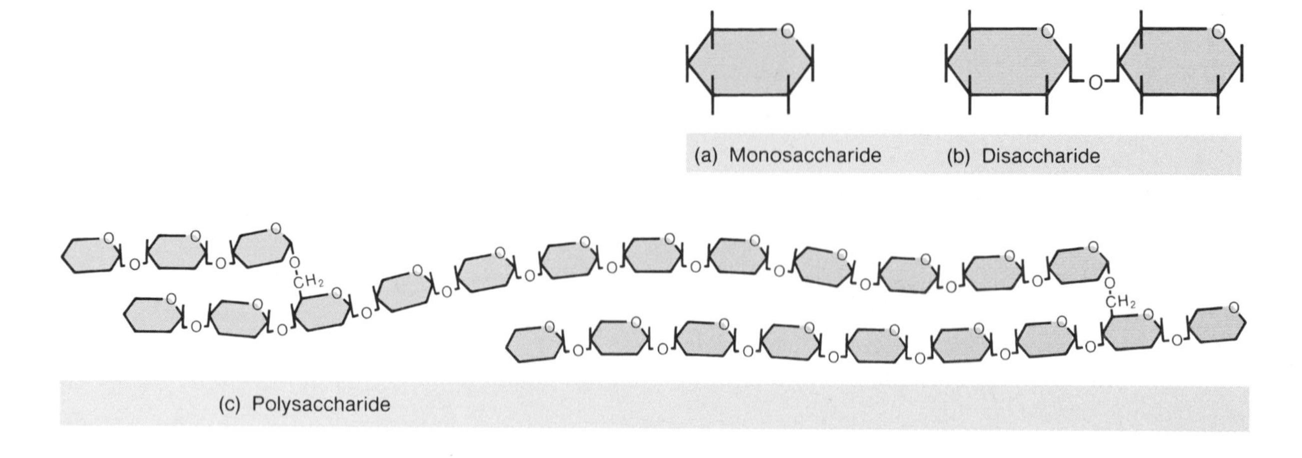

Figure 2.13

A triglyceride molecule consists of a glycerol portion and 3 fatty acid portions.

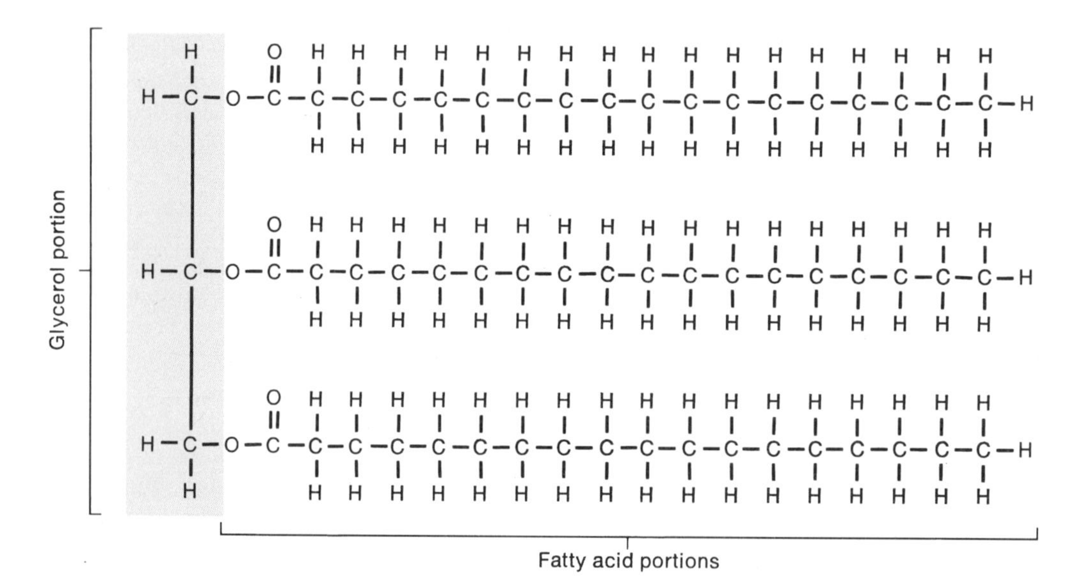

In complex carbohydrates, a number of simple sugar molecules are bound together to form molecules of varying sizes (fig. 2.12). Some, like sucrose (table sugar) and lactose (milk sugar), are *double sugars,* or **disaccharides,** whose molecules each contain two simple sugar building blocks. Others are made up of many simple sugar units joined together to form **polysaccharides,** such as plant starch.

Animals, including humans, synthesize a polysaccharide similar to starch, called *glycogen.*

Lipids

Lipids represent a group of organic substances that are insoluble in water, but are soluble in certain organic solvents, such as ether and chloroform. Lipids include a variety of compounds—fats, phospholipids, and steroids—that have vital functions in cells. The most common members of the group, however, are fats.

Fats are used to build cell parts and to supply energy for cellular activities. In fact, fat molecules can supply more energy, gram for gram, than carbohydrate molecules.

Like carbohydrates, fat molecules are composed of carbon, hydrogen, and oxygen atoms. They contain, however, a much smaller proportion of oxygen than do carbohydrates. This is illustrated by the formula for the fat, tristearin, $C_{57}H_{110}O_6$.

The building blocks of fat molecules are **fatty acids** and **glycerol.** These smaller molecules are united so that each glycerol molecule is combined with 3 fatty acid molecules. The result is a single fat, or *triglyceride,* molecule (fig. 2.13).

Chart 2.4 Important groups of lipids		
Group	**Basic molecular structure**	**Characteristics**
Triglycerides	Three fatty acid molecules bound to a glycerol molecule	Most common lipid in the body; stored in fat tissue as an energy supply. Fat tissue also provides insulation beneath the skin
Phospholipids	Two fatty acid molecules and a phosphate group bound to a glycerol molecule (may also include a nitrogen-containing molecule attached to the phosphate group)	Used as structural components in cell membranes; large amounts occur in the liver and parts of the nervous system
Steroids	Four interconnected rings of carbon atoms	Widely distributed in the body with a variety of functions; include cholesterol, hormones of the adrenal cortex, sex hormones, bile acids, and vitamin D

Although the glycerol portion of every fat molecule is the same, there are many kinds of fatty acids and, therefore, many kinds of fats. Fatty acid molecules differ in the lengths of their carbon atom chains, although such chains always contain an even number of carbon atoms. The chains also may vary in the way the carbon atoms are combined. In some cases, the carbon atoms are all joined by single carbon–carbon bonds. This type of fatty acid is said to be *saturated;* that is, each carbon atom is bound to as many hydrogen atoms as possible and is thus saturated with hydrogen atoms. Other fatty acid chains have one or more double bonds between carbon atoms. Those that contain such double bonds are said to be *unsaturated,* and fatty acid molecules with many double-bonded carbon atoms are called *polyunsaturated.* Similarly, fat molecules that contain only saturated fatty acids are called *saturated fats,* and those that include unsaturated fatty acids are called *unsaturated fats.*

A diet containing a high proportion of saturated fats seems to increase the chance of developing a serious disease (atherosclerosis) in which certain blood vessels (arteries) become obstructed. For this reason, many nutritionists recommend that polyunsaturated fats be substituted for some dietary saturated fats.

As a rule, saturated fats are more abundant in fatty foods that are solids at room temperature, such as butter, lard, and most other animal fats. Unsaturated fats, on the other hand, are likely to be plentiful in fatty foods that are liquids at room temperature, such as soft margarine and various seed oils, including corn, cottonseed, safflower, sesame, soybean, and sunflower oils. There are exceptions, however, because coconut and palm kernel oils are high in saturated fats.

Chart 2.4 lists three important groups of lipids and their characteristics.

Proteins

Proteins serve as structural materials, energy sources, hormones, and receptors on cell surfaces that are specialized to bond to particular kinds of molecules. Other proteins (antibodies) act against foreign substances that gain entrance into the body. Still others play vital roles in metabolic processes by acting as **enzymes.** Enzymes are proteins that serve as biological catalysts. That is, they have special properties and can speed up specific chemical reactions without being changed or used up themselves. (Enzymes are discussed in more detail in chapter 4.)

Like carbohydrates and lipids, proteins contain atoms of carbon, hydrogen, and oxygen. In addition, they always contain nitrogen atoms and sometimes contain sulfur atoms as well. The building blocks of proteins are smaller molecules called **amino acids** (fig. 2.14). About 20 different kinds of amino acids occur commonly in proteins. Within the protein molecules, the amino acid molecules are joined together in chains, varying in length from less than 100 to more than 50,000 amino acids (fig. 2.15).

Each type of protein molecule contains specific numbers and kinds of amino acids, arranged in a particular linear sequence. The amino acid chain is twisted to form a coil, and it, in turn, is folded up into a unique three-dimensional shape. Consequently, different kinds of protein molecules have different shapes, and their shapes are related to particular functions (fig. 2.16).

The special shapes of protein molecules can be altered by having bonds broken as a result of exposure to excessive heat, radiation, electricity, or various chemicals. When this occurs, the molecules lose their unique shapes, become disorganized, and are said to be *denatured.* At the same time, the proteins lose their special properties. For example, the protein in egg white (albumin) is denatured when it is heated. This treatment causes the egg white to change from a liquid to a solid—a change that cannot be reversed. Similarly, if cellular proteins are denatured, they are permanently changed and become nonfunctional.

Nucleic Acids

The most fundamental of the compounds in cells are the **nucleic acids.** These substances control cellular activities in very special ways.

Figure 2.14
Some representative amino acids and their structural formulas.

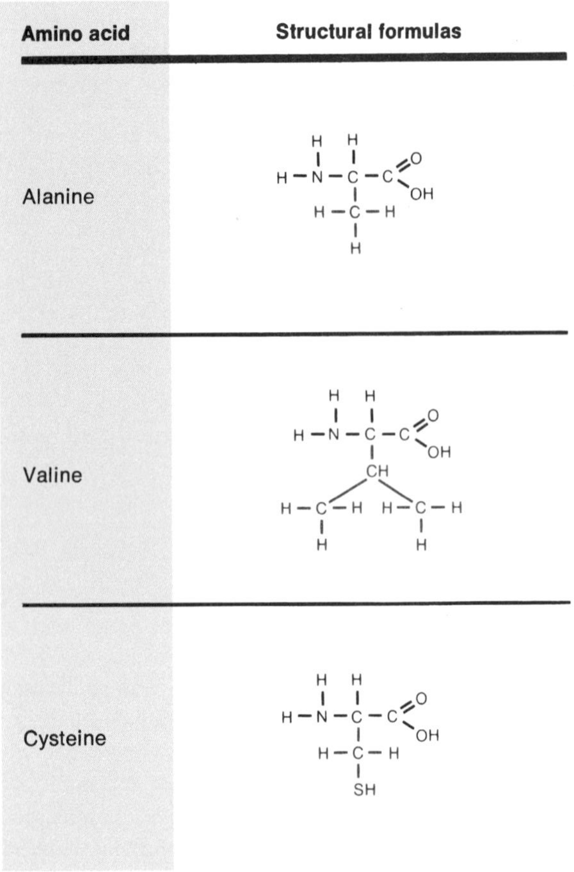

Amino acid	Structural formulas
Alanine	
Valine	
Cysteine	

Nucleic acid molecules are generally very large and complex. They contain atoms of carbon, hydrogen, oxygen, nitrogen, and phosphorus, which are bound into building blocks called **nucleotides.** Each nucleotide consists of a 5-carbon *sugar* (ribose or deoxyribose), a *phosphate group,* and one of several *organic bases* (fig. 2.17). A nucleic acid molecule consists of many nucleotides united in a chain (fig. 2.18).

There are two major types of nucleic acids. One type, composed of molecules whose nucleotides contain ribose sugar, is called **RNA** (ribonucleic acid). The nucleotides of the second type contain deoxyribose sugar, and this type is called **DNA** (deoxyribonucleic acid).

DNA molecules store information in a kind of molecular code. This information is used by cells in the process of constructing specific protein molecules, and these proteins, acting as enzymes, control all of the metabolic reactions that occur in cells. RNA molecules help to synthesize proteins. (Nucleic acids are discussed in more detail in chapter 4.)

Chart 2.5 summarizes these four groups of organic compounds.

1. Compare the chemical composition of carbohydrates, lipids, proteins, and nucleic acids.
2. How does an enzyme affect a chemical reaction?
3. What is likely to happen to a protein molecule that is exposed to excessive heat or radiation?
4. What is the function of the nucleic acids?

Figure 2.15
Each different shape in this chain represents a different amino acid molecule. The whole chain represents a portion of a protein molecule.

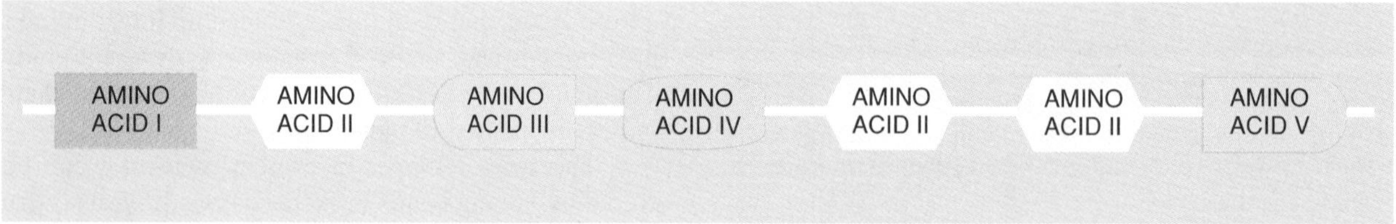

Chart 2.5 Organic compounds in cells

Compound	Elements present	Building blocks	Functions	Examples
Carbohydrates	C,H,O	Simple sugars	Provide energy, cell structure	Glucose, starch
Fats	C,H,O	Glycerol, fatty acids	Provide energy, cell structure	Tristearin, tripalmitin
Proteins	C,H,O,N (often S)	Amino acids	Provide cell structure, enzymes, energy	Albumins, hemoglobin
Nucleic acids	C,H,O,N,P	Nucleotides	Store information, control cell activities	RNA, DNA

Figure 2.16
A three-dimensional model of a portion of a protein molecule called collagen. The blue parts represent nitrogen atoms.

Figure 2.17
A nucleotide consists of a 5-carbon sugar (*S*), a phosphate group (*P*), and an organic base (*B*).

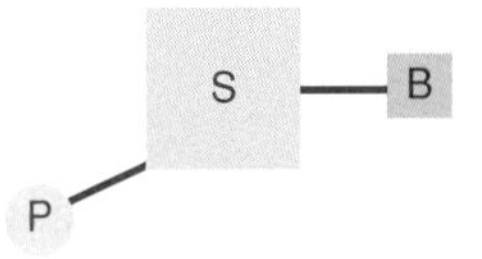

Figure 2.18
A nucleic acid molecule consists of nucleotides joined in a chain.

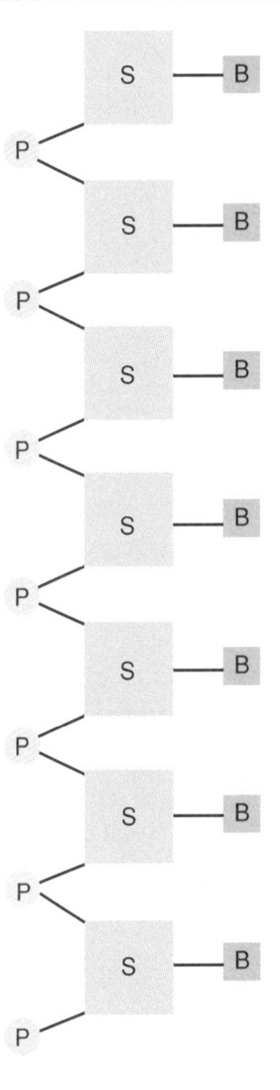

Chapter Summary

Introduction (page 36)

Chemistry deals with the composition of substances and various changes in their composition.

Structure of Matter (page 36)

1. Elements and atoms
   a. Naturally occurring matter on earth is composed of 91 elements.
   b. Some elements occur in pure form, but more frequently they are found in mixtures or are chemically combined.
   c. Elements are composed of atoms.
   d. Atoms of different elements vary in size and ways of interacting.
2. Atomic structure
   a. An atom consists of electrons surrounding a nucleus, which contains protons and neutrons.
   b. Electrons are negatively charged, protons positively charged, and neutrons uncharged.
   c. A complete atom is electrically neutral.
   d. The atomic number of an element is equal to the number of protons in each atom; the atomic weight is equal to the number of protons plus the number of neutrons in each atom.
3. Bonding of atoms
   a. When atoms combine they gain, lose, or share electrons.
   b. Electrons are arranged in shells around each nucleus.
   c. Atoms with completely filled outer shells are inactive, but atoms with incompletely filled outer shells tend to gain, lose, or share electrons and thus achieve stable structures.
   d. Atoms that lose electrons become positively charged; atoms that gain electrons become negatively charged.
   e. Ions with opposite charges are attracted to one another and become bound by electrovalent bonds; atoms that share electrons become bound by covalent bonds.

4. Molecules and compounds
   a. When two or more atoms of the same element are united, a molecule of that element is formed; when atoms of different elements are united, a molecule of a compound is formed.
   b. Molecules contain definite kinds and numbers of atoms.
   c. The numbers and kinds of atoms in a molecule can be represented by a molecular formula.
   d. The arrangement of atoms within a molecule can be represented by a structural formula.
5. Chemical reactions
   a. When a chemical reaction occurs, bonds between atoms, ions, or molecules are broken or created.
   b. Two kinds of chemical reaction are synthesis, in which larger molecules are formed from smaller particles, and decomposition, in which smaller particles are formed from larger molecules.
   c. Many reactions are reversible, and the direction of a reaction depends upon the proportions of reactants and end products present, the energy available, and the presence of catalysts.
6. Acids and bases
   a. Compounds that release ions when they dissolve in water are electrolytes.
   b. Electrolytes that release hydrogen ions are acids, and those that release hydroxyl or other ions that react with hydrogen ions are bases.
   c. The concentration of hydrogen ions (H^+) and hydroxyl ions (OH^-) in a solution can be represented by pH.
   d. A solution with equal numbers of H^+ and OH^- is neutral and has a pH of 7.0; a solution with more H^+ than OH^- is acidic and has a pH less than 7.0; a solution with fewer H^+ than OH^- is basic and has a pH greater than 7.0.
   e. There is a tenfold difference in the hydrogen ion concentration between each whole number on the pH scale.

Chemical Constituents of Cells (page 42)

Molecules lacking carbon atoms are inorganic and usually are electrolytes; those containing carbon and hydrogen atoms are organic and usually are nonelectrolytes.

1. Inorganic substances
   a. Water is the most abundant compound in cells and serves as a substance in which chemical reactions occur; it also transports chemicals and heat.
   b. Oxygen is used in releasing energy, needed to drive metabolic activities, from glucose and other molecules.
   c. Carbon dioxide is produced when energy is released during metabolic processes.
   d. Inorganic salts provide a variety of ions needed in metabolic processes.
2. Organic substances
   a. Carbohydrates provide much of the energy needed by cells; their basic building blocks are simple sugar molecules.
   b. Lipids supply energy also; the basic building blocks of fats—the most abundant lipid—are molecules of glycerol and fatty acids.
   c. Proteins serve as structural materials, energy sources, hormones, surface receptors, and enzymes.
      (1) Enzymes speed up chemical reactions.
      (2) The building blocks of proteins are amino acids.
      (3) Proteins vary in the numbers and kinds of amino acids they contain and in the sequence in which these amino acids are arranged.
      (4) Protein molecules can be denatured by exposure to excessive heat, radiation, electricity, or various chemicals.
   d. Nucleic acids are the most fundamental compounds in cells because they control cell activities.
      (1) Two major kinds of nucleic acids are RNA and DNA.
      (2) Nucleic acid molecules are composed of nucleotides.
      (3) DNA molecules store information that is used by cell parts to construct specific kinds of protein molecules, including enzymes; RNA molecules help to synthesize proteins.

Clinical Application of Knowledge

1. What acidic and alkaline substances do you encounter in your everyday activities? What acidic foods do you eat regularly? What alkaline foods do you eat?
2. Read the nutritional information on the packages of several food products. Which of these foods contain relatively high concentrations of carbohydrates? Of lipids? Of proteins?
3. How would you explain the importance of amino acids and proteins in a diet to a person who is following a diet composed primarily of carbohydrates?

Review Activities

1. Define *chemistry.*
2. Define *matter.*
3. Explain the relationship between elements and atoms.
4. List the four most abundant elements in the human body.
5. Describe the major parts of an atom.
6. Explain why a complete atom is electrically neutral.
7. Define *atomic number* and *atomic weight.*
8. Explain how electrons are arranged within an atom.
9. Distinguish between an electrovalent bond and a covalent bond.
10. Explain the relationship between molecules and compounds.
11. Distinguish between a molecular formula and a structural formula.
12. Explain what is meant by the formula $C_6H_{12}O_6$.
13. Describe two major types of chemical reactions.
14. Explain what is meant by a reversible reaction.
15. Define *catalyst.*
16. Define *acid* and *base.*
17. Explain what is meant by pH, and describe the pH scale.
18. Distinguish between inorganic and organic substances.
19. Distinguish between electrolyte and nonelectrolyte.
20. Describe the roles played by water and oxygen in the human body.
21. List several of the ions found in body fluids.
22. Describe the general characteristics of carbohydrates.
23. Distinguish between simple and complex carbohydrates.
24. Describe the general characteristics of lipids.
25. Define *triglyceride.*
26. Distinguish between saturated and unsaturated fats.
27. Describe the general characteristics of proteins.
28. Define *enzyme.*
29. Explain how protein molecules may become denatured.
30. Describe the general characteristics of nucleic acids.
31. Explain the major functions of nucleic acids.

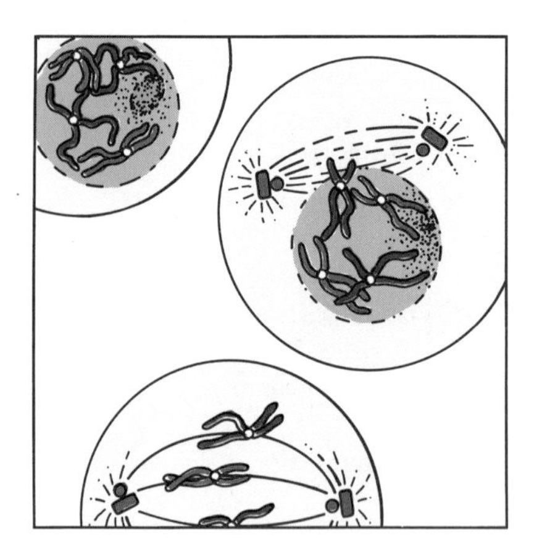

3
Cells

The human body is composed entirely of cells, products of cells, and various fluids. These cells represent the structural units of the body; they are the building blocks from which all larger parts are formed. They are also the functional units of the body, because whatever a body part can do is the result of activities within its cells.

Cells account for the shape, organization, and construction of the body and carry on its life processes. In addition, they can reproduce and thus provide the new cells, needed for growth, development, and the replacement of worn and injured tissues.

Chapter Objectives

After you have studied this chapter, you should be able to

1. Explain how cells vary from one another.
2. Describe the general characteristics of a composite cell.
3. Explain how the structure of a cell membrane is related to its function.
4. Describe each kind of cytoplasmic organelle and explain its function.
5. Describe the cell nucleus and its parts.
6. Explain how substances move through cell membranes.
7. Describe the life cycle of a cell.
8. Explain how a cell reproduces.
9. Complete the review activities at the end of this chapter. Note that the items are worded in the form of specific learning objectives. You may want to refer to them before reading the chapter.

Key Terms

active transport (ak′tiv trans′port)

centrosome (sent′tro-sōm)

chromosome (kro′mo-sōm)

cytoplasm (si′to-plazm)

differentiation (dif″er-en″she-a′shun)

diffusion (dĭ-fu′zhun)

endocytosis (en″do-si-to′sis)

endoplasmic reticulum (en′do-plaz′mik rē-tik′u-lum)

equilibrium (e″kwĭ-lib′re-um)

facilitated diffusion (fah-sil″ĭ-tat′ed dĭ-fu′zhun)

filtration (fil-tra′shun)

Golgi apparatus (gol′je ap″ah-ra′tus)

lysosome (li′so-sōm)

mitochondrion (mi″to-kon′dre-un); plural, mitochondria (mi-″to-kon′dre-ah)

mitosis (mi-to′sis)

nucleolus (nu-kle′o-lus)

nucleus (nu′kle-us)

osmosis (oz-mo′sis)

permeable (per′me-ah-bl)

phagocytosis (fag″o-si-to′sis)

pinocytosis (pi″no-si-to′sis)

ribosome (ri′bo-sōm)

vesicle (ves′ĭ-k'l)

Aids to Understanding Words

cyt-, cell: *cyt*oplasm—the fluid that occupies the space between the cell membrane and nuclear envelope.

endo-, within: *endo*plasmic reticulum—a complex of membranous structures within the cytoplasm.

hyper-, above: *hyper*tonic—a solution that has a greater concentration of dissolved particles than another solution.

hypo-, below: *hypo*tonic—a solution that has a lesser concentration of dissolved particles than another solution.

inter-, between: *inter*phase—the stage that occurs between mitotic divisions of a cell.

iso-, equal: *iso*tonic—a solution that has a concentration of dissolved particles equal to that of another solution.

mit-, thread: *mit*osis—the process of cell division during which threadlike chromosomes appear within a cell.

phag-, to eat: *phag*ocytosis—the process by which a cell takes in solid particles.

pino-, to drink: *pino*cytosis—the process by which a cell takes in tiny droplets of liquid.

-som, body: ribo*some*—a tiny, spherical organelle that contains protein and RNA.

Figure 3.1
Cells vary in structure and function. (*a*) A nerve cell transmits impulses from one body part to another; (*b*) epithelial cells protect underlying cells; and (*c*) muscle cells pull parts closer together.

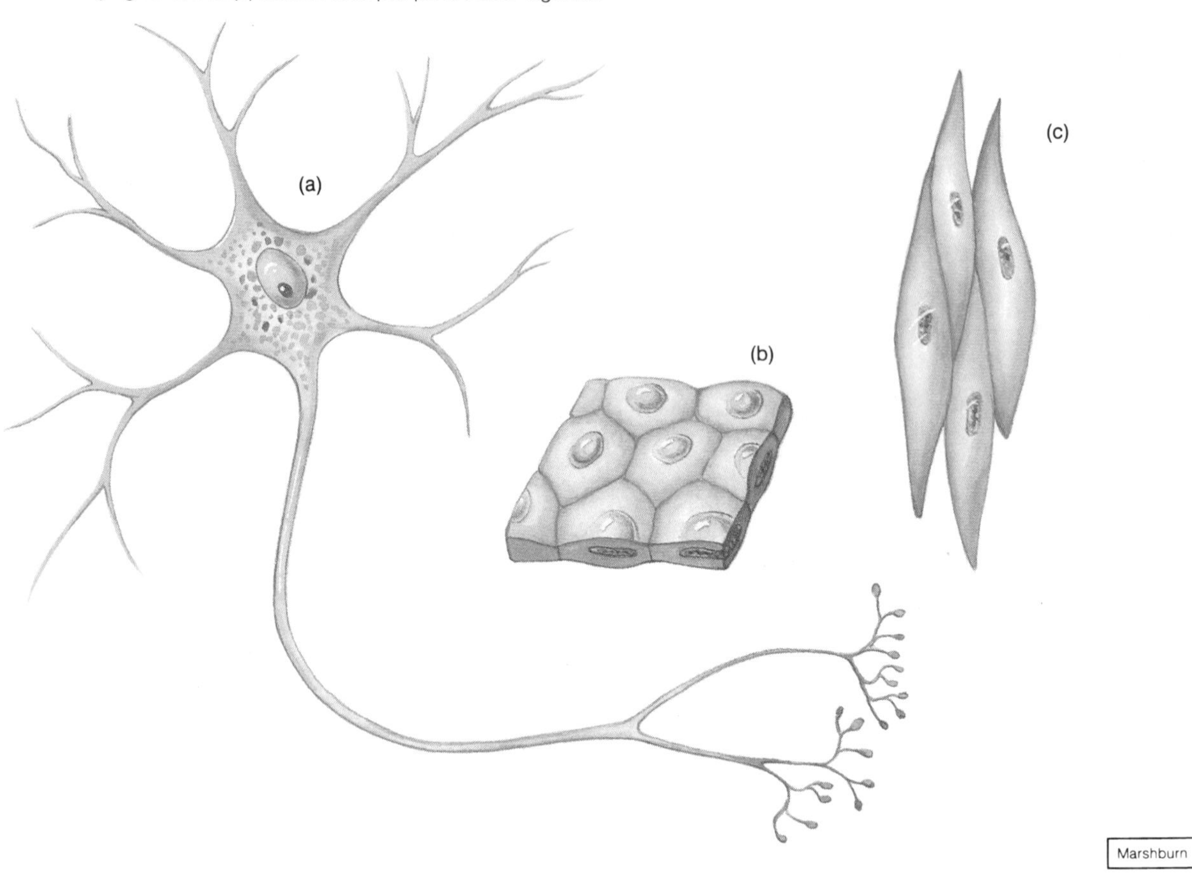

Introduction

THE ESTIMATED 75 trillion cells that make up an adult human body have much in common, yet those in different tissues vary in a number of ways. For example, cells vary considerably in size.

Cells also vary in shape, and typically their shapes are closely related to their functions. For instance, nerve cells often have long, threadlike extensions that transmit nerve impulses from one part of the body to another. Epithelial cells that line the inside of the mouth are thin, flattened, and tightly packed, somewhat like floor tiles. They are protective cells that serve to shield underlying cells. Muscle cells, which function to pull parts closer together, are slender and rodlike, with their ends attached to the parts they move (fig. 3.1).

Composite Cell

It is not possible to describe a typical cell because cells vary so greatly. For purposes of discussion, however, it is convenient to imagine that one exists. Thus, the cell shown in figure 3.2 and described in this chapter is not real. Instead, it is a *composite cell*—one that includes many known cell structures.

Commonly a cell consists of two major parts—the nucleus and the cytoplasm. The *nucleus* is the innermost part and is enclosed by a thin, double-layered membrane called a *nuclear envelope*. The *cytoplasm* is a mass of fluid that surrounds the nucleus and is itself encircled by a *cell membrane* (or cytoplasmic membrane).

Within the cytoplasm are specialized cellular parts called *cytoplasmic organelles*. These organelles perform specific metabolic functions. The nucleus, on the other hand, directs the overall activities of the cell.

1. Give two examples to illustrate that the shape of a cell is related to its function.
2. Name the two major parts of a cell.
3. What are the general functions of these two parts?

Cell Membrane

The **cell membrane** is the outermost limit of a cell, but it is more than a simple envelope surrounding the cellular contents; it is an actively functioning part of the living material, and many important metabolic reactions take place on its surfaces.

Figure 3.2
A composite cell.

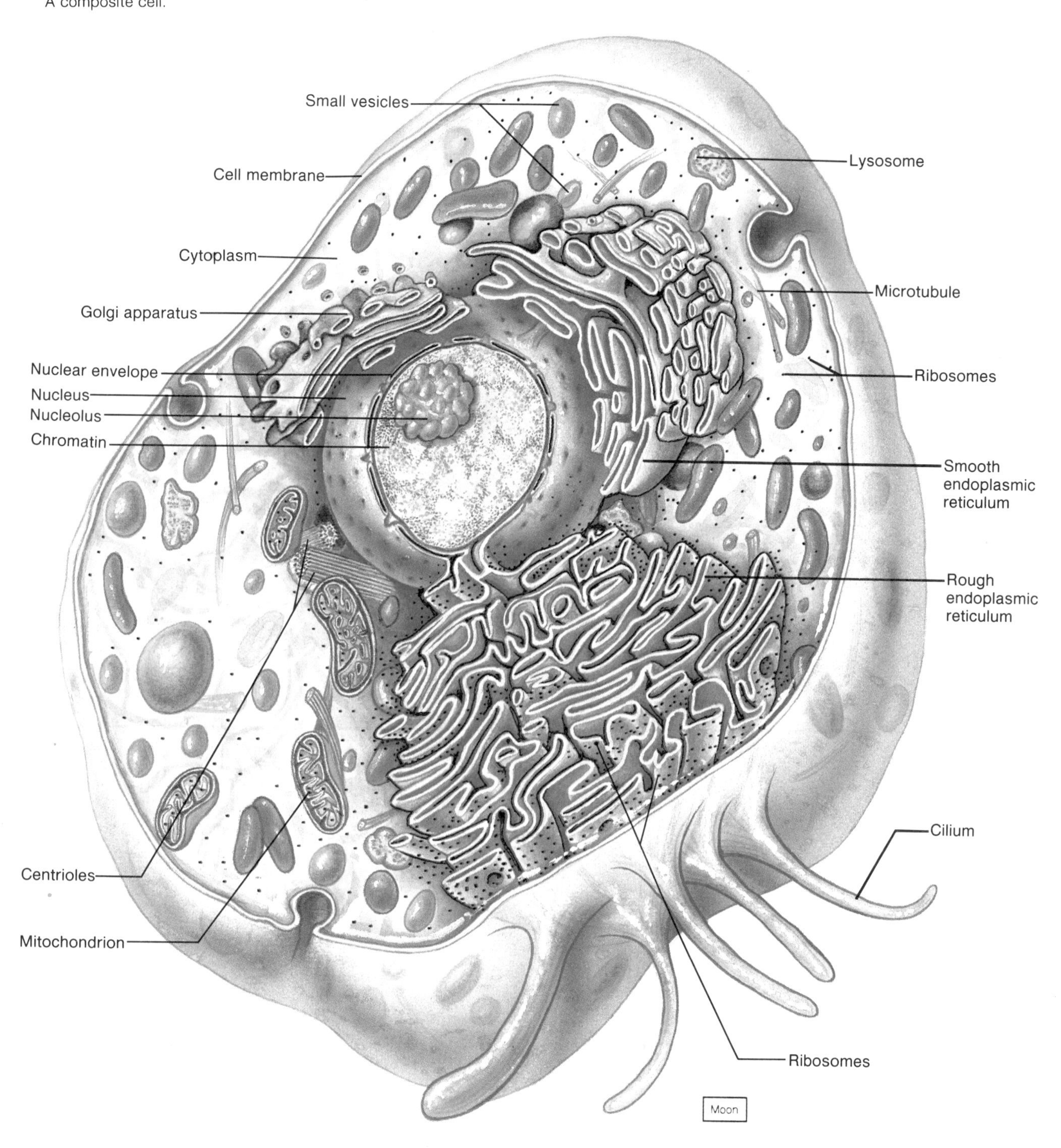

General Characteristics

The cell membrane is extremely thin—visible only with the aid of an electron microscope—but it is flexible and somewhat elastic. It typically has complex surface features with many outpouchings and infoldings that provide extra surface area (fig. 3.2).

In addition to maintaining the wholeness of the cell, the membrane controls the entrance and exit of substances. That is, it allows some substances to pass through, and it excludes others. When a membrane functions in this way, it is called *selectively permeable* (also known as semipermeable or differentially permeable). A *permeable* membrane, on the other hand, allows the free passage of such substances.

Figure 3.3
(*a*) The cell membrane is composed primarily of phospholipids, with proteins scattered throughout the lipid layer and associated with its surfaces. (*b*) Electron micrograph of a cell membrane (×250,000).

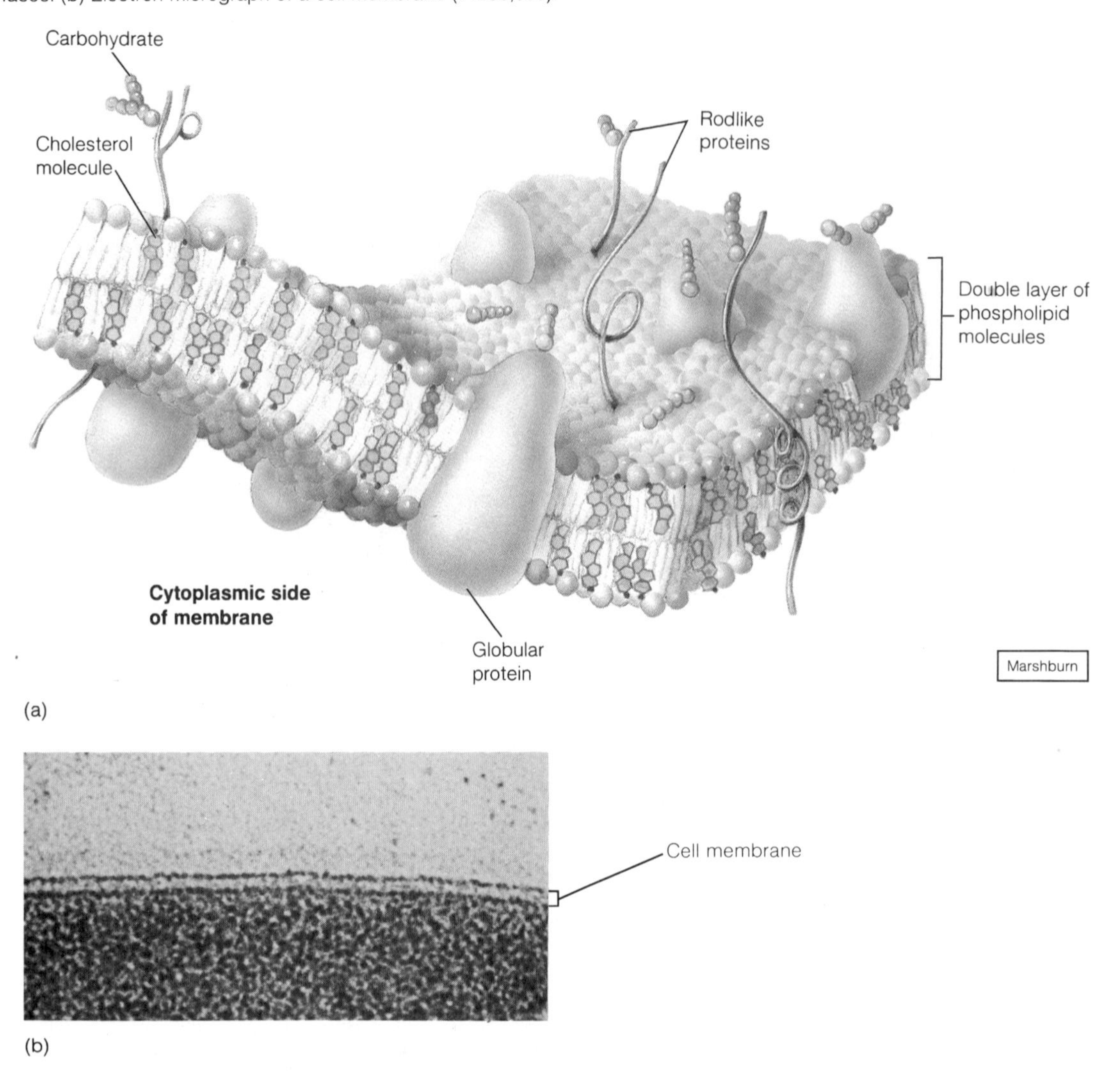

Membrane Structure

Chemically, the cell membrane is composed mainly of lipids and proteins, although it contains a small quantity of carbohydrate. Its basic framework consists of a double layer (bilayer) of phospholipid molecules. Each of these molecules includes a phosphate group and two fatty acids bound to a glycerol molecule (see chapter 2). In the membrane, the fatty acid portions of the phospholipids are arranged so that they form the interior of the structure (fig. 3.3).

The lipid molecules are relatively free to move sideways within the plane of the membrane, and together they create a thin, but stable liquid film. The film is also soft and flexible.

Because the interior of the membrane consists largely of the fatty acid portions of the phospholipid molecules, it has an oily characteristic. Molecules that are soluble in lipids, such as oxygen and carbon dioxide molecules, can pass through this layer easily; however, the layer is impermeable to water soluble molecules, such as amino acids, sugars, proteins, nucleic acid molecules, and various ions. Many cholesterol molecules embedded in the interior of the membrane also help make the membrane less permeable to water soluble substances. In addition, the relatively rigid structure of the cholesterol molecules helps make the membrane more stable than it would be otherwise.

Although the cell membrane includes only a few types of lipid molecules, it contains many kinds of proteins. These proteins are responsible for special functions, and they can be classified according to their shapes. One group, for example, consists of tightly coiled rodlike molecules embedded in the bilayer of phospholipids. These rodlike proteins may completely span the membrane; that is, they may extend outward from the surface on one side, while their opposite ends communicate with the cell's interior. Such proteins often function as *receptors* that are specialized to combine with specific kinds of molecules, such as hormones.

Another group of membrane proteins have molecules with more compact, globular shapes. These proteins also are embedded in the interior of the phospholipid bilayer. Typically they span the membrane and create narrow passageways, or *channels,* through which various ions and molecules can cross the bilayer. For example, some globular proteins form "pores" in the membrane that allow the movement of water molecules. Others serve as selective channels that only allow particular substances to pass through. In muscle and nerve cells, for example, such selective channels control the movements of sodium and potassium ions, which play important roles in muscle contraction and nerve impulse conduction. The processes of muscle contraction and nerve impulse conduction are described in chapters 8 and 9, respectively.

Cytoplasm

When viewed through a laboratory microscope, **cytoplasm** usually appears as a clear liquid with specks scattered throughout. An electron microscope, which produces much greater magnification and resolution, however, reveals the cytoplasm to be highly structured and filled with networks of membranes and other organelles (fig. 3.2).

The activities of a cell occur largely in its cytoplasm. It is there that food molecules are received, processed, and used. In other words, cytoplasm is a site of metabolic reactions in which the following **cytoplasmic organelles** play specific roles:

1. **Endoplasmic reticulum.** The endoplasmic reticulum (ER) is a complex organelle composed of membrane-bound flattened sacs and elongated canals. These membranous parts are interconnected, and they communicate with the cell membrane, nuclear envelope, and certain cytoplasmic organelles. The endoplasmic reticulum functions as a tubular communication system through which molecules can be transported from one cell part to another.

   The endoplasmic reticulum also plays roles in the synthesis of certain types of molecules. Commonly, for example, its outer membranous surface has numerous tiny, spherical organelles called *ribosomes* attached to it. These particles cause the endoplasmic reticulum to have a more textured appearance when viewed with an electron microscope. Such endoplasmic reticulum is termed *rough ER.*

Figure 3.4
A transmission electron micrograph of rough endoplasmic reticulum. Each tiny black dot represents a ribosome (×100,000).

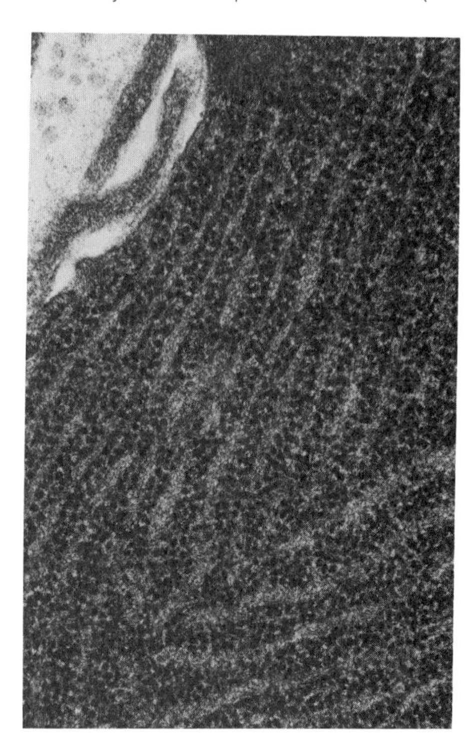

   Endoplasmic reticulum that lacks ribosomes is called *smooth ER* (figs. 3.2 and 3.4).

   Ribosomes synthesize proteins, and the resulting molecules may be transported through the tubules of the endoplasmic reticulum to the Golgi apparatus for further processing. Smooth endoplasmic reticulum, on the other hand, contains enzymes involved in the manufacturing of various lipid molecules, including certain steroid hormones.
2. **Ribosomes.** Although many ribosomes are attached to membranes of the endoplasmic reticulum, others occur as free particles scattered throughout the cytoplasm. In both cases, these particles are composed of protein and RNA molecules, and they function in the synthesis of proteins. (This function is described in chapter 4.)
3. **Golgi apparatus.** The Golgi apparatus is usually located near the nucleus. It consists of a stack of about six flattened, membranous sacs whose membranes are continuous with those of the endoplasmic reticulum.

   The Golgi apparatus is involved with refining and "packaging" the proteins which were synthesized by the ribosomes associated with the endoplasmic reticulum. Proteins arrive at the Golgi apparatus enclosed in tiny sacs

Figure 3.5
The Golgi apparatus functions in packaging glycoproteins, which then may be moved to the cell membrane and released as a secretion.

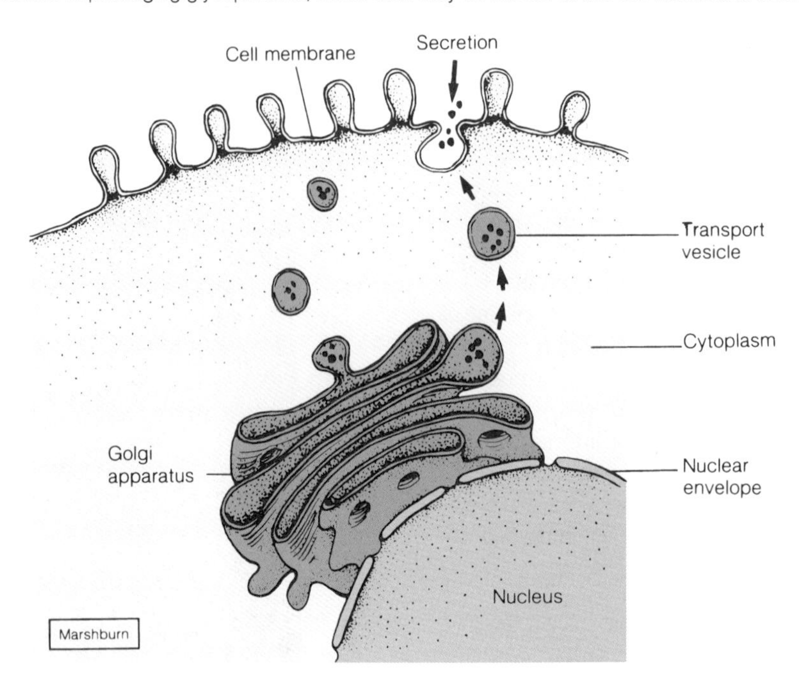

composed of membrane from the endoplasmic reticulum. These sacs fuse to the membrane at the end of the Golgi apparatus, which is specialized to receive proteins. Previously these protein molecules were combined with sugar molecules, and thus, they are called *glycoproteins.*

Glycoproteins pass from layer to layer through the stacks of Golgi membrane and are modified chemically. For example, some sugar molecules may be added or removed from them. When they reach the outermost layer, the altered glycoproteins are packaged in bits of Golgi membrane, which bud off and form transport vesicles. Such a vesicle may then move to the cell membrane where it may fuse with the membrane and release its contents to the outside of the cell as a secretion (figs. 3.2 and 3.5). Other vesicles may transport their glycoproteins to various cytoplasmic organelles within the cell.

1. What is meant by a selectively permeable membrane?
2. Describe the chemical structure of a cell membrane.
3. What are the functions of the endoplasmic reticulum?
4. Describe the functions of the Golgi apparatus.

Figure 3.6
A transmission electron micrograph of a mitochondrion (×79,000). What other cellular features can you identify in this micrograph?

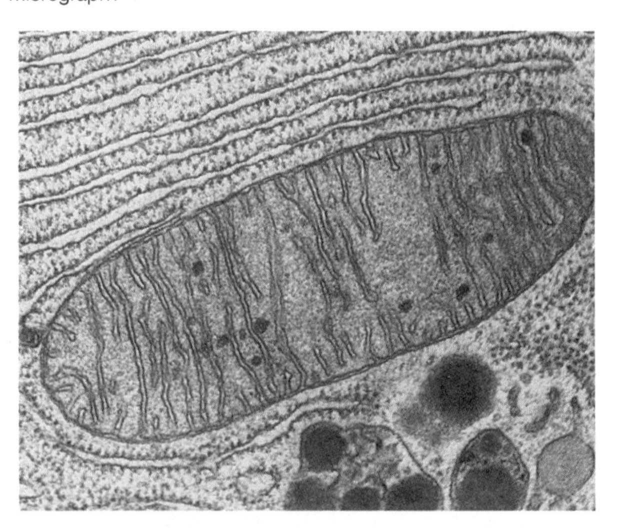

4. **Mitochondria.** Mitochondria are elongated fluid-filled sacs, which vary in size and shape. They often move about slowly in the cytoplasm and can reproduce themselves by dividing.

   The membrane surrounding a mitochondrion has an outer and an inner layer (figs. 3.2 and 3.6). The inner layer is folded

Figure 3.7
(*a*) A transmission electron micrograph of the two centrioles in a centrosome (×142,000). (*b*) Note that the centrioles lie at a right angle to one another.

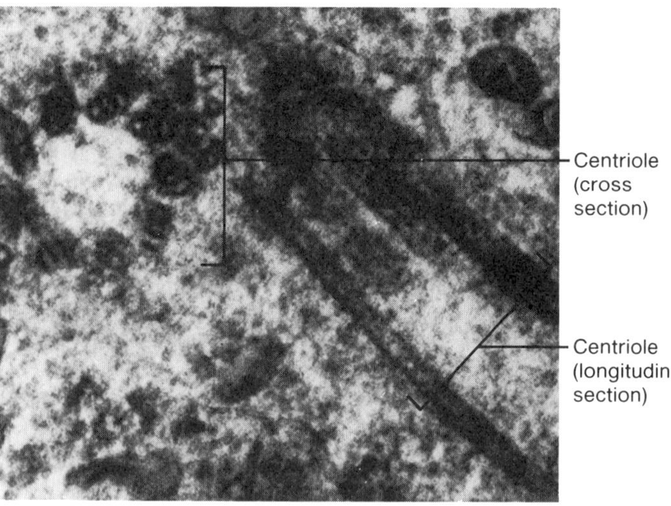

(a)

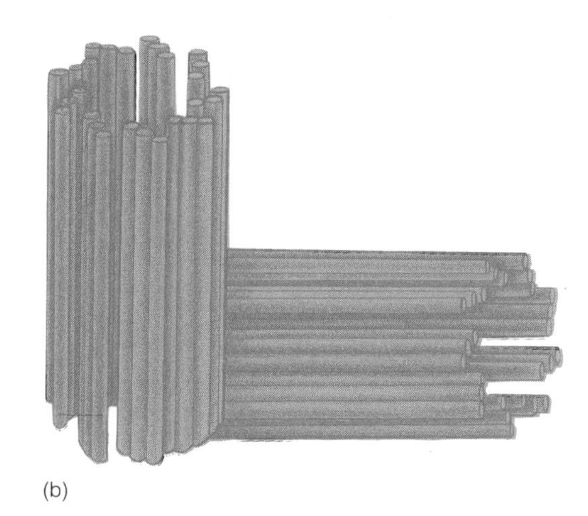
(b)

extensively to form partitions called *cristae*. Connected to the cristae are enzymes that control some of the chemical reactions by which energy is released from glucose and other organic molecules. Mitochondria also function in transforming this energy into a chemical form that is usable by various cell parts. (This energy-releasing function is described in more detail in chapter 4.)

5. **Lysosomes.** Lysosomes appear as tiny, membranous sacs (fig. 3.2). They contain powerful enzymes that are capable of breaking down molecules of nutrients or foreign particles that may enter cells. Certain white blood cells, for example, can engulf bacteria, which are then digested by the lysosomal enzymes. Consequently, white blood cells help prevent bacterial infections. Lysosomes also function in the destruction of worn cellular parts. In fact, the lysosomes of some scavenger cells may digest entire cells that have been engulfed.

Lysosomal digestive activity seems to be responsible for decreasing the size of body tissues at certain times. Such a regression in size occurs in the maternal uterus following birth of an infant, in the breasts after the weaning of an infant, and in skeletal muscles during periods of prolonged inactivity.

6. **Centrosome.** The centrosome is located in the cytoplasm near the Golgi apparatus and nucleus. It is nonmembranous and consists of two hollow cylinders, called *centrioles*. The centrioles lie at right angles to each other and function in cell reproduction (figs. 3.2 and 3.7). During this process, they aid in the distribution of parts in the nucleus (chromosomes) to the newly forming cells.
7. **Vesicles.** Vesicles (vacuoles) are membranous sacs formed by an action of the cell membrane in which a portion of the membrane folds inward and pinches off. As a result, a tiny, bubblelike vesicle, containing some liquid or solid material that was outside the cell a moment before, appears in the cytoplasm (fig. 3.2).
8. **Cilia** and **flagella.** Cilia and flagella are motile processes that extend outward from the surfaces of certain cells. They are similar structures and differ mainly in terms of their length and the number of them present.

    Cilia occur in large numbers on the free surfaces of some epithelial cells. Each cilium is a tiny, hairlike structure that is attached beneath the cell membrane (fig. 3.2). Cilia are arranged in precise patterns, and they have a "to-and-fro" type of movement. This movement is coordinated so that rows of cilia beat one after the other, creating a wave of motion that sweeps over the ciliated surface. This action serves to move fluids, such as mucus, over the surface of certain tissues, such as those that form the inner linings of the respiratory tubes (fig. 3.8).

Figure 3.8
Cilia, such as these (arrow) are common on the surface of certain cells that form the inner lining of respiratory tubes (×10,000).

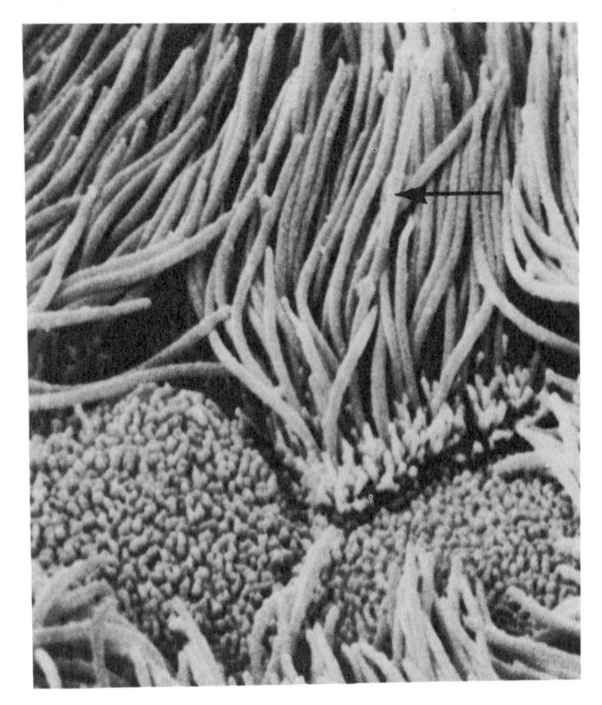

Flagella are considerably longer than cilia, and usually there is a single flagellum on a cell. These projections have an undulating wavelike motion, which begins at the base of a flagellum. The tail of a sperm cell, for example, is a flagellum that causes the swimming movement of this cell. Flagella also occur in certain cells of the kidneys, testes, and various glands.

9. **Microfilaments** and **microtubules.** Two types of thin, threadlike processes found within cytoplasm are microfilaments and microtubules. Microfilaments are tiny rods of protein, arranged in meshworks or bundles. They function to cause various kinds of cellular movement. In muscle cells, for example, they are highly developed as *myofibrils,* which help these cells to shorten or contract (see chapter 8).

   Microtubules are long, slender tubes with diameters two or three times greater than microfilaments and are composed of globular proteins. They are usually stiff and form an "internal skeleton" that helps to maintain the shape of a cell. They also provide strength to the structure of various cell parts, such as cilia and flagella, and they aid in moving organelles. For instance, microtubules appear during cellular reproduction and are involved in the distribution of chromosomes to the newly forming cells. (Cell reproduction is described in a subsequent section of this chapter.)

1. Describe a mitochondrion.
2. What is the function of a lysosome?
3. Distinguish between a centrosome and a centriole.
4. How do microfilaments and microtubules differ?

Cell Nucleus

The **nucleus** (figs. 3.2 and 3.9) is a cellular organelle that is usually located near the center of a cell. It is a relatively large, spherical structure enclosed in a double-layered **nuclear envelope,** consisting of inner and outer membranes. The envelope is porous and allows substances to move between the nucleus and the cytoplasm. The nucleus contains a fluid in which other structures float. These structures include the following:

1. **Nucleolus.** A nucleolus ("little nucleus") is a small, dense body composed largely of RNA and protein. It has no surrounding membrane and is formed in specialized regions of certain chromosomes. It functions in the production of ribosomes. Once the ribosomes are formed, they migrate through the pores in the nuclear envelope and enter the cytoplasm.
2. **Chromatin.** Chromatin consists of loosely coiled fibers. When the cell begins to undergo the reproductive process, these fibers become more tightly coiled and are transformed into tiny, rodlike *chromosomes.* Chromatin fibers are composed of protein and DNA molecules, which in turn contain the information for synthesizing proteins that promote cellular life processes.

Chart 3.1 summarizes the structure and function of the cellular organelles.

1. How are the nuclear contents separated from the cytoplasm of a cell?
2. What is the function of the nucleolus?
3. What is chromatin?

Figure 3.9
(*a*) A transmission electron micrograph of a cell nucleus (×8,000). It contains a nucleolus and masses of chromatin.
(*b*) The nuclear envelope is porous and allows substances to pass between the nucleus and the cytoplasm.

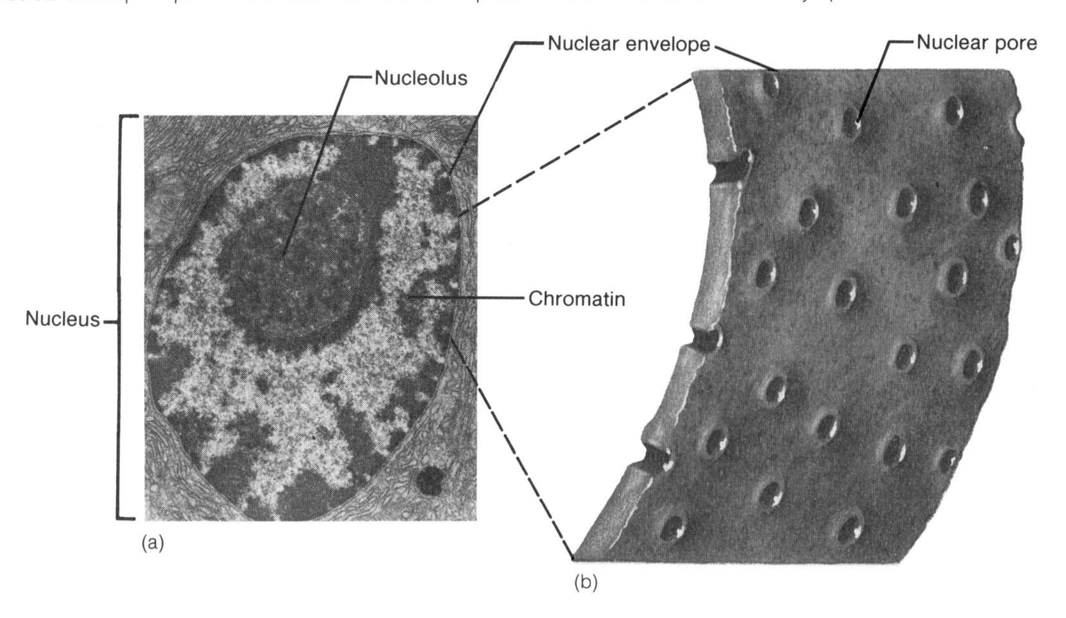

Chart 3.1 Structure and function of cellular organelles		
Organelles	**Structure**	**Function**
Cell membrane	Membrane composed of protein and lipid molecules	Maintains wholeness of the cell and controls the passage of materials into and out of the cell
Endoplasmic reticulum	Complex of interconnected membrane-bound sacs and canals	Transports materials within the cell, provides attachment for ribosomes, and synthesizes lipids
Ribosomes	Particles composed of protein and RNA molecules	Synthesis of proteins
Golgi apparatus	Group of flattened, membranous sacs	Packages protein molecules for secretion
Mitochondria	Membranous sacs with inner partitions	Release energy from food molecules and transform energy into usable form
Lysosomes	Membranous sacs	Digest worn cellular parts or substances that enter cells
Centrosome	Nonmembranous structure composed of two rodlike centrioles	Helps distribute chromosomes to new cells during cell reproduction
Vesicles	Membranous sacs	Contain various substances that recently entered the cell
Cilia and flagella	Motile projections attached beneath the cell membrane	Propel fluid over cellular surfaces and enable sperm cells to move
Microfilaments and microtubules	Thin rods and tubules	Provide support to the cytoplasm and help move substances and organelles in the cytoplasm
Nuclear envelope	Porous membrane that separates the nuclear contents from the cytoplasm	Maintains the wholeness of the nucleus and controls the passage of materials between the nucleus and cytoplasm
Nucleolus	Dense, nonmembranous body composed of protein and RNA molecules	Forms ribosomes
Chromatin	Fibers composed of protein and DNA molecules	Contains cellular information for synthesizing proteins needed in carrying on life processes

Figure 3.10
An example of diffusion. (*a, b, c*) If a sugar cube is placed in water, it slowly disappears. As this happens, the sugar molecules diffuse from regions where they are more concentrated toward regions where they are less concentrated. (*d*) Eventually, the sugar molecules are distributed evenly throughout the water.

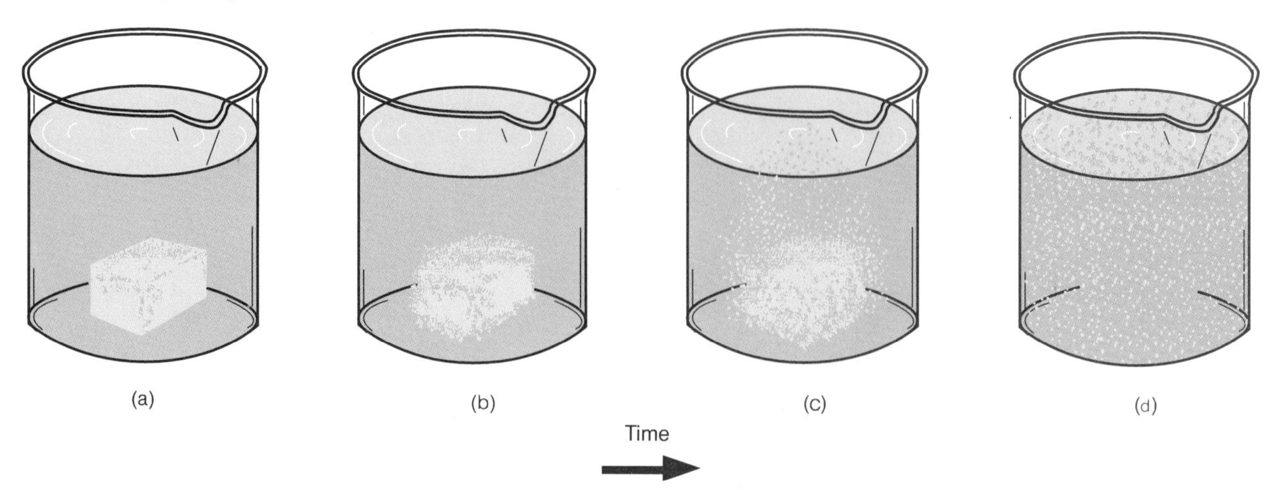

Movements through Cell Membranes

The cell membrane provides a surface through which various substances enter and leave the cell. These movements involve diffusion, facilitated diffusion, osmosis, filtration, active transport, and endocytosis.

Diffusion

Diffusion is the process by which molecules or ions become scattered or are spread spontaneously from regions where they are in higher concentrations toward regions where they are in lower concentrations.

Under natural conditions these molecules and ions are constantly moving at high speeds. Each particle travels in a separate path along a straight line until it collides and bounces off some other particle. Then it moves in another direction, only to collide again and change direction once more. Such random motion accounts for the mixing of molecules that commonly occurs when different kinds of substances are put together.

For example, when some sugar (a solute) is put into a glass of water (a solvent), as illustrated in figure 3.10, the sugar seems to remain in high concentration at the bottom of the glass. Then it slowly disappears into solution. As this is happening, the moving water and sugar molecules are colliding with one another, and in time the sugar and water molecules will be evenly mixed. This mixing process by which the sugar molecules spread from regions where they are in higher concentration toward the regions where they are less concentrated is diffusion. As a result of the process, the sugar molecules eventually become uniformly distributed in the water. This state of uniform distribution of molecules is called *equilibrium*. Although molecules continue to move after equilibrium is achieved, their concentrations no longer change.

To better understand how diffusion accounts for the movement of molecules through a cell membrane, imagine a container of water that is separated into two compartments by a permeable membrane (fig. 3.11). This membrane has numerous pores that are large enough for water and sugar molecules to pass through. The sugar molecules are placed in one compartment (*A*) but not in the other (*B*). Although the sugar molecules move in all directions, more diffuse from compartment *A* (where they are in greater concentration) through the pores in the membrane and into compartment *B* (where they are in lesser concentration) than move in the other direction. At the same time, the water molecules tend to diffuse from compartment *B* (where they are in greater concentration) through the pores into compartment *A* (where they are in lesser concentration). Eventually, equilibrium will be achieved with equal numbers of water and sugar molecules in each compartment.

Similarly, oxygen molecules diffuse through cell membranes and enter cells if these molecules are more highly concentrated on the outside than on the inside. Carbon dioxide molecules, too, diffuse through cell membranes and leave cells if they are more concentrated on the inside than on the outside of the membrane. Thus, it is by diffusion that oxygen and carbon dioxide molecules are exchanged between the air and the blood in the lungs and between the blood and the cells of various tissues.

Figure 3.11
(*1*) The container is separated into two compartments by a permeable membrane. Compartment *A* contains water molecules and sugar molecules, but compartment *B* contains only water molecules. (*2*) As a result of molecular motions, the sugar molecules tend to diffuse from compartment *A* into compartment *B*. The water molecules tend to diffuse from compartment *B* into compartment *A*. (*3*) Eventually, equilibrium is achieved.

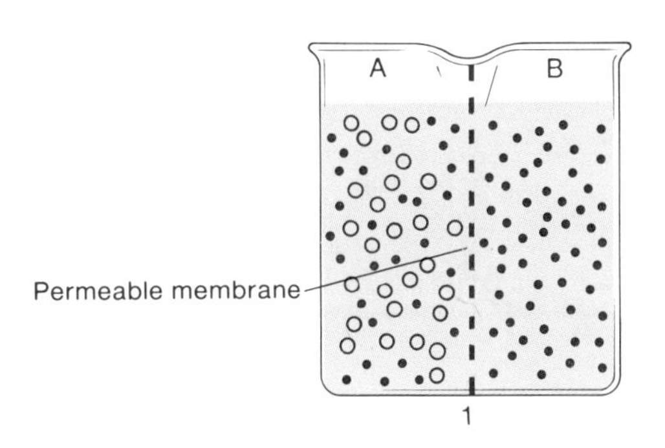

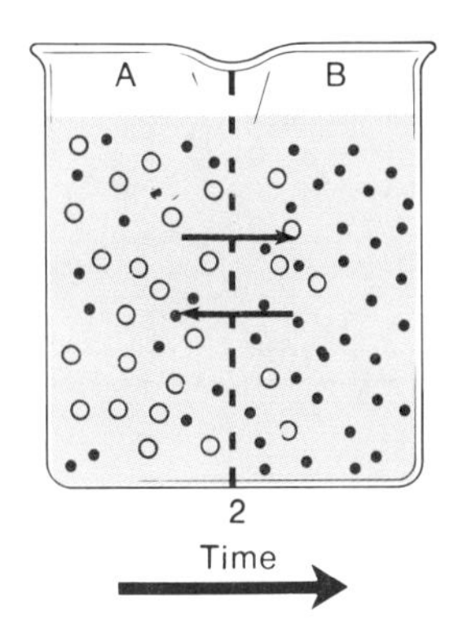

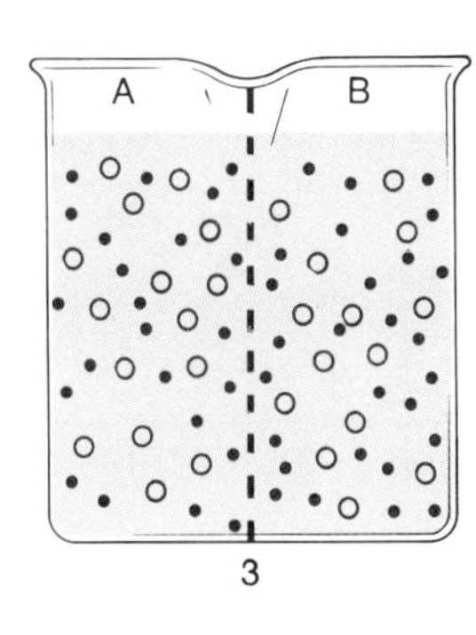

Dialysis is the process of separating smaller molecules from larger ones in a liquid by means of diffusion. This process is employed in the clinical procedure called *hemodialysis* when an artificial kidney is used to treat patients suffering from kidney damage or failure. When an artificial kidney (dialyzer) is operating, blood from a patient is passed through a long, coiled tubing composed of porous cellophane. The size of the pores allows the smaller molecules carried in the blood, such as urea, to pass out through the cellophane, while the larger molecules, such as those of blood proteins, remain inside the tubing. The tubing is submerged in a tank of dialyzing fluid (wash solution), which contains varying concentrations of chemicals. The solution is low, for instance, in concentrations of substances that should leave the blood and higher in concentrations of those that should remain in the blood.

Because it is desirable for an artificial kidney to remove blood urea, the dialyzing fluid must have a lower concentration of urea than the blood does; it is also desirable to maintain the blood glucose concentration, so the concentration of glucose in the wash solution must be kept at least equal to that of the blood. Thus, by altering the concentrations of molecules in the dialyzing fluid, it is possible to control those molecules that will diffuse out of the blood and those that will remain in it.

Facilitated Diffusion

Most sugars are insoluble in lipids and have molecular sizes that prevent them from passing through membrane pores. However, some of these substances, including glucose, may still enter through the lipid portion of the membrane by a process called **facilitated diffusion** (fig. 3.12). In this process, which occurs in most cells, the glucose molecule combines with a special protein carrier molecule at the surface of the membrane. This union of the glucose and carrier molecules creates a compound that is soluble in lipid, and this compound can diffuse to the other side. There the glucose portion is released, and the carrier molecule then can return to the opposite side of the membrane and pick up another glucose molecule. The hormone called *insulin* promotes facilitated diffusion of glucose through the membranes of certain cells (see chapter 11).

Facilitated diffusion is similar to simple diffusion in that it only can cause the movement of molecules from regions of higher concentration toward regions of lower concentration. The rate at which facilitated diffusion can occur, however, is limited by the number of carrier molecules in the cell membrane.

Osmosis

Osmosis is a special case of diffusion. It occurs whenever *water* molecules diffuse from a region of higher water concentration (where the solute concentration is lower) toward a region of lower water concentration (where the solute concentration is higher) through a selectively permeable membrane (fig. 3.13).

Ordinarily, the concentrations of water molecules are equal on either side of a cell membrane. Sometimes, however, the water on one side has more solute dissolved in it than the water on the other side. For example, if there is a greater concentration of glucose in the water outside a cell, there must be a lesser concentration of water there, because glucose molecules occupy space that would otherwise contain water molecules. Under such conditions, water molecules diffuse

Figure 3.12
Some substances are moved through a cell membrane by facilitated diffusion, a process by which carrier molecules transport a substance from a region of higher concentration to one of lower concentration.

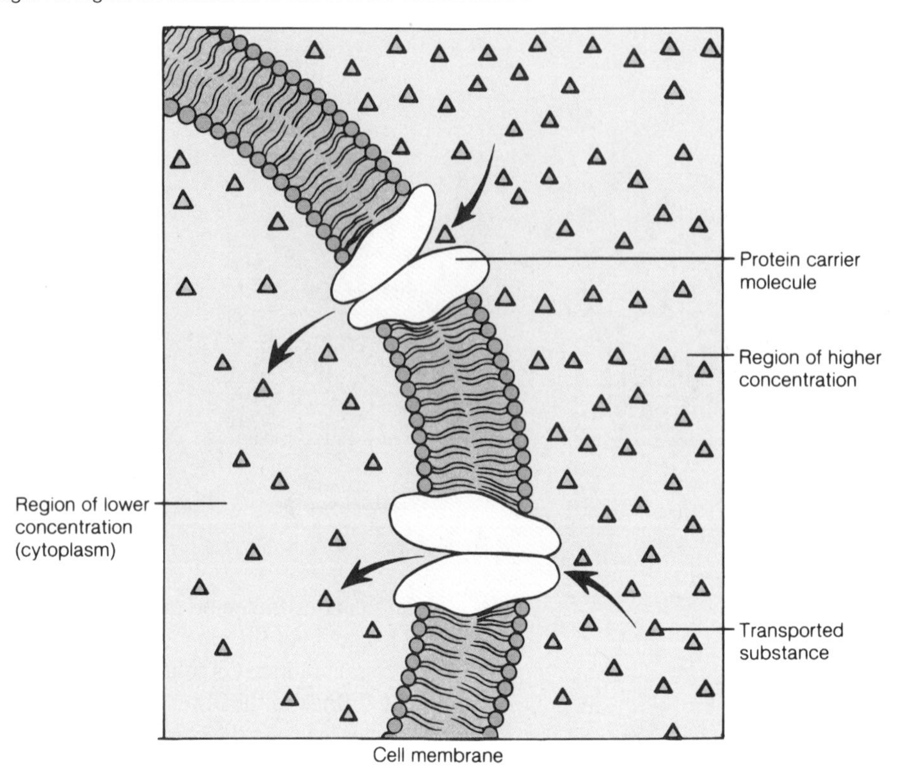

Figure 3.13
An example of osmosis. (*1*) The container is separated into two compartments by a selectively permeable membrane. Compartment *A* contains water and sugar molecules, but compartment *B* contains only water molecules. As a result of molecular motion, the water molecules tend to diffuse by osmosis from compartment *B* into compartment *A*. The sugar molecules remain in compartment *A* because they are too large to pass through the pores of the membrane. (*2*) Because more water enters compartment *A* than leaves it, water accumulates in this compartment, and the level of the liquid in compartment *A* rises above that of compartment *B*.

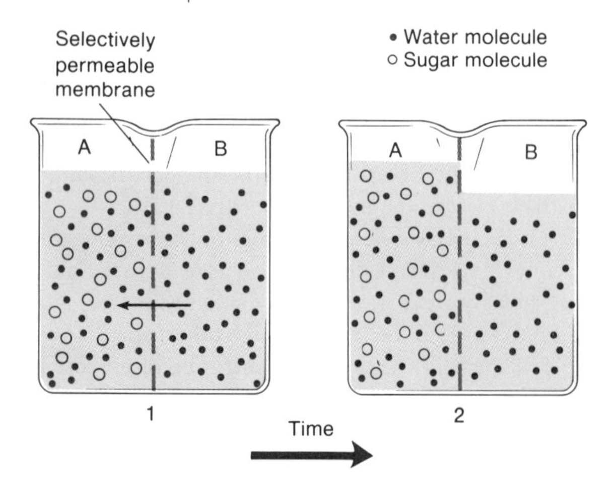

from inside the cell (where they are in higher concentration) toward the outside (where they are in lesser concentration).

Osmosis (fig. 3.13) is similar to diffusion (fig. 3.11). In the process of osmosis, however, the membrane is *selectively permeable,* that is, water molecules pass through readily, but glucose molecules do not; in the process of diffusion, the membrane is *permeable* and allows both water and glucose molecules to pass through.

Note in figure 3.13 that as osmosis occurs, the volume of water on side *A* increases. This increase in volume would be resisted if pressure were applied to the surface of the liquid on side *A*. The amount of pressure needed to stop osmosis is called *osmotic pressure.* Thus, the osmotic pressure of a solution is potential pressure and is due to the presence of nondiffusable solute particles in that solution. Furthermore, the greater the number of solute particles in the solution, the greater the osmotic pressure of that solution.

When osmosis occurs, water tends to move toward the region of greater osmotic pressure. Because the amount of osmotic pressure depends upon the difference in concentration of solute particles on opposite sides of the membrane, the greater the difference, the greater the tendency for water to move toward the region of higher solute concentration.

If some cells were put into a water solution that had a greater concentration of solute particles (higher osmotic pressure) than the cells, there would be a net movement of water out of the cells, and they would begin to shrink. Such cells are said to become *crenated.* A solution of this type, in which more water leaves a cell than enters it (because the concentration of solute particles is greater outside the cell) is called **hypertonic.**

Conversely, if there is a greater concentration of solute particles inside a cell than in the water around it, water will diffuse into the cell. As this happens, water accumulates within the cell, and it begins to swell. A solution in which more water enters a cell than leaves it (because of a lesser concentration of solute particles outside the cell) is called **hypotonic.**

A solution that contains the same concentration of solute particles as a cell is said to be **isotonic** to that cell. In such a solution, water enters and leaves the cell in equal amounts, and the cell size remains unchanged.

It is important to control the concentration of solute in solutions that are infused into body tissues or blood. Otherwise, osmosis may cause cells to swell or shrink, and they may be damaged. For instance, if red blood cells are placed in distilled water (which is hypotonic to them), water will diffuse into the cells and they will burst (hemolyze). On the other hand, if red blood cells are exposed to 0.9% NaCl solution (normal saline), the cells will remain unchanged because this solution is isotonic to human cells.

Filtration

The passage of substances through membranes by diffusion or osmosis is due to movements of the molecules of those substances. In other instances, molecules are forced through membranes by the hydrostatic pressure, called blood pressure, that is greater on one side of the membrane than on the other. This process by which molecules are forced through membranes is called **filtration.**

In the laboratory, filtration is commonly used to separate solids from a liquid. One method is to pour a mixture onto filter paper in a funnel. The paper serves as a porous membrane through which the liquid can pass, leaving the solids behind. Hydrostatic pressure, which is created by the weight of water on the paper due to gravity, forces the water through to the other side.

Figure 3.14
In this example of filtration, the smaller molecules are forced through tiny openings in the wall of a capillary by blood pressure. The larger molecules remain inside.

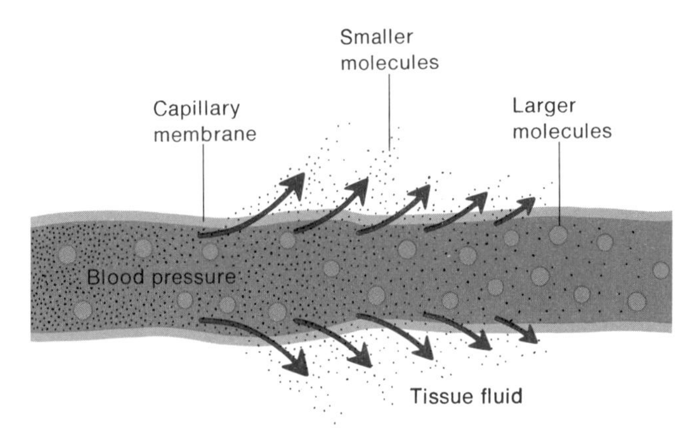

Similarly, in the body, tissue fluid is formed when water and dissolved substances are forced out through the thin, porous walls of blood capillaries, but larger particles such as blood protein molecules are left inside (fig. 3.14). The force for this movement comes from blood pressure, created largely by heart action, which is greater within the vessel than outside.

1. What kinds of substances move most readily through a cell membrane by diffusion?
2. Explain the differences between diffusion and osmosis.
3. Distinguish between hypertonic, hypotonic, and isotonic solutions.
4. Explain how filtration occurs within the body.

Active Transport

When molecules or ions pass through cell membranes by diffusion, facilitated diffusion, or osmosis, their net movements are from regions of higher concentration toward regions of lower concentration. Sometimes, however, the net movement of particles passing through membranes is in the opposite direction; that is, the particles move from a region of lower concentration to one of higher concentration.

It is known, for example, that sodium ions can diffuse through cell membranes. Yet the concentration of these ions typically remains many times greater on the outside of cells than on the inside. Furthermore, sodium ions are continually moved through cell membranes from the regions of lower concentration (inside) to regions of higher concentration (outside). Movement of this type is called **active transport.** It depends

Figure 3.15
(*a*) During active transport, a molecule or an ion combines with a carrier protein, whose shape is altered as a result. (*b*) This change in shape helps to move the transported particle through the cell membrane.

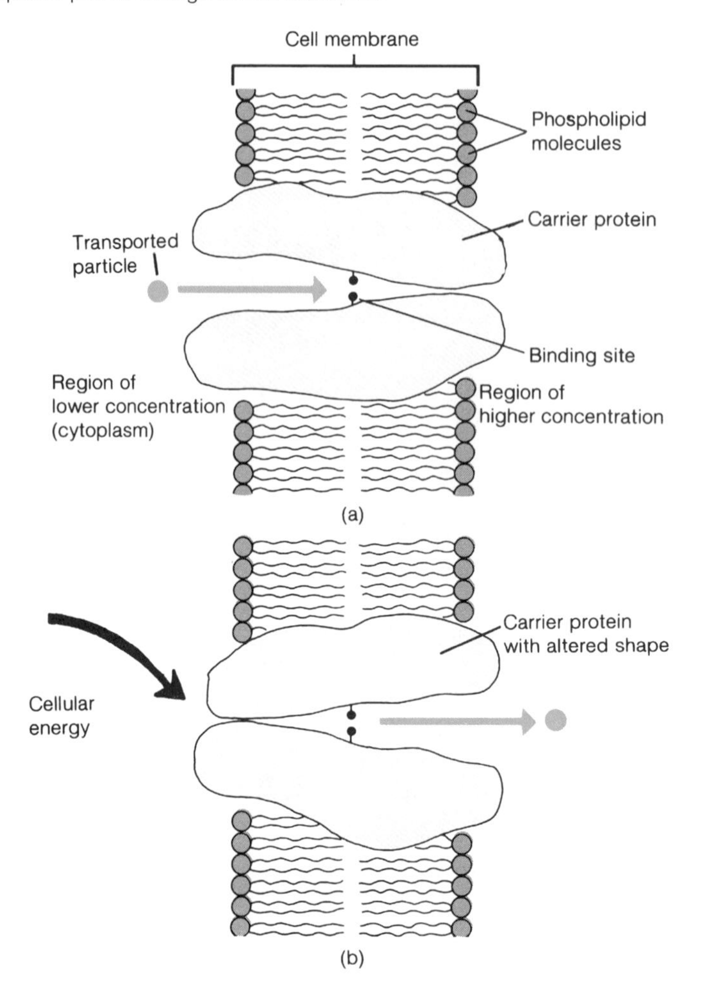

on life processes within cells and requires energy that is released by cellular metabolism. In fact, it is estimated that up to 40% of a cell's energy supply may be used for active transport of particles through its membranes.

The mechanism of active transport involves specific carrier molecules found within cell membranes (fig. 3.15). These carrier molecules are proteins which have binding sites that combine with the particles being transported. Such a union triggers the release of some cellular energy, and this causes the shape of a carrier molecule to be altered. As a result a "passenger" molecule is moved through the membrane. Once on the other side, the transported molecules are released, and the carriers can accept other passenger molecules at their binding sites.

Particles that are carried across cell membranes by active transport include various sugars and amino acids, as well as a variety of ions, such as those of sodium, potassium, calcium, and hydrogen. Active transport is also involved in the absorption of food molecules into cells that line intestinal walls.

Endocytosis

Another process by which substances may move through cell membranes is called **endocytosis.** In this case, molecules or other particles that are too large to enter a cell by diffusion or active transport may be conveyed by means of a vesicle formed from a section of the cell membrane. Two forms of endocytosis are pinocytosis and phagocytosis.

Pinocytosis refers to the process by which cells take in tiny droplets of liquid from their surroundings. When this happens, a small portion of a cell membrane becomes indented. The open end of the tubelike part thus formed seals off and creates a small vesicle, which becomes detached from the surface and moves into the cytoplasm.

Eventually the vesicular membrane breaks down, and the liquid inside becomes part of the cytoplasm. In this way, a cell is able to take in water and the particles dissolved in it, such as proteins, that otherwise might be unable to enter because of their large molecular size.

Figure 3.16
A cell takes in a solid particle from its surroundings by phagocytosis.

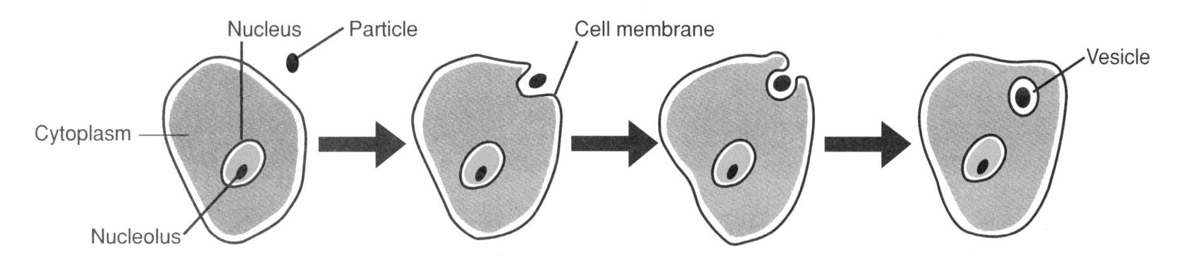

Chart 3.2 Movements through cell membranes

Process	Characteristics	Source of energy	Example
Diffusion	Molecules or ions move from regions of higher concentration toward regions of lower concentration	Molecular motion	Exchange of oxygen and carbon dioxide in lungs
Facilitated diffusion	Molecules are moved through a membrane by carrier molecules from a region of higher concentration to one of lower concentration	Molecular motion	Movement of glucose through a cell membrane
Osmosis	Water molecules move from regions of higher concentration toward regions of lower concentration through a selectively permeable membrane	Molecular motion	Distilled water entering a cell
Filtration	Molecules are forced from regions of higher pressure to regions of lower pressure	Hydrostatic pressure	Molecules leaving blood capillaries
Active transport	Molecules or ions are carried through membranes by other molecules from regions of lower concentration toward regions of higher concentration	Cellular energy	Movement of various ions and amino acids through membranes
Pinocytosis	Membrane acts to engulf minute droplets of liquid from surroundings	Cellular energy	Membrane forming vesicles containing liquid and dissolved particles
Phagocytosis	Membrane acts to engulf solid particles from surroundings	Cellular energy	White blood cell engulfing bacterial cell

Phagocytosis describes a process that is essentially the same as pinocytosis. In phagocytosis, however, the material taken into the cell is solid rather than liquid.

Certain kinds of white blood cells are called *phagocytes* because they can take in tiny solid particles, such as bacterial cells, by phagocytosis. When a phagocyte first encounters a particle, the particle becomes attached to the phagocyte's cell membrane. This stimulates a portion of the membrane to project outward, surround the particle, and slowly draw it inside. The part of the membrane surrounding the solid detaches from the cell's surface, and a vesicle containing the particle is formed (fig. 3.16).

Commonly, a lysosome then combines with such a newly formed vesicle, and the lysosomal digestive enzymes cause the vesicular contents to be decomposed. The products of this decomposition may diffuse out of the lysosome and into the cell's cytoplasm. Any remaining residue usually is expelled from the cell (exocytosis).

Chart 3.2 summarizes the various types of movements through cell membranes.

1. What type of mechanism is responsible for maintaining unequal concentrations of ions on opposite sides of a cell membrane?
2. How are the processes of facilitated diffusion and active transport similar? How are they different?
3. What is the difference between pinocytosis and phagocytosis?

Life Cycle of a Cell

The series of changes that a cell undergoes from the time it is formed until it reproduces is called its *life cycle*. Superficially, this cycle seems rather simple—a newly formed cell grows for a time and then divides to form two new cells, which in turn may grow and divide. Yet the details of the cycle are quite complex, involving mitosis, cytoplasmic division, interphase, and differentiation.

Mitosis

Cell reproduction involves the dividing of a cell into two portions and includes two separate processes: (1) the division of the nuclear parts, which is called **mitosis;** and (2) the division of the cytoplasm. Another type of cell division, called *meiosis,* occurs during maturation of sex cells. It is described in chapter 19.

The division of the nuclear parts by mitosis is, of necessity, very precise, because the nucleus contains information, in the form of DNA molecules, that directs cell parts to carry on life processes. Each new cell resulting from mitosis must have a copy of this information in order to survive.

Although mitosis is often described in terms of stages, the process is really a continuous one without marked changes between one step and the next (fig. 3.17). The idea of stages is useful, however, to indicate the sequence in which major events occur.

The stages of mitosis include the following:

1. **Prophase.** One of the first indications that a cell is going to reproduce is the appearance of *chromosomes* scattered throughout the nucleus. These structures form from chromatin in the nucleus as fibers of chromatin condense into tightly coiled, rodlike parts. The resulting chromosomes contain DNA and protein molecules. Sometime earlier (during interphase, which is described on page 68), the DNA molecules became replicated (duplicated), and consequently each chromosome is composed of two identical portions (chromatids). These parts are temporarily attached by a region on each called the *centromere.*

   The centrioles of the centrosome replicate just before the onset of mitosis, and during prophase the two newly formed pairs move to opposite sides of the cytoplasm. Soon the nuclear envelope and the nucleolus breakup and disappear. Microtubules are assembled from proteins in the cytoplasm, and these structures become associated with the centrioles and chromosomes. A spindle-shaped group of microtubules (spindle fibers) forms between the centrioles as they move apart.
2. **Metaphase.** The chromosomes are lined up in an orderly fashion about midway between the centrioles, as a result of microtubule activity. Spindle fibers become attached to the centromeres of the chromosomes so that a fiber accompanying one pair of centrioles is attached to one side of a centromere, and a microtubule accompanying the other pair of centrioles is attached to the other side.
3. **Anaphase.** Soon the centromeres of the chromosome parts separate, and these identical pairs become individual chromosomes. The separated chromosomes now move in opposite directions. Once again, the movement results from microtubule activity. In this case, the spindle fibers shorten and their attached chromosomes move toward the centrioles at opposite sides of the cell.
4. **Telophase.** The final stage of mitosis is said to begin when the chromosomes complete their migration toward the centrioles. It is much like prophase, but in reverse. As the chromosomes approach the centrioles, the chromosomes begin to elongate and change from rodlike into threadlike structures. A nuclear envelope forms around each chromosome set, and nucleoli appear within the newly formed nuclei. Finally, the microtubules disappear (see fig. 3.18).

Cytoplasmic Division

Cytoplasmic division begins during anaphase, when the cell membrane starts to constrict. The membrane continues constricting through telophase. This process involves the musclelike contraction of a ring of microfilaments, which are assembled in the cytoplasm and attached to the inner surface of the cell membrane. This contractile ring is positioned at right angles to the microtubules that pulled the chromosomes to opposite sides of the cell during mitosis. As it pinches inward, the ring separates the two newly formed nuclei and divides about half of the cytoplasmic organelles into each of the new cells.

Although the newly formed cells may differ slightly in size and number of cytoplasmic parts, they have identical chromosomes and thus contain identical DNA information. Except for size, they are copies of the parent cell.

Figure 3.17
Mitosis is a continuous process during which the nuclear parts of a cell are divided into two equal portions. After reading about mitosis, identify the phases of the process and the cell parts shown in this diagram.

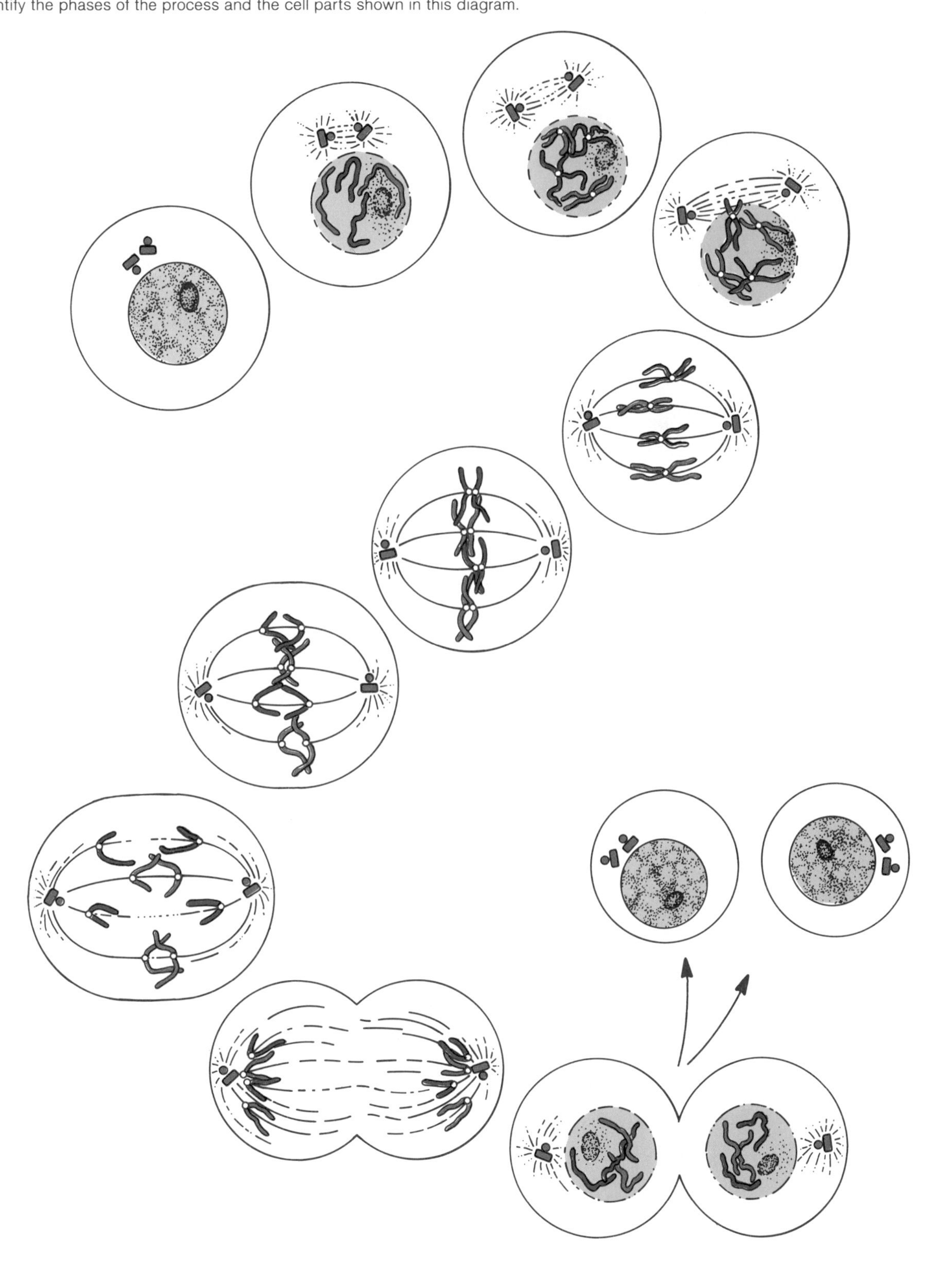

Figure 3.18
Note the two groups of darkly stained chromosomes within the cell near the center of this micrograph (×250). What stage in the process of mitosis does this illustrate?

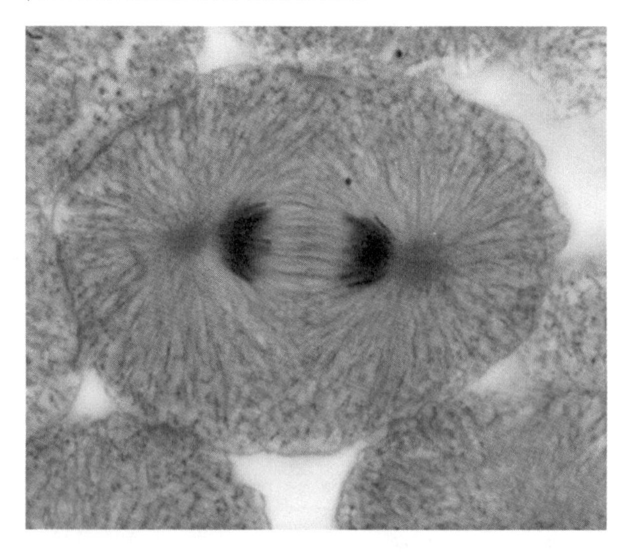

Interphase

Once formed, the new cells usually begin growing. This requires that the young cells obtain nutrients, utilize them in the manufacture of new living material, and synthesize many vital compounds. At the same time, various cell parts become duplicated. In the nucleus, the chromosomes may be doubling (at least in cells that will soon reproduce); and in the cytoplasm the centrosomes replicate, and new ribosomes, lysosomes, mitochondria, and various membranes appear. This stage in the life cycle is called **interphase,** and it lasts until the cell begins to undergo mitosis.

Many kinds of body cells are constantly growing and reproducing, thus increasing the number of cells present. Such activity is responsible for the growth and development of an embryo into a child and of a child into an adult. It also is necessary for replacing cells with relatively short life spans, such as those that form the skin and stomach lining, as well as cells that are worn out or lost due to injury or disease.

Cell Differentiation

Because all body cells are formed by mitosis and contain the same DNA information, they might be expected to look and act alike; obviously, they do not.

A human begins life as a single cell—a fertilized egg cell. This cell reproduces to form 2 new cells; they, in turn, divide into 4 cells, the 4 become 8, and so forth. Then, sometime during development, the cells begin to *specialize* (fig. 3.19). That is, they develop special structures or begin to function in different ways. Some become skin cells, others become bone cells, and still others become nerve cells.

The process by which cells develop different characteristics in structure and function is called **differentiation.** The mechanism responsible for this phenomenon is not well understood, but it involves the expression of some of the DNA information and the repression of other DNA information. Thus, the DNA information needed for general cell activities may be "switched on" in both nerve and bone cells. The information related to specific bone cell functions, however, may be "switched off" in nerve cells. Similarly, the information related to specific nerve cell functions may be "switched off" in bone cells.

Although the mechanism of differentiation is obscure, the results are obvious. Cells of many kinds are produced, and each kind carries on specialized functions. For example, skin cells protect underlying tissues, red blood cells carry oxygen, and nerve cells transmit impulses. Each type of cell somehow helps the others and aids in the survival of the organism.

1. Why is it important that the division of nuclear materials during mitosis be so precise?
2. Describe the events that occur during mitosis.
3. Name the process by which some cells become muscle cells and others become nerve cells.

Figure 3.19
During development, numerous body cells are produced from a single fertilized egg cell by mitosis. As these cells undergo differentiation, they become different kinds of cells that carry on specialized functions.

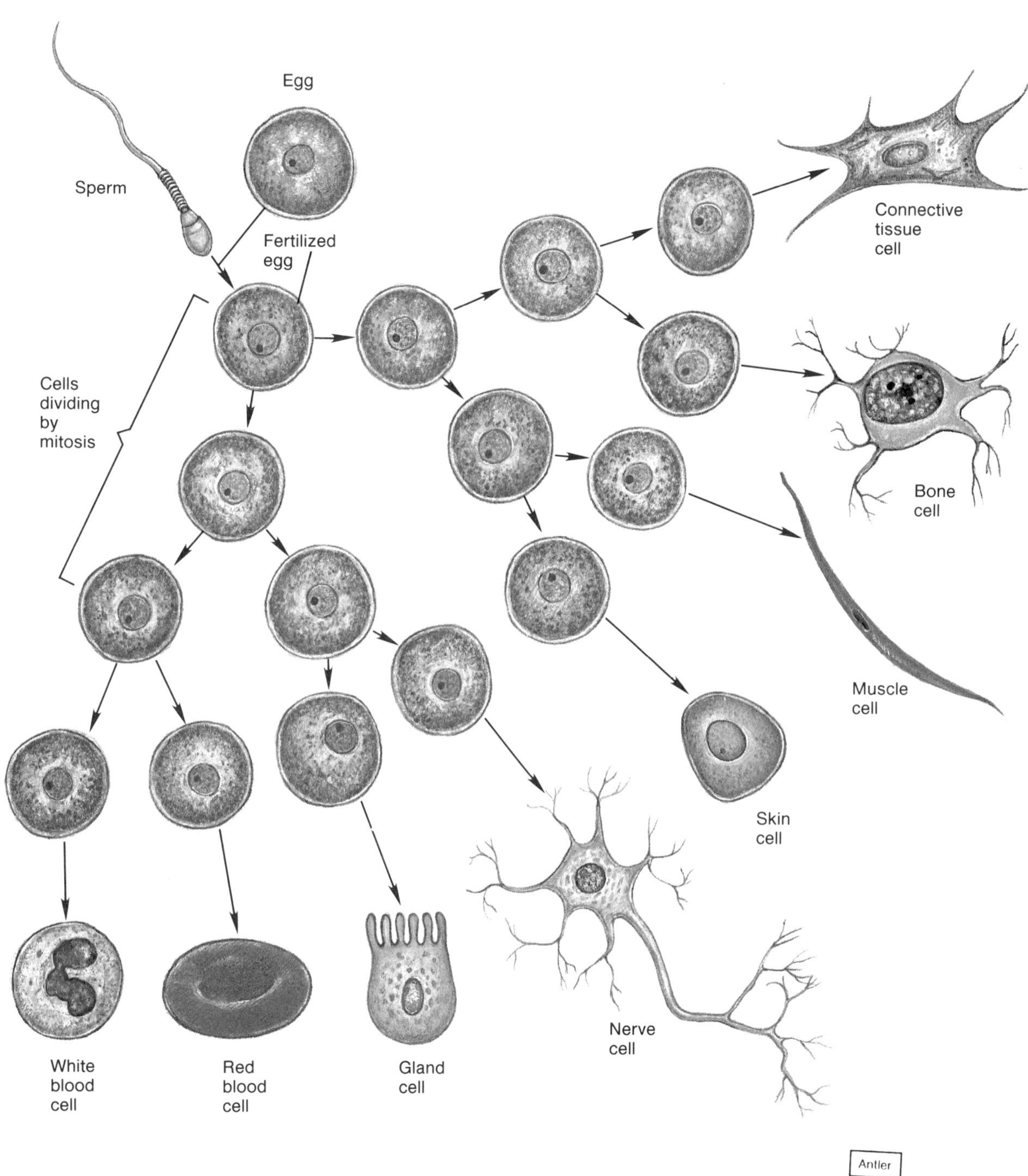

Antler

A Current Topic
Cancer

Cancer is a group of closely related diseases. It can occur in many different tissues, and it results from changes in cells that inactivates some of the mechanisms that regulate cellular activities. Cancers have certain common characteristics, including the following:

1. **Hyperplasia.** Hyperplasia is the uncontrolled reproduction of cells. Although the rate of reproduction among cancer cells is usually unchanged, they are not responsive to normal controls on cell numbers. As a result, cancer cells tend to give rise to very large cell populations.
2. **Anaplasia.** The word anaplasia refers to the appearance of abnormalities in cellular structure. Typically, cancer cells resemble undifferentiated cells. That is, they fail to develop the specialized structure of the kind of cell they represent. Also, they fail to function in the expected ways. Certain white blood cells, for example, normally function in resisting bacterial infections. When such cells become cancerous, they do not function effectively, and the cancer patient becomes more subject to infectious diseases. Cancer cells are also likely to form disorganized masses rather than to become arranged in orderly groups as normal cells do.
3. **Metastasis.** Metastasis is a tendency to spread. Normal cells are usually cohesive; that is, they stick together in groups of similar kinds. Cancer cells often become detached from their cellular mass and may then be carried away from their place of origin to establish new cancerous growths in other body parts. Metastasis is the characteristic most closely associated with the word *malignant,* which suggests a power to threaten life. This characteristic is often used to distinguish a cancerous growth from a noncancerous (benign) one.

The cause or causes of cancer remain poorly understood, although increasing evidence indicates the cause or causes are complex. Among the factors that may be involved are exposure to various chemicals or to harmful radiation, viral infections, changes in DNA structure, the presence of altered genes (oncogenes) that encode abnormal proteins (see chapter 4), or deficiencies in the body's immune system (see chapter 16). Two or more of these factors may be needed to cause a cancer, or perhaps different factors or combinations of factors cause different kinds of cancer.

When untreated, cancer cells eventually accumulate in large numbers. They may damage normal cells by effectively competing with them for nutrients. At the same time, cancer cells may invade vital organs and interfere with their functions, obstruct important passageways, or penetrate blood vessels and cause internal bleeding.

If detected in its early stage, a cancerous growth may be removed surgically. At other times, cancers are treated with radiation or drugs (chemotherapy), or with some combination of surgery and treatments. The drugs most commonly used act on the structure of DNA molecules, affecting the DNA of rapidly reproducing normal cells as well as the DNA of cancer cells. However, the normal cells seem able to repair DNA damage more effectively than cancer cells can. Thus, the cancer cells are more likely to be destroyed by the drug's action.

Chapter Summary

Introduction (page 52)

The shapes of cells are closely related to their functions.

Composite Cell (page 52)

A cell includes a nucleus, cytoplasm, and a cell membrane. Cytoplasmic organelles perform specific vital functions, but the nucleus controls the overall activities of the cell.

1. Cell membrane
   a. The cell membrane forms the outermost limit of the living material.
   b. It acts as a selectively permeable passageway, which controls the entrance and exit of substances.
   c. It includes protein, lipid, and carbohydrate molecules.
   d. The cell membrane's framework consists mainly of a double layer of phospholipid molecules.
   e. Molecules that are soluble in lipids pass through the membrane easily, but the membrane creates a barrier to the passage of water soluble molecules.
2. Cytoplasm
   a. The endoplasmic reticulum provides a tubular communication system in the cytoplasm and functions in the synthesis of lipids and proteins.
   b. Ribosomes function in protein synthesis.
   c. The Golgi apparatus packages glycoproteins for secretion.
   d. Mitochondria contain enzymes involved with releasing energy from food molecules.
   e. Lysosomes contain digestive enzymes, which can destroy substances that enter cells.
   f. The centrosome aids in the distribution of chromosomes during cell reproduction.
   g. Vesicles contain substances that recently entered the cell.
   h. Cilia and flagella are motile processes that extend outward from cell surfaces.
   i. Microfilaments and microtubules aid cellular movements and provide support and stability to the cytoplasm and various cellular parts.
3. Cell nucleus
   a. The nucleus is enclosed in a nuclear envelope.
   b. It contains a nucleolus, which functions in the production of ribosomes.
   c. It contains chromatin, which is composed of loosely coiled fibers of protein and DNA; chromatin fibers become chromosomes during cellular reproduction.

Movements through Cell Membranes (page 60)

The cell membrane provides a surface through which substances enter and leave a cell.

1. Diffusion
   a. Diffusion is a scattering of molecules or ions from regions of higher concentration toward regions of lower concentration.
   b. It is responsible for exchanges of oxygen and carbon dioxide within the body.
2. Facilitated diffusion
   a. In facilitated diffusion, special carrier molecules move substances through the cell membrane.
   b. This process can only move substances from regions of higher concentration toward regions of lower concentration.
3. Osmosis
   a. Osmosis is a case of diffusion in which water molecules diffuse from regions of higher water concentration toward regions of lower water concentration through a selectively permeable membrane.
   b. Osmotic pressure increases as the number of particles dissolved in a solution increases.
   c. Cells lose water when placed in hypertonic solutions and gain water when placed in hypotonic solutions.
   d. A solution is isotonic when it contains the same concentration of solute particles as a cell.
4. Filtration
   a. Filtration involves the movement of molecules from regions of higher hydrostatic pressure toward regions of lower hydrostatic pressure.
   b. Blood pressure causes filtration through porous capillary walls.
5. Active transport
   a. Active transport is responsible for the movement of molecules or ions from regions of lower concentration toward regions of higher concentration.
   b. It requires cellular energy and involves the action of carrier molecules.
6. Endocytosis
   a. Endocytosis is a process by which relatively large particles may be conveyed through a cell membrane.
   b. A cell membrane may engulf tiny droplets of liquid by pinocytosis.
   c. Solid particles may be engulfed by phagocytosis.

Life Cycle of a Cell (page 65)

The life cycle of a cell includes mitosis, cytoplasmic division, interphase, and differentiation.

1. Mitosis
   a. Mitosis involves the division and distribution of nuclear parts to new cells during cellular reproduction.
   b. The stages of mitosis include prophase, metaphase, anaphase, and telophase.
2. Cytoplasmic division is a process by which the cytoplasm is divided into two portions following mitosis.
3. Interphase
   a. Interphase is the stage in the life cycle when a cell grows and forms new organelles.
   b. It terminates when the cell begins to undergo mitosis.
4. Cell differentiation involves the development of specialized structures and functions.

Clinical Application of Knowledge

1. Which process—diffusion, osmosis, or filtration—is most closely related to each of the following situations?
   a. A person is given an injection of isotonic glucose solution.
   b. The concentration of urea in the wash solution of an artificial kidney is decreased.
2. What characteristic of cell membranes may account for the observation that fat-soluble substances like chloroform and ether cause rapid effects upon cells?
3. A person who has been exposed to excessive amounts of X ray may develop a decrease in white blood cell number and an increase in susceptibility to infections. In what way are these effects related?
4. Exposure to tobacco smoke causes cilia to become immobile and eventually to disappear. How would you relate this to the fact that tobacco smokers have an increased incidence of respiratory infections?

Review Activities

1. Describe how the shapes of nerve and muscle cells are related to their functions.
2. Name the two major portions of a cell, and describe their relationship to one another.
3. Define *selectively permeable.*
4. Describe the chemical structure of a cell membrane.
5. Explain how the structure of a cell membrane is related to its permeability.
6. Describe the structure and functions of each of the following:
   a. endoplasmic reticulum
   b. ribosome
   c. Golgi apparatus
   d. mitochondrion
   e. lysosome
   f. centrosome
   g. vesicle
   h. cilium and flagellum
   i. microfilament and microtubule
7. Describe the structure of the nucleus and the functions of its parts.
8. Define *diffusion.*
9. Explain how diffusion aids in the exchange of gases within the body.
10. Distinguish between diffusion and facilitated diffusion.
11. Define *osmosis.*
12. Define *osmotic pressure.*
13. Distinguish between solutions that are hypertonic, hypotonic, and isotonic.
14. Define *filtration.*
15. Explain how filtration aids in the movement of substances through capillary walls.
16. Explain the function of carrier molecules in active transport.
17. Distinguish between pinocytosis and phagocytosis.
18. Explain how the cytoplasm is divided during cell reproduction.
19. List the phases in the life cycle of a cell.
20. Name the two processes included in cell reproduction.
21. Describe the major events of mitosis.
22. Explain what happens during interphase.
23. Define *differentiation.*

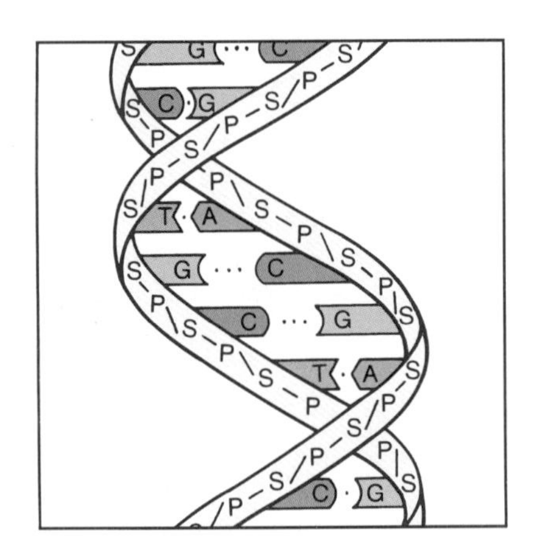

4 Cellular Metabolism

Cellular metabolism includes all the chemical reactions that occur within cells. These reactions, for the most part, involve the utilization of food substances and are of two major types. In one type of reaction, molecules of nutrients are converted into simpler forms through changes accompanied by the release of energy. In the other type, nutrient molecules are used in constructive processes that produce cell structural and functional materials or molecules that store energy.

In either case, metabolic reactions are controlled by proteins called enzymes, which are synthesized in cells. The production of these enzymes is, in turn, controlled by genetic information held within molecules of DNA, which instructs a cell how to synthesize specific kinds of enzyme molecules.

Chapter Objectives

After you have studied this chapter, you should be able to

1. Define *anabolic* and *catabolic metabolism.*
2. Explain how enzymes control metabolic processes.
3. Explain how chemical energy is released by respiratory processes.
4. Describe how energy is made available for cellular activities.
5. Describe the general metabolic pathways of carbohydrates, lipids, and proteins.
6. Explain how genetic information is stored within nucleic acid molecules.
7. Explain how genetic information is used in the control of cellular processes.
8. Describe how DNA molecules are replicated.
9. Complete the review activities at the end of this chapter. Note that the items are worded in the form of specific learning objectives. You may want to refer to them before reading the chapter.

Key Terms

aerobic respiration (a″er-o′bik res″pĭ-ra′shun)

anabolic metabolism (an″ah-bol′ik mĕ-tab′o-lizm)

anaerobic respiration (an″a-er-o′bik res″pĭ-ra′shun)

anticodon (an″tĭ-ko′don)

catabolic metabolism (kat″ah-bol′ik mĕ-tab′o-lizm)

codon (ko′don)

dehydration synthesis (de″hi-dra′shun sin′thĕ-sis)

enzyme (en′zīm)

gene (jēn)

hydrolysis (hi-drol′ĭ-sis)

oxidation (ok″sĭ-da′shun)

replication (re″pli-ka′shun)

substrate (sub′strāt)

Aids to Understanding Words

an-, without: *an*aerobic respiration—a respiratory process that proceeds without oxygen.

ana-, up: *ana*bolic metabolism—cellular processes in which smaller molecules are used to build up larger ones.

cata-, down: *cata*bolic metabolism—cellular processes in which larger molecules are broken down into smaller ones.

de-, undoing: *de*amination—a process by which the nitrogen-containing portions of amino acid molecules are removed.

mut-, change: *mut*ation—a change in the genetic information of a cell.

-zym, causing to ferment: en*zyme*—a protein that initiates or speeds up a chemical reaction without itself being changed.

Figure 4.1
What type of reaction is represented in this diagram?

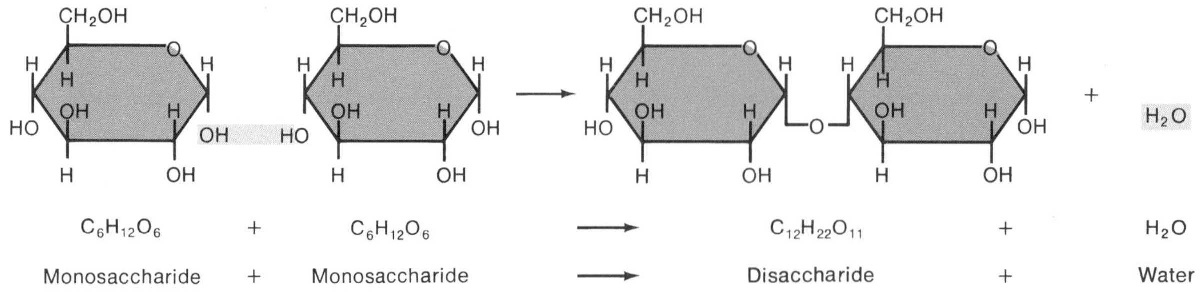

$C_6H_{12}O_6$ + $C_6H_{12}O_6$ → $C_{12}H_{22}O_{11}$ + H_2O

Monosaccharide + Monosaccharide → Disaccharide + Water

Introduction

ALTHOUGH A LIVING cell may appear to be idle, it is actually very active because the numerous metabolic processes needed to maintain its life are occurring all the time. These processes involve numerous chemical reactions, each of which is controlled by a specific protein and depends upon the release of cellular energy.

Metabolic Processes

Metabolic processes can be divided into two major types. One type, called *anabolic metabolism,* involves the buildup of larger molecules from smaller ones and utilizes energy. The other, called *catabolic metabolism,* involves the breakdown of larger molecules into smaller ones and releases energy.

Anabolic Metabolism

Anabolic metabolism includes all the constructive processes used to manufacture the substances needed for cellular growth and repair. For example, cells often join many simple sugar molecules (monosaccharides) to form larger molecules of glycogen by an anabolic process called **dehydration synthesis.** In this process, the larger carbohydrate molecule is built by bonding monosaccharide molecules together into a complex chain. As adjacent monosaccharide units are joined, an $-OH$ (hydroxyl group) from one monosaccharide molecule and an $-H$ (hydrogen atom) from an $-OH$ group of another are removed. These particles react to produce a water molecule, and the monosaccharides are united by a shared oxygen atom (fig. 4.1). As this process is repeated, the molecular chain becomes longer.

Similarly, glycerol and fatty acid molecules are joined by dehydration synthesis in fat (adipose) tissue cells to form fat molecules. In this case, 3 hydrogen atoms are removed from a glycerol molecule, and an $-OH$ group is removed from each of the 3 fatty acid molecules (fig. 4.2). The result is 3 water molecules and a single fat molecule, whose glycerol and fatty acid portions are bound by shared oxygen atoms.

Cells also build protein molecules by joining amino acids by means of dehydration synthesis (fig. 4.3). When two amino acid molecules are united, an $-OH$ from one and a hydrogen atom from the $-NH_2$ group of another are removed. A water molecule is formed, and the amino acid molecules are joined by a bond between a carbon atom and a nitrogen atom. This type of bond, which is called a *peptide bond,* holds the amino acids together. Two amino acids bound together form a *dipeptide molecule,* and many joined in a chain form a *polypeptide molecule.* Generally, a polypeptide consisting of 100 or more amino acid molecules is called a *protein,* although the boundary distinguishing polypeptides and proteins is not defined precisely.

Anabolic steroids are a group of lipids that stimulate anabolic metabolism and, thus, promote the growth of certain tissues. These substances are sometimes prescribed by physicians to treat various disease conditions; however, anabolic steroids are also used illicitly by some individuals to increase muscle mass, often with the hope of enhancing athletic performance. Such nonmedical use of steroids is associated with a variety of undesirable and dangerous side effects, including adverse changes in liver functions, increased risk of heart and blood vessel diseases, upsets in normal hormonal balances, and severe psychological disorders.

Catabolic Metabolism

Physiological processes in which larger molecules are broken down into smaller ones constitute **catabolic metabolism.** An example of such a process is **hydrolysis,** which can bring about the decomposition of carbohydrates, lipids, and proteins.

When a molecule of one of these substances is hydrolyzed, a water molecule is used, and the original molecule is split into two simpler parts. The hydrolysis

Figure 4.2
A glycerol molecule and 3 fatty acid molecules join by dehydration synthesis to form a fat molecule.

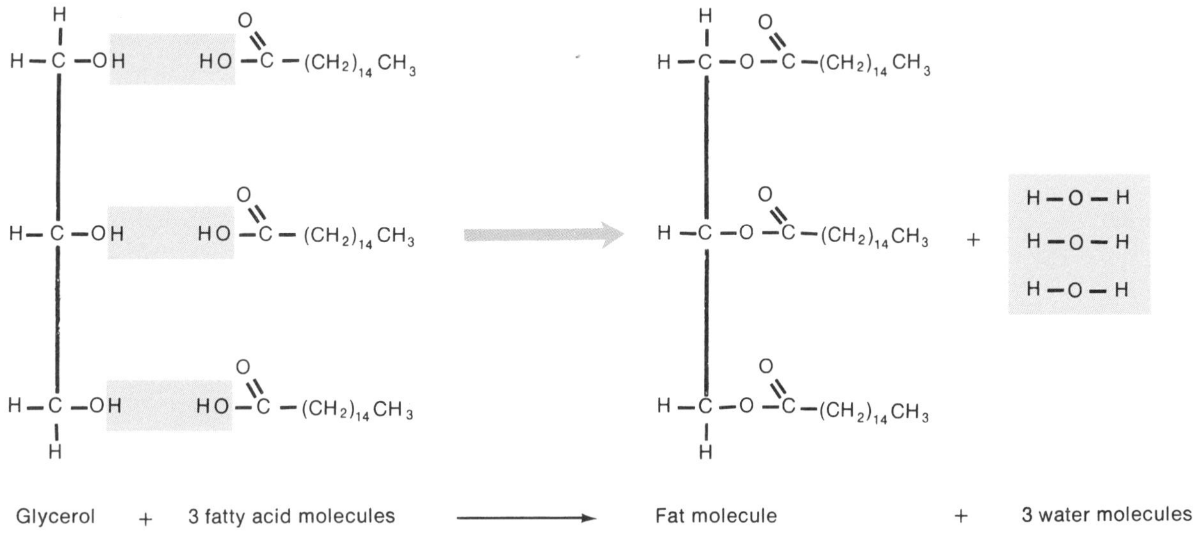

Glycerol + 3 fatty acid molecules ⟶ Fat molecule + 3 water molecules

Figure 4.3
When 2 amino acid molecules are united by dehydration synthesis, a peptide bond is formed between a carbon atom and a nitrogen atom.

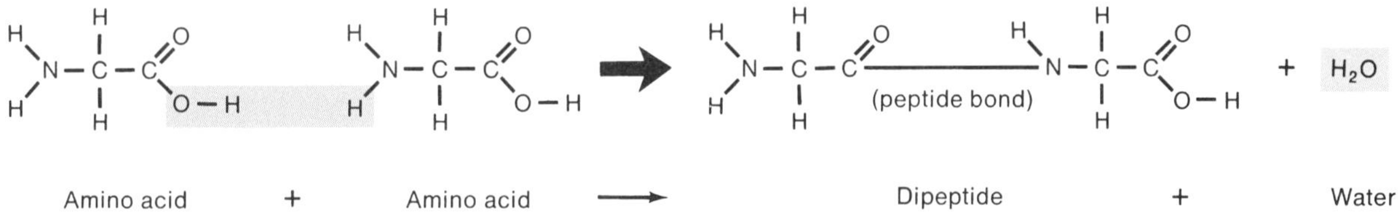

Amino acid + Amino acid ⟶ Dipeptide (peptide bond) + Water (H_2O)

of a disaccharide such as sucrose, for instance, results in the molecules of two monosaccharides, glucose and fructose.

$$\underset{\text{(sucrose)}}{C_{12}H_{22}O_{11}} + \underset{\text{(water)}}{H_2O} \rightarrow \underset{\text{(glucose)}}{C_6H_{12}O_6} + \underset{\text{(fructose)}}{C_6H_{12}O_6}$$

In this case, the bond between the simple sugars within the sucrose molecule is broken, and the water molecule supplies a hydrogen atom to one sugar molecule and a hydroxyl group to the other. Thus, hydrolysis is the reverse of dehydration synthesis, illustrated in figures 4.1, 4.2, and 4.3.

$$\text{Disaccharide} + \text{Water} \underset{\text{Dehydration synthesis}}{\overset{\text{Hydrolysis}}{\rightleftarrows}} \text{Monosaccharide} + \text{Monosaccharide}$$

Hydrolysis, more commonly called *digestion,* occurs in various regions of the digestive tract and is responsible for the breakdown of carbohydrates into monosaccharides, fats into glycerol and fatty acids, proteins into amino acids, and nucleic acids into nucleotides. (Digestion is discussed in more detail in chapter 12.)

1. What general functions does anabolic metabolism serve? Catabolic metabolism?
2. What substance is formed by the anabolic metabolism of monosaccharides? Of amino acids? Of glycerol and fatty acids?
3. Distinguish between dehydration synthesis and hydrolysis.

Control of Metabolic Reactions

Although different kinds of cells may conduct specialized metabolic processes, all cells perform certain basic reactions, such as the buildup and breakdown of carbohydrates, lipids, proteins, and nucleic acids. These reactions actually include hundreds of very specific chemical changes, which must occur in orderly fashion.

Enzymes and Their Actions

Like other chemical reactions, metabolic reactions generally require a certain amount of energy, such as heat, before they occur. The temperature conditions that exist in cells are usually too mild to promote the reactions needed to support life, but this is not a problem because cells contain enzymes.

Enzymes are proteins that promote chemical reactions within cells by lowering the amount of energy (activation energy) needed to start these reactions. Thus, in the presence of enzymes, metabolic reactions are speeded up. Enzymes are needed in very small quantities, because as they function, they are not used up and can, therefore, function again and again.

The reaction promoted by a particular enzyme is very specific. Each enzyme acts only on a particular substance, which is called its **substrate.** For example, the substrate of one enzyme, called *catalase,* is hydrogen peroxide, a toxic by-product of certain metabolic reactions. This enzyme's only function is to cause the decomposition of hydrogen peroxide into water and oxygen. In this way, it helps prevent an accumulation of hydrogen peroxide, which might damage cells.

Cellular metabolism includes hundreds of different chemical reactions, and each of these reactions is controlled by a specific kind of enzyme. Thus, hundreds of different kinds of enzymes must be present in every cell, and each enzyme must be able to "recognize" its specific substrate. This ability to identify its substrate depends upon the shape of the enzyme molecule. That is, each enzyme's polypeptide chain is twisted and coiled into a unique three-dimensional form that fits the special shape of its substrate molecule. In short, an enzyme molecule fits a molecule of its substrate much as a key fits a particular lock.

During an enzyme-controlled reaction, particular regions of the enzyme molecule called *active sites* temporarily combine with portions of the substrate, creating a substrate-enzyme complex (fig. 4.4). At the same time, the interaction between the molecules seems to distort or strain chemical bonds within the substrate, increasing the likelihood of a change in the substrate molecule. When the substrate is changed, the product of the reaction appears, and the enzyme is released in its original form.

Often the protein portion of an enzyme molecule is inactive until it is combined with an additional substance. This nonprotein part is needed to complete the proper shape of the active site of the enzyme molecule or to help bind the enzyme to its substrate. Such a substance is called a **cofactor**. A cofactor may be an ion of an element, such as copper, iron, or zinc, or it may be a relatively small organic molecule, in which case it is called a **coenzyme**. Most coenzymes are obtained from vitamins. (Vitamins are discussed in more detail in chapter 12.)

Figure 4.4
(*a*) The shape of a substrate molecule fits (*b*) the shape of the enzyme's active site. (*c*) When the substrate molecule becomes temporarily combined with the enzyme, a chemical reaction occurs. The result is (*d*) product molecules and (*e*) an unaltered enzyme.

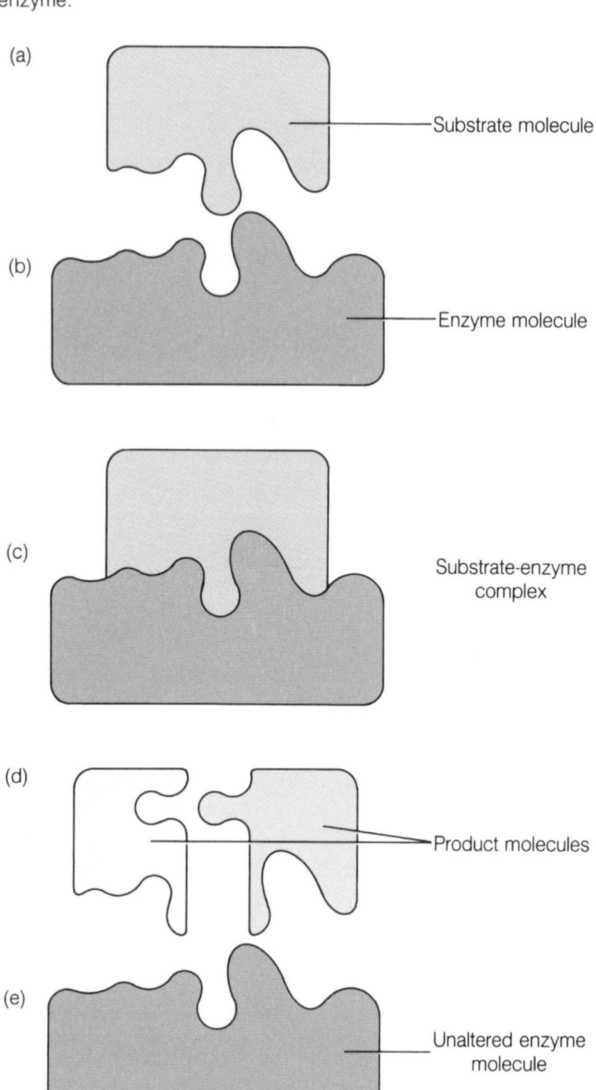

Factors That Alter Enzymes

Enzymes are proteins, and like other proteins they can be denatured by exposure to excessive heat, radiation, electricity, certain chemicals, or fluids with extreme pH values. For example, many enzymes become inactive at 45° C, and nearly all of them are denatured at 55° C. Some poisons are substances that cause enzymes to be denatured. Cyanides, for instance, can interfere with respiratory enzymes and damage cells by halting their energy-releasing processes.

1. What is an enzyme?
2. How does an enzyme recognize its substrate?
3. What factors are likely to denature enzymes?

Energy for Metabolic Reactions

Energy is the capacity to produce changes in matter or to move something; that is, energy is the ability to do work. Therefore, energy is recognized by what it can do; it is utilized whenever changes take place. Common forms of energy include heat, light, sound, electrical energy, mechanical energy, and chemical energy.

Release of Chemical Energy

Most metabolic processes use chemical energy. This form of energy is held in the bonds between the atoms of molecules and is released when these bonds are broken. For example, the chemical energy of many substances can be released by burning. Such a reaction must be started, and this is usually accomplished by applying heat to activate the burning process. As the substance burns, molecular bonds are broken, and energy escapes as heat and light.

Similarly, glucose molecules are "burned" in cells; this process is more correctly called **oxidation.** The energy released by the oxidation of glucose is used to promote cellular metabolism. There are, however, some important differences between the oxidation of substances inside cells and the burning of substances outside them.

Burning usually requires a relatively large amount of energy to activate the process, and most of the energy released escapes as heat or light. In cells, the oxidation process is initiated by enzymes that reduce the amount of energy (activation energy) needed. Also, by transferring energy to special energy-carrying molecules, cells are able to capture in the form of chemical energy about half of the energy released. The rest escapes as heat, which helps maintain body temperature.

1. What is energy?
2. How does cellular oxidation differ from burning?

Anaerobic Respiration

When a 6-carbon glucose molecule is decomposed in the process called *cellular respiration,* enzymes control a series of reactions that break glucose into two 3-carbon pyruvic acid molecules (fig. 4.5). This phase of the process occurs in the cytoplasm, and because it takes place in the absence of oxygen, it is called **anaerobic respiration** (glycolysis).

Figure 4.5
Anaerobic respiration occurs in the cytoplasm in the absence of oxygen, but aerobic respiration occurs in mitochondria in the presence of oxygen.

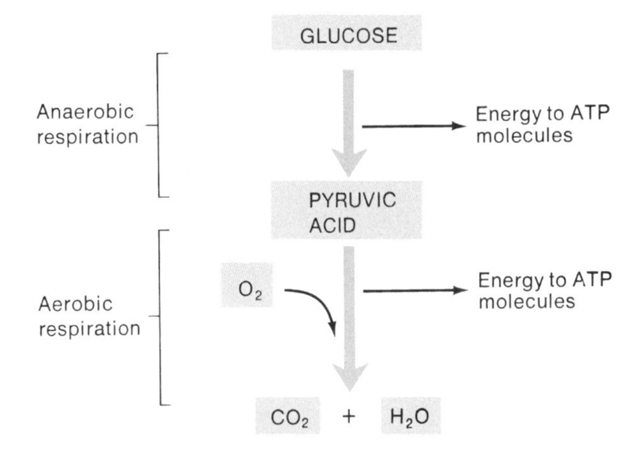

Although some energy is needed to activate the reactions of anaerobic respiration, more energy is released than is used. The excess is transferred to molecules of an energy-carrying substance called **ATP** (adenosine triphosphate).

Aerobic Respiration

Following the anaerobic phase of cellular respiration, oxygen must be available in order for the process to continue. For this reason, the second phase is called **aerobic respiration.** It takes place within mitochondria, and as a result, considerably more energy is transferred to ATP molecules.

When the decomposition of a glucose molecule is complete, carbon dioxide molecules and hydrogen atoms remain. The carbon dioxide diffuses out of the cell as a waste, and the hydrogen atoms combine with oxygen to form water molecules. Thus, the final products of glucose oxidation are carbon dioxide, water, and energy (fig. 4.5).

ATP Molecules

For each glucose molecule that is decomposed completely, 38 molecules of ATP can be produced. Two of these ATP molecules are the result of anaerobic respiration, but the rest are formed during the aerobic phase.

Each ATP molecule contains 3 phosphates in a chain (fig. 4.6). As energy is released during cellular respiration, some of it is captured in the bond of the end phosphate. When energy is needed for some metabolic process, the terminal phosphate bond of an ATP molecule is broken, and the energy stored in this bond is released. Such energy is used for a variety of cellular

Figure 4.6
What is the significance of this cyclic process?

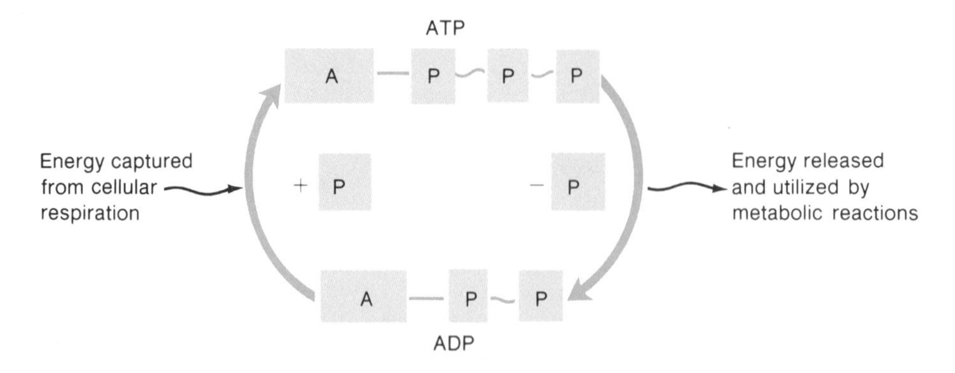

Figure 4.7
A metabolic pathway consists of a series of enzyme-controlled reactions leading to a product.

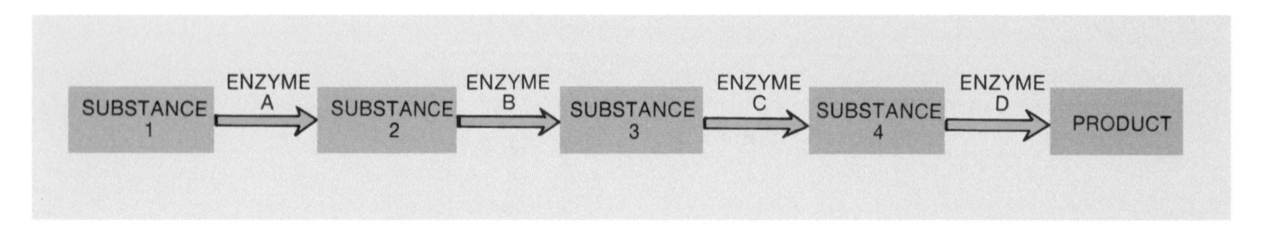

functions including muscle contractions, active transport mechanisms, and the manufacture of various compounds.

An ATP molecule that has lost its terminal phosphate becomes an **ADP** (adenosine diphosphate) molecule. The ADP molecule can convert back into an ATP, however, by capturing some energy and a phosphate. Thus, as figure 4.6 shows, ATP and ADP molecules shuttle back and forth between the energy-releasing reactions of cellular respiration and the energy-utilizing reactions of the cell.

1. What is meant by anaerobic respiration? By aerobic respiration?
2. What are the final products of cellular respiratory reactions?
3. What is the function of ATP molecules?

Metabolic Pathways

Anabolic and catabolic reactions that occur in cells usually involve a number of different steps that must occur in a particular sequence. For example, the anaerobic phase of cellular respiration, by which glucose is converted to pyruvic acid, involves ten separate reactions. Since each reaction is controlled by a specific kind of enzyme, these enzymes must act in proper order. Such a precision of activity suggests that the enzymes are positioned in the exact sequence as that of the reactions they control. The enzymes responsible for aerobic respiration seem to be located in tiny stalked particles on the membranes (cristae) within the mitochondria.

Such a sequence of enzyme-controlled reactions that leads to the production of particular products is called a **metabolic pathway** (fig. 4.7).

Carbohydrate Pathways

The average human diet consists largely of carbohydrates, which are changed by digestion into monosaccharides such as glucose. These substances are used primarily as cellular energy sources, which means they usually enter the catabolic pathways of cellular respiration. As was discussed previously, the first phase of this process occurs in the cytoplasm and is anaerobic (fig. 4.8). It involves the conversion of glucose into molecules of *pyruvic acid.* During the second phase of the process, the pyruvic acid is transformed into a 2-carbon acetyl group, which combines with a molecule of coenzyme A to form a substance called *acetyl coenzyme A.* It, in turn, is transported into a mitochondrion and changed within the mitochondrion into a number of intermediate products by a complex series of chemical reactions known as the **citric acid cycle** (Krebs cycle). As these changes occur, energy is released, and some of it is transferred to molecules of ATP, but the rest is

Figure 4.8
Aerobic respiration, the second phase of cellular respiration, includes a complex series of chemical reactions called the citric acid cycle.

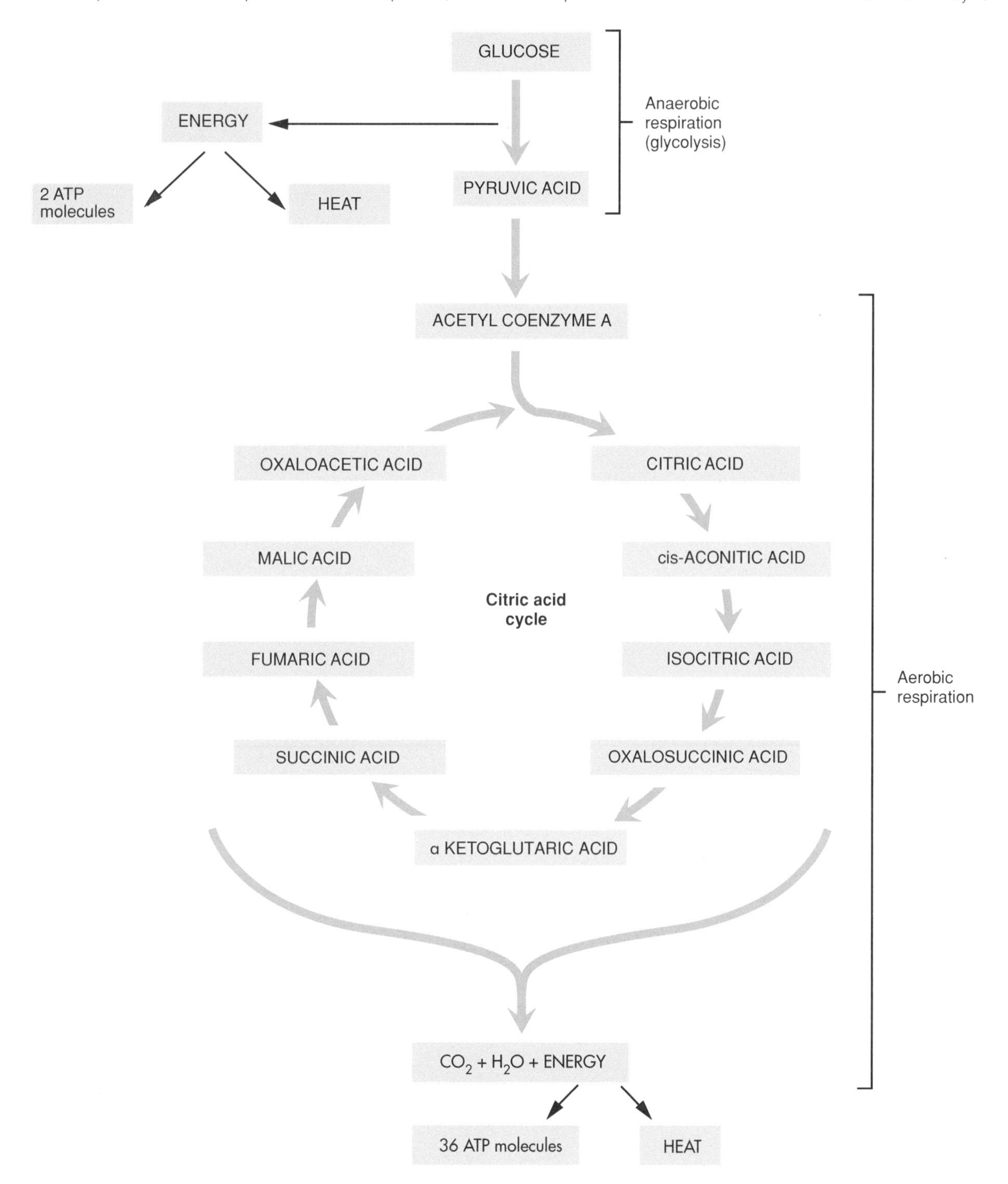

lost as heat. Any excess glucose may enter anabolic pathways and be converted into storage forms such as glycogen or fat.

Lipid Pathways

Although foods may contain lipids in the form of phospholipids or cholesterol, the most common dietary lipids are fats called *triglycerides*. As was explained in chapter 2, the molecule of such a fat consists of a glycerol portion and 3 fatty acids.

Lipids provide for a variety of physiological functions; however, fats are used mainly to supply energy. Gram for gram, fats contain more than twice as much chemical energy as carbohydrates or proteins.

Figure 4.9
Fats from foods are digested into glycerol and fatty acids. These molecules may enter catabolic pathways and be used as energy sources.

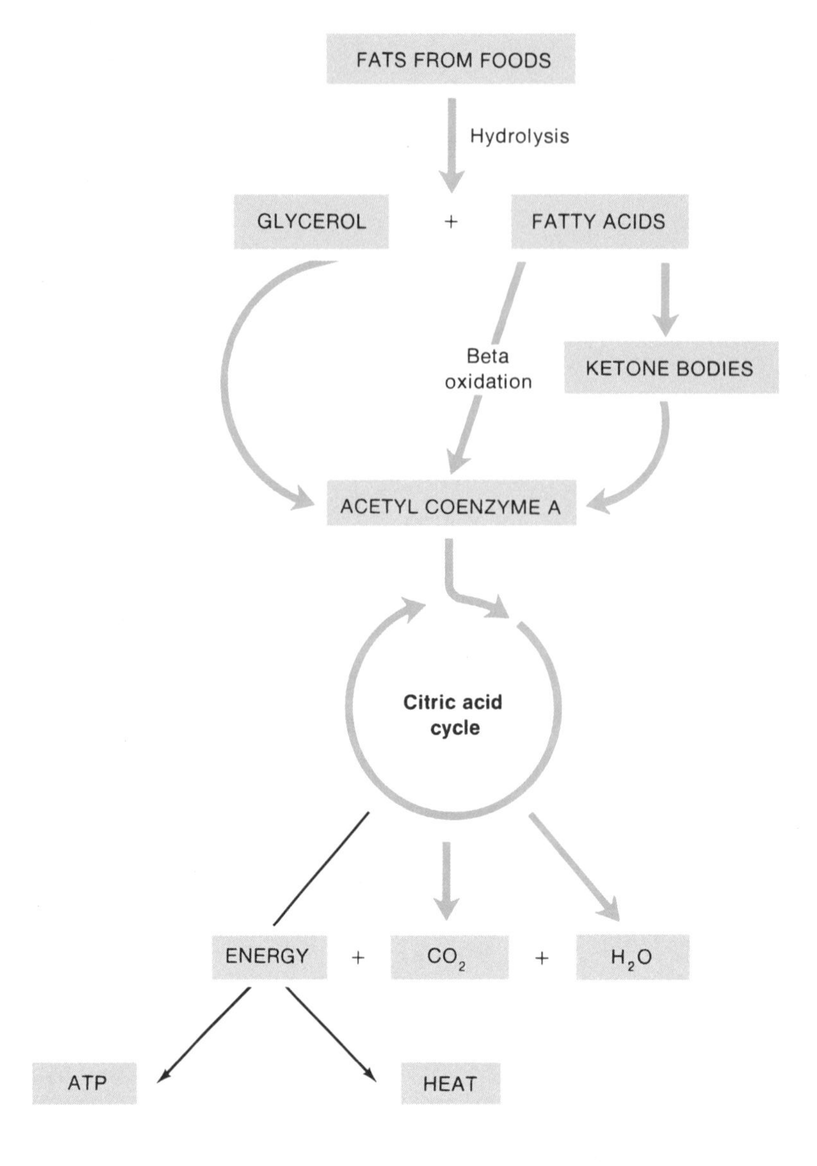

Before energy can be released from a fat molecule it must undergo hydrolysis. As shown in figure 4.9, some of the resulting fatty acid portions can then be converted into molecules of acetyl coenzyme A by a series of reactions called *beta oxidation.* Others can be converted into compounds called *ketone bodies,* such as acetone. Later these may be changed to acetyl coenzyme A, which in turn is oxidized by means of the citric acid cycle. The glycerol portions of the triglyceride molecules can also enter metabolic pathways leading to the citric acid cycle, or they can be used to synthesize glucose.

Another possibility for fatty acid molecules resulting from the hydrolysis of fats is to be changed back into fat molecules by anabolic processes and stored in fat tissue.

When ketone bodies are formed faster than they can be decomposed, some of them are eliminated through the lungs and kidneys. Consequently, the breath and urine may develop a fruity odor due to the presence of the ketone, *acetone.* This sometimes happens when a person fasts or diets to lose weight, thus forcing the cells to metabolize body fat. Persons suffering from *diabetes mellitus* are also likely to metabolize excessive amounts of fats, and they too may have acetone in the breath and urine. At the same time, they may develop a serious imbalance in the pH called *acidosis,* which is due to an accumulation of acidic ketone bodies.

Figure 4.10
Proteins from foods are digested into amino acids, but before these smaller molecules can be used as energy sources, they must be deaminated.

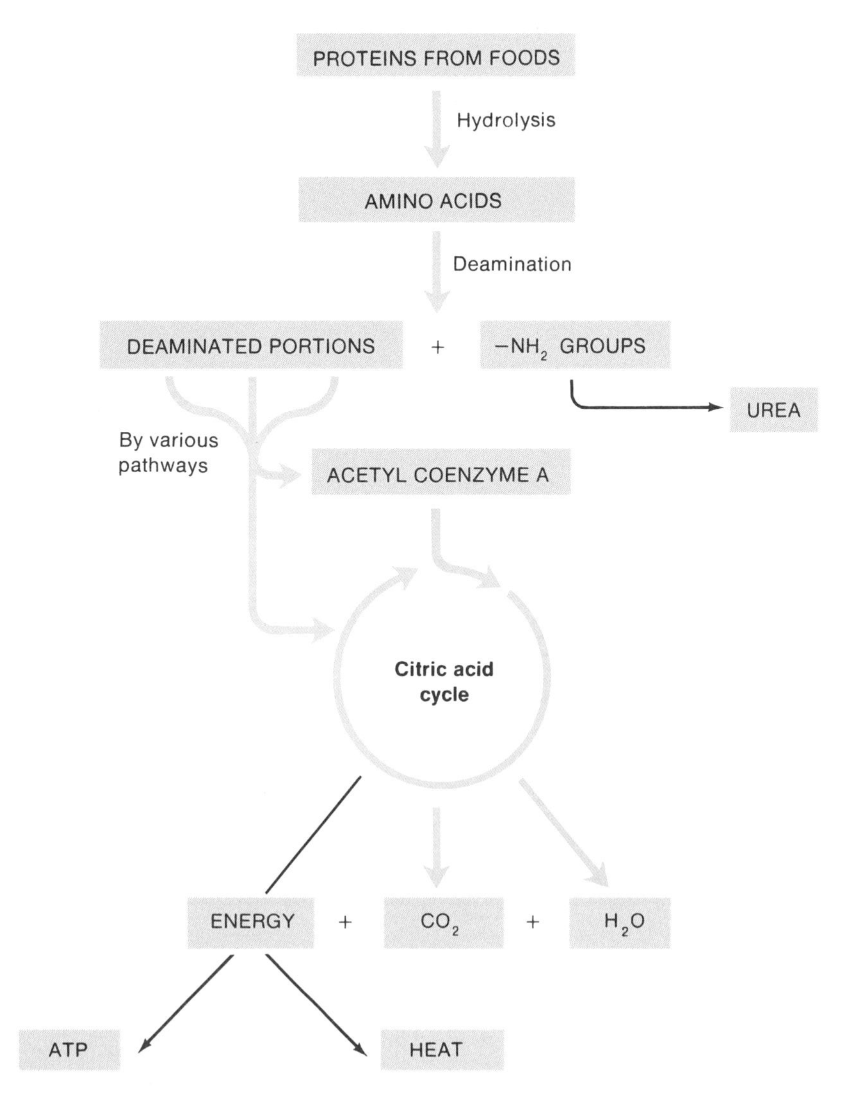

Protein Pathways

When dietary proteins are digested, the resulting amino acids are absorbed and transported by the blood to various body cells. Many of these amino acids are joined to form protein molecules again, which may be incorporated into cell parts or used as enzymes. Others may be decomposed and used to supply energy.

When protein molecules are used as energy sources, they first must be broken down into amino acids. The amino acids then undergo *deamination,* a process that occurs in the liver and involves removing the nitrogen-containing portions ($-NH_2$ groups) from the amino acids (see fig. 2.14). These $-NH_2$ groups are converted later into a waste substance called *urea,* which is excreted in urine.

Depending upon the amino acids involved, the remaining deaminated portions of the amino acid molecules are decomposed by one of several pathways (fig. 4.10). Some of these pathways lead to the formation of acetyl coenzyme A, and others lead more directly to various steps of the citric acid cycle. As energy is released from the cycle, some of it is captured in molecules of ATP. If energy is not needed immediately, the deaminated portions of the amino acids may be changed into glucose or fat molecules by still other metabolic pathways.

1. What is meant by a metabolic pathway?
2. How are carbohydrates used within cells?
3. What must happen to fat molecules before they can be used as energy sources?
4. How do cells use proteins?

Figure 4.11
Nucleotides of a DNA strand are joined to form a sugar-phosphate backbone (*S-P*). Organic bases of the nucleotides (*B*) extend from this backbone and are weakly held to the bases of the second strand by hydrogen bonds (*dashed lines*).

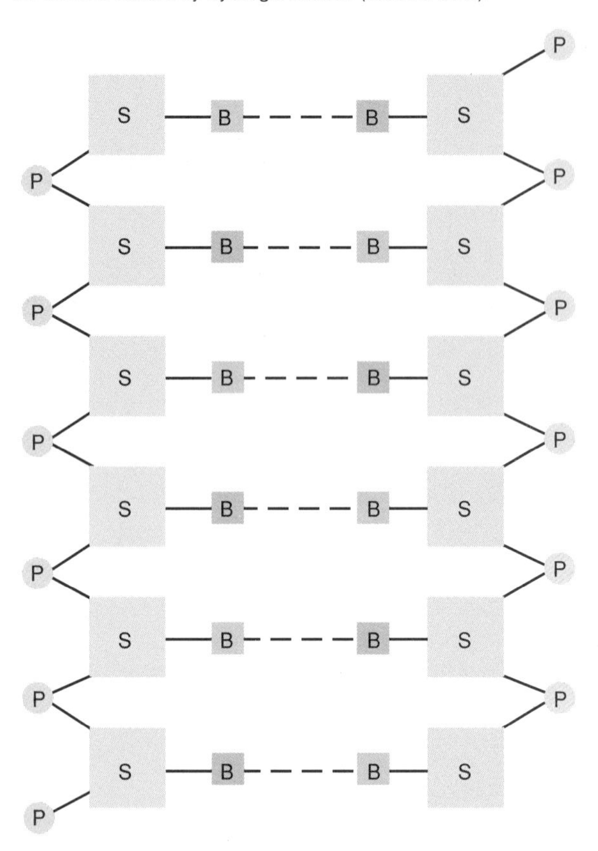

Figure 4.12
The molecular ladder of a double-stranded DNA molecule is twisted into the form of a double helix. Each strand contains four kinds of organic bases—adenine (*A*), thymine (*T*), cytosine (*C*), and guanine (*G*).

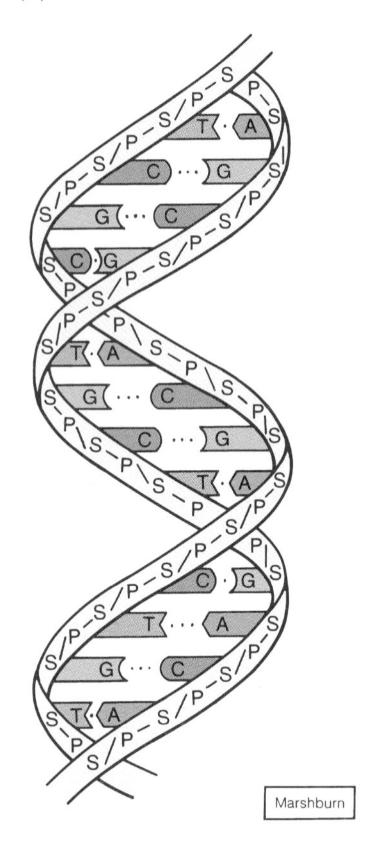

Nucleic Acids and Protein Synthesis

Because enzymes control the metabolic processes that enable cells to survive, cells must possess information for producing these specialized proteins. Such information is held in **DNA** (deoxyribonucleic acid) molecules in the form of a genetic code. This code "instructs" cells how to synthesize specific protein molecules.

Genetic Information

Children resemble their parents because of inherited traits, but what is actually passed from parents to a child is *genetic information.* This information is received in the form of DNA molecules from the parents' sex cells, and as an offspring develops, the information is passed from cell to cell by mitosis. Genetic information "tells" the cells of the developing body how to construct specific protein molecules, which in turn function as structural materials, enzymes, or other vital substances.

The portion of a DNA molecule that contains the genetic information for making one kind of protein is called a **gene.** Thus, inherited traits are determined by the genes contained in the parents' sex cells, which fuse to form the first cell of the offspring's body. These genes instruct cells to synthesize the particular enzymes needed to control metabolic pathways.

DNA Molecules

As described in chapter 2, the building blocks of nucleic acids (nucleotides) are joined together so that the sugar and phosphate portions alternate and form a long chain, or "backbone" (see fig. 2.18).

In a DNA molecule, the organic bases project out from this backbone and are bound weakly to those of the second strand (fig. 4.11). The resulting structure is something like a ladder in which the uprights represent the sugar and phosphate backbones of the two strands, and the crossbars represent the organic bases. In addition, the molecular ladder is twisted to form a *double helix* (figs. 4.12 and 4.13).

The organic base of a DNA nucleotide can be one of four kinds: *adenine* (A), *thymine* (T), *cytosine* (C), or *guanine* (G). Therefore, there are only four kinds of DNA nucleotides: adenine nucleotide, thymine nucleotide, cytosine nucleotide, and guanine nucleotide.

Figure 4.13
A three-dimensional model of a portion of a DNA molecule.

Figure 4.14
A strand of DNA consists of a chain of nucleotides arranged in a particular sequence. Within the chain are four kinds of nucleotides: adenine (*A*) nucleotide; thymine (*T*) nucleotide; cytosine (*C*) nucleotide; and guanine (*G*) nucleotide.

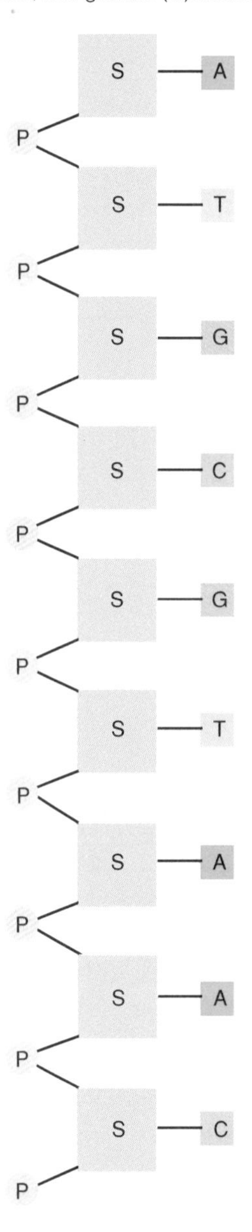

A strand of DNA consists of nucleotides arranged in a particular sequence (fig. 4.14). Moreover, the nucleotides of one strand are paired in a special way with those of the other strand. Only certain organic bases have the molecular shapes needed to fit together so that their nucleotides can bond with one another. Specifically, because of their molecular shapes an adenine will bond only to a thymine, and a cytosine will bond only to a guanine. As a consequence of such base pairing, a DNA strand possessing the base sequence T, C, G, C would have to be joined to a second strand with the complementary base sequence A, G, C, G (see the upper region of DNA in fig. 4.12). It is the particular sequence of base pairs that encodes the genetic information held in a DNA molecule.

Genetic Code

Genetic information contains the instructions for synthesizing proteins, and because proteins consist of twenty different amino acids joined in particular sequences, the genetic information must tell how to position the amino acids correctly in a polypeptide chain.

Each of the twenty different amino acids is represented in a DNA molecule by a particular sequence of three nucleotides. That is, the sequence G, G, T in a DNA strand represents one kind of amino acid; the sequence G, C, A represents another kind, and T, T, A still another kind (chart 4.1). Other nucleotide sequences represent the instructions for beginning or ending the synthesis of a protein molecule. Thus, the sequence in which the nucleotide groups are arranged within a DNA molecule can denote the arrangement of amino acids within a protein molecule as well as indicate where to start or stop the synthesis of a molecule. This method of storing information for the synthesis of protein molecules is termed the **genetic code.**

Although DNA molecules are located in the chromatin within a cell's nucleus, protein synthesis

Chart 4.1 Some nucleotide sequences of the genetic code		
Amino acids	**DNA sequence**	**RNA sequence**
Alanine	CGT	GCA
Arginine	GCA	CGU
Asparagine	TTA	AAU
Aspartic acid	CTA	GAU
Cysteine	ACA	UGU
Glutamic acid	CTT	GAA
Glutamine	GTT	CAA
Glycine	CCG	GGC
Histidine	GTA	CAU
Isoleucine	TAG	AUC
Leucine	GAA	CUU
Lysine	TTT	AAA
Methionine	TAC	AUG
Phenylalanine	AAA	UUU
Proline	GGA	CCU
Serine	AGG	UCC
Threonine	TGC	ACG
Tryptophan	ACC	UGG
Tyrosine	ATA	UAU
Valine	CAA	GUU
Instructions		
Start protein synthesis	TAC	AUG
Stop protein synthesis	ATT	UAA

occurs in the cytoplasm. Therefore the genetic information must be transferred from the nucleus into the cytoplasm. This transfer of information is the function of certain RNA molecules.

RNA Molecules

RNA (ribonucleic acid) molecules differ from DNA molecules in several ways. For example, RNA molecules are usually single-stranded, and their nucleotides contain ribose rather than deoxyribose sugar. Like DNA, RNA nucleotides each contain one of four organic bases; but while adenine, cytosine, and guanine nucleotides occur in both DNA and RNA, thymine nucleotides are found only in DNA. In place of thymine nucleotides, RNA molecules contain *uracil* (U) nucleotides.

One step in the transfer of information from the nucleus to the cytoplasm involves the synthesis of a type of RNA called **messenger RNA** (mRNA). In the process of producing messenger RNA, an enzyme (RNA polymerase) becomes associated with the DNA base sequence that represents the beginning of a gene—that is, the instructions for synthesizing a particular protein. As a result of the enzyme's action, a part of the double-stranded DNA molecule unwinds and pulls apart, exposing a portion of the gene. The enzyme then moves along the strand, causing other portions of the gene to be exposed. At the same time, a molecule of messenger RNA is formed of nucleotides complementary to those arranged along the DNA strand. For example, if the sequence of DNA bases was A,T,G,C,G,T,A,A,C, then the complementary bases in the developing messenger RNA molecule would be U,A,C,G,C,A,U,U,G (fig. 4.15).

The enzyme continues to move along the DNA strand exposing the gene until it reaches a special DNA

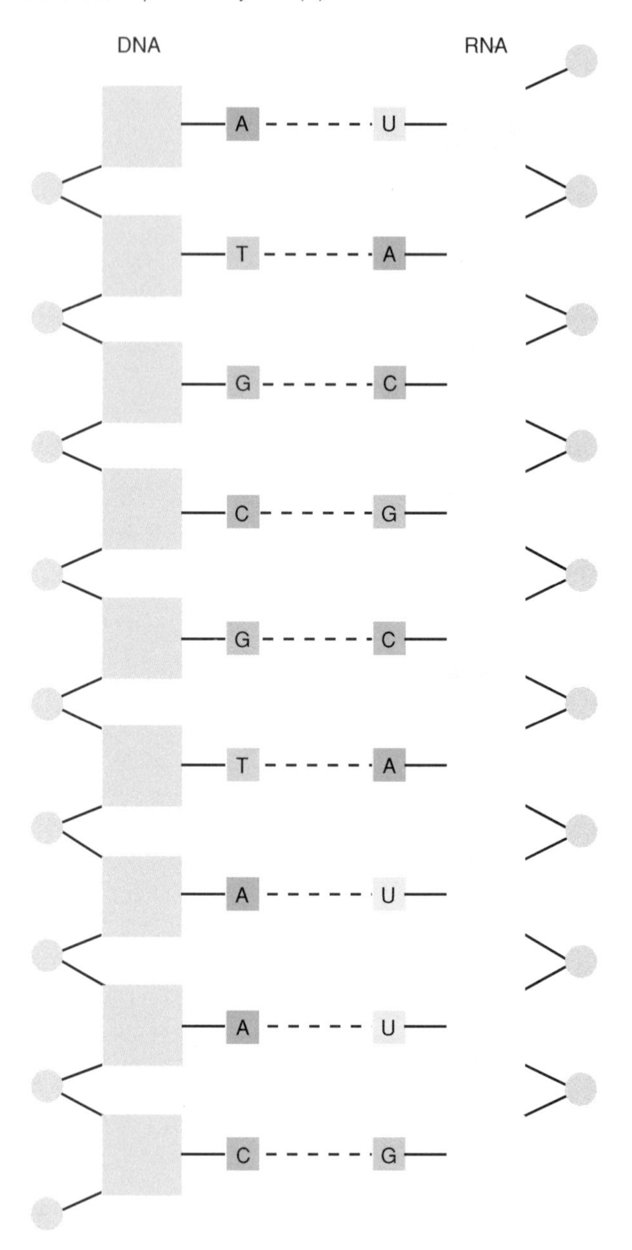

Figure 4.15
When an RNA molecule is synthesized beside a strand of DNA, complementary nucleotides bond as in a double-stranded molecule of DNA with one exception: RNA contains uracil (*U*) nucleotides in place of thymine (*T*) nucleotides.

Figure 4.16
After transcribing a section of DNA information, a messenger RNA (*mRNA*) molecule moves out of the nucleus and enters the cytoplasm. There it becomes associated with a ribosome.

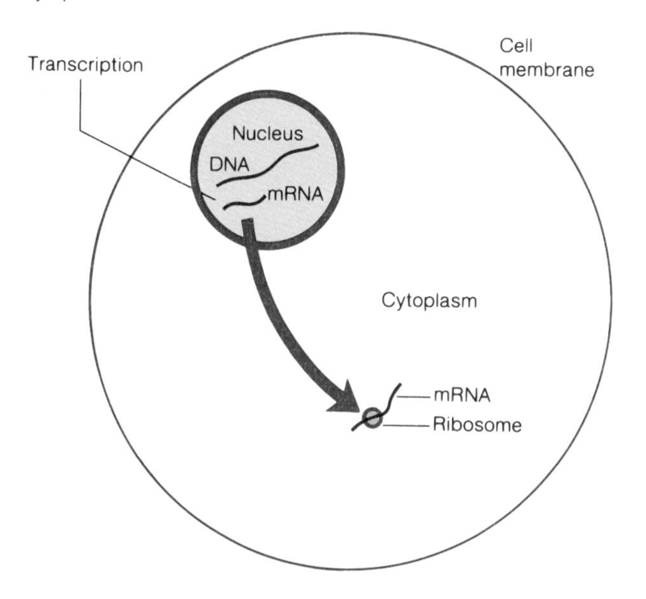

base sequence (termination signal) that represents the end of the gene. At this point, the enzyme releases the newly formed messenger RNA molecule and leaves the DNA. This process of copying DNA information into the structure of messenger RNA is called *transcription.*

Because an amino acid is represented by a sequence of three nucleotides in a DNA molecule, the same amino acid will be represented in the transcribed messenger RNA by the complementary set of three nucleotides. Such a set of nucleotides in messenger RNA is called a **codon.**

Once formed, messenger RNA molecules (each of which consists of hundreds or even thousands of nucleotides) can move out of the nucleus through the tiny pores in the nuclear envelope and enter the cytoplasm (fig. 4.16). There they become associated with ribosomes and act as patterns or templates for the synthesis of protein molecules—a process called *translation* (fig. 4.17).

Figure 4.17
Transcription is the process by which DNA information is copied into the structure of messenger RNA. Translation is the process by which messenger RNA is used in the synthesis of protein.

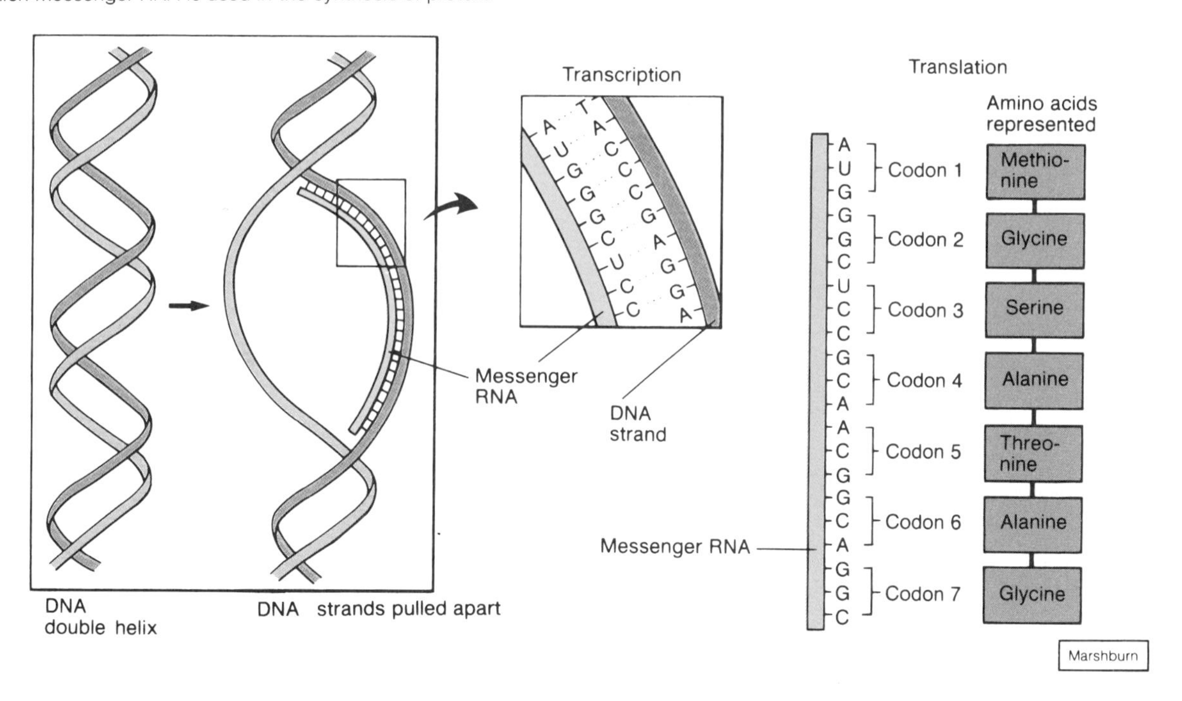

Figure 4.18
Molecules of transfer RNA (tRNA) bring specific amino acids and place them in the sequence determined by the codons of the messenger RNA molecule. These amino acids bond and form the polypeptide chain of a protein molecule.

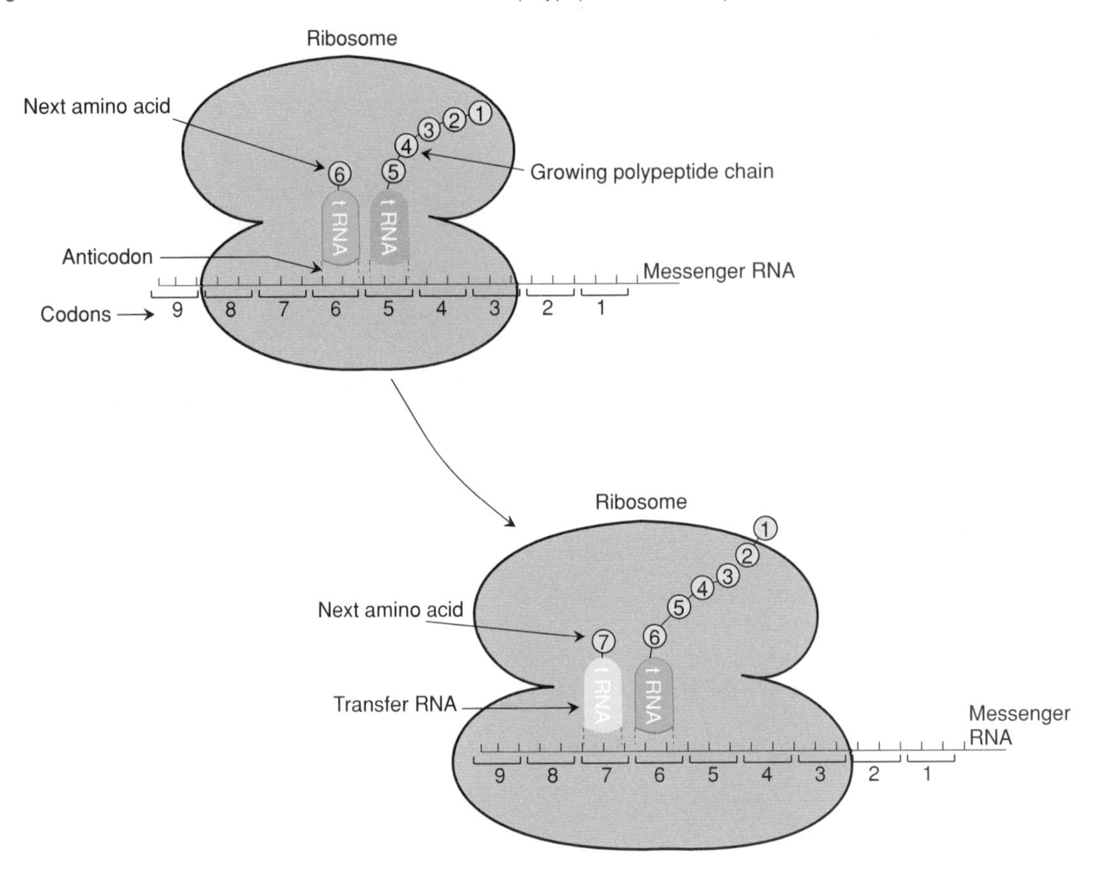

Protein Synthesis

Before a protein molecule can be synthesized, the correct amino acids must be present in the cytoplasm to serve as building blocks. Furthermore, these amino acids must be positioned in the proper locations along a strand of messenger RNA. The positioning of amino acid molecules is the function of a second kind of RNA molecule, which is synthesized in the nucleus and called **transfer RNA** (tRNA).

Since twenty different kinds of amino acids are involved in protein synthesis, there must be at least twenty different kinds of transfer RNA molecules to serve as guides.

Each type of transfer RNA, which is a relatively small molecule, contains three nucleotides in a particular sequence. These nucleotides can only bond to the complementary set of three nucleotides of a messenger RNA molecule—a codon. Thus, the set of three nucleotides in the transfer RNA is called an **anticodon.** In this way, the transfer RNA carries its amino acid to a correct position on a messenger RNA strand. This action occurs in close association with a ribosome.

A ribosome binds to a molecule of messenger RNA near a codon at the beginning of the messenger strand, allowing a transfer RNA molecule with the complementary anticodon to bring the amino acid it carries into position and become temporarily joined to the ribosome. Then the transfer RNA molecule releases its amino acid and returns to the cytoplasm (fig. 4.18).

This process is repeated again and again as the ribosome moves along messenger RNA. The amino acids, which are released by the transfer RNA molecules, are added one at a time to a developing protein molecule under the influence of enzymes associated with the ribosome.

As the protein molecule forms, it folds into its unique shape, and when it is completed, it is released to become a separate functional molecule. The transfer RNA molecules can pick up other amino acids from the cytoplasm, and the messenger RNA molecules can function again and again.

The process of protein synthesis is summarized in chart 4.2.

1. What is the function of DNA?
2. How is information carried from the nucleus to the cytoplasm?
3. How are protein molecules synthesized?

Mutations and Mutagenic Factors

A Current Topic

The amount of genetic information contained within a set of human chromosomes is very large, and since each of the trillions of cells in an adult body results from mitosis, this large amount of information must be replicated many times during development. Although the replication process usually is very precise, occasionally a mistake occurs and the information is altered. For example, during DNA replication, an organic base may be paired incorrectly within the newly forming strand, or some extra organic bases may be built into its structure. In other cases, sections of DNA strands may be deleted, moved to other regions of the molecule, or attached to other chromosomes. Such a change in genetic information is called a *mutation.*

The effects of mutations vary greatly. At one extreme, there is little or no effect; and at the other extreme, the mutation results in cell death. The mutations that cause the most concern are those between the two extremes—they bring about a decrease in cell efficiency, and although the cell with the mutation may live, it has difficulty functioning normally. In time, an increasing number of such mutated cells may produce abnormal effects in structure and function of tissues and organs.

Well over one hundred human diseases are known to be related to mutations. The majority of these involve an inability of cells to produce specific enzymes. As an example, the disease called *phenylketonuria* (PKU) results from a defective enzyme. The normal enzyme functions in the metabolic breakdown of the amino acid phenylalanine. When the enzyme is defective, phenylalanine is only partially metabolized, and a toxic substance accumulates. This toxin interferes with the normal activity of the developing nervous system. If this condition is untreated in an affected newborn, it usually results in severe mental retardation.

Although some mutations arise spontaneously, without known cause, certain factors are known to increase the likelihood of mutations. These factors are called *mutagens,* and they include ionizing radiation and various chemicals.

Some common forms of ionizing radiation are ultraviolet light and X ray. Ultraviolet light is not very penetrating, so its damaging effect is usually confined to the surfaces of the skin and eyes. X ray, on the other hand, is very penetrating and can affect cells located in deep tissues and organs.

Mutagenic chemicals occur in a variety of commonly used substances such as tobacco and petroleum products. Although it is difficult to completely avoid contact with mutagenic factors in everyday living, the wise course is to limit exposure to them whenever possible.

Chart 4.2 Protein synthesis

Transcription (occurring within the nucleus)

1. RNA polymerase becomes associated with the base sequence of a gene within a DNA molecule.
2. This enzyme causes a part of the DNA molecule to unwind and its strands to pull apart, thus exposing the gene.
3. The RNA polymerase moves along the exposed gene and causes the production of an mRNA molecule, whose nucleotides are complementary to those of the gene.
4. When the RNA polymerase reaches the end of the gene, the newly formed mRNA molecule is released.
5. The DNA molecule rewinds and assumes its previous structure.
6. The mRNA molecule passes through a pore in the nuclear envelope and enters the cytoplasm.

Translation (occurring within the cytoplasm)

1. A ribosome binds to the mRNA molecule near the codon at the beginning of the messenger strand.
2. A tRNA molecule that has the complementary anticodon becomes associated with the ribosome, and the tRNA molecule releases the amino acid it carries.
3. This process is repeated for each codon in the mRNA sequence as the ribosome moves along its length.
4. Enzymes associated with the ribosome cause the amino acids that are released from the tRNA molecules to bind together, thus forming a chain of amino acids.
5. As the chain of amino acids develops, it folds into the unique shape of a functional protein molecule.
6. The completed protein molecule is released. The mRNA molecule, ribosome, and tRNA molecules can function again and again to synthesize other protein molecules.

Replication of DNA

When a cell reproduces, each newly formed cell needs a copy of the parent cell's genetic information so that it will be able to synthesize the proteins necessary to build cellular parts and carry on metabolism.

DNA molecules can be replicated (duplicated), and this replication takes place during the interphase of the cell's life cycle.

As the replication process begins, the bonds are broken between complementary base pairs of the double

Figure 4.19
When a DNA molecule replicates, its strands pull apart (*center of figure*), and a new strand of complementary nucleotides forms along each of the old strands.

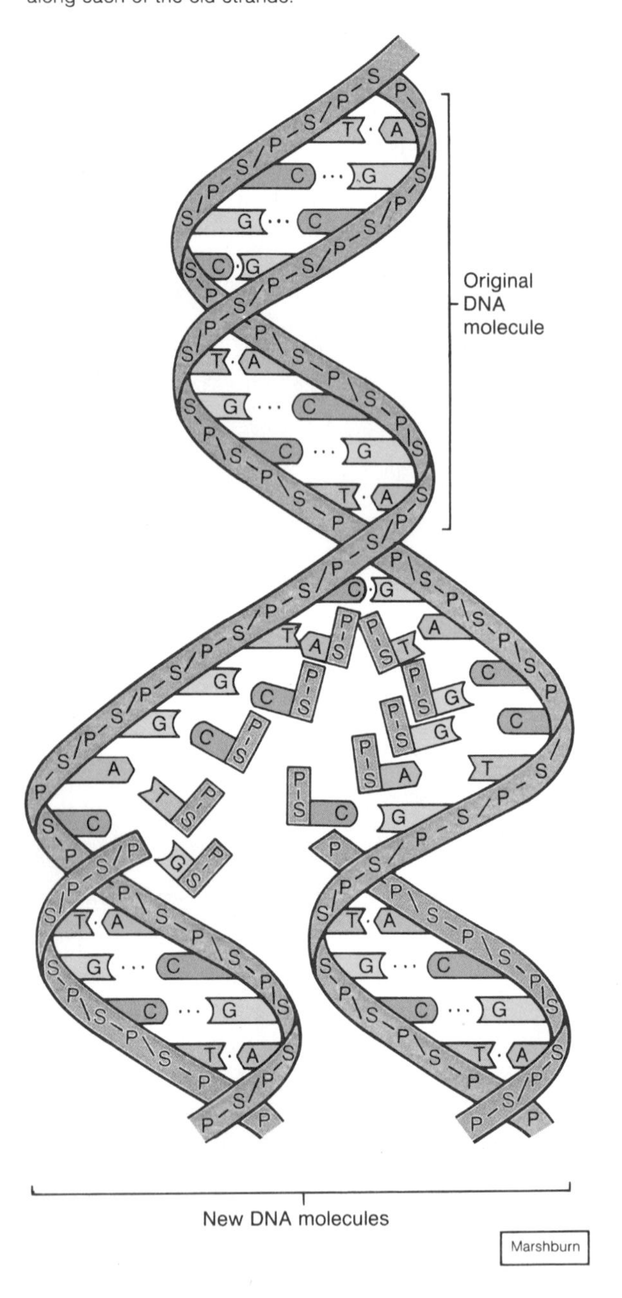

strands in each DNA molecule (fig. 4.19). Then the double-stranded structure pulls apart and unwinds, exposing the organic bases of its nucleotides. New nucleotides of the four types found in DNA become paired with the exposed bases, and enzymes cause the complementary bases to join together. In this way, a new strand of complementary nucleotides is constructed along each of the old strands. As a result, two complete DNA molecules are produced, each with one old strand of the original molecule and one new strand.

These two DNA molecules become incorporated into chromosomes and are separated during mitosis so that one passes to each of the newly forming cells.

1. Why must DNA molecules be replicated?
2. How is this replication accomplished?

Chapter Summary

Introduction (page 76)

A cell continuously carries on metabolic processes.

Metabolic Processes (page 76)

1. Anabolic metabolism
   a. Anabolic metabolism consists of constructive processes in which smaller molecules are used to build larger ones.
   b. In dehydration synthesis, water is formed, and smaller molecules become bound by shared atoms.
   c. Complex carbohydrates are synthesized from monosaccharides, fats are synthesized from glycerol and fatty acids, and proteins are synthesized from amino acids.
2. Catabolic metabolism
   a. Catabolic metabolism consists of decomposition processes in which larger molecules are broken down into smaller ones.
   b. In hydrolysis, a water molecule is used, and the bond between these two portions of the molecule is broken.
   c. Complex carbohydrates are decomposed into monosaccharides, fats are decomposed into glycerol and fatty acids, and proteins are decomposed into amino acids.

Control of Metabolic Reactions (page 77)

Metabolic reactions usually involve many specific chemical changes.

1. Enzymes and their actions
   a. Enzymes are proteins that promote metabolic reactions without being used up themselves.
   b. An enzyme acts upon a specific substrate.
   c. The shape of an enzyme molecule fits the shape of its substrate molecule.
   d. When an enzyme combines with its substrate, the substrate is changed, resulting in a product, but the enzyme is unaltered.
2. Factors that alter enzymes
   a. Enzymes are proteins and can be denatured.
   b. Factors that may denature enzymes include excessive heat, radiation, electricity, certain chemicals, and extreme pH values.

Energy for Metabolic Reactions (page 79)

Energy is a capacity to produce change or to do work. Common forms of energy include heat, light, sound, electrical energy, mechanical energy, and chemical energy.

1. Release of chemical energy
   a. Most metabolic processes utilize chemical energy that is released when molecular bonds are broken.
   b. The energy released from glucose during cellular respiration is used to promote cellular metabolism.
2. Anaerobic respiration
   a. The first phase of glucose decomposition does not require oxygen.
   b. Some of the energy released is transferred to molecules of ATP.
3. Aerobic respiration
   a. The second phase of glucose decomposition requires oxygen.
   b. Considerably more energy is transferred to ATP molecules during this phase than during the anaerobic phase.
   c. The final products of glucose decomposition are carbon dioxide, water, and energy.
4. ATP molecules
   a. Energy is captured in the bond of the terminal phosphate of each ATP molecule.
   b. When energy is needed by a cellular process, the terminal phosphate bond of an ATP molecule is broken, and the stored energy is released.
   c. An ATP molecule that loses its terminal phosphate becomes an ADP molecule.
   d. An ADP molecule can be converted to an ATP molecule by capturing some energy and a phosphate.

Metabolic Pathways (page 80)

Metabolic processes usually involve a number of steps that must occur in correct sequence. A sequence of enzyme-controlled reactions is called a metabolic pathway.

1. Carbohydrate pathways
   a. Carbohydrates may enter catabolic pathways and be used as energy sources.
   b. Carbohydrates may enter anabolic pathways and be converted into glycogen or fat.
2. Lipid pathways
   a. Before fats can be used as an energy source they must be converted into glycerol and fatty acids.
   b. Fatty acids can be changed to acetyl coenzyme A, which in turn can be oxidized by the citric acid cycle.
3. Protein pathways
   a. Proteins are used as building materials for cellular parts, as enzymes, and as energy sources.
   b. Before proteins can be used as energy sources, they must be decomposed into amino acids, and the amino acids must be deaminated.
   c. The deaminated portions of amino acids can be broken down into carbon dioxide and water or converted into glucose or fat.

Nucleic Acids and Protein Synthesis (page 84)

DNA molecules contain information that instructs a cell how to synthesize proteins.

1. Genetic information
   a. Inherited traits result from DNA information that is passed from parents to offspring.
   b. A gene is a portion of a DNA molecule that contains the genetic information for making one kind of protein.
2. DNA molecules
   a. A DNA molecule consists of two strands of nucleotides twisted into a double helix.
   b. The nucleotides of a DNA strand are arranged in a particular sequence.
   c. The nucleotides of each strand are paired with those of the other strand in a complementary fashion.
3. Genetic code
   a. The sequence of nucleotides in a DNA molecule represents the sequence of amino acids in a protein molecule.
   b. Genetic information is transferred from the nucleus to the cytoplasm by RNA molecules.
4. RNA molecules
   a. RNA molecules are usually single strands, contain ribose instead of deoxyribose, and contain uracil nucleotides in place of thymine nucleotides.
   b. Messenger RNA molecules contain a nucleotide sequence that is complementary to that of an exposed strand of DNA.
   c. Messenger RNA molecules become associated with ribosomes and act as patterns for the synthesis of protein molecules.
5. Protein synthesis
   a. Molecules of transfer RNA serve to position amino acids along a strand of messenger RNA.
   b. A ribosome binds to a messenger RNA molecule and allows a transfer RNA molecule to recognize its correct position on the messenger RNA.
   c. Amino acids released from the transfer RNA molecules become joined together and form a protein molecule with a unique shape.
6. Replication of DNA
   a. Each new cell needs a copy of the parent cell's genetic information.
   b. DNA molecules are replicated during interphase of the cell's life cycle.
   c. Each new DNA molecule contains one old strand and one new strand.

Clinical Application of Knowledge

1. Because enzymes are proteins, they may become denatured. Relate this to the fact that changes in the pH of body fluids accompanying an illness may be life-threatening.

2. Some reducing diets drastically limit the dieter's intake of carbohydrates but allow the liberal use of fat and protein foods. What changes would such a diet cause in the cellular metabolism of the dieter? What changes might be noted in the urine of such a person?
3. Why do you think it is of particular importance for a pregnant woman to avoid exposure to forms of ionizing radiation or to chemicals that are known to cause mutations?
4. What mutagenic factors might health care workers encounter more frequently than other workers?

Review Activities

1. Distinguish between anabolic and catabolic metabolism.
2. Distinguish between dehydration synthesis and hydrolysis.
3. Define *enzyme.*
4. Describe how an enzyme interacts with its substrate.
5. Explain how an enzyme may be denatured.
6. Explain how the oxidation of molecules inside cells differs from the burning of substances outside cells.
7. Distinguish between anaerobic and aerobic respiration.
8. Explain the importance of ATP to cellular processes.
9. Describe the relationship between ATP and ADP molecules.
10. Explain what is meant by a *metabolic pathway.*
11. Describe what happens to carbohydrates that enter catabolic pathways.
12. Explain how fats may serve as energy sources.
13. Define *deamination,* and explain its importance.
14. Explain what is meant by *genetic information.*
15. Describe the relationship between a DNA molecule and a gene.
16. Explain how genetic information is stored in a DNA molecule.
17. Distinguish between messenger RNA and transfer RNA.
18. Define *transcription.*
19. Define *translation.*
20. Distinguish between a codon and an anticodon.
21. Explain the function of a ribosome in protein synthesis.
22. Explain why a new cell needs a copy of the parent cell's genetic information.
23. Describe the process by which DNA molecules are replicated.

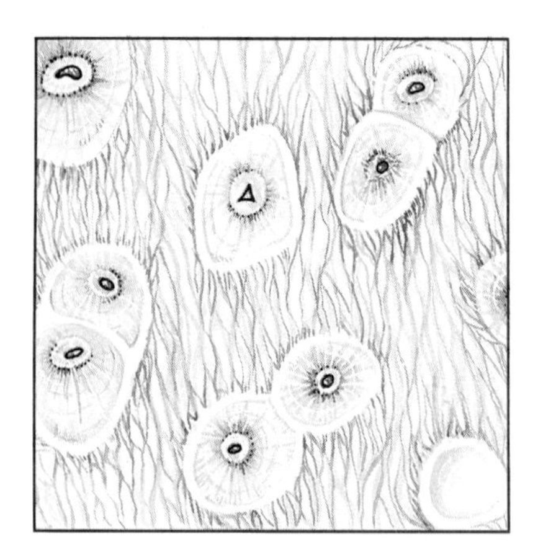

5 Tissues

Cells, the basic units of structure and function within the human organism, are organized into groups and layers called *tissues.*

Each type of tissue is composed of similar cells that are specialized to carry on particular functions. For example, epithelial tissues form protective coverings and function in secretion and absorption. Connective tissues provide support for softer body parts and bind structures together. Muscle tissues are responsible for producing body movements, and nervous tissue is specialized to conduct impulses that help to control and coordinate body activities.

Chapter Objectives

After you have studied this chapter, you should be able to

1. Describe the general characteristics and functions of epithelial tissue.
2. Name the types of epithelium, and identify an organ in which each is found.
3. Explain how glands can be classified.
4. Describe the general characteristics of connective tissue.
5. Describe the major cell types and fibers of connective tissue.
6. List the types of connective tissue that occur within the body.
7. Describe the major functions of each type of connective tissue.
8. Distinguish between the three types of muscle tissue.
9. Describe the general characteristics and functions of nervous tissue.
10. Complete the review activities at the end of this chapter. Note that the items are worded in the form of specific learning objectives. You may want to refer to them before reading the chapter.

Key Terms

adipose tissue (ad′ĭ-pōs tish′u)

cartilage (kar′tĭ-lij)

chondrocyte (kon′dro-sīt)

connective tissue (kŏ-nek′tiv tish′u)

epithelial tissue (ep″ĭ-the′le-al tish′u)

fibroblast (fi′bro-blast)

fibrous tissue (fi′brus tish′u)

macrophage (mak′ro-fāj)

muscle tissue (mus′el tish′u)

nervous tissue (ner′vus tish′u)

neuroglia (nu-rog′le-ah)

neuron (nu′ron)

osteocyte (os″te-o-sīt″)

osteon (os′te-on)

reticuloendothelial tissue (rĕ-tik″u-lo-en″do-the′le-al tish′u)

Aids to Understanding Words

adip-, fat: *adip*ose tissue—a tissue that stores fat.

chondr-, cartilage: *chondr*ocyte—a cartilage cell.

-cyt, cell: osteo*cyte*—a bone cell.

epi-, upon: *epi*thelial tissue—a tissue that covers all the free body surfaces.

-glia, glue: neuro*glia*—the cells that bind nervous tissue together.

inter-, between: *inter*calated disk—a band located between the ends of adjacent cardiac muscle cells.

macro-, large: *macro*phage—a large phagocytic cell.

oss-, bone: *oss*eous tissue—bone tissue.

pseudo-, false: *pseudo*stratified epithelium—a tissue whose cells appear to be arranged in layers, but are not.

squam-, scale: *squam*ous epithelium—a tissue whose cells appear flattened or scalelike.

strat-, layer: *strat*ified epithelium—a tissue whose cells occur in layers.

Introduction

WITHIN THE BODY, groups of cells perform specialized structural and functional roles. Such a group constitutes a **tissue.** Although the cells of different tissues vary in size, shape, arrangement, and function, those within each kind of tissue are quite similar.

The tissues of the human body include four major types: *epithelial tissue, connective tissue, muscle tissue,* and *nervous tissue.*

---

1. What is a tissue?
2. List the four major types of tissues.

Epithelial Tissue

General Characteristics

Epithelial tissues are widespread throughout the body. They cover all the body surfaces—inside and out—and are the major tissues of glands.

Because epithelium covers organs, forms the inner lining of body cavities, and lines hollow organs, it always has a free surface—one that is exposed to the outside or to an open space internally. The underside of this tissue is always anchored to connective tissue by a thin, nonliving layer, called the *basement membrane.*

As a rule, epithelial tissues lack blood vessels; however, epithelial cells are nourished by substances that diffuse from underlying connective tissues, which are well supplied with blood vessels.

Although the cells of some tissues have limited abilities to reproduce, those of epithelium reproduce readily. Injuries to epithelium are likely to heal rapidly as new cells replace lost or damaged ones. For example, skin cells and cells that line the stomach and intestines are continually being damaged and replaced.

Epithelial cells are tightly packed, and there is little intercellular material between them. Consequently, these cells are effective protective barriers in such structures as the outer layer of the skin and the lining of the mouth. Other epithelial functions include secretion, absorption, excretion, and sensory reception. In the following descriptions, note that the free surfaces of various epithelial cells are modified in ways that reflect their specialized functions.

Simple Squamous Epithelium

Simple squamous epithelium consists of a single layer of thin, flattened cells. These cells fit tightly together, somewhat like floor tiles, and their nuclei are usually broad and thin (fig. 5.1).

As a rule, substances pass rather easily through this type of tissue, and it occurs commonly where diffusion and filtration are taking place. For instance, simple squamous epithelium lines the air sacs of the lungs where oxygen and carbon dioxide are exchanged. It also forms the walls of capillaries, lines the insides of blood vessels, and covers the membranes that line body cavities. However, because it is so thin and delicate, simple squamous epithelium can be damaged relatively easily.

Simple Cuboidal Epithelium

Simple cuboidal epithelium consists of a single layer of cube-shaped cells. These cells usually have centrally located, spherical nuclei (fig. 5.2).

This tissue covers the ovaries and lines the kidney tubules and the ducts of glands, such as the salivary glands, thyroid gland, pancreas, and liver. In the kidneys, it functions in secretion and absorption; in the glands, it is concerned with the secretion of glandular products.

Simple Columnar Epithelium

The cells of **simple columnar epithelium** are elongated; that is, they are longer than they are wide (fig. 5.3). This tissue is composed of a single layer whose nuclei are usually located at about the same level near the basement membrane.

Simple columnar epithelium occurs in the linings of the uterus and various organs of the digestive tract, including the stomach and intestines. Because its cells are elongated, this tissue is relatively thick, providing protection for underlying tissues. It also functions in the secretion of digestive fluids and in the absorption of nutrient molecules resulting from the digestion of foods.

Columnar cells, whose principal function is absorption, often have numerous minute, cylindrical processes extending outward from their cell surfaces. These processes, called *microvilli,* function to increase the surface of the cell membrane where it is exposed to the substances being absorbed.

Typically, specialized, flask-shaped glandular cells are scattered among the columnar cells of this tissue. These cells, called *goblet cells,* secrete a protective fluid (*mucus*) onto the free surface of the tissue.

Pseudostratified Columnar Epithelium

The cells of **pseudostratified columnar epithelium** appear stratified or layered, but they are not (fig. 5.4). The layered effect occurs because their nuclei are located at two or more levels within the cells of the tissue.

Figure 5.1
(*a*) Simple squamous epithelium consists of a single layer of tightly packed flattened cells. (*b*) What features of the tissue, viewed on the flat surface, can you identify in this micrograph (×250)? (*c*) Simple squamous epithelium viewed from the edge (×250).

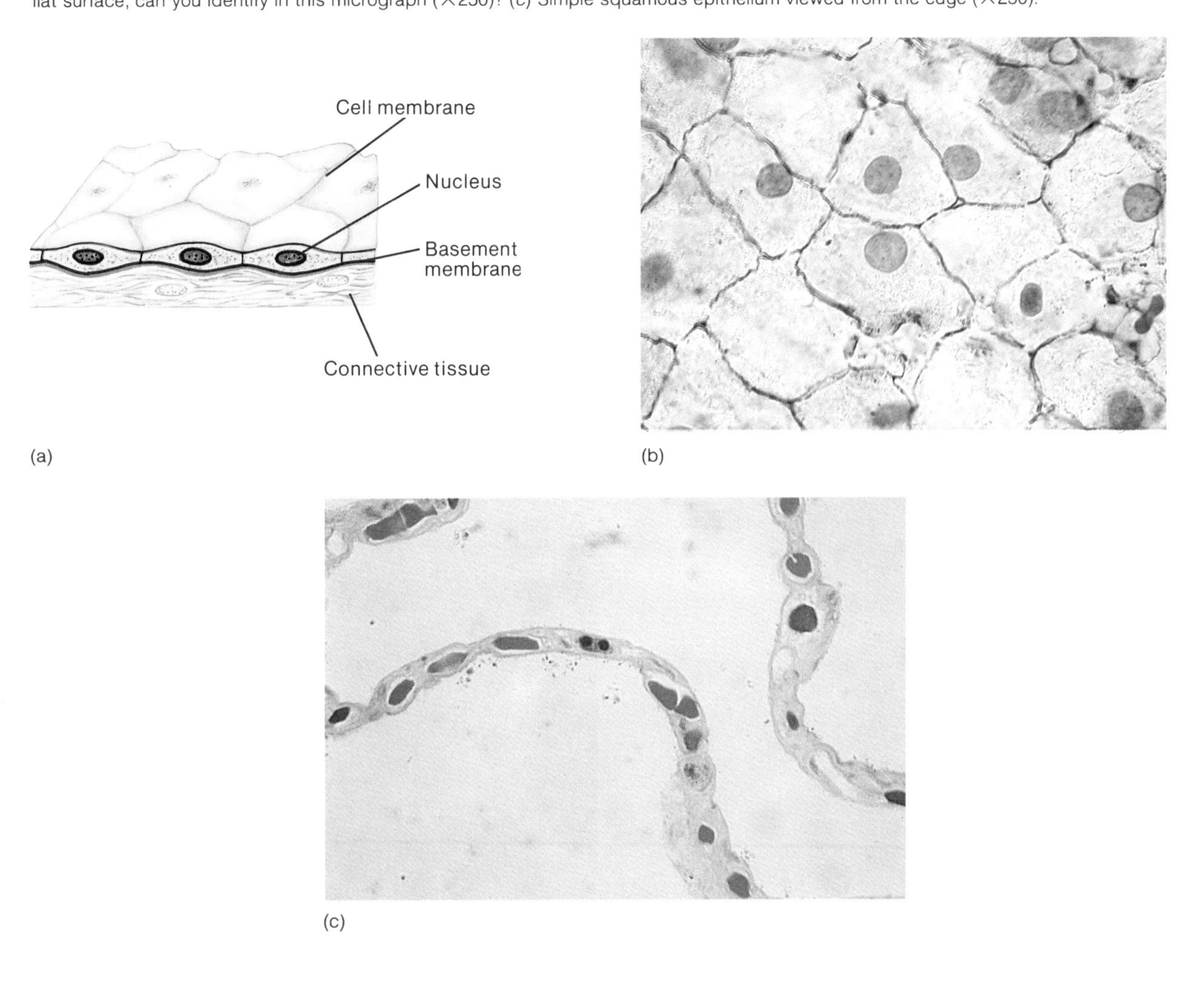

Figure 5.2
(*a*) Simple cuboidal epithelium consists of a single layer of tightly packed cube-shaped cells. (*b*) Note the single layer of simple cuboidal cells indicated by the arrow in the micrograph (×250).

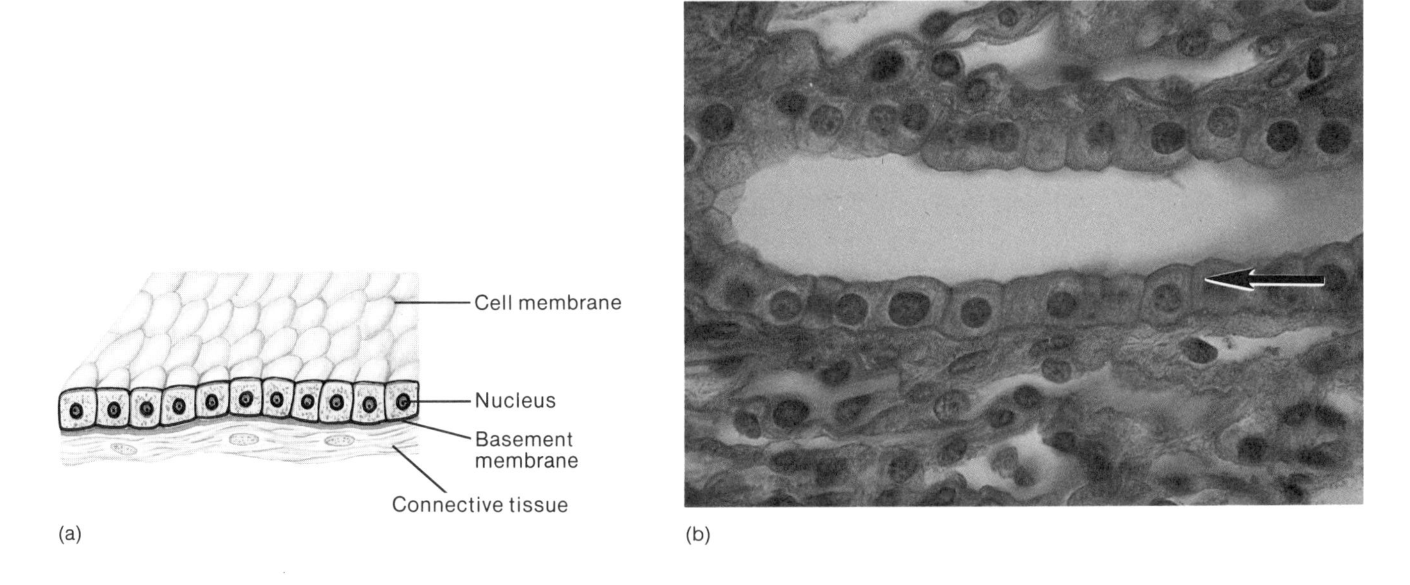

Figure 5.3
(*a*) Simple columnar epithelium consists of a single layer of elongated cells. (*b*) Note the goblet cell (*arrow*) in this micrograph (×400).

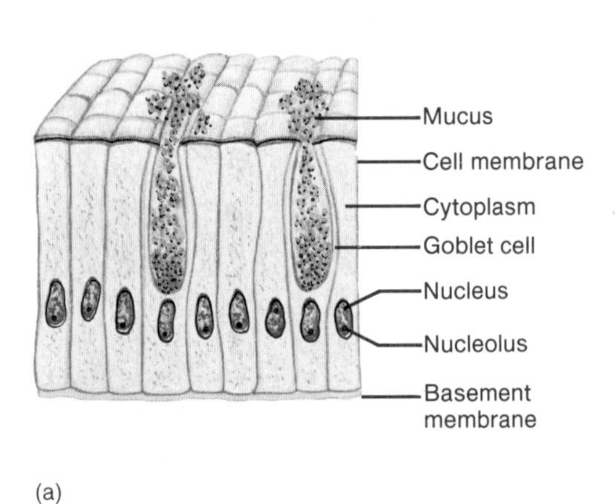

(a)

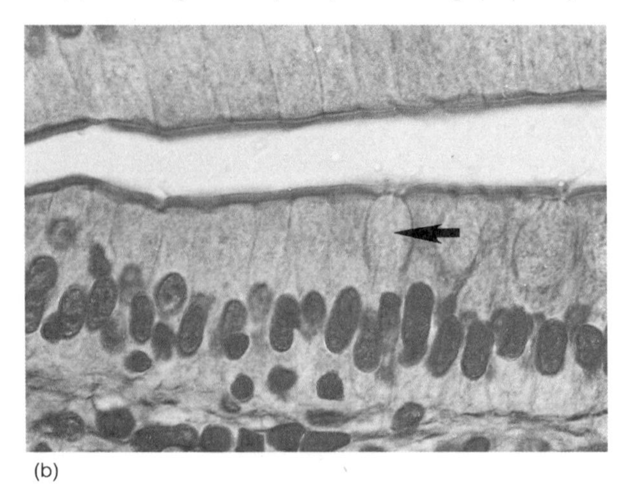
(b)

Figure 5.4
(*a*) Pseudostratified columnar epithelium appears stratified, because the nuclei of various cells are located at different levels. (*b*) What features of the tissue can you identify in this micrograph (×500)?

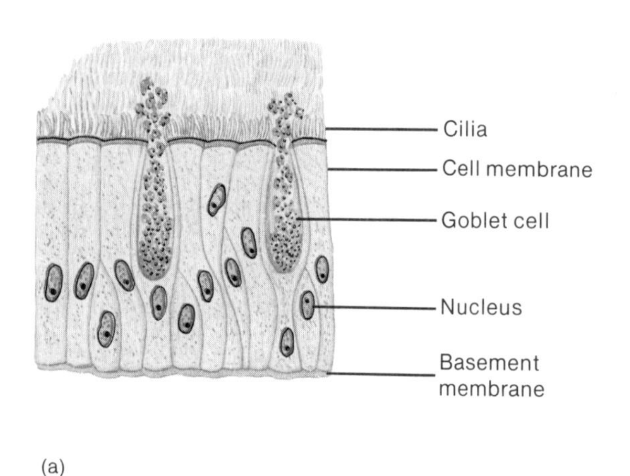

(a)

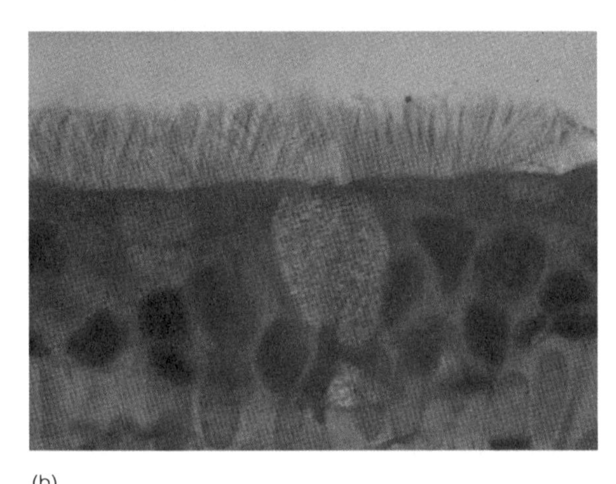
(b)

These cells commonly possess cilia, which extend from their free surfaces and move constantly (see chapter 3). Goblet cells are also scattered throughout this tissue, and the mucus they secrete is swept along by the cilia.

Pseudostratified columnar epithelium is found lining the passages of the respiratory system and in various tubes of the reproductive systems. In the respiratory passages, the mucus-covered linings are sticky and tend to trap particles of dust and microorganisms, which enter with the air. The cilia move the mucus and its captured particles upward and out of the airways. In the reproductive tubes, the cilia aid in moving the sex cells from one region to another.

Stratified Squamous Epithelium

Stratified squamous epithelium consists of many layers of cells, making this tissue relatively thick (fig. 5.5). Cellular reproduction occurs in the deeper layers, and as the newer cells grow, the older ones are pushed farther and farther outward.

This tissue forms the outer layer of the skin (epidermis). As the skin cells age, they accumulate a protein, called *keratin,* and become hardened and die. This action produces a covering of dry, tough, protective material, which prevents water from escaping underlying tissues and various microorganisms from entering.

Stratified squamous epithelium also lines the mouth cavity, throat, vagina, and anal canal. In these parts, the tissue is not keratinized; it stays moist, and the cells on the free surfaces remain alive.

Transitional Epithelium

Transitional epithelium is specialized to undergo changes in response to increased tension (fig. 5.6). It forms the inner lining of the urinary bladder and the

Figure 5.5
(*a*) Stratified squamous epithelium consists of many layers of cells. (*b*) In this micrograph, note how the cells near the surface of the tissue have become flattened (×35).

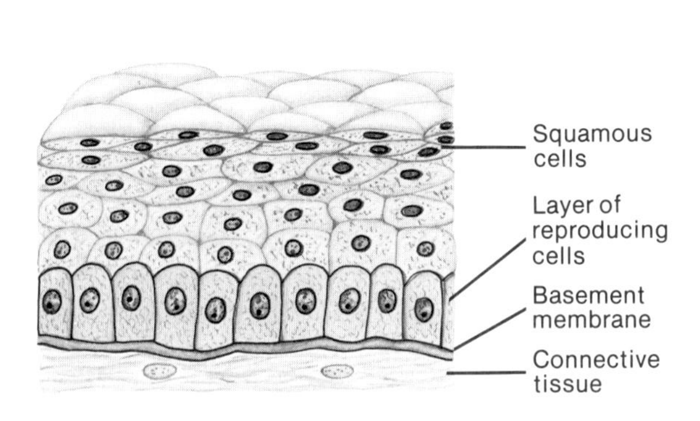

(a)

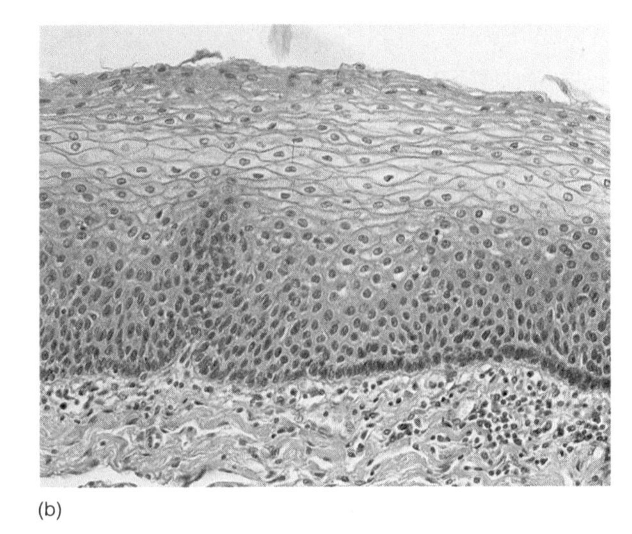

(b)

Figure 5.6
(*a*) A micrograph of transitional epithelium (×100). (*b*) This tissue consists of many layers when the organ wall is contracted. (*c*) The tissue is thinner when the wall is stretched.

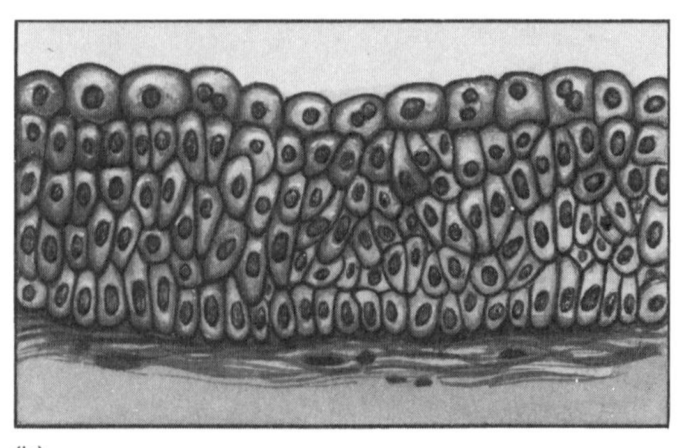

(b)

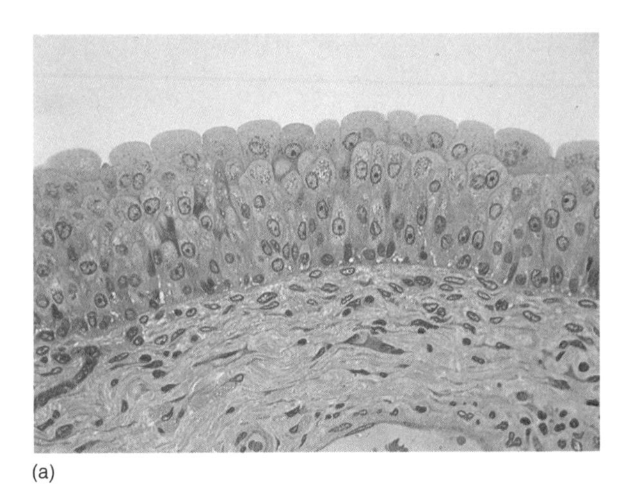

(a)

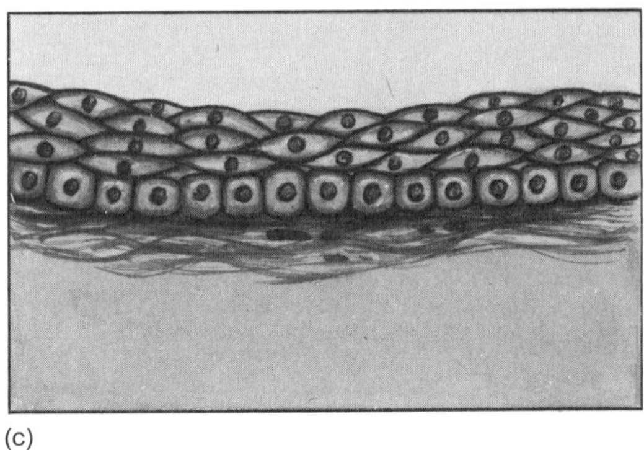

(c)

passageways of the urinary system. When the wall of one of these organs is contracted, the tissue consists of several layers of cuboidal cells; however, when the organ is distended, the tissue is stretched and may have only a few layers.

In addition to providing an expandable lining, transitional epithelium is thought to form a barrier that helps to prevent the contents of the urinary tract from diffusing out of its passageways.

Chart 5.1 summarizes the characteristics of epithelial tissues.

Chart 5.1 Epithelial tissues		
Type	Function	Location
Simple squamous epithelium	Filtration, diffusion, osmosis	Air sacs of the lungs, walls of capillaries, linings of blood vessels
Simple cuboidal epithelium	Secretion, absorption	Surface of the ovaries, linings of kidney tubules, and linings of the ducts of various glands
Simple columnar epithelium	Protection, secretion, absorption	Linings of the uterus and tubes of the digestive tract
Pseudostratified columnar epithelium	Protection, secretion, the movement of mucus and sex cells	Linings of respiratory passages and various tubes of the reproductive systems
Stratified squamous epithelium	Protection	Outer layers of the skin, linings of the mouth cavity, throat, vagina, and anal canal
Transitional epithelium	Distensibility, protection	Inner lining of the urinary bladder and passageways of the urinary tract

Figure 5.7
(*a*) Merocrine glands release fluid without a loss of cytoplasm; (*b*) apocrine glands lose small portions of their cell bodies during secretion; (*c*) holocrine glands release entire cells filled with secretory products.

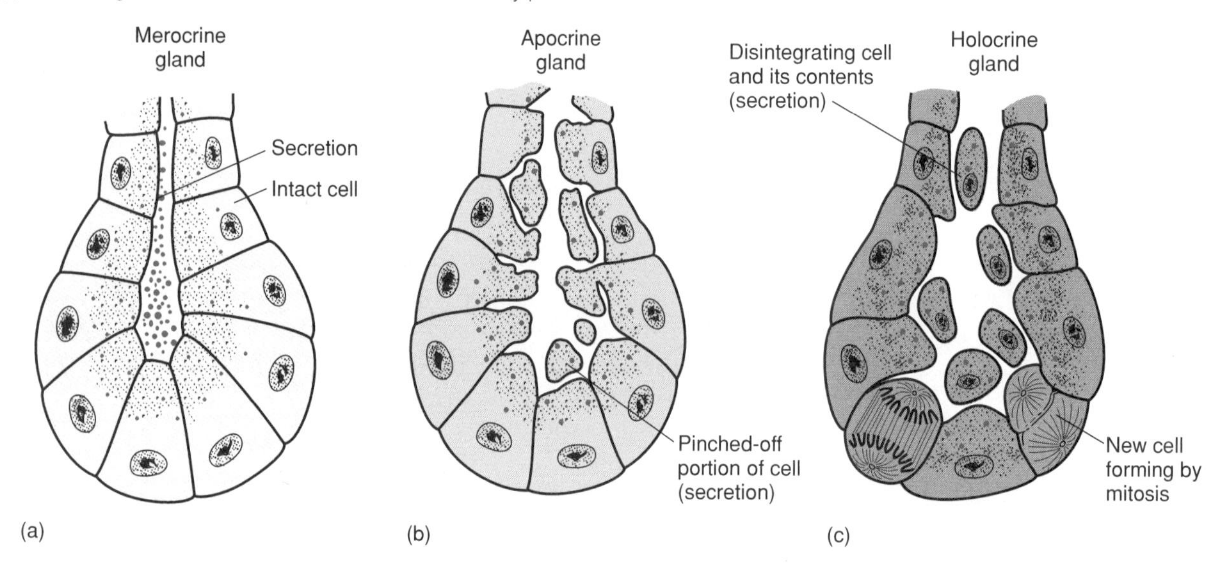

A cancer originating in epithelium is called a *carcinoma.* It is estimated that up to 90% of all human cancers are of this type. Most carcinomas begin on surfaces that contact the external environment, such as the skin, linings of the airways in the respiratory tract, or linings of the stomach or intestines in the digestive tract. This observation suggests that the more common cancer-causing agents may not penetrate tissues very deeply.

1. List the general characteristics of epithelial tissue.
2. Describe the structure of each type of epithelium.
3. Describe the special functions of each type of epithelium.

Glandular Epithelium

Glandular epithelium is composed of cells that are specialized to produce and secrete substances. Such cells occur most commonly within columnar and cuboidal epithelia, and one or more of these cells constitute a *gland.* Glands that secrete their products into ducts opening onto an internal or external surface are called *exocrine glands,* and those that secrete into tissue fluid or blood are called *endocrine glands.* (Endocrine glands are discussed in chapter 11.)

Exocrine glands are classified according to whether their secretions consist of cellular products or portions of glandular cells (fig. 5.7). Thus, glands that release fluid products through cell membranes without any loss of cytoplasm are known as *merocrine glands,*

Chart 5.2 Types of glandular secretions		
Type	Description of secretion	Example
Merocrine glands	Fluid cellular product that is released through the cell membrane	Salivary glands, pancreatic glands, certain sweat glands
Apocrine glands	Cellular product and portions of the free ends of glandular cells that are pinched off during secretion	Mammary glands, certain sweat glands
Holocrine glands	Entire cells that are laden with secretory products	Sebaceous glands

those that lose small portions of cells during secretion are called *apocrine glands,* and those that release entire cells filled with secretory products are called *holocrine glands.* Examples of these glands are provided in chart 5.2.

Most glandular cells are merocrine, and they can be further subdivided into *serous cells* and *mucous cells.* The secretion of serous cells is usually watery and is called *serous fluid.* It has a relatively high concentration of enzymes. Such cells are common in the glands of the digestive tract. Mucous cells, on the other hand, secrete a thicker fluid, mucus. Mucus is rich in a glycoprotein called *mucin* and is secreted abundantly by glands found in the inner linings of digestive and respiratory tubes.

1. Distinguish between exocrine glands and endocrine glands.
2. Explain how exocrine glands are classified.
3. Distinguish between a serous cell and a mucous cell.

Connective Tissue

General Characteristics

Connective tissue occurs throughout the body and represents the most abundant type of tissue by weight. It binds structures together, provides support, serves as framework, fills spaces, stores fat, produces blood cells, provides protection against infections, and helps to repair tissue damage. Connective tissue cells are usually farther apart than epithelial cells, and they have an abundance of intercellular material, called *matrix,* between them. This material consists of *fibers* and a *ground substance,* whose consistency varies from fluid to semisolid to solid.

Connective tissue cells are able to reproduce. In most cases, they have good blood supplies and are well nourished. Although some connective tissues, such as bone and cartilage, are quite rigid, loose connective tissue, adipose connective tissue, and fibrous connective tissue are more flexible.

Figure 5.8
A scanning electron micrograph of a fibroblast (×6,000).

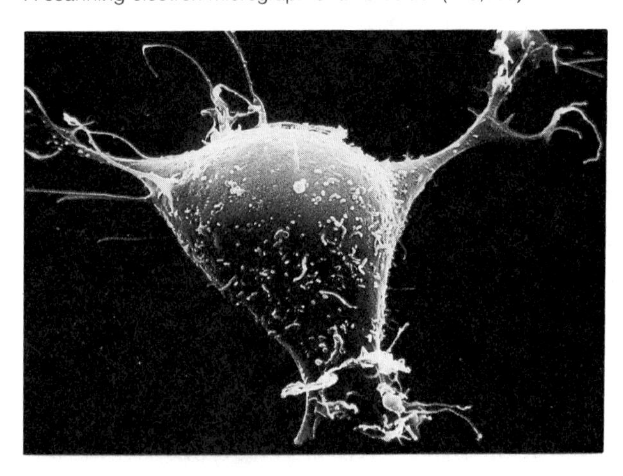

Major Cell Types

Connective tissues contain a variety of cell types. Some of them are called *resident cells* because they are usually present in relatively stable numbers. These include **fibroblasts, macrophages,** and **mast cells.** Another group, known as *wandering cells,* appears temporarily in the tissues, usually in response to an injury or infection. Wandering cells include several types of white blood cells.

The fibroblast is the most common kind of resident cell in connective tissue (fig. 5.8). It is a relatively large cell and usually is star shaped. Fibroblasts produce fibers by secreting proteins into the matrix of connective tissue.

Macrophages, which are almost as numerous as fibroblasts in some connective tissues, are specialized to carry on phagocytosis. They can move about and function as scavenger and defensive cells that clear foreign particles from tissues (fig. 5.9).

Mast cells are relatively large cells that are widely distributed in connective tissues and usually are located near blood vessels (fig. 5.10). Although their function is not fully understood, they release a compound (heparin) that prevents blood clotting. They also release a substance (histamine) that promotes reactions associated with inflammation and allergies (see chapter 16).

Figure 5.9
Macrophages are common in connective tissues where they function as scavenger cells. In this scanning electron micrograph, a macrophage can be seen engulfing two cells (about ×5,600).

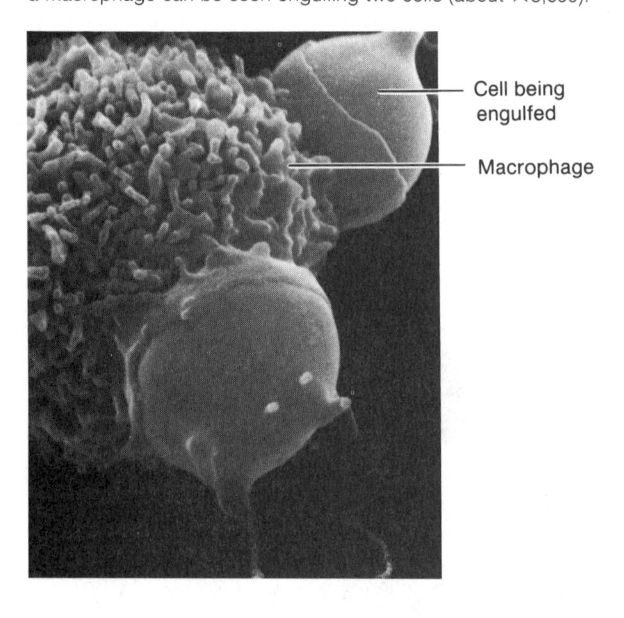

Figure 5.10
A transmission electron micrograph of a mast cell (×4,000).

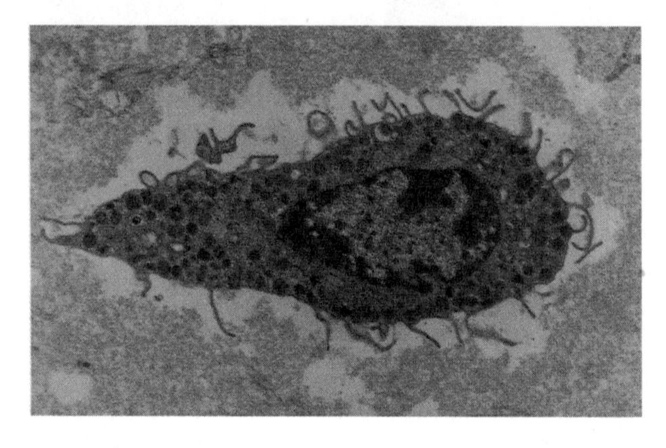

Connective Tissue Fibers

Three types of fibers are produced by fibroblasts. They are called **collagenous fibers, elastic fibers,** and **reticular fibers.** Of these, collagenous and elastic fibers are the most abundant (fig. 5.11). Collagenous fibers are relatively thick, threadlike parts composed of the protein *collagen.* These fibers are grouped in long, parallel bundles, and they are flexible but only slightly elastic. More importantly, they have great tensile strength—that is, they are capable of resisting considerable pulling force. Thus, collagenous fibers are important components of parts that hold structures together, such as cordlike tendons, which connect muscles to bones.

When collagenous fibers are abundant, the tissue containing them is known as *dense connective tissue.* Such tissue appears white, and for this reason, collagenous fibers are sometimes called *white fibers.*

Elastic fibers are composed of microfibrils embedded in a protein, called *elastin.* These fibers tend to be branched and to form complex networks. They have less strength than collagenous fibers, but they are very elastic. That is, they are easily stretched and can resume their original lengths and shapes. They are common in parts that are often stretched, such as the vocal cords. Elastic fibers are sometimes called *yellow fibers* because tissues well supplied with them appear yellowish.

Reticular fibers are very thin fibers and are composed of collagen. They are highly branched and form delicate supporting networks in a variety of tissues.

Loose Connective Tissue

Loose connective tissue forms delicate, thin membranes throughout the body. Cells of this tissue are mainly fibroblasts (fig. 5.11). They are located some

Figure 5.11
(*a*) Loose connective tissue contains numerous fibroblasts, which produce collagenous and elastic fibers. (*b*) A micrograph of loose connective tissue (×250).

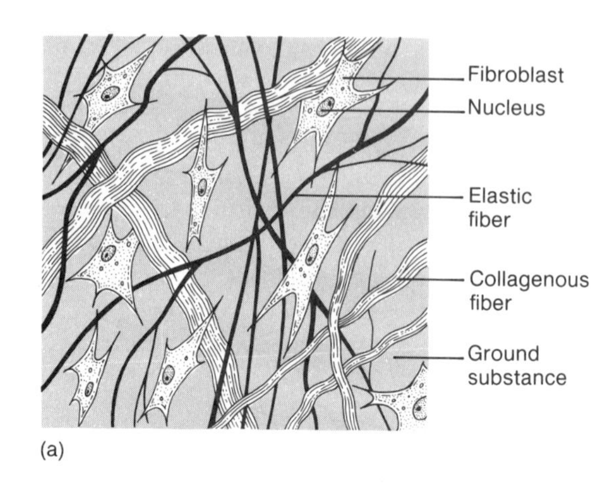

(a)

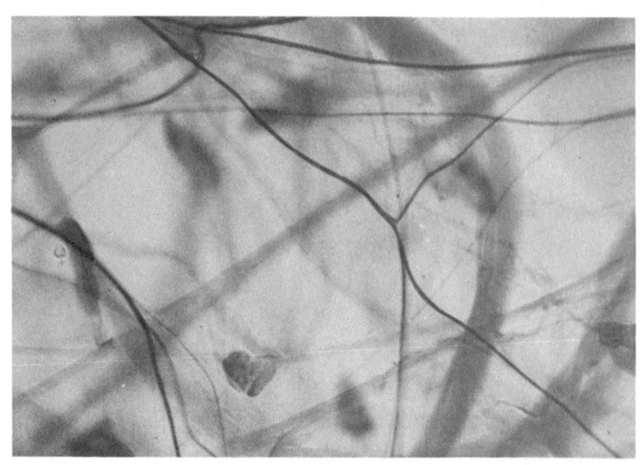

(b)

Figure 5.12
(*a*) Adipose tissue cells contain large fat droplets, which cause their nuclei to be pushed close to their cell membranes. (*b*) Note the nucleus (*arrow*) in this micrograph (×250).

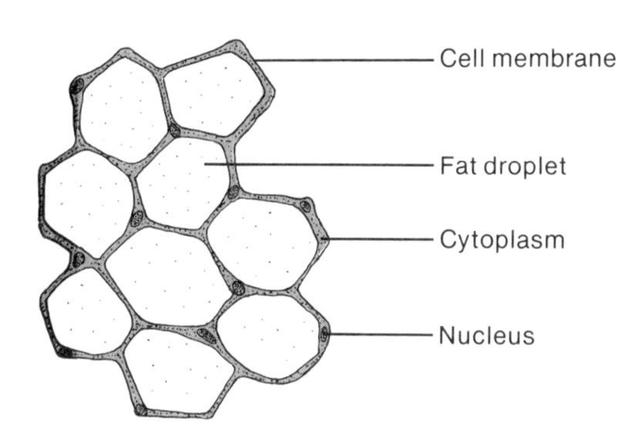

(a)

(b)

distance apart and are separated by a gellike ground substance that contains many collagenous and elastic fibers.

This tissue binds the skin to the underlying organs and fills the spaces between muscles. It lies beneath most layers of epithelium, where its numerous blood vessels provide nourishment for nearby epithelial cells.

Adipose Tissue

Adipose tissue, commonly called fat, is a specialized form of loose connective tissue (fig. 5.12). Certain cells within loose connective tissue store fat in droplets within their cytoplasm and become enlarged. When they occur in such large numbers that other cell types are crowded out, they form adipose tissue.

Adipose tissue is found beneath the skin and in the spaces between muscles. It also occurs around the kidneys, behind the eyeballs, in certain abdominal membranes, on the surface of the heart, and around various joints.

Adipose tissue serves as a protective cushion for joints and some organs, such as the kidneys. It also functions as an insulator beneath the skin, and it stores energy in fat molecules.

The amount of adipose tissue present in the body is usually related to a person's diet. If a person overeats, for example, excess food substances are likely to be converted to fat and stored in adipose tissue. During periods of fasting, however, fat is used to supply energy, and adipose cells lose their fat droplets, shrink in size, and become more like fibroblasts in appearance.

Fibrous Connective Tissue

Fibrous connective tissue is a dense tissue that contains many, closely packed, thick, collagenous fibers and a fine network of elastic fibers. It has relatively few cells, almost all of which are fibroblasts (fig. 5.13).

Since collagenous fibers are very strong, this type of tissue can withstand pulling forces, and it often functions to bind body parts together. For example, **tendons,** which connect muscles to bones and muscles to other muscles, and **ligaments,** which connect bones to bones at joints, are largely composed of fibrous connective tissue. This type of tissue also occurs in the protective white layer of the eyeball and in the deeper portions of the skin.

The blood supply to fibrous connective tissue is relatively poor, so tissue repair occurs slowly.

1. What are the general characteristics of connective tissue?
2. What are the characteristics of collagen and elastin?
3. How is loose connective tissue related to adipose tissue?
4. Distinguish between a tendon and a ligament.

Cartilage

Cartilage is one of the rigid connective tissues. It supports parts, provides frameworks and attachments, protects underlying tissues, and forms structural models for many developing bones.

The intercellular material of cartilage is abundant and is composed largely of collagenous fibers embedded in a gellike ground substance. Cartilage cells,

Figure 5.13

(*a*) Fibrous connective tissue consists largely of tightly packed collagenous fibers. (*b*) The collagenous fibers are stained red in this micrograph (×100).

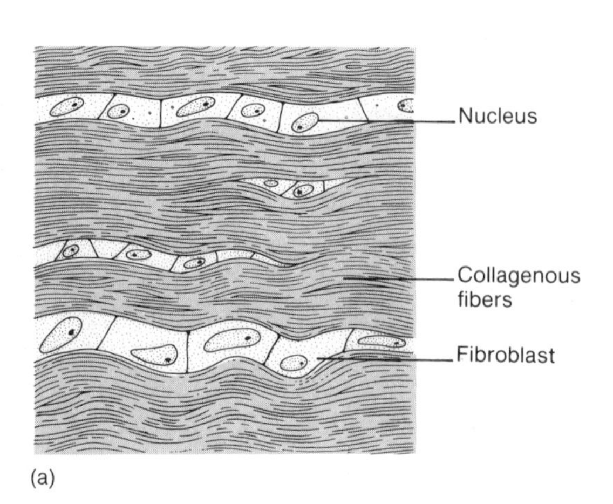

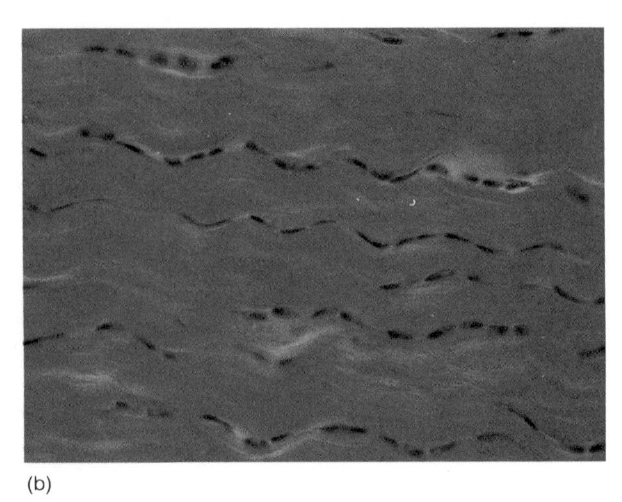

Figure 5.14

(*a*) Hyaline cartilage cells (chondrocytes) are located in lacunae, which are surrounded by intercellular material. (*b*) A micrograph of hyaline cartilage (×250).

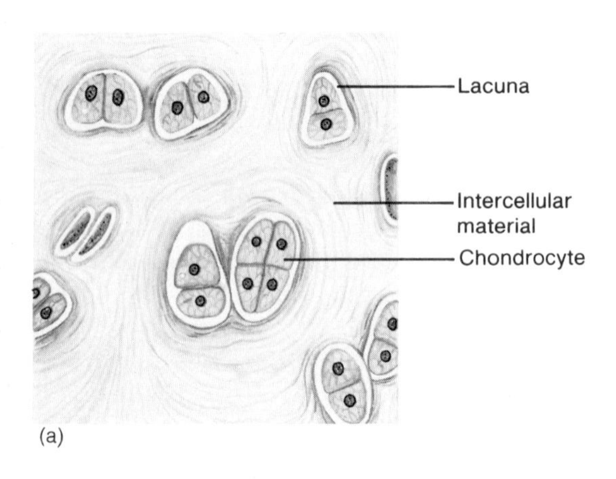

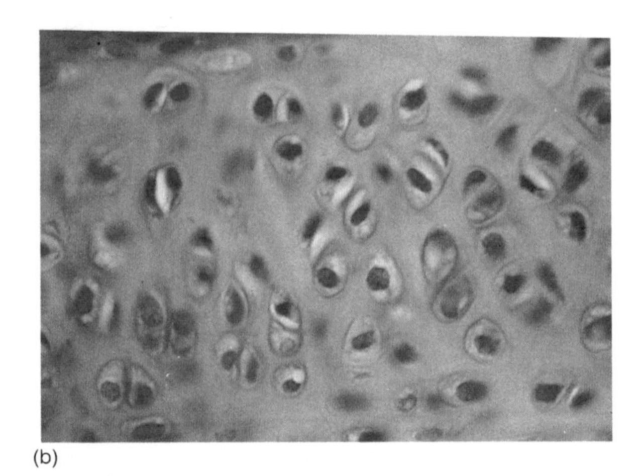

or **chondrocytes,** occupy small chambers, called *lacunae,* and thus are completely surrounded by matrix (fig. 5.14).

A cartilaginous structure is enclosed in a covering of fibrous connective tissue, called the *perichondrium.* Although cartilage tissue lacks a direct blood supply, there are blood vessels in the perichondrium surrounding it. Cartilage cells obtain nutrients from these vessels by diffusion. This lack of a direct blood supply is related to the slow rate of cellular reproduction and repair, which is characteristic of cartilage.

There are three kinds of cartilage, and each kind contains a different type of intercellular material. Hyaline cartilage has very fine collagenous fibers in its matrix, elastic cartilage contains a dense network of elastic fibers, and fibrocartilage contains many large collagenous fibers.

Hyaline cartilage, the most common type of cartilage, looks somewhat like milk glass (fig. 5.14). It occurs on the ends of bones in many joints, in the soft part of the nose, and in the supporting rings of respiratory passages. Hyaline cartilage also plays an important role in the development of many bones (see chapter 7).

Elastic cartilage is more flexible than hyaline cartilage because its matrix contains numerous elastic fibers (fig. 5.15). It provides the framework for the external ears and parts of the larynx.

Fibrocartilage, a very tough tissue, contains many collagenous fibers (fig. 5.16). It often serves as a shock absorber for structures that are subjected to pressure. For example, fibrocartilage forms pads (intervertebral disks) between the individual parts of the backbone. It also forms protective cushions between certain bones in the knees and between bones in the pelvic girdle.

Figure 5.15
(*a*) Elastic cartilage contains many elastic fibers in its intercellular material. (*b*) A micrograph of elastic cartilage (×250).

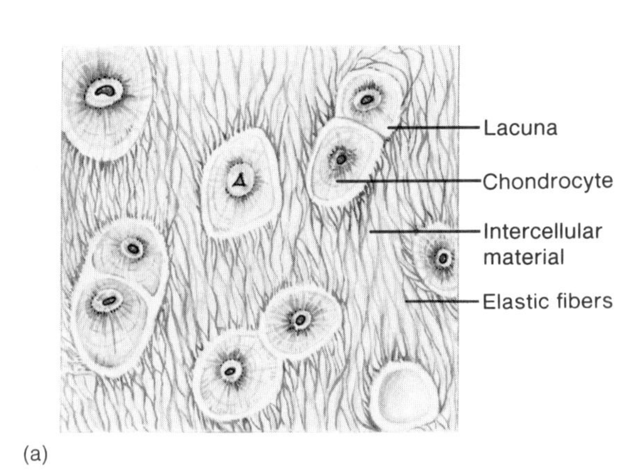

(a)

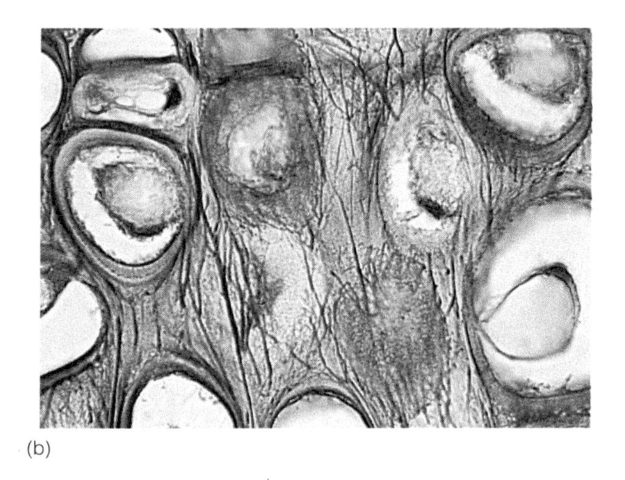

(b)

Figure 5.16
(*a*) Fibrocartilage contains many large, collagenous fibers in its intercellular material. (*b*) A micrograph of fibrocartilage (×195).

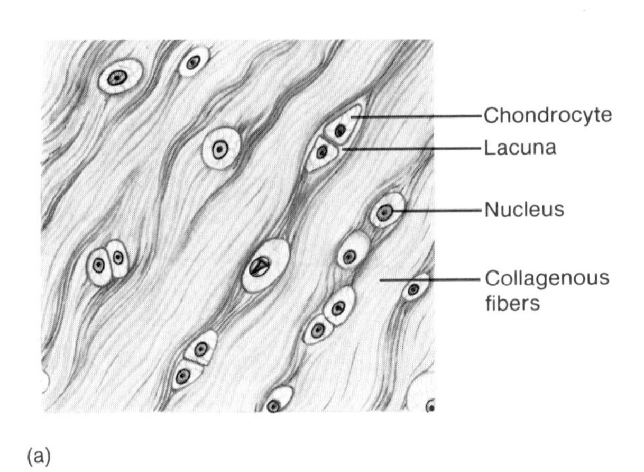

(a)

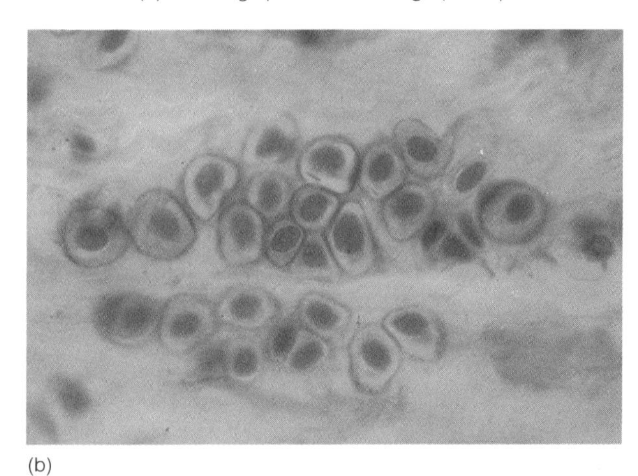

(b)

Bone

Bone is the most rigid of the connective tissues. Its hardness is due largely to the presence of mineral salts in the intercellular material. This matrix also contains a great amount of collagen in fibers that provide flexible reinforcement for the mineral components of the bone tissue.

Bone provides an internal support for body structures. It protects vital parts in various cavities and serves as an attachment for muscles. It also functions in forming blood cells and in storing certain inorganic salts.

Bone matrix is deposited in thin layers called *lamellae,* which most commonly are arranged in concentric patterns around tiny longitudinal tubes, called *osteonic canals* (haversian canals) (fig. 5.17). Bone cells, or **osteocytes,** are located in lacunae, rather evenly spaced between the lamellae. Consequently, they too are arranged in patterns of concentric circles.

In a bone, the osteocytes and layers of intercellular material, which are clustered concentrically around an osteonic canal, form a cylinder-shaped unit, called an **osteon** (haversian system). Many of these units cemented together form the substance of bone.

Each osteonic canal contains a blood vessel, so that every bone cell is fairly close to a nutrient supply. In addition, bone cells have numerous cytoplasmic processes, which extend outward and pass through minute tubes in the matrix called *canaliculi.* These cellular processes are attached to the membranes of nearby cells or connect to the osteonic canals. As a result, materials can move rapidly between blood vessels and bone cells. Thus, in spite of its inert appearance, bone is a very active tissue, which heals much more rapidly than cartilage when it is injured.

Chart 5.3 lists the characteristics of the major types of connective tissue.

Figure 5.17
(*a*) Bone matrix is deposited in concentric layers around osteonic canals. (*b*) What features of the tissue can you identify in this micrograph (×160)? (*c*) Transmission electron micrograph of an osteocyte.

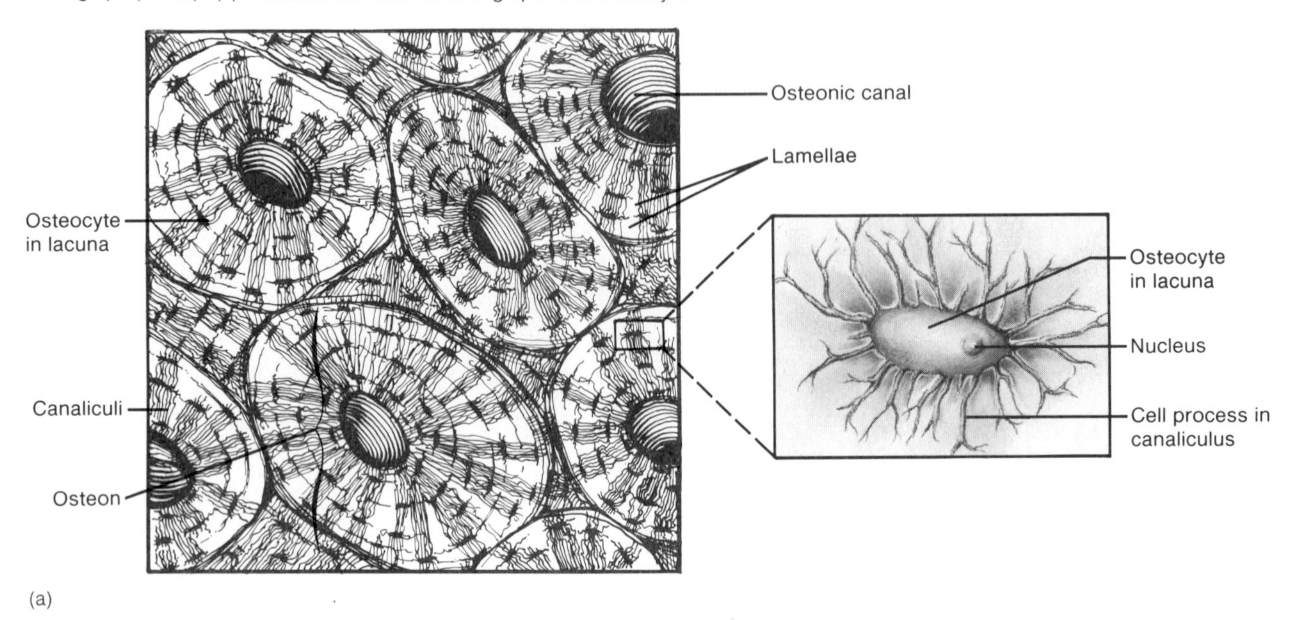

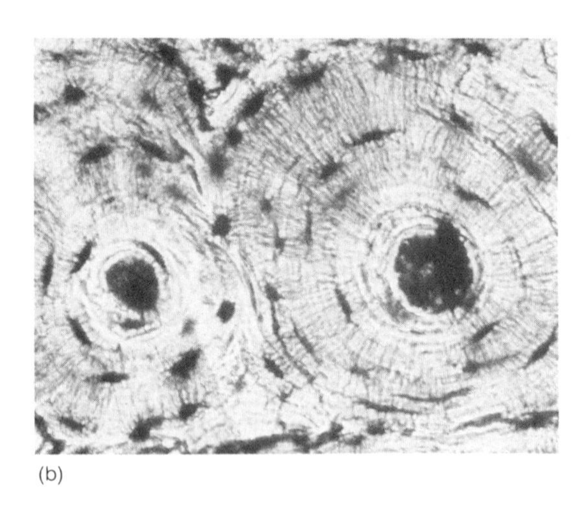

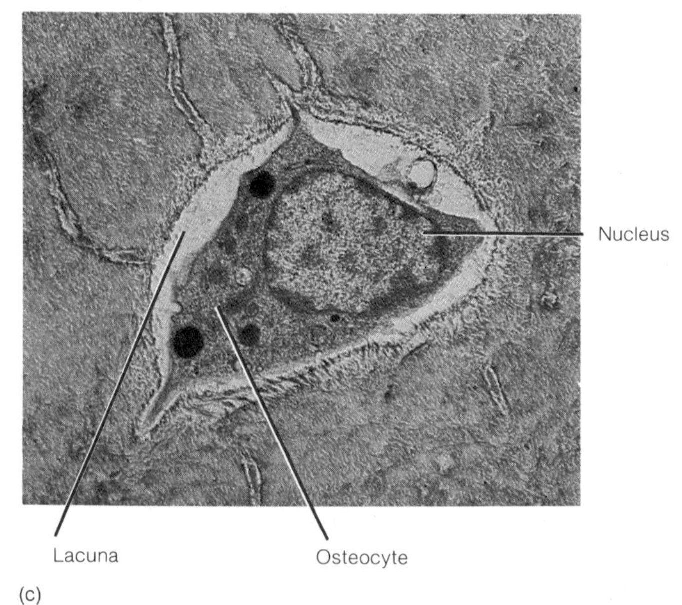

Other Connective Tissues

Other important connective tissues include the blood (vascular tissue) and reticuloendothelial tissue.

The **blood** is composed of cells that are suspended in the fluid intercellular matrix called *blood plasma* (fig. 5.18). These cells include *red blood cells, white blood cells,* and certain cellular fragments, called *platelets.* Most blood cells are formed by special tissues found in the red marrow within the hollow parts of certain bones. The blood is described in chapter 14.

Reticuloendothelial tissue is composed of specialized cells that are widely scattered throughout the body. As a group, these cells are phagocytic, and they function to ingest and destroy foreign particles such as microorganisms. Thus, they are particularly important in defending the body against infection.

The reticuloendothelial cells include types found in the blood, lungs, brain, bone marrow, spleen, liver, and lymph glands. The most common ones, however, are *macrophages.*

Reticuloendothelial tissue is discussed in chapter 16.

Chart 5.3 Connective tissues		
Type	**Function**	**Location**
Loose connective tissue	Binds organs together, holds tissue fluids	Beneath the skin, between muscles, beneath epithelial tissues
Adipose tissue	Protection, insulation, and the storage of fat	Beneath the skin, around the kidneys, behind the eyeballs, on the surface of the heart
Fibrous connective tissue	Binds organs together	Tendons, ligaments, and deep layer of the skin
Hyaline cartilage	Support, protection, provides a framework	Ends of bones, nose, and rings in the walls of respiratory passages
Elastic cartilage	Support, protection, provides a flexible framework	Framework of the external ear and part of the larynx
Fibrocartilage	Support, protection, shock absorption	Between the bony parts of the backbone and knee
Bone	Support, protection, provides a framework	Bones of the skeleton

Figure 5.18
(*a*) Blood tissue consists of an intercellular fluid in which red blood cells, white blood cells, and platelets are suspended. (*b*) A micrograph of human blood (×640).

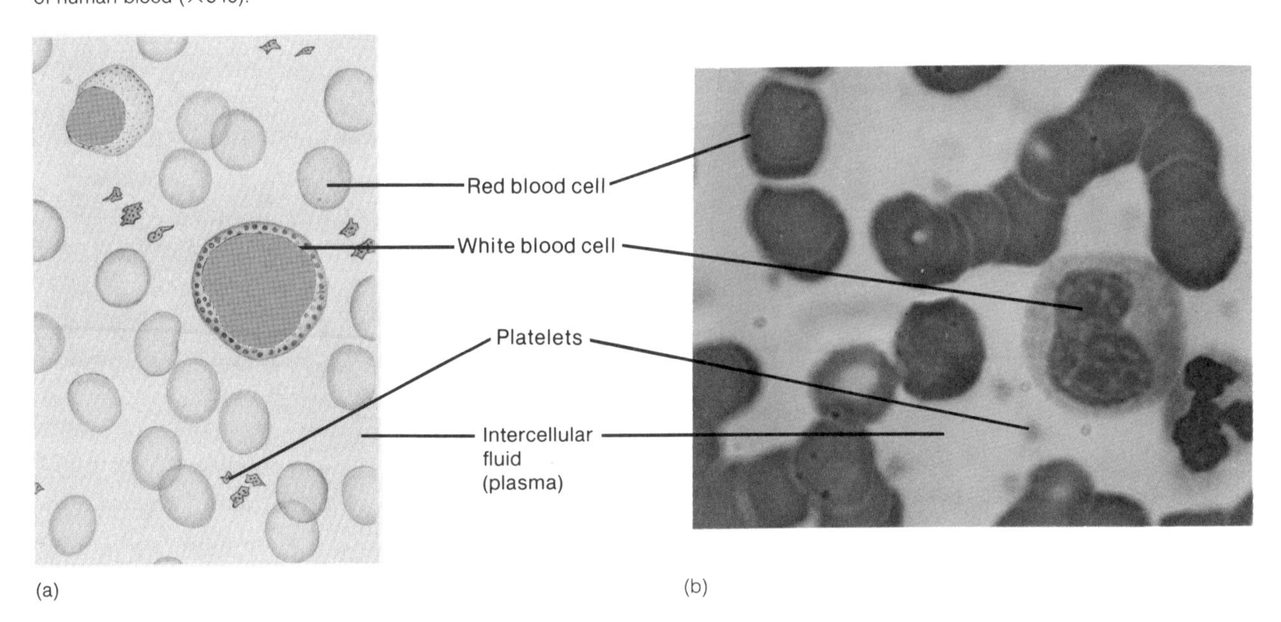

1. Describe the general characteristics of cartilage.
2. Explain why injured bone heals more rapidly than injured cartilage.
3. Name the cells found in the blood.
4. What do the cells of reticuloendothelial tissue have in common?

Muscle Tissue

General Characteristics

Muscle tissues are contractile—that is, their elongated cells, or *muscle fibers,* can change shape by becoming shorter and thicker. As they contract, muscle fibers pull at their attached ends and cause body parts to move. The three types of muscle tissue are skeletal muscle, smooth muscle, and cardiac muscle.

Skeletal Muscle Tissue

Skeletal muscle tissue (fig. 5.19) is found in muscles that are usually attached to bones and can be controlled by conscious effort. For this reason, the tissue is sometimes called *voluntary* muscle. The cells, or muscle fibers, are long and threadlike, with alternating light and dark cross-markings, called *striations.* Each fiber has many nuclei, located just beneath its cell membrane. When the muscle fiber is stimulated by an action of a nerve fiber, it contracts and then relaxes.

Figure 5.19
(*a*) Skeletal muscle tissue is composed of striated muscle fibers, which contain many nuclei. (*b*) A micrograph of skeletal muscle tissue (×250).

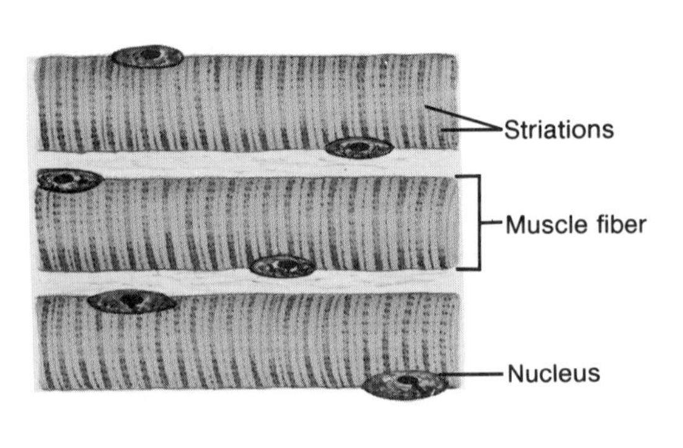

(a)

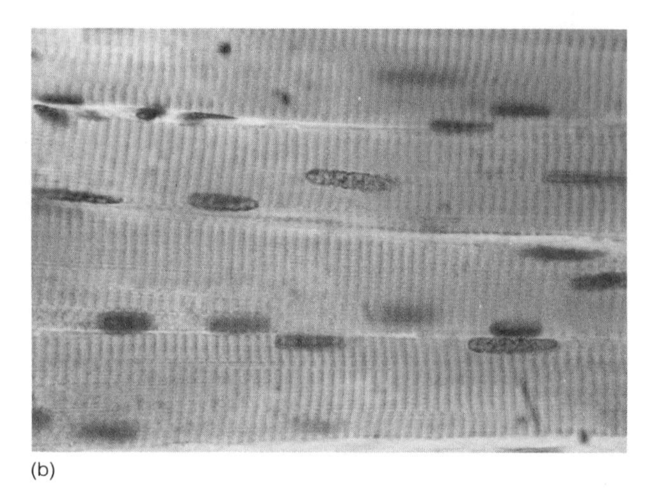

(b)

Figure 5.20
(*a*) Smooth muscle tissue is formed of spindle-shaped cells, each containing a single nucleus. (*b*) Note the elongated nucleus (*arrow*) in this micrograph (×250).

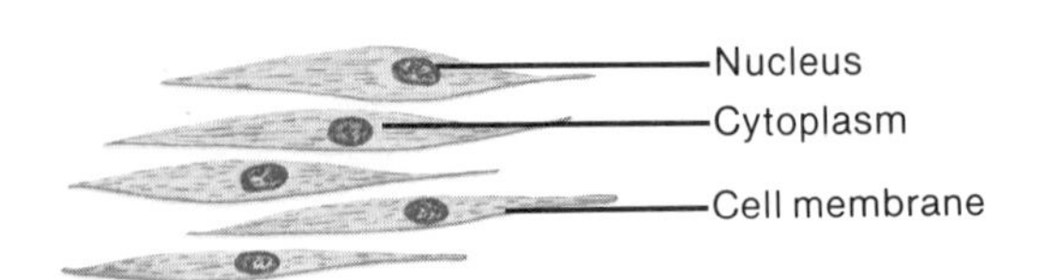

(a)

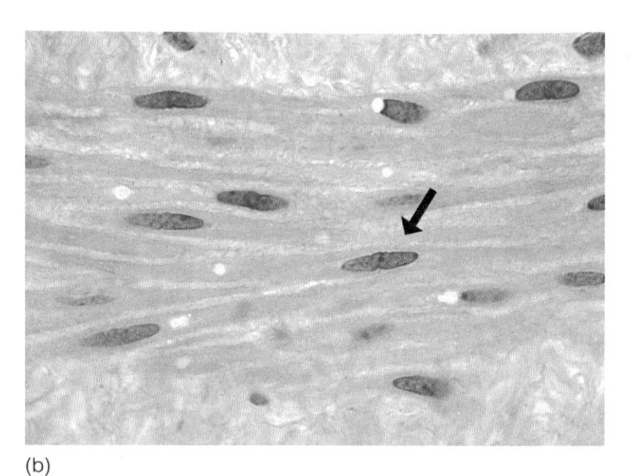

(b)

The muscles containing skeletal muscle tissue are responsible for moving the head, trunk, and limbs. They also produce the movements involved with making facial expressions, writing, talking, and singing, and with such actions as chewing, swallowing, and breathing.

Smooth Muscle Tissue

Smooth muscle tissue is called smooth because its cells lack striations (fig. 5.20). This tissue is found in the walls of hollow internal organs, such as the stomach, intestines, urinary bladder, uterus, and blood vessels. Unlike skeletal muscle, smooth muscle usually cannot be stimulated to contract by conscious effort; therefore, it is a type of *involuntary* muscle tissue. Smooth muscle cells are shorter than those of skeletal muscle, and each has a single, centrally located nucleus. Smooth muscle tissue is responsible for moving food through the digestive tube, constricting blood vessels, and emptying the urinary bladder.

Cardiac Muscle Tissue

Cardiac muscle tissue occurs only in the heart (fig. 5.21). Its cells, which are striated, are joined end to end. The resulting muscle fibers are branched and interconnected in complex networks. Each cell within a fiber usually has a single nucleus. At its end, where it touches another cell, there is a specialized intercellular connection, called an *intercalated disk.*

Cardiac muscle, like smooth muscle, is controlled involuntarily. This tissue makes up the bulk of the heart and is responsible for pumping the blood through the heart chambers and into certain blood vessels.

Figure 5.21
(*a*) How is cardiac muscle tissue similar to skeletal muscle? How is it similar to smooth muscle? (*b*) Note the intercalated disk (*arrow*) in this micrograph (×400).

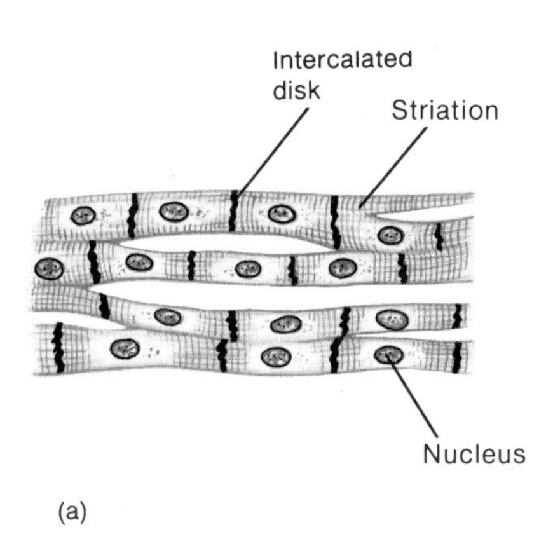

(a)

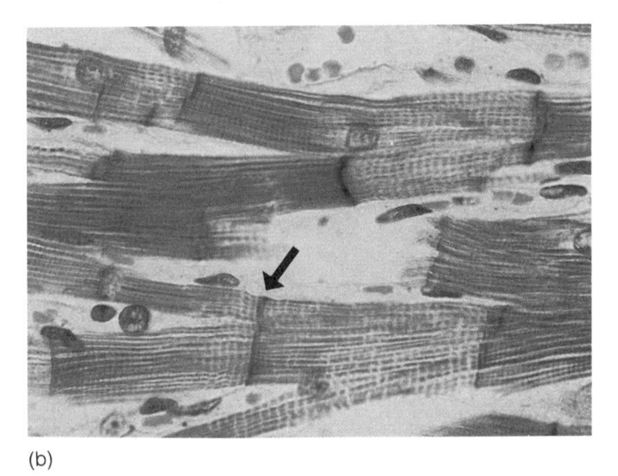

(b)

Figure 5.22
(*a*) Neurons function to transmit impulses to other neurons or to muscles or glands. (*b*) What features of a neuron can you identify in this micrograph (×450)?

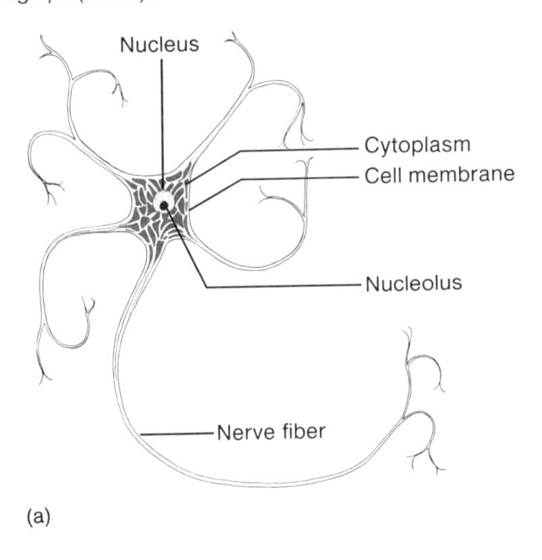

(a)

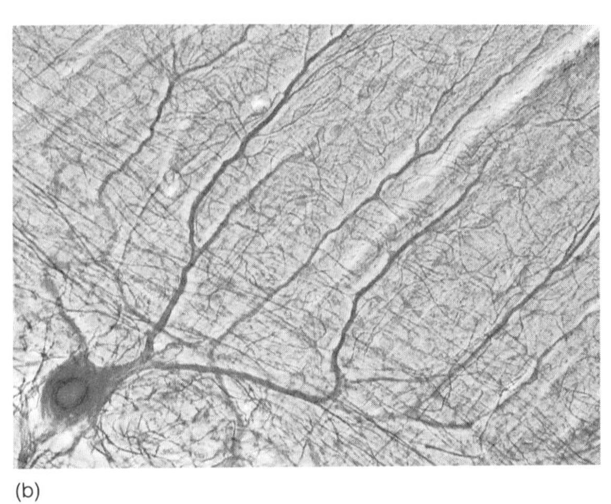

(b)

1. List the general characteristics of muscle tissue.
2. Distinguish between skeletal, smooth, and cardiac muscle tissue.

Nervous Tissue

Nervous tissue is found in the brain, spinal cord, and peripheral nerves. The basic cells of this tissue are called *nerve cells,* or **neurons** (fig. 5.22). Neurons are sensitive to certain types of changes in their surroundings, and they respond by transmitting nerve impulses along cytoplasmic extensions (nerve fibers) to other neurons or to muscles or glands.

Because neurons communicate with each other and with various body parts, they are able to coordinate, regulate, and integrate many body functions.

In addition to neurons, nerve tissue contains **neuroglial cells** (see fig. 9.5). These cells support and bind the components of nervous tissue together, carry on phagocytosis, and supply nutrients to neurons by connecting them to blood vessels. Nervous tissue is discussed in more detail in chapter 9.

Chart 5.4 summarizes the general characteristics of muscle and nervous tissues.

1. Describe the general characteristics of nervous tissue.
2. Distinguish between neurons and neuroglial cells.

Chart 5.4 Muscle and nervous tissues		
Type	Function	Location
Skeletal muscle tissue	Voluntary movements of skeletal parts	Muscles attached to bones
Smooth muscle tissue	Involuntary movements of internal organs	Walls of hollow internal organs
Cardiac muscle tissue	Heart movements	Heart muscle
Nervous tissue	Sensitivity and conduction of nerve impulses	Brain, spinal cord, and peripheral nerves

Chapter Summary

Introduction (page 96)

Cells are arranged in tissues that perform specialized structural and functional roles. The four major types of human tissue are epithelial tissue, connective tissue, muscle tissue, and nervous tissue.

Epithelial Tissue (page 96)

1. General characteristics
   a. Epithelial tissues cover all the free body surfaces and are the major tissues of glands.
   b. Epithelium is anchored to connective tissue by a basement membrane, lacks blood vessels, contains little intercellular material, and is replaced continuously.
   c. It functions in protection, secretion, absorption, excretion, and sensory reception.
2. Simple squamous epithelium
   a. This tissue consists of a single layer of flattened cells.
   b. It functions in the exchange of gases in the lungs and lines the blood vessels and various membranes.
3. Simple cuboidal epithelium
   a. This tissue consists of a single layer of cube-shaped cells.
   b. It carries on secretion and absorption in the kidneys and various glands.
4. Simple columnar epithelium
   a. This tissue is composed of elongated cells whose nuclei are located near the basement membrane.
   b. It lines the uterus and digestive tract.
   c. Absorbing cells often possess microvilli.
   d. This tissue contains goblet cells, which secrete mucus.
5. Pseudostratified columnar epithelium
   a. This tissue appears stratified because the nuclei are located at two or more levels within the cells.
   b. Its cells may have cilia, which function to move mucus or sex cells.
   c. It lines various tubes of the respiratory and reproductive systems.
6. Stratified squamous epithelium
   a. This tissue is composed of many cell layers.
   b. It protects underlying cells.
   c. It covers the skin and lines the mouth, throat, vagina, and anal canal.
7. Transitional epithelium
   a. This tissue is specialized to undergo changes in tension.
   b. It occurs in the walls of various organs of the urinary system.
8. Glandular epithelium
   a. Glandular epithelium is composed of cells that are specialized to secrete substances.
   b. One or more cells constitute a gland.
      (1) Exocrine glands secrete into ducts.
      (2) Endocrine glands secrete into body fluids or the blood.
   c. Exocrine glands are classified according to the composition of their secretions.
      (1) Merocrine glands secrete fluid without the loss of cytoplasm.
      (2) Apocrine glands lose portions of cells during secretion.
      (3) Holocrine glands release entire cells during secretion.
   d. Serous cells secrete watery fluid with a high enzyme concentration; mucous cells secrete mucus.

Connective Tissue (page 101)

1. General characteristics
   a. Connective tissues connect, support, protect, provide frameworks, fill spaces, store fat, and produce blood cells.
   b. Connective tissue cells are usually some distance apart, and they have a considerable amount of intercellular material between them.
   c. This intercellular matrix consists of fibers and a ground substance.
   d. Major cell types
      (1) Connective tissue cells include resident and wandering cells.
      (2) Fibroblasts produce collagenous and elastic fibers.
      (3) Macrophages function as phagocytes.
      (4) Mast cells may release heparin and histamine, and are usually found near blood vessels.

e. Connective tissue fibers
    (1) Collagenous fibers are composed of collagen and have great tensile strength.
    (2) Tissue containing an abundance of collagenous fibers is called dense connective tissue.
    (3) Elastic fibers are composed of microfibrils embedded in elastin and are very elastic.
    (4) Reticular fibers are very fine collagenous fibers.

2. Loose connective tissue
    a. This tissue forms thin membranes between organs.
    b. It is found beneath the skin and between muscles.
3. Adipose tissue
    a. Adipose tissue is a specialized form of loose connective tissue that stores fat.
    b. It is found beneath the skin, in certain abdominal membranes, and around the kidneys, heart, and various joints.
4. Fibrous connective tissue
    a. This tissue is composed largely of strong, collagenous fibers.
    b. It is found in the tendons, ligaments, white portions of the eyes, and the deep layer of the skin.
5. Cartilage
    a. Cartilage provides support and a framework for various parts.
    b. Its intercellular material is largely composed of fibers and a gellike ground substance.
    c. Cartilaginous structures are enclosed in a perichondrium.
    d. Cartilage lacks a direct blood supply and is slow to heal following an injury.
    e. The major types are hyaline cartilage, elastic cartilage, and fibrocartilage.
6. Bone
    a. The intercellular matrix of bone contains mineral salts and collagen.
    b. Its cells are arranged in concentric circles around osteonic canals and are interconnected by canaliculi.
    c. It is an active tissue that heals rapidly following an injury.
7. Other connective tissues
    a. Blood
        (1) The blood is composed of red cells, white cells, and platelets suspended in plasma.
        (2) The cells are formed by special tissue in the hollow parts of certain bones.
    b. Reticuloendothelial tissue
        (1) This tissue is composed of phagocytic cells that are widely distributed throughout the body.
        (2) It defends the body against invasion by microorganisms.

Muscle Tissue (page 107)

1. General characteristics
    a. Muscle tissue is contractile tissue that moves parts that are attached to it.
    b. Three types are skeletal, smooth, and cardiac muscle tissues.
2. Skeletal muscle tissue
    a. The muscles containing this tissue are usually attached to bones and controlled by conscious effort.
    b. The cells, or muscle fibers, are long and threadlike.
    c. Muscle fibers contract when stimulated by nerve action and then relax immediately.
3. Smooth muscle tissue
    a. This tissue is found in the walls of hollow internal organs.
    b. It is usually controlled by involuntary activity.
4. Cardiac muscle tissue
    a. This tissue is found only in the heart.
    b. Its cells are joined by intercalated disks and arranged in branched, interconnecting networks.

Nervous Tissue (page 109)

1. Nervous tissue is found in the brain, spinal cord, and peripheral nerves.
2. Neurons
    a. Neurons are sensitive to changes and respond by transmitting nerve impulses to other neurons or body parts.
    b. They function in coordinating, regulating, and integrating body activities.
3. Neuroglial cells
    a. Some forms bind and support nervous tissue.
    b. Others carry on phagocytosis.
    c. Still others connect neurons to blood vessels.

Clinical Application of Knowledge

1. Joints, such as the elbow, shoulder, and knee, contain considerable amounts of cartilage and fibrous connective tissue. How is this related to the fact that joint injuries are often very slow to heal?
2. There is a group of disorders called collagenous diseases that are characterized by the deterioration of connective tissues. Why would you expect such diseases to produce widely varying symptoms?
3. Sometimes, in response to the presence of irritants, mucous cells secrete excessive amounts of mucus. What symptoms might this produce if it occurred in (a) the respiratory passageways; (b) the digestive tract?

Review Activities

1. Define *tissue.*
2. Name the four major types of tissue found in the human body.
3. Describe the general characteristics of epithelial tissues.
4. Explain how the structure of simple squamous epithelium is related to its function.
5. Name an organ in which each of the following tissues is found, and give the function of the tissue in each case.
   a. Simple squamous epithelium.
   b. Simple cuboidal epithelium.
   c. Simple columnar epithelium.
   d. Pseudostratified columnar epithelium.
   e. Stratified squamous epithelium.
   f. Transitional epithelium.
6. Define *gland.*
7. Distinguish between exocrine and endocrine gland.
8. Explain how glandular secretions differ.
9. Define *mucus.*
10. Describe the general characteristics of connective tissues.
11. Distinguish between resident cells and wandering cells.
12. Distinguish between collagen and elastin.
13. Define *dense connective tissue.*
14. Explain the relationship between loose connective tissue and adipose tissue.
15. Explain why injured fibrous connective tissue and cartilage are usually slow to heal.
16. Name the types of cartilage, and describe their differences and similarities.
17. Describe how bone cells are arranged in bone tissue.
18. Describe the composition of the blood.
19. Define *reticuloendothelial tissue.*
20. Describe the general characteristics of muscle tissues.
21. Distinguish between skeletal, smooth, and cardiac muscle tissues.
22. Describe the general characteristics of nervous tissue.
23. Distinguish between neurons and neuroglial cells.

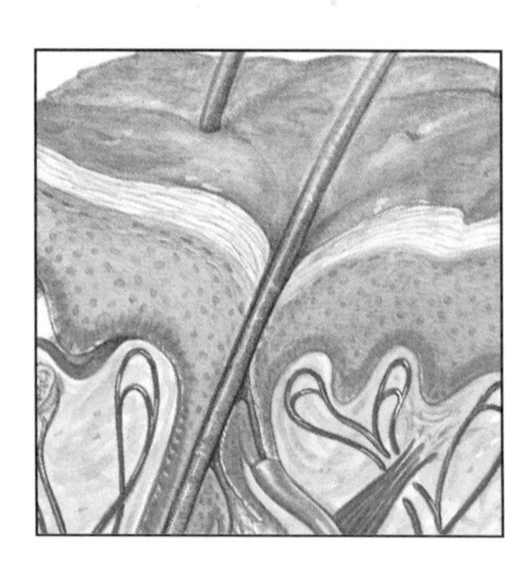

6
Skin and the Integumentary System

The previous chapters dealt with the lower levels of organization within the human organism—the tissues, cells, cellular organelles, and chemical substances that form these parts.

Chapter 6 explains how tissues are grouped to form organs and how organs comprise organ systems. In this explanation, various membranes, including the skin, are used as examples of organs. Since the skin acts with the hair follicles, sebaceous glands, and sweat glands to provide a variety of functions, these organs together constitute the integumentary organ system.

Chapter Objectives

After you have studied this chapter, you should be able to

1. Describe the four major types of membranes.
2. Describe the structure of the various layers of the skin.
3. List the general functions of each layer of skin.
4. Summarize the factors that determine skin color.
5. Describe the accessory organs associated with the skin.
6. Explain how the skin functions in regulating body temperature.
7. Complete the review activities at the end of this chapter. Note that the items are worded in the form of specific learning objectives. You may want to refer to them before reading the chapter.

Key Terms

cutaneous membrane (ku-ta′ne-us mem′brān)

dermis (der′mis)

epidermis (ep″ĭ-der′mis)

hair follicle (hār fol′ĭ-kl)

integumentary (in-teg-u-men′tar-e)

keratinization (ker″ah-tin″ĭ-za′shun)

melanin (mel′ah-nin)

mucous membrane (mu′kus mem′brān)

sebaceous gland (se-ba′shus gland)

serous membrane (se′rus mem′brān)

subcutaneous (sub″ku-ta′ne-us)

sweat gland (swet gland)

synovial membrane (sĭ-no′ve-al mem′brān)

Aids to Understanding Words

cut-, skin: sub*cut*aneous—beneath the skin.

derm-, skin: *derm*is—the inner layer of the skin.

epi-, upon: *epi*dermis—the outer layer of the skin.

follic-, small bag: hair *follic*le—a tubelike depression in which a hair develops.

kerat-, horn: *kerat*in—a protein produced as epidermal cells die and harden.

melan-, black: *melan*in—a dark pigment produced by certain cells.

seb-, grease: *seb*aceous gland—a gland that secretes an oily substance.

Introduction

TWO OR MORE kinds of tissues grouped together and performing specialized functions constitute an organ. Thus, the thin, sheetlike membranes, composed of epithelium and connective tissue, that cover body surfaces and line body cavities are organs.

One of these membranes, the cutaneous membrane, together with certain accessory organs, makes up the **integumentary system.**

Types of Membranes

The four major types of membranes are: *serous, mucous, cutaneous,* and *synovial.*

Serous membranes line body cavities that lack openings to the outside. They form the inner lining of the thorax (parietal pleura) and abdomen (parietal peritoneum), and they cover the organs within these cavities (visceral pleura and visceral peritoneum, respectively). A serous membrane consists of a layer of simple squamous epithelium and a thin layer of loose connective tissue. It secretes watery *serous fluid,* which helps to lubricate the surfaces of the membrane.

Mucous membranes line cavities and tubes that open to the outside of the body. These include the oral and nasal cavities and the tubes of the digestive, respiratory, urinary, and reproductive systems. A mucous membrane consists of epithelium overlying a layer of loose connective tissue. Specialized cells within a mucous membrane secrete *mucus.*

The **cutaneous membrane** is an organ of the integumentary system and is more commonly called *skin.* It is described in detail in the following section of this chapter.

Synovial membranes form the inner linings of the joint cavities between the ends of bones at freely movable joints (synovial joints). These membranes usually include fibrous connective tissue overlying loose connective tissue and adipose tissue. They secrete a thick, colorless *synovial fluid* into the joint cavity, which lubricates the ends of the bones within the joint.

Skin and Its Tissues

The skin is one of the larger and more versatile organs of the body, and it plays vital roles in the maintenance of homeostasis. For example, the skin functions as a protective covering, aids in the regulation of body temperature, retards water loss from the deeper tissues, houses sensory receptors, synthesizes various chemicals and excretes small quantities of waste substances.

The skin includes two distinct layers of tissues (fig. 6.1). The outer layer, called the **epidermis,** is composed of stratified squamous epithelium. The inner layer, or **dermis,** is thicker than the epidermis and includes a variety of tissues, such as fibrous connective tissue, epithelial tissue, smooth muscle tissue, nervous tissue, and blood. These two layers of the skin are separated by a basement membrane, which is anchored to the dermis by short fibrils.

Beneath the dermis are masses of loose connective and adipose tissues which bind the skin to the underlying organs. These tissues form the **subcutaneous layer** (hypodermis).

Subcutaneous injections are administered into the subcutaneous layer beneath the skin. Intradermal injections, on the other hand, are injected into the layers of tissues within the skin. Subcutaneous injections and intramuscular injections, which are administered into muscles, are sometimes called hypodermic injections.

1. Name the four types of membranes, and explain how they differ.
2. List the general functions of the skin.
3. Name the tissue found in the outer layer of the skin.
4. Name the tissues found in the inner layer of the skin.

Epidermis

Since the epidermis is composed of stratified squamous epithelium, it lacks blood vessels; however, the deepest cells of the epidermis, which are next to the dermis, are nourished by the dermal blood vessels and are capable of reproducing (fig. 6.1). These deep cells form a layer called *stratum basale* (stratum germinativum). As the cells of this layer divide and grow, the older epidermal cells are pushed slowly away from the dermis toward the surface. The farther the cells travel, the poorer their nutrient supply becomes, and in time they die.

Meanwhile, the maturing cells (keratinocytes) undergo a hardening process, called **keratinization,** during which the cytoplasm develops strands of a tough, fibrous, waterproof protein, called *keratin.* As a result, many layers of tough, dead cells accumulate in the outer portion of the epidermis. This outermost layer is called *stratum corneum,* and the dead cells that compose it are easily rubbed away.

In healthy skin, the production of epidermal cells is closely balanced with the loss of stratum corneum, so that the skin seldom wears away completely. In fact, the rate of cellular reproduction tends to increase in regions where the skin is being rubbed or pressed regularly. This response causes thickened regions, called *calluses,* to develop on the palms and soles.

Figure 6.1
A section of skin.

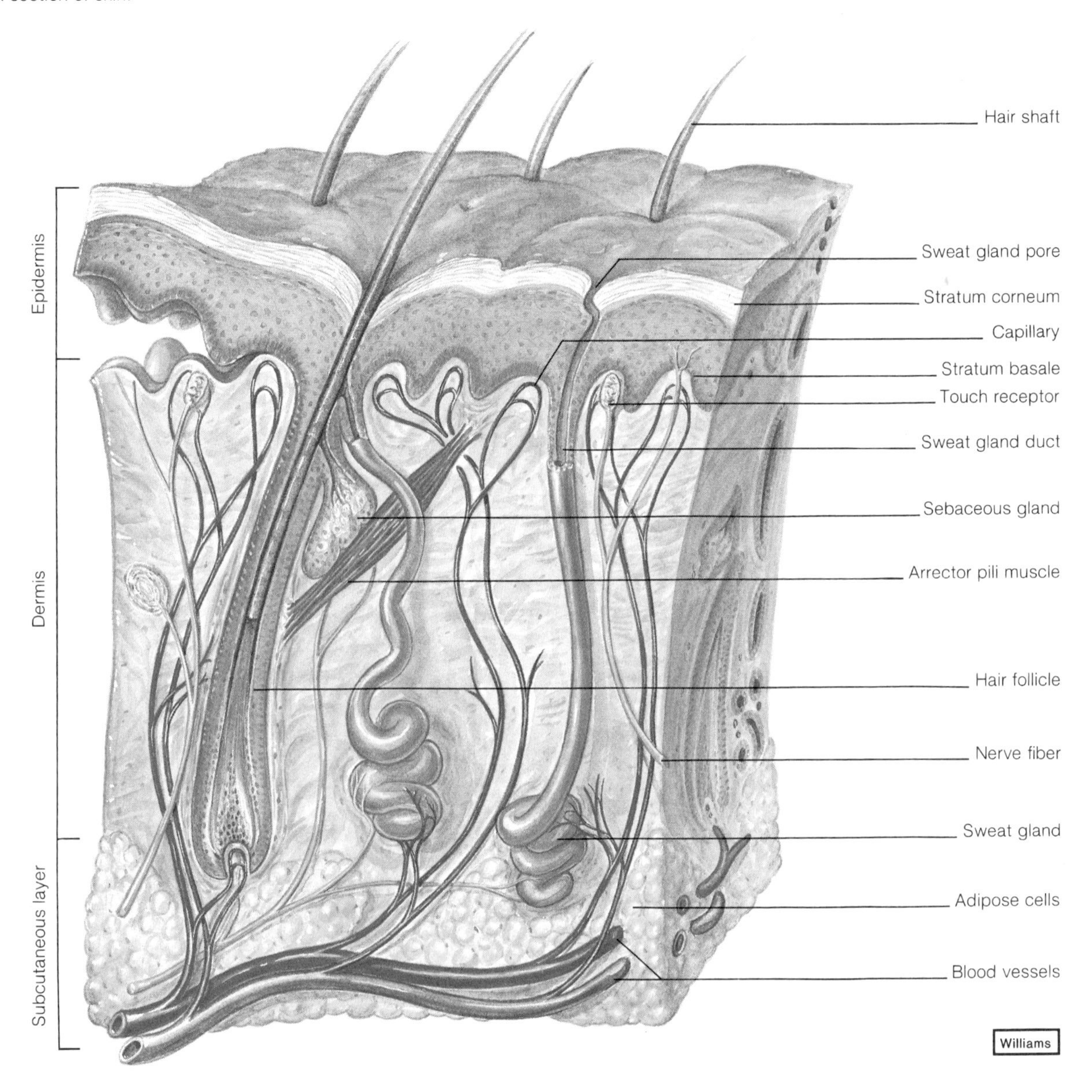

The epidermis has important protective functions. It shields the moist underlying tissues against excessive water loss, mechanical injury, and the effects of harmful chemicals. When it is unbroken, the epidermis also prevents the entrance of many disease-causing microorganisms. **Melanin** is a dark pigment that occurs in the deeper layers of the epidermis and is produced by specialized cells known as *melanocytes* (fig. 6.2). Melanin absorbs light energy and, in this way, helps protect still deeper cells from the damaging effects of sunlight.

Skin Color

Skin color is due largely to the presence of the melanin produced by epidermal melanocytes. Regardless of racial origin, all humans have about the same concentration of melanocytes in their skins. Humans vary, however, in the amount of melanin that their melanocytes produce and in the size of the pigment granules within these cells. Thus, persons whose genes cause the production of a relatively large amount of pigment and relatively large pigment granules have darker skins; whereas, those whose genes direct the formation of less pigment and smaller granules have lighter complexions.

Skin color is also influenced by various environmental and physiological factors. For example, environmental factors such as sunlight, ultraviolet light from sunlamps, and X rays stimulate additional pigment to be produced. The blood in the dermal vessels

Figure 6.2
(*a*) Melanocytes (*arrows*), which occur mainly in the deeper epidermal layers, produce a pigment called melanin (×160).
(*b*) Transmission electron micrograph of a melanocyte with melanin-containing granules (×10,000).

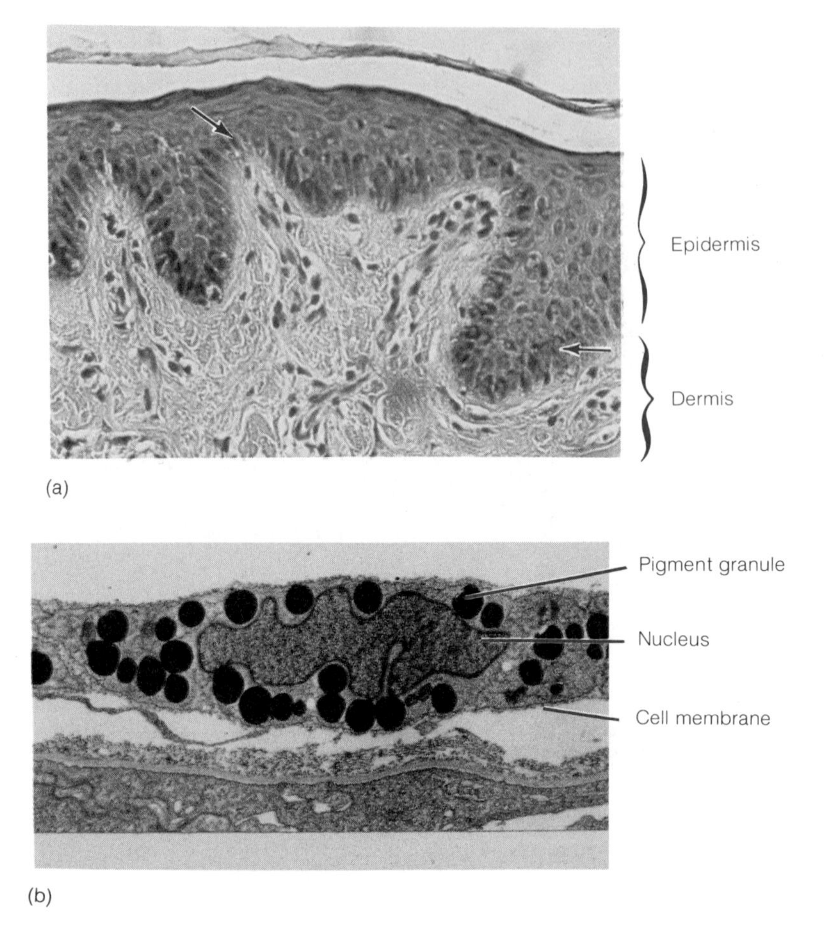

may affect skin color as physiological changes occur. Thus, when the blood is well oxygenated, the blood pigment (hemoglobin) is bright red, making the skin of light-complexioned persons appear pinkish. On the other hand, when the blood oxygen concentration is relatively low, the blood pigment is darker, and the skin may appear bluish—a condition called *cyanosis*.

1. Explain how the epidermis is formed.
2. Distinguish between stratum basale and stratum corneum.
3. What is the function of melanin?
4. What factors affect skin color?

Dermis

The dermis binds the epidermis to the underlying tissues (fig. 6.1). It is composed of fibrous connective tissue, which includes tough collagenous fibers and elastin fibers, surrounded by a gellike ground substance. Networks of these fibers give the dermis its strength and elasticity.

The blood vessels in the dermis supply nutrients to all the skin cells, including those of the epidermis. These dermal blood vessels also aid in regulating body temperature, as is explained in a subsequent section of this chapter.

Because the blood vessels in the dermis supply nutrients to the epidermis, any interference with blood flow is likely to result in the death of epidermal cells. For example, when a person lies in one position for a prolonged period of time, the weight of the body pressing against the bed interferes with the skin's blood supply. If the cells die, the tissues begin to break down and a *pressure ulcer* (also called a decubitus ulcer, or bedsore) may appear.

Pressure ulcers usually occur in the skin overlying bony projections, such as on the hip, heel, elbow, or shoulder. These ulcers can often be prevented by changing the body position frequently or by massaging the skin to stimulate blood flow in the regions associated with bony prominences.

Numerous nerve fibers are scattered throughout the dermis. Some of them (motor fibers) carry impulses to the dermal muscles and glands, causing these

Skin Cancer

A Current Topic

Skin cancer is most likely to arise from the nonpigmented epithelial cells within the deep layer of the epidermis or from pigmented melanocytes. Cancers originating from epithelial cells are more common and are called *cutaneous carcinomas;* those arising from melanocytes are known as *cutaneous melanomas* (melanocarcinomas).

Cutaneous carcinomas occur most frequently in light-skinned people who are over forty years of age. These cancers seem to result from the effect of sunlight on the DNA of the epithelial cells and usually appear in skin that has been exposed to sunlight regularly. Thus, the incidence of cutaneous carcinomas is increased in persons who have spent a considerable amount of time outdoors—farmers, sailors, athletes, sunbathers, and so forth.

Cutaneous carcinoma often develops from hard, dry, scaly growths (lesions) that have reddish bases. Such growths may be either flat or raised above the surface, and they adhere firmly to the skin. Fortunately, this type of cancer is typically slow-growing and usually can be cured completely through surgical removal or radiation treatment.

Because melanomas develop from melanocytes, they are pigmented with melanin. Such growths often have a variety of colored areas—variegated brown, black, gray, or blue—that are arranged haphazardly. They also have irregular margins, rather than smooth, regular outlines.

Cutaneous melanomas occur in young adults as well as in older ones and seem to be caused by relatively short, intermittent exposures to high intensity sunlight. Thus, the incidence of melanoma is increased in persons who generally stay indoors, but occasionally sustain blistering sunburns during weekend activities or vacations.

Cutaneous melanomas appear most often in light-skinned persons, particularly those whose skins tend to burn rather than tanning. Such a growth may arise from normal-appearing skin or from a mole (nevus). Typically the growth enlarges by spreading through the skin horizontally, but eventually it may thicken and invade deeper tissues. If a melanoma is removed while it is in its horizontal growth phase, it may be cured. Once it has thickened and grown downward, however, it becomes difficult to treat, and the survival rate for persons with this form of cancer is very low.

To reduce the chance of developing skin cancer, it is advisable to avoid exposing the skin to high intensity sunlight or other sources of ultraviolet radiation, to wear protective clothing when outdoors, or to use sun-screening lotions with high protective factors. In addition, the skin should be examined regularly, and any unusual growths—particularly those undergoing changes in color, shape, or surface texture—should be checked by a physician.

structures to react. Others (sensory fibers) carry impulses away from specialized sensory receptors, such as touch receptors, located within the dermis. Hair follicles, sebaceous glands, and sweat glands also occur within the dermis (fig. 6.1). These parts are discussed in subsequent sections of this chapter.

Subcutaneous Layer

As was mentioned previously, the subcutaneous layer (hypodermis) lies beneath the dermis and consists largely of loose connective and adipose tissues (fig. 6.1). The collagenous and elastic fibers in this layer are continuous with those of the dermis, and although most of the fibers run parallel to the surface of the skin, they travel in all directions. As a result, no sharp boundary exists between the dermis and the subcutaneous layer.

The adipose tissue of the subcutaneous layer functions as an insulator—helping to conserve body heat and impeding the entrance of heat from the outside. The subcutaneous layer also contains the major blood vessels that supply the skin and underlying adipose tissue.

1. Describe the dermis.
2. What are the functions of the dermis?
3. What are the functions of the subcutaneous layer?

Accessory Organs of the Skin

Hair Follicles

Hair is present on all the skin surfaces except the palms, soles, lips, nipples, and various parts of the external reproductive organs.

Each hair develops from a group of epidermal cells at the base of a tubelike depression, called a **hair follicle** (figs. 6.1 and 6.3). This follicle extends from the surface into the dermis and is occupied by the *root* of the hair. The epidermal cells at its base receive nourishment from the dermal blood vessels, which occur in a projection of connective tissue at the deep end of the follicle. As these cells divide and grow, older cells are pushed toward the surface. The cells that move upward and away from their nutrient supply become keratinized and die. Their remains constitute the structure of

Figure 6.3

(*a*) A hair grows from the base of a hair follicle when epidermal cells undergo cell division and older cells move outward and become keratinized. (*b*) What features can you identify in this micrograph of a hair follicle?

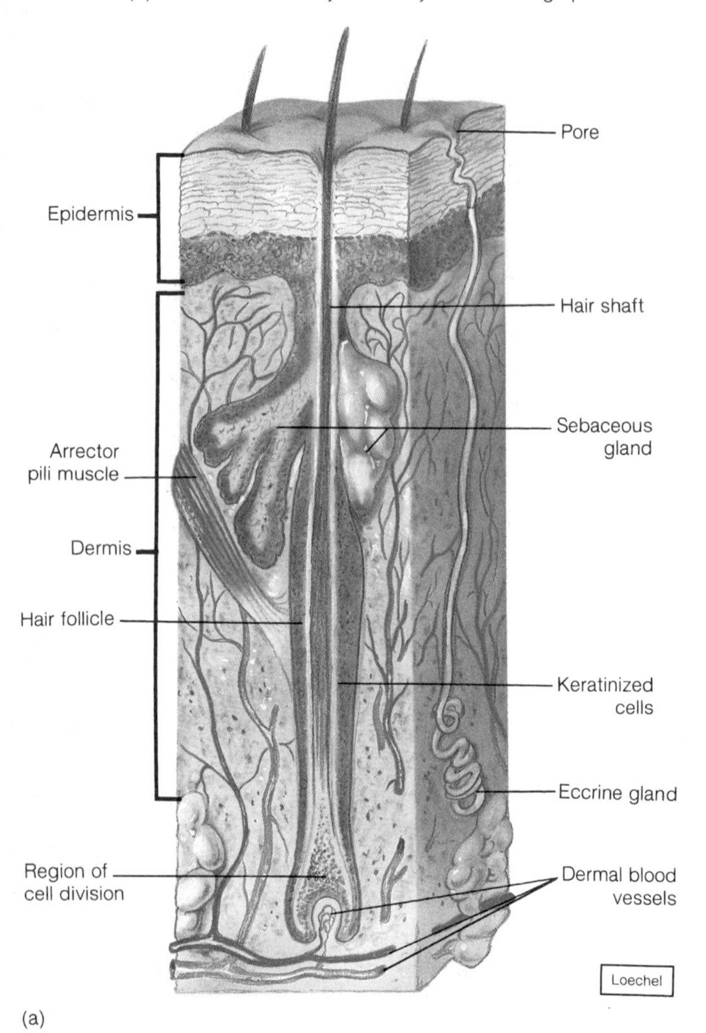

(a)

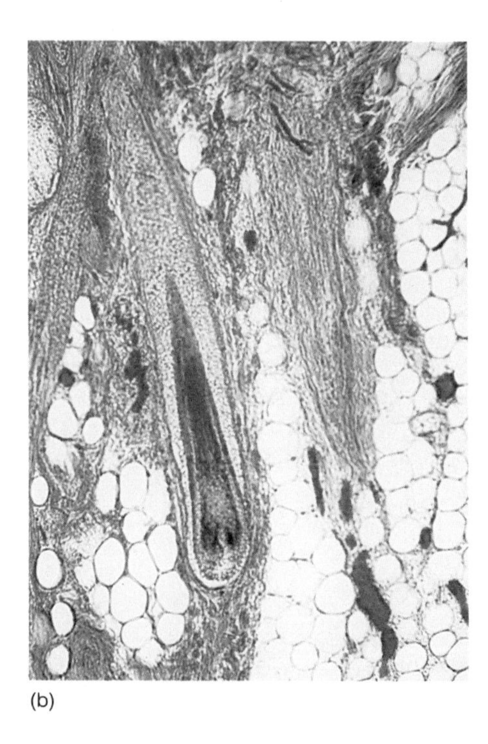

(b)

a developing hair, whose *shaft* extends away from the skin surface. In other words, a hair is composed of dead epidermal cells.

A bundle of smooth muscle cells, forming the *arrector pili muscle,* is attached to each hair follicle. This muscle is positioned so that the hair within the follicle tends to stand on end when the muscle contracts. If a person is emotionally upset or very cold, nerve impulses may stimulate the arrector pili muscles to contract, causing gooseflesh or goose bumps.

Hair color, like skin color, is determined by the genes that direct the type and amount of pigment produced by the epidermal melanocytes. If these cells, which are located at the deep end of a follicle, produce an abundance of melanin, the hair is dark; if an intermediate quantity of pigment is produced, the hair is blond; if no pigment appears, the hair is white. A mixture of pigmented and unpigmented hair appears gray.

Sebaceous Glands

Sebaceous glands (figs. 6.3 and 6.4) contain groups of specialized epithelial cells and are usually associated with hair follicles. They are holocrine glands (see chapter 5) that produce an oily secretion called *sebum,* which is a mixture of fatty material and cellular debris. Sebum is secreted through small ducts into the hair follicles and helps keep the hair and skin soft, pliable, and relatively waterproof.

Sebaceous glands are responsible for acne, a common adolescent disorder. In this condition, the sebaceous glands in certain body regions become overactive in response to hormonal changes, and their excessive secretions cause the openings of hair follicles to become dilated and plugged. The result is blackheads; if pus-forming bacteria are present, pimples or boils may develop in the follicles.

Figure 6.4
A sebaceous gland secretes sebum into a hair follicle (shown here in oblique section; ×175).

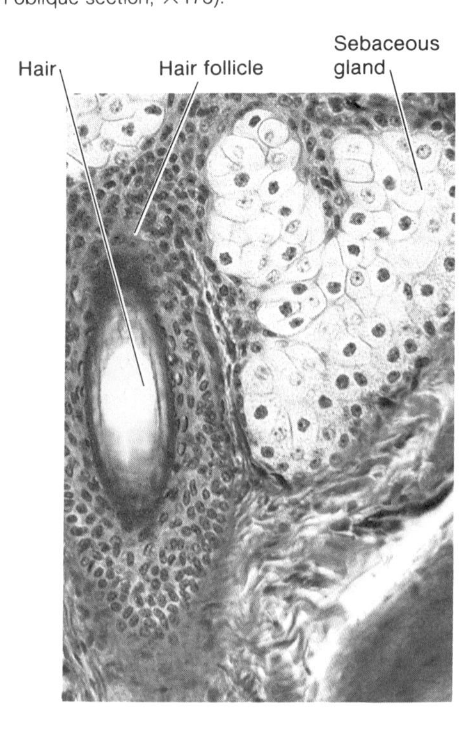

Nails

Nails are protective coverings on the ends of fingers and toes (fig. 6.5). Each nail is produced by layers of specialized epithelial cells that are continuous with the epithelium of the skin. The whitish, half-moon-shaped region (lunula) at the base of a nail is its most active growing portion. The epithelial cells formed in this region undergo keratinization and become part of the nail. The keratin they contain, however, is harder than that formed in the epidermis and hair follicles.

Sweat Glands

Sweat glands occur in nearly all regions of the skin, but are most numerous in the palms and soles. Each gland consists of a tiny tube that originates as a ball-shaped coil in the deeper dermis or superficial subcutaneous layer. The coiled portion of the gland is closed at its end and lined with sweat-secreting epithelial cells.

Certain sweat glands, known as *apocrine glands,* respond to emotional stress and become active when a person is emotionally upset, frightened, or experiencing pain. They are most numerous in the armpits and groin, and are usually connected with hair follicles. The development of these glands is stimulated by sex hormones, so they begin to function as an individual becomes sexually mature (puberty).

Figure 6.5
A nail is produced by epithelial cells that reproduce and undergo keratinization in the lunula of the nail.

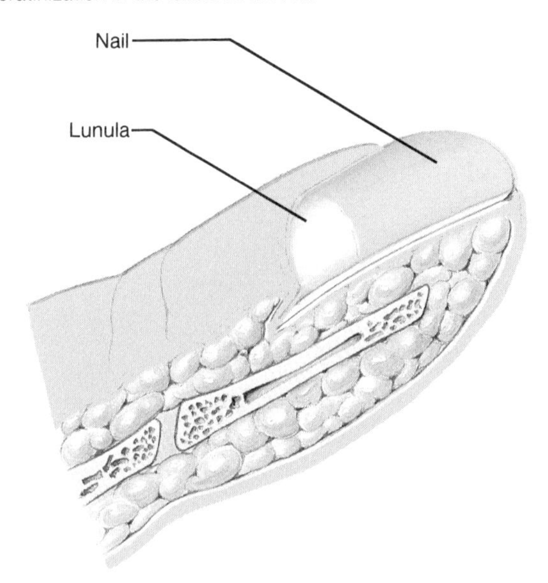

Other sweat glands, the *eccrine glands,* are not associated with hair follicles (fig. 6.6). They function throughout life and respond primarily to elevated body temperatures associated with environmental heat or physical exercise. These glands are common on the forehead, neck, and back, where they produce profuse sweating on hot days and during physical exertion.

The fluid secreted by eccrine glands (sweat) is carried away by a tubular part, which opens at the surface as a *pore.* Although it is mostly water, sweat contains small quantities of salt and certain wastes, such as urea and uric acid. Thus, the secretion of sweat is, to a limited degree, an excretory function.

1. Explain how a hair forms.
2. What is the function of the sebaceous glands?
3. Distinguish between the apocrine and eccrine glands.

Regulation of Body Temperature

The regulation of body temperature is of vital importance because metabolic processes involve chemical reactions that are speeded or slowed by changing heat conditions. As a result, even slight shifts in body temperature can disrupt the normal rates of such reactions and produce metabolic disorders.

Normally, the temperature of the deeper body parts remains close to 37° C (98.6° F), and the maintenance of a stable temperature requires that the amount of heat lost by the body be balanced by the

Figure 6.6
How do the functions of eccrine sweat glands and apocrine sweat glands differ?

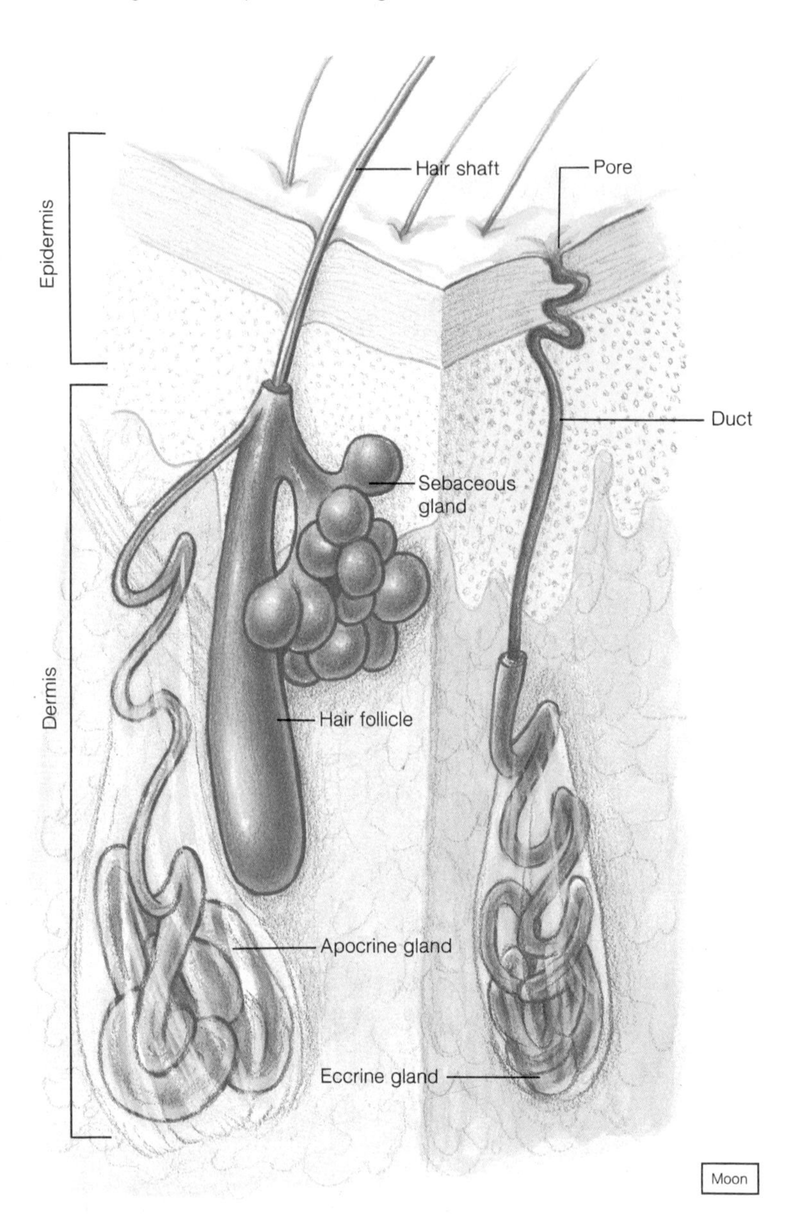

amount it produces. The skin plays a key role in the homeostatic mechanism that regulates body temperature.

Heat Production and Loss

Heat is a product of cellular metabolism; thus, the more active cells of the body are the major heat producers. They include the skeletal and cardiac muscle cells and the cells of certain glands, such as the liver.

When heat production is excessive, nerve impulses stimulate structures in the skin and other organs to react in ways that promote heat loss. For example, during physical exercise, the active muscles release heat, and the blood carries away the excess. The brain's temperature control center senses the increasing blood temperature, and signals from the brain cause the muscles in the walls of the dermal blood vessels to relax. As these blood vessels dilate, more heat-carrying blood enters the dermis, and some of the heat is likely to escape to the outside (see figure 1.4).

At the same time, the nervous system stimulates the eccrine glands to become active and to release fluid onto the surface of the skin. As this fluid evaporates (changes from a liquid to a gas), it carries heat away from the surface, cooling the skin.

If heat is lost excessively, as may occur in a very cold environment, the brain triggers different responses

Elevated Body Temperature

A Current Topic

Elevated body temperature (fever or pyrexia) may be due to an infectious disease or to failure of the body's temperature-regulating mechanism. In the case of disease, substances from infectious agents, such as bacteria, or from diseased tissues cause phagocytic cells to release yet another substance, called *interleukin*.

Interleukin is carried by the blood to the brain and causes the set point of the brain's thermostat to be raised. At the same time, signals from the brain cause the skeletal muscles to increase heat production, the flow of blood into the skin to be reduced, and the sweat glands to decrease their secretion of fluid. As a consequence, the body temperature rises toward the new set point, and the person develops a fever.

If the temperature-regulating mechanism fails, body heat may accumulate faster than it is given off, and the body temperature may become elevated even though the set point of the thermostat is normal. This sometimes happens to long distance runners, whose active muscles release great amounts of heat, especially in warm weather or when they become dehydrated.

In any case, a body temperature above 40° C (104° F) is of particular concern, expecially in very young or very old persons. For example, an extremely elevated body temperature (hyperthermia) may cause seizures in a child under six years of age. Similarly, it may cause mental disturbances, low blood pressure, or heart problems in adults.

Treatment of hyperthermia may involve providing liquids to replace lost body fluids and electrolytes, administering agents (antipyretic drugs) that act to normalize the set point of the body's thermostat, sponging the skin with water to increase cooling by evaporation, or covering the person with a refrigerated blanket.

in the skin structures. For example, the muscles in the walls of dermal blood vessels are stimulated to contract; this decreases the flow of heat-carrying blood through the skin and helps to reduce heat loss. Also, the sweat glands remain inactive, decreasing heat loss by evaporation. If the body temperature continues to drop, the nervous system may stimulate muscle fibers in the skeletal muscles throughout the body to contract slightly. This action requires an increase in the rate of cellular respiration and produces heat as a by-product. If this response does not raise the body temperature to normal, small groups of muscles may be stimulated to contract rhythmically with still greater force, and the person begins to shiver, generating more heat.

In addition to the skin, the circulatory and respiratory systems function in body temperature regulation. For example, if the body temperature is rising above normal, the heart is stimulated to beat faster. This causes more blood to be moved from the deeper, heat-generating tissues into the skin. At the same time, the breathing rate is increased, and more heat is lost from the lungs as an increased volume of air is moved in and out.

1. Why is the regulation of body temperature so important?
2. How can the body lose heat when heat production is excessive?
3. What actions help the body to conserve heat?

Common Skin Disorders

acne (ak′ne)—a disease of the sebaceous glands, accompanied by blackheads and pimples.

alopecia (al″o-pe′she-ah)—loss of hair.

athlete's foot (ath′-lētz foot) (tinea pedis)—a fungus infection usually involving the skin of the toes and soles.

birthmark (berth′mark)—a vascular tumor involving the skin and subcutaneous tissues that is visible at birth or soon after.

boil (boil) (furuncle)—a bacterial infection involving a hair follicle and/or sebaceous glands.

carbuncle (kar′bung-kl)—a bacterial infection, similar to a boil, that spreads into the subcutaneous tissues.

cyst (sist)—a liquid-filled sac or capsule.

dermatitis (der″mah-ti′tis)—an inflammation of the skin.

eczema (ek′zĕ-mah)—a noncontagious skin rash, often accompanied by itching, blistering, and scaling.

erythema (er″ĭ-the′mah)—reddening of the skin, due to dilation of the dermal blood vessels, in response to injury or inflammation.

herpes (her′pēz)—an infectious disease of the skin, usually caused by the virus called *herpes simplex* and characterized by recurring formations of small clusters of vesicles.

impetigo (im″pĕ-ti′go)—a contagious disease of bacterial origin, characterized by pustules that rupture and become covered with loosely held crusts.

keloids (ke′loidz)—an elevated, enlarging fibrous scar usually initiated by an injury.

mole (mōl) (nevus)—a fleshy skin tumor that usually is pigmented; colors range from brown to black.

pediculosis (pē-dik″u-lo′sis)—a disease produced by an infestation of lice.
pruritus (proo-ri′tus)—itching of the skin.
psoriasis (so-ri′ah-sis)—a chronic skin disease, characterized by red patches covered with silvery scales.
pustule (pus′tūl)—elevated, pus-filled area on the skin.
scabies (ska′bēz)—a disease resulting from an infestation of mites.
seborrhea (seb″o-re′ah)—a disease characterized by hyperactivity of the sebaceous glands and accompanied by greasy skin and dandruff.
shingles (shing′gelz) (Varicella-zoster)—a disease caused by a viral infection of certain spinal nerves; it is characterized by severe localized pain and blisters in the skin areas supplied by the affected nerves.
ulcer (ul′ser)—an open sore.
urticaria (ur″ti-ka′re-ah)—an allergic reaction of the skin that produces reddish, elevated patches (hives).
wart (wort)—a flesh-colored, raised area caused by a viral infection.

Chapter Summary

Introduction (page 116)

Organs, such as membranes, are composed of two or more kinds of tissues. The skin is an organ and together with its accessory organs constitute the integumentary system.

Types of Membranes (page 116)

1. Serous membranes
    a. Serous membranes line body cavities that lack openings to the outside.
    b. They secrete watery serous fluid, which lubricates membrane surfaces.
2. Mucous membranes
    a. Mucous membranes line cavities and tubes that open to the outside.
    b. They secrete mucus.
3. The cutaneous membrane is the external body covering, commonly called the skin.
4. Synovial membranes
    a. Synovial membranes line joint cavities.
    b. They secrete synovial fluid that lubricates the ends of the bones at joints.

Skin and Its Tissues (page 116)

Skin functions as a protective covering, aids in regulating the body temperature, retards water loss, houses sensory receptors, synthesizes various chemicals, and excretes wastes. It is composed of an epidermis and a dermis, separated by a basement membrane.

1. Epidermis
    a. The deepest layer of the epidermis contains cells undergoing mitosis.
    b. Epidermal cells undergo keratinization as they mature and are pushed toward the surface.
    c. The outermost layer is composed of dead epidermal cells.
    d. The epidermis functions to protect the underlying tissues against water loss, mechanical injury, and the effects of harmful chemicals.
    e. All humans possess about the same concentration of melanocytes.
    f. Skin color is due largely to the amount of melanin and the size of the pigment granules in the epidermis.
    g. Skin color is influenced by environmental and physiological factors, as well as by the genes.
2. Dermis
    a. The dermis is a layer that binds the epidermis to underlying tissues.
    b. The dermal blood vessels supply nutrients to all the skin cells and aid in regulating the body temperature.
    c. Nervous tissue is scattered throughout the dermis.
        (1) Some dermal nerve fibers carry impulses to the muscles and glands of the skin.
        (2) Other dermal nerve fibers are associated with various sensory receptors in the skin.
    d. The dermis also contains hair follicles, sebaceous glands, and sweat glands.
3. Subcutaneous layer
    a. The subcutaneous layer is composed of loose connective tissue and adipose tissue.
    b. Adipose tissue helps to conserve body heat.
    c. This layer contains blood vessels that supply the skin and underlying adipose tissue.

Accessory Organs of the Skin (page 119)

1. Hair follicles
   a. Each hair develops from the epidermal cells at the base of a hair follicle.
   b. As the newly formed cells develop and grow, the older cells are pushed toward the surface and undergo keratinization.
   c. A bundle of smooth muscle cells is attached to each hair follicle.
   d. Hair color is determined by genes that direct the amount of melanin produced by the melanocytes associated with hair follicles.
2. Sebaceous glands
   a. Sebaceous glands are usually associated with hair follicles.
   b. They secrete sebum, which helps keep the skin and hair soft and waterproof.
3. Nails
   a. Nails are produced by epidermal cells that undergo keratinization.
   b. The keratin of nails is harder than that produced by epidermal cells of the skin.
4. Sweat glands
   a. Each sweat gland consists of a coiled tube.
   b. Apocrine glands respond to emotional stress, but eccrine glands respond to an elevated body temperature.
   c. Sweat is primarily water, but also contains salts and waste products.

Regulation of Body Temperature (page 121)

The regulation of body temperature is vital because heat affects the rates of metabolic reactions.

The normal temperature of the deeper body parts is about 37° C (98.6° F).

1. When the body temperature rises above normal, the dermal blood vessels dilate and sweat glands secrete sweat.
2. If the body temperature drops below normal, the dermal blood vessels constrict and sweat glands become inactive.
3. When heat is lost excessively, the skeletal muscles are stimulated to contract involuntarily.

Clinical Application of Knowledge

1. What special problems would result from a loss of 50% of a person's functional skin surface? How might this person's environment be modified to compensate partially for such a loss?
2. A premature infant typically lacks subcutaneous adipose tissue. Also, the surface area of an infant's small body is relatively large compared to its volume. How do you think these factors influence an infant's ability to regulate its body temperature?
3. Which of the following would result in the more rapid absorption of a drug: a subcutaneous injection or an intradermal injection? Why do you think so?
4. How would you explain to an athlete the importance of keeping the body hydrated when exercising in warm weather?

Review Activities

1. Explain why a membrane is an organ.
2. Define *integumentary system.*
3. Distinguish between the serous and mucous membranes.
4. Define *synovial membrane.*
5. List six functions of the skin.
6. Distinguish between the epidermis and the dermis.
7. Explain what happens to epidermal cells as they undergo keratinization.
8. Describe the function of melanocytes.
9. List three factors that affect skin color.
10. Describe the subcutaneous layer and its functions.
11. Explain how blood is supplied to various skin layers.
12. Review the functions of nervous tissue.
13. Distinguish between a hair and a hair follicle.
14. Explain the function of the sebaceous glands.
15. Describe how the nails are formed.
16. Distinguish between the apocrine and eccrine glands.
17. Explain how body heat is produced.
18. Explain how the sweat glands function in regulating the body temperature.
19. Describe the body's responses to a decreasing body temperature.

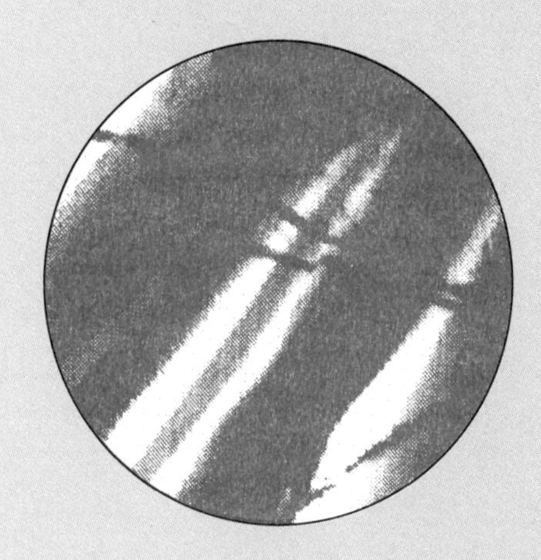

Unit 2
Support and Movement

The chapters of unit 2 deal with the structures and functions of the skeletal and muscular systems. They describe how the organs of the skeletal system support and protect other body parts and how they function with organs of the muscular system to enable body parts to move. They also describe how skeletal structures participate in the formation of blood and in the storage of inorganic salts, and how muscular tissues act to produce body heat and to move body fluids. This unit includes:

These circular computer-manipulated images appear on the front cover and are described on the back cover of this text.

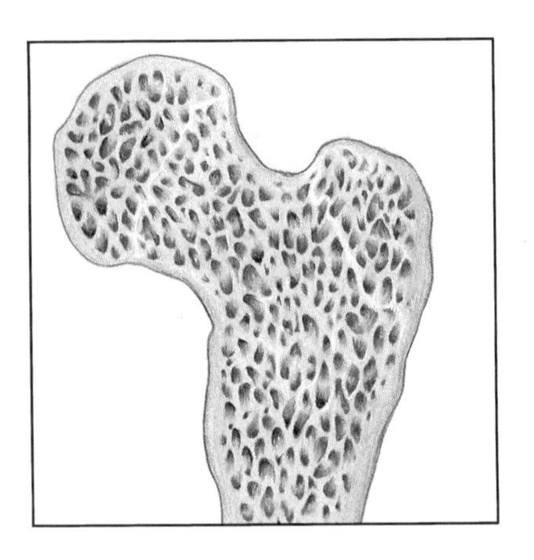

7 Skeletal System

Bones are composed of several kinds of tissues; thus, they are the organs of the *skeletal system*. Because bones are relatively rigid structures, they provide support and protection for softer tissues, and they act together with skeletal muscles to make body movements possible. They also contain tissue that produces blood cells, and they store inorganic salts.

The shapes of individual bones are closely related to their functions. Projections provide places for the attachment of muscles, tendons, and ligaments; openings serve as passageways for blood vessels and nerves; and the ends of bones are modified to form joints with other bones.

Chapter Objectives

After you have studied this chapter, you should be able to

1. Describe the general structure of a bone, and list the functions of its parts.
2. Distinguish between intramembranous and endochondral bones, and explain how such bones develop and grow.
3. Discuss the major functions of bones.
4. Distinguish between the axial and appendicular skeletons, and name the major parts of each.
5. Locate and identify the bones and the major features of the bones that comprise the skull, vertebral column, thoracic cage, pectoral girdle, upper limb, pelvic girdle, and lower limb.
6. List three types of joints, describe their characteristics, and name an example of each.
7. List six types of freely movable joints, and describe the actions of each.
8. Explain how skeletal muscles produce movements at joints, and identify several types of such movements.
9. Complete the review activities at the end of this chapter. Note that the items are worded in the form of specific learning objectives. You may want to refer to them before reading the chapter.

Key Terms

articular cartilage (ar-tik′u-lar kar′tĭ-lij)

compact bone (kom′pakt bōn)

diaphysis (di-af′ĭ-sis)

endochondral bone (en″do-kon′dral bōn)

epiphyseal disk (ep″ĭ-fiz′e-al disk)

epiphysis (e-pif′ĭ-sis)

hematopoiesis (hem″ah-to-poi-e′sis)

intramembranous bone (in″trah-mem′brah-nus bōn)

lever (lev′er)

marrow (mar′o)

medullary cavity (med′u-lăr″e kav′ĭ-te)

osteoblast (os′te-o-blast)

osteoclast (os′te-o-klast)

osteocyte (os′te-o-sīt)

periosteum (per″e-os′te-um)

spongy bone (spun′je bōn)

Aids to Understanding Words

acetabul-, vinegar cup: *acetabul*um—a depression in the coxal bone that articulates with the head of a femur.

ax-, axis: *ax*ial skeleton—the upright portion of the skeleton that supports the head, neck, and trunk.

-blast, budding: osteo*blast*—a cell that forms bone tissue.

carp-, wrist: *carp*als—wrist bones.

-clast, broken: osteo*clast*—a cell that breaks down bone tissue.

condyl-, knob: *condyle*—a rounded, bony process.

corac-, beaklike: *corac*oid process—a beaklike process of the scapula.

cribr-, sievelike: *cribr*iform plate—a portion of the ethmoid bone with many small openings.

crist-, ridge: *crist*a galli—a bony ridge that projects upward into the cranial cavity.

fov-, pit: *fov*ea capitis—the pit in the head of a femur.

glen-, joint socket: *glen*oid cavity—a depression in the scapula that articulates with the head of the humerus.

inter-, between: *inter*vertebral disk—a structure located between adjacent vertebra.

intra-, inside: *intra*membranous bone—bone that forms within sheetlike masses of connective tissue.

hema-, blood: *hema*toma—a blood clot.

meat-, passage: auditory *meat*us—a canal of the temporal bone that leads inward to parts of the ear.

odont-, tooth: *odont*oid process—a toothlike process of the second cervical vertebra.

poie-, making: hemato*poie*sis—a process by which blood cells are formed.

Introduction

AN INDIVIDUAL BONE is composed of a variety of tissues, including bone tissue, cartilage, fibrous connective tissue, blood, and nerve tissue. Because much nonliving material is present in the matrix of bone tissue, the whole organ may appear to be inert. A bone, however, contains very active, living tissues.

Bone Structure

Although the various bones of the skeletal system differ greatly in size and shape, they are similar in their structure, development, and functions.

Parts of a Long Bone

In describing the structure of bone, a long bone, such as one found in an arm or a leg, will be used as an example (fig. 7.1). At each end of the bone there is an expanded portion called an **epiphysis** (pl. *epiphyses*), which articulates (forms a joint) with another bone. On its outer surface, the articulating portion of the epiphysis is coated with a layer of hyaline cartilage, called **articular cartilage.** The shaft of the bone, which is located between the epiphyses, is called the **diaphysis.**

Except for the articular cartilage on its ends, the bone is completely enclosed by a tough, vascular covering of fibrous tissue, called the **periosteum.** This membrane is firmly attached to the bone, and the periosteal fibers are continuous with various ligaments and tendons that are connected to the membrane. The periosteum also functions in the formation and repair of bone tissue.

Each bone has a shape closely related to its functions. Bony projections called *processes,* for example, provide sites for the attachment of ligaments and tendons; grooves and openings serve as passageways for blood vessels and nerves; a depression of one bone might articulate with a process of another.

The wall of the diaphysis is composed mainly of tightly packed tissue called **compact bone** (cortical bone). This type of bone is solid, strong, and resistant to bending. The epiphyses, on the other hand, are composed largely of **spongy bone** (cancellous bone) with thin layers of compact bone on their surfaces. Spongy bone consists of numerous branching bony plates. Irregular interconnected spaces occur between these plates and help reduce the weight of the bone. Spongy bone provides strength, and its bony plates are most highly developed in the regions of the epiphyses that are subjected to the forces of compression.

The compact bone in the diaphysis of a long bone forms a semirigid tube with a hollow chamber, called the **medullary cavity.** This cavity is continuous with the spaces of spongy bone. All of these areas are lined with a thin layer of cells, called **endosteum,** and filled with a specialized type of soft connective tissue, called **marrow.**

Figure 7.1
Major parts of a long bone.

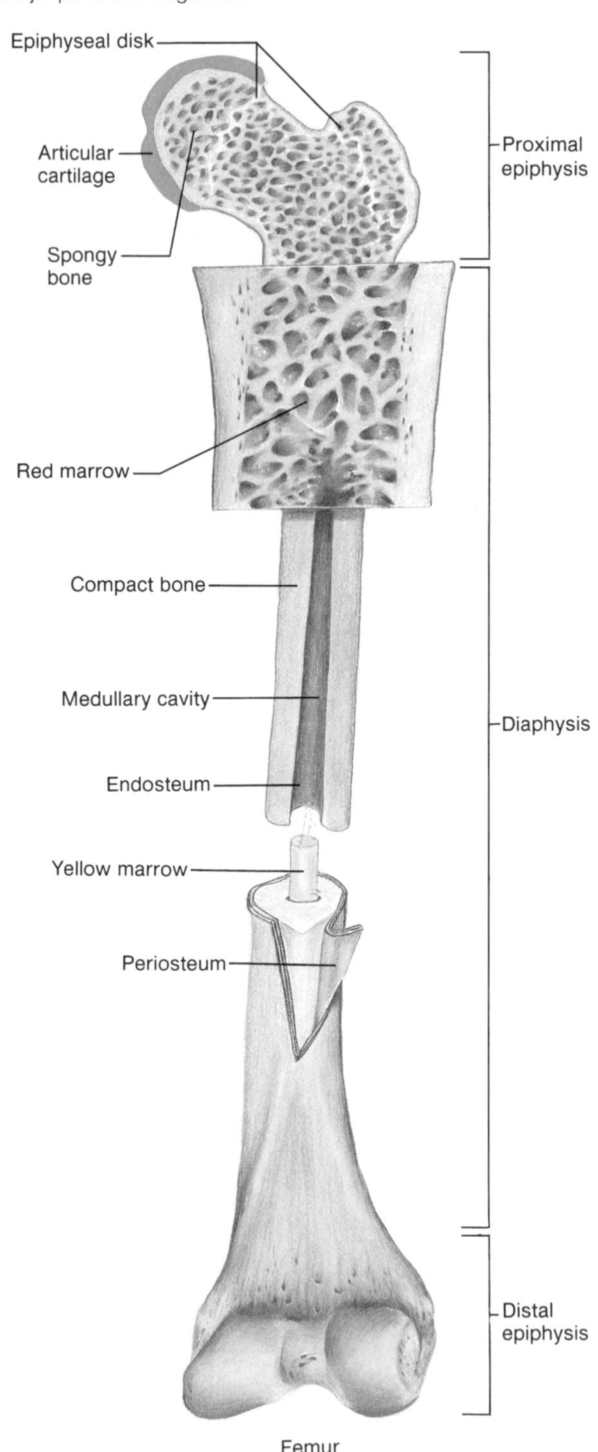

Figure 7.2
Compact bone is composed of osteons cemented together.

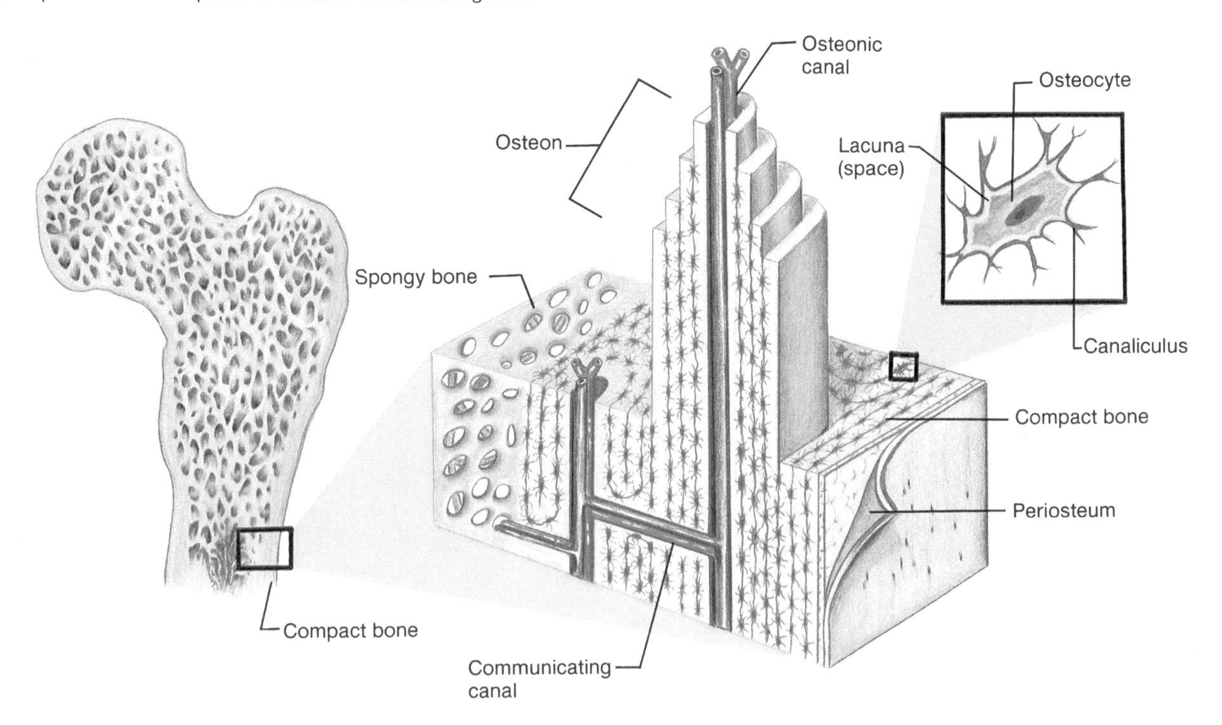

Microscopic Structure

As was discussed in chapter 5, bone cells (osteocytes) are located in minute, bony chambers, called *lacunae,* which are arranged in concentric circles around *osteonic canals* (figs. 5.17 and 7.2). These cells communicate with nearby cells by means of cellular processes passing through canaliculi. The intercellular material of bone tissue is composed largely of collagen and inorganic salts. The collagen gives the bone strength and resilience, and the inorganic salts make it hard and resistant to crushing.

In compact bone, the osteocytes and layers of intercellular material clustered, concentrically around an osteonic canal, form a cylinder-shaped unit called an *osteon* (haversian system). Many of these units cemented together form the substance of compact bone.

Each osteonic canal (haversian canal) contains one or two small blood vessels (usually capillaries), surrounded by some loose connective tissue. The blood in these vessels provides nourishment for the bone cells associated with the osteonic canal.

Osteonic canals travel longitudinally through bone tissue. They are interconnected by transverse *communicating canals* (Volkmann's canals). These canals contain larger blood vessels by which the vessels in osteonic canals communicate with the surface of the bone and the medullary cavity (fig. 7.2).

Spongy bone also is composed of osteocytes and intercellular material. However, its bony plates are usually very thin, and the bone cells are not arranged around osteonic canals. Instead, the cells are nourished by the diffusion of substances into canaliculi that lead from the bone cells to the surface of the bony plates.

1. List five major parts of a long bone.
2. How do compact and spongy bone differ in structure?
3. Describe the microscopic structure of compact bone.

Bone Development and Growth

Parts of the skeletal system begin to form during the first few weeks of life, and bony structures continue to develop and grow into adulthood. Bones form by the replacement of existing connective tissues, in one of two ways. Some appear between sheetlike layers of connective tissues and are called intramembranous bones. Others begin as masses of cartilage, which are later replaced by bone tissue. They are called endochondral bones.

Intramembranous Bones

Examples of **intramembranous bones** are the broad, flat bones of the skull. During their development, sheetlike masses of connective tissues appear at the sites of the future bones. Then some primitive connective tissue

Figure 7.3
The tissues of this miscarried fetus (about fourteen weeks old) have been cleared, and the developing bones have been selectively stained.

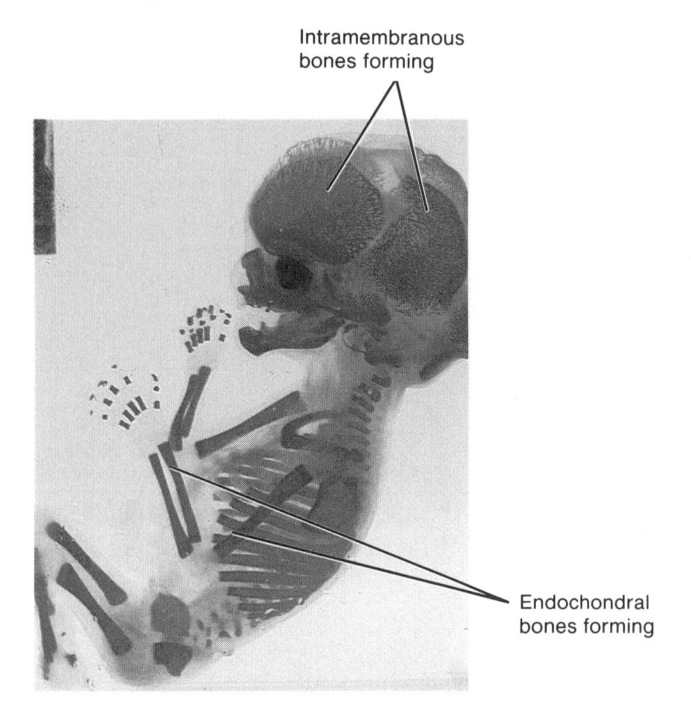

cells that are present enlarge and differentiate into bone-forming cells, called **osteoblasts.** The osteoblasts become active within the membranes and deposit bony intercellular materials around themselves. As a result, spongy bone tissue is produced in all directions within the membranes. Eventually cells of the membranous tissues that persist outside the developing bone give rise to the periosteum. At the same time osteoblasts on the inside of the periosteum form a layer of compact bone over the surface of the newly built spongy bone. When osteoblasts are completely surrounded by matrix, they are called **osteocytes.**

Endochondral Bones

Most of the bones of the skeleton are **endochondral bones.** They develop from masses of hyaline cartilage with shapes similar to future bony structures (fig. 7.3). These cartilaginous models grow rapidly for a while, and then begin to undergo extensive changes. In a long bone, for example, the changes begin in the center of the diaphysis where the cartilage slowly breaks down and disappears (fig. 7.4). At about the same time, a periosteum forms from connective tissues which encircles the developing diaphysis. Blood vessels and osteoblasts from the periosteum invade the disintegrating cartilage, and spongy bone is formed in its place. This region of bone formation is called the *primary ossification center,* and bone tissue develops from it toward the ends of the cartilaginous structure.

Figure 7.4
Major stages (*a–f*) in the development of an endochrondral bone.

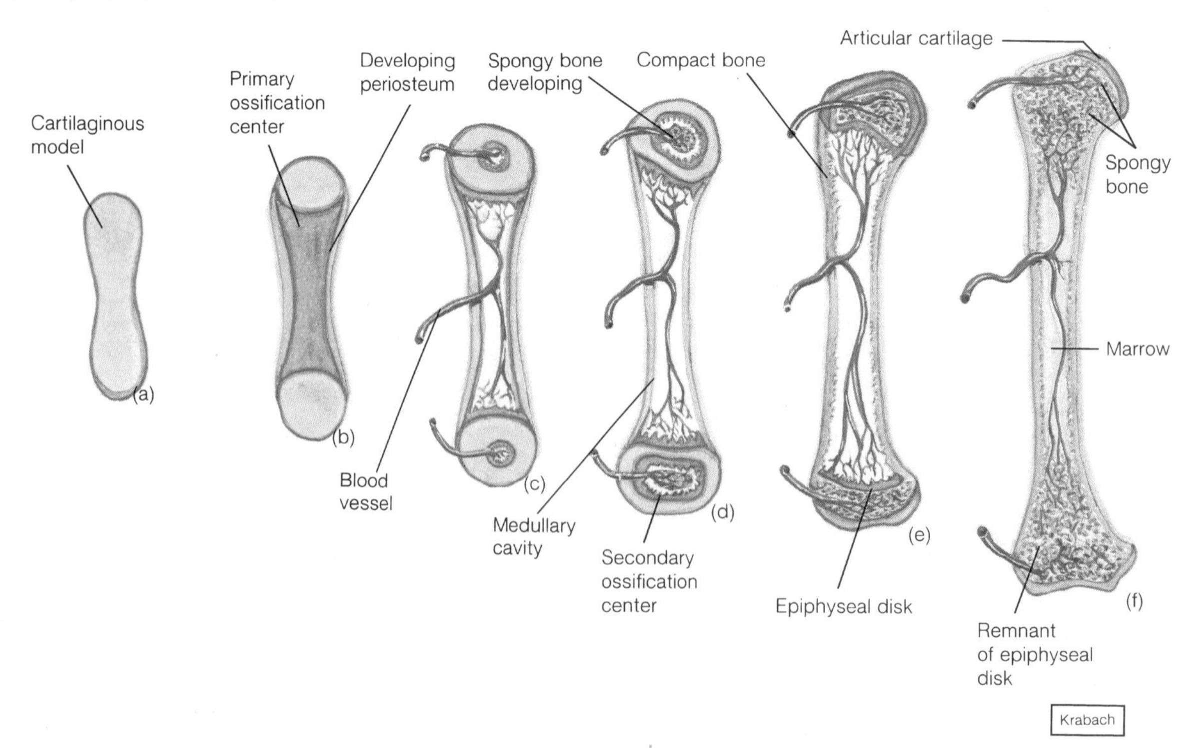

A Current Topic

Repair of a Bone Fracture

A *fracture* is a break in a bone. Whenever a bone is broken, blood vessels within the bone and its periosteum are ruptured, and the periosteum is likely to be torn. Blood escaping from the broken vessels spreads through the damaged area and soon forms a blood clot, or *hematoma*. In response to the injury, blood vessels in surrounding tissues dilate, and those tissues become swollen and inflamed.

Within days or weeks, the hematoma is invaded by developing blood vessels and large numbers of osteoblasts originating from the periosteum. The osteoblasts multiply rapidly in the regions close to the new blood vessels, building spongy bone nearby. In regions further from the blood supply, fibroblasts produce masses of fibrocartilage.

Meanwhile, phagocytic cells begin to remove the blood clot as well as any dead or damaged cells in the affected area. Osteoclasts also appear and resorb bone fragments, thus aiding in the clean up process.

In time, a large amount of fibrocartilage fills the gap between the ends of the broken bone, and this mass is termed a *cartilaginous callus*. The callus is later replaced by bone tissue in much the same way that the hyaline cartilage of a developing endochrondral bone is replaced. That is, the cartilaginous callus is broken down, the area is invaded by blood vessels and osteoblasts, and the space is filled with a *bony callus*.

Usually more bone is produced at the site of a healing fracture than is needed to replace the damaged tissues. However, osteoclasts eventually remove the excess, and the final result of the repair process is a bone shaped very much like the original one (fig. 7.5).

Figure 7.5
Major steps in the repair of a fracture.

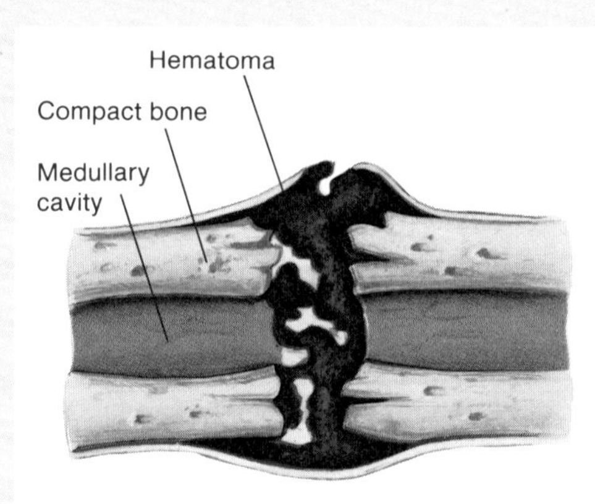

(a) Blood escapes from ruptured blood vessels and forms a hematoma

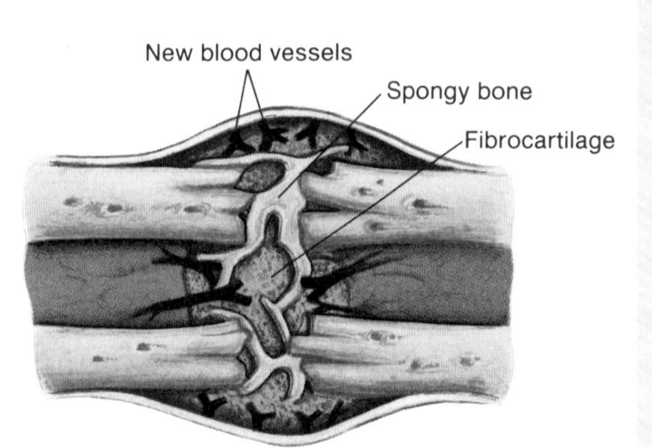

(b) Spongy bone forms in regions close to developing blood vessels, and fibrocartilage forms in more distant regions

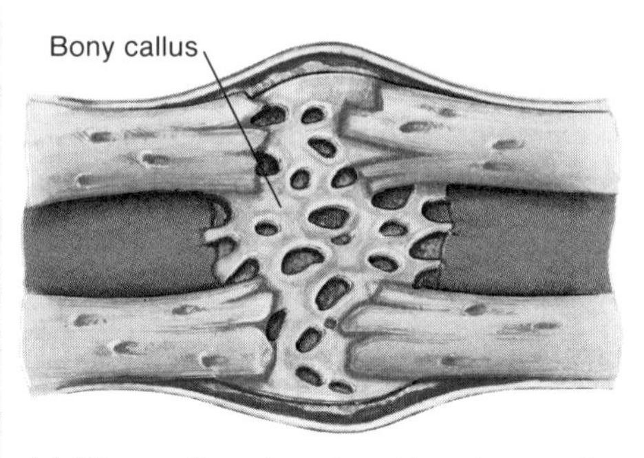

(c) Fibrocartilage is replaced by a bony callus

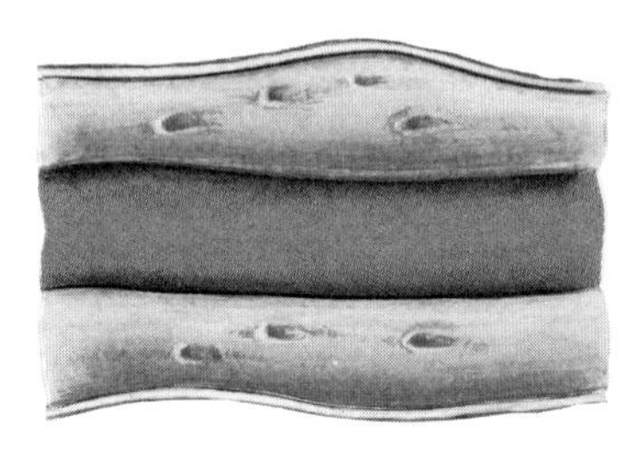

(d) Osteoclasts remove excess bony tissue, making new bone structure much like the original

Meanwhile, osteoblasts from the periosteum deposit a thin layer of compact bone around the primary ossification center. The epiphyses of the developing bone remain cartilaginous and continue to grow. Later *secondary ossification centers* appear in the epiphyses, and spongy bone forms in all directions from them. As bone is deposited in the diaphysis and epiphysis, a band of cartilage, called the **epiphyseal disk** (physis), is left between these two ossification centers.

The cartilaginous cells of an epiphyseal disk include rows of young cells that are undergoing mitosis and producing new cells. As these cells enlarge and intercellular material is formed around them, the cartilaginous disk thickens. Consequently, the length of the bone increases. At the same time, calcium salts accumulate in the matrix adjacent to the oldest cartilaginous cells, and as the matrix becomes calcified, the cells begin to die.

Later, this calcified intercellular substance is broken down by large multinucleated cells called **osteoclasts.** These large cells originate by the the fusion of certain single-nucleated white blood cells (monocytes) (see chapter 14).

Osteoclasts secrete an acid that dissolves the inorganic component of the calcified substance, and at the same time, their lysosomal enzymes digest the organic components of the substance. After the matrix is removed, bone-building osteoblasts invade the region and deposit bone tissue in place of the calcified cartilage.

A long bone will continue to grow in length while the cartilaginous cells of the epiphyseal disks are active. However, once the ossification centers of the diaphysis and epiphyses come together and the epiphyseal disks become ossified, growth in length is no longer possible.

If an epiphyseal disk is damaged before it becomes ossified, growth of the long bone may cease prematurely, or if growth continues, it may be uneven. For this reason, injuries to the epiphyses of a young person's bones are of special concern. On the other hand, an epiphysis is sometimes altered surgically in order to equalize the growth of bones that are developing at very different rates.

A developing long bone grows in thickness as compact bone tissue is deposited on the outside, just beneath the periosteum. As this is occurring, other bone tissue is being eroded away on the inside by the activity of the osteoclasts. The space that is created becomes the medullary cavity of the diaphysis, which later fills with marrow.

Bone in the central regions of the epiphyses and diaphysis remains spongy, and hyaline cartilage on the ends of the epiphyses persists throughout life as articular cartilage.

Homeostasis of Bone Tissue

After the intramembranous and endochondral bones have been formed, they are continually remodeled by the activities of osteoclasts and osteoblasts. Thus, throughout life, osteoclasts are being stimulated to resorb bone tissue at specific sites, and osteoblasts are being activated to replace the bone. These opposing processes of resorption and replacement, however, are well regulated. Consequently, the total mass of bone tissue within an adult skeleton normally remains nearly constant, even though 3% to 5% of bone calcium is exchanged each year.

1. Describe the development of an intramembranous bone.
2. Explain how an endochondral bone develops.
3. Explain how osteoclasts and osteoblasts remodel bone from time to time.

Functions of Bones

Skeletal parts provide shape, support, and protection for body structures. They also aid body movements, house tissues that produce blood cells, and store various inorganic salts.

Support and Protection

Bones give shape to structures such as the head, face, thorax, and limbs. They also provide support and protection. For example, the bones of the feet, legs, pelvis, and backbone support the weight of the body. The bones of the skull protect the eyes, ears, and brain. Those of the rib cage and shoulder girdle protect the heart and lungs, and the bones of the pelvic girdle protect the lower abdominal and internal reproductive organs.

Body Movement

Whenever limbs or other body parts are moved, bones and muscles function together as simple mechanical devices called *levers*. Such a lever has four basic components: (*a*) a rigid bar or rod; (*b*) a pivot, or fulcrum, on which the bar turns; (*c*) an object or weight (resistance) that is moved; (*d*) a force that supplies energy for the movement of the bar.

The actions of bending and straightening the arm at the elbow, for example, involve bones and muscles functioning together as levers, as illustrated in figure 7.6. When the arm is bent, the lower arm bones represent the rigid bar, the elbow joint is the pivot, the hand is the weight that is moved, and the force is supplied by muscles on the anterior side of the upper arm. One of these muscles, the *biceps brachii,* is attached

Figure 7.6
When the arm is (*a*) bent or (*b*) straightened at the elbow, muscles and bones work together as a lever.

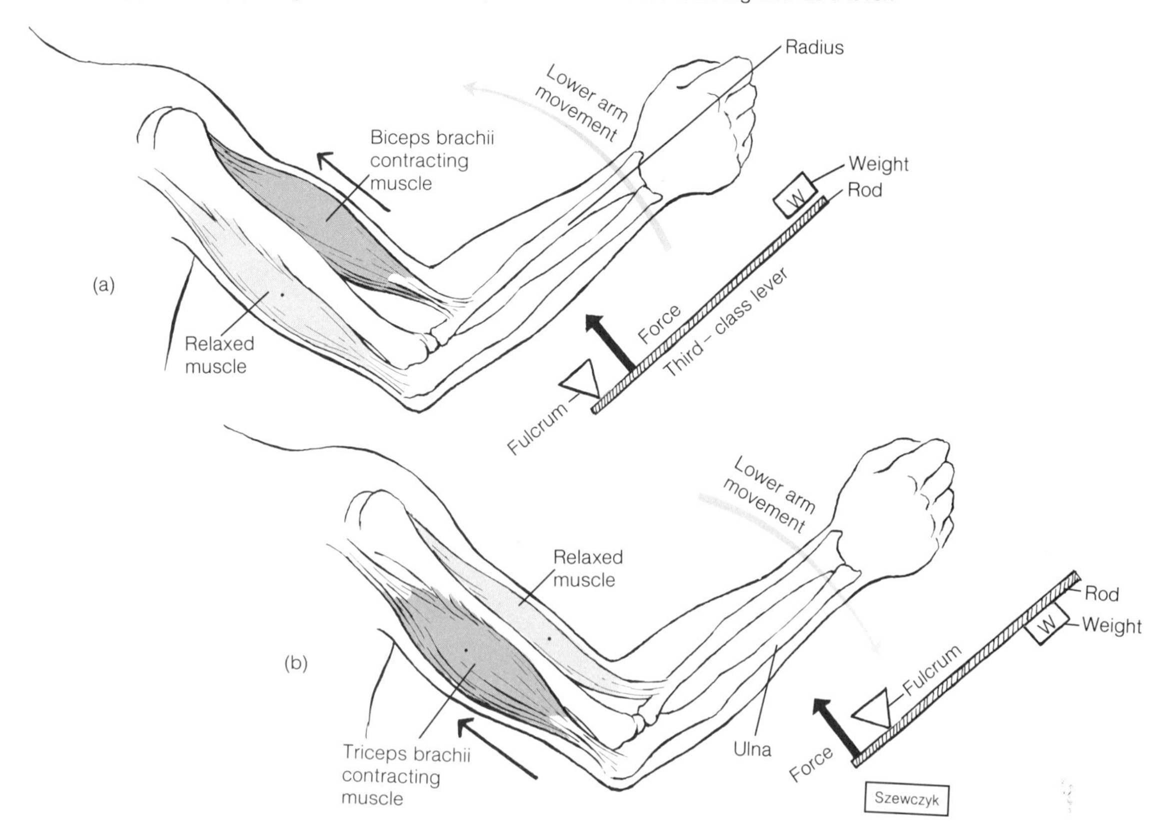

by a tendon to a process on the *radius* bone in the lower arm, a short distance below the elbow.

When the arm is straightened at the elbow, the lower arm bones again serve as a rigid bar and the elbow joint as the pivot. However, this time the force is supplied by the *triceps brachii,* a muscle located on the posterior side of the upper arm. A tendon of this muscle is attached to a process of the *ulna* bone at the point of the elbow.

Blood Cell Formation

The process of blood cell formation is called **hematopoiesis.** Very early in life it occurs in a structure called the *yolk sac,* which lies outside the body of a human embryo (see chapter 19). Later in development, blood cells are manufactured in the liver and spleen, and still later, they are formed in the marrow of various bones.

Marrow is a soft mass of connective tissue found within the medullary cavities of long bones, in the irregular spaces of spongy bone, and in the larger osteonic canals of compact bone.

There are two kinds of marrow—red and yellow. *Red marrow* functions in the formation of red blood cells, white blood cells, and blood platelets. It is red because of the red, oxygen-carrying pigment, called **hemoglobin,** that is contained within the red blood cells.

Red marrow occupies the cavities of most bones in an infant. With increasing age, however, more and more of it is replaced by *yellow marrow,* which functions as fat storage tissue and is inactive in blood cell production.

In an adult, red marrow is found primarily in spongy bone of the ribs, sternum, vertebrae, and pelvis, and in the epiphyses of the humerus and femur. Blood cell formation is described in more detail in chapter 14.

Red marrow may be damaged or destroyed by excessive exposure to X ray, adverse drug reactions, or the presence of cancerous tissues. The treatment of this condition sometimes involves a *marrow transplant.*

In this procedure, normal red marrow cells are removed from the spongy bone of a donor by means of a hollow needle and syringe. These cells are then injected into the recipient's blood with the hope that they will lodge in the spaces normally inhabited by red marrow and will, in time, replace the damaged tissue.

Storage of Inorganic Salts

The intercellular matrix of bone tissue contains large quantities of calcium salts—mostly in the form of calcium phosphate.

Osteoporosis

A Current Topic

Osteoporosis is a disorder of the skeletal system in which there is an excessive loss of bone volume and mineral content. The affected bones develop spaces and canals that become enlarged and filled with fibrous and fatty tissues. Such bones are easily fractured and may break spontaneously as they become unable to support body weight. For example, a person with osteoporosis may suffer a spontaneous fracture of the upper leg bone (femur) at the hip joint or a collapse of sections of backbone (vertebrae).

Osteoporosis is associated with aging and is responsible for a large proportion of fractures occurring in persons over forty-five years of age. Although it may affect persons of either gender, it is most common in light-skinned females after menopause (see chapter 19).

Factors that increase the risk of developing osteoporosis include a low intake of dietary calcium, a lack of physical exercise (particularly during the early growing years), and in females, a decrease in blood estrogen concentration. (Estrogen is a hormone produced by the ovaries; however, the ovaries cease to secrete estrogen at menopause.) The heavy use of beverage alcohol or smoking tobacco also seem to increase risk.

Osteoporosis may be prevented if steps are taken early enough. It is known, for example, that bone mass reaches a maximum at about age thirty-five. Thereafter, bone loss may exceed bone formation in both males and females. To reduce such loss, persons in their mid-twenties and older are advised to ensure that their dietary calcium intake reaches at least the recommended daily allowance of 800 milligrams. (Some nutritionists believe that 1000–1500 milligrams of calcium are needed daily to help control bone loss.) It is also recommended that adults engage in some type of regular physical exercise in which their bones support their body weight. Additionally, because women typically lose bone mass more rapidly following menopause, older women may require estrogen replacement therapy to prevent osteoporosis.

Calcium is needed for vital metabolic processes. When a low blood calcium ion concentration exists, osteoclasts are stimulated to breakdown bone tissue, and calcium ions are released into the blood. On the other hand, if the blood calcium ion concentration is high, osteoclast activity is inhibited, and osteoblasts are stimulated to form bone tissue. As a result, the excessive calcium is stored in the intercellular matrix of bone. (The details of this mechanism are described in chapter 11.)

In addition to storing calcium and phosphorus (as calcium phosphate), bone tissue contains lesser amounts of magnesium, sodium, potassium, and carbonate ions. Bones also tend to accumulate certain metallic elements such as lead, radium, or strontium, which are not normally present in the body but are sometimes ingested accidentally.

1. Name three functions of bones.
2. Distinguish between red marrow and yellow marrow.
3. List the substances normally stored in bone.

Organization of the Skeleton

For purposes of study, it is convenient to divide the skeleton into two major portions—an axial skeleton and an appendicular skeleton (fig. 7.7).

The **axial skeleton** consists of the bony and cartilaginous parts that support and protect the organs of the head, neck, and trunk. These parts include the following:

1. **Skull.** The skull is composed of the *cranium* (braincase) and facial bones.
2. **Hyoid** (hi′oid) **bone.** The hyoid bone is located in the neck between the lower jaw and larynx. It supports the tongue and serves as an attachment for certain muscles that help to move the tongue and function in swallowing.
3. **Vertebral column.** The vertebral column (backbone) consists of many vertebrae, separated by cartilaginous *intervertebral disks.* Near its distal end, several vertebrae are fused to form the **sacrum** (sa′krum), which is part of the pelvis. A small, rudimentary tailbone, called the **coccyx** (kok′siks), is attached to the end of the sacrum.
4. **Thoracic cage.** The thoracic cage protects the organs of the thorax and the upper abdomen. It is composed of twelve pairs of **ribs,** which articulate posteriorly with the thoracic vertebrae. It also includes the **sternum** (ster′num), to which most of the ribs are attached anteriorly.

The **appendicular skeleton** consists of the bones of the limbs and the bones that anchor the limbs to the axial skeleton. It includes the following:

1. **Pectoral girdle.** The pectoral girdle is formed by a **scapula** (scap′u-lah) and a **clavicle** (klav′ĭ-k′l) on both sides of the body. The pectoral girdle connects the bones of the arms to the axial skeleton and aids in arm movements.

Figure 7.7

Major bones of the skeleton: (*a*) an anterior view; (*b*) a posterior view. (Note that the axial and appendicular portions are distinguished by color.)

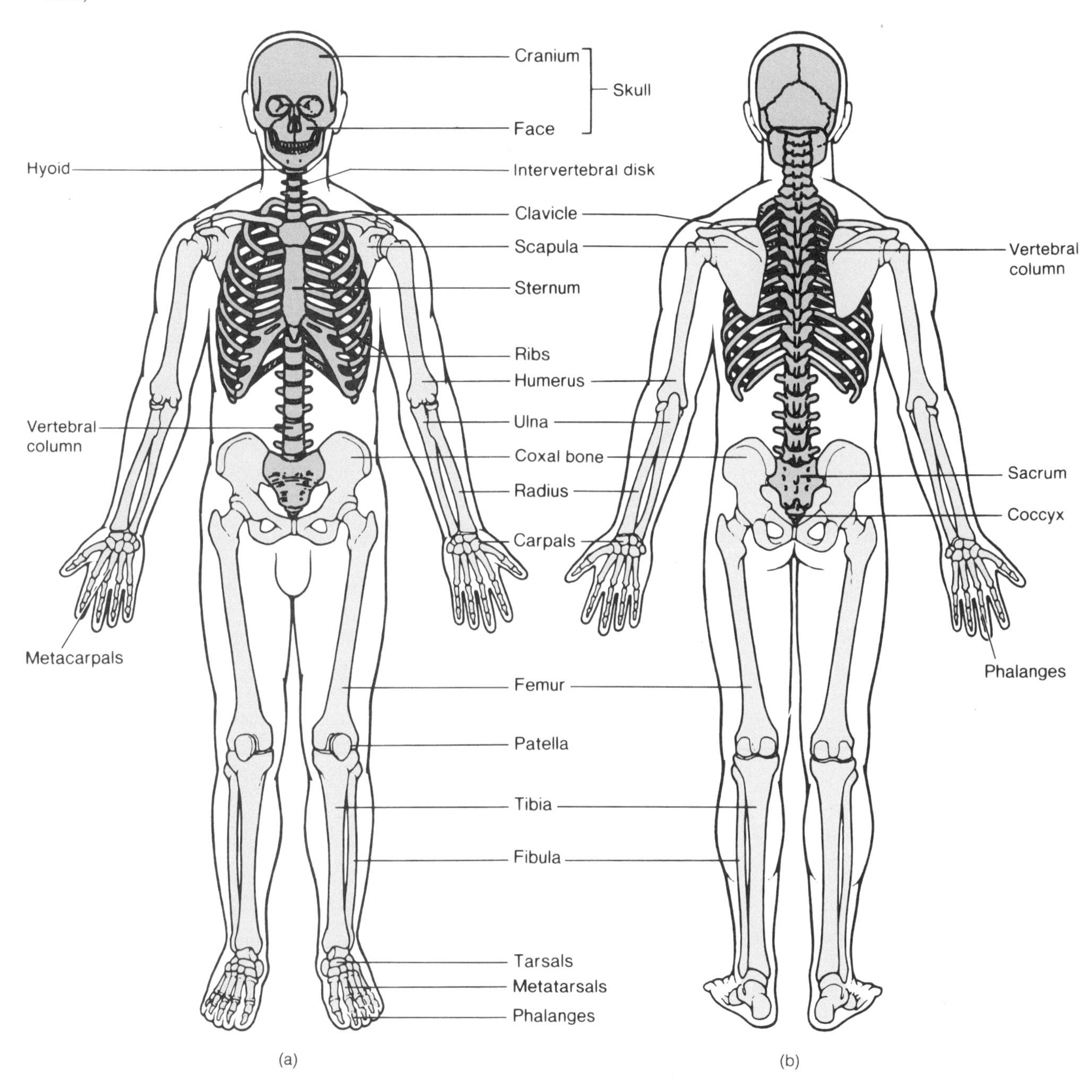

2. **Upper limbs** (arms). Each upper limb consists of a **humerus** (hu′mer-us) and two lower arm bones—a **radius** (ra′de-us) and an **ulna** (ul′nah). These three bones articulate at the elbow joint. At the distal end of the radius and ulna, there are eight **carpals** (kar′pals). The bones of the palm are called **metacarpals** (met″ah-kar′pals), and the finger bones are called **phalanges** (fah-lan′jēz).
3. **Pelvic girdle.** The pelvic girdle is formed by two **coxal** (kok′sal) **bones** (hipbones), which are attached to each other anteriorly and to the sacrum posteriorly. They connect the bones of the legs to the axial skeleton and, with the sacrum and coccyx, form the **pelvis.**
4. **Lower limbs** (legs). Each lower limb consists of a **femur** (fe′mur) and two lower leg bones—a large **tibia** (tib′e-ah) and a slender **fibula** (fib′u-lah). These three bones articulate at the knee

Chart 7.1 Terms used to describe skeletal structures		
Term	**Definition**	**Example**
Condyle (kon′dil)	A rounded process, which usually articulates with another bone	Occipital condyle of the occipital bone (fig. 7.11)
Crest (krest)	A narrow, ridgelike projection	Iliac crest of the ilium (fig. 7.27)
Epicondyle (ep″ĭ-kon′dil)	A projection situated above a condyle	Medial epicondyle of the humerus (fig. 7.23)
Facet (fas′et)	A small, nearly flat surface	Rib facet of a thoracic vertebra (fig. 7.15)
Fontanel (fon″tah-nel′)	A soft spot in the skull where membranes cover the space between bones	Anterior fontanel between the frontal and parietal bones (fig. 7.14)
Foramen (fo-ra′men)	An opening through a bone, which usually serves as a passageway for blood vessels, nerves, or ligaments	Foramen magnum of the occipital bone (fig. 7.11)
Fossa (fos′ah)	A relatively deep pit or depression	Olecranon fossa of the humerus (fig. 7.23)
Fovea (fo′ve-ah)	A tiny pit or depression	Fovea capitis of the femur (fig. 7.29)
Head (hed)	An enlargement of the end of a bone	Head of the humerus (fig. 7.23)
Meatus (me-a′tus)	A tubelike passageway within a bone	External auditory meatus of the ear (fig. 7.10)
Process (pros′es)	A prominent projection on a bone	Mastoid process of the temporal bone (fig. 7.10)
Sinus (si′nus)	A cavity within a bone	Frontal sinus of the frontal bone (fig. 7.13)
Spine (spīn)	A thornlike projection	Spine of the scapula (fig. 7.22)
Suture (soo′cher)	An interlocking line of union between bones	Lambdoidal suture between the occipital and parietal bones (fig. 7.10)
Trochanter (tro-kan′ter)	A relatively large process	Greater trochanter of the femur (fig. 7.29)
Tubercle (tu′ber-kl)	A small, knoblike process	Greater tubercle of the humerus (fig. 7.23)
Tuberosity (tu″bĕ-ros′ĭ-te)	A knoblike process usually larger than a tubercle	Radial tuberosity of the radius (fig. 7.24)

joint, where the **patella** (pah-tel′ah) covers the anterior surface. At the distal ends of the tibia and fibula, there are seven **tarsals** (tahr′sals). The bones of the foot are called **metatarsals** (met″ah-tahr′sals), and those of the toes (like the fingers) are called **phalanges.**

Chart 7.1 includes some terms that are used to describe skeletal structures.

1. Distinguish between the axial and the appendicular skeletons.
2. Name the bones of each portion of the skeleton.

Skull

A human skull usually consists of twenty-two bones, which except for the lower jaw, are firmly interlocked along lines, called *sutures* (su'churz) (fig. 7.8). Eight of these interlocked bones make up the cranium, and thirteen form the facial skeleton. The **mandible** (man′dĭ-b'l) is a movable bone held to the cranium by ligaments.

Cranium

The **cranium** (kra′ne-um) encloses and protects the brain, and its surface provides attachments for various muscles that make chewing and head movements possible. Some cranial bones contain air-filled cavities, called *sinuses,* which are lined with mucous membranes and connected by passageways to the nasal cavity (fig. 7.9). Sinuses reduce the weight of the skull and increase the intensity of the voice by serving as resonant sound chambers.

The eight bones of the cranium (figs. 7.8 and 7.10) are as follows:

1. **Frontal bone.** The frontal (frun′tal) bone forms the anterior portion of the skull above the eyes. On the upper margin of each orbit (the bony socket of the eye), the frontal bone is marked by a *supraorbital foramen* (or *supraorbital notch* in some skulls), through which blood vessels and nerves pass to the tissues of the forehead. Within the frontal bone are two *frontal sinuses,* one above each eye near the midline (fig. 7.9).
2. **Parietal bones.** One parietal (pah-ri′ĕ-tal) bone is located on each side of the skull, just behind the frontal bone (fig. 7.10). Together, the parietal bones form the bulging sides and roof of the cranium. They are fused at the midline along the *sagittal suture,* and they meet the frontal bone along the *coronal suture.*
3. **Occipital bone.** The occipital (ok-sip′ĭ-tal) bone (figs. 7.10 and 7.11) joins the parietal bones

Figure 7.8
An anterior view of the skull.

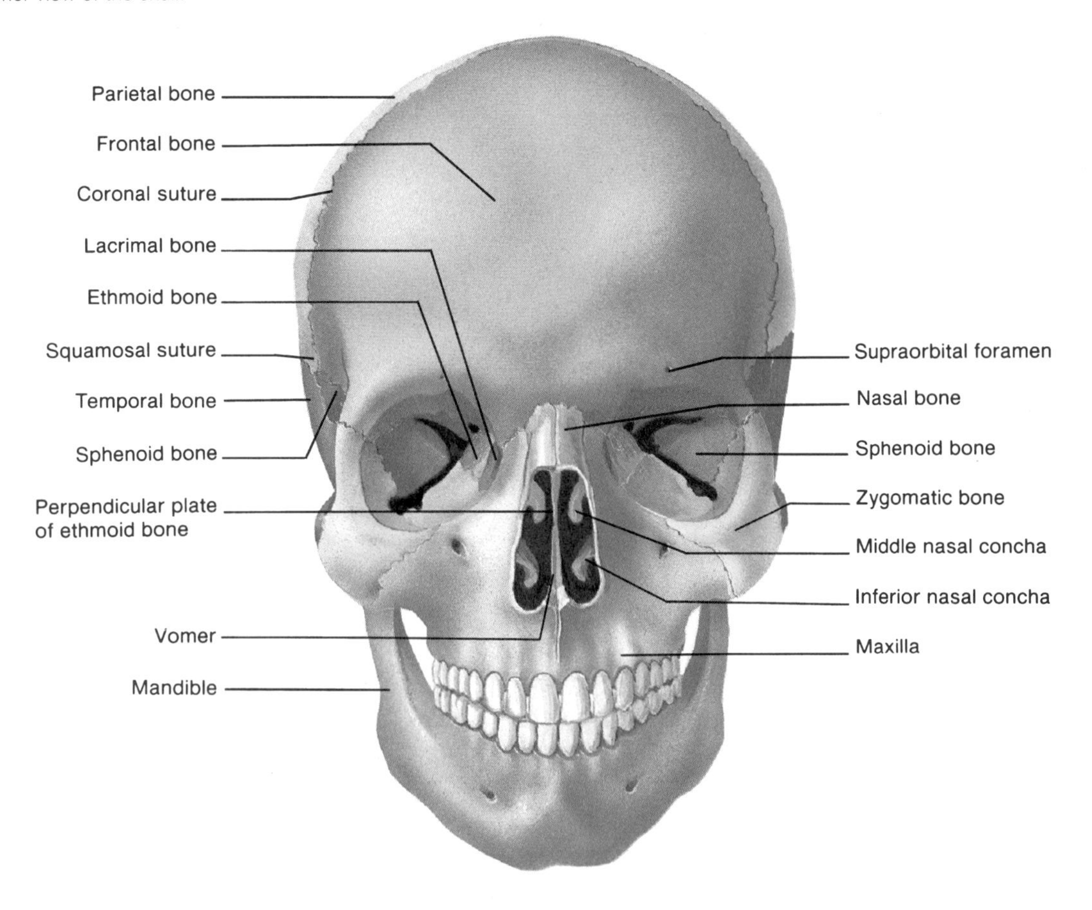

Figure 7.9
Locations of the sinuses.

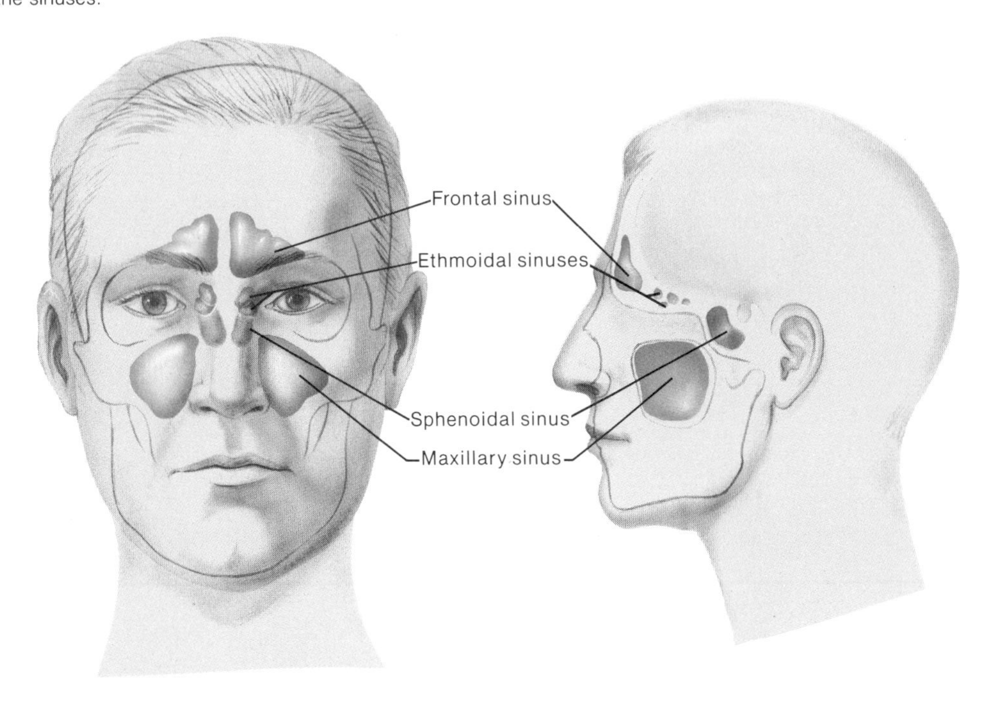

Figure 7.10
A lateral view of the skull.

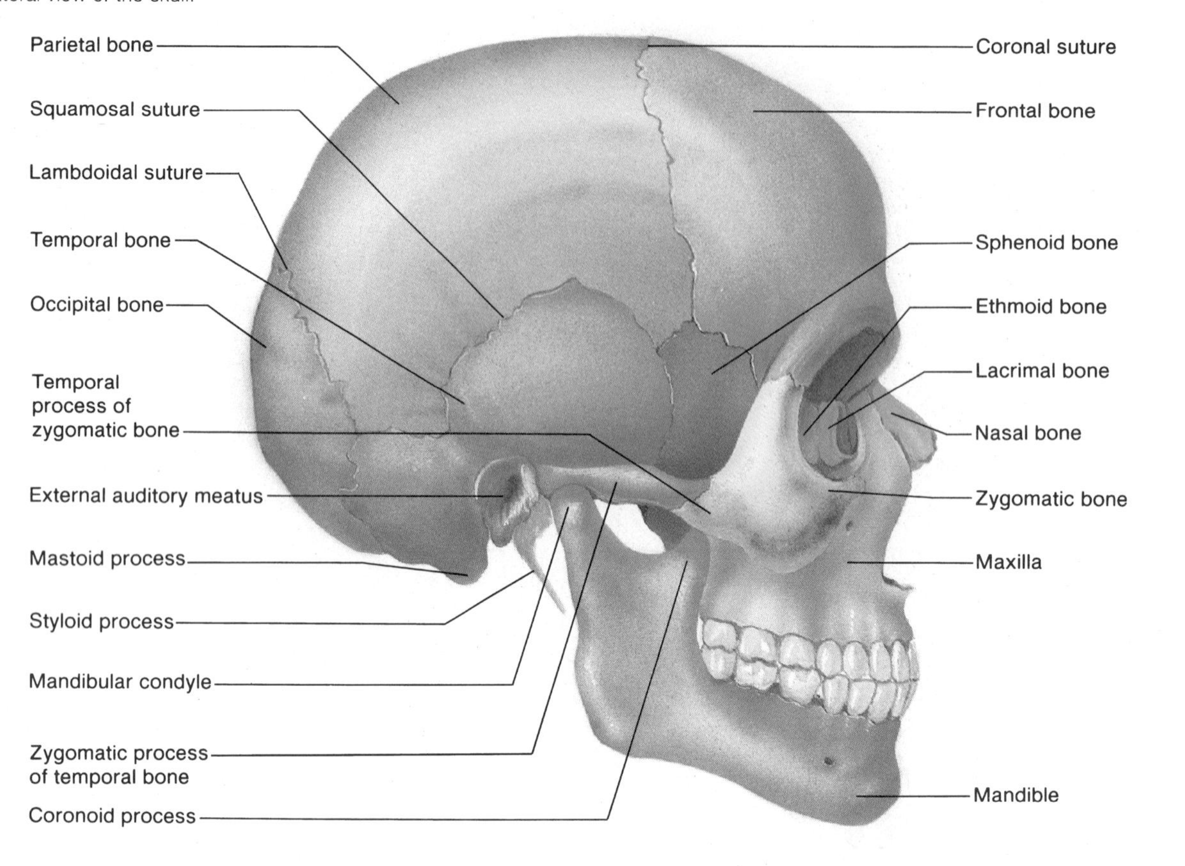

Figure 7.11
An inferior view of the skull.

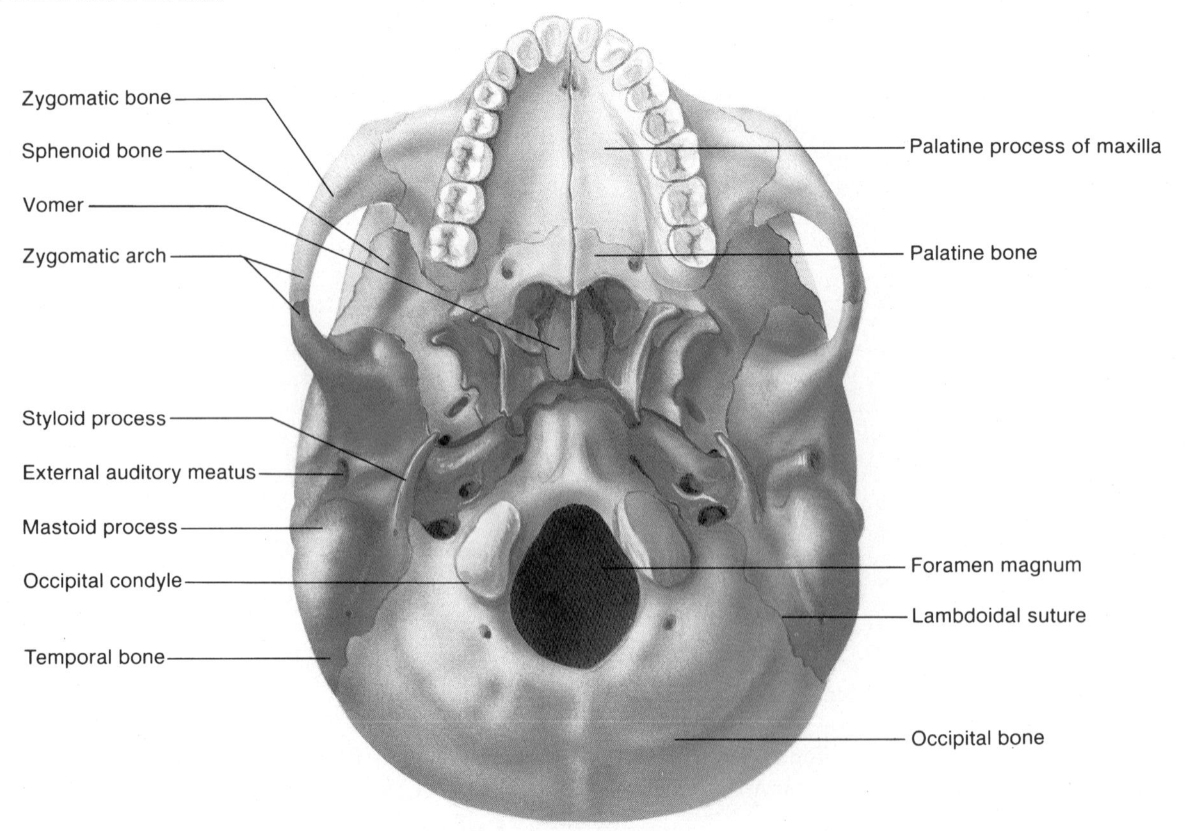

Figure 7.12
Floor of the cranial cavity viewed from above.

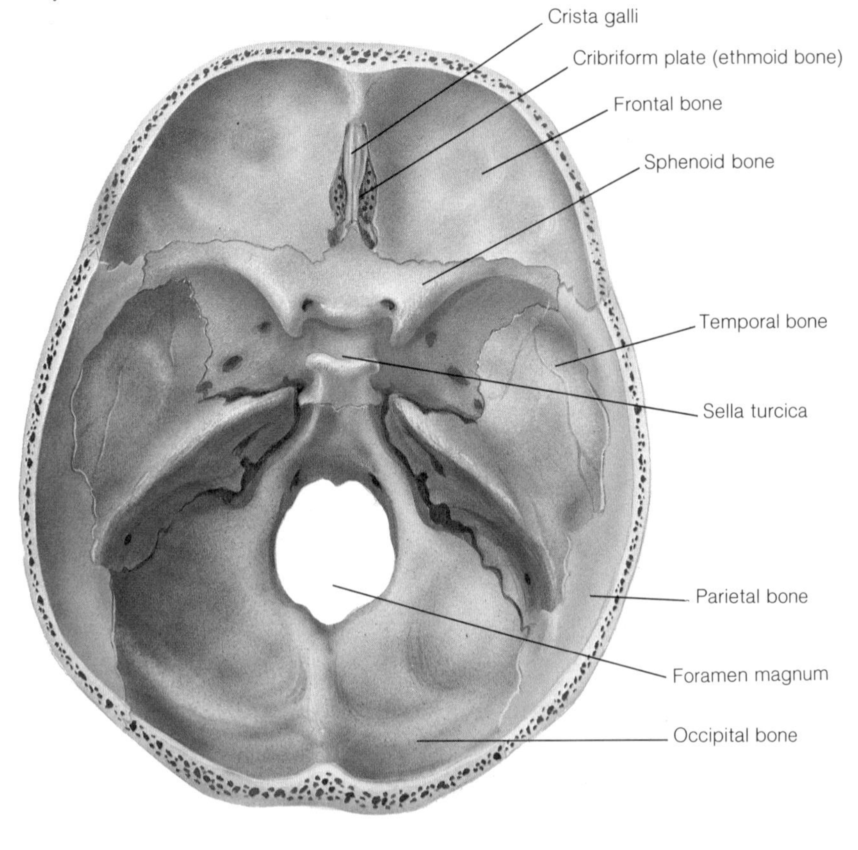

along the *lambdoidal* (lam′doid-al) *suture.* It forms the back of the skull and the base of the cranium. There is a large opening on its lower surface called the *foramen magnum,* through which nerve fibers from the brain pass and enter the vertebral canal. Rounded processes called *occipital condyles,* which are located on each side of the foramen magnum, articulate with the first vertebra (atlas) of the vertebral column.

4. **Temporal bones.** A temporal (tem′po-ral) bone (fig. 7.10) on each side of the skull joins the parietal bone along a *squamosal* (skwa-mo′sal) *suture.* The temporal bones form parts of the sides and the base of the cranium. Located near the inferior margin is an opening, the *external auditory meatus,* which leads inward to parts of the ear. The temporal bones have depressions, called the *mandibular fossae,* that articulate with processes of the mandible. Below each external auditory meatus, there are two projections—a rounded *mastoid process* and a long, pointed *styloid process.* The mastoid process provides an attachment for certain muscles of the neck, and the styloid process serves as an anchorage for muscles associated with the tongue and pharynx.
   A *zygomatic process* projects anteriorly from the temporal bone, joins the *zygomatic bone,* and helps form the prominence of the cheek.
5. **Sphenoid bone.** The sphenoid (sfe′noid) bone (figs. 7.10 and 7.11) is wedged between several other bones in the anterior portion of the cranium. It consists of a central part and two winglike structures, which extend laterally toward each side of the skull. This bone helps form the base of the cranium, sides of the skull, and floors and sides of the orbits. Along the midline within the cranial cavity, a portion of the sphenoid bone rises up and forms a saddle-shaped mass called the *sella turcica* (sel′ah tur′si-ka) (Turk's saddle). This depression is occupied by the pituitary gland.
   The sphenoid bone also contains two *sphenoidal sinuses* (fig. 7.9).
6. **Ethmoid bone.** The ethmoid (eth′moid) bone (figs. 7.10 and 7.12) is located in front of the sphenoid bone. It consists of two masses, one on each side of the nasal cavity, which are joined horizontally by thin *cribriform* (krib′ri-form) *plates.* These plates form part of the roof of the nasal cavity (fig. 7.12).
   Projecting upward into the cranial cavity between the cribriform plates is a triangular process of the ethmoid bone, called the *crista galli* (kris′ta gal′li) (cock's comb). This process

Figure 7.13
A sagittal section of the skull.

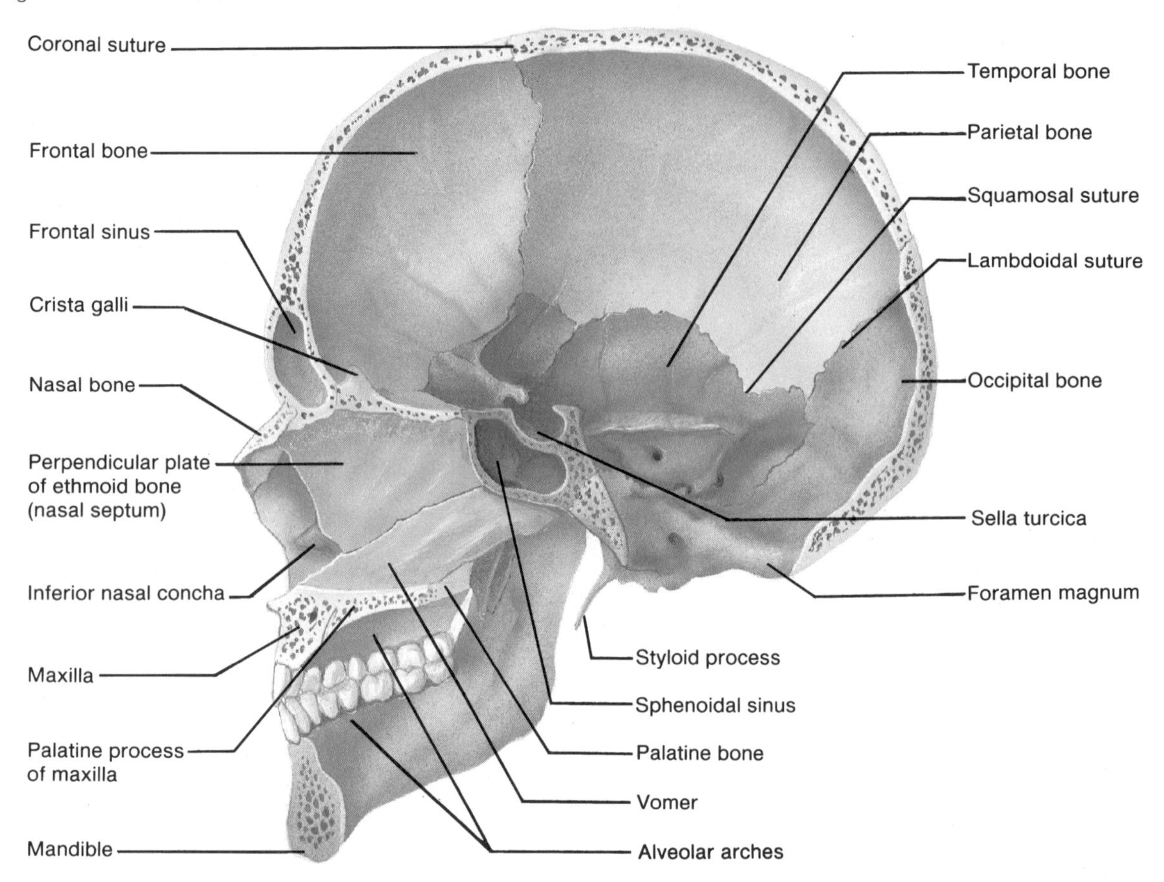

serves as an attachment for membranes that enclose the brain (figs. 7.12 and 7.13). Portions of the ethmoid bone also form sections of the cranial floor, orbital walls, and nasal cavity walls. A *perpendicular plate* projects downward in the midline from the cribriform plates and forms most of the nasal septum (fig. 7.13).

Delicate scroll-shaped plates, called the *superior nasal concha* (kong′kah) and *middle nasal concha,* project inward from the lateral portions of the ethmoid bone toward the perpendicular plate (fig. 7.13). The lateral portions of the ethmoid bone contain many small air spaces, the *ethmoidal sinuses* (fig. 7.9).

Facial Skeleton

The **facial skeleton** consists of thirteen immovable bones and a movable lower jawbone. In addition to forming the basic shape of the face, these bones provide attachments for various muscles that move the jaw and control facial expressions.

The bones of the facial skeleton are as follows:

1. **Maxillary bones.** The maxillary (mak′sĭ-ler″e) bones (pl. maxillae, mak-si-l′e) (figs. 7.10 and 7.11) form the upper jaw. Portions of these bones comprise the anterior roof of the mouth (*hard palate*), floors of the orbits, and sides and floor of the nasal cavity. They also contain the sockets of the upper teeth. Inside the maxillae, lateral to the nasal cavity, are *maxillary sinuses,* which are the largest of the sinuses (fig. 7.9).

   During development, portions of the maxillae, called *palatine processes,* grow together and fuse along the midline to form the anterior section of the hard palate.

   The inferior border of each maxillary bone projects downward forming an *alveolar* (al-ve′o-lar) *process.* Together these processes form a horseshoe-shaped *alveolar arch* (dental arch) (fig. 7.13). Cavities in this arch (dental alveoli) are occupied by the teeth, which are attached to these bony sockets by fibrous connective tissues.

Sometimes the fusion of the palatine processes of the maxillae is incomplete at the time of birth; the result is called a *cleft palate*. Infants with this deformity may have trouble suckling because of the opening that remains between the oral and nasal cavities.

2. **Palatine bones.** The palatine (pal′ah-tīn) bones (figs. 7.11 and 7.13) are located behind the maxillae. Each bone is roughly L-shaped. The horizontal portions serve as both the posterior section of the hard palate and the floor of the nasal cavity. The perpendicular portions help form the lateral walls of the nasal cavity.
3. **Zygomatic bones.** The zygomatic (zi″go-mat′ik) bones (figs. 7.10 and 7.11) are responsible for the prominences of the cheeks below and to the sides of the eyes. These bones also help form the lateral walls and floors of the orbits. Each bone has a *temporal process*, which extends posteriorly to join the zygomatic process of a temporal bone. Together these processes form a *zygomatic arch*.
4. **Lacrimal bones.** A lacrimal (lak′rĭ-mal) bone (figs. 7.8 and 7.10) is a thin, scalelike structure located in the medial wall of each orbit between the ethmoid bone and maxilla.
5. **Nasal bones.** The nasal (na′zal) bones (figs. 7.8 and 7.10) are long, thin, and nearly rectangular. They lie side by side and are fused at the midline, where they form the bridge of the nose.
6. **Vomer.** The thin, flat vomer (vo′mer) (figs. 7.8 and 7.13) is located along the midline within the nasal cavity. Posteriorly it joins the perpendicular plate of the ethmoid bone, and together they form the nasal septum.
7. **Inferior nasal conchae.** The inferior nasal conchae (kong′ke) are fragile, scroll-shaped bones attached to the lateral walls of the nasal cavity (figs. 7.8 and 7.13). Like the superior and middle conchae, the inferior conchae provide support for mucous membranes within the nasal cavity.
8. **Mandible.** The mandible (man′dĭ-b′l) consists of a horizontal, horseshoe-shaped body with a flat portion projecting upward at each end (figs. 7.8 and 7.10). This projection is divided into two processes—a posterior *mandibular condyle* and an anterior *coronoid process*. The mandibular condyles articulate with the mandibular fossae of the temporal bones, and the coronoid processes serve as attachments for muscles used in chewing. Other large chewing muscles are inserted on the lateral surface of the mandible. A curved bar of bone on the superior border of the mandible, the *alveolar arch*, contains the hollow sockets (dental alveoli) that bear the lower teeth (fig. 7.13).

Infantile Skull

At birth, the skull is incompletely developed, and the cranial bones are separated by fibrous membranes. These membranous areas are called **fontanels** (fon″tah-nels), or more commonly, soft spots (fig. 7.14). They permit some movement between the bones, so that the developing skull is partially compressible and can change shape slightly. This action enables an infant's skull to pass more easily through the birth canal. Eventually the fontanels close as the cranial bones grow together.

Other characteristics of an infantile skull include a relatively small face with a prominent forehead and large orbits. The jaw and nasal cavity are small, the sinuses are incompletely formed, and the frontal bone is in two parts. The skull bones are thin, but they are also somewhat flexible and thus are less easily fractured than adult bones.

1. Locate and name each bone of the cranium.
2. Locate and name each facial bone.
3. Explain how an adult skull differs from that of an infant.

Vertebral Column

The **vertebral column** extends from the skull to the pelvis and forms the vertical axis of the skeleton. It is composed of many bony parts called **vertebrae** (ver′tĕ-bre). These are separated by masses of fibrocartilage, called *intervertebral disks*, and connected to one another by ligaments. The vertebral column supports the head and trunk of the body. It also protects the spinal cord, which passes through a *vertebral canal*, formed by openings in the vertebrae.

Normally the vertebral column has four curvatures, which give it a degree of resiliency. The names of the curves correspond to the regions in which they occur, as indicated in figure 7.15.

Typical Vertebra

Although the vertebrae in different regions of the vertebral column have special characteristics, they also have features in common. Thus, a typical vertebra has a drum-shaped *body*, which forms the thick, anterior

Figure 7.14
(*a*) Lateral view and (*b*) superior view of the infantile skull.

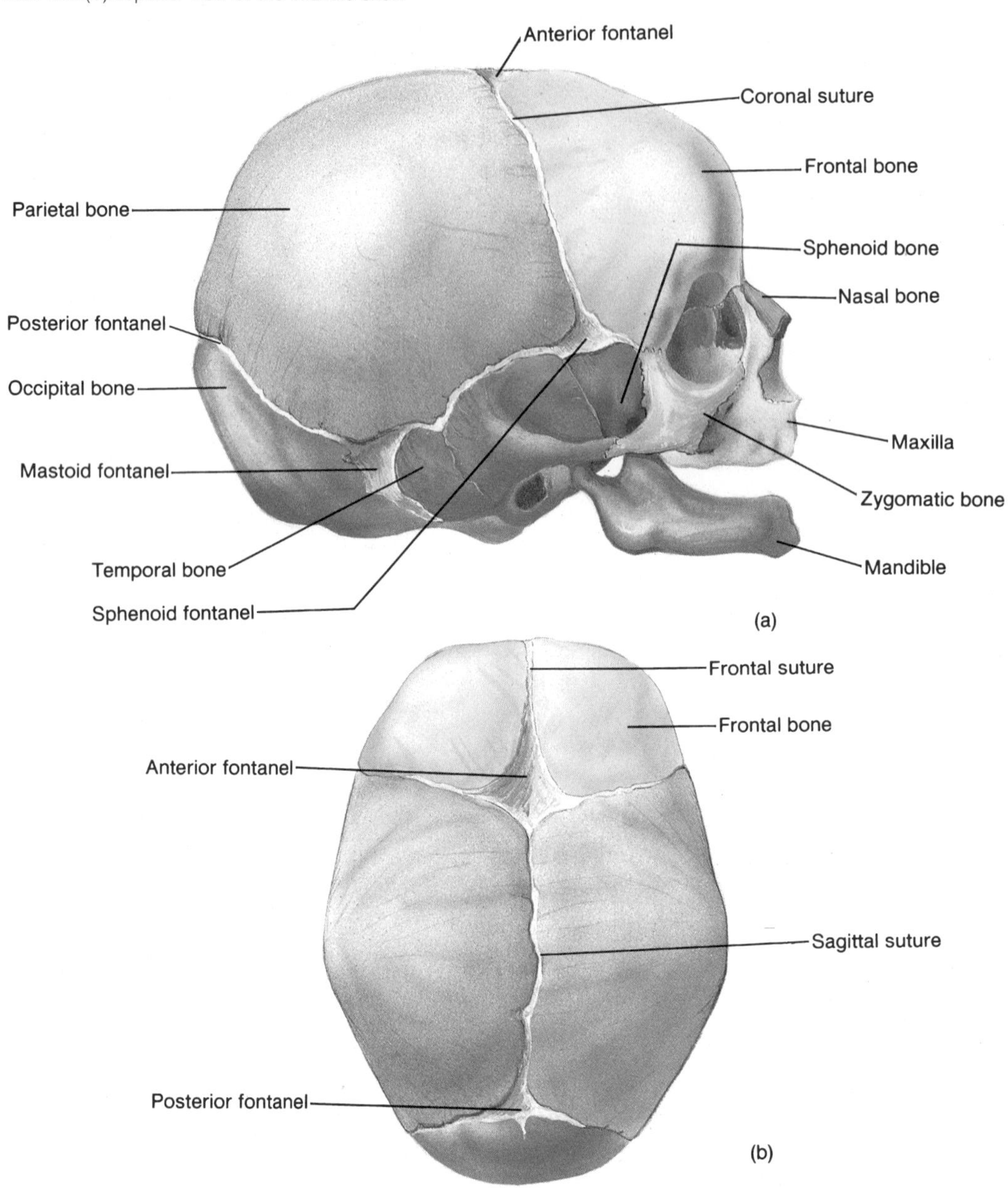

portion of the bone (fig. 7.16). A longitudinal row of these bodies supports the weight of the head and trunk. Intervertebral disks, which separate adjacent vertebral bodies, cushion and soften the forces produced by such movements as walking and jumping (fig. 7.15).

Projecting posteriorly from each vertebral body are two short stalks, called *pedicles* (ped'i-k'lz). Two plates, called *laminae* (lam'i-ne) arise from the pedicles and fuse in the back to become a *spinous process.* The pedicles, laminae, and spinous process together complete a bony *vertebral arch* around a *vertebral foramen,* through which the spinal cord passes.

If the laminae of the vertebrae fail to unite during development, the vertebral arch remains incomplete. This condition is called *spina bifida.* As a result of it, the contents of the vertebral canal may protrude outward. This problem occurs most frequently in the lumbosacral region.

Figure 7.15
The curved vertebral column consists of many vertebrae, separated by intervertebral disks.

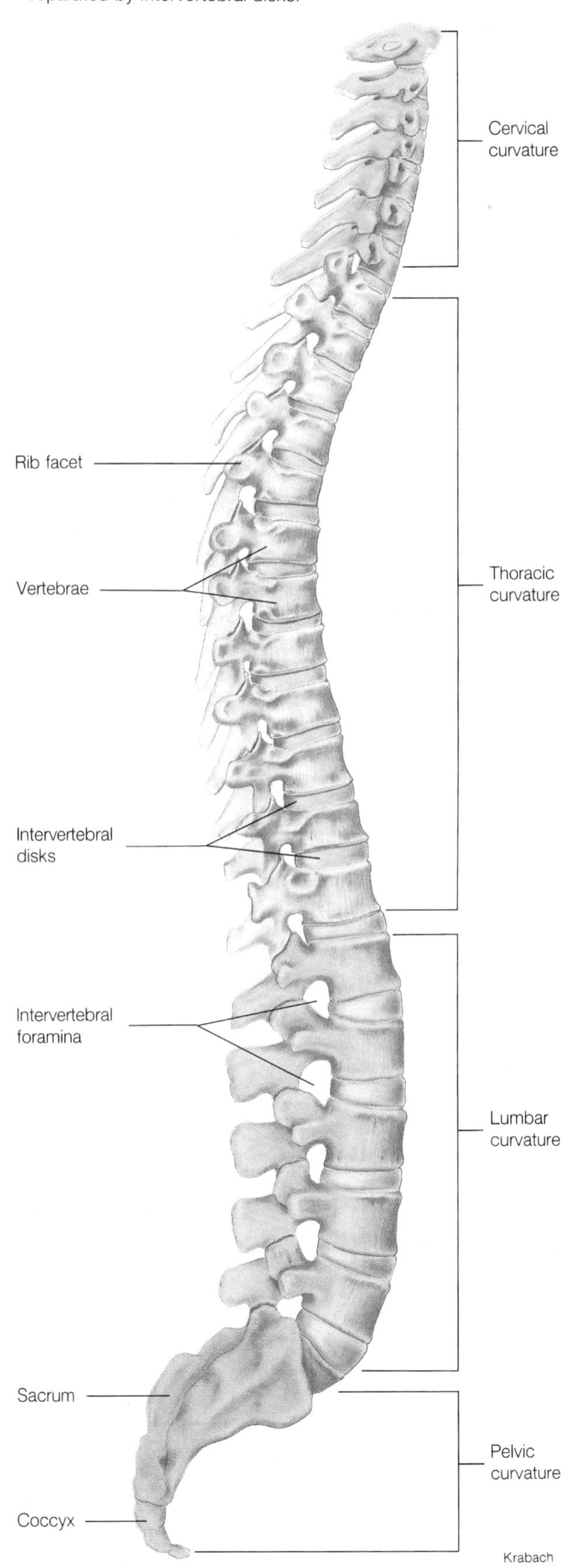

Between the pedicles and laminae of a typical vertebra is a *transverse process,* which projects laterally and toward the back. Various ligaments and muscles are attached to the dorsal spinous process and the transverse processes. Projecting upward and downward from each vertebral arch are *superior* and *inferior articulating processes.* These processes bear cartilage-covered facets, by which each vertebra is joined to the one above and the one below it.

On the lower surfaces of the vertebral pedicles are notches that align to form openings, called *intervertebral foramina* (in″ter-ver′te-bral fo-ram′i-nah). These openings provide passageways for spinal nerves that proceed between adjacent vertebrae and connect to the spinal cord (fig. 7.15).

Cervical Vertebrae

Seven **cervical vertebrae** comprise the bony axis of the neck (fig. 7.15). The transverse processes of these vertebrae are distinctive because they have *transverse foramina,* which serve as passageways for arteries leading to the brain (fig. 7.17).

Two of the cervical vertebrae, shown in figure 7.17, are of special interest. The first vertebra, or **atlas** (at′las), supports and balances the head. On its upper surface, are two kidney-shaped *facets,* which articulate with the occipital condyles of the skull.

The second cervical vertebra, or **axis** (ak′sis), bears a toothlike *odontoid process* (dens) on its body. This process projects upward and lies in the ring of the atlas. As the head is turned from side to side, the atlas pivots around the odontoid process.

Thoracic Vertebrae

The twelve **thoracic vertebrae** are larger than those in the cervical region (fig. 7.15). Each vertebra has a long, pointed spinous process, which slopes downward, and facets on the sides of its body, which articulate with a rib.

Beginning with the third thoracic vertebra and downward, the bodies of these bones increase in size. Thus, they are adapted to the stress placed on them by the increasing amounts of body weight they bear.

Lumbar Vertebrae

There are five **lumbar vertebrae** (fig. 7.15) in the small of the back (loin). The lumbars are adapted to support more weight than the vertebrae above them and have larger and stronger bodies.

Figure 7.16
A superior view of (*a*) a cervical vertebra, (*b*) a thoracic vertebra, and (*c*) a lumbar vertebra.

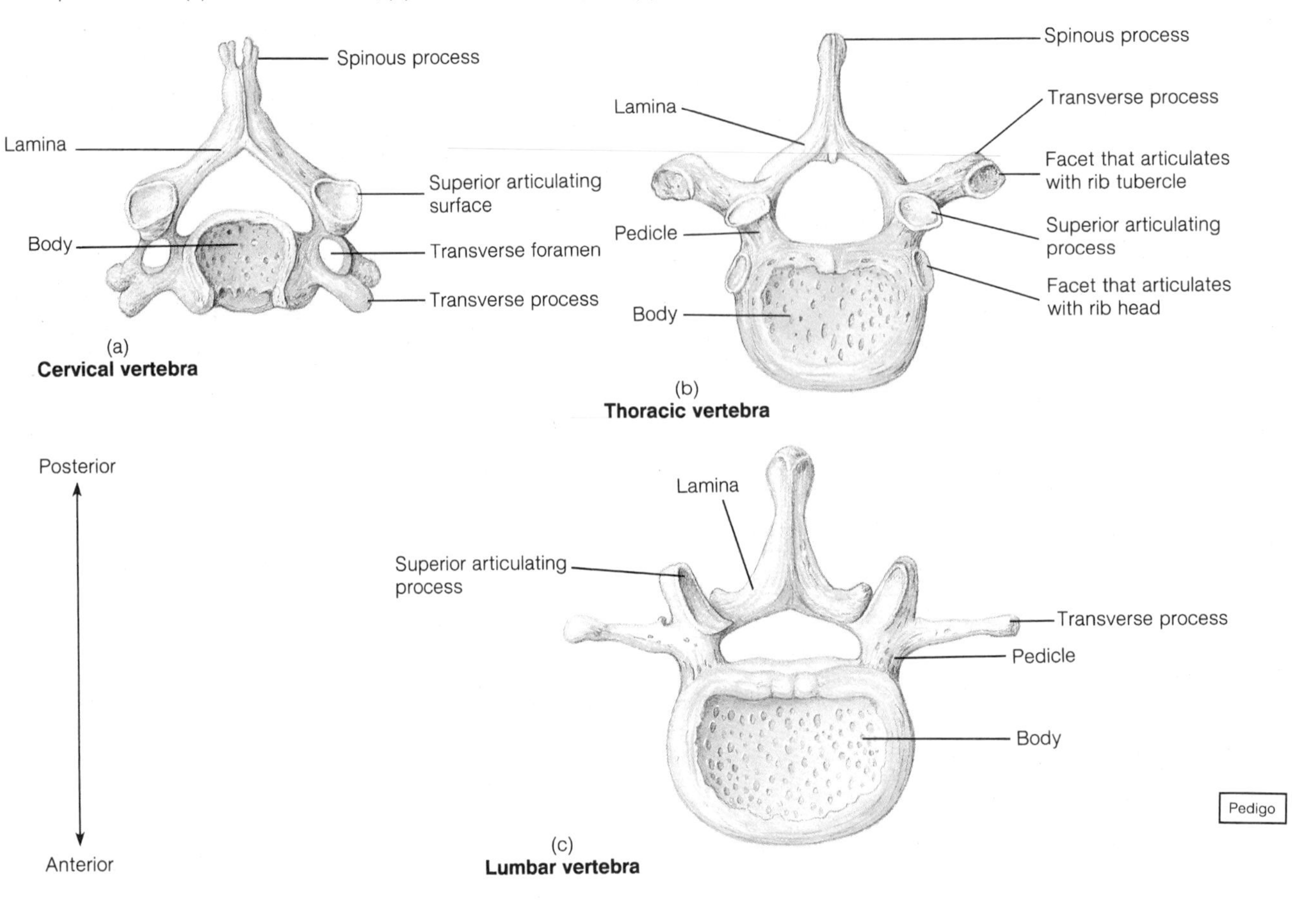

Figure 7.17
How do the (*a*) atlas and (*b*) axis function together to allow movement of the head?

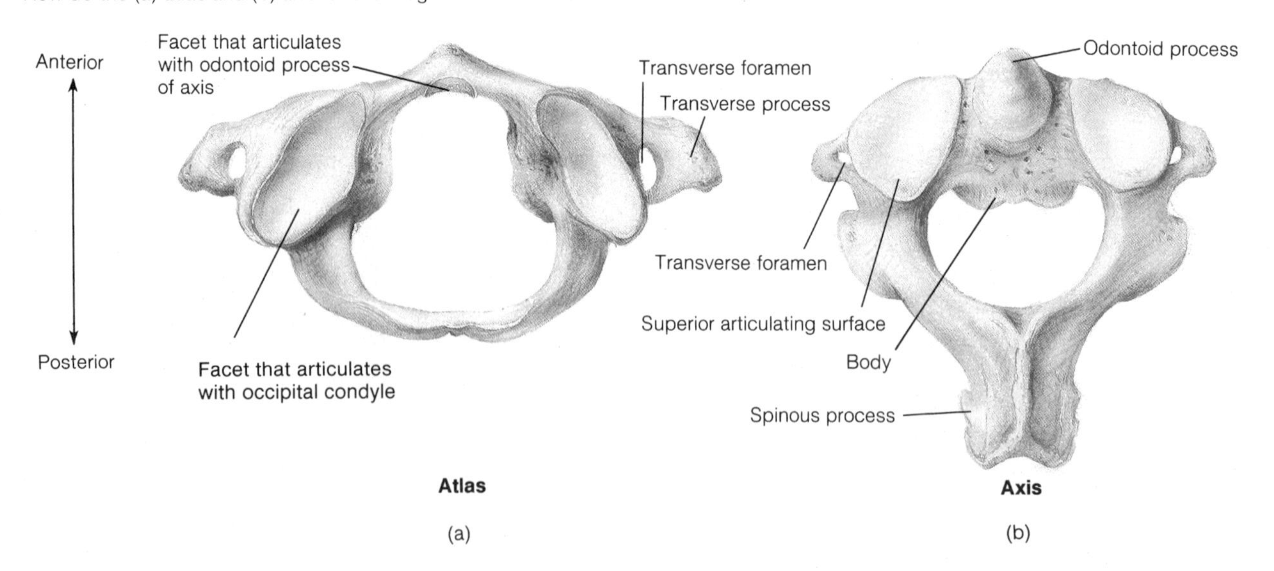

Sacrum

The **sacrum** is a triangular structure, composed of five fused vertebrae, that forms the base of the vertebral column (fig. 7.18). The spinous processes of these fused bones are represented by a ridge of *tubercles*. To the sides of the tubercles are rows of openings, the *dorsal sacral foramina,* through which nerves and blood vessels pass.

The vertebral foramina of the sacral vertebrae form the *sacral canal,* which continues through the

Figure 7.18
(*a*) Anterior view of the sacrum and coccyx; (*b*) posterior view.

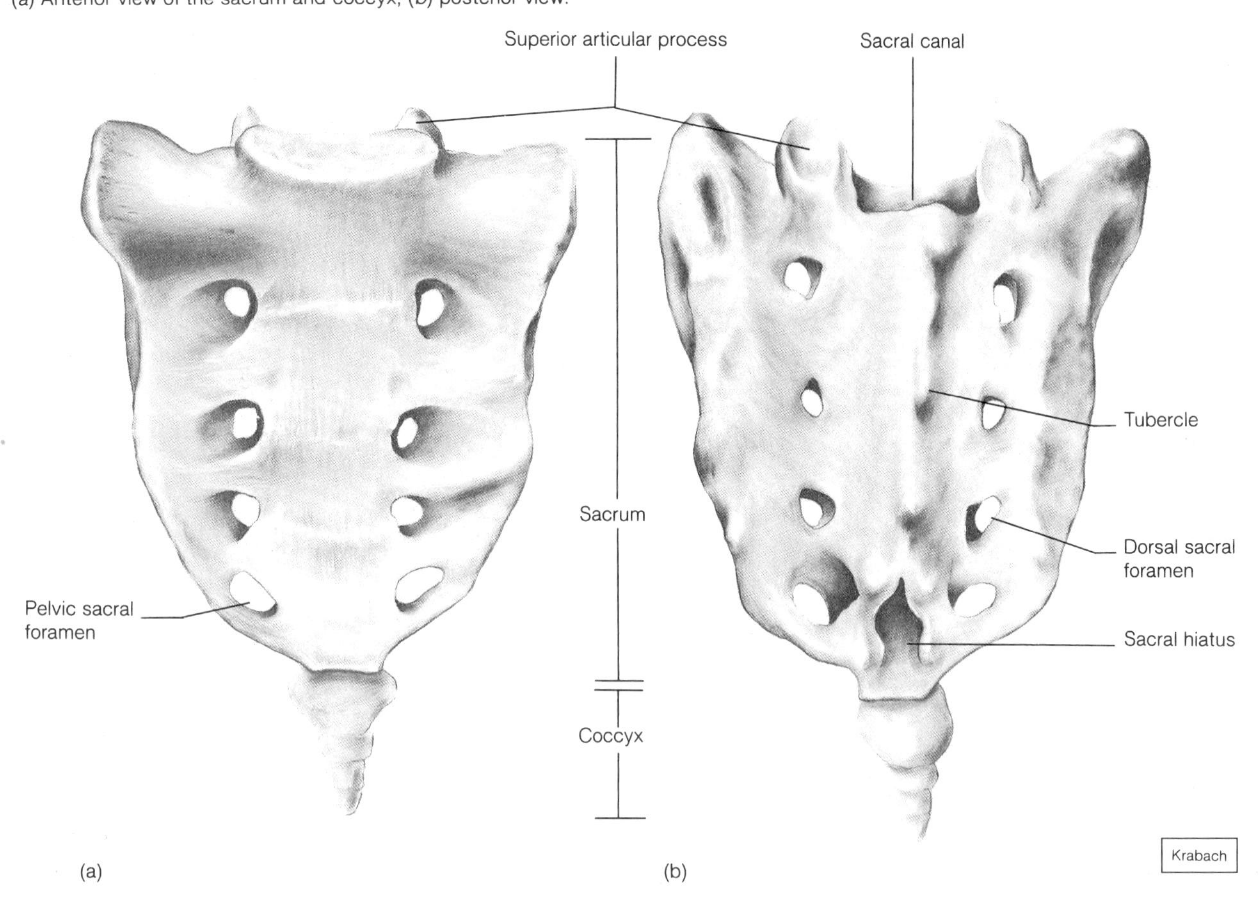

sacrum to an opening of variable size at the tip called the *sacral hiatus* (sa′kral hi-a′tus). On the anterior surface of the sacrum, four pairs of *pelvic sacral foramina* provide passageways for nerves and blood vessels.

Coccyx

The **coccyx** is the lowest part of the vertebral column and is composed of four fused vertebrae (fig. 7.18). It is attached by ligaments to the margins of the sacral hiatus.

A common back problem involves changes in the intervertebral disks. Each disk is composed of a tough, outer layer of fibrocartilage and an elastic central mass. As a person ages, these disks tend to undergo degenerative changes in which the central masses lose their firmness, and the outer layers become thinner and weaker and develop cracks. Extra pressure, as produced when a person falls or lifts a heavy object, may break the outer layer of a disk and allow the central mass to squeeze out. Such a rupture may cause pressure on the spinal cord or on a spinal nerve that branches from it. This condition—a ruptured or herniated disk—may cause back pain and numbness or the loss of muscular function in the parts innervated by the affected spinal nerve.

1. Describe the structure of the vertebral column.
2. Describe a typical vertebra.
3. How do the structures of a cervical, thoracic, and lumbar vertebra differ?

Thoracic Cage

The **thoracic cage** includes the ribs, thoracic vertebrae, sternum (breastbone), and costal cartilages, by which the ribs are attached to the sternum (fig. 7.19). These parts support the shoulder girdle and arms, protect the visceral organs in the thoracic and upper abdominal cavities, and play a role in breathing.

Ribs

Usually each person has twelve pairs of ribs—one pair attached to each of the twelve thoracic vertebrae.

The first seven rib pairs, *true ribs* (vertebrosternal ribs), join the sternum directly by their costal cartilages. The remaining five pairs are called *false ribs* because their cartilages do not reach the sternum directly. Instead, the cartilages of the upper three false ribs (vertebrochondral ribs) join the cartilages of the ribs next above. The last two rib pairs (or sometimes the last three pairs) are called *floating ribs* (vertebral ribs) because they have no cartilaginous attachments to the sternum.

Figure 7.19
The thoracic cage includes the thoracic vertebrae, sternum, ribs, and costal cartilages that attach the ribs to the sternum.

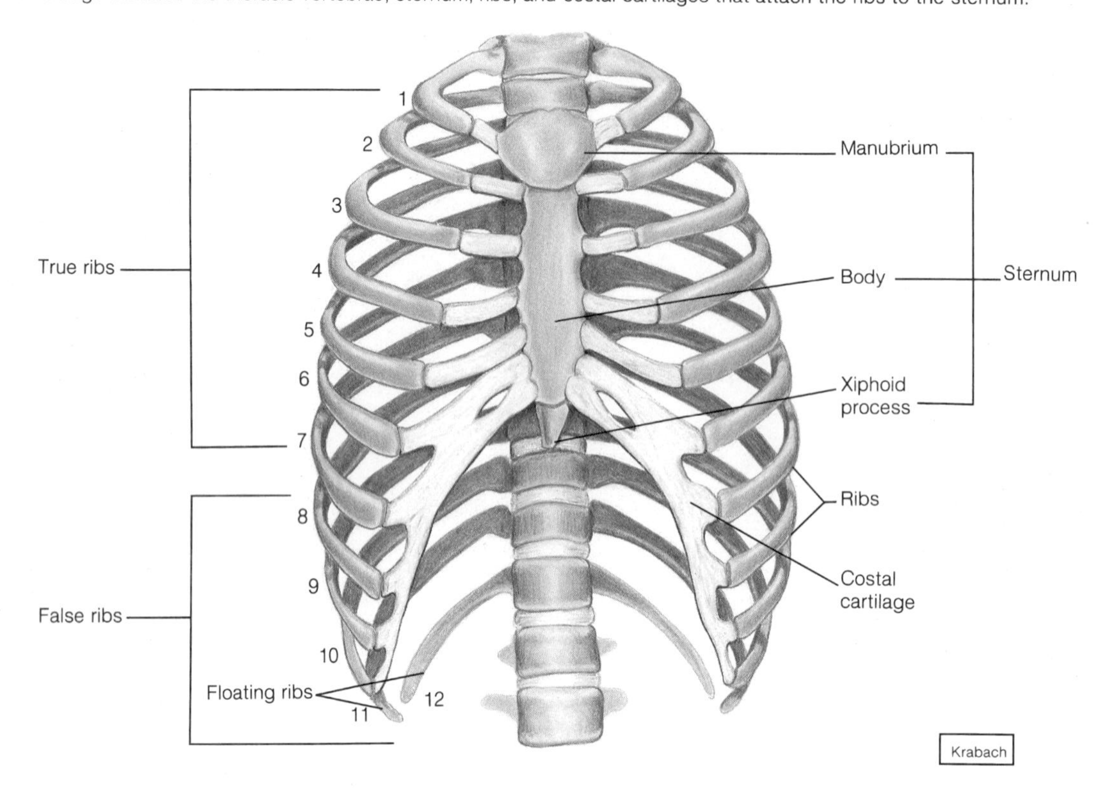

A typical rib has a long, slender shaft, which curves around the chest and slopes downward (fig. 7.19). On the posterior end is an enlarged *head* by which the rib articulates with a *facet* on the body of its own vertebra and with the body of the next higher vertebra. Also near the head is a *tubercle,* which articulates with the transverse process of the vertebra.

1. What bones make up the thoracic cage?
2. Describe a typical rib.
3. What are the differences between vertebrosternal, vertebrochondral, and vertebral ribs.

Sternum

The **sternum** (ster′num) or breastbone is located along the midline in the anterior portion of the thoracic cage (fig. 7.19). It is a flat, elongated bone that develops in three parts—an upper *manubrium* (mah-nu′bre-um), a middle *body,* and a lower *xiphoid* (zif′oid) *process,* which projects downward. The manubrium articulates with the clavicles by facets on its superior border.

Red marrow within the spongy bone of the sternum functions in blood-cell formation into adulthood. Since the sternum has a thin covering of compact bone and is easy to reach, samples of its blood-cell-forming tissue may be removed for use in diagnosing diseases. This procedure, a *sternal puncture,* involves suctioning (aspirating) some marrow through a hollow needle. Marrow may also be removed from the iliac crest of a coxal bone.

Pectoral Girdle

The **pectoral** (pek′to-ral) **girdle** or shoulder girdle is composed of four parts—two *clavicles* and two *scapulae* (figs. 7.20 and 7.21). Although the word *girdle* suggests a ring-shaped structure, the pectoral girdle is an incomplete ring. It is open in the back between the scapulae, and its bones are separated in the front by the sternum. The pectoral girdle supports the arms and serves as an attachment for several muscles that move the arms.

Clavicles

The **clavicles** (klav′i-k′lz) or collarbones are slender, rodlike bones with elongated S-shapes (fig. 7.20). They are located at the base of the neck and run horizontally between the manubrium and scapulae.

The clavicles act as braces for the freely movable scapulae and thus help to hold the shoulders in place. They also provide attachments for muscles of the arms, chest, and back.

Figure 7.20
The pectoral girdle, to which the arms are attached, consists of a clavicle and scapula on each side.

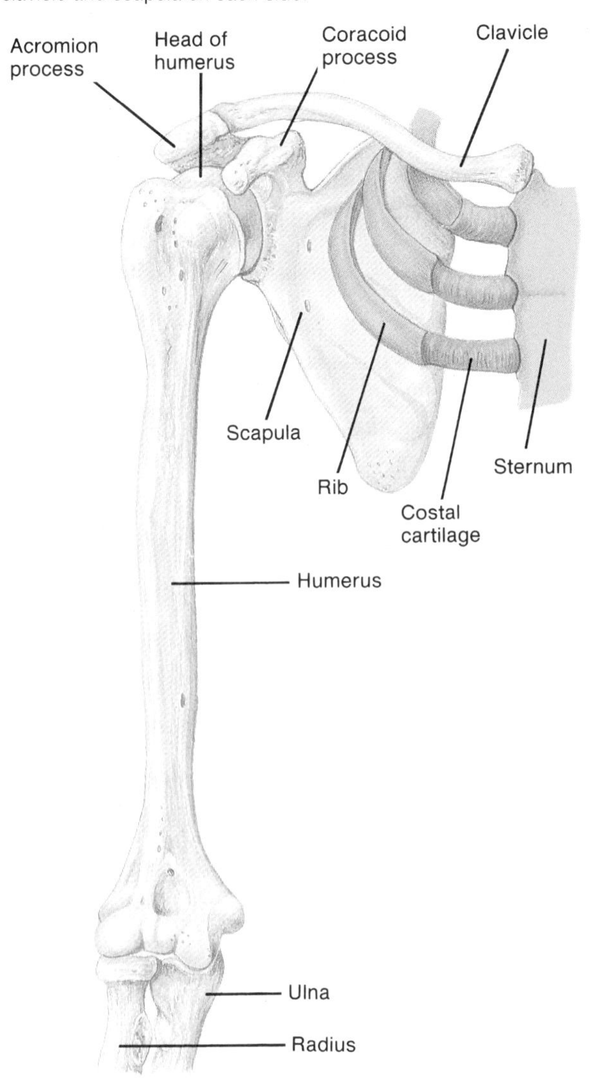

Figure 7.21
An X-ray film of the right shoulder region viewed from the front. What features can you identify?

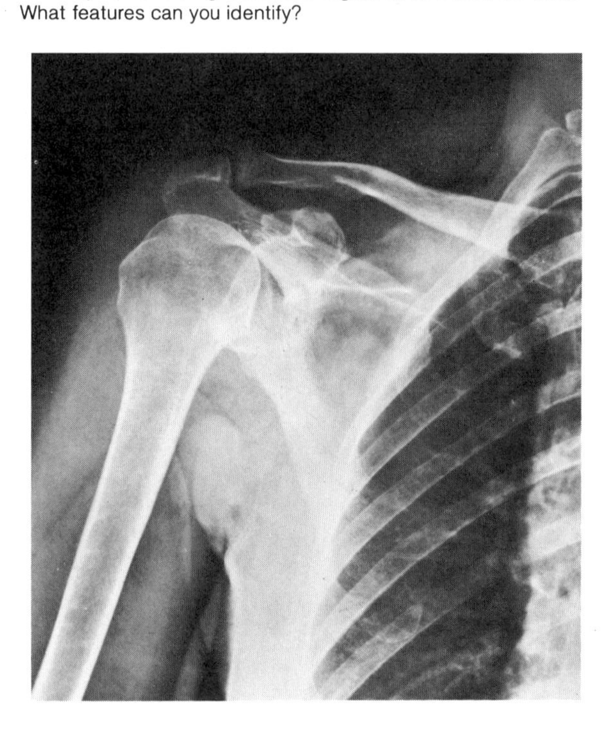

Figure 7.22
(*a*) The posterior surface of the scapula; (*b*) a lateral view showing the glenoid cavity, which articulates with the head of the humerus.

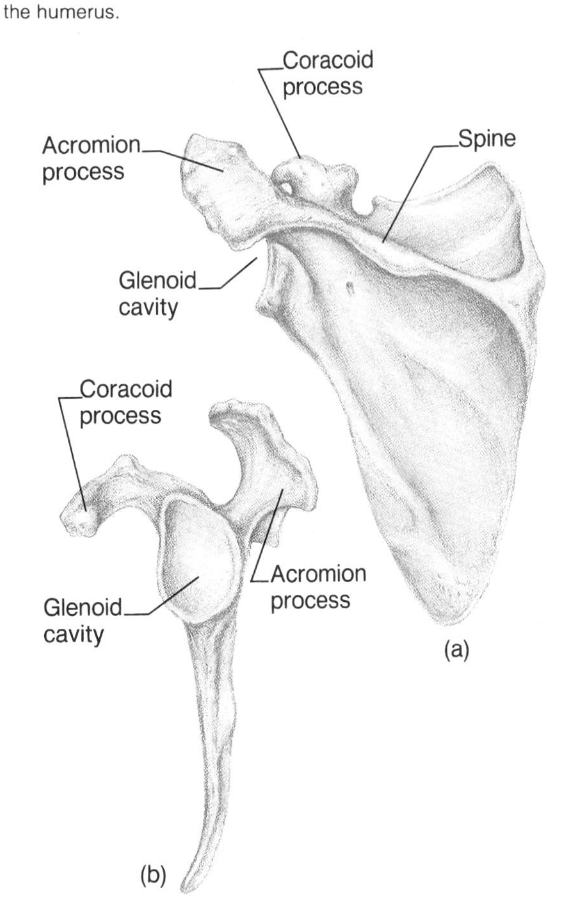

Scapulae

The **scapulae** (skap′u-le) or shoulder blades are broad, somewhat triangular bones located on either side of the upper back (figs. 7.20 and 7.22). The posterior surface of each scapula is divided into unequal portions by a *spine.* This spine leads to two processes—an *acromion* (ah-kro′me-on) *process,* which forms the tip of the shoulder, and a *coracoid* (kor′ah-koid) *process,* which curves forward and downward below the clavicle. The acromion process articulates with a clavicle and provides attachments for muscles of the arm and chest. The coracoid process also provides attachments for arm and chest muscles.

Between the processes is a depression called the *glenoid cavity.* It articulates with the head of the upper arm bone (humerus).

1. What bones form the pectoral girdle?
2. What is the function of the pectoral girdle?

Upper Limb

The bones of the upper limb form the framework of the arm, wrist, palm, and fingers. They also provide attachments for muscles, and they function as levers that move limb parts. These bones include a humerus, a radius, an ulna, and several carpals, metacarpals, and phalanges (fig. 7.7).

Humerus

The **humerus** (hu′mer-us) is a heavy bone that extends from the scapula to the elbow (fig. 7.23). At its upper end, it has a smooth, rounded *head* that fits into the glenoid cavity of the scapula. Just below the head, there are two processes—a *greater tubercle* on the lateral side and a *lesser tubercle* anteriorly. These tubercles provide attachments for muscles that move the arm at the shoulder. Between them is a narrow furrow, the *intertubercular groove.*

A narrow depression along the lower margin of the humerus head, which separates it from the tubercles, is called the *anatomical neck*. Just below the head and tubercles is a region called the *surgical neck,* so named because fractures commonly occur there. Near the middle of the bony shaft on the lateral side, there is a rough V-shaped area, called the *deltoid tuberosity.* It provides an attachment for the muscle (deltoid) that raises the arm horizontally to the side.

At the lower end of the humerus, there are two smooth *condyles* (a lateral capitulum and a medial trochlea), which articulate with the radius on the lateral side and the ulna on the medial side.

Above the condyles on either side are *epicondyles,* which provide attachments for muscles and ligaments of the elbow. Between the epicondyles anteriorly there is a depression, the *coronoid* (kor′o-noid) *fossa,* that receives a process of the ulna (coronoid process) when the elbow is bent. Another depression on the posterior surface, the *olecranon* (o″lek′ra-non) *fossa,* receives an ulnar process (olecranon process) when the arm is straightened at the elbow.

Figure 7.23
(*a*) The posterior surface and (*b*) the anterior surface of the left humerus.

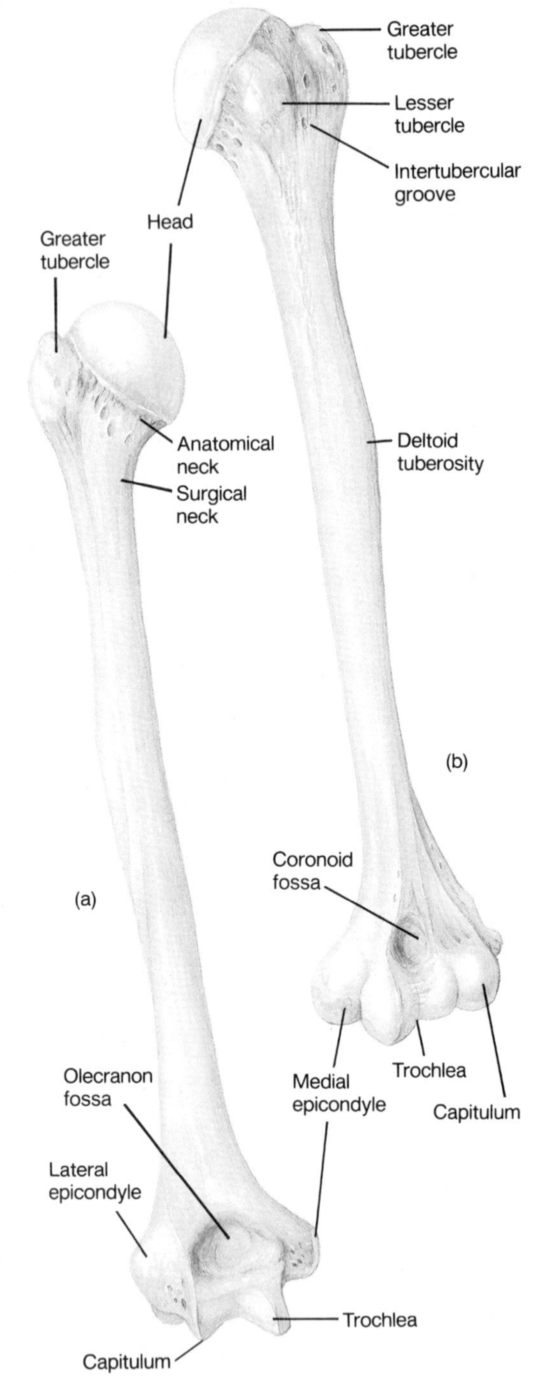

Radius

The **radius** (ra′de-us) extends from the elbow to the wrist and crosses over the ulna when the hand is turned so that the palm faces backward (fig. 7.24).

A thick, disklike *head* at the upper end of the radius articulates with the humerus and a notch of the ulna (radial notch). This arrangement allows the radius to rotate freely.

On the radial shaft, just below the head, is a process called the *radial tuberosity.* It serves as an attachment for a muscle (biceps brachii). At the lower end of the radius, a lateral *styloid* (sti′loid) *process* provides attachments for ligaments of the wrist.

Figure 7.24
The head of the radius articulates with the radial notch of the ulna, and the head of the ulna articulates with the ulnar notch of the radius.

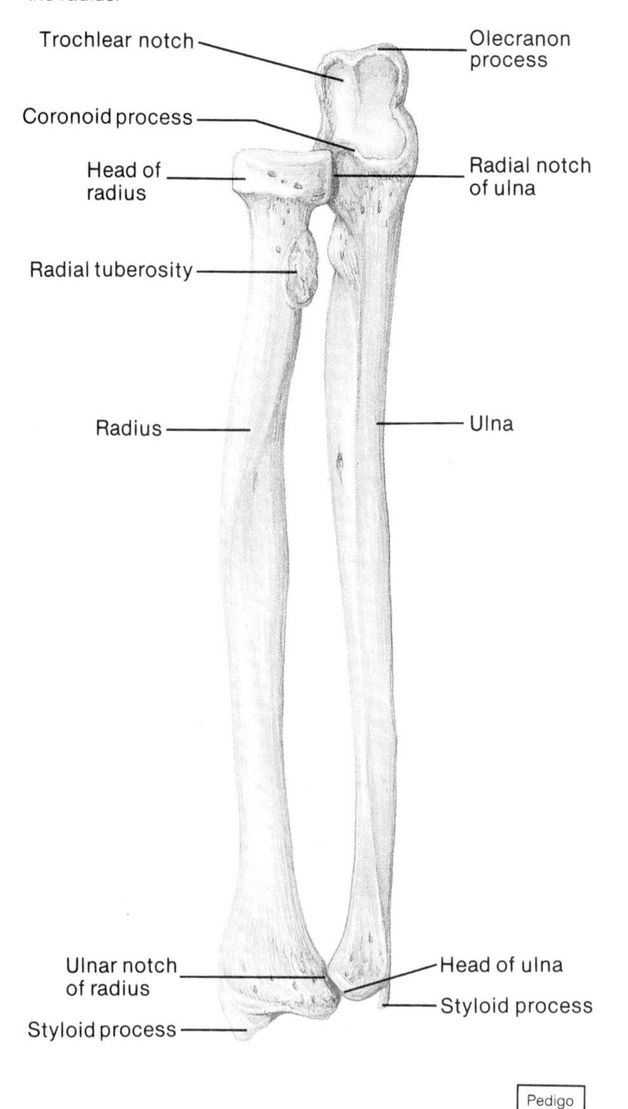

Ulna

The **ulna** (ul′nah) overlaps the end of the humerus posteriorly (fig. 7.24). At its upper end, the ulna has a wrenchlike opening, the *trochlear notch,* which articulates with the humerus. There are two processes, one on either side of this notch, the *olecranon process* and the *coronoid process,* which provide attachments for muscles.

At the lower end, the knoblike *head* of the ulna articulates with a notch of the radius (ulnar notch) laterally and with a disk of fibrocartilage inferiorly. This disk, in turn, joins a wrist bone (triangular). A medial *styloid process* at the distal end of the ulna provides attachments for ligaments of the wrist.

Figure 7.25
The left hand viewed from the back.

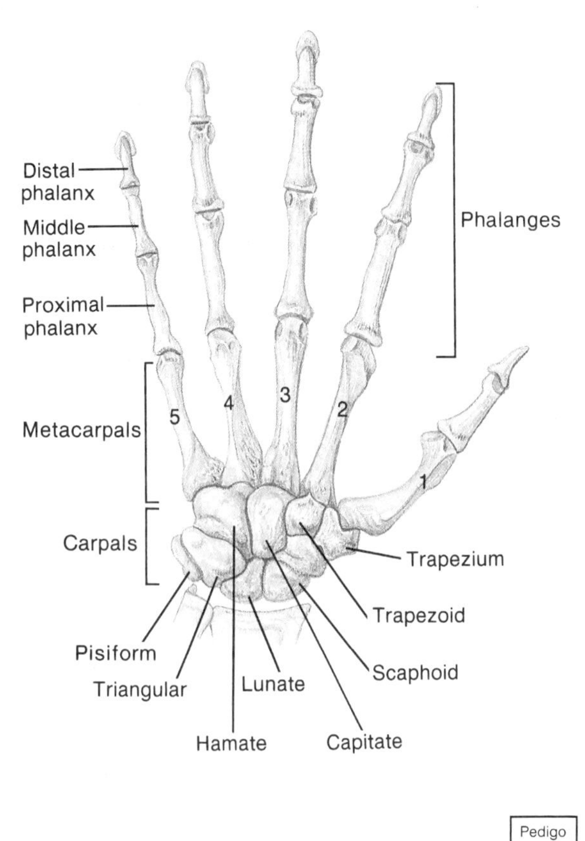

Hand

The **hand** is composed of a wrist, a palm, and five fingers (fig. 7.25). The skeleton of the wrist consists of eight small **carpal** (kar′pel) **bones,** which are firmly joined in two rows of four bones each. The resulting compact mass is called a *carpus* (kar′pus). The carpus articulates with the radius and with the fibrocartilaginous disk on the ulnar side. Its distal surface articulates with the metacarpal bones. The individual bones of the carpus are named in figure 7.25.

Five **metacarpal** (met″ah-kar′pal) **bones,** one in line with each finger, form the framework of the palm. These bones are cylindrical, with rounded distal ends that form the knuckles on a clenched fist. They are numbered 1 to 5, beginning with the metacarpal of the thumb (fig. 7.25). The metacarpals articulate proximally with the carpals and distally with the phalanges.

The **phalanges** (fah-lan′jez) are the bones of the fingers. There are three in each finger—a proximal, a middle, and a distal phalanx—and two in the thumb (it lacks a middle phalanx).

Figure 7.26
Anterior views of the (*a*) male pelvis and (*b*) female pelvis.

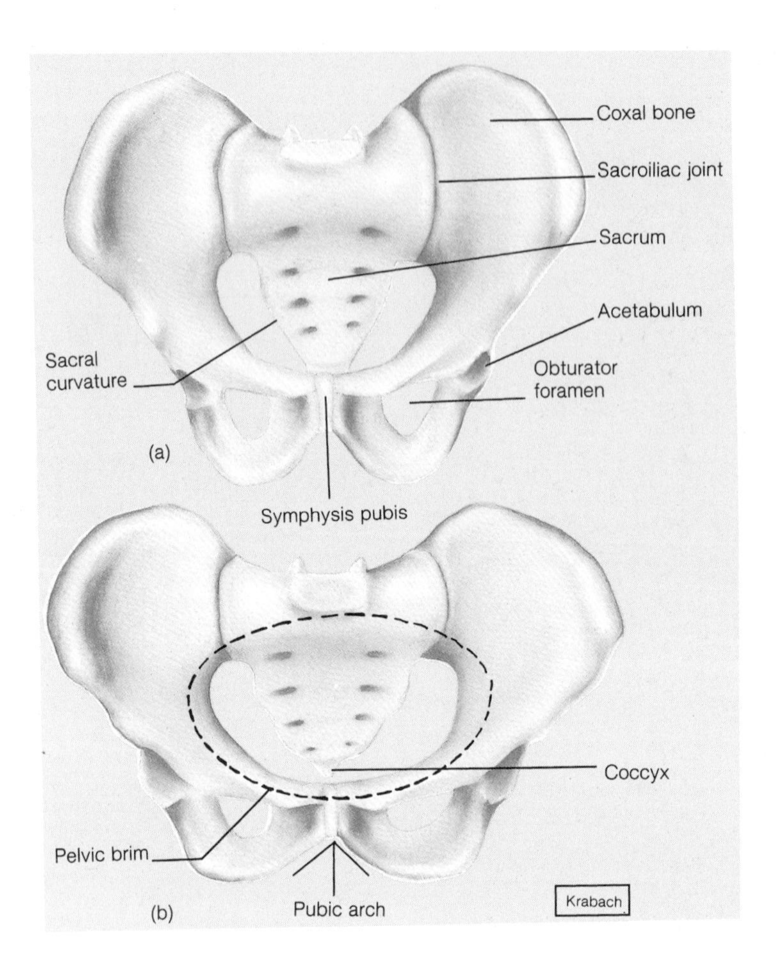

1. Locate and name each bone of the upper limb.
2. Explain how the bones of the upper limb articulate with one another.

Pelvic Girdle

The **pelvic girdle** consists of two coxal bones, which articulate with each other anteriorly and with the sacrum posteriorly (fig. 7.26). The sacrum, coccyx, and pelvic girdle together form the ringlike *pelvis,* which provides support for the trunk of the body and attachments for the legs.

Coxal Bones

Each **coxal** (kok′sal) **bone** (os coxa) develops from three parts—an ilium, an ischium, and a pubis (figs. 7.26 and 7.27). These parts fuse in the region of a cup-shaped cavity, called the *acetabulum* (as″e-tab′u-lum). This depression is on the lateral surface of the hipbone, and it receives the rounded head of the femur (thighbone).

The **ilium** (il′e-um), which is the largest and uppermost portion of the coxal bone, flares outward to form the prominence of the hip. The margin of this prominence is called the *iliac crest.*

Posteriorly, the ilium joins the sacrum at the *sacroiliac* (sa″kro-il′e-ak) *joint.* A projection of the ilium, the *anterior superior iliac spine,* which can be felt lateral to the groin, provides attachments for ligaments and muscles.

The **ischium** (is′ke-um), which forms the lowest portion of the coxal bone, is L-shaped with its angle, the *ischial tuberosity,* pointing posteriorly and downward. This tuberosity has a rough surface, which provides attachments for ligaments and leg muscles. It also supports the weight of the body when a person is sitting. Above the ischial tuberosity, near the junction of the ilium and ischium, is a sharp projection, called the *ischial spine.*

The **pubis** (pu′bis) constitutes the anterior portion of the coxal bone. The two pubic bones come together in the midline to form a joint called the *symphysis pubis* (sim′fi-sis pu′bis). The angle formed

Figure 7.27
The lateral surface of the right coxal bone.

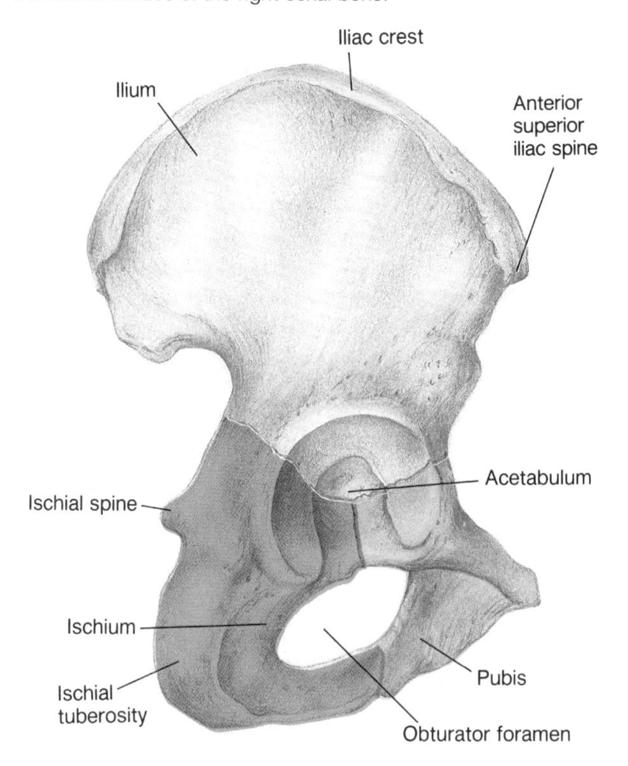

Figure 7.28
What features can you identify in this X-ray film of the pelvic girdle?

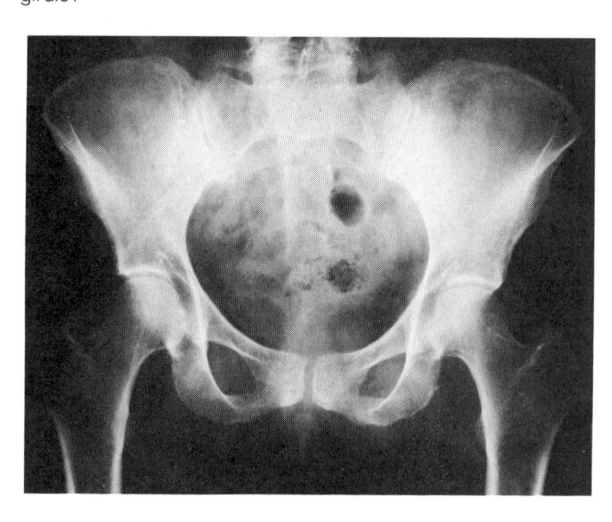

by these bones below the symphysis is the *pubic arch* (fig. 7.26).

A portion of each pubis passes posteriorly and downward to join an ischium. Between the bodies of these bones on either side there is a large opening, the *obturator foramen,* which is the largest foramen in the skeleton (figs. 7.27 and 7.28).

If a line were drawn along each side of the pelvis from the upper anterior margin of the sacrum (sacral promontory) downward and anteriorly to the upper margin of the symphysis pubis, it would mark the *pelvic brim* (linea terminalis). This margin separates the lower, or lesser (true) pelvis from the upper, or greater (false) pelvis (fig. 7.26).

Some differences in the male and female pelves and other skeletal structures are summarized in chart 7.2.

1. Locate and name each pelvic bone.
2. Describe a coxal bone.

Lower Limb

The bones of the lower limb form the frameworks of the leg, ankle, instep, and toes. They include a femur, a tibia, a fibula, and several tarsals, metatarsals, and phalanges (fig. 7.7).

Chart 7.2 Differences between the male and female skeletons

Part	Differences
Skull	Female skull is relatively smaller and lighter, and its muscular attachments are less conspicuous. The female forehead is longer vertically, the facial area is rounder, the jaw is smaller, and the mastoid process is less prominent than those of a male
Pelvis	Female pelvic bones are lighter, thinner, and have less obvious muscular attachments. The obturator foramina and acetabula are smaller and farther apart than those of a male
Pelvic cavity	Female pelvic cavity is wider in all diameters and is shorter, roomier, and less funnel-shaped. The distances between the ischial spines and ischial tuberosities are greater than in a male
Sacrum	Female sacrum is relatively wider, the first sacral vertebra projects forward to a lesser degree, and the sacral curvature is bent more sharply posteriorly than in a male
Coccyx	Female coccyx is more movable than that of a male

Femur

The **femur** (fe′mur) is the longest bone in the body and extends from the hip to the knee (fig. 7.29). A large, rounded *head* at its upper end projects medially into the acetabulum of the coxal bone. On the head, a pit, called the *fovea capitis,* marks the attachment of a ligament (ligamentum capitis). Just below the head, there are a constriction, or *neck,* and two large processes—an upper, lateral *greater trochanter* and a lower, medial *lesser trochanter.* These processes provide attachments for muscles of the legs and buttocks.

Figure 7.29
(*a*) The anterior surface and (*b*) the posterior surface of the left femur.

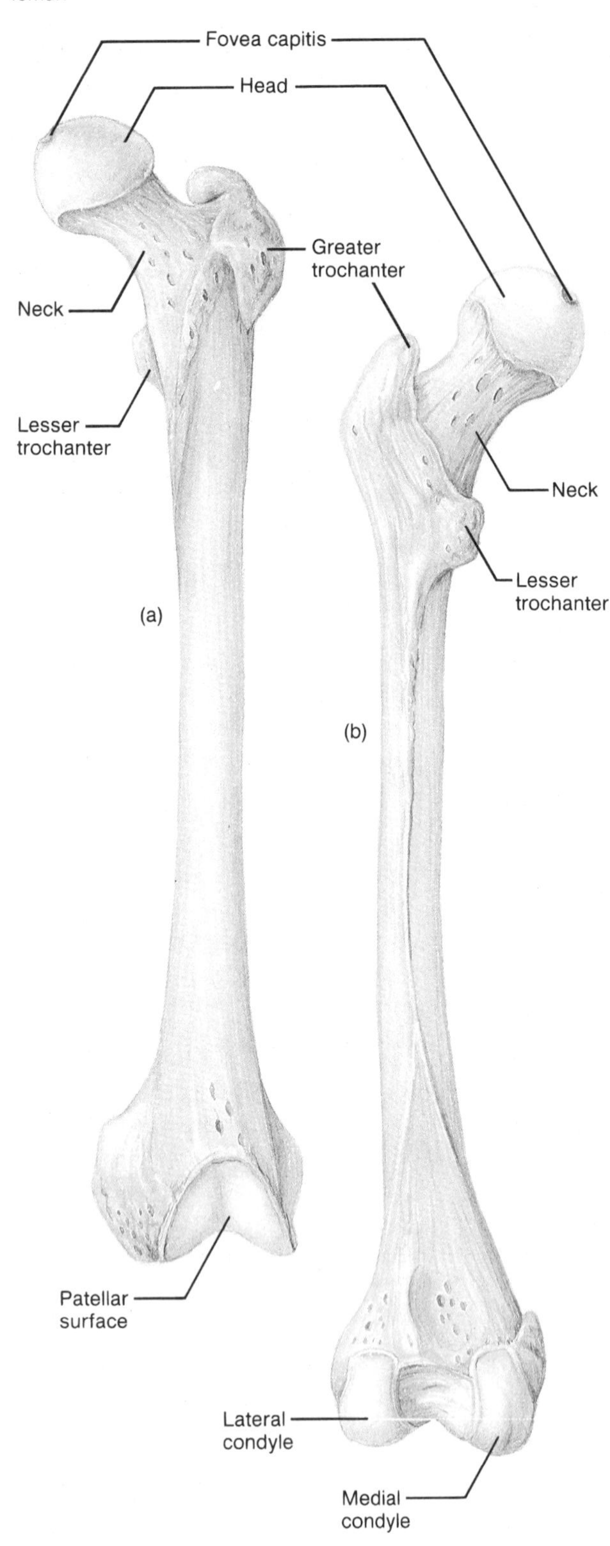

At the lower end of the femur, two rounded processes, the *lateral* and *medial condyles,* articulate with the tibia of the lower leg. A **patella** (pah-tel'ah), or kneecap, also articulates with the femur on its distal anterior surface (fig. 7.7). It is located in a tendon that passes anteriorly over the knee.

Hip fracture is one of the more serious causes of hospitalization among elderly persons. The site of hip fracture is most commonly the neck of a femur or the region between the trochanters of a femur. Such a fracture is usually caused by a fall or by a bone disease in which calcified bone tissue has been lost excessively.

Tibia

The **tibia** (tib'e-ah) or shinbone is the larger of the two lower leg bones and is located on the medial side (fig. 7.30). Its upper end is expanded into the *medial* and *lateral condyles,* which have concave surfaces and articulate with the condyles of the femur. Below the condyles, on the anterior surface, is a process called the *tibial tuberosity,* which provides an attachment for the *patellar ligament*—a continuation of the patella-bearing tendon.

At its lower end, the tibia expands to form a prominence on the inner ankle called the *medial malleolus* (mah-le'o-lus), which serves as an attachment for ligaments. On its lateral side is a depression that articulates with the fibula. The inferior surface of the tibia's distal end articulates with a large bone (the talus) in the foot.

During a life-threatening medical emergency, if it is necessary to administer fluids or drugs to a young child, red bone marrow is sometimes used as an entrance into the vascular system. In this procedure, called *intraosseous infusion,* substances are introduced through a hollow needle that has been inserted into the marrow cavity of a long bone, such as the tibia. Once inside the marrow, the substances quickly enter the general circulation.

Fibula

The **fibula** (fib'u-lah) (fig. 7.30) is a long, slender bone located on the lateral side of the tibia. Its ends are slightly enlarged into an upper *head* and a lower *lateral malleolus.* The head articulates with the tibia just below the lateral condyle; however, it does not enter into the knee joint and does not bear any body weight. The lateral malleolus articulates with the ankle and forms a prominence on the lateral side.

Figure 7.30
Bones of the left lower leg viewed from the front.

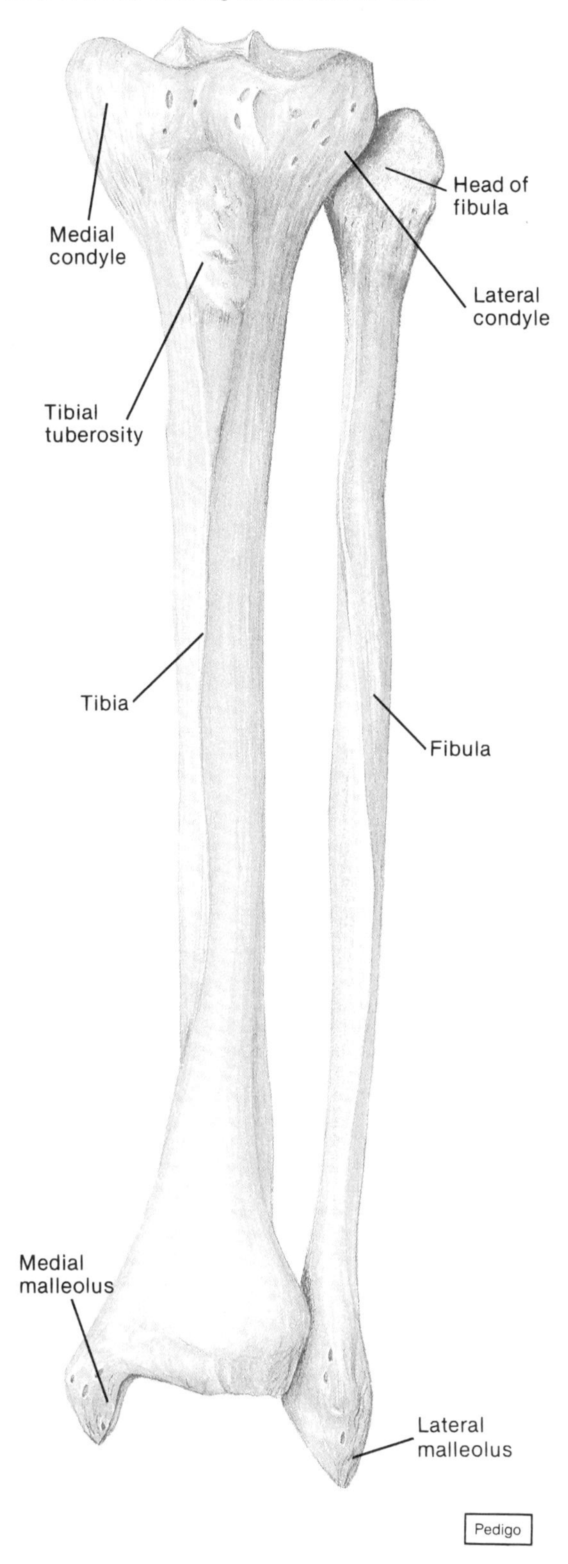

Foot

The **foot** consists of an ankle, an instep, and five toes. The ankle is composed of seven **tarsal** (tahr-sal) **bones,** forming a group called the *tarsus* (tahr′sus) (figs. 7.31 and 7.32). These bones are arranged so that one of them, the **talus** (ta′lus), can move freely where it joins the tibia and fibula. The remaining tarsal bones are bound firmly together, forming a mass on which the talus rests.

The individual bones of the tarsus are named in figure 7.32.

The largest of the ankle bones, the **calcaneus** (kal-ka′ne-us), is located below the talus, where it projects backward to form the base of the heel. The calcaneus helps support the weight of the body and provides an attachment for muscles that move the foot.

The instep consists of five, elongated **metatarsal** (met″ah-tar′sal) **bones,** which articulate with the tarsus. They are numbered 1 to 5, beginning on the medial side (fig. 7.32). The heads at the distal ends of these bones form the ball of the foot. The tarsals and metatarsals are arranged and bound by ligaments to form the arches of the foot. A longitudinal arch extends from the heel to the toe, and a transverse arch stretches across the foot. These arches provide a stable, springy base for the body.

The **phalanges** of the toes are similar to those of the fingers. They are in line with the metatarsals and articulate with them. There are three phalanges in each toe—a proximal, a middle, and a distal phalanx—except the great toe which lacks a middle phalanx.

1. Locate and name each bone of the lower limb.
2. Explain how the bones of the lower limb articulate with one another.
3. Describe how the foot is adapted to support the body.

Joints

Joints (articulations) are functional junctions between bones. Although they vary considerably in structure, they can be classified according to the amount of movement they make possible. On this basis, three general groups can be identified—immovable joints, slightly movable joints, and freely movable joints.

Immovable Joints

Immovable joints (synarthroses) occur between bones that come into close contact with one another. The bones at such joints are separated by a thin layer of fibrous connective tissue or cartilage, as in the case of a *suture* between a pair of flat bones of the cranium (fig. 7.33). No active movement takes place at an immovable joint.

Figure 7.31
The talus moves freely where it articulates with the tibia and fibula.

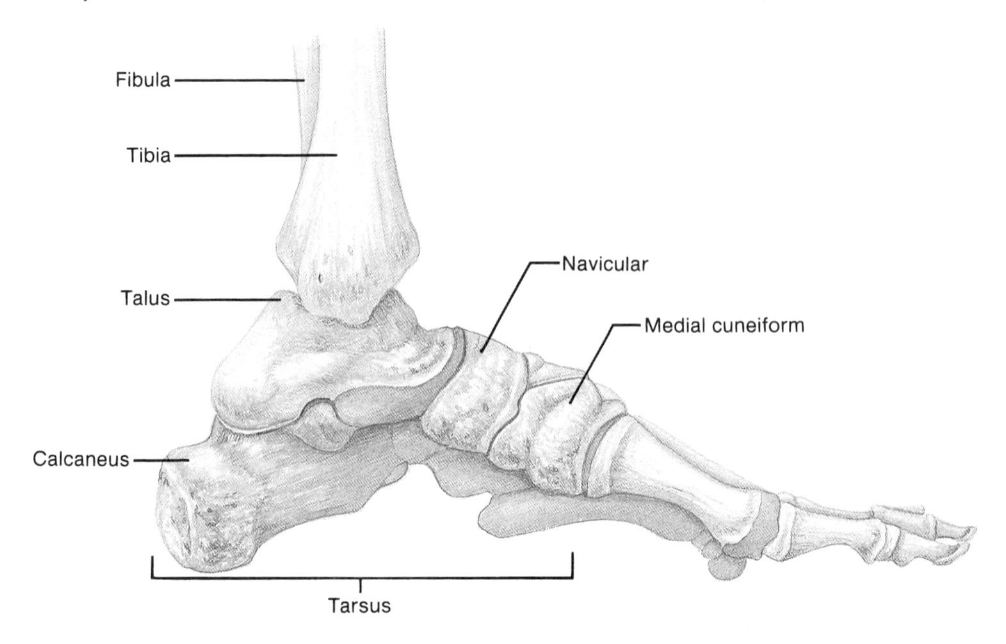

Figure 7.32
The left foot viewed from above.

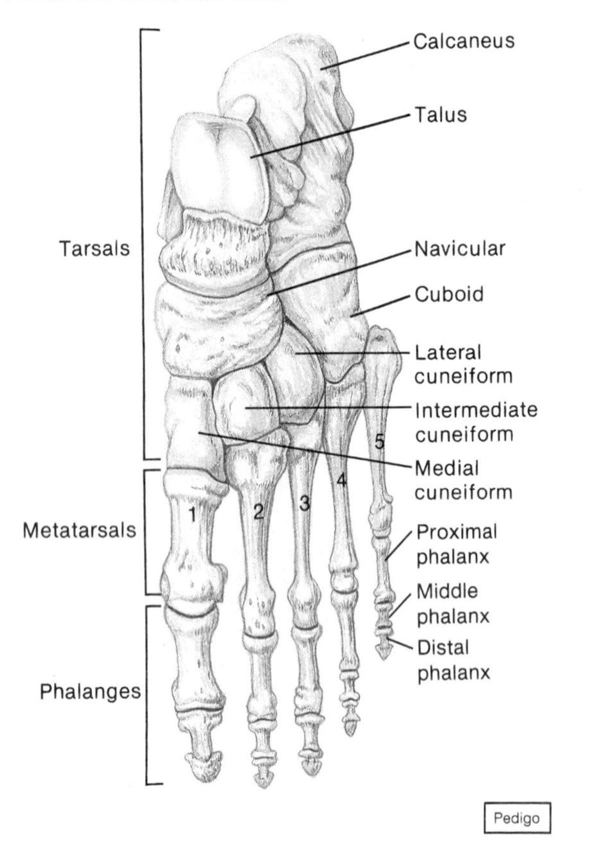

Figure 7.33
(*a*) The joints between the bones of the cranium are immovable and are called sutures; (*b*) the bones at a suture are separated by a thin layer of connective tissue.

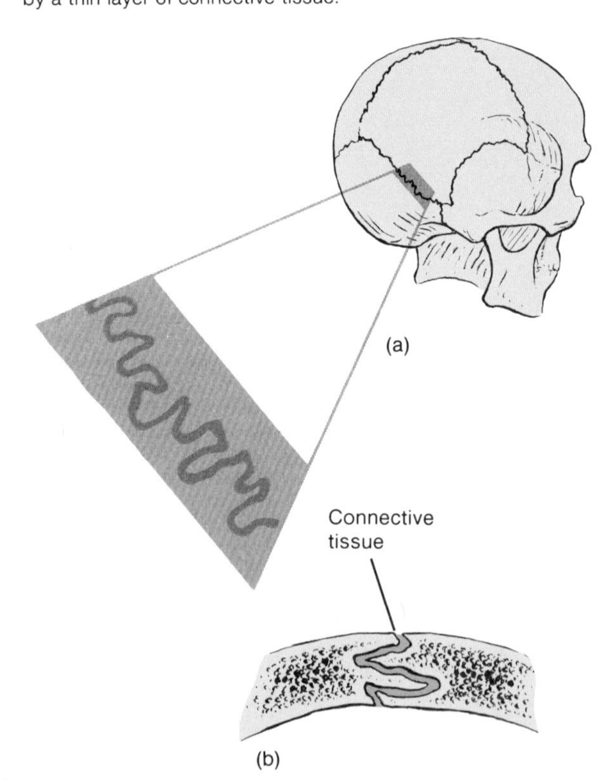

Figure 7.34
The generalized structure of a synovial joint. What is the function of the synovial fluid within this type of joint?

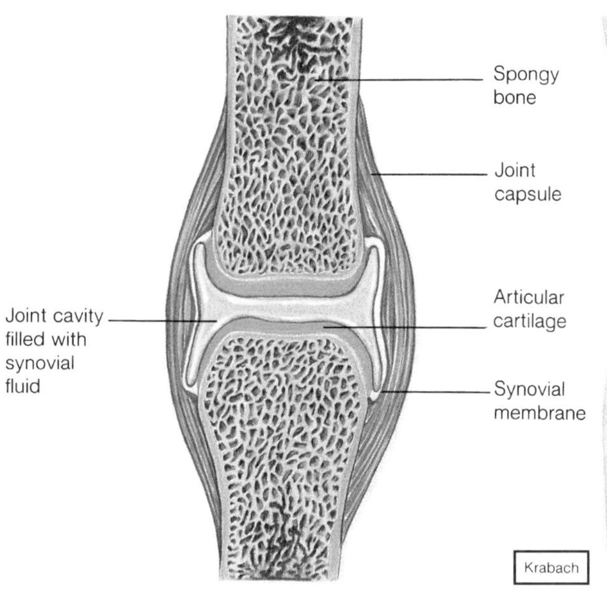

Figure 7.35
The knee joint contains semilunar cartilages. (Note the bursae associated with this joint.)

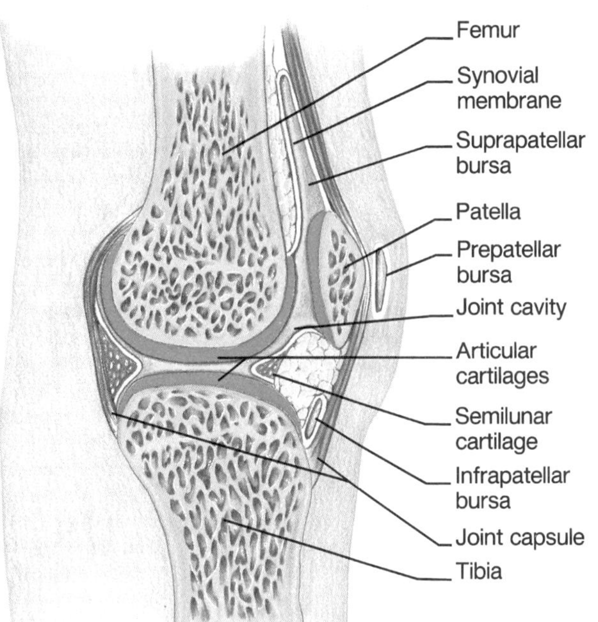

Slightly Movable Joints

The bones of **slightly movable joints** (amphiarthroses) are connected by disks of fibrocartilage or by ligaments. The vertebrae of the vertebral column, for instance, are separated by joints of this type. The articulating surfaces of the vertebrae are covered by thin layers of hyaline cartilage. This cartilage, in turn, is attached to the intervertebral disk, which separates adjacent vertebral bodies.

Each intervertebral disk is composed of a band of fibrous fibrocartilage surrounding a pulpy or gelatinous core (nucleus pulposus). The disk acts as a shock absorber and helps to equalize pressures between adjacent vertebral bodies during body movements (fig. 7.15).

Due to the slight flexibility of the disks, these joints allow a limited amount of movement, as when the back is bent forward or to the side or is twisted. Other examples of slightly movable joints include the symphysis pubis, the sacroiliac joint, and the joint in the lower leg between the distal ends of the tibia and fibula.

Freely Movable Joints

Most joints within the skeletal system are freely movable, and they have more complex structures than immovable or slightly movable ones.

The ends of bones at **freely movable joints** (diarthroses) are covered with hyaline cartilage (articular cartilage) and held together by a surrounding, tubelike capsule of dense fibrous tissue (fig. 7.34). This *joint capsule* is composed of an outer layer of ligaments and an inner lining of *synovial membrane,* which secretes synovial fluid. For this reason, freely movable joints are often called *synovial joints.* Synovial fluid has a consistency somewhat like egg white, and it acts as a joint lubricant.

Some freely movable joints have flattened, shock-absorbing pads of fibrocartilage between the articulating surfaces of the bones. The knee joint, for example, contains pads called *semilunar cartilages* (menisci) (fig. 7.35). Such joints may also have closed, fluid-filled sacs, called **bursae,** associated with them. Bursae are lined with synovial membrane, which may be continuous with the synovial membranes of nearby joint cavities.

Bursae are commonly located between the skin and underlying bony prominences, as in the case of the patella of the knee or the olecranon process of the elbow. They aid in the movement of tendons that pass over these bony parts or over other tendons. Figures 7.35 and 7.36 show some of the bursae associated with the knee and shoulder.

Based on the shapes of their parts and the movements they permit, joints can be classified as follows:

1. **Ball-and-socket joints.** A ball-and-socket joint consists of a bone with a ball-shaped head that articulates with a cup-shaped socket of another bone. Such a joint allows for a wider range of motion than does any other kind. Movements in all planes, as well as rotational movement

Figure 7.36
The shoulder joint allows movements in all directions.

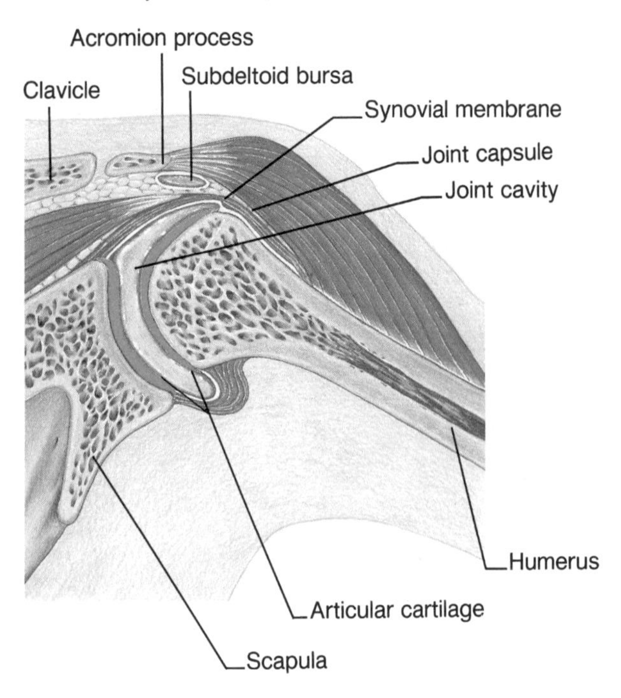

around a central axis, are possible. The shoulder and hip contain joints of this type (figs. 7.36 and 7.37).

2. **Condyloid joints.** In a condyloid joint, an oval-shaped condyle of one bone fits into an elliptical cavity of another bone, as in the case of the joints between the metacarpals and phalanges (fig. 7.25). This type of joint allows a variety of movements in different planes; rotational movements, however, are not possible.
3. **Gliding joints.** The articulating surfaces of gliding joints are nearly flat or only slightly curved. Such joints are found between some wrist bones (fig. 7.25), some ankle bones, and between the articular processes of adjacent vertebrae. They allow sliding and twisting movements.
4. **Hinge joints.** In a hinge joint, the convex surface of one bone fits into the concave surface of another, as in the case of the elbow (fig. 7.38) and the joints of the phalanges. This type of joint allows movement in one plane only, like the motion of a single-hinged door.
5. **Pivot joints.** In a pivot joint, a cylindrical surface of one bone rotates within a ring formed of bone and fibrous tissue. The movement at such a joint is limited to rotation about a central axis. The joint between the proximal ends of the radius and the ulna is of this type (fig. 7.24).
6. **Saddle joints.** A saddle joint is formed between bones whose articulating surfaces have both concave and convex regions. The surfaces of one bone fit the complementary surfaces of the other. This arrangement allows a variety of movements, as in the case of the joint between a carpal (trapezium) and the metacarpal of the thumb (fig. 7.25).

Figure 7.37
The hip is a ball-and-socket joint.

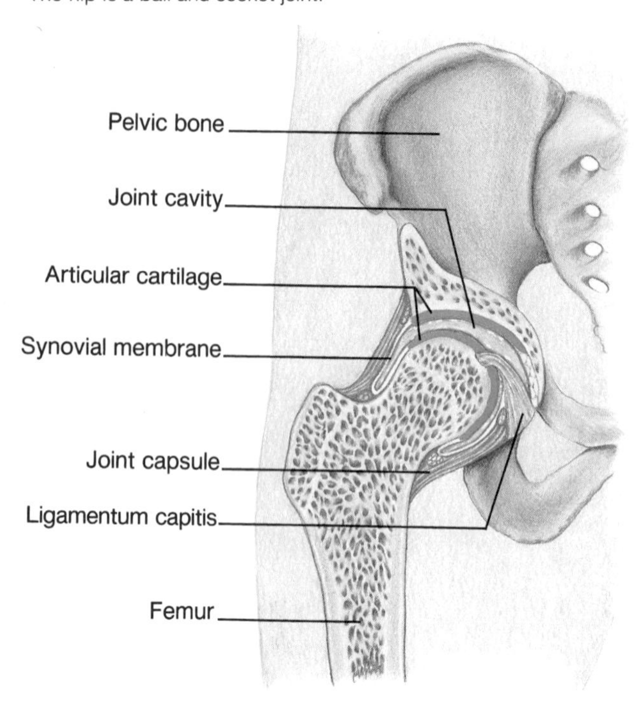

Figure 7.38
The elbow is a hinge joint.

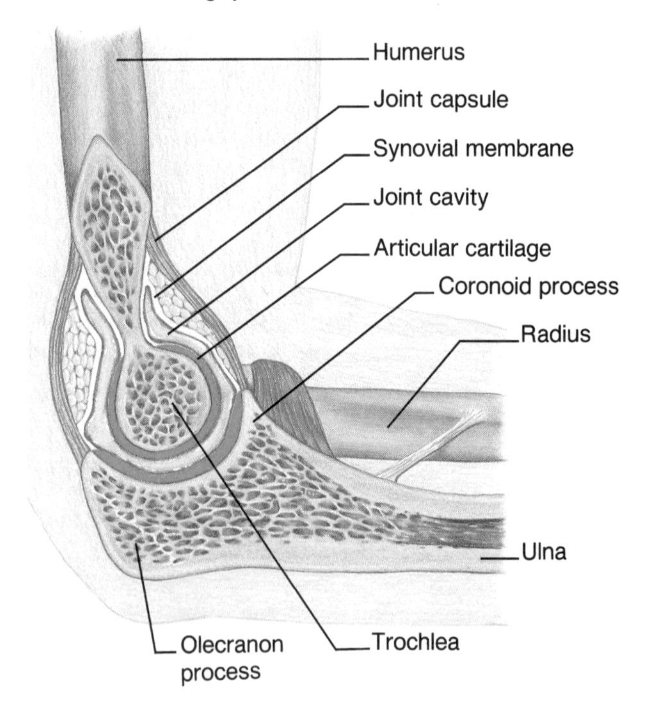

Chart 7.3 Types of joints			
Type	**Description**	**Possible movements**	**Example**
Immovable	Articulating bones in close contact and separated by a thin layer of fibrous tissue or cartilage	No active movement	Suture between bones of the cranium
Slightly movable	Articulating bones separated by disks of fibrocartilage	Limited movements as when the back is bent or twisted	Joints between the vertebrae, symphysis pubis, sacroiliac joint
Freely movable	Articulating bones surrounded by a joint capsule of ligaments and synovial membranes; ends of articulating bones covered by hyaline cartilage and separated by synovial fluid		
1. Ball-and-socket	Ball-shaped head of one bone articulates with the cup-shaped socket of another	Movements in all planes and rotation	Shoulder, hip
2. Condyloid	Oval-shaped condyle of one bone articulates with an elliptical cavity of another	Variety of movements in different planes, but no rotation	Joints between the metacarpals and phalanges
3. Gliding	Articulating surfaces are nearly flat or slightly curved	Sliding or twisting	Joints between various bones of the wrist and ankle
4. Hinge	Convex surface of one bone articulates with the concave surface of another	Flexion and extension	Elbow, joints of phalanges
5. Pivot	Cylindrical surface of one bone articulates with a ring of bone and fibrous tissue	Rotation	Joint between the proximal ends of radius and ulna
6. Saddle	Articulating surfaces have both concave and convex regions; the surface of one bone fits the complementary surface of another	Variety of movements	Joint between the carpal and metacarpal of the thumb

Chart 7.3 summarizes the characteristics of the various types of joints.

Arthritis is a condition that causes inflamed, swollen, and painful joints. Although there are several different types of arthritis, the most common forms are *rheumatoid arthritis* and *osteoarthritis.*

In rheumatoid arthritis, which is the most painful and crippling of the arthritic diseases, the synovial membrane of a freely movable joint becomes inflamed and grows thicker. This change is usually followed by damage to the articular cartilages on the ends of the bones and an invasion of the joint by fibrous tissues. These fibrous tissues increasingly interfere with joint movements, and, in time, the tissues may become ossified so that the articulating bones are fused together. The cause of rheumatoid arthritis is unknown.

Osteoarthritis is a degenerative disease that occurs as a result of aging, as well as other factors, and affects a large percentage of persons over sixty years of age. In this condition, the articular cartilages soften and disintegrate gradually so that the articular surfaces become roughened. Consequently, the joints are sore and less movement is possible. Osteoarthritis is most likely to affect joints that have received the greatest use over the years, such as those in the knees and the lower regions of the vertebral column.

Types of Joint Movements

Movements at synovial joints are produced by the actions of skeletal muscles. Typically, one end of a muscle is attached to a relatively immovable or fixed part on one side of a joint, and the other end of the muscle is fastened to a movable part on the other side. When the muscle contracts, fibers within the muscle pull its movable end (insertion) toward its fixed end (origin), and a movement occurs at the joint.

The following terms are used to describe various movements at joints (figs. 7.39, 7.40, and 7.41).

flexion (flek′shun)—bending a joint so that the angle between its parts is decreased and the parts come closer together (bending the leg at the knee).

extension (ek-sten′shun)—straightening a joint so that the angle between its parts is increased and the parts move farther apart (straightening the leg at the knee).

dorsiflexion (dor″sĭ-flek′shun)—flexing the foot at the ankle (bending the foot upward).

plantar flexion (plan′tar flek′shun)—extending the foot at the ankle (bending the foot downward).

hyperextension (hi″per-ek-sten′shun)—excessive extension of the parts at a joint, beyond the anatomical position (bending the head back beyond the upright position).

Figure 7.39
Joint movements illustrating flexion, extension, hyperextension, dorsiflexion, plantar flexion, abduction, and adduction.

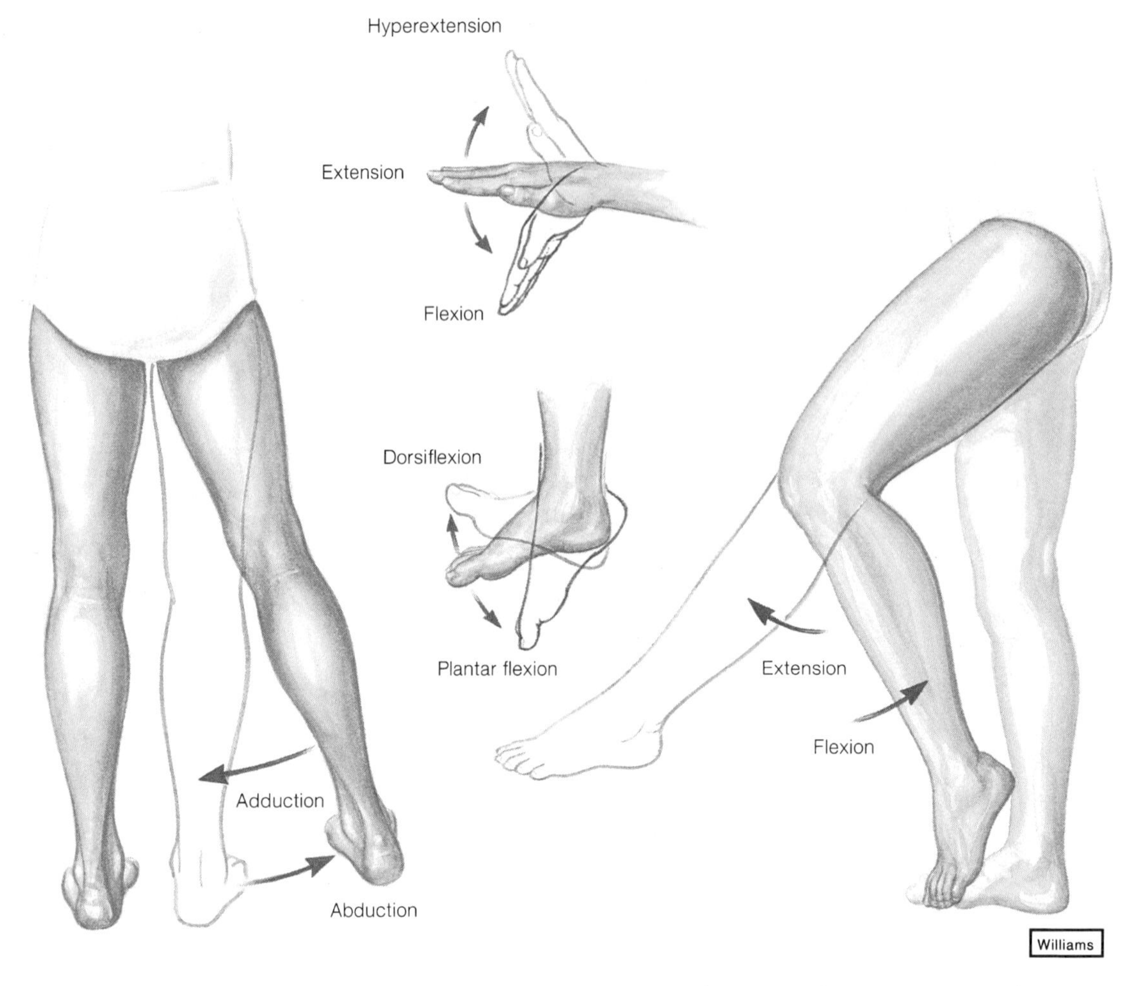

abduction (ab-duk'shun)—moving a part away from the midline (lifting the arm horizontally to form a right angle with the side of the body).
adduction (ah-duk'shun)—moving a part toward the midline (returning the arm from the horizontal position to the side of the body).
rotation (ro-ta'shun)—moving a part around an axis (twisting the head from side to side).
circumduction (ser"kum-duk'shun)—moving a part so that its end follows a circular path (moving the finger in a circular motion without moving the hand).
pronation (pro-na'shun)—turning the hand so the palm is downward.
supination (soo"pĭ-na'shun)—turning the hand so the palm is upward.
eversion (e-ver'zhun)—turning the foot so the sole is outward.
inversion (in-ver'zhun)—turning the foot so the sole is inward.
retraction (re-trak'shun)—moving a part backward (pulling the chin backward).
protraction (pro-trak'shun)—moving a part forward (thrusting the chin forward).
elevation (el"ĕ-va'shun)—raising a part (shrugging the shoulders).
depression (de-presh'un)—lowering a part (drooping the shoulders).

1. Describe the characteristics of the three major types of joints.
2. List six different types of freely movable joints.
3. Describe the movements each type of joint makes possible.
4. What terms are used to describe various movements at joints?

Clinical Terms Related to the Skeletal System

acromegaly (ak"ro-meg'ah-le)—a condition caused by an overproduction of growth hormone in adults and characterized by an abnormal enlargement of the facial features, hands, and feet.
ankylosis (ang"kĭ-lo'sis)—an abnormal stiffness of a joint or fusion of bones at a joint, often due to damage of the joint membranes from chronic rheumatoid arthritis.

Figure 7.40
Joint movements illustrating rotation, circumduction, pronation, and supination.

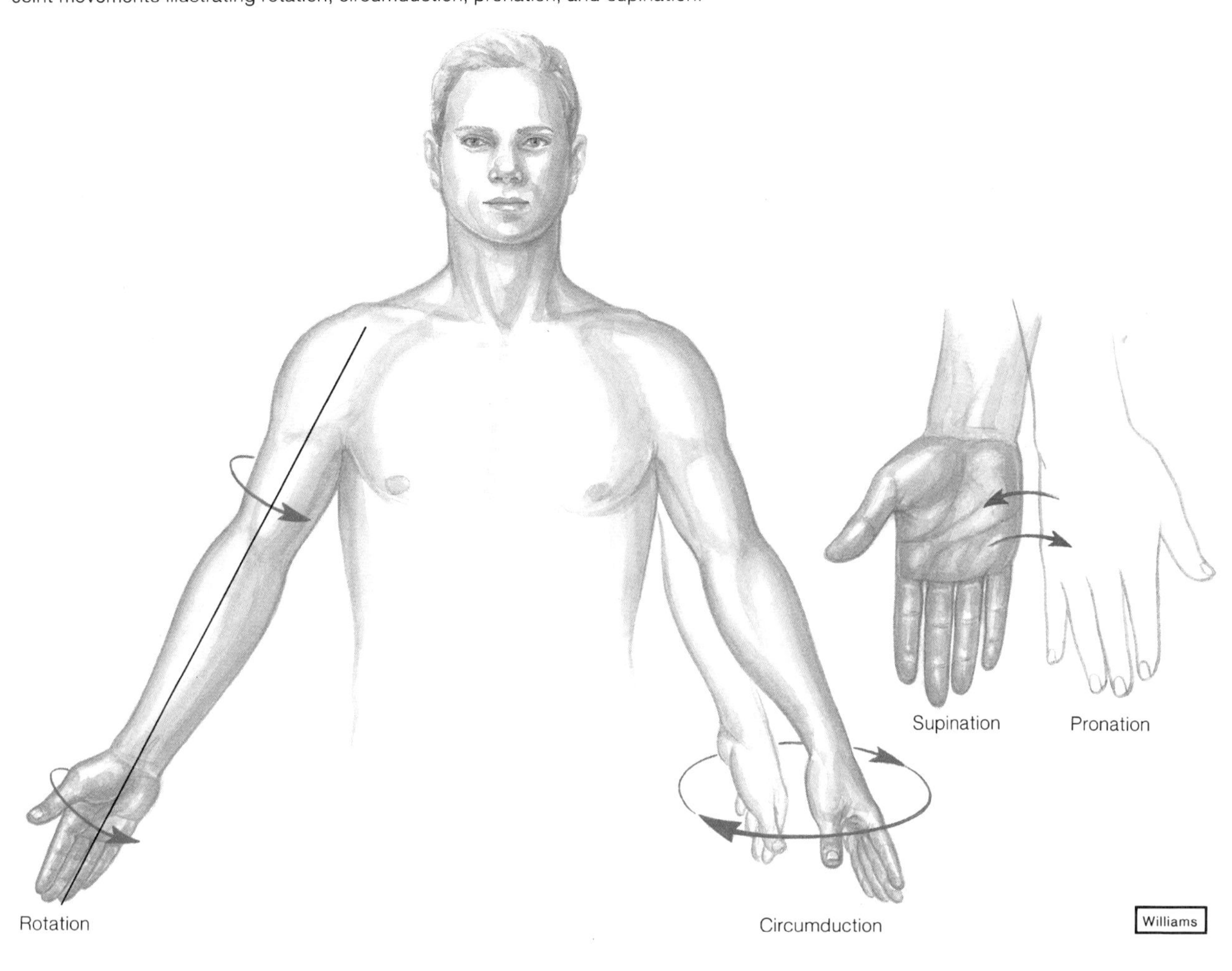

Figure 7.41
Joint movements illustrating eversion, inversion, retraction, protraction, elevation, and depression.

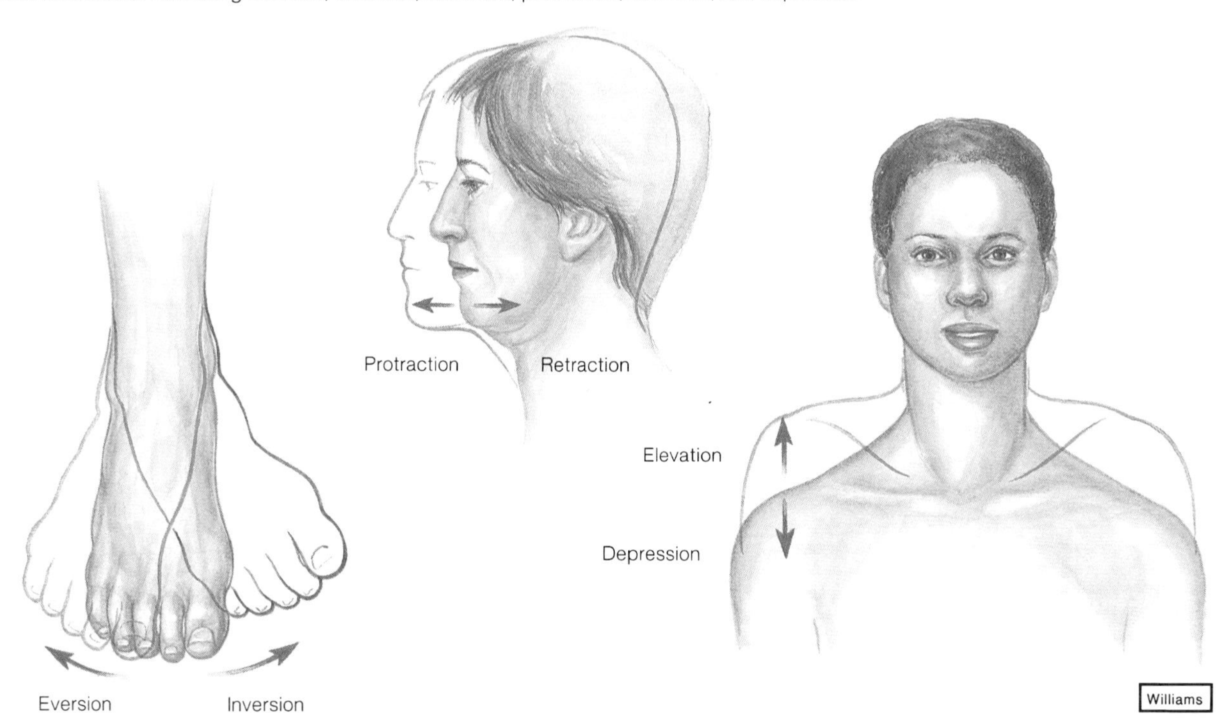

arthralgia (ar-thral'je-ah)—pain in a joint.
arthrocentesis (ar"thro-sen-te'sis)—the puncture of and removal of fluid from a joint cavity.
arthrodesis (ar"thro-de'sis)—surgery performed to fuse the bones at a joint.
arthroplasty (ar'thro-plas"te)—surgery performed to make a joint movable.
arthroscopy (ar-thros'ko-pe)—a procedure by which the interior of a joint is examined using a tubular instrument called an arthroscope.
Colles' fracture (kol'ēz frak'ture)—a fracture at the distal end of the radius in which the smaller fragment is displaced posteriorly.
epiphysiolysis (ep"ĭ-fiz"e-ol'ĭ-sis)—a separation or loosening of the epiphysis from the diaphysis of a bone.
gout (gowt)—a metabolic disease in which excessive uric acid in the blood may be deposited in the joints, causing them to become inflamed, swollen, and painful.
hemarthrosis (hem"ar-thro'sis)—blood in a joint cavity.
laminectomy (lam"ĭ-nek'to-me)—the surgical removal of the posterior arch of a vertebra, usually to relieve the symptoms of a ruptured intervertebral disk.
lumbago (lum-ba'go)—a dull ache in the lumbar region of the back.
orthopedics (or"tho-pe'diks)—the science of the prevention, diagnosis, and treatment of diseases and abnormalities involving the skeletal and muscular systems.
ostealgia (os"te-al'je-ah)—pain in a bone.
ostectomy (os-tek'to-me)—the surgical removal of a bone.
osteitis (os"te-i'tis)—inflammation of bone tissue.
osteochondritis (os"te-o-kon-dri'tis)—inflammation of bone and cartilage tissues.
osteogenesis (os"te-o-jen'ĕ-sis)—the development of bone.
osteogenesis imperfecta (os"te-o-jen'ĕ-sis im-per-fek'ta)—a congenital condition characterized by the development of deformed and abnormally brittle bones.
osteoma (os"te-o'mah)—a tumor composed of bone tissue.
osteomalacia (os"te-o-mah-la'she-ah)—a softening of adult bone due to a disorder in calcium and phosphorus metabolism, usually caused by a deficiency of vitamin D.
osteomyelitis (os"te-o-mi"ĕ-li'tis)—inflammation of bone caused by the action of bacteria or fungi.
osteonecrosis (os"te-o-ne-kro'sis)—the death of bone tissue. This condition occurs most commonly in the head of the femur in elderly persons and may be due to obstructions in the arteries that supply the bone.
osteopathology (os"te-o-pah-thol'o-je)—the study of bone diseases.
osteotomy (os"te-ot'o-me)—the cutting of a bone.
roentgenogram (rent-gen'o-gram")—an image obtained on film by using X rays.
synovectomy (sin"o-vek'to-me) the surgical removal of the synovial membrane of a joint.

Chapter Summary

Introduction (page 130)

Individual bones are the organs of the skeletal system.
Bone contains very active, living tissues.

Bone Structure (page 130)

Bone structure reflects its function.

1. Parts of a long bone
   a. Epiphyses are covered with articular cartilage and articulate with other bones.
   b. The shaft of a bone is called the diaphysis.
   c. Except for the articular cartilage, a bone is covered by a periosteum.
   d. Compact bone provides strength and resistance to bending.
   e. Spongy bone provides strength and reduces the weight of bone.
   f. The diaphysis contains a medullary cavity filled with marrow.
2. Microscopic structure
   a. Compact bone contains osteons cemented together.
   b. Osteonic canals contain blood vessels that nourish the cells of the osteons.
   c. The cells of spongy bone are nourished by diffusion from the surface of the thin bony plates.

Bone Development and Growth (page 131)

1. Intramembranous bones
   a. Intramembranous bones develop from layers of connective tissues.
   b. Bone tissue is formed by osteoblasts within the membranous layers.
   c. Mature bone cells are called osteocytes.
2. Endochondral bones
   a. Endochondral bones develop first as hyaline cartilage, which is later replaced by bone tissue.
   b. The primary ossification center appears in the diaphysis, and secondary ossification centers appear in the epiphyses.
   c. An epiphyseal disk remains between the primary and secondary ossification centers.
   d. The epiphyseal disks are responsible for growth in length.
   e. Long bones continue to grow in length until the epiphyseal disks are ossified.
   f. Growth in thickness is due to intramembranous ossification occurring beneath the periosteum.

Functions of Bones (page 134)

1. Support and protection
   a. Skeletal parts provide shape and form of body structures.
   b. They support and protect softer, underlying tissues.
2. Body movement
   a. Bones and muscles function together as levers.
   b. A lever consists of a rod, pivot (fulcrum), weight that is moved, and a force that supplies energy.

3. Blood cell formation
    a. At different ages, hematopoiesis occurs in the yolk sac, liver and spleen, and red bone marrow.
    b. Red marrow functions in the production of red blood cells, white blood cells, and blood platelets; yellow marrow stores fat.
4. Storage of inorganic salts
    a. The intercellular material of bone tissue contains large quantities of calcium phosphate.
    b. When blood calcium is low, osteoclasts break down bone; when blood calcium is high, osteoblasts build bone.
    c. Bone also stores lesser amounts of magnesium, sodium, potassium, and carbonate ions.

Organization of the Skeleton (page 136)

1. The skeleton can be divided into axial and appendicular portions.
2. The axial skeleton consists of the skull, hyoid bone, vertebral column, and thoracic cage.
3. The appendicular skeleton consists of the pectoral girdle, upper limbs, pelvic girdle, and lower limbs.

Skull (page 138)

The skull consists of 22 bones, which include 8 cranial bones, 13 facial bones, and 1 mandible.

1. Cranium
    a. The cranium encloses and protects the brain.
    b. Some cranial bones contain air-filled sinuses.
    c. Cranial bones include the frontal bone, parietal bones, occipital bone, temporal bones, sphenoid bone, and ethmoid bone.
2. Facial skeleton
    a. Facial bones form the basic shape of the face and provide attachments for muscles.
    b. Facial bones include the maxillary bones, palatine bones, zygomatic bones, lacrimal bones, nasal bones, vomer, inferior nasal conchae, and mandible.
3. Infantile skull
    a. Incompletely developed bones are separated by fontanels.
    b. Proportions of the infantile skull are different from those of the adult skull.

Vertebral Column (page 143)

The vertebral column extends from the skull to the pelvis and protects the spinal cord.

It is composed of vertebrae, separated by intervertebral disks.

It has four curvatures, which give it resiliency.

1. Typical vertebra
    a. A typical vertebra consists of a body and a bony arch, which surrounds the spinal cord.
    b. Notches on the lower surfaces provide intervertebral foramina, through which spinal nerves pass.
2. Cervical vertebrae
    a. Transverse processes bear transverse foramina.
    b. The atlas (first vertebra) supports and balances the head.
    c. The odontoid process of the axis (second vertebra) provides a pivot for the atlas.
3. Thoracic vertebrae
    a. Thoracic vertebrae are larger than cervical vertebrae.
    b. Facets on the sides articulate with the ribs.
4. Lumbar vertebrae
    a. Vertebral bodies are large and strong.
    b. They support more body weight than other vertebrae.
5. Sacrum
    a. The sacrum is a triangular structure formed of five fused vertebrae.
    b. Vertebral foramina form the sacral canal.
6. Coccyx
    a. The coccyx forms the lowest part of the vertebral column.
    b. It is composed of four fused vertebrae.

Thoracic Cage (page 147)

The thoracic cage includes the ribs, thoracic vertebrae, sternum, and costal cartilages. It supports the shoulder girdle and arms, protects visceral organs, and functions in breathing.

1. Ribs
    a. The ribs are attached to the thoracic vertebrae.
    b. The costal cartilages of the true ribs join the sternum directly; those of the false ribs join it indirectly.
    c. A typical rib bears a shaft, a head, and tubercles, which articulate with the vertebrae.
2. Sternum
    a. The sternum consists of a manubrium, body, and xiphoid process.
    b. It articulates with the clavicles.

Pectoral Girdle (page 148)

The pectoral girdle is composed of two clavicles and two scapulae. It forms an incomplete ring that supports the arms and provides attachments for muscles.

1. Clavicles
    a. The clavicles are located between the manubrium and scapulae.
    b. They function to hold the shoulders in place and provide attachments for muscles.
2. Scapulae
    a. The scapulae are broad, triangular bones.
    b. They articulate with the humerus and provide attachments for muscles.

Upper Limb (page 150)

Bones of the upper limb provide frameworks of arms, wrists, palms, and fingers. They also provide attachments for muscles and function in levers that move the limb and its parts.

1. Humerus
   a. The humerus extends from the glenoid cavity of the scapula to the elbow.
   b. It articulates with the radius and ulna at the elbow.
2. Radius
   a. The radius extends from the elbow to the wrist.
   b. It articulates with the humerus, ulna, and wrist.
3. Ulna
   a. The ulna overlaps the humerus posteriorly.
   b. It articulates with the radius laterally and with a disk of fibrocartilage inferiorly.
4. Hand
   a. The hand is composed of a wrist, palm, and 5 fingers.
   b. It includes 8 carpals that form a carpus, 5 metacarpals, and 14 phalanges.

Pelvic Girdle (page 152)

The pelvic girdle consists of two coxal bones that articulate with each other anteriorly and with the sacrum posteriorly. The sacrum, coccyx, and pelvic girdle form the pelvis.

Coxal bones
Each coxal bone consists of three bones, which are fused in the region of the acetabulum.

a. The ilium
   (1) The ilium is the largest portion of the coxal bone.
   (2) It joins the sacrum at the sacroiliac joint.
b. The ischium
   (1) The ischium is the lowest portion of the coxal bone.
   (2) It supports body weight when sitting.
c. The pubis
   (1) The pubis is the anterior portion of the coxal bone.
   (2) The pubic bones are fused anteriorly at the symphysis pubis.

Lower Limb (page 153)

Bones of the lower limb provide frameworks of the leg, ankle, instep, and toes.

1. Femur
   a. The femur extends from the knee to the hip.
   b. The patella articulates with its anterior surface.
2. Tibia
   a. The tibia is located on the medial side of the lower leg.
   b. It articulates with the talus of the ankle.
3. Fibula
   a. The fibula is located on the lateral side of the tibia.
   b. It articulates with the ankle, but does not bear body weight.
4. Foot
   a. The foot consists of an ankle, instep, and 5 toes.
   b. It includes 7 tarsals that form the tarsus, 5 metatarsals, and 14 phalanges.

Joints (page 155)

Joints can be classified on the basis of the amount of movement they make possible.

1. Immovable joints
   a. The bones of immovable joints are in close contact, separated by a thin layer of fibrous tissue or cartilage, as in a suture.
   b. No active movements are possible at these joints.
2. Slightly movable joints
   a. The bones of slightly movable joints are connected by disks of fibrocartilage or by ligaments, as in the vertebrae.
   b. Such a joint allows a limited amount of movement.
3. Freely movable joints
   a. The bones of a freely movable joint are covered with hyaline cartilage and held together by a fibrous capsule.
   b. The joint capsule consists of an outer layer of ligaments and an inner lining of synovial membrane.
   c. Bursae are often located between the skin and underlying bony prominences.
   d. Freely movable joints include several types: ball-and-socket, condyloid, gliding, hinge, pivot, and saddle.
4. Types of joint movements
   a. The movements of synovial joints are produced by muscles that are fastened on either side of the joint.
   b. The movements include flexion, extension, dorsiflexion, plantar flexion, hyperextension, abduction, adduction, rotation, circumduction, pronation, supination, eversion, inversion, retraction, protraction, elevation, and depression.

Clinical Application of Knowledge

1. What steps do you think should be taken to reduce the chances of persons accumulating foreign metallic elements, such as lead, radium, and strontium, in their bones?
2. When a child's bone is fractured, growth may be stimulated at the epiphyseal disk of that bone. What problems might this extra growth create in an arm or a leg before the growth of the other limb compensates for the difference in length?
3. How would you explain to an athlete the reason damaged ligaments and cartilages of joints are so slow to heal following an injury?
4. Why are women more likely to develop osteoporosis than men? What steps might be taken to reduce the risk of developing this condition?

Review Activities

Part A

1. Sketch a typical long bone, and label its epiphyses, diaphysis, medullary cavity, periosteum, and articular cartilages.
2. Distinguish between spongy and compact bone.
3. Explain how osteonic and communicating canals are related.
4. Explain how the development of intramembranous bone differs from that of endochondral bone.
5. Distinguish between osteoblasts and osteocytes.
6. Explain the function of an epiphyseal disk.
7. Explain how a long bone grows in thickness.
8. Provide several examples to illustrate how bones support and protect body parts.
9. Describe a lever.
10. Explain how arm movements involve levers.
11. Explain the functions of red and yellow bone marrow.
12. Describe how bone cells help control blood calcium concentration.
13. Distinguish between the axial and appendicular skeletons.
14. List the bones that form the pectoral and pelvic girdles.
15. Name the bones of the cranium and facial skeleton.
16. Explain the importance of fontanels.
17. Describe a typical vertebra.
18. Explain the differences between the cervical, thoracic, and lumbar vertebrae.
19. Name the bones that comprise the thoracic cage.
20. Name the bones of the upper limb.
21. Define *coxal bone*.
22. List the bones of the lower limb.
23. Define *joint*.
24. Describe an immovable joint, a slightly movable joint, and a freely movable joint.
25. Define *bursa*.
26. List six types of freely movable joints, and name an example of each type.

Part B

Match the parts listed in column I with the bones listed in column II.

I		II	
1.	Coronoid process	A.	Ethmoid bone
2.	Cribriform plate	B.	Frontal bone
3.	Foramen magnum	C.	Mandible
4.	Mastoid process	D.	Maxillary bone
5.	Palatine process	E.	Occipital bone
6.	Sella turcica	F.	Temporal bone
7.	Supraorbital foramen	G.	Sphenoid bone
8.	Temporal process	H.	Zygomatic bone
9.	Acromion process	I.	Femur
10.	Deltoid tuberosity	J.	Fibula
11.	Greater trochanter	K.	Humerus
12.	Lateral malleolus	L.	Radius
13.	Medial malleolus	M.	Scapula
14.	Olecranon process	N.	Sternum
15.	Radial tuberosity	O.	Tibia
16.	Xiphoid process	P.	Ulna

Part C

Match the terms in column I with the movements in column II.

I		II	
17.	Rotation	Q.	Turning the palm upward
18.	Supination	R.	Decreasing the angle between the parts
19.	Extension	S.	Moving the part forward
20.	Eversion	T.	Moving the part around an axis
21.	Protraction	U.	Turning the sole of the foot outward
22.	Flexion	V.	Increasing the angle between the parts
23.	Pronation	W.	Lowering a part
24.	Abduction	X.	Turning the palm downward
25.	Depression	Y.	Moving the part away from the midline

Part D

What features can you identify in these photographs of a human skull?

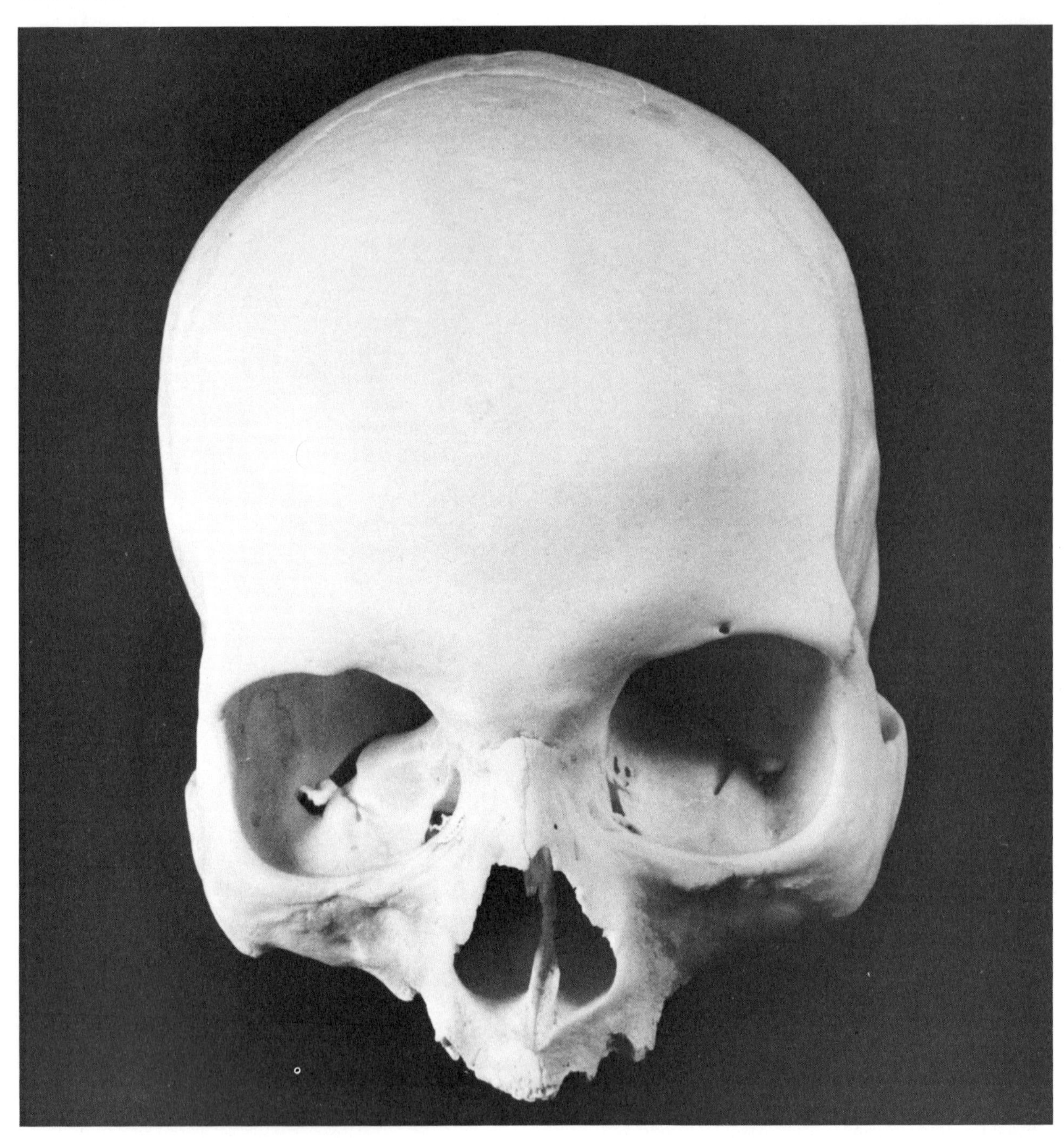

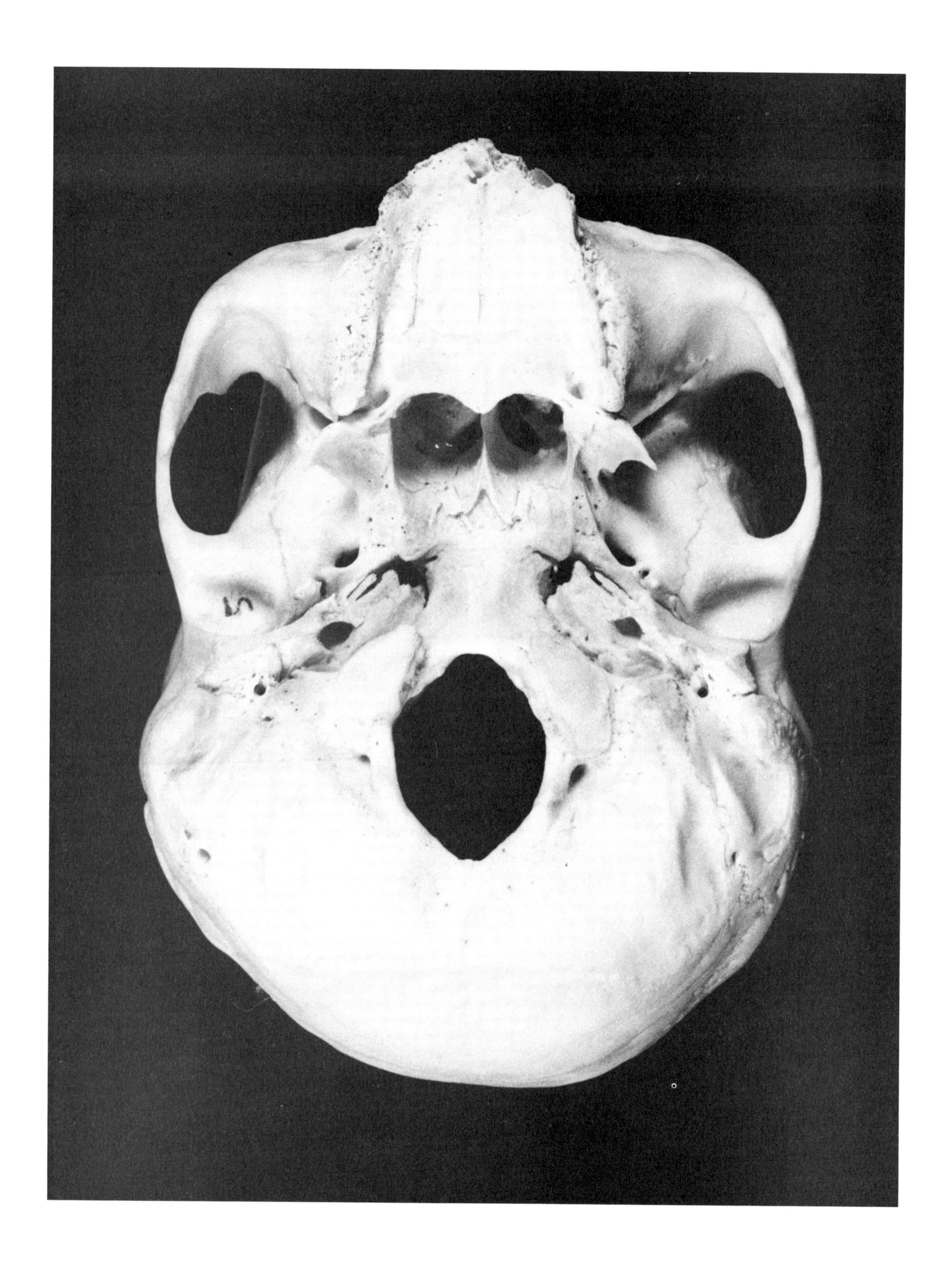

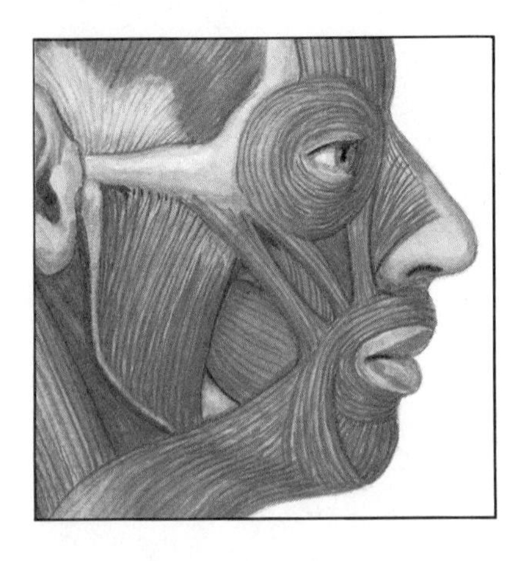

8 Muscular System

Muscles, the organs of the *muscular system,* consist largely of muscle cells. They are specialized to undergo muscular contractions during which the chemical energy from nutrients is converted into mechanical energy or movement.

When muscle cells contract, they pull on the parts to which they are attached. This action usually causes movement, as when the joints of the legs are flexed and extended during walking. At other times, muscular contractions resist motion, as when they help to hold body parts in postural positions. Muscles are also responsible for the movement of body fluids, such as blood and urine. In addition, they function in heat production, which aids in maintaining body temperature.

Chapter Objectives

After you have studied this chapter, you should be able to

1. Describe how connective tissue is included in the structure of a skeletal muscle.
2. Name the major parts of a skeletal muscle fiber, and describe the function of each part.
3. Explain the major events that occur during muscle fiber contraction.
4. Explain how energy is supplied to the muscle fiber contraction mechanism.
5. Describe how oxygen debt develops and how a muscle may become fatigued.
6. Distinguish between a twitch and a sustained contraction.
7. Explain how various types of muscular contractions produce body movements and help maintain posture.
8. Describe how skeletal muscles are affected by exercise.
9. Distinguish between the structures and functions of a multiunit smooth muscle and a visceral smooth muscle.
10. Compare the fiber contraction mechanisms of skeletal, smooth, and cardiac muscles.
11. Explain how the locations of skeletal muscles are related to the movements they produce and how muscles interact to produce such movements.
12. Identify and describe the locations of the major skeletal muscles of each body region, and describe the action of each muscle.
13. Complete the review activities at the end of this chapter. Note that the items are worded in the form of specific learning objectives. You may want to refer to them before reading the chapter.

Key Terms

actin (ak′tin)

antagonist (an-tag′o-nist)

aponeurosis (ap″o-nu-ro′sēz)

fascia (fash′e-ah)

insertion (in-ser′shun)

motor neuron (mo′tor nu′ron)

motor unit (mo′tor u′nit)

muscle impulse (mus′el im′puls)

myofibril (mi″o-fi′bril)

myosin (mi′o-sin)

neurotransmitter (nu″ro-trans′mit-er)

origin (or′ĭ-jin)

oxygen debt (ok′sĭ-jen det)

prime mover (prim mo͞ov′er)

recruitment (re-kro͞ot′ment)

synergist (sin′er-jist)

threshold stimulus (thresh′old stim′u-lus)

Aids to Understanding Words

calat-, something inserted: inter*calat*ed disk—a membranous band that separates adjacent cardiac muscle cells.

erg-, work: syn*erg*ist—a muscle that works together with a prime mover to produce a movement.

hyper-, over, more: muscular *hyper*trophy—the enlargement of muscle fibers.

inter-, between: *inter*calated disk—a membranous band that separates adjacent cardiac muscle cells.

laten-, hidden: *laten*t period—the period between the time a stimulus is applied and the beginning of a muscle contraction.

myo-, muscle: *myo*fibril—a contractile fiber of a muscle cell.

sarco-, flesh: *sarco*plasm—the substance (cytoplasm) within a muscle fiber.

syn-, together: *syn*ergist—a muscle that works together with a prime mover to produce a movement.

tetan-, stiff: *tetan*ic contraction—a sustained muscular contraction.

-troph, well fed: muscular hyper*trophy*—the enlargement of muscle fibers.

Introduction

AS IS DESCRIBED in chapter 5, there are three types of muscle tissue within the body—skeletal muscle, smooth muscle, and cardiac muscle. This chapter, however, is primarily concerned with skeletal muscle, the type found in muscles that are attached to bones and under conscious control.

Structure of a Skeletal Muscle

A skeletal muscle is an organ of the muscular system and is composed of several kinds of tissue. These include skeletal muscle tissue, nervous tissue, blood, and various connective tissues.

Connective Tissue Coverings

An individual skeletal muscle is separated from adjacent muscles and held in position by layers of fibrous connective tissue, called **fascia** (fig. 8.1).

This connective tissue surrounds each muscle and may project beyond the end of its muscle fibers to form a cordlike **tendon.** Fibers in a tendon may intertwine with those in the periosteum of a bone, thus attaching the muscle to the bone. In other cases, the connective

Figure 8.1

(*a*) A skeletal muscle is composed of a variety of tissues, including layers of connective tissue. (*b*) Fascia covers the surface of the muscle, epimysium lies beneath the fascia, and perimysium extends into the structure of the muscle where it separates muscle fibers into fasciculi. (*c*) Individual muscle fibers are separated by endomysium. (*d*) A single muscle fiber.

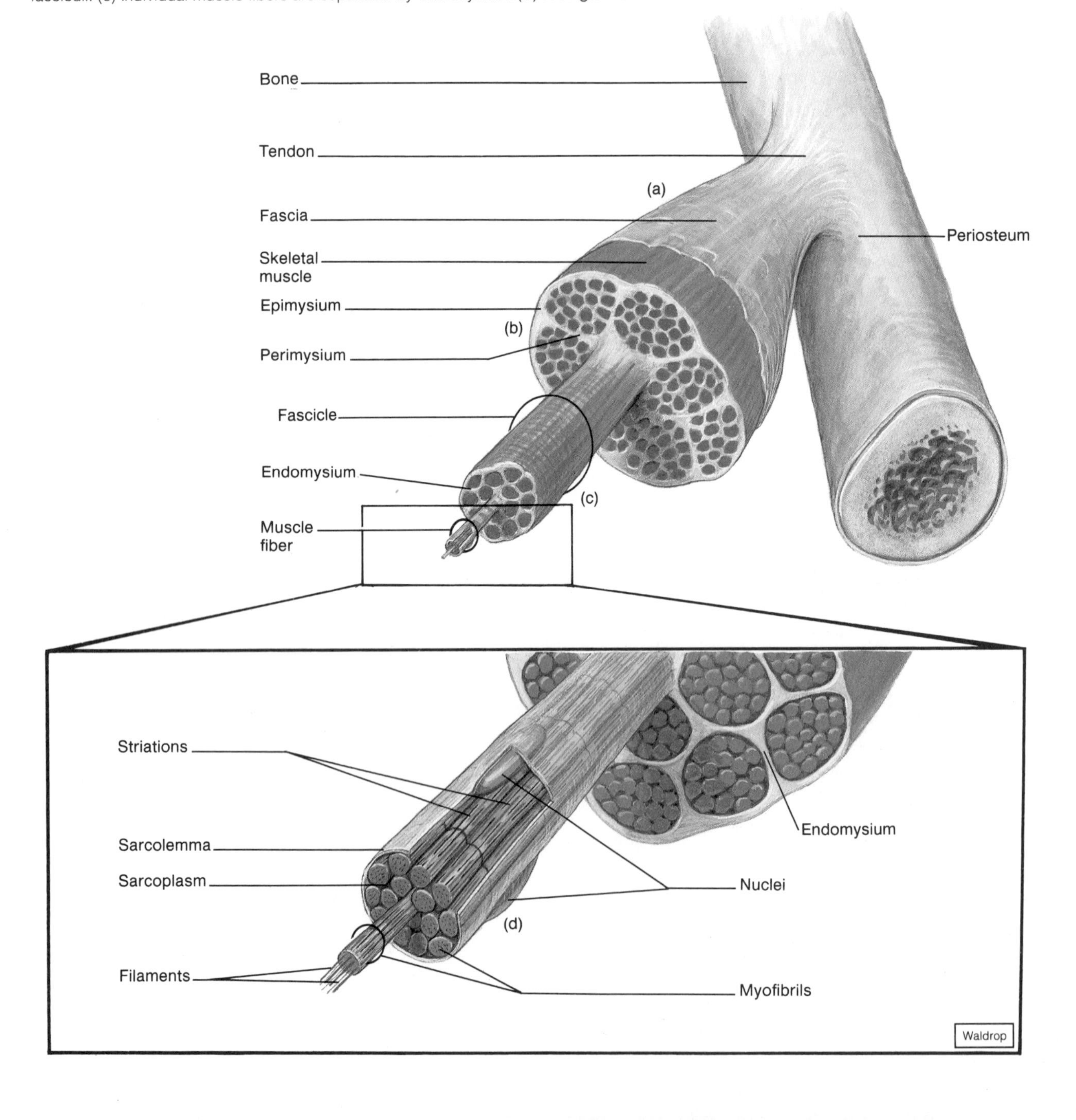

Figure 8.2

(*a*) A skeletal muscle fiber contains numerous myofibrils, each consisting of (*b*) units called sarcomeres. (*c*) The characteristic striations of a sarcomere are due to the arrangement of actin and myosin filaments.

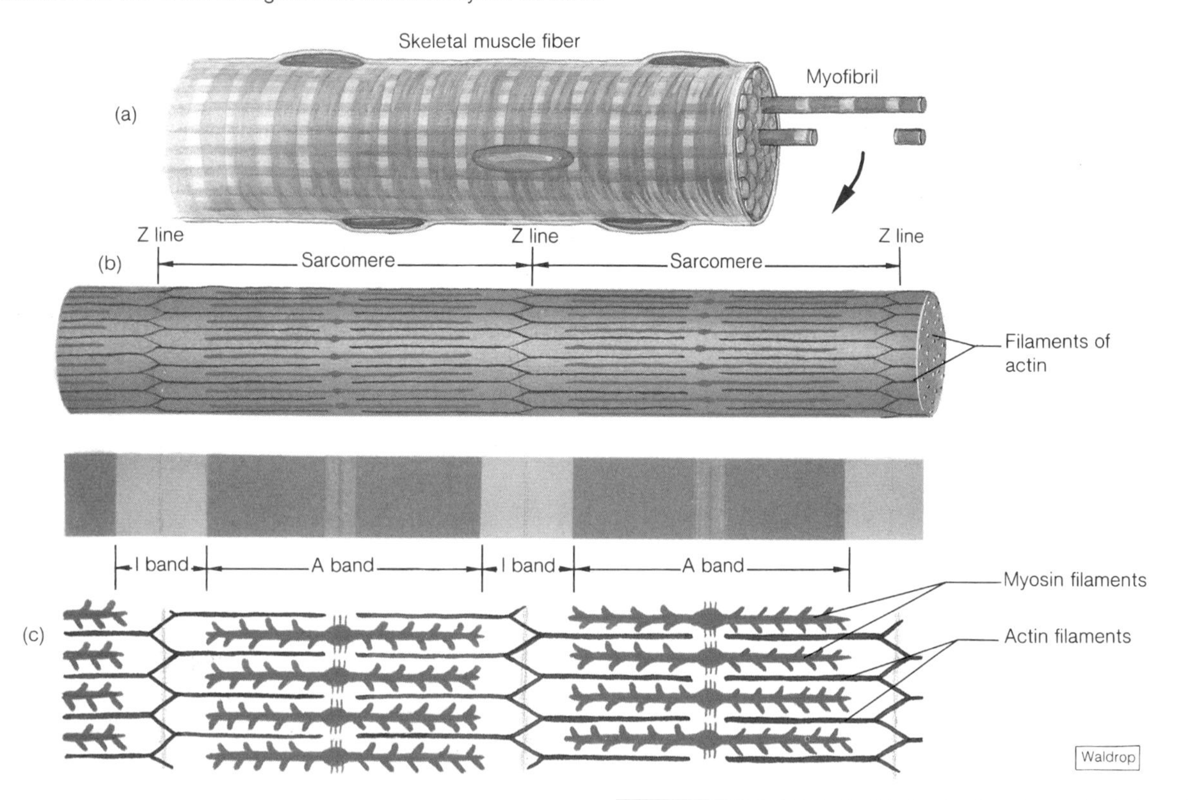

tissue forms broad fibrous sheets, called **aponeuroses,** which may be attached to the coverings of adjacent muscles (fig. 8.18).

The layer of connective tissue that closely surrounds a skeletal muscle is called *epimysium* (fig. 8.1). Other layers of connective tissue, called *perimysium,* extend inward from the epimysium and separate the muscle tissue into small compartments. These compartments contain bundles of skeletal muscle fibers, called *fascicles* (fasciculi). Each muscle fiber within a fascicle (fasciculus) is surrounded by a layer of connective tissue in the form of a thin covering, called *endomysium.*

Thus, all parts of a skeletal muscle are enclosed in layers of connective tissue, which form a network extending throughout the muscular system.

Figure 8.3

Identify the bands of the striations in this electron micrograph of myofibrils (×20,000).

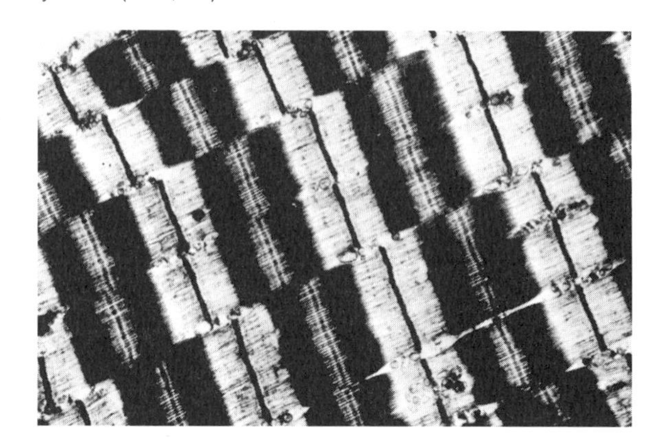

Skeletal Muscle Fibers

A skeletal muscle fiber represents a single cell of a muscle. This fiber responds to stimulation by contracting and then relaxing. Each skeletal muscle fiber is a thin, elongated cylinder with rounded ends, and it may extend the full length of the muscle. Just beneath its cell membrane (or *sarcolemma*), the cytoplasm (or *sarcoplasm)* of the fiber contains many small, oval nuclei and mitochondria (fig. 8.1). Also within the sarcoplasm are numerous threadlike **myofibrils,** which lie parallel to one another.

Myofibrils play a fundamental role in the muscle contraction mechanism. They contain two kinds of protein filaments—thick ones composed of the protein **myosin** and thin ones composed of the protein **actin** (figs. 8.2 and 8.3). The arrangement of these filaments produces the characteristic alternating light and dark *striations* of skeletal muscle fiber.

Figure 8.4
Within the sarcoplasm of a skeletal muscle fiber are a network of sarcoplasmic reticulum and a system of transverse tubules.

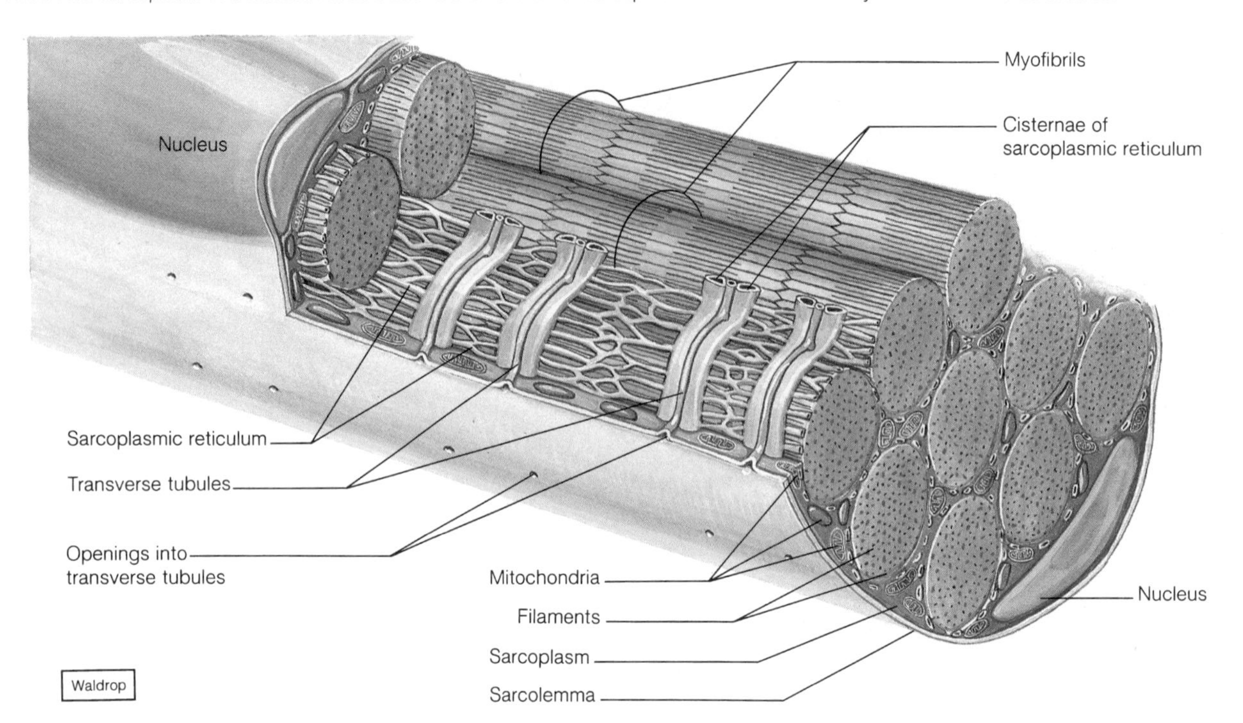

The myosin filaments are located within the dark portions *(A bands)* of the striations, and the actin filaments occur primarily in the light areas *(I bands)*. The actin filaments, however, also extend into the A bands, and when the muscle fiber contracts, the actin filaments slide farther into these bands.

The actin filaments are attached to the Z lines at the end of the I bands. These Z lines extend across the muscle fiber so that those of adjacent myofibrils lie side by side. The segment of the myofibril between two successive Z lines is called a **sarcomere,** and the consequent regular arrangement of the sarcomeres causes the muscle fiber to appear striated.

Within the cytoplasm of a muscle fiber is a network of membranous channels, which surrounds each myofibril and runs parallel to it (fig. 8.4). These membranes form the **sarcoplasmic reticulum,** which corresponds to the endoplasmic reticulum of other cells. Another set of membranous channels, called **transverse tubules** (T-tubules), extends inward as invaginations from the fiber's membrane and passes all the way through the fiber. Thus, each tubule opens to the outside of the muscle fiber and contains extracellular fluid. Furthermore, each transverse tubule lies between two enlarged portions of the sarcoplasmic reticulum, called *cisternae,* near the region where the actin and myosin filaments overlap. The sarcoplasmic reticulum and transverse tubules activate the muscle contraction mechanism when the fiber is stimulated.

Although muscle fibers and the connective tissues associated with them are flexible, they can be torn if overstretched. This type of injury, which is common in athletes, is called *muscle strain* or *muscle pull.* The seriousness of the injury depends on the degree of damage sustained by the tissues. In a mild strain, for example, only a few muscle fibers are injured, the fascia remains intact, and there is little loss of function. In a severe strain, however, many muscle fibers as well as the fascia are torn, and muscle function may be lost completely. Such a severe strain is painful and accompanied by discoloration and swelling of tissues.

1. Describe how connective tissue is associated with a skeletal muscle.
2. Describe the general structure of a skeletal muscle fiber.
3. Explain why skeletal muscle fibers appear striated.
4. Explain the relationship between the sarcoplasmic reticulum and transverse tubules.

Neuromuscular Junction

Each skeletal muscle fiber is connected to a fiber from a nerve cell called a **motor neuron.** This nerve fiber extends outward from the brain or spinal cord, and a muscle fiber contracts only when it is stimulated by such a motor neuron.

Figure 8.5
A neuromuscular junction includes the end of a motor neuron and the motor end plate of a muscle fiber.

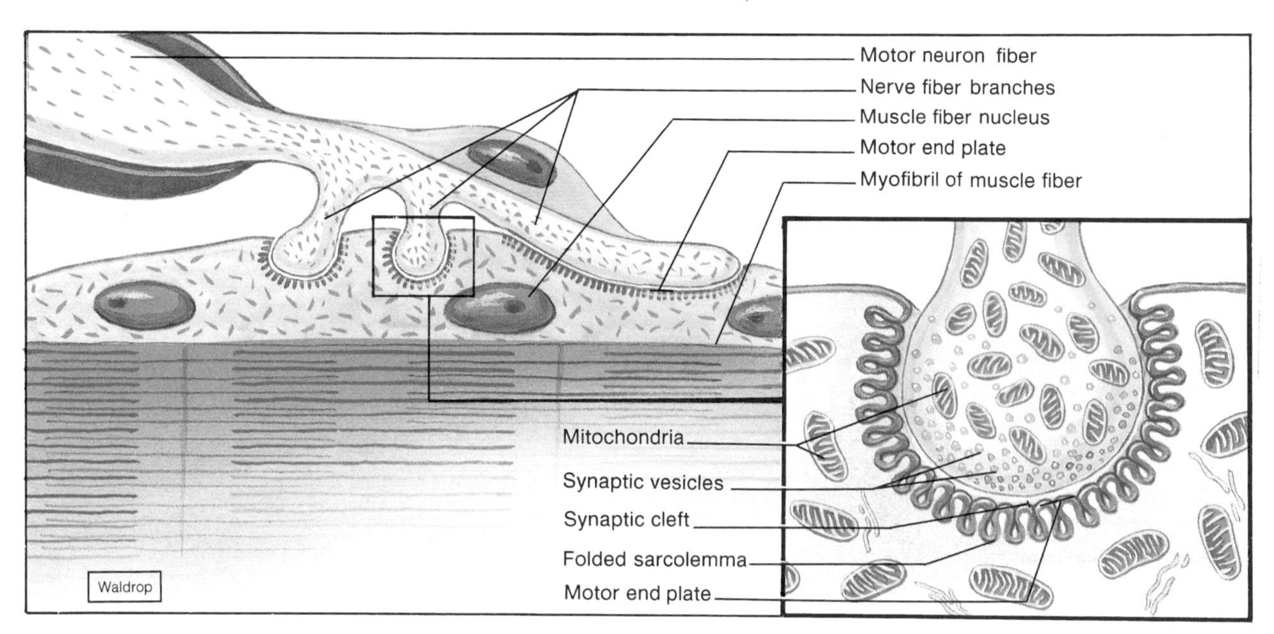

The site where the nerve fiber and muscle fiber meet is called a **neuromuscular junction.** At this junction the muscle fiber membrane is specialized to form a **motor end plate.** In this region of the muscle fiber, mitochondria are abundant, and the cell membrane is extensively folded (fig. 8.5).

The end of the motor nerve fiber is branched, and the ends of these branches project into recesses (synaptic clefts) of the muscle fiber membrane. The cytoplasm at the distal ends of the nerve fibers is rich in mitochondria and contains many tiny vesicles (synaptic vesicles) that store chemicals, called **neurotransmitters.**

When a nerve impulse traveling from the brain or spinal cord reaches the end of a motor nerve fiber, some of the vesicles release a neurotransmitter into the gap between the nerve and the motor end plate of the muscle fiber. This action stimulates the muscle fiber to contract.

Motor Units

Although a muscle fiber usually has a single motor end plate, the nerve fibers of motor neurons are densely branched. By means of these branches, one motor fiber may be connected to many muscle fibers. Furthermore, when the motor nerve fiber transmits an impulse, all of the muscle fibers it is connected to are stimulated to contract simultaneously. Together, a motor neuron and the muscle fibers that it controls constitute a **motor unit** (fig. 8.6).

Figure 8.6
A motor unit consists of one motor neuron and all the muscle fibers with which it communicates.

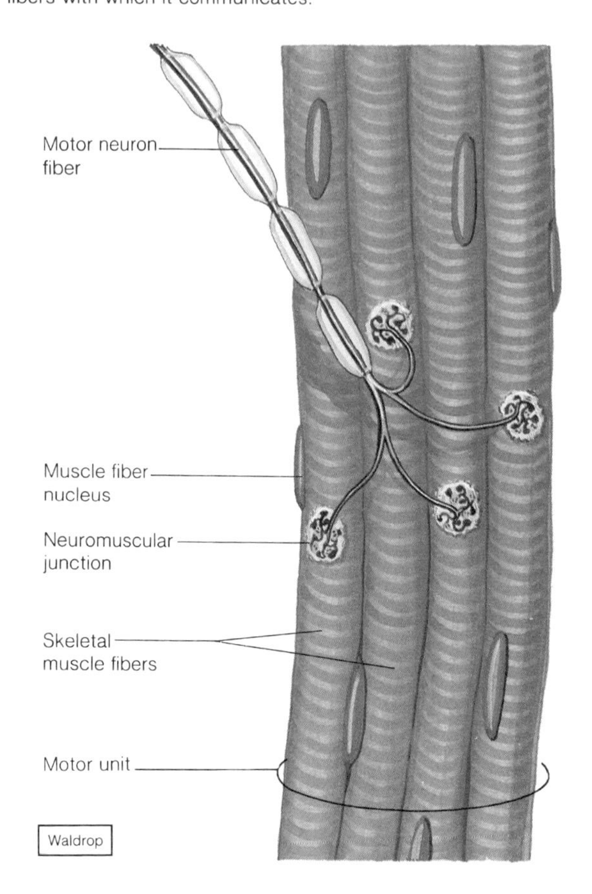

Skeletal Muscle Contraction

A muscle fiber contraction is a complex action involving a number of cell parts and chemical substances. The final result is a sliding movement within the myofibrils in which the filaments of actin and myosin merge. When this happens, the muscle fiber is shortened, and it pulls on its attachments.

Role of Myosin and Actin

A myosin molecule is composed of protein strands with globular parts, called *crossbridges,* projecting outward along their lengths. In the presence of calcium ions, these crossbridges can react with the actin filaments and form linkages with them. This reaction between the myosin and actin filaments generates the force involved in shortening the myofibrils during a muscle contraction.

An actin molecule also consists of strands of protein, and it has ADP molecules attached to its surface. These ADP molecules serve as active sites for the formation of linkages with the crossbridges of the myosin molecules.

It is not completely understood how the formation of linkages results in the shortening of the myofibrils. One theory (the ratchet theory) suggests that the end of a myosin crossbridge can attach to an actin active site and bend slightly, pulling the actin filament with it. Then the end may release, straighten itself, and combine with another active site further down the actin filament. Presumably, this cycle can be repeated again and again, and as the actin filament is moved toward the center of the sarcomere, the sarcomere shortens (fig. 8.7).

Figure 8.7
According to the ratchet theory, (*a*) when calcium ions are present, active sites on an actin filament are exposed. (*b*) Cross-bridges on a myosin filament form linkages at the active sites. (*c*) A myosin cross-bridge bends slightly, pulling an actin filament. (*d*) The linkage is broken, and (*e*) the myosin cross-bridge forms a linkage with the next active site.

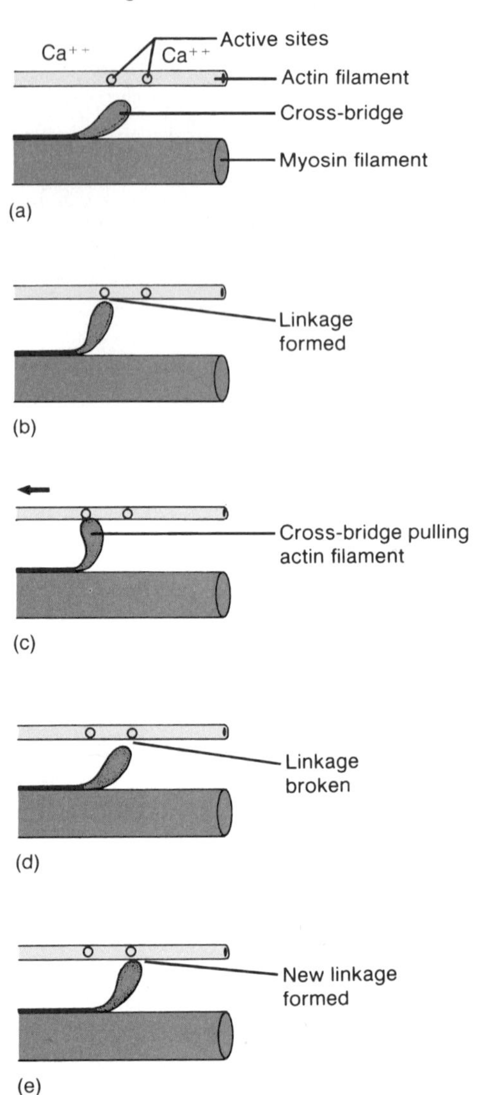

Stimulus for Contraction

A skeletal muscle fiber normally does not contract until it is stimulated by a neurotransmitter. In skeletal muscle, the neurotransmitter is a compound called **acetylcholine.** This substance is synthesized in the cytoplasm of the motor neuron and stored in vesicles. When a nerve impulse reaches the end of the nerve fiber, some of the vesicles release their acetylcholine into the gap between the nerve fiber and motor end plate (fig. 8.5).

The acetylcholine diffuses rapidly across the gap and combines with certain protein molecules (receptors) in the muscle fiber membrane, thus stimulating it. As a result of this stimulus, a **muscle impulse,** very much like a nerve impulse (see chapter 9), is generated, and the impulse passes in all directions over the surface of the muscle fiber membrane. It also travels through the transverse tubules deep into the fiber (fig. 8.4).

The sarcoplasmic reticulum contains a high concentration of calcium ions. In response to a muscle impulse, the membranes of the cisternae become more permeable to these ions, and the ions diffuse into the sarcoplasm of the muscle fiber.

When a relatively high concentration of calcium ions is present in the sarcoplasm, linkages form between the actin and myosin filaments, and a muscle contraction occurs (fig. 8.8). The contraction continues while the calcium ions are present, but the ions are moved quickly back into the sarcoplasmic reticulum by an active transport mechanism (calcium pump). Consequently, when the calcium concentration of the sar-

Chart 8.1 Major events of muscle contraction and relaxation

Muscle fiber contraction	Muscle fiber relaxation
1. Acetylcholine is released from the distal end of a motor neuron.	1. Cholinesterase causes acetylcholine to decompose, and the muscle fiber membrane is no longer stimulated.
2. Acetylcholine diffuses across the gap at the neuromuscular junction.	2. Calcium ions are actively transported into the sarcoplasmic reticulum.
3. The muscle fiber membrane is stimulated, and a muscle impulse travels deep into the fiber through the transverse tubules and reaches the sarcoplasmic reticulum.	3. Linkages between actin and myosin filaments are broken.
4. Calcium ions diffuse from the sarcoplasmic reticulum into the sarcoplasm.	4. Actin and myosin filaments slide apart.
5. Linkages form between actin and myosin filaments.	5. Muscle fiber relaxes.
6. Actin filaments slide inward along the myosin filaments.	
7. Muscle fiber shortens as a contraction occurs.	

Figure 8.8
(*a*) As linkages form between filaments of actin and myosin, (*b and c*) the actin filaments are pulled toward the center of the A band, causing the myofibril to contract.

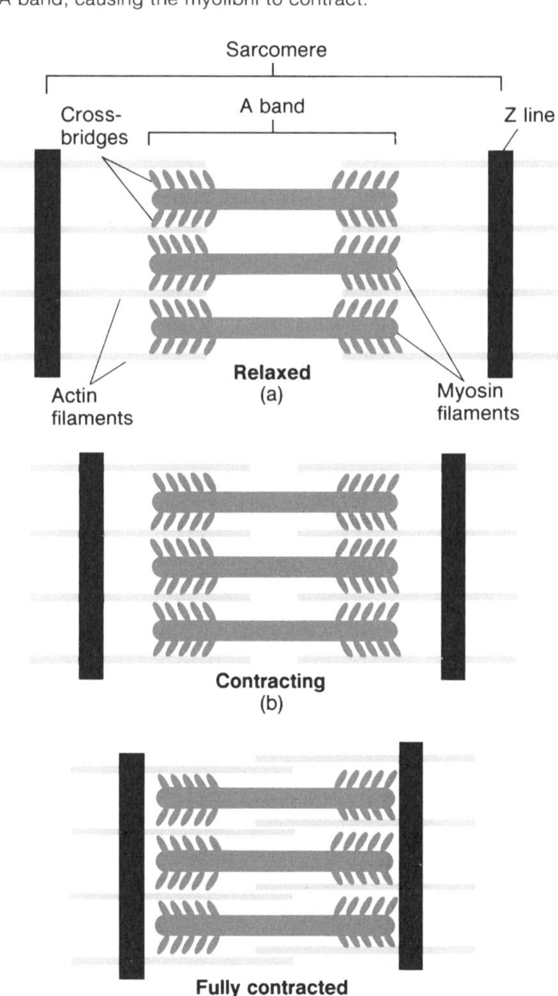

coplasm is decreased, the linkages are broken. Within a fraction of a second, the actin and myosin filaments slide apart, and muscle fiber relaxes.

Meanwhile, the acetylcholine that stimulated the muscle fiber in the first place is rapidly decomposed by the action of an enzyme called **cholinesterase.** This enzyme is present at the neuromuscular junction within the membranes of the motor end plate, and its action prevents a single nerve impulse from causing a continued stimulation of the muscle fiber.

A substance called botulinus toxin, produced by a bacterium *(Clostridium botulinum)*, can prevent the release of acetylcholine from motor nerve fibers at the neuromuscular junctions. This bacterium is responsible for a very serious form of food poisoning called *botulism*. This condition is most likely to result from eating home-processed food that has not been heated enough to kill the bacteria present in it or to inactivate the toxin.

When botulinus toxin affects the body, muscle fibers fail to be stimulated, and muscles, including those responsible for breathing, may be paralyzed. As a consequence, without prompt medical treatment, the fatality rate for botulism is relatively high.

Chart 8.1 summarizes the major events leading to muscle contraction and relaxation.

1. Describe a neuromuscular junction.
2. Define *motor unit.*
3. Explain how the filaments of a myofibril interact during muscle contraction.
4. Explain how a motor nerve impulse can trigger a muscle contraction.

Figure 8.9
Energy released by cellular respiration may be used to promote the synthesis of ATP or to synthesize creatine phosphate. Later, energy from creatine phosphate may be used to promote ATP synthesis.

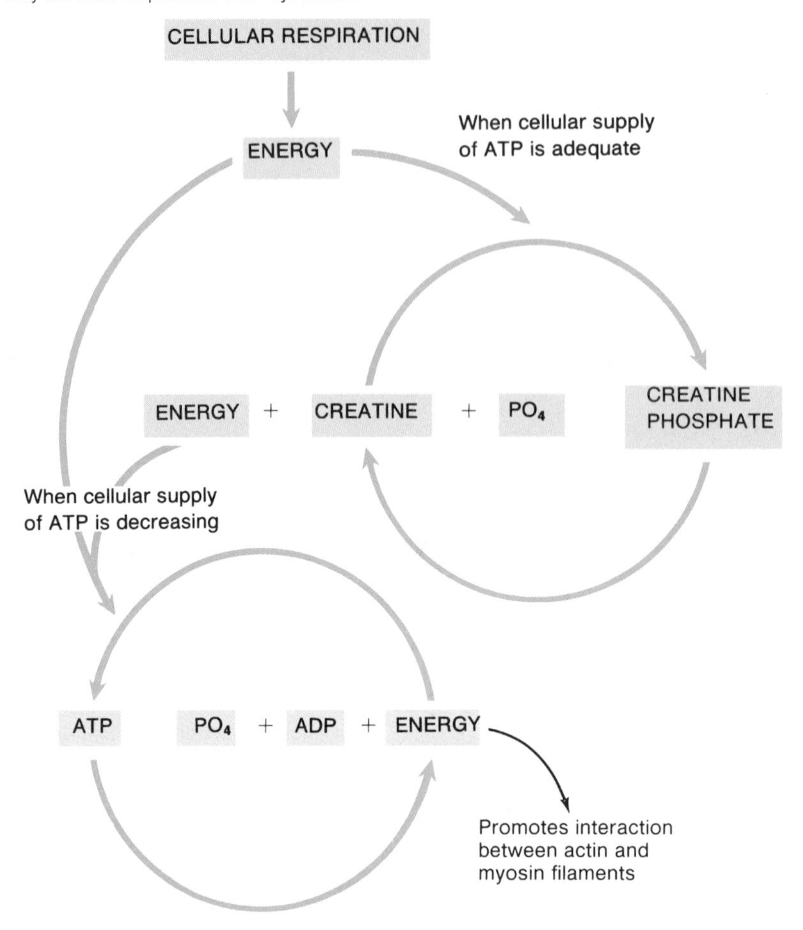

Energy Sources for Contraction

The energy used during muscle fiber contraction comes from ATP molecules, supplied by numerous mitochondria that are positioned close to the myofibrils. The crossbridges of the myosin filaments contain an enzyme called **ATPase.** This enzyme causes ATP to decompose into ADP and phosphate and, at the same time, to release energy. This energy makes possible the reaction between the actin and myosin filaments. There is only enough ATP in a muscle fiber to operate the contraction mechanism for a very short time, however, so when a fiber is active, ATP must be regenerated.

The primary source of energy available to regenerate ATP from ADP and phosphate is a substance called **creatine phosphate.** Like ATP, creatine phosphate contains high-energy phosphate bonds, and it is actually four to six times more abundant in the muscle fibers than ATP. Creatine phosphate, however, cannot directly supply energy to a cell's energy-utilizing reactions. Instead, it acts to store energy released from the mitochondria. Thus, whenever sufficient ATP is present, an enzyme in the mitochondria (creatine phosphokinase) promotes the synthesis of creatine phosphate, and energy is stored in its phosphate bonds (fig. 8.9). Then, at those times when ATP is being decomposed, the energy from creatine phosphate can be transferred to ADP molecules, and they, in turn, are quickly converted back into ATP.

In active muscle, the supply of creatine phosphate is exhausted rapidly. When this happens, the muscle fibers become dependent upon the cellular respiration of glucose as a source of energy for synthesizing ATP.

Oxygen Supply and Cellular Respiration

As is described in chapter 4, the early phase of cellular respiration takes place in the absence of oxygen. The more complete breakdown of glucose, however, occurs in the mitochondria and requires oxygen (fig. 8.10).

The oxygen needed to support this aerobic respiration is carried from the lungs to the body cells by the blood. Oxygen is transported within the red blood cells loosely bonded to molecules of **hemoglobin,** the pigment responsible for the red color of the blood.

Figure 8.10
The oxygen needed to support aerobic respiration is carried in the blood and stored in myoglobin.

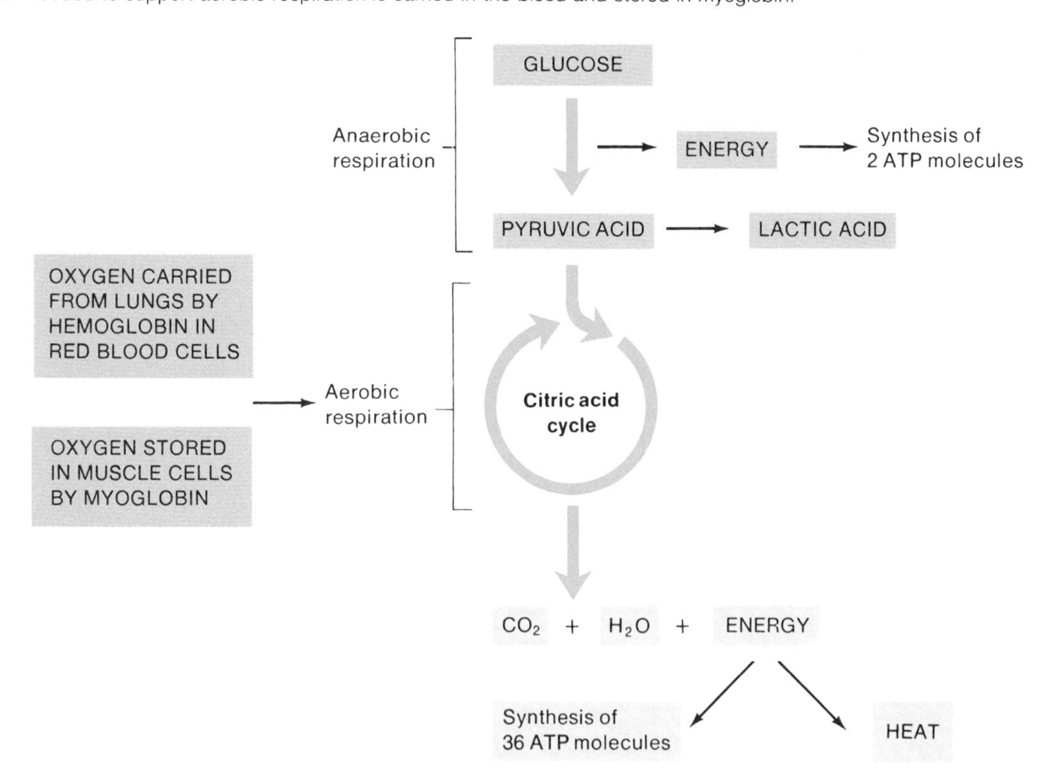

Another pigment, **myoglobin,** is synthesized in the muscle cells and is responsible for the reddish brown color of skeletal muscle tissue. Myoglobin has properties similar to hemoglobin in that it can combine loosely with oxygen. This ability to store oxygen temporarily reduces a muscle's need for a continuous blood supply during muscular contraction.

Oxygen Debt

When a person is resting or moderately active, the respiratory and circulatory systems can usually supply sufficient oxygen to the skeletal muscles to support aerobic respiration. However, when the skeletal muscles are used strenuously for even a minute or two, these systems usually cannot supply enough oxygen to meet the needs of aerobic respiration. Consequently, the muscle fibers must depend more and more on the anaerobic phase of respiration for their energy.

In anaerobic respiration, glucose molecules are changed to *pyruvic acid* (see chapter 4). If the oxygen supply is low, however, the pyruvic acid is quickly converted to *lactic acid,* which diffuses out of the muscle fibers and is carried to the liver by the blood (fig. 8.10).

The liver cells can change lactic acid into *glucose;* however, this conversion also requires energy from ATP. During strenuous exercise, the available oxygen is used primarily to synthesize the ATP needed for the muscle fiber contraction mechanism rather than to make ATP for changing lactic acid into glucose. Consequently, as lactic acid accumulates, a person develops an **oxygen debt** that must be repaid later. The amount of oxygen debt acquired is equal to the amount of oxygen needed by the liver cells to convert the accumulated lactic acid into glucose plus the amount needed by the muscle cells to resynthesize ATP and creatine phosphate and return them to their original concentrations.

Because the conversion of lactic acid back into glucose is a relatively slow process, it may require several hours to repay an oxygen debt following vigorous exercise.

Muscle Fatigue

If a muscle is exercised strenuously for a prolonged period, it may lose its ability to contract, a condition called *fatigue.* This condition may result from an interruption in the muscle's blood supply or, rarely, from an exhaustion of the supply of acetylcholine in its motor nerve fibers. Muscle fatigue, however, is most likely to arise from an accumulation of lactic acid in the muscle as a result of anaerobic respiration. The lactic acid causes factors, such as pH, to change so that the muscle fibers are no longer responsive to stimulation.

Occasionally a muscle becomes fatigued and develops cramps at the same time. A cramp is a painful condition in which the muscle contracts spasmodically, but does not relax completely. This condition seems to be due to a lack of ATP, which is needed to move the calcium ions back into the sarcoplasmic reticulum and to break the linkages between the actin and myosin filaments before the muscle fibers can relax.

A few hours after death, the skeletal muscles undergo a partial contraction that causes the joints to become fixed. This condition, *rigor mortis,* may continue for seventy-two hours or more. It seems to result from an increase in membrane permeability to calcium ions and a decrease in ATP in the muscle fibers, which prevents relaxation. Thus, the actin and myosin filaments of the muscle fibers remain linked together until the muscles begin to decompose.

Heat Production

Only about 25% of the energy released by cellular respiration is available for use in metabolic processes; the rest becomes heat. Although heat is generated by all active cells, muscle tissue is a major source of heat because muscle represents such a large proportion of the total body mass. Thus, whenever muscles are active, large amounts of heat are released. This heat is transported to other tissues by the blood and helps to maintain body temperature.

1. What substances provide the energy used to regenerate ATP?
2. What are the sources of oxygen needed for aerobic respiration?
3. How are lactic acid, the oxygen debt, and muscle fatigue related?
4. What is the relationship between cellular respiration and heat production?

Muscular Responses

One way to observe muscle contraction is to remove a single muscle fiber from a skeletal muscle and connect it to a device that records changes in the fiber's length. In such experiments, the muscle fiber is usually stimulated by an electrical stimulator that is capable of producing stimuli of varying strengths and frequencies.

Threshold Stimulus

By exposing an isolated muscle fiber to a series of stimuli of increasing strength, it can be shown that the fiber remains unresponsive until a certain strength of stimulation is applied. This minimal strength needed to cause a contraction is called the **threshold stimulus.**

All-or-None Response

When a muscle fiber is exposed to a stimulus of threshold strength (or above), it responds to its fullest extent. Increasing the strength of the stimulus does not affect the degree to which the fiber contracts. In other words, there are no partial contractions of a muscle fiber—if it contracts at all, it contracts completely, even though in some instances it may not shorten completely. This phenomenon is called the **all-or-none response.**

Recruitment of Motor Units

Since the muscle fibers within a muscle are organized into motor units, and each motor unit is controlled by a single motor neuron, all the muscle fibers in a motor unit are stimulated at the same time. Consequently, a motor unit also responds in an all-or-none manner. A whole muscle, however, does not behave like this, because it is composed of many motor units, controlled by different motor neurons, which respond to different thresholds of stimulation. Thus, if only the motor neurons with low thresholds of stimulation are stimulated, a relatively small number of motor units contract. At higher intensities of stimulation, other motor neurons respond, and more motor units are activated. Such an increase in the number of motor units being activated is called **recruitment.** As the intensity of stimulation increases, the recruitment of motor units continues, until finally all possible motor units are activated, and the muscle is contracting with maximal tension.

Twitch Contraction

To demonstrate how a whole muscle responds to stimulation, a skeletal muscle can be removed from a frog or another laboratory animal. The muscle is stimulated electrically, and when it contracts, its movement is recorded. The resulting pattern is called a **myogram.**

If a muscle is exposed to a single stimulus of sufficient strength to activate some of its motor units, the muscle will contract and then relax. This action—a single contraction that lasts only a fraction of a second—is called a **twitch.** A twitch produces a myogram like that in figure 8.11. It is apparent from this record that the muscle response did not begin immediately following stimulation. There is a delay between the time that the stimulus was applied and the time that the muscle responded. This time lag is called the **latent period.** In a frog muscle, the latent period lasts for about 0.01 second, and it is even shorter in a human muscle.

The latent period is followed by a **period of contraction,** during which the muscle pulls at its attachments, and a **period of relaxation,** during which it returns to its former length (fig. 8.11).

A Current Topic
Use and Disuse of Skeletal Muscles

Skeletal muscles are very responsive to use and disuse. For example, those that are forcefully exercised tend to enlarge. This phenomenon is called *muscular hypertrophy.* Conversely, a muscle that is not used undergoes *atrophy*—that is, it decreases in size and strength.

The way a muscle responds to use also depends on the type of exercise involved. For instance, when a muscle contracts relatively weakly, as during swimming and running, a specialized group of muscle fibers called *slow fibers,* which are fatigue-resistant, are activated. As a result of being used, these specialized muscle fibers develop more mitochondria, and more extensive capillary networks develop around them. Such changes increase the ability of the slow fibers to resist becoming fatigued during prolonged periods of exercise, although their sizes and strengths may remain unchanged.

Forceful excercise, such as weight lifting, in which a muscle exerts more than 75% of its maximum tension, involves another group of specialized muscle fibers called *fast fibers,* which are fatigable. In response to such exercise, these fibers develop new filaments of actin and myosin, the diameter of the muscle fibers increases, and the entire muscle enlarges. However, no new muscle fibers are produced as a result of the muscular hypertrophy.

The strength of a muscular contraction is directly related to the diameter of the muscle fibers being activated. Consequently, an enlarged muscle is capable of producing stronger contractions than before. Such a change, however, does not increase the muscle's ability to resist fatigue during activities like swimming or running.

If regular exercise is discontinued, there is a reduction in the capillary networks and in the number of mitochondria within the muscle fibers. Also, the size of the actin and myosin filaments decreases, and the entire muscle atrophies. Such atrophy commonly occurs when accidents or diseases interfere with motor nerve impulses and prevent them from reaching the muscle fibers. A muscle that cannot be exercised may decrease to less than one-half its usual size within a few months.

The fibers of muscles whose motor neurons are severed not only decrease in size, but also may become fragmented and, in time, be replaced by fat or fibrous connective tissue. However, if such a muscle is reinnervated within the first few months following an injury, its function may be restored. Meanwhile, atrophy may be delayed by treatments in which electrical stimulation is used to cause muscular contractions against loads.

Figure 8.11
A myogram of a single muscle twitch.

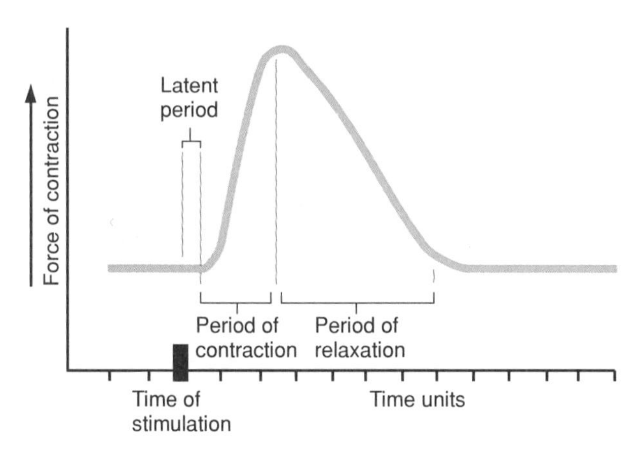

Sustained Contractions

If a muscle is exposed to a series of stimuli of increasing frequency, a point is reached when the muscle is unable to complete its relaxation period before the next stimulus in the series arrives. When this happens, the twitches begin to combine, and the muscle contraction becomes *sustained* (fig. 8.12).

At the same time that twitches are combining, the strength of the contractions may be increasing. This is due to the recruitment of motor units. The smaller motor units, which have finer fibers, are most easily stimulated and tend to respond earlier in the series of stimuli. The larger motor units, which contain thicker fibers, respond later and produce more forceful contractions. When a sustained, forceful contraction lacks even partial relaxation, it is termed a **tetanic contraction** (tetany).

Figure 8.12
Myograms of (*a*) a series of twitches, (*b*) a combining of twitches, and (*c*) a tetanic contraction. (Note: The frequency of stimulation increases from one myogram to the next.)

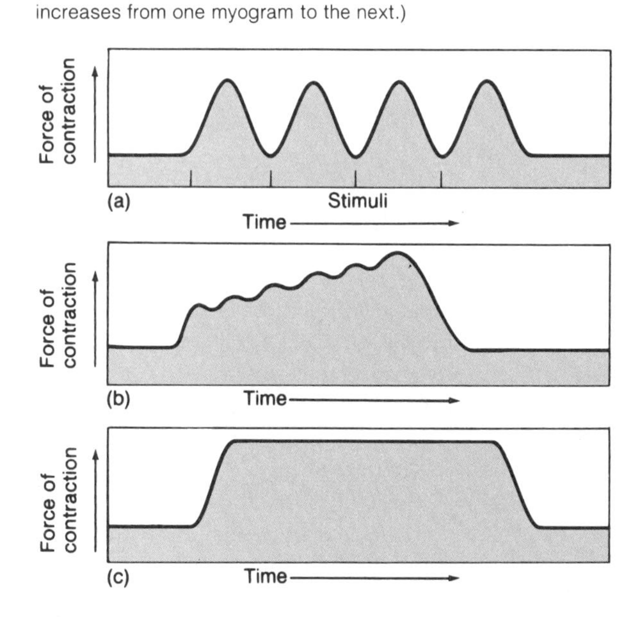

Although twitches may occur occasionally in human skeletal muscles, as when an eyelid twitches, such contractions are of limited use. More commonly, muscular contractions are sustained. When a person lifts a weight or walks, for example, sustained contractions are maintained in the arm or leg muscles for varying lengths of time. These contractions are responses to a rapid series of stimuli transmitted from the brain and spinal cord on motor neuron fibers.

Actually, even when a muscle appears to be at rest, a certain amount of sustained contraction is occurring in its fibers. This is called **muscle tone** (tonus). Muscle tone is a response to nerve impulses originating repeatedly from the spinal cord and traveling to small numbers of muscle fibers.

Muscle tone is particularly important in maintaining posture. If tone is suddenly lost, as happens when a person loses consciousness, the body will collapse.

1. Define *threshold stimulus.*
2. What is meant by an all-or-none response?
3. Distinguish between a twitch and a sustained contraction.
4. How is muscle tone maintained?

Smooth Muscles

The contractile mechanisms of smooth and cardiac muscles are essentially the same as those of skeletal muscles. The cells of these tissues, however, have some important structural and functional differences.

Smooth Muscle Fibers

As discussed in chapter 5, smooth muscle cells are elongated with tapering ends. They contain filaments of *actin* and *myosin* in myofibrils that extend the lengths of the cells. However, these filaments are very thin and more randomly arranged than those in skeletal muscle. Consequently, smooth muscle cells lack striations.

There are two major types of smooth muscles—multiunit and visceral. In the type called **multiunit smooth muscle,** the muscle fibers are less well organized and occur as separate fibers rather than in sheets. Smooth muscle of this type is found in the irises of the eyes and in the walls of blood vessels. Typically, this tissue contracts only after stimulation by motor nerve impulses.

Visceral smooth muscle is composed of sheets of spindle-shaped cells that are in close contact with one another (fig. 5.20). This type, which is more common, is found in the walls of hollow visceral organs, such as the stomach, intestines, urinary bladder, and uterus.

The fibers of visceral smooth muscles are capable of stimulating each other. Consequently, when one fiber is stimulated, the impulse moving over its surface may excite adjacent fibers, which, in turn, stimulate still others. Visceral smooth muscles also display *rhythmicity*—a pattern of repeated contractions. This phenomenon is due to the presence of self-exciting fibers, from which spontaneous impulses travel periodically into the surrounding muscle tissue.

These two features—the transmission of impulses from cell to cell and rhythmicity—are largely responsible for the wavelike motion, called **peristalsis,** that occurs in various tubular organs, such as the intestines, and helps force the contents of these organs along their lengths.

Smooth Muscle Contraction

Smooth muscle contraction resembles skeletal muscle contraction in a number of ways. Both mechanisms involve reactions of actin and myosin, both are triggered by membrane impulses and the release of calcium ions, and both use energy from ATP molecules. There are, however, significant differences between these two types of muscle tissue.

For example, although acetylcholine is the neurotransmitter substance in skeletal muscle, two neurotransmitters affect smooth muscle—*acetylcholine* and *norepinephrine.* Each of these substances stimulates contractions in some smooth muscles and inhibits contractions in others (see chapter 9). Also, smooth muscles are affected by a number of hormones, which stimulate contractions in some cases and alter the amount of response to neurotransmitters in others.

Smooth muscle is slower to contract and slower to relax than skeletal muscle. On the other hand, smooth muscle can maintain a forceful contraction for a longer time with a given amount of ATP. Also, unlike skeletal muscle, smooth muscle fibers can change length without changing tautness; therefore, smooth muscles in the stomach and intestinal walls can stretch as these organs become filled, but the pressure inside the organs remains unchanged.

1. Describe two major types of smooth muscle.
2. What special characteristics of visceral smooth muscle make peristalsis possible?
3. How does smooth muscle contraction differ from that of skeletal muscle?

Cardiac Muscle

Cardiac muscle occurs only in the heart. It is composed of striated cells joined end-to-end, forming fibers (fig. 5.21). These fibers are interconnected in branching,

Chart 8.2 Types of muscle tissue			
	Skeletal	Smooth	Cardiac
Major location	Skeletal muscles	Walls of hollow visceral organs	Wall of the heart
Major function	Movement of bones at joints, maintenance of posture	Movement of visceral organs, peristalsis	Pumping action of the heart
Cellular characteristics			
Striations	Present	Absent	Present
Nucleus	Many nuclei	Single nucleus	Single nucleus
Special features	Transverse tubule system is well developed	Lacks transverse tubules	Transverse tubule system is well developed, adjacent cells are separated by intercalated disks
Mode of control	Voluntary	Involuntary	Involuntary
Contraction characteristics	Contracts and relaxes relatively rapidly	Contracts and relaxes relatively slowly, self-exciting, rhythmic	Network of fibers contracts as a unit, self-exciting, rhythmic

three-dimensional networks. Each cell contains numerous filaments of actin and myosin, similar to those in skeletal muscle. A cardiac muscle cell also has a well-developed sarcoplasmic reticulum, many mitochondria, and a system of transverse tubules. The cisternae of cardiac muscle fibers, however, are less well developed and store less calcium than those of skeletal muscle. On the other hand, the transverse tubules of cardiac muscle are larger, and they release large quantities of calcium ions into the sarcoplasm in response to muscle impulses. This extra calcium from the transverse tubules is obtained from fluid outside the muscle fibers, and it enables cardiac muscle fibers to maintain a contraction for a longer time than skeletal muscle fibers.

The opposing ends of cardiac muscle cells are separated by cross-bands called *intercalated disks.* These bands are the result of elaborate junctions of cell membranes. They help to hold adjacent cells together and to transmit the force of contraction from cell to cell. Intercalated disks also have a very low resistance to the passage of impulses, so that cardiac muscle fibers transmit impulses relatively rapidly.

When one portion of the cardiac muscle network is stimulated, the resulting impulse passes to the other fibers of the network and the whole structure contracts as a unit; that is, the network responds to stimulation in an all-or-none manner. Cardiac muscle is also self-exciting and rhythmic. Consequently, a pattern of contraction and relaxation is repeated again and again and is responsible for the rhythmic contractions of the heart.

Chart 8.2 summarizes the characteristics of the three types of muscle tissues.

1. How is cardiac muscle similar to smooth muscle?
2. What is the function of intercalated disks?
3. What characteristic of cardiac muscle is responsible for contraction of the heart as a unit?

Skeletal Muscle Actions

Skeletal muscles are responsible for a variety of body movements. The movement caused by each muscle depends largely upon the kind of joint it is associated with and the way it is attached on either side of that joint.

Origin and Insertion

One end of a skeletal muscle is usually fastened to a relatively immovable or fixed part, and the other end is connected to a movable part on the other side of a joint. The immovable end of the muscle is called its **origin,** and the movable one is its **insertion.** When a muscle contracts, its insertion is pulled toward its origin (fig. 8.13).

Some muscles have more than one origin or insertion. The *biceps brachii* in the upper arm, for example, has two origins. This is reflected in its name, *biceps,* which means *two heads.* (Note: the head of a muscle is the part nearest its origin.) One head of the muscle is attached to the coracoid process of the scapula, and the other head arises from a tubercle above the glenoid cavity of the scapula. The muscle extends along the front surface of the humerus and is inserted by means of a tendon on the radial tuberosity of the radius. When the biceps brachii contracts, its insertion is pulled toward its origin, and the arm bends at the elbow.

Interaction of Skeletal Muscles

Skeletal muscles almost always function in groups rather than singly. Consequently, for a particular body movement to occur, a person must do more than cause a single muscle to contract; instead, the person wills the movement to occur, and the appropriate group of muscles responds to the decision.

Figure 8.13
The biceps brachii has 2 heads that originate on the scapula. This muscle is inserted on the radius by means of a tendon. What movement results as this muscle contracts?

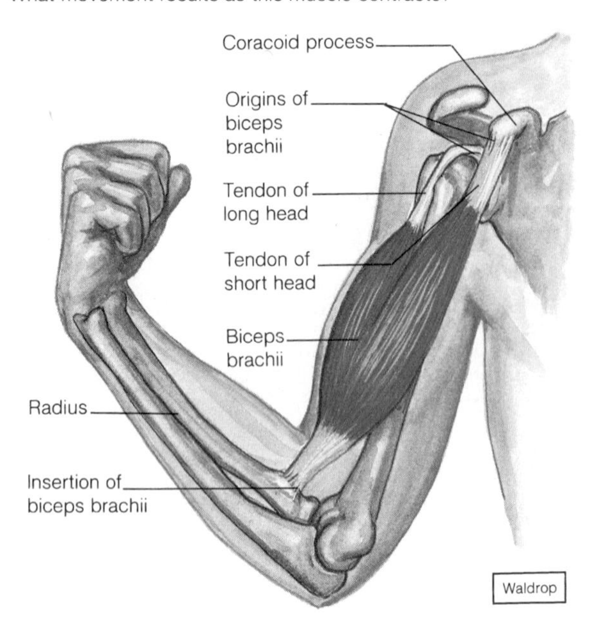

By carefully observing body movements, it is possible to determine the particular roles of various muscles. For instance, when the arm is lifted horizontally away from the side, a contracting *deltoid* muscle (fig. 8.14) is responsible for most of the movement, and so is said to be the **prime mover.** However, while a prime mover is acting, certain nearby muscles are also contracting. In the case of the contracting deltoid muscle, nearby muscles help to hold the shoulder steady and, in this way, make the action of the prime mover more effective. Muscles that contract and assist the prime mover are called **synergists.**

Still other muscles act as **antagonists** to prime movers. These muscles are capable of resisting a prime mover's action and are responsible for movement in the opposite direction—the antagonist of the prime mover that raises the arm can lower the arm, or the antagonist of the prime mover that bends the arm can straighten it (see figure 7.6). If both a prime mover and its antagonist contract simultaneously, the part they act upon remains rigid. Consequently, smooth body movements depend upon the antagonists relaxing and, thus, giving way to the prime movers whenever the prime movers are contracting. These complex actions are controlled by the nervous system, as is described in chapter 9.

1. Distinguish between the origin and insertion of a muscle.
2. Define *prime mover.*
3. What is the function of a synergist? An antagonist?

Major Skeletal Muscles

The following section concerns the locations, actions, and attachments of some of the major skeletal muscles. (Figures 8.14 and 8.15 show the locations of the superficial skeletal muscles—those near the surface.)

Note that the names of these muscles often describe them in some way. A name may indicate a muscle's relative size, shape, location, action, number of attachments, or the direction of its fibers, as in the following examples:

pectoralis major—of large size (major) located in the pectoral region (chest)
deltoid—shaped like a delta or triangle
extensor digitorum—acts to extend the digits (fingers or toes)
biceps brachii—with two heads (biceps) or points of origin and located in the brachium (arm)
sternocleidomastoid—attached to the sternum, clavicle, and mastoid process
external oblique—located near the outside with fibers that run obliquely (in a slanting direction)

Muscles of Facial Expression

A number of small muscles that lie beneath the skin of the face and scalp enable us to communicate feelings through facial expression (fig. 8.16). Many of these muscles are located around the eyes and mouth, and they are responsible for such expressions as surprise, sadness, anger, fear, disgust, and pain. As a group, the muscles of facial expression join the bones of the skull to connective tissue in various regions of the overlying skin. They include the following:

epicranius (ep″ĭ-kra′ne-us)
orbicularis oculi (or-bik′u-la-rus ok′u-li)
orbicularis oris (or-bik′u-la-rus o′ris)
buccinator (buk′sĭ-na″tor)
zygomaticus (zi″go-mat′ik-us)
platysma (plah-tiz′mah)

Chart 8.3 lists the origins, insertions, and actions of the muscles of facial expression. (The muscles that move the eyes are listed in chapter 10.)

Muscles of Mastication

Chewing movements are produced by the muscles that are attached to the mandible. Two pairs of these muscles act to close the lower jaw, as in biting. They are the following (figs. 8.14 and 8.15):

masseter (mas-se′ter)
temporalis (tem-po-ra′lis)

Chart 8.4 lists the origins, insertions, and actions of the muscles of mastication.

Figure 8.14
An anterior view of the superficial skeletal muscles.

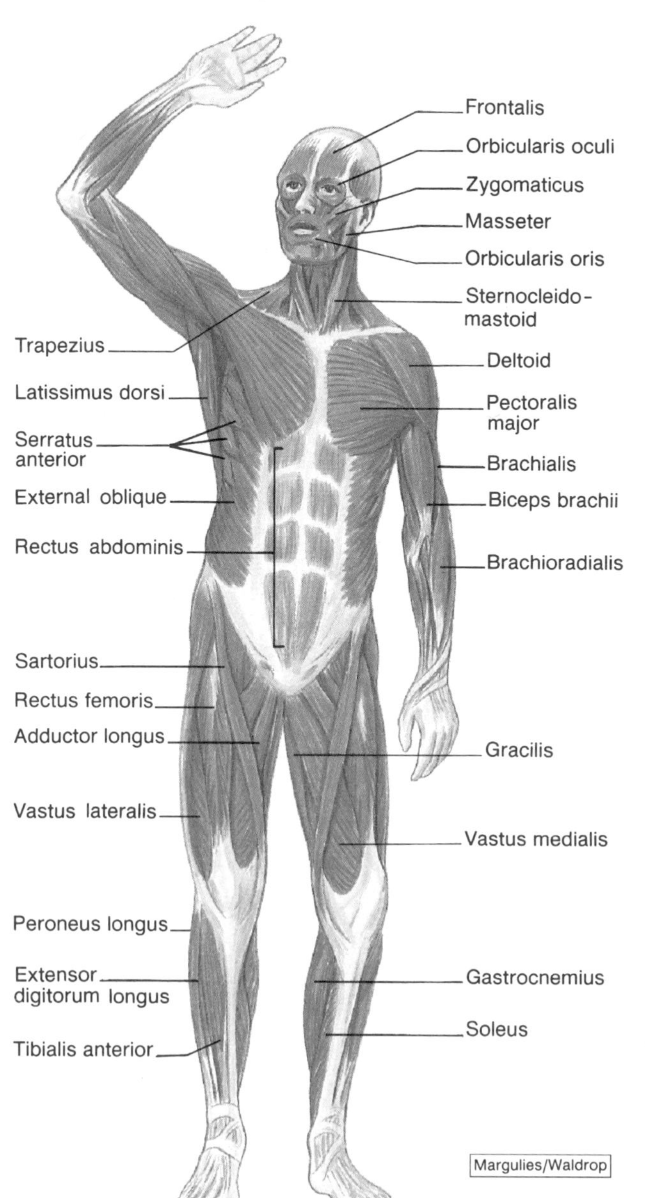

Figure 8.15
A posterior view of the superficial skeletal muscles.

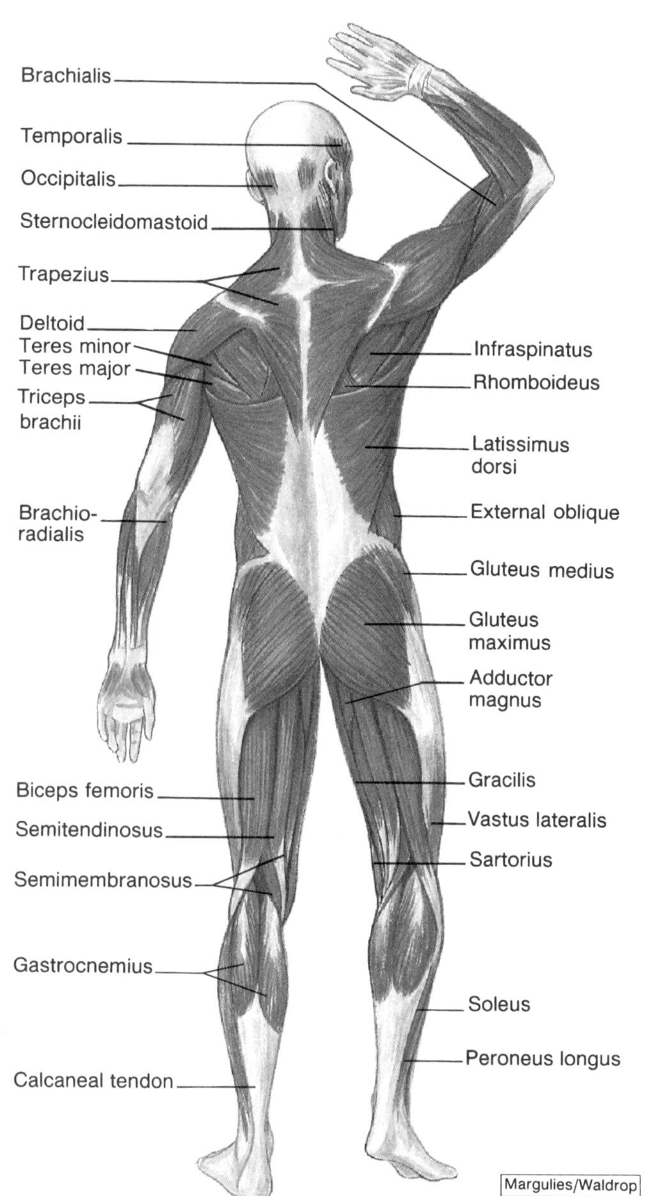

Figure 8.16
(*a*) Muscles of facial expression and mastication; (*b*) posterior view of muscles that move the head.

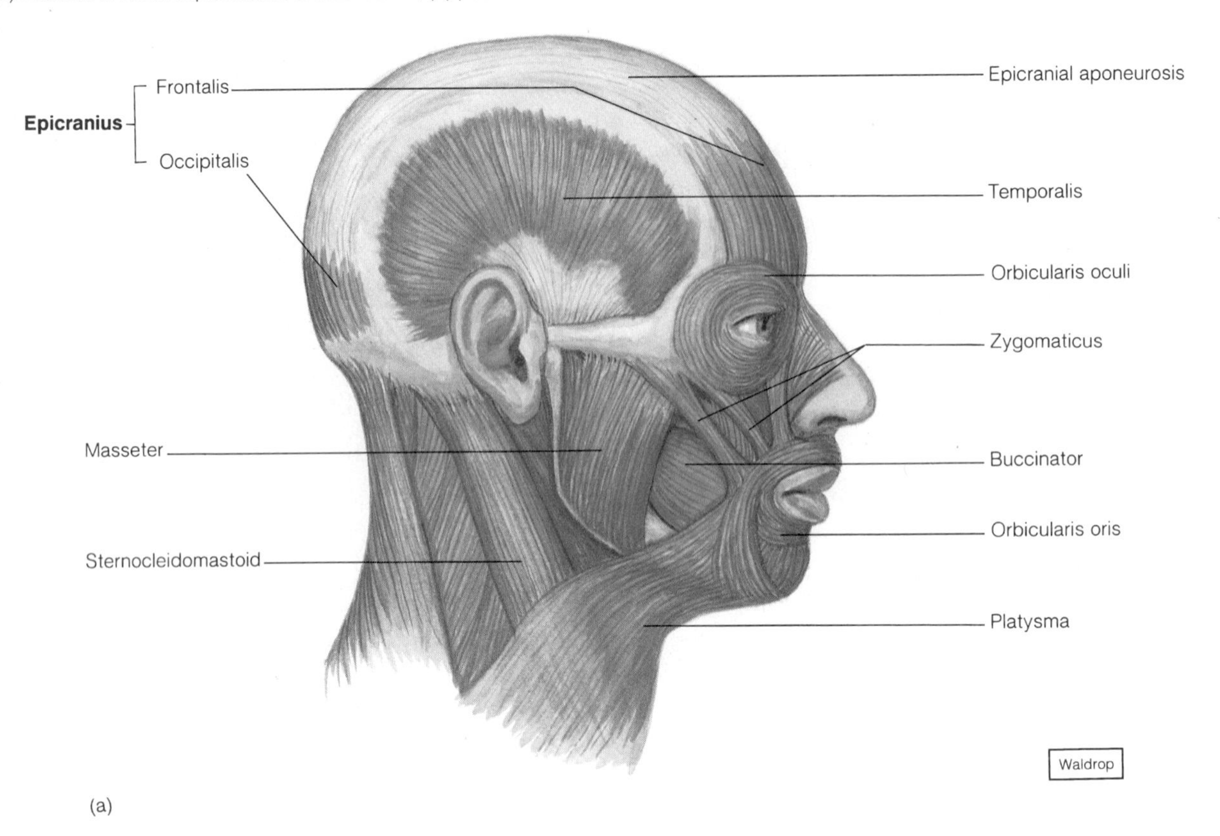

(a)

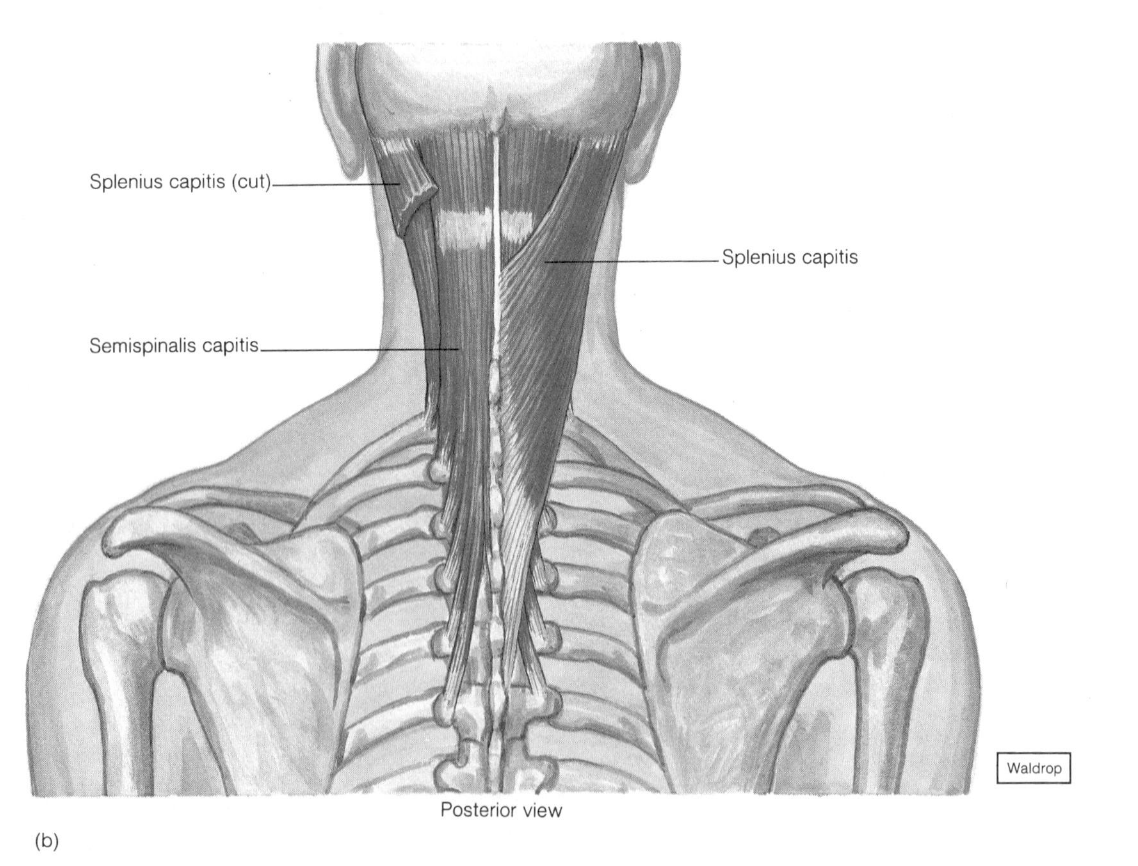

(b)

Chart 8.3 Muscles of facial expression			
Muscle	**Origin**	**Insertion**	**Action**
Epicranius	Occipital bone	Skin and muscles around the eye	Raises the eyebrow
Orbicularis oculi	Maxillary and frontal bones	Skin around the eye	Closes the eye
Orbicularis oris	Muscles near the mouth	Skin of the lips	Closes and protrudes the lips
Buccinator	Outer surfaces of the maxilla and mandible	Orbicularis oris	Compresses the cheeks inward
Zygomaticus	Zygomatic bone	Orbicularis oris	Raises the corner of the mouth
Platysma	Fascia in the upper chest	Lower border of the mandible	Draws the angle of the mouth downward

Chart 8.4 Muscles of mastication			
Muscle	**Origin**	**Insertion**	**Action**
Masseter	Lower border of the zygomatic arch	Lateral surface of the mandible	Closes the jaw
Temporalis	Temporal bone	Coronoid process and lateral surface of the mandible	Closes the jaw

Chart 8.5 Muscles that move the head			
Muscle	**Origin**	**Insertion**	**Action**
Sternocleidomastoid	Anterior of the sternum and upper surface of the clavicle	Mastoid process of the temporal bone	Pulls the head to one side, pulls the head toward the chest, or raises the sternum
Splenius capitis	Spinous processes of the lower cervical and upper thoracic vertebrae	Mastoid process of the temporal bone	Rotates the head, bends the head to one side, or brings the head into an upright position
Semispinalis capitis	Processes of the lower cervical and upper thoracic vertebrae	Occipital bone	Extends the head, bends the head to one side, or rotates the head

Muscles That Move the Head

Head movements result from the actions of paired muscles in the neck and upper back. These muscles are responsible for flexing, extending, and rotating the head. They include the following (fig. 8.16):

sternocleidomastoid (ster″no-kli″do-mas′toid)
splenius capitis (sple′ne-us kap′ĭ-tis)
semispinalis capitis (sem″e-spi-na′lis kap′ĭ-tis)

Chart 8.5 lists the origins, insertions, and actions of the muscles that move the head.

Muscles That Move the Pectoral Girdle

The muscles that move the pectoral girdle are closely associated with those that move the upper arm. A number of these chest and shoulder muscles connect the scapula to nearby bones and act to move the scapula upward, downward, forward, and backward. They include the following (figs. 8.17 and 8.18):

trapezius (trah-pe′ze-us)
rhomboideus major (rom′boid-e-us)
levator scapulae (le-va′tor scap′u-lē)
serratus anterior (ser-ra′tus an-te′re-or)
pectoralis minor (pek″to-ra′lis)

Chart 8.6 lists the origins, insertions, and actions of the muscles that move the pectoral girdle.

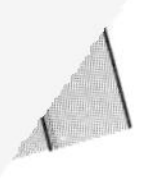

rior shoulder. (The right trapezius is removed to show underlying muscles.)

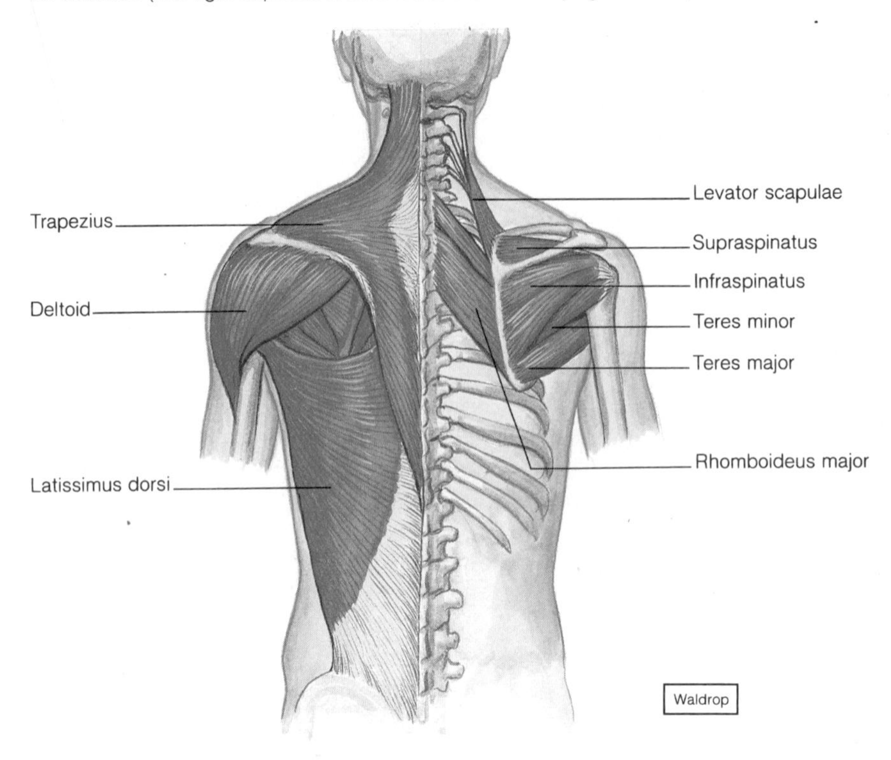

Figure 8.18

Muscles of the anterior chest and abdominal wall. (The left pectoralis major is removed to show the pectoralis minor.)

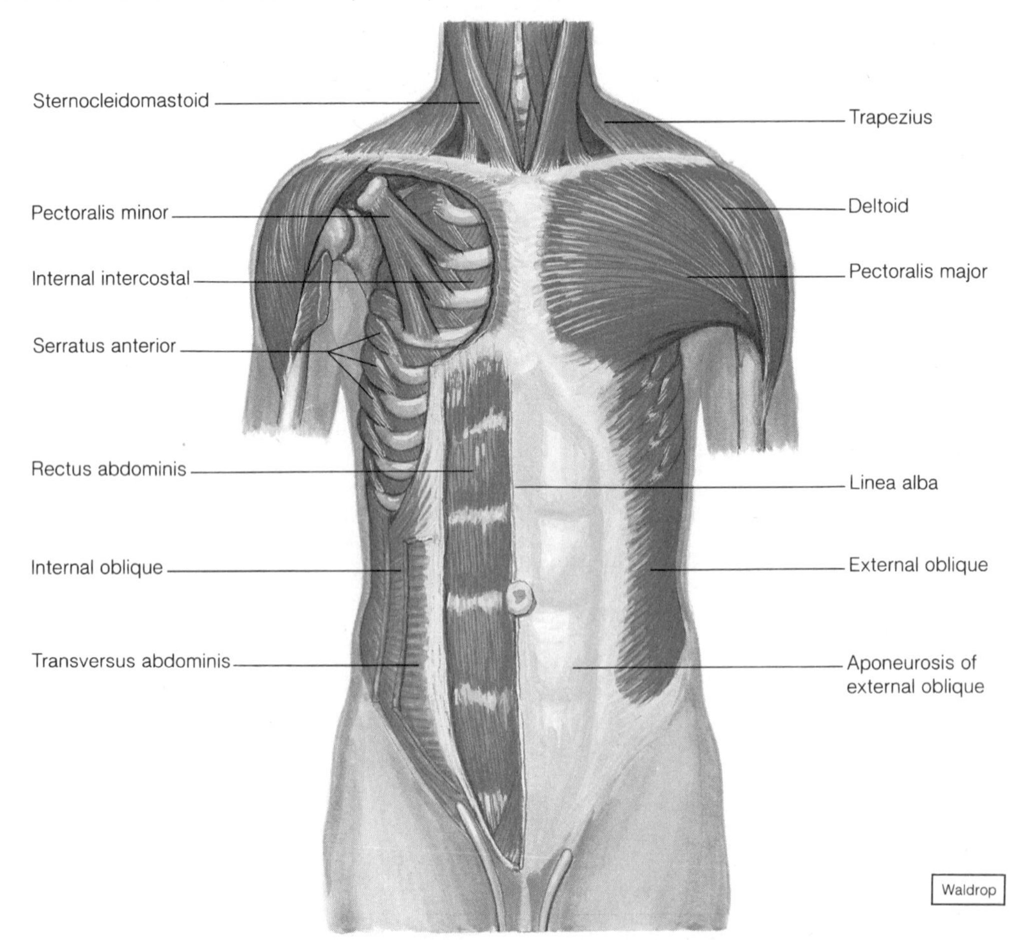

Chart 8.6 Muscles that move the pectoral girdle			
Muscle	**Origin**	**Insertion**	**Action**
Trapezius	Occipital bone and spines of the cervical and thoracic vertebrae	Clavicle; spine and acromion process of the scapula	Rotates the scapula and flexes the arm; raises the scapula; pulls the scapula medially; or pulls the scapula and shoulder downward
Rhomboideus major	Spines of the upper thoracic vertebrae	Medial border of the scapula	Raises and adducts the scapula
Levator scapulae	Transverse processes of the cervical vertebrae	Medial margin of the scapula	Elevates the scapula
Serratus anterior	Outer surfaces of the upper ribs	Ventral surface of the scapula	Pulls the scapula anteriorly and downward
Pectoralis minor	Sternal ends of the upper ribs	Coracoid process of the scapula	Pulls the scapula forward and downward or raises the ribs

Figure 8.19
Muscles of the posterior surface of the scapula and upper arm.

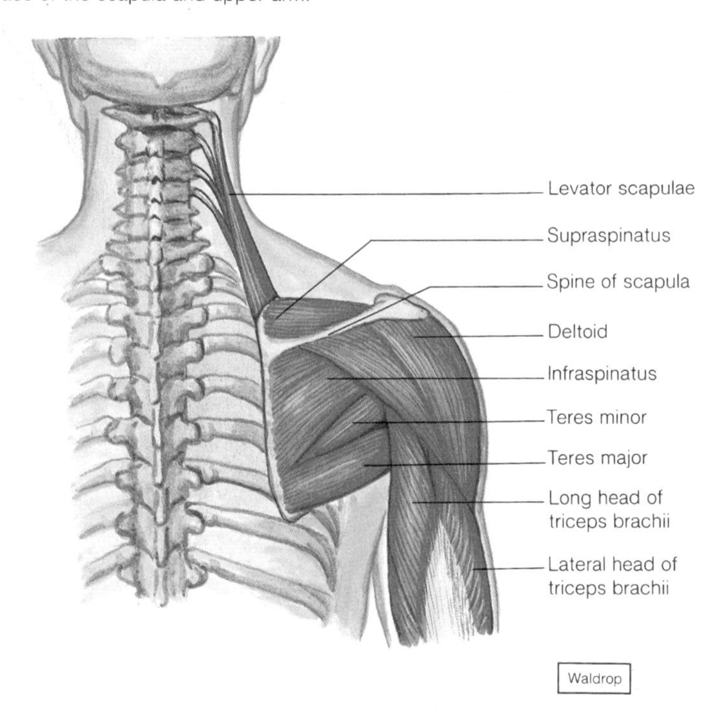

Muscles That Move the Upper Arm

The upper arm is one of the more freely movable parts of the body. Its many movements are made possible by the muscles that connect the humerus to various regions of the pectoral girdle, ribs, and vertebral column (figs. 8.17, 8.18, 8.19, and 8.20). These muscles can be grouped according to their primary actions—flexion, extension, abduction, and rotation—as follows:

Flexors
- *coracobrachialis* (kor″ah-ko-bra′ke-al-is)
- *pectoralis major* (pek″to-ra′lis)

Extensors
- *teres major* (te′rez)
- *latissimus dorsi* (lah-tis′ĭ-mus dor′si)

Abductors
- *supraspinatus* (su″prah-spi′na-tus)
- *deltoid* (del′toid)

Rotators
- *subscapularis* (sub-scap′u-lar-is)
- *infraspinatus* (in″frah-spi′na-tus)
- *teres minor* (te′rēz)

Chart 8.7 lists the origins, insertions, and actions of the muscles that move the upper arm.

Figure 8.20
Muscles of the anterior shoulder and upper arm, with the rib cage removed.

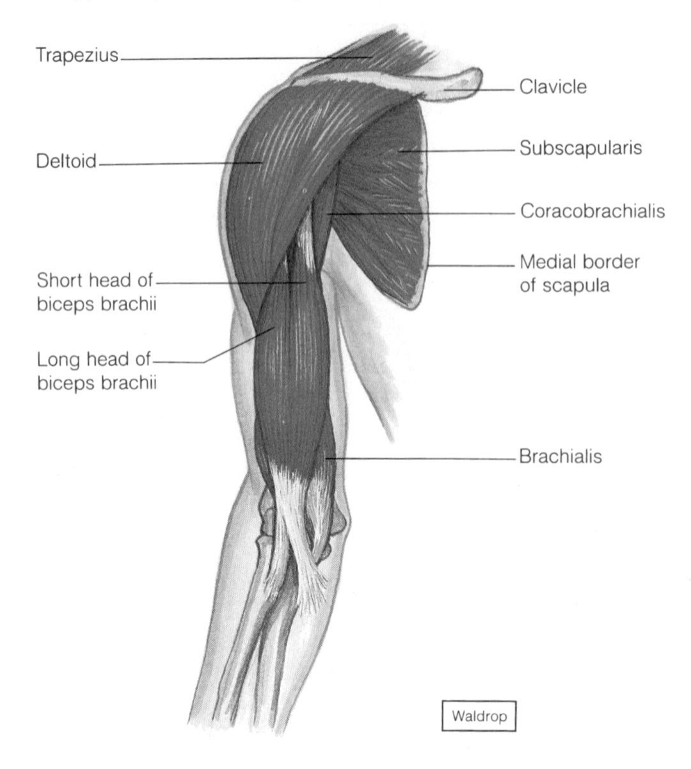

Chart 8.7 Muscles that move the upper arm			
Muscle	**Origin**	**Insertion**	**Action**
Coracobrachialis	Coracoid process of the scapula	Shaft of the humerus	Flexes and adducts the upper arm
Pectoralis major	Clavicle, sternum, and costal cartilages of the upper ribs	Intertubercular groove of the humerus	Pulls the arm forward and across the chest, rotates the humerus, or adducts the arm
Teres major	Lateral border of the scapula	Intertubercular groove of the humerus	Extends the humerus, or adducts and rotates the arm medially
Latissimus dorsi	Spines of the sacral, lumbar, and lower thoracic vertebrae, the iliac crest, and the lower ribs	Intertubercular groove of the humerus	Extends and adducts the arm and rotates the humerus inwardly, or pulls the shoulder downward and back
Supraspinatus	Posterior surface of the scapula	Greater tubercle of the humerus	Abducts the upper arm
Deltoid	Acromion process, spine of the scapula, and clavicle	Deltoid tuberosity of the humerus	Abducts the upper arm, extends the humerus, or flexes the humerus
Subscapularis	Anterior surface of the scapula	Lesser tubercle of the humerus	Rotates the arm medially
Infraspinatus	Posterior surface of the scapula	Greater tubercle of the humerus	Rotates the arm laterally
Teres minor	Lateral border of the scapula	Greater tubercle of the humerus	Rotates the arm laterally

Figure 8.21
Muscles of the anterior forearm.

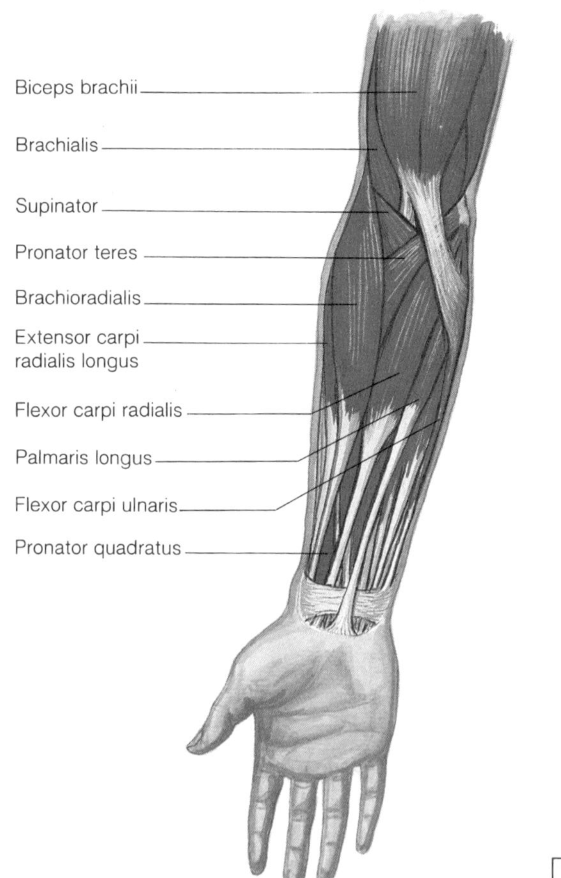

Muscles That Move the Forearm

Most forearm movements are produced by the muscles that connect the radius or ulna to the humerus or pectoral girdle. A group of muscles located along the anterior surface of the humerus act to flex the elbow, and a single posterior muscle serves to extend this joint. Other muscles cause movements at the radioulnar joint and are responsible for rotating the forearm.

The muscles that move the forearm include the following (figs. 8.19, 8.20, and 8.21):

Flexors
- *biceps brachii* (bi′seps bra′ke-i)
- *brachialis* (bra′ke-al-is)
- *brachioradialis* (bra″ke-o-ra″de-a′lis)

Extensor
- *triceps brachii* (tri′seps bra′ke-i)

Rotators
- *supinator* (su′pi̅-na-tor)
- *pronator teres* (pro-na′tor te′rēz)
- *pronator quadratus* (pro-na′tor kwod-ra′tus)

Chart 8.8 lists the origins, insertions, and actions of the muscles that move the forearm.

Chart 8.8 Muscles that move the forearm

Muscle	Origin	Insertion	Action
Biceps brachii	Coracoid process and tubercle above the glenoid cavity of the scapula	Radial tuberosity of the radius	Flexes the arm at the elbow and rotates the hand laterally
Brachialis	Anterior shaft of the humerus	Coronoid process of the ulna	Flexes the arm at the elbow
Brachioradialis	Distal lateral end of the humerus	Lateral surface of the radius above the styloid process	Flexes the arm at the elbow
Triceps brachii	Tubercle below the glenoid cavity and lateral and medial surfaces of the humerus	Olecranon process of the ulna	Extends the arm at the elbow
Supinator	Lateral epicondyle of the humerus and crest of the ulna	Lateral surface of the radius	Rotates the forearm laterally
Pronator teres	Medial epicondyle of the humerus and coronoid process of the ulna	Lateral surface of the radius	Rotates the arm medially
Pronator quadratus	Anterior distal end of the ulna	Anterior distal end of the radius	Rotates the arm medially

Figure 8.22
Muscles of the posterior forearm

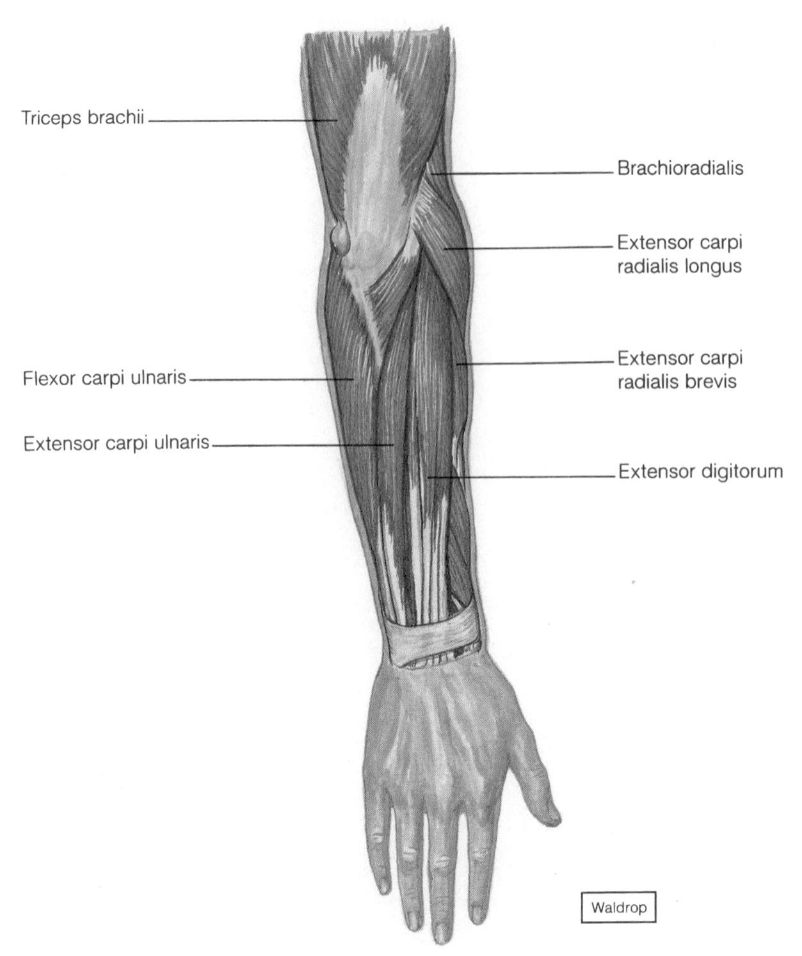

Muscles That Move the Wrist, Hand, and Fingers

Many muscles are responsible for wrist, hand, and finger movements. They originate from the distal end of the humerus and from the radius and ulna. The two major groups of these muscles are flexors on the anterior side of the forearm and extensors on the posterior side. These muscles include the following (figs. 8.21 and 8.22):

Flexors
- *flexor carpi radialis* (flek′sor kar-pi′ ra″de-a′lis)
- *flexor carpi ulnaris* (flex′sor kar-pi′ ul-na′ris)
- *palmaris longus* (pal-ma′ris long′gus)
- *flexor digitorum profundus* (flex′sor dij″ĭ-to′rum pro-fun′dus)

Extensors
- *extensor carpi radialis longus* (eks-ten′sor kar-pi′ ra″de-a′lis long′gus)
- *extensor carpi radialis brevis* (eks-ten′sor kar-pi′ ra″de-a′lis bre′ĭs)
- *extensor carpi ulnaris* (eks-ten′sor kar-pi′ ul-na′ris)
- *extensor digitorum* (eks-ten′sor dij″ĭ-to′rum)

Chart 8.9 lists the origins, insertions, and actions of the muscles that move the wrist, hand, and fingers.

Muscles of the Abdominal Wall

Although the walls of the chest and pelvic regions are supported directly by bone, those of the abdomen are not. Instead, the anterior and lateral walls of the abdomen are composed of broad, flattened muscles arranged in layers. These muscles connect the rib cage and vertebral column to the pelvic girdle. A band of tough connective tissue, called the **linea alba,** extends from the xiphoid process of the sternum to the symphysis pubis (fig. 8.18). It serves as an attachment for some of the abdominal wall muscles.

Contraction of these muscles decreases the size of the abdominal cavity and increases the pressure inside. This action helps to press air out of the lungs

Chart 8.9 Muscles that move the wrist, hand, and fingers			
Muscle	**Origin**	**Insertion**	**Action**
Flexor carpi radialis	Medial epicondyle of the humerus	Base of the second and third metacarpals	Flexes and abducts the wrist
Flexor carpi ulnaris	Medial epicondyle of the humerus and olecranon process	Carpal and metacarpal bones	Flexes and adducts the wrist
Palmaris longus	Medial epicondyle of the humerus	Fascia of the palm	Flexes the wrist
Flexor digitorum profundus	Anterior surface of the ulna	Bases of the distal phalanges in fingers two through five	Flexes the distal joints of the fingers
Extensor carpi radialis longus	Distal end of the humerus	Base of the second metacarpal	Extends the wrist and abducts the hand
Extensor carpi radialis brevis	Lateral epicondyle of the humerus	Base of the second and third metacarpals	Extends the wrist and abducts the hand
Extensor carpi ulnaris	Lateral epicondyle of the humerus	Base of the fifth metacarpal	Extends and adducts the wrist
Extensor digitorum	Lateral epicondyle of the humerus	Posterior surface of the phalanges in fingers two through five	Extends the fingers

Chart 8.10 Muscles of the abdominal wall			
Muscle	**Origin**	**Insertion**	**Action**
External oblique	Outer surfaces of the lower ribs	Outer lip of the iliac crest and linea alba	Tenses the abdominal wall and compresses the abdominal contents
Internal oblique	Crest of the ilium and inguinal ligament	Cartilages of the lower ribs, linea alba, and crest of the pubis	Same as above
Transversus abdominis	Costal cartilages of the lower ribs, processes of the lumbar vertebrae, lip of the iliac crest, and inguinal ligament	Linea alba and crest of the pubis	Same as above
Rectus abdominis	Crest of the pubis and symphysis pubis	Xiphoid process of the sternum and costal cartilages	Same as above, also flexes the vertebral column

during forceful exhalation and also aids in defecation, urination, vomiting, and childbirth.

The abdominal wall muscles include the following (fig. 8.18):

external oblique (eks-ter′nal ō-blēk)
internal oblique (in-ter′nal ō-blēk)
transversus abdominis (trans-ver′sus ab-dom′ĭ-nis)
rectus abdominis (rek′tus ab-dom′ĭ-nis)

Chart 8.10 lists the origins, insertions, and actions of the muscles of the abdominal wall.

Muscles of the Pelvic Outlet

The outlet of the pelvis is spanned by two muscular sheets—a deeper **pelvic diaphragm** and a more superficial **urogenital diaphragm.** The pelvic diaphragm forms the floor of the pelvic cavity, and the urogenital diaphragm fills the space within the pubic arch (fig. 7.26). The muscles of the male and female pelvic outlets include the following (fig. 8.23):

Pelvic diaphragm
levator ani (le-va′tor ah-ni′)
Urogenital diaphragm
superficial transversus perinei (su″per-fish′al trans-ver′sus per″ĭ-ne′i)
bulbospongiosus (bul″bo-spon″je-o′sus)
ischiocavernosus (is″ke-o-kav″er-no′sus)

Chart 8.11 lists the origins, insertions, and actions of the pelvic outlet muscles.

Figure 8.23
External view of muscles of (*a*) the male pelvic outlet and (*b*) the female pelvic outlet.

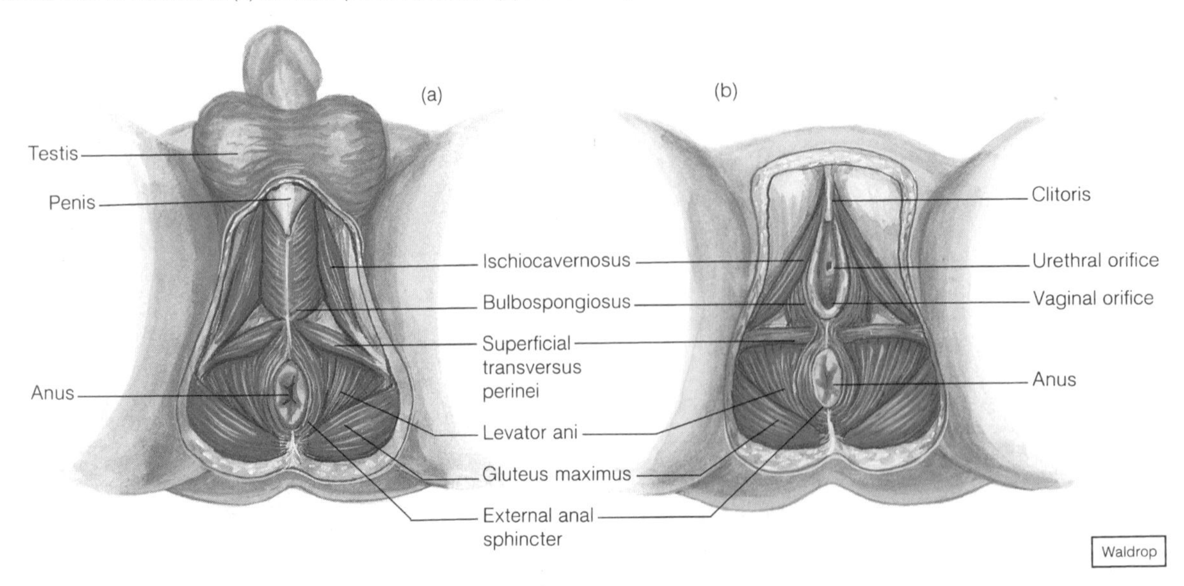

Chart 8.11 Muscles of the pelvic outlet

Muscle	Origin	Insertion	Action
Levator ani	Pubic bone and ischial spine	Coccyx	Supports the pelvic viscera, and provides a sphincterlike action in the anal canal and vagina
Superficial transversus perinei	Ischial tuberosity	Central tendon	Supports the pelvic viscera
Bulbospongiosus	Central tendon	Males: the urogenital diaphragm and fascia of the penis	Males: assists in emptying the urethra
		Females: the pubic arch and root of the clitoris	Females: constricts the vagina
Ischiocavernosus	Ischial tuberosity	Pubic arch	Assists the function of the bulbospongiosus

Muscles That Move the Thigh

The muscles that move the thigh are attached to the femur and to some part of the pelvic girdle. They can be separated into anterior and posterior groups. The muscles of the anterior group act primarily to flex the thigh; those of the posterior group extend, abduct, or rotate it. The muscles in these groups include the following (figs. 8.24, 8.25, and 8.26):

Anterior group
- *psoas major* (so′as)
- *iliacus* (il′e-ak-us)

Posterior group
- *gluteus maximus* (gloo′te-us mak′si-mus)
- *gluteus medius* (gloo′te-us me′de-us)
- *gluteus minimus* (gloo′te-us min′ĭ-mus)
- *tensor fasciae latae* (ten′sor fash′e-e lah-tē)

Still another group of muscles attached to the femur and pelvic girdle function to adduct the thigh. They include the following:

- *adductor longus* (ah-duk′tor long′gus)
- *adductor magnus* (ah-duk′tor mag′nus)
- *gracilis* (gras′il-is)

Chart 8.12 lists the origins, insertions, and actions of the muscles that move the thigh.

Figure 8.24
Muscles of the anterior right thigh.

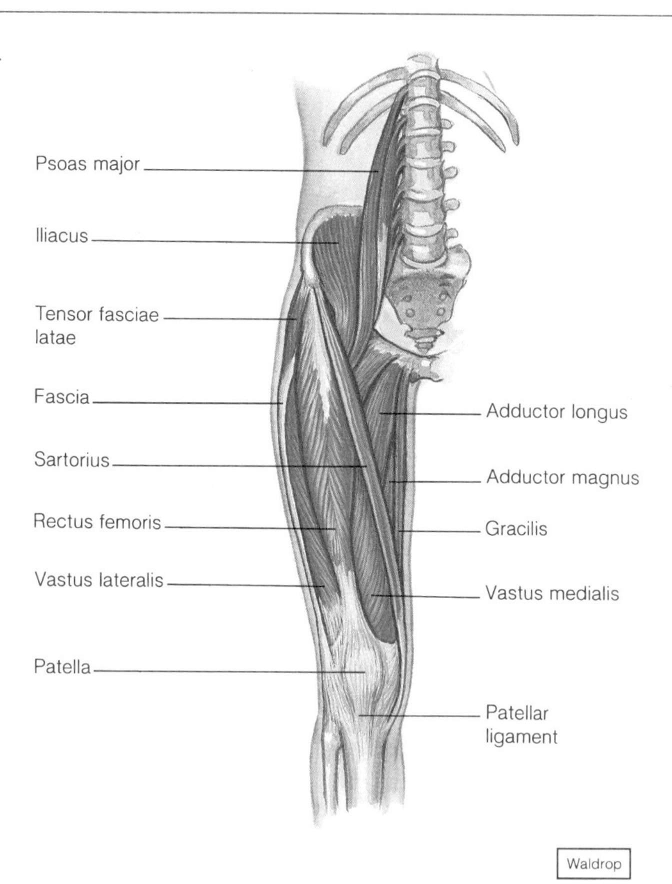

Figure 8.25
Muscles of the lateral right thigh.

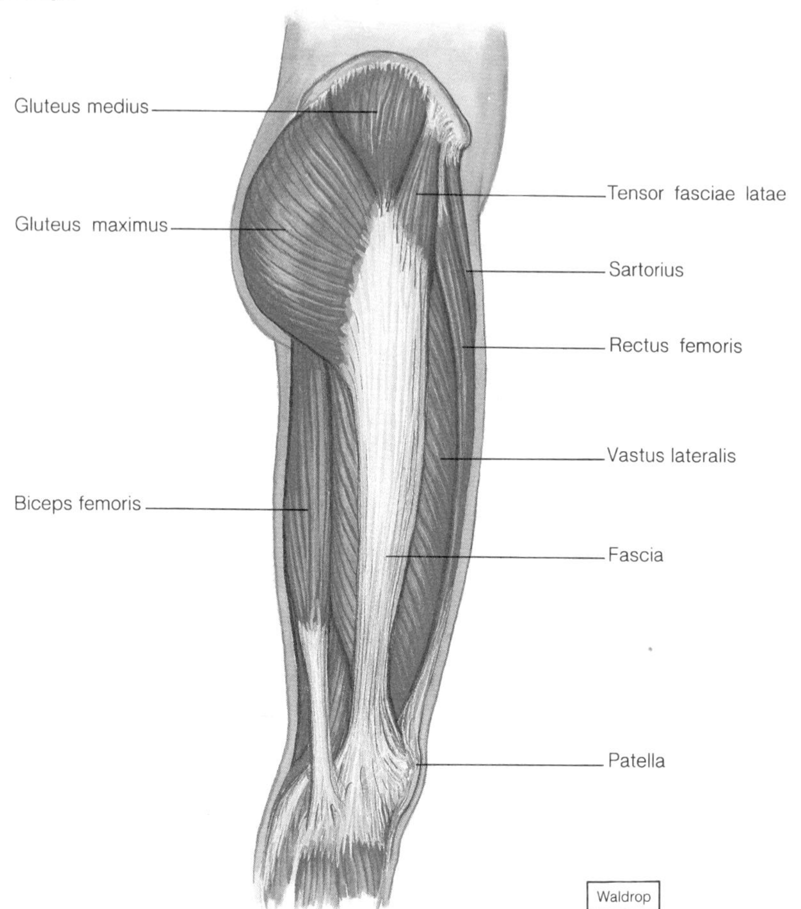

Figure 8.26
Muscles of the posterior right thigh.

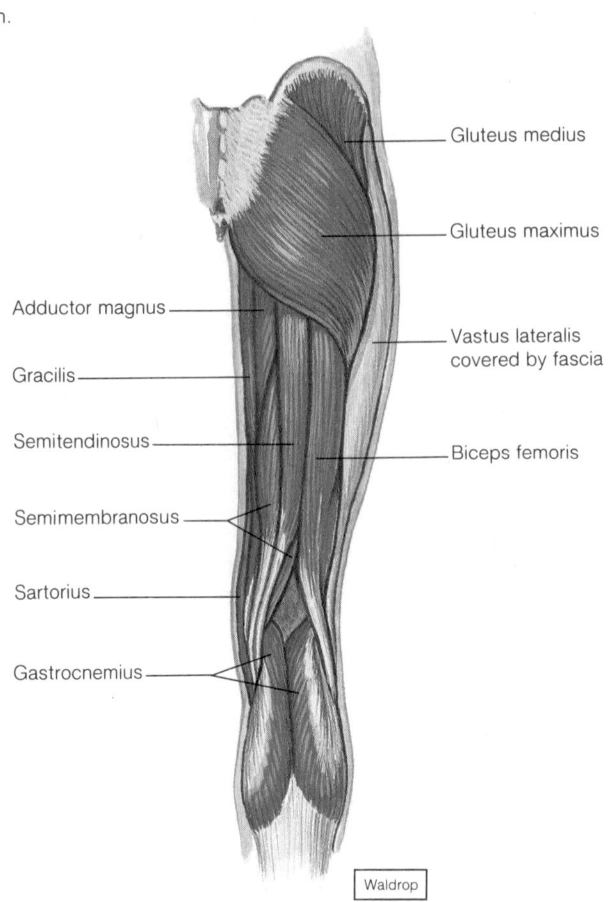

Chart 8.12 Muscles that move the thigh			
Muscle	**Origin**	**Insertion**	**Action**
Psoas major	Lumbar intervertebral disks, bodies and transverse processes of the lumbar vertebrae	Lesser trochanter of the femur	Flexes the thigh
Iliacus	Iliac fossa of the ilium	Lesser trochanter of the femur	Flexes the thigh
Gluteus maximus	Sacrum, coccyx, and posterior surface of the ilium	Posterior surface of the femur and fascia of the thigh	Extends the leg at the hip
Gluteus medius	Lateral surface of the ilium	Greater trochanter of the femur	Abducts and rotates the thigh medially
Gluteus minimus	Lateral surface of the ilium	Greater trochanter of the femur	Same as the gluteus medius
Tensor fasciae latae	Anterior iliac crest	Fascia of the thigh	Abducts, flexes, and rotates the thigh medially
Adductor longus	Pubic bone near the symphysis pubis	Posterior surface of the femur	Adducts, flexes, and rotates the thigh laterally
Adductor magnus	Ischial tuberosity	Posterior surface of the femur	Adducts, extends, and rotates the thigh laterally
Gracilis	Lower edge of the symphysis pubis	Medial surface of the tibia	Adducts the thigh, flexes and rotates the leg medially at the knee

Chart 8.13 Muscles that move the lower leg			
Muscle	**Origin**	**Insertion**	**Action**
Hamstring group			
Biceps femoris	Ischial tuberosity and posterior surface of the femur	Head of the fibula and lateral condyle of the tibia	Flexes and rotates the leg laterally and extends the thigh
Semitendinosus	Ischial tuberosity	Medial surface of the tibia	Flexes and rotates the leg medially and extends the thigh
Semimembranosus	Ischial tuberosity	Medial condyle of the tibia	Flexes and rotates the leg medially and extends the thigh
Sartorius	Anterior superior iliac spine	Medial surface of the tibia	Flexes the leg and thigh, abducts the thigh, rotates the thigh laterally, and rotates the leg medially
Quadriceps femoris group			
Rectus femoris	Spine of the ilium and margin of the acetabulum	Patella by the tendon, which continues as the patellar ligament to the tibial tuberosity	Extends the leg at the knee
Vastus lateralis	Greater trochanter and posterior surface of the femur	Patella by the tendon, which continues as the patellar ligament to the tibial tuberosity	Extends the leg at the knee
Vastus medialis	Medial surface of the femur	Patella by the tendon, which continues as the patellar ligament to the tibial tuberosity	Extends the leg at the knee
Vastus intermedius	Anterior and lateral surfaces of the femur	Patella by the tendon, which continues as the patellar ligament to the tibial tuberosity	Extends the leg at the knee

Muscles That Move the Lower Leg

The muscles that move the lower leg connect the tibia or fibula to the femur or to the pelvic girdle. They can be separated into two major groups—those that cause flexion at the knee and those that cause extension at the knee. The muscles of these groups include the following (figs. 8.24, 8.25, and 8.26):

Flexors
- *biceps femoris* (bi′seps fem′or-is)
- *semitendinosus* (sem″e-ten′dĭ-no-sus)
- *semimembranosus* (sem″e-mem′brah-no-sus)
- *sartorius* (sar-to′re-us)

Extensor
- *quadriceps femoris group* (kwod′rĭ-seps fem′or-is)

Chart 8.13 lists the origins, insertions, and actions of the muscles that move the lower leg.

Muscles That Move the Ankle, Foot, and Toes

A number of muscles that function to move the ankle, foot, and toes are located in the lower leg. They attach the femur, tibia, and fibula to various bones of the foot and are responsible for a variety of movements—moving the foot upward (dorsal flexion) or downward (plantar flexion), and turning the sole of the foot inward (inversion) or outward (eversion). These muscles include the following (figs. 8.27, 8.28, and 8.29):

Dorsal flexors
- *tibialis anterior* (tib″e-a′lis an-te′re-or)
- *peroneus tertius* (per″o-ne′us ter′shus)
- *extensor digitorum longus* (eks-ten′sor dij″ĭ-to′rum long′gus)

Plantar flexors
- *gastrocnemius* (gas″trok-ne′me-us)
- *soleus* (so′le-us)
- *flexor digitorum longus* (flek′sor dij″ĭ-to′rum long′gus)

Evertor
- *peroneus longus* (per″o-ne′us long′gus)

Chart 8.14 lists the origins, insertions, and actions of the muscles that move the ankle, foot, and toes.

Figure 8.27
Muscles of the anterior right lower leg.

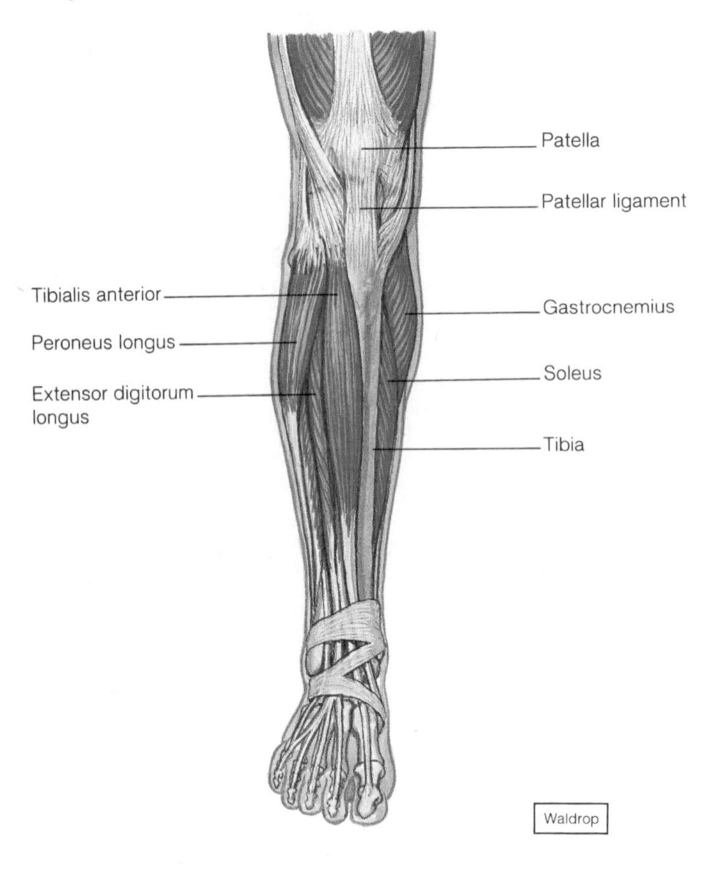

Figure 8.28
Muscles of the lateral right lower leg.

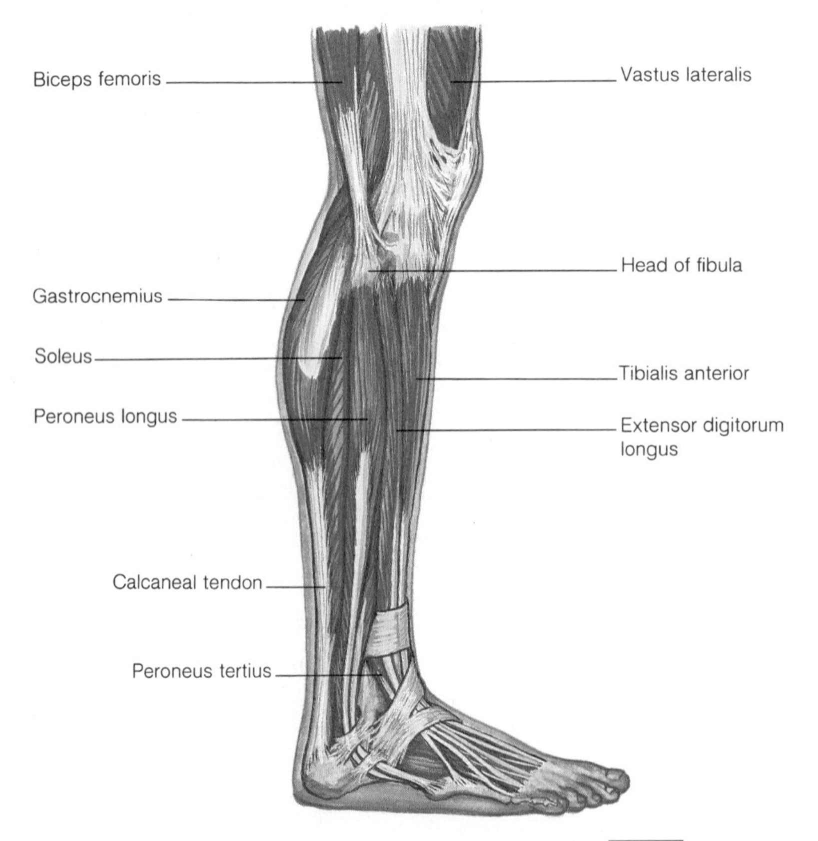

Figure 8.29
Muscles of the posterior right lower leg.

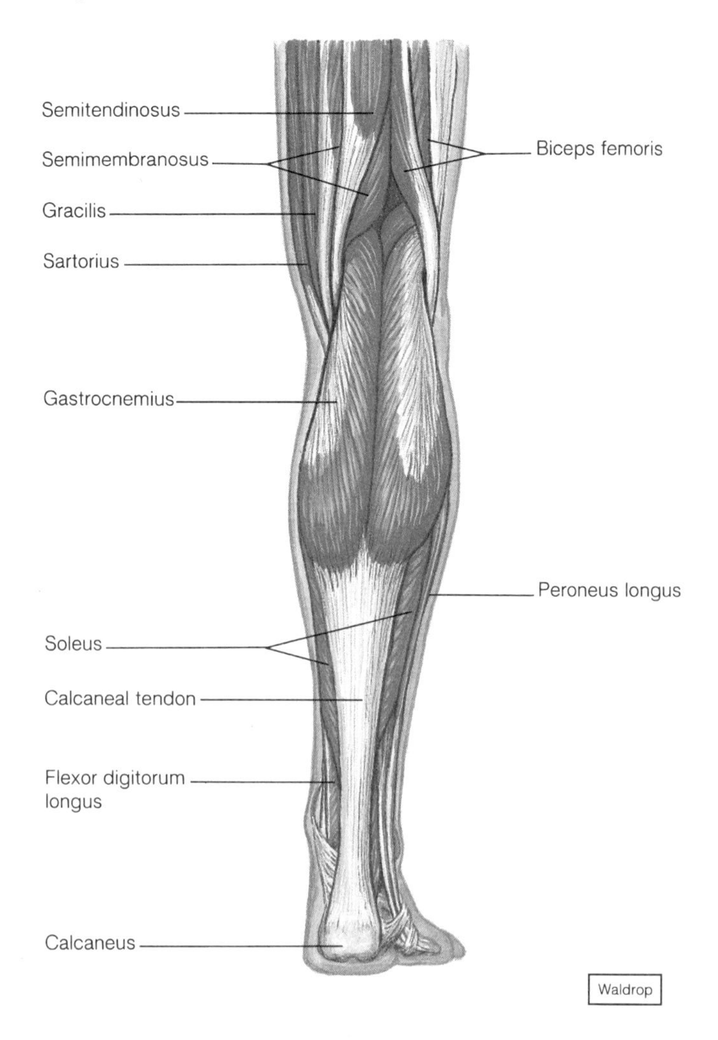

Chart 8.14 Muscles that move the ankle, foot, and toes

Muscle	Origin	Insertion	Action
Tibialis anterior	Lateral condyle and lateral surface of the tibia	Tarsal bone (cuneiform) and first metatarsal	Dorsal flexion and inversion of the foot
Peroneus tertius	Anterior surface of the fibula	Dorsal surface of the fifth metatarsal	Dorsal flexion and eversion of the foot
Extensor digitorum longus	Lateral condyle of the tibia and anterior surface of the fibula	Dorsal surfaces of the second and third phalanges of the four lateral toes	Dorsal flexion and eversion of the foot and extension of the toes
Gastrocnemius	Lateral and medial condyles of the femur	Posterior surface of the calcaneus	Plantar flexion of the foot and flexion of the leg at the knee
Soleus	Head and shaft of the fibula and posterior surface of the tibia	Posterior surface of the calcaneus	Plantar flexion of the foot
Flexor digitorum longus	Posterior surface of the tibia	Distal phalanges of the four lateral toes	Plantar flexion and inversion of the foot and flexion of the four lateral toes
Peroneus longus	Lateral condyle of the tibia and head and shaft of the fibula	Tarsal and metatarsal bones	Plantar flexion and eversion of the foot; also supports the arch

Clinical Terms Related to the Muscular System

contracture (kon-trak'tur) a condition in which there is great resistance to the stretch of a muscle.
convulsion (kun-vul'shun)—an involuntary contraction of muscles.
electromyography (e-lek"tro-mi-og'rah-fe)—a technique for recording the electrical changes that occur in muscle tissues.
fibrillation (fi"bri-la'shun)—spontaneous contractions of individual muscle fibers, producing rapid and uncoordinated activity within a muscle.
fibrosis (fi-bro'sis)—a degenerative disease in which skeletal muscle tissue is replaced by fibrous connective tissue.
fibrositis (fi"bro-si'tis)—an inflammatory condition of fibrous connective tissues, especially in the muscle fascia. This disease is also called muscular rheumatism.
muscular dystrophy (mus'ku-lar dis'tro-fe)—a progressively crippling disease of unknown cause in which the muscles gradually weaken and atrophy.
myalgia (mi-al'je-ah)—pain resulting from any muscular disease or disorder.
myasthenia gravis (mi"as-the'ne-ah gra'vis)—a chronic disease characterized by muscles that are weak and easily fatigued. It results from a disorder at some of the neuromuscular junctions so that a stimulus is not transmitted from the motor neuron to the muscle fiber.
myokymia (mi"o-ki'me-ah)—persistent quivering of a muscle.
myology (mi-ol'o-je)—the study of muscles.
myoma (mi-o'mah)—a tumor composed of muscle tissue.
myopathy (mi-op'ah-the)—any muscular disease.
myositis (mi"o-si'tis)—an inflammation of skeletal muscle tissue.
myotomy (mi-ot'o-me)—the cutting of muscle tissue.
myotonia (mi"o-to'ne-ah)—a prolonged muscular spasm.
paralysis (pah-ral'i-sis)—the loss of ability to move a body part.
paresis (pah-re'sis) a partial or slight paralysis of muscles.
shin splints (shin splints)—a soreness on the front of the lower leg due to straining the flexor digitorum longus, often as a result of walking up and down hills.
torticollis (tor"ti-kol'is)—a condition in which the neck muscles, such as the sternocleidomastoids, contract involuntarily. It is more commonly called wryneck.

Chapter Summary

Introduction (page 170)

There are three types of muscle tissue—skeletal, smooth, and cardiac.

Structure of a Skeletal Muscle (page 170)

Individual muscles are the organs of the muscular system.

They contain skeletal muscle tissue, nervous tissue, blood, and various connective tissues.

1. Connective tissue coverings
   a. Skeletal muscles and their parts are covered with fascia.
   b. Other connective tissues attach muscles to bones or to other muscles.
   c. A network of connective tissue extends throughout the muscular system.
2. Skeletal muscle fibers
   a. Each skeletal muscle fiber represents a single muscle cell, which is the unit of contraction.
   b. The cytoplasm contains mitochondria, a sarcoplasmic reticulum, and myofibrils, composed of actin and myosin.
   c. Striations are produced by the arrangement of the actin and myosin filaments.
   d. Transverse tubules extend inward from the cell membrane and are associated with the sarcoplasmic reticulum.
3. Neuromuscular junction
   a. Muscle fibers are stimulated to contract by motor neurons.
   b. In response to a nerve impulse, the end of a motor nerve fiber secretes a neurotransmitter, which stimulates the muscle fiber.
4. Motor units
   a. One motor neuron and the muscle fibers associated with it constitute a motor unit.
   b. All the fibers of a motor unit contract together.

Skeletal Muscle Contraction (page 174)

Muscle fiber contraction results from a sliding movement in which the actin and myosin filaments merge.

1. Role of myosin and actin
   a. The crossbridges of the myosin filaments can form linkages with the actin filaments.
   b. The reaction between the actin and myosin filaments generates the force of contraction.

2. Stimulus for contraction
   a. A skeletal muscle fiber is stimulated by acetylcholine, released from a motor nerve fiber.
   b. Acetylcholine causes the muscle fiber to conduct an impulse that reaches the deep parts of the fiber by means of the transverse tubules.
   c. A muscle impulse signals the sarcoplasmic reticulum to release calcium ions.
   d. Linkages form between actin and myosin, and the actin filaments move inward.
   e. The muscle fiber relaxes when the calcium ions are transported back into the sarcoplasmic reticulum.
   f. Cholinesterase decomposes acetylcholine.
3. Energy sources for contraction
   a. ATP supplies the energy for muscle fiber contraction.
   b. Creatine phosphate stores energy that can be used to synthesize ATP.
4. Oxygen supply and cellular respiration
   a. Aerobic respiration requires the presence of oxygen.
   b. Oxygen is carried to the body cells by the red blood cells.
   c. Myoglobin in the muscle cells stores oxygen temporarily.
5. Oxygen debt
   a. During rest or moderate exercise, oxygen is supplied to the muscles in sufficient concentration to support aerobic respiration.
   b. During strenuous exercise, an oxygen deficiency may develop, and lactic acid may accumulate.
   c. The amount of oxygen needed to convert accumulated lactic acid to glucose and to restore the supplies of ATP and creatine phosphate is called the oxygen debt.
6. Muscle fatigue
   a. A fatigued muscle loses its ability to contract.
   b. Muscle fatigue is usually due to an accumulation of lactic acid.
7. Heat production
   a. Most of the energy released by cellular respiration is lost as heat.
   b. Muscles represent an important source of body heat.

Muscular Responses (page 178)

1. The threshold stimulus is the minimal stimulus needed to elicit a muscular contraction.
2. All-or-none response
   a. If a muscle fiber contracts at all, it will contract completely.
   b. Motor units respond in an all-or-none manner.
3. Recruitment of motor units
   a. At a low intensity of stimulation, relatively small numbers of motor units contract.
   b. At an increasing intensity of stimulation, other motor units are recruited until the muscle contracts with maximal tension.
4. Twitch contraction
   a. A myogram is a recording of a muscular contraction.
   b. The myogram of a twitch includes a latent period, a period of contraction, and a period of relaxation.
5. Sustained contractions
   a. A rapid series of stimuli may produce a combining of twitches and a sustained contraction.
   b. A tetanic contraction is forceful and sustained.
   c. Even when a muscle is at rest, its fibers usually remain partially contracted.

Smooth Muscles (page 179)

The contractile mechanisms of smooth and cardiac muscles are similar to those of skeletal muscle.

1. Smooth muscle fibers
   a. Smooth muscle cells contain filaments of actin and myosin.
   b. The types include multiunit smooth muscle and visceral smooth muscle.
   c. Visceral smooth muscle displays rhythmicity and is self-exciting.
2. Smooth muscle contraction
   a. Smooth muscles are affected by two neurotransmitters—acetylcholine and norepinephrine—and are influenced by a number of hormones.
   b. Smooth muscle can maintain a contraction for a longer time with a given amount of energy than can skeletal muscle.
   c. Smooth muscles can change their lengths without changing their tautness.

Cardiac Muscle (page 180)

1. Cardiac muscle can maintain a contraction for a longer time than skeletal muscle.
2. The ends of adjacent cardiac muscle cells are separated by intercalated disks.
3. A network of fibers contracts as a unit and responds to stimulation in an all-or-none manner.
4. Cardiac muscle is self-exciting and rhythmic.

Skeletal Muscle Actions (page 180)

The type of movement produced by a skeletal muscle depends on the way it is attached on either side of a joint.

1. Origin and insertion
   a. The movable end of a skeletal muscle is its insertion, and the immovable end is its origin.
   b. Some muscles have more than one origin.
2. Interaction of skeletal muscles
   a. Skeletal muscles function in groups.
   b. A prime mover is responsible for most of a movement; synergists aid prime movers; antagonists can resist the movement of a prime mover.
   c. Smooth movements depend upon antagonists giving way to the actions of prime movers.

Major Skeletal Muscles (page 182)

1. Muscles of facial expression
   a. These muscles lie beneath the skin of the face and scalp and are used to communicate feelings through facial expression.
   b. They include the epicranius, orbicularis oculi, orbicularis oris, buccinator, zygomaticus, and platysma.
2. Muscles of mastication
   a. These muscles are attached to the mandible and are used in chewing.
   b. They include the masseter and temporalis.
3. Muscles that move the head
   a. Head movements are produced by muscles in the neck and upper back.
   b. They include the sternocleidomastoid, splenius capitis, and semispinalis capitis.
4. Muscles that move the pectoral girdle
   a. Most of these muscles connect the scapula to nearby bones and are closely associated with muscles that move the upper arm.
   b. They include the trapezius, rhomboideus major, levator scapulae, serratus anterior, and pectoralis minor.
5. Muscles that move the upper arm
   a. These muscles connect the humerus to various regions of the pectoral girdle, ribs, and vertebral column.
   b. They include the coracobrachialis, pectoralis major, teres major, latissimus dorsi, supraspinatus, deltoid, subscapularis, infraspinatus, and teres minor.
6. Muscles that move the forearm
   a. These muscles connect the radius and ulna to the humerus or pectoral girdle.
   b. They include the biceps brachii, brachialis, brachioradialis, triceps brachii, supinator, pronator teres, and pronator quadratus.
7. Muscles that move the wrist, hand, and fingers
   a. These muscles arise from the distal end of the humerus and from the radius and ulna.
   b. They include the flexor carpi radialis, flexor carpi ulnaris, palmaris longus, flexor digitorum profundus, extensor carpi radialis longus, extensor carpi radialis brevis, extensor carpi ulnaris, and extensor digitorum.
8. Muscles of the abdominal wall
   a. These muscles connect the rib cage and vertebral column to the pelvic girdle.
   b. They include the external oblique, internal oblique, transversus abdominis, and rectus abdominis.
9. Muscles of the pelvic outlet
   a. These muscles form the floor of the pelvic cavity and fill the space of the pubic arch.
   b. They include the levator ani, superficial transversus perinei, bulbospongiosus, and ischiocavernosus.
10. Muscles that move the thigh
    a. These muscles are attached to the femur and to some part of the pelvic girdle.
    b. They include the psoas major, iliacus, gluteus maximus, gluteus medius, gluteus minimus, tensor fasciae latae, adductor longus, adductor magnus, and gracilis.
11. Muscles that move the lower leg
    a. These muscles connect the tibia or fibula to the femur or pelvic girdle.
    b. They include the biceps femoris, semitendinosus, semimembranosus, sartorius, and the quadriceps femoris group.
12. Muscles that move the ankle, foot, and toes
    a. These muscles attach the femur, tibia, and fibula to various bones of the foot.
    b. They include the tibialis anterior, peroneus tertius, extensor digitorum longus, gastrocnemius, soleus, flexor digitorum longus, and peroneus longus.

Clinical Application of Knowledge

1. Why do you think athletes generally perform better if they warm up by exercising before a competitive event?
2. What steps might be taken to minimize atrophy of the skeletal muscles in patients who are confined to bed for prolonged times?
3. As lactic and other substances accumulate in an active muscle, they tend to stimulate pain receptors, and the muscle may feel sore. How might the application of heat or substances that cause blood vessels to dilate help to relieve such soreness?
4. Following an injury to a nerve, the muscles it supplies with motor fibers may be paralyzed. How would you explain to a patient the importance of having the disabled muscles moved passively or causing them to contract by using electrical stimulation?

Review Activities

Part A

1. List the three types of muscle tissue.
2. Distinguish between a tendon and an aponeurosis.
3. Describe how connective tissue is associated with skeletal muscle?
4. List the major parts of a skeletal muscle fiber, and describe the function of each part.
5. Describe a neuromuscular junction.
6. Explain the function of a neurotransmitter.
7. Define *motor unit.*
8. Describe the major events that occur when a muscle fiber contracts.

9. Explain how ATP and creatine phosphate are related.
10. Describe how oxygen is supplied to muscles.
11. Describe how an oxygen debt may develop.
12. Explain how muscles may become fatigued.
13. Explain how the maintenance of body temperature is related to the actions of skeletal muscles.
14. Define *threshold stimulus*.
15. Explain what is meant by an *all-or-none response*.
16. Explain what is meant by *motor unit recruitment*.
17. Sketch a myogram of a single muscular twitch, and identify the latent period, period of contraction, and period of relaxation.
18. Explain how a skeletal muscle can be stimulated to produce a sustained contraction.
19. Distinguish between tetanic contraction and muscle tone.
20. Distinguish between multiunit and visceral smooth muscle fibers.
21. Compare the characteristics of smooth and skeletal muscle contractions.
22. Compare the structure of cardiac and skeletal muscle fibers.
23. Distinguish between a muscle's origin and its insertion.
24. Define *prime mover, synergist,* and *antagonist*.

Part B

Match the muscles in column I with the descriptions and functions in column II.

I	II
1. Buccinator	A. Inserted on the coronoid process of the mandible.
2. Epicranius	B. Draws the corner of the mouth upward.
3. Orbicularis oris	C. Can raise and adduct the scapula.
4. Platysma	D. Can pull the head into an upright position.
5. Rhomboideus major	E. Raises the eyebrow.
6. Splenius capitis	F. Compresses the cheeks.
7. Temporalis	G. Extends over the neck from the chest to the face.
8. Zygomaticus	H. Closes the lips.

I	II
9. Biceps brachii	I. Extends the arm at the elbow.
10. Brachialis	J. Pulls the shoulder back and downward.
11. Deltoid	K. Abducts the arm.
12. Latissimus dorsi	L. Inserted on the radial tuberosity.
13. Pectoralis major	M. Pulls the arm forward and across the chest.
14. Pronator teres	N. Rotates the arm medially.
15. Teres minor	O. Inserted on the coronoid process of the ulna.
16. Triceps brachii	P. Rotates the arm laterally.

I	II
17. Biceps femoris	Q. Inverts the foot.
18. External oblique	R. Member of the quadriceps femoris group.
19. Gastrocnemius	S. Plantar flexor of the foot.
20. Gluteus maximus	T. Compresses the contents of the abdominal cavity.
21. Gluteus medius	U. Extends the leg at the hip.
22. Gracilis	V. Hamstring muscle.
23. Rectus femoris	W. Adducts the thigh.
24. Tibialis anterior	X. Abducts the thigh.

Part C

What muscles can you identify in the bodies of these models?

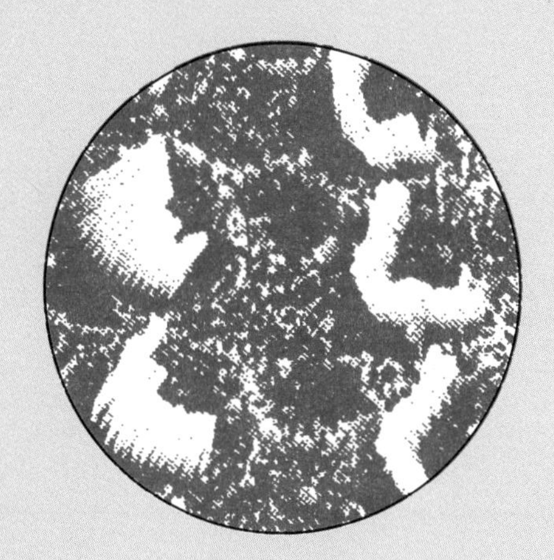

Unit 3
Integration and Coordination

The chapters of unit 3 are concerned with the structures and functions of the nervous and endocrine systems. They describe how the organs of these systems keep the parts of the human body functioning together as a whole and how these organs help maintain a stable internal environment, which is vital to the survival of the organism. This unit includes:

These circular images are a computer-manipulated version of figure 10.11.

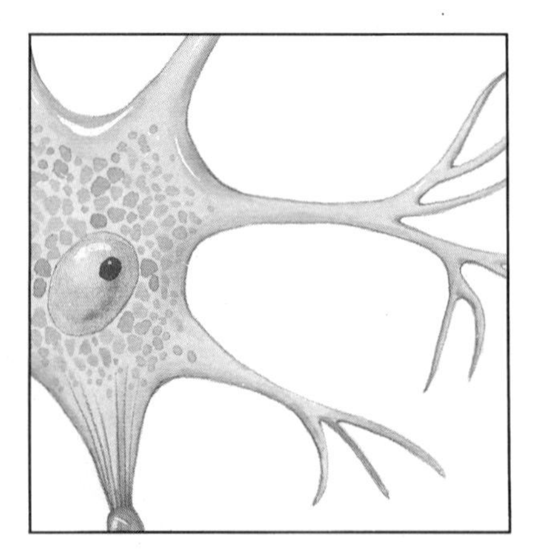

9 Nervous System

If a human organism is to survive, the actions of its cells, tissues, organs, and systems must be directed toward a single goal—the maintenance of homeostasis. To accomplish this, the functions of all the body parts must be controlled and coordinated so that the parts work as a unit and respond to changes in ways that help maintain a stable internal environment.

The general task of controlling and coordinating body activities is handled by the nervous and endocrine systems. Of these, the *nervous system* provides a more rapid and precise mode of action. It controls muscular contractions and glandular secretions by means of nerve fibers, which extend from the brain and spinal cord to various body parts. Some of these nerve fibers bring information concerning changes occurring inside and outside the body to the brain and spinal cord from sensory receptors. Other fibers carry impulses away from the brain or spinal cord and stimulate muscles or glands to respond.

Chapter Objectives

After you have studied this chapter, you should be able to

1. Describe the general structure of a neuron.
2. Name four types of neuroglial cells, and describe the functions of each.
3. Describe the events that lead to the conduction of a nerve impulse.
4. Explain how a nerve impulse is transmitted from one neuron to another.
5. Explain how differences in structure and function are used to classify neurons.
6. Name the parts of a reflex arc, and describe the function of each part.
7. Describe the structure of the spinal cord and its major functions.
8. Name the major parts of the brain, and describe the functions of each part.
9. Distinguish between motor, sensory, and association areas of the cerebral cortex.
10. Describe the formation and function of cerebrospinal fluid.
11. List the major parts of the peripheral nervous system.
12. Name the cranial nerves, and list their major functions.
13. Describe the structure of a spinal nerve.
14. Describe the functions of the autonomic nervous system.
15. Distinguish between the sympathetic and parasympathetic divisions of the autonomic nervous system.
16. Describe a sympathetic and a parasympathetic nerve pathway.
17. Complete the review activities at the end of this chapter. Note that the items are worded in the form of specific learning objectives. You may want to refer to them before reading the chapter.

Key Terms

action potential (ak′shun po-ten′shal)

autonomic nervous system (aw″to-nom′ik ner′vus sis′tem)

axon (ak′son)

central nervous system (sen′tral ner′vus sis′tem)

convergence (kon-ver′jens)

dendrite (den′drit)

divergence (di-ver′jens)

effector (ĕ-fek′tor)

facilitation (fah-sil″ĭ-ta′shun)

ganglion (gang′gle-on)

meninges (mĕ-nin′jēz)

myelin (mi′ĕ-lin)

neurilemma (nu″rĭ-lem′mah)

neurotransmitter (nu″ro-trans-mit′er)

nissl body (nis′l bod′e)

parasympathetic nervous system (par″ah-sim″pah-thet′ik ner′vus sis′tem)

peripheral nervous system (pĕ-rif′er-al ner′vus sis′tem)

plexus (plek′sus)

receptor (re-sep′tor)

reflex (re′fleks)

sympathetic nervous system (sim″pah-thet′ik ner′vus sis′tem)

synapse (sin′aps)

Aids to Understanding Words

ax-, axle: *ax*on—a cylindrical nerve fiber that carries impulses away from a neuron cell body.

dendr-, tree: *dendr*ite—a branched nerve fiber that serves as a receptor surface of a neuron.

funi-, small cord or fiber: *funi*culus—a major nerve tract or bundle of myelinated nerve fibers within the spinal cord.

gangli-, a swelling: *gangli*on—a mass of neuron cell bodies.

-lemm, rind or peel: neuri*lemma*—a sheath that surrounds the myelin of a nerve fiber.

mening-, membrane: *mening*es—membranous coverings of the brain and spinal cord.

moto-, moving: *moto*r neuron—a neuron that stimulates a muscle to contract or a gland to release a secretion.

peri, all around: *peri*pheral nervous system—the portion of the nervous system that consists of nerves branching from the brain and spinal cord.

plex-, interweaving: choroid *plex*us—a mass of specialized capillaries associated with spaces in the brain.

sens-, feeling: *sens*ory neuron—a neuron that can be stimulated by a sensory receptor and conducts impulses into the brain or spinal cord.

syn-, together: *syn*apse—the junction between two neurons.

ventr-, belly or stomach: *ventr*icle—a fluid-filled space within the brain.

Introduction

THE ORGANS OF the nervous system, like other organs, are composed of various kinds of tissues, including nervous tissue, connective tissues, and blood. These organs can be divided into two groups. One group, consisting of the brain and spinal cord, forms the **central nervous system (CNS),** and the other, composed of the nerves (peripheral nerves) that connect the central nervous system to other body parts, is called the **peripheral nervous system (PNS).** Together these systems provide three general functions—a sensory function, an integrative function, and a motor function.

General Functions of the Nervous System

The **sensory function** of the nervous system involves *sensory receptors* at the ends of peripheral nerves (see chapter 10). These receptors are specialized to gather information by detecting changes that occur inside and outside the body. They monitor such external environmental factors as light and sound intensities as well as the temperature, oxygen concentration, and conditions of the body's internal fluids.

The information gathered by the sensory receptors is converted into signals in the form of *nerve impulses,* which are then transmitted over peripheral nerves to the central nervous system. There the signals are integrated; that is, they are brought together, creating sensations (perceptions), adding to memory, or helping to produce thoughts. As a result of this **integrative function,** conscious or subconscious decisions are made and then acted upon by means of motor functions.

The **motor functions** of the nervous system employ peripheral nerves, which carry impulses from the central nervous system to responsive parts called *effectors.* These effectors are outside the nervous system and include muscles, that contract when stimulated by nerve impulses, and glands, that produce secretions when stimulated.

Thus, the nervous system can detect changes occurring outside and within the body, make decisions on the basis of the information received, and cause muscles or glands to respond. Typically, these responses are directed toward counteracting the effects of the changes that were detected, and in this way the nervous system helps to maintain homeostasis.

Nervous Tissue

As is described in chapter 5, nervous tissue contains masses of nerve cells, or **neurons** (fig. 9.1). These cells are the structural and functional units of the nervous system and are specialized to react to physical and chemical changes occurring in their surroundings. They also conduct nerve impulses to other neurons and to cells outside the nervous system.

Figure 9.1
Neurons are the structural and functional units of the nervous system (×50).

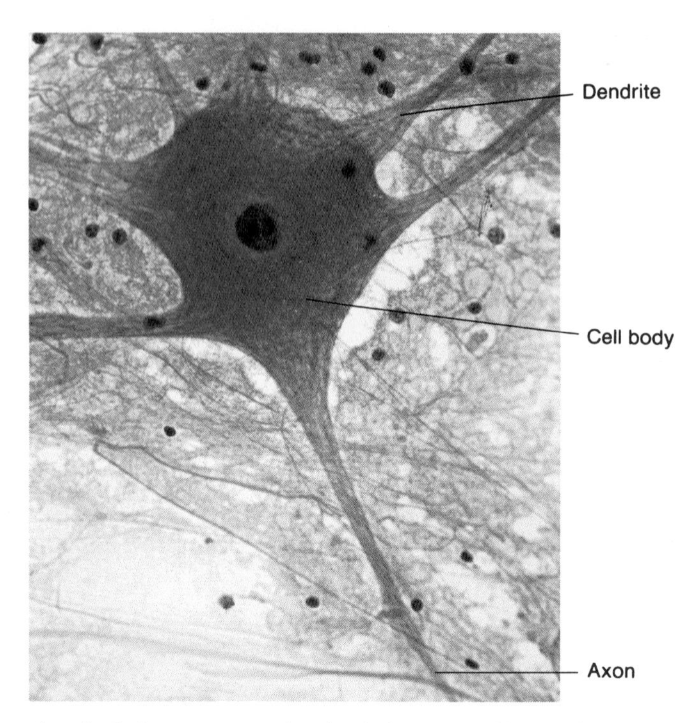

Between the neurons, there are *neuroglial cells,* which function like the connective tissue cells in other systems.

Neuroglial cells are described in a subsequent section of this chapter.

Neuron Structure

Although neurons vary considerably in size and shape, they have certain features in common. For example, every neuron has a cell body and tubular processes filled with cytoplasm called *nerve fibers,* which conduct nerve impulses to or from the cell body (fig. 9.2).

The neuron **cell body** contains a mass of granular cytoplasm, a cell membrane, and various other organelles usually found in cells. Inside the cell body, for example, are mitochondria, lysosomes, a Golgi apparatus, and a network of fine threads called **neurofibrils,** which extend into the nerve fibers. Scattered throughout the cytoplasm are many membranous sacs called **nissl bodies,** which are similar to rough endoplasmic reticulum in other cells. Ribosomes attached to the surfaces of these parts function in the synthesis of vital protein molecules.

Near the center of the neuron cell body is a large, spherical nucleus with a conspicuous nucleolus. This

Figure 9.2
(*a*) Motor neuron and (*b*) sensory neuron.

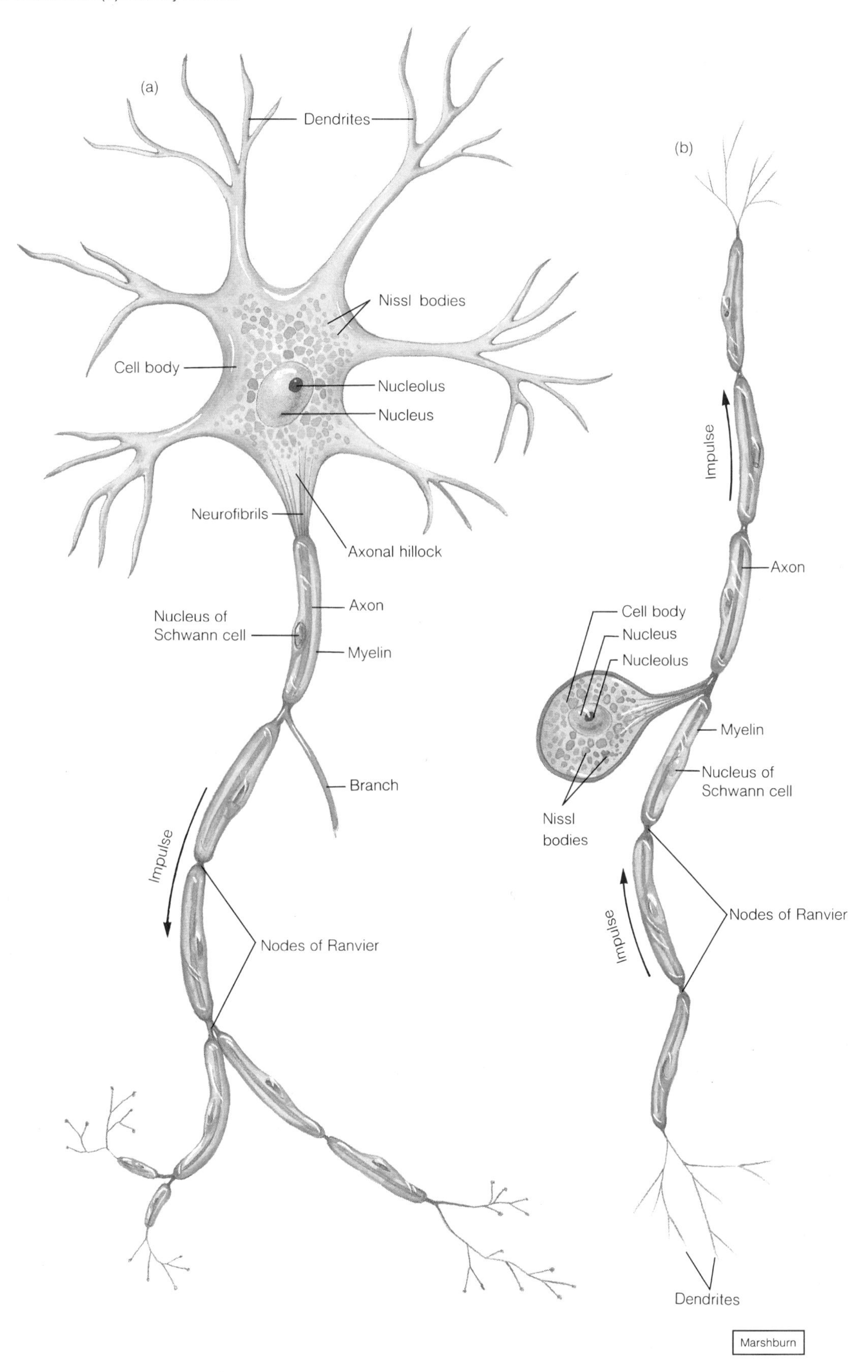

Figure 9.3
The portion of a Schwann cell that is tightly wound around an axon forms a myelin sheath, and the cytoplasm and nucleus of the Schwann cell, remaining outside, form a neurilemmal sheath.

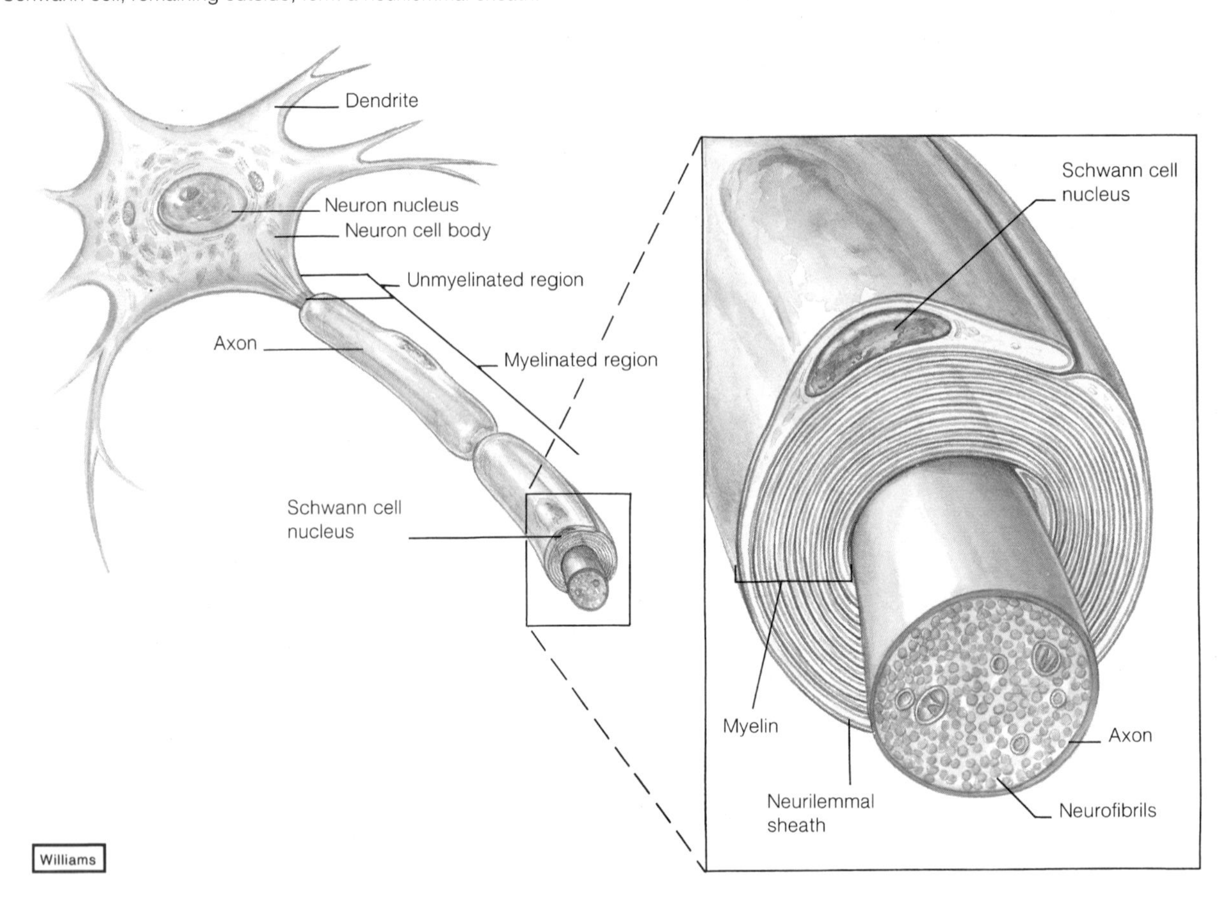

nucleus apparently does not undergo mitosis after the nervous system is developed, and consequently, mature neurons seem to be incapable of reproduction.

Two kinds of nerve fibers, called **dendrites** and **axons,** extend from the cell bodies of most neurons. Although a neuron usually has many dendrites, it has only one axon.

In most neurons, the dendrites are relatively short and highly branched. These fibers, together with the membrane of the cell body, provide the main receptive surfaces of the neuron to which fibers from other neurons communicate.

The axon usually arises from a slight elevation of the cell body called the *axonal hillock.* It is specialized to conduct nerve impulses away from the cell body. Many mitochondria, microtubules, and neurofibrils occur within the cytoplasm of the axon. Although it begins as a single fiber, an axon may give off side branches, and at its terminal end, it may have many fine extensions, each with a specialized ending which contacts the receptive surface of another cell.

Larger axons of peripheral nerves commonly are enclosed in sheaths composed of neuroglial cells, called **Schwann cells** (figs. 9.3 and 9.4). These cells are tightly wound around the axons, somewhat like a bandage wrapped many times around an injured arm. As a result, such axons are coated with many layers of cell membrane that have little or no cytoplasm between them. These membrane layers are composed largely of a lipid-protein (lipoprotein) that has a higher proportion of lipid than other surface membranes. This lipoprotein is called **myelin,** and it forms a *myelin sheath* on the outside of an axon. In addition, the portions of the Schwann cells that contain most of the cytoplasm and the nuclei remain outside the myelin sheath and comprise a **neurilemma** (neurilemmal sheath), which surrounds the myelin sheath. Narrow gaps in the myelin sheath between adjacent Schwann cells are called *nodes of Ranvier.*

Smaller axons are also enclosed by Schwann cells, but the Schwann cells may not be wound around these axons. Consequently, such axons lack myelin sheaths.

Axons that possess myelin sheaths are called *myelinated* nerve fibers, and those that lack sheaths are *unmyelinated.* Groups of myelinated fibers appear white, and masses of such fibers are responsible for the *white matter* in the nervous system. Unmyelinated nerve fibers and neuron cell bodies appear as *gray matter* within the nervous system.

Figure 9.4
A transmission electron micrograph of myelinated and unmyelinated axons in cross section.

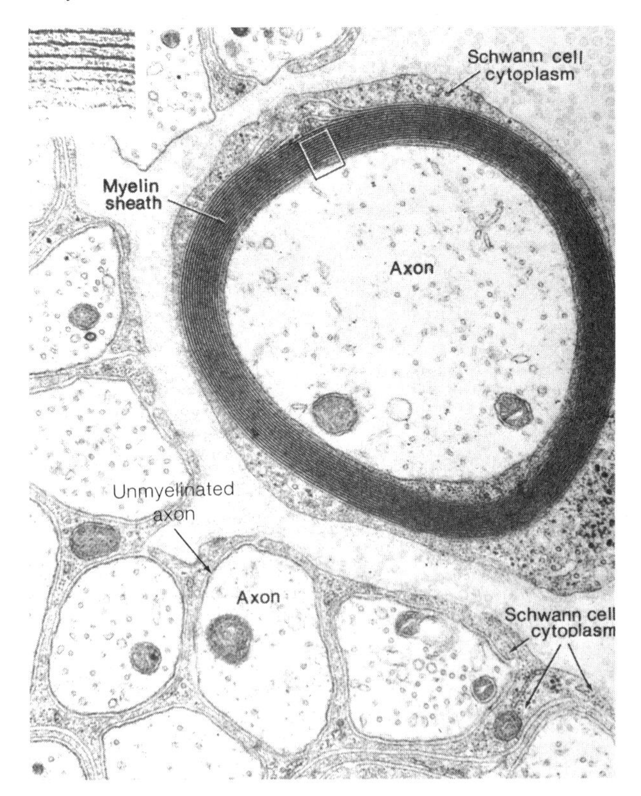

When neurons are deprived of oxygen, they undergo a series of irreversible structural changes in which their shapes are altered and their nuclei shrink. This phenomenon is called *ischemic cell change*, and, in time, the affected cells disintegrate. Such an oxygen deficiency can result from a lack of blood flow through nerve tissue (ischemia), an abnormally low blood oxygen concentration (hypoxemia), or the presence of toxins that block aerobic respiration.

Neuroglial Cells

Neuroglial cells occur within the organs of the nervous system, where they fill spaces, support neurons, provide structural frameworks, produce myelin, and carry on phagocytosis.

Within the peripheral nervous system, neuroglial cells include the *Schwann cells,* previously described. In the central nervous system, where neuroglial cells greatly outnumber neurons, the following types are present (fig. 9.5):

1. **Astrocytes.** These cells are commonly found between nerve tissue and blood vessels. They provide structural support, hold parts together by means of numerous cellular processes, and help regulate the concentrations of nutrients and ions within nervous tissue. Astrocytes also are responsible for the formation of scar tissue, which fills spaces following an injury to the nervous system.
2. **Oligodendrocytes.** These cells are commonly arranged in rows along nerve fibers, and they function in the formation of myelin within the brain and spinal cord. However, unlike the Schwann cells in the peripheral nervous system, oligodendrocytes fail to form neurilemmal sheaths.
3. **Microglia.** These cells are scattered throughout the central nervous system, where they help support neurons and phagocytize bacterial cells and cellular debris.
4. **Ependyma.** These cells form an epitheliallike membrane, which covers specialized brain parts *(choroid plexuses)* and forms the inner linings that enclose spaces within the brain (ventricles) and spinal cord (central canal).

1. Describe a neuron.
2. Distinguish between an axon and a dendrite.
3. Describe how a myelin sheath is formed.
4. Distinguish between a neuron and a neuroglial cell.

Cell Membrane Potential

The surface of a cell membrane is usually electrically charged, or *polarized,* with respect to the inside. This polarization is due to an unequal distribution of ions on either side of the membrane, and it is particularly important in the conduction of muscle and nerve impulses.

Distribution of Ions

The distribution of ions associated with membranes is determined in part by the presence of pores or channels in those membranes (see chapter 3). Some channels are always open, and others can be opened or closed. Furthermore, channels can be selective; that is, a channel may allow one kind of ion to pass through and exclude other kinds (fig. 9.6).

As a consequence of such factors, potassium ions tend to pass through cell membranes much more easily than sodium ions, and sodium ions pass through more easily than calcium ions. The relative ease with which potassium ions diffuse through membranes makes them a major contributor to membrane polarization.

Resting Potential

When nerve cells are at rest (that is, not conducting impulses), there is a relatively greater concentration of

Figure 9.5
Types of neuroglial cells found within the central nervous system include (*a*) microglial cell, (*b*) oligodendrocyte, (*c*) astrocyte, and (*d*) ependymal cell.

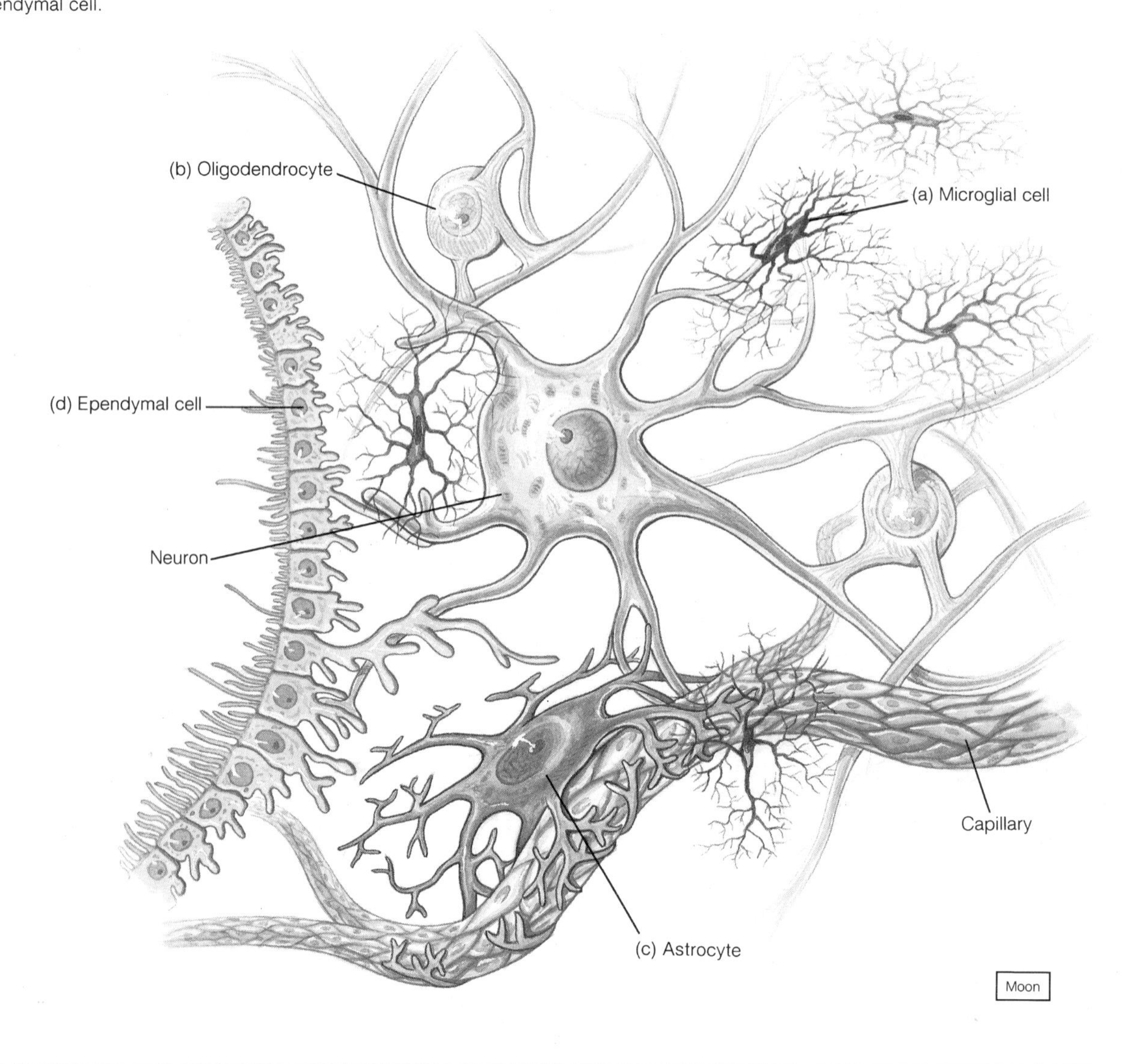

Figure 9.6
Some of the channels in cell membranes through which ions pass can be (*a*) closed or (*b*) opened by a gatelike mechanism.

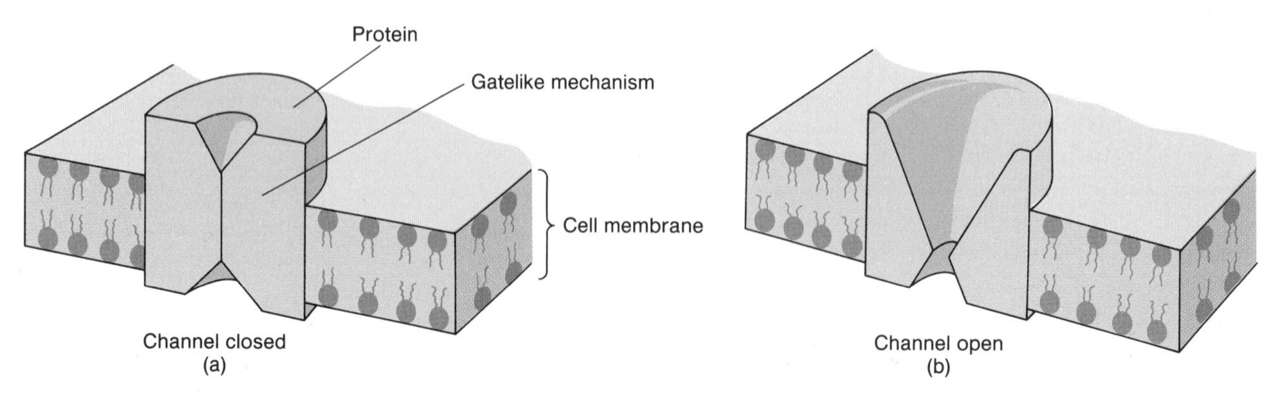

Figure 9.7
A nerve fiber at rest is polarized as a result of an unequal distribution of ions on either side of its membrane.

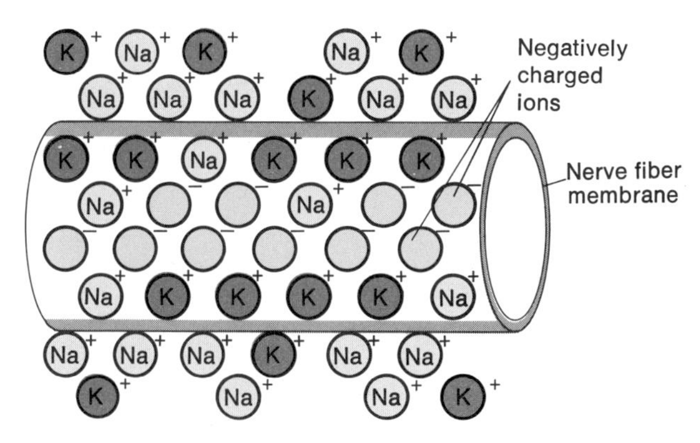

Figure 9.8
An active transport mechanism in the nerve fiber membrane moves sodium ions outward and potassium ions inward.

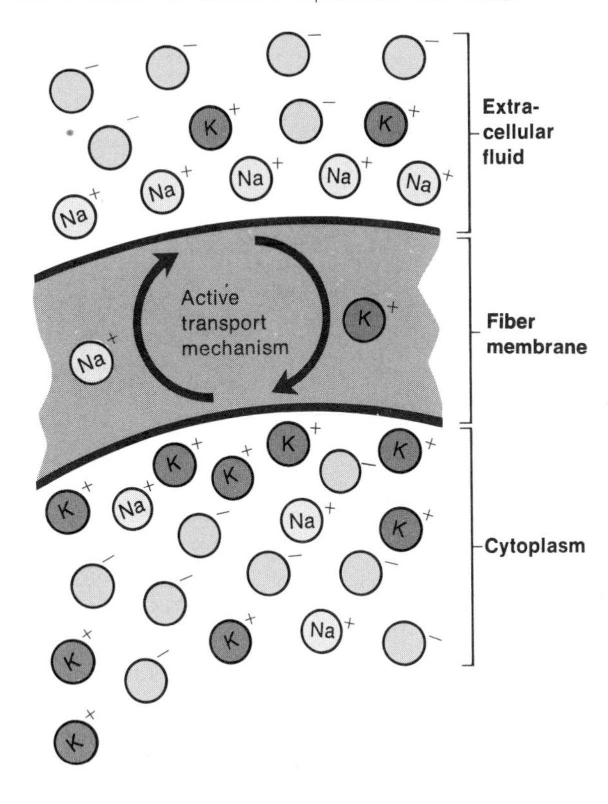

sodium ions (Na^+) outside their membranes and a relatively greater concentration of potassium ions (K^+) inside their membranes (fig. 9.7). In the cytoplasm of these cells, there are large numbers of negatively charged ions, including those of phosphate, sulfate, and protein, that cannot diffuse through cell membranes.

However, because a cell membrane is very permeable to potassium ions and only slightly permeable to sodium ions, potassium ions tend to diffuse freely through the membrane to the outside, and sodium ions diffuse inward more slowly. At the same time, the membrane expends energy to actively transport these ions in the opposite directions, which prevents them from reaching equilibrium by diffusion. Therefore, sodium ions are actively transported outward, and potassium ions are actively transported inward. Because potassium ions can readily diffuse out again and sodium ions can enter only slowly, the net effect is for more positively charged ions to leave a cell than to enter it. Consequently, the outside of a cell membrane becomes positively charged with respect to the inside, which is negatively charged (fig. 9.8).

The difference in electrical charge between two regions is called a *potential difference.* In the case of a resting nerve cell, the potential difference between the region inside and the region outside of the membrane is called a **resting potential.** As long as a nerve cell membrane is undisturbed, the membrane remains in this polarized state.

Potential Changes

Nerve cells are excitable; that is, they can respond to changes in their surroundings. Some nerve cells, for example, are specialized to detect changes in temperature, light, or pressure occurring outside the body. Other neurons are responsive to signals coming from nearby nerve fibers. In any case, such changes or stimuli usually affect the resting potential in a particular region

Figure 9.9
When a polarized nerve fiber is stimulated, sodium channels open, some sodium ions diffuse inward, and the membrane is depolarized.

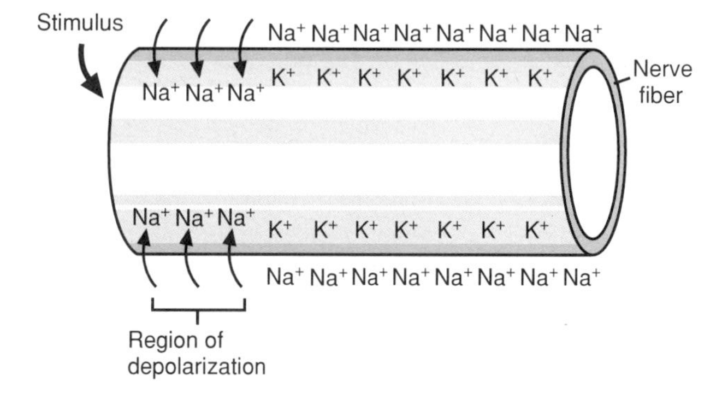

of a nerve cell membrane, and if the membrane's resting potential becomes decreased, the membrane is said to be *depolarizing* (fig. 9.9).

Changes that occur in the resting potential of a membrane are graded. This means the amount of change in potential is directly related to the intensity of stimulation received. Furthermore, if additional stimulation is received before the effect of some previous stimulation subsides, the change in potential is still greater. This additive phenomena is called *summation,* and as a result of summated potentials, a level called **threshold potential** may be reached. Thus, many subthreshold potential changes may be combined to reach the threshold level, and once threshold is achieved, an *action potential* occurs.

Action Potential

When threshold potential is reached, the region of cell membrane being stimulated undergoes a sudden change in permeability. Channels that are highly selective for sodium ions open, and they allow sodium to diffuse freely inward (fig. 9.9). This movement is aided by the negative electrical condition on the opposite side of the membrane, which attracts the positively charged sodium ions.

As sodium ions diffuse inward, the membrane loses its electrical charge and becomes depolarized. At the same time, however, membrane channels open that allow potassium ions to pass through, and as these ions diffuse outward, the outside of the membrane becomes positively charged once more (fig. 9.10). Thus, the membrane becomes *repolarized,* and it remains in this state until it is stimulated again.

This rapid sequence of events, involving depolarization and repolarization, takes about one-thousandth of a second and is called an **action potential.** Because only a small proportion of the sodium and potassium ions present move through a membrane during an action potential, many action potentials can occur before the original concentrations of these ions change significantly. Eventually, however, the active transport mechanism within the membrane soon reestablishes the original concentrations of sodium and potassium ions on either side, and the resting potential returns.

Nerve Impulse

When an action potential occurs in one region of a nerve fiber membrane, it causes an electric current to flow to adjacent portions of the membrane (fig. 9.11). This local current stimulates the membrane to its threshold level and triggers other action potentials. These, in turn, stimulate still other regions, and a wave of action potentials moves away from the point of stimulation in all directions, traveling to the end of the fiber. This propagation of action potentials along a nerve fiber constitutes a **nerve impulse.**

Chart 9.1 summarizes the events leading to the conduction of a nerve impulse.

Impulse Conduction

An unmyelinated nerve fiber conducts an impulse over its entire surface. A myelinated fiber functions differently, because myelin serves as an insulator that prevents almost all of the flow of ions through the membrane.

Figure 9.10
When the potassium channels open, potassium ions diffuse outward, and the membrane is repolarized.

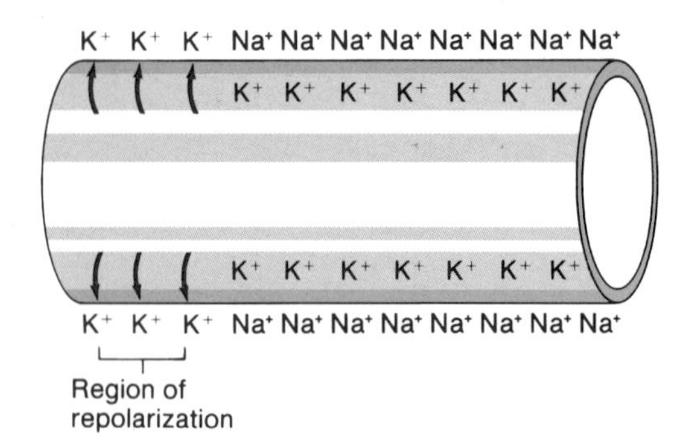

Considering this, it might seem that the myelin sheath would prevent the conduction of a nerve impulse altogether, and this would be true if the sheath were continuous. It is, however, interrupted by the constrictions called **nodes of Ranvier,** which occur between adjacent Schwann cells (fig. 9.2). At these nodes, the fiber membrane can become especially permeable to sodium and potassium ions, and a nerve impulse traveling along a myelinated fiber appears to jump from node to node. This type of impulse conduction (saltatory conduction) is many times faster than conduction on an unmyelinated fiber.

All-or-None Response

Like muscle fiber contraction, nerve impulse conduction is an *all-or-none response.* In other words, if a nerve fiber responds at all, it responds completely. Thus, a nerve impulse is conducted whenever a stimulus of threshold intensity or above is applied to a nerve fiber, and all impulses carried on that fiber will be of the same strength. A greater intensity of stimulation does not produce a stronger nerve impulse.

Certain drugs, such as procaine, produce effects by decreasing membrane permeability to sodium ions. When one of these drugs is present in the fluids surrounding a nerve fiber, impulses are prevented from passing through the affected region. Consequently, the drugs are useful as local anesthetics because they help keep impulses from reaching the brain and, thus, prevent the sensations of touch and pain.

1. Summarize how a nerve fiber becomes polarized.
2. List the major events that occur during an action potential.
3. Explain how impulse conduction differs in myelinated and unmyelinated nerve fibers.
4. Define *"all-or-none" response.*

Figure 9.11
(*a*) An action potential in one region stimulates the adjacent region, and (*b, c*) a wave of action potentials (nerve impulse) moves along the fiber.

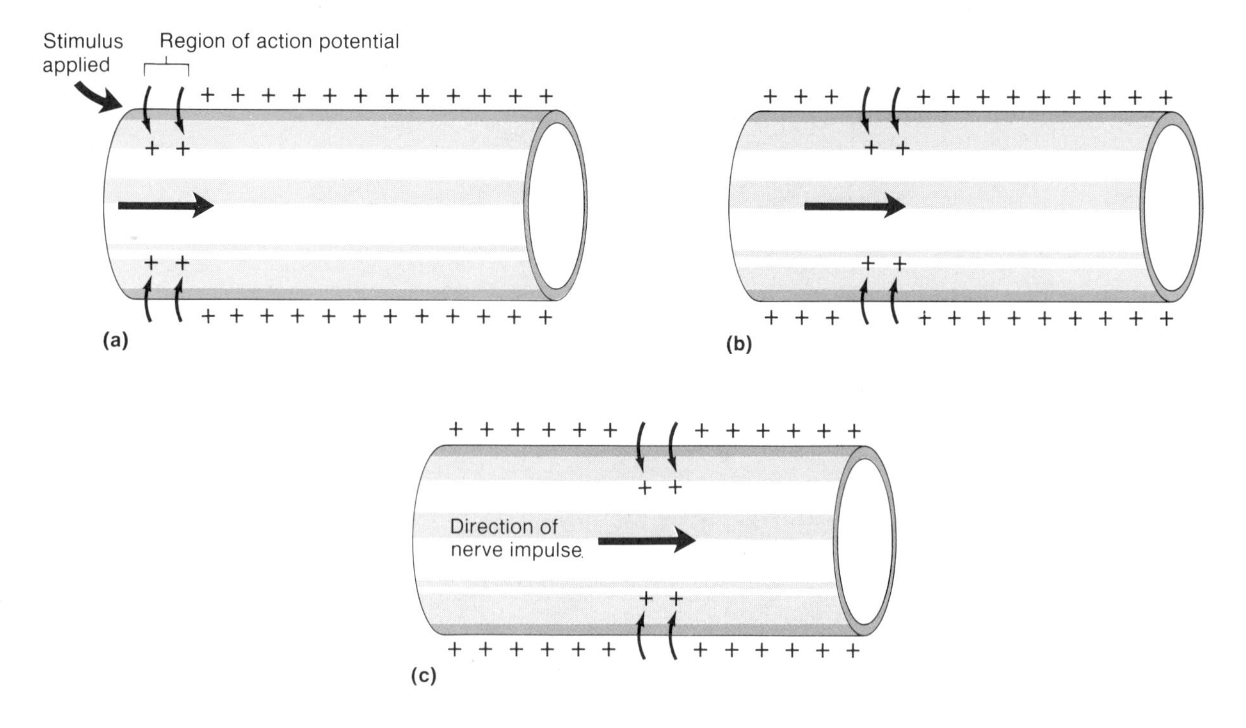

Chart 9.1 Events leading to the conduction of a nerve impulse	
1. Nerve fiber membrane develops resting potential. 2. Threshold stimulus is received. 3. Sodium channels in a local region of the membrane open. 4. Sodium ions diffuse inward, causing the membrane to depolarize. 5. Potassium channels in membrane open.	6. Potassium ions diffuse outward, causing the membrane to repolarize. 7. The resulting action potential causes a local electric current that stimulates adjacent portions of the membrane. 8. Wave of action potentials travels the length of the nerve fiber as a nerve impulse.

Synapse

Within the nervous system, nerve impulses travel from neuron to neuron along complex nerve pathways. The junction between two neurons is called a **synapse.** Actually, the neurons are not in direct contact at a synapse. There is a gap, called a *synaptic cleft,* between them. For an impulse to continue along a nerve pathway, it must cross this gap (fig. 9.12).

Synaptic Transmission

As was mentioned, a nerve impulse travels in both directions away from the point of stimulation. Within a neuron (presynaptic neuron), however, an impulse travels from a dendrite to its cell body and then moves along the axon to its end. There the impulse crosses a synapse and continues to a dendrite or cell body of another neuron (postsynaptic neuron). The process of crossing the gap at a synapse is called *synaptic transmission.*

The typical one-way transmission from an axon to a dendrite or cell body is due to the fact that axons usually have several rounded *synaptic knobs* at their distal ends, which dendrites lack (fig. 9.13). These knobs contain numerous membranous sacs, called *synaptic vesicles,* and when a nerve impulse reaches a knob, some of the vesicles respond by releasing a **neurotransmitter** (fig. 9.14).

The neurotransmitter diffuses across the synaptic cleft and reacts with specific receptors of the postsynaptic neuron membrane. If enough neurotransmitter is released, the postsynaptic membrane is stimulated to threshold level, and a nerve impulse is triggered.

Neurotransmitter Substances

Many different neurotransmitter substances are produced in the nervous system. Each neuron, however, seems to release only one or two kinds. The neurotransmitters include *acetylcholine,* which stimulates skeletal muscle contractions (see chapter 8); a group of compounds called *monoamines* (such as epinephrine, norepinephrine, dopamine, and serotonin), which are formed from modified amino acid molecules; several

Figure 9.12
For an impulse to continue from one neuron to another, it must cross the synaptic cleft at a synapse. A synapse may occur (*a*) between an axon and a dendrite or (*b*) between an axon and a cell body.

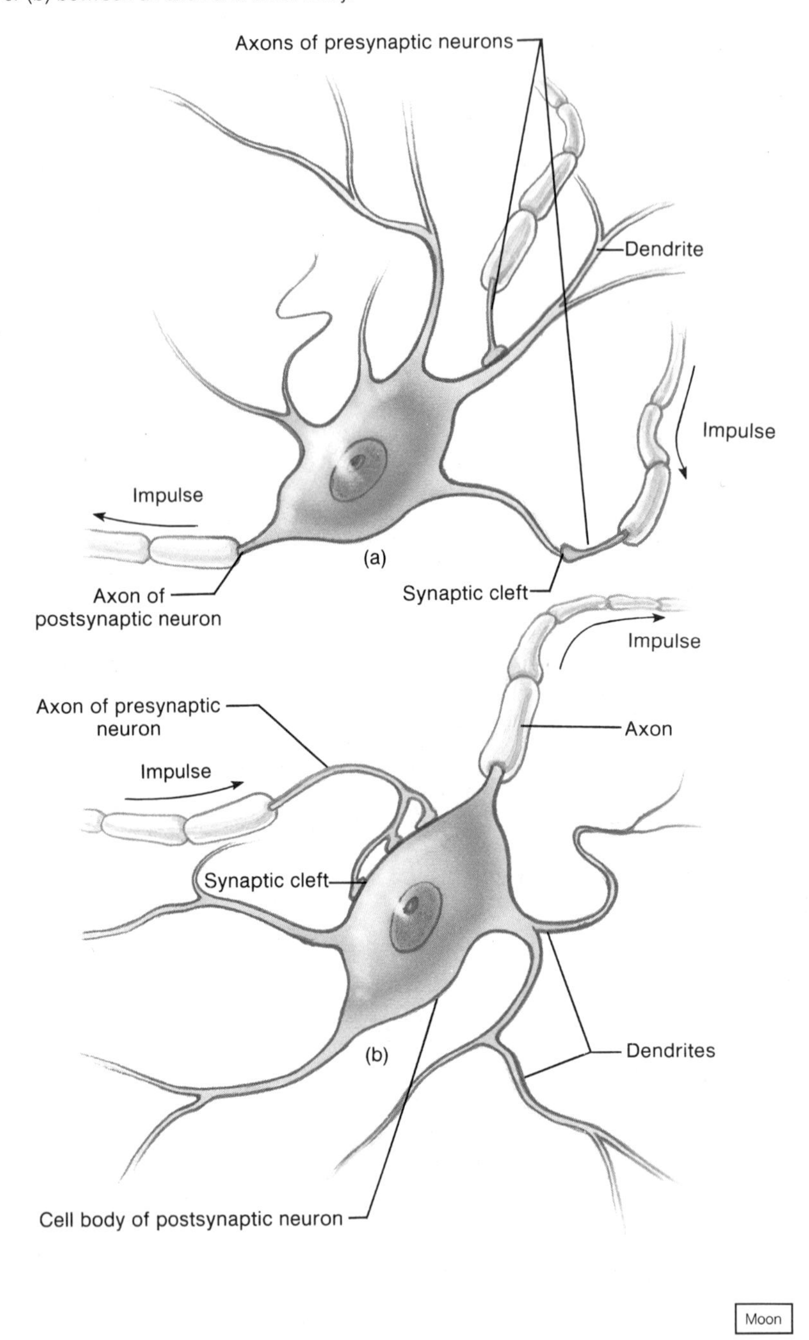

amino acids (such as glycine, glutamic acid, aspartic acid, and gamma-aminobutyric acid—GABA); and a large group of *peptides,* each of which consists of a relatively short chain of amino acids. These substances are usually synthesized in the cytoplasm of the synaptic knobs and stored in the synaptic vesicles.

When an action potential passes over the membrane of a synaptic knob, it produces an increase in the membrane's permeability to calcium ions by causing its calcium ion channels to open. Consequently, calcium ions diffuse inward, and in response to their presence, some synaptic vesicles fuse with the membrane and release their contents into the synaptic cleft. A vesicle that has released its neurotransmitter breaks away from the membrane and reenters the cytoplasm, where it is quickly resupplied with neurotransmitters.

After being released, some neurotransmitters are decomposed by enzymes present in the synaptic cleft. Other neurotransmitters are removed from the cleft by being transported back into the synaptic knob that re-

Figure 9.13
(*a*) When a nerve impulse reaches the synaptic knob at the end of an axon, (*b*) synaptic vesicles release a neurotransmitter substance that diffuses across the synaptic cleft.

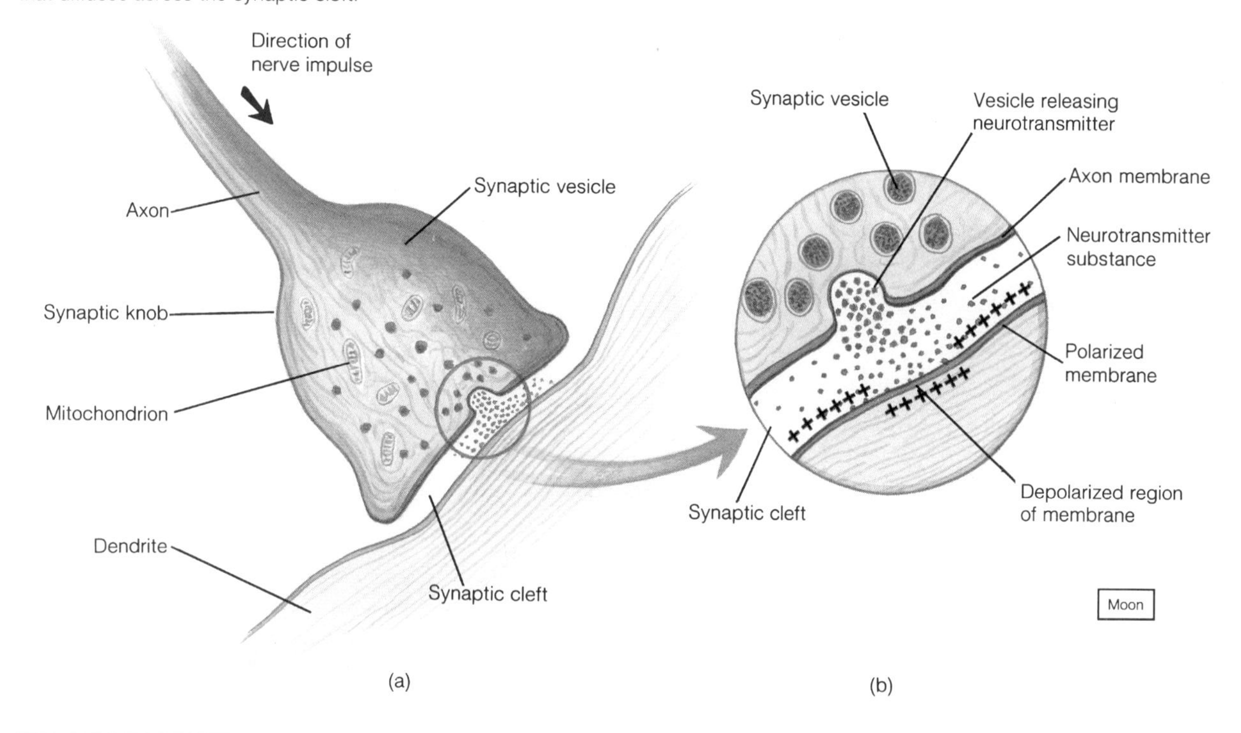

Figure 9.14
A transmission electron micrograph of a synaptic knob filled with synaptic vesicles.

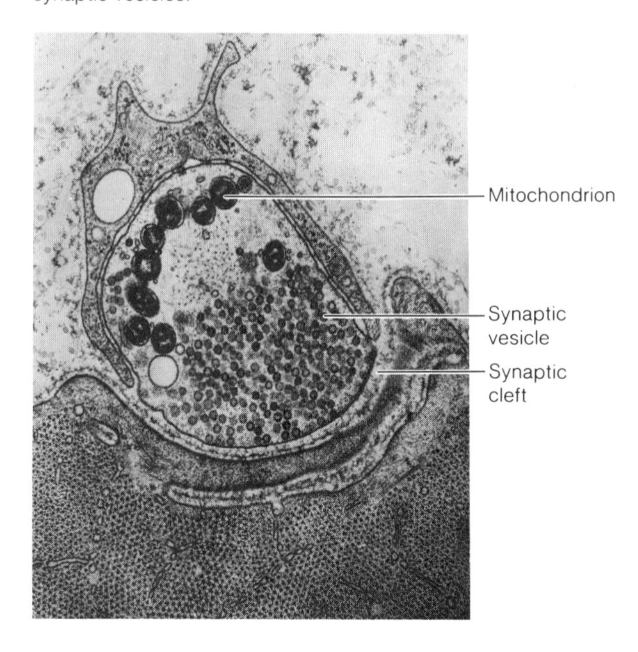

leased them or into nearby neurons or neuroglial cells. Acetylcholine, for example, is decomposed by the enzyme *cholinesterase,* which is present at synapses in the membranes that employ this neurotransmitter. Similarly, monoamine neurotransmitters are inactivated by the enzyme *monoamine oxidase,* found in the mitochondria. Such decomposition or removal of neurotransmitter substances prevents a continuous stimulation of the postsynaptic neurons.

Chart 9.2 summarized the events leading to the release of a neurotransmitter.

Excitatory and Inhibitory Actions

Neurotransmitters that cause an increased membrane permeability to sodium ions and, thus, trigger nerve impulses are said to be *excitatory.* Substances of this type include serotonin, dopamine, and norepinephrine.

Other neurotransmitters cause a decreased membrane permeability to sodium ions, thus causing the threshold of stimulation to be raised. This action is called *inhibitory,* because it lessens the chance that a

Chart 9.2 Events leading to the release of a neurotransmitter

1. Action potential passes along a nerve fiber and over the surface of its synaptic knob.
2. Synaptic knob membrane becomes more permeable to calcium ions, and they diffuse inward.
3. In the presence of calcium ions, synaptic vessels fuse to synaptic knob membrane.
4. Synaptic vessels release their neurotransmitter into the synaptic cleft.
5. Synaptic vesicles reenter the cytoplasm of the nerve fiber and are refilled with neurotransmitter.

nerve impulse will be transferred to an adjoining neuron. Inhibitory substances include the amino acids GABA and glycine.

The synaptic knobs of a thousand or more neurons may communicate with the dendrites and cell body of a particular neuron. Neurotransmitters released by some of these knobs probably have an excitatory action, but those from other knobs very likely have an inhibitory action. The effect on the postsynaptic neuron will depend on which knobs are activated from moment to moment. In other words, if more excitatory than inhibitory neurotransmitters are released, the postsynaptic neuron's threshold may be exceeded, and a nerve impulse will be triggered to pass over its surface. Conversely, if most of the neurotransmitters released are inhibitory, no impulse will be conducted.

Processing of Impulses

The way the nervous system processes nerve impulses and acts upon them reflects, in part, the organization of neurons and their nerve fibers within the brain and spinal cord.

Neuronal Pools

The neurons within the central nervous system are organized into groups called *neuronal pools* that have varying numbers of cells. Each neuronal pool receives impulses from input nerve fibers. These impulses are processed according to the special characteristics of the pool, and any resulting impulses are conducted away on output fibers.

Each input fiber divides many times as it enters, and its branches spread over a certain region of the neuronal pool. The branches give off smaller branches and their terminals form hundreds of synapses with the dendrites and cell bodies of the neurons in the pool.

Facilitation

As a result of incoming impulses, and the release of neurotransmitters, a particular neuron of a neuronal pool may receive excitatory and inhibitory stimulation. As mentioned before, if the net effect of the stimulation is excitatory, threshold may be reached, and an outgoing impulse will be triggered. If the net effect is excitatory but subthreshold, an impulse will not be triggered. However, in this case, the neuron becomes more excitable to incoming stimulation than before and is said to be *facilitated.*

Convergence

Any single neuron in a neuronal pool may receive impulses from two or more incoming fibers. Furthermore, these fibers may originate from different parts of the nervous system, and they are said to *converge* when they lead to the same neuron (fig. 9.15*a*).

Convergence makes it possible for impulses arriving from different sources to create an additive effect upon a neuron. For example, if a neuron is facilitated by receiving subthreshold stimulation from one input fiber, its threshold may be reached if it receives additional stimulation from a second input fiber. As a result, an output impulse may travel to a particular effector and cause a response.

Incoming impulses often represent information from various sensory receptors that have detected changes taking place. Convergence allows the nervous system to bring a variety of kinds of information together, to process it, and to respond to it in a special way.

Divergence

Impulses leaving a neuron of a neuronal pool often *diverge* by passing into several other output fibers (fig. 9.15*b*). For example, an impulse from one neuron may stimulate two others; each of these, in turn, may stimulate several others, and so forth. Such an arrangement of diverging nerve fibers can cause an impulse to be *amplified*—that is, to be spread to increasing numbers of neurons within the pool.

As a result of divergence, an impulse originating from a single neuron in the central nervous system may be amplified so that enough impulses reach the motor units within a skeletal muscle to cause a forceful contraction (chapter 8).

Similarly, an impulse originating from a sensory receptor may diverge and reach several different regions of the central nervous system, where the resulting impulses can be processed and acted upon.

1. Describe the function of a neurotransmitter.
2. Distinguish between the excitatory and inhibitory actions of neurotransmitters.
3. Define *neuronal pool.*
4. Distinguish between convergence and divergence.

Types of Neurons and Nerves

Neurons differ in the structure, size, and shape of their cell bodies. Likewise, they vary in the length and size of their axons and dendrites and in the number of synaptic knobs, by which they communicate with other neurons.

Neurons also vary in function. Some carry impulses into the brain or spinal cord, others transmit impulses out of the brain or spinal cord, and still others conduct impulses from neuron to neuron within the brain or spinal cord.

A Current Topic

Factors Affecting Synaptic Transmission

If nerve impulses reach synaptic knobs at too rapid a rate, the supplies of neurotransmitters may become exhausted. Impulses cannot be transferred between the neurons involved until more neurotransmitters are synthesized.

Such a condition seems to occur during an epileptic seizure, when abnormal and excessive discharges of impulses originate from certain brain cells. Some of these impulses reach skeletal muscle fibers and stimulate violent contractions. In time, the synaptic knobs seem to run out of neurotransmitters, and the seizure subsides.

A drug called Dilantin (diphenylhydantoin) is sometimes used to treat epilepsy. It seems to have a stabilizing effect upon excitable neuron membranes, by increasing the effectiveness of the sodium active transport mechanism. As sodium ions are moved from inside the neurons, the membrane thresholds are stabilized against excessive stimulation.

Other factors that affect synaptic transmission include various drugs. For example, caffeine, which is found in coffee, tea, and cola drinks, stimulates activity in the nervous system. It does this by lowering the thresholds at synapses. Consequently, when caffeine is present, certain neurons are more easily excited than usual.

A number of drugs produce their special effects by interfering with the normal actions of neurotransmitters. In fact, nearly all of the drugs used to treat functional disorders of the nervous system act by blocking or promoting the effects of some neurotransmitter substance. For example, the stimulants called *amphetamines* cause a person to feel more alert and energetic by enhancing the release of the neurotransmitters norepinephrine and dopamine from the axonal ends of certain neurons. The *tricyclic antidepressant drugs* (as well as cocaine) produce similar results by blocking the normal uptake of norepinephrine and dopamine, thus leaving these neurotransmitters in the synaptic clefts where they continue to produce effects longer than usual.

Figure 9.15
(*a*) Nerve fibers of neurons *1* and *2* converge to the cell body of neuron *3;* (*b*) the nerve fiber of neuron *4* diverges to the cell bodies of neurons *5* and *6.*

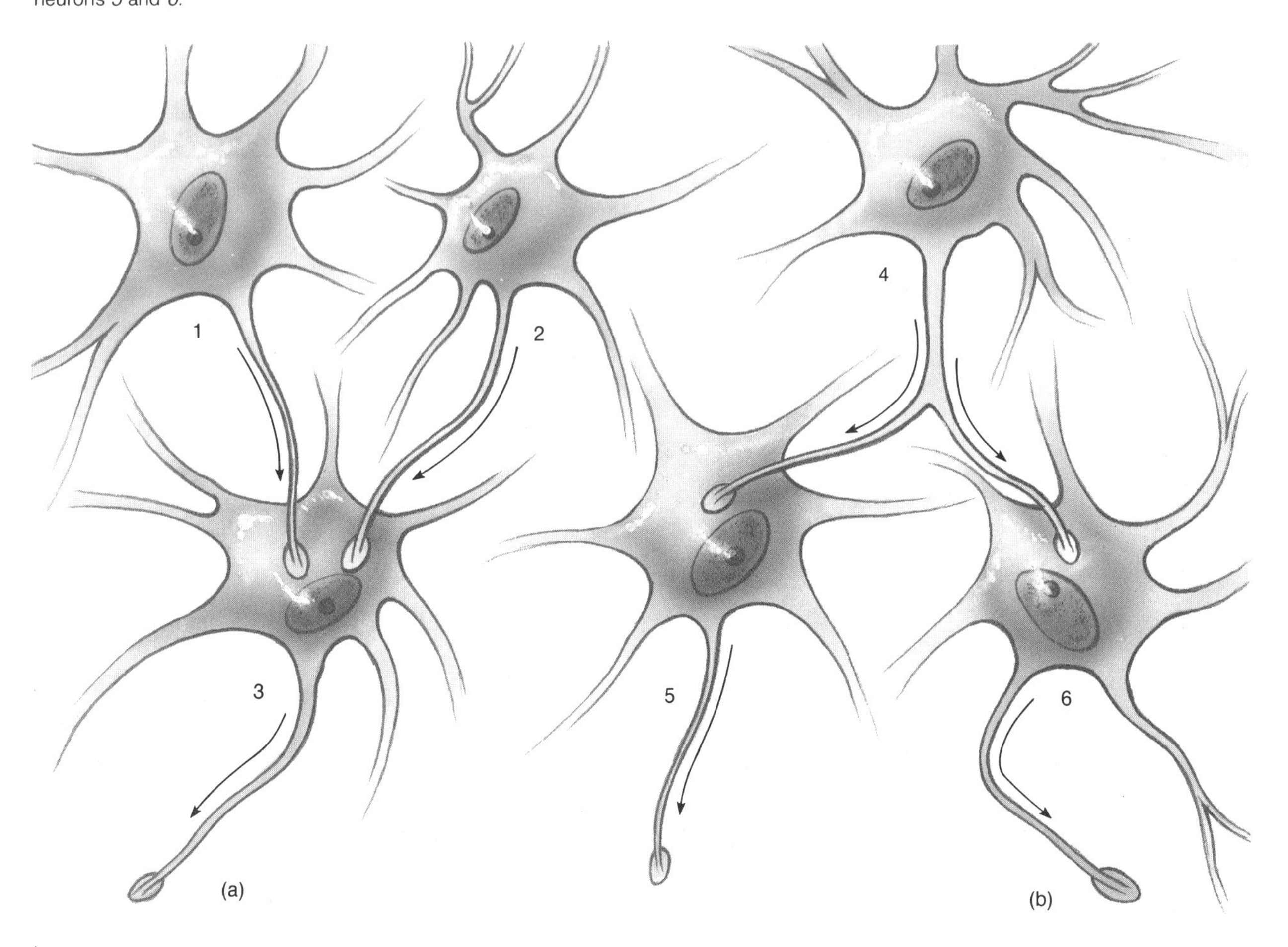

Classification of Neurons

On the basis of *structural differences,* neurons can be classified into three major groups as follows:

1. **Multipolar neurons.** Multipolar neurons have many nerve fibers arising from their cell bodies. Only one fiber of each neuron is an axon; the rest are dendrites. Most of the neurons whose cell bodies lie within the brain or spinal cord are of this type. The motor neuron in figure 9.2 is multipolar.
2. **Bipolar neurons.** The cell body of a bipolar neuron has only two nerve fibers, one arising from each end. Although these fibers have similar structural characteristics, one serves as an axon and the other as a dendrite. Such neurons are found within specialized parts of the eyes, nose, and ears.
3. **Unipolar neurons.** Each unipolar neuron has a single nerve fiber extending from its cell body. A short distance from the cell body, this fiber divides into two branches: one branch is connected to a peripheral body part and serves as a dendrite, and the other enters the brain or spinal cord and serves as an axon. Unipolar neurons occur in specialized masses of nerve tissue called *ganglia,* which are located outside the brain and spinal cord. (The sensory neuron in figure 9.2 is unipolar.)

Based upon *functional differences,* neurons can be grouped as follows:

1. **Sensory neurons** (afferent neurons) are those that carry nerve impulses from peripheral body parts into the brain or spinal cord (fig. 9.2*b*). These neurons either have specialized *receptor ends* at the tips of their dendrites, or they have dendrites that are closely associated with *receptor cells* located in the skin or in various sensory organs.

   Changes that occur inside or outside the body are likely to stimulate receptor ends or receptor cells, triggering sensory nerve impulses. The impulses travel along the sensory neuron fibers, which lead to the brain or spinal cord, and they are then processed in these parts by other neurons.
2. **Interneurons** (also called intercalated, internuncial, or association neurons) lie within the brain or spinal cord. They form links between other neurons (fig. 9.15). Interneurons transmit impulses from one part of the brain or spinal cord to another. That is, they may direct incoming sensory impulses to appropriate parts for processing and interpreting. Other incoming impulses are transferred to motor neurons.
3. **Motor neurons** (efferent neurons) carry nerve impulses out of the brain or spinal cord to *effectors*—parts of the body capable of responding, such as muscles or glands (fig. 9.2*a*). For example, when motor impulses reach muscles, these effectors are stimulated to contract; when motor impulses reach glands, the glands are stimulated to release secretions.

Types of Nerves

Although a nerve fiber is an extension of a neuron, a **nerve** is a cordlike bundle (or group of bundles) of nerve fibers held together by layers of connective tissue.

Like nerve fibers, nerves that conduct impulses into the brain or spinal cord are called **sensory nerves,** and those that carry impulses to muscles or glands are termed **motor nerves.** Most nerves, however, include both sensory and motor fibers, and they are called **mixed nerves.**

---

1. Explain how neurons are classified according to structure or function.
2. How is a neuron related to a nerve?
3. What is a mixed nerve?

Nerve Pathways

The routes followed by nerve impulses as they travel through the nervous system are called *nerve pathways.* The simplest of these pathways includes only a few neurons and is called a reflex arc.

Reflex Arcs

A **reflex arc** begins with a receptor at the end of a sensory nerve fiber. This fiber usually leads to several interneurons within the central nervous system, which serve as a processing center, or *reflex center.* Fibers from the interneurons may connect with interneurons in other parts of the nervous system. They also communicate with motor neurons, whose fibers pass outward to effectors. Such a reflex arc represents the *behavioral unit* of the nervous system. That is, it constitutes the structural and functional basis for the simplest acts—reflexes.

Reflex Behavior

Reflexes are automatic, unconscious responses to changes occurring within or outside the body. They are mechanisms that help to maintain homeostasis by controlling many involuntary processes such as heart rate, breathing rate, blood pressure, and digestive activities. Reflexes also are involved in the automatic actions of swallowing, sneezing, coughing, and vomiting.

Figure 9.16
The knee jerk reflex involves two neurons—a sensory neuron and a motor neuron.

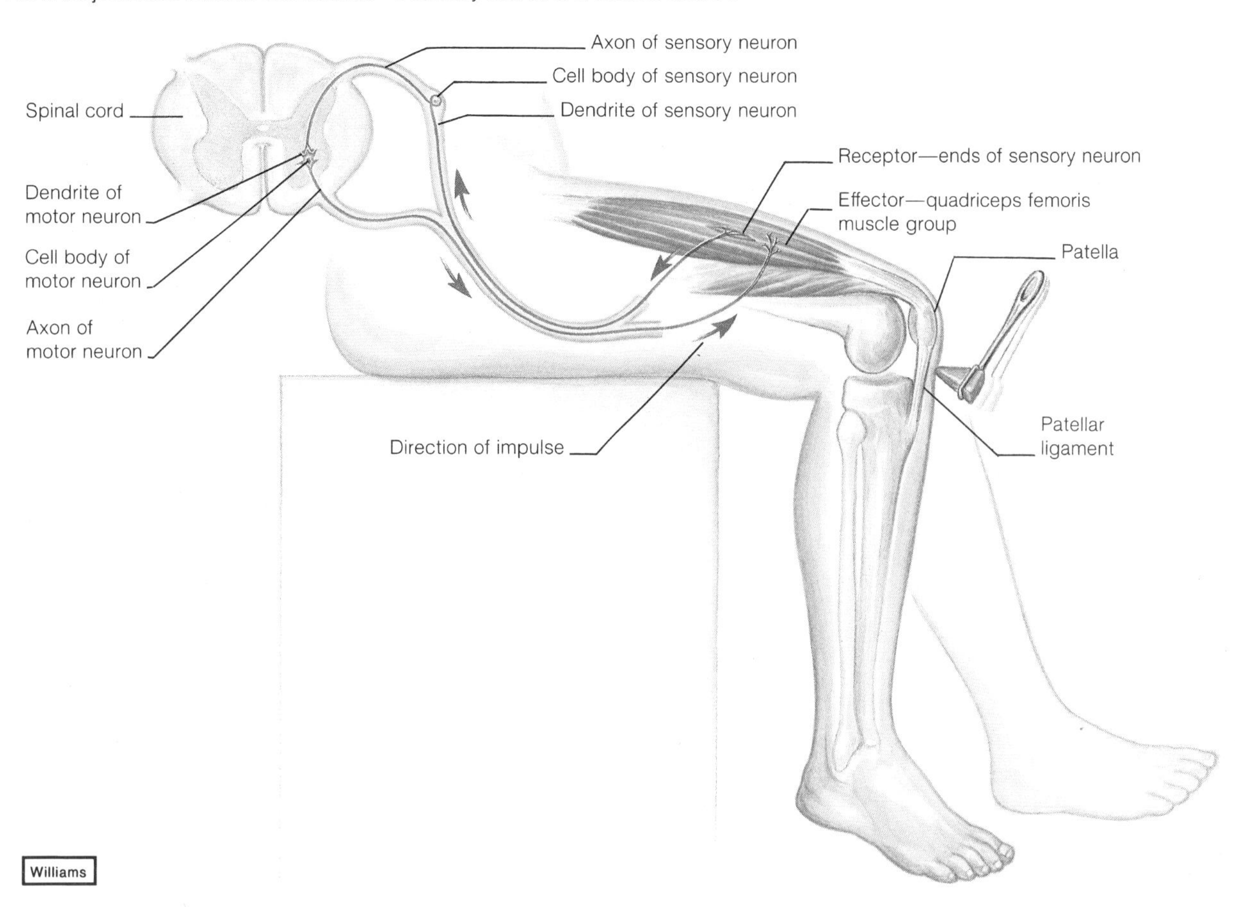

The *knee jerk reflex* (patellar reflex) is an example of a simple reflex that employs only two neurons—a sensory neuron connected directly to a motor neuron (fig. 9.16). This reflex is initiated by striking the patellar ligament just below the patella. As a result, the quadriceps femoris group of muscles, which is attached to the patella by a tendon, is pulled slightly, and stretch receptors located within the muscle group are stimulated. These receptors, in turn, trigger impulses that pass along the fibers of a sensory neuron into the spinal cord. Within the spinal cord, the sensory axon forms a synapse with a dendrite of a motor neuron. The impulse then continues along the axon of the motor neuron and travels back to the quadriceps femoris. The muscle group responds by contracting, and the reflex is completed as the lower leg extends.

This reflex is helpful in maintaining an upright posture. For example, if the knee begins to bend as a result of gravity when a person is standing still, the quadriceps femoris is stretched, the reflex is triggered, and the leg straightens again.

Another type of reflex, called a *withdrawal reflex,* occurs when a person unexpectedly touches a finger to something painful (fig. 9.17). As this happens, the skin receptors are activated, and sensory impulses travel to the spinal cord. There the impulses pass on to interneurons of a reflex center and are directed to motor neurons. The motor neurons transmit the signals to flexor muscles in the arm, and the muscles contract in response. At the same time, the antagonistic extensor muscles are inhibited, and the hand is rapidly and unconsciously withdrawn from the painful stimulation.

Concurrent with the withdrawal reflex, other interneurons in the spinal cord carry sensory impulses to the brain, and the person becomes aware of the experience and may feel pain.

A withdrawal reflex is, of course, protective because it prevents excessive tissue damage when a body part touches something that is potentially harmful.

Chart 9.3 summarizes the parts of a reflex arc.

Because normal reflexes depend on normal neuron functions, reflexes are commonly used to obtain information concerning the condition of the nervous system. An anesthesiologist, for instance, may try to initiate a reflex in a patient who is being anesthetized in order to determine how well the anesthetic drug is affecting nerve functions. Also, in the case of injury to some part of the nervous system, various reflexes may be tested to discover the location and extent of the damage.

Figure 9.17
A withdrawal reflex involves a sensory neuron, an interneuron, and a motor neuron.

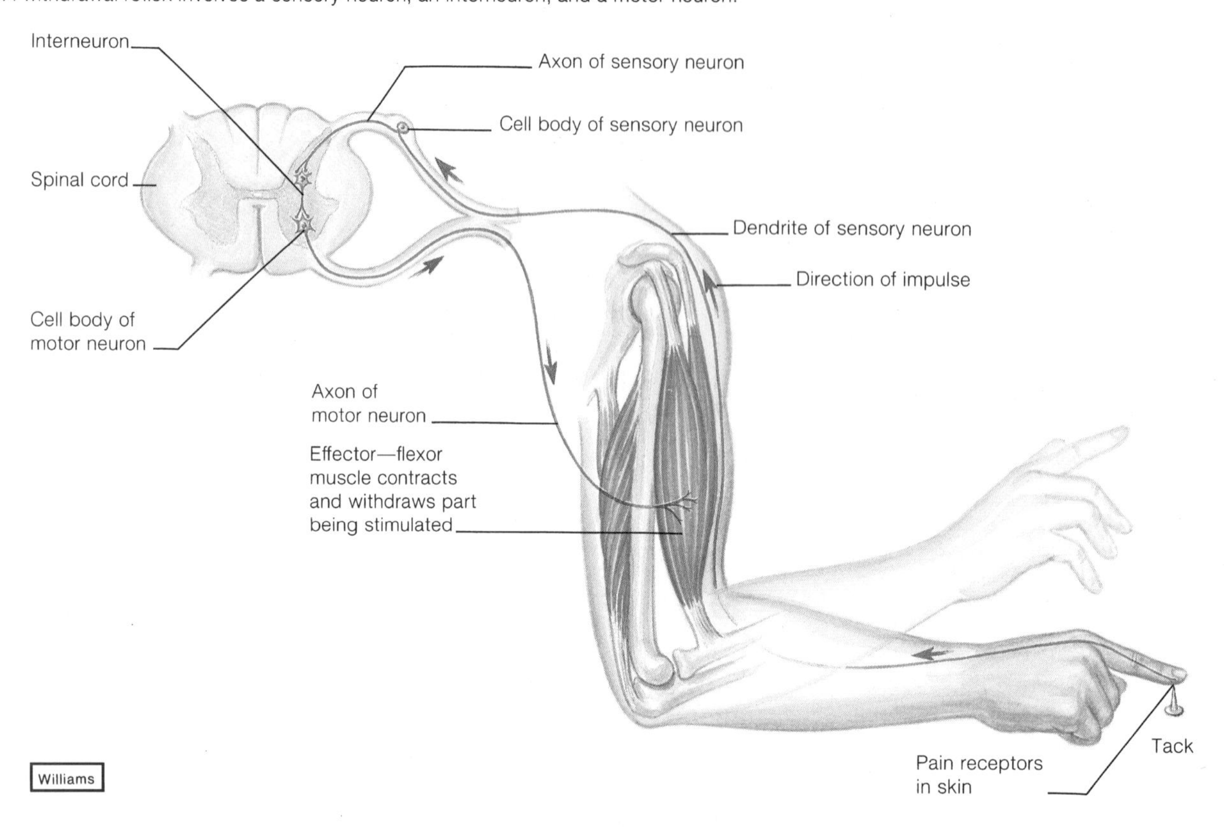

Chart 9.3 Parts of a reflex arc

Part	Description	Function
Receptor	Receptor end of a dendrite or a specialized receptor cell in a sensory organ	Sensitive to a specific type of internal or external change
Sensory neuron	Dendrite, cell body, and axon of a sensory neuron	Transmits a nerve impulse from the receptor into the brain or spinal cord
Interneuron	Dendrite, cell body, and axon of a neuron within the brain or spinal cord	Conducts a nerve impulse from the sensory neuron to a motor neuron
Motor neuron	Dendrite, cell body, and axon of a motor neuron	Transmits a nerve impulse from the brain or spinal cord out to an effector
Effector	Muscle or gland outside the nervous system	Responds to stimulation by the motor neuron and produces the reflex or behavioral action

1. What is a nerve pathway?
2. List the parts of a reflex arc.
3. Define *reflex*.
4. Review the actions that occur during a withdrawal reflex.

Coverings of the Central Nervous System

The organs of the central nervous system (CNS) are surrounded by bones, membranes, and fluid. More specifically, the brain lies within the cranial cavity of the skull, and the spinal cord occupies the vertebral canal within the vertebral column (fig. 9.18*a*). Beneath these bony coverings, the brain and spinal cord are protected by membranes called *meninges,* which are located between the bone and soft tissues of the nervous system.

Meninges

The **meninges** have three layers—dura mater, arachnoid mater, and pia mater (fig. 9.18*b*).

The **dura mater** is the outermost layer. It is composed primarily of tough, white fibrous connective tissue and contains many blood vessels and nerves. It is attached to the inside of the cranial cavity and forms the internal periosteum of the surrounding skull bones.

Figure 9.18
(*a*) The brain and spinal cord are enclosed by bone and by membranes called meninges. (*b*) The meninges include three layers—dura mater, arachnoid mater, and pia mater.

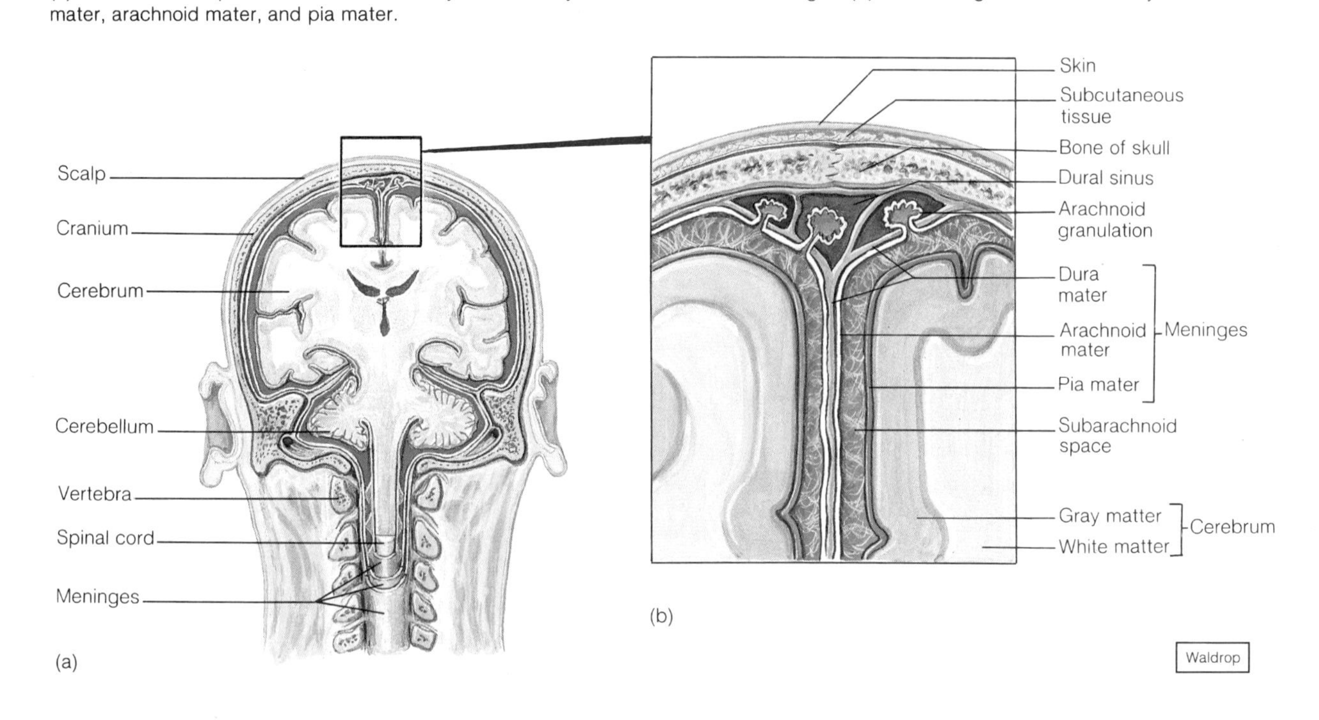

Figure 9.19
Tissues in the epidural space provide a protective pad around the spinal cord.

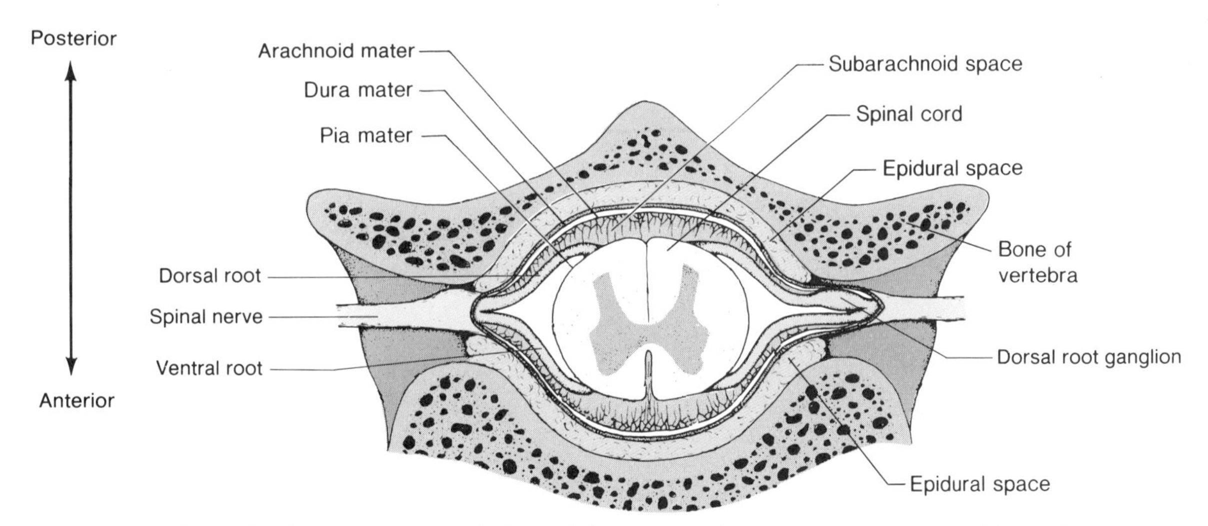

In some regions, the dura mater extends inward between lobes of the brain and forms partitions that support and protect these parts.

The dura mater continues into the vertebral canal as a strong, tubular sheath that surrounds the spinal cord. It terminates as a blind sac below the end of the cord. The membrane around the spinal cord is not attached directly to the vertebrae, but is separated by an *epidural space,* which lies between the dural sheath and the bony walls (fig. 9.19). This space contains loose connective and fat tissues, which provide a protective pad around the spinal cord.

The **arachnoid mater** is a thin, netlike membrane that lacks blood vessels and is located between the dura and pia maters. It spreads over the brain and spinal cord, but generally does not dip into the grooves and depressions on their surfaces.

Between the arachnoid and pia maters is a *subarachnoid space,* which contains the clear, watery **cerebrospinal fluid.**

The **pia mater** is very thin and contains many nerves as well as blood vessels, which aid in nourishing the underlying cells of the brain and spinal cord. This layer is attached to the surfaces of these organs and follows their irregular contours, passing over the high areas and dipping into the depressions.

A blow to the head, may cause some blood vessels associated with the brain to be broken, and blood may collect in the space beneath the dura mater. This condition, called *subdural hematoma*, causes an increasing pressure to develop between the rigid bones of the skull and the soft tissues of the brain. Unless the accumulating blood is evacuated, the resulting compression of the brain may lead to functional losses or even death.

1. Describe the meninges.
2. Name the layers of the meninges.
3. Explain where cerebrospinal fluid occurs.

Spinal Cord

The **spinal cord** is a slender nerve column that passes downward from the brain into the vertebral canal (fig. 9.20). Although it is continuous with the brain, the spinal cord is said to begin where nerve tissue leaves the cranial cavity at the level of the foramen magnum. The cord tapers to a point and terminates near the intervertebral disk that separates the first and second lumbar vertebrae.

Structure of the Spinal Cord

The spinal cord consists of thirty-one segments, each of which gives rise to a pair of **spinal nerves.** These nerves branch out to various body parts and connect them with the central nervous system.

In the neck region, a thickening in the spinal cord, called the *cervical enlargement*, gives off nerves to the arms. A similar thickening in the lower back, the *lumbar enlargement*, gives off nerves to the legs (fig. 9.20).

Two grooves, a deep *anterior median fissure* and a shallow *posterior median sulcus*, extend the length of the spinal cord, dividing it into right and left halves (fig. 9.21). A cross section of the cord reveals that it consists of a core of gray matter surrounded by white matter. The pattern produced by the gray matter roughly resembles a butterfly with its wings outspread. The upper and lower wings of gray matter are called the *posterior horns* and *anterior horns*, respectively. Between them on either side there is a protrusion of gray matter called the *lateral horn*.

A horizontal bar of gray matter in the middle of the spinal cord, the *gray commissure*, connects the wings of the gray matter on the right and left sides. This bar surrounds the **central canal,** which contains cerebrospinal fluid.

Neurons with large cell bodies located in the anterior horns give rise to motor fibers, which pass out through spinal nerves and lead to various skeletal muscles. The majority of neurons in the gray matter of the spinal cord, however, are interneurons.

The white matter of the spinal cord is divided by gray matter into three regions on each side. These regions are known as the *anterior*, *lateral*, and *posterior funiculi* (fig. 9.21). Each funiculus consists of longitudinal bundles of myelinated nerve fibers, which comprise major nerve pathways, called **nerve tracts.**

Figure 9.20
The spinal cord begins at the level of the foramen magnum. At what level does it terminate?

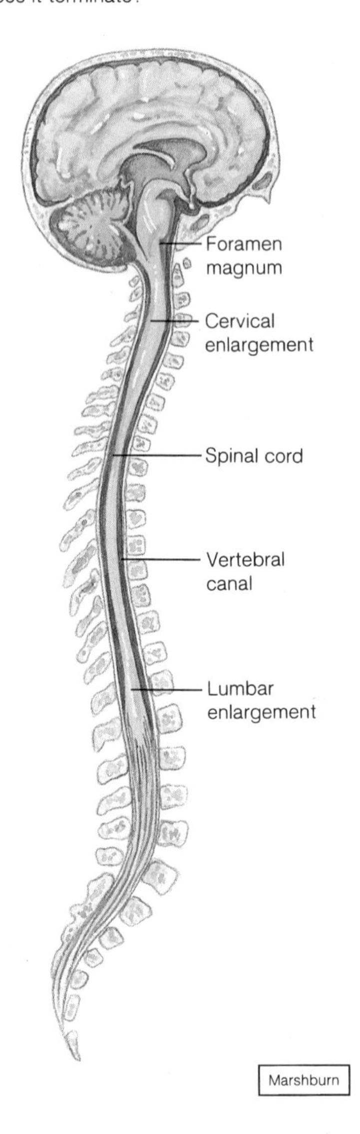

Figure 9.21
(*a*) A cross section of the spinal cord. (*b*) What parts can you identify in this micrograph of the cord?

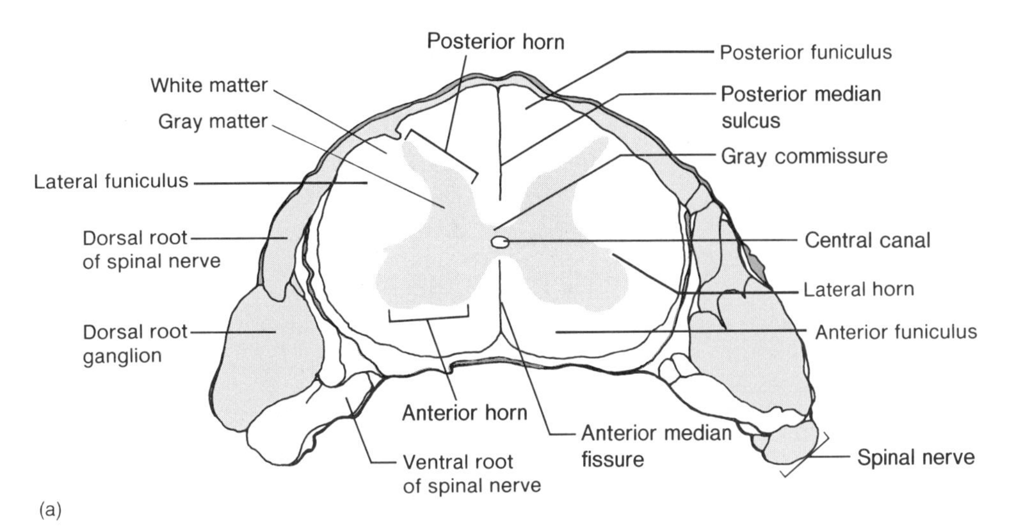

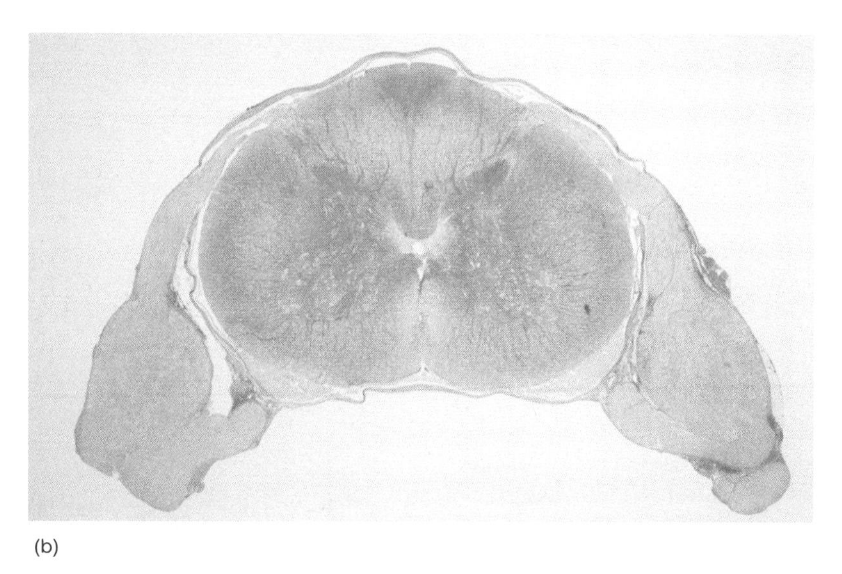

(b)

Functions of the Spinal Cord

The spinal cord has two major functions—to conduct nerve impulses and to serve as a center for spinal reflexes.

The nerve tracts of the spinal cord provide a two-way system of communication between the brain and parts outside the nervous system. The tracts that conduct impulses from body parts and carry sensory information to the brain are called **ascending tracts** (fig. 9.26); those that conduct motor impulses from the brain to muscles and glands are called **descending tracts** (fig. 9.25).

The nerve fibers within these tracts are axons, and usually all the axons within a given tract originate from neuron cell bodies located in the same part of the nervous system, and they terminate together in some other part. The names that identify nerve tracts often reflect these common origins and terminations. For example, a *spinothalamic* tract begins in the spinal cord and carries the sensory impulses associated with the sensations of pain and touch to the thalamus of the brain; a *corticospinal* tract originates in the cortex of the brain and carries motor impulses downward through the spinal cord and spinal nerves. These impulses function in the control of skeletal muscle movements.

Corticospinal tracts are also called *pyramidal tracts* after the pyramid-shaped areas within a region of the brain (medulla oblongata) through which they

Figure 9.22
The major portions of the brain include the cerebrum, cerebellum, and brain stem.

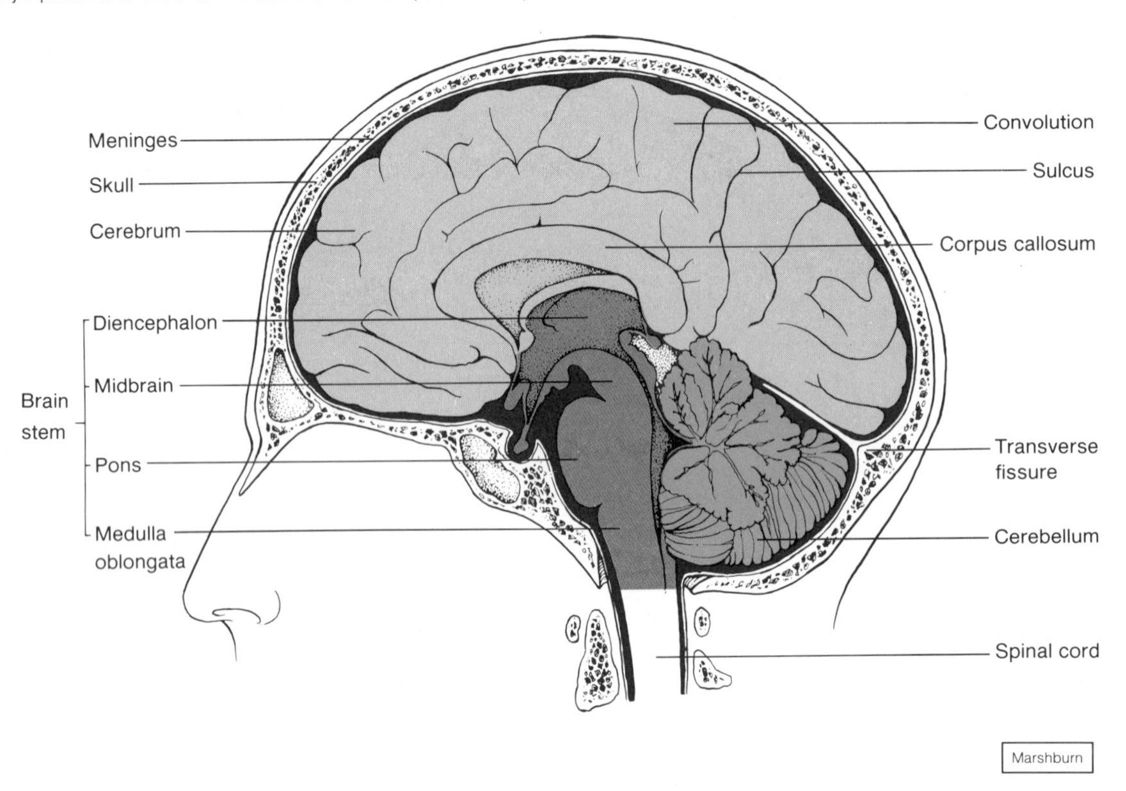

pass. Other descending tracts are called *extrapyramidal tracts.* Extrapyramidal tracts function in the control of the motor activities associated with the maintenance of balance and posture.

In addition to serving as a pathway for various nerve tracts, the spinal cord functions in many reflexes like the knee jerk and withdrawal reflexes, described previously. Such reflexes are called **spinal reflexes** because their reflex arcs pass through the cord.

Injuries to the spinal cord may be caused indirectly, as by a blow to the head or by a fall, or they may be due to forces applied directly to the cord. The consequences will depend on the amount of damage sustained by the cord. The spinal cord may, for example, be compressed or distorted by a minor injury, and its functions may be disturbed only temporarily. If nerve fibers are severed, however, some of the cord's functions are likely to be lost permanently.

1. Describe the structure of the spinal cord.
2. Distinguish between an ascending and a descending tract.
3. Describe the general functions of the spinal cord.

Brain

The **brain** is composed of about one hundred billion (10^{11}) multipolar neurons and innumerable nerve fibers by which these neurons communicate with one another and with neurons in other parts of the system.

As figure 9.22 shows, the brain can be divided into three major portions—the cerebrum, cerebellum, and brain stem. The **cerebrum,** which is the largest part, contains nerve centers associated with sensory and motor functions. It is also concerned with higher mental functions, including memory and reasoning. The cerebellum includes centers associated with the coordination of voluntary muscular movements. The **brain stem** contains nerve pathways by which various parts of the nervous system are interconnected and also contains nerve centers involved in the regulation of various visceral activities.

Structure of the Cerebrum

The cerebrum consists of two large masses called **cerebral hemispheres,** which are essentially mirror images of each other. These hemispheres are connected by a deep bridge of nerve fibers called the **corpus callosum** and separated by a layer of dura mater (falx cerebri).

Figure 9.23
Some motor, sensory, and association areas of the cerebral cortex.

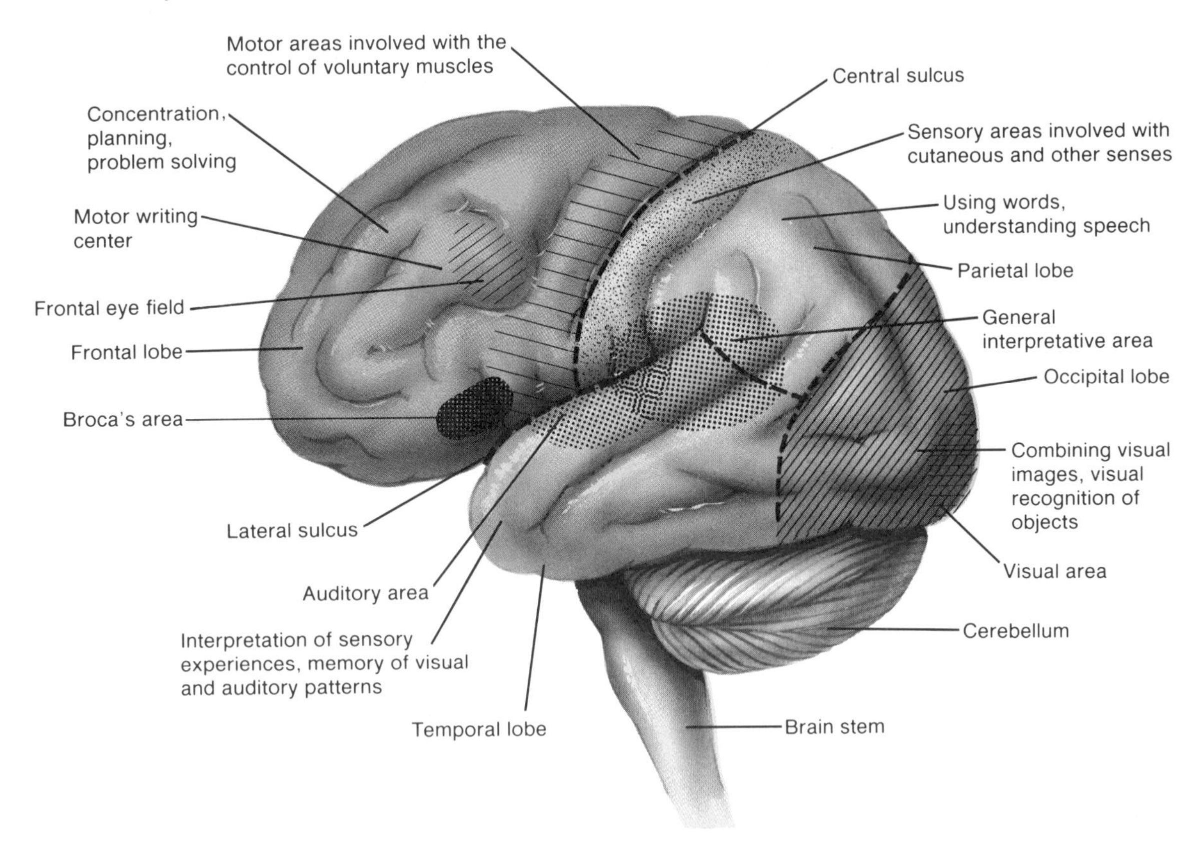

The surface of the cerebrum is marked by numerous ridges, or **convolutions** (gyri), which are separated by grooves. A shallow groove is called a **sulcus,** and a deep groove is called a **fissure.** Although the arrangement of these elevations and depressions is complex, they form fairly distinct patterns in all normal brains. For example, a *longitudinal fissure* separates the right and left cerebral hemispheres, a *transverse fissure* separates the cerebrum from the cerebellum, and various sulci divide each hemisphere into lobes.

The lobes of the cerebral hemispheres are named after the skull bones that they underlie (fig. 9.23). They include the following:

1. **Frontal lobe.** The frontal lobe forms the anterior portion of each cerebral hemisphere. It is bordered posteriorly by a *central sulcus,* which extends out from the longitudinal fissure at a right angle, and inferiorly by a *lateral sulcus,* which extends out from the undersurface of the brain along its sides.
2. **Parietal lobe.** The parietal lobe is posterior to the frontal lobe and separated from it by the central sulcus.
3. **Temporal lobe.** The temporal lobe lies below the frontal lobe and is separated from it by the lateral sulcus.
4. **Occipital lobe.** The occipital lobe forms the posterior portion of each cerebral hemisphere and is separated from the cerebellum by a shelflike extension of dura mater (tentorium cerebelli). There is no distinct boundary between the occipital lobe and the parietal and temporal lobes.
5. **The insula.** The insula is located deep within the lateral sulcus and is covered by parts of the frontal, parietal, and temporal lobes. It is separated from them by a *circular sulcus.*

   A thin layer of gray matter, called the **cerebral cortex,** constitutes the outermost portion of the cerebrum. This layer covers the convolutions and dips into the sulci and fissures. It is estimated to contain nearly 75% of the neuron cell bodies in the nervous system.

Just beneath the cerebral cortex are masses of white matter, making up the bulk of the cerebrum. These masses contain bundles of myelinated nerve fibers

Figure 9.24
A frontal section of the left cerebral hemisphere reveals the basal ganglia.

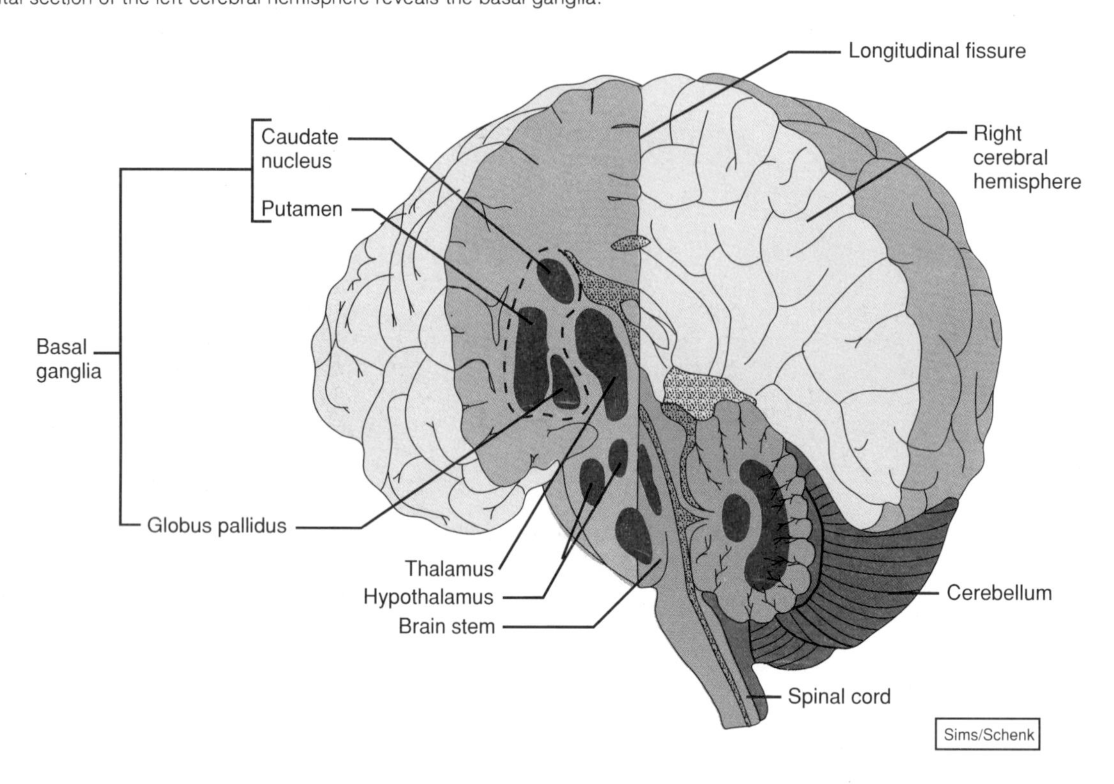

that connect the neuron cell bodies of the cortex with other parts of the nervous system. Some of these fibers pass from one cerebral hemisphere to the other by way of the corpus callosum, and others carry sensory or motor impulses from portions of the cortex to nerve centers in the brain or spinal cord.

Deep within each cerebral hemisphere are several masses of gray matter called *basal ganglia* (fig. 9.24). Although the precise function of these parts is not completely understood, the neuron cell bodies they contain are known to serve as relay stations for motor impulses originating in the cerebral cortex and passing into the brain stem and spinal cord. It is also known that most of the inhibitory neurotransmitter dopamine is produced in the basal ganglia. The impulses from these parts normally inhibit motor functions and thus aid in the control of various skeletal muscle activities.

Because the basal ganglia function to inhibit muscular activity, disorders in this portion of the brain may be accompanied by a reduction in mobility (hypokinesia) due to an excessive discharge of impulses from the ganglia or by involuntary movements (hyperkinesia) due to a lack of inhibiting impulses from them. Parkinson's disease and Huntington's chorea are conditions resulting from lesions in the basal ganglia.

Functions of the Cerebrum

The cerebrum is concerned with higher brain functions in that it contains centers for interpreting sensory impulses arriving from various sense organs as well as centers for initiating voluntary muscular movements. It stores the information of memory and utilizes this information in reasoning processes. It also functions in determining a person's intelligence and personality.

Functional Regions of the Cortex

The regions of the cerebral cortex that perform specific functions have been identified. Although there is considerable overlap in these areas, the cortex can be divided into sections known as the *motor, sensory,* and *association areas.*

The primary **motor areas** of the cerebral cortex lie in the frontal lobes, just in front of the central sulcus (fig. 9.23). The nervous tissue in these regions contains numerous, large *pyramidal cells,* so named because of their pyramid-shaped cell bodies.

The impulses from the pyramidal cells travel downward through the brain stem and into the spinal cord on the *corticospinal tracts* (fig. 9.25). Most of the nerve fibers in these tracts cross over from one side of the brain to the other within the brain stem. As a result, the motor area of the right cerebral hemisphere gen-

Figure 9.25
Motor fibers of the corticospinal tract begin in the cerebral cortex, cross over in the medulla, and descend in the spinal cord. There they synapse with neurons whose fibers lead to the spinal nerves that supply skeletal muscles.

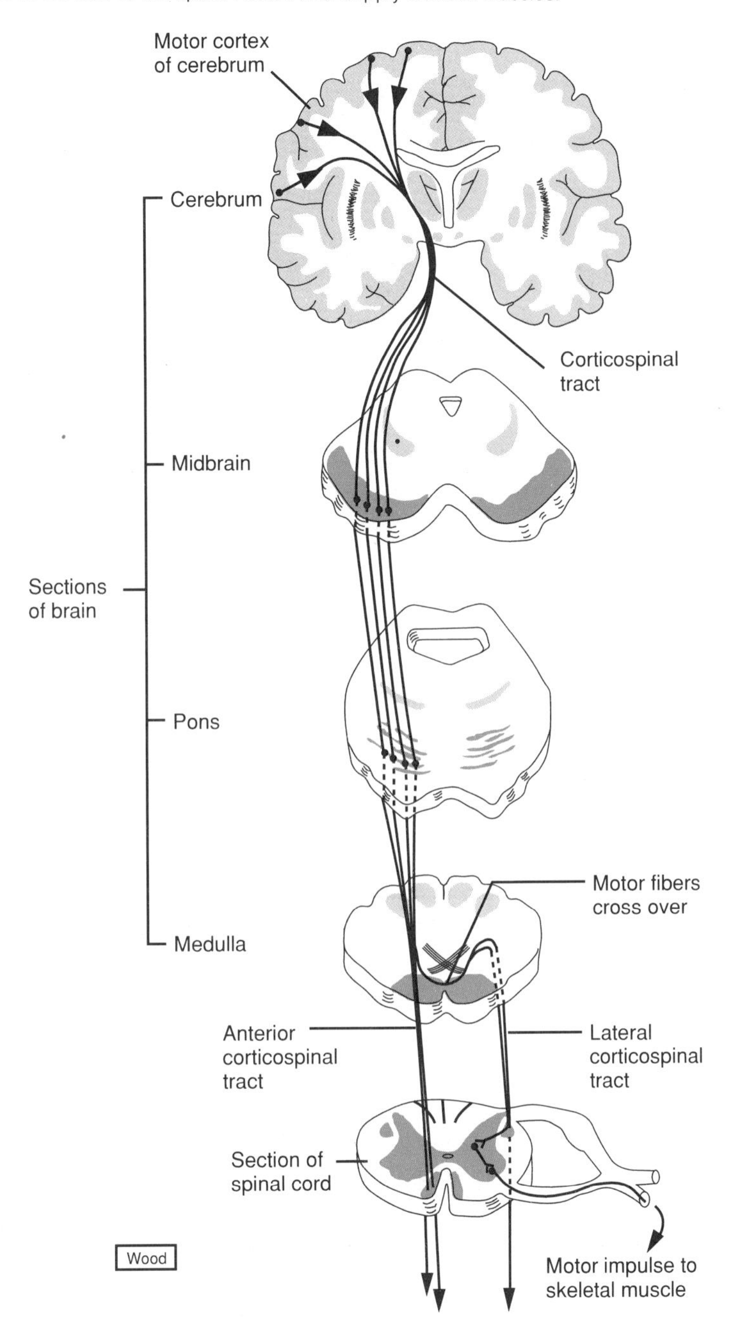

erally controls skeletal muscles on the left side of the body and vice versa.

In addition to the primary motor areas, certain other regions of the frontal lobe are involved with motor functions. For example, a region called *Broca's area* is just anterior to the primary motor cortex and above the lateral sulcus (fig. 9.23). It coordinates the complex muscular actions of the mouth, tongue, and larynx, which make speech possible. Above Broca's area is a region called the *frontal eye field.* The motor cortex in this area controls the voluntary movements of the eyes and eyelids. Another region just in front of the primary motor area controls the muscular movements of the hands and fingers that make skills such as writing possible.

Figure 9.26
Sensory impulses originating in skin receptors ascend in the spinal cord and cross over in the medulla of the brain.

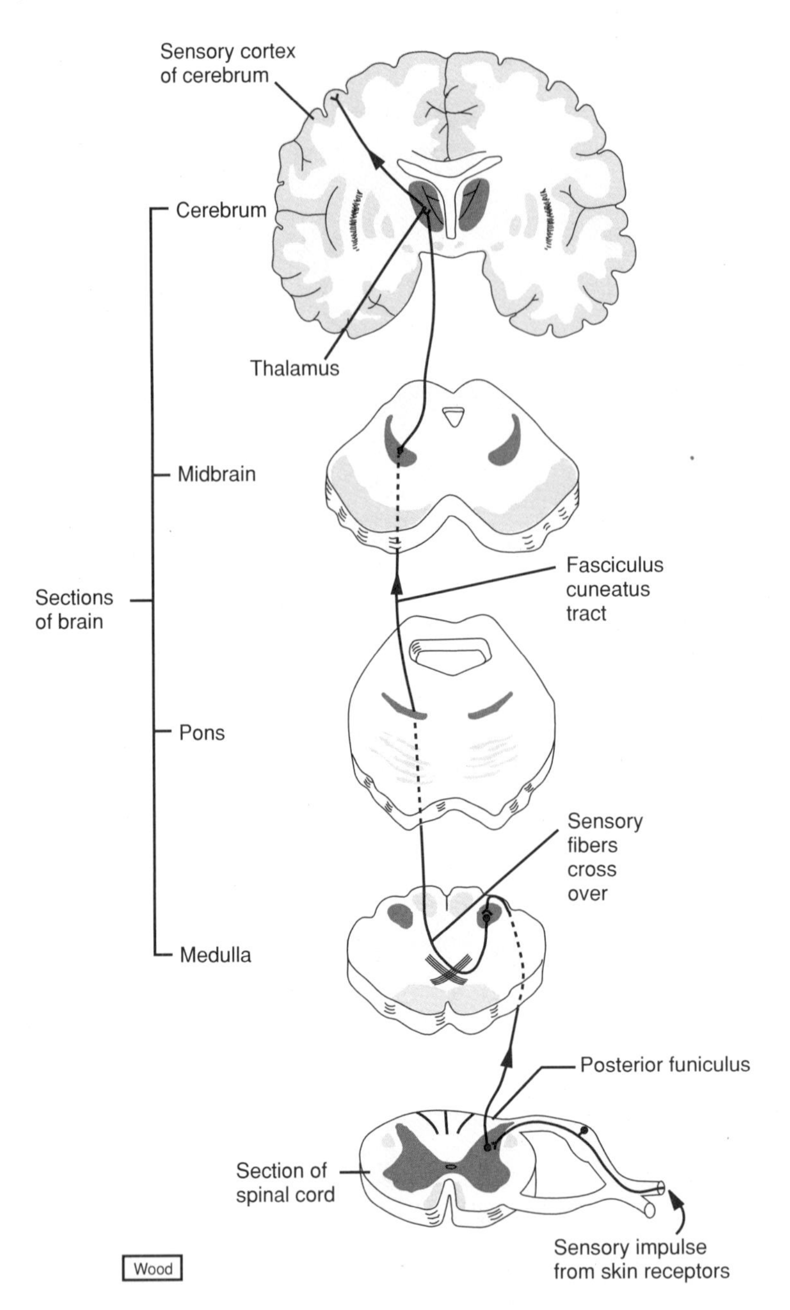

Sensory areas, which occur in several lobes of the cerebrum, function in interpreting impulses that arrive from various sensory receptors. These interpretations give rise to feelings or sensations. For example, sensations from all parts of the skin arise in the anterior portions of the parietal lobes along the central sulcus (fig. 9.23). The posterior parts of the occipital lobes are concerned with vision, and the temporal lobes contain the centers for hearing. The sensory areas for taste are located near the bases of the central sulci along the lateral sulci, and the sense of smell arises from centers deep within the cerebrum.

Like motor fibers, sensory fibers, such as those of the fasciculus cuneatus tract, cross over (fig. 9.26). Thus, the centers in the right cerebral hemisphere interpret impulses originating from the left side of the body and vice versa.

Association areas function in the analysis and interpretation of sensory experiences and are involved with memory, reasoning, verbalizing, judgment, and emotional feelings. These areas occupy the anterior portions of the frontal lobes and are widespread in the lateral portions of the parietal, temporal, and occipital lobes (fig. 9.23).

The association areas of the frontal lobes are concerned with a number of higher intellectual processes, including those necessary for concentrating, planning, complex problem solving, and judging the possible consequences of behavior.

The association areas of the parietal lobes aid in understanding speech and choosing the words needed to express thoughts and feelings.

The association areas of the temporal lobes and regions at the posterior ends of the lateral fissures are concerned with the interpretation of complex sensory experiences, such as those needed to understand speech and read printed words. These regions are also involved with the memory of visual scenes, music, and other complex sensory patterns.

The association areas of the occipital lobes that are adjacent to the visual centers are important in analyzing visual patterns and combining visual images with other sensory experiences, as when one recognizes another person or an object.

Of particular importance is the region where the parietal, temporal, and occipital association areas come together, near the posterior end of the lateral sulcus. This region is called the *general interpretative area,* and it plays the primary role in complex thought processing (fig. 9.23).

The effects of injuries to the cerebral cortex depend on which areas are damaged and to what extent. When portions of the cortex are injured, the special functions of these portions are likely to be lost or at least depressed.

It is often possible to deduce the location and extent of a brain injury by determining what abilities the patient is missing. For example, if the motor areas of one frontal lobe have been damaged, the person is likely to be partially or completely paralyzed on the opposite side of the body.

A person with damage to the association areas of the frontal lobes may have difficulty in concentrating on complex mental tasks. Such an individual usually appears disorganized and is easily distracted. A person who suffers damage to association areas of the temporal lobes may have difficulty recognizing printed words or arranging words into meaningful thoughts.

1. List the major divisions of the brain.
2. Describe the cerebral cortex.
3. What are the major functions of the cerebrum?
4. Locate the major functional regions of the cerebral cortex.

Hemisphere Dominance

Both cerebral hemispheres participate in basic functions, such as receiving and analyzing sensory impulses, controlling skeletal muscles, and storing memory. However, in most persons one side acts as a **dominant hemisphere** for certain other functions.

In over 90% of the population, for example, the left hemisphere is dominant for the language-related activities of speech, writing, and reading. It is also dominant for complex intellectual functions requiring verbal, analytical, and computational skills. In other persons, the right hemisphere is dominant, and in some, the hemispheres are equally dominant.

In addition to carrying on basic functions, the nondominant hemisphere seems to specialize in nonverbal functions, such as those involving motor tasks that require orientation of the body in the surrounding space, understanding and interpreting musical patterns, and nonverbal visual experiences. It is also concerned with emotional and intuitive thought processes.

Nerve fibers of the *corpus callosum,* which connect the cerebral hemispheres, make it possible for the dominant side to control the motor cortex of the nondominant hemisphere. These fibers also allow sensory information reaching the nondominant hemisphere to be transferred to the dominant one, where the information can be used in decision making.

Ventricles and Cerebrospinal Fluid

Within the cerebral hemispheres and brain stem is a series of interconnected cavities called **ventricles** (fig. 9.27). These spaces are continuous with the central canal of the spinal cord, and like it, they are filled with cerebrospinal fluid.

The largest ventricles are the *lateral ventricles* (first and second ventricles), which extend into the cerebral hemispheres and occupy portions of the frontal, temporal, and occipital lobes.

A narrow space that constitutes the *third ventricle* is located in the midline of the brain, beneath the corpus callosum. This ventricle communicates with the lateral ventricles through openings (interventricular foramina) in its anterior end.

The *fourth ventricle* is located in the brain stem just in front of the cerebellum. It is connected to the third ventricle by a narrow canal, the *cerebral aqueduct,* which passes lengthwise through the brain stem. This ventricle is continous with the central canal of the spinal cord and has openings in its roof that lead into the subarachnoid space of the meninges.

Cerebrospinal Fluid

Cerebrospinal fluid (CSF) is a clear liquid that is secreted by tiny cauliflowerlike masses of specialized capillaries from the pia mater, called **choroid plexuses** (fig. 9.28). These structures project into the cavities of the ventricles. Most of the cerebrospinal fluid seems to arise in the lateral ventricles. From there, it circulates

Figure 9.27
(*a*) An anterior and (*b*) a lateral view of the ventricles within the cerebral hemispheres and brain stem.

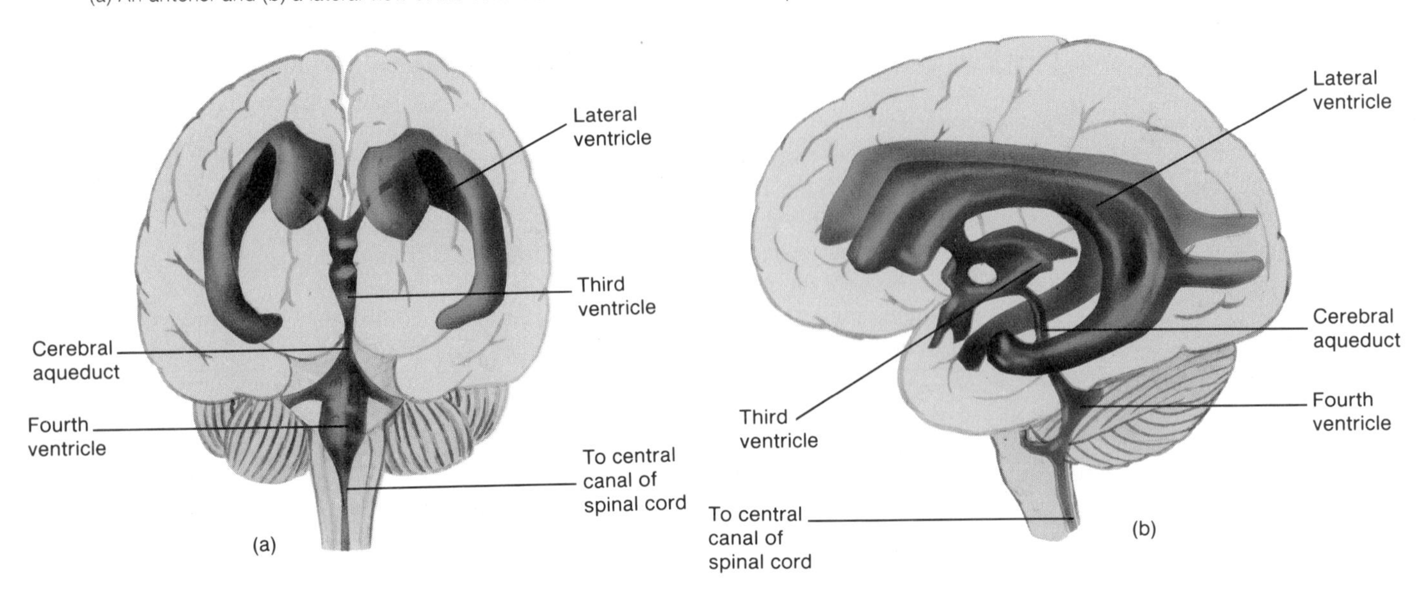

Figure 9.28
Cerebrospinal fluid is secreted by the choroid plexuses in the walls of the ventricles. The fluid circulates through the ventricles and central canal, enters the subarachnoid space, and is reabsorbed into the blood.

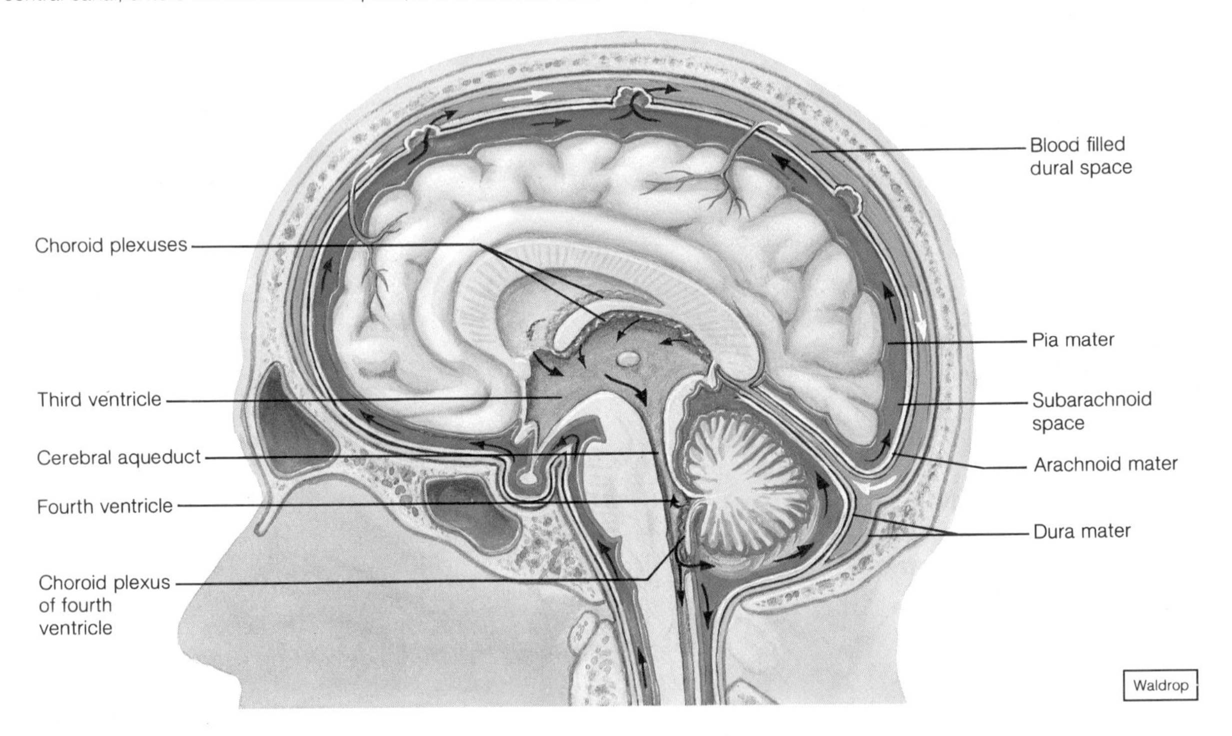

slowly into the third and fourth ventricles and into the central canal of the spinal cord. It also enters the subarachnoid space of the meninges by passing through the wall of the fourth ventricle near the cerebellum and completes its circuit by being reabsorbed into the blood.

Because it occupies the subarachnoid space of the meninges, cerebrospinal fluid completely surrounds the brain and spinal cord. In effect, these organs float in the fluid, which supports and protects them by absorbing forces that might otherwise jar and damage their delicate tissues. Cerebrospinal fluid also aids in maintaining a stable ionic concentration in the central nervous system and provides a pathway to the blood for waste substances.

Because cerebrospinal fluid is secreted and reabsorbed continuously, the fluid pressure in the ventricles normally remains relatively constant. Sometimes, however, an infection, a tumor, or a blood clot interferes with the circulation of this fluid, and the pressure within the ventricles increases. When this happens, there is a danger that the brain may be injured by being forced against the rigid skull.

The pressure of cerebrospinal fluid can be determined by performing a *lumbar puncture* (spinal tap). This procedure involves inserting a fine, hollow needle into the subarachnoid space between the third and fourth or between the fourth and fifth lumbar vertebrae and measuring the pressure by using an instrument called a *manometer.*

1. What is meant by hemisphere dominance?
2. What are the major functions of the dominant hemisphere? The nondominant one?
3. Where are the ventricles of the brain located?
4. Describe the pattern of cerebrospinal fluid circulation.

Brain Stem

The **brain stem** is a bundle of nervous tissue that connects the cerebrum to the spinal cord. It consists of numerous tracts of nerve fibers and several masses of gray matter called *nuclei.* The parts of the brain stem include the diencephalon, midbrain, pons, and medulla oblongata (fig. 9.22).

Diencephalon

The **diencephalon** is located between the cerebral hemispheres and above the midbrain. It surrounds the third ventricle and is composed largely of gray matter. Within the diencephalon, a dense mass, called the *thalamus,* bulges into the third ventricle from each side (fig. 9.24). Another region of the diencephalon that includes many nuclei is called the *hypothalamus.* It lies below the thalamus and forms the lower walls and floor of the third ventricle.

The **thalamus** serves as a central relay station for sensory impulses ascending from other parts of the nervous system to the cerebral cortex. It receives all sensory impulses (except those associated with the sense of smell) and channels them to the appropriate regions of the cortex for interpretation. In addition, all regions of the cerebral cortex can communicate with the thalamus by means of descending fibers.

Although the cerebral cortex pinpoints the origin of sensory stimulation, the thalamus produces a general awareness of certain sensations such as pain, touch, and temperature.

The **hypothalamus** is interconnected by nerve fibers to the cerebral cortex, thalamus, and other parts of the brain stem so that it can receive impulses from them and send impulses to them. The hypothalamus plays key roles in maintaining homeostasis by regulating a variety of visceral activities and by serving as a link between the nervous and endocrine systems.

Among the many important functions of the hypothalamus are the following:

1. Regulation of heart rate and arterial blood pressure.
2. Regulation of body temperature.
3. Regulation of water and electrolyte balance.
4. Control of hunger and regulation of body weight.
5. Control of movements and glandular secretions of the stomach and intestines.
6. Production of neurosecretory substances that stimulate the pituitary gland to release various hormones.
7. Regulation of sleep and wakefulness.

Structures in the general region of the diencephalon also play important roles in the control of emotional responses. For example, portions of the cerebral cortex in the medial parts of the frontal and temporal lobes are interconnected with the hypothalamus, thalamus, basal ganglia, and other deep nuclei. Together these structures comprise a complex called the limbic system.

The **limbic system** is involved in emotional experience and expression. It can modify the way a person acts because it functions to produce such feelings as fear, anger, pleasure, and sorrow. More specifically, the limbic system seems to recognize upsets in a person's physical or psychological condition that might threaten life. By causing pleasant or unpleasant feelings about experiences, the limbic system guides a person into behavior that is likely to increase the chance of survival.

The diencephalon also includes the **pineal gland,** which is a small, cone-shaped structure attached to the upper portion of the thalamus. Its function as an endocrine gland is discussed in chapter 11.

Midbrain

The **midbrain** is a short section of the brain stem located between the diencephalon and pons (fig. 9.22). It contains bundles of myelinated nerve fibers, which join lower parts of the brain stem and spinal cord with higher parts of the brain. The midbrain includes several masses of gray matter that serve as reflex centers. For example, it contains the centers for certain visual reflexes, such as those responsible for moving the eyes to view something as the head is turned. It also contains the auditory reflex centers that operate when it is necessary to move the head so that sounds can be heard more distinctly.

Two prominent bundles of nerve fibers on the underside of the midbrain include the corticospinal tracts and are the main motor pathways between the cerebrum and lower parts of the nervous system.

Pons

The **pons** appears as a rounded bulge on the underside of the brain stem, where it separates the midbrain from the medulla oblongata (fig. 9.22). The dorsal portion of the pons consists largely of longitudinal nerve fibers, which relay impulses to and from the medulla oblongata and the cerebrum. Its ventral portion contains large bundles of transverse nerve fibers, which transmit impulses from the cerebrum to centers within the cerebellum.

Several nuclei of the pons relay sensory impulses from peripheral nerves to higher brain centers. Other nuclei function with centers of the medulla oblongata to regulate the rate and depth of breathing. (See chapter 13.)

Medulla Oblongata

The **medulla oblongata** (fig. 9.22) is an enlarged continuation of the spinal cord extending from the pons to the foramen magnum of the skull. Its dorsal surface is flattened to form the floor of the fourth ventricle, and its ventral surface is marked by the corticospinal tracts, most of whose fibers cross over at this level (fig. 9.25).

Because of its location, all the ascending and descending nerve fibers connecting the brain and spinal cord must pass through the medulla oblongata. As in the spinal cord, the white matter of the medulla surrounds a central mass of gray matter. Here, however, the gray matter is broken up into nuclei that are separated by nerve fibers. Some of these nuclei relay ascending impulses to the other side of the brain stem and then onto higher brain centers.

Other nuclei within the medulla oblongata function as control centers for vital visceral activities. These centers include the following:

1. **Cardiac center.** Impulses originating in the cardiac center are transmitted to the heart on peripheral nerves. Impulses on these nerves can cause the heart rate to decrease or increase.
2. **Vasomotor center.** Certain cells of the vasomotor center initiate impulses that travel to smooth muscles in the walls of certain blood vessels and stimulate them to contract. This action causes constriction of the blood vessels (vasoconstriction) and a rise in blood pressure. Other cells of the vasomotor center produce the opposite effect—a dilation of the blood vessels (vasodilation) and a consequent drop in the blood pressure.
3. **Respiratory center.** The respiratory center acts with centers in the pons to regulate the rate, rhythm, and depth of breathing.

Still other nuclei within the medulla oblongata function as centers for the reflexes associated with coughing, sneezing, swallowing, and vomiting.

Reticular Formation

Scattered throughout the medulla oblongata, pons, and midbrain is a complex network of nerve fibers associated with tiny islands of gray matter. This network, the **reticular formation** (reticular activating system), extends from the upper portion of the spinal cord into the diencephalon. Its intricate system of nerve fibers interconnects centers of the hypothalamus, basal ganglia, cerebellum, and cerebrum with fibers in all the major ascending and descending tracts.

When sensory impulses reach the reticular formation, it responds by activating the cerebral cortex into a state of wakefulness. Without this arousal, the cortex remains unaware of stimulation and cannot interpret sensory information or carry on thought processes. Thus, decreased activity in the reticular formation results in sleep. If the reticular formation ceases to function, as in certain injuries, the person remains unconscious and cannot be aroused, even with strong stimulation (a comatose state).

---

1. List the structures of the brain stem.
2. What are the major functions of the thalamus? The hypothalamus?
3. How may the limbic system influence a person's behavior?
4. What vital reflex centers are located in the brain stem?
5. What is the function of the reticular formation?

Cerebellum

The **cerebellum** is a large mass of tissue located below the occipital lobes of the cerebrum and posterior to the pons and medulla oblongata (fig. 9.22). It consists of two lateral hemispheres partially separated by a layer of dura mater (falx cerebelli) and connected in the midline by a structure called the *vermis.*

Like the cerebrum, the cerebellum is composed primarily of white matter with a thin layer of gray matter, the **cerebellar cortex,** on its surface.

The cerebellum communicates with other parts of the central nervous system by means of three pairs of nerve tracts, called *cerebellar peduncles.* One pair (the inferior peduncles) brings sensory information concerning the position of the limbs, joints, and other body parts to the cerebellum. Another pair (the middle peduncles) transmits signals from the cerebral cortex to the cerebellum concerning the desired positions of these parts. After integrating and analyzing this information, the cerebellum sends impulses via a third pair (the superior peduncles) to the midbrain. In response, motor impulses travel from the midbrain downward through the pons, medulla oblongata, and spinal cord. These impulses appropriately stimulate or inhibit the skeletal muscles to cause desired body movements. Thus, the cerebellum functions as a reflex center in the control and coordination of complex skeletal muscle movements. It also helps to maintain posture.

Damage to the cerebellum is likely to result in tremors, inaccurate movements of voluntary muscles, the loss of muscle tone, a reeling walk, and the loss of equilibrium.

1. Where is the cerebellum located?
2. What are the major functions of the cerebellum?

Peripheral Nervous System

The **peripheral nervous system (PNS)** consists of the nerves that branch out from the central nervous system (CNS) and connect it to other body parts. The peripheral nervous system includes the *cranial nerves,* which arise from the brain, and the *spinal nerves,* which arise from the spinal cord.

The peripheral nervous system can also be subdivided into the somatic and autonomic nervous systems. Generally, the **somatic system** consists of the cranial and spinal nerve fibers that connect the CNS to the skin and skeletal muscles; it is involved in conscious activities. The **autonomic system** includes those fibers that connect the CNS to the visceral organs, such as the heart, stomach, intestines, and various glands; it is concerned with unconscious activities. Chart 9.4 outlines the subdivisions of the nervous system.

Cranial Nerves

Twelve pairs of **cranial nerves** arise from various locations on the underside of the brain (fig. 9.29). With the exception of the first pair, which begins within the cerebrum, these nerves originate from the brain stem.

Chart 9.4 Subdivisions of the nervous system

1. **Central nervous system (CNS)**
   a. Brain
   b. Spinal cord
2. **Peripheral nervous system (PNS)**
   a. Cranial nerves arising from the brain
      (1) Somatic fibers connecting to the skin and skeletal muscles
      (2) Autonomic fibers connecting to the visceral organs
   b. Spinal nerves arising from the spinal cord
      (1) Somatic fibers connecting to the skin and skeletal muscles
      (2) Autonomic fibers connecting to the visceral organs

Figure 9.29
Except for the first pair, the cranial nerves arise from the brain stem. They are identified either by numbers, indicating their order, or by names, describing their function or the general distribution of their fibers.

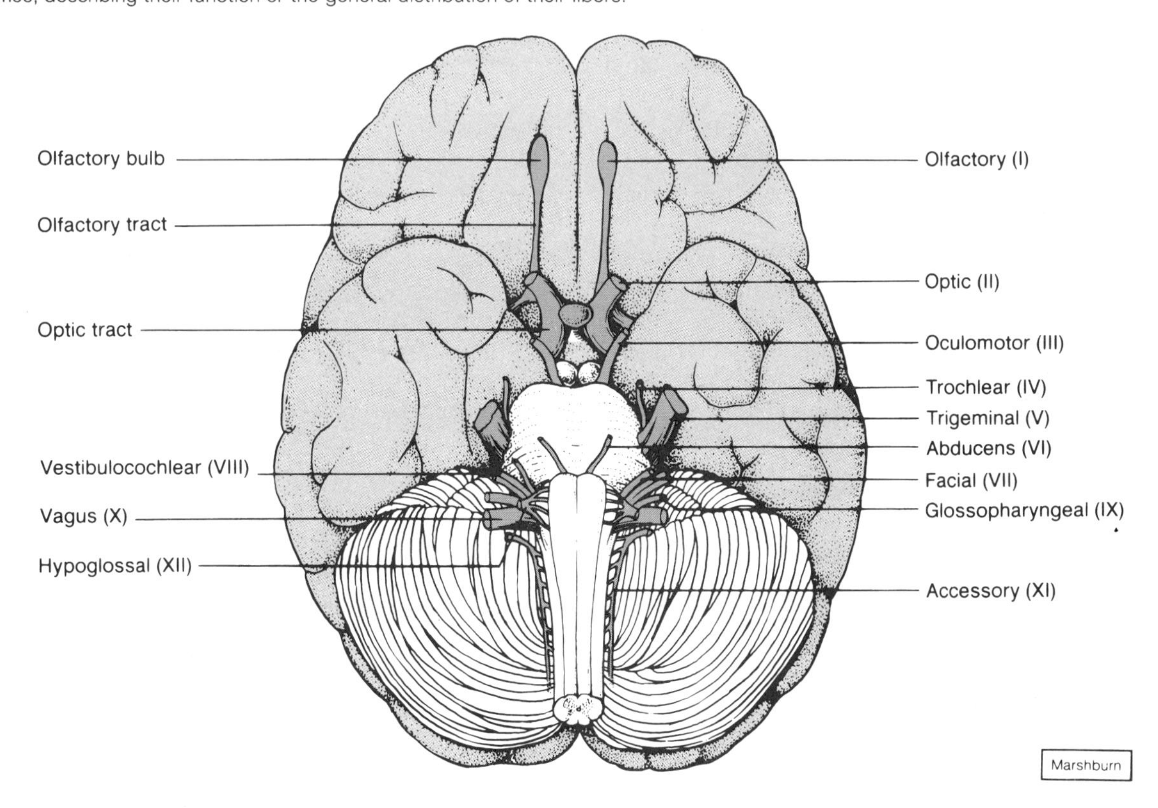

They pass from their sites of origin through various foramina of the skull and lead to parts of the head, neck, and trunk.

Although most of the cranial nerves are mixed nerves, some of those associated with special senses, such as smell and vision, contain only sensory fibers. Others that are closely involved with the activities of muscles and glands are composed primarily of motor fibers.

When sensory fibers are present in the cranial nerves, the neuron cell bodies to which the fibers are attached are located outside the brain and are usually in groups called *ganglia* (sing. *ganglion*). On the other hand, the motor neuron cell bodies are typically located within the gray matter of the brain.

The cranial nerves are designated either by numbers or names. The numbers indicate the order in which the nerves arise from the front to the back of the brain, and the names describe their primary functions or the general distribution of their fibers.

The first pair of cranial nerves, the **olfactory nerves (I)**, are associated with the sense of smell and contain only sensory neurons. These neurons are located in the lining of the upper nasal cavity, where they serve as the *olfactory receptor cells*. Axons from these receptors pass upward through the cribriform plates of the ethmoid bone and into *olfactory bulbs* that lie just beneath the frontal lobes of the cerebrum. Sensory impulses travel from these bulbs along *olfactory tracts* to the cerebral centers where they are interpreted.

The second pair, the **optic nerves (II)**, lead from the eyes to the brain and are associated with the sense of sight. The sensory nerve cell bodies of these nerve fibers occur in ganglion cell layers within the eyes, and their axons pass through the *optic foramina* of the orbits and continue into the visual nerve pathways of the brain (see chapter 10). Sensory impulses transmitted on the optic nerves are interpreted in the visual cortices of the occipital lobes.

The third pair, the **oculomotor nerves (III)**, arise from the midbrain and pass into the orbits of the eyes. One component of each nerve connects to the voluntary muscles that raise the eyelid and certain muscles that move the eye. A second portion of each oculomotor nerve is part of the autonomic nervous system and supplies involuntary muscles within the eyes that adjust the amount of light entering the eyes and focus the lenses of the eyes.

The fourth pair, the **trochlear nerves (IV)**, are the smallest of the cranial nerves. They arise from the midbrain and carry motor impulses to certain voluntary muscles that move the eyes but are not supplied by the oculomotor nerves.

The fifth pair, the **trigeminal nerves (V)**, are the largest of the cranial nerves and arise from the pons. They are mixed nerves, but their sensory portions are more extensive than their motor portions. Each sensory component includes three large branches, called the ophthalmic, maxillary, and mandibular divisions.

The *ophthalmic divisions* of the trigeminal nerves consist of sensory fibers that bring impulses to the brain from the surface of the eyes, the tear glands, and the skin of the anterior scalp, forehead, and upper eyelids. The fibers of the *maxillary divisions* carry sensory impulses from the upper teeth, upper gum, and upper lip, as well as from the mucous lining of the palate and the skin of the face. The *mandibular division* includes both motor and sensory fibers. The sensory branches transmit impulses from the scalp behind the ears, the skin of the jaw, the lower teeth, the lower gum, and the lower lip. The motor branches supply the muscles of mastication and certain muscles in the floor of the mouth.

The sixth pair, the **abducens nerves (VI)**, are quite small and originate from the pons near the medulla oblongata. They enter the orbits of the eyes and supply motor impulses to a pair of muscles that move the eyes.

The seventh pair, the **facial nerves (VII)**, arise from the lower part of the pons and emerge on the sides of the face. Their sensory branches are associated with taste receptors on the anterior two-thirds of the tongue, and some of their motor fibers transmit impulses to muscles of facial expression. Still other motor fibers of these nerves function in the autonomic nervous system and stimulate secretions from tear glands and salivary glands.

The eighth pair, the **vestibulocochlear nerves (VIII)**, are sensory nerves that arise from the medulla oblongata. Each of these nerves has two distinct parts—a vestibular branch and a cochlear branch.

The neuron cell bodies of the *vestibular branch* fibers are located in ganglia associated with parts of the inner ear. These parts contain the receptors involved with reflexes that help to maintain equilibrium. The neuron cell bodies of the *cochlear branch* fibers are located in the parts of the inner ear that house the hearing receptors. Impulses from these branches pass through the pons and medulla oblongata on their way to the temporal lobes, where they are interpreted.

The ninth pair, the **glossopharyngeal nerves (IX)**, are associated with the tongue and pharynx. These nerves arise from the medulla oblongata, and although they are mixed nerves, their predominant fibers are sensory. These sensory fibers carry impulses from the linings of the pharynx, tonsils, and posterior third of the tongue to the brain. Fibers of the motor component innervate muscles of the pharynx that function in swallowing. Other branches function in the autonomic nervous system, helping to regulate blood pressure and controlling the secretion of saliva.

The tenth pair, the **vagus nerves (X)**, originate in the medulla oblongata and extend downward through the neck into the chest and abdomen. These nerves are

Chart 9.5 Functions of cranial nerves		
Nerve	**Type**	**Function**
I. Olfactory	Sensory	Sensory fibers transmit impulses associated with the sense of smell
II. Optic	Sensory	Sensory fibers transmit impulses associated with the sense of vision
III. Oculomotor	Primarily motor	Motor fibers transmit impulses to muscles that raise the eyelids, move the eyes, adjust the amount of light entering the eyes, and focus the lenses Some sensory fibers transmit impulses associated with the condition of the muscles
IV. Trochlear	Primarily motor	Motor fibers transmit impulses to muscles that move the eyes Some sensory fibers transmit impulses associated with the condition of the muscles
V. Trigeminal	Mixed	
Opthalmic division		Sensory fibers transmit impulses from the surface of the eyes, tear glands, scalp, forehead, and upper eyelids
Maxillary division		Sensory fibers transmit impulses from the upper teeth, upper gum, upper lip, lining of the palate, and skin of the face
Mandibular division		Sensory fibers transmit impulses from the scalp, skin of the jaw, lower teeth, lower gum, and lower lip Motor fibers transmit impulses to the muscles of mastication and some muscles in the floor of the mouth
VI. Abducens	Primarily motor	Motor fibers transmit impulses to the muscles that move the eyes Some sensory fibers transmit impulses associated with the condition of the muscles
VII. Facial	Mixed	Sensory fibers transmit impulses associated with the taste receptors of the anterior tongue Motor fibers transmit impulses to the muscles of facial expression, tear glands, and salivary glands
VIII. Vestibulocochlear	Sensory	
Vestibular branch		Sensory fibers transmit impulses associated with the sense of equilibrium
Cochlear branch		Sensory fibers transmit impulses associated with the sense of hearing
IX. Glossopharyngeal	Mixed	Sensory fibers transmit impulses from the pharynx, tonsils, posterior tongue, and carotid arteries Motor fibers transmit impulses to the muscles of the pharynx used in swallowing and to the salivary glands
X. Vagus	Mixed	Somatic motor fibers transmit impulses to the muscles associated with speech and swallowing; autonomic motor fibers transmit impulses to the heart, smooth muscles, and glands in the thorax and abdomen Sensory fibers transmit impulses from the pharynx, larynx, esophagus, and visceral organs of the thorax and abdomen
XI. Accessory	Motor	
Cranial branch		Motor fibers transmit impulses to the muscles of the soft palate, pharynx, and larynx
Spinal branch		Motor fibers transmit impulses to the muscles of the neck and back
XII. Hypoglossal	Motor	Motor fibers transmit impulses to the muscles that move the tongue

mixed, and although they contain both somatic and autonomic branches, the autonomic fibers are the predominant ones. Certain of their somatic motor fibers carry impulses to muscles of the larynx that are associated with speech and swallowing. Autonomic vagal motor fibers supply the heart and a variety of smooth muscles and glands in the thorax and abdomen.

The eleventh pair, the **accessory nerves (XI),** originate in the medulla oblongata and the spinal cord; thus, they have both cranial and spinal branches. Each *cranial branch* joins a vagus nerve and carries impulses to muscles of the soft palate, pharynx, and larynx. The *spinal branch* descends into the neck and supplies motor fibers to the trapezius and sternocleidomastoid muscles.

The twelfth pair of cranial nerves, the **hypoglossal nerves (XII),** arise from the medulla oblongata and pass into the tongue. They include motor fibers that carry impulses to muscles that move the tongue in speaking, chewing, and swallowing.

The functions of the cranial nerves are summarized in chart 9.5.

1. Define the peripheral nervous system.
2. Distinguish between the somatic and autonomic nerve fibers.
3. Name the cranial nerves, and list the major functions of each.

Spinal Nerves

Thirty-one pairs of **spinal nerves** originate from the spinal cord (fig. 9.30). They are mixed nerves, and they

Figure 9.30
Spinal nerves and plexuses.

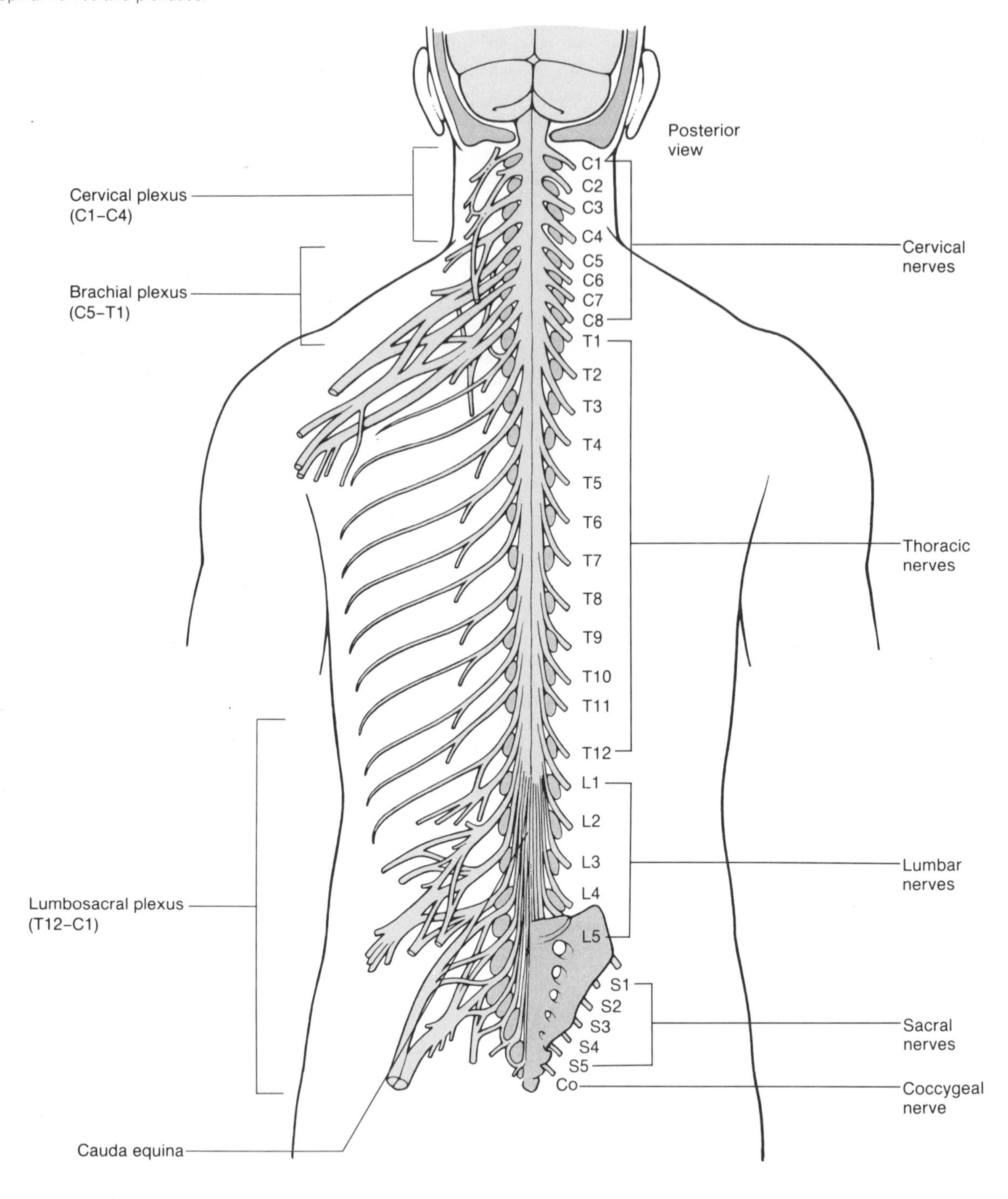

provide two-way communication between the spinal cord and parts of the arms, legs, neck, and trunk.

Although the spinal nerves are not named individually, they are grouped according to the level from which they arise, and each nerve is numbered in sequence. Thus, there are eight pairs of *cervical nerves* (numbered C1 to C8), twelve pairs of *thoracic nerves* (numbered T1 to T12), five pairs of *lumbar nerves* (numbered L1 to L5), five pairs of *sacral nerves* (numbered S1 to S5), and one pair of *coccygeal nerves* (Co).

The adult spinal cord ends at the level between the first and second lumbar vertebrae, so the lumbar, sacral, and coccygeal nerves descend to their exits beyond the end of the cord. These descending nerves form a structure called the *cauda equina* (horse's tail).

Each spinal nerve emerges from the cord by two short branches, or *roots,* which lie within the vertebral column. The **dorsal root** (sensory root) can be identified by an enlargement called the *dorsal root ganglion* (fig. 9.21). This ganglion contains the cell bodies of the sensory neurons whose dendrites conduct impulses inward from the peripheral body parts. The axons of these neurons extend through the dorsal root and into the spinal cord, where they form synapses with the dendrites of other neurons.

The **ventral root** (motor root) of each spinal nerve consists of axons from the motor neurons, whose cell bodies are located within the gray matter of the cord.

A ventral root and a dorsal root unite to form a spinal nerve, which extends outward from the vertebral canal through an *intervertebral foramen* (fig. 7.15). Just beyond its foramen, each spinal nerve divides into several parts.

Except in the thoracic region, the main portions of the spinal nerves combine to form complex networks, called **plexuses,** instead of continuing directly to the peripheral body parts (fig. 9.30). In a plexus, the fibers of various spinal nerves are sorted and recombined, so that the fibers associated with a particular peripheral part reach it in the same nerve, even though the fibers originate from different spinal nerves.

Cervical Plexuses

The **cervical plexuses** lie deep in the neck on either side. They are formed by the branches of the first four cervical nerves. Fibers from these plexuses supply the muscles and skin of the neck. In addition, fibers from the third, fourth, and fifth cervical nerves pass into the right and left **phrenic nerves,** which conduct motor impulses to the muscle fibers of the diaphragm.

Brachial Plexuses

Branches of the lower four cervical nerves and the first thoracic nerve give rise to the **brachial plexuses.** These networks of nerve fibers are located deep within the shoulders between the neck and axillae (armpits). The major branches emerging from the brachial plexuses supply the muscles and skin of the arm, forearm, and hand, and include the **musculocutaneous, ulnar, median, radial,** and **axillary nerves.**

Lumbosacral Plexuses

The **lumbosacral plexuses** are formed on either side by the last thoracic nerve and the lumbar, sacral, and coccygeal nerves. These networks of nerve fibers extend from the lumbar region of the back into the pelvic cavity, giving rise to a number of motor and sensory fibers associated with the muscles and skin of the lower abdominal wall, external genitalia, buttocks, thighs, legs, and feet. The major branches of these plexuses include the **obturator, femoral,** and **sciatic nerves.**

1. How are the spinal nerves grouped?
2. Describe the way a spinal nerve joins the spinal cord.
3. Name and locate the major nerve plexuses.

Spinal nerves may be injured in a variety of ways including stabs, gunshot wounds, birth injuries, dislocations and fractures of the vertebrae, and pressure from tumors in surrounding tissues. The nerves of the cervical plexuses, for example, are sometimes compressed by a sudden bending of the neck, called *whiplash,* which may occur during rear-end automobile collisions. A victim of such an injury may suffer continuing headaches and pain in the neck and skin, which are supplied by the cervical nerves.

Autonomic Nervous System

The **autonomic nervous system** is the portion of the peripheral nervous system that functions independently (autonomously) and continuously without conscious effort. This system controls visceral functions by regulating the actions of smooth muscle, cardiac muscle, and glands. It is concerned with regulating heart rate, blood pressure, breathing rate, body temperature, and other visceral activities that aid in maintaining homeostasis. Portions of the autonomic nervous system are also responsive during times of emotional stress, and they prepare the body to meet the demands of strenuous physical activity.

General Characteristics

Autonomic activities are regulated largely by reflexes in which the sensory signals originate from receptors within the visceral organs and skin. These signals are received by nerve centers within the hypothalamus, brain stem, or spinal cord. In response, motor impulses travel out from these centers on the peripheral nerve fibers within the cranial and spinal nerves.

Typically these fibers lead to ganglia outside the central nervous system. The impulses they carry are integrated within these ganglia and relayed to various visceral organs—muscles and glands—which respond by contracting, releasing secretions, or being inhibited. The integrative function of the ganglia provides the autonomic system with a degree of independence from the brain and spinal cord.

The autonomic nervous system includes two sections, called the **sympathetic** and **parasympathetic divisions.** Some visceral organs are supplied by nerve fibers from each of the subdivisions. In such cases, impulses on one set of fibers may activate an organ, while impulses on the other set inhibit it. Thus, the divisions may act antagonistically, so that the actions of some visceral organs are regulated by alternately being activated or inhibited.

The functions of the autonomic divisions are mixed—that is, each activates some organs and inhibits others. However, the divisions have important

Figure 9.31
(*a*) Autonomic neuron pathways involve two interneurons between the central nervous system and an effector. (*b*) Somatic neuron pathways usually have a single neuron between the central nervous system and an effector.

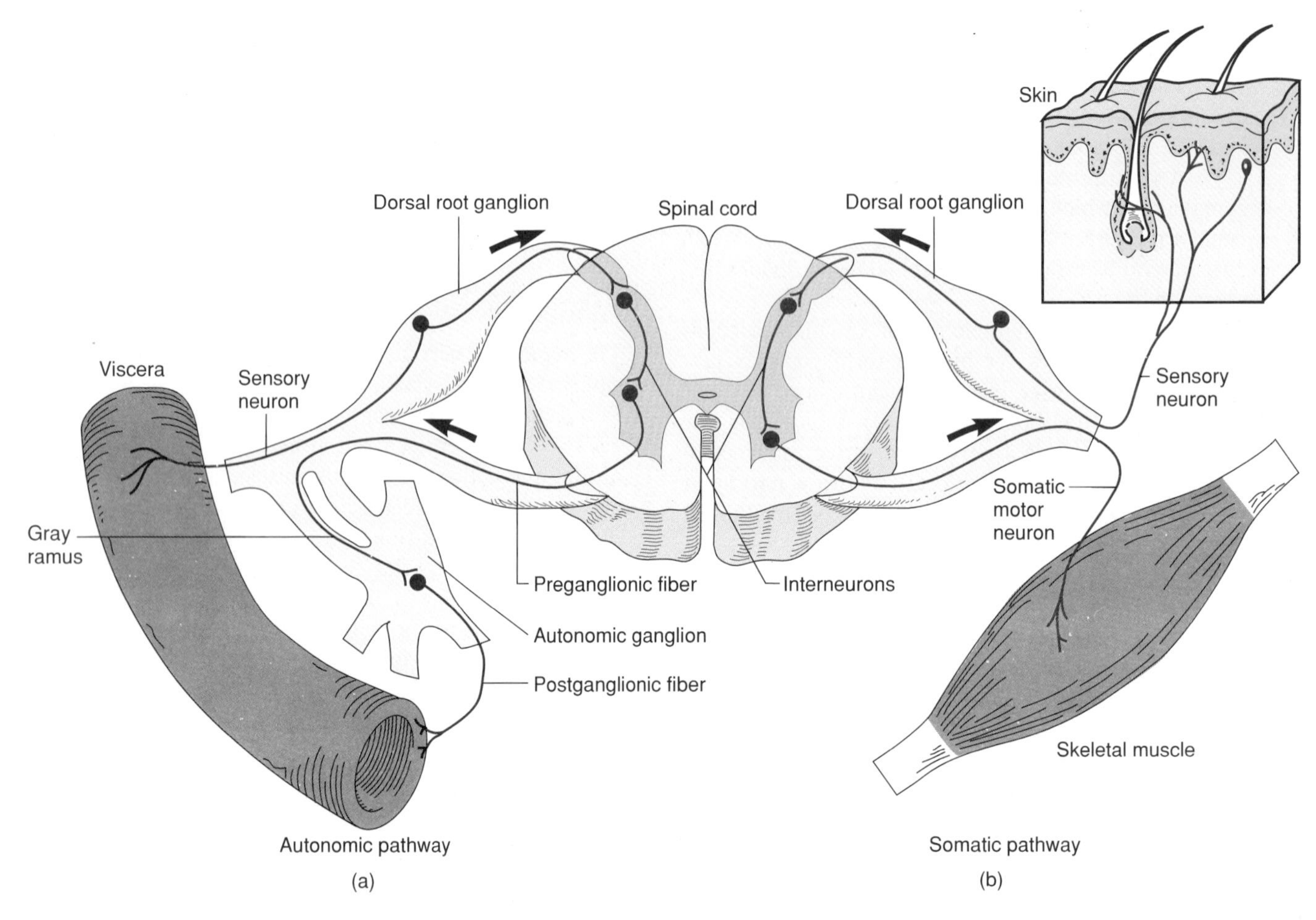

functional differences. The sympathetic division is concerned primarily with preparing the body for energy-expending, stressful, or emergency situations. Conversely, the parasympathetic division is most active under ordinary, restful conditions. It also counterbalances the effects of the sympathetic division and restores the body to a resting state following a stressful experience. For example, during an emergency, the sympathetic division will cause the heart and breathing rates to increase, and following an emergency, the parasympathetic division will slow these activities.

Autonomic Nerve Fibers

The nerve fibers of the autonomic nervous system are motor fibers. Unlike the motor pathways of the somatic nervous system, however, which usually include a single neuron between the brain or spinal cord and a skeletal muscle, those of the autonomic system involve two neurons (fig. 9.31). The cell body of one neuron is located in the brain or spinal cord. Its axon, the **preganglionic fiber,** leaves the CNS and forms a synapse with one or more nerve fibers whose cell bodies are housed within an autonomic ganglion. The axon of such a second neuron is called a **postganglionic fiber,** and it extends to a visceral effector.

Within the sympathetic division, the preganglionic fibers begin from neurons in the gray matter of the spinal cord (fig. 9.32). Their axons leave the cord through the ventral roots of spinal nerves in the first thoracic through the second lumbar segments. After traveling a short distance, these fibers leave the spinal nerves, and each enters a member of a chain of *paravertebral ganglia.* One of these chains extends longitudinally along each side of the vertebral column.

Within a paravertebral ganglion, a preganglionic fiber forms a synapse with a second neuron. The axon of this neuron, the postganglionic fiber, typically returns to a spinal nerve and extends with it to a visceral effector.

The preganglionic fibers of the parasympathetic division arise from the *brain stem* and *sacral region* of the spinal cord (fig. 9.33). From there, they lead outward on the cranial or sacral nerves to the ganglia located near or within various visceral organs. The relatively short postganglionic fibers continue from the ganglia to specific muscles or glands within these visceral organs.

Figure 9.32
The preganglionic fibers of the sympathetic division arise from the thoracic and lumbar regions of the spinal cord.

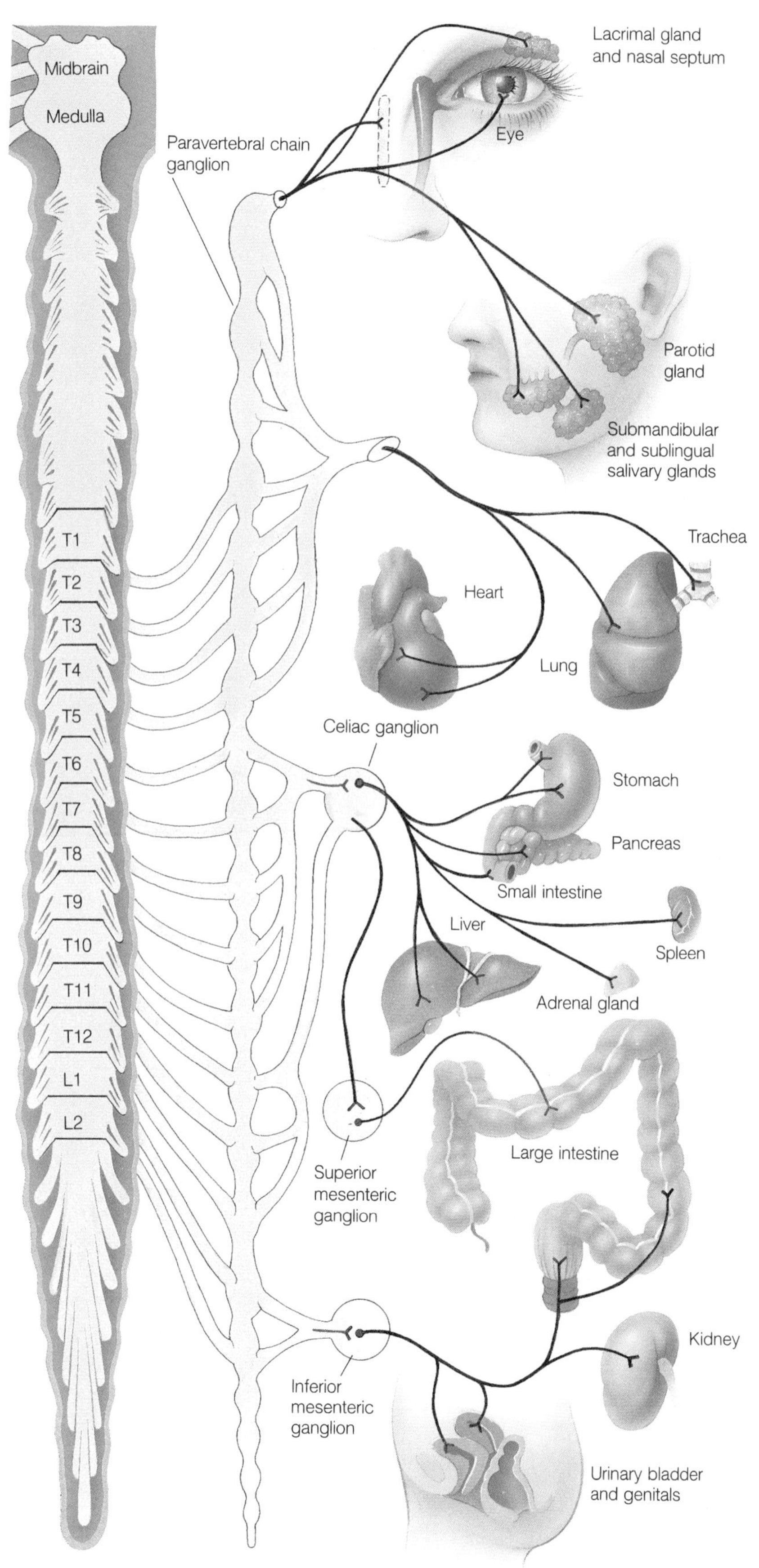

Figure 9.33
The preganglionic fibers of the parasympathetic division of the autonomic nervous system arise from the brain and sacral region of the spinal cord.

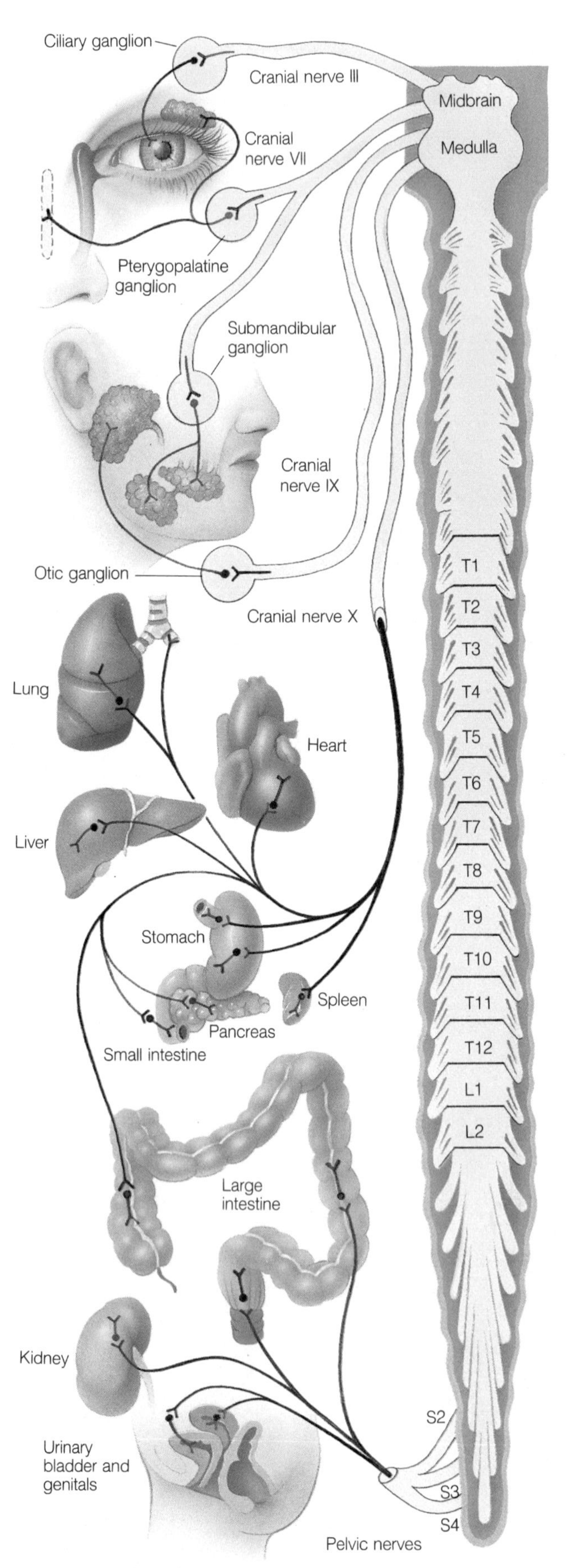

Figure 9.34
Most sympathetic fibers are adrenergic and secrete norepinephrine at the ends of the postganglionic fibers; parasympathetic fibers are cholinergic and secrete acetylcholine at the ends of the postganglionic fibers.

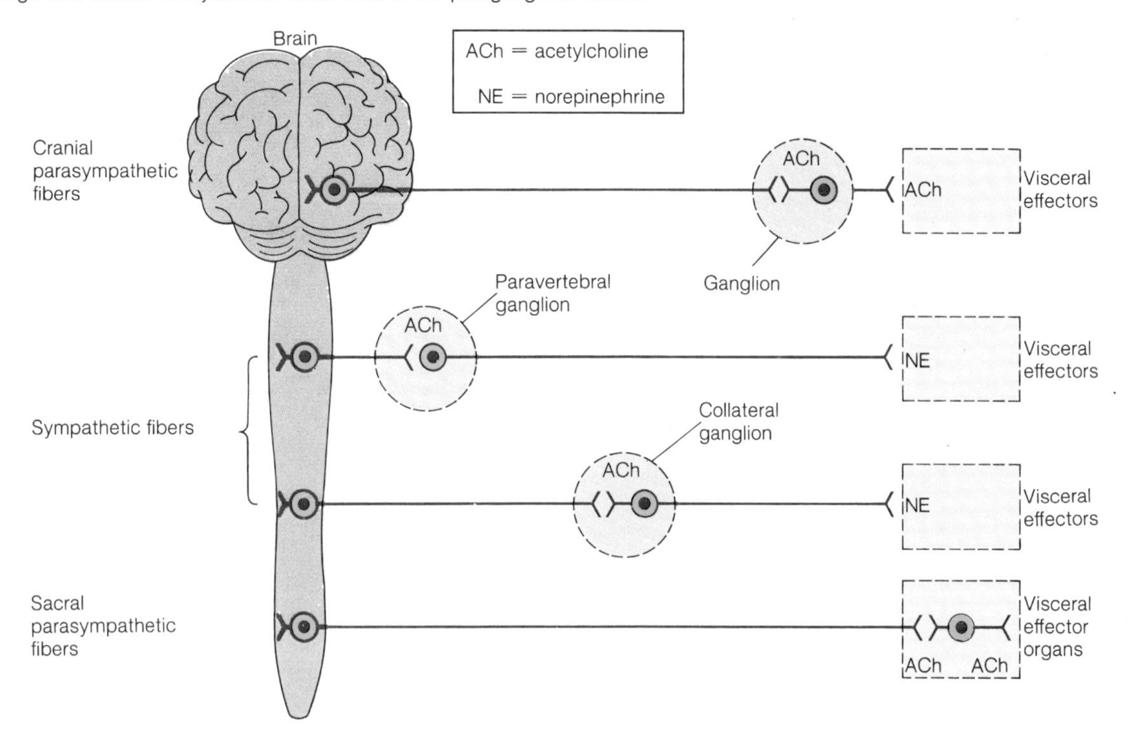

1. What parts of the nervous system are included in the autonomic nervous system?
2. How are the divisions of the autonomic system distinguished?
3. Describe a sympathetic nerve pathway. A parasympathetic nerve pathway.

Autonomic Neurotransmitters

The preganglionic fibers of the sympathetic and parasympathetic divisions secrete *acetylcholine.* The parasympathetic postganglionic fibers also secrete *acetylcholine* and for this reason are called **cholinergic fibers** (fig. 9.34). Most sympathetic postganglionic fibers, however, secrete *norepinephrine* (noradrenalin) and are called **adrenergic fibers.** The different postganglionic neurotransmitters are responsible for the different effects that the sympathetic and parasympathetic divisions have on the visceral organs.

Although each divison can activate some effectors and inhibit others, most visceral organs are controlled primarily by one division. In other words, the divisions are usually not actively antagonistic. For example, the diameter of most blood vessels, which lack parasympathetic innervation, is regulated by the sympathetic division. Smooth muscles in the walls of these vessels are continuously stimulated and thus maintained in a state of partial contraction (tone). The diameter of a vessel can be increased (dilated) by a decrease of sympathetic stimulation, which allows the muscular wall to relax. Conversely, the vessel can be constricted by an increasing amount of sympathetic stimulation.

Similarly, the parasympathetic division is dominant in controlling movements in the digestive system. Parasympathetic impulses stimulate stomach and intestinal motility, and when these impulses decrease, such movement is reduced.

The effects of adrenergic and cholinergic fibers on some visceral effectors are summarized in chart 9.6.

Control of Autonomic Activity

Although the autonomic nervous system has some degree of independence resulting from the integrative function of its ganglia, it is controlled largely by the brain and spinal cord. For example, as discussed previously, there are control centers in the medulla oblongata for cardiac, vasomotor, and respiratory activities. These reflex centers receive sensory impulses from the visceral organs by means of vagus nerve fibers, and they employ autonomic nerve pathways to stimulate motor responses in various muscles and glands. Similarly, the hypothalamus helps regulate body temperature, hunger, thirst, and water and electrolyte balance by using and, thus, controlling the autonomic pathways.

Chart 9.6 Some effects of neurotransmitter substances upon visceral effectors or actions

Visceral effector or action	Response to adrenergic stimulation (sympathetic)	Response to cholinergic stimulation (parasympathetic)
Pupil of the eye	Dilation	Constriction
Heart rate	Increases	Decreases
Bronchioles of the lungs	Dilation	Constriction
Muscles of the intestinal wall	Slows peristaltic action	Speeds peristaltic action
Intestinal glands	Secretion decreases	Secretion increases
Blood distribution	More blood to the skeletal muscles; less blood to the digestive organs	More blood to the digestive organs; less blood to the skeletal muscles
Blood glucose concentration	Increases	Decreases
Salivary glands	Secretion decreases	Secretion increases
Tear glands	No action	Secretion
Muscles of the gallbladder	Relaxation	Contraction
Muscles of the urinary bladder	Relaxation	Contraction

Still higher levels within the brain, including the limbic system and cerebral cortex, control the autonomic nervous system when a person is emotionally stressed. These parts use the autonomic pathways to regulate a person's emotional expression and behavior.

1. What neurotransmitters are used in the autonomic nervous system?
2. How do the divisions of the autonomic system function to regulate visceral activities?
3. How are autonomic activities controlled?

Clinical Terms Related to the Nervous System

analgesia (an″al-je′ze-ah)—a loss or reduction in the ability to sense pain, but without a loss of consciousness.

analgesic (an″al-je′sik)—a pain-relieving drug.

anesthesia (an″es-the′ze-ah)—a loss of feeling.

aphasia (ah-fa′ze-ah)—a disturbance or loss in the ability to use words or to understand them, usually due to damage to the cerebral association areas.

apraxia (ah-prak′se-ah)—an impairment in a person's ability to make correct use of objects.

ataxia (ah-tak′se-ah)—a partial or complete inability to coordinate voluntary movements.

cerebral palsy (ser′ĕ-bral pawl′ze)—a condition characterized by partial paralysis and the lack of muscular coordination.

coma (ko′mah)—an unconscious condition in which a person does not respond to stimulation.

cordotomy (kor-dot′o-me)—a surgical procedure in which a nerve tract within the spinal cord is severed, usually to relieve intractable pain.

craniotomy (kra″ne-ot′o-me)—a surgical procedure in which part of the skull is opened.

electroencephalogram (EEG) (e-lek″tro-en-sef′ah-lo-gram″)—a recording of the electrical activity of the brain.

encephalitis (en″sef-ah-li′tis)—an inflammation of the brain and meninges, characterized by drowsiness and apathy.

epilepsy (ep′ĭ-lep″se)—a disorder of the central nervous system that is characterized by temporary disturbances in normal brain impulses; it may be accompanied by convulsive seizures and a loss of consciousness.

hemiplegia (hem″ĭ-ple′je-ah)—paralysis on one side of the body and of the limbs on that side.

Huntington's chorea (hunt′ing-tunz ko-re′ah)—a rare hereditary disorder of the brain, characterized by involuntary convulsive movements and mental deterioration.

laminectomy (lam″ĭ-nek′to-me)—the surgical removal of the posterior arch of a vertebra, usually to relieve the symptoms of a ruptured intervertebral disk that is pressing on a spinal nerve.

monoplegia (mon″o-ple′je-ah)—paralysis of a single limb.

multiple sclerosis (mul′tĭ-pl skle-ro′sis)—a disease of the central nervous system, characterized by loss of myelin and the appearance of scarlike patches throughout either the brain and spinal cord or both.

neuralgia (nu-ral′je-ah)—a sharp, recurring pain associated with a nerve, usually caused by inflammation or injury.

neuritis (nu-ri′tis)—an inflammation of a nerve.

paraplegia (par″ah-ple′je-ah)—paralysis of both legs.

quadriplegia (kwod″rĭ-ple′je-ah)—paralysis of all four limbs.

vagotomy (va-got′o-me)—the surgical severing of a vagus nerve.

Chapter Summary

Introduction (page 208)

The organs of the nervous system are divided into the central and peripheral nervous systems. These parts provide sensory, integrative, and motor functions.

General Functions of the Nervous System (page 208)

1. Sensory functions employ receptors that detect internal and external changes in body conditions.

2. The integrative functions bring sensory information together and result in decisions that are acted upon by means of motor functions.
3. The motor functions use effectors that respond when they are stimulated by motor impulses.

Nervous Tissue (page 208)

1. Nervous tissue includes neurons, which are the structural and functional units of the nervous system, and neuroglial cells.
2. Neuron structure
    a. A neuron includes a cell body, nerve fibers, and the organelles usually found in cells.
    b. The dendrites and the cell body provide receptive surfaces.
    c. A single axon arises from the cell body and may be enclosed in a myelin sheath and a neurilemma.
3. Neuroglial cells
    a. Neuroglial cells fill spaces, support neurons, hold nervous tissue together, produce myelin, help regulate the concentrations of nutrients and ions, and carry on phagocytosis.
    b. They include Schwann cells, astrocytes, oligodendrocytes, microglia, and ependyma.

Cell Membrane Potential (page 211)

1. A cell membrane is usually polarized as a result of an unequal distribution of ions.
2. Distribution of ions
    a. The distribution of ions is due to the presence of pores and channels in the membranes, which allow the passage of some ions, but not others.
    b. Potassium ions pass more easily through cell membranes than do sodium ions.
3. Resting potential
    a. There is a high concentration of sodium ions outside a membrane and a high concentration of potassium ions inside.
    b. There are large numbers of negatively charged ions inside a cell.
    c. In a resting cell, more positive ions leave than enter, so the outside of the cell membrane develops a positive charge.
4. Potential changes
    a. Stimulation of a membrane affects its resting potential.
    b. When its resting potential decreases, a membrane becomes depolarized.
    c. Potential changes are subject to summation.
    d. If threshold potential is achieved, an action potential is triggered.
5. Action potential
    a. At threshold, the sodium channels open, and sodium ions diffuse inward, causing depolarization.
    b. At the same time, the potassium channels open, and potassium ions diffuse outward, causing repolarization.
    c. This rapid change in potential is an action potential.
    d. Many action potentials can occur before an active transport mechanism reestablishes the original resting potential.

Nerve Impulse (page 214)

A wave of action potentials is a nerve impulse.

1. Impulse conduction
    a. Unmyelinated fibers conduct impulses that travel over their entire surfaces.
    b. Myelinated fibers conduct impulses more rapidly.
2. All-or-none response
    a. A nerve impulse is conducted in an all-or-none manner whenever a stimulus of threshold intensity is applied to a fiber.
    b. All the impulses conducted on a fiber are of the same strength.

Synapse (page 215)

A synapse is a junction between two neurons.

1. Synaptic transmission
    a. Impulses usually travel from a dendrite to a cell body, then along the axon to a synapse.
    b. Axons have synaptic knobs at their distal ends, which secrete neurotransmitters.
    c. A neurotransmitter is released when a nerve impulse reaches the end of an axon.
    d. When the neurotransmitter reaches the nerve fiber on the distal side of the cleft, a nerve impulse is triggered.
2. Neurotransmitter substances
    a. Many different neurotransmitters are produced in the nervous system.
    b. They include acetylcholine, monoamines, amino acids, and peptides.
    c. Neurotransmitters are released from a synaptic knob when an action potential causes the membrane permeability to sodium to increase.
    d. After being released, neurotransmitters are decomposed or removed from the synaptic clefts.
3. Excitatory and inhibitory actions
    a. Neurotransmitters that trigger nerve impulses are excitatory; those that inhibit impulses are inhibitory.
    b. The effect of the synaptic knobs communicating with a neuron will depend upon which knobs are activated from moment to moment.

Processing of Impulses (page 218)

The way impulses are processed reflects the organization of the neurons in the brain and spinal cord.

1. Neuronal pools
    a. Neurons are organized into pools within the central nervous system.
    b. Each pool receives impulses, processes them, and conducts impulses away.
2. Facilitation
    a. Each neuron in a pool may receive excitatory and inhibitory stimuli.
    b. A neuron is facilitated when it receives subthreshold stimuli and becomes more excitable.

3. Convergence
   a. Impulses from two or more incoming fibers may converge on a single neuron.
   b. Convergence makes it possible for impulses from different sources to create an additive effect on a neuron.
4. Divergence
   a. Impulses leaving a pool may diverge by passing onto several output fibers.
   b. Divergence allows impulses to be amplified.

Types of Neurons and Nerves (page 218)

1. On the basis of structure, neurons can be classified as multipolar, bipolar, or unipolar.
2. On the basis of function, neurons can be classified as sensory neurons, interneurons, or motor neurons.
3. Types of nerves
   a. Nerves are cordlike bundles of nerve fibers.
   b. Nerves can be classified as sensory, motor, or mixed.

Nerve Pathways (page 220)

A nerve pathway is a route followed by an impulse as it travels through the nervous system.

1. Reflex arcs
   a. A reflex arc includes a sensory neuron, a reflex center composed of interneurons, and a motor neuron.
   b. The reflex arc is the behavioral unit of the nervous system.
2. Reflex behavior
   a. Reflexes are automatic, unconscious responses to changes.
   b. They help maintain homeostasis.
   c. The knee jerk reflex employs only two neurons.
   d. Withdrawal reflexes are protective actions.

Coverings of the Central Nervous System (page 222)

1. The brain and spinal cord are surrounded by bone and protective membranes called meninges.
2. Meninges
   a. The meninges consist of the dura mater, arachnoid mater, and pia mater.
   b. Cerebrospinal fluid occupies the space between the arachnoid and pia maters.

Spinal Cord (page 224)

The spinal cord is a nerve column that extends from the brain into the vertebral canal.

1. Structure of the spinal cord
   a. The spinal cord is composed of thirty-one segments, each of which gives rise to a pair of spinal nerves.
   b. It is characterized by a cervical enlargement and a lumbar enlargement.
   c. It has a central core of gray matter, which is surrounded by white matter.
   d. The white matter is composed of bundles of myelinated nerve fibers.
2. Functions of the spinal cord
   a. The cord provides a two-way communication system between the brain and other body parts.
   b. Ascending tracts carry sensory impulses to the brain; descending tracts carry motor impulses to muscles and glands.

Brain (page 226)

The brain is subdivided into the cerebrum, cerebellum, and brain stem.

1. Structure of the cerebrum
   a. The cerebrum consists of two cerebral hemispheres connected by the corpus callosum.
   b. The cerebral cortex is a thin layer of gray matter near the surface.
   c. White matter consists of myelinated nerve fibers that interconnect neurons within the nervous system and communicate with other body parts.
2. Functions of the cerebrum
   a. The cerebrum is concerned with higher brain functions.
   b. The cerebral cortex can be subdivided into the sensory, motor, and association areas.
   c. In most persons, one cerebral hemisphere is dominant for certain intellectual functions.
3. Ventricles and cerebrospinal fluid
   a. Ventricles are interconnected cavities within the cerebral hemispheres and brain stem.
   b. These spaces are filled with cerebrospinal fluid.
   c. Cerebrospinal fluid is secreted by the choroid plexuses in the walls of the ventricles.
4. Brain stem
   a. The brain stem consists of the diencephalon, midbrain, pons, and medulla oblongata.
   b. The diencephalon contains the thalamus, which serves as a central relay station for incoming sensory impulses, and the hypothalamus, which plays important roles in maintaining homeostasis.
   c. The limbic system functions to produce emotional feelings and to modify behavior.
   d. The midbrain contains reflex centers associated with eye and head movements.
   e. The pons transmits impulses between the cerebrum and other parts of the nervous system and contains centers that help regulate the rate and depth of breathing.
   f. The medulla oblongata transmits all ascending and descending impulses and contains several vital and nonvital reflex centers.
   g. The reticular formation filters incoming sensory impulses, arousing the cerebral cortex into wakefulness whenever significant impulses are received.
5. Cerebellum
   a. The cerebellum consists of two hemispheres.
   b. It functions primarily as a reflex center in the coordination of skeletal muscle movements and the maintenance of equilibrium.

Peripheral Nervous System (page 235)

The peripheral nervous system consists of cranial and spinal nerves that branch out from the brain and spinal cord to all body parts. It can be subdivided into the somatic and autonomic portions.

1. Cranial nerves
   a. Twelve pairs of cranial nerves connect the brain to parts in the head, neck, and trunk.
   b. Although most cranial nerves are mixed, some are purely sensory, and others are primarily motor.
   c. The names of the cranial nerves indicate their primary functions or the general distributions of their fibers.
   d. Some cranial nerves are somatic and others are autonomic.
2. Spinal nerves
   a. Thirty-one pairs of spinal nerves originate from the spinal cord.
   b. These mixed nerves provide a two-way communication system between the spinal cord and parts of the arms, legs, neck, and trunk.
   c. Spinal nerves are grouped according to the levels from which they arise, and they are numbered in sequence.
   d. Each nerve emerges by a dorsal and a ventral root.
   e. Just beyond its foramen, each spinal nerve divides into several branches.
   f. Most spinal nerves combine to form plexuses in which nerve fibers are sorted and recombined so that those fibers associated with a particular part reach it together.

Autonomic Nervous System (page 239)

The autonomic nervous system consists of the portions of the nervous system that function without conscious effort. It is concerned primarily with the regulation of the visceral activities that aid in maintaining homeostasis.

1. General characteristics
   a. Autonomic functions operate as reflex actions controlled from centers in the hypothalamus, brain stem, and spinal cord.
   b. The autonomic nervous system consists of two divisions—sympathetic and parasympathetic.
   c. The sympathetic division responds to stressful and emergency conditions.
   d. The parasympathetic division is most active under ordinary conditions.
2. Autonomic nerve fibers
   a. The autonomic nerve fibers are motor fibers.
   b. Sympathetic fibers leave the spinal cord and synapse in paravertebral ganglia.
   c. Parasympathetic fibers begin in the brain stem and sacral region of the spinal cord and synapse in ganglia near various visceral organs.
3. Autonomic neurotransmitters
   a. Preganglionic sympathetic and parasympathetic fibers secrete acetylcholine.
   b. Postganglionic parasympathetic fibers secrete acetylcholine; postganglionic sympathetic fibers secrete norepinephrine.
   c. The different effects of the autonomic divisions are due to different neurotransmitter substances released by the postganglionic fibers.
   d. Most visceral organs are controlled mainly by one division.
4. Control of autonomic activity
   a. The autonomic nervous system has some degree of independence.
   b. Control centers in the medulla oblongata and hypothalamus employ autonomic nerve pathways.
   c. The limbic system and cerebral cortex control the autonomic system when a person is emotionally stressed.

Clinical Application of Knowledge

1. What functional losses would you expect to observe in a patient who has suffered injury to the right occipital lobe of the cerebral cortex? The right temporal lobe?
2. Based on your knowledge of the cranial nerves, devise a set of tests to assess the normal functions of each of these nerves.
3. Multiple sclerosis is a disease in which nerve fibers in the central nervous system lose their myelin. Why would this loss be likely to affect the person's ability to control skeletal muscles?
4. Why are rapidly growing cancers that originate in nervous tissue most likely to be composed of neuroglial cells rather than neurons?
5. Substances used by intravenous drug abusers are sometimes obtained in tablet form and must be crushed and dissolved before they can be injected. Such tablets may contain fillers, such as talc or cornstarch, that were added during the manufacturing process. Either talc or cornstarch may cause obstructions in tiny blood vessels of the cerebrum. What problems might such obstructions create?

Review Activities

1. Explain the relationship between the central nervous system and the peripheral nervous system.
2. List three general functions of the nervous system.
3. Distinguish between neurons and neuroglial cells.
4. Describe the generalized structure of a neuron, and explain the functions of its parts.
5. Distinguish between myelinated and unmyelinated nerve fibers.
6. Discuss the functions of each type of neuroglial cell.
7. Explain how a membrane becomes polarized.
8. Describe how ions associated with nerve cell membranes are distributed.

9. Define *resting potential.*
10. Explain how threshold potential may be achieved.
11. List the events occurring during an action potential.
12. Explain how nerve impulses are related to action potentials.
13. Explain how impulses are conducted on myelinated and unmyelinated nerve fibers.
14. Define *synapse.*
15. Explain how a nerve impulse passes from one neuron to another.
16. List four types of neurotransmitters.
17. Explain what happens to neurotransmitters after they are released.
18. Distinguish between excitatory and inhibitory actions of neurotransmitters.
19. Describe a neuronal pool.
20. Distinguish between convergence and divergence.
21. Explain how neurons can be classified on the basis of their structure.
22. Distinguish between sensory, motor, and mixed nerves.
23. Define *reflex.*
24. Describe a reflex arc that consists of two neurons.
25. Name the layers of the meninges, and explain their functions.
26. Describe the structure of the spinal cord.
27. Distinguish between the ascending and descending tracts of the spinal cord.
28. Name the three major portions of the brain, and describe the general functions of each.
29. Describe the general structure of the cerebrum.
30. Distinguish between the *cerebral cortex* and *basal ganglia.*
31. Describe the location of the motor, sensory, and association areas of the cerebral cortex, and describe the general functions of each.
32. Define *hemisphere dominance.*
33. Explain the function of the corpus callosum.
34. Describe the location of the ventricles of the brain.
35. Explain how cerebrospinal fluid is produced and how it functions.
36. Name the parts of the brain stem, and describe the general functions of each part.
37. Define *limbic system,* and explain its functions.
38. Name the parts of the midbrain, and describe the general functions of each part.
39. Describe the pons and its functions.
40. Describe the medulla oblongata and its functions.
41. Describe the functions of the cerebellum.
42. Name, locate, and describe the major functions of each pair of cranial nerves.
43. Explain how the spinal nerves are grouped and numbered.
44. Describe the structure of a spinal nerve.
45. Define *plexus,* and locate the major plexuses of the spinal nerves.
46. Describe the general functions of the autonomic nervous system.
47. Distinguish between the sympathetic and the parasympathetic divisions of the autonomic nervous system.
48. Distinguish between a preganglionic and a postganglionic nerve fiber.
49. Explain why the effects of the sympathetic and parasympathetic divisions differ.
50. Describe how portions of the central nervous system control autonomic activities.

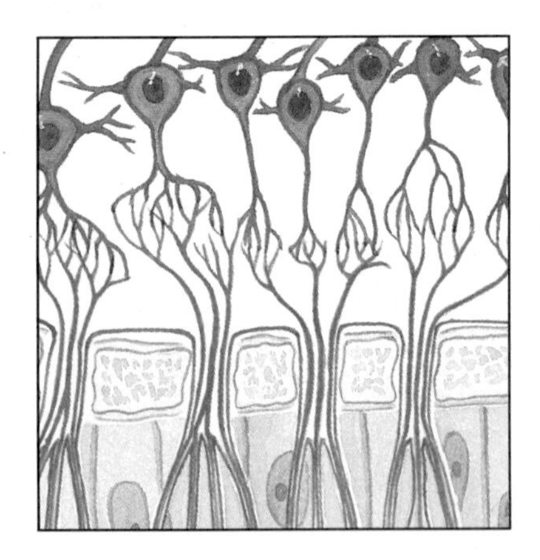

10

Somatic and Special Senses

Before the nervous system can act to control body functions, it must detect what is occurring inside and outside the body. This information is gathered by sensory receptors, which are sensitive to changes taking place in their surroundings.

Although receptors vary greatly in their individual characteristics, they can be grouped into two major categories. The receptors of one group are widely distributed throughout the skin and deeper tissues and generally have simple forms. These receptors are associated with the *somatic senses* of touch, pressure, temperature, and pain. The receptors of the second group are parts of complex, specialized sensory organs that are responsible for the *special senses* of smell, taste, hearing, equilibrium, and vision.

Chapter Objectives

After you have studied this chapter, you should be able to

1. Name five kinds of receptors, and explain the function of each kind.
2. Explain how a sensation is produced.
3. Describe the somatic senses.
4. Describe the receptors associated with the senses of touch, pressure, temperature, and pain.
5. Describe how the sense of pain is produced.
6. Explain the relationship between the senses of smell and taste.
7. Name the parts of the ear, and explain the function of each part.
8. Distinguish between static and dynamic equilibrium.
9. Name the parts of the eye, and explain the function of each part.
10. Explain how light is refracted by the eye.
11. Describe the visual nerve pathway.
12. Complete the review activities at the end of this chapter. Note that the items are worded in the form of specific learning objectives. You may want to refer to them before reading the chapter.

Key Terms

accommodation (ah-kom″o-da′shun)

ampulla (am-pul′lah)

chemoreceptor (ke″mo-re-sep′tor)

cochlea (kok′le-ah)

cornea (kor′ne-ah)

dynamic equilibrium (di-nam′ik e″kwi-lib′re-um)

labyrinth (lab′i-rinth)

macula (mak′u-lah)

mechanoreceptor (mek″ah-no-re-sep′tor)

olfactory (ol-fak′to-re)

optic (op′tik)

photoreceptor (fo″to-re-sep′tor)

projection (pro-jek′shun)

referred pain (re-furd′ pān)

refraction (re-frak′shun)

retina (ret′i-nah)

rhodopsin (ro-dop′sin)

sclera (skle′rah)

sensory adaptation (sen′so-re ad″ap-ta′shun)

static equilibrium (stat′ik e″kwi-lib′re-um)

Aids to Understanding Words

choroid, skinlike: *choroid* coat—the middle, vascular layer of the eye.

cochlea, snail: *cochlea*—the coiled tube within the inner ear.

iris, rainbow: *iris*—the colored, muscular part of the eye.

labyrinth, maze: *labyrinth*—a complex system of interconnecting chambers and tubes of the inner ear.

lacri-, tears: *lacri*mal gland—a tear gland.

macula, spot: *macula* lutea—a yellowish spot on the retina.

olfact-, to smell: *olfact*ory—pertaining to the sense of smell.

scler-, hard: *scler*a—the tough, outer protective layer of the eye.

tympan-, drum: *tympan*ic membrane—the eardrum.

vitre-, glass: *vitre*ous humor—a clear, jellylike substance within the eye.

Introduction

AS CHANGES OCCUR within the body and in its surroundings, *sensory receptors* are stimulated, and they, in turn, trigger nerve impulses. These impulses travel on sensory pathways into the central nervous system to be processed and interpreted. As a result, a person experiences or perceives a particular feeling or sensation.

Receptors and Sensations

Although there are many kinds of sensory receptors, they have features in common. For example, each type of receptor is particularly sensitive (that is, has a low threshold) to a distinct kind of environmental change and is much less sensitive to other forms of stimulation.

Types of Receptors

Five general groups of sensory receptors have been identified based upon their sensitivities. They are those stimulated by changes in the chemical concentration of substances (chemoreceptors); those stimulated by tissue damage (pain receptors); those stimulated by changes in temperature (thermoreceptors); those stimulated by changes in pressure or movement in fluids (mechanoreceptors); and those stimulated by light energy (photoreceptors).

Sensations

A **sensation** (perception) is a feeling that occurs when sensory impulses are interpreted by the brain. Because all the nerve impulses that travel from sensory receptors into the central nervous system are alike, different kinds of sensations must be due to the way the brain interprets the impulses rather than to differences in the receptors. In other words, when a receptor is stimulated, the resulting sensation depends on what region of the brain receives the impulse. For example, the impulses reaching one region are always interpreted as sounds, and those reaching another portion are always sensed as touch.

At the same time that a sensation is created, the cerebral cortex causes the feeling to seem to come from the receptors being stimulated. This process is called **projection,** because the brain projects the sensation back to its apparent source. Projection allows a person to pinpoint the region of stimulation; thus, the eyes seem to see and the ears seem to hear.

Sensory Adaptation

When receptors are subjected to continuous stimulation, many of them undergo an adjustment called **sensory adaptation.** As receptors adapt, impulses leave them at decreasing rates, until finally these receptors may fail completely to send signals. Once receptors have adapted, impulses can be triggered only if the strength of the stimulus is changed.

Sensory adaptation is experienced when a person enters a room where there is a strong odor. At first the scent seems intense, but it becomes less and less noticeable as the smell (olfactory) receptors adapt.

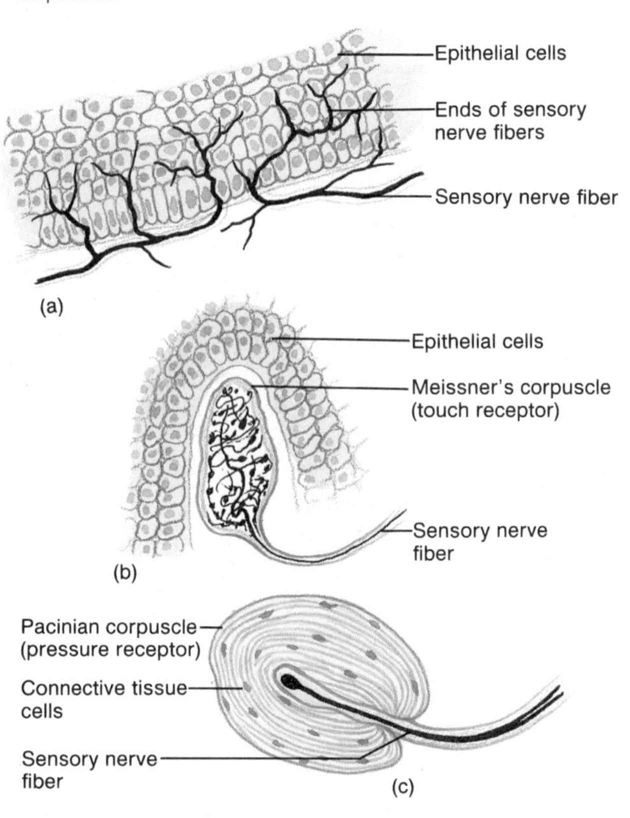

Figure 10.1
Touch and pressure receptors include (*a*) free ends of sensory nerve fibers, (*b*) Meissner's corpuscles, and (*c*) pacinian corpuscles.

1. List five general types of sensory receptors.
2. Explain how a sensation occurs.
3. What is meant by sensory adaptation?

Somatic Senses

The somatic senses include the senses associated with the skin as well as those associated with the muscles, joints, and visceral organs.

Touch and Pressure Senses

The senses of touch and pressure employ three kinds of receptors (fig. 10.1). As a group, these receptors are sensitive to the mechanical forces that cause tissues to be deformed or displaced. Touch and pressure receptors include the following:

Headache
A Current Topic

One of the more commonly experienced forms of pain is a *headache*. Although the brain tissue itself lacks pain receptors, nearly all the other tissues of the head, including the meninges and blood vessels, are well supplied with them.

Most headaches seem to be related to stressful life situations that result in fatigue, emotional tension, anxiety, or frustration. These conditions are reflected in various physiological changes. For example, they may cause prolonged contraction of the skeletal muscles in the forehead, sides of the head, or back of the neck. Such contractions may stimulate pain receptors and produce what is often called a *tension headache*. In other cases, people suffer more severe *vascular headaches,* which accompany the constriction or dilation of the cranial blood vessels. The throbbing headache of a "hangover," following an excessive consumption of alcohol, for example, may be due to blood pulsating through such cranial vessels.

Still another form of vascular headache is called *migraine*. In this disorder, certain cranial blood vessels seem to constrict, producing a localized cerebral blood deficiency. As a result, the person may experience a variety of symptoms, such as seeing patterns of bright light, which may interfere with vision, or feeling numbness in the limbs or face. Typically, vasoconstriction is followed by vasodilation of the affected vessels and a severe headache, which is usually limited to one side of the head and may last for several hours.

Even though the pain of some headaches originates inside the cranium, the sensation often seems to come from the surface. Other causes of headaches include sensitivity to certain food additives, high blood pressure, increased intracranial pressure due to a tumor or to blood escaping from a ruptured vessel, decreased cerebrospinal fluid pressure following a lumbar puncture, or sensitivity to or withdrawal from various drugs.

1. **Sensory nerve fibers.** These receptors are common in epithelial tissues, where their free ends occur between epithelial cells. They are associated with the sensations of touch and pressure.
2. **Meissner's corpuscles.** These structures consist of small, oval masses of flattened connective tissue cells surrounded by connective tissue sheaths. Two or more sensory nerve fibers branch into each corpuscle and end as tiny knobs.

   Meissner's corpuscles are especially numerous in the hairless portions of the skin, such as the lips, fingertips, palms, soles, nipples, and external genital organs. They are sensitive to the motion of objects that barely contact the skin, and impulses from them are related to the sensation of light touch.
3. **Pacinian corpuscles.** These sensory bodies are relatively large structures composed of connective tissue fibers and cells. They are common in the deeper subcutaneous tissues and occur in the tendons of muscles and the ligaments of joints.

   Pacinian corpuscles are stimulated by heavy pressure and associated with the sensation of deep pressure.

Temperature Senses

The temperature senses employ two kinds of skin receptors. Although there is some question about the identity of these receptors, they seem to include two types of *free nerve endings,* called *heat receptors* and *cold receptors*. The heat receptors are most sensitive to temperatures above 25° C (77° F) and become unresponsive at temperatures above 45° C (113° F). As 45° C is approached, the pain receptors are also triggered, producing a *burning* sensation.

The cold receptors are most sensitive to temperatures between 10° C (50° F) and 20° C (68° F). If the temperature drops below 10° C, the pain receptors are stimulated, and the person feels a *freezing* sensation.

Both the heat and cold receptors demonstrate rapid adaptation, so that within about a minute of continuous stimulation, the sensation of heat or cold begins to fade.

Sense of Pain

The sense of pain also involves receptors that consist of *free nerve endings*. These receptors are widely distributed throughout the skin and internal tissues, except in the nervous tissue of the brain, which lacks pain receptors.

The pain receptors have a protective function in that they are stimulated whenever tissues are being damaged. The pain sensation is usually perceived as unpleasant, and it serves as a signal that something should be done to remove the source of stimulation.

The pain receptors adapt poorly, if at all, and once such a receptor has been activated, even by a single stimulus, it may continue to send impulses into the central nervous system for some time.

The way in which tissue damage excites the pain receptors is poorly understood. It is thought, however, that injuries promote the release of certain chemicals and that sufficient quantities of these chemicals may stimulate pain receptors. A deficiency of oxygen-rich blood (ischemia) in a tissue or the stimulation of certain mechanical-sensitive receptors also triggers pain sensations. The pain elicited during a muscle cramp, for example, seems to be related to an interruption of blood flow, which occurs as the sustained contraction squeezes the capillaries and reduces blood flow, as well as to the stimulation of the mechanical-sensitive pain receptors.

The stimulation of pain receptors associated with injuries to bones, tendons, or ligaments may also cause nearby skeletal muscles to undergo contraction. As the muscles contract, they may become ischemic, and this condition may trigger still other pain receptors within the muscle tissue. This additional stimulation of pain receptors may lead to increased muscular contraction, thus establishing a "vicious circle."

Visceral Pain

As a rule, the pain receptors are the only receptors in the visceral organs whose stimulation produces sensations. The pain receptors in these organs seem to respond differently to stimulation than those associated with the surface tissues. For example, localized damage to the intestinal tissue, as may occur during surgical procedures, may not elicit any pain sensations even in a conscious person. When the visceral tissues are subjected to a more widespread stimulation, however, as when the intestinal tissues are stretched or when the smooth muscles in the intestinal walls undergo spasms, a strong pain sensation may follow. Once again, the resulting pain seems to be related to the stimulation of mechanical-sensitive receptors and to a decreased blood flow, accompanied by a lower tissue oxygen concentration and an accumulation of pain-stimulating chemicals.

Another characteristic of visceral pain is that it may feel as if it is coming from some part of the body other than the part being stimulated—a phenomenon called **referred pain.** For example, pain originating from the heart may be referred to the left shoulder or left arm (fig. 10.2).

The occurrence of referred pain seems to be related to *common nerve pathways* that carry sensory impulses from skin areas as well as visceral organs. In other words, pain impulses from the heart seem to be conducted over the same nerve pathways as those conducting impulses from the skin of the left shoulder and left arm (fig. 10.3). Consequently, the cerebral cortex may incorrectly interpret the source of pain impulses during a heart attack as the left shoulder or arm rather than the heart.

Figure 10.2
Surface regions to which visceral pain may be referred.

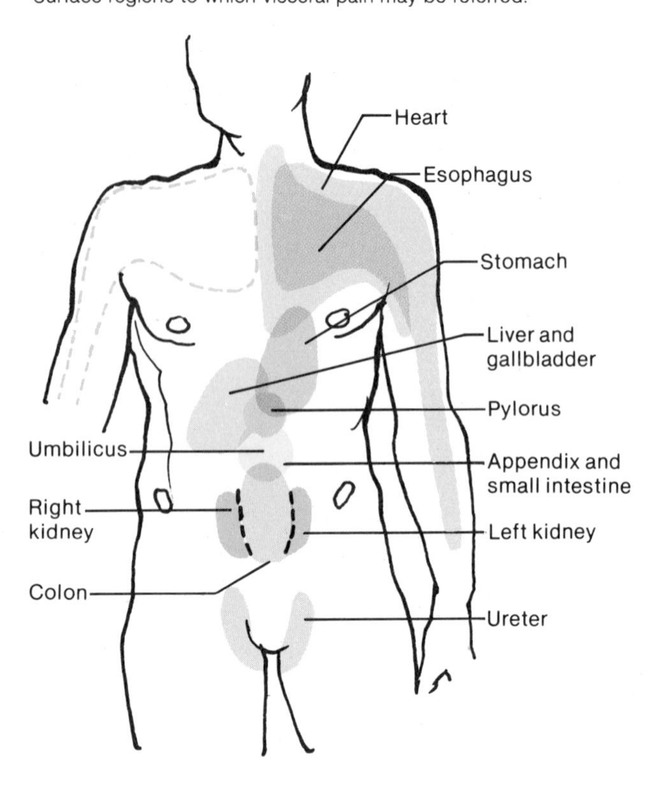

1. Describe the three types of touch and pressure receptors.
2. Describe the receptors involved in the temperature senses.
3. What types of stimuli excite the pain receptors?
4. What is referred pain?

Pain Nerve Fibers

The nerve fibers that conduct impulses away from pain receptors include two main types: acute pain fibers and chronic pain fibers.

The *acute pain fibers* are relatively thin, myelinated nerve fibers. They conduct nerve impulses rapidly and are associated with the sensation of sharp pain. This pain typically seems to originate from a restricted area of the skin, and the pain seldom continues after the pain-producing stimulus is discontinued.

The *chronic pain fibers* are thin, unmyelinated nerve fibers. They conduct impulses more slowly and are related to a dull, aching pain sensation that may be widespread and difficult to pinpoint. Such pain may continue for some time after the original stimulus has been eliminated. Although acute pain is usually sensed as coming from the skin, chronic pain is likely to be felt in deeper tissues, as well as in the skin.

Figure 10.3
Pain originating in the heart may feel as if it is coming from the skin, because sensory impulses from these two regions follow common nerve pathways to the brain.

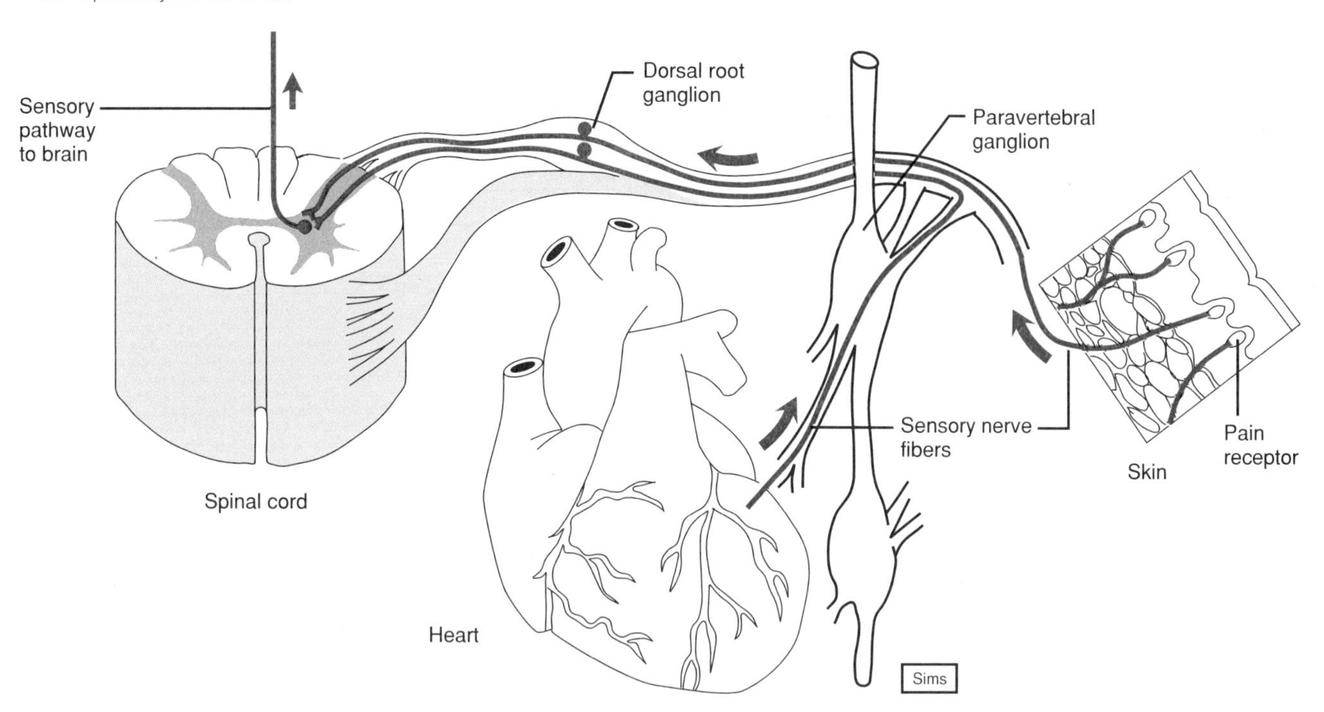

Commonly, an event that stimulates pain receptors will trigger impulses on both types of pain fibers. This causes a dual sensation—a sharp, pricking pain, followed shortly by a dull, aching one. The aching pain is usually more intense and may become even more intense with time. Such chronic pain sometimes gives rise to prolonged suffering that is very resistant to relief and control.

Pain impulses that originate from tissues of the head reach the brain on sensory fibers of various cranial nerves. All other pain impulses travel on the sensory fibers of spinal nerves, and they pass into the spinal cord by way of the dorsal roots of these spinal nerves. Within the spinal cord, pain impulses are processed in the gray matter of the dorsal horn and are transmitted to the brain.

Within the brain, most pain fibers terminate in the reticular formation (see chapter 9), and from there the impulses are conducted on still other neurons to the thalamus, hypothalamus, and cerebral cortex.

Regulation of Pain Impulses

The awareness of pain seems to occur when the pain impulses reach the thalamus—that is, even before they reach the cerebral cortex. However, the cerebral cortex is needed to judge the intensity of the pain and locate its source. The cerebral cortex is also responsible for various emotional and motor responses to pain.

Still other parts of the brain regulate the flow of pain impulses up from the spinal cord. These parts include areas of gray matter in the midbrain, pons, and medulla oblongata. Impulses from special neurons in these areas descend in the lateral funiculus (see chapter 9) to various levels of the spinal cord. These impulses cause the ends of certain nerve fibers to release substances that can block pain signals by inhibiting presynaptic nerve fibers in the dorsal horn of the spinal cord.

Among the inhibiting substances released in the dorsal horn are a group of neuropeptides called *enkephalins* and the monoamine called *serotonin* (see chapter 9). The enkephalins can suppress both acute and chronic pain impulses. Thus, they can relieve relatively strong pain sensations, much as morphine and other opiate drugs do. In fact, enkephalins seem to bind to the same receptor sites on neuron membranes as does morphine. Serotonin is thought to stimulate other neurons to release enkephalins.

Another group of neuropeptides that has pain-suppressing, morphinelike actions are called *endorphins.* These substances occur in the pituitary gland and various regions of the nervous system, such as the hypothalamus. It is believed that enkephalins and endorphins are released when pain impulses are triggered excessively, and that in this way they provide a natural pain control. Such pain-relieving substances may also

Figure 10.4
(*a*) The olfactory receptor cells, which have cilia at their distal ends, are supported by columnar epithelial cells. (*b*) The olfactory area is associated with the superior nasal concha.

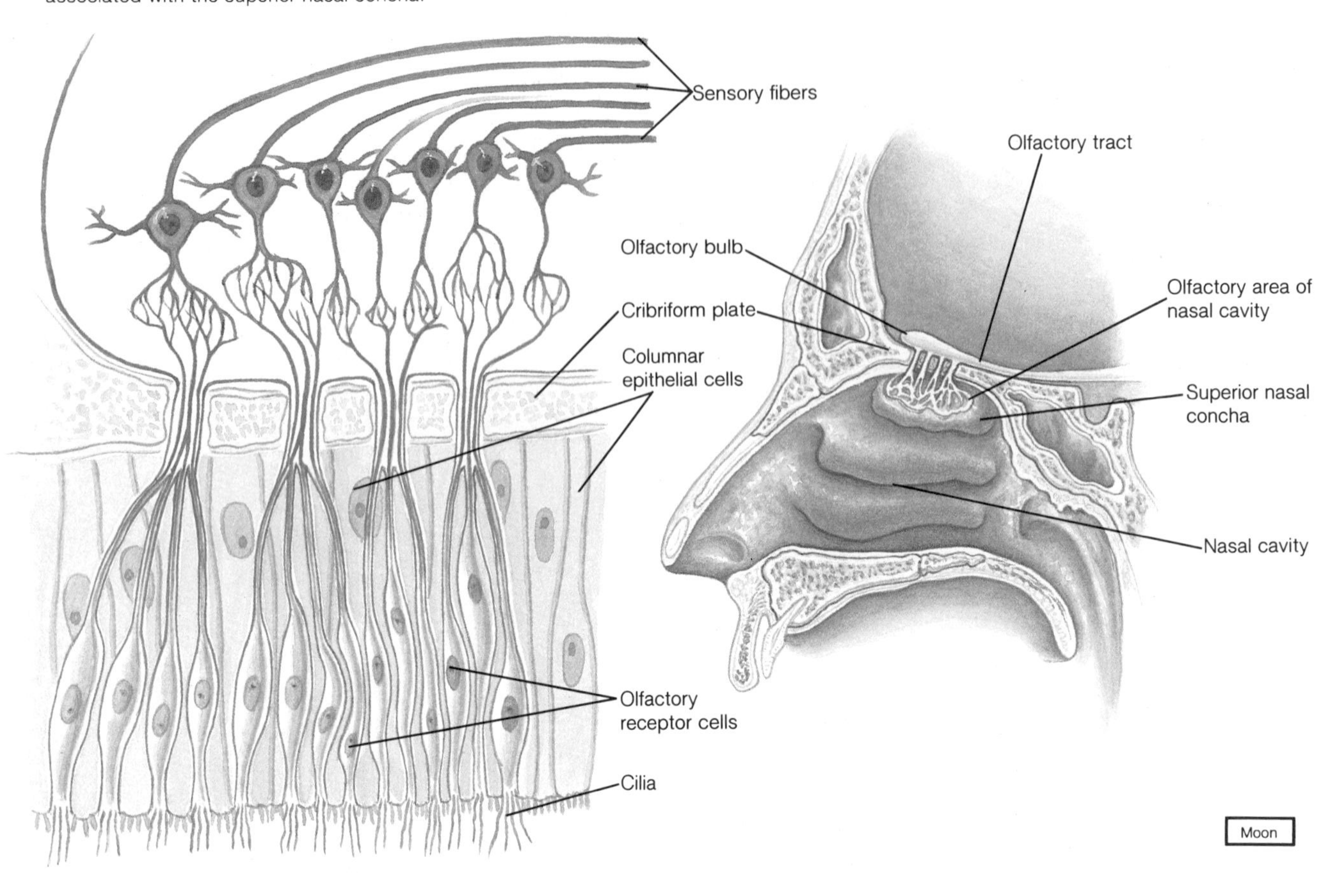

be released in response to indirect stimulation by electrical impulses or acupuncture needles.

1. Describe two types of pain fibers.
2. How do acute pain and chronic pain differ?
3. What parts of the brain are involved with interpreting pain impulses?
4. How do neuropeptides help to control pain?

Special Senses

Special senses are those whose sensory receptors occur within relatively large, complex sensory organs in the head. These organs include the olfactory organs, taste buds, ears, organs of equilibrium, and eyes, which are associated with the senses of smell, taste, hearing, static equilibrium, dynamic equilibrium, and sight, respectively.

Sense of Smell

The sense of smell is associated with complex sensory structures in the upper region of the nasal cavity.

Olfactory Receptors

The smell, or olfactory, receptors are similar to those for taste (described in a subsequent section) in that they are chemoreceptors, stimulated by chemicals dissolved in liquids. These two senses function closely together and aid in food selection, since food is usually smelled at the same time it is tasted.

Olfactory Organs

The **olfactory organs,** which contain the olfactory receptors, appear as yellowish-brown masses that cover the upper parts of the nasal cavity, the superior nasal conchae, and a portion of the nasal septum.

The **olfactory receptor cells** are neurons surrounded by columnar epithelial cells (fig. 10.4). These neurons have tiny knobs at the distal ends of their dendrites that are covered by hairlike cilia. The cilia project into the nasal cavity and are thought to be the sensitive portions of the receptors.

Chemicals that stimulate the olfactory receptors enter the nasal cavity as gases, but they must dissolve at least partially in the watery fluids that surround the cilia before they can be detected.

Olfactory Nerve Pathway

Once the olfactory receptors have been stimulated, nerve impulses are triggered and travel along the axons of the receptor cells that are the fibers of the olfactory nerves. These fibers lead to neurons located in the enlargements called the **olfactory bulbs,** which lie on either side of the crista galli of the ethmoid bone (fig. 7.12). Within the olfactory bulbs, the impulses are analyzed, and as a result, additional impulses travel along the **olfactory tracts** to a portion of the limbic system (see chapter 9). The major interpreting areas (olfactory cortex) for these impulses are located within the temporal lobes and at the base of the frontal lobes just anterior to the hypothalamus.

Olfactory Stimulation

The mechanism by which various substances stimulate the olfactory receptors is poorly understood. One hypothesis suggests that the shapes of gaseous molecules may fit complementary shapes of receptor sites on the cilia. A nerve impulse, according to this idea, is triggered when a molecule binds to its particular receptor site.

Because the olfactory organs are located high in the nasal cavity above the usual pathway of inhaled air, a person may have to sniff and force air over the receptor areas to smell something that has a faint odor. Also, the olfactory receptors undergo sensory adaptation rather rapidly, but even though they have adapted to one scent, their sensitivity to other odors remains unchanged.

Partial or complete loss of smell is called *anosmia.* This condition may be caused by a variety of factors including inflammation of the nasal cavity lining, as occurs during a head cold. It may also result from excessive tobacco smoking, or from the use of certain drugs, such as epinephrine or cocaine.

1. Where are the olfactory receptors located?
2. Trace the pathway of an olfactory impulse from a receptor to the cerebrum.

Sense of Taste

Taste buds are the special organs of taste (fig. 10.5). They occur primarily on the surface of the tongue and are associated with tiny elevations called *papillae.* They are also found in smaller numbers in the roof of the mouth and walls of the pharynx.

Taste Receptors

Each taste bud includes a group of modified epithelial cells, the **taste cells** (gustatory cells), which function as receptors. The taste bud also includes a number of epithelial supporting cells. The entire structure is somewhat spherical with an opening, the **taste pore,** on its free surface. Tiny projections, called **taste hairs,** protrude from the outer ends of the taste cells and jut out through the taste pore. It is believed that these taste hairs are the sensitive parts of the receptor cells.

Interwoven among the taste cells and wrapped around them is a network of nerve fibers. When a receptor cell is stimulated, an impulse is triggered on a nearby nerve fiber and carried into the brain.

Before the taste of a particular chemical can be detected, the chemical must be dissolved in the watery fluid surrounding the taste buds. This fluid is supplied by the salivary glands.

The mechanism by which various substances stimulate taste cells is not well understood. One possible explanation holds that various substances combine with specific receptor sites on the taste hair surfaces. Such a combination is thought to be responsible for the generation of sensory impulses on nearby nerve fibers.

Although the taste cells in all taste buds appear very much alike microscopically, there are at least four types. Each type is most sensitive to a particular kind of chemical stimulus; consequently, there are at least four primary taste (gustatory) sensations.

Taste Sensations

The four *primary taste sensations* are as follows:

1. *sweet,* as produced by table sugar
2. *sour,* as produced by vinegar
3. *salty,* as produced by table salt
4. *bitter,* as produced by caffeine or quinine

In addition to these, some investigators recognize two other taste sensations, which they call *alkaline* and *metallic.*

Each of the four major types of taste receptors is most highly concentrated in certain regions of the tongue's surface. Figure 10.6 illustrates this special distribution of the receptors.

Each of the many flavors we experience daily is believed to result from one of the primary sensations or from some combination of two or more of them. The way we experience flavors also involves the concentration of chemicals as well as the sensations of odor, texture (touch), and temperature. Furthermore, the chemicals in some foods—chili peppers and ginger, for instance—may stimulate pain receptors, which cause the tongue to burn.

Taste receptors, like olfactory receptors, undergo sensory adaptation relatively rapidly. The resulting loss of taste can be avoided by moving bits of food over the surface of the tongue to stimulate different receptors at different moments.

Figure 10.5
(*a*) Taste buds on the surface of the tongue are associated with nipplelike elevations called papillae. (*b*) A taste bud contains taste cells and has an opening, the taste pore, at its free surface.

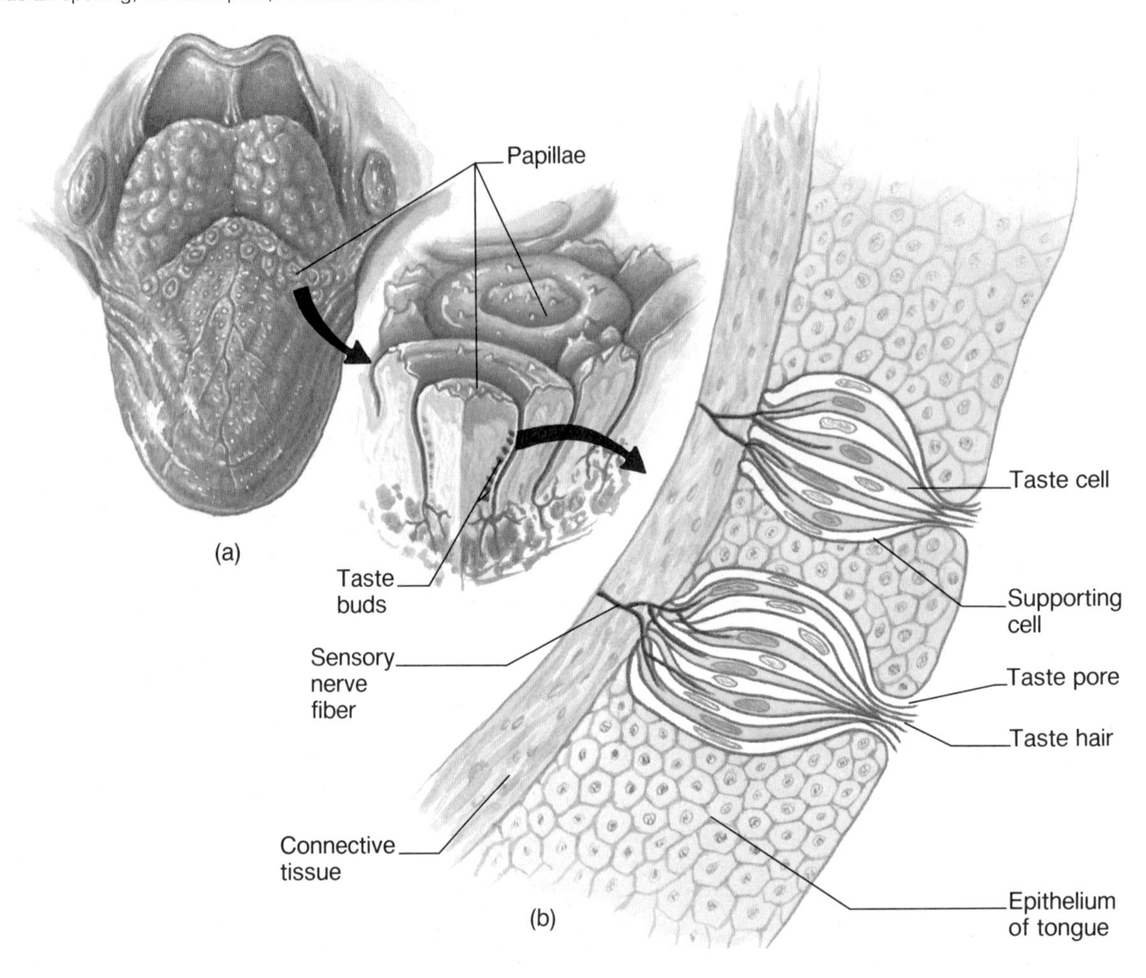

Figure 10.6
Patterns of taste receptor distribution are indicated by color in these diagrams: (*a*) sweet receptors; (*b*) sour receptors; (*c*) salt receptors; (*d*) bitter receptors.

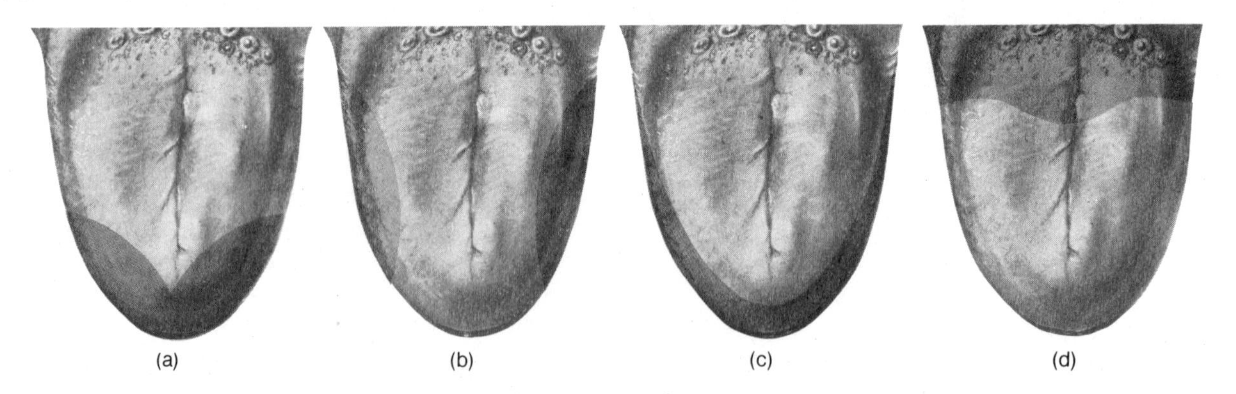

Taste Nerve Pathway

Sensory impulses from the taste receptors located in various regions of the tongue travel on fibers of the facial, glossopharyngeal, and vagus nerves into the medulla oblongata. From there, the impulses ascend to the thalamus and are directed to the gustatory cortex, which is located in the parietal lobe of the cerebrum along a deep portion of the lateral sulcus (fig. 9.23).

1. Why is saliva necessary for the sense of taste?
2. Name the four primary taste sensations.
3. Trace a sensory impulse from a taste receptor to the cerebral cortex.

Figure 10.7
Major parts of the ear.

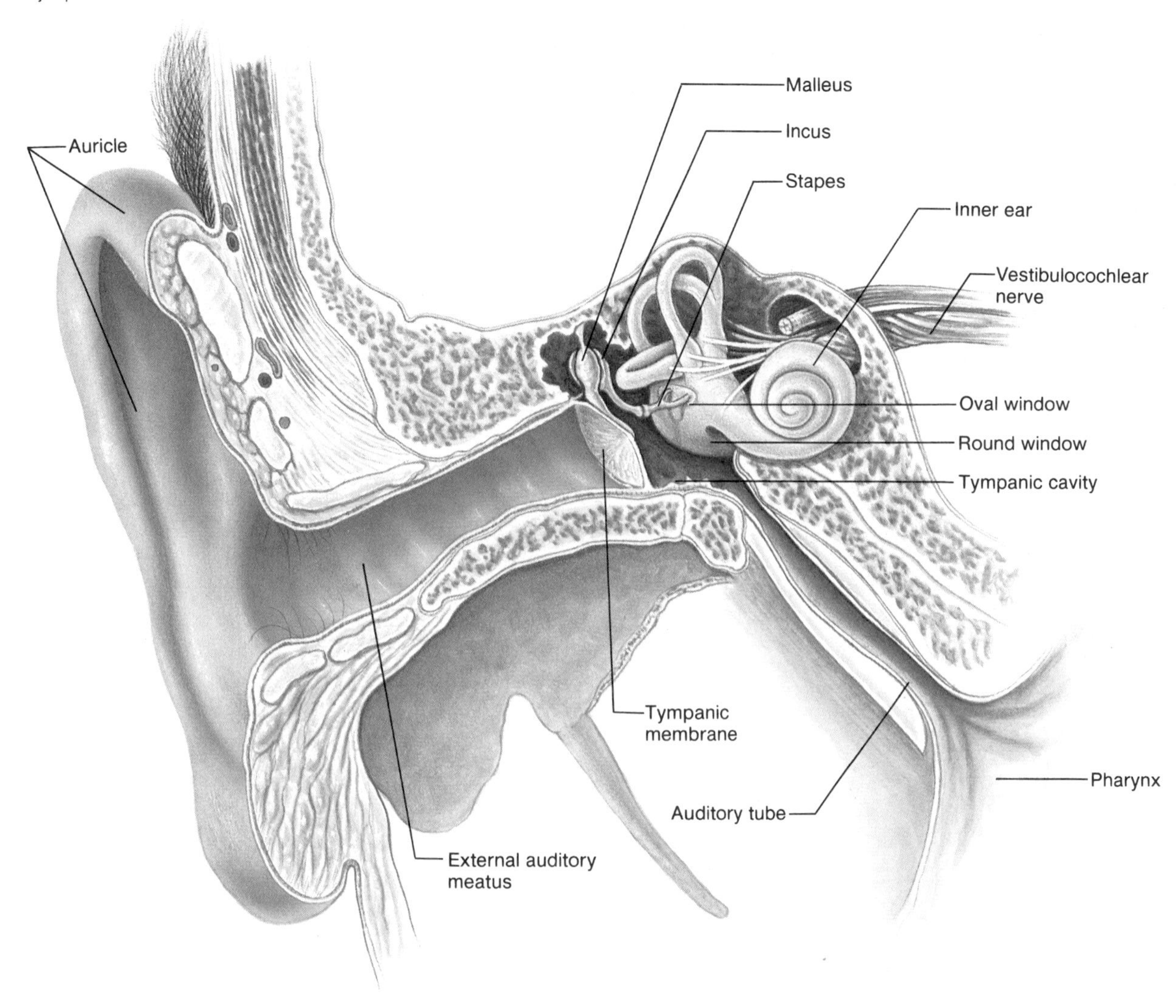

Sense of Hearing

The organ of hearing, the **ear,** has external, middle, and inner parts. In addition to making hearing possible, the ear functions in the sense of equilibrium, which is discussed in a subsequent section of this chapter.

External Ear

The external ear consists of two parts: an outer, funnellike structure, called the **auricle,** and an S-shaped tube, called the **external auditory meatus,** which leads into the temporal bone for about 2.5 centimeters (fig. 10.7).

Sounds are created by vibrating objects, and the vibrations are transmitted through matter in the form of sound waves. For example, the sounds of some musical instruments are produced by vibrating strings or reeds, and the sounds of the voice are created by vibrating vocal folds in the larynx. The auricle of the ear helps collect sound waves traveling through air and directs them into the auditory meatus.

Middle Ear

The middle ear includes an air-filled space in the temporal bone, called the tympanic cavity, an eardrum, or tympanic membrane, and three small bones called auditory ossicles.

The **tympanic membrane** (eardrum) is a semitransparent membrane, covered by a thin layer of skin on its outer surface and by mucous membrane on the inside. It has an oval margin and is cone-shaped with the apex of the cone directed inward. Its cone shape is maintained by the attachment of one of the auditory ossicles (malleus).

Sound waves that enter the auditory meatus cause pressure changes on the eardrum, which moves back and forth in response and, thus, reproduces the vibrations of the sound wave source.

The three **auditory ossicles** are called the *malleus* (hammer), *incus* (anvil), and *stapes* (stirrup). They are attached to the wall of the tympanic cavity by tiny ligaments and are covered by mucous membrane. These bones form a bridge connecting the eardrum to the inner ear and function to transmit vibrations between these parts. Specifically, the malleus is attached to the eardrum, and when the eardrum vibrates, the malleus vibrates in unison. The malleus causes the incus to vibrate, and the incus passes the movement onto the stapes. The stapes is held by ligaments to an opening in the wall of the tympanic cavity. This opening, called the **oval window,** leads into the inner ear. Vibration of the stapes at the oval window causes motion of a fluid within the inner ear, which is responsible for stimulating the hearing receptors.

In addition to transmitting vibrations, the auditory ossicles help to increase the force of the vibrations as they are passed from the eardrum to the oval window. For example, because the ossicles transmit vibrations from the relatively large surface of the eardrum to a much smaller area at the oval window, the vibrational force becomes concentrated as it travels from the external to the inner ear. As a result, the pressure (per mm^2) applied by the stapes at the oval window is many times greater than that exerted on the eardrum by sound waves.

Auditory Tube

An **auditory tube** (eustachian tube) connects each middle ear to the throat. This tube allows air to pass between the tympanic cavity and the outside of the body by way of the throat (nasopharynx) and mouth. It helps maintain equal air pressure on both sides of the eardrum, which is necessary for normal hearing.

The function of the auditory tube can be experienced during rapid change in altitude. For example, as a person moves from a high altitude to a lower one, the air pressure on the outside of the eardrum becomes greater and greater. As a result the eardrum may be pushed inward, out of its normal position, and hearing may be impaired.

When the air pressure difference is great enough, some air may force its way through the auditory tube into the middle ear. This allows the pressure on both sides of the eardrum to equalize, and the membrane moves back into its regular position. The person usually hears a popping sound at this moment, and normal hearing is restored.

The mucous membranes that line the auditory tubes are continuous with the linings of the middle ears. Consequently, the tubes provide a route by which mucous membrane infections of the throat may spread and cause an infection of the middle ear. For this reason, it is poor practice to pinch one nostril when blowing the nose, because the pressure in the nasal cavity may force material from the throat up the auditory tube and into the middle ear.

Inner Ear

The inner ear consists of a complex system of intercommunicating chambers and tubes, called a **labyrinth;** in fact, there are two such structures in each ear—the osseous labyrinth and the membranous labyrinth (fig. 10.8).

The *osseous labyrinth* is a bony canal in the temporal bone; the *membranous labyrinth* is a tube that lies within the osseous labyrinth and has a similar shape. Between the osseous and membranous labyrinths is a fluid, called *perilymph,* that is secreted by cells in the wall of the bony canal. The membranous labyrinth contains another fluid, called *endolymph.*

The inner ear includes three **semicircular canals,** which function in providing a sense of equilibrium (to be discussed in a subsequent section), and a **cochlea,** which functions in hearing.

The cochlea contains a bony core and a thin bony shelf that winds around the core like the threads of a screw. The shelf divides the bony labyrinth of the cochlea into upper and lower compartments. The upper compartment, called the *scala vestibuli,* leads from the oval window to the apex of the spiral. The lower compartment, the *scala tympani,* extends from the apex of the cochlea to a membrane-covered opening in the wall of the inner ear called the **round window** (fig. 10.9).

That portion of the membranous labyrinth within the cochlea is called the *cochlear duct.* It lies between the two bony compartments and ends as a closed sac at the apex of the cochlea. The cochlear duct is separated from the scala vestibuli by a *vestibular membrane* (Reissner's membrane) and from the scala tympani by a *basilar membrane* (fig. 10.10).

The basilar membrane contains many thousands of stiff, elastic fibers, whose lengths vary, becoming progressively longer from the base of the cochlea to its apex. Sound vibrations entering the perilymph at the oval window travel along the scala vestibuli and pass through the vestibular membrane and into the endolymph of the cochlear duct, where they cause movements in the basilar membrane.

After passing through the basilar membrane, the vibrations enter the perilymph of the scala tympani, and their forces are dissipated into the air of the tympanic cavity by movements of the membrane covering the round window.

Figure 10.8
The osseous labyrinth of the inner ear is separated from the membranous labyrinth by perilymph. The membranous labyrinth contains endolymph.

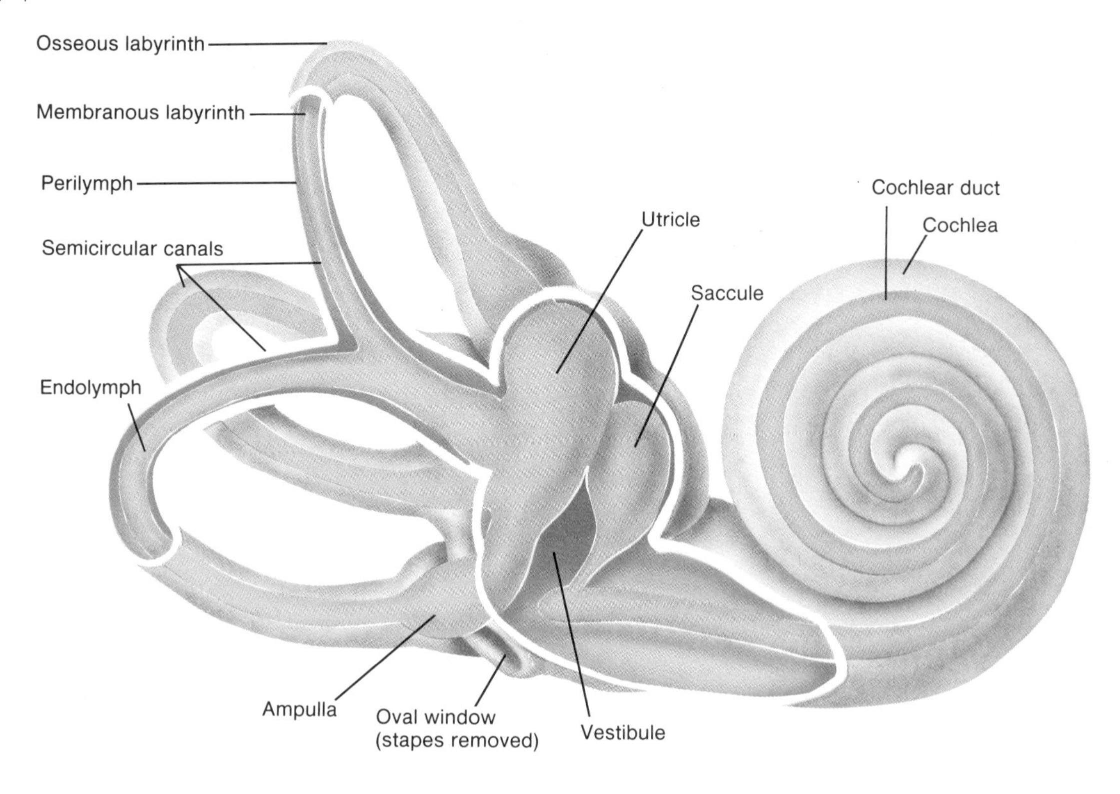

Figure 10.9
(*a*) The cochlea consists of a coiled, bony canal with a portion of the membranous labyrinth inside. (*b*) If the cochlea could be unwound, the membranous tube, also called the cochlear duct, would be seen ending as a closed sac at the apex of the bony canal.

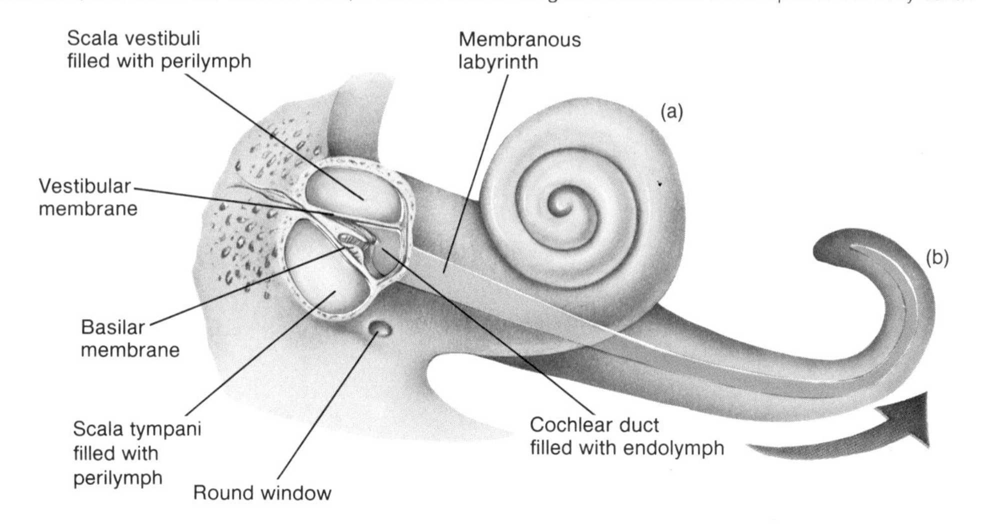

An **organ of Corti,** which contains the hearing receptors, is located on the upper surface of the basilar membrane and stretches from the apex to the base of the cochlea (fig. 10.10). Its receptor cells are arranged in rows, and they possess numerous hairlike processes that project into the endolymph of the cochlear duct. Above these hair cells is a *tectorial membrane,* which is attached to the bony shelf of the cochlea. It passes over the receptor cells and makes contact with the tips of their hairs.

As sound vibrations pass through the inner ear, the hairs shear back and forth against the tectorial membrane, and the resulting mechanical deformation of the hairs stimulates the receptor cells (figs. 10.10

Figure 10.10
(a) A cross section of the cochlea; (b) the organ of Corti.

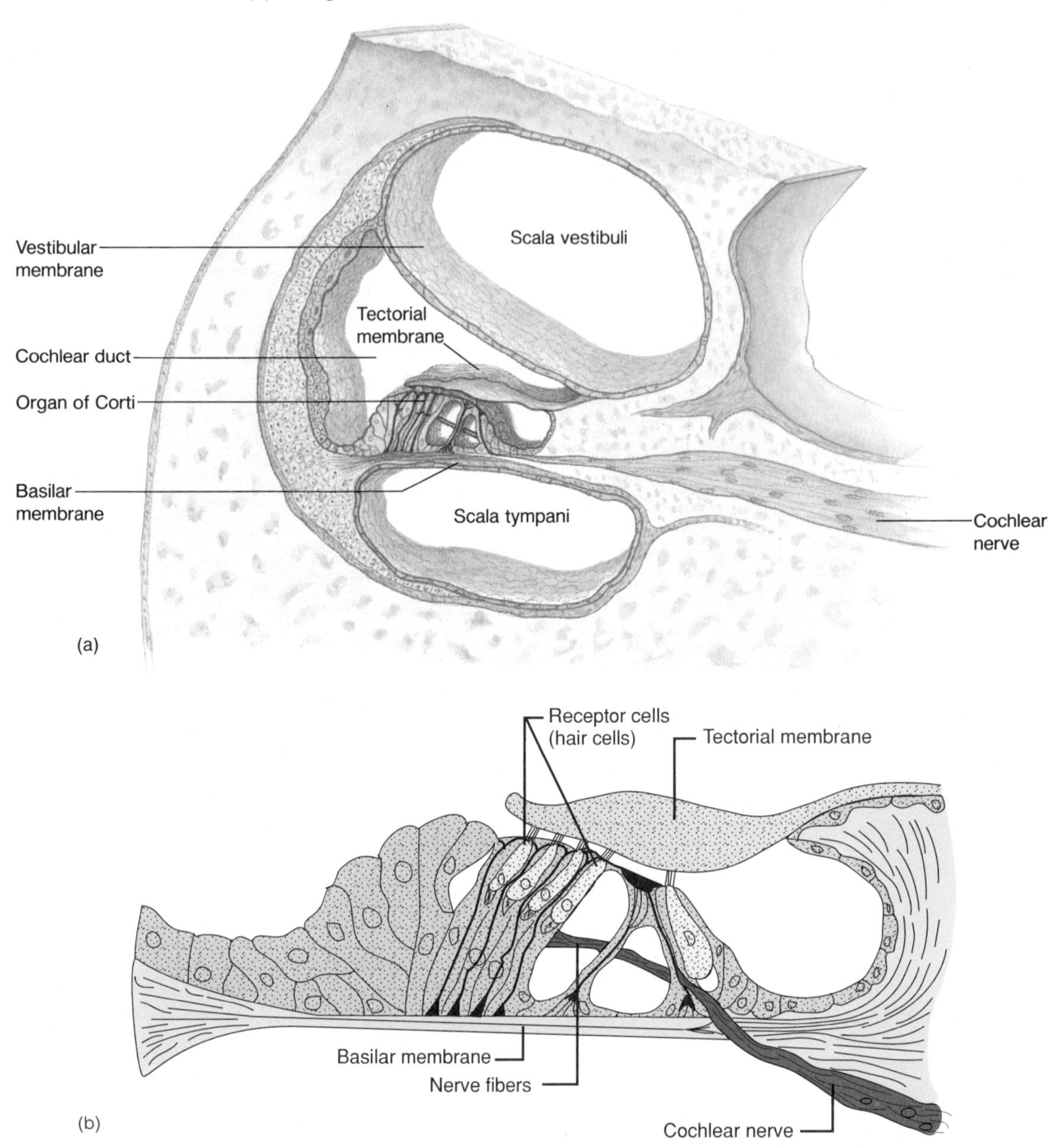

and 10.11). Various receptor cells, however, have slightly different sensitivities to such deformation of their hairs. Thus, a sound that produces a particular frequency of vibration excites certain receptor cells, and a sound involving another frequency stimulates a different set of hair cells.

Although the hearing receptor cells are epithelial cells, they act somewhat like neurons. For example, when such a cell is at rest, its membrane is polarized. When it is stimulated, its membrane becomes depolarized, its ion channels open, and the membrane becomes more permeable to calcium ions. The receptor cell has no axon or dendrites; however, it has neurotransmitter-containing vesicles in the cytoplasm near its base. In the presence of calcium ions, some of these vesicles fuse with the cell membrane and release neurotransmitter substance. The neurotransmitter stimulates the nearby ends of sensory nerve fibers, and in response they transmit impulses along the cochlear branch of the vestibulocochlear nerve to the auditory cortex of the temporal lobe of the brain.

Although the human ear is able to detect sound waves with frequencies varying from about 20 to 20,000 vibrations per second, the range of greatest sensitivity is between 2,000 and 3,000 vibrations per second.

Chart 10.1 summarizes the steps involved with the hearing mechanism.

Chart 10.1 Steps in the generation of sensory impulses from the ear

1. Sound waves enter the external auditory meatus.
2. Waves of changing pressures cause the eardrum to reproduce the vibrations coming from the sound wave source.
3. Auditory ossicles amplify and transmit vibrations to the end of the stapes.
4. Movement of the stapes at the oval window transmits vibrations to the perilymph in the scala vestibuli.
5. Vibrations pass through the vestibular membrane and enter the endolymph of the cochlear duct.
6. Different frequencies of vibration in endolymph stimulate different sets of receptor cells.
7. A receptor cell becomes depolarized; its membrane becomes more permeable to calcium ions.
8. In the presence of calcium ions, vesicles at the base of the receptor cell release neurotransmitter.
9. Neurotransmitter stimulates the ends of nearby sensory neurons.
10. Sensory impulses are triggered on fibers of the cochlear branch of the vestibulocochlear nerve.
11. The auditory cortex of the temporal lobe interprets the sensory impulses.

Figure 10.11
A scanning electron micrograph of hair cells in the organ of Corti (×13,000).

Auditory Nerve Pathway

The nerve fibers associated with the hearing receptors enter the auditory nerve pathways, which pass into the auditory cortices of the temporal lobes of the cerebrum, where they are interpreted. On the way, some of these fibers cross over, so that impulses arising from each ear are interpreted on both sides of the brain. Consequently, damage to a temporal lobe on one side of the brain is not necessarily accompanied by complete hearing loss in the ear on that side.

1. How are sound waves transmitted through the external, middle, and inner ears?
2. Distinguish between the osseous and membranous labyrinths.
3. Describe the organ of Corti.

Partial or complete hearing loss can be caused by a variety of factors, including interference with the transmission of vibrations to the inner ear (conductive deafness) or damage to the cochlea, auditory nerve, or auditory nerve pathways (sensorineural deafness).

Conductive deafness may be due to plugging of the external auditory meatus or to changes in the eardrum or auditory ossicles. The eardrum, for example, may harden as a result of disease and thus be less responsive to sound waves, or it may be torn or perforated by disease or injury.

Sensorineural deafness can be caused by exposure to excessively loud sounds, by tumors in the central nervous system, by brain damage as a result of vascular accidents, or by the use of certain drugs.

Sense of Equilibrium

The sense of equilibrium actually involves two senses—a sense of *static equilibrium* and a sense of *dynamic equilibrium*—that result from the actions of different sensory organs.

Static Equilibrium

The organs of **static equilibrium** sense the position of the head. They help in maintaining the stability and posture of the head and body when these parts are motionless. They are located within the **vestibule,** a bony chamber between the semicircular canals and cochlea. The membranous labyrinth inside the vestibule consists of two expanded chambers—an **utricle** and a **saccule** (fig. 10.8).

On the anterior wall of the utricle is a tiny structure, called a **macula,** that contains numerous hair cells, which serve as sensory receptors, and supporting cells (fig. 10.12). When the head is upright, the hairs of the hair cells project upward into a mass of gelatinous material, which has grains of calcium carbonate (otoliths) embedded in it. These particles increase the weight of the gelatinous structure.

Figure 10.12
The macula is responsive to changes in the position of the head. (*a*) Macula with the head in an upright position; (*b*) macula with the head bent forward.

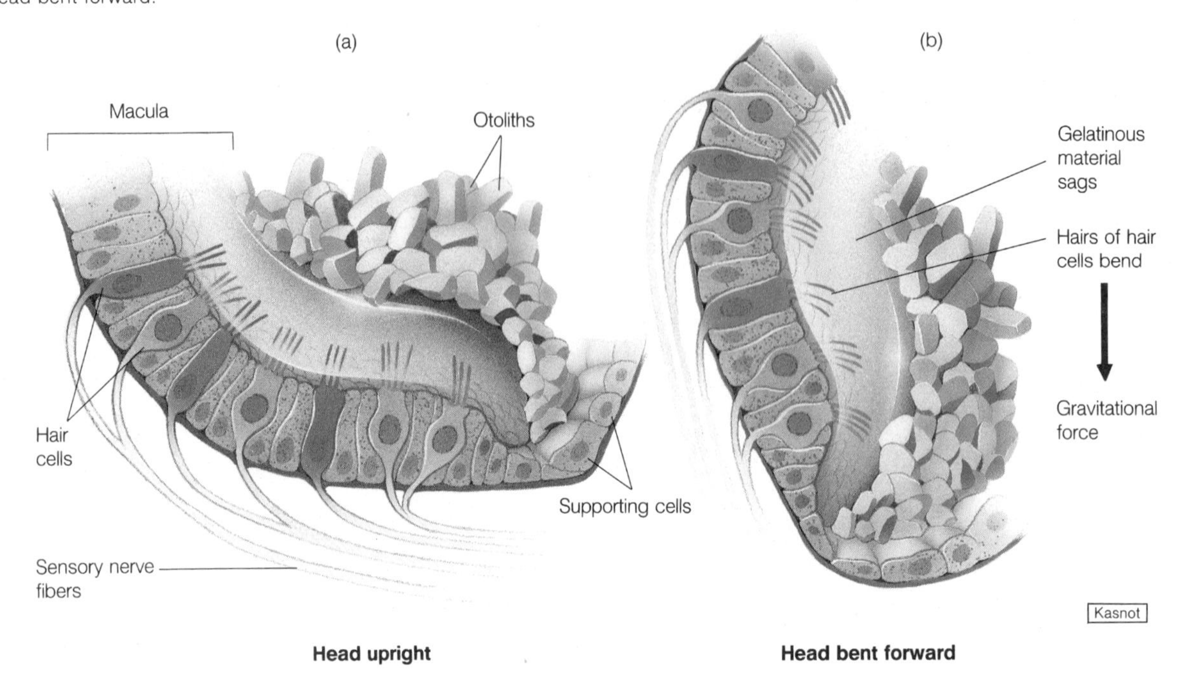

The usual stimulus to the hair cells occurs when the head is bent forward, backward, or to one side. Such movements cause the gelatinous masses of the maculae to be tilted, and as the gelatinous masses sag in response to gravity, the hairs projecting into them are bent. This action stimulates the hair cells, and they signal the nerve fibers associated with them in a manner similar to that of hearing receptors. The resulting nerve impulses travel into the central nervous system, by means of the vestibular branch of the vestibulocochlear nerve. These impulses inform the brain as to the position of the head. The brain acts on this information by sending motor impulses to skeletal muscles, and they may contract or relax appropriately so that balance is maintained.

Dynamic Equilibrium

The three **semicircular canals** function to detect motion of the head, and they aid in balancing the head and body when they are moved suddenly. These canals lie at right angles to each other, and each occupies a different plane in space (fig. 10.8).

Suspended in the perilymph of the bony portion of each semicircular canal is a membranous canal that ends in a swelling, called an **ampulla.**

The sensory organs of the semicircular canals are located within the ampullae. Each of these organs, called a **crista ampullaris,** contains a number of sensory hair cells and supporting cells (fig. 10.13). As in a macula, the hair cells have hairs that extend upward into a dome-shaped gelatinous mass, called the *cupula.*

Figure 10.13
A crista ampullaris is located within the ampulla of each semicircular canal.

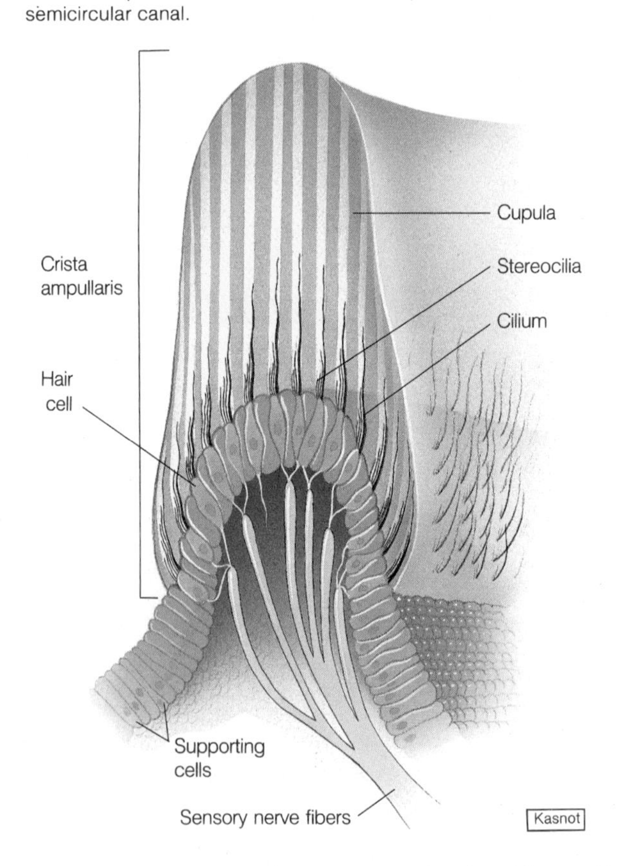

The hair cells of the crista are stimulated by rapid turns of the head or body. At such times, the semicircular canals move with the head or body, but the fluid

Figure 10.14
A sagittal section of the eyelids and anterior portion of the eye.

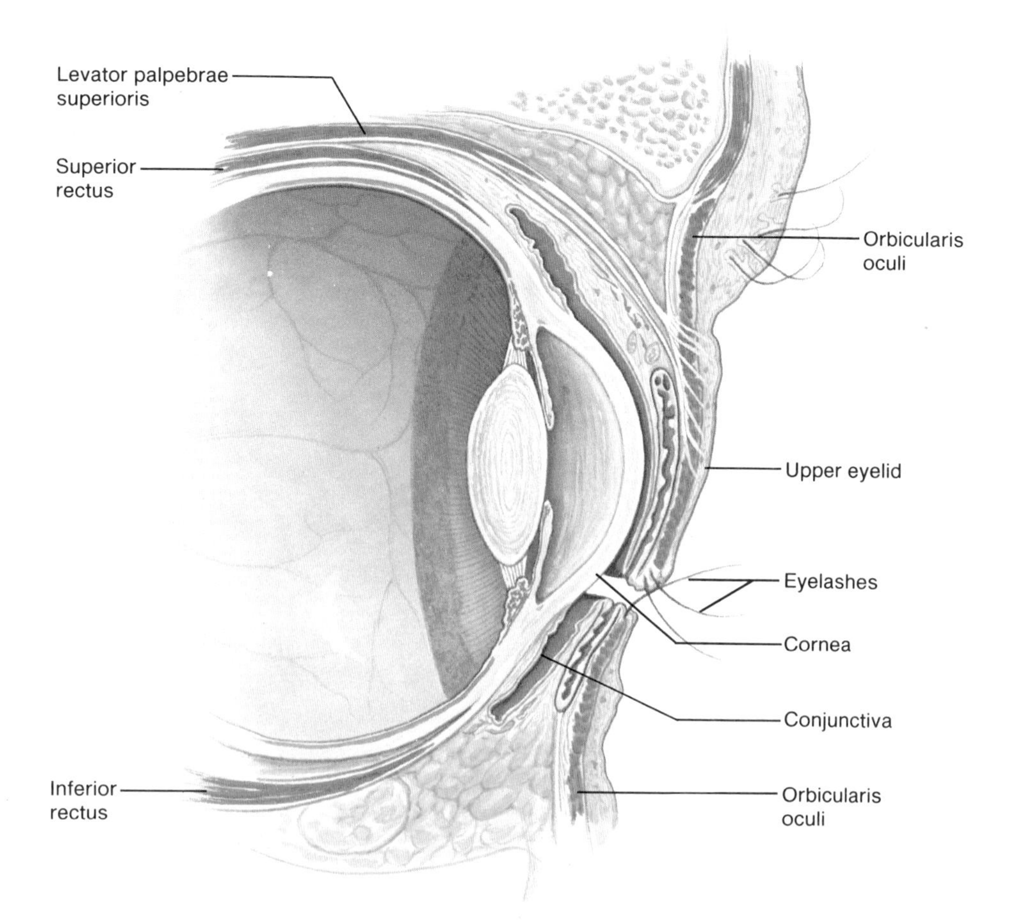

inside the membranous canals tends to remain stationary. This causes the cupula in one or more of the canals to bend in a direction opposite to that of the head or body movement, and the hairs embedded in it also bend. This stimulates the hair cells to signal their associated nerve fibers, and as a result, impulses travel to the brain.

Parts of the cerebellum are particularly important in interpreting impulses from the semicircular canals. Analysis of such information allows the brain to predict the consequences of rapid body movements, and by stimulating skeletal muscles appropriately, the brain can prevent loss of balance.

Other sensory structures also aid in maintaining equilibrium. For example, certain mechanoreceptors (proprioceptors), particularly those associated with the joints of the neck, supply the brain with information concerning the position of the body parts. In addition, the eyes can detect changes in posture that result from body movements. Such visual information is so important that even though a person has suffered damage to the organs of equilibrium, he or she may be able to maintain normal balance by keeping the eyes open and performing body movements slowly.

1. Distinguish between the senses of static and dynamic equilibrium.
2. What structures provide the sense of static equilibrium? Of dynamic equilibrium?
3. How does sensory information from other receptors help maintain equilibrium?

Sense of Sight

Although the eye contains the visual receptors, its functions are assisted by a number of *accessory organs*. These include the eyelids and lacrimal apparatus, which help protect the eye, and a set of extrinsic muscles, which move it.

Visual Accessory Organs

The eye, lacrimal gland, and extrinsic muscles of the eye are housed within the pear-shaped orbital cavity of the skull (fig. 7.8). This orbit, which is lined with the periosteums of various bones, also contains fat, blood vessels, nerves, and a variety of connective tissues.

Each **eyelid** is composed of four layers—skin, muscle, connective tissue, and conjunctiva (fig. 10.14).

The skin of the eyelid, which is the thinnest skin of the body, covers the lid's outer surface and fuses with its inner lining near the margin of the lid.

The eyelids are moved by the *orbicularis oculi* muscle (fig. 8.16), which acts as a sphincter and closes the lids when it contracts, and by the *levator palpebrae superioris* muscle, which raises the upper lid and thus helps open the eye (fig. 10.14).

The **conjunctiva** is a mucous membrane that lines the inner surfaces of the eyelids and folds back to cover the anterior surface of the eyeball, except for its central portion (cornea).

The *lacrimal apparatus* consists of the **lacrimal gland,** which secretes tears, and a series of *ducts,* which carry the tears into the nasal cavity (fig. 10.15). The gland is located in the orbit and secretes tears continuously. The tears pass out through tiny tubules and flow downward and medially across the eye.

The tears are collected by two small ducts (the superior and inferior canaliculi) and flow first into the *lacrimal sac,* which lies in a deep groove of the lacrimal bone, and then into the *nasolacrimal duct,* which empties into the nasal cavity.

The secretion of the lacrimal gland keeps the surface of the eye and the lining of the lids moist and lubricated. Also, tears contain an enzyme *(lysozyme)* that functions as an antibacterial agent, reducing the chance of eye infections.

The **extrinsic muscles** of the eye arise from the bones of the orbit and are inserted by broad tendons on the eye's tough outer surface. There are six such muscles that function to move the eye in various directions. Although any given eye movement may involve more than one of them, each muscle is associated with one primary action. Figure 10.16 illustrates the locations of these extrinsic muscles, and chart 10.2 lists their functions.

When one eye deviates from the line of vision, the person may have double vision (diplopia). If this condition persists, there is danger that changes will occur in the brain to suppress the image from the deviated eye. As a result, the turning eye may become blind (suppression amblyopia). Such monocular blindness often can be prevented if the eye deviation is treated early in life with exercises, glasses, and surgery.

1. Explain how the eyelid is moved.
2. Describe the conjunctiva.
3. What is the function of the lacrimal apparatus?

Structure of the Eye

The eye is a hollow, spherical structure about 2.5 centimeters in diameter. Its wall has three distinct layers—

Figure 10.15
The lacrimal apparatus consists of a tear-secreting gland and a series of ducts.

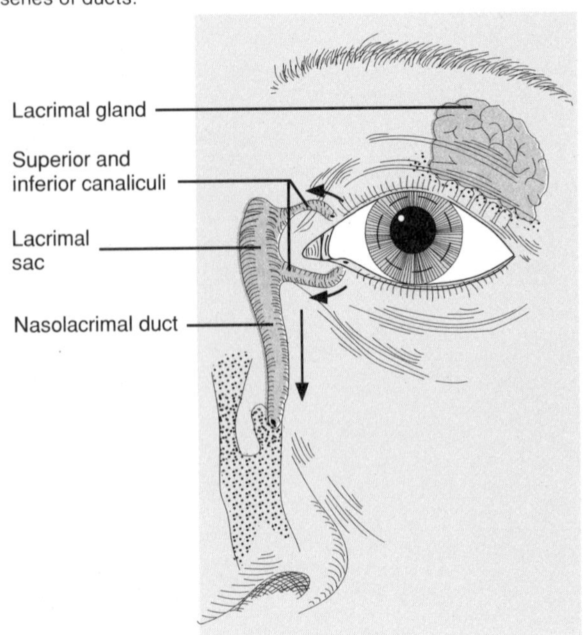

Chart 10.2 Muscles associated with the eyelids and eyes

Name	Innervation	Function
Muscles of the eyelids		
Orbicularis oculi	Facial nerve (VII)	Closes the eye
Levator palpebrae superioris	Oculomotor nerve (III)	Opens the eye
Extrinsic muscles of the eyes		
Superior rectus	Oculomotor nerve (III)	Rotates the eye upward and toward the midline
Inferior rectus	Oculomotor nerve (III)	Rotates the eye downward and toward the midline
Medial rectus	Oculomotor nerve (III)	Rotates the eye toward the midline
Lateral rectus	Abducens nerve (VI)	Rotates the eye away from the midline
Superior oblique	Trochlear nerve (IV)	Rotates the eye downward and away from the midline
Inferior oblique	Oculomotor nerve (III)	Rotates the eye upward and away from the midline

Figure 10.16
The extrinsic muscles of the eye.

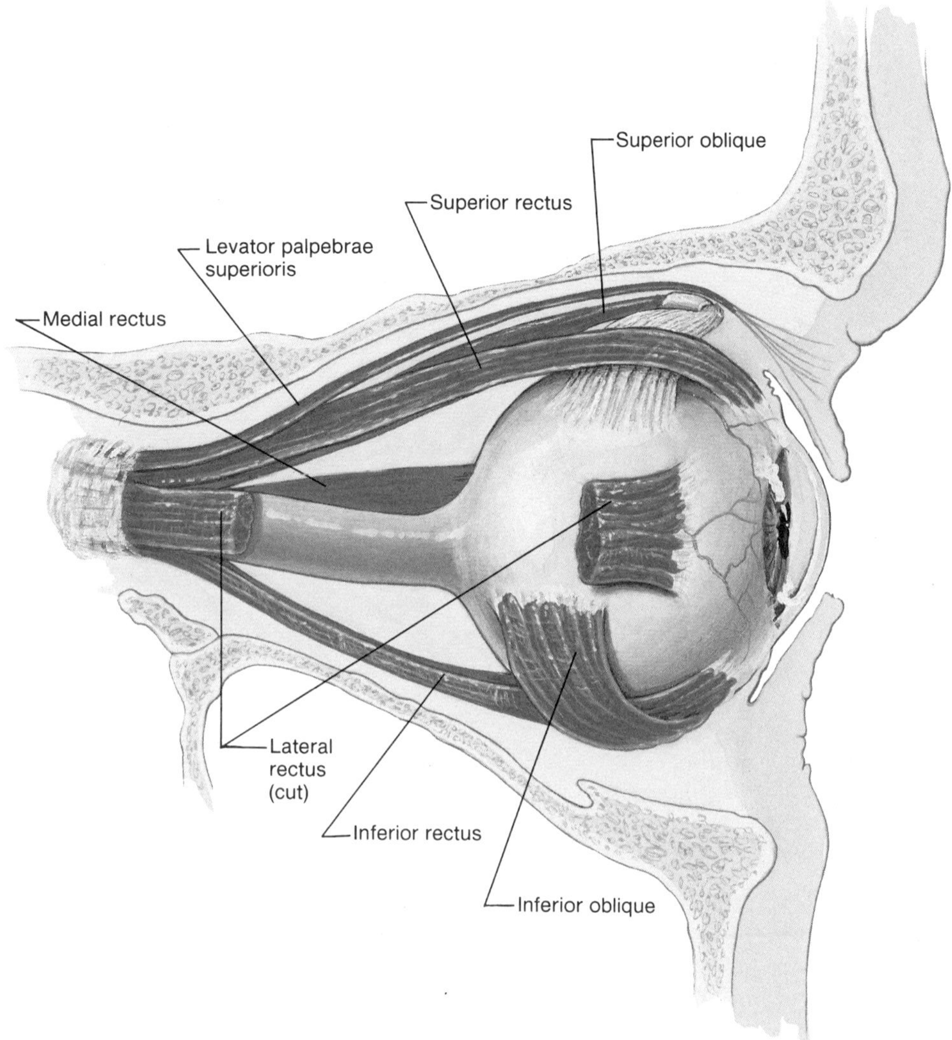

an outer *fibrous tunic,* a middle *vascular tunic,* and an inner *nervous tunic.* The spaces within the eye are filled with fluids that provide support for its wall and internal parts and help maintain its shape. Figure 10.17 shows the major parts of the eye.

Outer Tunic

The anterior sixth of the outer tunic bulges forward as the transparent **cornea,** which serves as the window of the eye and helps focus entering light rays. It is composed largely of connective tissue with a thin layer of epithelium on its surface. The transparency of the cornea is due to the fact that it contains relatively few cells and no blood vessels, and its cells and collagenous fibers are arranged in unusually regular patterns.

Along its circumference, the cornea is continuous with the **sclera,** the white portion of the eye. This part makes up the posterior five-sixths of the outer tunic and is opaque due to the presence of many large, haphazardly arranged, collagenous and elastic fibers. The sclera provides protection and serves as an attachment for the extrinsic muscles. In the back of the eye, the sclera is pierced by the **optic nerve** and certain blood vessels.

Worldwide, the most common cause of blindness is corneal disease in which the transparency of the cornea is lost. Treatment of this condition may involve *corneal transplantation* (penetrating keratoplasty). In this procedure, the central two-thirds of the defective cornea is removed and replaced by a similar-sized portion of cornea from a donor eye. Because corneal tissues lack blood vessels, the transplanted tissue is usually not rejected by the recipient's immune system (see chapter 16), and the success rate of the procedure is very high. Unfortunately for many persons in need of corneal transplantation, the supply of donor eyes remains inadequate.

Middle Tunic

The middle tunic includes the choroid coat, ciliary body, and iris (fig. 10.17).

Figure 10.17
Transverse section of the eye.

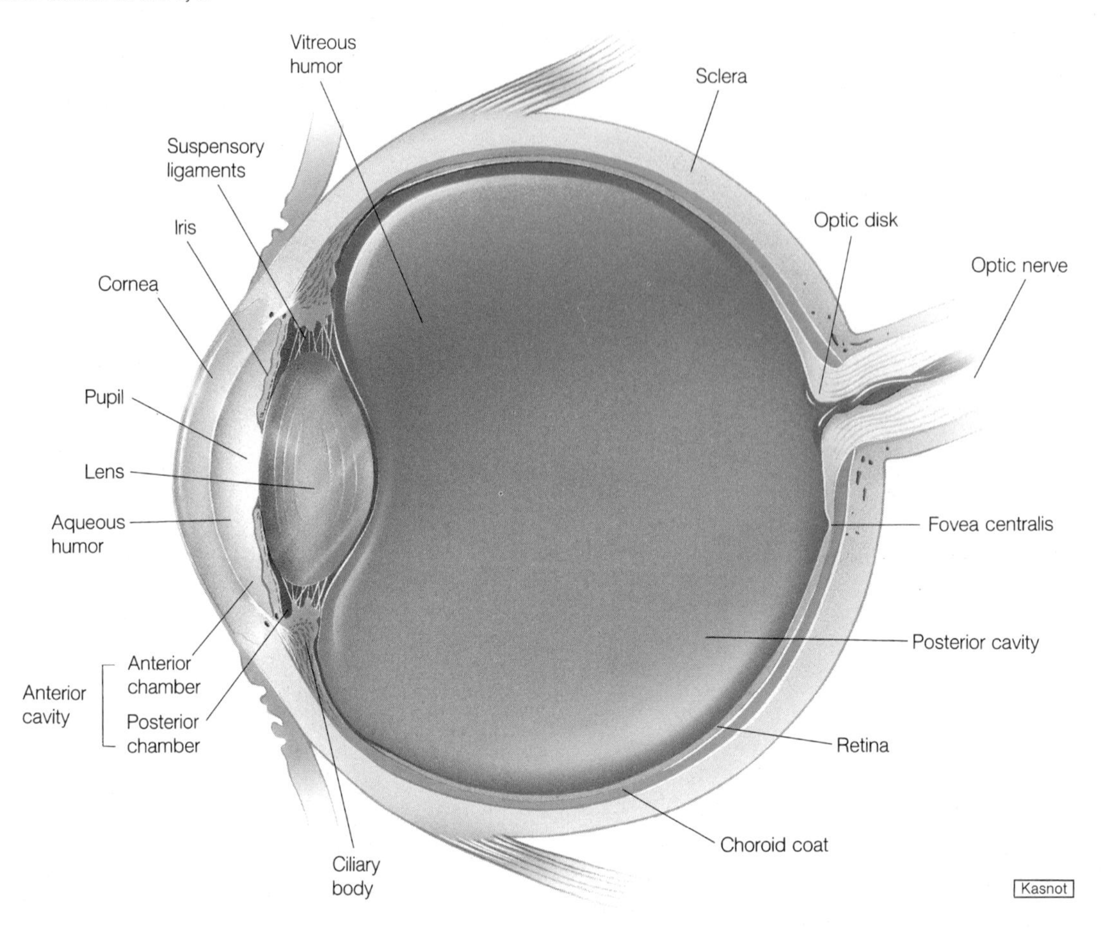

The **choroid coat,** in the posterior five-sixths of the globe of the eye, is loosely joined to the sclera and is honeycombed with blood vessels, which provide nourishment to the surrounding tissues. The choroid also contains numerous pigment-producing melanocytes. The melanin of these cells absorbs excess light and, thus, helps to keep the inside of the eye dark.

The **ciliary body,** which is the thickest part of the middle tunic, extends forward from the choroid and forms an internal ring around the front of the eye. Within the ciliary body, there are many radiating folds, called *ciliary processes,* and groups of muscle fibers, which constitute the *ciliary muscles.*

The transparent **lens** is held in position by a large number of strong but delicate fibers, called *suspensory ligaments,* that extend inward from the ciliary processes (fig. 10.18). The distal ends of these fibers are attached along the margin of a thin capsule that surrounds the lens. The body of the lens lies directly behind the iris and is composed of "fibers" that arise from epithelial cells. In fact, the cytoplasm of these cells makes up the transparent substance of the lens.

Figure 10.18
The lens and ciliary body viewed from behind.

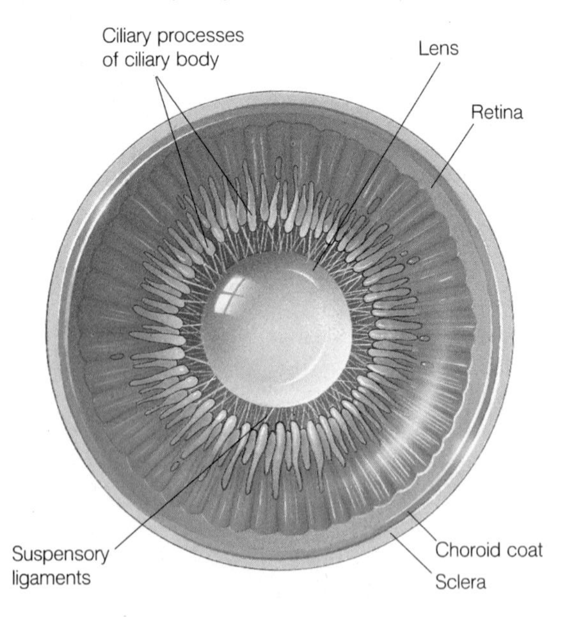

A relatively common eye disorder, particularly in older people, is called *cataract.* In this condition, the lens or its capsule slowly loses its transparency and becomes cloudy and opaque. In time, the person may become blind.

Cataract often is treated by surgically removing the lens and replacing it with an artificial one (intraocular lens implant). Sometimes the loss of refractive power following removal of the lens may be corrected with eyeglasses or contact lenses.

The lens capsule is a clear, membranelike structure composed largely of intercellular material. It is quite elastic, and this quality keeps it under constant tension. As a result, the lens can assume a globular shape. The suspensory ligaments attached to the margin of the capsule are also under tension, however, and as they pull outward, the capsule and the lens inside are kept somewhat flattened (fig. 10.19).

If the tension on the suspensory ligaments is relaxed, the elastic capsule rebounds, and the lens surface becomes more convex. This change occurs in the lens when it is focused to view a close object. This adjustment is called **accommodation.**

Relaxation of the suspensory ligaments during accommodation is a function of the ciliary muscles. For example, one set of these muscle fibers extends back from fixed points in the sclera to the choroid coat. When the fibers contract, the choroid is pulled forward and the ciliary body is shortened. This action causes the suspensory ligaments to become relaxed, and the lens thickens in response (fig. 10.19). In this thickened state, the lens is focused for viewing objects closer than before. To focus on more distant objects, the ciliary muscles are relaxed, tension on the suspensory ligaments increases, and the lens becomes thinner again.

1. Describe the outer and middle tunics of the eye.
2. What factors contribute to the transparency of the cornea?
3. How does the shape of the lens change during accommodation?

The **iris** is a thin diaphragm, composed largely of connective tissue and smooth muscle fibers. It is seen from the outside as the colored portion of the eye. The iris extends forward from the periphery of the ciliary body and lies between the cornea and lens (fig. 10.17). The iris divides the space separating these parts into an *anterior chamber* (between the cornea and iris) and a *posterior chamber* (between the iris and vitreous body and occupied by the lens).

The epithelium on the inner surface of the ciliary body secretes a watery fluid, called **aqueous humor,** into the posterior chamber. The fluid circulates from this chamber through the **pupil,** a circular opening in the center of the iris, and into the anterior chamber. Aqueous humor fills the space between the cornea and lens, helps to nourish these parts, and aids in maintaining the shape of the front of the eye. It subsequently leaves the anterior chamber through veins and a special drainage canal (canal of Schlemm) located in its wall.

Figure 10.19
(*a*) How is focus of the eye affected when the lens becomes thin as the ciliary muscle fibers relax? (*b*) How is focus affected when the lens thickens as the ciliary muscle fibers contract?

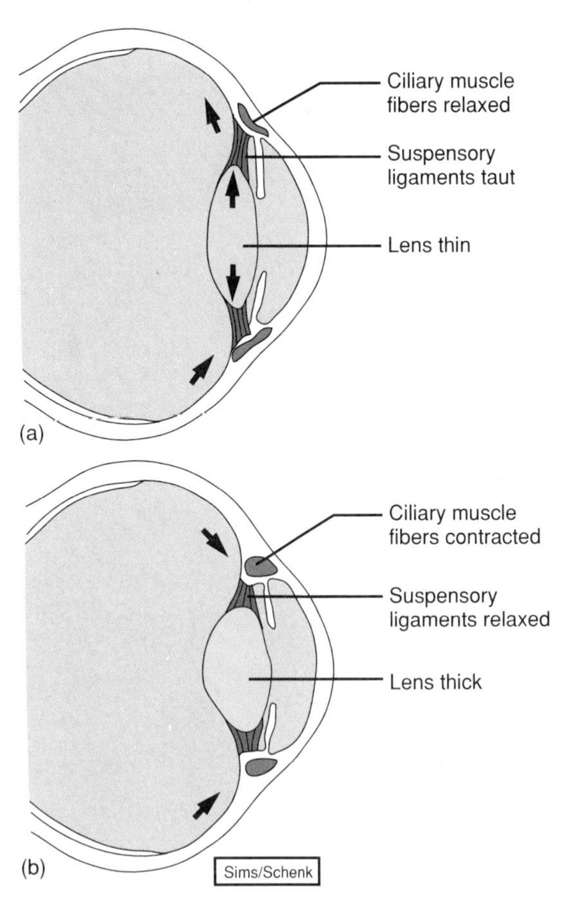

A disorder called *glaucoma* sometimes develops in the eyes as a person ages. This condition occurs when the rate of aqueous humor formation exceeds the rate of its removal. As a result, fluid accumulates in the anterior chamber of the eye, and the fluid pressure rises.

Because liquids cannot be compressed, increasing pressure from the anterior chamber is transmitted to all parts of the eye, and in time the blood vessels that supply the receptor cells of the retina may be squeezed closed. If this happens, cells that fail to receive needed nutrients and oxygen may die, and a form of permanent blindness may result.

When it is diagnosed early, glaucoma can usually be treated successfully with drugs, laser therapy, or surgery that promote the outflow of aqueous humor. However, since glaucoma in its early stages typically produces no symptoms, discovery of the condition depends largely on measurement of the intraocular pressure using an instrument called a *tonometer.*

Figure 10.20
The retina consists of several cell layers.

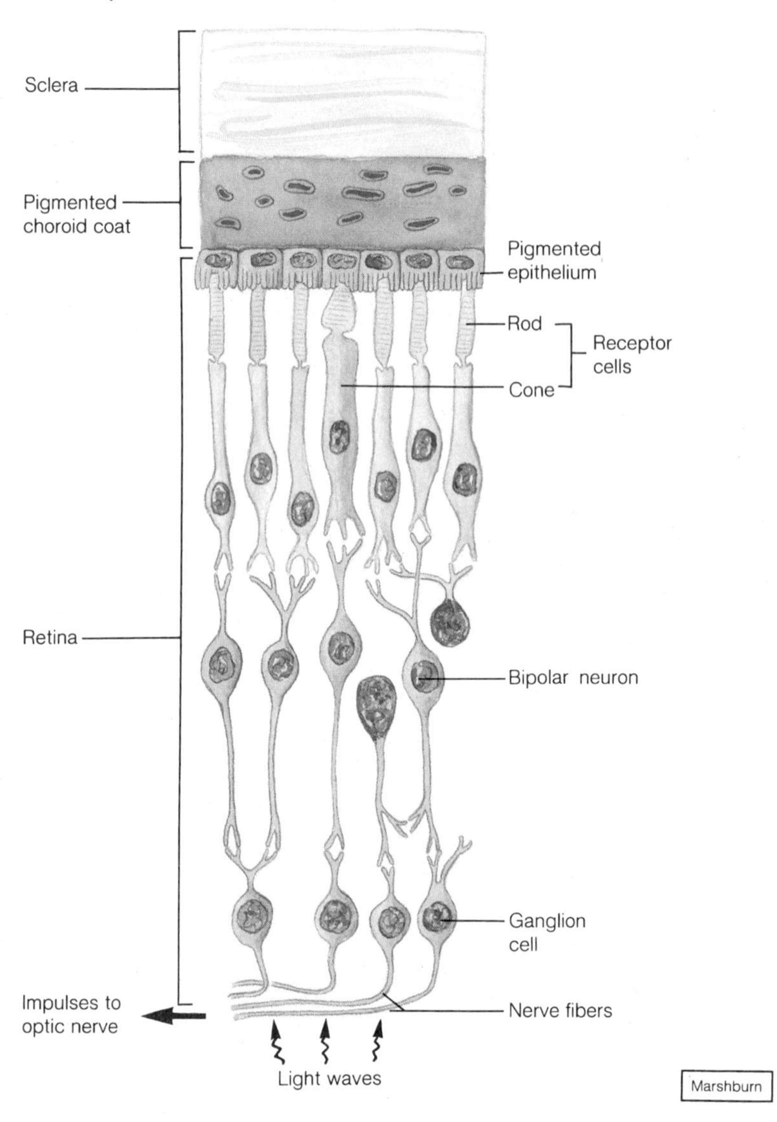

The smooth muscle fibers of the iris are arranged into two groups, a circular set and a radial set. These muscles control the size of the pupil, which is the opening that light passes through as it enters the eye. The circular set of muscle fibers acts as a sphincter. When it contracts, the pupil gets smaller, and the intensity of the light entering decreases. When the radial muscle fibers contract, the diameter of the pupil increases, and the intensity of the light entering increases.

Inner Tunic

The inner tunic consists of the **retina,** which contains the visual receptor cells (photoreceptors). This nearly transparent sheet of tissue is continuous with the optic nerve in the back of the eye and extends forward as the inner lining of the eyeball. It ends just behind the margin of the ciliary body.

Although the retina is thin and delicate, its structure is quite complex. It has a number of distinct layers, as figures 10.20 and 10.21 illustrate.

In the central region of the retina there is a yellowish spot (macula lutea), which has a depression in its center, called **fovea centralis** (fig. 10.17). This depression is in the region of the retina that produces the sharpest vision.

Just medial to the fovea centralis is an area called the **optic disk** (fig. 10.22). Here nerve fibers from the retina leave the eye and become parts of the optic nerve. A central artery and vein also pass through the optic disk. These vessels are continuous with the capillary networks of the retina, and together with vessels in the underlying choroid coat, they supply blood to the cells of the inner tunic. Because there are no receptor cells in the region of the optic disk, it is commonly referred to as the *blind spot* of the eye.

Figure 10.21
Note the layers of cells and nerve fibers in this light micrograph of the retina (×80).

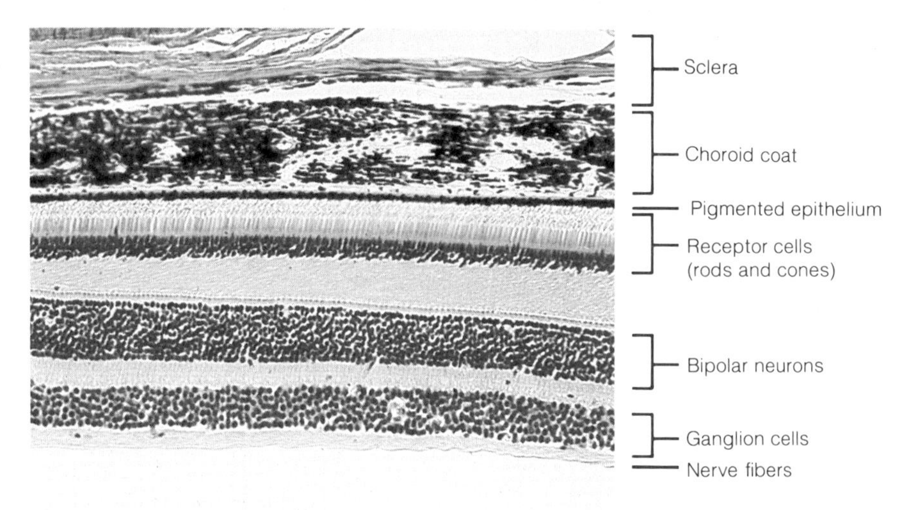

Figure 10.22
Nerve fibers leave the eye in the area of the optic disk (*arrow*) to form the optic nerve.

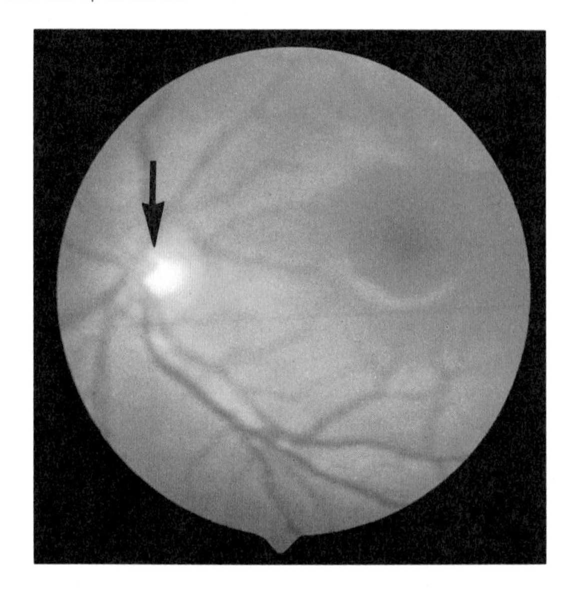

The space bounded by the lens, ciliary body, and retina is the largest compartment of the eye and is called the *posterior cavity* (fig. 10.17). It is filled with a transparent, jellylike fluid, called **vitreous humor,** which together with some collagenous fibers comprise the *vitreous body*. The vitreous body supports the internal parts of the eye and helps maintain its shape.

1. Explain the origin of aqueous humor, and trace its path through the eye.
2. How is the size of the pupil regulated?
3. Describe the structure of the retina.

Figure 10.23
A lens with a convex surface causes light waves to converge.

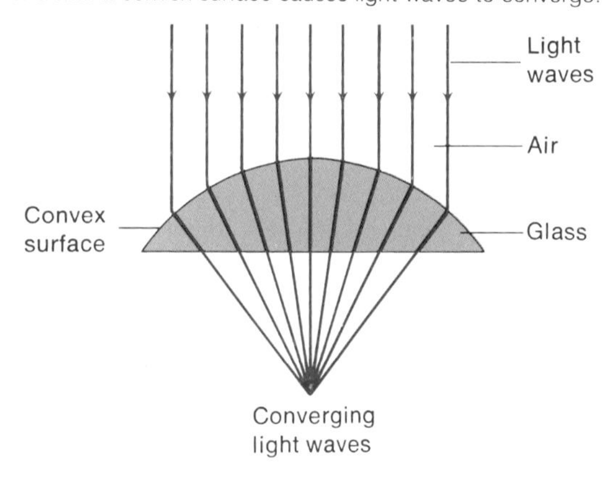

Refraction of Light

When a person sees something, either the object is giving off light or light waves are being reflected from it. These waves enter the eye, and an image of what is seen becomes focused upon the retina. This focusing process involves a bending of light waves—a phenomenon called **refraction.**

Refraction occurs when light waves pass at an oblique angle from a medium of one optical density into a medium of a different optical density. For example, when light passes obliquely from a less dense medium such as air into a denser medium such as glass or from air into the cornea of the eye, light is bent toward a line perpendicular to the surface between these substances. When the surface between such refracting media is curved, a lens is formed. A lens with a *convex* surface causes light waves to converge (fig. 10.23).

When light arrives from objects outside the eye, light waves are refracted primarily by the convex surface of the cornea. Then, light is refracted again by the convex surface of the lens and, to a lesser extent, by the surfaces of the fluids within the chambers of the eye.

If the shape of the eye is normal, light waves are focused sharply upon the retina, much as a motion picture image is focused on a screen for viewing. Unlike the motion picture image, however, the one formed on the retina is upside down and reversed from left to right. When the visual cortex of the cerebrum interprets such an image, it somehow corrects this, and things are seen in their proper positions.

1. What is meant by refraction?
2. What parts of the eye provide refracting surfaces?

Visual Receptors

Visual receptor cells are modified neurons, and there are two distinct kinds, as illustrated in figure 10.20. One group of receptor cells have long, thin projections at their terminal ends and are called **rods.** The cells of the other group have short, blunt projections and are called **cones.**

Instead of being located in the surface layer of the retina, the rods and cones are found in a deep portion, closely associated with a layer of pigmented epithelium (fig. 10.21). Projections from these receptors, which are loaded with visual pigments, extend into the pigmented layer.

The epithelial pigment of the retina absorbs light waves that are not absorbed by the receptor cells, and together with the pigment of the choroid coat, it keeps light from reflecting off the surfaces inside the eye.

Visual receptors are stimulated only when light reaches them. Thus, when a light image is focused on an area of the retina, some receptors are stimulated, and impulses travel away from them to the brain. The impulse leaving each activated receptor, however, provides only a fragment of the information needed for the brain to interpret a total scene.

Rods and cones function differently. For example, the rods are hundreds of times more sensitive to light than the cones, and as a result, rods enable persons to see in relatively dim light. In addition, rods produce colorless vision, but the cones can detect colors.

Still another difference involves visual acuity—the sharpness of the images perceived. The cones allow persons to see sharp images, but the rods enable them to see more general outlines of objects. This characteristic is related to the fact that nerve fibers from many rods may converge, and their impulses may be transmitted to the brain on the same nerve fiber (fig. 10.24). Thus, if a point of light stimulates a rod, the brain cannot tell which one of many receptors has been stimulated. Such a convergence of impulses occurs to a much lesser degree among the cones, so that when a cone is stimulated, the brain is able to pinpoint the stimulation more accurately (fig. 10.24).

As was mentioned, the area of sharpest vision is the fovea centralis (fig. 10.17). This area lacks rods, but contains densely packed cones with few or no converging fibers. Also, the overlying layers of the retina, as well as the retinal blood vessels, are displaced to the sides in the fovea. This displacement more fully exposes the receptors to incoming light. Consequently, to view something in detail a person moves the eye so that the important part of an image falls upon the fovea centralis.

Visual Pigments

Both the rods and cones contain light-sensitive pigments, which decompose when they absorb light energy. The light-sensitive substance in the rods is called **rhodopsin** (visual purple). In the presence of light, rhodopsin molecules break down into molecules of a colorless protein, called *opsin,* and a yellowish substance, called *retinal* (retinene), which is synthesized from vitamin A.

When rhodopsin molecules decompose, an enzyme is activated that initiates a complex series of reactions. As a result of these reactions, the permeability of the rod cell membrane is altered, and a nerve impulse is triggered. The impulse travels away from the retina, along the optic nerve, and into the brain.

In bright light, nearly all of the rhodopsin in the rods of the retina is decomposed, and the sensitivity of these receptors is greatly reduced. In dim light, however, rhodopsin can be regenerated from opsin and retinal faster than it is broken down. This regeneration process requires cellular energy, which is provided by the energy-carrying molecules of ATP (see chapter 4).

Many people, particularly children, suffer from vitamin A deficiency due to improper diet. In such cases, the quantity of retinal available for manufacture of rhodopsin may be reduced, and consequently, the sensitivity of the rods may be low. This condition, called *night blindness,* is characterized by poor vision in dim light. Fortunately, the problem is usually easy to correct by adding vitamin A to the diet or providing vitamin A by injection.

The light-sensitive pigments of the cones are similar to rhodopsin in that they are composed of retinal combined with a protein; the protein, however, differs from the protein in the rods. In fact, there are three different sets of cones within the retina, each containing an abundance of one of three different visual pigments.

Figure 10.24
(*a*) Impulses from several rods may be transmitted to the brain on a single sensory nerve fiber. (*b*) Impulses from cones are often transmitted to the brain on separate nerve fibers. (*c*) A scanning electron micrograph of a rod and a cone.

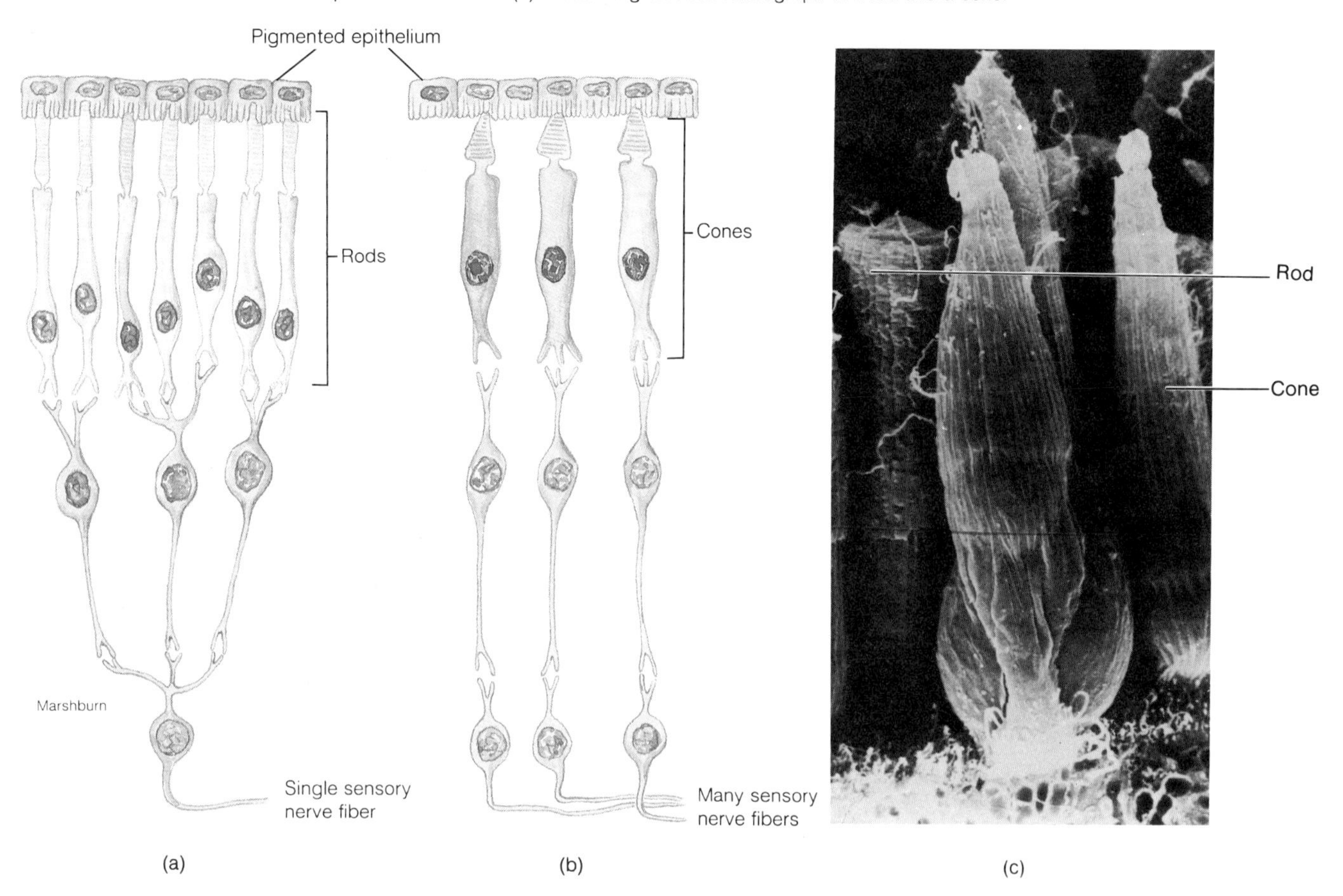

The wavelength of a particular kind of light determines the color perceived from it. For example, the shortest wavelengths of visible light are perceived as violet, and the longest wavelengths of visible light are perceived as red. As far as the cone pigments are concerned, one type (erythrolabe) is most sensitive to red light waves, another (chlorolabe) to green light waves, and a third (cyanolabe) to blue light waves. The color a person perceives depends upon which set of cones or combination of sets is stimulated by the light in a given image. If all three sets of cones are stimulated, the person senses the light as white, and if none are stimulated, the person senses black.

Visual Nerve Pathway

The axons of the retinal neurons leave the eyes to form the *optic nerves* (fig. 10.25). Just anterior to the pituitary gland, these nerves give rise to the X-shaped *optic chiasma,* and within the chiasma, some of the fibers cross over. More specifically, the fibers from the nasal (medial) half of each retina cross over, but those from the temporal (lateral) sides do not. Thus, the fibers from the nasal half of the left eye and the temporal half of the right eye form the *right optic tract;* and the fibers from the nasal half of the right eye and the temporal half of the left eye form the *left optic tract.*

The nerve fibers continue in the optic tracts, and just before they reach the thalamus, a few of them leave to enter the nuclei that function in various visual reflexes. Most of the fibers, however, enter the thalamus and synapse in its posterior portion (lateral geniculate body). From this region, the visual impulses enter nerve pathways called *optic radiations,* which lead to the visual cortex of the occipital lobes.

1. Distinguish between the rods and cones of the retina.
2. Explain the roles of visual pigments.
3. Trace a nerve impulse from the retina to the visual cortex.

Clinical Terms Related to the Senses

amblyopia (am″ble-o′pe-ah)—dimness of vision due to a cause other than a refractive disorder or a lesion.
anopia (an-o′pe-ah)—absence of the eye.

Figure 10.25
The visual pathway includes the optic nerve, optic chiasma, optic tract, and optic radiations.

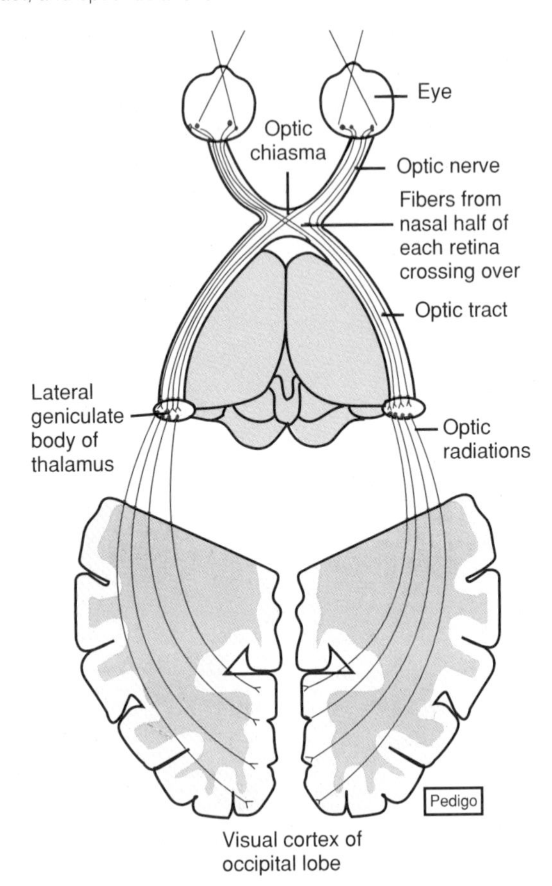

audiometry (aw″de-om′ĕ-tre)—the measurement of auditory acuity for various frequencies of sound waves.
blepharitis (blef″ah-ri′tis)—an inflammation of the margins of the eyelids.
causalgia (kaw-zal′je-ah)—a persistent, burning pain usually associated with injury to a limb.
conjunctivitis (kon-junk″tĭ-vi′tis)—an inflammation of the conjunctiva.
diplopia (dĭ-plo′pe-ah)—double vision, or the sensation of seeing two objects when only one is viewed.
emmetropia (em″ĕ-tro′pe-ah)—the normal condition of the eyes; eyes with no refractive defects.
enucleation (e-nu″kle-a′shun)—removal of the eyeball.
exophthalmos (ek″sof-thal′mos)—a condition in which the eyes protrude abnormally.
hemianopsia (hem″e-an-op′se-ah)—defective vision affecting half of the visual field.
hyperalgesia (hi″per-al-je′ze-ah)—an abnormally increased sensitivity to pain.
iridectomy (ir″ĭ-dek′to-me)—the surgical removal of part of the iris.
iritis (i-ri′tis)—an inflammation of the iris.
keratitis (ker″ah-ti′tis)—an inflammation of the cornea.
labyrinthectomy (lab″ĭ-rin-thek′to-me)—the surgical removal of the labyrinth.
labyrinthitis (lab″ĭ-rin-thi′tis)—an inflammation of the labyrinth.
Meniere's disease (men″e-erz′ dĭ-zēz)—an inner ear disorder, characterized by ringing in the ears, increased sensitivity to sounds, dizziness, and loss of hearing.
neuralgia (nu-ral′je-ah)—pain resulting from an inflammation of a nerve or a group of nerves.
neuritis (nu-ri′tis)—an inflammation of a nerve.
nystagmus (nis-tag′mus)—an involuntary oscillation of the eyes.
otitis media (o-ti′tis me′de-ah)—an inflammation of the middle ear.
otosclerosis (o″to-skle-ro′sis)—a formation of spongy bone in the inner ear, which often causes deafness by fixing the stapes to the oval window.
pterygium (tĕ-rij′e-um)—an abnormally thickened patch of conjunctiva that extends over part of the cornea.
retinitis pigmentosa (ret″ĭ-ni′tis pig″men-to′sa)—a progressive retinal sclerosis characterized by deposits of pigment in the retina and by atrophy of the retina.
retinoblastoma (ret″ĭ-no-blas-to′mah)—an inherited, highly malignant tumor arising from immature retinal cells.
tinnitus (tĭ-ni′tus)—a ringing or buzzing noise in the ears.
tonometry (to-nom′ĕ-tre)—the measurement of fluid pressure within the eyeball.
trachoma (trah-ko′mah)—a bacterial disease of the eye, characterized by conjunctivitis, which may lead to blindness.
tympanoplasty (tim″pah-no-plas′te)—the surgical reconstruction of the middle ear bones and the establishment of continuity from the tympanic membrane to the oval window.
uveitis (u″ve-i′tis)—an inflammation of the uvea, the region of the eye that includes the iris, ciliary body, and the choroid coat.
vertigo (ver′tĭ-go)—a sensation of dizziness.

Chapter Summary

Introduction (page 252)

Sensory receptors are sensitive to changes occurring in their surroundings.

Receptors and Sensations (page 252)

1. Types of receptors
   a. Each type of receptor is most sensitive to a distinct type of stimulus.
   b. The major types of receptors include chemoreceptors, pain receptors, thermoreceptors, mechanoreceptors, and photoreceptors.
2. Sensations
   a. Sensations are feelings resulting from sensory stimulation.
   b. A particular part of the sensory cortex interprets every impulse reaching it in the same way.
   c. The cerebral cortex projects a sensation back to the region of stimulation.
3. Sensory adaptations are adjustments made by sensory receptors to continuous stimulation in which impulses leave them at lower and lower rates.

Somatic Senses (page 252)

Somatic senses are those involved with receptors in the skin, muscles, joints, and visceral organs.

1. Touch and pressure senses
   a. The free ends of sensory nerve fibers are receptors for the sensations of touch and pressure.
   b. Meissner's corpuscles are receptors for the sensation of light touch.
   c. Pacinian corpuscles are receptors for the sensation of heavy pressure.
2. The temperature receptors include two sets of free nerve endings that serve as heat and cold receptors.
3. Sense of pain
   a. The pain receptors are free nerve endings that are stimulated by tissue damage.
   b. The pain receptors are the only receptors in the visceral organs that provide sensations.
   c. The sensations produced from the visceral receptors are likely to feel as if they were coming from some other part.
   d. Visceral pain may be referred because the sensory impulses from the skin and visceral organs travel on common nerve pathways.
   e. Pain nerve fibers
      (1) The two main types of pain fibers are acute pain fibers and chronic pain fibers.
      (2) Acute pain fibers are fast conducting; chronic pain fibers are slower conducting.
      (3) Pain impulses are processed in the gray matter of the spinal cord and ascend to the brain.
      (4) Within the brain, pain impulses pass through the reticular formation before being conducted to the cerebral cortex.
   f. Regulation of pain impulses
      (1) Awareness of pain occurs when impulses reach the thalamus.
      (2) The cerebral cortex judges the intensity of pain and locates its source.
      (3) Impulses descending from the brain cause neurons to release pain-relieving neuropeptides, such as enkephalins.

Special Senses (page 256)

Special senses are those whose receptors occur in relatively large, complex sensory organs of the head.

Sense of Smell (page 256)

1. Olfactory receptors
   a. The olfactory receptors are chemoreceptors that are stimulated by chemicals dissolved in liquid.
   b. Olfactory receptors function together with taste receptors and aid in food selection.
2. Olfactory organs
   a. The olfactory organs consist of the receptors and supporting cells in the nasal cavity.
   b. The olfactory receptors are neurons with cilia.
3. Nerve impulses travel from the olfactory receptors through the olfactory nerves, olfactory bulbs, and olfactory tracts to interpreting centers in the temporal and frontal lobes of the cerebrum.
4. Olfactory stimulation
   a. Olfactory impulses may result when various gaseous molecules combine with specific sites on the cilia of the receptor cells.
   b. The olfactory receptors adapt rapidly.

Sense of Taste (page 257)

1. Taste receptors
   a. Taste buds consist of receptor cells and supporting cells.
   b. The taste cells have taste hairs.
   c. Taste hair surfaces have receptor sites to which chemicals combine.
2. Taste sensations
   a. The four primary taste sensations are sweet, sour, salty, and bitter.
   b. Various taste sensations result from the stimulation of one or more sets of taste receptors.
3. Taste nerve pathway
   a. Sensory impulses from the taste receptors travel on fibers of the facial, glossopharyngeal, and vagus nerves.
   b. These impulses are carried to the medulla and then ascend to the thalamus, from which they travel to the gustatory cortex in the parietal lobes.

Sense of Hearing (page 259)

1. The external ear collects sound waves created by vibrating objects.
2. Middle ear
   a. The auditory ossicles of the middle ear conduct sound waves from the tympanic membrane to the oval window of the inner ear.
   b. The auditory tubes connect the middle ears to the throat and help maintain equal air pressure on both sides of the eardrums.
3. Inner ear
   a. The inner ear consists of a complex system of interconnected tubes and chambers—the osseous and membranous labyrinths.
   b. The organ of Corti contains the hearing receptors, which are stimulated by vibrations in the fluids of the inner ear.
   c. Different frequencies of vibrations stimulate different sets of receptor cells.
4. Auditory nerve pathway
   a. The auditory nerves carry impulses to the auditory cortices of the temporal lobes.
   b. Some auditory nerve fibers cross over, so that impulses arising from each ear are interpreted on both sides of the brain.

Sense of Equilibrium (page 263)

1. Static equilibrium is concerned with maintaining the stability of the head and body when they are motionless.
2. Dynamic equilibrium is concerned with balancing the head and body when they are moved or rotated suddenly.

3. Other parts that help maintain equilibrium include the eyes and mechanoreceptors associated with certain joints.

Sense of Sight (page 265)

1. The visual accessory organs include the eyelids, lacrimal apparatus, and extrinsic muscles of the eyes.
2. Structure of the eye
   a. The wall of the eye has an outer, a middle, and an inner layer, which function as follows:
      (1) The outer layer (sclera) is protective, and its transparent anterior portion (cornea) refracts light entering the eye.
      (2) The middle layer (choroid coat) is vascular and contains pigments.
      (3) The inner layer (retina) contains the visual receptor cells.
   b. The lens is a transparent, elastic structure whose shape is controlled by the action of the ciliary muscles.
   c. The iris is a muscular diaphragm that controls the amount of light entering the eye.
   d. Spaces within the eye are filled with fluids that help to maintain its shape.
3. Refraction of light
   a. Light waves are refracted primarily by the cornea and lens.
   b. The lens must be thickened to focus on close objects.
4. Visual receptors
   a. The visual receptors are called rods and cones.
   b. The rods are responsible for colorless vision in relatively dim light, and the cones are responsible for color vision.
   c. Visual pigments
      (1) A light-sensitive pigment in the rods decomposes in the presence of light and triggers a complex series of reactions that initiate nerve impulses.
      (2) Color vision is related to the presence of three sets of cones containing different light-sensitive pigments.
5. Visual nerve pathway
   a. Nerve fibers from the retina form the optic nerves.
   b. Some fibers cross over in the optic chiasma.
   c. Most of the fibers enter the thalamus and synapse with others that continue to the visual cortex in the occipital lobes.

Clinical Application of Knowledge

1. How would you explain the following observation? A person enters a tub of water and reports that it is uncomfortably warm, yet a few moments later says the water feels comfortable, even though the water temperature remains unchanged.
2. How would you explain the fact that some serious injuries, such as those produced by a bullet entering the abdomen, may be relatively painless, but others, such as those involving crushing of the skin, may produce considerable discomfort?
3. Labyrinthitis is a condition in which the tissues of the inner ear are inflamed. What symptoms would you expect to observe in a patient with this disorder?
4. A patient with heart disease experiences pain at the base of the neck and in the left shoulder and arm during exercise. How would you explain the probable origin of this pain to the patient?

Review Activities

1. List five groups of sensory receptors, and name the kind of change to which each is sensitive.
2. Define *sensation.*
3. Explain what is meant by the projection of a sensation.
4. Define *sensory adaptation,* and provide an example of this phenomenon.
5. Describe the functions of free nerve endings, Meissner's corpuscles, and pacinian corpuscles.
6. Define *referred pain,* and provide an example of this phenomenon.
7. Explain why pain may be referred.
8. Describe the olfactory organ and its function.
9. Trace a nerve impulse from an olfactory receptor to the interpreting center of the cerebrum.
10. Explain how the salivary glands aid the function of the taste receptors.
11. Name the four primary taste sensations.
12. Trace the pathway of a taste impulse from a receptor to the cerebral cortex.
13. Distinguish between the external, middle, and inner ears.
14. Trace the path of a sound wave from the tympanic membrane to the hearing receptors.
15. Describe the functions of the auditory ossicles.
16. Explain the function of the auditory tube.
17. Distinguish between the osseous and membranous labyrinths.
18. Describe the cochlea and its function.
19. Describe a hearing receptor.
20. Explain how a hearing receptor stimulates a sensory neuron.
21. Trace a nerve impulse from the organ of Corti to the interpreting centers of the cerebrum.
22. Describe the organs of static and dynamic equilibrium and their functions.
23. List the visual accessory organs, and describe the functions of each organ.
24. Name the three layers of the eye wall, and describe the functions of each layer.

25. Describe how accommodation is accomplished.
26. Explain how the iris functions.
27. Distinguish between the aqueous humor and vitreous humor.
28. Distinguish between the fovea centralis and optic disk.
29. Explain how light waves are focused on the retina.
30. Distinguish between the rods and cones.
31. Explain why cone vision is generally more acute than rod vision.
32. Describe the function of rhodopsin.
33. Describe the relationship between light wavelengths and color vision.
34. Trace a nerve impulse from the retina to the visual cortex.

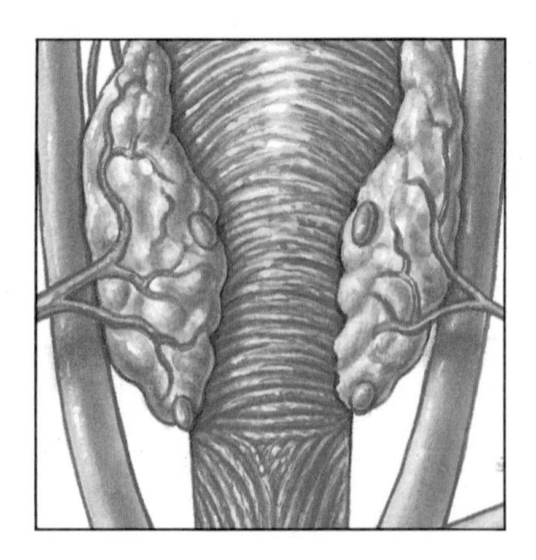

11 Endocrine System

The endocrine system consists of a variety of loosely related cells, tissues, and organs that act together with parts of the nervous system to control body activities and maintain homeostasis. The endocrine and nervous systems each provide a means by which the body parts can communicate with one another and adjust to changing needs. Whereas the parts of the nervous system communicate with various cells by means of nerve impulses carried on nerve fibers, the parts of the endocrine system use hormones, which are carried in the body fluids and act as chemical messengers to their target cells.

Chapter Objectives

After you have studied this chapter, you should be able to

1. Distinguish between endocrine and exocrine glands.
2. Explain how steroid and nonsteroid hormones produce effects on target cells.
3. Discuss how hormonal secretions are regulated by negative feedback mechanisms.
4. Explain how hormonal secretions may be controlled by the nervous system.
5. Name and describe the location of the major endocrine glands, and list the hormones they secrete.
6. Describe the general functions of the hormones secreted by endocrine glands.
7. Explain how the secretion of each hormone is regulated.
8. Complete the review activities at the end of this chapter. Note that the items are worded in the form of specific learning objectives. You may want to refer to them before reading the chapter.

Key Terms

adrenal cortex (ah-dre′nal kor′teks)

adrenal medulla (ah-dre′nal me-dul′ah)

anterior pituitary (an-ter′e-or pi̇-tu′i-tar″e)

hormone (hor′mōn)

negative feedback (neg′ah-tiv fēd′bak)

pancreas (pan′kre-as)

parathyroid gland (par″ah-thi′roid gland)

pineal gland (pin′e-al gland)

posterior pituitary (pos-ter′e-or pi̇-tu′i-tar″e)

prostaglandin (pros″tah-glan′din)

target cell (tar′get sel)

thymus gland (thi′mus gland)

thyroid gland (thi′roid gland)

Aids to Understanding Words

-crin, to secrete: endo*crine*—pertaining to internal secretions.

diuret-, to pass urine: *diuret*ic—a substance that promotes the production of urine.

endo-, within: *endo*crine gland—a gland that releases its secretion internally into a body fluid.

exo-, outside: *exo*crine gland—a gland that releases its secretion to the outside through a duct.

hyper-, above: *hyper*thyroidism—a condition resulting from an above normal secretion of thyroid hormone.

hypo-, below: *hypo*thyroidism—a condition resulting from a below normal secretion of thyroid hormone.

para-, beside: *para*thyroid glands—a set of glands located on the surface of the thyroid gland.

toc-, birth: oxy*toc*in—a hormone that stimulates the uterine muscles to contract during childbirth.

-tropic, influencing: adrenocortico*tropic* hormone—a hormone secreted by the anterior pituitary gland that stimulates the adrenal cortex.

Introduction

THE TERM *ENDOCRINE* describes the cells, tissues, and organs that secrete hormones into the body fluids. By contrast, the term *exocrine* refers to those parts whose secretions are carried by tubes or ducts to some internal or external body surface. Thus, the thyroid and parathyroid glands, which secrete hormones into the blood are endocrine glands (ductless glands), and the sweat glands and salivary glands are exocrine glands (see chapter 6).

General Characteristics of the Endocrine System

As a group, endocrine glands and their hormones help regulate metabolic processes. They control the rates of certain chemical reactions, aid in the transport of substances through membranes, and help regulate water and electrolyte balances. They also play vital roles in the reproductive processes and in development and growth.

Although many hormones have localized effects, those described in this chapter have widespread actions and are produced by the larger endocrine glands. These glands include the pituitary gland, thyroid gland, parathyroid glands, adrenal glands, and pancreas (fig. 11.1). Several other hormone-secreting glands and tissues, such as those involved in the processes of digestion and reproduction, are discussed in subsequent chapters.

Hormones and Their Actions

A **hormone** is an organic substance secreted by a cell that has an effect on the functions of another cell. Such substances are released into the extracellular spaces surrounding the hormone-secreting cells. Some hormones travel only short distances and produce their effects in nearby cells (paracrine secretion). Others are transported in the blood to all parts of the body and may produce general effects. In either case, the physiological action of a particular hormone is restricted to its *target cells*—those cells that possess specific receptors for the hormone molecules. In other words, a hormone's target cells possess receptors that other cells lack.

Hormones are very potent and, thus, can stimulate changes in target cells even though they are present in extremely low concentrations. Chemically, most hormones are either amines, peptides, proteins, or glycoproteins that are synthesized from amino acids, or they are steroids or steroidlike substances that are synthesized from cholesterol (chart 11.1).

Figure 11.1
Locations of the major endocrine glands.

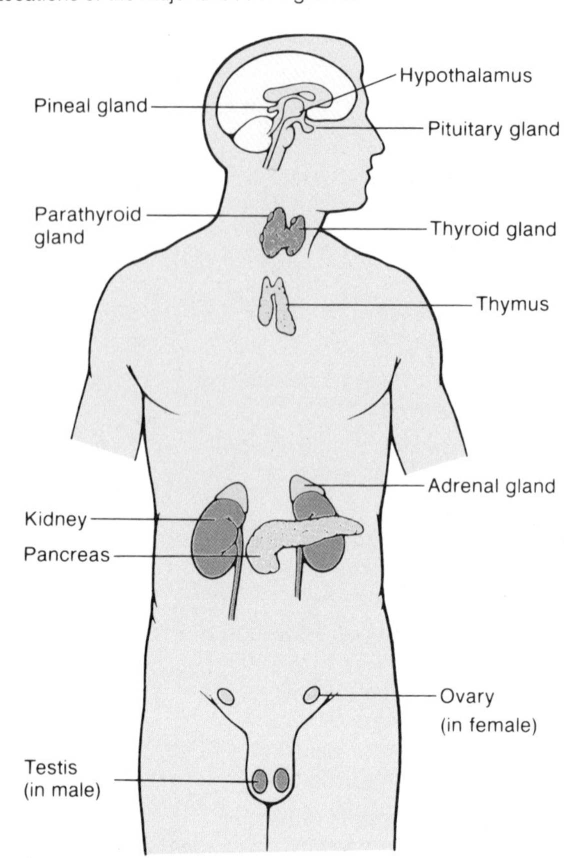

Chart 11.1 Types of hormones

Type of compound	Formed from	Examples
Amines	Amino acids	Norepinephrine, epinephrine
Peptides	Amino acids	Antidiuretic hormone, oxytocin, thyrotropin-releasing hormone
Protein	Amino acids	Parathyroid hormone, growth hormone, prolactin
Glycoprotein	Protein and carbohydrate	Follicle-stimulating hormone, luteinizing hormone, thyroid-stimulating hormone
Steroid	Cholesterol	Estrogen, testosterone, aldosterone, cortisol

1. Distinguish between an endocrine and an exocrine gland.
2. Describe the general function of the endocrine system.
3. What is a hormone?

Figure 11.2
(*a*) A steroid hormone passes through a cell membrane and (*b*) combines with a protein receptor in the nucleus. (*c*) The steroid-protein complex activates the synthesis of messenger RNA. (*d*) The messenger RNA leaves the nucleus and (*e*) functions in manufacture of protein molecules.

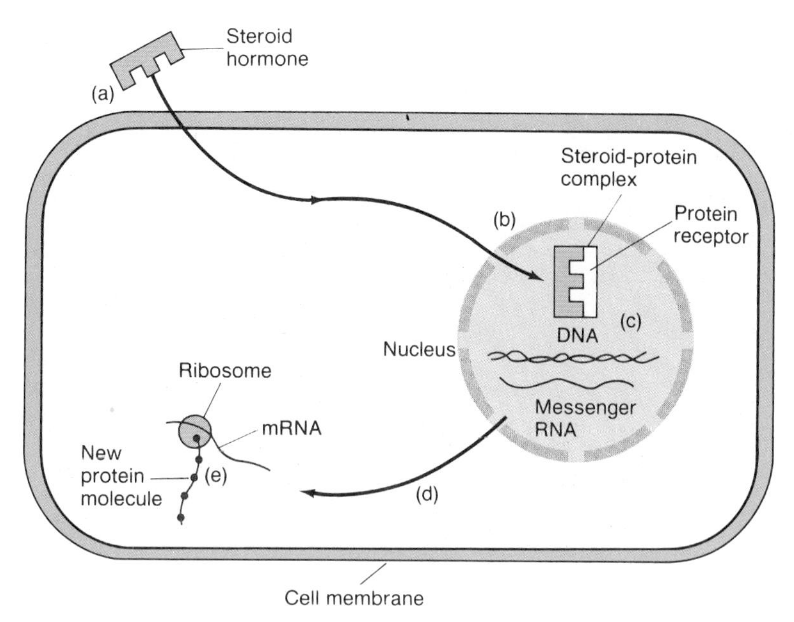

Actions of Hormones

Hormones exert their effects by altering various metabolic processes. For example, they may change the rate at which proteins or enzymes are synthesized, the rates of enzyme activities, or the rates at which molecules are transported through cell membranes.

Steroid Hormones

Steroids are compounds whose molecules contain complex rings of carbon and hydrogen atoms. The difference between one type of steroid and another is due to the kinds and numbers of atoms attached to these rings and the ways they are joined.

Steroid hormones, unlike amines, peptides, and proteins, are soluble in the lipids that make up the bulk of cell membranes (fig. 11.2). For this reason, steroid molecules can enter cells relatively easily by diffusion. Once inside of a target cell, they may combine with specific nuclear protein molecules—the receptors. The steroid-protein complex thus formed binds to particular regions of the target cell's DNA molecules, and activates certain genes. The activated genes, in turn, cause the synthesis of particular kinds of messenger RNA (mRNA) molecules.

As discussed in chapter 4, a messenger RNA molecule can leave the nucleus and enter the cytoplasm where it functions in the manufacture of specific proteins. Thus, steroid hormones influence their target cells by causing special proteins to be synthesized—proteins that act as enzymes and alter the rates of cellular processes, act as parts of membrane transport systems, or cause enzymes to be activated or inhibited.

Nonsteroid Hormones

Nonsteroid hormones, such as amines, peptides, and proteins, usually act by combining with specific receptor molecules located in target cell membranes (fig. 11.3). Each receptor molecule is a protein that has a *binding site* and an *activity site*. A hormone delivers its message to its target cell by uniting with the binding site of a receptor. This combination causes the receptor's activity site to interact with other membrane proteins. As a result, the actions of membrane-bound enzymes or membrane transport mechanisms are altered, causing changes in the concentrations of other cellular substances. These substances serve as "second messengers" in the hormonal stimulation mechanism because they, in turn, induce cellular changes, which appear as responses to the original hormone.

The second messenger associated with one group of hormones is *cyclic adenosine monophosphate* (cAMP). In this mechanism, a hormone binds to its receptor, and the resulting hormone-receptor complex activates an enzyme called *adenylate cyclase,* which is bound to the inside of the cell membrane. Activated adenylate cyclase causes ATP molecules within the cytoplasm to become molecules of cAMP. The cAMP, in turn, activates another set of enzymes called *protein kinases*. Protein kinases bring about the transfer of phosphate groups from ATP molecules to various protein substrate molecules. This action (phosphorylation) alters the shapes of the substrate molecules and converts some of them from inactive forms into active ones. The activated protein molecules then induce changes in various cellular processes (fig. 11.3). Thus,

Figure 11.3
(*a*) Nonsteroid hormone molecules reach the target cell by means of body fluids and (*b*) combine with receptor sites on the cell membrane. (*c*) As a result, molecules of adenylate cyclase are activated and (*d*) cause the change of ATP into cyclic AMP. (*e*) Cyclic AMP promotes a series of reactions leading to various cellular changes.

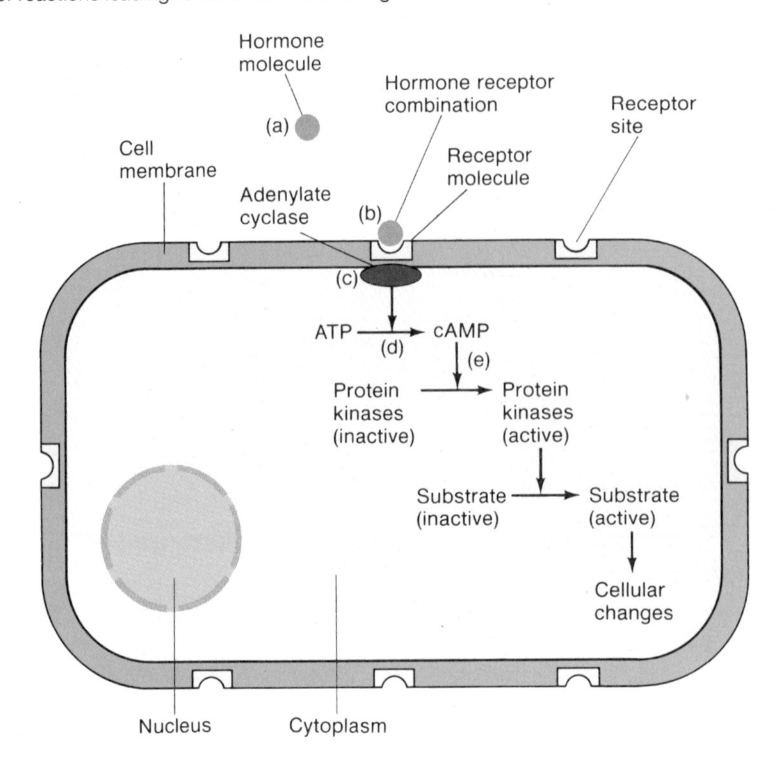

the response of any particular cell to such a hormone is determined not only by the type of membrane receptors present, but also by the kinds of protein substrate molecules that the cell contains.

Cellular responses to this second messenger mechanism include altering membrane permeabilities, activating various enzymes, promoting the synthesis of certain proteins, stimulating or inhibiting specific metabolic pathways, promoting cellular movements, and initiating the secretion of hormones or other substances.

Cyclic AMP produced by the mechanism is quickly inactivated by still another enzyme (phosphodiesterase) so that the action of cAMP is short-lived. For this reason, a continuing response within a target cell depends upon a continuing signal produced by the hormone molecules combining with the target cell's membrane receptors.

Prostaglandins

Another group of substances, called **prostaglandins,** also have regulating effects on cells. These substances are lipids and are synthesized from the fatty acids found in the cell membranes. They occur in a variety of cells, including those of the liver, kidneys, heart, lungs, thymus gland, pancreas, brain, and various reproductive organs.

Like hormones, prostaglandins are very potent compounds and are present in very small quantities. They are not stored in cells, but instead are synthesized just before they are released and are then inactivated rapidly.

Prostaglandins produce a variety of effects. For example, some can cause smooth muscles in the airways of the lungs and in the blood vessels to relax. Prostaglandins also can cause smooth muscles in the walls of the uterus and intestines to contract. They stimulate the secretions of hormones from the adrenal cortex and inhibit the secretion of hydrochloric acid from the wall of the stomach. They also influence the movements of sodium ions and water molecules in the kidneys, help regulate blood pressure, and have a powerful effect on both male and female reproductive physiology.

1. How does a steroid hormone promote cellular changes? A nonsteroid hormone?
2. Explain what is meant by a "second messenger."
3. What are prostaglandins?
4. What kinds of effects do prostaglandins produce?

Figure 11.4
An example of a negative feedback system: (*1*) gland *A* secretes a hormone that stimulates gland *B* to increase the secretion of another hormone; (*2*) the hormone from gland *B* inhibits the activity of gland *A*.

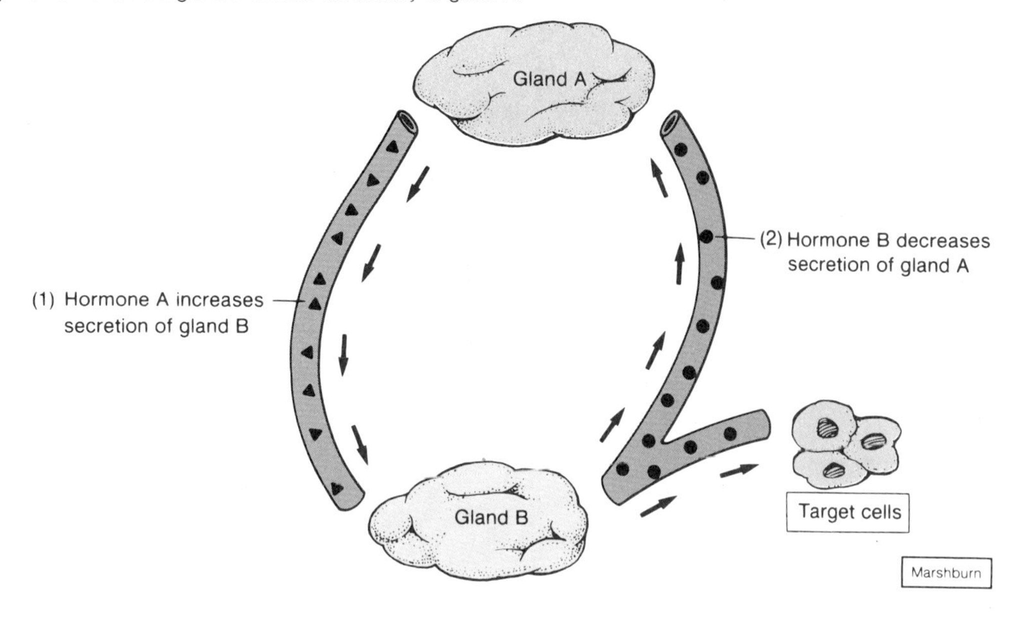

Figure 11.5
As a result of negative feedback systems, some hormone concentrations remain relatively stable, although they may fluctuate slightly above and below the average concentration.

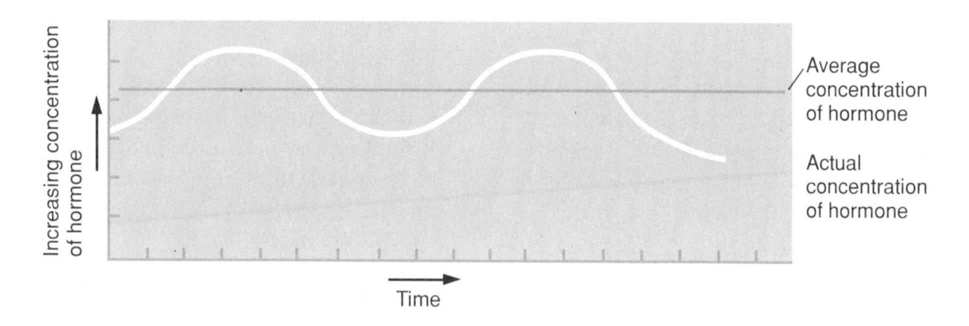

Control of Hormonal Secretions

Since hormones are very potent substances, their release by endocrine cells must be regulated precisely so that the amounts released are balanced with the amounts used by other body cells. One mechanism that functions to control hormonal secretion is a *feedback system.*

Negative Feedback Systems

In a feedback system, a body part continuously receives information (feedback) concerning the cellular processes it controls. The part receives this information in the form of signals, and if conditions change, the part can act upon the information and adjust its activity.

Commonly, the control of hormonal secretions involves a **negative feedback system** (fig. 11.4). In such a system, an endocrine gland is sensitive either to the concentration of a substance it regulates or to the concentration of a product from a process it controls. Whenever this concentration reaches a certain level, the endocrine gland is inhibited (a negative effect) and its secretory activity decreases. Then, as the concentration of the gland's hormone drops, the concentration of the regulated substances drops also, and inhibition of the gland ceases. When the gland is no longer inhibited, it begins to secrete its hormone again.

As a result of such negative feedback systems, the concentrations of some hormones remain relatively stable, although they may fluctuate slightly within normal ranges (fig. 11.5).

Nerve Control

Another type of control mechanism involves the nervous system. For example, some endocrine glands, such

Figure 11.6
The pituitary gland is attached to the hypothalamus and lies in the sella turcica of the sphenoid bone.

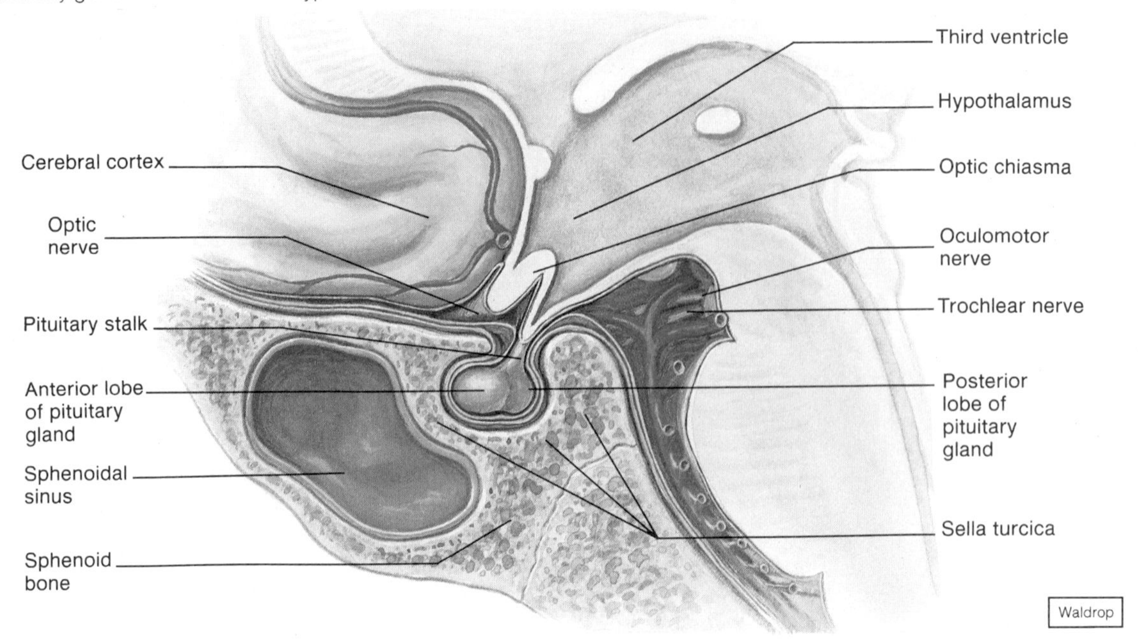

as the adrenal medulla, secrete hormones in response to nerve impulses coming from various parts of the nervous system.

Still another control system involves an interaction between an endocrine gland and the hypothalamus of the brain. In this system, neurosecretory cells in the hypothalamus secrete substances, called **releasing** (or inhibiting) **hormones.** The target cells of these releasing hormones are in the anterior pituitary gland. The anterior pituitary gland responds to a releasing hormone by secreting its own hormone. Then, as the gland's hormone reaches a certain concentration in the body fluids, a negative feedback system inhibits the hypothalamus, and its secretion of the releasing hormone decreases.

1. Describe a negative feedback system.
2. Explain two mechanisms involving the nervous system that help to control hormonal secretions.

Pituitary Gland

The **pituitary gland** (hypophysis) is located at the base of the brain, where it is attached to the hypothalamus by a pituitary stalk (infundibulum). The gland is about 1 centimeter in diameter and consists of two distinct portions—an anterior lobe and a posterior lobe (fig. 11.6).

Most of the pituitary gland's activities are controlled by the brain. For example, the release of hormones from its posterior lobe occurs when nerve impulses from the hypothalamus signal the axon ends of neurosecretory cells in the posterior lobe (fig. 11.7). On the other hand, secretions from the anterior lobe are controlled by releasing hormones produced by the hypothalamus. These releasing hormones are transmitted by blood in the vessels of a capillary net associated with the hypothalamus. These vessels merge to form the **hypophyseal portal veins,** which pass downward along the pituitary stalk and give rise to a capillary net in the anterior lobe. Thus, substances released into the blood from the hypothalamus are carried directly to the anterior lobe.

1. Where is the pituitary gland located?
2. Explain how the hypothalamus controls the actions of the posterior lobe of the pituitary gland. Of the anterior lobe.

Anterior Pituitary Hormones

The anterior lobe of the pituitary gland is enclosed by a capsule of dense collagenous connective tissue and consists largely of epithelial tissue arranged in blocks around many thin-walled blood vessels. Within the epithelium five types of secretory cells have been identified. Four of these each secrete a different hormone—growth hormone (GH), prolactin (PRL), thyroid-stimulating hormone (TSH), and adrenocorticotropic hormone (ACTH). The fifth type of cell secretes follicle-stimulating hormone (FSH) and luteinizing hormone (LH). (Note: In males, luteinizing hormone is known as interstitial stimulating hormone or ISCH.)

Figure 11.7
Axons in the posterior lobe of the pituitary gland are stimulated to release hormones by nerve impulses originating in the hypothalamus; cells of the anterior lobe are stimulated by releasing hormones secreted from hypothalamic neurons.

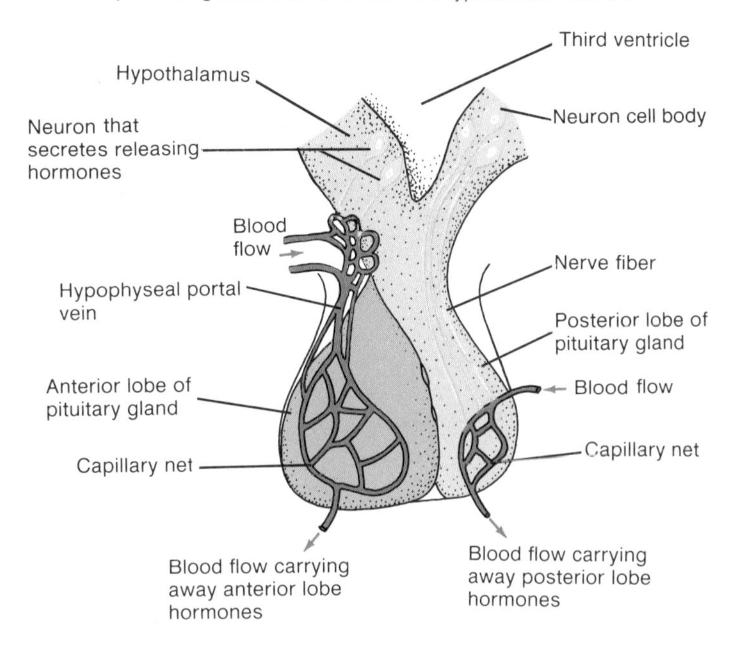

Growth hormone (GH) stimulates the body cells to increase in size and undergo rapid cell division. It also enhances the movement of amino acids through the cell membranes and causes an increase in the rate at which the cells utilize carbohydrates and fats. The hormone's effect on amino acids, however, seems to be the more important one.

The mechanism for controlling growth hormone secretion involves two substances from the hypothalamus (growth hormone-releasing hormone and growth hormone release-inhibiting hormone). A person's nutritional state also seems to play a role in the control of GH. For example, more GH is released during periods of protein deficiency and abnormally low blood glucose concentration. Conversely, when blood protein and glucose concentrations increase, growth hormone secretion decreases.

Prolactin (PRL) promotes milk production. More specifically, it stimulates and sustains the mother's milk production following the birth of an infant. This action is discussed in chapter 19. Prolactin has little effect in males, although excessive secretion can cause a deficiency of male sex hormones.

Thyroid-stimulating hormone (TSH) has as its major function the control of secretions from the thyroid gland, which are described in a subsequent section of this chapter.

TSH secretion is partially regulated by the hypothalamus, which produces *thyrotropin-releasing hormone* (TRH) (fig. 11.8). TSH secretion also is regulated by circulating thyroid hormones that exert an inhibiting effect on the release of TRH and TSH; therefore, as the blood concentration of thyroid hormones increases, the secretions of TRH and TSH are reduced.

If an insufficient amount of growth hormone is secreted during childhood, body growth is limited, and a type of *dwarfism* (hypopituitary dwarfism) results. In this condition, the body parts are usually correctly proportioned and mental development is normal. However, an abnormally low secretion of growth hormone is usually accompanied by lessened secretions of other anterior lobe hormones, leading to additional hormone deficiency symptoms. For example, a hypopituitary dwarf often fails to develop adult sexual features unless hormone therapy is provided.

An oversecretion of growth hormone during childhood may result in *gigantism*—a condition in which the person's height may exceed 8 feet. Gigantism, which is relatively rare, is usually accompanied by a tumor of the pituitary gland. In such cases, various pituitary hormones in addition to GH are likely to be secreted excessively, so that a giant often suffers from a variety of metabolic disturbances.

1. How does growth hormone affect the synthesis of proteins?
2. What is the function of prolactin?
3. How is TSH secretion regulated?

Adrenocorticotropic hormone (ACTH) controls the manufacture and secretion of certain hormones from the outer layer, or *cortex*, of the adrenal gland. These adrenal cortical hormones are discussed in a subsequent section of this chapter.

Figure 11.8
TRH from the hypothalamus stimulates the anterior pituitary gland to release TSH. TSH stimulates the thyroid gland to release hormones, which in turn cause the hypothalamus to reduce its secretion of TRH.

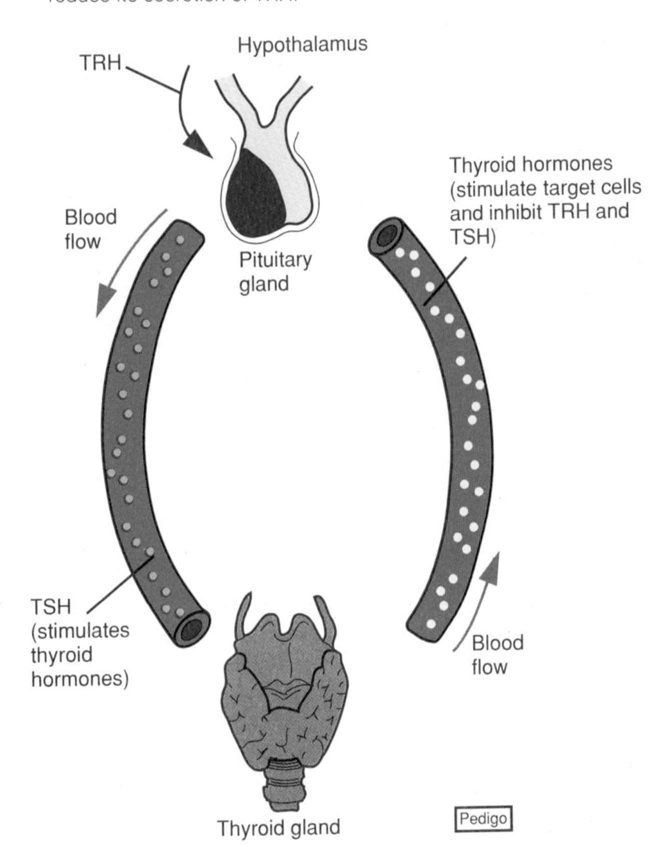

The secretion of ACTH is regulated in part by *corticotropin-releasing hormone* (CRH), which is released from the hypothalamus in response to decreased concentrations of adrenal cortical hormones. Also, various forms of stress result in an increased secretion of ACTH by stimulating the release of CRH.

Follicle-stimulating hormone (FSH) and **luteinizing hormone** (LH) are called *gonadotropins,* which means they exert their actions on the gonads or reproductive organs. The functions of these gonadotropins and the ways they interact with each other are discussed in chapter 19.

1. What is the function of ACTH?
2. What is a gonadotropin?

Posterior Pituitary Hormones

Unlike the anterior lobe of the pituitary gland, which is composed primarily of glandular epithelial cells, the posterior lobe consists largely of nerve fibers and neuroglial cells. The neuroglial cells function to support the nerve fibers, which originate in the hypothalamus.

The two hormones associated with the posterior lobe—antidiuretic hormone (ADH) and oxytocin (OT)—are produced by specialized neurons in the hypothalamus (fig. 11.7). These hormones travel down axons through the pituitary stalk to the posterior lobe and are stored in vesicles (secretory granules) near the ends of the axons. The hormones are then released into the blood in response to nerve impulses coming from the hypothalamus.

A *diuretic* is a substance that increases urine production. An *antidiuretic,* then, is a chemical that decreases urine formation. **Antidiuretic hormone** (ADH) produces an antidiuretic effect by acting on the kidneys and causing them to reduce the amount of water they excrete. In this way, ADH plays an important role in regulating the water concentration of the body fluids.

The secretion of ADH is regulated by the hypothalamus. Certain neurons in this part of the brain, called *osmoreceptors,* are sensitive to changes in the osmotic pressure of body fluids. For example, if a person is dehydrating due to a lack of water intake, the solutes in blood become more and more concentrated. The osmoreceptors can sense the resulting increase in the osmotic pressure, and their signals cause the posterior lobe to release ADH, which is transmitted by the blood to the kidneys. As a result of the effects of ADH on kidney function, less urine is produced. This action conserves water.

If, on the other hand, a person drinks an excessive amount of water, the body fluids become more dilute and the release of ADH is inhibited. Consequently, the kidneys excrete more dilute urine until the water concentration of the body fluids returns to normal.

If any parts of the ADH-regulating mechanism are damaged due to an injury or a tumor, insufficient quantities of the hormone may be synthesized or released. The resulting ADH deficiency produces a condition called *diabetes insipidus.* This condition is characterized by an output of as much as 25 to 30 liters of very dilute urine per day and a rise in the concentration of solutes in the body fluids.

Oxytocin (OT) also has an antidiuretic action, but it is weaker in this respect than ADH. In addition, oxytocin causes contractions of smooth muscles in the uterine wall and plays a role in the later stages of childbirth by stimulating uterine contractions. The mechanism that triggers the release of oxytocin during childbirth is not clearly understood.

Oxytocin has an effect on the breasts, causing contractions in certain cells associated with the milk-producing glands and their ducts. In lactating breasts, this action forces liquid from the milk glands into the milk ducts—an effect that is necessary for breastfeeding.

Chart 11.2 Hormones of the pituitary gland

Anterior lobe		
Hormone	*Action*	*Source of control*
Growth hormone (GH)	Stimulates an increase in the size and rate of reproduction of body cells; enhances the movement of amino acids through membranes	Growth hormone-releasing hormone and growth hormone release-inhibiting hormone from the hypothalamus
Prolactin (PRL)	Sustains milk production after birth	Secretion is restrained by prolactin release-inhibiting factor and stimulated by prolactin-releasing factor from the hypothalamus
Thyroid-stimulating hormone (TSH)	Controls the secretion of hormones from the thyroid gland	Thyrotropin-releasing hormone (TRH) from the hypothalamus
Adrenocorticotropic hormone (ACTH)	Controls the secretion of certain hormones from the adrenal cortex	Corticotropin-releasing hormone (CRH) from the hypothalamus
Follicle-stimulating hormone (FSH)	Responsible for the development of the egg-containing follicles in the ovaries; stimulates the follicular cells to secrete estrogen; in males stimulates the production of sperm cells	Gonadotropin-releasing hormone from the hypothalamus
Luteinizing hormone (LH)	Promotes the secretion of sex hormones; plays a role in the release of an egg cell in females	Gonadotropin-releasing hormone from the hypothalamus
Posterior lobe		
Hormone	*Action*	*Source of control*
Antidiuretic hormone (ADH)	Causes the kidneys to reduce water excretion; in a high concentration, causes the blood pressure to rise	Hypothalamus in response to changes in the blood water concentration
Oxytocin (OT)	Causes contractions of the muscles in the uterine wall; causes the muscles associated with the milk-secreting glands to contract	Hypothalamus in response to stretch in the uterine and vaginal walls and stimulation of the breasts

Chart 11.2 reviews the hormones of the pituitary gland.

1. What is the function of ADH?
2. How is the secretion of ADH controlled?
3. What effects does oxytocin produce in females?

Thyroid Gland

The **thyroid gland** is a very vascular structure that consists of two large lobes connected by a broad isthmus (fig. 11.9). It is located just below the larynx on either side and in front of the trachea.

Structure of the Gland

The thyroid gland is covered by a capsule of connective tissue and is made up of many secretory parts called *follicles.* The cavities within these follicles are lined with a single layer of cuboidal epithelial cells and filled with a clear, viscous substance, called *colloid.* The follicular cells produce and secrete hormones that may either be stored in the colloid or released into the blood of nearby capillaries.

Thyroid Hormones

Two hormones are synthesized by the follicular cells of the thyroid gland. They are **thyroxine** (tetraiodothyronine), which is also known as T4 because its molecule contains four atoms of iodine, and **triiodothyronine,** which is known as T3 because its molecule includes three atoms of iodine. A third hormone, **calcitonin,** is produced by the extrafollicular cells of the gland.

Thyroxine and triiodothyronine have similar actions, although triiodothyronine is three to five times more potent. Actually, thyroxine is thought to act by being converted into triiodothyronine within the liver. These hormones help regulate the metabolism of carbohydrates, lipids, and proteins. For example, they increase the rate at which the cells release energy from carbohydrates; they enhance the rate of protein synthesis; and they stimulate the breakdown and mobilization of lipids. As a result of their actions, these hormones are needed for normal growth and development. They also are essential to the maturation of the nervous system.

Before follicular cells can produce thyroxine and triiodothyronine, they must be supplied with iodine salts (iodides). Such salts are normally obtained from foods, and after they have been absorbed from the intestine,

Figure 11.9
(*a*) The thyroid gland consists of two lobes connected anteriorly by an isthmus. (*b*) Thyroid hormones are secreted by follicular cells.

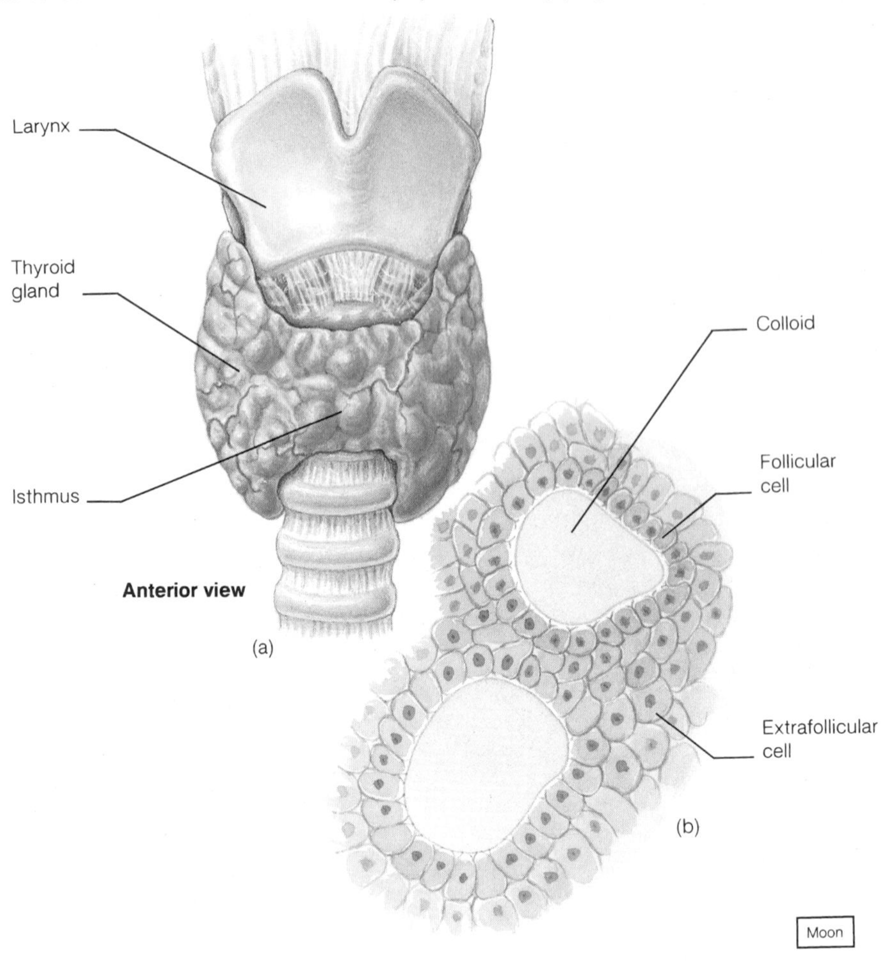

some are carried by the blood to the thyroid gland. An efficient active transport mechanism moves the iodides into the follicular cells, where they are used in the synthesis of the hormones. As was explained previously, the release of these hormones is controlled by the hypothalamus and pituitary gland. Once they are in the blood, thyroxine and triiodothronine combine with blood proteins (plasma proteins) and are transported to body cells.

Calcitonin influences the blood calcium and phosphate concentrations. It functions with parathyroid hormone (PTH) from the parathyroid glands to regulate the concentrations of calcium and phosphate ions.

The release of calcitonin is triggered by an increasing blood concentration of calcium ions or by the release of certain hormones from the digestive system in response to the intake of dietary calcium. The effect of calcitonin is to inhibit the bone-resorbing activity of osteoclasts (see chapter 7) and to increase the excretion of calcium and phosphate ions by the kidneys—actions that cause a lowering of the blood calcium and phosphate concentrations.

Many functional disorders of the thyroid gland are characterized by overactivity (hyperthyroidism) or underactivity (hypothyroidism) of the glandular cells. One form of *hypothyroidism* appears in infants when their thyroid glands fail to function normally. An affected child may appear normal at birth, because it has received an adequate supply of thyroid hormones from its mother during pregnancy. When its own thyroid gland fails to produce sufficient quantities of these hormones, the child soon develops a condition called *cretinism.* Cretinism is characterized by severe symptoms including stunted growth, abnormal bone formation, retarded mental development, low body temperature, and sluggishness. Without treatment within a month or so following birth, the child is likely to suffer from permanent mental retardation.

Hyperthyroidism is characterized by an elevated metabolic rate, restlessness, and overeating. Also, the person's eyes are likely to protrude (exophthalmos) because of edematous swelling in the tissues behind them. At the same time, the thyroid gland is likely to enlarge, producing a bulge in the neck called a *goiter.*

Chart 11.3 Hormones of the thyroid gland

Hormone	Action	Source of control
Thyroxine	Increases the rate at which energy is released from carbohydrates; increases the rate of protein synthesis; accelerates growth; stimulates activity in the nervous system	TSH from the anterior pituitary gland
Triiodothyronine	Same as above, but 3 to 5 times more potent than thyroxine	Same as above
Calcitonin	Lowers the blood calcium and phosphate concentrations by inhibiting the release of calcium and phosphate ions from bones and by increasing the excretion of these ions by the kidneys	Blood calcium concentration

Figure 11.10
The parathyroid glands are embedded in the posterior surface of the thyroid gland.

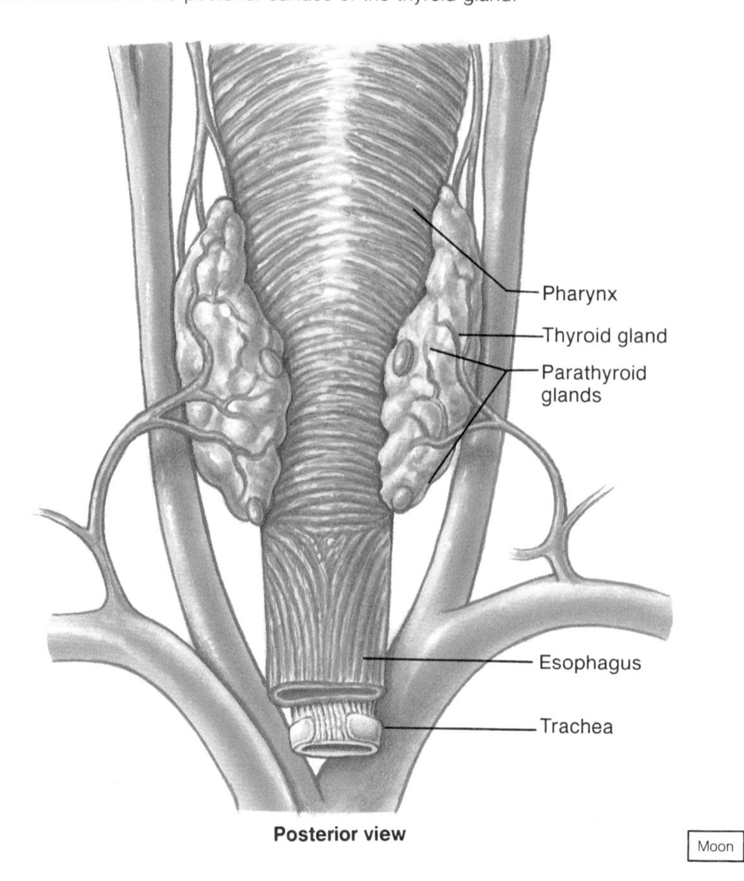

The secretion of calcitonin is thought to be controlled directly by the blood calcium concentration. As this concentration increases, so does the secretion of calcitonin.

Chart 11.3 reviews the actions and controls of the thyroid hormones.

1. Where is the thyroid gland located?
2. What hormones of the thyroid gland affect carbohydrate metabolism and protein synthesis?
3. How does the thyroid gland influence the concentrations of blood calcium and phosphate ions?

Parathyroid Glands

The **parathyroid glands** are located on the posterior surface of the thyroid gland, as shown in figure 11.10. Usually there are four of them—a superior and an inferior gland associated with each of the thyroid's lateral lobes.

Structure of the Glands

Each parathyroid gland is a small, yellowish brown structure, covered by a thin capsule of connective tissue.

The body of the gland consists of numerous tightly packed secretory cells that are closely associated with capillary networks.

Parathyroid Hormone

The hormone secreted by the parathyroid glands is called **parathyroid hormone (PTH).** This substance causes an increase in the blood calcium concentration and a decrease in the blood phosphate concentration. It does this through actions in the bones, kidneys, and intestine.

The intercellular matrix of bone tissue contains a considerable amount of mineral salts, including calcium phosphate (see chapter 7). PTH stimulates bone resorption by osteocytes and osteoclasts and inhibits the activity of osteoblasts. As a result of this increased resorption of bone, calcium and phosphate ions are released, and the blood concentrations of these substances increase (fig. 11.11). At the same time, PTH causes the kidneys to conserve blood calcium and to excrete more phosphate ions in the urine. It also stimulates the absorption of calcium from food in the intestine so that the blood calcium concentration increases still more.

As in the case of calcitonin, secretion of PTH is regulated by a negative feedback mechanism operating between the glands and the blood calcium concentration. As the concentration of blood calcium rises, less PTH is secreted; as the concentration of blood calcium drops, more PTH is released.

To summarize, a stable concentration of blood calcium is maintained by the action of calcitonin, which tends to decrease the blood calcium concentration when it is relatively high, and by the action of PTH, which tends to increase the blood calcium concentration, when it is relatively low.

Hyperparathyroidism may be caused by a tumor associated with a parathyroid gland. The resulting increase in PTH secretion stimulates excessive osteoclastic activity, and as bone tissue is resorbed, the bones become soft, deformed, and subject to spontaneous fractures.

The excessive calcium and phosphate released into the body fluids as a result of increased PTH activity may be deposited in abnormal places, causing new problems such as kidney stones.

Hypoparathyroidism can result from an injury to the parathyroids or from the surgical removal of these glands. In either case, decreased PTH secretion is reflected in reduced osteoclastic activity, and although the bones remain strong, the blood calcium concentration decreases. At the same time, the nervous system may become abnormally excitable, and impulses may be triggered spontaneously. As a result, muscles may undergo tetanic contractions, and the person may die due to a failure of respiratory movements.

Figure 11.11
Parathyroid hormone causes the release of calcium ions from bone and the conservation of calcium ions by the kidneys. It indirectly stimulates the absorption of calcium ions by the intestine. The resulting increase in blood calcium ion concentration inhibits the secretion of parathyroid hormone.

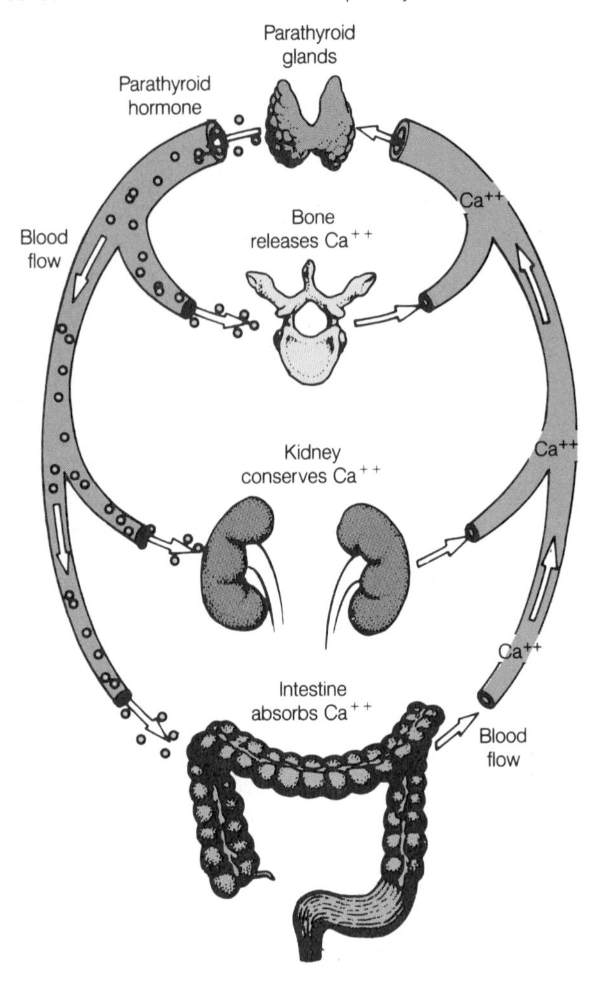

1. Where are the parathyroid glands located?
2. How does parathyroid hormone help regulate the concentrations of blood calcium and phosphate ions?

Adrenal Glands

The **adrenal glands** are closely associated with the kidneys (fig. 11.12). A gland sits atop each kidney like a cap and is embedded in the mass of fat that encloses the kidney.

Structure of the Glands

Each adrenal gland is very vascular and consists of two parts: the central portion is called the adrenal medulla, and the outer part is called the adrenal cortex. Although these regions are not sharply divided, they represent distinct glands that secrete different hormones.

Figure 11.12
An adrenal gland consists of an outer cortex and an inner medulla.

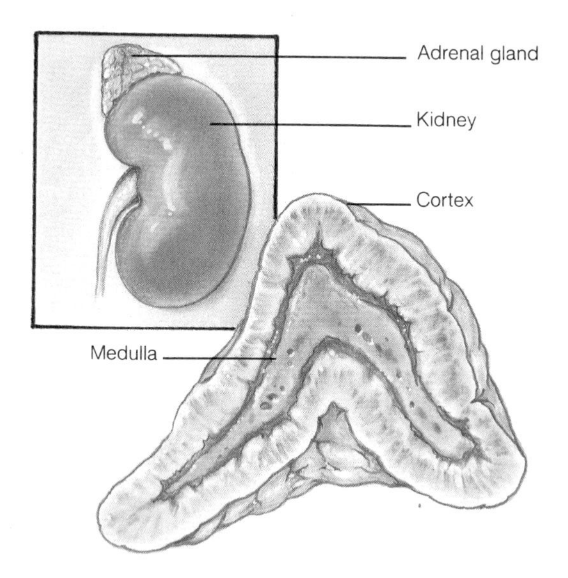

The **adrenal medulla** consists of irregularly shaped cells that are arranged in groups around blood vessels. These cells are intimately connected with the sympathetic division of the autonomic nervous system. In fact, the adrenal medullary cells are modified postganglionic neurons, and preganglionic autonomic nerve fibers lead to them from the central nervous system (see chapter 9).

The **adrenal cortex,** which makes up the bulk of the adrenal gland, is composed of closely packed masses of epithelial cells, arranged in layers. These layers form an outer, a middle, and an inner zone of the cortex. As in the case of the cells of the adrenal medulla, the cells of the adrenal cortex are well supplied with blood vessels.

1. Where are the adrenal glands located?
2. Describe the two portions of an adrenal gland.

Hormones of the Adrenal Medulla

The cells of the adrenal medulla secrete two closely related hormones, **epinephrine** (adrenaline) and **norepinephrine** (noradrenaline). These substances have similar molecular structures and physiological functions. In fact, epinephrine, which makes up 80% of the adrenal medullary secretion, is produced from norepinephrine.

The synthesis of these hormones begins with the amino acid tyrosine. In the first step of the process, tyrosine is converted into a substance called *dopa* by an enzyme in the medullary secretory cells. Dopa is changed to *dopamine* by a second enzyme, and dopamine is changed to norepinephrine by a third enzyme. In about 10% of the adrenal medullary cells the process ends with norepinephrine; in the remainder, still another enzyme converts norepinephrine to epinephrine.

The effects of the adrenal medullary hormones generally resemble those occurring when sympathetic nerve fibers stimulate their effectors. The hormonal effects, however, last up to ten times longer because the hormones are removed from the tissues relatively slowly. These effects include increased heart rate and increased force of cardiac muscle contraction, elevated blood pressure, increased breathing rate, and decreased activity in the digestive system.

Impulses arriving by way of sympathetic nerve fibers stimulate the adrenal medulla to release its hormones at the same time other effectors are being stimulated by sympathetic impulses. As a rule, these impulses originate in the hypothalamus in response to various types of stress. Thus, the adrenal medullary secretions function together with the sympathetic division of the autonomic nervous system in preparing the body for energy-expending action—fight or flight. Chart 11.4 compares some of the effects of the adrenal medullary hormones.

Although a lack of adrenal medullary hormones produces no significant disorder, tumors in the adrenal medulla are sometimes accompanied by excessive hormonal secretions. In such cases, the secretion of norepinephrine usually predominates, and affected persons show signs of prolonged sympathetic responses—high blood pressure, increased heart rate, elevated blood sugar, and so forth. The treatment of this condition generally involves a surgical removal of the tumorous growth.

1. Name the hormones secreted by the adrenal medulla.
2. What effects are produced by hormones from the adrenal medulla?
3. What stimulates the release of hormones from the adrenal medulla?

Hormones of the Adrenal Cortex

The cells of the adrenal cortex produce over thirty different steroids, among which are several hormones. Unlike the adrenal medullary hormones, which a person can survive without, some of those released by the cortex are vital. In fact, in the absence of adrenal cortical secretions, a person usually dies within a week unless extensive electrolyte therapy is provided.

The most important adrenal cortical hormones are aldosterone, cortisol, and sex hormones.

Part or function affected	Epinephrine	Norepinephrine
Heart	Rate increases	Less effect (rate may slow)
	Force of contraction increases	Force of contraction increases
Blood vessels	Vessels in the skeletal muscle dilate; a decreasing resistance to blood flow	Increased blood flow to skeletal muscles, resulting from constriction of blood vessels in skin and viscera
Systemic blood pressure	Some increase due to increased cardiac output	Great increase due to vasoconstriction
Airways	Dilation	Less effect
Reticular formation of brain	Activated	Little effect
Liver	Promotes the change of glycogen to glucose, increasing blood sugar	Little effect on blood sugar
Metabolic rate	Increases	Increases

Chart 11.4 Comparative effects of epinephrine and norepinephrine

Aldosterone

Aldosterone is synthesized by cells in the outer zone of the adrenal cortex. It is called a *mineralocorticoid* because it helps regulate the concentration of mineral electrolytes, such as sodium and potassium. More specifically, aldosterone causes sodium ions to be reabsorbed and conserved and causes potassium ions to be excreted. As a consequence of the sodium reabsorption and an increasing blood sodium ion concentration, ADH is released from the posterior pituitary gland. The action of ADH promotes the conservation of water and reduces urine output. By indirectly promoting water conservation, aldosterone causes the blood volume to increase. This increase in blood volume is accompanied by an increase in blood pressure.

The cells that secrete aldosterone are stimulated by a decrease in the blood concentration of sodium ions or an increase in the blood concentration of potassium ions.

Cortisol

Cortisol (hydrocortisone) is a *glucocorticoid,* which means it affects glucose metabolism. It is produced in the middle zone of the adrenal cortex and has a molecular structure similar to aldosterone. In addition to affecting glucose, cortisol influences protein and fat metabolism.

Among the more important actions of this hormone are the following:

1. It inhibits the synthesis of protein in various tissues, thus causing an increase in the blood concentration of amino acids.
2. It promotes the release of fatty acids from adipose tissue, thus causing an increase in the use of fatty acids as an energy source and a decrease in the use of glucose for this purpose.
3. It stimulates the liver cells to form glucose from noncarbohydrates, such as circulating amino acids and glycerol, thus promoting an increase in the blood glucose concentration.

These actions help to keep the blood glucose concentration within the normal range between meals, because the supply of glycogen stored within the liver (that is used to provide glucose) can be exhausted in a few hours without food.

The release of cortisol is controlled by a negative feedback mechanism, which involves the hypothalamus, anterior pituitary gland, and adrenal cortex. More specifically, the hypothalamus secretes CRH (corticotropin-releasing hormone) into the hypophyseal portal veins, which were described previously in this chapter. These vessels carry CRH to the anterior pituitary gland, stimulating it to secrete ACTH. In turn, ACTH causes the adrenal cortex to release cortisol. Cortisol has an inhibiting effect on the release of both CRH and ACTH, and as these substances decrease in concentration, less cortisol is produced (fig. 11.13).

Such a mechanism might act to maintain a relatively stable blood concentration of a hormone, but in the case of cortisol, the set point of the mechanism is changed from time to time (see chapter 1). In this way, the output of hormone can be altered to meet the demands of changing conditions. For example, when the body is subjected to stressful conditions—injury, disease, extreme temperatures, extreme emotional feelings, and so forth—nerve impulses provide the brain with information concerning the stressful factors. In response, brain centers acting through the hypothalamus can cause the release of more cortisol and maintain a higher concentration of the hormone until the stress is reduced.

Stress and Its Effects

A Current Topic

Because survival depends upon the maintenance of homeostasis, factors that cause changes in the body's internal environment are potentially life threatening. When such dangers are sensed, nerve impulses are directed to the hypothalamus, and physiological responses are triggered that tend to resist a loss of homeostasis. These responses include increased activity in the sympathetic division of the autonomic nervous system and an increased secretion of adrenal and other hormones. A factor capable of stimulating such a response is called a *stressor,* and the condition it produces in the body is called *stress.*

Stressors include physical factors, such as exposure to extreme heat or cold, decreased oxygen concentration, infections, injuries, prolonged heavy exercise, and loud sounds. Stressors also include psychological factors, such as thoughts about real or imagined dangers, personal losses, and unpleasant social interactions. Psychological stress can also result from feelings of anger, fear, grief, anxiety, depression, or guilt. In other instances, pleasant stimuli, such as friendly social contact, feelings of joy and happiness, or sexual arousal, may be stressful.

Regardless of its cause, physiological responses to stress are directed toward maintaining homeostasis, and they usually involve a set of reactions called the *general stress syndrome,* which is controlled largely by the hypothalamus.

Typically the hypothalamus responds to stress by activating mechanisms that prepare the body for "fight or flight." More specifically, the hypothalamus causes a rise in the blood glucose concentration, an increase in the concentrations of blood glycerol and fatty acids, an increase in heart rate, a rise in blood pressure, an increase in breathing rate, dilation of the air passages, a shunting of blood from the skin and digestive organs into the skeletal muscles, and an increased secretion of epinephrine from the adrenal medulla.

At the same time, the hypothalamus releases corticotropin-releasing hormone (CRH), which, in turn, stimulates the anterior pituitary gland to secrete ACTH. ACTH causes the adrenal cortex to increase its secretion of cortisol. Cortisol promotes an increase in the blood amino acid concentration, release of fatty acids, and formation of glucose from noncarbohydrates. Thus, while the body is being prepared for physical activity, the actions of cortisol supply the cells with substances that may be needed during times of stress.

Other hormones whose secretions increase with stress include glucagon, growth hormone, and antidiuretic hormone. The general effect of glucagon and growth hormone is to aid in mobilizing energy sources, such glucose, glycerol, fatty acids, and amino acids. ADH promotes the retention of sodium ions and water by the kidneys, which increases the blood volume—an important action if a person is bleeding excessively or sweating heavily.

An increased secretion of cortisol may be accompanied by a decrease in the number of certain white blood cells (lymphocytes). Because such cells function in defending the body against infections, a person experiencing prolonged stress may have a lowered resistance to infectious diseases and to the growth of some cancers. Also, exposure to excessive amounts of cortisol may promote the development of high blood pressure, atherosclerosis, and gastrointestinal ulcers.

The hyposecretion of cortical hormones leads to *Addison's disease,* a condition characterized by decreased blood sodium, increased blood potassium, low blood glucose concentration (hypoglycemia), dehydration, low blood pressure, and increased skin pigmentation. The treatment of Addison's disease involves the use of mineralocorticoids and glucocorticoids. An untreated person is likely to live only a few days because of severe disturbances in electrolyte balance.

Hypersecretion of cortical hormones, which may be associated with an adrenal tumor or with an oversecretion of ACTH by the anterior pituitary gland, results in *Cushing's syndrome.* A person with this condition has changes in carbohydrate and protein metabolism and in electrolyte balance. For example, when mineralocorticoids and glucocorticoids are excessive, the blood glucose concentration remains high, and there is a great decrease in tissue protein. Also, sodium is retained abnormally, and as a result, the tissue fluids tend to increase, and the skin becomes puffy. At the same time, an increase in the adrenal sex hormones may produce masculinizing effects in a female, such as the growth of a beard or development of a deeper voice.

Sex Hormones

Adrenal sex hormones are produced by cells in the inner zone of the cortex. Although the hormones of this group are male types (adrenal androgens), some of them are converted to female hormones (estrogens) by the skin, liver, and adipose tissue. The normal functions of these hormones are not clear; however, they may supplement the supply of sex hormones from the gonads and stimulate early development of the reproductive organs. Also, there is some evidence that the adrenal androgens play a role in controlling the female sex drive.

Chart 11.5 summarizes the characteristics of the hormones produced by the adrenal cortex.

1. Name the most important hormones of the adrenal cortex.
2. What is the function of aldosterone?
3. What actions does cortisol produce?
4. How are the blood concentrations of aldosterone and cortisol regulated?

Chart 11.5 Hormones of the adrenal cortex		
Hormone	**Action**	**Factor regulating secretion**
Aldosterone	Helps regulate the concentration of extracellular electrolytes by causing sodium ions to be conserved and potassium ions to be excreted	Electrolyte concentrations in the body fluids
Cortisol	Decreases protein synthesis, increases fatty acid release, and stimulates the formation of glucose from noncarbohydrates	CRH from the hypothalamus and ACTH from the anterior pituitary gland
Adrenal androgens	Supplement the sex hormones from the gonads; converted to estrogens in females	

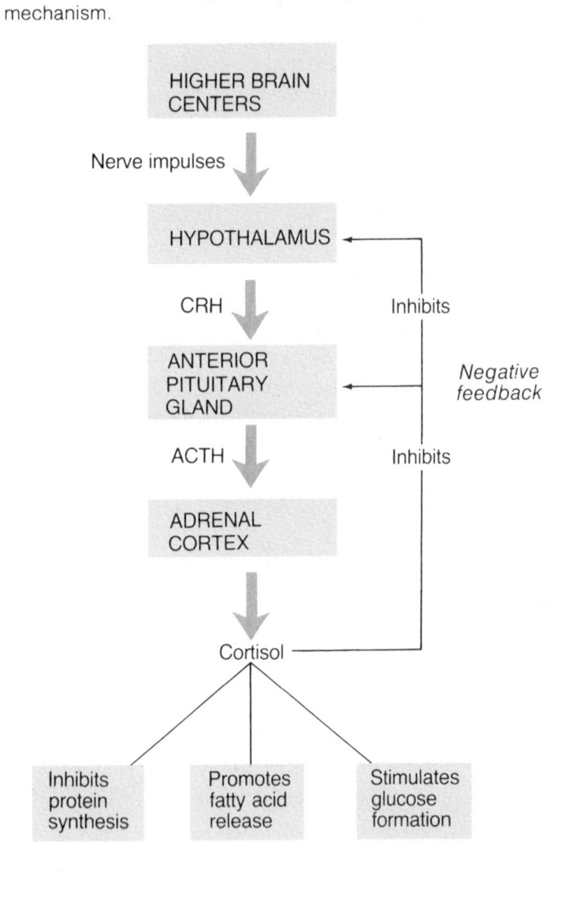

Figure 11.13
Cortisol secretion is regulated by a negative feedback mechanism.

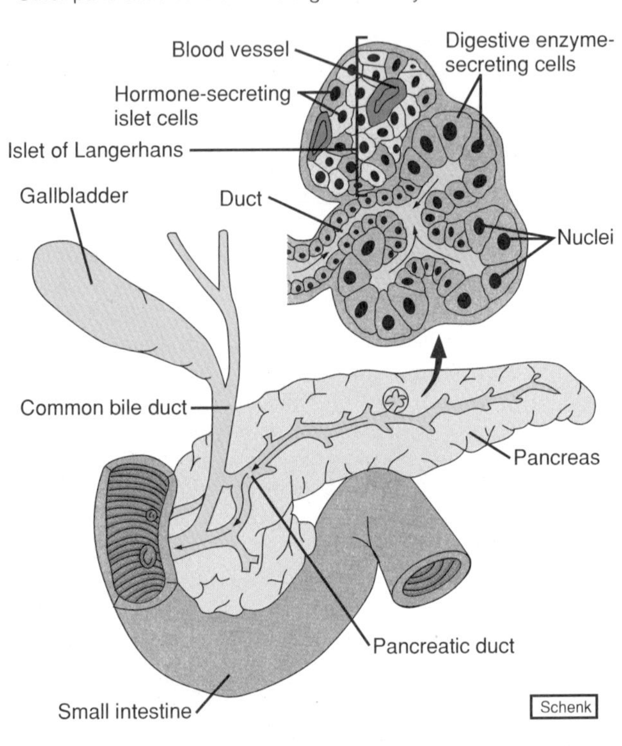

Figure 11.14
The hormone-secreting cells of the pancreas are arranged in clusters or islets that are closely associated with blood vessels. Other pancreatic cells secrete digestive enzymes into ducts.

Pancreas

The **pancreas** contains two major types of secretory tissues, reflecting its dual function as an exocrine gland that secretes digestive juice, and an endocrine gland that releases hormones (fig. 11.14).

Structure of the Gland

The pancreas is an elongated, somewhat flattened organ that is posterior to the stomach and behind the parietal peritoneum. It is attached to the first section of the small intestine (duodenum) by a duct, which transports its digestive juice into the intestine.

The endocrine portion of the pancreas consists of cells arranged in groups, closely associated with blood vessels. These groups, called *islets of Langerhans,* include two distinct types of cells—alpha cells, which secrete the hormone glucagon, and beta cells, which secrete the hormone insulin (fig. 11.15).

The digestive functions of the pancreas are discussed in chapter 12.

Figure 11.15
Light micrograph of an islet of Langerhans within the pancreas (×50).

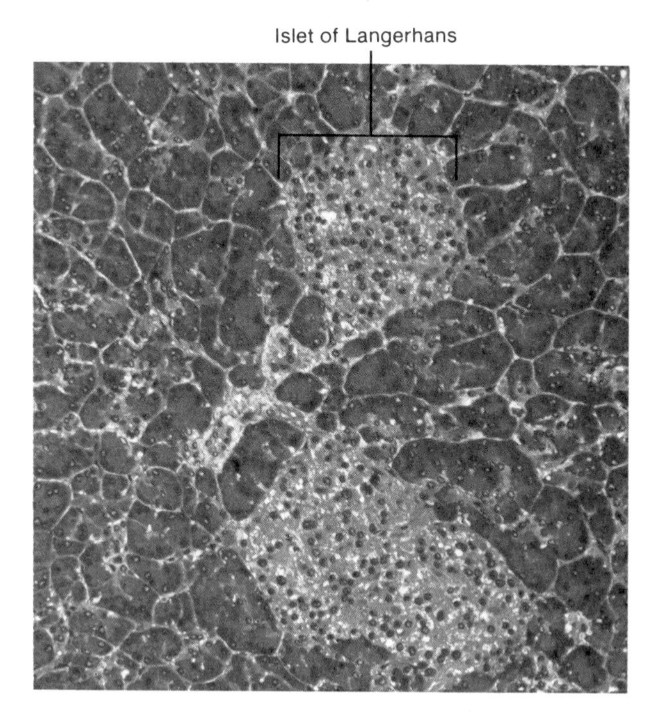

Hormones of the Islets of Langerhans

Glucagon stimulates the liver to convert glycogen and certain noncarbohydrates, such as amino acids, into glucose. This causes the blood glucose concentration to rise. Although its effect is similar to that of epinephrine, glucagon is many times more effective in elevating blood sugar.

The secretion of glucagon is regulated by a negative feedback system in which a low concentration of blood sugar stimulates the release of the hormone from the alpha cells. When the blood sugar concentration rises, glucagon is no longer secreted. This mechanism prevents hypoglycemia from occurring when the glucose concentration is relatively low, such as between meals, or when glucose is being used rapidly—during periods of exercise, for example.

The main effect of **insulin** is exactly opposite to that of glucagon. Insulin acts on the liver to stimulate the formation of glycogen from glucose and to inhibit the conversion of noncarbohydrates into glucose. Insulin also has the special effect of promoting the facilitated diffusion (see chapter 3) of glucose through the membranes of cells that possess insulin receptors on their cell membranes. These cells include those of skeletal muscle, cardiac muscle, and adipose tissue. As a result of these actions, insulin causes a decrease in the concentration of blood glucose. In addition, it promotes the transport of amino acids into the cells, increases the synthesis of proteins, and stimulates the adipose cells to synthesize and store fat.

As in the case of glucagon secretion, insulin secretion is regulated by a negative feedback system which is sensitive to the concentration of blood glucose. When glucose is relatively high, as may occur following a meal, insulin is released from the beta cells. By promoting the formation of glycogen in the liver and the entrance of glucose into adipose and muscle cells, the hormone helps to prevent an excessive rise in the blood glucose concentration. Then, when the glucose concentration is relatively low, between meals or during the night, less insulin is released.

As insulin decreases, less and less glucose enters the adipose and muscle cells, and the glucose remaining in the blood is available for use by cells that lack insulin receptors, such as nerve cells.

At the same time that insulin is decreasing, glucagon secretion is increasing, so that these hormones function together to maintain a relatively stable blood glucose concentration, despite great variation in the amount of carbohydrates eaten (fig. 11.16).

Because the nerve cells, including those of the brain, lack insulin receptors, they must obtain glucose by simple diffusion. For this reason, the nerve cells are particularly sensitive to changes in the blood glucose concentration, and conditions that cause such changes—excessive insulin secretion, for example—are likely to alter brain functions.

1. What is the name of the endocrine portion of the pancreas?
2. What is the function of glucagon?
3. What is the function of insulin?
4. How are the secretions of glucagon and insulin controlled?
5. Why are nerve cells particularly sensitive to changes in the blood glucose concentration?

Other Endocrine Glands

Other organs that produce hormones and thus are parts of the endocrine system include the pineal gland, thymus gland, reproductive glands, and certain glands of the digestive tract.

Pineal Gland

The **pineal gland** is a small structure located deep between the cerebral hemispheres, where it is attached to

Diabetes Mellitus

A Current Topic

Diabetes mellitus is characterized by a deficiency of insulin and by severe disturbances in carbohydrate metabolism as well as disorders in protein and fat metabolism. More specifically, the movement of glucose into adipose and skeletal muscle cells decreases; glycogen formation decreases; and, as a result, the concentration of blood sugar rises (hyperglycemia). When the blood sugar reaches a certain high concentration, the kidneys begin to excrete the excess, and glucose appears in the urine. As a result of excessive urine output, the affected person becomes dehydrated and experiences great thirst.

At the same time, the synthesis of proteins and fats decreases, and there is increased use of proteins as an energy source by the glucose-starved cells. Consequently, tissues tend to waste away, the person loses weight, and the ability to grow and repair tissues decreases. Changes in fat metabolism lead to an accumulation of fatty acids and ketone bodies in the blood, and this, in turn, causes a decrease in pH (metabolic acidosis). Brain cells may be affected adversely by the dehydration and acidosis, and the person may become disoriented, comatose, and die.

The two common forms of diabetes mellitus are called *type I* (insulin-dependent) and *type II* (noninsulin-dependent). Type I usually appears before the age of twenty and is an autoimmune disease. This means that it occurs when the body's immune system abnormally destroys the beta cells of the pancreas (see chapter 16). The usual treatment for type I diabetes mellitus involves administering by injection enough insulin to control carbohydrate metabolism. However, such therapy almost never achieves the blood glucose stability found in a healthy person. Also, the amount and type of insulin needed varies with individuals, depending on such factors as diet, physical activity, and health.

Type II diabetes mellitus is found in 70% to 80% of diabetic patients. It usually develops gradually after the age of forty and produces milder symptoms than type I. Type II disease is thought to be caused by an inherited disorder, although most affected persons are overweight when they first experience symptoms. In type II diabetes mellitus the beta cells continue to function; in fact, the patient may have an excessive blood insulin concentration. Instead of resulting from a lack of insulin, the symptoms seem to involve a cellular loss of sensitivity to the hormone. Treatment for type II diabetes mellitus usually involves carefully controlling the diet, participation in an exercise program, avoiding substances that stimulate insulin secretion, and maintaining a desirable body weight.

Because a major goal in the treatment of a diabetic patient is to keep the blood glucose concentration under control, regular monitoring of blood glucose is necessary. Such testing may be done by the patient at home using blood glucose test paper or a blood glucose meter.

Figure 11.16
Insulin and glucagon function together to help maintain a relatively stable blood glucose concentration.

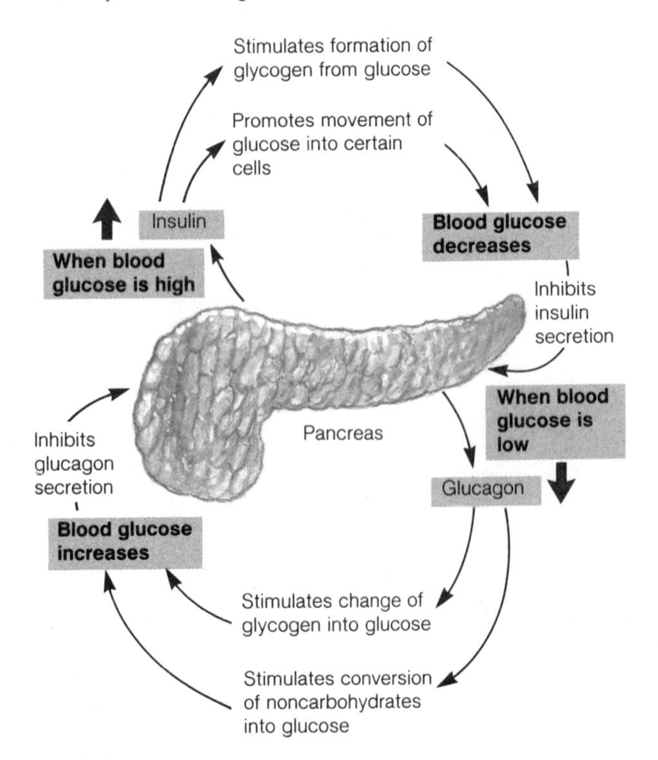

the upper portion of the thalamus near the roof of the third ventricle. Its body consists largely of specialized *pineal cells* and *neuroglial cells,* which provide support (fig. 11.1).

The pineal gland secretes the hormone **melatonin** in response to light conditions outside the body. Information concerning these conditions reaches the gland by means of nerve impulses that originate in the retinas of the eyes. More specifically, in the absence of light, nerve impulses from the eyes are decreased, and the secretion of melatonin increases.

This mechanism by which hormonal secretion is responsive to varying light conditions is sometimes involved in the regulation of **circadian rhythms.** Such rhythms are patterns of repeated activity that are associated with the environmental cycles of day and night, such as the sleep-wake rhythm and the seasonal cycles of fertility and infertility that occur in many mammals.

Although the process by which melatonin acts is not well understood, the hormone seems to inhibit the secretion of gonadotropins from the anterior pituitary gland and is thought to help regulate the female reproductive cycle (menstrual cycle).

Thymus Gland

The **thymus gland,** which lies in the mediastinum behind the sternum and between the lungs, is relatively large in young children but diminishes in size with age (fig. 11.1). This gland secretes a group of hormones, including one called **thymosin,** which affects the production of certain white blood cells (lymphocytes). In this way, the thymus plays an important role in the immune mechanism, which is discussed in chapter 16.

Reproductive Glands

The reproductive organs that secrete important hormones include the **ovaries,** which produce estrogens and progesterone; the **placenta,** which produces estrogens, progesterone, and gonadotropin; and the **testes,** which produce testosterone. These glands and their secretions are discussed in chapter 19.

Digestive Glands

The digestive glands that secrete hormones are generally associated with the linings of the stomach and small intestine. These structures and their secretions are described in chapter 12.

1. Where is the pineal gland located?
2. What seems to be the function of the pineal gland?
3. Where is the thymus gland located?
4. Which reproductive organs secrete hormones?

Clinical Terms Related to the Endocrine System

adrenalectomy (ah-dre″nah-lek′to-me)—the surgical removal of the adrenal glands.

adrenogenital syndrome (ah-dre″no-jen′ĭ-tal sin′drōm)—a group of symptoms associated with changes in sexual characteristics as a result of the increased secretion of adrenal androgens.

diabetes insipidus (di″ah-be′tēz in-sip′ĭdus)—a metabolic disorder, characterized by a large output of dilute urine containing no sugar and caused by a decreased secretion of ADH by the posterior pituitary gland.

exophthalmos (ek″sof-thal′mos)—an abnormal protrusion of the eyes.

hirsutism (her′sūt-izm)—the excessive growth of hair, especially in women.

hypercalcemia (hi″per-kal-se′me-ah)—an excess of blood calcium.

hyperglycemia (hi″per-gli-se′me-ah)—an excess of blood glucose.

hypocalcemia (hi″po-kal-se′me-ah)—a deficiency of blood calcium.

hypoglycemia (hi″po-gli-se′me-ah)—a deficiency of blood glucose.

hypophysectomy (hi-pof″i-sek′to-me)—the surgical removal of the pituitary gland.

parathyroidectomy (par″ah-thi″roi-dek′to-me)—the surgical removal of the parathyroid glands.

pheochromocytoma (fe-o-kro″mo-si-to′mah)—a type of tumor found in the adrenal medulla and usually accompanied by high blood pressure.

polyphagia (pol″e-fa′je-ah)—excessive eating.

thymectomy (thi-mek′to-me)—the surgical removal of the thymus gland.

thyroidectomy (thi″roi-dek′to-me)—the surgical removal of the thyroid gland.

thyroiditis (thi″roi-di′tis)—inflammation of the thyroid gland.

virilism (vir′ĭ-lizm)—masculinization of a female.

Chapter Summary

Introduction (page 280)

Endocrine glands secrete their products into the body fluids; exocrine glands secrete into ducts that lead to a body surface.

General Characteristics of the Endocrine System (page 280)

1. As a group, endocrine glands are concerned with the regulation of metabolic processes.
2. Some hormones have localized effects, and others produce more general actions.

Hormones and Their Actions (page 280)

Endocrine glands secrete hormones that affect target cells possessing specific receptors. Hormones are very potent substances.

1. Chemically hormones are amines, peptides, proteins, glycoproteins, or steroids.
2. Actions of hormones
   a. Steroid hormones
      (1) Steroid hormones enter a target cell and combine with receptors in the nucleus to form complexes.
      (2) These complexes activate specific genes, which cause special proteins to be synthesized.
   b. Nonsteroid hormones
      (1) Nonsteroid hormones combine with receptors in the target cell membrane.
      (2) The hormone-receptor combinations stimulate the membrane proteins, such as adenylate cyclase, to induce the formation of second messenger molecules.
      (3) A second messenger, such as cAMP, activates protein kinases.
      (4) Protein kinases activate protein substrate molecules, which, in turn, cause changes in the cellular processes.
3. Prostaglandins
   Prostaglandins are substances, present in small quantities, that have powerful hormonelike effects.

Control of Hormonal Secretions (page 283)

The concentration of each hormone in the body fluids is regulated.

1. Negative feedback systems
   a. When an imbalance in hormone concentration occurs, information is fed back to some part that acts to correct this imbalance.
   b. Negative feedback systems maintain relatively stable hormone concentrations.
2. Nerve control
   a. Some endocrine glands secrete their hormones in response to nerve impulses.
   b. Other glands secrete hormones in response to substances called releasing hormones.

Pituitary Gland (page 284)

The pituitary gland has an anterior lobe and a posterior lobe. Most pituitary secretions are controlled by the hypothalamus.

1. Anterior pituitary hormones
   a. The anterior lobe secretes GH, PRL, TSH, ACTH, FSH, and LH.
   b. Growth hormone (GH)
      (1) Growth hormone stimulates the body cells to increase in size and undergo an increased rate of reproduction.
      (2) The secretion of GH is controlled by growth hormone-releasing hormone and growth hormone release-inhibiting hormone from the hypothalamus.
   c. Prolactin promotes breast development and stimulates milk production.
   d. Thyroid-stimulating hormone (TSH)
      (1) TSH controls the secretion of hormones from the thyroid gland.
      (2) TSH secretion is regulated by the hypothalamus, which secretes TRH.
   e. Adrenocorticotropic hormone (ACTH)
      (1) ACTH controls the secretion of hormones from the adrenal cortex.
      (2) The secretion of ACTH is regulated by the hypothalamus, which secretes CRH.
   f. Follicle-stimulating hormone (FSH) and luteinizing hormone (LH) are gonadotropins.
2. Posterior pituitary hormones
   a. The posterior lobe of the pituitary gland consists largely of neuroglial cells and nerve fibers.
   b. The hormones of the posterior pituitary are produced in the hypothalamus.
   c. Antidiuretic hormone (ADH)
      (1) ADH causes the kidneys to reduce the amount of water they excrete.
      (2) The secretion of ADH is regulated by the hypothalamus.
   d. Oxytocin (OT)
      (1) Oxytocin has an antidiuretic effect and can cause muscles in the uterine wall to contract.
      (2) Oxytocin also causes the contraction of certain cells associated with the production and ejection of milk.

Thyroid Gland (page 287)

The thyroid gland is located in the neck and consists of two lobes.

1. Structure of the gland
   a. The thyroid gland consists of many follicles.
   b. The follicles are fluid-filled and store hormones.
2. Thyroid hormones
   a. Thyroxine and triiodothyronine cause the metabolic rate of cells to increase, enhance protein synthesis, and stimulate the utilization of lipids.
   b. Calcitonin helps to regulate the concentrations of blood calcium and phosphate ions.

Parathyroid Glands (page 289)

The parathyroid glands are located on the posterior surface of the thyroid.

1. Structure of the glands
   Each gland consists of secretory cells that are well supplied with capillaries.
2. Parathyroid hormone (PTH)
   a. PTH causes an increase in the blood calcium and a decrease in the blood phosphate concentrations.
   b. The parathyroid glands are regulated by a negative feedback mechanism that operates between the glands and the blood.

Adrenal Glands (page 290)

The adrenal glands are located atop the kidneys.

1. Structure of the glands
   a. Each gland consists of a medulla and a cortex.
   b. The adrenal medulla and adrenal cortex represent separate glands.
2. Hormones of the adrenal medulla
   a. The adrenal medulla secretes epinephrine and norepinephrine, which have similar effects.
   b. The secretion of these hormones is stimulated by sympathetic impulses.
   c. Adrenal medullary hormones are synthesized from tyrosine.
3. Hormones of the adrenal cortex
   a. The adrenal cortex produces several steroid hormones.
   b. Aldosterone is a mineralocorticoid, which causes the kidneys to conserve sodium and water and excrete potassium.
   c. Cortisol is a glucocorticoid, which affects carbohydrate, protein, and fat metabolism.
   d. Adrenal sex hormones
      (1) These hormones are of the male type, but may be converted to female hormones.
      (2) They are thought to supplement the sex hormones produced by the gonads.

Pancreas (page 293)

The pancreas secretes digestive juices as well as hormones.

1. Structure of the gland
   a. The pancreas is attached to the small intestine.
   b. The islets of Langerhans secrete glucagon and insulin.

2. Hormones of the islets of Langerhans
   a. Glucagon stimulates the liver to produce glucose from glycogen and noncarbohydrates.
   b. Insulin promotes the movement of glucose through some cell membranes, stimulates the storage of glucose, promotes the synthesis of proteins, and stimulates the storage of fats.
   c. Nerve cells lack insulin receptors and depend on simple diffusion for a glucose supply.

Other Endocrine Glands (page 295)

1. Pineal gland
   a. The pineal gland is attached to the thalamus.
   b. It secretes melatonin in response to varying light conditions.
   c. Melatonin may play a role in regulating the female reproductive cycle by inhibiting the secretion of gonadotropins from the anterior pituitary gland.
2. Thymus gland
   a. The thymus gland lies behind the sternum and between the lungs.
   b. It secretes thymosin, which affects the production of certain lymphocytes that play important roles in immunity.
3. Reproductive glands
   a. The ovaries secrete estrogen and progesterone.
   b. The placenta secretes estrogen, progesterone, and gonadotropin.
   c. The testes secrete testosterone.
4. Digestive glands
   Certain glands of the stomach and small intestine secrete hormones.

Clinical Application of Knowledge

1. What hormones would need to be administered to an adult whose anterior pituitary gland had been removed? Why?
2. How might the environment of a patient with hyperthyroidism be modified to minimize the drain on body energy resources?
3. A patient who has lost a relatively large volume of blood will have an increased secretion of aldosterone from the adrenal cortex. What effect will this increased secretion have on the patient's blood concentrations of sodium and potassium ions?
4. Both growth hormone and growth hormone-releasing hormone have been used successfully to promote growth in children with short statures. How would you explain the difference in the ways these hormones produce their effects?

Review Activities

1. Explain what is meant by an endocrine gland.
2. Define *hormone* and *target cell.*
3. Explain how steroid hormones produce their effects.
4. Explain how nonsteroid hormones employ second messenger molecules.
5. Describe how prostaglandins are similar to hormones.
6. Describe a negative feedback system.
7. Define *releasing hormone,* and provide an example of such a substance.
8. Describe the location and structure of the pituitary gland.
9. List the hormones secreted by the anterior lobe of the pituitary gland.
10. Explain how pituitary gland activity is controlled by the brain.
11. Explain how growth hormone produces its effects.
12. List the major factors that affect the secretion of growth hormone.
13. Summarize the function of prolactin.
14. Describe the mechanism that regulates the concentrations of circulating thyroid hormones.
15. Explain how the secretion of ACTH is controlled.
16. Compare the cellular structures of the anterior and posterior lobes of the pituitary gland.
17. Describe the functions of the posterior pituitary hormones.
18. Explain how the release of ADH is regulated.
19. Describe the location and structure of the thyroid gland.
20. Name the hormones secreted by the thyroid gland, and list the general functions of each hormone.
21. Describe the location and structure of the parathyroid glands.
22. Explain the general functions of parathyroid hormone.
23. Describe the mechanism that regulates the secretion of parathyroid hormone.
24. Distinguish between the adrenal medulla and the adrenal cortex.
25. List the hormones produced by the adrenal medulla, and describe their general functions.
26. List the steps in the synthesis of adrenal medullary hormones.
27. Name the most important hormones of the adrenal cortex, and describe the general functions of each.
28. Describe how the pituitary gland controls the secretion of adrenal cortical hormones.
29. Describe the location and structure of the pancreas.
30. List the hormones secreted by the islets of Langerhans, and describe the general functions of each.
31. Summarize how the secretion of hormones from the pancreas is regulated.
32. Describe the location and general function of the pineal gland.
33. Describe the location and general function of the thymus gland.
34. Name four additional hormone-secreting organs.

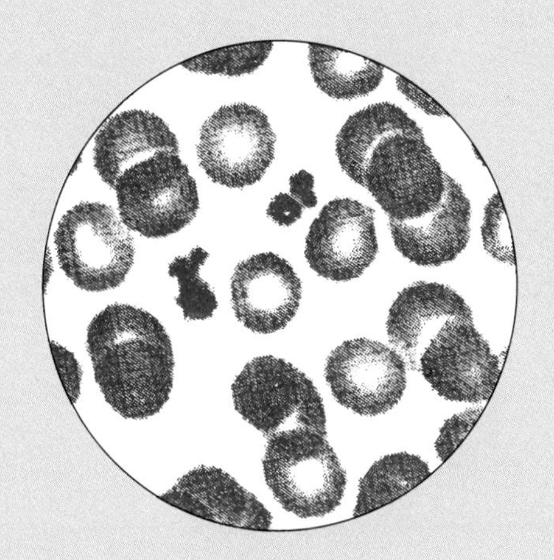

Unit 4
Processing and Transporting

The chapters of unit 4 are concerned with the digestive, respiratory, circulatory, lymphatic, and urinary systems. They describe how the organs of these systems obtain nutrients and oxygen from outside the body, how nutrients are altered chemically,and absorbed into the body fluids, and how nutrients and oxygen are transported to body cells. They also discuss how these substances are utilized by the cells, how the resulting wastes are transported and excreted, and how stable concentrations of various substances in the body fluids are maintained. This unit includes:

These circular computer-manipulated images appear on the front cover and are described on the back cover of this text.

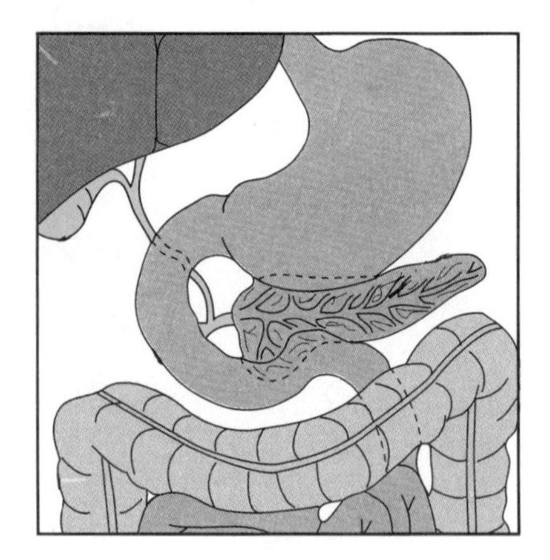

12 Digestion and Nutrition

Most food substances are composed of chemicals whose molecules cannot pass easily through membranes and so cannot be absorbed efficiently by cells. The *digestive system* solves this problem. Its parts are adapted to ingest foods, to break large particles into smaller ones, to secrete enzymes that decompose food molecules, to absorb the products of this digestive action, and to eliminate the unused residues.

The substances in foods that are needed for the maintenance of health are called *nutrients,* and they include various carbohydrates, lipids, proteins, vitamins, and minerals. Taken together, the processes involved in the ingestion, assimilation, and use of these nutrients is called *nutrition.*

Chapter Objectives

After you have studied this chapter, you should be able to

1. Name and describe the locations of the organs of the digestive system and their major parts.
2. Describe the general functions of each digestive organ and of the liver.
3. Describe the structure of the wall of the alimentary canal.
4. Explain how the contents of the alimentary canal are mixed and moved.
5. List the enzymes secreted by the various digestive organs, and describe the function of each enzyme.
6. Describe how the digestive secretions are regulated.
7. Explain how the products of digestion are absorbed.
8. List the major sources of carbohydrates, lipids, and proteins.
9. Describe how carbohydrates, lipids, and proteins are used by cells.
10. List the fat-soluble and water-soluble vitamins, and summarize the general functions of each vitamin.
11. List the major minerals and trace minerals, and summarize the general functions of each mineral.
12. Describe an adequate diet.
13. Complete the review activities at the end of this chapter. Note that the items are worded in the form of specific learning objectives. You may want to refer to them before reading the chapter.

Key Terms

absorption (ab-sorp′shun)

alimentary canal (al″ĭ-men′tar-e kah-nal′)

bile (bīl)

cellulose (sel′u-lōs)

chyme (kīm)

deciduous (de-sid′u-us)

feces (fe′sēz)

gastric juice (gas′trik jo͞os)

intestinal juice (in-tes′tĭ-nal jo͞os)

intrinsic (in-trin′sik)

mesentery (mes′en-ter″e)

mineral (min′er-al)

mucous membrane (mu′kus mem′brān)

nutrient (nu′tre-ent)

nutrition (nu-trish′un)

pancreatic juice (pan″kre-at′ik jo͞os)

peristalsis (per″ĭ-stal′sis)

sphincter muscle (sfingk′ter mus′el)

villi; singular, villus (vil′i, vil′us)

vitamin (vi′tah-min)

Aids to Understanding Words

aliment-, food: *aliment*ary canal—the tubelike portion of the digestive system.

chym-, juice: *chym*e—the semifluid paste of food particles and gastric juice formed in the stomach.

decidu-, falling off: *decidu*ous teeth—those that are shed during childhood.

gastr-, stomach: *gastr*ic gland—a portion of the stomach that secretes gastric juice.

hepat-, liver: *hepat*ic duct—the duct that carries bile from the liver to the common bile duct.

lingu-, tongue: *lingu*al tonsil—the mass of lymphatic tissue at the root of the tongue.

nutri-, nourishing: *nutri*ent—a chemical substance needed to nourish body cells.

peri-, around: *peri*stalsis—a wavelike ring of contraction that moves material along the alimentary canal.

pylor-, gatekeeper: *pylor*ic sphincter—a muscle that serves as a valve between the stomach and small intestine.

vill-, hairy: *vill*i—tiny projections of mucous membrane in the small intestine.

Introduction

Digestion is the process by which food substances are changed into forms that can be absorbed through cell membranes. The **digestive system** includes the organs that promote digestion and absorb the products of this process. It consists of an *alimentary canal,* which extends from the mouth to the anus and several *accessory organs,* which release secretions into the canal. The alimentary canal includes the mouth, pharynx, esophagus, stomach, small intestine, and large intestine, and the accessory organs include the salivary glands, liver, gallbladder, and pancreas. The major organs of this system are shown in figure 12.1.

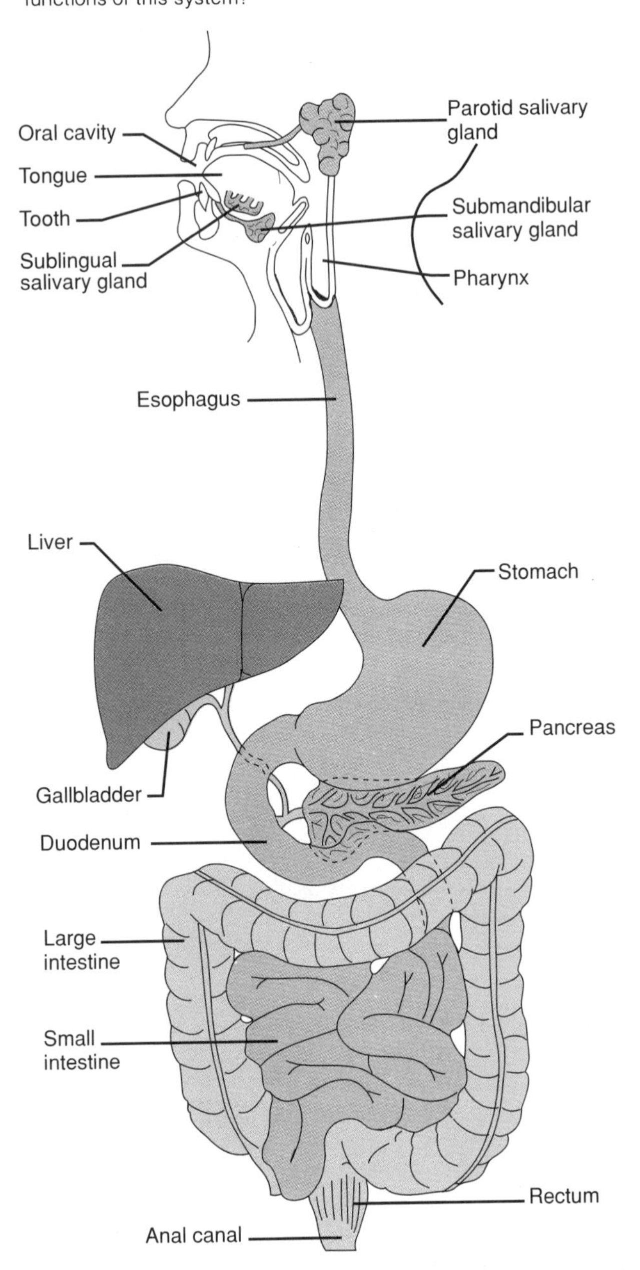

Figure 12.1
Major organs of the digestive system. What are the general functions of this system?

General Characteristics of the Alimentary Canal

The **alimentary canal** is a muscular tube about 9 meters long, which passes through the body's ventral cavity. Although it is specialized in various regions to carry on particular functions, the structure of its wall and the method by which it moves food are similar throughout its length.

Structure of the Wall

The wall of the alimentary canal consists of four distinct layers, although the degree to which they are developed varies from region to region. Beginning with the innermost tissues, these layers, shown in figure 12.2, include the following:

1. **Mucous membrane,** or **mucosa.** This layer is formed of surface epithelium, underlying connective tissue, and a small amount of smooth muscle. In some regions, it develops folds and tiny projections, which extend into the lumen of the digestive tube and increase its absorptive surface area. It also may contain glands, which are tubular invaginations into which lining cells secrete mucus and digestive enzymes. Thus, the mucosa protects the tissues beneath it and carries on absorption and secretion.
2. **Submucosa.** The submucosa contains considerable loose connective tissue as well as blood vessels, lymphatic vessels, and nerves. Its vessels nourish the surrounding tissues and carry away absorbed materials.
3. **Muscular layer.** This layer consists of two coats of smooth muscle tissue. The fibers of the inner coat are arranged so that they encircle the tube, and when these *circular fibers* contract, the diameter of the tube is decreased. The fibers of the outer muscular coat run lengthwise, and when these *longitudinal fibers* contract, the tube is shortened.
4. **Serous layer** or **serosa.** The serous, or outer, covering of the tube is composed of the *visceral peritoneum.* The cells of the serosa secrete serous fluid, which keeps the tube's outer surface moist and lubricates it so that the organs within the abdominal cavity slide freely against one another.

Movements of the Tube

The motor functions of the alimentary canal are of two basic types—mixing movements and propelling movements. Mixing occurs when smooth muscles in rela-

Figure 12.2
The wall of the small intestine, as in other portions of the alimentary canal, includes four layers: an inner mucous membrane, a submucosa, a muscular layer, and an outer serous layer.

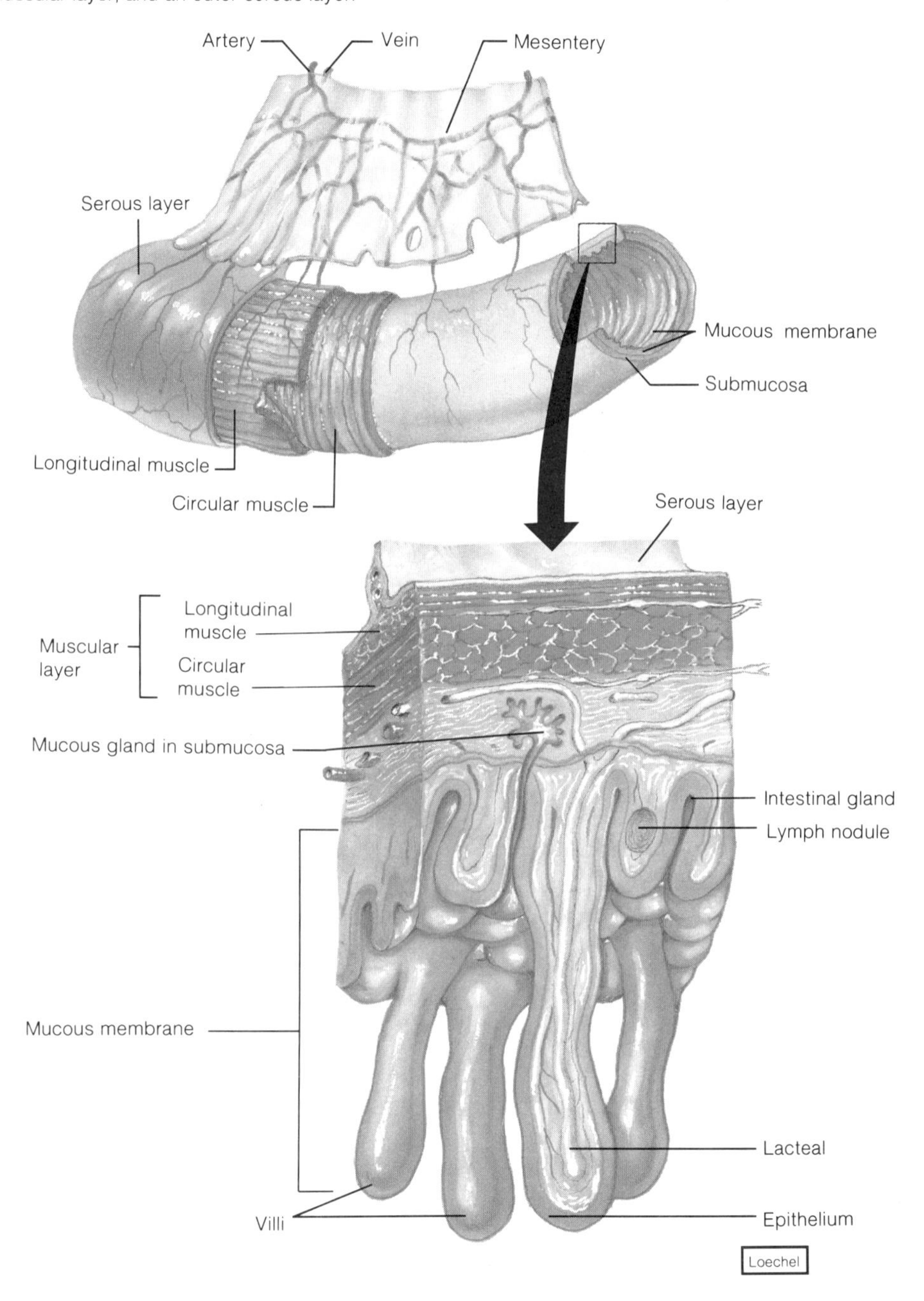

tively small segments of the tube undergo rhythmic contractions. For example, when the stomach is full, waves of muscular contractions move along its walls from one end to the other. These waves mix food substances with the digestive juices secreted by the mucosa.

Propelling movements include a wavelike motion called **peristalsis.** When peristalsis occurs, a ring of contraction appears in the wall of the tube. At the same time, the muscular wall just ahead of the ring relaxes. As the peristaltic wave moves along, it pushes the tubular contents ahead of it.

1. What organs constitute the digestive system?
2. Describe the wall of the alimentary canal.
3. Describe the propelling movement that occurs in the alimentary canal.

Mouth

The **mouth** is adapted to receive food and to prepare it for digestion by mechanically reducing the size of solid particles and mixing them with saliva. The mouth is

Figure 12.3
The mouth is adapted for ingesting food and preparing it for digestion.

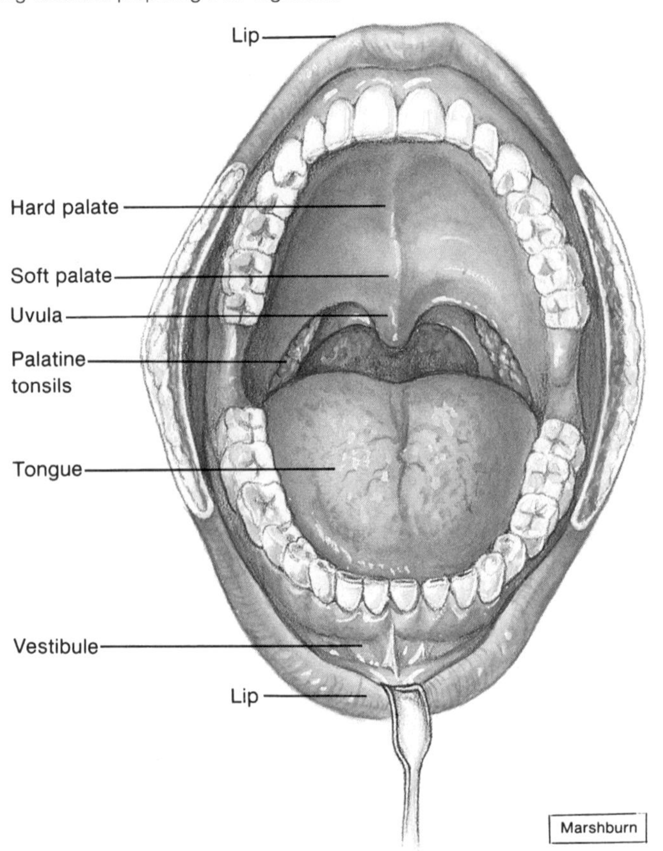

surrounded by the lips, cheeks, tongue, and palate, and includes a chamber between the palate and tongue, called the *oral cavity,* as well as a narrow space between the teeth, cheeks, and lips, called the *vestibule* (fig. 12.3).

Cheeks and Lips

The **cheeks** consist of outer layers of skin, pads of subcutaneous fat, muscles associated with expression and chewing, and inner linings of moist stratified squamous epithelium.

The **lips** are highly mobile structures. They contain skeletal muscles and a variety of sensory receptors, which are useful in judging the temperature and texture of foods. Their normal reddish color is due to an abundance of blood vessels near their surfaces.

Tongue

The **tongue** nearly fills the oral cavity when the mouth is closed. It is covered by mucous membrane and is connected in the midline to the floor of the mouth by a membranous fold called the **frenulum.**

The *body* of the tongue is composed largely of skeletal muscle. These muscles aid in mixing food particles with saliva during chewing and in moving food toward the pharynx during swallowing. Rough projections, called **papillae,** on the surface of the tongue provide friction, which is useful in handling food. These papillae also contain taste buds (see chapter 10).

The posterior region, or *root,* of the tongue is anchored to the hyoid bone and is covered with rounded masses of lymphatic tissue, called **lingual tonsils** (fig. 12.4).

Palate

The **palate** forms the roof of the oral cavity and consists of a hard anterior part (*hard palate*) and a soft posterior part (*soft palate*). The soft palate forms a muscular arch, which extends posteriorly and downward as a cone-shaped projection, called the **uvula.**

During swallowing, muscles draw the soft palate and uvula upward. This action closes the opening between the nasal cavity and pharynx, preventing food from entering the nasal cavity.

In the back of the mouth, on either side of the tongue and closely associated with the palate, are masses of lymphatic tissue, called **palatine tonsils** (fig. 12.4). These structures lie beneath the epithelial lining of the mouth and, like other lymphatic tissues, help protect the body against infections.

Figure 12.4
A sagittal section of the mouth, nasal cavity, and pharynx.

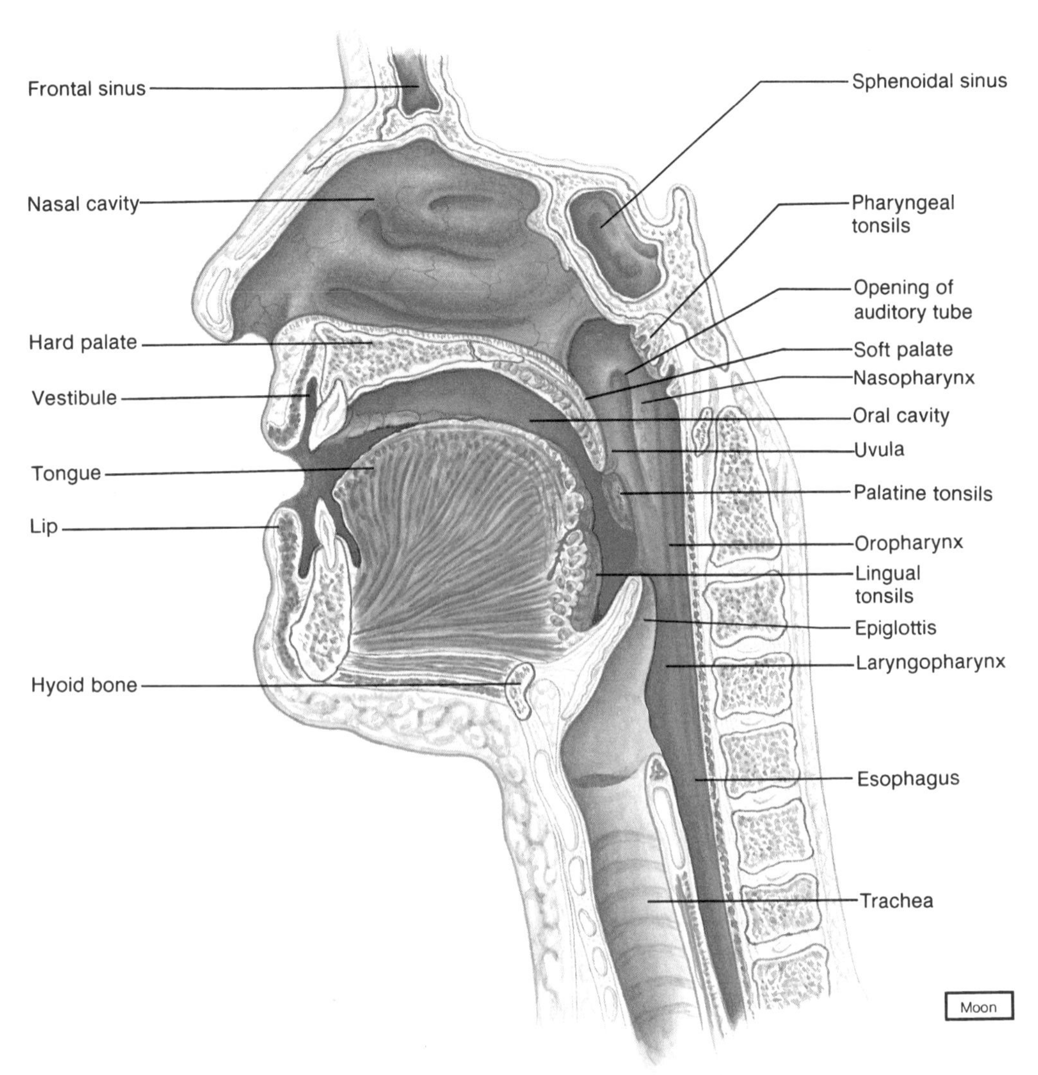

The palatine tonsils are common sites of infections, and if they become inflamed, the condition is termed *tonsillitis*. Infected tonsils may become so swollen that they block the passageways of the pharynx and interfere with breathing and swallowing. Because the mucous membranes of the pharynx, auditory tubes, and middle ears are continuous, there is danger that such an infection may travel from the throat into the middle ears.

Still other masses of lymphatic tissue, called **pharyngeal tonsils,** or *adenoids* (fig. 12.4), occur on the posterior wall of the pharynx, above the border of the soft palate.

1. How does the tongue function as part of the digestive system?
2. What is the role of the soft palate in swallowing?
3. Where are the various tonsils located?

Teeth

Teeth are unique structures in that two different sets form during development. The members of the first set, the *primary teeth* (deciduous teeth) usually erupt through the gums at regular intervals between the ages of six months and two and one-half years (fig. 12.5). There are twenty deciduous teeth—ten in each jaw.

The primary teeth are usually shed in the same order they appeared. Before this happens, their roots are resorbed. The teeth are then pushed out of their sockets by pressure from the developing *secondary teeth* (permanent teeth). This secondary set consists of thirty-two teeth—sixteen in each jaw (fig. 12.6).

The secondary teeth usually begin to appear at about age six years, but the set may not be completed until the third molars appear between seventeen and twenty-five years of age.

Figure 12.5
The primary and secondary teeth developing in the maxilla and mandible are revealed in this skull of a child.

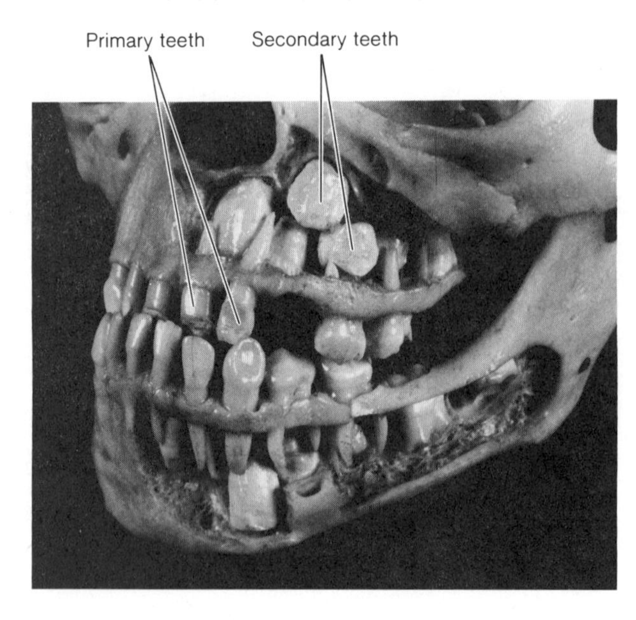

Figure 12.6
The secondary teeth of the upper and lower jaws.

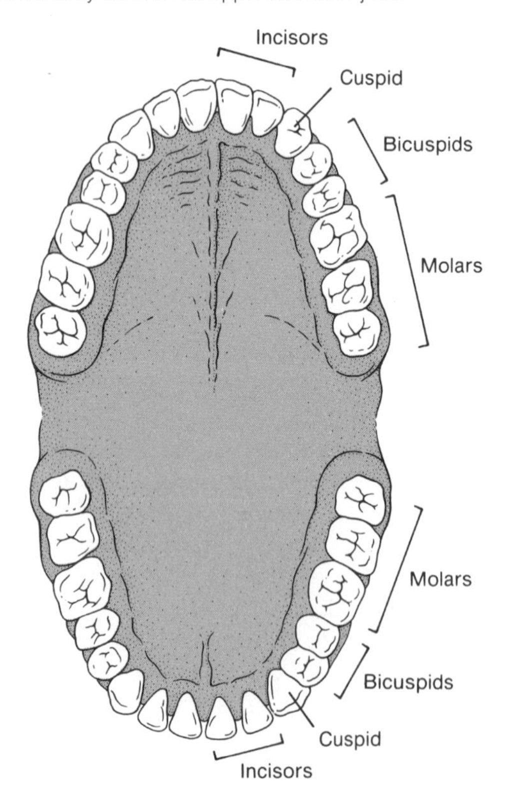

The teeth break pieces of food into smaller pieces. This action increases the surface area of food particles and thus makes it possible for digestive enzymes to react more effectively with the food molecules.

Different teeth are adapted to handle food in different ways. The *incisors* (front teeth) are chisel-shaped, and their sharp edges are used to bite off relatively large pieces of food. The *cuspids* (canine teeth) are cone-shaped, and they are useful in grasping or tearing food. The *bicuspids* (premolars) and *molars* have somewhat flattened surfaces and are specialized for grinding food particles (fig. 12.6).

Chart 12.1 Primary and secondary teeth

Primary teeth (deciduous)		Secondary teeth (permanent)	
Type	*Number*	*Type*	*Number*
Incisor		Incisor	
central	4	central	4
lateral	4	lateral	4
Cuspid	4	Cuspid	4
		Bicuspid	
		first	4
		second	4
Molar		Molar	
first	4	first	4
second	4	second	4
		third	4
Total	20	Total	32

Chart 12.1 summarizes the number and kinds of teeth that appear during development.

Each tooth consists of two main portions, called the *crown,* which projects beyond the gum (gingiva), and the *root,* which is anchored to the alveolar process of the jaw (fig. 12.7). The region where these portions meet is called the *neck* of the tooth.

The crown is covered by glossy, white *enamel,* which consists mainly of calcium salts and is the hardest substance in the body. Unfortunately, if damaged by abrasive action or injury, the enamel is not replaced.

The bulk of a tooth beneath the enamel is composed of *dentin,* a substance much like bone but somewhat harder. Dentin, in turn, surrounds the tooth's central cavity (pulp cavity), which contains blood vessels, nerves, and connective tissue (pulp). Blood vessels and nerves reach this cavity through tubular *root canals,* which extend into the root.

The root is enclosed by a thin layer of bonelike material called *cementum,* which is surrounded by a *periodontal ligament.* This ligament contains bundles of thick collagenous fibers, which pass between the cementum and the bone of the alveolar process, thus firmly attaching the tooth to the jaw. It also contains blood vessels and nerves.

1. How do the primary teeth differ from the secondary teeth?
2. Describe the structure of a tooth.
3. Explain how a tooth is attached to the bone of the jaw.

Dental Caries

A Current Topic

Dental caries (decay) involves the decalcification of tooth enamel and is usually followed by the destruction of the enamel and its underlying dentin. The result is a cavity, which must be cleaned and filled to prevent further erosion of the tooth.

Although the cause or causes of dental caries are not well understood, a lack of dental cleanliness and a diet high in sugar and starch seem to promote the problem. Accumulations of food particles on the surfaces and between the teeth are thought to aid the growth of certain kinds of bacteria. These microorganisms utilize the carbohydrates in food particles and produce acid by-products. The acids then begin the process of destroying tooth enamel.

Preventing dental caries requires brushing the teeth at least once a day, using dental floss or tape regularly to remove debris from between the teeth, and limiting the intake of sugar and starch, especially between meals. The use of fluoridated drinking water or the application of fluoride solution to children's teeth also helps prevent dental decay.

Loss of teeth is most commonly associated with diseases of the gums (gingivitis) and the dental pulp (endodontitis). Such diseases can usually be avoided by practicing good oral hygiene and obtaining regular dental treatment.

Figure 12.7
A section of a cuspid tooth.

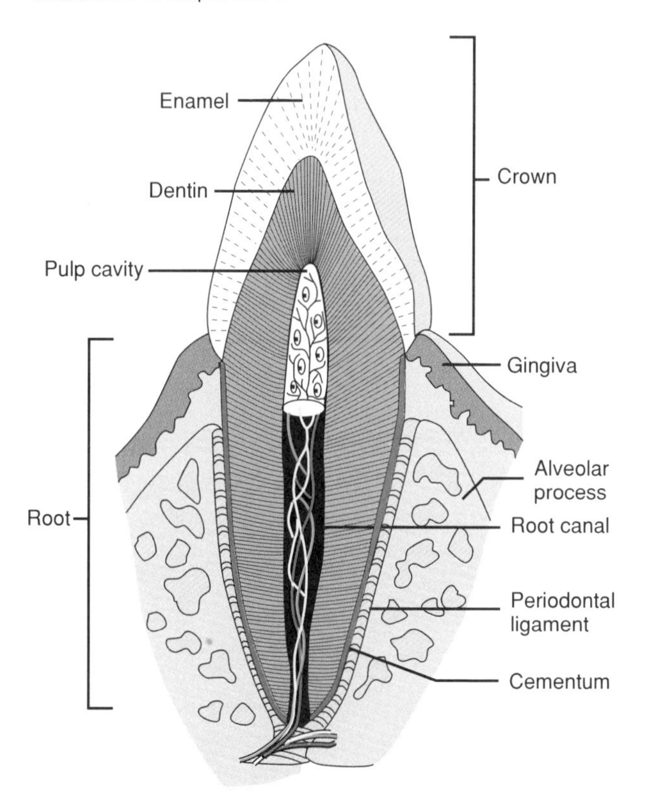

Salivary Glands

The **salivary glands** secrete saliva. This fluid moistens food particles, helps bind them together, and begins the digestion of carbohydrates. Saliva also acts as a solvent by dissolving various food chemicals—a process that is necessary before they can be tasted—and by helping to cleanse the mouth and teeth.

Salivary Secretions

Within a salivary gland are two types of secretory cells, *serous cells* and *mucous cells*. These cells occur in varying proportions within different glands. The serous cells produce a watery fluid that contains a digestive enzyme, called **amylase.** This enzyme splits starch and glycogen molecules into disaccharides—the first step in the digestion of carbohydrates. The mucous cells secrete the thick, stringy liquid called **mucus,** which binds food particles together and acts as a lubricant during swallowing.

When a person sees, smells, tastes, or even thinks about pleasant food, parasympathetic nerve impulses elicit the secretion of a large volume of watery saliva. Conversely, if food looks, smells, or tastes unpleasant, parasympathetic activity is inhibited so that less saliva is produced, and swallowing may become difficult (see chapter 10).

Major Salivary Glands

There are three pairs of major salivary glands—the parotid, submandibular, and sublingual glands—and many minor ones, which are associated with the mucous membrane of the tongue, palate, and cheeks (fig. 12.8).

The **parotid glands** are the largest of the major salivary glands. One gland lies in front of and somewhat below each ear, between the skin of the cheek and the masseter muscle. The parotid glands secrete a clear, watery fluid that is rich in amylase.

The **submandibular glands** are located in the floor of the mouth on the inside surface of the lower jaw. The secretory cells of these glands are predominantly serous, but some mucous cells are present. Consequently, the submandibular glands secrete a more viscous fluid than the parotid glands.

The **sublingual glands** are the smallest of the major salivary glands. They are on the floor of the

Figure 12.8
Locations of the major salivary glands.

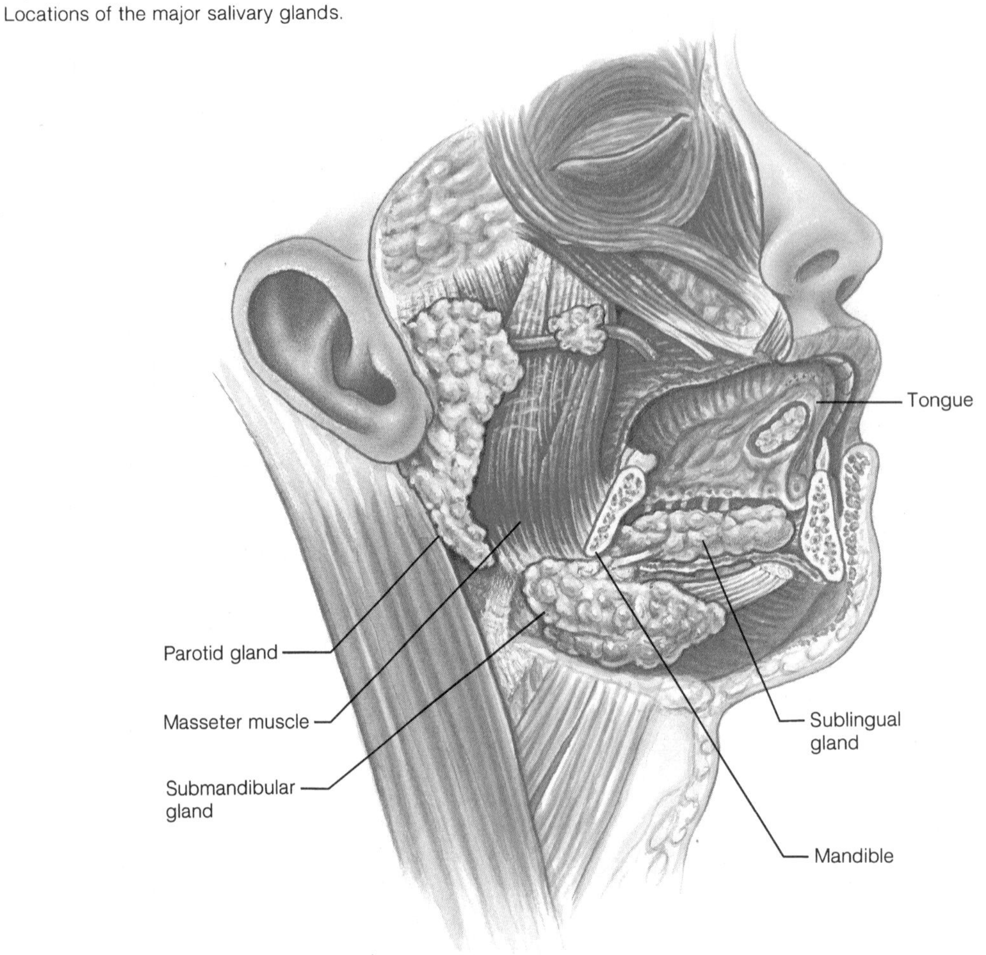

mouth under the tongue. Because their cells are primarily the mucous type, their secretions tend to be thick and stringy.

1. What is the function of saliva?
2. What stimulates the salivary glands to secrete saliva?
3. Where are the major salivary glands located?

Pharynx and Esophagus

The pharynx is a cavity behind the mouth from which the tubular esophagus leads to the stomach (fig. 12.1). Although neither of these organs contributes to the digestive process, both are important passageways, and their muscular walls function in swallowing.

Structure of the Pharynx

The **pharynx** connects the nasal and oral cavities with the larynx and esophagus. It can be divided into the nasopharynx, oropharynx, and laryngopharynx (fig. 12.4).

The **nasopharynx** communicates with the nasal cavity and provides a passageway for air during breathing.

The **oropharynx** opens behind the soft palate into the nasopharynx. It functions as a passageway for food moving downward from the mouth and for air moving to and from the nasal cavity.

The **laryngopharynx** is located just below the oropharynx. It opens into the larynx and esophagus.

Swallowing Mechanism

The act of swallowing involves a set of complex reflexes and can be divided into three stages. In the first stage, which is initiated voluntarily, food is chewed and mixed with saliva. Then this mixture is rolled into a mass (bolus) and forced into the pharynx by the tongue.

The second stage begins as the food reaches the pharynx and stimulates sensory receptors located around the pharyngeal opening. This stimulation triggers the swallowing reflex, which includes the following actions:

1. The soft palate is raised, preventing food from entering the nasal cavity.
2. The hyoid bone and the larynx are elevated, so that food is less likely to enter the trachea; a flaplike structure attached to the larynx, called the *epiglottis,* is pressed downward by the base of the tongue, and this action also helps to prevent food from entering the trachea.
3. Muscles in the lower portion of the pharynx relax, opening the esophagus.
4. A peristaltic wave begins in the pharyngeal muscles, and this wave forces the food into the esophagus.

As the swallowing reflex occurs, breathing is momentarily inhibited. Then, during the third stage of swallowing, the food is transported by the esophagus to the stomach by peristalsis.

Esophagus

The **esophagus** is a straight, collapsible tube about 25 centimeters long, which provides a passageway for substances between the pharynx and stomach (figs. 12.1 and 12.4). It begins at the base of the pharynx and descends behind the trachea, passing through the mediastinum. The esophagus penetrates the diaphragm and is continuous with the stomach on the abdominal side of the diaphragm.

Just above the point where the esophagus joins the stomach, some circular muscle fibers in its wall are thickened. These fibers usually are contracted and function to close the entrance to the stomach. In this way, they help prevent regurgitation of the stomach contents into the esophagus. When peristaltic waves reach the stomach, these muscle fibers relax and allow food to enter.

Mucous glands are scattered throughout the mucosa of the esophagus, and their secretions keep the inner lining of the tube moist and lubricated.

If regurgitation (reflux) of the stomach contents occurs, gastric secretions may irritate the esophageal wall, producing discomfort or pain. This condition is commonly called *heartburn* because it seems to originate from behind the sternum in the region of the heart.

1. Describe the regions of the pharynx.
2. List the major events that occur during swallowing.
3. What is the function of the esophagus?

Stomach

The **stomach** (figs. 12.1 and 12.9) is a J-shaped, pouchlike organ, which hangs under the diaphragm in the upper left portion of the abdominal cavity. It has a capacity of about 1 liter or more, and its inner lining is marked by thick folds (rugae), which tend to disappear as its wall is distended. The stomach receives food from the esophagus, mixes it with gastric juice, initiates the digestion of proteins, carries on a limited amount of absorption, and moves food into the small intestine.

Parts of the Stomach

The stomach, shown in figure 12.9, is divided into the cardiac, fundic, body, and pyloric regions. The *cardiac region* is a small area near the esophageal opening. The *fundic region,* which balloons above the cardiac portion, acts as a temporary storage area. The dilated *body region,* which is the main part of the stomach, is located between the fundic and pyloric portions. The *pyloric region* narrows and becomes the *pyloric canal* as it approaches the small intestine.

At the end of the pyloric canal, the muscular wall is thickened, forming a powerful circular muscle, called the **pyloric sphincter** (pylorus). This muscle serves as a valve that prevents the regurgitation of food from the intestine back into the stomach.

Gastric Secretions

The mucous membrane that forms the inner lining of the stomach is relatively thick, and its surface is studded with many small openings. These openings, called *gastric pits,* are located at the ends of tubular **gastric glands** (figs. 12.10 and 12.11).

The gastric glands generally contain three types of secretory cells. One type, the *mucous cell,* occurs in the necks of the glands near the openings of the gastric

Figure 12.9
Major regions of the stomach.

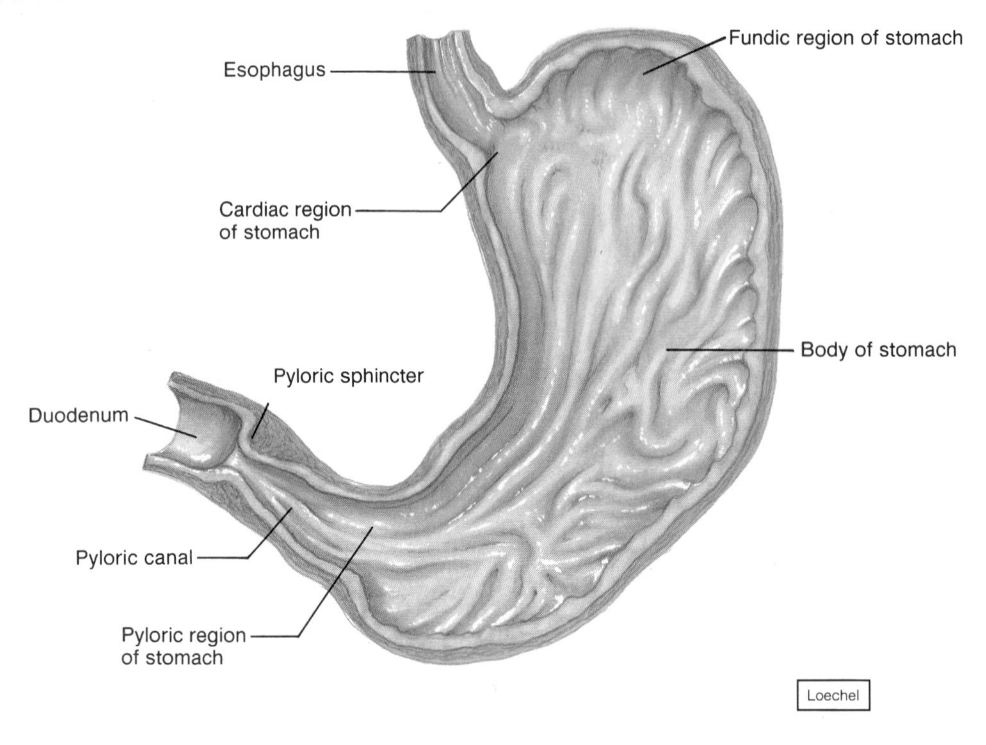

Figure 12.10
(*a*) The mucosa of the stomach is studded with gastric pits, which are the openings of gastric glands; (*b*) gastric glands include mucous cells, parietal cells, and chief cells.

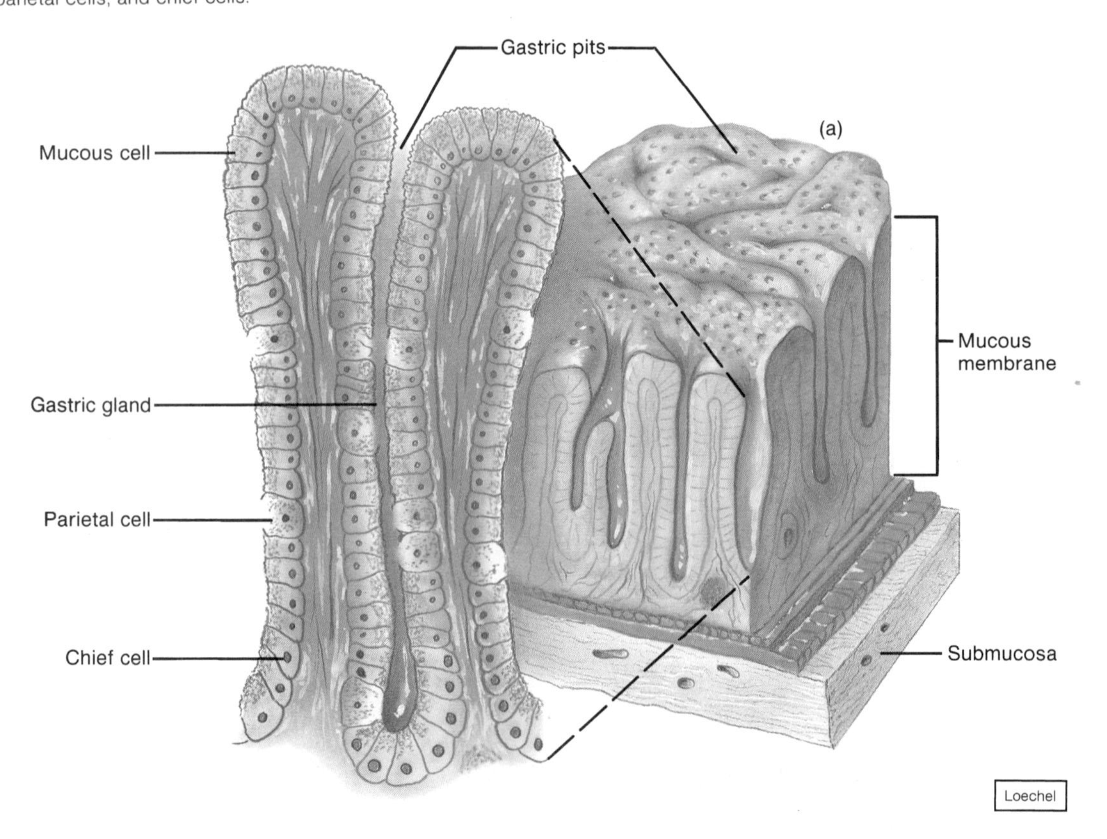

Chart 12.2 Composition of gastric juice		
Component	**Source**	**Function**
Pepsinogen	Chief cells of the gastric glands	Inactive form of pepsin
Pepsin	Formed from pepsinogen in the presence of gastric juice	Protein-splitting enzyme capable of digesting nearly all types of protein
Hydrochloric acid	Parietal cells of the gastric glands	Changes pepsinogen into pepsin; provides an acid environment needed for the action of pepsin
Mucus	Goblet cells and mucous glands	Provides a viscous, alkaline protective layer on the stomach wall
Intrinsic factor	Parietal cells of the gastric glands	Aids in the absorption of vitamin B_{12}

Figure 12.11
A light micrograph of the gastric mucosa (×100).

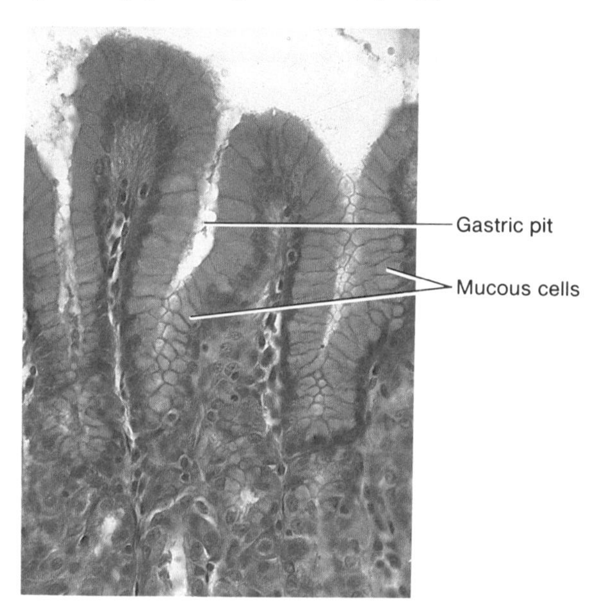

pits. The other types, *chief cells* and *parietal cells,* are found in the deeper parts of the glands. The chief cells secrete *digestive enzymes,* and the parietal cells release *hydrochloric acid.* The products of the mucous cells, chief cells, and parietal cells together form **gastric juice.**

Although gastric juice contains several digestive enzymes, **pepsin** is by far the most important of them. It is secreted by the chief cells as an inactive enzyme precursor called **pepsinogen.** When pepsinogen contacts the hydrochloric acid of gastric juice, however, it changes rapidly into pepsin.

Pepsin is a protein-splitting enzyme capable of beginning the digestion of nearly all types of dietary protein. This enzyme is most active in an acid environment, and the hydrochloric acid in gastric juice provides such an environment.

The mucous cells of the gastric glands secrete large quantities of thin mucus. In addition, the cells of the mucous membrane, associated with the inner lining of the stomach and between these glands, release a more viscous and alkaline secretion, which forms a protective coating on the inside of the stomach wall. This coating is especially important because pepsin is capable of digesting the proteins of the stomach tissues as well as those in foods. Thus the coating normally prevents the stomach from digesting itself.

Still another component of gastric juice is **intrinsic factor.** This substance, which is secreted by the parietal cells of the gastric glands, aids in the absorption of vitamin B_{12} from the small intestine.

The substances in gastric juice are summarized in chart 12.2.

1. What is secreted by the chief cells of the gastric glands? By the parietal cells?
2. What is the most important digestive enzyme in gastric juice? Why is it important?
3. How is the stomach prevented from digesting itself?

Regulation of Gastric Secretions

Although gastric juice is produced continuously, the rate of its production varies considerably from time to time and is under the control of neural and hormonal mechanisms (fig. 12.12). More specifically, when a person tastes, smells, or even sees pleasant food, or when food enters the stomach, parasympathetic impulses cause acetylcholine to be released from nerve fibers. It, in turn, stimulates the gastric glands to secrete large amounts of gastric juice, which is rich in hydrochloric acid and pepsin. These impulses also stimulate certain stomach cells to release a peptide hormone, called **gastrin,** that causes the gastric glands to increase their secretory activity.

As food moves into the small intestine, the secretion of gastric juice from the stomach wall is inhibited. This inhibition is due to sympathetic nerve impulses triggered by the presence of acid substances in the upper part of the small intestine. Also, the presence of proteins and fats in this region causes the release of a hormone, called *cholecystokinin,* from the intestinal wall. Its action decreases gastric motility as the small intestine fills with food.

Figure 12.12
The secretion of gastric juice is regulated in part by parasympathetic nerve impulses that stimulate the release of gastric juice and gastrin.

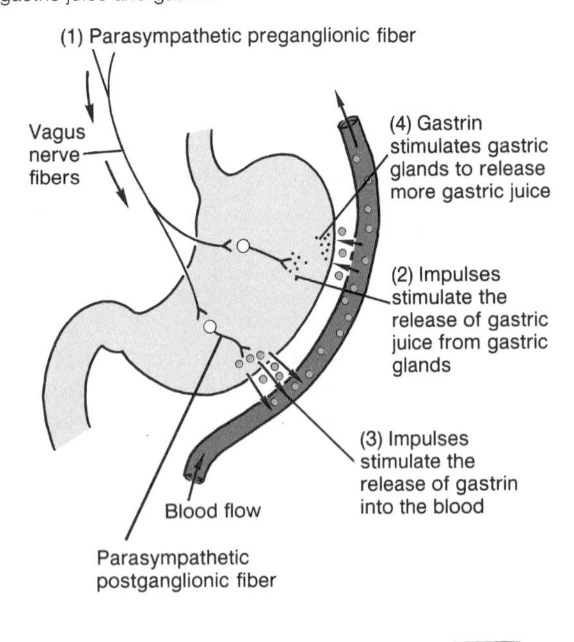

An *ulcer* is an open sore in the mucous membrane resulting from a localized breakdown of the tissues. *Gastric ulcers* are most likely to develop in the wall of the stomach near the liver. Ulcers are also common in the first portion of the small intestine (duodenum). *Duodenal ulcers* occur in regions that are exposed to gastric juice as the contents of the stomach enter the intestine. Both of these types of ulcers are caused by the digestive action of pepsin, and for this reason, are known as *peptic ulcers.*

Gastric Absorption

Although gastric enzymes begin the breakdown of proteins, the stomach wall is not well adapted to absorb digestive products. However, small quantities of water, glucose, certain salts, alcohol, and various lipid-soluble drugs may be absorbed by the stomach.

1. How is the secretion of gastric juice stimulated?
2. How is the secretion of gastric juice inhibited?
3. What substances may be absorbed in the stomach?

Mixing and Emptying Actions

Following a meal, the mixing movements of the stomach wall aid in producing a semifluid paste of food particles and gastric juice, called **chyme.** Peristaltic waves push the chyme toward the pyloric region of the stomach, and as it accumulates near the pyloric sphincter, this muscle begins to relax. The muscular pyloric region then pumps the chyme a little at a time into the small intestine.

The rate at which the stomach empties depends on several factors, including the fluidity of the chyme and the type of food present. For example, liquids usually pass through the stomach quite rapidly, but solids remain until they are well mixed with gastric juice. Fatty foods may remain in the stomach from three to six hours; foods high in proteins tend to be moved through more quickly; and carbohydrates usually pass through more rapidly than either fats or proteins.

As the stomach contents enter the duodenum, accessory organs add their secretions to the chyme. These organs include the pancreas, liver, and gallbladder.

Vomiting is a complex reflex that empties the stomach in another way. This action is usually triggered by irritation or distension in some part of the alimentary canal, such as the stomach or intestines. Sensory impulses travel from the site of stimulation to the *vomiting center* in the medulla oblongata, and a number of motor responses follow. These include taking a deep breath, raising the soft palate and thus closing the nasal cavity, closing the opening to the trachea (glottis), relaxing the circular muscle fibers at the base of the esophagus, contracting the diaphragm so it moves downward over the stomach, and contracting the abdominal wall muscles so that pressure inside the abdominal cavity is increased. As a result, the stomach is squeezed from all sides, and its contents are forced upward and out through the esophagus, pharynx, and mouth.

1. How is chyme produced?
2. What factors influence the speed with which the chyme leaves the stomach?

Pancreas

The shape of the **pancreas** and its general location are described in chapter 11, as are its endocrine functions (see figure 11.14). The pancreas also has an exocrine function—secretion of digestive juice.

Structure of the Pancreas

The pancreas is closely associated with the small intestine. It extends horizontally across the posterior abdominal wall in the C-shaped curve of the duodenum (figs. 12.1 and 12.13).

The cells that produce pancreatic juice make up the bulk of the pancreas. These cells (pancreatic acinar cells) are clustered around tiny tubes into which they release their secretions. The smaller tubes unite to form larger ones, which, in turn, give rise to a *pancreatic*

Figure 12.13
The pancreas is closely associated with the duodenum.

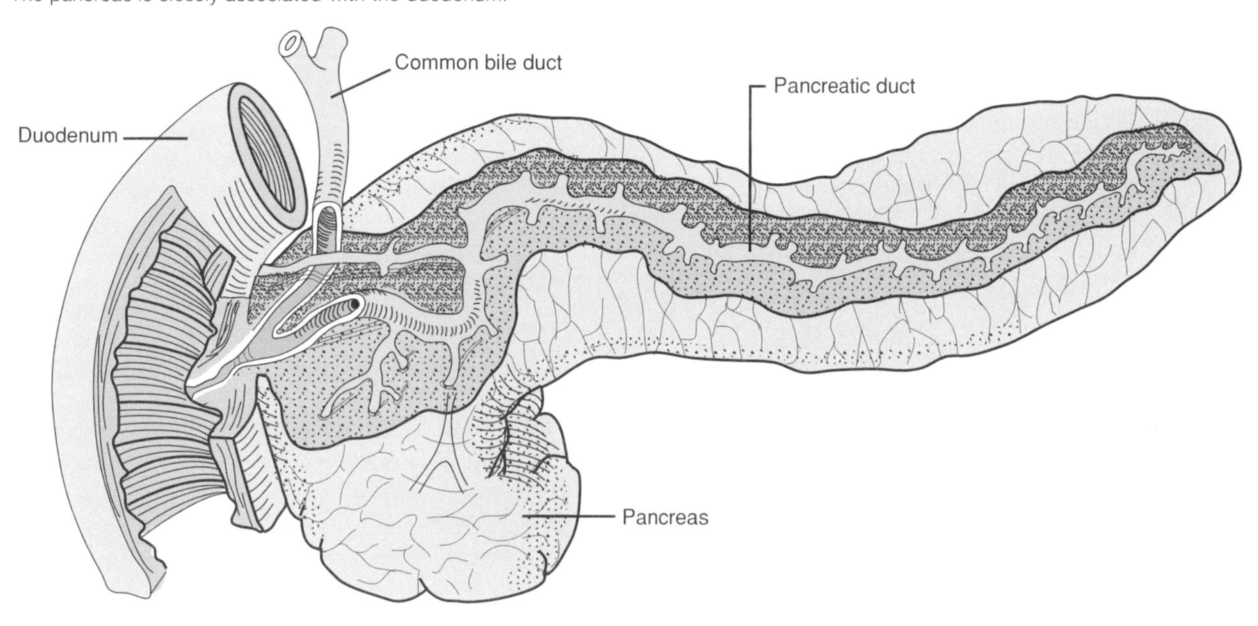

duct, extending the length of the pancreas. This duct usually connects with the duodenum at the place where the bile duct from the liver and gallbladder joins the duodenum (fig. 12.13).

Pancreatic Juice

Pancreatic juice contains enzymes capable of digesting carbohydrates, fats, proteins, and nucleic acids.

The carbohydrate-digesting enzyme is called *pancreatic amylase.* It splits molecules of starch or glycogen into double sugars (disaccharides); the fat-digesting enzyme, *pancreatic lipase,* breaks fat molecules (triglycerides) into fatty acids and glycerol.

The protein-splitting enzymes (proteinases) are *trypsin, chymotrypsin,* and *carboxypeptidase.* Each of these enzymes acts to split the bonds between particular combinations of amino acids in proteins. Since no single enzyme can split all the possible combinations, the presence of several enzymes is necessary for the complete digestion of protein molecules.

These protein-splitting (proteolytic) enzymes are stored in inactive forms within tiny cellular structures called *zymogen granules.* These enzymes, like gastric pepsin, are secreted in inactive forms and must be activated by other enzymes after they reach the small intestine. For example, the pancreatic cells release inactive *trypsinogen.* This substance becomes active trypsin when it contacts an enzyme called *enterokinase,* which is secreted by the mucosa of the small intestine.

If the release of pancreatic juice is blocked, it may accumulate in the duct system of the pancreas, and the trypsinogen may become activated. As a result, portions of the pancreas may become digested, causing a painful condition called *acute pancreatitis.*

Pancreatic juice also contains two **nucleases,** enzymes that break nucleic acid molecules into nucleotides.

Regulation of Pancreatic Secretion

As with gastric and small intestinal secretions, the release of pancreatic juice is regulated by nerve actions as well as hormones. For example, when parasympathetic impulses stimulate the secretion of gastric juice, other parasympathetic impulses travel to the pancreas and stimulate it to release digestive enzymes. Also, as the acidic chyme enters the duodenum, a peptide hormone called **secretin** is released into the blood from the duodenal mucous membrane (fig. 12.14). This hormone stimulates the secretion of pancreatic juice, which has a high concentration of bicarbonate ions. These ions neutralize the acid of chyme and provide a favorable environment for digestive enzymes in the intestine.

The presence of proteins and fats in the chyme in the duodenum also stimulates the release of **cholecystokinin** from the intestinal wall. As is the case with secretin, cholecystokinin reaches the pancreas by way of the blood; however, cholecystokinin causes the secretion of pancreatic juice with a high concentration of digestive enzymes.

Figure 12.14
Acidic chyme entering the duodenum from the stomach stimulates the release of secretin, which in turn stimulates the release of pancreatic juice.

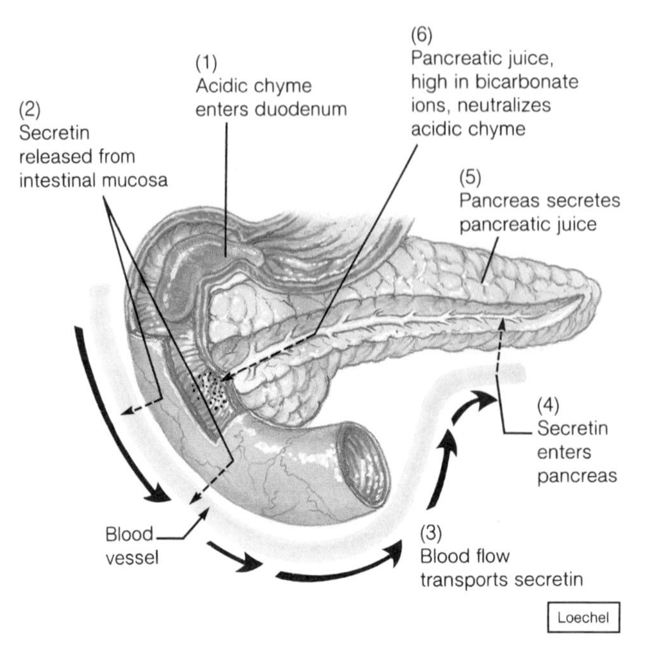

1. List the enzymes found in pancreatic juice.
2. What are the functions of the enzymes in pancreatic juice?
3. How is the secretion of pancreatic juice regulated?

Liver

The **liver** is located in the upper right and central portions of the abdominal cavity, just below the diaphragm. It is partially surrounded by the ribs and extends from the level of the fifth intercostal space to the lower margin of the ribs. It is reddish brown in color and well supplied with blood vessels (fig. 12.1).

Functions of the Liver

The liver carries on many important metabolic activities. For example, it plays a key role in carbohydrate metabolism by helping to maintain the normal concentration of blood glucose. As described in chapter 11, liver cells responding to various hormones can decrease blood glucose by converting glucose to glycogen; they also can increase blood glucose by changing glycogen to glucose or by converting noncarbohydrates into glucose.

The liver's effects on lipid metabolism include oxidizing fatty acids at an especially high rate (see chapter 4); synthesizing lipoproteins, phospholipids, and cholesterol; and converting carbohydrates into proteins and fats. Fats synthesized in the liver are transported by the blood to adipose tissue for storage.

The most vital liver functions are probably those related to protein metabolism. They include deaminating amino acids; forming urea (see chapter 4); synthesizing various blood proteins, including several that are necessary for blood clotting (see chapter 14); and converting various amino acids to other types of amino acids.

The liver also stores a variety of substances including glycogen; iron; and vitamins A, D, and B_{12}. In addition, various liver cells (macrophages) help to destroy damaged red blood cells and foreign substances by phagocytosis. In addition the liver alters the composition of toxic substances (detoxification), such as alcohol, and it secretes bile.

Because many of these liver functions are not directly related to the digestive system, they are discussed in other chapters. Bile secretion, however, is important to digestion and is explained in a subsequent section of this chapter.

Chart 12.3 summarizes the major functions of the liver.

Hepatitis
A Current Topic

Hepatitis is an inflammation of the liver. Most commonly, it is caused by a viral infection.

One form of viral hepatitis, called *type A hepatitis,* usually occurs in children or young adults. It is spread by contact with food or objects, such as eating utensils or toys, that have been contaminated with virus-containing feces. This form of hepatitis is often mild, although it may be accompanied by weakness, abdominal discomfort, nausea, and jaundice. Usually the patient recovers completely with no lasting damage to the liver.

Type B hepatitis produces symptoms similar to those of type A, but the effects may last for a much longer time. This form of the disease is spread by contact with virus-containing body fluids, such as blood, saliva, or semen. Thus, it may be transmitted by means of blood transfusions, hypodermic needle injections, or sexual activities. Most patients recover completely from type B hepatitis; however, some patients continue to harbor live viruses and become "carriers," who may seem healthy but can transmit the disease to others.

Another type of viral hepatitis is called *non-A, non-B hepatitis.* This form of hepatitis is thought to be caused by at least two viruses, one resembling the hepatitis A virus and the other resembling the hepatitis B virus. Non-A, non-B hepatitis is usually transmitted by means of blood or blood products and is responsible for 80% to 90% of the cases of hepatitis that develop following blood transfusions.

Patients who exhibit the symptoms of hepatitis for six months or more are said to have *chronic hepatitis.* In such cases, there is danger that the liver will be permanently damaged and have its functions impaired. In addition to being caused by viral infections, chronic hepatitis may be caused by the effects of certain drugs or alcohol, or an abnormal reaction of the immune system.

Chart 12.3 Major functions of the liver

General function	Specific function
Carbohydrate metabolism	Conversion of glucose to glycogen, of glycogen to glucose, and of noncarbohydrates to glucose
Lipid metabolism	Oxidation of fatty acids; the synthesis of lipoproteins, phospholipids, and cholesterol; the conversion of carbohydrates and proteins into fats
Protein metabolism	Deamination of amino acids; the formation of urea; the synthesis of blood proteins; the interconversion of amino acids
Storage	Stores glycogen, vitamins A, D, and B_{12}, and iron
Blood filtering	Removes damaged red blood cells and foreign substances by phagocytosis
Detoxification	Alters the composition of toxic substances
Secretion	Secretes bile

Structure of the Liver

The liver is enclosed in a fibrous capsule and divided by connective tissue into *lobes*—a large right lobe and a smaller left lobe (fig. 12.15). Each lobe is separated into numerous tiny **hepatic lobules,** which are the functional units of the gland (figs. 12.16 and 12.17). A lobule consists of numerous hepatic cells that radiate outward from a *central vein.* Platelike groups of these cells are separated from each other by vascular channels, called **hepatic sinusoids.** Blood from the digestive tract, which is carried in the portal veins (see chapter 15), brings newly absorbed nutrients into the sinusoids and nourishes the hepatic cells.

Large *Kupffer cells,* which are phagocytic cells, are fixed to the inner linings of the hepatic sinusoids. They remove bacteria or other foreign particles that sometimes gain entrance into the blood of the portal veins through the intestinal wall. The blood passes from these sinusoids into the central veins of the hepatic lobules and moves out of the liver.

Within the liver lobules are many fine *bile canals,* which receive secretions from the hepatic cells. The canals of neighboring lobules unite to form larger ducts, and these converge to become the **hepatic ducts.** These ducts, in turn, merge to form the **common bile duct.**

1. Describe the location of the liver.
2. Review the functions of the liver.
3. Describe a hepatic lobule.

Composition of Bile

Bile is a yellowish green liquid that is secreted continuously by the hepatic cells. In addition to water, it contains *bile salts, bile pigments* (bilirubin and biliverdin), cholesterol, and various electrolytes. Of these, the bile salts are the most abundant, and they are the only substances in bile that have a digestive function.

Figure 12.15
The lobes of the liver as viewed (*a*) from the front and (*b*) from below.

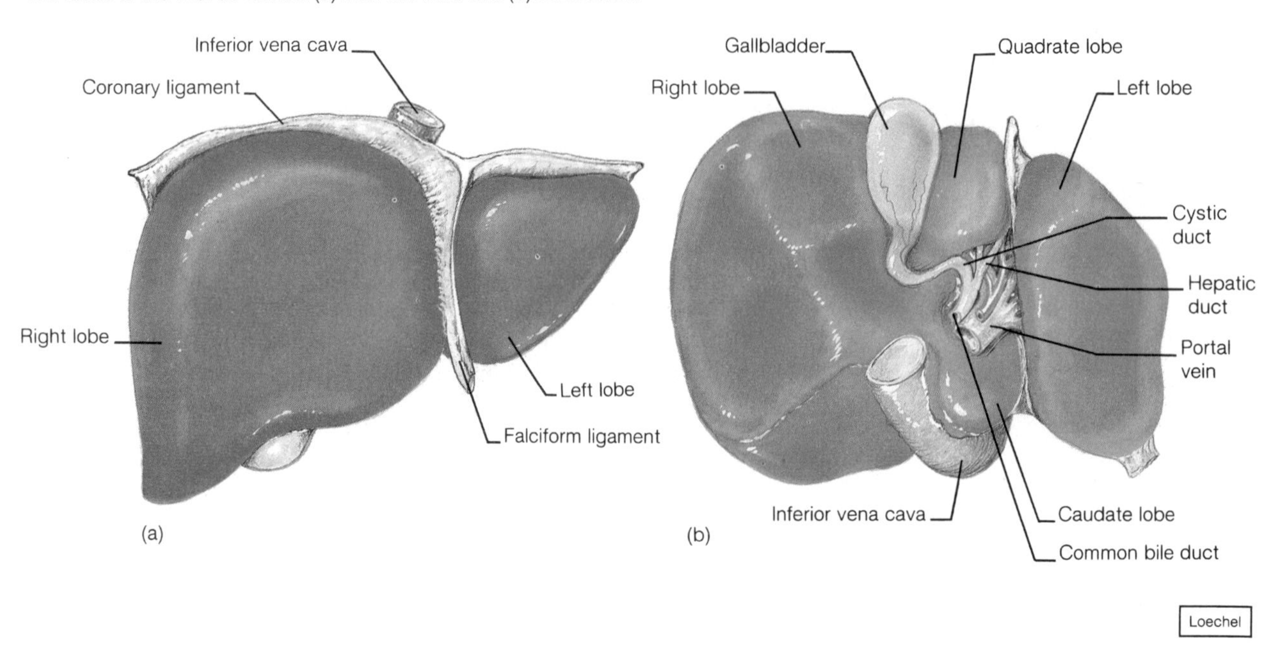

Figure 12.16
A cross section of a part of a hepatic lobule, the functional unit of the liver.

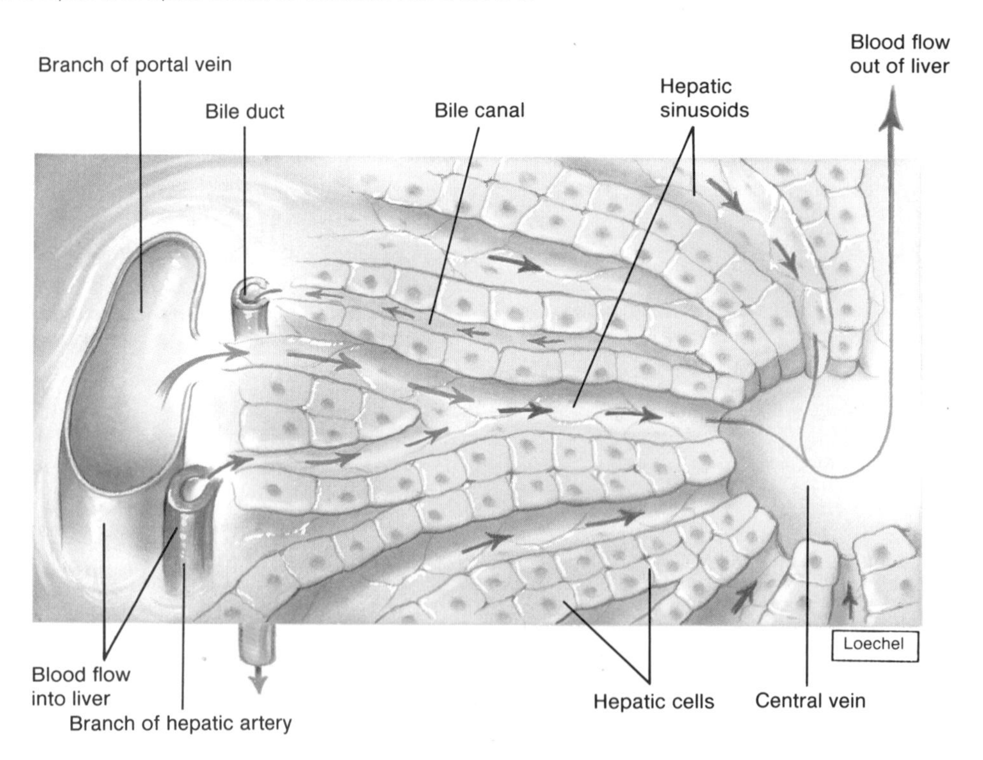

Figure 12.17
A light micrograph of hepatic lobules. What features can you identify?

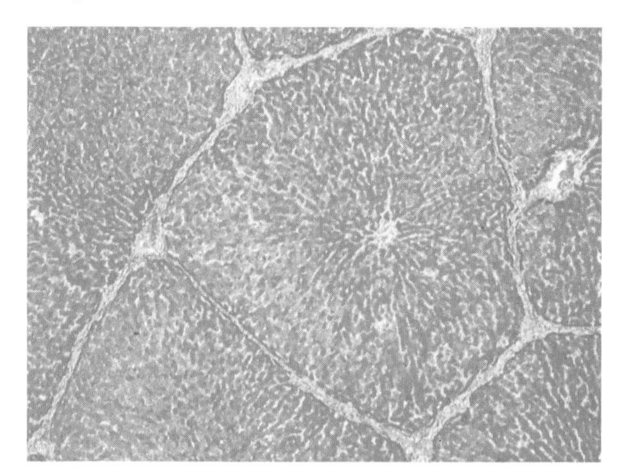

The bile pigments are products of red blood cell breakdown and are normally excreted in the bile (see chapter 14).

If the excretion of bile pigments is prevented by the obstruction of bile ducts, the pigments accumulate in the blood and tissues, causing a yellowish tinge in the skin and other body parts. This condition is called *obstructive jaundice.*

The skin begins to appear yellow when the concentration of bile pigments in the body fluid reaches a level about three times normal. In addition to being caused by obstructions, jaundice may accompany liver diseases in which liver cells are unable to secrete ordinary amounts of pigment (hepatocellular jaundice) or it may occur when an excessive destruction of red blood cells is accompanied by a rapid release of pigments (hemolytic jaundice).

Gallbladder and Its Functions

The **gallbladder** is a pear-shaped sac attached to the ventral surface of the liver by the **cystic duct,** which, in turn, joins the hepatic duct (figs. 12.1 and 12.15). The gallbladder is lined with epithelial cells and has a strong muscular layer in its wall. It stores bile between meals, concentrates bile by reabsorbing water, and releases bile into the small intestine.

The **common bile duct** is formed by the union of the hepatic and cystic ducts. It leads to the duodenum (fig. 12.13), where its exit is guarded by a sphincter muscle (sphincter of Oddi). This sphincter normally remains contracted, so that bile collects in the duct and backs up into the cystic duct. When this happens, the bile flows into the gallbladder and is stored there.

Figure 12.18
An X-ray film of a gallbladder that contains gallstones (*arrow*).

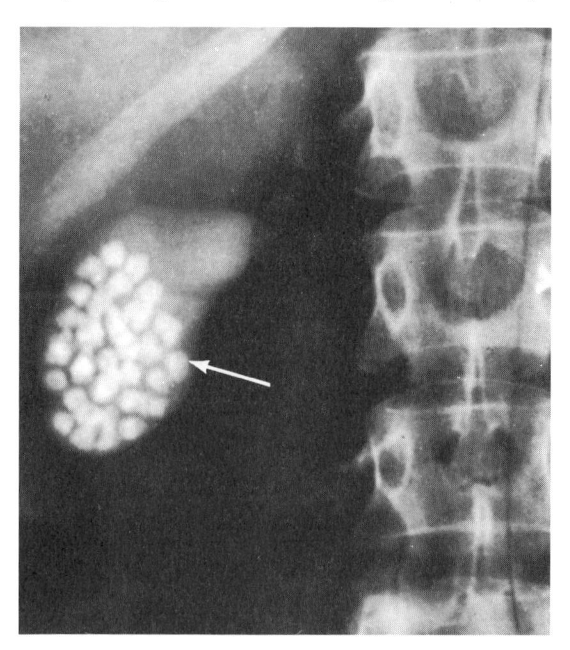

Although the cholesterol in bile normally remains in solution, it may precipitate and form crystals under certain conditions. The resulting solids are called *gallstones,* and if they get into the bile duct, they may block the flow of bile into the small intestine and cause considerable pain.

Generally gallstones that cause obstructions are surgically removed (fig. 12.18). At the same time, the gallbladder is removed by a surgical procedure called *cholecystectomy.*

Regulation of Bile Release

Normally bile does not enter the duodenum until the gallbladder is stimulated to contract by the hormone *cholecystokinin* (fig. 12.19). The usual stimulus involves the presence of protein and fat in the small intestine, which triggers the release of the hormone from the mucosa of the small intestine. The sphincter at the base of the common bile duct remains contracted until a peristaltic wave in the duodenal wall passes by. Then the sphincter relaxes slightly, and a squirt of bile enters the small intestine.

Functions of Bile Salts

Although they do not act as digestive enzymes, bile salts aid the actions of digestive enzymes and enhance the absorption of fatty acids and certain fat-soluble vitamins.

Molecules of fats tend to clump together, forming masses called *fat globules.* Bile salts affect fat globules

Figure 12.19
The gallbladder is stimulated to release bile when fat-containing chyme enters the duodenum.

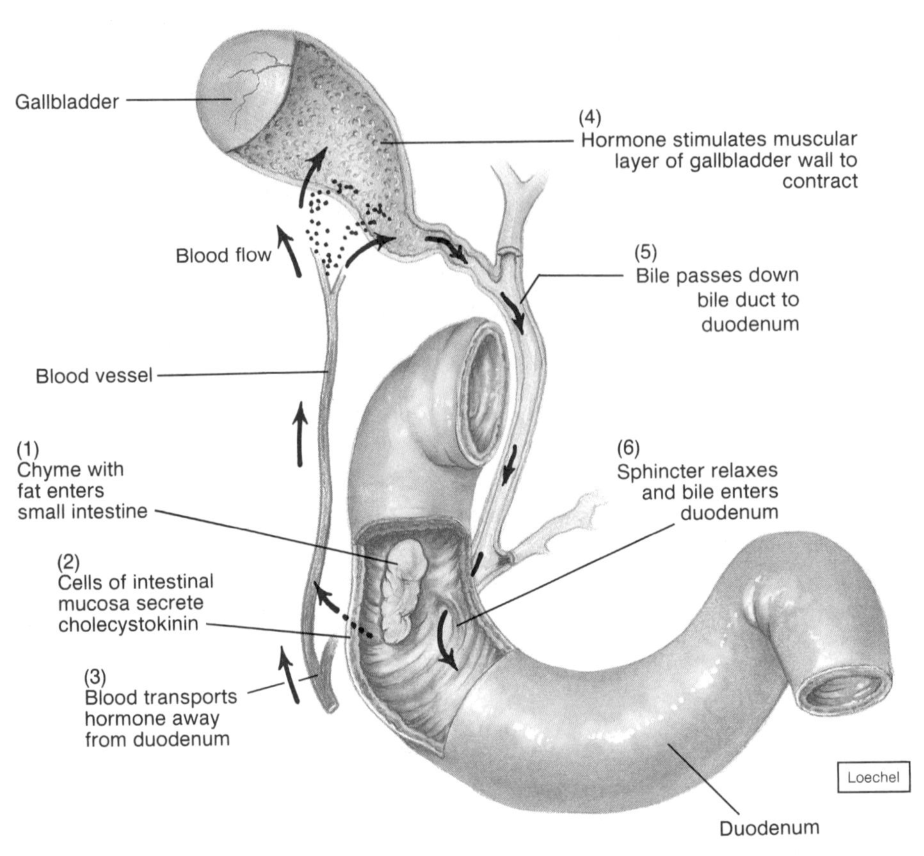

Chart 12.4 Hormones of the digestive tract

Hormone	Source	Function
Gastrin	Gastric cells, in response to the presence of food	Causes the gastric glands to increase their secretory activity
Cholecystokinin	Intestinal wall cells, in response to the presence of proteins and fats in the small intestine	Causes the gastric glands to decrease their secretory activity and inhibits gastric motility; stimulates the pancreas to secrete fluid with a high digestive enzyme concentration; stimulates the gallbladder to contract and release bile
Secretin	Cells in the duodenal wall, in response to acidic chyme entering the small intestine	Stimulates the pancreas to secrete fluid with a high bicarbonate ion concentration

much as a soap or detergent would affect them. That is, bile salts cause fat globules to break up into smaller droplets, an action called **emulsification.** As a result of emulsification, the total surface area of the fatty substance is greatly increased, and the tiny droplets can mix with water. The fat-splitting enzymes (lipases) can then digest the fat molecules more effectively. Bile salts also aid in the absorption of fatty acids and cholesterol. Along with these lipids, various fat-soluble vitamins, such as vitamins A, D, E, and K, also are absorbed. Thus, if bile salts are lacking, lipids may be poorly absorbed, and the person is likely to develop vitamin deficiencies.

The hormones that help to control digestive functions are summarized in chart 12.4.

1. Explain how bile originates.
2. Describe the functions of the gallbladder.
3. How is the secretion of bile regulated?
4. How does bile function in digestion?

Small Intestine

The **small intestine** is a tubular organ that extends from the pyloric sphincter to the beginning of the large in-

Figure 12.20
The three parts of the small intestine are the duodenum, the jejunum, and the ileum.

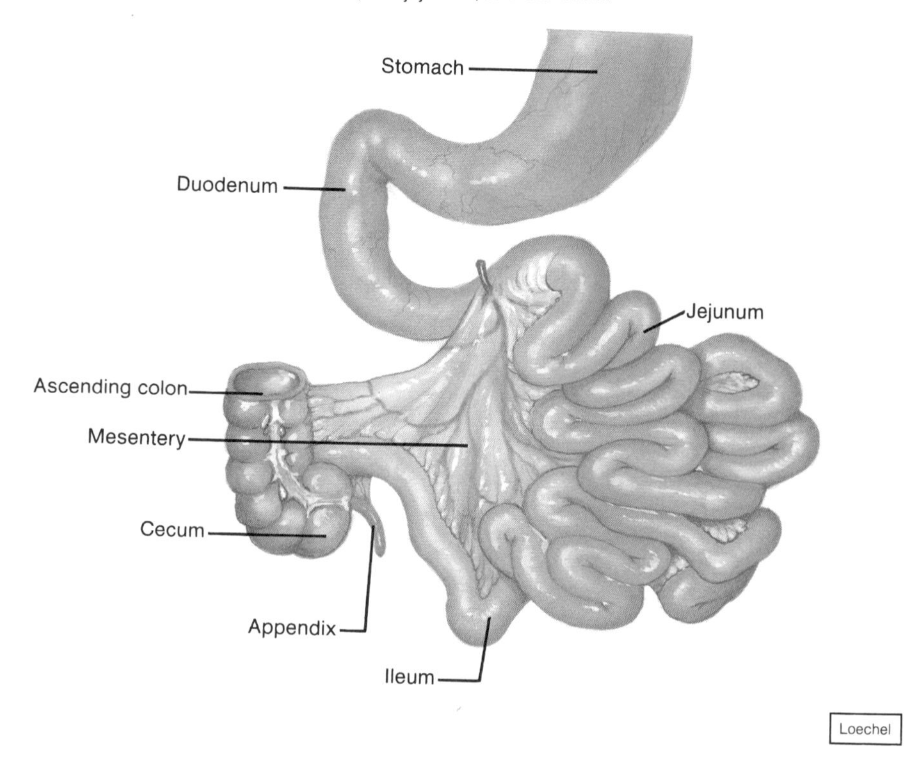

testine. With its many loops and coils, it fills much of the abdominal cavity (fig. 12.1).

As was mentioned, this portion of the alimentary canal receives secretions from the pancreas and liver. It also completes the digestion of the nutrients in chyme, absorbs the various products of digestion, and transports the remaining residues to the large intestine.

Parts of the Small Intestine

The small intestine consists of three portions: the duodenum, jejunum, and ileum (figs. 12.20 and 12.21).

The **duodenum,** which is about 25 centimeters long and 5 centimeters in diameter, lies behind the parietal peritoneum and is the most fixed portion of the small intestine. It follows a C-shaped path as it passes in front of the right kidney and upper three lumbar vertebrae.

The remainder of the small intestine is mobile and lies free in the peritoneal cavity. The proximal two-fifths of this portion is called the **jejunum,** and the remainder is the **ileum.** These portions are suspended from the posterior abdominal wall by a double-layered fold of peritoneum, called **mesentery** (fig. 12.20). This supporting tissue contains the blood vessels, nerves, and lymphatic vessels that supply the intestinal wall.

Although there is no distinct separation between the jejunum and ileum, the diameter of the jejunum tends to be greater, and its wall is thicker, more vascular, and more active than that of the ileum.

Figure 12.21
X-ray film showing a normal small intestine.

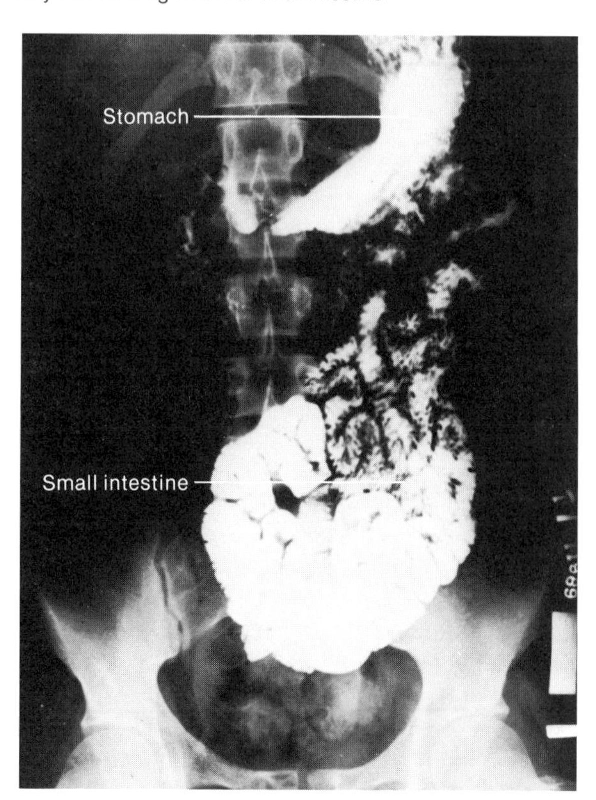

Figure 12.22
The structure of a single intestinal villus. What is the function of a villus?

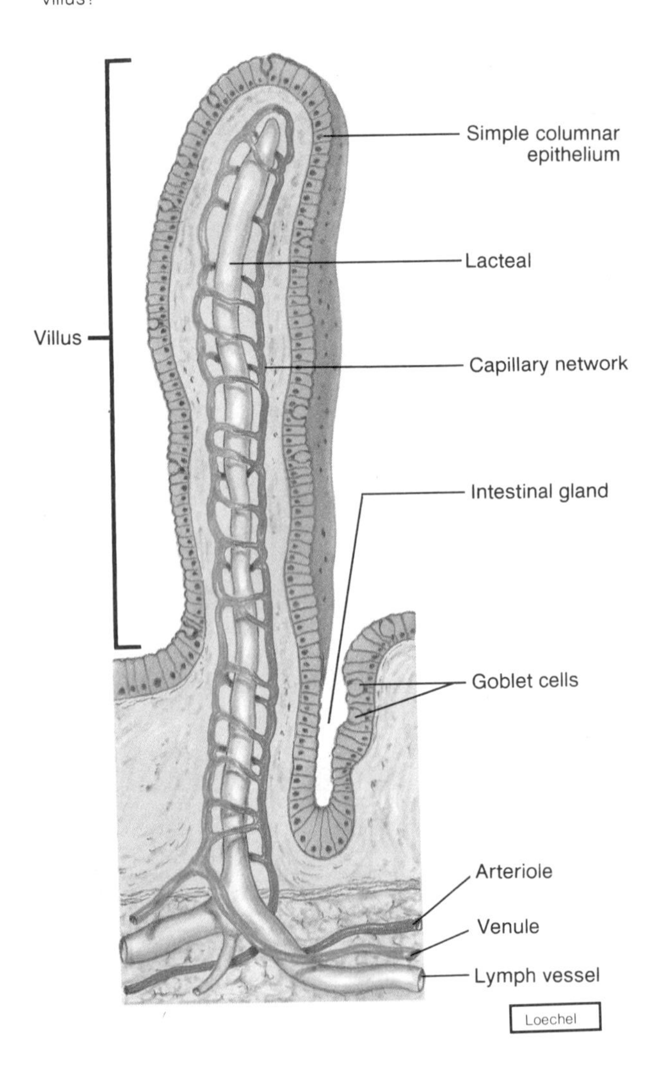

Figure 12.23
Light micrograph of intestinal villi from the wall of the duodenum.

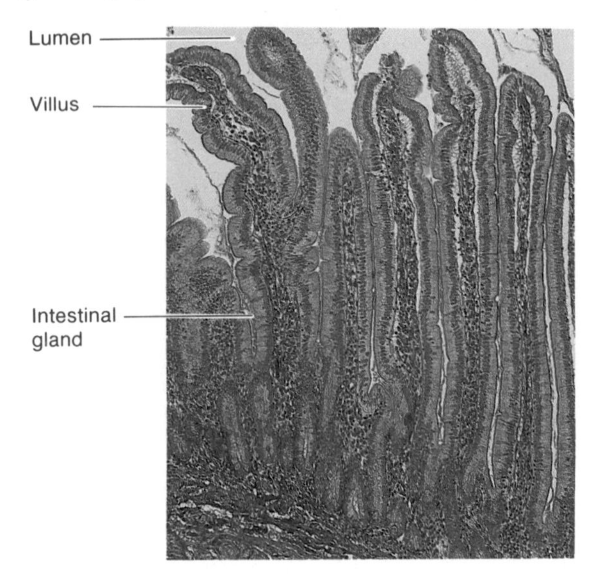

Structure of the Small Intestinal Wall

Throughout its length, the inner wall of the small intestine has a velvety appearance. This is due to the presence of innumerable tiny projections of the mucous membrane, called **intestinal villi** (figs. 12.2, 12.22, and 12.23). These structures are most numerous in the duodenum and the proximal portion of the jejunum. They project into the passageway, or **lumen,** of the alimentary canal, where they come in contact with the intestinal contents. The villi increase the surface area of the intestinal lining, and they play an important role in the absorption of digestive products.

Each villus consists of a layer of simple columnar epithelium and a core of connective tissue containing blood capillaries, a lymphatic capillary (called a *lacteal*), and nerve fibers.

The blood and lymph capillaries function to carry away substances absorbed by a villus, and the nerve fibers act to stimulate or inhibit its activities.

Between the bases of adjacent villi are tubular **intestinal glands,** which extend downward into the mucous membrane (figs. 12.2 and 12.22).

The epithelial cells that form the lining of the small intestine are continually being replaced. The new cells are formed within the intestinal glands by mitosis, and these cells migrate outward onto the surface of a villus. When the cells reach the tip of the villus, they are shed. As a result of this process, which is called *cellular turnover,* the epithelial lining of the small intestine is renewed every three to six days.

Secretions of the Small Intestine

In addition to the mucus-secreting goblet cells, which occur extensively throughout the mucosa of the small intestine, there are many specialized *mucus-secreting glands* within the submucosa of the proximal duodenum. These glands secrete large quantities of thick, alkaline mucus in response to various stimuli.

The intestinal glands at the bases of the villi secrete great amounts of a watery fluid. The villi rapidly reabsorb this fluid, and it provides a vehicle for moving digestive products into the villi. The fluid secreted by the intestinal glands has a pH that is nearly neutral (6.5–7.5), and it seems to lack digestive enzymes. The epithelial cells of the intestinal mucosa, however, have digestive enzymes embedded in the surfaces of their microvilli. These enzymes break down food molecules just before absorption takes place. The enzymes include **peptidases,** which split peptides into amino acids; **sucrase, maltase,** and **lactase,** which split the double sugars (disaccharides) sucrose, maltose, and lactose into

Chart 12.5 Summary of the major digestive enzymes		
Enzyme	**Source**	**Digestive action**
Salivary enzyme		
Amylase	Salivary glands	Begins carbohydrate digestion by converting starch and glycogen to disaccharides
Gastric enzyme		
Pepsin	Gastric glands	Begins the digestion of proteins
Pancreatic enzymes		
Amylase	Pancreas	Converts starch and glycogen into disaccharides
Lipase	Pancreas	Converts fats into fatty acids and glycerol
Proteinases	Pancreas	Converts proteins or partially digested proteins into peptides
a. Trypsin		
b. Chymotrypsin		
c. Carboxypeptidase		
Nucleases	Pancreas	Convert nucleic acids into nucleotides
Intestinal enzymes		
Peptidase	Mucosal cells	Converts peptides into amino acids
Sucrase, maltase, lactase	Mucosal cells	Converts disaccharides into monosaccharides
Lipase	Mucosal cells	Converts fats into fatty acids and glycerol
Enterokinase	Mucosal cells	Converts trypsinogen into trypsin

the simple sugars (monosaccharides) glucose, fructose, and galactose, respectively; and **intestinal lipase,** which splits fats into fatty acids and glycerol.

Chart 12.5 summarizes the sources and actions of the major digestive enzymes.

The amount of lactase produced in the small intestine usually reaches a maximum shortly after birth and thereafter tends to decrease. As a result, some adults produce insufficient quantities of lactase to break down the lactose (milk sugar) in their diets—a condition called *lactose intolerance.* When this happens, the lactose from milk and various milk products remains undigested and causes an increase in the osmotic pressure of the intestinal contents. Consequently, water is drawn from the tissues into the intestine. At the same time, intestinal bacteria may act upon the undigested sugar and produce organic acids and gases. As a result, the person may feel bloated and suffer from intestinal cramps and diarrhea.

Regulation of Small Intestinal Secretions

Secretions from the goblet cells and intestinal glands are stimulated by direct contact with chyme, which provides both chemical and mechanical stimuli, and by reflexes triggered by distension of the intestinal wall. The reflex actions involve parasympathetic motor impulses that cause secretory cells to increase their activities.

1. Describe the parts of the small intestine.
2. Define *intestinal villi.*
3. What is the function of the intestinal glands?
4. List the digestive enzymes formed by the intestinal cells.

Absorption in the Small Intestine

Because the villi greatly increase the surface area of the intestinal mucosa, the small intestine is the most important absorbing organ of the alimentary canal. In fact, the small intestine is so effective in absorbing digestive products, water, and electrolytes that very little absorbable material reaches its distal end.

Carbohydrate digestion begins in the mouth with the activity of salivary amylase. It is completed in the small intestine by enzymes from the intestinal mucosa and pancreas. The resulting monosaccharides are absorbed by the villi and enter the blood capillaries. Even though small quantities of these simple sugars may pass into the villi by diffusion, most of them are absorbed by active transport or facilitated diffusion (see chapter 3).

Protein digestion begins in the stomach as a result of pepsin activity and is completed in the small intestine by enzymes from the intestinal mucosa and pancreas. During this process, large protein molecules are converted into amino acids. These smaller particles then enter the villi by active transport and are carried away by the blood.

Fat molecules are digested almost entirely by enzymes from the intestinal mucosa and pancreas. The resulting fatty acids and glycerol molecules diffuse into the epithelial cells of the villi (fig. 12.24). They are resynthesized into fat molecules by action of the endoplasmic reticulum. The resynthesized fat molecules are similar to those previously digested. These fats are encased in protein to form tiny droplets, called chylomicrons, which make their way to the lacteals of the villi. Lymph in the lacteals and other lymphatic vessels carries the chylomicrons to the blood (see chapter 16).

Figure 12.24
Fatty acid absorption involves several steps.

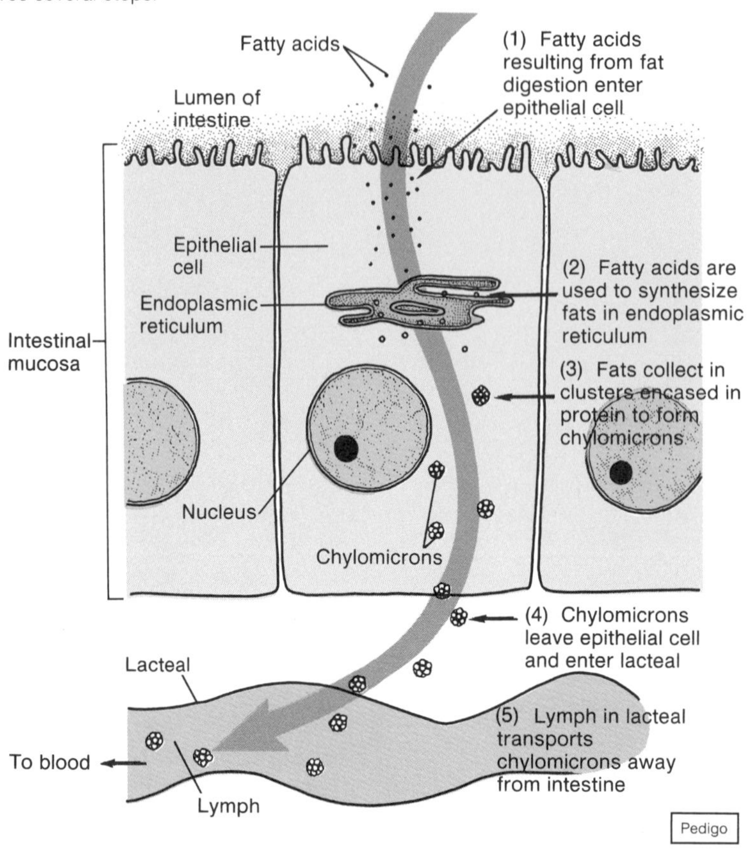

On the other hand, some fatty acids with relatively short carbon chains may be absorbed directly into the blood capillary of a villus without being converted back into fat.

In addition to absorbing the products of carbohydrate, protein, and fat digestion, the intestinal villi absorb various electrolytes and water. Chart 12.6 summarizes how nutrients are absorbed.

1. What substances resulting from the digestion of carbohydrate, protein, and fat molecules are absorbed by the small intestine?
2. Describe how fatty acids are absorbed.

Movements of the Small Intestine

Like the stomach, the small intestine carries on mixing movements and peristalsis. The mixing movements include small, ringlike contractions that occur periodically, cutting the chyme into segments and moving it to and fro.

The chyme is propelled through the small intestine by peristaltic waves. These waves are usually weak, and they stop after pushing the chyme a short distance.

Chart 12.6 Intestinal absorption of nutrients

Nutrient	Absorption mechanism	Means of transport
Monosaccharides	Diffusion and active transport	Blood in the capillaries
Amino acids	Active transport	Blood in the capillaries
Fatty acids and glycerol	Diffusion into cells (a) Most fatty acids are converted back into fats and incorporated in chylomicrons for transport	Lymph in the lacteals
	(b) Some fatty acids with relatively short carbon chains are transported without being converted back into fats	Blood in the capillaries
Electrolytes	Diffusion and active transport	Blood in the capillaries
Water	Osmosis	Blood in the capillaries

Figure 12.25
Major parts of the large intestine (*anterior view*).

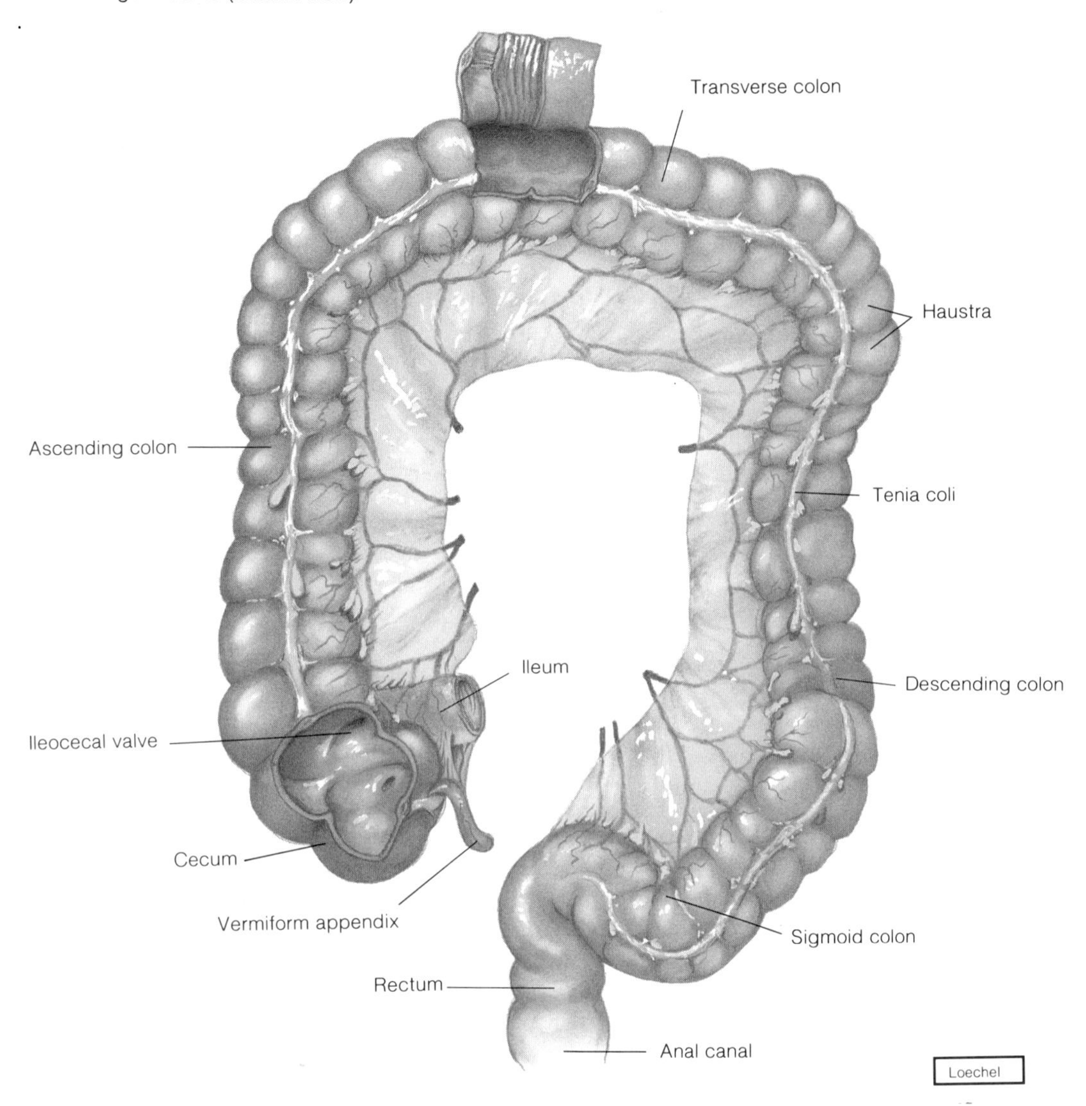

Consequently, food materials move relatively slowly through the small intestine, taking from three to ten hours to travel its length.

Sometimes, however, stimulation of the small intestinal wall by overdistension or by irritation may elicit a strong *peristaltic rush* that passes along the entire length of the small intestine. This type of movement serves to sweep the contents of the small intestine into the large intestine relatively rapidly.

The rapid movement of chyme through the intestine may prevent the normal absorption of water, nutrients, and electrolytes from the intestinal contents. The result is *diarrhea,* a condition in which defecation becomes more frequent and the stools are watery. If diarrhea continues for a prolonged time, imbalances in water and electrolyte concentrations are likely to develop.

At the distal end of the small intestine, where the ileum joins the cecum of the large intestine, is a sphincter muscle called the **ileocecal valve** (fig. 12.25). Normally this sphincter remains constricted, preventing the contents of the small intestine from entering. At the same time, it prevents the contents of the large intestine from backing up into the ileum. After a meal however, a reflex is elicited, and peristalsis in the ileum is increased. This action forces some of the contents of the small intestine into the cecum.

1. Describe the movements of the small intestine.
2. What stimulus causes the ileocecal valve to relax?

Large Intestine

The **large intestine** is so named because its diameter is greater than that of the small intestine. This portion of the alimentary canal is about 1.5 meters long, and it begins in the lower right side of the abdominal cavity, where the ileum joins the cecum. From there, the large intestine travels upward on the right side, crosses obliquely to the left, and descends into the pelvis. At its distal end, it opens to the outside as the anus (fig. 12.1).

The large intestine reabsorbs water and electrolytes from the chyme remaining in the alimentary canal. It also forms and stores the feces until defecation occurs.

Parts of the Large Intestine

The large intestine, shown in figures 12.25 and 12.26, consists of the cecum, colon, rectum, and anal canal.

The **cecum,** which represents the beginning of the large intestine, is a dilated, pouchlike structure that hangs slightly below the ileocecal opening. Projecting downward from it is a narrow tube with a closed end called the **vermiform appendix.** Although the human appendix has no digestive function, it does contain lymphatic tissue that can provide immune functions (chapter 16).

Occasionally the appendix may become infected and inflamed, causing the condition called *appendicitis.* When this happens, the appendix is often removed surgically to prevent it from rupturing. If it breaks open, the contents of the large intestine may enter the abdominal cavity and cause a serious infection of the peritoneum, called *peritonitis.*

The **colon** is divided into four portions—the ascending, transverse, descending, and sigmoid colons. The **ascending colon** begins at the cecum and travels upward against the posterior abdominal wall to a point just below the liver. There it turns sharply to the left and becomes the **transverse colon.** The transverse colon is the longest and the most mobile part of the large intestine. It is suspended by a fold of peritoneum and tends to sag in the middle below the stomach. As the transverse colon approaches the spleen, it turns abruptly downward and becomes the **descending colon.** At the brim of the pelvis, the descending colon makes an S-shaped curve, called the **sigmoid colon,** and then becomes the rectum.

The **rectum** lies next to the sacrum and follows its curvature. It is firmly attached to the sacrum by connective tissue, and it ends about 5 centimeters below the tip of the coccyx, where it becomes the anal canal (fig. 12.25).

Figure 12.26
An X-ray film of the large intestine. What portions can you identify?

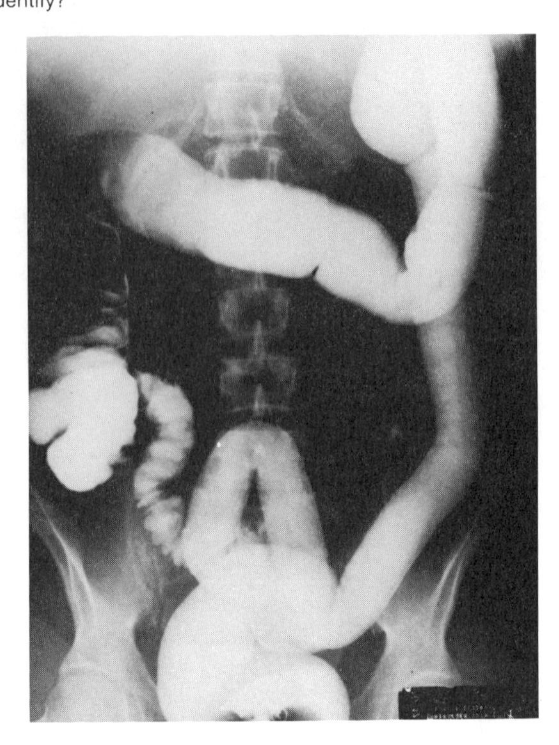

The **anal canal** is formed by the last 2.5 to 4.0 centimeters of the large intestine (fig. 12.27). The mucous membrane in the anal canal is folded into a series of six to eight longitudinal *anal columns.* At its distal end, the canal opens to the outside as the **anus.** This opening is guarded by two sphincter muscles, an *internal anal sphincter,* composed of smooth muscle, under involuntary control, and an *external anal sphincter,* composed of skeletal muscle, under voluntary control.

1. What is the general function of the large intestine?
2. Describe the parts of the large intestine.

Each anal column contains a branch of the rectal vein, and if something interferes with the blood flow in these vessels, the anal columns may become enlarged and inflamed. This condition, called *hemorrhoids,* may be aggravated by bowel movements and may be accompanied by discomfort and bleeding.

Structure of the Large Intestinal Wall

Although the wall of the large intestine includes the same types of tissues found in other parts of the alimentary canal, it has some unique features. For ex-

Figure 12.27
The rectum and anal canal are located at the distal end of the alimentary canal.

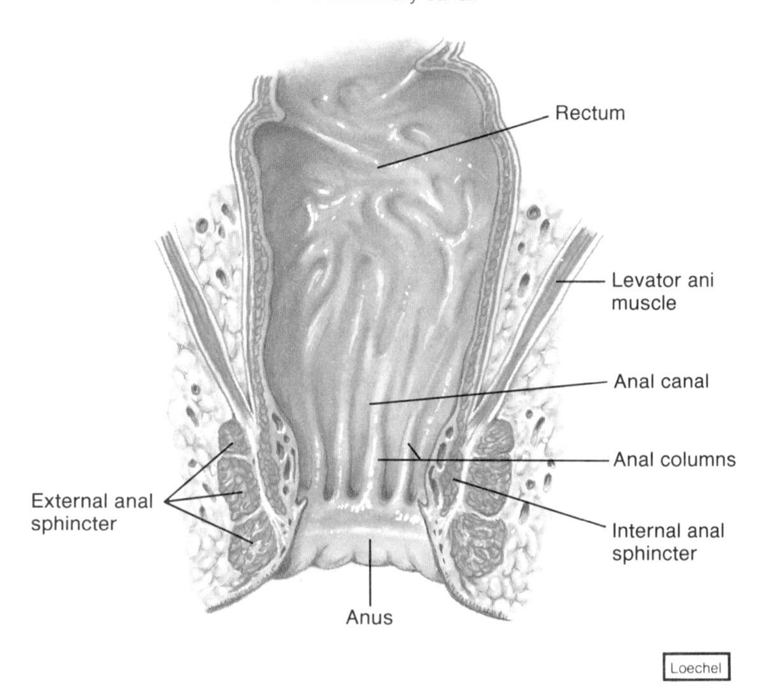

ample, it lacks the villi that are characteristic of the small intestine. Also, the layer of longitudinal muscle fibers does not cover its wall uniformly. Instead, the fibers are arranged in three distinct bands (teniae coli) that extend the entire length of the colon (fig. 12.25). These bands exert tension on the wall, creating a series of pouches (haustra).

Functions of the Large Intestine

Unlike the small intestine, which secretes digestive enzymes and absorbs the products of digestion, the large intestine has little or no digestive function. On the other hand, the mucous membrane that forms the inner lining of the large intestine contains many tubular glands. Structurally these glands are similar to those in the small intestine, but they are composed almost entirely of goblet cells. Consequently, mucus is the only significant secretion of the large intestine.

The mucus secreted into the large intestine protects the intestinal wall against the abrasive action of the material passing through. It also aids in holding particles of fecal matter together, and because it is alkaline, mucus helps to control the pH of the large intestinal contents.

The chyme entering the large intestine contains materials that were not digested or absorbed by the small intestine. It also contains water, various electrolytes, and bacteria. The proximal half of the large intestine functions to reabsorb some of the water and electrolytes. The substances that remain in the tube become feces, which are stored for a time in the distal portion of the large intestine.

1. How does the structure of the large intestine differ from that of the small intestine?
2. What substances can be absorbed by the large intestine?

Movements of the Large Intestine

The movements of the large intestine—mixing and peristalsis—are similar to those of the small intestine, although they are usually more sluggish. Also, instead of occurring frequently, peristaltic waves in the large intestine come only two or three times each day. These waves produce *mass movements* in which a relatively large section of the colon constricts vigorously, forcing its contents to move toward the rectum. Typically, mass movements occur following a meal, as a result of a reflex that is initiated in the small intestine. Abnormal irritations of the mucosa also can trigger such movements. For instance, a person suffering from an inflamed colon (colitis) may experience frequent mass movements.

When it is appropriate to defecate, a person usually can initiate a *defecation reflex* by holding a deep breath and contracting the abdominal wall muscles. This action increases the internal abdominal pressure and forces the feces into the rectum. As the rectum fills, its wall is distended and the defecation reflex is triggered. As a result, peristaltic waves in the descending colon are stimulated, and the internal anal sphincter relaxes. At the same time, other reflexes involving the sacral region of the spinal cord cause the peristaltic waves to strengthen, the diaphragm to lower, the glottis to close, and the abdominal wall muscles to contract. These actions cause an additional increase in the internal abdominal pressure and assist in squeezing the rectum. The external anal sphincter is signaled to relax, and the feces are forced to the outside. A person usually can inhibit defecation voluntarily by keeping the external anal sphincter contracted.

The action of taking a deep breath, closing the glottis, and forcibly contracting the abdominal wall muscles is called the *Valsalva maneuver.* This action causes the internal abdominal pressure to increase and aids defecation. It also causes the internal thoracic pressure to rise and may interfere with the return of blood to the heart. For this reason, the Valsalva maneuver may present a hazard to persons with heart disease.

Feces

As was mentioned, the **feces** are composed largely of materials that were not digested or absorbed, together with water, electrolytes, mucus, and bacteria. Usually the feces are about 75% water, and their color is normally due to the presence of bile pigments that have been altered somewhat by bacterial actions.

The pungent odor of the feces results from a variety of compounds produced by bacteria acting upon the residues.

1. How does peristalsis in the large intestine differ from peristalsis in the small intestine?
2. List the major events that occur during defecation.
3. Describe the composition of the feces.

Nutrition and Nutrients

Nutrition is the process by which necessary food substances are taken in and utilized by the body. These food substances, or **nutrients,** include carbohydrates, lipids, proteins, vitamins, and water (see chapters 2 and 4). Some of them, such as certain amino acids and fatty acids, are of particular importance because they cannot be synthesized in adequate amounts by human cells. Because it is essential for health that they be provided in the diet, they are known as *essential nutrients.*

Carbohydrates

Carbohydrates are organic compounds that are used primarily to supply energy for cellular processes.

Sources of Carbohydrates

Carbohydrates are ingested in a variety of forms including starch from grains and certain vegetables; glycogen from meats and seafoods; disaccharides from cane sugar, beet sugar, and molasses; and simple sugars from honey and various fruits. During digestion, the more complex carbohydrates, such as starch and glycogen, are changed to monosaccharides—a form in which they can be absorbed and transported by the blood to the body cells.

An exception to this general process is the polysaccharide called *cellulose,* which is very abundant in plant foods. Although cellulose is composed of glucose units arranged in chains, somewhat like plant starch, it cannot be broken down by human digestive enzymes. Consequently, it passes through the alimentary canal largely unchanged. The presence of cellulose in foods, however, is important to the function of the digestive system. It provides bulk (sometimes called fiber or roughage) in the digestive tube against which the muscular wall can push. Thus, cellulose facilitates the movement of food substances through the digestive tract.

Utilization of Carbohydrates

The monosaccharides that are absorbed from the digestive tract include *fructose, galactose,* and *glucose.* Fructose and galactose are normally converted into glucose by enzymes of the liver, and glucose is the form of carbohydrate that is most commonly oxidized by the cells as fuel.

Some of the glucose present usually is converted to *glycogen* and stored in the liver and muscles (fig. 12.28). Glucose can be mobilized rapidly from glycogen when it is needed to supply energy. Only a certain amount of glycogen can be stored, however, and excess glucose is usually converted into fat and stored in adipose tissue.

Although most carbohydrates are used to supply cellular energy, some are used by the body to synthesize vital substances. These substances include the 5-carbon sugars *ribose* and *deoxyribose,* which are needed for the synthesis of the nucleic acids RNA and DNA, as well as the disaccharide *lactose* (milk sugar), which is synthesized when the breasts are actively secreting milk.

Many cells can also obtain energy by oxidizing fatty acids. Some cells, however, such as neurons are dependent on a continuous supply of glucose for sur-

Figure 12.28
Monosaccharides from foods are used for energy, stored as glycogen, or converted to fat.

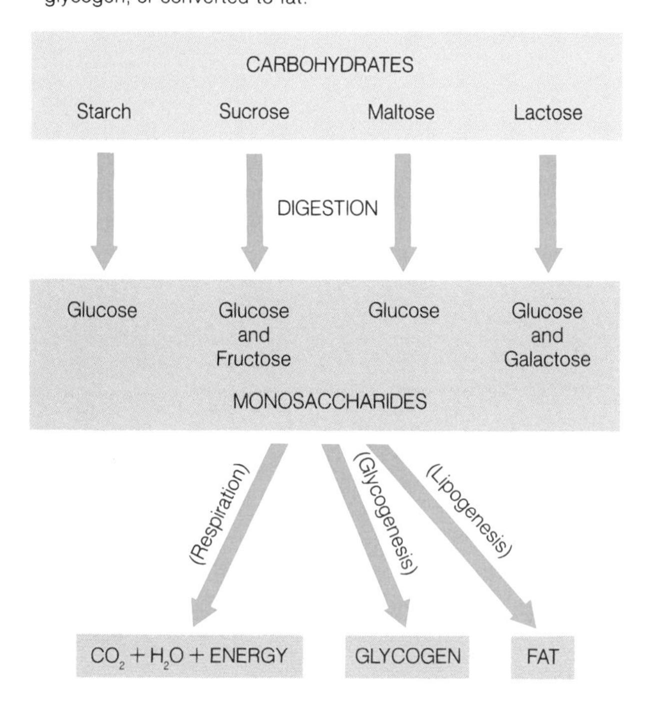

vival. Even a temporary decrease in the glucose concentration may result in a serious functional disorder of the nervous system or the death of nerve cells. Consequently, the presence of a minimum amount of carbohydrate in the body is essential. If an adequate supply of carbohydrates is not received from foods, the liver may convert noncarbohydrates, such as amino acids from proteins, into glucose. Thus, the need for glucose has physiological priority over the need to synthesize proteins from available amino acids.

Carbohydrate Requirement

Because carbohydrates provide the primary source of fuel for cellular processes, the need for carbohydrates varies with individual energy requirements. Thus, persons who are physically active require more fuel than those who are sedentary. The minimal requirement for carbohydrates in the human diet is unknown. It is estimated, however, that an intake of at least 100–125 grams daily is necessary to spare protein (that is, to avoid an excessive breakdown of protein) and to avoid metabolic disorders that sometimes accompany an excessive utilization of fats.

1. List several common sources of carbohydrates.
2. Explain the importance of cellulose in a diet.
3. Explain why the need for glucose has priority over protein synthesis.

Lipids

Lipids comprise the group of organic compounds that includes fats, oils, and fatlike substances (see chapter 2). They are used to supply energy for cellular processes and to build structures, such as cell membranes. Although lipids include fats, phospholipids, and cholesterol, the most common dietary lipids are the fats called *triglycerides.*

Sources of Lipids

Triglycerides are contained in foods of both plant and animal origin. They occur in meats, eggs, milk, and lard, as well as in various nuts and plant oils, such as corn oil, peanut oil, and olive oil.

Cholesterol is obtained in relatively high concentrations from such foods as liver, egg yolk, and brain and is present in lesser amounts in whole milk, butter, cheese, and meats. It does not occur in foods of plant origin.

Utilization of Lipids

During digestion, triglycerides are broken down into fatty acids and glycerol. After being absorbed, these products are transported by the lymph and blood to various tissues. The metabolism of these substances is controlled mainly by the liver and adipose tissues.

The liver can convert fatty acids from one form to another, but it cannot synthesize the type of fatty acid called *linoleic acid.* Thus, this substance is an **essential fatty acid.** It is needed for the synthesis of certain phospholipids, which, in turn, are necessary for the formation of cell membranes and the transport of circulating lipids. Good sources of linoleic acid include corn oil, cottonseed oil, and soy oil.

The liver uses free fatty acids to synthesize triglycerides, phospholipids, and lipoproteins that may then be released into the blood (fig. 12.29). Thus, the liver is largely responsible for the control of circulating lipids. In addition, it regulates the total amount of cholesterol in the body by synthesizing cholesterol and releasing it into the blood, or by removing cholesterol from the blood and excreting it into the bile. The liver also uses cholesterol in the production of bile salts.

Cholesterol is not used as an energy source, but it does provide structural material for a variety of cell parts, including cell membranes, and it furnishes molecular components for the synthesis of various sex hormones and hormones produced by the adrenal cortex.

Excessive triglycerides are stored in adipose tissue, and if the blood lipid concentration drops (in response to fasting, for example), some of these triglycerides are hydrolyzed into free fatty acids and glycerol and then released into the blood.

Figure 12.29
Fatty acids are used by the liver to synthesize a variety of lipids.

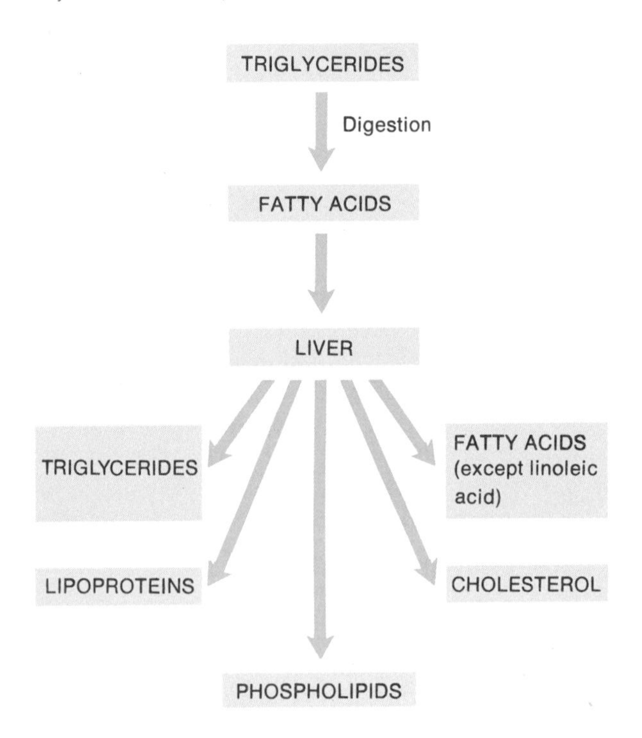

Lipid Requirements

The amounts and types of fats needed for health are unknown. Because linoleic acid is an essential fatty acid, however, nutritionists recommend that infants receive formulas in which 3% of the energy intake is in the form of linoleic acid. Fatty acid deficiencies have not been observed in adults, so it has been concluded that a typical adult diet consisting of a variety of foods provides an adequate supply of this essential nutrient.

Because fats contain fat-soluble vitamins, the intake of fats must also be sufficient to supply the needed amounts of these nutrients.

A report from the Surgeon General of the United States recommends that to improve health, most people should reduce their intake of saturated fats by selecting leaner meats and fish for their meals and by using low-fat dairy products. The report also suggests that people should increase their intake of fruits, vegetables, and whole grains.

The American Heart Association recommends that fats make up no more than 30% of the calories in a person's diet.

1. Which fatty acid is an essential nutrient?
2. What is the role of the liver in the utilization of lipids?
3. What is the function of cholesterol?

Chart 12.7 Amino acids found in foods

Alanine	Leucine (e)
Arginine (ch)	Lysine (e)
Aspartic acid	Methionine (e)
Asparagine	Phenylalanine (e)
Cysteine	Proline
Glutamic acid	Serine
Glutamine	Threonine (e)
Glycine	Tryptophan (e)
Histidine (ch)	Tyrosine
Isoleucine (e)	Valine (e)
Hydroxyproline	

Eight essential amino acids (e) cannot be synthesized by human cells and must be provided in the diet. Two additional amino acids (ch) are essential in growing children.

Proteins

Proteins are organic compounds that serve as structural materials in cells, act as enzymes that regulate metabolic reactions, and are often used to supply energy (see chapter 2).

Sources of Proteins

Foods that are rich in proteins include meats, fish, poultry, cheese, nuts, milk, eggs, and cereals. Various legumes, such as beans and peas, contain lesser amounts.

During digestion, proteins are broken down into their component amino acids, and these smaller molecules are absorbed by intestinal tissues and transported to the body cells by the blood.

Although human body cells can synthesize many amino acids (nonessential amino acids), eight amino acids needed by the adult body (ten needed by growing children) cannot be synthesized or are not synthesized in adequate amounts. For this reason they are called **essential amino acids.**

All essential amino acids must be present in the body at the same time if growth and the repair of tissues are to occur. In other words, if one essential molecule is missing, the process of protein synthesis cannot take place.

Chart 12.7 lists the amino acids found in foods and indicates those that are essential.

On the basis of the kinds of amino acids that they provide, proteins can be classified as complete or incomplete. **Complete proteins,** which include those available in milk, meats, fish, poultry, and eggs, contain adequate amounts of the essential amino acids. **Incomplete proteins,** such as *zein* in corn, which lacks adequate amounts of the essential amino acids tryptophan

Figure 12.30
Amino acids resulting from the digestion of proteins are used to synthesize proteins and to supply energy.

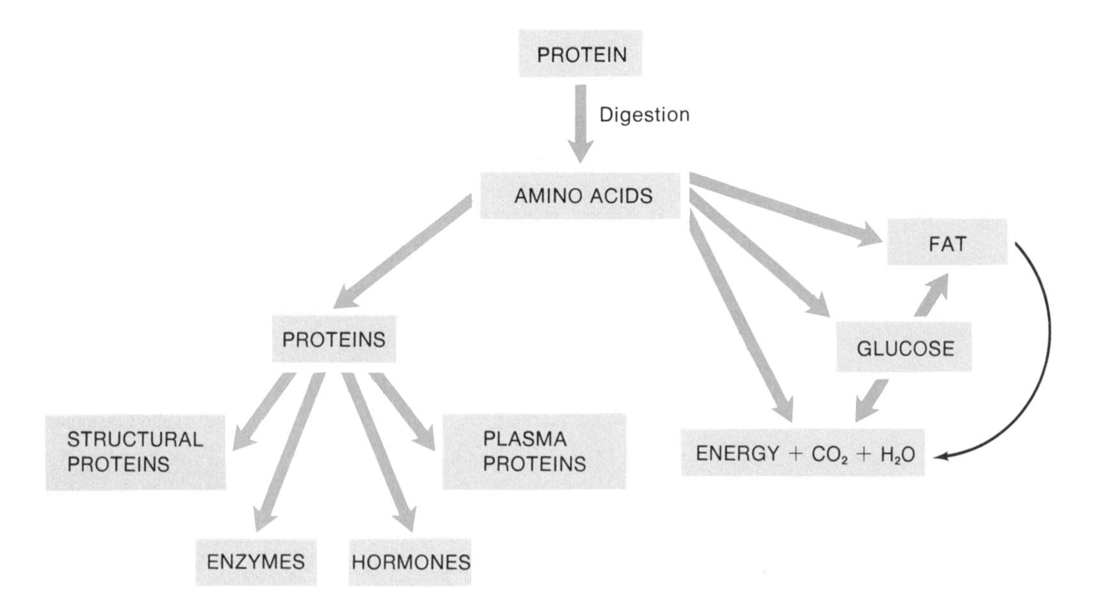

and lysine, and *gelatin,* which lacks adequate amounts of tryptophan, are unable to support tissue maintenance or normal growth and development.

A protein called *gliadin,* which occurs in wheat, is an example of a **partially complete protein.** It is deficient in the essential amino acid lysine, and although it does not contain enough lysine to promote growth, it contains enough to maintain life.

Plant proteins typically contain less than adequate amounts of one or more of the essential amino acids. Protein-containing plant foods, however, can be combined in a meal so that one supplies the amino acids another lacks. For example, beans are deficient in the essential amino acid *methionine,* but they contain adequate amounts of *lysine.* Rice is deficient in *lysine* but contains adequate amounts of *methionine.* Thus, if beans and rice are eaten together, they complement one another, and the meal provides a suitable combination of essential amino acids.

1. What foods provide rich sources of proteins?
2. Why are some amino acids called essential?
3. Distinguish between a complete protein and an incomplete protein.

Utilization of Amino Acids

Amino acids are used by the body cells in a variety of ways (fig. 12.30). For example, some amino acids are incorporated into protein molecules that provide cellular structure as in the *actin* and *myosin* of muscle fibers, the *collagen* of connective tissue fibers, and the *keratin* of the skin. Other amino acids are used to synthesize proteins that function in various body processes: hemoglobin, which transports oxygen; plasma proteins, which help to regulate water balance and control pH; enzymes, which catalyze chemical reactions; and hormones, which help to coordinate metabolic processes.

Amino acids also represent potential sources of energy. When they are present in excess or when the supplies of carbohydrates and fats are insufficient to provide needed energy, amino acids are oxidized using various metabolic pathways.

Protein Requirements

In addition to supplying essential amino acids, proteins are needed to provide molecular parts and nitrogen for the synthesis of nonessential amino acids. Consequently, the amount of protein required by individuals varies according to body size, metabolic rate, and nitrogen needs.

For an average adult, nutritionists recommend a daily protein intake of about 0.8 grams per kilogram of body weight. For a pregnant woman, the recommendation is increased by an additional 30 grams of protein per day. Similarly, a nursing mother requires an additional 20 grams of protein per day to maintain a high level of milk production.

Chart 12.8 Fat-soluble vitamins			
Vitamin	Characteristics	Functions	Sources
Vitamin A	Occurs in several forms; synthesized from carotenes; stored in the liver; stable in heat, acids, and alkalis; unstable in light	Necessary for the synthesis of visual pigments, mucoproteins, and mucopolysaccharides; for the normal development of bones and teeth; and for the maintenance of epithelial cells	Liver, whole milk, butter, eggs, leafy green vegetables, and yellow and orange vegetables and fruits
Vitamin D	Group of sterols; resistant to heat, oxidation, acids, and alkalis; stored in the liver, skin, brain, spleen, and bones	Promotes the absorption of calcium and phosphorus	Produced in skin exposed to ultraviolet light; milk, egg yolk, fish liver oils, fortified foods
Vitamin E	Group of compounds; resistant to heat and visible light; unstable in the presence of oxygen and ultraviolet light; stored in muscles and adipose tissue	Antioxidant; prevents the oxidation of vitamin A and polyunsaturated fatty acids; may help maintain the stability of cell membranes	Oils from cereal seeds, salad oils, margarine, shortenings, fruits, and vegetables
Vitamin K	Occurs in several forms; resistant to heat but destroyed by acids, alkalis, and light; stored in the liver	Needed for the synthesis of prothrombin, which functions in blood clotting	Leafy green vegetables, egg yolk, pork liver, soy oil, tomatoes, cauliflower

The amount of energy contained in foods consisting of carbohydrates, lipids, or proteins can be expressed in units of heat energy called *calories*. Although a calorie is defined as the amount of heat needed to raise the temperature of a gram of water by 1 degree Celsius (°C), the calorie used in the measurement of food energy is 1,000 times greater. This larger calorie (Cal) is called a *kilocalorie;* but in nutritional studies, it is usually referred to simply as a Calorie, or a food calorie.

As a result of cellular oxidation, 1 gram of carbohydrate or 1 gram of protein yields about 4.1 food calories, but 1 gram of fat yields 9.5 food calories.

1. What are the physiological functions of proteins?
2. How much protein is recommended for an adult diet?

Vitamins

Vitamins are organic compounds (other than carbohydrates, lipids, and proteins) that must be present in small amounts for normal metabolic processes, but cannot be synthesized in adequate amounts by the body cells. Thus, they are essential nutrients that must be supplied in foods.

Vitamins can be classified on the basis of their solubilities, since some are soluble in fats (or fat solvents) and others are soluble in water. Those that are *fat-soluble* include vitamins A, D, E, and K; the *water-soluble* group includes the B vitamins and vitamin C.

Fat-Soluble Vitamins

Since the fat-soluble vitamins dissolve in fats, they occur in association with lipids and are influenced by the same factors that affect lipid absorption. For example, the presence of bile salts in the intestine promotes the absorption of these vitamins. As a group, the fat-soluble vitamins are stored in moderate quantities within various tissues, and because they are fairly resistant to the effects of heat, they are usually not destroyed by cooking and food processing.

Chart 12.8 lists the characteristics, functions, and sources of the fat-soluble vitamins.

1. What are vitamins?
2. How do bile salts affect absorption of fat-soluble vitamins?

Water-Soluble Vitamins

The water-soluble vitamins include the B vitamins and vitamin C. The **B vitamins** consist of several compounds that are essential for normal cellular metabolism and are involved with the oxidation of carbohydrates, lipids, and proteins. Since the B vitamins often occur together in foods, they usually are referred to as a group called the *vitamin B complex.* Members of this group differ chemically, and they have unique functions. Some of them are easily destroyed by cooking and food processing.

Vitamin C (ascorbic acid), which is one of the least stable of the vitamins, is fairly widespread in plant foods. Although its action is poorly understood, it is needed for the synthesis of collagen and a variety of metabolic processes.

Chart 12.9 lists the characteristics, functions, and sources of the water-soluble vitamins.

1. Name the water-soluble vitamins.
2. What is meant by the vitamin B complex?
3. Distinguish between the fat-soluble and water-soluble vitamins.

Chart 12.9 Water-soluble vitamins			
Vitamin	**Characteristics**	**Functions**	**Sources**
Thiamine (Vitamin B_1)	Destroyed by heat and oxygen, especially in an alkaline environment	Part of the coenzyme needed for the oxidation of carbohydrates, and the coenzyme needed in the synthesis of ribose	Lean meats, liver, eggs, whole grain cereals, leafy green vegetables, legumes
Riboflavin (Vitamin B_2)	Stable in heat, acids, and oxidation; destroyed by alkalis and light	Part of the enzymes and coenzymes needed for the oxidation of glucose and fatty acids and for cellular growth	Meats, dairy products, leafy green vegetables, whole grain cereals
Niacin (Nicotinic acid)	Stable in heat, acids, and alkalis; converted to niacinamide by the cells; synthesized from tryptophan	Part of the coenzymes needed for the oxidation of glucose and the synthesis of proteins, fats, and nucleic acids	Liver, lean meats, poultry, peanuts, legumes
Vitamin B_6	Group of three compounds; stable in heat and acids; destroyed by oxidation, alkalis, and ultraviolet light	Coenzyme needed for the synthesis of proteins and various amino acids, for the conversion of tryptophan to niacin, for the production of antibodies, and for the synthesis of nucleic acids	Liver, meats, fish, poultry, bananas, avocados, beans, peanuts, whole grain cereals, egg yolk
Pantothenic acid	Destroyed by heat, acids, and alkalis	Part of the coenzyme needed for the oxidation of carbohydrates and fats	Meats, fish, whole grain cereals, legumes, milk, fruits, vegetables
Cyanocobalamin (Vitamin B_{12})	Complex, cobalt-containing compound; stable in heat; inactivated by light, strong acids, and strong alkalis; absorption regulated by intrinsic factor from gastric glands; stored in the liver	Part of the coenzyme needed for the synthesis of nucleic acids and for the metabolism of carbohydrates; plays a role in the synthesis of myelin	Liver, meats, poultry, fish, milk, cheese, eggs
Folacin (Folic acid)	Occurs in several forms; destroyed by oxidation in an acid environment or by heat in an alkaline environment; stored in the liver where it is converted into folinic acid	Coenzyme needed for the metabolism of certain amino acids and for the synthesis of DNA; promotes the production of normal red blood cells	Liver, leafy green vegetables, whole grain cereals, legumes
Biotin	Stable in heat, acids, and light; destroyed by oxidation and alkalis	Coenzyme needed for the metabolism of amino acids and fatty acids and for the synthesis of nucleic acids	Liver, egg yolk, nuts, legumes, mushrooms
Ascorbic acid (Vitamin C)	Closely related to monosaccharides; stable in acids, but destroyed by oxidation, heat, light, and alkalis	Needed for the production of collagen, the conversion of folacin to folinic acid, and the metabolism of certain amino acids; promotes the absorption of iron and the synthesis of hormones from cholesterol	Citrus fruits, citrus juices, tomatoes, cabbage, potatoes, leafy green vegetables, fresh fruits

Minerals

Dietary **minerals** are inorganic elements that play essential roles in human metabolism. They are usually extracted from the soil by plants. Humans, in turn, obtain them from plant foods or from animals that have eaten plants.

Characteristics of Minerals

Minerals are responsible for about 4% of the body weight and are most concentrated in the bones and teeth. In fact, the minerals *calcium* and *phosphorus,* which are very abundant in these tissues, account for nearly 75% of the body's minerals.

Minerals are usually incorporated into organic molecules. For example, phosphorous occurs in phospholipids, iron in hemoglobin, and iodine in thyroxine. However, some occur in inorganic compounds, such as the calcium phosphate of bone; others occur as free ions, such as the sodium, chloride, and calcium ions in the blood.

Minerals are present in all body cells, where they comprise parts of the structural materials. They function as portions of enzyme molecules, help create the osmotic pressure of body fluids, and play vital roles in the conduction of nerve impulses, contraction of muscle fibers, coagulation of blood, and maintenance of pH.

Mineral	Distribution	Functions	Sources
Calcium (Ca)	Mostly in the inorganic salts of the bones and teeth	Structure of the bones and teeth; essential for nerve impulse conduction, muscle fiber contraction and blood coagulation; increases the permeability of cell membranes; activates certain enzymes	Milk, milk products, leafy green vegetables
Phosphorus (P)	Mostly in the inorganic salts of the bones and teeth	Structure of the bones and teeth; component in nearly all metabolic reactions; constituent of nucleic acids, many proteins, some enzymes, and some vitamins; occurs in the cell membrane, ATP, and phosphates of body fluids	Meats, poultry, fish, cheese, nuts, whole grain cereals, milk, legumes
Potassium (K)	Widely distributed; tends to be concentrated inside the cells	Helps maintain intracellular osmotic pressure and regulate pH; promotes metabolism; needed for nerve impulse conduction and muscle fiber contraction	Avocados, dried apricots, meats, nuts, potatoes, bananas
Sulfur (S)	Widely distributed	Essential part of various amino acids, thiamine, insulin, biotin, and mucopolysaccharides	Meats, milk, eggs, legumes
Sodium (Na)	Widely distributed; a large proportion occurs in extracellular fluids and is bonded to the inorganic salts of bone	Helps maintain the osmotic pressure of extracellular fluids and regulate water balance; needed for the conduction of nerve impulses and contraction of muscle fibers; aids in the regulation of pH and in the transport of substances across cell membranes	Table salt, cured ham, sauerkraut, cheese, graham crackers
Chlorine (Cl)	Closely associated with sodium; most highly concentrated in the cerebrospinal fluid and gastric juice	Helps maintain the osmotic pressure of extracellular fluids, regulate pH, and maintain electrolyte balance; essential in the formation of hydrochloric acid; aids in the transport of carbon dioxide by red blood cells	Same as for sodium
Magnesium (Mg)	Abundant in bones	Needed in metabolic reactions that occur in the mitochondria and are associated with the production of ATP; plays a role in the conversion of ATP to ADP	Milk, dairy products, legumes, nuts, leafy green vegetables

Chart 12.10 Major minerals

The concentrations of various minerals in the body fluids are regulated by homeostatic mechanisms, which ensure that the excretion of these substances are balanced with their dietary intake. Thus, toxic excesses are avoided, and minerals present in limited amounts are conserved.

1. How are minerals obtained?
2. What are the major functions of minerals?

Major Minerals

As mentioned, calcium and phosphorus account for nearly 75% of the mineral elements in the body; thus, they are **major minerals.** Other major minerals, each of which accounts for 0.05% or more of the body weight, include potassium, sulfur, sodium, chlorine, and magnesium.

Chart 12.10 lists the distribution, functions, and sources of major minerals.

Chart 12.11 Trace elements			
Trace Element	**Distribution**	**Functions**	**Sources**
Iron (Fe)	Primarily in the blood; stored in the liver, spleen, and bone marrow	Part of the hemoglobin molecule; catalyzes the formation of vitamin A; incorporated into a number of enzymes	Liver, lean meats, dried apricots, raisins, enriched whole grain cereals, legumes, molasses
Manganese (Mn)	Most concentrated in the liver, kidneys, and pancreas	Occurs in the enzymes needed for the synthesis of fatty acids and cholesterol, the formation of urea, and normal functioning of the nervous system	Nuts, legumes, whole grain cereals, leafy green vegetables, fruits
Copper (Cu)	Most highly concentrated in the liver, heart, and brain	Essential for the synthesis of hemoglobin, the development of bone, the production of melanin, and the formation of myelin	Liver, oysters, crabmeat, nuts, whole grain cereals, legumes
Iodine (I)	Concentrated in the thyroid gland	Essential component for the synthesis of thyroid hormones	Food content varies with soil content in different geographic regions; iodized table salt
Cobalt (Co)	Widely distributed	Component of cyanocobalamin; needed for the synthesis of several enzymes	Liver, lean meats, poultry, fish, milk
Zinc (Zn)	Most concentrated in the liver, kidneys, and brain	Constituent of several enzymes involved in digestion, respiration, bone metabolism, liver metabolism; necessary for normal wound healing and maintaining the integrity of the skin	Seafoods, meats, cereals, legumes, nuts, vegetables

Trace Elements

Trace elements are essential minerals that occur in minute amounts, each making up less than 0.005% of the adult body weight. They include iron, manganese, copper, iodine, cobalt, and zinc.

Chart 12.11 lists the distribution, functions, and sources of the trace elements.

1. Distinguish between a major mineral and a trace element.
2. Name the major minerals and trace elements.

Adequate Diets

An adequate diet is one that provides sufficient *energy, essential fatty acids, essential amino acids, vitamins,* and *minerals* to support optimal growth and to maintain and repair body tissues. Since individual needs for nutrients vary greatly with age, sex, growth rate, and amount of physical activity, as well as with genetic and environmental factors, it is not possible to design a diet that is adequate for everyone.

On the other hand, nutrients are so widely distributed in foods that satisfactory amounts and combinations of essential substances can usually be obtained in spite of individual food preferences that are related to cultural backgrounds, life-styles, and emotional attitudes.

However, if essential nutrients are lacking in the diet or a person fails to use available foods to best advantage, the result is called *malnutrition.* This condition may involve *undernutrition* and include the symptoms of deficiency diseases, or it may be due to *overnutrition,* arising from an excessive intake of nutrients.

The factors leading to malnutrition are varied. A deficiency condition may stem, for example, from the lack of availability or poor quality of food. On the other hand, malnutrition may result from the excessive intake of vitamin supplements or from excessive caloric intake.

Chart 12.12 Summary of Recommended Dietary Allowances (RDA), 1989*

Age (years)	Weight kg	Weight lb.	Height cm	Height in.	Protein (g)	(RE) Vitamin A	(µg) Vitamin D	(mg) Vitamin E	(mg) Vitamin K	(mg) Vitamin C	(mg) Thiamin	(mg) Riboflavin	(mg equiv.) Niacin	(mg) Vitamin B_6	(µg) Folate	(µg) Vitamin B_{12}	(mg) Calcium	(mg) Phosphorus	(mg) Magnesium	(mg) Iron	(mg) Zinc	(µg) Iodine	(µg) Selenium
Infants																							
0.0–0.5	6	13	60	24	13	375	7.5	3	5	30	0.3	0.4	5	0.3	25	0.3	400	300	40	6	5	40	10
0.5–1.0	9	20	71	28	14	375	10	4	10	35	0.4	0.5	6	0.6	35	0.5	600	500	60	10	5	50	15
Children																							
1–3	13	29	90	35	16	400	10	6	15	40	0.7	0.8	9	1.0	50	0.7	800	800	80	10	10	70	20
4–6	20	44	112	44	24	500	10	7	20	45	0.9	1.1	12	1.1	75	1.0	800	800	120	10	10	90	20
7–10	28	62	132	52	28	700	10	7	30	45	1.0	1.2	13	1.4	100	1.4	800	800	170	10	10	120	30
Males																							
11–14	45	99	157	62	45	1,000	10	10	45	50	1.3	1.5	17	1.7	150	2.0	1,200	1,200	270	12	15	150	40
15–18	66	145	176	69	59	1,000	10	10	65	60	1.5	1.8	20	2.0	200	2.0	1,200	1,200	400	12	15	150	50
19–24	72	160	177	70	58	1,000	10	10	70	60	1.5	1.7	19	2.0	200	2.0	1,200	1,200	350	10	15	150	70
25–50	79	174	176	70	63	1,000	5	10	80	60	1.5	1.7	19	2.0	200	2.0	800	800	350	10	15	150	70
51+	77	170	173	68	63	1,000	5	10	80	60	1.2	1.4	15	2.0	200	2.0	800	800	350	10	15	150	70
Females																							
11–14	46	101	157	62	46	800	10	8	45	50	1.1	1.3	15	1.4	150	2.0	1,200	1,200	280	15	12	150	45
15–18	55	120	163	64	44	800	10	8	55	60	1.1	1.3	15	1.5	180	2.0	1,200	1,200	300	15	12	150	50
19–24	58	128	164	65	46	800	10	8	60	60	1.1	1.3	15	1.6	180	2.0	1,200	1,200	280	15	12	150	55
25–50	63	138	163	64	50	800	5	8	65	60	1.1	1.3	15	1.6	180	2.0	800	800	280	15	12	150	55
51+	65	143	160	63	50	800	5	8	65	60	1.0	1.2	13	1.6	180	2.0	800	800	280	10	12	150	55
Pregnant					60	800	10	10	65	70	1.5	1.6	17	2.2	400	2.2	1,200	1,200	320	30	15	175	65
Lactating																							
1st 6 mo					65	1,300	10	12	65	95	1.6	1.8	20	2.1	280	2.6	1,200	1,200	355	15	19	200	75
2nd 6 mo					62	1,200	10	11	65	90	1.6	1.7	20	2.1	260	2.6	1,200	1,200	340	15	16	200	75

In response to emotional problems and an intense fear of becoming overweight, teenagers (particularly girls) sometimes develop a disorder called *anorexia nervosa.* In this condition, which occurs less commonly in adults, the person severely restricts food intake and may exercise excessively. As a result, 25% or more of the body weight is lost. The person also may engage in episodes of secretive, unrestrained eating (bulimia) followed by self-induced vomiting and the use of laxatives and diuretics (binge-purge behavior). Symptoms of anorexia nervosa that accompany the malnutrition associated with chronic starvation include cessation of menstruation, decreased heart rate, inability to maintain normal body temperature, impaired judgment, and hallucinations. In 3% or more of the cases, anorexics die suddenly, usually as a result of electrolyte imbalances and cardiac disorders.

Chart 12.12 lists the recommended daily allowance for each of the major nutrients.

1. What is meant by an adequate diet?
2. What factors influence individual needs for nutrients?
3. What causes malnutrition?

Clinical Terms Related to the Digestive System

achalasia (ak″ah-la′ze-ah)—a failure of the smooth muscle to relax at some junction in the digestive tube, such as that between the esophagus and stomach.

achlorhydria (ah″klor-hi′dre-ah)—a lack of hydrochloric acid in the gastric secretions.

aphagia (ah-fa′je-ah)—an inability to swallow.

cachexia (kah-kek′se-ah)—a state of chronic malnutrition and physical wasting.

cholecystitis (ko″le-sis-ti′tis)—an inflammation of the gallbladder.

Food Selection

A Current Topic

Although recommended daily allowances of essential nutrients are often used by nutritionists and dietitians as guides in planning adequate diets, these values are usually of limited use to the average person. A more useful aid for most persons is the *basic food groups.*

A basic food group is a class of foods that will supply sufficient amounts of certain essential nutrients when a given quantity of food from the group is included in the diet. For example, one basic food plan includes four food groups, each of which provides a unique contribution toward achieving an adequate diet. In this plan, the food groups and the quantities of food required from each are as follows:

Group 1: Milk and Dairy Products. This group includes milk, cottage cheese, cream cheese, natural and processed cheese, and ice cream. It provides calcium, phosphorus, magnesium, protein, vitamin B_6, vitamin B_{12}, and riboflavin. It is recommended that an adult have two 8-ounce glasses of milk or the equivalent in other dairy foods each day.

Group 2: Fruits and Vegetables. This group includes all fruits and vegetables, and it provides vitamin C, vitamin A, iron, magnesium, and vitamin B_6. It is recommended that an adult diet contain four servings from this group each day and that one of these be citrus fruits or some other fruit or vegetable that is particularly high in vitamin C content. At least every other day, one of the servings should be a dark green, a yellow, or an orange vegetable.

Group 3: Meat, Poultry, and Fish. This group includes all meats, poultry, and fish as well as meat substitutes such as eggs, beans, peas, and nuts. It provides protein, vitamin A, thiamine, niacin, vitamin B_6, vitamin B_{12}, riboflavin, phosphorus, magnesium, and iron. For an adult, two or more servings from this group are recommended each day.

Group 4: Breads and Cereals. This group includes all whole grain or enriched breads and cereals. It provides iron, thiamine, niacin, riboflavin, protein, phosphorus, and magnesium. Four or more servings per day from this group are recommended for adults.

Chart 12.13 provides a food guide based on these basic food groups.

Chart 12.13 A daily food guide

Milk group (8-ounce cups)

2 to 3 cups for children under 9 years

3 cups or more for children 9 to 12 years

4 cups or more for teenagers

2 cups or more for adults

3 cups or more for pregnant women

4 cups or more for nursing mothers

Meat group

2 servings or more. Count as one serving:

2 to 3 ounces lean, cooked beef, veal, pork, lamb, poultry, fish—without bone

2 eggs

1 cup of cooked dry beans, dry peas, lentils

4 tablespoons of peanut butter

Vegetable and fruit group (½ cup serving, or 1 piece of fruit, etc.)

4 servings or more per day, including:

1 serving of citrus fruit or another fruit or vegetable as a good source of vitamin C, or 2 servings of a fair source

1 serving, at least every other day, of a dark green or deep yellow vegetable for vitamin A

2 servings or more of other vegetables and fruits, including potatoes

Bread and cereals group

4 servings or more daily (whole grain, enriched, or restored). Count as one serving:

1 slice of bread

1 ounce of ready-to-eat cereal

½to ¾ cup of cooked cereal, cornmeal, grits, macaroni, noodles, rice, or spaghetti

Source: "A Daily Food Guide" in *Consumers All.* Yearbook of Agriculture, 1965, U.S. Department of Agriculture, Washington, D.C., 1965, p. 394.

cholelithiasis (ko″le-lĭ-thi′ah-sis)—the presence of stones in the gallbladder.

cirrhosis (si-ro′sis)—a liver condition in which the hepatic cells degenerate and the surrounding connective tissues thicken.

diverticulitis (di″ver-tik″u-li′tis)—an inflammation of the small pouches (diverticula) that sometimes form in the lining and wall of the colon.

dumping syndrome (dum′ping sin′drōm)—a set of symptoms, including diarrhea, that often occur following a gastrectomy.

dysentery (dis′en-ter″e)—an intestinal infection caused by viruses, bacteria, or protozoans, which is accompanied by diarrhea and cramps.

dyspepsia (dis-pep′se-ah)—indigestion; difficulty in digesting a meal.

dysphagia (dis-fa′je-ah)—difficulty in swallowing.

enteritis (en″tĕ-ri′tis)—inflammation of the intestine.

esophagitis (e-sof″ah-ji′tis)—inflammation of the esophagus.

gastrectomy (gas-trek′to-me)—the partial or complete removal of the stomach.

gastritis (gas-tri′tis)—an inflammation of the stomach lining.
gastrostomy (gas-tros′to-me)—the creation of an opening in the stomach wall through which food and liquids may be administered when swallowing is not possible.
gingivitis (jin″jĭ-vi′tis)—an inflammation of the gums.
glossitis (glŏ-si′tis)—an inflammation of the tongue.
hemorrhoidectomy (hem″ŏ-roi-dek′to-me)—the removal of hemorrhoids.
hepatitis (hep″ah-ti′tis)—an inflammation of the liver.
ileitis (il″e-i′tis)—an inflammation of the ileum.
pharyngitis (far″in-ji′tis)—an inflammation of the pharynx.
pyloric stenosis (pi-lor′ik stĕ-no′sis)—a congenital obstruction at the pyloric sphincter due to an enlarged pyloric muscle.
pylorospasm (pi-lor′o-spazm)—a spasm of the pyloric portion of the stomach or of the pyloric sphincter.
pyorrhea (pi″o-re′ah)—an inflammation of the dental periosteum accompanied by the formation of pus.
stomatitis (sto″mah-ti′tis)—an inflammation of the lining of the mouth.
vagotomy (va-got′o-me)—sectioning of the vagus nerve fibers.

Chapter Summary

Introduction (page 304)
Digestion is the process of changing food substances into forms that can be absorbed. The digestive system consists of an alimentary canal and several accessory organs.

General Characteristics of the Alimentary Canal (page 304)
Various regions of the canal are specialized to perform specific functions.

1. Structure of the wall
   The wall consists of four layers—the mucosa, submucosa, muscular layer, and serosa.
2. Movements of the tube
   Motor functions include mixing and propelling movements.

Mouth (page 305)
The mouth is adapted to receive food and begin preparing it for digestion.

1. Cheeks and lips
   a. The cheeks consist of outer layers of skin, pads of fat, muscles associated with expression and chewing, and inner linings of epithelium.
   b. The lips are highly mobile and possess a variety of sensory receptors.
2. Tongue
   a. The tongue's rough surface contains taste buds and aids in handling food.
   b. The lingual tonsils are located on the root of the tongue.
3. Palate
   a. The palate includes the hard and soft portions.
   b. The soft palate closes the opening to the nasal cavity during swallowing.
   c. The palatine tonsils are located on either side of the tongue in the back of the mouth.
4. Teeth
   a. There are twenty primary and thirty-two secondary teeth.
   b. The teeth function to break food into smaller pieces, increasing the surface area of the food.
   c. Each tooth consists of a crown and root and is composed of enamel, dentin, pulp, nerves, and blood vessels.
   d. A tooth is attached to the alveolar process by a periodontal ligament.

Salivary Glands (page 309)
The salivary glands secrete saliva, which moistens food, helps bind food particles together, begins the digestion of carbohydrates, makes taste possible, and helps cleanse the mouth.

1. Salivary secretions
   A salivary gland includes serous cells which secrete digestive enzymes, and mucous cells, which secrete mucus.
2. Major salivary glands
   a. The parotid glands secrete saliva rich in amylase, which begins the digestion of carbohydrates.
   b. The submandibular glands produce saliva that is more viscous than that of the parotid glands.
   c. The sublingual glands primarily secrete mucus.

Pharynx and Esophagus (page 310)
The pharynx and esophagus serve as passageways.

1. Structure of the pharynx
   The pharynx is divided into the nasopharynx, oropharynx, and laryngopharynx.
2. Swallowing mechanism
   a. The act of swallowing occurs in three stages.
      (1) The food is mixed with saliva and forced into the pharynx.
      (2) Involuntary reflex actions move the food into the esophagus.
      (3) The food is transported to the stomach.
3. Esophagus
   a. The esophagus passes through the diaphragm and joins the stomach.
   b. Some circular muscle fibers at the distal end of the esophagus help to prevent the regurgitation of food from the stomach.

Stomach (page 311)
The stomach receives food, mixes it with gastric juice, carries on a limited amount of absorption, and moves food into the small intestine.

1. Parts of the stomach
   a. The stomach is divided into the cardiac, fundic, body, and pyloric regions.
   b. The pyloric sphincter serves as a valve between the stomach and small intestine.

2. Gastric secretions
   a. The gastric glands secrete gastric juice.
   b. Gastric juice contains pepsin, hydrochloric acid, and intrinsic factor.
   c. Regulation of the gastric secretions.
      (1) The gastric secretions are enhanced by parasympathetic impulses and the hormone gastrin.
      (2) The presence of food in the small intestine reflexly inhibits the gastric secretions.
3. Gastric absorption
   A few substances such as water and other small molecules may be absorbed through the stomach wall.
4. Mixing and emptying actions
   a. Mixing movements aid in producing the chyme; peristaltic waves move the chyme into the pyloric region of the stomach.
   b. The rate of emptying depends on the fluidity of the chyme and the type of food present.
   c. The muscular wall of the pyloric region pumps the chyme into the small intestine.

Pancreas (page 314)

1. Structure of the pancreas
   a. The pancreas produces pancreatic juice, which is secreted into a pancreatic duct.
   b. The pancreatic duct leads to the duodenum.
2. Pancreatic juice
   a. Pancreatic juice contains enzymes that can split carbohydrates, proteins, fats, and nucleic acids.
   b. Regulation of pancreatic secretion.
      (1) Secretin stimulates the release of pancreatic juice that has a high bicarbonate ion concentration.
      (2) Cholecystokinin stimulates the release of pancreatic juice that has a high concentration of digestive enzymes.

Liver (page 316)

1. Functions of the liver
   a. The liver carries on a variety of important functions involving the metabolism of carbohydrates, lipids, and proteins; the storage of substances; the filtering of the blood; the destruction of the toxic chemicals; and the secretion of bile.
   b. Bile is the only liver secretion that directly affects digestion.
2. Structure of the liver
   a. Each lobe contains hepatic lobules, the functional units of the liver.
   b. Bile from the lobules is carried by the bile canals to the hepatic ducts.
3. Composition of bile
   a. Bile contains bile salts, bile pigments, cholesterol, and various electrolytes.
   b. Only the bile salts have digestive functions.
4. Gallbladder and its functions
   a. The gallbladder stores bile between meals.
   b. The release of bile from the common bile duct is controlled by a sphincter muscle.
   c. The release is stimulated by cholecystokinin from the small intestine.
   d. A sphincter muscle at the base of the common bile duct relaxes as a peristaltic wave in the duodenal wall passes by.
   e. Bile salts emulsify fats and aid in the absorption of fatty acids, cholesterol, and certain vitamins.

Small Intestine (page 320)

The small intestine receives secretions from the pancreas and liver, completes the digestion of nutrients, absorbs the products of digestion, and transports the residues to the large intestine.

1. Parts of the small intestine
   The small intestine consists of the duodenum, jejunum, and ileum.
2. Structure of the small intestinal wall
   a. The wall is lined with villi, which aid in mixing and absorption.
   b. The intestinal glands are located between the villi.
3. Secretions of the small intestine
   a. The secretions include mucus and digestive enzymes.
   b. Digestive enzymes can split molecules of sugars, proteins, fats, and nucleic acids into simpler forms.
   c. Small intestinal secretions are enhanced by the presence of gastric juice and chyme and by the mechanical stimulation of distension.
4. Absorption in the small intestine
   a. The villi increase the surface area of the intestinal wall.
   b. Monosaccharides, amino acids, fatty acids, and glycerol are absorbed by the villi.
   c. Fat molecules with longer chains of carbon atoms enter the lacteals of the villi.
   d. Other digestive products enter the blood capillaries of the villi.
5. Movements of the small intestine
   a. These movements include mixing and peristalsis.
   b. The ileocecal valve controls movement of the intestinal contents from the small intestine into the large intestine.

Large Intestine (page 326)

The large intestine functions to reabsorb water and electrolytes and to form and store feces.

1. Parts of the large intestine
   a. The large intestine consists of the cecum, colon, rectum, and anal canal.
   b. The colon is divided into the ascending, transverse, descending, and sigmoid portions.

2. Structure of the large intestinal wall
   a. Basically the large intestinal wall is like the wall in other parts of the alimentary canal.
   b. Its unique features include a layer of longitudinal muscle fibers that are arranged in three distinct bands.
3. Functions of the large intestine
   a. The large intestine has little or no digestive function.
   b. The only significant secretion is mucus.
   c. Absorption in the large intestine is generally limited to water and electrolytes.
   d. Feces are formed and stored in the large intestine.
4. Movements of the large intestine
   a. Its movements are similar to those in the small intestine.
   b. Mass movements occur two to three times each day.
   c. Defecation involves a defecation reflex.
5. Feces
   a. Feces are formed and stored in the large intestine.
   b. The feces consist largely of water, undigested material, mucus, and bacteria.
   c. The color of the feces is due to bile salts altered by bacterial actions.

Nutrition and Nutrients (page 328)

Nutrition is the process of ingestion and utilization of the necessary food substances or nutrients.

1. Carbohydrates
   Carbohydrates are organic compounds that are used primarily to supply cellular energy.
   a. Sources of carbohydrates
      (1) Starch, glycogen, disaccharides, and monosaccharides are carbohydrates.
      (2) Cellulose is a polysaccharide that cannot be digested by human enzymes.
   b. Utilization of carbohydrates
      (1) Energy is released from glucose by oxidation.
      (2) Excessive glucose is stored as glycogen or converted to fat.
   c. Carbohydrate requirements
      (1) Most carbohydrates are utilized to supply energy.
      (2) Some cells depend on a continuous supply of glucose to survive.
      (3) Humans survive with a wide range of carbohydrate intakes.
2. Lipids
   Lipids are organic compounds that supply energy and are used to build cell structures.
   a. Sources of lipids
      (1) Triglycerides are obtained from foods of plant and animal origins.
      (2) Cholesterol is obtained from foods of animal origin only.
   b. Utilization of lipids
      (1) The metabolism of triglycerides is controlled mainly by the liver and adipose tissues.
      (2) Linoleic acid is an essential fatty acid.
   c. Lipid requirements
      (1) Humans survive with a wide range of lipid intakes.
      (2) Some fats contain fat-soluble vitamins.
3. Proteins
   Proteins are organic compounds that serve as structural materials, act as enzymes, and provide energy.
   a. Sources of proteins
      (1) Proteins are obtained mainly from meats, dairy products, cereals, and legumes.
      (2) Complete proteins contain adequate amounts of all the essential amino acids.
      (3) Incomplete proteins lack adequate amounts of one or more essential amino acids.
   b. Utilization of amino acids
      Amino acids are incorporated into various structural and functional proteins, including enzymes.
   c. Protein requirements
      Proteins and amino acids are needed to supply the essential amino acids and nitrogen for the synthesis of various molecules.
4. Vitamins
   Vitamins are organic compounds (other than carbohydrates, lipids, and proteins) that are essential for normal metabolic processes, but cannot be synthesized by the body cells in adequate amounts.
   a. Fat-soluble vitamins
      (1) These include vitamins A, D, E, and K.
      (2) They occur in association with lipids and are influenced by the same factors that affect lipid absorption.
      (3) They are fairly resistant to the effects of heat; thus, they are not destroyed by cooking or food processing.
   b. Water-soluble vitamins
      (1) This group includes the B vitamins and vitamin C.
      (2) B vitamins make up a group called the vitamin B complex and are generally involved with the oxidation of carbohydrates, lipids, and proteins.
      (3) Some are destroyed by cooking and food processing.
5. Minerals
   a. Characteristics of minerals
      (1) Most minerals are found in the bones and teeth.
      (2) Minerals are usually incorporated into organic molecules, although some occur in inorganic compounds or as free ions.
      (3) They comprise structural materials, function in enzymes, and play vital roles in various metabolic processes.

b. Major minerals
   They include calcium, phosphorus, potassium, sulfur, sodium, chlorine, and magnesium.
c. Trace elements
   They include iron, manganese, copper, iodine, cobalt, and zinc.

Adequate Diets (page 335)

1. An adequate diet provides sufficient energy and essential nutrients to support optimal growth as well as maintenance and repair of tissues.
2. Individual needs vary so greatly that it is not possible to design a diet that is adequate for everyone.
3. Malnutrition is poor nutrition due to lack of essential nutrients or failure to use available foods to best advantage.

Clinical Application of Knowledge

1. If a patient has 95% of the stomach removed (subtotal gastrectomy) in the treatment of severe ulcers or cancer, how would the patient's digestion and absorption of foods be affected? How would the patient's eating habits have to be altered? Why?
2. Why might a person with an inflammation of the gallbladder (cholecystitis) also develop an inflammation of the pancreas (pancreatitis)?
3. How would you explain the fact that the blood sugar concentration of a person whose diet is relatively low in carbohydrates remains stable.
4. Examine the label information on the packages of a variety of dry breakfast cereals. Which types of cereals provide adequate sources of vitamins and minerals?

Review Activities

1. List and describe the location of the major parts of the alimentary canal.
2. List and describe the location of the accessory organs of the digestive system.
3. Name the four layers of the wall of the alimentary canal.
4. Distinguish between mixing movements and propelling movements.
5. Define *peristalsis.*
6. Discuss the functions of the mouth and its parts.
7. Distinguish between the lingual, palatine, and pharyngeal tonsils.
8. Compare the primary and secondary teeth.
9. Describe the structure of a tooth.
10. Explain how a tooth is anchored to its socket.
11. List and describe the locations of the major salivary glands.
12. Explain how the secretions of the salivary glands differ.
13. Discuss the digestive functions of saliva.
14. Explain the function of the esophagus.
15. Describe the structure of the stomach.
16. List the enzymes in gastric juice, and explain the function of each enzyme.
17. Explain how gastric secretions are regulated.
18. Describe the location of the pancreas and the pancreatic duct.
19. List the enzymes found in pancreatic juice, and explain the function of each enzyme.
20. Explain how pancreatic secretions are regulated.
21. List the major functions of the liver.
22. Describe the structure of the liver.
23. Describe the composition of bile.
24. Define *cholecystokinin.*
25. Explain the functions of bile salts.
26. List and describe the locations of the parts of the small intestine.
27. Name the enzymes of the intestinal mucosa, and explain the function of each enzyme.
28. Explain how the secretions of the small intestine are regulated.
29. Describe the functions of the intestinal villi.
30. Summarize how digestive products are absorbed.
31. List and describe the locations of the parts of the large intestine.
32. Explain the general functions of the large intestine.
33. Describe the defecation reflex.
34. List some common sources of carbohydrates.
35. Summarize the importance of cellulose in the diet.
36. Explain why a temporary drop in the glucose concentration may produce functional disorders of the nervous system.
37. List some common sources of lipids.
38. Describe the role of the liver in fat metabolism.
39. List some common sources of proteins.
40. Distinguish between the essential and nonessential amino acids.
41. Distinguish between complete and incomplete proteins.
42. Discuss the general characteristics of fat-soluble vitamins.
43. List the fat-soluble vitamins, and describe the major functions of each vitamin.
44. List the water-soluble vitamins, and describe the major functions of each vitamin.
45. Discuss the general characteristics of the mineral nutrients.
46. List the major minerals, and describe the major functions of each mineral.
47. List the trace elements, and describe the major functions of each element.
48. Define *adequate diet.*
49. Define *malnutrition.*

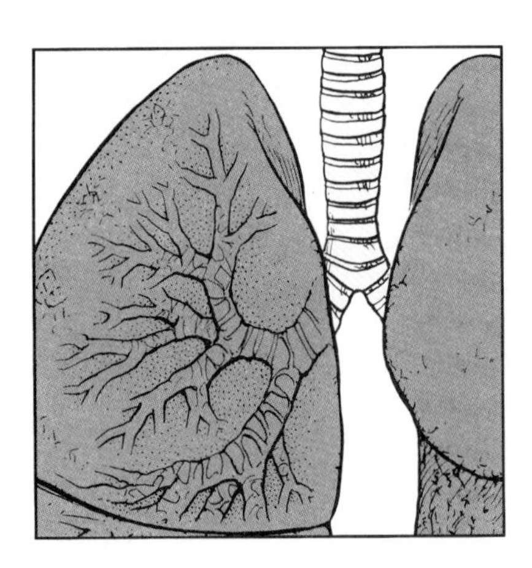

13 Respiratory System

Before the body cells can oxidize nutrients and release energy, they must be supplied with oxygen. Also, the carbon dioxide that results from oxidation must be excreted. These two general processes—obtaining oxygen and removing carbon dioxide—are the primary functions of the *respiratory system*.

In addition, the respiratory organs filter particles from the incoming air, help control the temperature and water content of the air, aid in producing vocal sounds, and play important roles in the sense of smell and in the regulation of the pH of the blood.

Chapter Objectives

After you have studied this chapter, you should be able to

1. List the general functions of the respiratory system.
2. Name and describe the locations of the organs of the respiratory system.
3. Describe the functions of each organ of the respiratory system.
4. Explain how inspiration and expiration are accomplished.
5. Name and define each of the respiratory air volumes.
6. Locate the respiratory center, and explain how it controls normal breathing.
7. Discuss how various factors affect the respiratory center.
8. Describe the structure and function of the respiratory membrane.
9. Explain how oxygen and carbon dioxide are transported in the blood.
10. Complete the review activities at the end of this chapter. Note that the items are worded in the form of specific learning objectives. You may want to refer to them before reading the chapter.

Key Terms

alveolus (al-ve′o-lus)

bronchial tree (brong′ke-al tre)

carbaminohemoglobin (kar-bam″ĭ-no-he″mo-glo′bin)

carbonic anhydrase (kar-bon′ik an-hi′drās)

cellular respiration (sel′u-lar res″pĭ-ra′shun)

expiration (ek″spĭ-ra′shun)

glottis (glot′is)

hemoglobin (he″mo-glo′bin)

hyperventilation (hi″per-ven″tĭ-la′shun)

inspiration (in″spĭ-ra′shun)

oxyhemoglobin (ok″se-he″mo-glo′bin)

partial pressure (par′shil presh′ur)

pleural cavity (ploo′ral kav′ĭ-te)

respiratory center (re-spi′rah-to″re sen′ter)

respiratory membrane (re-spi′rah-to″re mem′brān)

respiratory volume (re-spi′rah-to″re vol′ūm)

surface tension (ser′fas ten′shun)

surfactant (ser-fak′tant)

Aids to Understanding Words

alveol-, small cavity: *alveol*us—a microscopic air sac within a lung.

bronch-, windpipe: *bronch*us—a primary branch of the trachea.

cric-, ring: *cric*oid cartilage—a ring-shaped mass of cartilage at the base of the larynx.

epi-, upon: *epi*glottis—a flaplike structure that partially covers the opening into the larynx during swallowing.

hem-, blood: *hem*oglobin—the pigment in red blood cells that serves to transport oxygen and carbon dioxide.

Introduction

THE RESPIRATORY SYSTEM contains a group of passages that filter incoming air and transport it from outside the body into the lungs. It also includes numerous microscopic air sacs in which gas exchanges take place.

The entire process of exchanging gases between the atmosphere and the body cells is called **respiration,** and it involves several events. These include the movement of air in and out of the lungs—commonly called *breathing,* or *pulmonary ventilation;* the exchange of gases between the air in the lungs and the blood; the transport of gases by the blood between the lungs and body cells; and the exchange of gases between the blood and body cells. The utilization of oxygen and production of carbon dioxide by the cells is called **cellular respiration.**

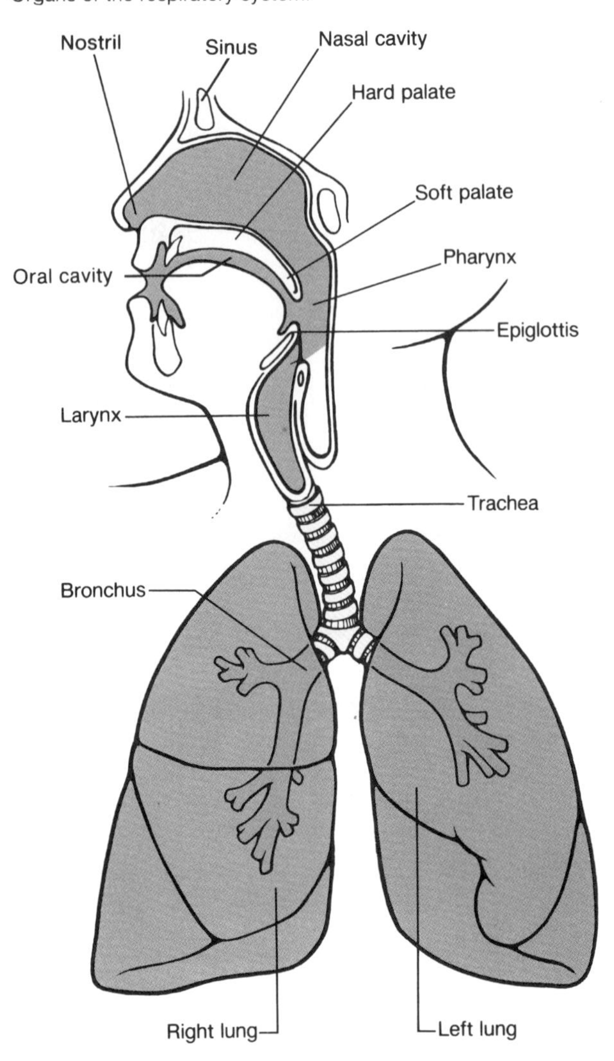

Figure 13.1
Organs of the respiratory system.

Organs of the Respiratory System

The organs of the respiratory system include the nose, nasal cavity, sinuses, pharynx, larynx, trachea, bronchial tree, and lungs. The parts of the respiratory system can be divided into two sets, or tracts. Those organs outside the thorax constitute the *upper respiratory tract,* and those within the thorax comprise the *lower respiratory tract* (fig. 13.1).

Nose

The **nose** is supported internally by bone and cartilage. Its two *nostrils* provide openings through which air can enter and leave the nasal cavity. These openings are guarded by numerous internal hairs, which help prevent the entrance of relatively large particles sometimes carried in the air.

Nasal Cavity

The **nasal cavity,** a hollow space behind the nose, is divided medially into right and left portions by the **nasal septum.**

Nasal conchae (fig. 7.8) curl out from the lateral walls of the nasal cavity on each side, dividing the cavity into passageways. They support the mucous membrane that lines the nasal cavity and help increase its surface area.

The mucous membrane contains pseudostratified ciliated epithelium that is rich in mucus-secreting goblet cells (see chapter 5). It also includes an extensive network of blood vessels, and as air passes over the membrane, heat leaves the blood and warms the air. In this way, the temperature of the incoming air quickly adjusts to that of the body. In addition, the incoming air tends to become moistened by the evaporation of water from the mucous lining. The sticky mucus secreted by the mucous membrane entraps dust and other small particles entering with the air.

As the cilia of the epithelial lining move, a thin layer of mucus and entrapped particles are pushed toward the pharynx (fig. 13.2). When the mucus reaches the pharynx, it is swallowed. In the stomach, any microorganisms in the mucus are likely to be destroyed by the action of gastric juice.

1. What is meant by respiration?
2. What organs constitute the respiratory system?
3. What are the functions of the mucous membrane that lines the nasal cavity?

Sinuses

As discussed in chapter 7, the **sinuses** are air-filled spaces located within (and named from) the *maxillary,*

Figure 13.2
Cilia move mucus and trapped particles from the nasal cavity to the pharynx.

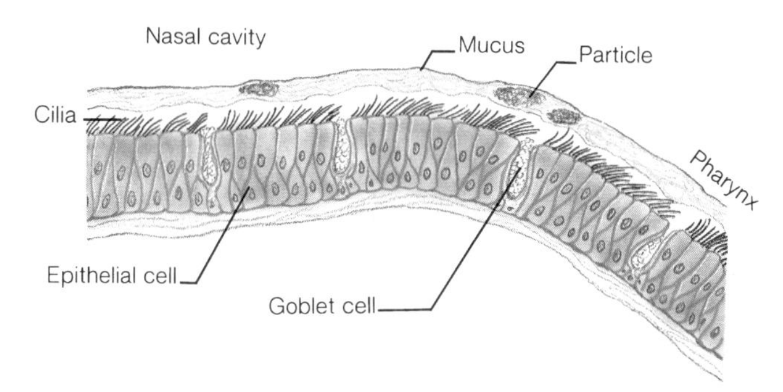

frontal, ethmoid, and *sphenoid bones* of the skull (fig. 7.9). These spaces open into the nasal cavity and are lined with mucous membranes that are continuous with the lining of the nasal cavity.

Although the sinuses function mainly to reduce the weight of the skull, they also serve as resonant chambers, which affect the quality of the voice.

Mucous secretions can drain from the sinuses into the nasal cavity. If this drainage is blocked by swollen membranes, which sometimes accompany nasal infections or allergic reactions, the accumulating fluids may cause increasing pressure within a sinus and produce a painful sinus headache.

1. Where are the sinuses located?
2. What are the functions of the sinuses?

Pharynx

The **pharynx** (throat) is located behind the oral cavity and between the nasal cavity and larynx (fig. 13.1). It functions as a passageway for food traveling from the oral cavity to the esophagus and for air passing between the nasal cavity and larynx. It also aids in producing the sounds of speech.

Larynx

The **larynx** is an enlargement in the airway at the top of the trachea and below the pharynx. It serves as a passageway for air moving in and out of the trachea and functions to prevent foreign objects from entering the trachea. In addition, it houses the *vocal cords.*

The larynx is composed primarily of muscles and cartilages, which form the framework of the larynx and are bound together by elastic tissue. The largest of the cartilages (fig. 13.3) are the *thyroid, cricoid,* and *epiglottic cartilages.*

Inside the larynx, two pairs of horizontal folds in the mucous membrane extend inward from the lateral walls (fig. 13.4). The upper folds are called *false vocal cords* because they do not function in the production of sounds. Muscle fibers within these folds help to close the larynx during swallowing.

Figure 13.3
(*a*) An anterior view and (*b*) a posterior view of the larynx.

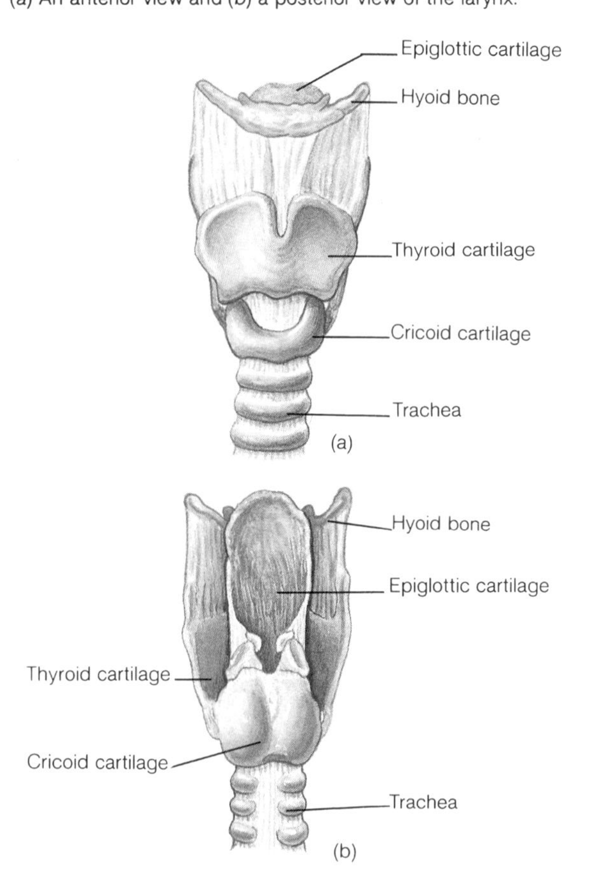

The lower folds are the *true vocal cords.* They contain elastic fibers and are responsible for vocal sounds, which are created when air is forced between the vocal cords, causing them to vibrate. This action generates sound waves, which can be formed into words by changing the shapes of the pharynx and oral cavity and by using the tongue and lips.

During normal breathing, the vocal cords remain relaxed, and the opening between them, called the **glottis,** appears as a triangular slit. When food or liquid is swallowed, however, the glottis is closed by muscles within the false vocal cords, and this prevents the food or liquid from entering the trachea.

The epiglottic cartilage supports a flaplike structure called the **epiglottis.** This structure usually stands upright and allows air to enter the larynx. During swallowing, however, the larynx is raised, and the epiglottis is pressed downward. As a result, the epiglottis partially covers the opening into the larynx and helps to prevent foods and liquids from entering the air passages (see chapter 12).

Figure 13.4

Vocal cords as viewed from above (*a*) with the glottis closed and (*b*) with the glottis open. (*c*) A photograph of the glottis and vocal folds.

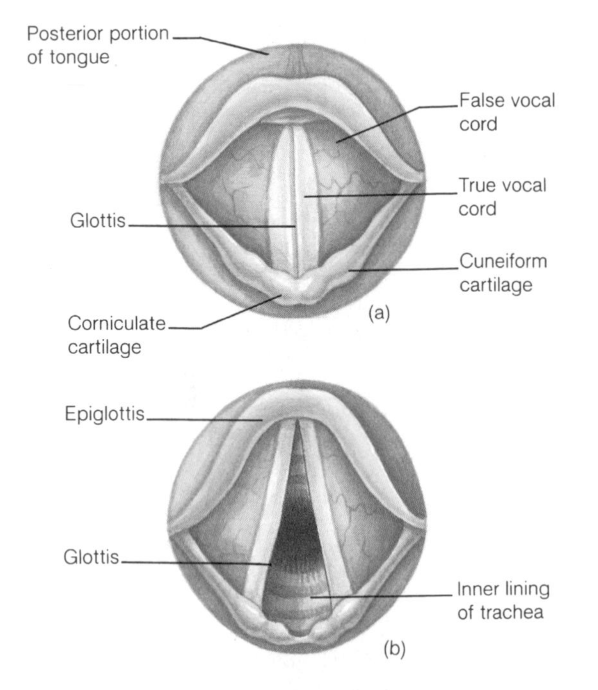

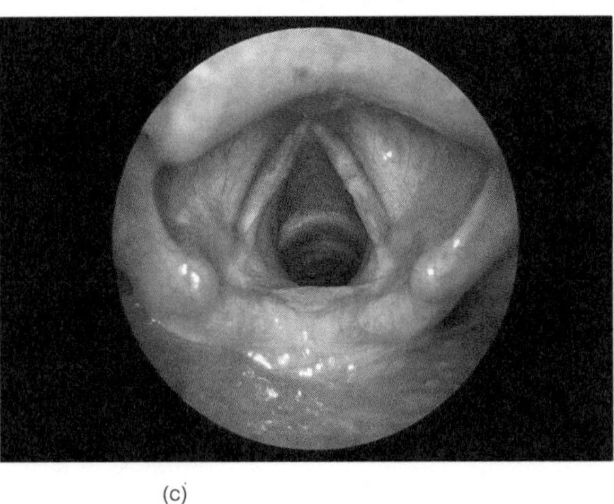

(c)

Occasionally the mucous membrane of the larynx becomes inflamed and swollen as a result of an infection or an irritation from inhaled vapors. When this happens, the vocal cords may not vibrate as freely as before, and the voice may sound harsh. This condition is called *laryngitis*, and although it is usually mild, laryngitis is potentially dangerous because the swollen tissues may obstruct the airway and interfere with breathing. In such cases it may be necessary to provide a passageway by inserting a tube (endotracheal tube) into the trachea through the nose or mouth.

1. Describe the structure of the larynx.
2. How do the vocal cords function to produce sounds?
3. What is the function of the glottis? The epiglottis?

Trachea

The **trachea** (windpipe) is a flexible cylindrical tube about 2.5 cm in diameter and 12.5 cm in length (fig. 13.5). It extends downward in front of the esophagus and into the thoracic cavity, where it splits into right and left bronchi.

The inner wall of the trachea is lined with a ciliated mucous membrane that contains many goblet cells. As mentioned before, this membrane continues to filter the incoming air and to move entrapped particles upward into the pharynx.

Within the tracheal wall are about twenty C-shaped pieces of hyaline cartilage, arranged one above the other. The open ends of these incomplete rings are directed posteriorly, and the gaps between their ends are filled with smooth muscle and connective tissues. These cartilaginous rings prevent the trachea from collapsing and blocking the airway. At the same time, the soft tissues that complete the rings in the back allow the nearby esophagus to expand as food moves through it on the way to the stomach.

Bronchial Tree

The **bronchial tree** consists of branched airways leading from the trachea to the microscopic air sacs in the lungs (fig. 13.6). It begins with the right and left **primary bronchi,** which arise from the trachea at the level of the fifth thoracic vertebra.

A short distance from its origin, each primary bronchus divides into secondary bronchi, which in turn branch again and again into finer and finer tubes. Among these smaller tubes, are some called **bronchioles.** They continue to divide, giving rise to very thin tubes called **alveolar ducts.** These ducts terminate in groups of microscopic air sacs called **alveoli,** which are surrounded by capillary nets (figs. 13.7 and 13.8).

Although, the structure of a bronchus is similar to that of the trachea, as finer and finer branches are given off, the amount of cartilage in the walls decreases and finally disappears in the bronchioles. As the cartilage decreases, however, a layer of smooth muscle surrounding the tube becomes more prominent. This muscular layer remains even in the smallest bronchioles, but only a few muscle fibers occur in the alveolar ducts.

Figure 13.5
The trachea transports air between the larynx and bronchi.

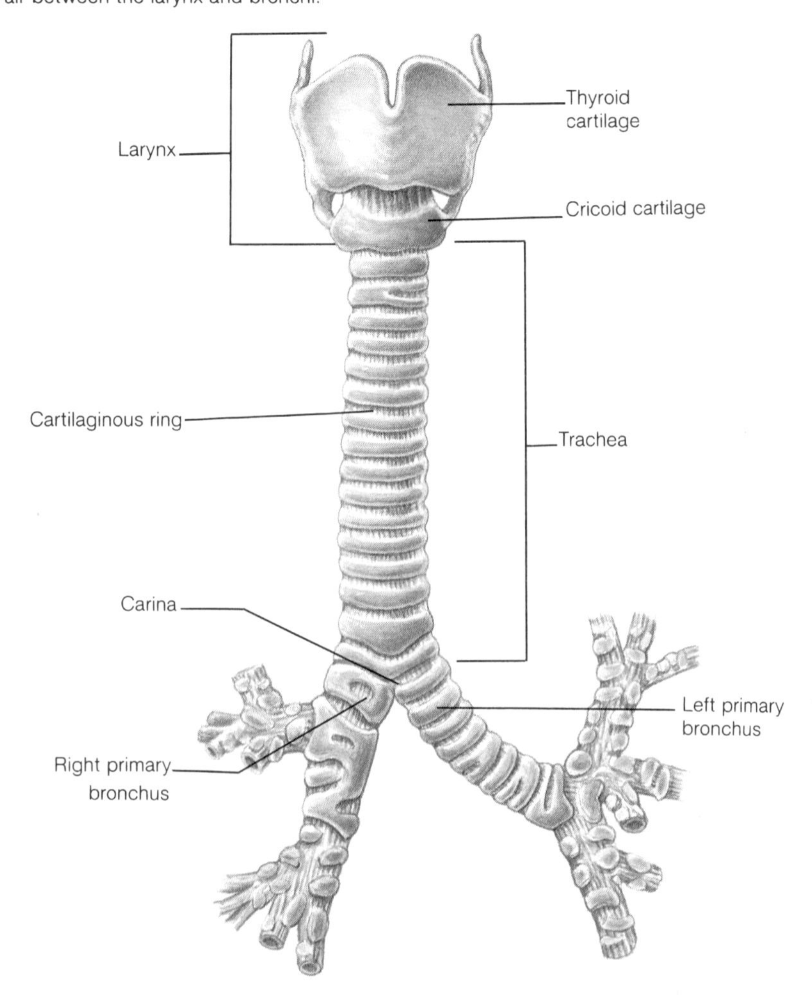

The branches of the bronchial tree serve as air passages, which continue to filter the incoming air and distribute it to the alveoli in all parts of the lungs. The alveoli, in turn, provide a large surface area of thin epithelial cells through which gas exchanges can occur. During these exchanges, oxygen diffuses through the alveolar walls and enters the blood in nearby capillaries, and carbon dioxide diffuses from the blood through the walls and enters the alveoli (fig. 13.9).

It is estimated that there are about 300 million alveoli in an adult lung and that these spaces have a total surface area between 70 and 80 square meters.

1. What is the function of the cartilaginous rings in the tracheal wall?
2. Describe the bronchial tree.
3. How are gases exchanged in the alveoli?

Lungs

The **lungs** (fig. 13.1) are soft, spongy, cone-shaped organs located in the thoracic cavity. The right and left lungs are separated medially by the heart and mediastinum, and they are enclosed by the diaphragm and thoracic cage.

Each lung occupies most of the thoracic space on its side and is suspended in the cavity by its attachments, which include a bronchus and some large blood vessels. These tubular parts enter the lung on its medial surface. A layer of serous membrane, the **visceral pleura,** is firmly attached to the surface on each lung, and this membrane folds back to become the **parietal pleura.** The parietal pleura, in turn, forms part of the mediastinum and lines the inner wall of the thoracic cavity (see fig. 1.8).

Figure 13.6
The bronchial tree consists of the passageways that connect the trachea and alveoli.

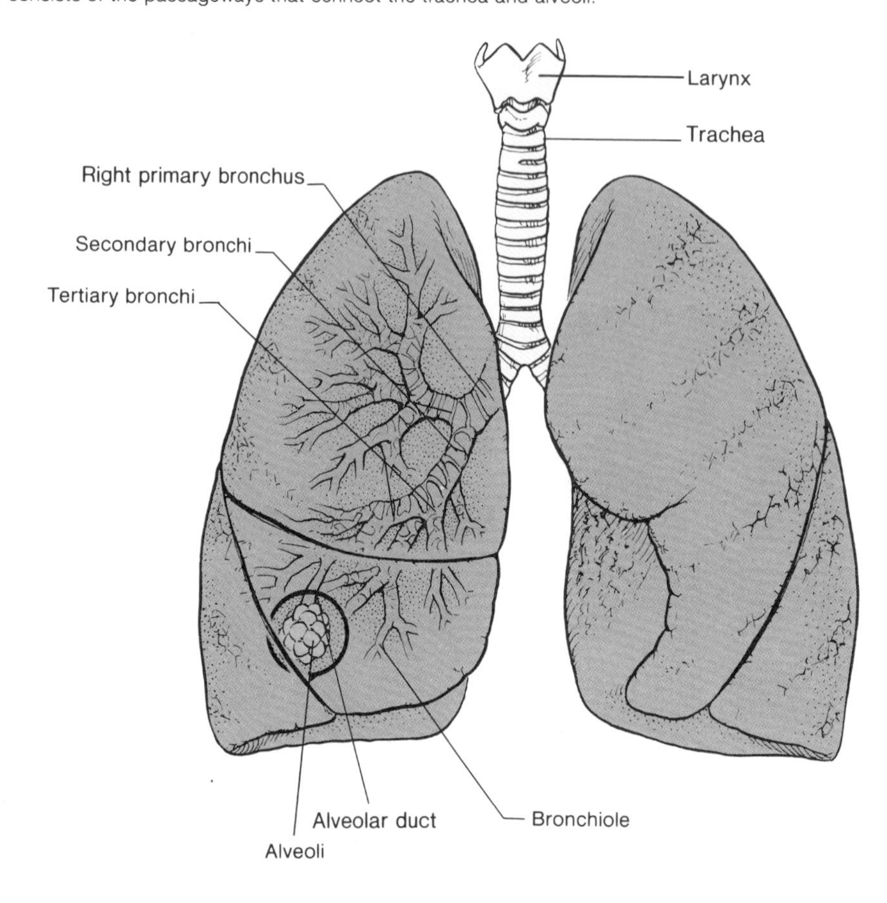

Figure 13.7
The respiratory tubes end in tiny alveoli, each of which is surrounded by a capillary network.

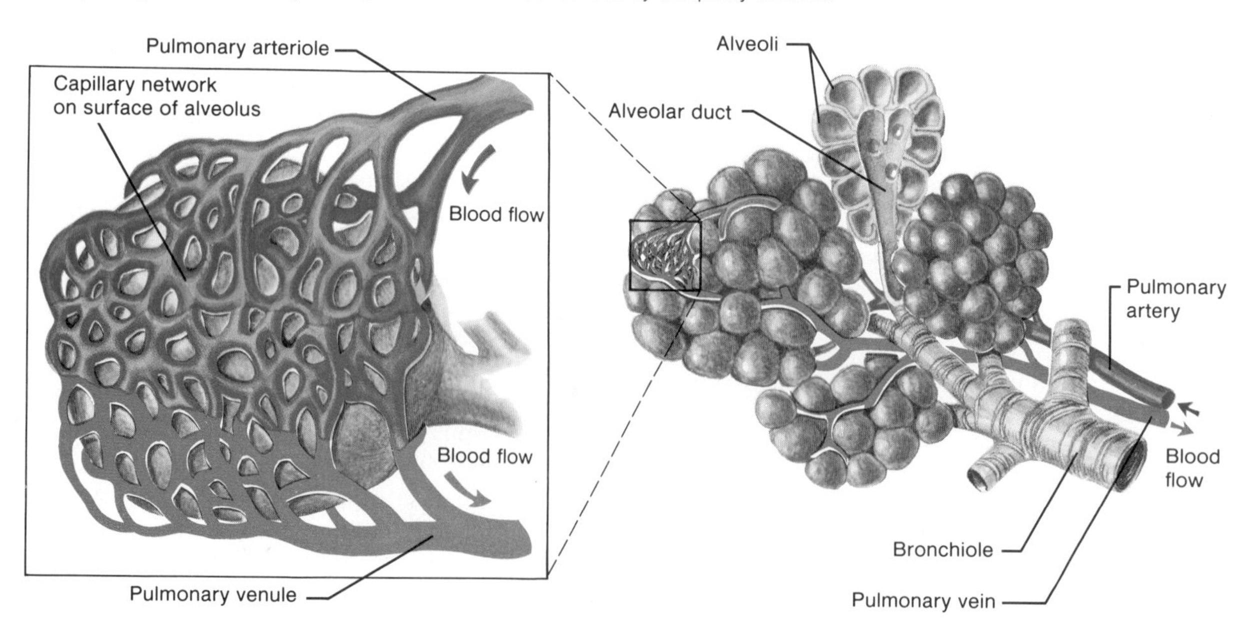

Figure 13.8
Alveoli appear as open spaces in this light micrograph of human lung tissue.

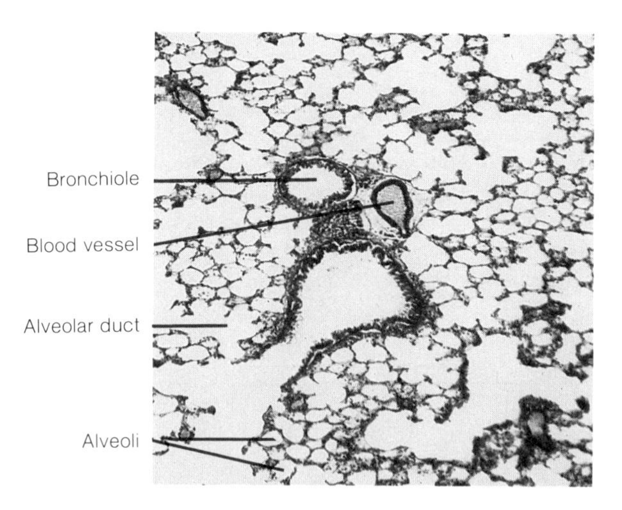

Figure 13.9
Oxygen diffuses from the air within the alveolus into the capillary, and carbon dioxide diffuses from the blood in the capillary into the alveolus.

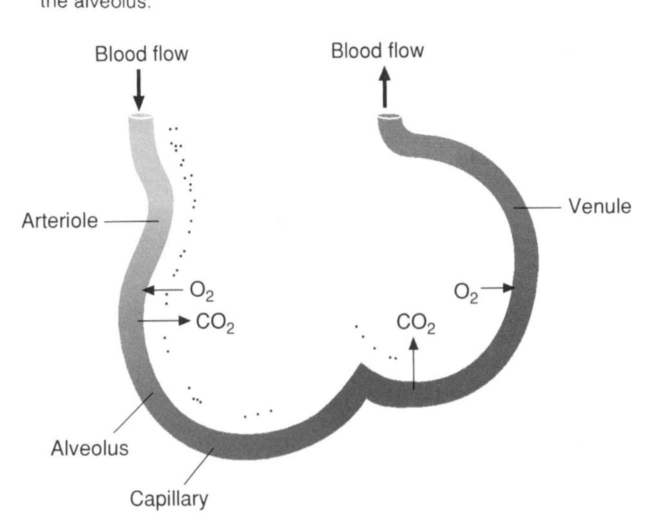

Chart 13.1 Parts of the respiratory system

Part	Description	Function
Nose	Part of face centered above the mouth and below the space between the eyes	Nostrils provide an entrance to the nasal cavity; internal hairs begin to filter the incoming air
Nasal cavity	Hollow space behind the nose	Conducts air to the pharynx; the mucous lining filters, warms, and moistens the air
Sinuses	Hollow spaces in various bones of the skull	Reduce the weight of the skull; serve as resonant chambers
Pharynx	Chamber behind the mouth cavity and between the nasal cavity and larynx	Passageway for air moving from the nasal cavity to larynx and for food moving from the mouth cavity to esophagus
Larynx	Enlargement at the top of the trachea	Passageway for air; prevents foreign objects from entering the trachea; houses the vocal cords
Trachea	Flexible tube that connects the larynx with the bronchial tree	Passageway for air; mucous lining continues to filter air
Bronchial tree	Branched tubes that lead from the trachea to the alveoli	Conducts air from the trachea to the alveoli; the mucous lining continues to filter air
Lungs	Soft, cone-shaped organs that occupy a large portion of the thoracic cavity	Contain the air passages, alveoli, blood vessels, connective tissues, lymphatic vessels, and nerves of the lower respiratory tract

The potential space between the visceral and parietal pleurae is called the **pleural cavity,** and it contains a thin film of serous fluid. This fluid lubricates the adjacent pleural surfaces, reducing friction as they move against one another during breathing. It also helps hold the pleural membranes together, as explained in the next section of this chapter.

The right lung is larger than the left one and is divided into three lobes. The left lung consists of two lobes (fig. 13.1).

Each lobe is supplied by a major branch of the bronchial tree. A lobe also has connections to blood and lymphatic vessels and is enclosed by connective tissues. Thus, the substance of a lung includes air passages, alveoli, blood vessels, connective tissues, lymphatic vessels, and nerves.

Chart 13.1 summarizes the characteristics of the major parts of the respiratory system.

1. Where are the lungs located?
2. What is the function of the serous fluid within the pleural cavity?
3. What kinds of structures make up a lung?

Emphysema and Lung Cancer

A Current Topic

Emphysema is a progressive, degenerative disease characterized by the destruction of many alveolar walls. As a result, clusters of small air sacs merge to form larger chambers, so that the total surface area of the respiratory membrane decreases. This decrease in respiratory membrane is accompanied by a decrease in the volume of gases that can be exchanged through the membrane. At the same time, the alveolar walls lose some of their elasticity, and the capillary networks associated with the alveoli become less abundant (fig. 13.10).

Because of the loss of tissue elasticity, a person with emphysema finds it increasingly difficult to force air out of the lungs because normal expiration involves the passive elastic recoil of inflated tissues. Consequently, with emphysema abnormal muscular effort is required to move the air.

The cause of emphysema is not well understood, however it seems to develop in response to prolonged exposure to respiratory irritants, such as those in tobacco smoke and polluted air.

Lung cancer, like other cancers, involves an uncontrolled growth of abnormal cells. These cells develop in and around normal tissues and deprive the normal tissues of their needs, such as nutrients and oxygen. In effect, the cancer cells cause the death of normal cells by crowding them out.

Some cancerous growths in the lungs result secondarily from cancer cells that have spread (metastasized) from other parts of the body, such as the breasts, intestines, liver, or kidneys. Other cancers begin in the lungs and are called *primary pulmonary cancers.*

Primary pulmonary cancer may arise from epithelial cells, connective tissue cells, or various blood cells. The most common form originates from epithelium and is called *bronchogenic carcinoma.* This type of cancer occurs in response to excessive irritation, such as that produced by prolonged exposure to tobacco smoke (fig. 13.11).

Once the cancer cells have appeared, they are likely to produce masses that obstruct air passages and reduce the amount of gas exchange. Furthermore, bronchogenic carcinoma is likely to spread relatively quickly and establish secondary cancers in the lymph nodes, liver, bones, brain, or kidneys.

Lung cancer is often difficult to control. Usually it is treated with surgery, ionizing radiation, and drugs (chemotherapy). However, despite treatment, the survival rate among lung cancer patients remains quite low.

Figure 13.10
(*a*) Normal lung tissue. (*b*) As emphysema develops, the alveoli tend to merge, forming larger chambers.

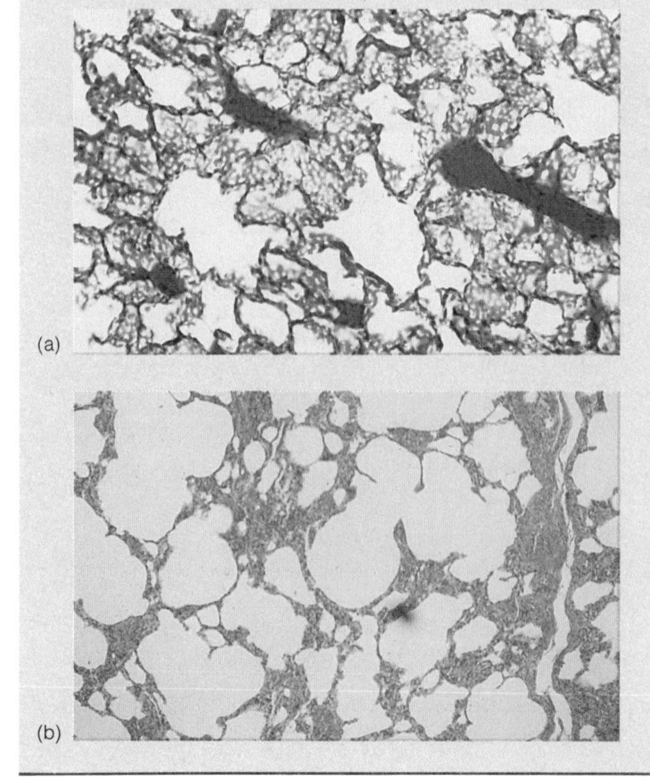

Figure 13.11
Lung cancer usually starts in the lining (epithelium) of a bronchus. (*a*) The normal lining shows (*4*) columnar cells with (*2*) hairlike cilia, (*3*) goblet cells that secrete (*1*) mucus, and (*5*) basal cells from which new columnar cells arise. (*6*) A basement membrane separates the epithelial cells from (*7*) the underlying connective tissue. (*b*) In the first stage of lung cancer, the basal cells divide repeatedly. The goblet cells secrete excessive mucus, and the cilia function less efficiently in moving the heavy mucus secretion. (*c*) With the continued multiplication of basal cells, the columnar and goblet cells are displaced. The basal cells penetrate the basement membrane and invade the deeper connective tissue.

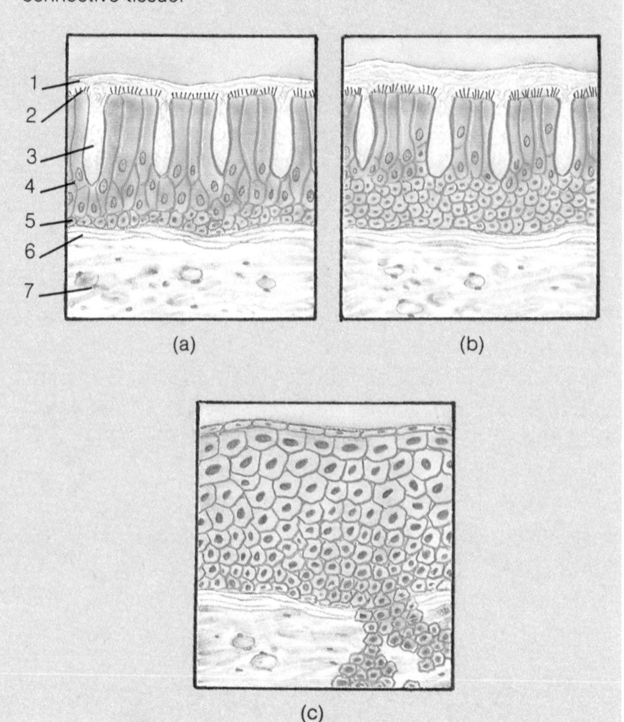

Figure 13.12

(*a*) Prior to inspiration, the intra-alveolar pressure is 760 mm Hg. (*b*) The intra-alveolar pressure decreases to about 758 mm Hg as the thoracic cavity enlarges, and air is forced into the airways by the atmospheric pressure.

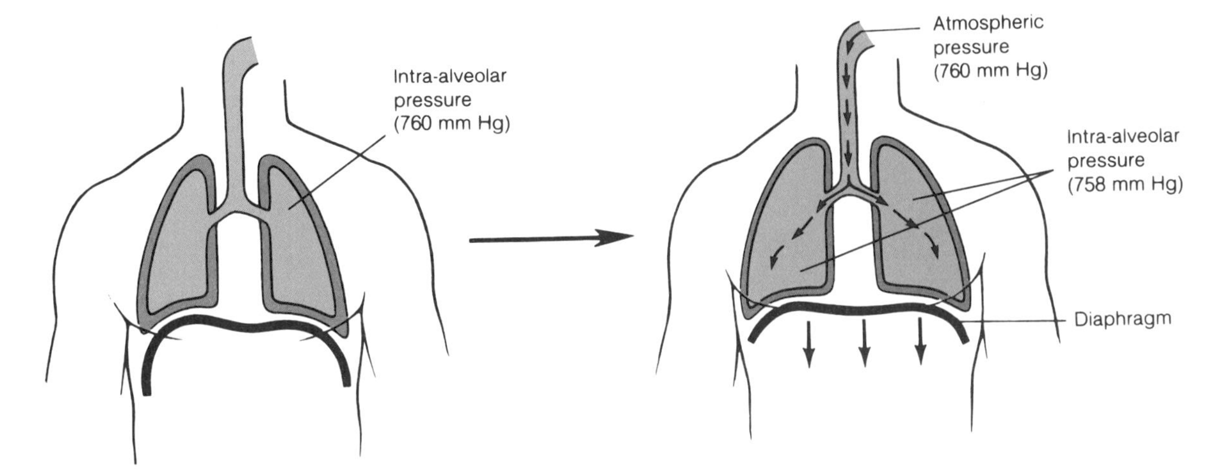

Breathing Mechanism

Breathing, or *pulmonary ventilation,* is the movement of air from outside the body into the bronchial tree and alveoli, followed by a reversal of this air movement. The actions responsible for these air movements are termed **inspiration** (inhalation) and **expiration** (exhalation).

Inspiration

Atmospheric pressure, due to the weight of the air, is the force that causes air to move into the lungs. At sea level, this pressure is sufficient to support a column of mercury about 760 millimeters (mm) high in a tube. Thus, normal air pressure is equal to 760 mm of mercury (Hg).

Air pressure is exerted on all surfaces in contact with the air, and since persons breathe air, the inside surfaces of their lungs also are subjected to pressure. In other words, the pressures on the inside of the lungs and alveoli and on the outside of the thoracic wall are about the same.

If the pressure inside the lungs and alveoli decreases, outside air will be pushed into the airways by atmospheric pressure. That is what happens during inspiration. Muscle fibers in the dome-shaped *diaphragm* below the lungs are stimulated to contract by impulses carried on the phrenic nerves, which are associated with the cervical plexuses (see chapter 9). As this happens, the diaphragm moves downward, the size of the thoracic cavity is enlarged, and the pressure within the alveoli is reduced about 2 mm Hg below that of atmospheric pressure. In response to this decreased pressure, air is forced into the airways by atmospheric pressure, and the lungs expand (fig. 13.12).

While the diaphragm is contracting and moving downward, the *external intercostal muscles* between the ribs may be stimulated to contract. This action raises the ribs and elevates the sternum, so that the size of the thoracic cavity increases even more. As a result, the pressure inside is further reduced and more air is forced into the airways by the greater atmospheric pressure.

The expansion of the lungs is aided by the fact that the parietal pleura, on the inner wall of the thoracic cavity, and the visceral pleura, attached to the surface of the lungs, are separated only by a thin film of serous fluid. The *water molecules* in this fluid have a great attraction to one another, creating a force called **surface tension.** This force holds the moist surfaces of the pleural membranes tightly together. Consequently, when the thoracic wall is moved upward and outward by the action of the external intercostal muscles, the parietal pleura is moved too, and the visceral pleura follows it. This action helps to expand the lungs in all directions.

The surface tension between the adjacent moist membranes is sufficient to cause the collapse of the alveoli, which have moist inner surfaces. Certain alveolar cells, however, synthesize a mixture of lipoproteins, called **surfactant.** Surfactant, which is secreted into the alveolar air spaces continuously, acts to reduce the surface tension and decreases the tendency of the alveoli to collapse.

Sometimes the lungs of a newborn fail to produce enough surfactant, and the newborn's breathing mechanism is unable to overcome the force of surface tension. Consequently, the lungs cannot be ventilated, and the newborn is likely to die of suffocation. This condition is called *respiratory distress syndrome* (hyaline membrane disease), and it is the primary cause of respiratory difficulty in premature newborns.

Figure 13.13
(*a*) The shape of the thorax at the end of normal inspiration. (*b*) The shape of the thorax at the end of maximal inspiration.

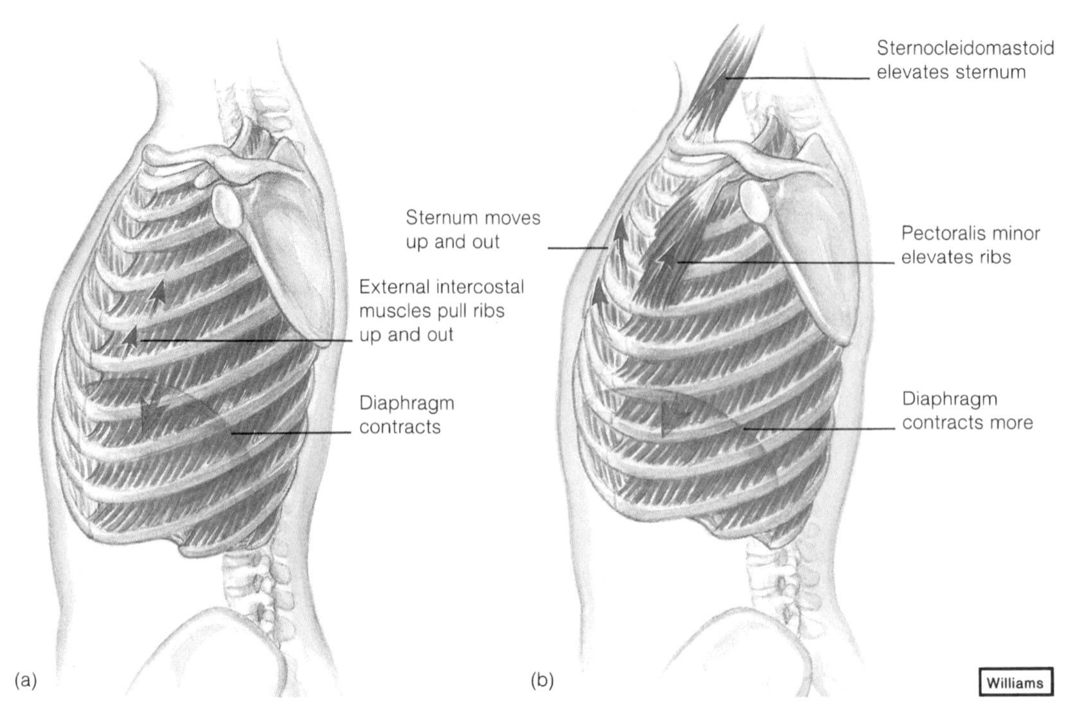

If a person needs to take a deeper than normal breath, the diaphragm and external intercostal muscles may be contracted to an even greater extent. Additional muscles, such as the pectoralis minors and sternocleidomastoids, can also be used to pull the thoracic cage farther upward and outward, enlarging the thoracic cavity and decreasing the internal pressure still more (fig. 13.13).

Because the visceral and parietal pleural membranes are held together by surface tension, no actual space exists normally in the pleural cavity between them. If the thoracic wall is punctured, however, atmospheric air may enter the pleural cavity and create a real space between the membranes. This condition is called *pneumothorax,* and when it occurs, the lung on the affected side may collapse because of its elasticity.

Expiration

The forces responsible for normal expiration come from the *elastic recoil* of tissues and from surface tension. The lungs and thoracic wall, for example, contain a considerable amount of elastic tissue, and as the lungs expand during inspiration, these tissues are stretched. As the diaphragm lowers, the abdominal organs beneath it are compressed. As the diaphragm and external intercostal muscles relax following inspiration, these elastic tissues cause the lungs and thoracic cage to recoil, and they return to their original shapes. Similarly, the abdominal organs spring back into their previous shapes, pushing the diaphragm upward (fig. 13.14*a*). At the same time, the surface tension that develops between the moist surfaces of the alveolar linings tends to cause a decrease in the diameter of the alveoli. Each of these factors tends to increase the alveolar pressure about 1 mm Hg above atmospheric pressure, so that the air inside the lungs is forced out through the respiratory passages. Thus, normal expiration is a passive process.

If a person needs to exhale more air than normal, the posterior *internal intercostal muscles* can be contracted (fig. 13.14*b*). These muscles pull the ribs and sternum downward and inward increasing the pressure in the lungs. Also, the *abdominal wall muscles,* including the external and internal obliques, transversus abdominis, and rectus abdominis, can be used to squeeze the abdominal organs inward (see fig. 8.18). Thus, the abdominal wall muscles can cause pressure in the abdominal cavity to increase and force the diaphragm still higher against the lungs. As a result of these actions, additional air can be squeezed out of the lungs.

1. Describe the events in inspiration.
2. How does surface tension aid in expanding the lungs during inspiration?
3. What forces are responsible for normal expiration?

Figure 13.14
(*a*) Normal expiration is due to the elastic recoil of the thoracic wall and abdominal organs; (*b*) maximal expiration is aided by contraction of the abdominal wall and posterior internal intercostal muscles.

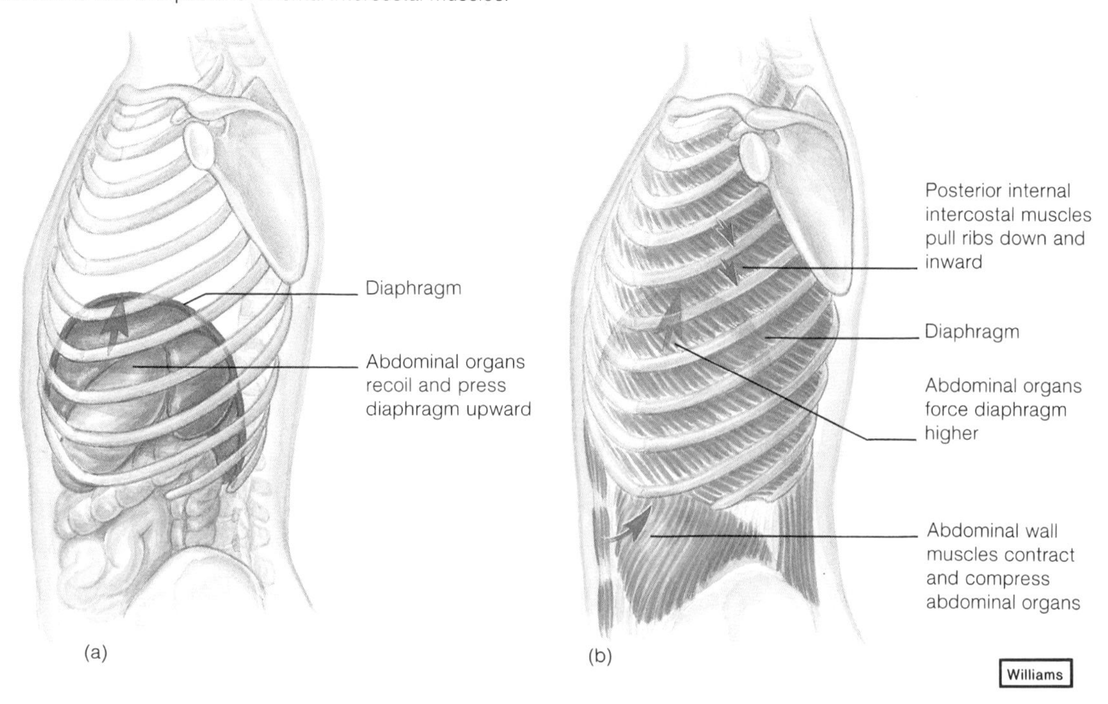

Figure 13.15
Respiratory air volumes.

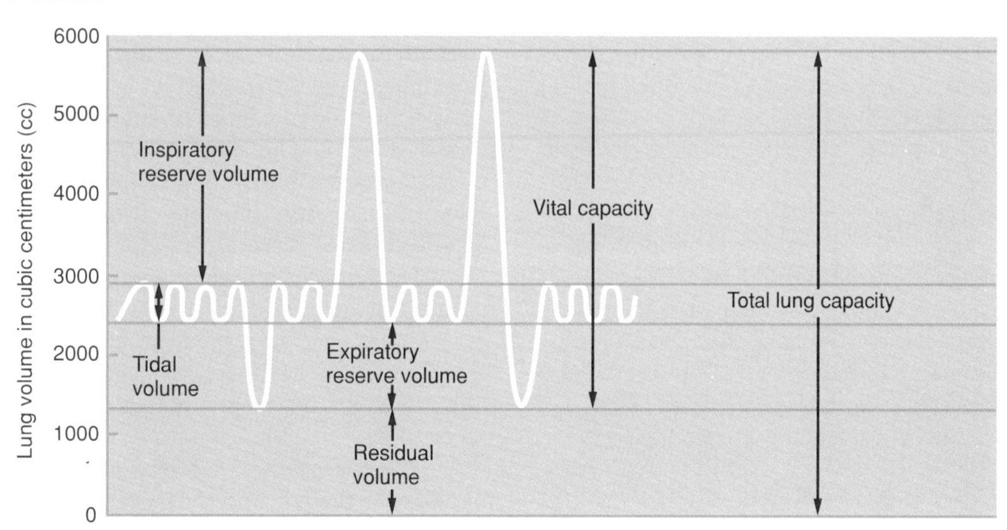

Respiratory Air Volumes

The amount of air that enters the lungs during a normal, quiet inspiration is about 500 cubic centimeters (cc). Approximately the same amount leaves during a normal expiration. This volume is termed the **tidal volume** (fig. 13.15).

During forced inspiration, a quantity of air in addition to the tidal volume enters the lungs. This additional volume is called the *inspiratory reserve volume* (complemental air), and it equals about 3000 cc.

During forced expiration, about 1100 cc of air in addition to the tidal volume can be expelled from the lungs. This quantity is called the *expiratory reserve volume* (supplemental air). Even after the most forceful expiration, however, about 1200 cc of air remains in the lungs. This volume is called the *residual volume.*

Residual air remains in the lungs at all times, and consequently, newly inhaled air is always mixed with the air that is already in the lungs. This prevents the oxygen and carbon dioxide concentrations in the lungs from fluctuating excessively with each breath.

Chart 13.2 Respiratory air volumes		
Name	Volume (average)	Description
Tidal volume (TV)	500 cc	Volume moved in or out of the lungs during quiet breathing
Inspiratory reserve volume (IRV)	3000 cc	Volume that can be inhaled during forced breathing in addition to the tidal volume
Expiratory reserve volume (ERV)	1100 cc	Volume that can be exhaled during forced breathing in addition to the tidal volume
Vital capacity (VC)	4600 cc	Maximum volume of air that can be exhaled after taking the deepest breath possible: VC = TV + IRV + ERV
Residual volume (RV)	1200 cc	Volume that remains in the lungs at all times
Total lung capacity (TLC)	5800 cc	Total volume of air that the lungs can hold: TLC = VC + RV

If the *inspiratory reserve volume* (3000 cc) is combined with the *tidal volume* (500 cc) and the *expiratory reserve volume* (1100 cc), the total is termed the **vital capacity** (4600 cc). This volume is the maximum amount of air a person can exhale after taking the deepest breath possible (fig. 13.15).

The *vital capacity* plus the *residual volume* equals the *total lung capacity* (about 5800 cc). This total varies with age, sex, and body size.

Some of the air that enters the respiratory tract during breathing fails to reach the alveoli. This volume (about 150 milliliters) remains in the passageways of the trachea, bronchi, and bronchioles. Since gas exchanges do not occur through the walls of these passages, this air is said to occupy *dead space*.

The respiratory air volumes are summarized in chart 13.2.

1. What is meant by the tidal volume?
2. Distinguish between the inspiratory and expiratory reserve volumes.
3. How is the vital capacity determined?
4. How is the total lung capacity calculated?

Nonrespiratory Air Movements

Air movements that occur in addition to breathing are called *nonrespiratory movements*. They are used to clear air passages, as in coughing and sneezing; or to express emotional feelings, as in laughing and crying; or to speak.

Chart 13.3 summarizes the characteristics of these movements.

Control of Breathing

Although the respiratory muscles can be controlled voluntarily, normal breathing is a rhythmic, involuntary act that continues even when a person is unconscious.

Respiratory Center

Breathing is controlled by a poorly defined group of neurons in the brain stem called the **respiratory center.** The components of this center are widely scattered throughout the pons and medulla oblongata (fig. 9.22). However, two areas of the respiratory center are of special interest. They are the rhythmicity area of the medulla and the pneumotaxic area of the pons.

The *medullary rhythmicity area* includes two groups of neurons that extend the length of the medulla. They are called the dorsal respiratory group and the ventral respiratory group.

The *dorsal respiratory group* is responsible for the basic rhythm of breathing. The neurons of this group emit bursts of impulses that signal the diaphragm and other inspiratory muscles to contract. The impulses of each burst begin weakly, increase in strength for about two seconds, and cease abruptly. The breathing muscles that contract in response to the impulses cause the volume of air entering the lungs to increase steadily. The neurons remain inactive while expiration occurs passively, and then they emit another burst of inspiratory impulses so that the inspiration-expiration cycle is repeated.

The *ventral respiratory group* is quiescent during normal breathing. However, when more forceful breathing is necessary, the neurons in this group generate impulses that increase inspiratory movements. Other neurons of the group activate the muscles associated with forceful expiration.

The neurons in the *pneumotaxic area* transmit impulses to the dorsal respiratory group continuously and regulate the duration of inspiratory bursts originating from the dorsal group. In this way, the pneumotaxic neurons control the rate of breathing. More specifically, when the pneumotaxic signals are strong, the inspiratory bursts have shorter durations, and the rate of breathing is increased; when the pneumotaxic signals are weak, the inspiratory bursts have longer durations, and the rate of breathing is decreased.

1. Where is the respiratory center located?
2. Describe how the respiratory center functions to maintain a normal breathing pattern.
3. Explain how the breathing pattern may be changed.

Chart 13.3 Nonrespiratory air movements		
Air movement	**Mechanism**	**Function**
Coughing	Deep breath is taken, the glottis is closed, and air is forced against the closure; suddenly the glottis is opened and a blast of air passes upward	Clears the lower respiratory passages
Sneezing	Same as coughing, except the air moving upward is directed into the nasal cavity by depressing the uvula	Clears the upper respiratory passages
Laughing	Deep breath is released in a series of short expirations	Expresses emotional happiness
Crying	Same as laughing	Expresses emotional sadness
Hiccuping	Diaphragm contracts spasmodically while the glottis is closed	No useful function
Yawning	Deep breath taken	Ventilates a large proportion of the alveoli and aids oxygenation of the blood
Speech	Air is forced through the larynx, causing the vocal cords to vibrate; words are formed by the lips, tongue, and soft palate	Communication

Factors Affecting Breathing

In addition to the controls exerted by the respiratory center, the breathing rate and depth are influenced by a variety of other factors. These include the presence of certain chemicals in the body fluids, the degree to which the lung tissues are stretched, and a person's emotional state. For example, there are *chemosensitive areas* within the respiratory center. These areas are located in the ventral portion of the medulla oblongata near the origins of the vagus nerves. They are very sensitive to changes in the blood concentrations of carbon dioxide and hydrogen ions. Thus, if the concentration of carbon dioxide or hydrogen ions rises, the chemosensitive areas signal the respiratory center, and the rate of breathing is increased. Then, as a result of increased breathing rate, more carbon dioxide is lost in exhaled air, the blood concentrations of these substances are reduced, and the breathing rate decreases.

Low blood oxygen seems to have little direct effect on the chemosensitive areas associated with the respiratory center. Instead, changes in the blood oxygen concentration are sensed by *chemoreceptors* in specialized structures called the *carotid* and *aortic bodies,* which are located in walls of certain large arteries (the carotid arteries and the aorta) in the neck and thorax (fig. 13.16). When these receptors are stimulated, impulses are transmitted to the respiratory center, and the breathing rate is increased. However, this mechanism is usually not triggered until the blood oxygen concentration reaches a very low level; thus, oxygen seems to play only a minor role in the control of normal respiration.

An *inflation reflex* helps to regulate the depth of breathing. This reflex occurs when stretch receptors in the visceral pleura, bronchioles, and alveoli are stimulated as a result of lung tissues being overstretched (fig. 13.17). The sensory impulses of this reflex travel via the vagus nerves to the pneumotaxic area of the respiratory center and cause the duration of inspiratory movements to shorten. This action prevents overinflation of the lungs during forceful breathing.

The normal breathing pattern may also be altered if a person is emotionally upset. Fear, for example, typically causes an increased breathing rate, as does pain. In addition, because the respiratory muscles are voluntary, breathing can be altered consciously. In fact, breathing can be stopped altogether if a person desires.

If breathing is stopped, the blood concentrations of carbon dioxide and hydrogen ions begin to rise, and the concentration of oxygen falls. These changes stimulate the respiratory center, and soon the need to inhale overpowers the desire to hold the breath. On the other hand, a person can increase the breath-holding time by breathing rapidly and deeply in advance. This action, called **hyperventilation,** causes a lowering of the blood carbon dioxide concentration, and following hyperventilation it takes longer than usual for the carbon dioxide concentration to reach the level needed to produce an overwhelming effect on the respiratory center. (Note: *Hyperventilation should never be used to help in holding the breath while swimming, because the person who has hyperventilated may lose consciousness under water and drown.*)

1. What chemical factors affect breathing?
2. Describe the inflation reflex.
3. How does hyperventilation result in a decreased respiratory rate?

Exercise and Breathing

A Current Topic

When a person engages in moderate to heavy physical exercise, the amount of oxygen used by the skeletal muscles increases greatly. For example, a young man at rest will utilize about 250 ml of oxygen per minute, but he may require 3600 ml per minute during maximal exercise.

While oxygen utilization is increasing, the volume of carbon dioxide produced increases also. Because decreased blood oxygen and increased blood carbon dioxide concentrations are stimulating to the respiratory center, it is not surprising that exercise is accompanied by an increased breathing rate. Studies have revealed, however, that the blood oxygen and carbon dioxide concentrations usually remain nearly unchanged during exercise—a reflection of the respiratory system's effectiveness in obtaining oxygen and releasing carbon dioxide to the outside.

The mechanism that seems to be responsible for most of the increase in the breathing rate during exercise involves the cerebral cortex and some sensory structures called *proprioceptors* that are associated with muscles and joints. Specifically, the cerebral cortex seems to transmit stimulating impulses to the respiratory center whenever it signals the skeletal muscles to contract. At the same time, muscular movements stimulate the proprioceptors, and a *joint reflex* is triggered. In this reflex, sensory impulses are transmitted from the proprioceptors to the respiratory center, and the breathing rate increases.

Whenever an increase in the breathing rate occurs during exercise, an increase in the blood flow is also needed in order to meet the needs of the skeletal muscles. Thus, physical exercise places a demand on the circulatory as well as on the respiratory system. If either of these systems fails to keep up with cellular demands, the person will begin to feel out of breath. This feeling, however, is usually due to the inability of the heart and circulatory system to move enough blood between the lungs and body cells, rather than the inability of the respiratory system to provide enough air.

Figure 13.16
Chemoreceptors in the carotid and aortic bodies are stimulated by decreased blood oxygen concentration.

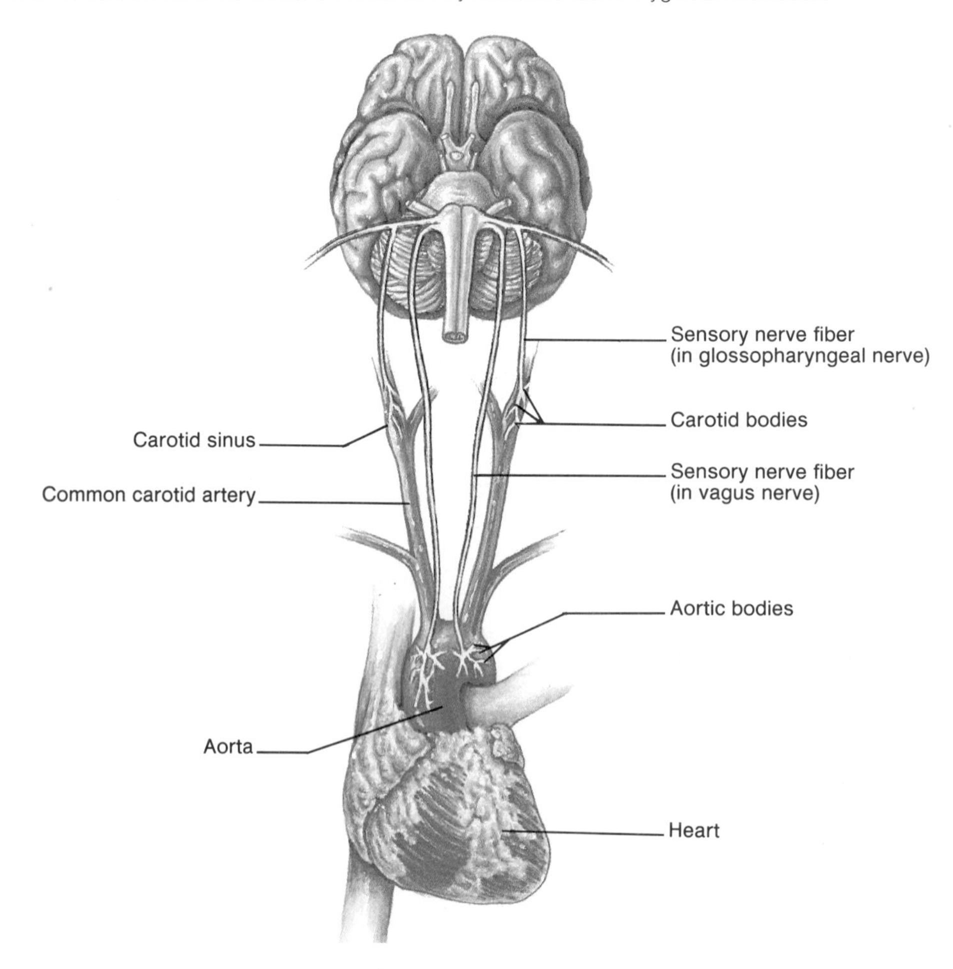

Figure 13.17
In the process of inspiration, motor impulses travel from the respiratory center to the diaphragm and external intercostal muscles, which contract and cause the lungs to expand. This expansion stimulates stretch receptors to send inhibiting impulses to the respiratory center.

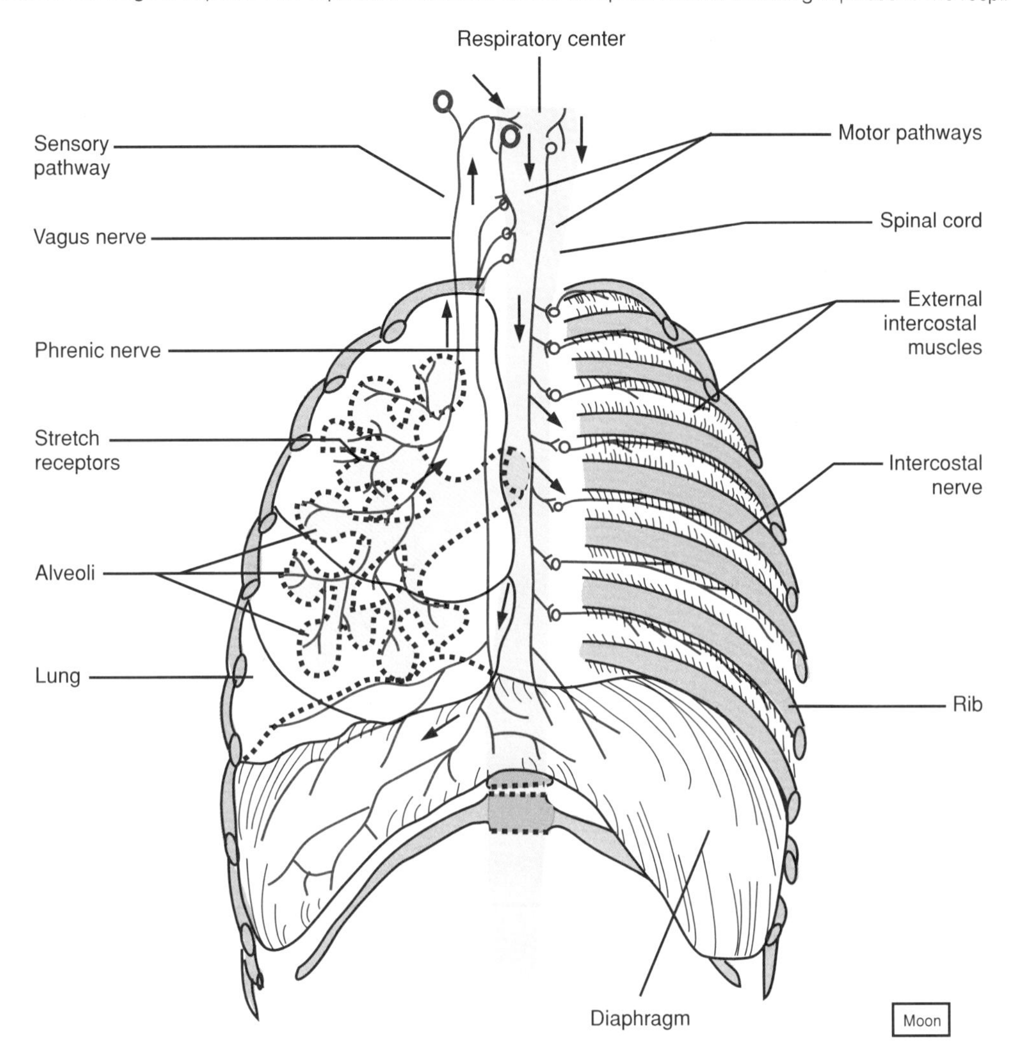

Alveolar Gas Exchanges

While other parts of the respiratory system move air and conduct it into and out of the air passages, the alveoli carry on the vital process of exchanging gases between the air and the blood.

Alveoli

The **alveoli** (fig. 13.7) are microscopic air sacs clustered at the distal ends of the finest respiratory tubes, the alveolar ducts. Each alveolus consists of a tiny space surrounded by a thin wall, which separates it from adjacent alveoli.

Respiratory Membrane

The wall of an alveolus consists of an inner lining of simple squamous epithelium and a dense network of capillaries, which are also lined with simple squamous epithelial cells. Thin basement membranes separate the layers of these flattened cells, and in the spaces between them, there are elastic and collagenous fibers, which help to support the wall. As figure 13.18 shows, there are at least two thicknesses of epithelial cells and basement membranes between the air in an alveolus and the blood in a capillary. These layers make up the **respiratory membrane,** which is of vital importance because it is through this membrane that gas exchanges occur between the blood and alveolar air.

Diffusion through the Respiratory Membrane

As is described in chapter 3, gas molecules diffuse from regions where they are in higher concentration toward regions where they are in lower concentration. Similarly, gases move from regions of higher pressure toward regions of lower pressure, and the *pressure* of a gas determines the rate at which it will diffuse from one region to another.

Figure 13.18
The respiratory membrane consists of the walls of the alveolus and the capillary.

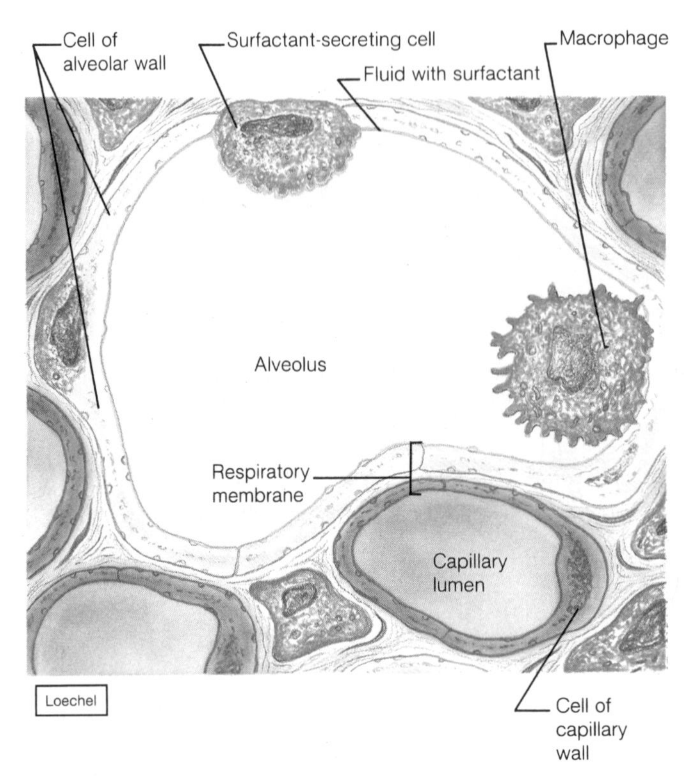

Measured by volume, ordinary air is about 78% nitrogen, 21% oxygen, and 0.04% carbon dioxide. Air also contains small amounts of other gases that have little or no physiological importance.

In a mixture of gases, such as the air, each gas is responsible for a portion of the total weight or pressure produced by the mixture. The amount of pressure each gas creates is called the **partial pressure,** and it is directly related to the concentration of the gas in the mixture. For example, because air is 21% oxygen, this gas is responsible for 21% of the atmospheric pressure. Since 21% of 760 mm Hg is equal to 160 mm Hg, it is said that the partial pressure of oxygen, symbolized PO_2, in atmospheric air is 160 mm Hg. Similarly, the partial pressure of carbon dioxide (PCO_2) in air can be calculated as 0.3 mm Hg.

When a mixture of gases dissolves in the blood, each gas exerts its own partial pressure in proportion to its dissolved concentration. Furthermore, each gas will diffuse between the blood and its surroundings, and this movement will tend to equalize its partial pressures in the two regions.

For example, the PCO_2 in capillary blood is 45 mm of Hg, but the PCO_2 in alveolar air is 40 mm of Hg. As a consequence of the difference between these partial pressures, carbon dioxide diffuses from the blood, where its pressure is higher, through the respiratory membrane and into the alveolar air (fig. 13.19). When the blood leaves the lungs, its PCO_2 is 40 mm Hg, which is about the same as the PCO_2 of the alveolar air.

Similarly, the PO_2 of capillary blood is 40 mm Hg, but that of alveolar air is 104 mm Hg. Thus, oxygen diffuses from the alveolar air into the blood, and the blood leaves the lungs with a PO_2 of 104 mm Hg.

If the lungs are exposed to an abnormally high oxygen concentration (hyperoxia) for a prolonged time, tissues may be damaged. The tissues most likely to suffer are those that form the capillary walls. As a result of such damage, excessive amounts of fluid may escape from the capillaries and flood the alveolar air spaces. Fluid in the alveoli will interfere with gas exchanges, and this may lead to death.

1. Describe the structure of the respiratory membrane.
2. What is meant by the partial pressure?
3. What causes oxygen and carbon dioxide to move across the respiratory membrane?

Transport of Gases

The transport of oxygen and carbon dioxide between the lungs and body cells is a function of the blood. As these gases enter the blood, they dissolve in the liquid portion (plasma). They combine chemically with various blood components, and most are carried in combination with other atoms or molecules.

Figure 13.19
Gas exchanges occur between the air of an alveolus and the blood of a capillary as a result of differences in partial pressures.

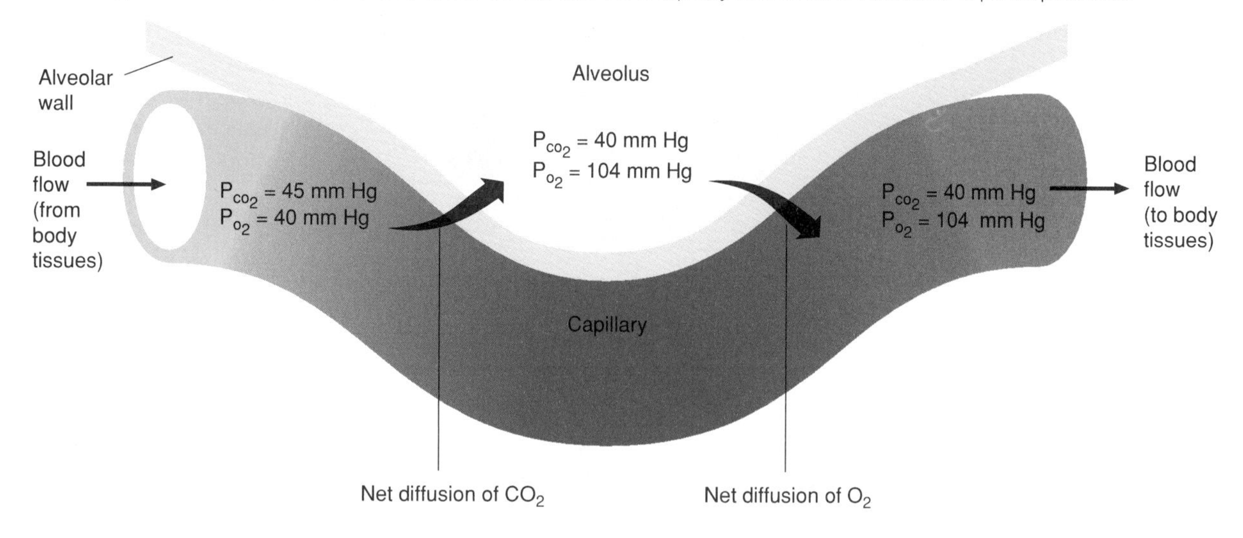

Figure 13.20
(*a*) Oxygen molecules, entering the blood from an alveolus, bond to hemoglobin and form oxyhemoglobin. (*b*) In regions of the body cells, oxyhemoglobin releases its oxygen.

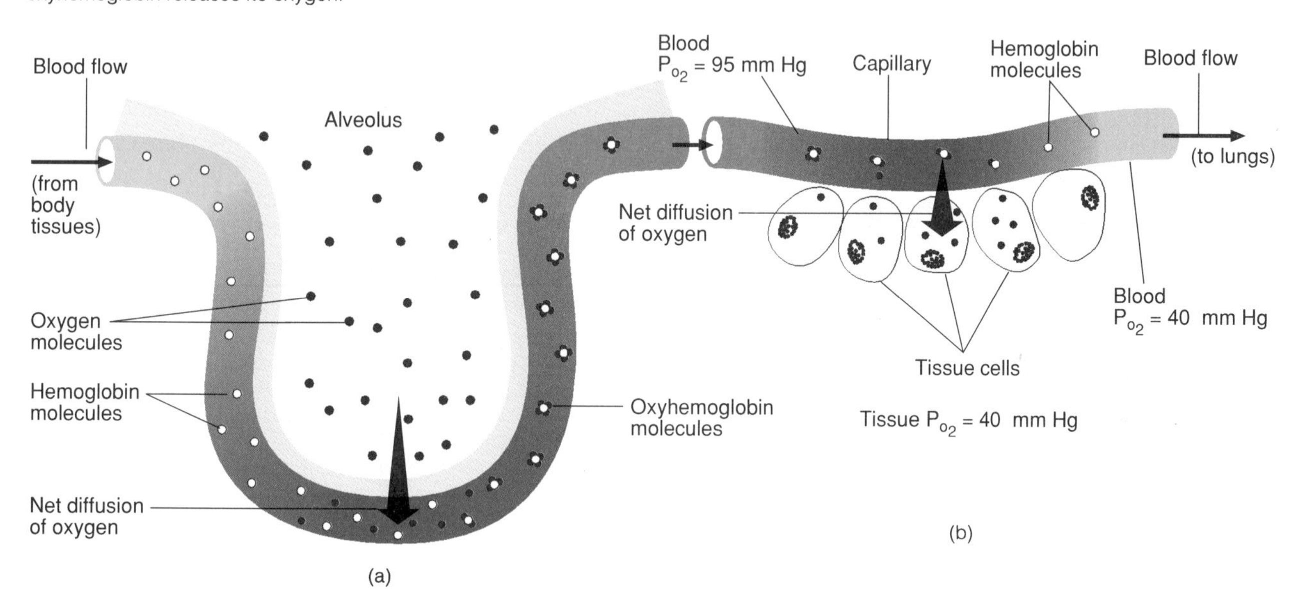

Oxygen Transport

Almost all the oxygen (over 98%) carried in the blood is combined with the iron-containing compound, **hemoglobin,** that occurs within the red blood cells. The remainder of the oxygen is dissolved in the blood plasma.

In the lungs, where the Po_2 is relatively high, oxygen dissolves in the blood and combines rapidly with the iron atoms of hemoglobin. The result of this chemical reaction is a new substance called **oxyhemoglobin** (fig. 13.20).

The chemical bonds that form between the oxygen and hemoglobin molecules are relatively unstable, and as the Po_2 decreases, oxygen is released from oxyhemoglobin molecules. This happens in tissues where the cells have used oxygen in their respiratory processes, and the free oxygen diffuses from the blood into nearby cells.

The amount of oxygen released from oxyhemoglobin is affected by several other factors, including the blood concentration of carbon dioxide, the blood pH, and the blood temperature. Thus, as the concentration of carbon dioxide increases, as the blood becomes more acidic, or as the blood temperature increases, more oxygen is released.

Due to these factors, more oxygen is released to the skeletal muscles during periods of physical exercise, because the increased muscular activity accompanied by an increased use of oxygen causes an increase in carbon dioxide concentration, a decrease in the pH,

Figure 13.21
The carbon dioxide produced by tissue cells is transported in the blood in a dissolved state, either combined with hemoglobin or in the form of bicarbonate ions (HCO_3^-).

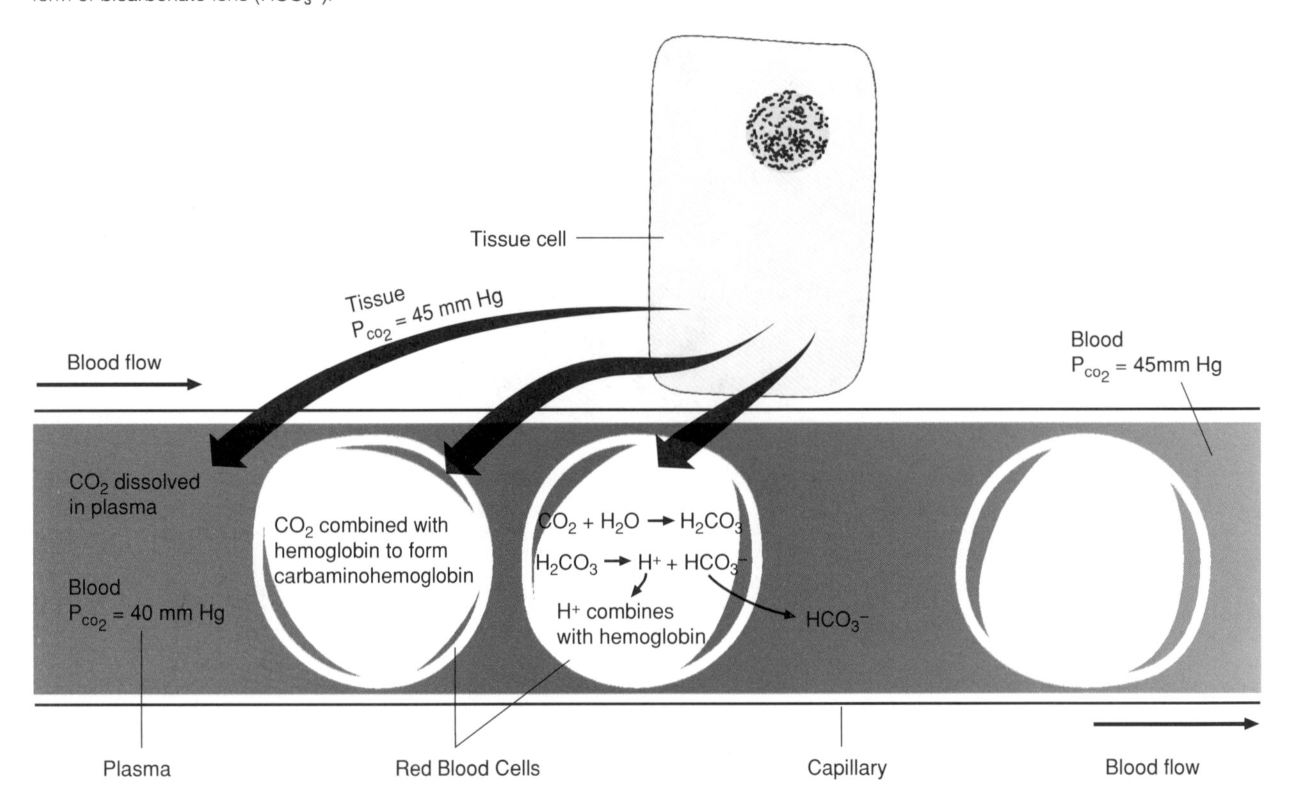

and a rise in the temperature. At the same time, less active cells receive relatively smaller amounts of oxygen.

Carbon monoxide (CO) is a toxic gas produced in gasoline engines as a result of incomplete combustion of the fuel. It is also a component of tobacco smoke. The toxic effect of carbon monoxide occurs because it combines with hemoglobin more effectively than oxygen does. Furthermore, carbon monoxide does not dissociate readily from hemoglobin. Thus, when a person breathes carbon monoxide, increasing quantities of hemoglobin become unavailable for oxygen transport, and the body cells soon begin to suffer from oxygen deficiencies.

A patient with carbon monoxide poisoning may be treated by administering oxygen in high concentration to replace some of the carbon monoxide bound to hemoglobin molecules. Carbon dioxide is usually administered simultaneously to stimulate the respiratory center, which, in turn, causes an increase in the breathing rate. Rapid breathing is desirable because it helps to reduce the concentration of carbon monoxide in the alveoli.

1. How is oxygen transported from the lungs to the body cells?
2. What stimulates oxygen to be released from the blood to the various tissues?

Carbon Dioxide Transport

Blood flowing through the capillaries of the body tissues gains carbon dioxide because the tissues have a relatively high P_{CO_2}. This carbon dioxide is transported to the lungs in one of three forms: as carbon dioxide dissolved in the blood, as part of a compound formed by bonding to hemoglobin, or as part of a bicarbonate ion (fig. 13.21).

The amount of carbon dioxide that dissolves in the blood is determined by its partial pressure. The higher the P_{CO_2} of the tissues, the more carbon dioxide will go into solution. However, only about 7% of the carbon dioxide is transported in this form.

Unlike oxygen, which combines with the iron atoms of hemoglobin molecules, carbon dioxide bonds with the amino groups ($-NH_2$) of these molecules. Consequently, oxygen and carbon dioxide do not compete for bonding sites, and both gases can be transported by a hemoglobin molecule at the same time.

When carbon dioxide combines with hemoglobin, a loosely bound compound called **carbaminohemoglobin** is formed. This substance decomposes readily in regions where the P_{CO_2} is low and, thus, releases its carbon dioxide. Although this method of transporting carbon dioxide is theoretically quite effective, carbaminohemoglobin forms relatively slowly. It is believed that only about 23% of the total carbon dioxide in the blood is carried this way.

Figure 13.22
In the lungs, carbon dioxide diffuses from the blood into the alveoli.

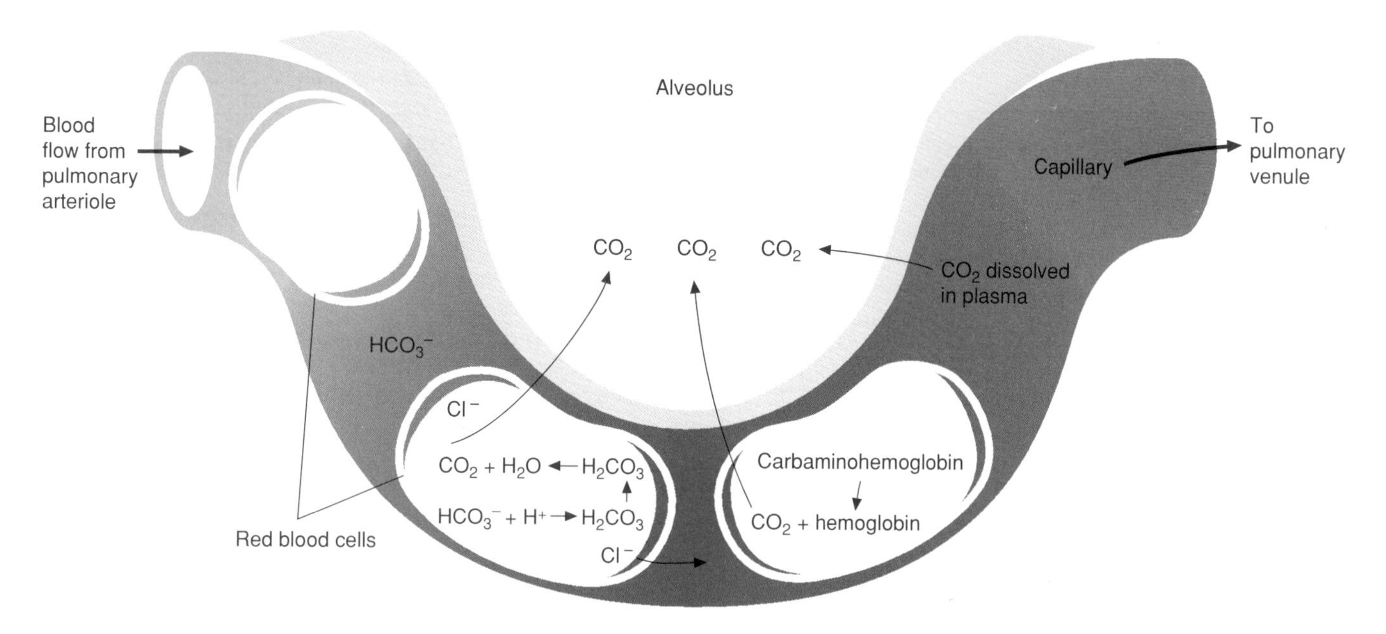

The most important carbon dioxide transport mechanism involves the formation of **bicarbonate ions** (HCO_3^-). Carbon dioxide reacts with water to form carbonic acid (H_2CO_3).

$$CO_2 + H_2O \rightarrow H_2CO_3$$

Although this reaction occurs slowly in the blood plasma, much of the carbon dioxide diffuses into the red blood cells, and these cells contain an enzyme, called **carbonic anhydrase,** that speeds the reaction between carbon dioxide and water.

The resulting carbonic acid then dissociates, releasing hydrogen ions (H^+) and bicarbonate ions (HCO_3^-).

$$H_2CO_3 \rightarrow H^+ + HCO_3^-$$

Most of the hydrogen ions combine quickly with hemoglobin molecules and, thus, are prevented from accumulating and causing a great change in the blood pH. The bicarbonate ions tend to diffuse out of the red blood cells and enter the blood plasma. It is estimated that nearly 70% of the carbon dioxide transported in the blood is carried in this form.

When the blood passes through the capillaries of the lungs, it loses its dissolved carbon dioxide by diffusion into the alveoli (fig. 13.22). This occurs in response to the relatively low P_{CO_2} of the alveolar air. At the same time, hydrogen ions and bicarbonate ions in the red blood cells recombine to form carbonic acid molecules, and under the influence of carbonic anhydrase, the carbonic acid gives rise to carbon dioxide and water.

$$H^+ + HCO_3^- \rightarrow H_2CO_3 \rightarrow CO_2 + H_2O$$

Chart 13.4 Gases transported in blood

Gas	Reaction involved	Substance transported
Oxygen	Combines with iron atoms of hemoglobin molecules.	Oxyhemoglobin
Carbon dioxide	About 7% dissolves in plasma.	Carbon dioxide
	About 23% combines with the amino groups of hemoglobin molecules.	Carbaminohemoglobin
	About 70% reacts with water to form carbonic acid; the carbonic acid then dissociates to release hydrogen ions and bicarbonate ions.	Bicarbonate ions

Carbaminohemoglobin also releases its carbon dioxide, and as carbon dioxide continues to diffuse out of the blood, an equilibrium is established between the P_{CO_2} of the blood and the P_{CO_2} of the alveolar air.

Chart 13.4 summarizes how blood gases are transported.

1. Describe three ways carbon dioxide can be carried from the body cells to the lungs.
2. How is it possible for hemoglobin to carry oxygen and carbon dioxide at the same time?
3. How is carbon dioxide released from the blood into the lungs?

Clinical Terms Related to the Respiratory System

anoxia (ah-nok'se-ah)—an absence or a deficiency of oxygen within tissues.

apnea (ap-ne'ah)—a temporary absence of breathing.

asphyxia (as-fik'se-ah)—a condition characterized by a deficiency of oxygen and an excess of carbon dioxide in the blood and tissues.

atelectasis (at"e-lek'tah-sis)—the collapse of a lung or some portion of it.

bradypnea (brad"e-ne'ah)—abnormally slow breathing.

bronchiolectasis (brong"ke-o-lek'tah-sis)—chronic dilation of the bronchioles.

bronchitis (brong-ki'tis)—an inflammation of the bronchial lining.

Cheyne-Stokes respiration (chān stōks res"pi-ra'shun)—irregular breathing, characterized by a series of shallow breaths that increase in depth and rate, followed by breaths that decrease in depth and rate.

dyspnea (disp'ne-ah)—difficulty in breathing.

eupnea (up-ne'ah)—normal breathing.

hemothorax (he"mo-tho'raks)—the presence of blood in the pleural cavity.

hypercapnia (hi"per-kap'ne-ah)—excessive carbon dioxide in the blood.

hyperoxia (hi"per-ok'-se-ah)—an excess in oxygenation of the blood.

hyperpnea (hi"perp-ne'ah)—an increase in the depth and rate of breathing.

hyperventilation (hi"per-ven"ti-la'shun)—prolonged, rapid, and deep breathing.

hypoxemia (hi"pok-se'me-ah)—a deficiency in the oxygenation of the blood.

hypoxia (hi-pok'se-ah)—a diminished availability of oxygen in the tissues.

lobar pneumonia (lo'ber nu-mo'ne-ah)—pneumonia that affects an entire lobe of a lung.

pleurisy (ploo'rĭ-se)—an inflammation of the pleural membranes.

pneumoconiosis (nu"mo-ko"ne-o'sis)—a condition characterized by the accumulation of particles from the environment in the lungs and the reaction of the tissues to their presence.

pneumothorax (nu"mo-tho'raks)—the entrance of air into the space between the pleural membranes, followed by collapse of the lung.

rhinitis (ri-ni'tis)—an inflammation of the nasal cavity lining.

sinusitis (si"nū-si'tis)—an inflammation of the sinus cavity lining.

tachypnea (tak"ip-ne'ah)—rapid, shallow breathing.

tracheotomy (tra"ke-ot'o-me)—an incision in the trachea for exploration or the removal of a foreign object.

Chapter Summary

Introduction (page 344)

The respiratory system includes the passages that transport air to and from the lungs and the air sacs, in which gas exchanges occur. Respiration is the entire process by which gases are exchanged between the atmosphere and body cells.

Organs of the Respiratory System (page 344)

The respiratory system includes the nose, nasal cavity, sinuses, pharynx, larynx, trachea, bronchial tree, and lungs.

The upper respiratory tract includes the respiratory organs outside the thorax; the lower respiratory tract includes those within the thorax.

1. Nose
   a. The nose is supported by bone and cartilage.
   b. The nostrils provide entrances for air.
2. Nasal cavity
   a. The nasal conchae divide the cavity into passageways and help increase the surface area of the mucous membrane.
   b. The mucous membrane filters, warms, and moistens the incoming air.
   c. Particles trapped in the mucus are carried to the pharynx by ciliary action and swallowed.
3. Sinuses
   a. The sinuses are spaces in the bones of the skull that open into the nasal cavity.
   b. They are lined with mucous membrane.
4. Pharynx
   a. The pharynx is located behind the mouth and between the nasal cavity and larynx.
   b. It functions as a common passage for air and food.
5. Larynx
   a. The larynx serves as a passageway for air and helps prevent foreign objects from entering the trachea.
   b. It is composed of muscles and cartilages.
   c. It contains the vocal cords, which produce sounds by vibrating as air passes over them.
   d. The glottis and epiglottis help prevent foods and liquids from entering the trachea.
6. Trachea
   a. The trachea extends into the thoracic cavity in front of the esophagus.
   b. It divides into right and left bronchi.
7. Bronchial tree
   a. The bronchial tree consists of branched air passages that lead from the trachea to the air sacs.
   b. The alveoli are located at the distal ends of the finest tubes.

8. Lungs
   a. The left and right lungs are separated by the mediastinum and enclosed by the diaphragm and thoracic cage.
   b. The visceral pleura is attached to the surface of the lungs; the parietal pleura lines the thoracic cavity.
   c. Each lobe is composed of alveoli, blood vessels, and supporting tissues.

Breathing Mechanism (page 351)

Inspiration and expiration movements are accompanied by changes in the size of the thoracic cavity.

1. Inspiration
   a. Air is forced into the lungs by atmospheric pressure.
   b. Inspiration occurs when the pressure inside the alveoli is reduced.
   c. Pressure within the alveoli is reduced when the diaphragm moves downward and the thoracic cage moves upward and outward.
   d. Expansion of the lungs is aided by surface tension.
2. Expiration
   a. The forces of expiration come from the elastic recoil of tissues and from the surface tension within the alveoli.
   b. Expiration can be aided by the thoracic and abdominal wall muscles.
3. Respiratory air volumes
   a. The amount of air that normally moves in and out during quiet breathing is the tidal volume.
   b. The additional air that can be inhaled is the inspiratory reserve volume; the additional air that can be exhaled is the expiratory reserve volume.
   c. Residual air remains in the lungs and is mixed with newly inhaled air.
   d. The vital capacity is the maximum amount of air a person can exhale after taking the deepest breath possible.
   e. The total lung capacity is equal to the vital capacity plus the residual air volume.
4. Nonrespiratory air movements
   Nonrespiratory air movements include coughing, sneezing, laughing, crying, hiccuping, yawning, and speaking.

Control of Breathing (page 354)

Normal breathing is rhythmic and involuntary.

1. Respiratory center
   a. The respiratory center is located in the brain stem and includes portions of the pons and medulla oblongata.
   b. The medullary rhythmicity area includes two groups of neurons.
      (1) The dorsal respiratory group is responsible for the basic rhythm of breathing.
      (2) The ventral respiratory group increases inspiratory and expiratory movements during forceful breathing.
   c. The pneumotaxic area regulates the rate of breathing.
2. Factors affecting breathing
   a. Breathing is affected by certain chemicals, stretching of the lung tissues, and emotional states.
   b. Chemosensitive areas are associated with the respiratory center.
      (1) Chemosensitive areas are influenced by the blood concentrations of carbon dioxide and hydrogen ions.
      (2) Stimulation of these areas causes the breathing rate to increase.
   c. There are chemoreceptors in the walls of certain large arteries.
      (1) These chemoreceptors are sensitive to a low oxygen concentration.
      (2) When the oxygen concentration is low, the breathing rate is increased.
   d. An inflation reflex is triggered by overstretching the lung tissues.
      (1) This reflex reduces the duration of inspiratory movements.
      (2) This prevents overinflation of the lungs during forceful breathing.
   e. Hyperventilation causes the blood carbon dioxide concentration to decrease, but *this is dangerous when associated with underwater swimming.*

Alveolar Gas Exchanges (page 357)

The alveoli carry on gas exchanges between the air and the blood.

1. Alveoli
   The alveoli are tiny air sacs clustered at the distal ends of the alveolar ducts.
2. Respiratory membrane
   a. This membrane consists of the alveolar and capillary walls.
   b. Gas exchanges take place through these walls.
   c. Diffusion through the respiratory membrane
      (1) The partial pressure of a gas is determined by the concentration of that gas in a mixture or the concentration dissolved in a liquid.
      (2) Gases diffuse from regions of higher partial pressure toward regions of lower partial pressure.
      (3) Oxygen diffuses from the alveolar air into the blood; carbon dioxide diffuses from the blood into the alveolar air.

Transport of Gases (page 364)

Blood transports gases between the lungs and body cells.

1. Oxygen transport
   a. Oxygen is mainly transported in combination with hemoglobin molecules.
   b. The resulting oxyhemoglobin is relatively unstable and releases its oxygen in regions where the PO_2 is low.
   c. More oxygen is released as the blood concentration of carbon dioxide increases, the blood becomes more acidic, or the blood temperature increases.
2. Carbon dioxide transport
   a. Carbon dioxide may be carried in solution either bound to hemoglobin or as a bicarbonate ion.
   b. Most carbon dioxide is transported in the form of bicarbonate ions.
   c. The enzyme, carbonic anhydrase, speeds the reaction between carbon dioxide and water.
   d. Carbonic acid dissociates to release hydrogen ions and bicarbonate ions.

Clinical Application of Knowledge

1. In certain respiratory disorders, such as emphysema, the capacity of the lungs to recoil elastically is reduced. Which respiratory air volumes will be affected by a condition of this type? Why?
2. What changes would you expect to occur in the relative concentrations of blood oxygen and carbon dioxide in a patient who breathes rapidly and deeply for a prolonged time? Why?
3. If a person has stopped breathing and is receiving pulmonary resuscitation, would it be better to administer pure oxygen or a mixture of oxygen and carbon dioxide? Why?

Review Activities

1. Describe the general functions of the respiratory system.
2. Distinguish between the upper and lower respiratory tracts.
3. Explain how the nose and nasal cavity function in filtering the incoming air.
4. Describe the locations of the major sinuses.
5. Distinguish between the pharynx and larynx.
6. Name and describe the locations of the larger cartilages of the larynx.
7. Distinguish between the false vocal cords and the true vocal cords.
8. Compare the structure of the trachea with the structure of the branches of the bronchial tree.
9. Distinguish between the visceral pleura and parietal pleura.
10. Explain how normal inspiration and forced inspiration are accomplished.
11. Define *surface tension,* and explain how it aids the breathing mechanism.
12. Define *surfactant,* and explain its function.
13. Explain how normal expiration and forced expiration are accomplished.
14. Distinguish between the vital capacity and total lung capacity.
15. Describe the location of the respiratory center, and name its major components.
16. Describe how the basic rhythm of breathing is controlled.
17. Explain the function of the pneumotaxic area of the respiratory center.
18. Describe the function of the chemoreceptors in the carotid and aortic bodies.
19. Describe the inflation reflex.
20. Define *hyperventilation,* and explain how it affects the respiratory center.
21. Define *respiratory membrane,* and explain its function.
22. Explain the relationship between the partial pressure of a gas and its rate of diffusion.
23. Summarize the gas exchanges that occur through the respiratory membrane.
24. Describe how oxygen is transported in the blood.
25. List three factors that cause an increased release of blood oxygen.
26. Explain how carbon dioxide is transported in the blood.

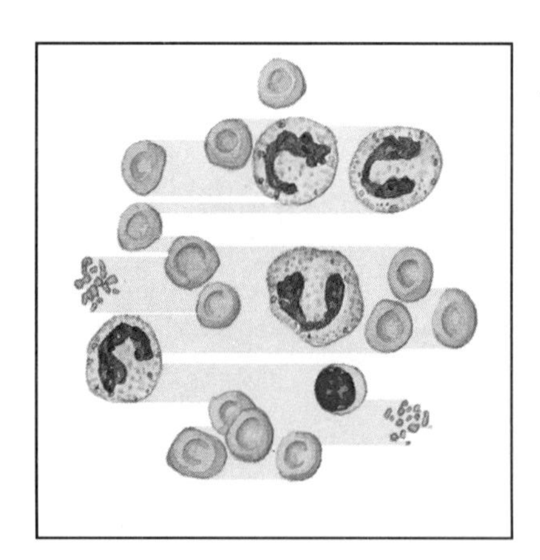

14 Blood

The blood, heart, and blood vessels constitute the *circulatory system* and provide a link between the body's internal parts and its external environment. More specifically, the blood transports nutrients and oxygen to the body cells and carries wastes from these cells to the respiratory and excretory organs. It transports hormones from the endocrine glands to target tissues and bathes the body cells in a liquid of relatively stable composition. It also aids in temperature control by distributing heat from the skeletal muscles and other active organs to all the body parts. Thus, the blood provides vital support for cellular activities and aids in maintaining a favorable cellular environment.

Chapter Objectives

After you have studied this chapter, you should be able to

1. Describe the general characteristics of the blood, and discuss its major functions.
2. Distinguish between the various types of cells found in blood.
3. Explain how red cell production is controlled.
4. List the major components of blood plasma, and describe the functions of each.
5. Define *hemostasis,* and explain the mechanisms that help to achieve it.
6. Review the major steps in blood coagulation.
7. Explain the basis for blood typing and how it is used to avoid adverse reactions following blood transfusions.
8. Describe how blood reactions may occur between the fetal and maternal tissues.
9. Complete the review activities at the end of this chapter. Note that the items are worded in the form of specific learning objectives. You may want to refer to them before reading the chapter.

Key Terms

agglutinin (ah-gloo′tĭ-nin)

agglutinogen (ag″loo-tin′o-jen)

albumin (al-bu′min)

basophil (ba′so-fil)

coagulation (ko-ag″u-la′shun)

eosinophil (e″o-sin′o-fil)

erythrocyte (ē-rith′ro-sīt)

erythropoietin (ē-rith″ro-poi′ē-tin)

fibrinogen (fi-brin′o-jen)

globulin (glob′u-lin)

hemostasis (he″mo-sta′sis)

leukocyte (lu′ko-sīt)

lymphocyte (lim′fo-sīt)

monocyte (mon′o-sīt)

neutrophil (nu′tro-fil)

plasma (plaz′mah)

platelet (plat′let)

Aids to Understanding Words

agglutin-, to glue together: *agglutin*ation—the clumping together of red blood cells.

bil-, bile: *bil*irubin—a pigment excreted in the bile.

embol-, stopper: *embol*us—an obstruction in a blood vessel.

erythr-, red: *erythr*ocyte—a red blood cell.

hemo-, blood: *hemo*globin—the red pigment responsible for the color of blood.

leuko-, white: *leuko*cyte—a white blood cell.

-osis, abnormal increase in production: leukocyt*osis*—condition in which white blood cells are produced in abnormally large numbers.

-poiet, making: erythro*poiet*in—a hormone that stimulates the production of red blood cells.

-stas, halting: hemo*stas*is—the arrest of bleeding from damaged blood vessels.

thromb-, clot: *thromb*ocyte—a blood platelet, which plays a role in the formation of a blood clot.

Introduction

As MENTIONED in chapter 5, blood is a type of connective tissue whose cells are suspended in a liquid intercellular material. It plays vital roles in transporting substances between the body cells and the external environment, and it aids in maintaining a stable cellular environment.

Blood and Blood Cells

Whole blood is slightly heavier and three to four times more viscous than water. Its cells, which are formed mostly in red bone marrow, include *red blood cells* and *white blood cells* (fig. 14.1). The blood also contains cellular fragments called *platelets*.

Volume and Composition of the Blood

The volume of the blood varies with body size. It also varies with changes in the fluid and electrolyte concentrations and the amount of fat tissue present. An average-size adult, however, will have a blood volume of about 5 liters.

A blood sample is usually about 45% red cells. This percentage is called the *hematocrit* (HCT). The remaining 55% of a blood sample consists of a clear liquid, called **plasma** (fig. 14.1). Plasma is a complex mixture that includes water, amino acids, proteins, carbohydrates, lipids, vitamins, hormones, electrolytes, and cellular wastes. The normal concentrations of the various plasma components are listed in the appendix.

1. What factors affect the blood volume?
2. What are the major components of the blood?

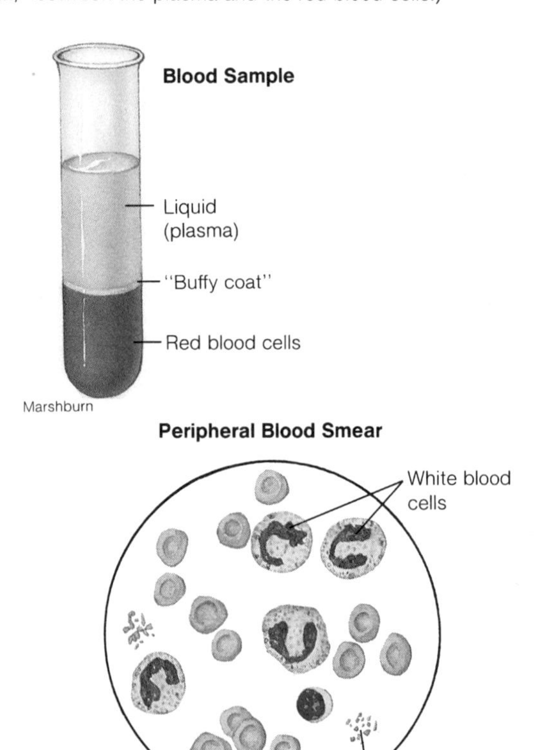

Figure 14.1
Blood consists of a liquid portion, called plasma, and a solid portion, which includes red blood cells, white blood cells, and platelets. (Note: When blood components are separated, the white blood cells and platelets form a thin layer, called the "buffy coat," between the plasma and the red blood cells.)

Characteristics of Red Blood Cells

Red blood cells, or **erythrocytes,** are tiny, biconcave disks, which are thin near their centers and thicker around their rims. This special shape is related to the red cell's function of transporting gases in that the shape provides an increased surface area through which gases can diffuse. The shape also places the cell membrane closer to various interior parts, where oxygen-carrying *hemoglobin* is found (fig. 14.2).

Each red blood cell is about one-third hemoglobin by volume, and this substance is responsible for the color of the blood. Specifically, when hemoglobin is combined with oxygen, the resulting *oxyhemoglobin* is bright red, and when oxygen is released, the resulting *deoxyhemoglobin* is darker.

A person experiencing a prolonged oxygen deficiency (hypoxia) may develop a symptom called *cyanosis*. In this condition, the skin and mucous membranes appear bluish due to an abnormally high concentration of deoxyhemoglobin in the blood. Cyanosis may also occur as a result of exposure to low temperature. In this case, the superficial blood vessels become constricted, the blood flow is slowed, and more oxygen than usual is removed from the blood flowing through the vessels.

Although the red blood cells have nuclei during their early stages of development, these nuclei are lost as the cells mature. This characteristic, like the shape of a red blood cell, seems to be related to the function of transporting oxygen, since the space previously occupied by the nucleus is available to hold hemoglobin.

1. Describe a red blood cell.
2. What is the function of hemoglobin?
3. What changes occur in a red blood cell as it matures?

Figure 14.2
(*a*) How is the biconcave shape of a red blood cell related to its function? (*b*) Scanning electron micrograph of human red blood cells.

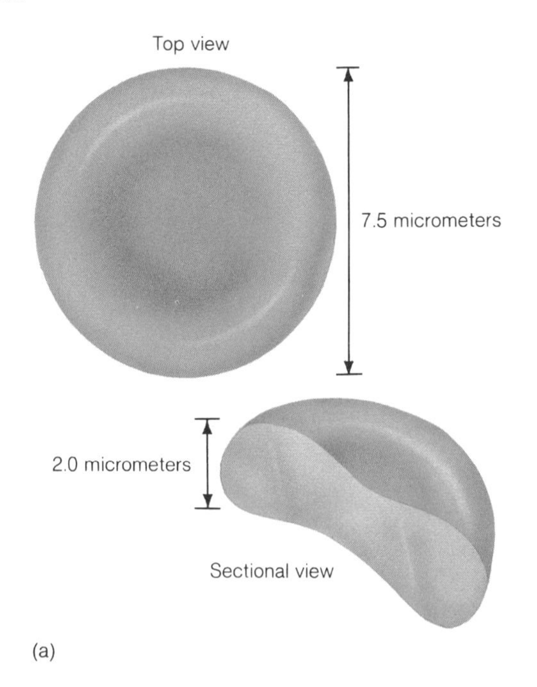

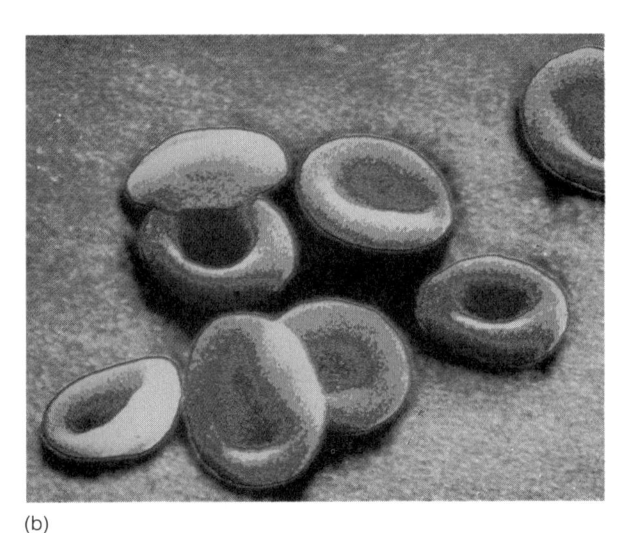
(b)

Red Blood Cell Counts

The number of red blood cells in a cubic millimeter (mm^3) of blood is called the *red blood cell count* (RBC count). Although this number varies from time to time, the normal range for adult males is 4,600,000–6,200,000 cells per mm^3, and that for adult females is 4,200,000–5,400,000 cells per mm^3.

Since the number of circulating red blood cells is closely related to the blood's *oxygen-carrying capacity,* any changes in this number may be significant. For this reason, red blood cell counts are routinely made to help diagnose and evaluate the courses of various diseases.

Destruction of Red Blood Cells

Red blood cells are quite elastic and flexible, and they readily change shape as they pass through small blood vessels. With time, however, these cells become more fragile, and they are often damaged or ruptured when they squeeze through capillaries, particularly those in active muscles.

Damaged red cells are phagocytized and destroyed, primarily in the liver and spleen, by **macrophages** (see chapter 5).

The hemoglobin molecules from the red cells undergoing destruction are broken down into molecular subunits of *heme,* an iron-containing portion, and *globin,* a protein. The heme is further decomposed into iron and a greenish pigment called **biliverdin.** The iron, combined with a protein, may be carried by the blood to the blood-cell-forming tissue in red bone marrow and reused in the synthesis of new hemoglobin. Otherwise, the iron may be stored in the liver cells in the form of an iron-protein complex. In time, the biliverdin is converted to an orange pigment called **bilirubin.** Biliverdin and bilirubin are excreted in the bile as bile pigments (see chapter 12).

About one-third of all newborns develop a mild disorder called *physiologic jaundice* within a few days following birth. In this condition, as in other forms of jaundice, the skin and eyes become yellowish due to an accumulation of bilirubin in the tissues.

Physiologic jaundice is thought to be the result of immature liver cells, which are somewhat ineffective in excreting bilirubin into the bile. The condition is treated by feedings that promote bowel movements and by exposure to fluorescent light, which reduces the concentration of bilirubin in the tissues.

1. What is the normal red blood cell count for an adult male? For an adult female?
2. What happens to damaged red blood cells?
3. What are the end products of hemoglobin breakdown?

Red Blood Cell Production

As mentioned in chapter 7, red blood cell formation (hematopoiesis) occurs initially in the yolk sac, liver, and spleen, but after an infant is born, these cells are produced almost exclusively by the tissue that lines the spaces within the red bone marrow.

The average life span of a red blood cell is about 120 days. Although a large number of these cells are removed from circulation each day, the number of cells in the circulating blood remains relatively stable. This observation suggests that a *homeostatic mechanism* regulates the rate of red cell production.

The mechanism involves *negative feedback,* which employs a hormone called **erythropoietin.** In response to prolonged oxygen deficiency, erythropoietin is released, primarily from the kidneys and to a lesser extent from the liver (fig. 14.3). At high altitudes, for example, where the Po_2 is reduced, the amount of oxygen delivered to the tissues decreases. This drop in oxygen concentration triggers the release of erythropoietin, which travels via the blood to the red bone marrow and stimulates increased red cell production.

After a few days, a large number of newly formed red cells begin to appear in the circulating blood, and the increased rate of production continues as long as the kidney and liver tissues experience an oxygen deficiency. Eventually, the number of red cells in circulation may be sufficient to supply these tissues with their oxygen needs. When this happens or when the Po_2 returns to normal, the release of erythropoietin ceases, and the rate of red cell production is reduced.

Figure 14.4 illustrates the stages in the development of red blood cells, white blood cells, and platelets.

Some individuals produce an abnormal type of hemoglobin molecule, which tends to form long chains when exposed to low oxygen concentrations. This causes the red blood cells containing such molecules to become distorted or sickle-shaped, and the person is said to have *sickle-cell anemia.*

In this condition, the distorted cells may block capillaries and, thus, interrupt the flow of blood to tissues. The resulting symptoms may include severe joint and abdominal pain, skin ulceration, and chronic kidney disease.

1. Where are red blood cells produced?
2. How is the production of red blood cells controlled?

Dietary Factors Affecting Red Blood Cell Production

Red blood cell production is significantly influenced by the availability of two of the B-complex vitamins—*vitamin* B_{12} and *folic acid.* These substances are required for the synthesis of DNA molecules and, thus, are needed by all cells for growth and reproduction. Since cellular reproduction occurs at a particularly high rate in red-blood-cell-forming tissue, this tissue is especially affected by a lack of either vitamin.

In addition, *iron* is needed for hemoglobin synthesis and normal red blood cell production. Iron can be absorbed from food passing through the small intestine; also, much of the iron released during the decomposition of hemoglobin from damaged red blood cells is available for reuse. Because of iron reuse, relatively small quantities of iron are needed in the daily diet.

Figure 14.3

Low Po_2 causes the release of erythropoietin from the kidneys and liver. This substance stimulates the production of red blood cells, and these cells carry additional oxygen to the tissues.

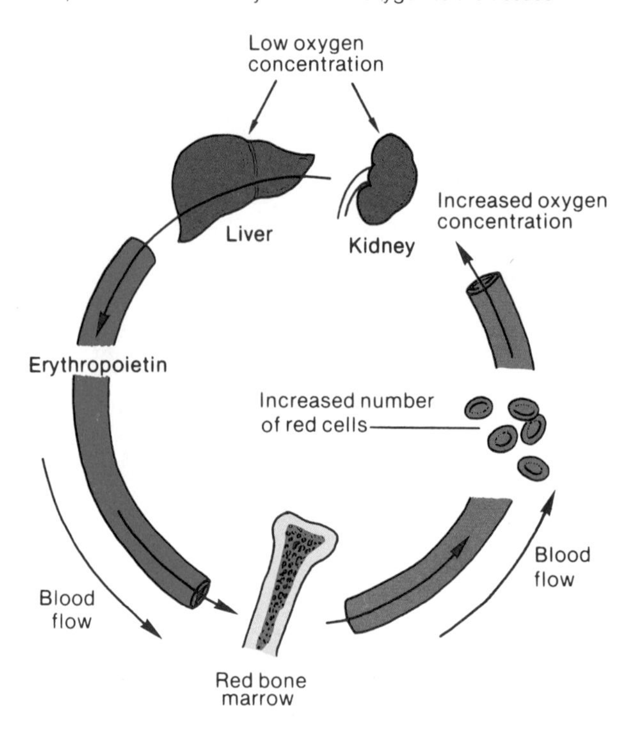

Figure 14.5 illustrates the life cycle of a red blood cell.

A deficiency of red blood cells or a reduction in the quantity of the hemoglobin they contain results in a condition called *anemia.* In disorders of this type, the oxygen-carrying capacity of the blood is reduced, and the person may appear pale and lack energy. Among the more common types of anemia are *hemorrhagic anemia,* caused by an abnormal loss of blood; *aplastic anemia,* due to a malfunction of red bone marrow, characterized by the decreased production of red blood cells; and *hemolytic anemia,* characterized by an abnormally high rate of red blood cell rupture (hemolysis).

In the absence of an adequate supply of iron, a person may develop a condition called *hypochromic anemia,* which is characterized by the presence of small, pale red blood cells with relatively low hemoglobin content.

1. What vitamins are necessary for red blood cell production?
2. Why is iron needed for the normal development of red blood cells?

Figure 14.4
Stages in the development of blood cells and platelets.

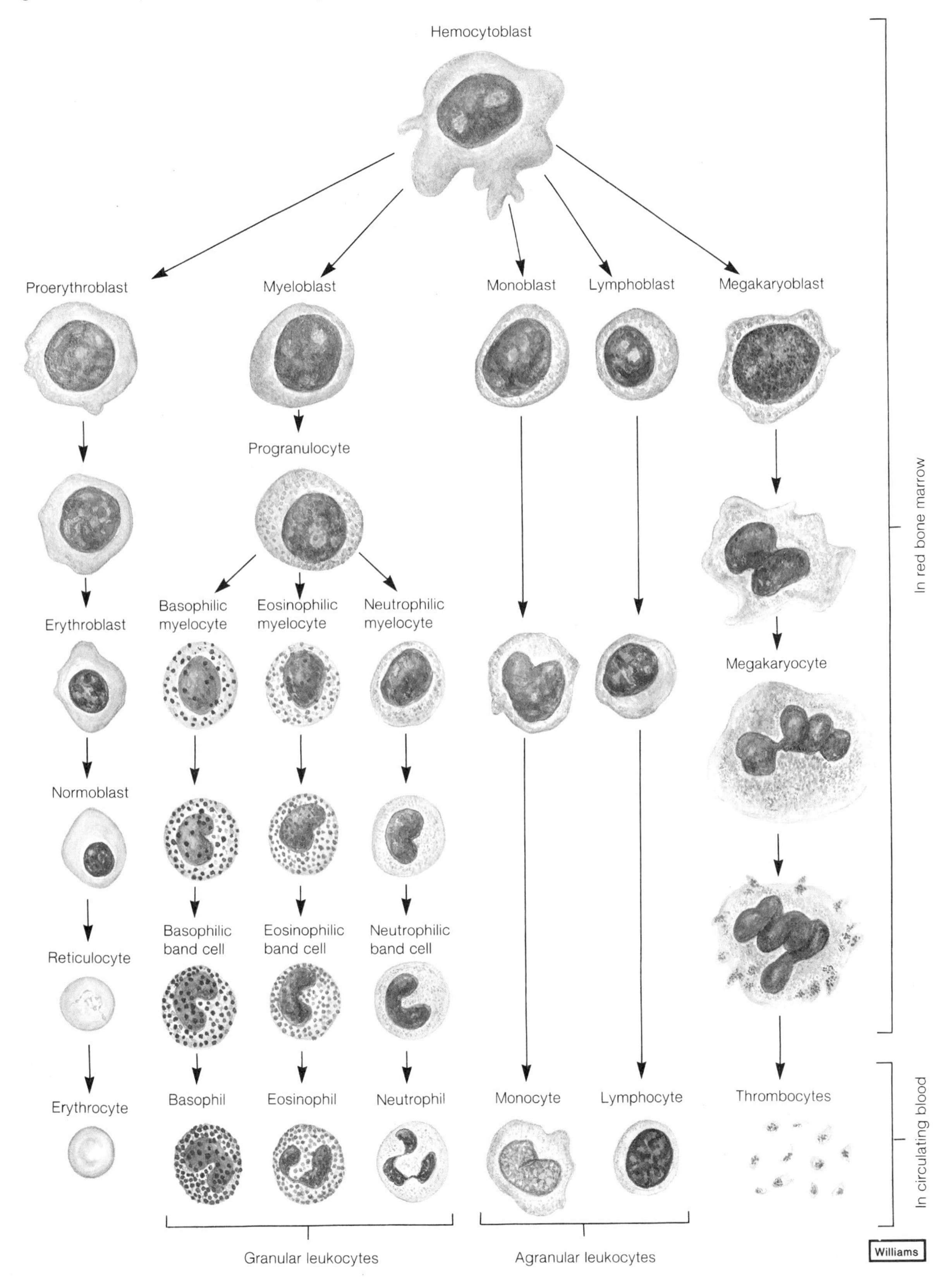

Figure 14.5
The life cycle of a red blood cell. (*1*) Essential nutrients are absorbed from the intestine; (*2*) nutrients are transported by the blood to red bone marrow; (*3*) red blood cells are produced in red bone marrow by mitosis; (*4*) mature red blood cells are released into the blood where they circulate for about 120 days; (*5*) damaged red blood cells are destroyed in the liver by macrophages; (*6*) hemoglobin from red blood cells is decomposed into heme and globin; (*7*) iron from heme is returned to red bone marrow and reused; (*8*) biliverdin is excreted in bile.

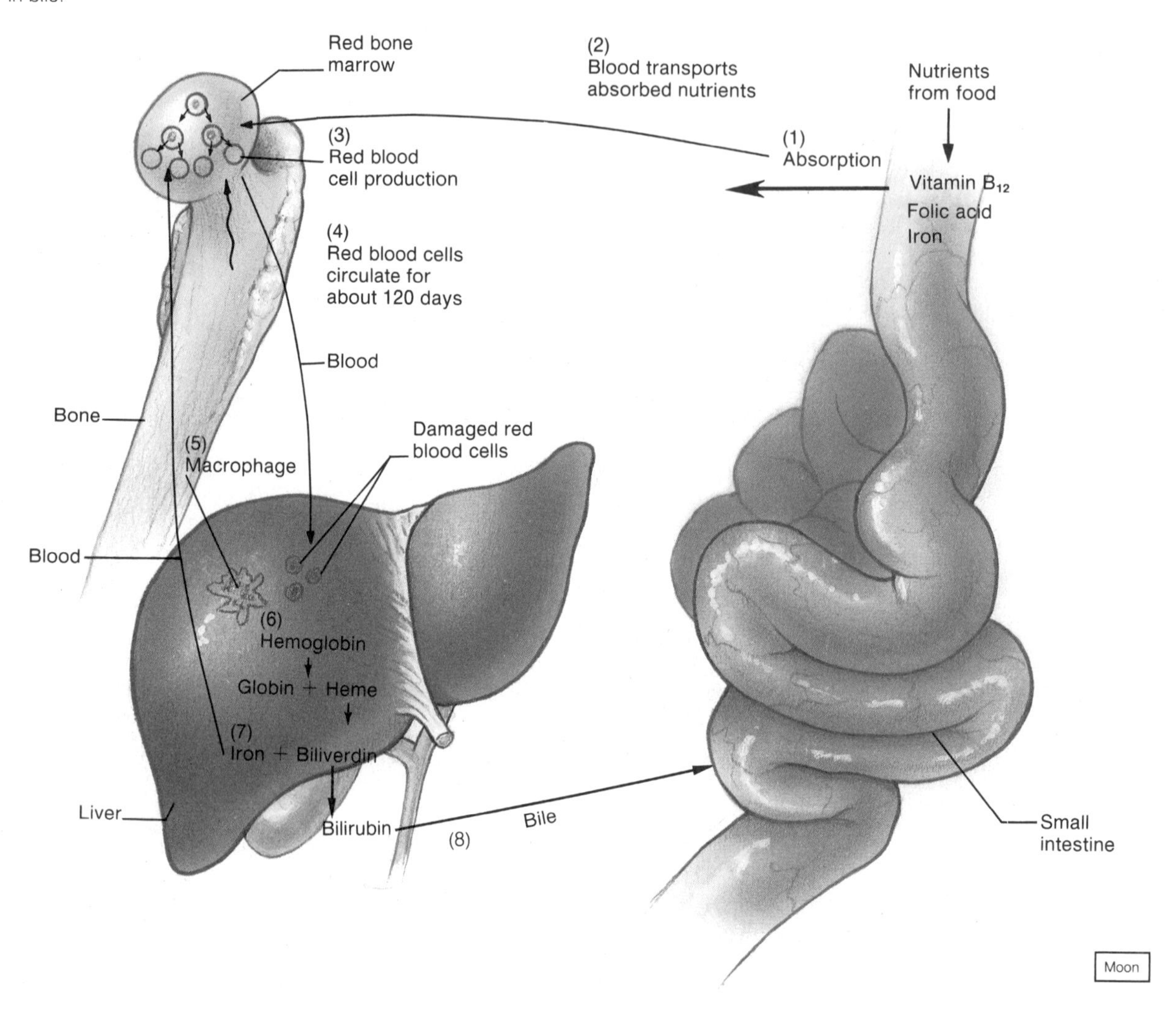

Types of White Blood Cells

White blood cells, or **leukocytes,** function primarily to control various disease conditions. Although these cells do most of their work outside the circulatory system, they use the blood for transportation.

Normally, five types of white cells can be found in the circulating blood. They are distinguished by their size, the nature of their cytoplasm, the shape of their nucleus, and their staining characteristics. For example, some types of leukocytes have granular cytoplasm and make up a group called *granulocytes,* while others lack cytoplasmic granules and are called *agranulocytes* (fig. 14.4).

A typical **granulocyte** is about twice the size of a red cell. The members of this group include three types of white cells—neutrophils, eosinophils, and basophils. These cells are produced in red bone marrow in much the same manner as red cells but have relatively short life spans, averaging about twelve hours.

Neutrophils are characterized by the presence of fine cytoplasmic granules that stain pinkish in neutral stain. The nucleus of a neutrophil is lobed and consists of two to five parts connected by thin strands of chromatin (fig. 14.6). Neutrophils account for 54% to 62% of the white cells in a normal blood sample.

Eosinophils contain coarse, uniformly sized cytoplasmic granules that stain deep red in acid stain (fig. 14.7). The nucleus usually has only two lobes. These cells make up 1% to 3% of the total number of circulating leukocytes.

Basophils are similar to eosinophils in size and in the shape of their nuclei, but they have fewer, more ir-

Figure 14.6
A neutrophil has a lobed nucleus with 2 to 5 parts (×400).

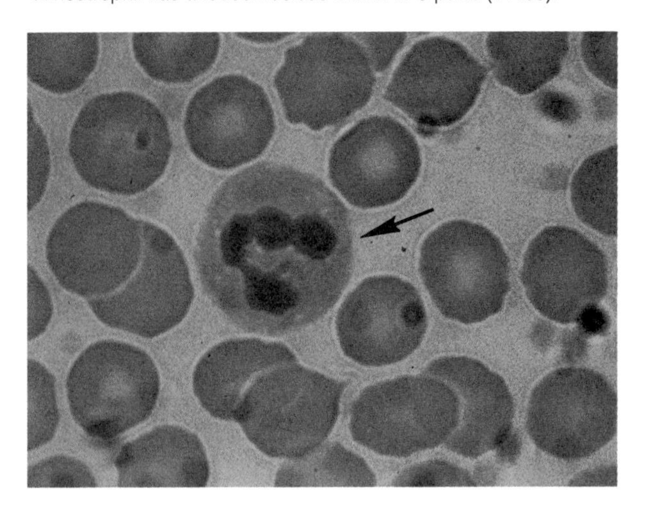

Figure 14.7
An eosinophil is characterized by the presence of red-staining cytoplasmic granules (×500).

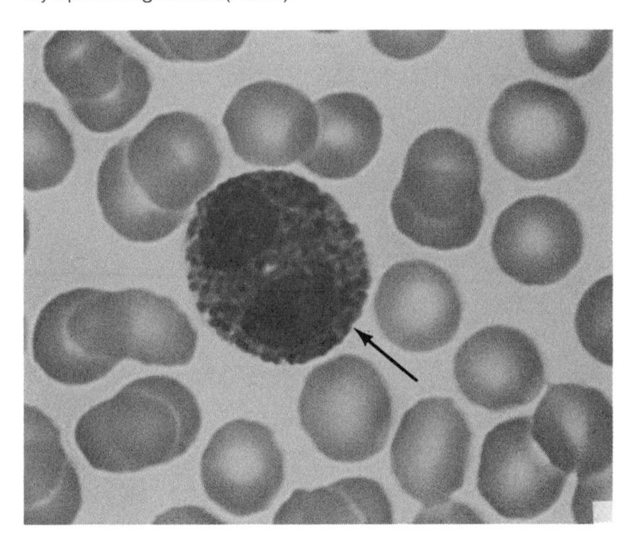

Figure 14.8
A basophil has cytoplasmic granules that stain deep blue (×563).

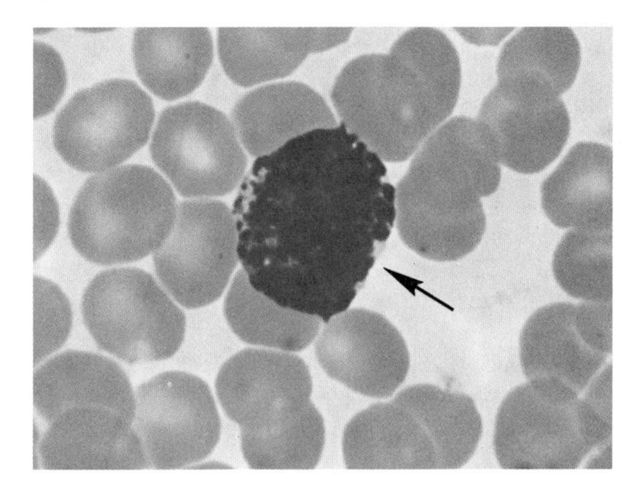

Figure 14.9
A monocyte is the largest type of blood cell (×640).

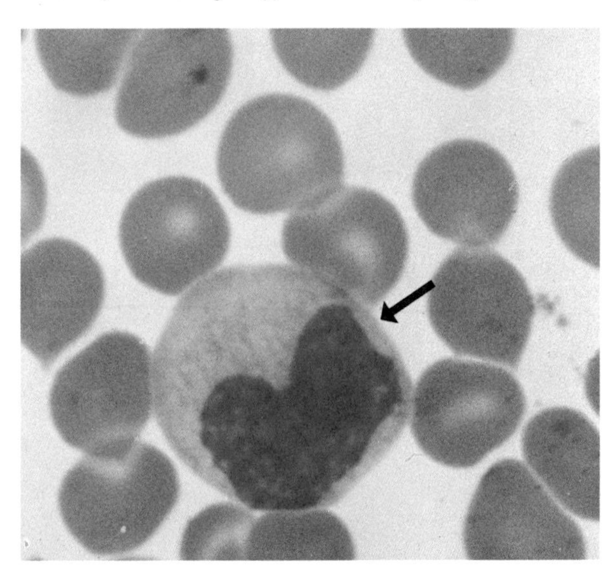

regularly shaped cytoplasmic granules that stain deep blue in basic stain (fig. 14.8). This type of leukocyte usually accounts for less than 1% of the white cells.

The leukocytes of the **agranulocyte** group include monocytes and lymphocytes. Both of these arise from red bone marrow; however, lymphocytes also are formed in the organs of the lymphatic system (chapter 16).

Monocytes are the largest cells found in the blood, having diameters two to three times greater than those of red cells (fig. 14.9). Their nuclei vary in shape and are described as round, kidney-shaped, oval, or lobed. They usually make up 3% to 9% of the leukocytes in a blood sample and seem to live for several weeks or even months.

Although large **lymphocytes** are sometimes found in the blood, usually they are only slightly larger than the red cells. A typical lymphocyte contains a relatively large, round nucleus surrounded by a thin rim of cytoplasm (fig. 14.10). These cells account for 25% to 33% of the circulating white blood cells. They seem to have relatively long life spans, which may extend for years.

1. Distinguish between granulocytes and agranulocytes.
2. List the five types of white blood cells, and explain how they differ from one another.

White Blood Cell Counts

The number of white blood cells in a cubic millimeter of human blood, called the *white blood cell count* (WBC), normally varies from 5,000 to 10,000. Because this number may change in response to abnormal conditions, white blood cell counts are of clinical interest. For example, some infectious diseases are accompanied by a rise in the number of circulating white cells, and if the total number of white cells exceeds 10,000 per mm^3 of blood, the person is said to have **leukocytosis.** This condition occurs during certain acute infections, such as appendicitis.

If the total white cell count drops below 5,000 per mm^3 of blood, the condition is called **leukopenia.** Such a deficiency may accompany typhoid fever, influenza, measles, mumps, chicken pox, and poliomyelitis.

Figure 14.10
The lymphocyte (*arrow*) contains a large, round nucleus; a neutrophil appears to the right of the lymphocyte (×640).

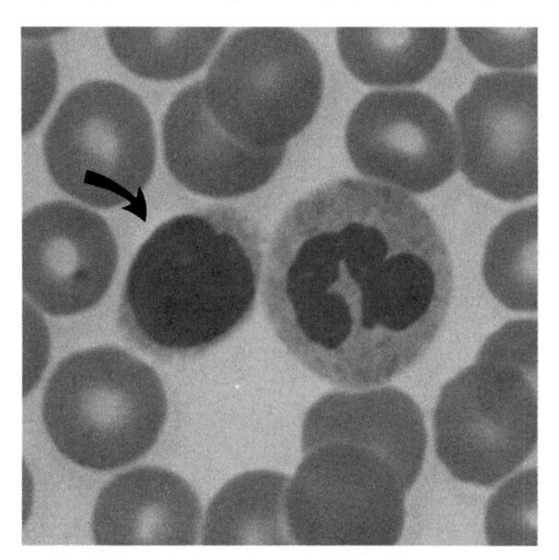

A *differential white blood cell count* (DIFF) is one in which the percentages of the various types of leukocytes in a blood sample are determined. This test is useful because the relative proportions of white cells may change in particular diseases. Neutrophils, for instance, usually increase during bacterial infections, and eosinophils may increase during certain parasitic infections and allergic reactions.

1. What is the normal human white blood cell count?
2. Distinguish between leukocytosis and leukopenia.
3. What is a differential white blood cell count?

Functions of White Blood Cells

As mentioned previously, white blood cells function to control disease conditions, such as infections caused by microorganisms. For example, some leukocytes phagocytize microorganisms that get into the body, and other leukocytes produce substances (antibodies) that react to destroy or disable such microorganisms.

Leukocytes can squeeze between the cells that form blood vessel walls. This movement, which is called *diapedesis,* allows the white cells to leave the circulation. Once outside the blood, they move through interstitial spaces with an amoebalike self-propulsion, called *amoeboid motion* (fig. 14.11).

The most mobile and active phagocytic leukocytes are *neutrophils* and *monocytes.* Although neutrophils are unable to ingest particles much larger than bacterial cells, monocytes can engulf relatively large objects. Both of these phagocytes contain numerous *lysosomes,* which are filled with digestive enzymes capable of breaking down various proteins and lipids, such as those in bacterial cell membranes (see chapter 3). As a result of their activities, neutrophils and monocytes often become so engorged with digestive products and bacterial toxins that they also die.

Figure 14.11
Leukocytes can squeeze between the cells of a capillary wall and enter the tissue space outside the blood system.

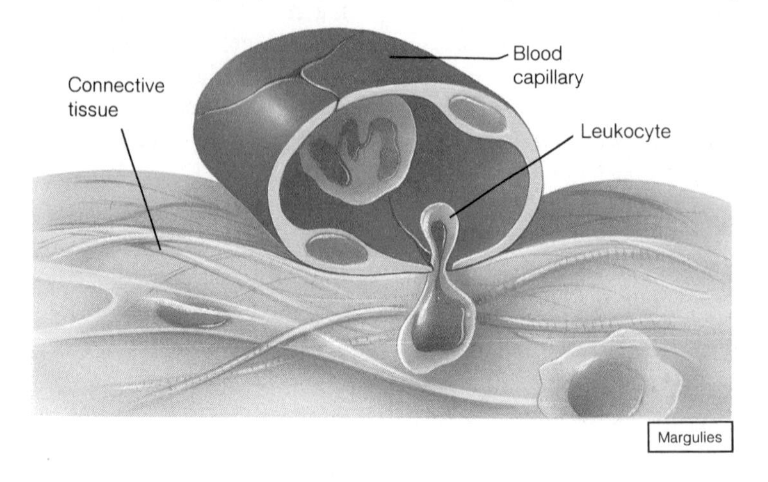

Leukemia

A Current Topic

Leukemia is a form of cancer characterized by uncontrolled production of leukocytes (fig. 14.12). There are two major types of leukemia: *myeloid leukemia*, resulting from abnormal production of granulocytes in red bone marrow, and *lymphoid leukemia*, due to increased formation of lymphocytes in lymph nodes. In both types, the newly formed leukocytes fail to mature into functional cells. Thus, even though large numbers of neutrophils may be produced in myeloid leukemia, the immature cells are unable to phagocytize bacteria, and the patient develops a lowered resistance to infections.

Eventually, the cells responsible for the overproduction of leukocytes tend to spread (metastasize) from the bone marrow or lymph nodes to other body parts. As with other forms of cancer, the leukemic cells eventually appear in such great numbers that they crowd out the normal, functional cells. Also, when normal red bone marrow is crowded out, the patient becomes anemic and develops a deficiency of blood platelets, accompanied by an increasing tendency to bleed.

Leukemias are also classified as *acute* or *chronic*. An acute condition appears suddenly, the symptoms progress rapidly, and death usually occurs within a few months if the disease is untreated. Chronic leukemia begins more slowly and may remain undetected for many months. Without treatment, however, life expectancy is about three years.

The greatest success in treatment has been achieved with acute lymphoid leukemia, which is the most common cancerous condition in children. This treatment usually involves counteracting the side effects of the disease, such as anemia, bleeding, and increased susceptibility to infections, as well as administering therapeutic drugs (chemotherapy).

Although acute lymphoid leukemia may occur at any age, the chronic form usually appears after the age of fifty. Acute myeloid leukemia also may occur at any age, but it is more common in adults; chronic myeloid leukemia is primarily a disease of adults between twenty and fifty years of age.

Figure 14.12
(*a*) Normal blood cells; (*b*) blood cells from a person with leukemia. (Note the increased number of leukocytes in *b*.)

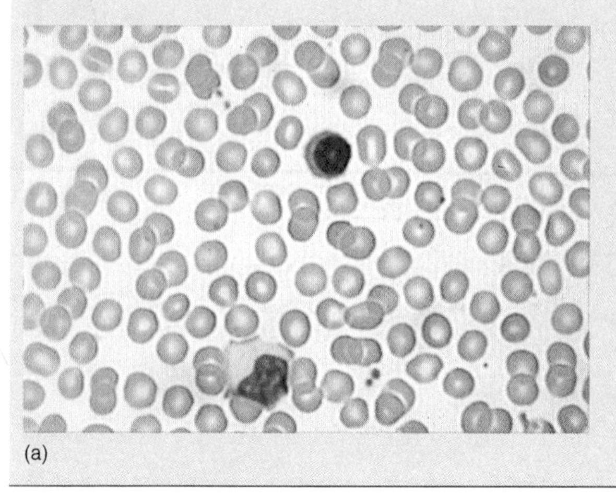

(a)

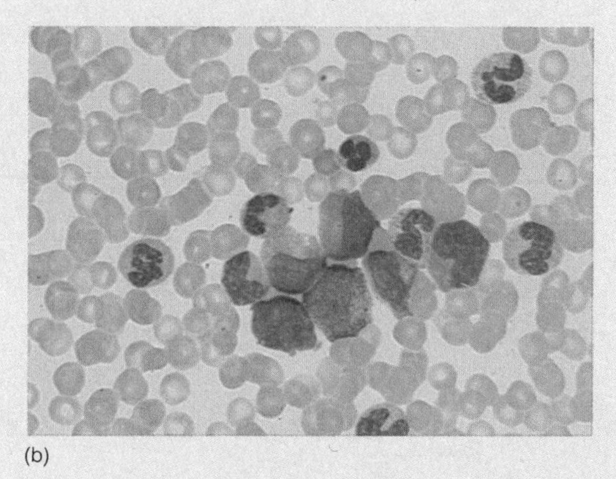

(b)

Eosinophils are only weakly phagocytic. However, they are attracted when various parasites enter tissues, and they can kill some of these invaders. Eosinophils also help to control allergic reactions by removing certain substances involved in such reactions.

Some of the cytoplasmic granules of *basophils* contain a blood-clot-inhibiting substance, called *heparin,* and other granules contain *histamine.* It is thought that basophils may aid in preventing intravascular blood clot formation by releasing heparin, and that they may cause an increase in blood flow to injured tissues by releasing histamine. Basophils are also involved in certain allergic reactions.

Lymphocytes play an important role in the process of *immunity.* Some, for example, form **antibodies** that act against specific foreign substances when they enter the body. This function is discussed in chapter 16.

1. What are the primary functions of white blood cells?
2. Which white blood cells are the most active phagocytes?
3. What are the functions of eosinophils and basophils?

Inflammation

A Current Topic

Inflammation is a localized response to factors that are damaging or potentially damaging to tissues. These factors include traumatic injuries, excessive heat, chemical irritants, bacterial infections, and other stressful conditions.

The inflammation reaction occurs mainly in connective tissues and involves various components of the circulatory system. It helps to prevent the spread of infections, brings about the destruction of foreign substances entering the tissues, and promotes the healing process.

As an example, imagine that some bacteria have entered connective tissue through a break in the skin (fig. 14.13). In response, chemicals, such as histamine, are released from the damaged and stressed cells. These substances, together with others from the invading bacteria, stimulate a variety of changes to take place. Nearby blood vessels become dilated and cause an increase in the blood supply to the affected tissues. This causes the region to appear redder and feel warmer. At the same time, the permeability of the smaller blood vessels increases, and extra fluid filters into the intercellular spaces of the tissue. This causes swelling (a condition called *edema*) and puts pressure on the local nerve endings, which causes the region to become painful.

The fluid that enters from the blood soon clots. This seals off the inflamed area by filling tissue spaces and lymphatic vessels, and it helps delay the spread of bacteria into surrounding tissues.

Macrophages residing in the connective tissue become mobilized and begin to phagocytize bacteria almost immediately. The inflamed tissue releases a substance (leukocytosis-inducing factor) that is carried away by the blood and stimulates the release of many white blood cells from bone marrow. Within a few hours, large numbers of these cells (neutrophils) migrate into the inflamed tissues from the blood by passing through capillary walls. They are attracted toward chemicals diffusing out from the affected tissues, and they also act as phagocytes.

Eventually, a fluid-filled cavity filled with a creamy, yellowish substance, called *pus,* may appear in the inflamed region. The pus represents an accumulation of tissue fluid, bacteria, white blood cells, and macrophages. Sometimes such a cavity opens itself to the surface or into a body cavity, and the pus drains out.

As the inflammation subsides, macrophages clean up the cellular debris by phagocytosis and the excess fluid is absorbed by the surrounding tissues. The gap is filled with new connective tissue, formed by fibroblasts that migrate into the area.

Figure 14.13

(*a*) When bacteria invade tissues, (*b*) neutrophils migrate into the region and destroy the bacteria by phagocytosis.

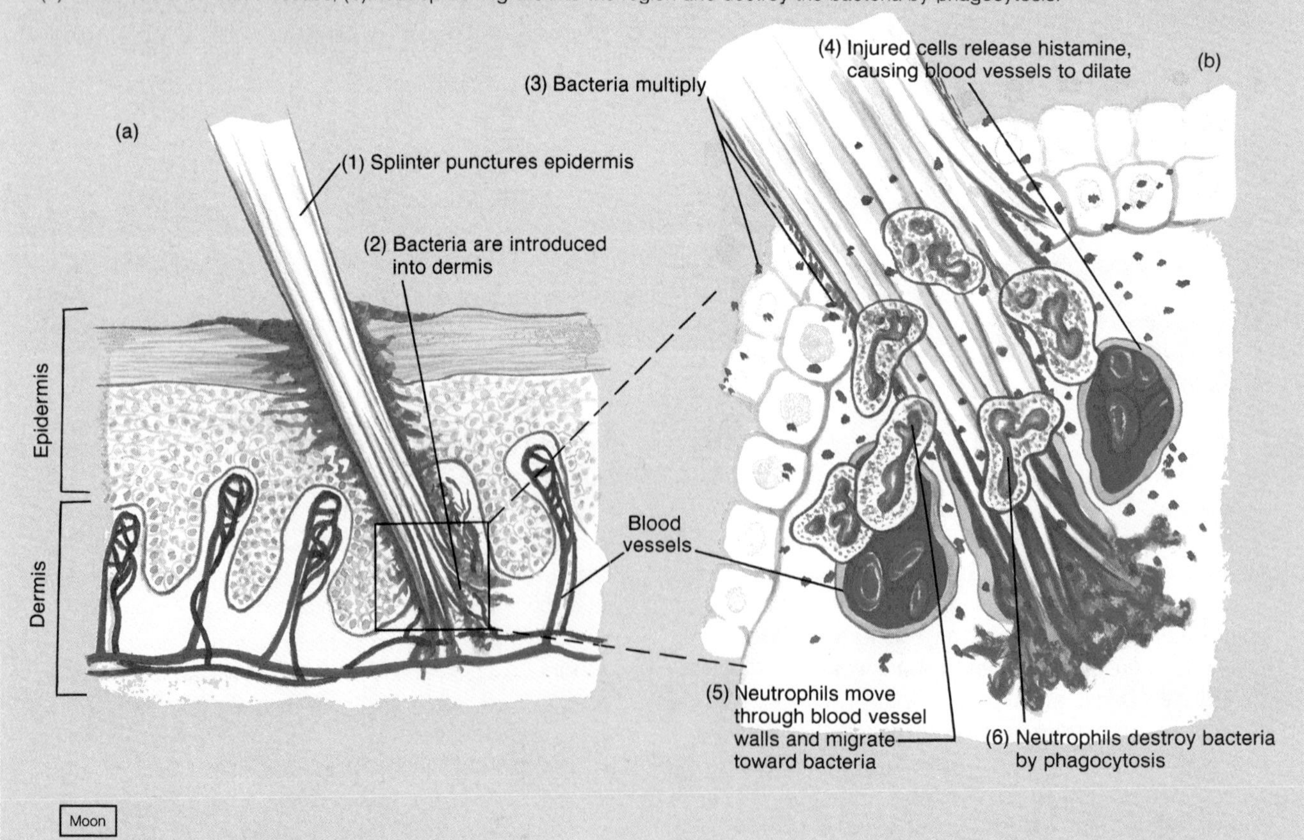

Chart 14.1 Cellular components of blood			
Component	**Description**	**Number present**	**Function**
Red blood cell (erythrocyte)	Biconcave disk without a nucleus, about one-third hemoglobin	4,000,000 to 6,000,000 per mm³	Transports oxygen and carbon dioxide
White blood cells (leukocytes)		5000 to 10,000 per mm³	Aids in the defense against infections by microorganisms
Granulocytes	About twice the size of red cells; cytoplasmic granules are present		
1. Neutrophil	Nucleus with two to five lobes; its cytoplasmic granules stain pink in neutral stain	54% to 62% of the white cells present	Destroys relatively small particles by phagocytosis
2. Eosinophil	Nucleus bilobed; its cytoplasmic granules stain red in acid stain	1% to 3% of the white cells present	Kills parasites and helps control allergic reactions
3. Basophil	Nucleus lobed; its cytoplasmic granules stain blue in basic stain	Less than 1% of the white cells present	Releases anticoagulant, heparin, and histamine
Agranulocytes	Cytoplasmic granules are absent		
1. Monocyte	Two to three times larger than a red cell; its nuclear shape varies from round to lobed	3% to 9% of the white cells present	Destroys relatively large particles by phagocytosis
2. Lymphocyte	Only slightly larger than a red cell; its nucleus nearly fills cell	25% to 33% of the white cells present	Responsible for immunity
Platelet (thrombocyte)	Cytoplasmic fragment	130,000 to 360,000 per mm³	Helps to control blood loss from broken vessels

Blood Platelets

Platelets, or **thrombocytes,** are not complete cells. They arise from very large cells in red bone marrow, called **megakaryocytes.** These cells release cytoplasmic fragments of themselves, and as the fragments detach and enter the circulation, the smaller ones become platelets. As they pass through the blood vessels of the lungs, the larger fragments break down to produce more platelets.

Each platelet is a round disk that lacks a nucleus and is less than half the size of a red blood cell. It is capable of amoeboid movement and may live for about ten days. In normal human blood, the *platelet count* will vary from 130,000 to 360,000 platelets per mm³.

Platelets help close the breaks in damaged blood vessels and function to initiate the formation of blood clots, as is explained in a subsequent section of this chapter.

Chart 14.1 summarizes the characteristics of blood cells and platelets.

1. What is the normal blood platelet count?
2. What is the function of blood platelets?

Blood Plasma

Plasma is the clear, straw-colored, liquid portion of the blood in which the cells and platelets are suspended. It is approximately 92% water and contains a complex mixture of organic and inorganic substances that function in a variety of ways. These functions include transporting nutrients, gases, and vitamins; regulating fluid and electrolyte balances; and maintaining a favorable pH.

Plasma Proteins

The most abundant of the dissolved substances (solutes) in plasma are the **plasma proteins.** These proteins remain in the blood and interstitial fluids and ordinarily are not used as energy sources. The three main groups are albumins, globulins, and fibrinogen. The members of each group differ in their chemical structures and in their physiological functions.

Albumins account for about 60% of the plasma proteins and their molecules are the smallest of these proteins. Albumins are synthesized in the liver, and because they are so plentiful, albumins, together with other solutes, help maintain the *osmotic pressure* of the blood.

As explained in chapter 3, whenever the concentration of solutes changes on either side of a cell membrane, water is likely to move through the membrane toward the region where the dissolved molecules are in higher concentration. For this reason, it is important that the concentration of solutes in the plasma remain relatively stable. Otherwise, water tends to leave the blood and enter the tissues or to leave the tissues and enter the blood by *osmosis.* Because albumins (and other plasma proteins) add to the osmotic pressure of the plasma, they aid in regulating the water balance between the blood and tissues. At the same time, they help control the blood volume, which, in turn, is directly related to the blood pressure (chapter 15).

The concentration of the plasma proteins may decrease significantly if a person is starving or has a protein-deficient diet and, thus, is forced to use body protein as an energy source. Similarly, the plasma protein concentration may drop if liver disease interferes with the synthesis of these proteins. In either case, as the blood protein concentration decreases, the osmotic pressure of the blood decreases, and water tends to accumulate in the intercellular spaces, causing the tissues to swell (edema).

Globulins, which make up about 36% of the plasma proteins, can be separated further into fractions called *alpha globulins, beta globulins,* and *gamma globulins.* Alpha and beta globulins are synthesized in the liver, and they have a variety of functions including the transport of lipids and fat-soluble vitamins. Gamma globulins are produced in the lymphatic tissues, and they include the proteins that function as *antibodies of immunity* (see chapter 16).

Fibrinogen, which constitutes about 4% of the plasma proteins, plays a primary role in the blood clotting mechanism. It is synthesized in the liver and has the largest molecules of the plasma proteins. Its function is described in a subsequent section of this chapter.

Chart 14.2 summarizes the characteristics of the plasma proteins.

1. List three types of plasma proteins.
2. How does albumin help to maintain the water balance between the blood and tissues?
3. What are the functions of globulins?

Nutrients and Gases

The *plasma nutrients* include amino acids, simple sugars, and various lipids that have been absorbed from the digestive tract and are being transported to organs and tissues by the blood.

The most important *blood gases* are oxygen and carbon dioxide. Although the plasma also contains a considerable amount of dissolved nitrogen, this gas ordinarily has no physiological function.

Nonprotein Nitrogenous Substances

Molecules that contain nitrogen atoms but are not proteins comprise a group called **nonprotein nitrogenous substances.** Within the plasma, this group includes amino acids, urea, and uric acid. The *amino acids* are present as a result of protein digestion and amino acid absorption. The *urea* and *uric acid* are the products of protein and nucleic acid catabolism, respectively, and are excreted in the urine.

Chart 14.2 Plasma proteins

Protein	Percentage of total	Origin	Function
Albumin	60%	Liver	Helps in the maintenance of the blood osmotic pressure
Globulin	36%		
Alpha globulins		Liver	Transports lipids and fat-soluble vitamins
Beta globulins		Liver	Same as above
Gamma globulins		Lymphatic tissues	Constitutes the antibodies of immunity
Fibrinogen	4%	Liver	Plays a key role in blood clot formation

Normally, the concentration of nonprotein nitrogenous (NPN) substances remains relatively stable because protein intake and utilization are balanced with the excretion of nitrogenous wastes. Because about half of the NPN is urea, which is ordinarily excreted by the kidneys, a rise in the plasma NPN concentration may suggest a kidney disorder. Such an increase may also occur as a result of excessive protein catabolism or the presence of an infection.

Plasma Electrolytes

The plasma contains a variety of *electrolytes,* which have been absorbed from the intestine or have been released as by-products of cellular metabolism. They include sodium, potassium, calcium, magnesium, chloride, bicarbonate, phosphate, and sulfate ions. Of these, sodium and chloride ions are the most abundant.

Such ions are important in maintaining the osmotic pressure and the pH of the plasma, and like other plasma constituents, they are regulated so that their blood concentrations remain relatively stable. These electrolytes are discussed in chapter 18 in connection with water and electrolyte balance.

1. What nutrients are found in the blood plasma?
2. What gases occur in the plasma?
3. What is meant by a nonprotein nitrogenous substance?
4. What are the sources of the plasma electrolytes?

Hemostasis

The term **hemostasis** refers to the stoppage of bleeding, which is vitally important when blood vessels are ruptured. Following an injury to blood vessels, several actions may occur that help prevent excessive blood loss. These include blood vessel spasm, platelet plug formation, and blood coagulation.

Blood Vessel Spasm

When a blood vessel is cut or broken, the smooth muscles in its wall are stimulated to contract, and blood loss is decreased almost immediately. In fact, the ends of a severed vessel may be closed completely by such a *spasm.*

Although this response may last only a few minutes, by then the platelet plug and blood coagulation mechanisms are normally operating. Also, as a platelet plug forms, the platelets release a substance called *serotonin,* which causes the smooth muscles in the blood vessel wall to contract. This vasoconstricting action helps to maintain a prolonged vascular spasm.

Platelet Plug Formation

Platelets tend to stick to any rough surface and to the *collagen* in connective tissue. Consequently, when a blood vessel is broken, the platelets adhere to the collagen that underlies the lining of the blood vessel. At the same time, they tend to stick to each other, creating a *platelet plug* in the vascular break. Such a plug may be able to control blood loss if the break is relatively small.

The steps in platelet plug formation are shown in figure 14.14.

1. What is meant by hemostasis?
2. How does a blood vessel spasm help control bleeding?
3. Describe the formation of a platelet plug.

Figure 14.14
Steps in platelet plug formation.

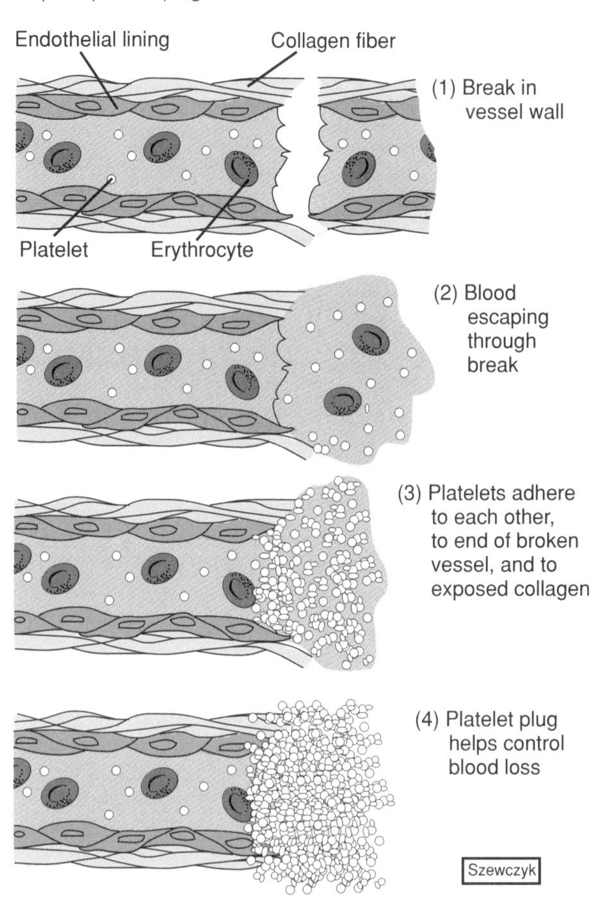

Blood Coagulation

Coagulation, which is the most effective of the hemostatic mechanisms, causes the formation of a *blood clot.*

The mechanism by which the blood coagulates is very complex and involves many substances called *clotting factors.* Some of these factors promote coagulation, and others inhibit it. Whether or not the blood coagulates depends on the balance that exists between these two groups of factors. Normally the anticoagulants prevail, and the blood does not clot. As a result of injury (trauma), however, substances that favor coagulation may increase in concentration, and the blood may coagulate.

The basic event in blood clot formation is the conversion of the soluble plasma protein *fibrinogen* into the relatively insoluble threads of the protein **fibrin.**

When tissues are damaged, the clotting mechanism initiates a series of reactions resulting in the production of a substance called *prothrombin activator.* This series of changes depends upon the presence of *calcium ions* as well as certain proteins and phospholipids for its completion.

Prothrombin is an alpha globulin that is continually produced by the liver and, thus, is normally present in the plasma. In the presence of calcium ions, prothrombin is converted into **thrombin** by the action of prothrombin activator. Thrombin, in turn, acts as an enzyme and causes a reaction in the molecules of fibrinogen. As a result, the fibrinogen molecules join, end to end, forming long threads of *fibrin.* The production of fibrin threads is also enhanced by the presence of calcium ions and certain proteins.

Once the threads of fibrin have formed, they tend to stick to the exposed surfaces of the damaged blood vessels and create a meshwork in which various blood cells and platelets become entangled (fig. 14.15). The resulting mass is a *blood clot,* which may effectively block a vascular break and prevent further loss of blood.

Figure 14.15
A scanning electron micrograph of fibrin threads (×5000). What factors serve to initiate the formation of fibrin?

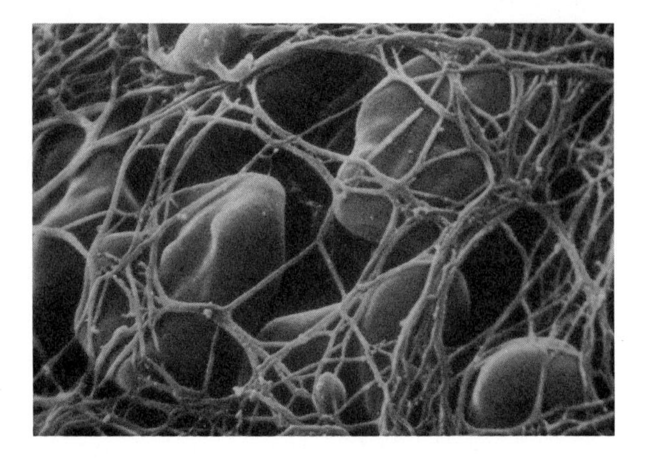

Figure 14.16
(*a*) A normal artery; (*b*) the inner wall of an artery that has changed as a result of atherosclerosis. How might this condition promote the formation of blood clots?

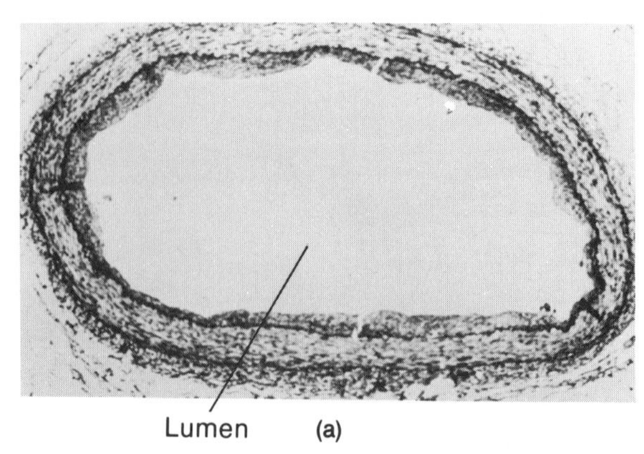

Fatty deposit (b) Lumen

Because the liver plays an important role in the synthesis of various plasma proteins, such as prothrombin, it is not surprising that liver diseases are often accompanied by a tendency to bleed. Also, bile salts from the liver are necessary for the efficient absorption of *vitamin K* from the intestine, and this vitamin is essential for the synthesis of prothrombin. If the liver fails to produce enough bile, or if the bile ducts become obstructed, a vitamin K deficiency is likely to develop, and the ability to form blood clots may be diminished. For this reason, vitamin K is often administered to patients with liver diseases or bile duct obstructions before they are treated surgically.

The amount of prothrombin activator that appears in the blood is directly proportional to the degree of tissue damage. Once a blood clot begins to form, it promotes still more clotting. This happens because thrombin also acts directly on blood-clotting factors other than fibrinogen, and it can cause prothrombin to form still more thrombin. This is an example of a **positive feedback system**, in which the original action stimulates more of the same type of action. Such a positive feedback mechanism produces very unstable conditions and can operate for only a short time in a living organism.

Normally, the formation of a massive clot throughout the blood system is prevented by the blood's movement, which rapidly carries excessive thrombin away and, thus, keeps its concentration too low to enhance further clotting. As a result, blood clot formation is usually limited to the blood that is standing still (or moving relatively slowly), and clotting ceases where a clot comes in contact with circulating blood.

Blood clots that form in ruptured vessels are soon invaded by *fibroblasts* (see chapter 5). These cells produce fibrous connective tissue throughout the clots, which helps strengthen and seal vascular breaks. Many clots, including those that form in tissues as a result of blood leakage (hematomas), disappear in time. This dissolution involves the activation of a plasma protein that can digest fibrin threads and other proteins associated with clots. Clots that fill large blood vessels, however, are seldom removed by natural processes.

If a blood clot forms in a vessel abnormally, it is termed a **thrombus.** If the clot becomes dislodged or if a fragment of it breaks loose and is carried away by the blood flow, it is called an **embolus.** Generally, emboli continue to move until they reach narrow places in vessels where they become lodged and may interfere with the blood flow.

Such abnormal clot formations are often associated with conditions that cause changes in the endothelial linings of vessels. In the disease called *atherosclerosis,* for example, arterial linings are changed by accumulations of fatty deposits. These changes may initiate the clotting mechanism (fig. 14.16).

Figure 14.17 summarizes the three primary hemostatic mechanisms.

Figure 14.17
Hemostasis following tissue damage is likely to involve blood vessel spasm, platelet plug formation, and blood clotting.

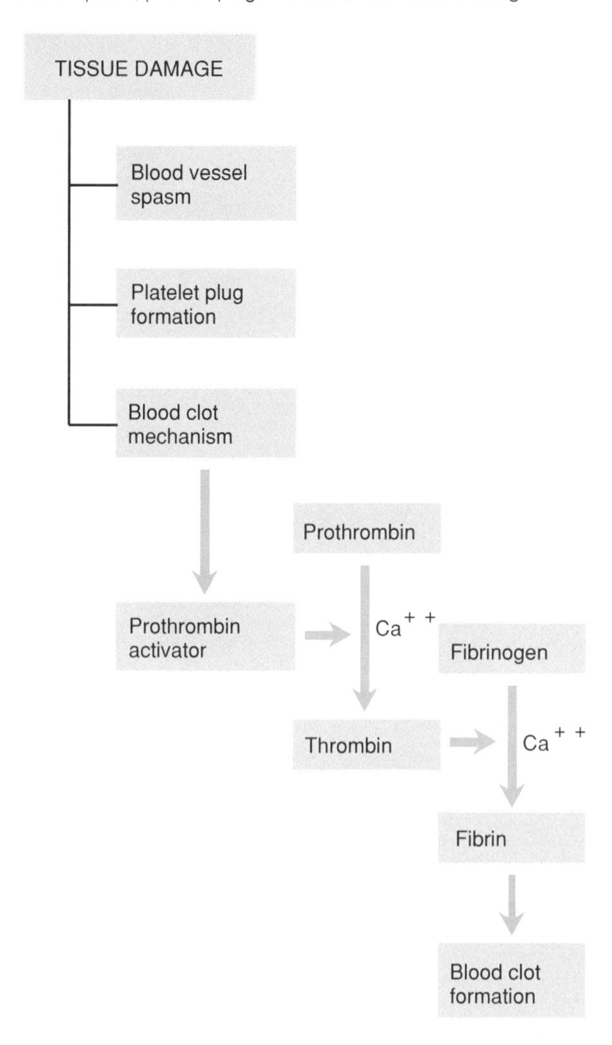

Hemophilia is a hereditary disease that appears almost exclusively in males. Hemophiliacs are deficient in a blood factor necessary for coagulation and, consequently, they usually experience repeated episodes of serious bleeding.

The treatment of hemophilia may involve applying pressure on accessible bleeding sites in an effort to control blood loss. Transfusions are often used to replace missing blood factors. For example, a common type of hemophilia caused by a deficiency of a substance called blood factor VIII may be treated with fresh plasma, fresh-frozen plasma, or plasma concentrates (cryoprecipitates) that contain the missing factor.

1. Review the major steps in the formation of a blood clot.
2. What prevents the formation of massive clots throughout the blood system?
3. Distinguish between a thrombus and an embolus.

Blood Groups and Transfusions

Early attempts to transfer blood from one person to another produced varied results. Sometimes the person receiving the transfusion was aided by the procedure. At other times, the recipient suffered a blood reaction in which the red blood cells clumped together, obstructing vessels and producing other serious consequences.

Eventually, it was discovered that each individual has a particular combination of substances in his or her blood. Some of these substances react with those in another person's blood. These discoveries led to the development of procedures for typing blood. It is now known that safe transfusions of whole blood depend upon properly matching the blood types of the donors and recipients.

Agglutinogens and Agglutinins

The clumping of red cells following a transfusion reaction is called **agglutination.** This phenomenon is due to the presence of substances called **agglutinogens** (antigens) in the red cell membranes and substances called **agglutinins** (antibodies) dissolved in the plasma.

Blood typing involves identifying the agglutinogens that are present in a person's red cells. Although many different agglutinogens are associated with human erythrocytes, only a few of them are likely to produce serious transfusion reactions. These include the agglutinogens of the ABO group and those of the Rh group.

Avoiding the mixture of certain kinds of agglutinogens and agglutinins prevents adverse transfusion reactions. Such reactions are described in a subsequent section of this chapter.

ABO Blood Group

The *ABO blood group* is based upon the presence (or absence) of two major agglutinogens in the red cell membranes—*agglutinogen A* and *agglutinogen B*—which are present at birth as a result of inheritance. The erythrocytes of each person contain one of the four following combinations of agglutinogens: only A, only B, both A and B, or neither A nor B.

A person with only agglutinogen A is said to have *type A blood;* a person with only agglutinogen B has *type B blood;* one with both agglutinogen A and B has *type AB blood;* and one with neither agglutinogen A nor B has *type O blood.* Thus, all humans have one of four possible blood types—A, B, AB, or O.

Certain agglutinins develop spontaneously in the plasma about two to eight months after birth. Specifically, whenever agglutinogen A is absent in the red blood cells, an agglutinin called *anti-A* develops; and

Figure 14.18
Each blood type is characterized by a different combination of agglutinogens and agglutinins.

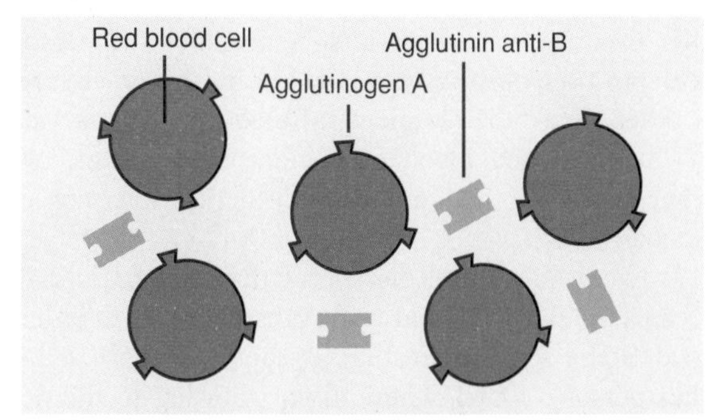

Type A blood

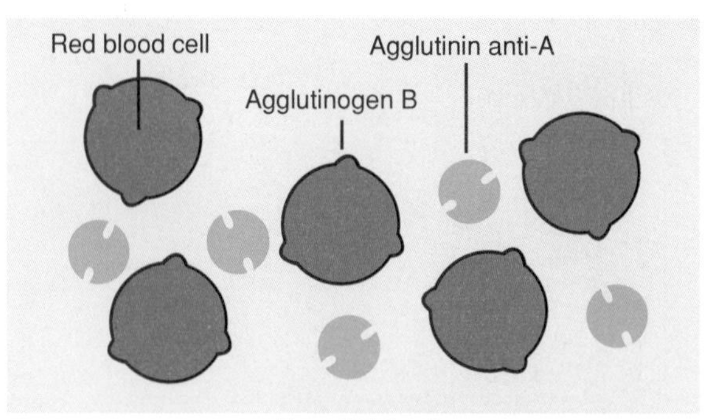

Type B blood

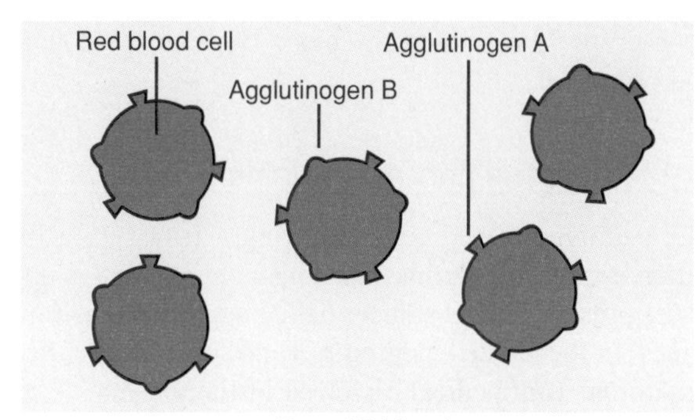

Type AB blood

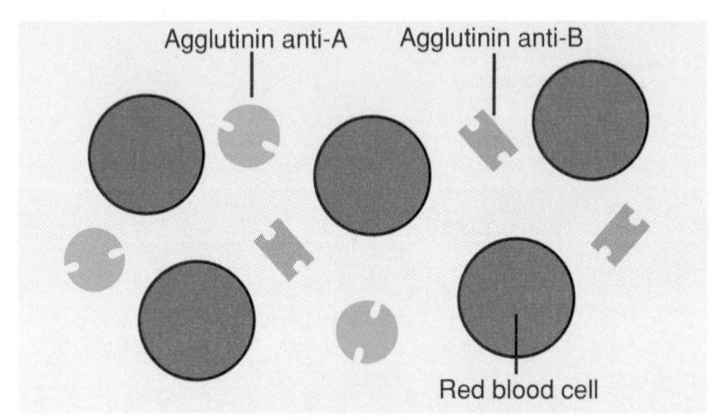

Type O blood

whenever agglutinogen B is absent, an agglutinin called *anti-B* develops. Therefore, persons with type A blood have agglutinin anti-B in their plasma; those with type B blood have agglutinin anti-A; those with type AB blood have neither agglutinin; and those with type O blood have both agglutinin anti-A and anti-B (fig. 14.18).

Chart 14.3 summarizes the agglutinogens and agglutinins of the ABO blood group.

Because an agglutinin of one kind will react with an agglutinogen of the same kind and cause red blood cells to clump together, such combinations must be avoided. The major concern in blood transfusion procedures is that the cells in the *transfused blood* not be agglutinated by the agglutinins in the recipient's plasma. For this reason, a person with type A (anti-B) blood must not be given blood of type B or AB, because the red cells of both types would be agglutinated by the anti-B in the recipient's type A blood. Likewise, a person with type B (anti-A) blood must not be given type A or AB blood, and a person with type O (anti-A and anti-B) blood must not be given type A, B, or AB blood (fig. 14.19).

Because type AB blood lacks both anti-A and anti-B agglutinins, it would appear that an AB person could receive a transfusion of blood of any other type. It should be noted, however, that type A (anti-B) blood, type B (anti-A) blood, and type O (anti-A and anti-B) blood still contain agglutinins (either anti-A or anti-B)

Chart 14.3 Agglutinogens and agglutinins of the ABO blood group

Blood type	Agglutinogen	Agglutinin
A	A	anti-B
B	B	anti-A
AB	A and B	Neither anti-A nor anti-B
O	Neither A nor B	Both anti-A and anti-B

that could cause agglutination of type AB cells. Similarly, because type O blood lacks agglutinogens A and B, it would seem that this type could be transfused into persons with blood of any other type. Type O blood, however, contains both anti-A and anti-B agglutinins, which can cause agglutination of types A, B, or AB cells. Consequently, even for AB individuals, it is always best to use donor blood of the same type as the recipient blood.

Although the agglutinogens of the ABO and Rh groups are the ones most likely to cause serious blood transfusion reactions, every person's blood contains a number of other factors that sometimes cause problems. For this reason, it is good practice to determine whether samples of the recipient and donor blood will cause agglutination of the red blood cells before administering a transfusion to a patient. This procedure,

Figure 14.19
(*a*) If red blood cells with agglutinogen A are added to blood containing agglutinin anti-A, (*b*) the agglutinins will react with the agglutinogens of the red blood cells and cause them to clump together.

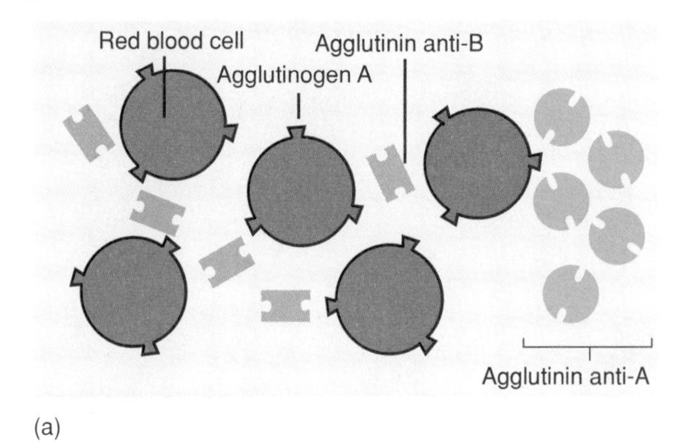

(a)

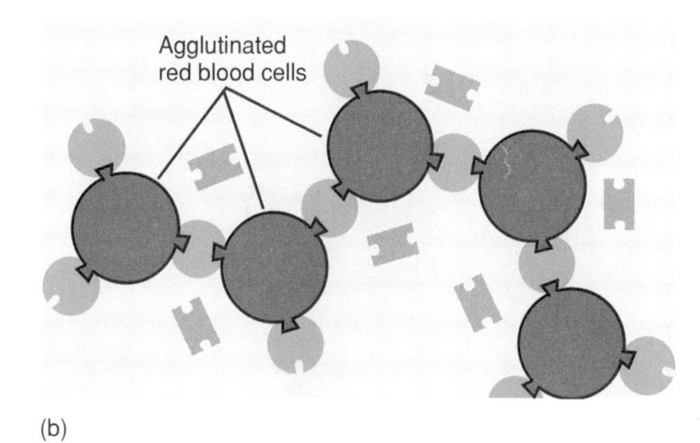

(b)

called *cross matching,* involves mixing a suspension of donor cells in the recipient serum and then mixing a suspension of recipient cells in the donor serum. If the red blood cells do not agglutinate in either case, it is probably safe to give the transfusion.

In an extreme emergency, when an AB person needs a transfusion and the matching blood type (AB) is not available, type A, B, or O blood is sometimes used. For this reason a type AB person is called a *universal recipient.*

Similarly, in an extreme emergency type O blood is sometimes given to persons with blood type A, B, or AB, so that type O persons have been called *universal donors.*

In such instances, the donor blood is transfused slowly so that it will be diluted by the recipient's larger blood volume. This precaution usually avoids serious reactions between donor agglutinins and recipient agglutinogens, providing only a small quantity of blood is transfused (see chart 14.4).

1. Distinguish between agglutinogens and agglutinins.
2. What is the main concern when blood is transfused from one individual to another?
3. Why is blood cross matched before a transfusion is given?

Rh Blood Group

The *Rh blood group* was named after the *rhesus monkey,* in which it was first studied. In humans this group includes several Rh agglutinogens (factors). The most important of these is *agglutinogen D;* however, if any Rh factors are present in the red cell membranes, the blood is said to be *Rh positive.* Conversely, if the red cells lack Rh agglutinogens, the blood is called *Rh negative.*

As in the case of agglutinogens A and B, the presence (or absence) of an Rh agglutinogen is an inherited trait. Unlike anti-A and anti-B, agglutinins for Rh (*anti-Rh*) do not appear spontaneously. Instead, they form only in Rh-negative persons in response to special stimulation.

Chart 14.4 Preferred and permissible blood types for transfusions

Blood type	Preferred transfusion	Permissible transfusion (in extreme emergency)
A	A	A, O
B	B	B, O
AB	AB	AB, A, B, O
O	O	O

If a person with Rh-negative blood receives a transfusion of Rh-positive blood, the recipient's antibody-producing cells are stimulated by the presence of the Rh agglutinogen, and they begin producing an *anti-Rh agglutinin.* Generally no serious consequences result from this initial transfusion, but if the Rh-negative person—now sensitized to Rh-positive blood—receives another transfusion of Rh-positive blood some months later, the donor's red cells are likely to agglutinate.

A related condition may occur when a woman with Rh-negative blood is pregnant with an Rh-positive fetus for the first time (fig. 14.20). Such a pregnancy may be uneventful; however, at the time of this infant's birth (or if a miscarriage occurs), the placental membrane, which separates the maternal blood from the fetal blood, may be broken, and some of the infant's Rh-positive blood cells may get into the maternal circulation. These Rh-positive cells may then stimulate the maternal tissues to begin producing anti-Rh agglutinins.

If the mother, who has already developed anti-Rh agglutinin, becomes pregnant with a second Rh-positive fetus, these anti-Rh agglutinins can pass slowly

Figure 14.20
(*a*) If an Rh-negative woman is pregnant with an Rh-positive fetus, (*b*) some of the fetal red blood cells with Rh agglutinogens may enter the maternal blood at the time of birth. (*c*) As a result, the woman's cells may produce anti-Rh agglutinins.

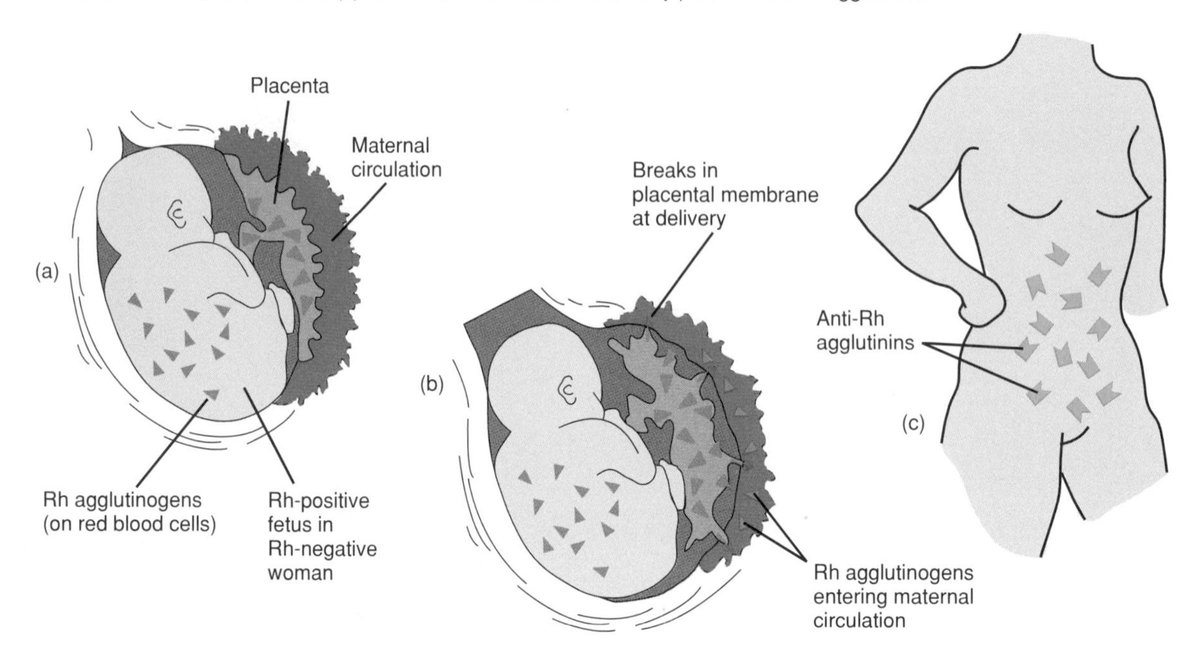

through the placental membrane and react with the fetal red cells, causing them to agglutinate. The fetus then develops a condition called **erythroblastosis fetalis** (fig. 14.21).

Erythroblastosis fetalis can be prevented in future offspring by treating Rh-negative mothers with a special blood serum within seventy-two hours following the birth of each Rh-positive child. This serum is obtained from the blood of another Rh-negative person who has formed anti-Rh agglutinin. This anti-Rh agglutinin inactivates any Rh-positive cells that may have entered the maternal blood at the time of the infant's birth. The maternal tissues, consequently, are not stimulated to manufacture anti-Rh agglutinins, and the anti-Rh agglutinins received during the treatment soon disappear.

Agglutination Reactions

In any blood reaction involving agglutinogens and agglutinins, the agglutinated red cells usually degenerate or are destroyed by phagocytic cells. At the same time, hemoglobin and other red cell contents are released, and the blood concentration of *free hemoglobin* increases greatly. Some of this hemoglobin may diffuse out of the vascular system and enter the tissues, where it is gradually converted into bilirubin. As a result, the tissues may develop a yellowish stain—a characteristic of the condition called *jaundice* (icterus).

Free hemoglobin also may pass into the kidneys and interfere with the vital functions of these organs, so that a person with a blood transfusion reaction may have kidney failure.

Infants with erythroblastosis fetalis are usually jaundiced and severely anemic. As their blood-cell-forming tissues respond to the need for more red cells, various immature erythrocytes, including *erythroblasts,* are released into the blood (fig. 14.4).

An affected infant may suffer permanent brain damage as a result of bilirubin precipitating in the brain tissues and injuring neurons. This condition is called *kernicterus,* and if the infant survives, it may have motor or sensory losses and exhibit mental deficiencies.

The treatment of erythroblastosis fetalis usually involves exposing the affected infant to bright blue or white *fluorescent light.* Bilirubin is a light-sensitive substance, and this exposure (phototherapy) causes a decrease in the blood bilirubin concentration.

In more severe cases, the infant's Rh-positive blood may be replaced slowly with Rh-negative blood. This procedure is called an *exchange transfusion,* and it reduces the concentration of bilirubin in the infant's tissues in addition to removing the agglutinating red cells, anti-Rh agglutinins, free hemoglobin, and other products of erythrocyte destruction. Such a transfusion also provides a temporary supply of red cells that will not be agglutinated by any remaining anti-Rh agglutinins. In time, the infant's blood-cell-forming tissues will replace the donor's blood cells with Rh-positive cells, but by then the maternal agglutinins will have disappeared.

Figure 14.21

(*a*) If a woman who has developed anti-Rh agglutinins is pregnant with an Rh-positive fetus, (*b*) agglutinins may pass through the placental membrane and cause the fetal red blood cells to agglutinate.

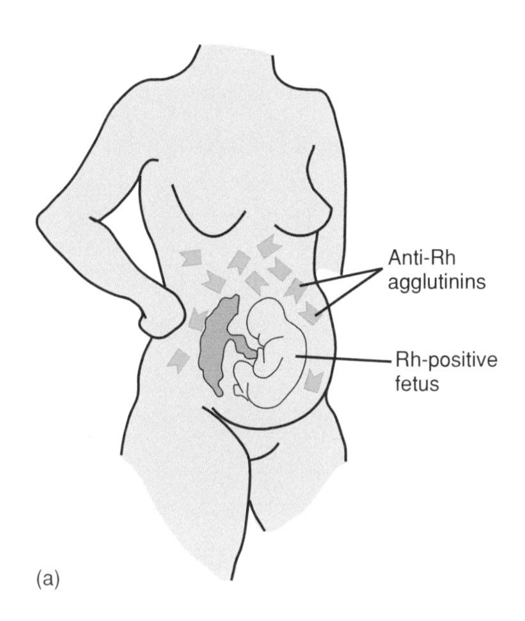

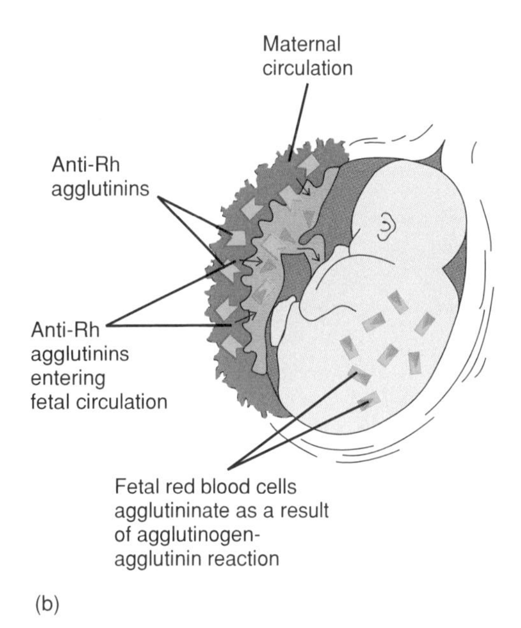

1. Under what conditions might a person with Rh-negative blood develop anti-Rh agglutinins?
2. What happens to red blood cells that are agglutinated?

Clinical Terms Related to the Blood

anisocytosis (an-i″so-si-to′sis)—condition characterized by an abnormal variation in the size of erythrocytes.

antihemophilic plasma (an″ti-he″mo-fil′ik plaz′mah)—normal blood plasma that has been processed to preserve an antihemophilic factor.

Christmas disease (kris′mas di-zēz′)—hereditary bleeding disease that is due to a deficiency in a specific clotting factor; also called *hemophilia B.*

citrated whole blood (sit′rāt-ed hōl blud)—normal blood to which a solution of acid citrate has been added to prevent coagulation.

dried plasma (drid plas′mah)—normal blood plasma that has been vacuum dried to prevent the growth of microorganisms.

hemorrhagic telangiectasia (hem″o-raj′ik tel-an″je-ek-ta′ze-ah)—hereditary disorder characterized by a tendency to bleed from localized lesions of the capillaries.

heparinized whole blood (hep′er-i-nizd″ hōl blud)—normal blood to which a solution of heparin has been added to prevent coagulation.

macrocytosis (mak″ro-si-to′sis)—condition characterized by the presence of abnormally large erythrocytes.

microcytosis (mi″kro-si-to′sis)—condition characterized by the presence of abnormally small erythrocytes.

neutrophilia (nu″tro-fil′e-ah)—condition in which the number of circulating neutrophils is increased.

normal plasma (nor′mal plaz′mah)—plasma from which the blood cells have been removed by centrifugation or sedimentation.

packed red cells —concentrated suspension of red blood cells from which the plasma has been removed.

pancytopenia (pan″si-to-pe′ne-ah)—condition characterized by an abnormal depression of all the cellular components of blood.

poikilocytosis (poi″ki-lo-si-to′sis)—condition in which the erythrocytes are irregularly shaped.

pseudoagglutination (su″do-ah-gloo″ti-na′shun)—the clumping of erythrocytes due to some factor other than an agglutinogen-agglutinin reaction.

purpura (per′pu-rah)—disease characterized by spontaneous bleeding into the tissues and through the mucous membrane.

spherocytosis (sfe″ro-si-to′sis)—hereditary form of hemolytic anemia characterized by the presence of spherical erythrocytes (spherocytes).

thalassemia (thal″ah-se′me-ah)—group of hereditary hemolytic anemias characterized by the presence of very thin, fragile erythrocytes.

von Willebrand's disease (fon vil′ĕ-brandz di-zēz′)—hereditary condition caused by a deficiency of an antihemophilic blood factor combined with capillary defects, which is characterized by bleeding from the nose, gums, and genitalia.

Chapter Summary

Introduction (page 368)

The blood is a type of connective tissue whose cells are suspended in liquid. It transports substances between the body cells and the external environment and helps maintain a stable cellular environment.

Blood and Blood Cells (page 368)

1. The blood contains red blood cells, white blood cells, and platelets.
2. Volume and composition of the blood.
   a. Blood volume varies with body size, fluid and electrolyte balance, and fat content.
   b. The blood can be separated into cellular and liquid portions.
   c. Plasma includes water, nutrients, hormones, electrolytes, and cellular wastes.
3. Characteristics of red blood cells
   a. Red blood cells are biconcave disks, whose shapes provide increased surface area.
   b. They contain hemoglobin, which combines with oxygen.
4. Red blood cell counts
   a. The red blood cell count equals the number of cells per mm^3 of blood.
   b. The normal red cell count varies from 4,600,000 to 6,200,000 cells per mm^3 in males and from 4,200,000 to 5,400,000 cells per mm^3 in females.
   c. The red cell count is related to the oxygen-carrying capacity of the blood.
5. Destruction of red blood cells
   a. Damaged red cells are phagocytized by macrophages in the liver and spleen.
   b. Hemoglobin molecules are decomposed and the iron they contain is conserved.
   c. Biliverdin and bilirubin are pigments resulting from hemoglobin breakdown.
6. Red blood cell production and its control
   a. Red cells are produced by the red bone marrow.
   b. The number of red blood cells remains relatively stable.
   c. The rate of red cell production is controlled by a negative feedback mechanism involving erythropoietin.
   d. Dietary factors affecting red blood cell production:
      (1) Production is affected by the availability of vitamin B_{12} and folic acid.
      (2) Iron is needed for hemoglobin synthesis.
7. Types of white blood cells
   a. White blood cells function to control disease conditions.
   b. Granulocytes include neutrophils, eosinophils, and basophils.
   c. Agranulocytes include monocytes and lymphocytes.
8. White blood cell counts
   a. Normal total white cell counts vary from 5,000 to 10,000 cells per mm^3.
   b. The number of white cells may change in abnormal conditions.
   c. A differential white cell count indicates the percentages of the various types of leukocytes present.
9. Functions of white blood cells
   a. Neutrophils and monocytes phagocytize foreign particles.
   b. Eosinophils kill parasites and help control allergic reactions.
   c. Basophils release heparin, which inhibits blood clotting.
   d. Lymphocytes produce antibodies that act against specific foreign substances.
10. Blood platelets
   a. Blood platelets are fragments of giant cells.
   b. The normal platelet count varies from 130,000 to 360,000 platelets per mm^3.
   c. They function to help close breaks in blood vessels.

Blood Plasma (page 377)

Plasma transports nutrients and gases, regulates fluid and electrolyte balance, and helps maintain stable pH.

1. Plasma proteins
   Three major groups exist.
   a. Albumins help maintain the osmotic pressure of the blood.
   b. Globulins transport lipids and fat-soluble vitamins, and they include the antibodies of immunity.
   c. Fibrinogen functions in blood clotting.
2. Nutrients and gases
   a. The plasma nutrients include amino acids, simple sugars, and lipids.
   b. The gases in plasma include oxygen, carbon dioxide, and nitrogen.
3. Nonprotein nitrogenous substances
   a. These are composed of molecules that contain nitrogen atoms but are not proteins.
   b. They include amino acids, urea, and uric acid.
4. Plasma electrolytes
   a. They include ions of sodium, potassium, calcium, magnesium, chlorine, bicarbonate, phosphate, and sulfate.
   b. They are important in the maintenance of osmotic pressure and pH.

Hemostasis (page 379)

Hemostasis refers to the stoppage of bleeding.

1. Blood vessel spasm
   a. The smooth muscles in the blood vessel walls contract following injury.
   b. Platelets release serotonin that stimulates vasoconstriction.
2. Platelet plug formation
   a. Platelets adhere to rough surfaces and exposed collagen.
   b. Platelets stick together at the sites of injuries and form platelet plugs in the broken vessels.

3. Blood coagulation
    a. Blood clotting is the most effective means of hemostasis.
    b. Clot formation depends on the balance between substances that promote clotting and those that inhibit clotting.
    c. The basic event of clotting is the conversion of soluble fibrinogen into insoluble fibrin.
    d. Factors that promote clotting include the presence of prothrombin activator, prothrombin, and calcium ions.
    e. A thrombus is a blood clot in a vessel; an embolus is a clot or fragment of a clot that has moved in a vessel.

Blood Groups and Transfusions (page 381)

The blood can be typed on the basis of the substances it contains. Blood substances of certain types will react adversely with other types.

1. Agglutinogens and agglutinins
    a. Red blood cell membranes may contain agglutinogens and blood plasma may contain agglutinins.
    b. Blood typing involves identifying the agglutinogens present in the red cell membranes.
2. ABO blood group
    a. The blood can be grouped according to the presence or absence of agglutinogens A and B.
    b. Adverse transfusion reactions are avoided by preventing the mixing of red cells that contain an agglutinogen with plasma that contains the corresponding agglutinin.
    c. Blood should be cross matched to determine if mixing the donor and recipient blood will cause agglutination.
3. Rh blood group
    a. Rh agglutinogens are present in the red cell membranes of Rh-positive blood; they are absent in Rh-negative blood.
    b. Mixing Rh-positive red cells with plasma that contains anti-Rh agglutinins results in agglutination of the positive cells.
    c. Anti-Rh agglutinins in maternal blood may pass through the placental tissues and react with the red cells of an Rh-positive fetus.
4. Agglutination reactions
    a. Agglutinated cells are destroyed by phagocytic cells.
    b. Free hemoglobin is likely to cause jaundice and interfere with kidney functions.

Clinical Application of Knowledge

1. If a patient with an inoperable cancer is treated by using a drug that reduces the rate of cell division, what changes might occur in the patient's white blood cell count? How might the patient's environment be modified to compensate for the effects of these changes?
2. Hypochromic (iron-deficiency) anemia is relatively common among aging persons who are admitted to hospitals for other conditions. What environmental and sociological factors might promote this form of anemia?
3. Why do patients with liver diseases commonly develop blood clotting disorders?
4. How would you explain to a patient with leukemia, who has an elevated white blood count, the importance of avoiding bacterial infections?

Review Activities

1. List the major components of the blood.
2. Describe a red blood cell.
3. Distinguish between oxyhemoglobin and deoxyhemoglobin.
4. Describe the life cycle of a red blood cell.
5. Distinguish between biliverdin and bilirubin.
6. Define *erythropoietin,* and explain its function.
7. Explain how vitamin B_{12} and folic acid deficiencies affect red blood cell production.
8. Distinguish between granulocytes and agranulocytes.
9. Name five types of leukocytes, and list the major functions of each type.
10. Explain the significance of white blood cell counts as aids to diagnosing diseases.
11. Describe a blood platelet, and explain its function.
12. Name three types of plasma proteins, and list the major functions of each type.
13. Define *nonprotein nitrogenous substances,* and name those commonly present in plasma.
14. Name several plasma electrolytes.
15. Define *hemostasis.*
16. Explain how blood vessel spasms are stimulated following an injury.
17. Explain how a platelet plug forms.
18. List the major steps leading to the formation of a blood clot.
19. Distinguish between fibrinogen and fibrin.
20. Provide an example of a positive feedback system that operates during blood clot formation.
21. Distinguish between thrombus and embolus.
22. Distinguish between agglutinogen and agglutinin.
23. Explain the basis of ABO blood types.
24. Explain why it is always best to match exactly the donor and recipient blood before giving a blood transfusion.
25. Explain how blood is cross matched.
26. Distinguish between Rh-positive and Rh-negative blood.
27. Describe how a person may become sensitized to Rh-positive blood.
28. Define *erythroblastosis fetalis,* and explain how this condition may develop.
29. Describe the consequences of an agglutination reaction.

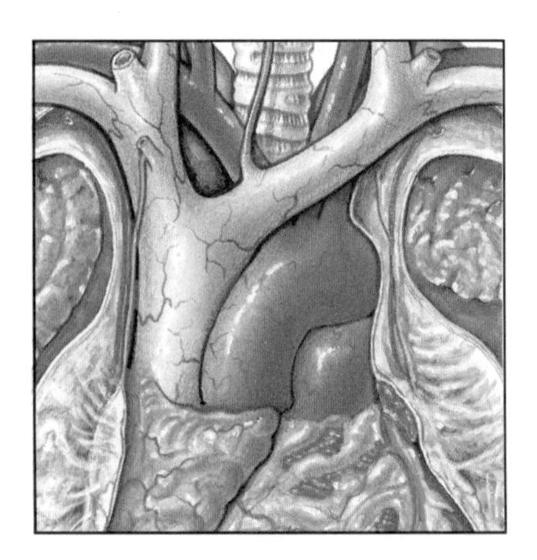

15 Cardiovascular System

The *cardiovascular system* is the portion of the circulatory system that includes the heart and blood vessels. It moves the blood between the body cells and the organs of the integumentary, digestive, respiratory, and urinary systems, which communicate with the external environment.

In performing this function, the heart acts as a pump that forces the blood through the blood vessels. The blood vessels, in turn, form a closed system of ducts, which transports the blood and allows exchanges of gases, nutrients, and wastes between the blood and the body cells.

Chapter Objectives

After you have studied this chapter, you should be able to

1. Name the organs of the cardiovascular system, and discuss their functions.
2. Name and describe the locations of the major parts of the heart, and discuss the function of each part.
3. Trace the pathway of the blood through the heart and vessels of the coronary circulation.
4. Discuss the cardiac cycle, and explain how it is controlled.
5. Identify the parts of a normal ECG pattern, and discuss the significance of this pattern.
6. Compare the structures and functions of the major types of blood vessels.
7. Describe the mechanism that aids in returning venous blood to the heart.
8. Explain how blood pressure is created and controlled.
9. Compare the pulmonary and systemic circuits of the cardiovascular system.
10. Identify and locate the major arteries and veins of the pulmonary and systemic circuits.
11. Complete the review activities at the end of this chapter. Note that the items are worded in the form of specific learning objectives. You may want to refer to them before reading the chapter.

Key Terms

arteriole (ar-te′re-ōl)

atrium (a′tre-um)

capillary (kap′i-lar″e)

cardiac cycle (kar′de-ak si′kl)

cardiac output (kar′de-ak owt′poot)

diastolic pressure (di″ah-stol′ik presh′ur)

electrocardiogram (e-lek″tro-kar′de-o-gram″)

endocardium (en″do-kar′de-um)

epicardium (ep″ĭ-kar′de-um)

functional syncytium (funk′shun-al sin-sish′e-um)

myocardium (mi″o-kar′de-um)

pericardium (per″ĭ-kar′de-um)

peripheral resistance (pĕ-rif′er-al re-zis′tans)

pulmonary circuit (pul′mo-ner″e sur′kit)

systemic circuit (sis-tem′ik sur′kit)

systolic pressure (sis-tol′ik presh′ur)

vasoconstriction (vas″o-kon-strik′shun)

vasodilation (vas″o-di-la′shun)

ventricle (ven′tri-kl)

venule (ven′ūl)

Aids to Understanding Words

brady-, slow: *brady*cardia—an abnormally slow heartbeat.

diastol-, dilation: *diastol*ic pressure—the blood pressure that occurs when the ventricle is relaxed (thus dilated).

-gram, something written: electrocardio*gram*—a recording of the electrical changes that occur in the heart muscle during a cardiac cycle.

papill-, nipple: *papill*ary muscle—a small mound of muscle within a chamber of the heart.

syn-, together: *syn*cytium—a mass of merging cells that act together.

systol-, contraction: *systol*ic pressure—the blood pressure that occurs during a ventricular contraction.

tachy-, rapid: *tachy*cardia—an abnormally fast heartbeat.

Introduction

A FUNCTIONAL CARDIOVASCULAR system is vital for survival because without circulation, the tissues lack a supply of oxygen and nutrients, and waste substances begin to accumulate. Under such conditions, the cells soon begin to undergo irreversible changes, which quickly lead to the death of the organism. The general pattern of the cardiovascular system is shown in figure 15.1.

Structure of the Heart

The heart is a hollow, cone-shaped, muscular pump, located within the thorax and resting upon the diaphragm (fig. 15.2).

Size and Location of the Heart

Although heart size varies with body size, the heart of an average adult is about 14 centimeters long and 9 centimeters wide.

The heart is within the mediastinum, which is bordered laterally by the lungs, posteriorly by the backbone, and anteriorly by the sternum. Its *base,* which is attached to several large blood vessels, lies beneath the second rib. Its distal end extends downward and to the left, terminating as a bluntly pointed *apex* at the level of the fifth intercostal space.

Coverings of the Heart

The heart and the proximal ends of the large blood vessels to which it is attached are enclosed by a **pericardium.** The pericardium consists of an outer fibrous bag, the *fibrous pericardium,* which surrounds a more delicate, double-layered sac. The inner layer of the sac, the *visceral pericardium* (epicardium), covers the heart. At the base of the heart, the visceral layer turns back upon itself to become the *parietal pericardium* (fig. 15.3). The parietal pericardium, in turn, forms the inner lining of the fibrous pericardium.

The fibrous pericardium is a tough, protective sac composed largely of white fibrous connective tissue. It is attached to the central portion of the diaphragm, behind the sternum, the vertebral column, and the large blood vessels emerging from the heart. Between the parietal and visceral layers of the pericardium is a potential space, the *pericardial cavity,* which contains a small amount of serous fluid (fig. 15.3). This fluid reduces friction between the pericardial membranes as the heart moves within them.

Figure 15.1
The cardiovascular system transports blood between body cells and the organs that communicate with the external environment.

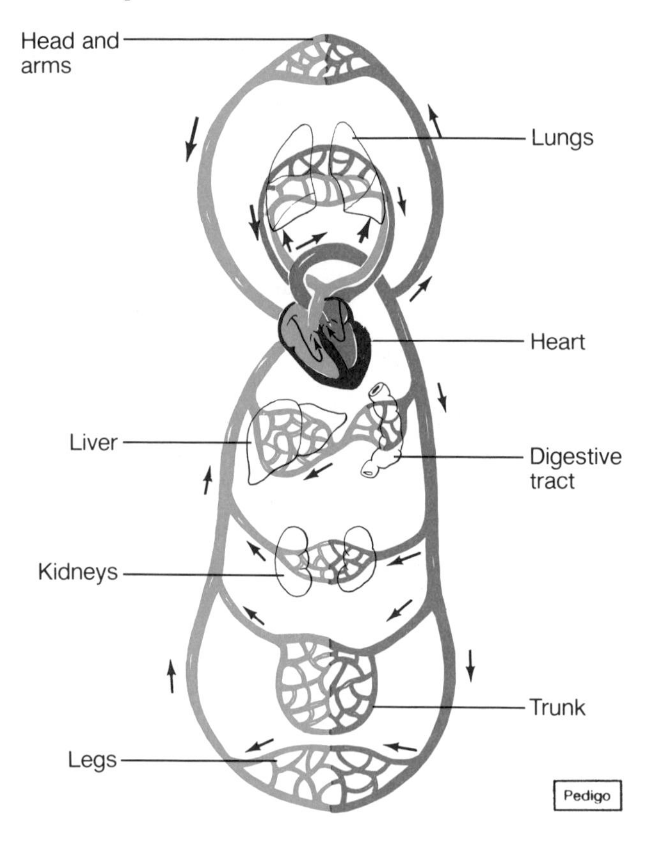

If the pericardium becomes inflamed due to a bacterial or viral infection, the condition is called *pericarditis.* As a result of this inflammation, the layers of the pericardium sometimes become stuck together by adhesions, and this may interfere with heart movements. If this happens, surgery may be required to separate the surfaces and make unrestricted heart actions possible again.

1. Where is the heart located?
2. Distinguish between the visceral pericardium and the parietal pericardium.

Wall of the Heart

The wall of the heart is composed of three distinct layers—an outer epicardium, a middle myocardium, and an inner endocardium (fig. 15.3).

The **epicardium,** which corresponds to the visceral pericardium, functions as a protective layer. This serous membrane consists of connective tissue covered

Figure 15.2
The heart is enclosed by a layered pericardium.

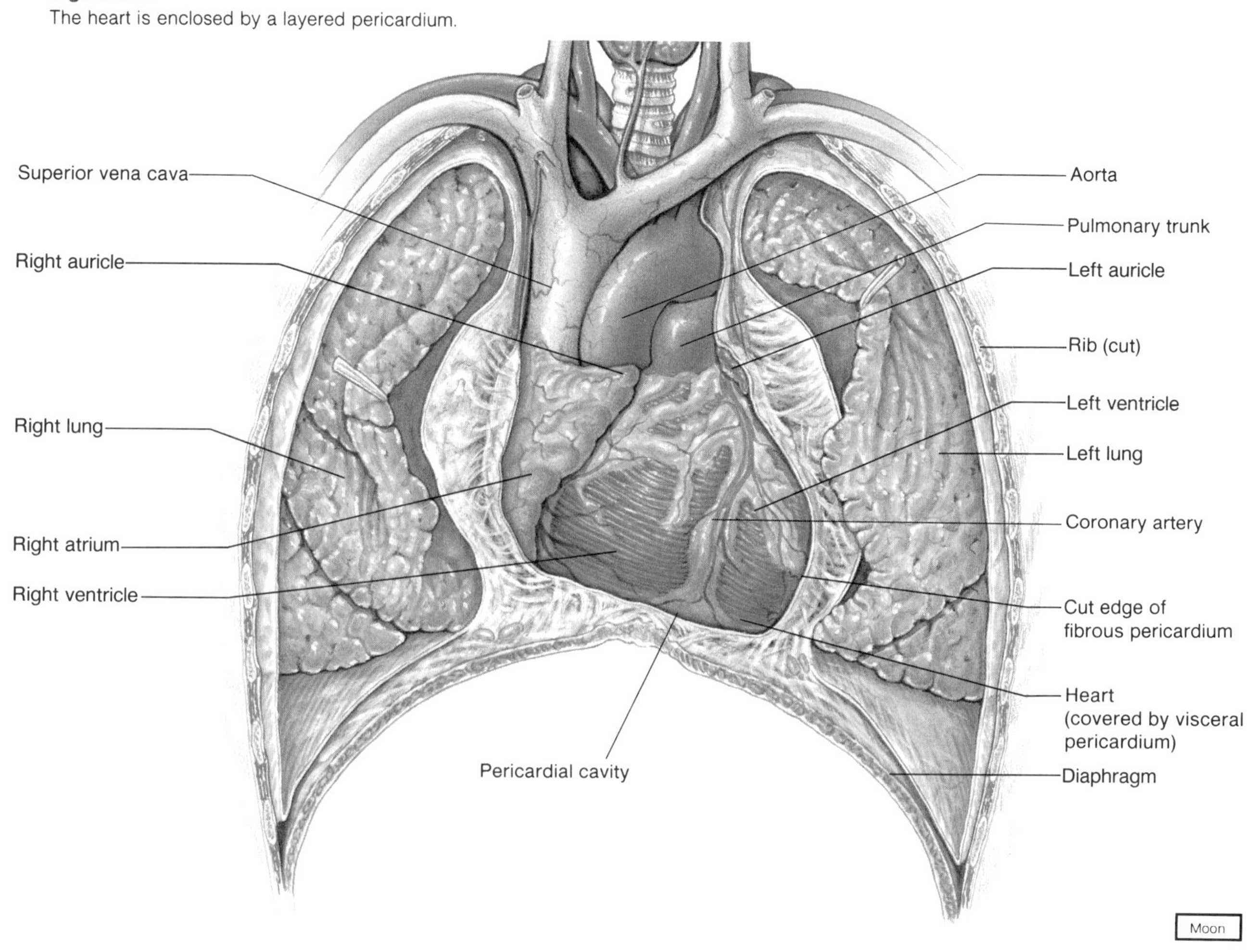

Figure 15.3
The wall of the heart consists of three layers: the endocardium, myocardium, and epicardium.

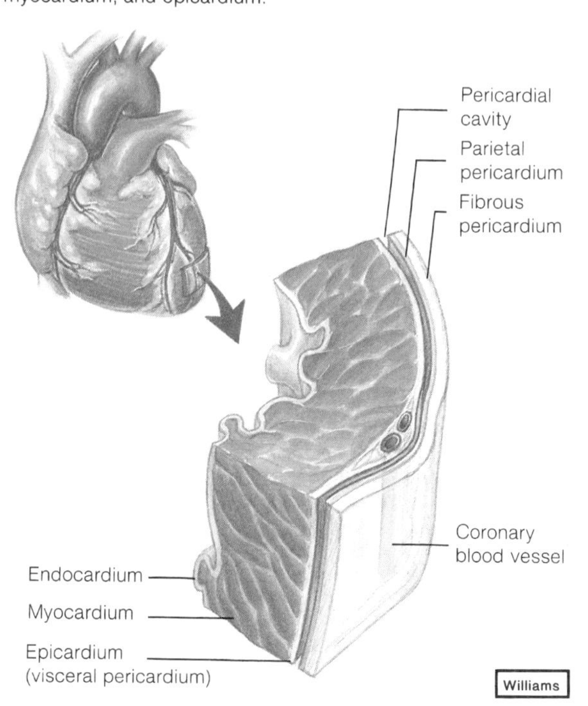

by epithelium. Its deeper portion often contains fat, particularly along the paths of the larger blood vessels.

The **myocardium** is relatively thick and consists largely of the cardiac muscle tissue responsible for forcing the blood out of the heart chambers. The muscle fibers are arranged in planes, separated by connective tissues, which are richly supplied with blood capillaries, lymph capillaries, and nerve fibers.

The **endocardium** consists of epithelium and connective tissue, which contains many elastic and collagenous fibers. The connective tissue also contains some specialized cardiac muscle fibers, called *Purkinje fibers,* whose function is described in a subsequent section of this chapter. This inner lining is continuous with the inner linings of the blood vessels attached to the heart.

Heart Chambers and Valves

Internally, the heart is divided into four chambers, two on the left and two on the right (fig. 15.4). The upper chambers, called **atria** (sing. *atrium*), have relatively thin walls and receive the blood from the veins. The lower chambers, the **ventricles,** force the blood out of

Figure 15.4
Frontal sections of the heart (*a*) showing the connection between the right ventricle and the pulmonary trunk, and (*b*) showing the connection between the left ventricle and the aorta.

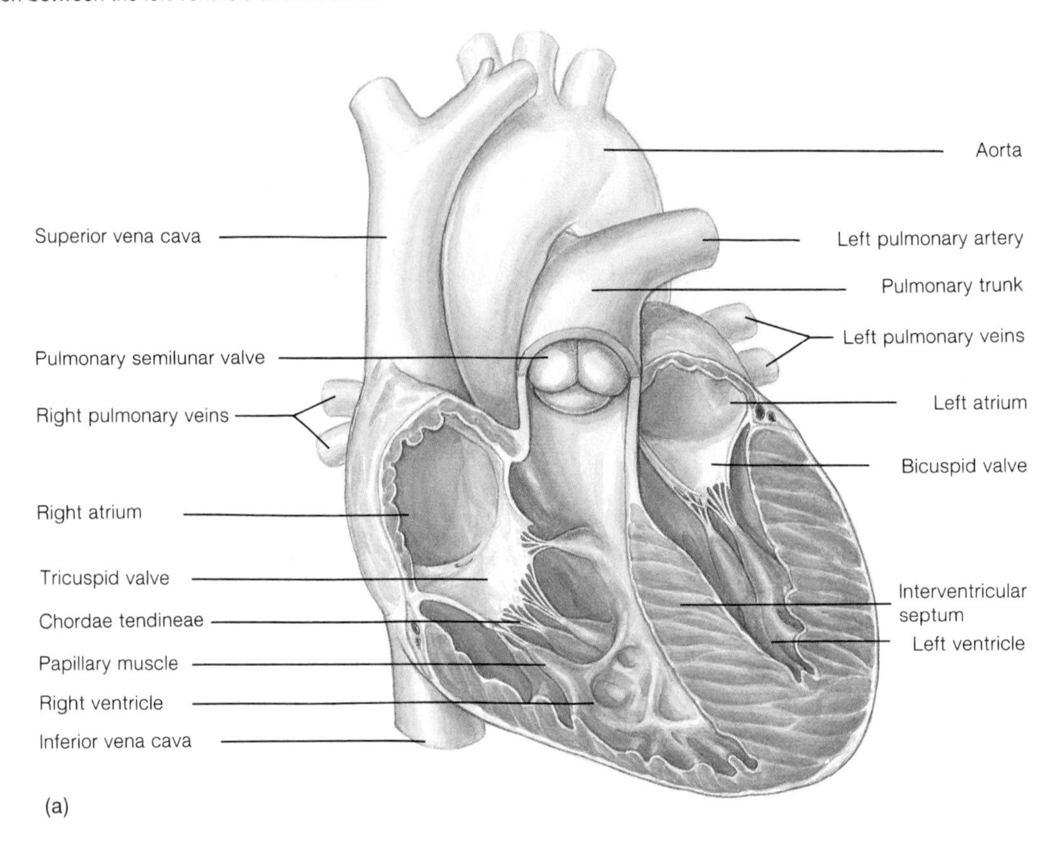

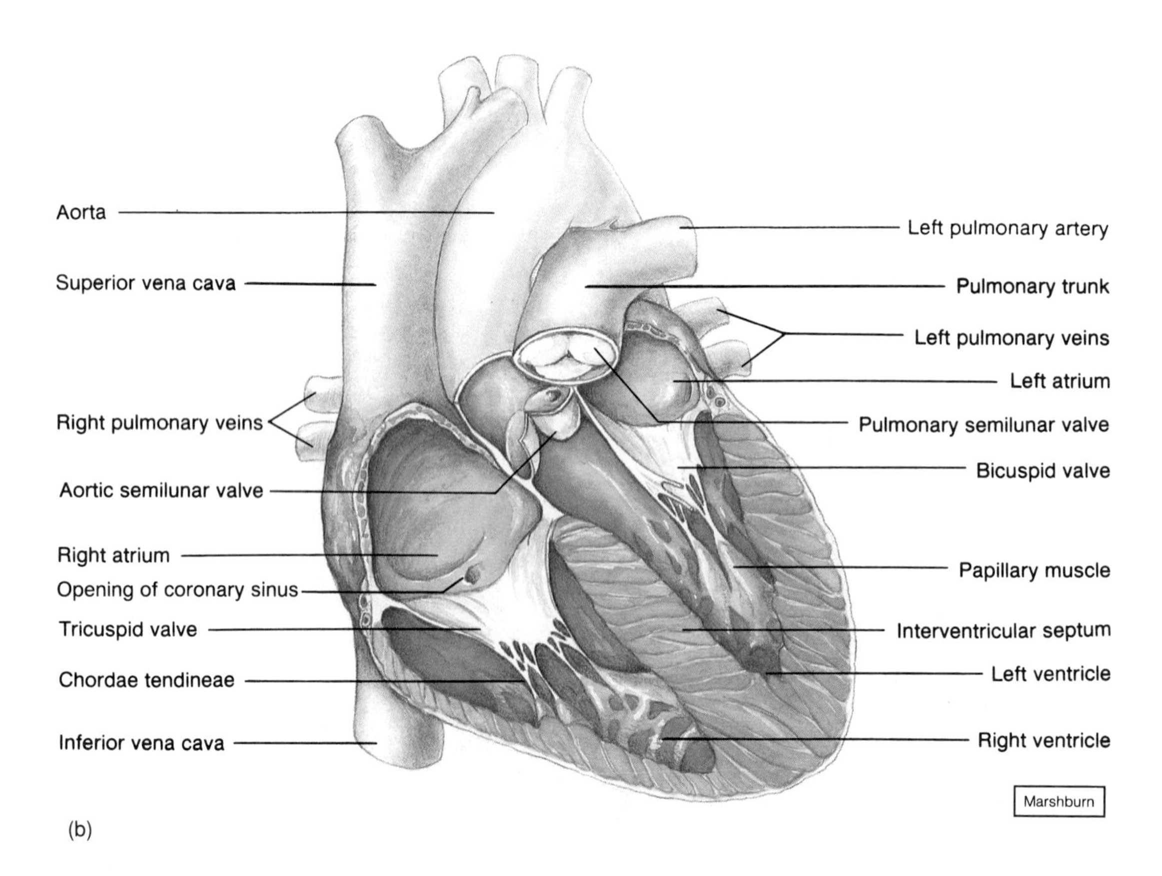

the heart into the arteries. (Note: Veins are blood vessels that carry the blood toward the heart; arteries carry the blood away from the heart.)

The atrium and ventricle on the right side are separated from those on the left by a *septum.* The atrium on each side communicates with its corresponding ventricle through an opening, which is guarded by a *valve.* Small earlike projections, called *auricles* extend outward from the atria (fig. 15.2).

The right atrium receives blood from two large veins—the *superior vena cava* and the *inferior vena cava.* A smaller vein, the *coronary sinus,* also drains blood into the right atrium from the wall of the heart.

The opening between the right atrium and the right ventricle is guarded by a large **tricuspid valve,** which is composed of three leaflets, or cusps (fig. 15.4). This valve permits the blood to move from the right atrium into the right ventricle and prevents it from passing in the opposite direction.

Strong, fibrous strings, called *chordae tendineae,* are attached to the cusps on the ventricular side. These strings originate from small mounds of muscle tissue, the **papillary muscles,** which project inward from the walls of the ventricle. When the tricuspid valve closes, the chordae tendineae and papillary muscles prevent the cusps from swinging back into the atrium.

The right ventricle has a much thinner muscular wall than the left ventricle (fig. 15.4). This right chamber pumps the blood a fairly short distance to the lungs against a relatively low resistance to blood flow. The left ventricle, on the other hand, must force the blood to all the other parts of the body against a much greater resistance to flow.

An exit from the right ventricle is provided by the *pulmonary trunk* that divides to form the left and right *pulmonary arteries,* which lead to the lungs. At the base of this trunk is a **pulmonary (semilunar) valve,** which consists of three cusps. This valve allows the blood to leave the right ventricle and prevents a return flow into the ventricular chamber.

The left atrium receives the blood from the lungs through four *pulmonary veins*—two from the right lung and two from the left lung. The blood passes from the left atrium into the left ventricle through the **bicuspid (mitral) valve,** which prevents the blood from flowing back into the left atrium from the ventricle. As with the tricuspid valve, the cusps of the bicuspid valve are prevented from swinging back into the left atrium by the chordae tendineae and papillary muscles.

The only exit from the left ventricle is through a large artery called the *aorta.* At the base of the aorta, there is an **aortic (semilunar) valve,** which consists of three cusps. It opens and allows the blood to leave the left ventricle, and when it closes, it prevents the blood from backing up into the ventricle (fig. 15.5).

Figure 15.5
A photograph of the semilunar valves (*superior view*). Which vessel is the aorta? Why do you think so?

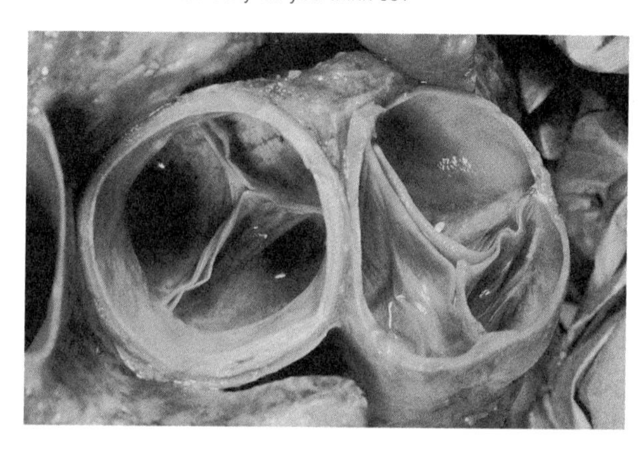

Cardiac muscle cells associated with the atria secrete a hormone called *atrial natriuretic factor* (ANF). This hormone seems to be released when the muscle cells are stretched abnormally, as may occur when the volume of blood within the atria is increased. The effect of ANF is to inhibit the release of a substance (renin) from the kidneys (chapter 17) and to inhibit the release of aldosterone from the adrenal cortex (chapter 11). As a result of these actions, the excretion of sodium and water from the kidneys is increased, leading to a decrease in blood volume and blood pressure.

1. Describe the layers of the heart wall.
2. Name and locate the four chambers and valves of the heart.

Skeleton of the Heart

At their proximal ends, the pulmonary trunk and aorta are surrounded by rings of dense fibrous connective tissue (fig. 15.6). The rings provide firm attachments for the heart valves and various muscle fibers. In addition, they prevent the outlets of the atria and ventricles from dilating during myocardial contraction. The fibrous rings together with other masses of dense fibrous tissue in the upper portion of the septum between the ventricles (interventricular septum) constitute the skeleton of the heart.

Path of Blood through the Heart

Blood that is relatively low in oxygen concentration and relatively high in carbon dioxide concentration enters the right atrium through the venae cavae and coronary sinus. As the right atrial wall contracts, the blood passes through the tricuspid valve and enters the chamber of the right ventricle (fig. 15.7).

Figure 15.6
The skeleton of the heart (*superior view*) consists of fibrous rings, to which the heart valves are attached.

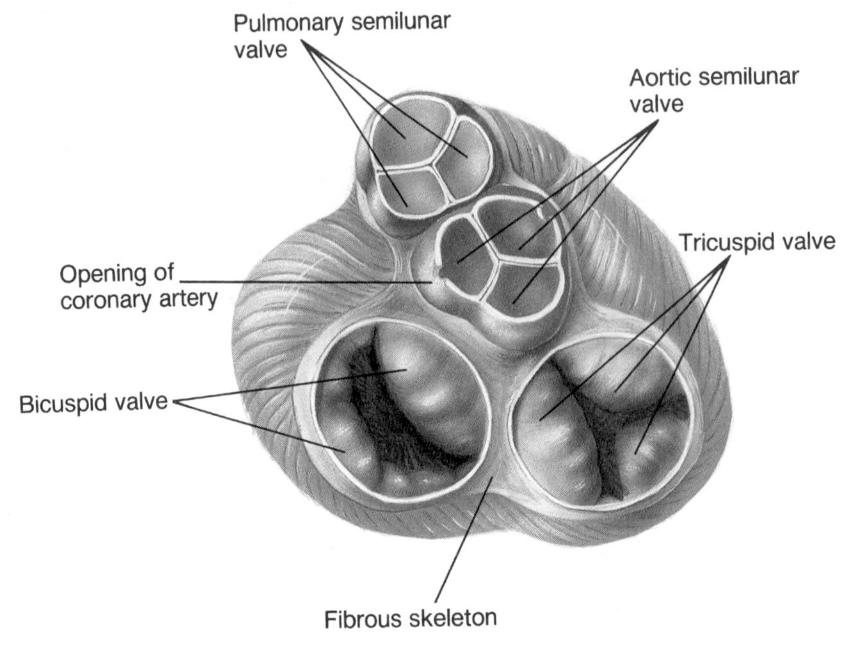

When the right ventricular wall contracts, the tricuspid valve closes, and the blood moves through the pulmonary semilunar valve and into the pulmonary trunk and its branches. From these vessels, the blood enters the capillaries associated with the alveoli of the lungs. Gas exchanges occur between the blood in the capillaries and the air in the alveoli. The freshly oxygenated blood, which is now relatively low in carbon dioxide concentration, returns to the heart through the pulmonary veins, which lead to the left atrium.

As the left atrial wall contracts, the blood moves through the bicuspid valve and into the chamber of the left ventricle. When the left ventricular wall contracts, the bicuspid valve closes and the blood passes through the aortic semilunar valve and into the aorta and its branches.

Blood Supply to the Heart

Blood is supplied to the tissues of the heart by the first two branches of the aorta, called the right and left **coronary arteries.** Their openings lie just beyond the aortic semilunar valve (fig. 15.8).

Because the heart must beat continually to supply blood to the body tissues, the myocardial cells require a constant supply of freshly oxygenated blood. The myocardium contains many capillaries fed by branches of the coronary arteries (fig. 15.9). The larger branches of these arteries usually have interconnections between vessels that provide alternate pathways for the blood.

If a branch of a coronary artery becomes abnormally constricted or obstructed, the myocardial cells it supplies may experience a blood deficiency, called *ischemia.* As a result of ischemia, the person may experience a painful condition called *angina pectoris.* The discomfort of angina pectoris usually occurs during physical activity or an emotional disturbance and is relieved by rest. It may take the form of pain or a sensation of heavy pressure, tightening, or squeezing in the chest. Although it is usually felt in the region behind the sternum or in the anterior portion of the upper thorax, the pain may radiate to other parts, including the neck, jaw, throat, arm, shoulder, elbow, back, or upper abdomen.

Sometimes a portion of the heart dies because a blood clot forms and completely obstructs a coronary artery or one of its branches. This condition, a *myocardial infarction,* which is more commonly called a heart attack, is one of the leading causes of death.

The blood that has passed through the capillaries of the myocardium is drained by branches of the **cardiac veins,** whose paths roughly parallel those of the coronary arteries. As figure 15.9*b* shows, these veins join the **coronary sinus,** an enlarged vein which is on the posterior surface of the heart and empties into the right atrium (fig. 15.4*b*).

1. What structures make up the skeleton of the heart?
2. Review the path of blood through the heart.
3. What vessels supply blood to the myocardium?
4. What vessels drain blood from the myocardium?

Figure 15.7
The right ventricle forces blood to the lungs, and the left ventricle forces blood to all other body parts. How does the composition of the blood in these two chambers differ?

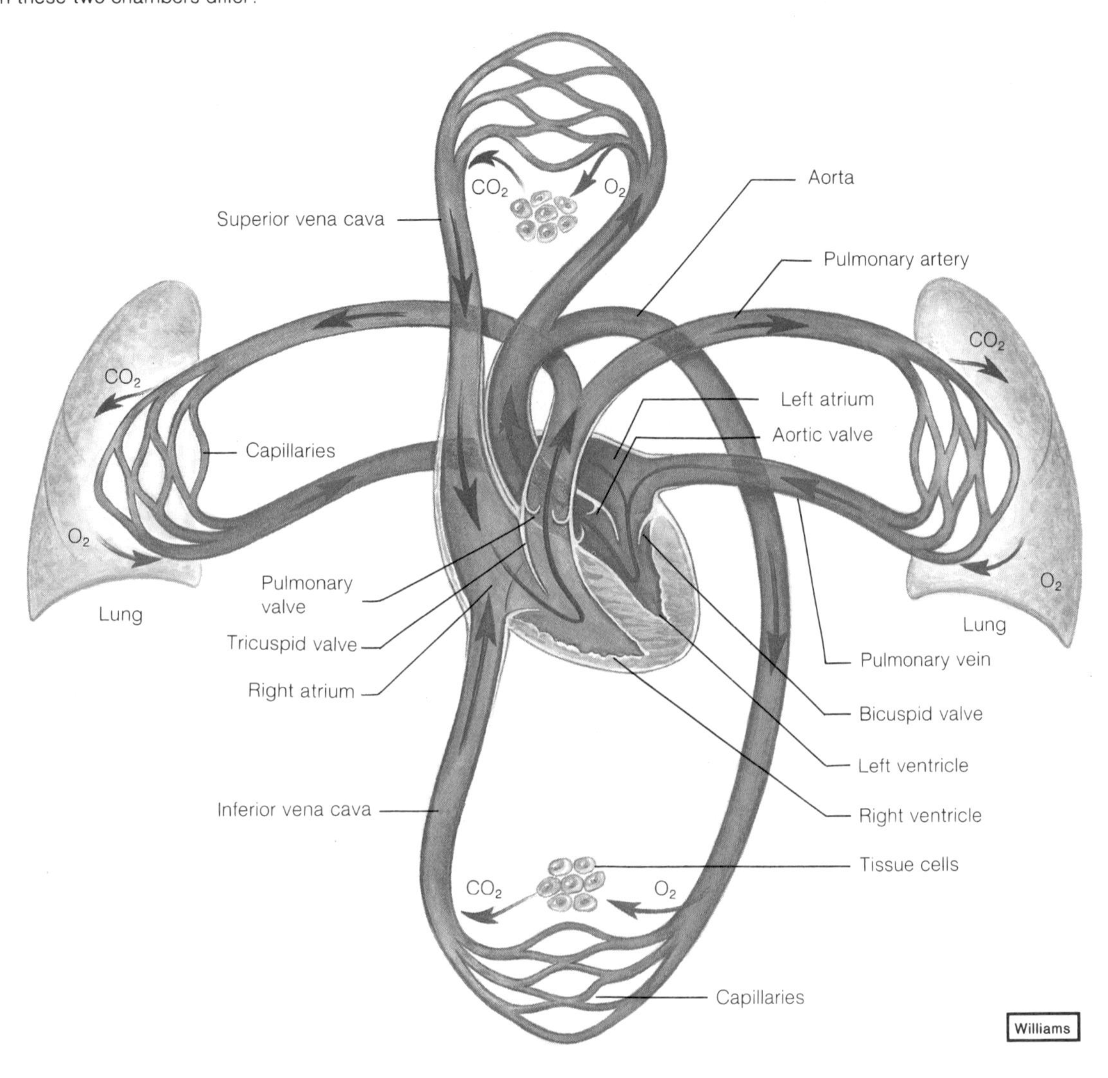

Figure 15.8
The openings of the coronary arteries lie just beyond the aortic semilunar valve.

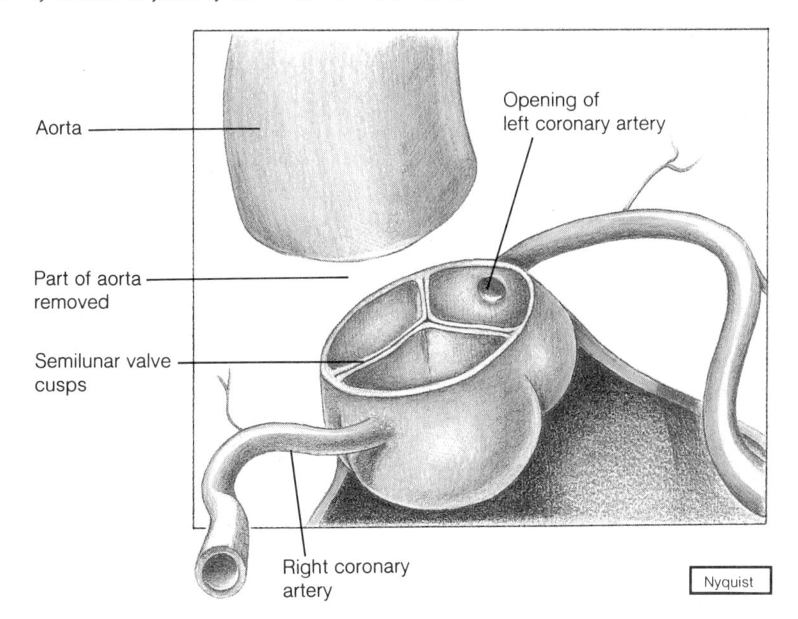

Figure 15.9
The blood vessels associated with the surface of the heart: (*a*) an anterior view; (*b*) a posterior view.

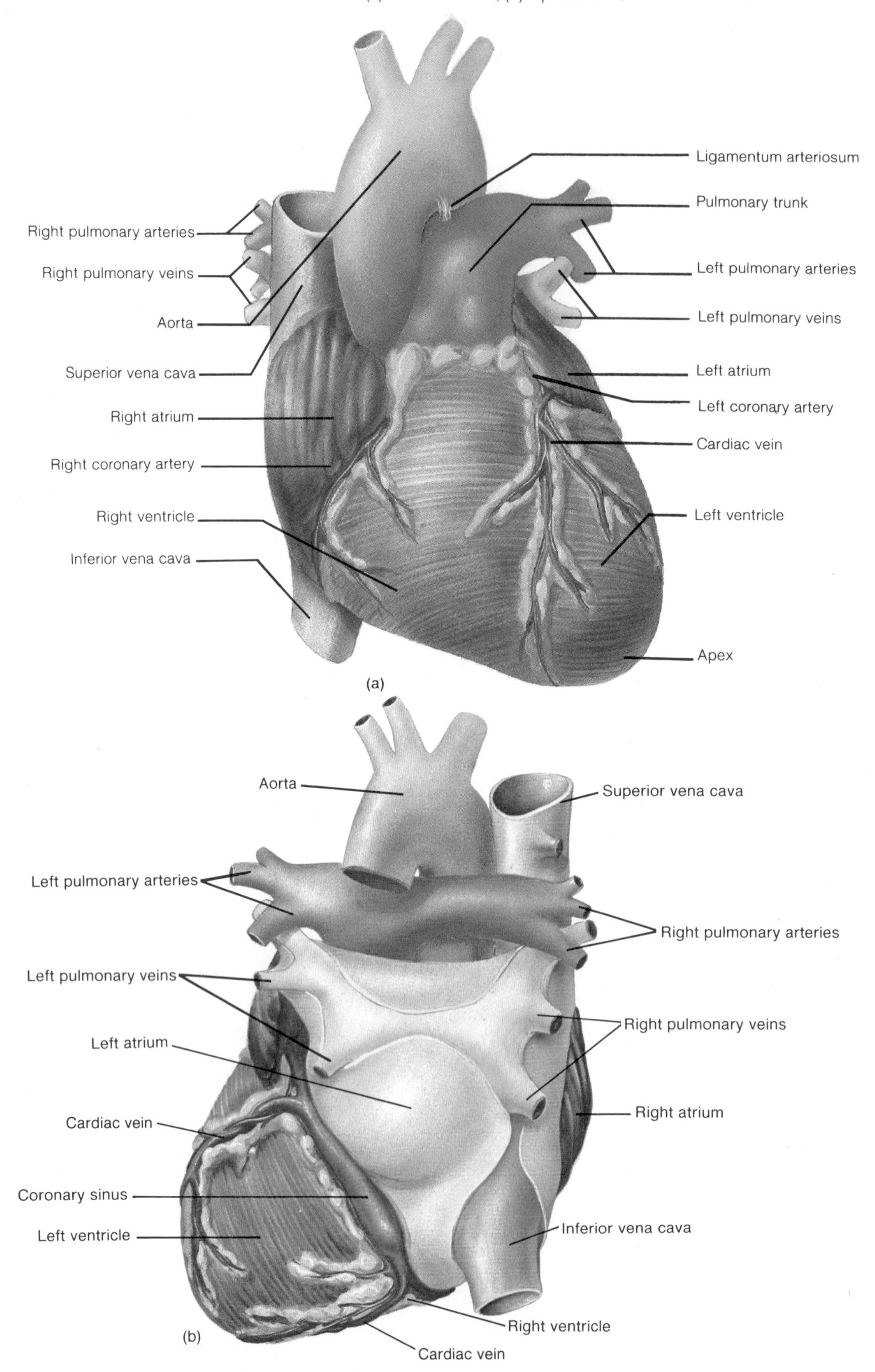

Figure 15.10
(*a*) The atria empty during atrial systole and (*b*) fill during atrial diastole.

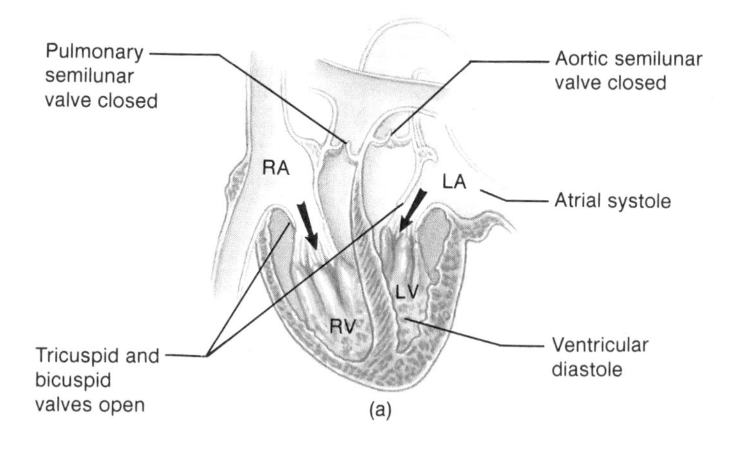

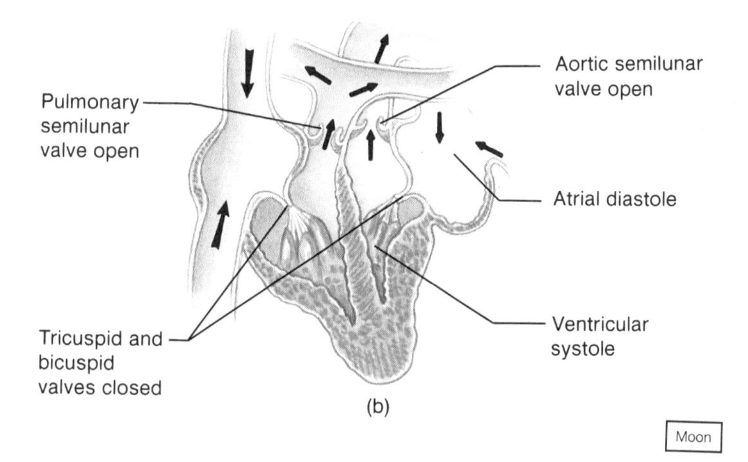

In a *heart transplantation* procedure, the recipient's failing heart is removed, except for the posterior walls of the right and left atria and their connections to the venae cavae and pulmonary veins. The donor heart is prepared similarly and is attached to the atrial cuffs remaining in the recipient's thorax. Finally, the recipient's aorta and pulmonary arteries are connected to those of the donor heart.

Actions of the Heart

Although the previous discussion described the actions of the heart chambers separately, they do not function independently. Instead, their actions are regulated so that the atrial walls contract while the ventricular walls are relaxed, and ventricular walls contract while the atrial walls are relaxed. Such a series of events constitutes a complete heartbeat, or **cardiac cycle** (fig. 15.10).

Cardiac Cycle

During a cardiac cycle, the pressure within the chambers of the heart rises and falls. For example, when the atria are relaxed, the blood flows into them from the large, attached veins. Then, the atrial walls contract (atrial systole), and the atrial pressure rises suddenly, forcing the atrial contents into the ventricles. This is followed by atrial relaxation (atrial diastole).

As the ventricles contract (ventricular systole), the tricuspid and bicuspid valves close, and the blood flows out of the ventricles into the arteries. When the ventricles relax (ventricular diastole), the tricuspid and bicuspid valves open, and the blood flows through them from the atria into the ventricles.

Heart Sounds

The sounds associated with a heartbeat can be heard with a stethoscope and are described as *lub-dup* sounds. These sounds are due to vibrations in the heart tissues that are created as the blood flow is suddenly speeded or slowed with the contraction and relaxation of the heart chambers and with the opening and closing of the valves.

The first part of a heart sound (*lub*) occurs during the ventricular contraction, when the tricuspid and bi-

Figure 15.11
The cardiac conduction system. What is the function of this system?

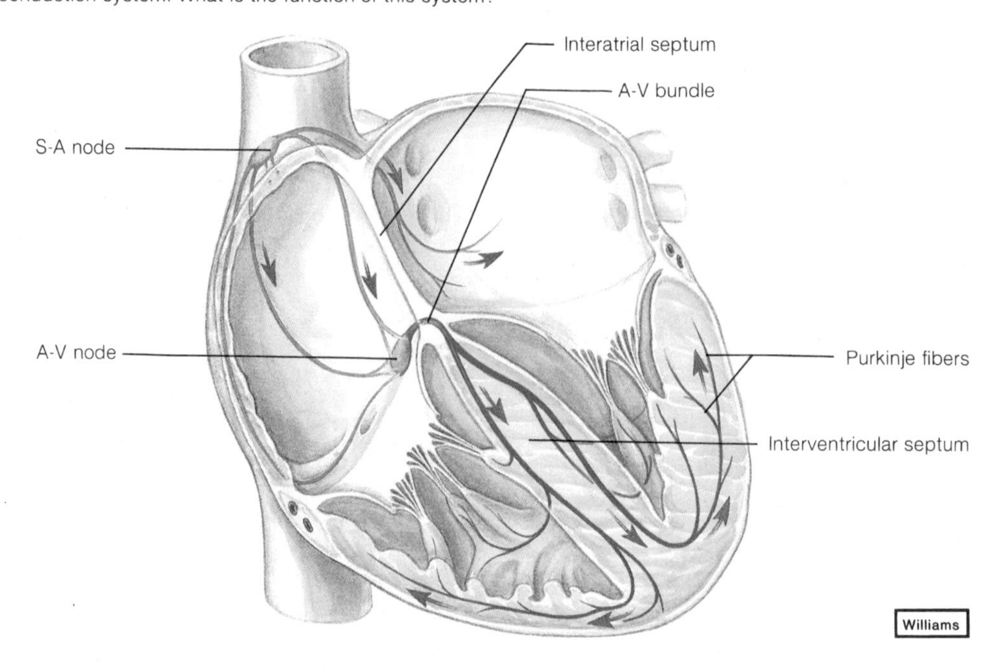

cuspid valves are closing. The second part (*dup*) occurs during ventricular relaxation, when the pulmonary and aortic semilunar valves are closing.

Heart sounds are of particular interest because they provide information concerning the condition of the heart valves. For example, an inflammation of the endocardium (endocarditis) may cause changes in the shapes of the valvular cusps (valvular stenosis). Then, when the cusps close, the closure may be incomplete, and some blood may leak back through the valve. If this happens, an abnormal sound called a *murmur* may be heard. The seriousness of a murmur depends on the amount of valvular damage. Fortunately for those who have serious problems, it may be possible to repair the damaged valves or to replace them by open heart surgery.

Cardiac Muscle Fibers

As mentioned in chapter 8, cardiac muscle fibers function much like those of skeletal muscles. In cardiac muscle, however, the fibers are interconnected in branching networks that spread in all directions through the heart. When any portion of this net is stimulated, an impulse travels to all of its parts, and the whole structure contracts as a unit.

A mass of merging cells that act as a unit is called a **functional syncytium.** There are two such structures in the heart—one in the atrial walls and another in the ventricular walls. These masses of muscle fibers are separated from each other by portions of the heart's fibrous skeleton, except for a small area in the right atrial floor. In this region, the ***atrial syncytium*** and the ***ventricular syncytium*** are connected by fibers of the cardiac conduction system.

1. Describe a cardiac cycle.
2. What causes heart sounds?
3. What is meant by a functional syncytium?

Cardiac Conduction System

Throughout the heart are clumps and strands of specialized cardiac muscle tissue whose fibers contain only a few myofibrils. Instead of contracting, these parts initiate and distribute impulses (cardiac impulses) throughout the myocardium. They comprise the **cardiac conduction system,** which functions to coordinate the events occurring during the cardiac cycle (fig. 15.11).

A key portion of this conduction system is called the **sinoatrial node** (S-A node). It consists of a small mass of specialized muscle tissue just beneath the epicardium. It is located in the posterior wall of the right atrium, below the opening of the superior vena cava, and its fibers are continuous with those of the atrial syncytium.

The cells of the S-A node have the ability to excite themselves. That is, without being stimulated by nerve fibers or any other outside agents, these cells initiate impulses that spread into the myocardium and stimulate the cardiac muscle fibers to contract. Furthermore, this activity is rhythmic. The S-A node initiates one impulse after another, seventy to eighty times a minute. Thus, it is responsible for the rhythmic contractions of the heart and is often called the **pacemaker.**

As a cardiac impulse travels from the S-A node into the atrial syncytium, the right and left atria contract almost simultaneously. Instead of passing directly into the ventricular syncytium, which is separated from the atrial syncytium by the fibrous skeleton of the heart, the cardiac impulse passes along fibers of the conduction system that are continuous with atrial muscle fibers. These conducting fibers lead to a mass of specialized muscle tissue called the **atrioventricular node** (A-V node). This node, located in the floor of the right atrium near the septum between the atria (interatrial septum) and just beneath the endocardium, provides the only normal conduction pathway between the atrial and ventricular syncytia.

The fibers that conduct the cardiac impulse into the A-V node (junctional fibers) have very small diameters, and because small fibers conduct impulses slowly, they cause the impulse to be delayed. The impulse is delayed still more as it travels through the A-V node, and this delay allows time for the atria to empty and the ventricles to fill with blood.

Once the cardiac impulse reaches the other side of the A-V node, it passes into a group of large fibers that make up the **A-V bundle** (bundle of His), and the impulse moves rapidly through them. The A-V bundle enters the upper part of the interventricular septum, and divides into right and left branches that lie just beneath the endocardium. About halfway down the septum, the branches give rise to the enlarged **Purkinje fibers.**

The Purkinje fibers spread from the interventricular septum, into the papillary muscles, which project inward from the ventricular walls, and then continue downward to the apex of the heart. There, they curve around the tips of the ventricles and pass upward over the lateral walls of these chambers. Along the way, the Purkinje fibers give off many small branches, which become continuous with cardiac muscle fibers.

The muscle fibers in the ventricular walls are arranged in irregular whorls, so that when they are stimulated by the impulses on the Purkinje fibers, the ventricular walls contract with a twisting motion (fig. 15.12). This action squeezes, or wrings, the blood out of the ventricular chambers and forces it into the arteries.

1. What kinds of tissues make up the cardiac conduction system?
2. How is a cardiac impulse initiated?
3. How is this impulse transmitted from the right atrium to the other heart chambers?

Figure 15.12
The muscle fibers within the ventricular walls are arranged in patterns of whorls.

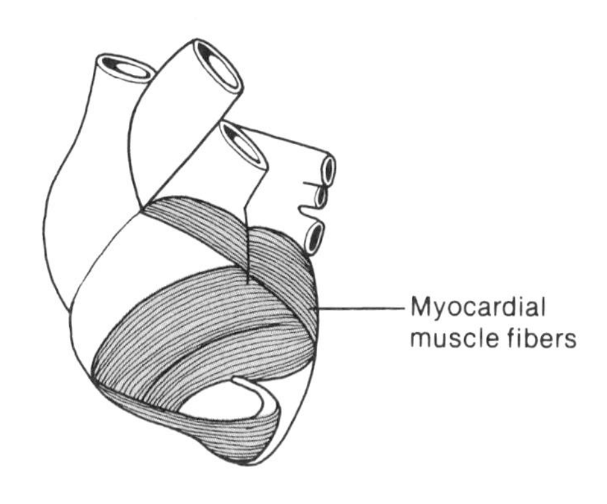

Certain cardiac tissue other than the S-A node can also function as a pacemaker. For example, if the S-A node is damaged, the impulses originating in the A-V node may travel upward into the atrial myocardium and downward into the ventricular walls and stimulate them to contract. Although the S-A node usually initiates 70–80 heartbeats per minute, under the influence of the A-V node, the rate may be 40–60 beats per minute. Similarly, the Purkinje fibers can initiate cardiac impulses, causing the heart to beat 15–40 times per minute.

A patient who has such a disorder of the cardiac conduction system might be treated with an *artificial pacemaker.* This device includes an electrical pulse generator and a lead wire that communicates with a portion of the myocardium. The pulse generator contains a permanent battery that serves as an energy source and a microprocessor that can sense the cardiac rhythm. The microprocessor can also signal the heart to increase or decrease its rate of contraction, as is needed.

An artificial pacemaker can be surgically implanted beneath the patient's skin, and its functions can be adjusted from the outside by means of an external programmer.

Electrocardiogram

A recording of the electrical changes that occur in the *myocardium* during a cardiac cycle is called an **electrocardiogram** (ECG). These changes result from the depolarization and repolarization associated with the contraction of muscle fibers. Because the body fluids can conduct electrical currents, such changes can be detected on the surface of the body.

To record an ECG, metal electrodes are placed at certain locations on the skin. These electrodes are connected by wires to an instrument that responds to very weak electrical changes by causing a pen or stylus to mark on a moving strip of paper. When the instrument is operating, up-and-down movements of the pen correspond to electrical changes occurring as a result of myocardial activity.

Figure 15.13
(*a*) A normal ECG; (*b*) In an ECG pattern, the P wave results from a depolarization of the atria, the QRS complex results from a depolarization of the ventricles, and the T wave results from a repolarization of the ventricles.

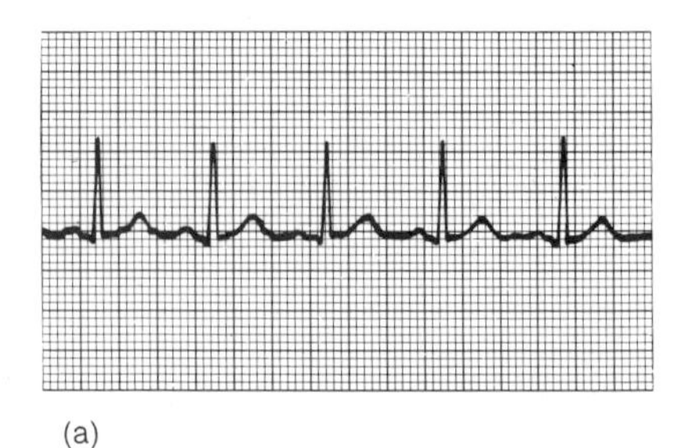

(a)

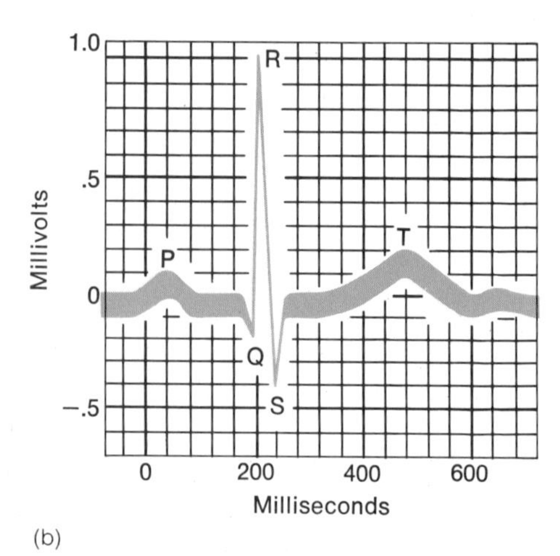

(b)

Because the paper moves past the pen at a known rate, the distance between pen deflections can be used to measure the time elapsing between the various phases of the cardiac cycle.

As figure 15.13*a* illustrates, the normal ECG pattern includes several deflections, or *waves*, during each cardiac cycle. Between cycles, the muscle fibers remain polarized, and no detectable electrical changes occur. Consequently, the pen does not move but creates a baseline as the paper passes through the instrument. When the S-A node triggers a cardiac impulse, however, the atrial fibers are stimulated to depolarize, and an electrical change occurs. As a result, the pen is deflected, and when this electrical change is completed, the pen returns to the base position. This first pen movement produces a *P wave* that is caused by a depolarization of the atrial fibers just before they contract (fig. 15.13*b*).

When the cardiac impulse reaches the ventricular fibers, they are stimulated to depolarize rapidly. Because the ventricular walls are much more extensive than those of the atria, the amount of electrical change is greater, and the pen is deflected to a greater degree than before. As before, when the electrical change is completed, the pen returns to the baseline, leaving a mark called the *QRS complex*, which usually consists of a *Q wave*, an *R wave*, and an *S wave*. This complex appears just prior to the contraction of the ventricular walls.

Near the end of the ECG pattern, the pen is deflected once again, producing a *T wave*. This wave is caused by the electrical changes occurring as the ventricular muscle fibers become repolarized relatively slowly. The record of the atrial repolarization is missing from the pattern because the atrial fibers repolarize at the same time that the ventricular fibers depolarize. The recording of the atrial repolarization is thus obscured by the QRS complex.

ECG patterns are especially important because they allow a physician to assess the heart's ability to conduct impulses and, thus, to judge its condition. For example, the time period between the beginning of a P wave and the beginning of a QRS complex (*P-Q* or *P-R interval*) indicates how long it takes for the cardiac impulse to travel from the S-A node through the A-V node and into the ventricular walls. If ischemia or other problems involving the fibers of the A-V conduction pathways are present, this P-Q interval may increase. Similarly, if the Purkinje fibers are injured, the duration of the QRS complex may increase, because it may take longer for an impulse to spread throughout the ventricular walls.

1. What is an electrocardiogram?
2. What cardiac event is represented by the P wave? By the QRS complex? By the T wave?

Regulation of the Cardiac Cycle

The primary function of the heart is to pump the blood to the body cells, and when the needs of these cells change, the quantity of the blood pumped must change also. For example, during strenuous exercise, the amount of blood required by the skeletal muscles increases greatly, and the rate of the heartbeat increases in response to this need. Since the S-A node normally controls the heart rate, changes in this rate often involve factors that affect the pacemaker. These include the motor impulses carried on the parasympathetic and sympathetic nerve fibers (see chapter 9).

The parasympathetic fibers that supply the heart arise from neurons in the medulla oblongata (fig. 15.14). Most of these fibers branch to the S-A and A-V nodes. When the nerve impulses reach their endings, these fibers secrete acetylcholine, which causes a decrease in S-A and A-V nodal activity. As a result, the rate of heartbeat decreases.

The parasympathetic fibers seem to carry impulses continually to the S-A and A-V nodes, and these impulses impose a braking action on the heart. Con-

Figure 15.14
The activities of the S-A and A-V nodes can be altered by autonomic nerve impulses.

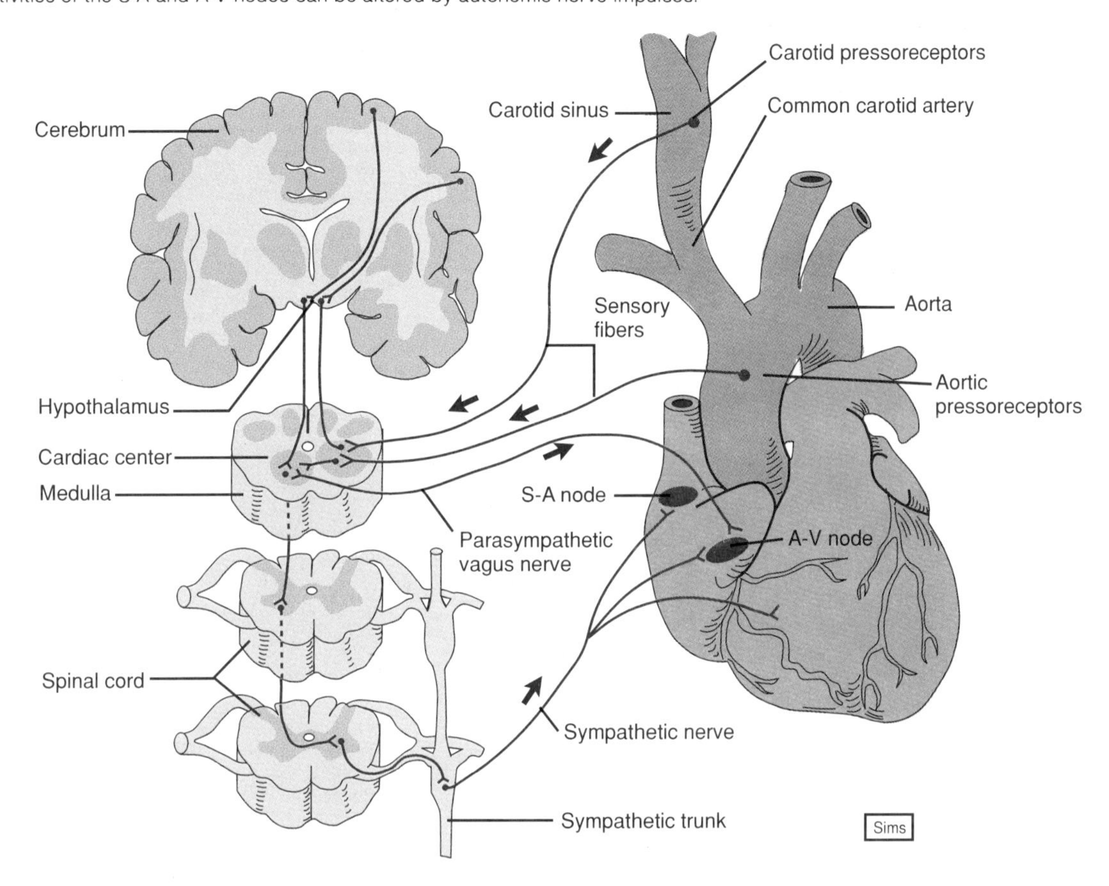

sequently, parasympathetic activity can cause the heart rate to change in either direction. An increase in the impulses causes a slowing of the heart, and a decrease in the impulses releases the parasympathetic brake and allows the heartbeat to increase.

Sympathetic fibers also reach the heart and join the S-A and A-V nodes as well as other areas of the atrial and ventricular myocardium. The endings of these fibers secrete norepinephrine, and this substance causes an increase in the rate and force of myocardial contractions.

A normal balance between the inhibitory effects of the parasympathetic fibers and the excitatory effects of the sympathetic fibers is controlled by the *cardiac center* of the medulla oblongata. This center receives sensory impulses from various parts of the circulatory system and relays motor impulses to the heart in response.

For example, receptors that are sensitive to being stretched are located in certain regions of the aorta (aortic sinus and aortic arch) and in the carotid arteries (carotid sinuses) (fig. 15.14). These receptors, called *pressoreceptors* (baroreceptors), can detect changes in the blood pressure. For example, if the pressure rises, the pressoreceptors are stretched, and they signal the cardiac center in the medulla. In response, the medulla sends *parasympathetic* motor impulses to the heart, causing the heart rate and force of contraction to decrease. This response also causes the blood pressure to drop toward the normal level.

The cardiac control center can also be influenced by impulses from the cerebrum or hypothalamus. Such impulses may cause the heart rate to decrease, as occurs when a person faints following an emotional upset; or they may cause the heart rate to increase during a period of anxiety.

Two other factors that influence the heart rate are temperature change and the presence of various ions. Heart action is increased by a rising body temperature and is decreased by abnormally low body temperature. Consequently, a patient's body temperature is sometimes deliberately lowered (hypothermia) to slow the heart during surgery.

Of the ions that influence heart action, the most important are potassium (K^+) and calcium (Ca^{++}) ions. An excess of *potassium ions* (hyperkalemia), for example, results in a decrease in the rate and force of contractions. If the potassium concentration drops below normal (hypokalemia), the heart may develop a serious abnormal rhythm (arrhythmia).

Figure 15.15
(*a*) The wall of an artery and (*b*) the wall of a vein.

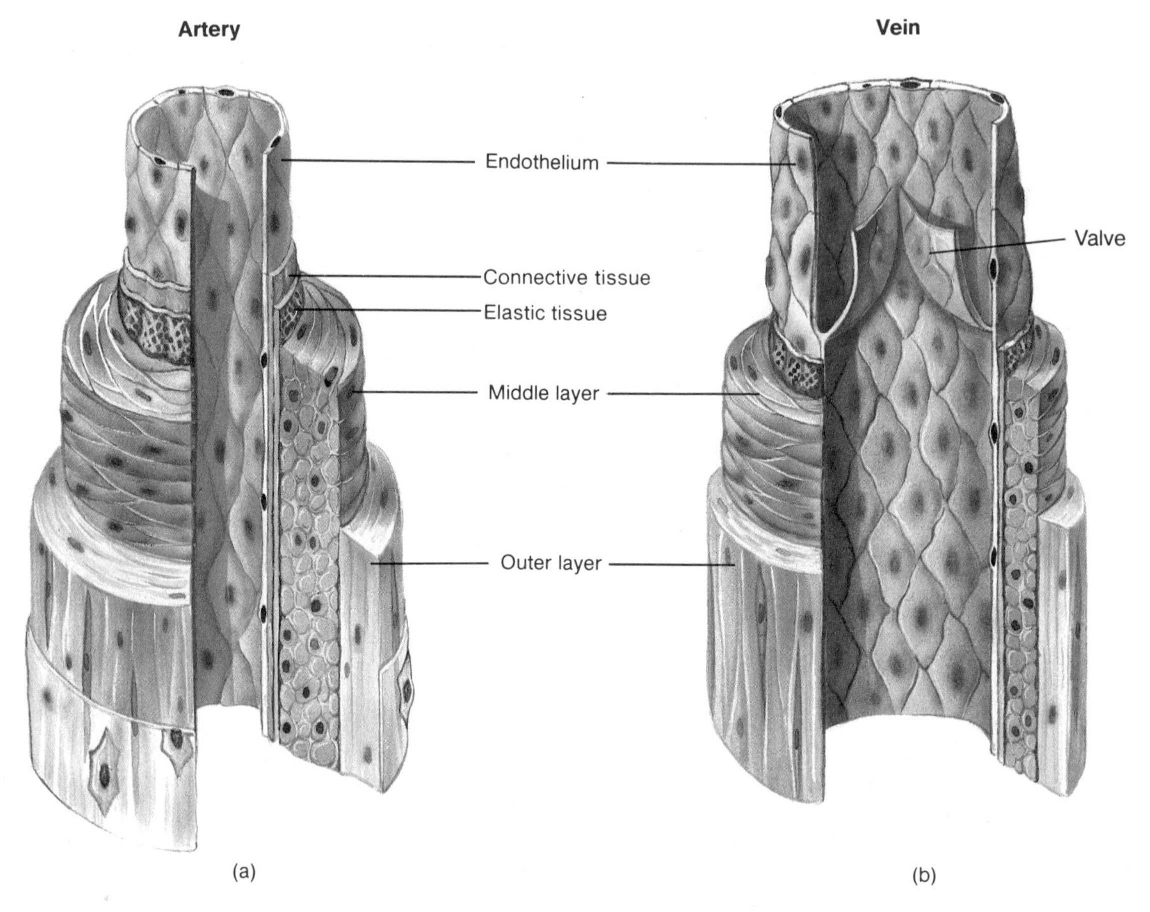

Excessive *calcium ions* (hypercalcemia) cause increased heart actions, and there is a danger that the heart will undergo a prolonged contraction. Conversely, low calcium concentration (hypocalcemia) depresses heart action.

Although slight variations in heart actions sometimes occur normally, marked changes in the usual rate or rhythm may suggest cardiovascular disease. Such abnormal actions, termed cardiac *arrhythmias,* include the following:

1. *Tachycardia*—an abnormally fast heartbeat, usually over 100 beats per minute.
2. *Bradycardia*—a slow heart rate, usually less than 60 beats per minute.
3. *Flutter*—a very rapid heart rate, such as 250–350 contractions per minute.
4. *Fibrillation*—characterized by rapid heart actions but, unlike flutter, the accompanying contractions are *uncoordinated.* In fibrillation, small regions of the myocardium contract and relax independently of all the other areas. As a result, the myocardium fails to contract as a whole, and the walls of the fibrillating chambers are completely ineffective in pumping the blood. Although a person may survive atrial fibrillation, because the venous blood pressure may continue to force the blood into the ventricles, ventricular fibrillation is very likely to cause death unless the heart can be defibrillated within a few minutes.

1. How do parasympathetic and sympathetic impulses help to control the heart rate?
2. How do changes in body temperature affect the heart rate?
3. Describe the effects of abnormal concentrations of potassium and calcium ions on the heart.

Blood Vessels

The vessels of the cardiovascular system form a closed circuit of tubes that carry the blood from the heart to the body cells and back again. These vessels include arteries, arterioles, capillaries, venules, and veins.

Arteries and Arterioles

Arteries are strong, elastic vessels that are adapted for carrying the blood away from the heart under relatively high pressure. These vessels subdivide into progressively thinner tubes and eventually give rise to fine branches called **arterioles.**

The wall of an artery consists of three distinct layers, shown in figure 15.15. The innermost layer is composed of simple squamous epithelium, called *endothelium,* resting on a connective tissue membrane, which is rich in elastic and collagenous fibers.

Atherosclerosis

A Current Topic

It is estimated that nearly half of all deaths in the United States are due to the arterial disease called *atherosclerosis.* In this condition, fatty materials, particularly cholesterol, accumulate on the inner walls of certain arteries (fig. 14.16). Such deposits are called *plaque,* and as they develop, they protrude into the lumens of the blood vessels and interfere with blood flow. Furthermore, plaque creates a surface that can initiate blood clot formation. As a result, persons with atherosclerosis may develop thrombi or emboli that cause blood deficiencies (ischemia) or tissue death (necrosis) downstream from the obstructions.

The walls of affected arteries tend to undergo degenerative changes during which they lose their elasticity and become hardened (sclerotic). This stage of the disease is called *arteriosclerosis,* and when it occurs, there is danger that a sclerotic blood vessel will rupture under the force of blood pressure.

Although the cause of atherosclerosis is not completely understood, it is often associated with diets containing excessive amounts of saturated fats, elevated blood pressure, tobacco smoking, obesity, and lack of physical exercise. Emotional and genetic factors may also increase susceptibility to the development of atherosclerosis.

Arteries that are obstructed with atherosclerotic plaque are sometimes treated using *percutaneous transluminal angioplasty.* In this procedure, a thin plastic catheter is passed through a tiny incision in the skin and into the lumen of the affected blood vessel. The catheter has a tiny deflated balloon at its tip, and it is pushed along the vessel and into the region of the obstruction. When it is in position, the balloon is inflated with relatively high pressure for a few minutes. The inflating balloon compresses the atherosclerotic plaque against the arterial wall, increasing the diameter of the arterial lumen, and allowing an increased volume of blood flow.

Another procedure for treating arterial obstruction is *bypass graft surgery.* In this case, a portion of a vein is removed from some part of the patient's body and is surgically connected between a healthy artery and the affected artery at a point beyond its obstruction. This allows blood from the healthy artery to bypass the narrowed region of the affected artery and supply the tissues downstream.

More recently, laser energy has been used to destroy atherosclerotic plaque and form channels through arterial obstructions, thus increasing blood flow. In this procedure, called *laser angioplasty,* the light energy of a laser is transmitted through a bundle of optical fibers that has been passed through a small incision in the skin and into the lumen of an obstructed artery.

In addition to separating the flowing blood from the blood vessel wall, endothelium helps prevent blood clotting by secreting substances that inhibit platelet aggregation. Endothelium may also play a role in regulating local blood pressure by secreting substances that can cause blood vessel dilation and other substances that can cause blood vessel constriction.

The middle layer makes up the bulk of the arterial wall. It includes smooth muscle fibers, which encircle the tube, and a thick layer of elastic connective tissue.

The outer layer is relatively thin and consists chiefly of connective tissue with irregularly arranged elastic and collagenous fibers. This layer attaches the artery to the surrounding tissues.

The smooth muscles in the walls of arteries and arterioles are innervated by the sympathetic branches of the autonomic nervous system. Impulses on these *vasomotor* fibers cause the smooth muscles to contract, reducing the diameter of the vessels. This action is called **vasoconstriction.** If such vasomotor impulses are inhibited, the muscle fibers relax and the diameter of the vessels increases. In this case, the vessels are said to undergo **vasodilation.** Changes in the diameters of arteries and arterioles greatly influence the flow and pressure of the blood.

Figure 15.16
Small arterioles have some smooth muscle fibers in their walls; capillaries lack these fibers.

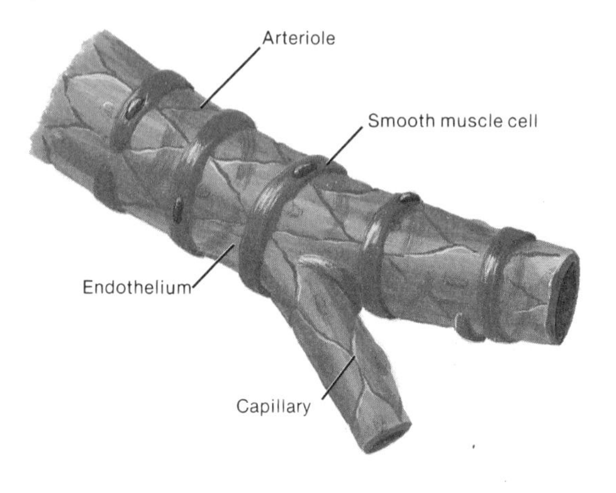

Arterioles, which are microscopic continuations of arteries, join capillaries. Although the walls of the larger arterioles have three layers, similar to those of arteries, these walls become thinner and thinner as the arterioles approach the capillaries. The wall of a very small arteriole consists only of an endothelial lining and some smooth muscle fibers, surrounded by a small amount of connective tissue (fig. 15.16).

1. Describe the wall of an artery.
2. What is the function of the smooth muscle in the arterial wall?
3. How is the structure of an arteriole different from that of an artery?

Capillaries

Capillaries are the smallest blood vessels. They form connections between the smallest arterioles and the smallest venules. Capillaries are essentially extensions of the inner linings of these larger vessels in that their walls consist of endothelium—a single layer of squamous epithelial cells (fig. 15.16). These thin walls form the semipermeable membranes through which substances in the blood are exchanged for substances in the tissue fluid surrounding body cells.

The openings (pores) in capillary walls are thin slits that occur where two adjacent endothelial cells overlap. The sizes of such openings, and consequently the permeability of the capillary walls, vary from tissue to tissue. For example, the capillaries in muscle tissues are less permeable than those of the liver, spleen, or bone marrow.

The density of the capillaries within the tissues varies directly with the tissues' rates of metabolism. Thus, muscle and nerve tissues, which utilize relatively large quantities of oxygen and nutrients, are richly supplied with capillaries; while tissues of cartilage, the epidermis, and the cornea, whose metabolic rates are very slow, lack capillaries.

The patterns of capillary arrangement also differ in various body parts. For example, some capillaries pass directly from arterioles to venules, but others lead to highly branched networks (fig. 15.17). Such arrangements make it possible for the blood to follow different pathways through a tissue and to meet the varying demands of its cells. During periods of exercise, for example, the blood can be directed into the capillary networks of the skeletal muscles, where the cells are experiencing an increasing need for oxygen and nutrients. At the same time, the blood can bypass some of the capillary nets in the tissues of the digestive tract, where the demand for blood is less critical.

The distribution of blood in the various capillary pathways is regulated mainly by the smooth muscles that encircle the capillary entrances. These muscles form *precapillary sphincters,* which may close a capillary by contracting or open it by relaxing. How precapillary sphincters are controlled is not clear, but they seem to respond to the demands of the cells supplied by their individual capillaries. When the cells are low in concentrations of oxygen and nutrients, the sphincter relaxes; when the cellular needs are met, the sphincter may contract again.

Figure 15.17
Light micrograph of a capillary network.

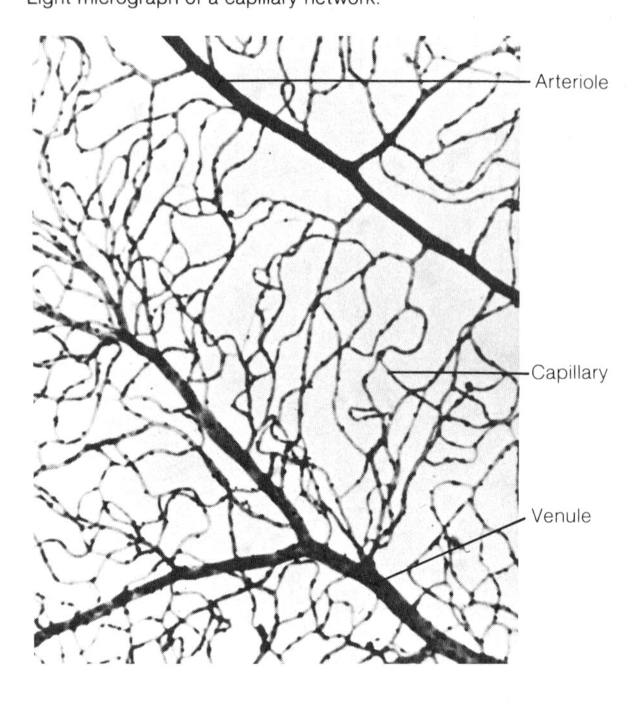

1. Describe the wall of a capillary.
2. What is the function of a capillary?
3. How is blood flow into capillaries controlled?

Exchanges in the Capillaries

The vital function of exchanging gases, nutrients, and metabolic by-products between the blood and the tissue fluid surrounding the body cells occurs in the capillaries. The substances exchanged move through the capillary walls primarily by the processes of diffusion, filtration, and osmosis, described in chapter 3. Of these processes, diffusion provides the most important means of transfer.

It is by diffusion that molecules and ions move from regions where they are more highly concentrated toward regions where they are in lower concentration. Because the blood entering the capillaries of the tissues outside the lungs generally carries relatively high concentrations of oxygen and nutrients, these substances diffuse through the capillary walls and enter the tissue fluid. Conversely, the concentrations of carbon dioxide and other wastes are generally greater in these tissues, and the wastes tend to diffuse into the capillary blood.

The plasma proteins generally remain in the blood because their molecular size is too great to permit diffusion through the membrane pores between the endothelial cells of most capillaries.

Filtration involves the forcing of molecules through a membrane by *hydrostatic pressure.* In the capillaries, the force is provided by the blood pressure generated by contractions of the ventricular walls.

Figure 15.18
Substances leave capillaries because of a net outward filtration pressure; substances enter capillaries because of a net inward force of osmotic pressure.

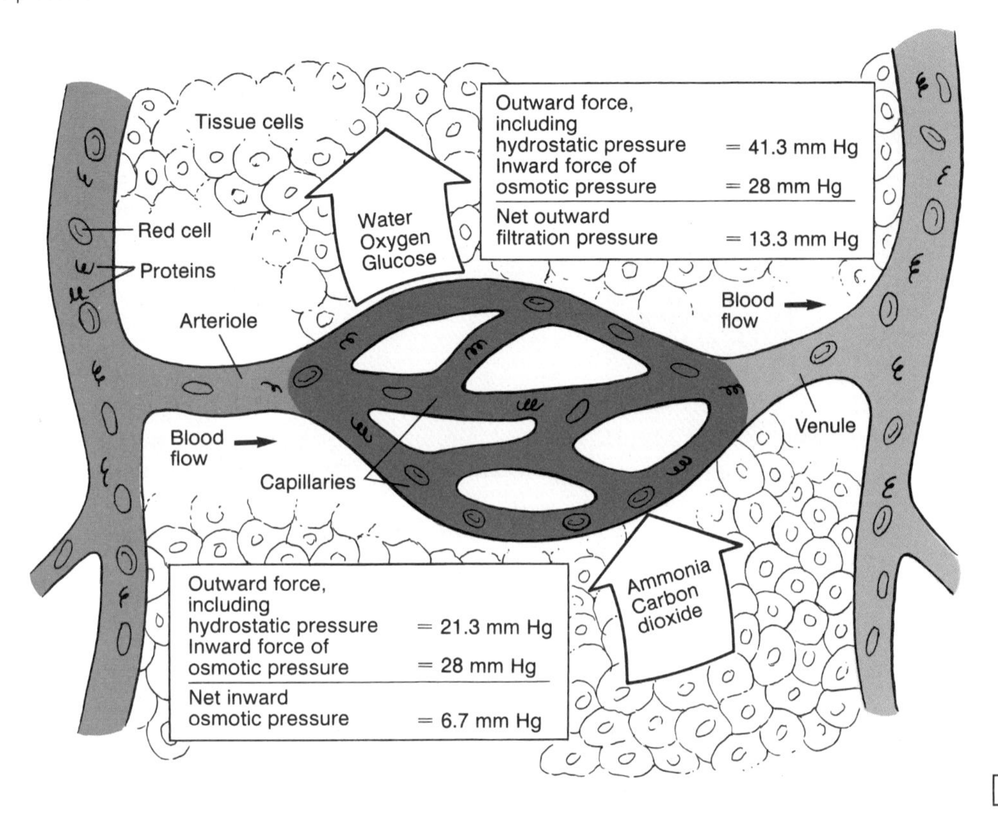

In the brain, the endothelial cells of the capillary walls are more tightly fused than those in other body regions. Consequently, some substances that readily leave the capillaries in other tissues enter the brain tissues only slightly or not at all. This resistance to movement is called the *blood-brain barrier,* and it is of particular interest because it prevents various drugs from entering the brain tissues or cerebrospinal fluid in sufficient concentrations to effectively treat certain diseases.

Blood pressure is also responsible for moving blood through the arteries and arterioles. Pressure tends to decrease, however, as the distance from the heart increases, because friction (peripheral resistance) between the blood and the vessel walls slows the flow. For this reason, blood pressure is greater in the arteries than in the arterioles and greater in the arterioles than in the capillaries. It is similarly greater at the arteriole end of a capillary than at the venule end. Therefore, the filtration effect occurs primarily at the arteriole ends of capillaries.

The plasma proteins, which remain in the capillaries, help to make the *osmotic pressure* of the blood greater (hypertonic) than that of the tissue fluid. Although the capillary blood has a greater osmotic attraction for water than does the tissue fluid, this attraction is overcome by the greater force of the blood pressure. As a result, the net movement of water and dissolved substances is outward at the arteriole end of the capillary by filtration (fig. 15.18).

Since the blood pressure decreases as the blood moves through the capillary, however, the outward filtration force is less than the osmotic pressure of the blood at the venule end. Consequently, there is a net movement of water and dissolved materials into the venule end of the capillary by osmosis.

Normally, more fluid leaves the capillaries than returns to them, and the excess is collected and returned to the venous circulation by *lymphatic vessels.* This mechanism is discussed in chapter 16.

Sometimes unusual events cause an increase in the permeability of the capillaries, and an excessive amount of fluid is likely to enter the spaces between the tissue cells. This may occur, for instance, following a traumatic injury to the tissues or in response to certain chemicals such as *histamine,* either of which increases membrane permeability. In any case, so much fluid may leak out of the capillaries that the lymphatic drainage is overwhelmed, and the affected tissues become swollen (edematous) and painful.

Figure 15.19
Note the structural differences in the cross sections of (*a*) this artery (×100) and (*b*) this vein (×160). What layers in their walls can you identify?

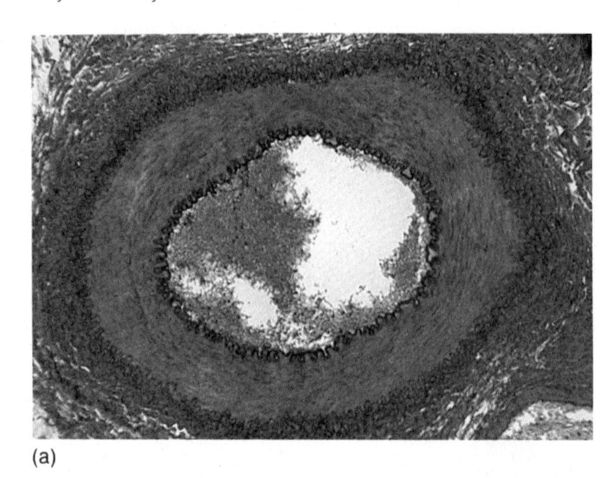
(a)

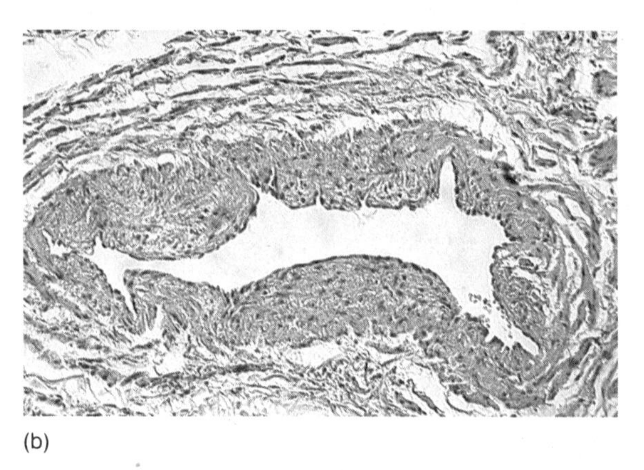
(b)

1. What forces are responsible for the exchange of substances between the blood and tissue fluid?
2. Why is the fluid movement out of a capillary greater at its arteriole end than at its venule end?

Venules and Veins

Venules are microscopic vessels that continue from the capillaries and merge to form **veins.** The veins, which carry the blood back to the atria, follow pathways that roughly parallel those of the arteries.

The walls of veins are similar to those of arteries in that they are composed of three distinct layers (fig. 15.15). Because the middle layer of the venous wall is poorly developed, however, veins have thinner walls and contain less smooth muscle and less elastic tissue than comparable arteries (fig. 15.19).

Many veins, particularly those in the arms and legs, contain flaplike *valves,* which project inward from their linings. These valves, shown in figure 15.20, are usually composed of two leaflets that close if the blood begins to back up in a vein. In other words, the valves aid in returning the blood to the heart, since the valves open as long as the flow is toward the heart, but close if flow is in the opposite direction.

Figure 15.20
(*a*) Venous valves allow blood to move toward the heart, but (*b*) prevent blood from moving away from the heart.

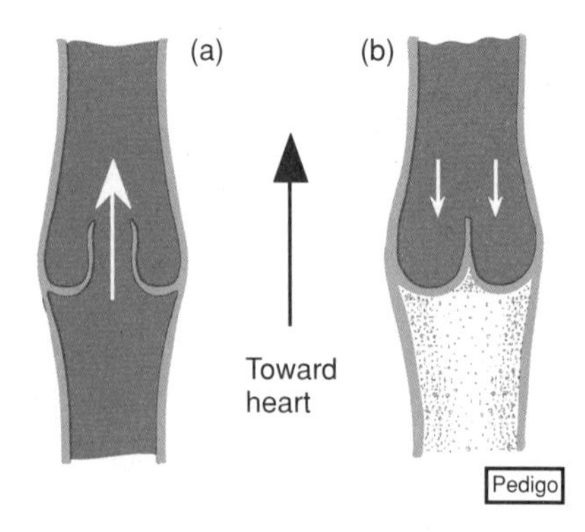

In addition to providing pathways for the blood returning to the heart, the veins function as blood reservoirs, which can be drawn on in times of need. For example, if a hemorrhage accompanied by a drop in arterial blood pressure occurs, the muscular walls of the veins are stimulated reflexly by sympathetic nerve impulses. The resulting venous constrictions help raise the blood pressure. This mechanism ensures a nearly normal blood flow even when as much as 25% of the blood volume has been lost.

The characteristics of the blood vessels are summarized in chart 15.1.

1. How does the structure of a vein differ from that of an artery?
2. How does the venous circulation help to maintain the blood pressure when some blood is lost by a hemorrhage?

Blood Pressure

Blood pressure is the force exerted by the blood against the inner walls of the blood vessels. Although such a force occurs throughout the vascular system, the term *blood pressure* most commonly refers to the arterial pressure in various branches of the aorta.

Arterial Blood Pressure

The arterial blood pressure rises and falls in a pattern corresponding to the phases of the cardiac cycle. That is, when the ventricles contract (ventricular systole), their walls squeeze the blood inside their chambers and force it into the pulmonary trunk and aorta. As a result, the pressures in these arteries and their branches rise sharply. The maximum pressure achieved during ven-

Chart 15.1 Characteristics of blood vessels		
Vessel	**Type of wall**	**Function**
Artery	Thick, strong wall with three layers—an endothelial lining, a middle layer of smooth muscle and elastic tissue, and an outer layer of connective tissue	Carries high pressure blood from the heart to arterioles
Arteriole	Thinner wall than an artery but with three layers; smaller arterioles have an endothelial lining, some smooth muscle tissue, and a small amount of connective tissue	Connects an artery to a capillary; helps to control the blood flow into a capillary by undergoing vasoconstriction or vasodilation
Capillary	Single layer of squamous epithelium	Provides a membrane through which nutrients, gases, and wastes are exchanged between the blood and tissue cells; connects an arteriole to a venule
Venule	Thin wall, less smooth muscle and elastic tissue than in an arteriole	Connects a capillary to a vein
Vein	Thinner wall than an artery but with similar layers; the middle layer is more poorly developed; some with flaplike valves	Carries low pressure blood from a venule to the heart; the valves prevent a backflow of blood; serves as blood reservoir

Figure 15.21
Blood pressure decreases as the distance from the left ventricle increases.

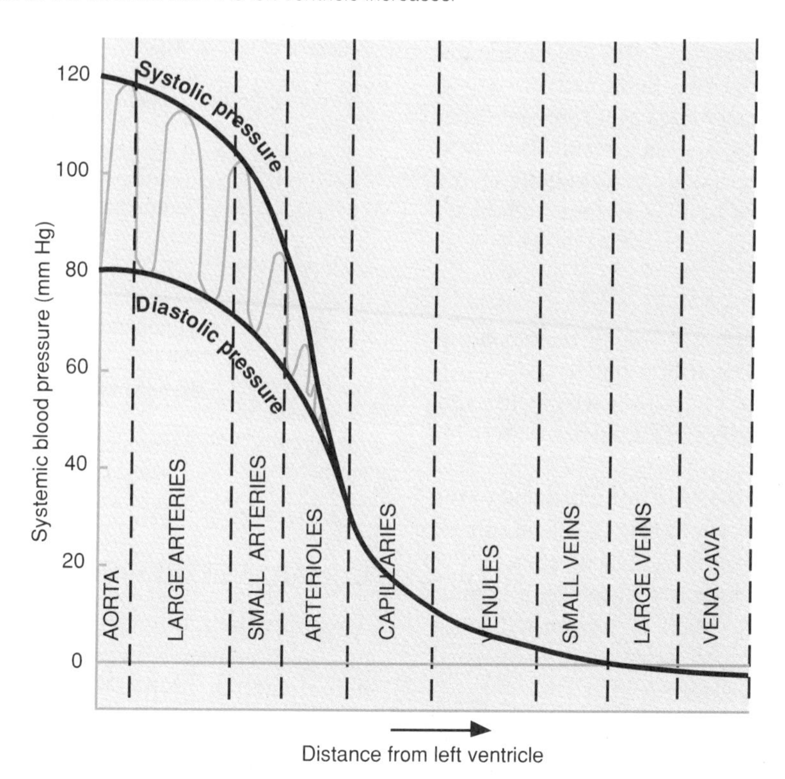

tricular contraction is called the **systolic pressure** (fig. 15.21). When the ventricles relax (ventricular diastole), the arterial pressure drops, and the lowest pressure that remains in the arteries before the next ventricular contraction is termed the **diastolic pressure.**

The surge of blood entering the arterial system during a ventricular contraction causes the elastic walls of the arteries to swell, but the pressure drops almost immediately as the contraction is completed, and the arterial walls recoil. This alternate expanding and recoiling of an arterial wall can be felt as a *pulse* in an artery that runs close to the surface.

1. What is meant by *blood pressure?*
2. Distinguish between systolic and diastolic blood pressure.
3. What causes a pulse in an artery?

Factors That Influence Arterial Blood Pressure

The arterial pressure depends on a variety of factors, including the heart action, blood volume, resistance to flow, and viscosity of the blood.

Heart Action

In addition to creating the blood pressure by forcing the blood into the arteries, the heart action determines how much blood will enter the arterial system with each ventricular contraction. The heart also determines the rate of this fluid output.

The volume of blood discharged from the ventricle with each contraction is called the **stroke volume** and equals about 70 milliliters (ml). The volume discharged from the ventricle per minute is called the **cardiac output.** It is calculated by multiplying the stroke volume by the heart rate in beats per minute. (Cardiac output = stroke volume × heart rate.) Thus, if the stroke volume is 70 ml, and the heart rate is 72 beats per minute, the cardiac output is 5,040 ml per minute.

The blood pressure varies with the cardiac output. If either the stroke volume or the heart rate increases, so does the cardiac output, and as a result, the blood pressure rises. Conversely, if the stroke volume or the heart rate decreases, the cardiac output and blood pressure decrease also.

Blood Volume

The **blood volume** is equal to the sum of the blood cell and plasma volumes in the cardiovascular system. Although the blood volume varies somewhat with age, body size, and sex, it usually remains about 5 liters for adults.

The blood pressure is directly proportional to the volume of blood. Thus, any changes in the blood volume are accompanied by changes in the blood pressure. For example, if the blood volume is reduced by a hemorrhage, the blood pressure drops. If the normal blood volume is restored by a blood transfusion, the normal pressure may be reestablished.

Peripheral Resistance

Friction between the blood and the walls of the blood vessels creates a force called the **peripheral resistance,** which hinders the blood flow. This force must be overcome by the blood pressure if the blood is to continue flowing. Consequently, factors that alter the peripheral resistance cause changes in the blood pressure.

For example, if the smooth muscles in the walls of the arterioles contract, the peripheral resistance of these constricted vessels increases. The blood tends to back up into the arteries supplying the arterioles, and the arterial pressure rises. Dilation of the arterioles has the opposite effect—the peripheral resistance lessens, and the arterial blood pressure drops in response (fig. 15.22).

Viscosity

Viscosity is a physical property of a fluid and is related to the ease with which its molecules slide past one another. A fluid with a high viscosity tends to be sticky like syrup, and one with a low viscosity flows easily like water.

The presence of blood cells and plasma proteins increases the viscosity of the blood. Since the greater the blood's resistance to flowing, the greater the force needed to move it through the vascular system, it is not surprising that the blood pressure rises as the blood viscosity increases and drops as its viscosity decreases.

Although the viscosity of blood normally remains relatively stable, any condition that alters the concentrations of blood cells or plasma proteins may cause changes in viscosity. For example, anemia and hemorrhage may be accompanied by a decreasing viscosity and a consequent drop in the blood pressure. If red blood cells are present in abnormally high numbers (polycythemia), the viscosity increases, and a rise in the blood pressure is likely.

1. What is the relationship between the cardiac output and blood pressure?
2. How does the blood volume affect the blood pressure?
3. What is the relationship between the peripheral resistance and blood pressure?

Control of Blood Pressure

Two important mechanisms for maintaining normal arterial pressure involve the regulation of the cardiac output and the peripheral resistance.

As was mentioned, the cardiac output depends on the volume of blood discharged from the ventricle with each contraction (stroke volume) and the rate of the heartbeat.

For example, the volume of blood entering the ventricle affects the stroke volume. As the blood enters, myocardial fibers in the ventricular wall are mechanically stretched. Within limits, the greater the length of these fibers, the greater the force with which they contract. This relationship between fiber length and force of contraction is called *Starling's law of the heart.* Because of it, the heart can respond to the immediate

Figure 15.22
(*a*) Relaxation of the smooth muscle in the arteriole wall produces vasodilation, while (*b*) contraction of the smooth muscle causes vasoconstriction.

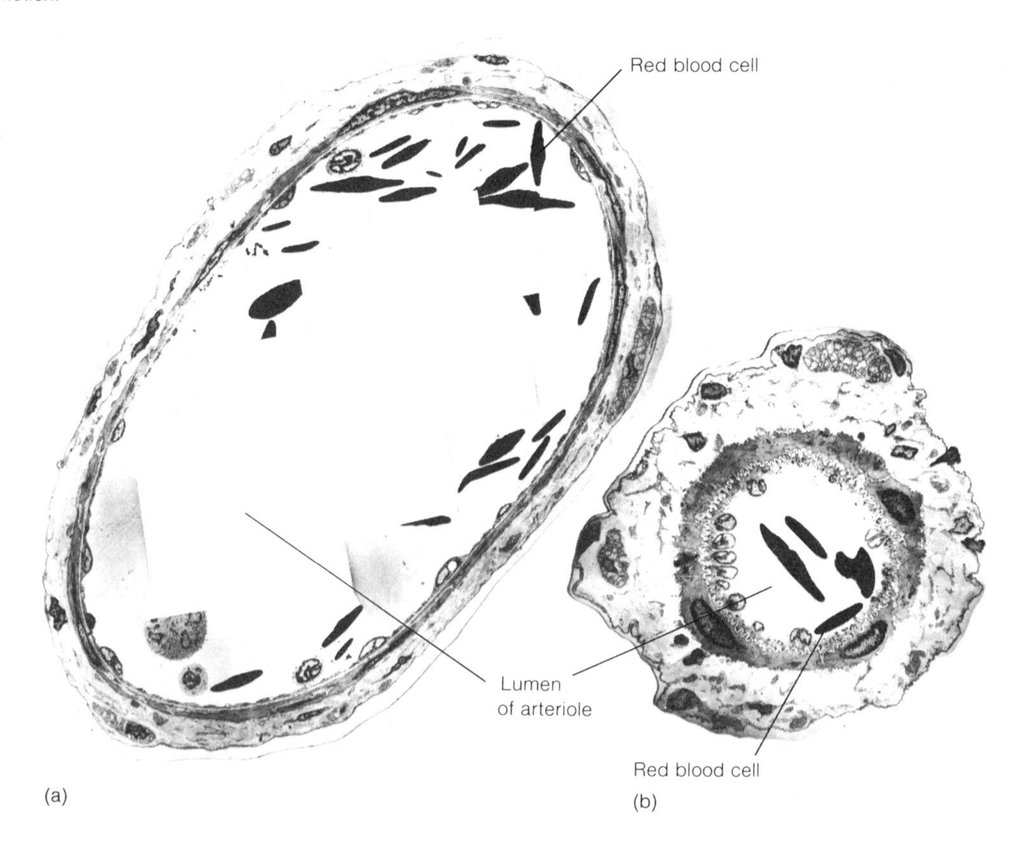

demands placed on it by the varying quantities of blood that return from the venous system. In other words, the more blood that enters the heart from the veins, the stronger the ventricular contraction, the greater the stroke volume, and the greater the cardiac output. This mechanism ensures that the volume of blood discharged from the heart is equal to the volume entering its chambers.

The neural regulation of heart rate was described previously in this chapter. To review, pressoreceptors located in the walls of the aorta and carotid arteries are sensitive to changes in the blood pressure. If the arterial pressure rises, nerve impulses travel from the receptors to the cardiac center of the medulla oblongata. This center relays parasympathetic impulses to the S-A node in the heart, and the rate of heartbeat decreases in response. As a result of this reflex, the cardiac output is reduced, and the blood pressure falls toward the normal level.

Conversely, if the arterial blood pressure drops, a reflex involving sympathetic impulses to the S-A node is initiated, and the heart beats faster. This response increases the cardiac output, and the arterial pressure rises.

Other factors that cause an increase in heart rate and a rise in blood pressure include the presence of chemicals such as epinephrine, emotional responses such as fear and anger, physical exercise, and an increase in body temperature.

Peripheral resistance is regulated primarily by changes in the diameters of arterioles. Because blood vessels with smaller diameters offer a greater resistance to blood flow, factors that cause arteriole vasoconstriction bring about an increase in peripheral resistance, and factors causing vasodilation produce a decrease in resistance.

The *vasomotor center* of the medulla oblongata continually sends sympathetic impulses to the smooth muscles in the arteriole walls. As a result, these muscles are kept in a state of tonic contraction, which helps to maintain the peripheral resistance associated with a normal blood pressure. Because the vasomotor center is responsive to changes in the blood pressure, it can cause an increase in the peripheral resistance by increasing its outflow of sympathetic impulses, or it can cause a decrease in such resistance by decreasing its sympathetic outflow. In the latter case, the vessels undergo vasodilation as sympathetic stimulation is decreased.

Hypertension

A Current Topic

Hypertension, or high blood pressure, is characterized by a persistently elevated arterial blood pressure, and is one of the more common diseases of the cardiovascular system.

When the cause of hypertension is unknown, as is often the case, it may be called *essential hypertension* (also primary or idiopathic hypertension). Sometimes the elevated pressure is related to another problem, such as arteriosclerosis. Arteriosclerosis is accompanied by decreasing elasticity of the arterial walls and narrowing of the lumens of these vessels. Both of these effects promote an increase in blood pressure.

The consequences of prolonged, uncontrolled hypertension can be very serious. For example, the left ventricle must contract with greater force than usual in order to discharge a normal volume of blood against increased arterial pressure. As a result of the increased work load, the myocardium tends to thicken, and the heart enlarges. At first, such hypertrophy of the myocardium may benefit cardiac function. However, as the muscle tissue increases, the coronary blood vessels may not increase to the same extent. Consequently, the coronary vessels may become less and less able to supply the myocardium with sufficient blood. This change may be accompanied by degeneration of muscle fibers and a replacement of muscle with fibrous tissue. In time, the enlarged heart is weakened and may fail.

Hypertension also enhances the development of atherosclerosis. As a result, various arteries, including those of the coronary circuit, tend to accumulate plaque that may cause arterial occlusions. The patient may then suffer a *coronary thrombosis* or *coronary embolism.* Similar changes in the arteries of the brain increase the chances of a *cerebral vascular accident* (CVA), which is due to a cerebral thrombosis, embolism, or hemorrhage, and is more commonly called a stroke.

The treatment of hypertension may include exercising regularly, controlling body weight, reducing stress, and using foods that are low in sodium and high in potassium. It also may involve the use of drugs, such as diuretics that reduce the volume of body fluids, and/or substances that inhibit the activity of sympathetic neurotransmitters (see chapter 9).

For instance, whenever the arterial blood pressure suddenly rises, the pressoreceptors in the aorta and carotid arteries signal the vasomotor center, and the sympathetic outflow to the arteriole walls is reduced. The resulting vasodilation causes a decrease in the peripheral resistance, and the blood pressure falls toward the normal level.

Chemical substances, including carbon dioxide, oxygen, and hydrogen ions, also influence peripheral resistance by affecting the smooth muscles in the walls of arterioles and the actions of precapillary sphincters. For example, an increasing P_{CO_2}, a decreasing P_{O_2}, and a decreasing pH cause relaxation of these muscles and a consequent drop in the blood pressure.

1. What factors affect cardiac output?
2. What is the function of the pressoreceptors in the walls of the aorta and carotid arteries?
3. How does the vasomotor center control the diameter of the arterioles?

Venous Blood Flow

Blood pressure decreases as the blood moves through the arterial system and into the capillary networks. In fact, little pressure remains at the venule ends of capillaries (fig. 15.21). Therefore, blood flow through the venous system is not the direct result of heart action, but depends on other factors, such as skeletal muscle contraction and breathing movements.

For example, when skeletal muscles contract, they thicken and press on nearby vessels, squeezing the blood inside. As was mentioned, many veins, particularly those in the arms and legs, contain flaplike valves. These valves offer little resistance to blood flowing toward the heart, but they close if the blood moves in the opposite direction (fig. 15.20). Consequently, as *skeletal muscles* exert pressure on veins with valves, some blood is moved from one valve section to another. This massaging action of contracting skeletal muscles helps push the blood through the venous system toward the heart.

Respiratory movements provide another means of moving venous blood. During inspiration, the pressure within the thoracic cavity is reduced, as the diaphragm contracts and the rib cage moves upward and outward (see chapter 13). At the same time, the pressure within the abdominal cavity is increased, as the diaphragm presses downward on the abdominal viscera. Consequently, the blood tends to be squeezed out of the abdominal veins and forced into the thoracic veins. Backflow into the legs is prevented by the valves in the veins of the legs.

During exercise, respiratory movements act together with skeletal muscle contractions to increase the return of venous blood to the heart.

Cardiovascular Adjustment to Exercise
A Current Topic

When a person engages in strenuous exercise, a number of changes occur in the circulatory system that help increase the blood flow, and therefore oxygen availability, to the skeletal muscles. For example, when the skeletal muscles are resting, only about one-fifth of their capillaries are open. As exercise begins, oxygen concentration in the active skeletal muscles decreases and, in response, essentially all of the capillaries in the muscles open. At the same time, blood flow into the active muscles greatly increases due to vasodilation stimulated by such factors as sympathetic nerve impulses, decreasing oxygen concentration, increasing carbon dioxide concentration, and decreasing pH. The blood flow to the myocardium and skin also increases.

In certain other parts of the body, sympathetic nerve impulses cause vasoconstriction and, as a result, the blood flow into the abdominal viscera and inactive skeletal muscles is reduced. The blood flow to the brain and kidneys, however, remains nearly unchanged, so that these vital organs continue to receive oxygen and nutrients at the rate required to support their needs.

When the walls of the veins in visceral organs constrict, blood is moved out of these blood reservoirs, adding to the venous return to the heart. This increased venous return is also aided by an increased respiratory rate during exercise.

The heart rate increases in response to decreased parasympathetic stimulation of the S-A and A-V nodes and to sympathetic reflexes triggered by the stimulation of sensory receptors (proprioceptors) in the skeletal muscles and stretch receptors in the lungs. As the volume of blood returning to the heart increases, the ventricular walls are stretched, and this stimulates them to contract with greater force. Consequently, with each heartbeat, the stroke volume increases and the blood pressure rises.

A number of physiological changes occur in response to physical conditioning that may improve athletic performance. These include increases in heart-pumping efficiency, blood volume, blood hemoglobin concentration, and mitochondrial content of muscle fibers. As a result of such changes, oxygen can be delivered to and utilized by muscle tissue more effectively. In the case of athletes such as marathoners, who train for endurance-type performance, the size of the heart and its chambers may enlarge 40% or more, providing for greater cardiac output.

Another mechanism that promotes the return of venous blood involves constriction of the veins. When venous pressure is low, sympathetic reflexes stimulate smooth muscles in the walls of veins to contract. Such *venoconstriction* increases the venous pressure and forces more blood toward the heart.

Also, as mentioned previously, the veins provide a blood reservoir that can adapt its capacity to changes in the blood volume. If some blood is lost and the blood pressure decreases, venoconstriction can help return the blood pressure to normal by forcing blood out of this reservoir.

1. What is the function of the venous valves?
2. How do skeletal muscles and respiratory movements affect venous blood flow?
3. What factors stimulate venoconstriction?

Paths of Circulation

The blood vessels of the cardiovascular system can be divided into two major pathways—a pulmonary circuit and a systemic circuit. The **pulmonary circuit** consists of those vessels that carry the blood from the heart to the lungs and back to the heart. The **systemic circuit** is responsible for carrying the blood from the heart to all other parts of the body and back again (fig. 15.1).

The circulatory pathways described in the following sections are those of an adult. The fetal pathways are somewhat different and are described in chapter 20.

Pulmonary Circuit

The blood enters the pulmonary circuit as it leaves the right ventricle through the *pulmonary trunk*. The pulmonary trunk extends upward and posteriorly from the heart, and at about 5 cm above its origin, it divides into the right and left *pulmonary arteries* (fig. 15.4). These branches penetrate the right and left lungs, respectively. After repeated divisions, they give rise to the arterioles that continue into the capillary networks associated with the walls of the alveoli, where gas exchanges occur between the blood and the air (see chapter 13).

From the pulmonary capillaries, the blood enters the venules, which merge to form small veins. They, in turn, converge to form still larger ones. Four *pulmonary veins,* two from each lung, return the blood to the left atrium, and this completes the vascular loop of the pulmonary circuit.

Systemic Circuit

The freshly oxygenated blood received by the left atrium is forced into the systemic circuit by contraction of the left ventricle. This circuit includes the aorta

Figure 15.23
Principal branches of the aorta (*a.* stands for *artery*).

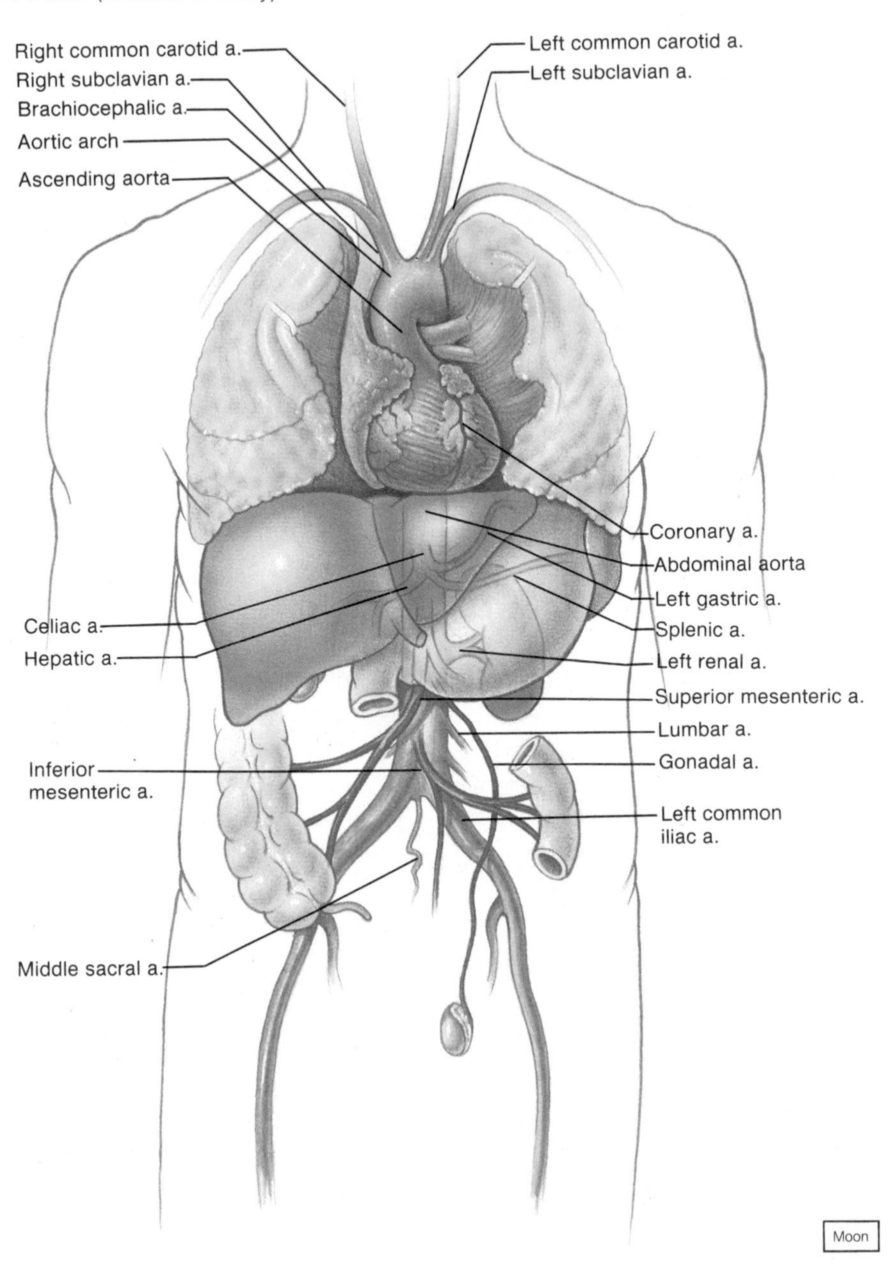

and its branches that lead to all the body tissues, as well as the companion system of veins, which returns the blood to the right atrium.

1. Distinguish between the pulmonary and systemic circuits of the cardiovascular system.
2. Trace a drop of blood through the pulmonary circuit from the right ventricle.

Arterial System

The **aorta** is the largest artery in the body. It extends upward from the left ventricle, arches over the heart to the left, and descends just in front of the vertebral column. Figure 15.23 shows the aorta and its main branches.

Principal Branches of the Aorta

The first portion of the aorta is called the *ascending aorta.* At its base are located the three cusps of the aortic valve, and opposite each cusp is a swelling in the aortic wall called an **aortic sinus.** The right and left *coronary arteries* spring from two of these sinuses (fig. 15.8).

Three major arteries originate from the *arch of the aorta* (aortic arch). They are the **brachiocephalic** (brak″e-o-sĕ-fal′ik) **artery,** the left **common carotid** (kah-rot′id) **artery,** and the left **subclavian** (sub-kla′ve-an) **artery.**

Although the upper part of the *descending aorta* is positioned left of the midline, it gradually moves medially and finally lies directly in front of the vertebral column at the level of the twelfth thoracic vertebra.

The portion of the descending aorta above the diaphragm is known as the **thoracic** (tho-ras′ik) **aorta,** and it gives off numerous small branches to the thoracic wall and thoracic visceral organs.

Below the diaphragm, the descending aorta becomes the **abdominal aorta,** and it gives off branches to the abdominal wall and various abdominal visceral organs.

The abdominal aorta terminates near the brim of the pelvis, where it divides into the right and left *common iliac arteries.* These vessels supply blood to the lower regions of the abdominal wall, the pelvic organs, and the lower extremities.

The names of the major branches of the aorta and the general regions or organs they supply with blood are listed in chart 15.2.

If the aortic wall becomes weakened as a result of injury or disease, a blood-filled dilation or sac may appear in the vessel. This condition is called an *aortic aneurysm,* and it occurs most frequently in the abdominal aorta below the level of the renal arteries. An aneurysm usually increases in size with time, and the swelling is likely to rupture unless the affected portion of the artery is removed and replaced with a vessel graft.

1. Name the portions of the aorta.
2. List the major branches of the aorta.

Chart 15.2 Aorta and its principal branches

Portion of aorta	Major branch	General regions or organs supplied
Ascending aorta	Right and left coronary arteries	Heart
Arch of aorta	Brachiocephalic artery	Right arm, right side of head
	Left common carotid artery	Left side of head
	Left subclavian artery	Left arm
Descending aorta		
Thoracic aorta	Bronchial artery	Bronchi
	Pericardial artery	Pericardium
	Esophageal artery	Esophagus
	Mediastinal artery	Mediastinum
	Posterior intercostal artery	Thoracic wall
Abdominal aorta	Celiac artery	Organs of upper digestive system
	Phrenic artery	Diaphragm
	Superior mesenteric artery	Portions of small and large intestines
	Suprarenal artery	Adrenal gland
	Renal artery	Kidney
	Gonadal artery	Ovary or testis
	Inferior mesenteric artery	Lower portions of large intestine
	Lumbar artery	Posterior abdominal wall
	Middle sacral artery	Sacrum and coccyx
	Common iliac artery	Lower abdominal wall, pelvic organs, and leg

Arteries to the Neck, Head, and Brain

Blood is supplied to parts within the neck, head, and brain through branches of the subclavian and common carotid arteries (fig. 15.24). The main divisions of the subclavian artery to these regions are the vertebral and thyrocervical arteries. The common carotid artery communicates to parts of the head and neck by means of the internal and external carotid arteries.

The **vertebral** (ver′te-bral) **arteries** pass upward through the foramina of the transverse processes of the cervical vertebrae and enter the skull by way of the foramen magnum. Along their paths, these vessels supply blood to the vertebrae and to the ligaments and muscles associated with them.

Within the cranial cavity, the vertebral arteries unite to form a single *basilar artery.* This vessel passes along the ventral brain stem and gives rise to branches leading to the pons, midbrain, and cerebellum.

The **thyrocervical** (thi″ro-ser′vĭ-kal) **arteries** are short vessels that give off branches to the thyroid glands, parathyroid glands, larynx, trachea, esophagus, and pharynx, as well as to various muscles in the neck, shoulder, and back.

The left and right **common carotid arteries** divide to form the internal and external carotid arteries.

The *external carotid artery* courses upward on the side of the head, giving off branches to various structures in the neck, face, jaw, scalp, and base of the skull.

The *internal carotid artery* follows a deep course upward along the pharynx to the base of the skull. Entering the cranial cavity, it provides the major blood supply to the brain.

Near the base of each internal carotid artery is an enlargement called a **carotid sinus.** Like the aortic sinuses, these structures contain pressoreceptors that function in the reflex control of the blood pressure.

Figure 15.24
Main arteries of the head, neck, and brain.

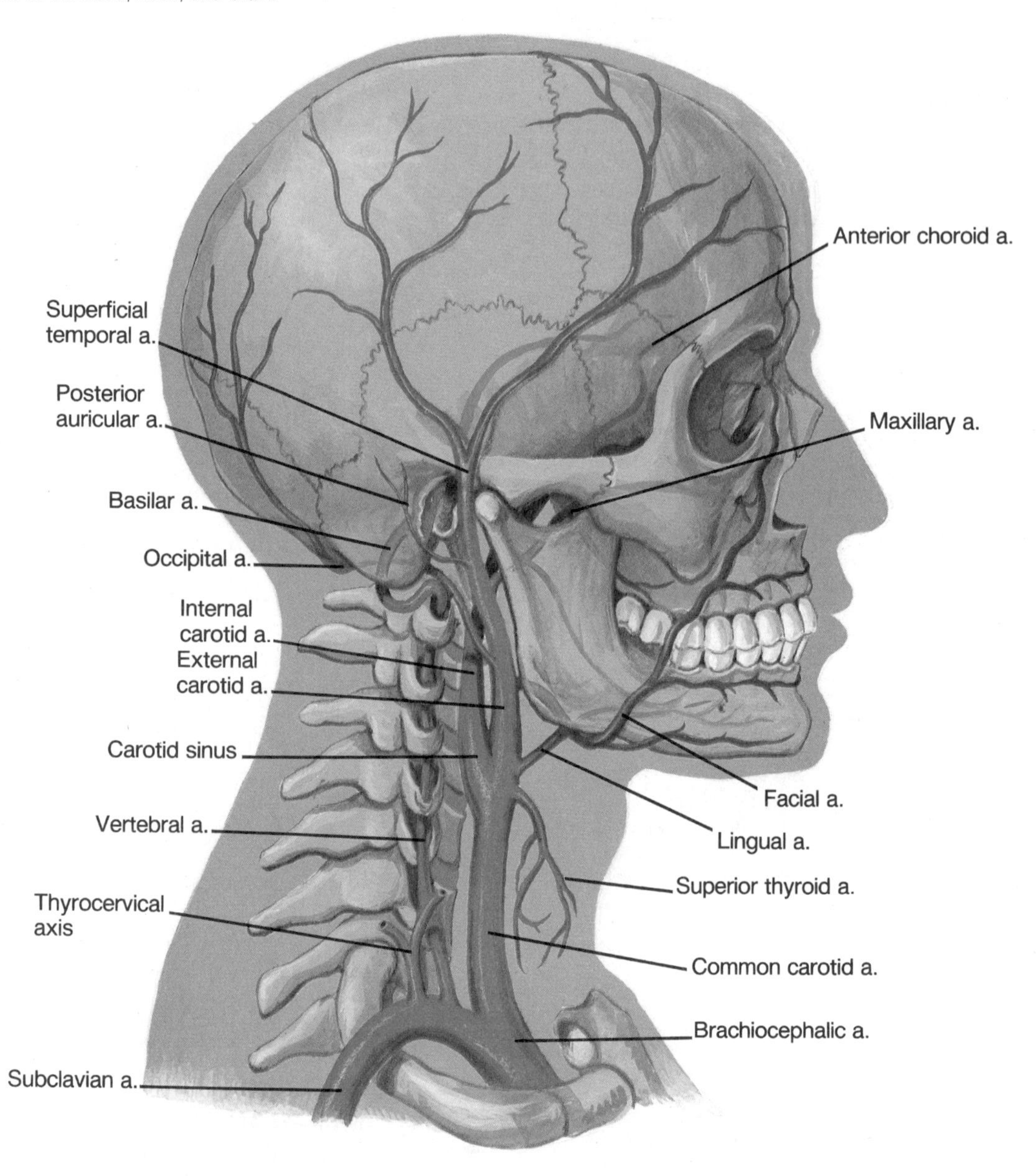

The major branches of the external and internal carotid arteries and the general regions and organs they supply with blood are listed in chart 15.3.

Arteries to the Thoracic and Abdominal Walls

The blood reaches the thoracic wall through several vessels including the following (fig. 15.25):

1. The **internal thoracic artery,** a branch of the subclavian artery, which gives off two *anterior intercostal arteries.* They, in turn, supply the intercostal muscles and mammary glands.
2. The **posterior intercostal** (in″ter-kos′tal) **arteries,** which arise from the thoracic aorta and enter the intercostal spaces. They supply the intercostal muscles, vertebrae, spinal cord, and various deep muscles of the back.

The blood supply to the anterior abdominal wall is provided primarily by branches of the *internal thoracic* and *external iliac arteries.* Structures in the posterior and lateral abdominal wall are supplied by paired vessels originating from the abdominal aorta, including the *phrenic* and *lumbar arteries.*

Arteries to the Shoulder and Arm

The subclavian artery, after giving off branches to the neck, continues into the upper arm (fig. 15.26). It passes between the clavicle and first rib and becomes the axillary artery.

Chart 15.3 Major branches of the external and internal carotid arteries

Artery	Major branch	General region or organs supplied
External carotid artery	Superior thyroid artery	Larynx and thyroid gland
	Lingual artery	Tongue and salivary glands
	Facial artery	Pharynx, chin, lips, and nose
	Occipital artery	Posterior scalp, meninges, and neck muscles
	Posterior auricular artery	Ear and lateral scalp
	Maxillary artery	Teeth, jaw, cheek, and eyelids
	Superficial temporal artery	Parotid salivary gland and surface of the face and scalp
Internal carotid artery	Ophthalmic artery	Eye and eye muscles
	Anterior choroid artery	Choroid plexus and brain

The **axillary** (ak′sĭ-ler′e) **artery** supplies branches to various muscles and other structures in the axilla and chest wall and becomes the brachial artery.

The **brachial** (bra′ke-al) **artery,** which courses along the humerus to the elbow, gives rise to a *deep brachial artery.* Within the elbow, the brachial artery divides into an ulnar and a radial artery.

The **ulnar** (ul′nar) **artery** leads downward on the ulnar side to the forearm and wrist. Some of its branches supply the elbow joint, while others supply blood to muscles in the lower arm.

The **radial** (ra′de-al) **artery** travels along the radial side of the forearm to the wrist, supplying the muscles of the forearm.

At the wrist, the branches of the ulnar and radial arteries join to form an interconnecting network of vessels. Arteries arising from this network supply blood to structures in the wrist, hand, and fingers.

Arteries to the Pelvis and Leg

The abdominal aorta divides to form the **common iliac** (il′e-ak) **arteries** at the level of the pelvic brim, and these vessels provide blood to the pelvic organs, gluteal region, and legs (fig. 15.26). Each common iliac artery divides into an internal and external branch.

Figure 15.25
Arteries that supply the thoracic wall (*m.* stands for *muscle*).

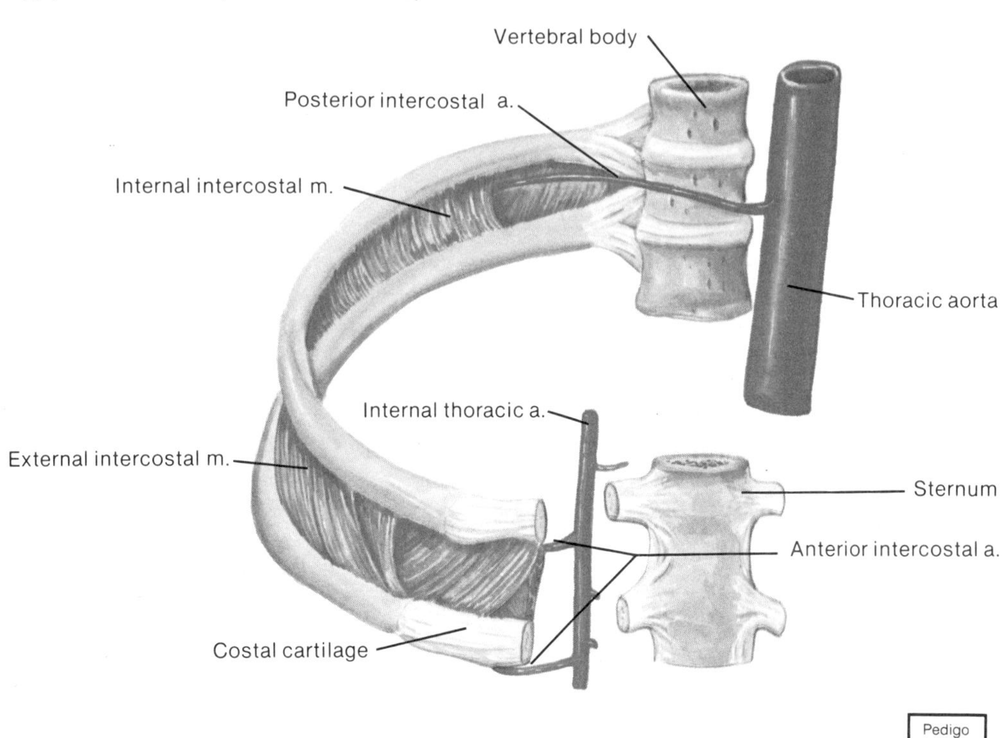

Figure 15.26
Major vessels of the arterial system.

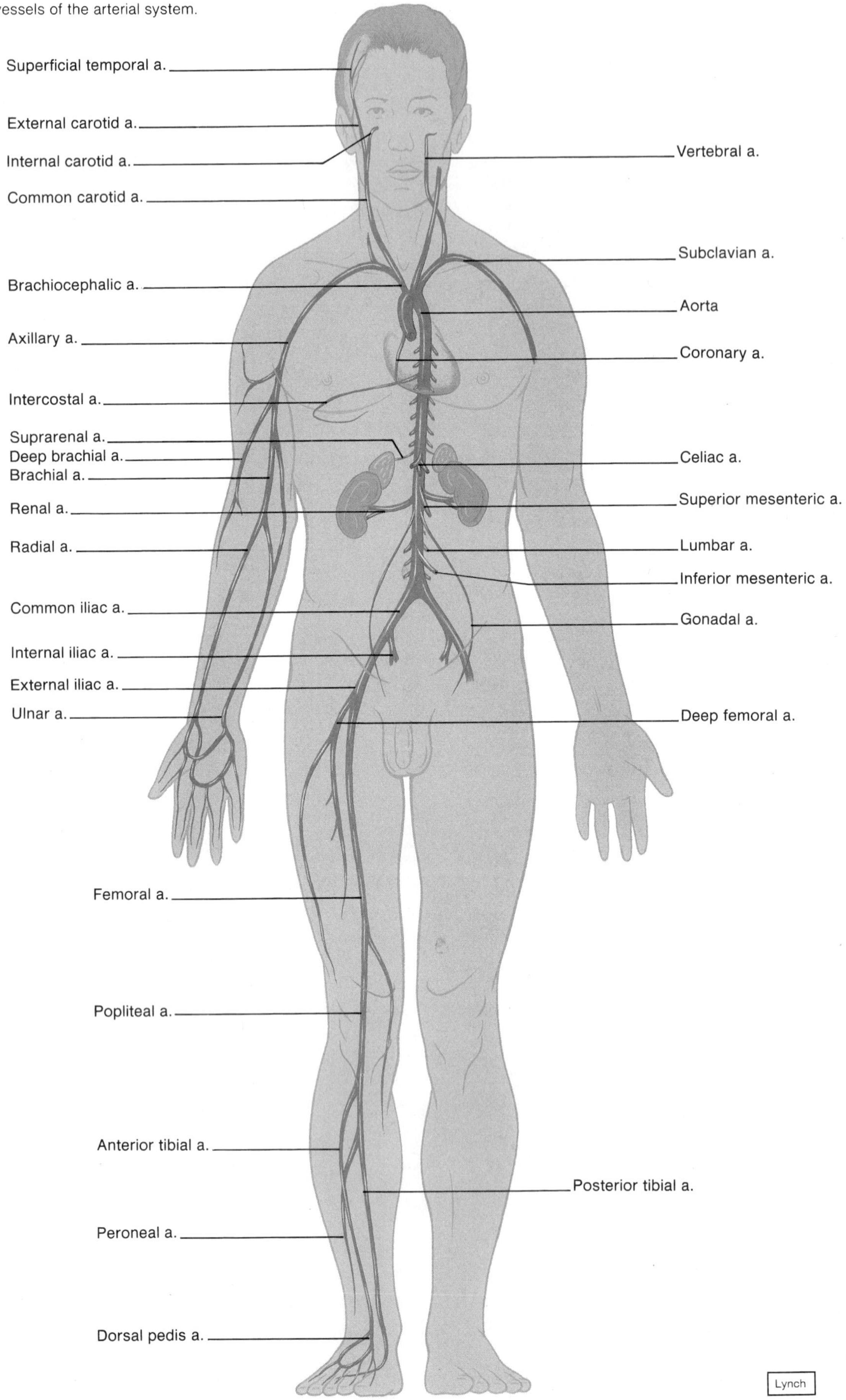

Superficial temporal a.
External carotid a.
Internal carotid a.
Common carotid a.
Brachiocephalic a.
Axillary a.
Intercostal a.
Suprarenal a.
Deep brachial a.
Brachial a.
Renal a.
Radial a.
Common iliac a.
Internal iliac a.
External iliac a.
Ulnar a.
Femoral a.
Popliteal a.
Anterior tibial a.
Peroneal a.
Dorsal pedis a.
Vertebral a.
Subclavian a.
Aorta
Coronary a.
Celiac a.
Superior mesenteric a.
Lumbar a.
Inferior mesenteric a.
Gonadal a.
Deep femoral a.
Posterior tibial a.

Lynch

The **internal iliac artery** gives off numerous branches to various pelvic muscles and visceral structures, as well as to the gluteal muscles and the external genitalia.

The **external iliac artery** provides the main blood supply to the legs. It passes downward along the brim of the pelvis and gives off branches that supply the muscles and skin in the lower abdominal wall.

Midway between the symphysis pubis and the anterior superior iliac spine of the ilium, the external iliac artery becomes the femoral artery.

The **femoral** (fem′or-al) **artery,** which passes fairly close to the anterior surface of the upper thigh, gives off many branches to the muscles and superficial tissues of the thigh. These branches also supply the skin of the groin and lower abdominal wall.

As the femoral artery reaches the proximal border of the space behind the knee, it becomes the **popliteal** (pop-lit′e-al) **artery.** The branches of this artery supply blood to the knee joint and to certain muscles in the thigh and calf. The popliteal artery divides into the anterior and posterior tibial arteries.

The **anterior tibial** (tib′e-al) **artery** passes downward between the tibia and fibula, giving off branches to the skin and muscles in the anterior and lateral regions of the lower leg. This vessel continues into the foot as the *dorsalis pedis artery,* which supplies blood to the foot and toes.

The **posterior tibial artery,** the larger of the two popliteal branches, descends beneath the calf muscles, giving off branches to the skin, muscles, and other tissues of the lower leg along the way.

1. Name the branches of the thoracic and abdominal aorta.
2. Which vessels supply blood to the head? The arm? The abdominal wall? The leg?

Venous System

Venous circulation is responsible for returning the blood to the heart after exchanges of gases, nutrients, and wastes have occurred between the blood and body cells.

Characteristics of Venous Pathways

The vessels of the venous system begin with the merging of capillaries into venules, venules into small veins, and small veins into larger ones. Unlike the arterial pathways, however, those of the venous system are difficult to follow. This is because the vessels are commonly interconnected in irregular networks, so that many unnamed tributaries may join to form a relatively large vein.

Figure 15.27
Major veins of the brain, head, and neck (*v.* stands for *vein*).

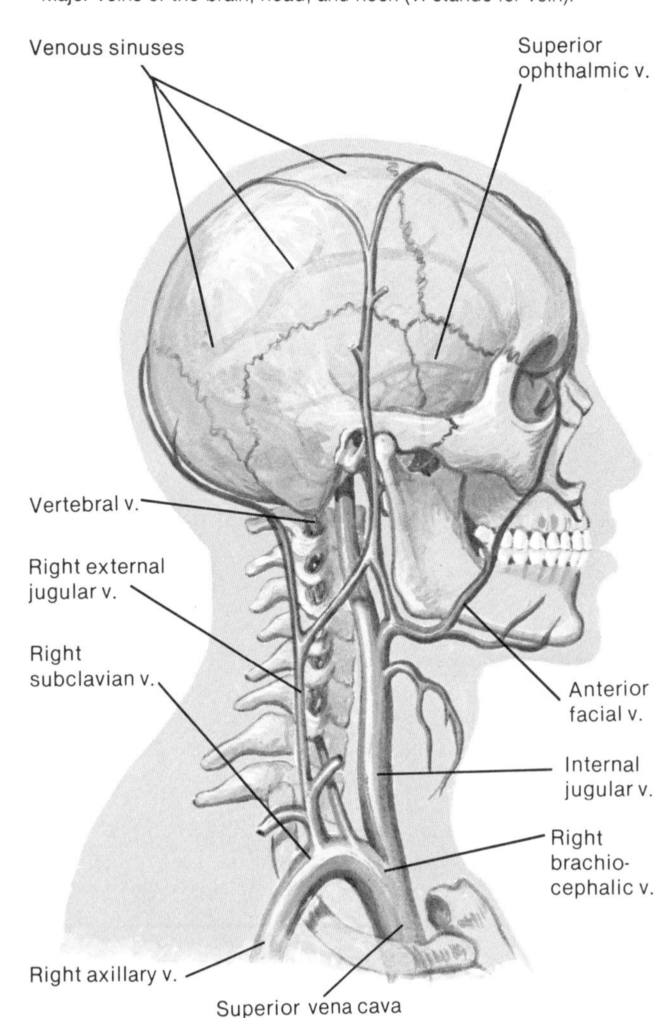

On the other hand, the larger veins typically parallel the courses taken by named arteries, and these veins often have the same names as their companions in the arterial system. Thus, with some exceptions, the name of a major artery also provides the name of the vein next to it. For example, the renal vein parallels the renal artery, the common iliac vein accompanies the common iliac artery and so forth.

The veins that carry the blood from the lungs and myocardium back to the heart have already been described. The veins from all the other parts of the body converge into two major pathways, which lead to the right atrium. They are the *superior* and *inferior venae cavae.*

Veins from the Brain, Head, and Neck

The blood from the face, scalp, and superficial regions of the neck is drained by the **external jugular** (jug′u-lar) **veins.** These vessels descend on either side of the neck and empty into the *subclavian veins* (fig. 15.27).

Figure 15.28
Veins that drain the thoracic wall.

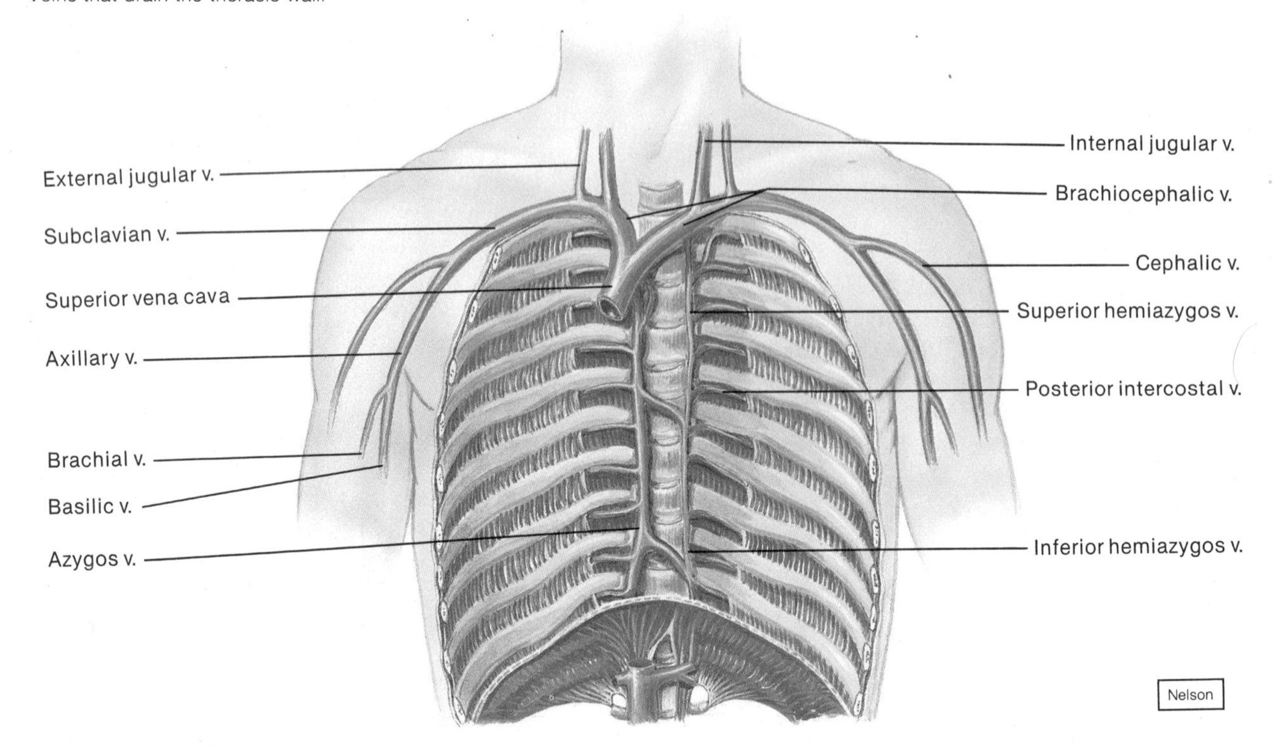

The **internal jugular veins,** which are somewhat larger than the external jugular veins, arise from numerous veins and venous sinuses of the brain and from deep veins in various parts of the face and neck. They pass downward through the neck and join the subclavian veins. These unions of the internal jugular and subclavian veins form large **brachiocephalic veins** on each side. These vessels then merge and give rise to the **superior vena cava** (ven′ah kav′ah), which enters the right atrium.

Veins from the Abdominal and Thoracic Walls

The abdominal and thoracic walls are drained mainly by tributaries of the brachiocephalic and azygos veins. For example, the *brachiocephalic vein* receives blood from the *internal thoracic vein,* which generally drains the tissues supplied by the internal thoracic artery. Some *intercostal veins* also empty into the brachiocephalic vein (fig. 15.28).

The **azygos** (az′ĭ-gos) **vein** originates in the dorsal abdominal wall and ascends through the mediastinum on the right side of the vertebral column to join the superior vena cava. It drains most of the muscular tissue in the abdominal and thoracic walls.

Tributaries of the azygos vein include the *posterior intercostal veins* on the right side, which drain the intercostal spaces, and the *superior* and *inferior hemiazygos veins,* which receive blood from the posterior intercostal veins on the left. The right and left *ascending lumbar veins,* whose tributaries include vessels from the lumbar and sacral regions, also connect to the azygos system.

Veins from the Abdominal Viscera

Although the veins usually carry the blood directly to the atria of the heart, those that drain the abdominal viscera are exceptions (fig. 15.29). They originate in the capillary networks of the stomach, intestines, pancreas, and spleen and carry the blood from these organs through a **portal** (por′tal) **vein** to the liver. This unique venous pathway is called the **hepatic portal system.**

The tributaries of the portal vein include the following vessels:

1. The right and left *gastric veins* from the stomach.
2. The *superior mesenteric vein* from the small intestine, ascending colon, and transverse colon.
3. The *splenic vein* from a convergence of several veins draining the spleen, the pancreas, and a portion of the stomach. Its largest tributary, the *inferior mesenteric vein,* brings blood upward from the descending colon, sigmoid colon, and rectum.

About 80% of the blood flowing to the liver in the hepatic portal system comes from the capillaries in the stomach and intestines.

Figure 15.29
Veins that drain the abdominal viscera.

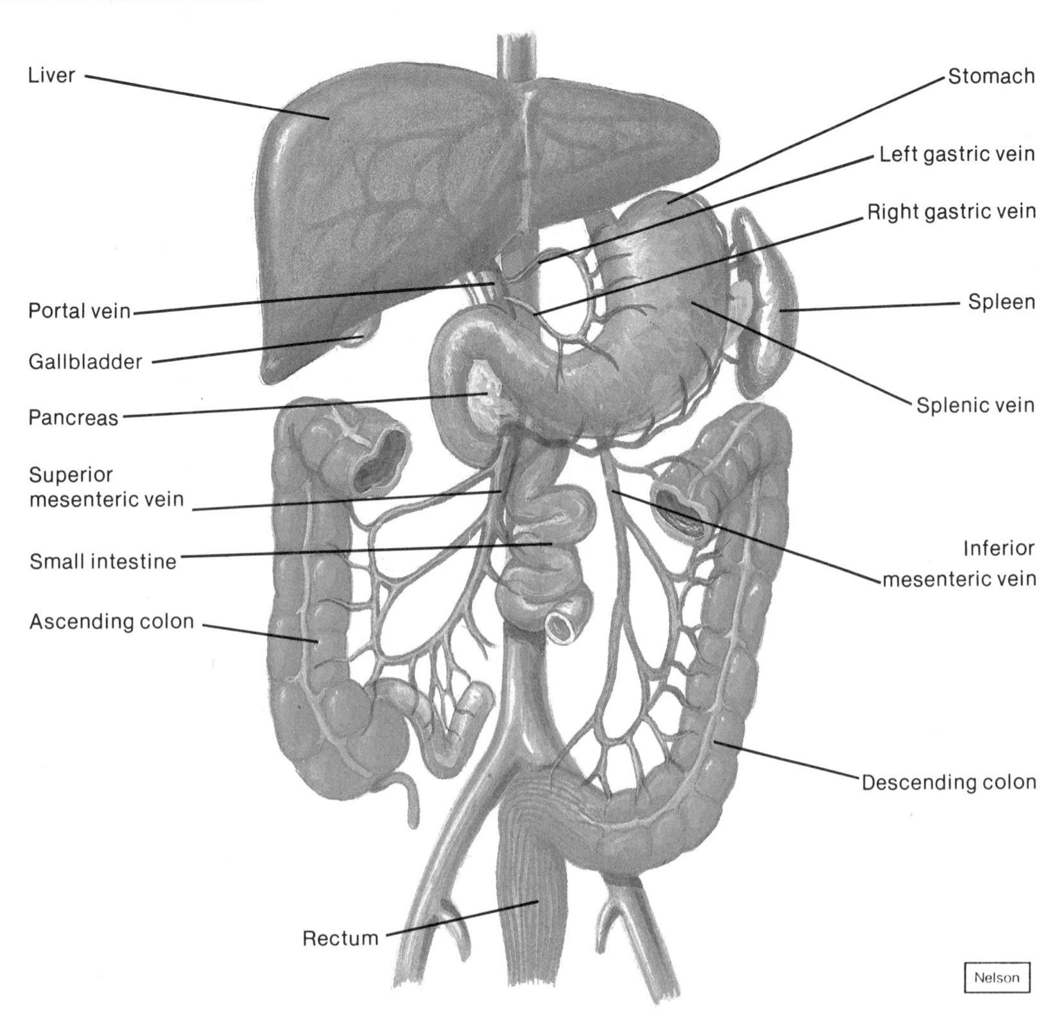

After passing through the hepatic sinusoids of the liver, the blood in the hepatic portal system is carried through a series of merging vessels into **hepatic veins.** These veins empty into the *inferior vena cava;* thus, the blood is returned to the general circulation.

Veins from the Arm and Shoulder

The arm is drained by a set of deep veins and a set of superficial ones. The deep veins generally parallel the arteries in each region and are given similar names, such as the *radial vein, ulnar vein, brachial vein,* and *axillary vein.* The superficial veins are interconnected in complex networks just beneath the skin. They also communicate with the deep vessels of the arm, providing many alternate pathways through which the blood can leave the tissues (fig. 15.30). The main vessels of the superficial network are the basilic and cephalic veins.

The **basilic** (bah-sil′ik) **vein** ascends from the forearm to the middle of the upper arm. There it penetrates the tissues deeply and joins the *brachial vein.* As the basilic and brachial veins merge, they form the *axillary vein.*

The **cephalic** (sĕ-fal′ik) **vein** courses upward from the hand to the shoulder. In the shoulder, it pierces the tissues and empties into the axillary vein. Beyond the axilla, the axillary vein becomes the *subclavian vein.*

In the bend of the elbow, a *median cubital vein* ascends from the cephalic vein on the lateral side of the arm to the basilic vein on the medial side. This vein is often used as a site for *venipuncture,* when it is necessary to remove a sample of blood for examination or to add fluids to the blood.

Veins from the Leg and Pelvis

As in the arm, the veins that drain the blood from the leg can be divided into deep and superficial groups (fig. 15.30).

Figure 15.30
Major vessels of the venous system.

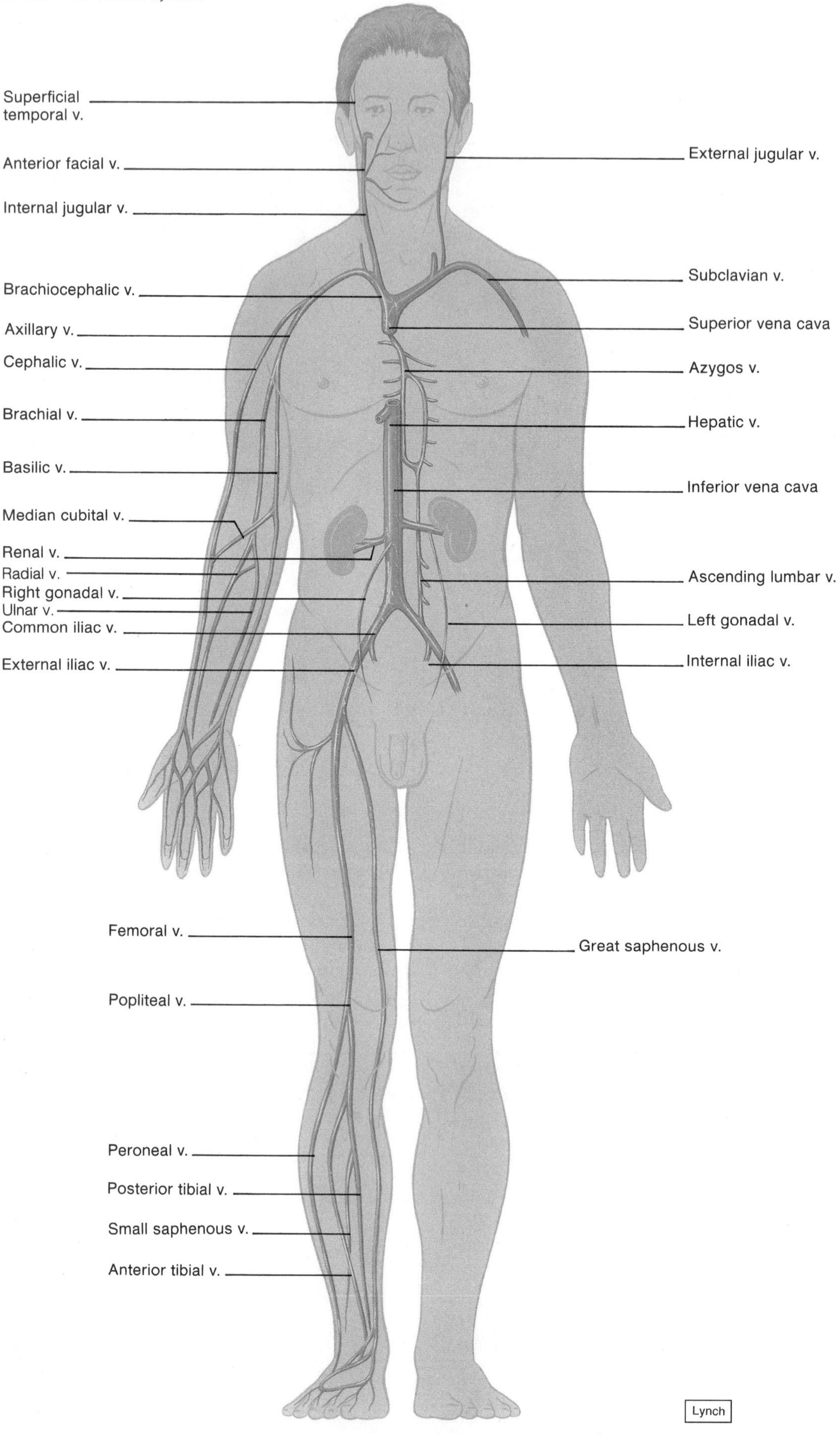

Superficial temporal v.
Anterior facial v.
Internal jugular v.
Brachiocephalic v.
Axillary v.
Cephalic v.
Brachial v.
Basilic v.
Median cubital v.
Renal v.
Radial v.
Right gonadal v.
Ulnar v.
Common iliac v.
External iliac v.
External jugular v.
Subclavian v.
Superior vena cava
Azygos v.
Hepatic v.
Inferior vena cava
Ascending lumbar v.
Left gonadal v.
Internal iliac v.
Femoral v.
Great saphenous v.
Popliteal v.
Peroneal v.
Posterior tibial v.
Small saphenous v.
Anterior tibial v.

Lynch

Figure 15.31
A schematic drawing of the circulatory system.

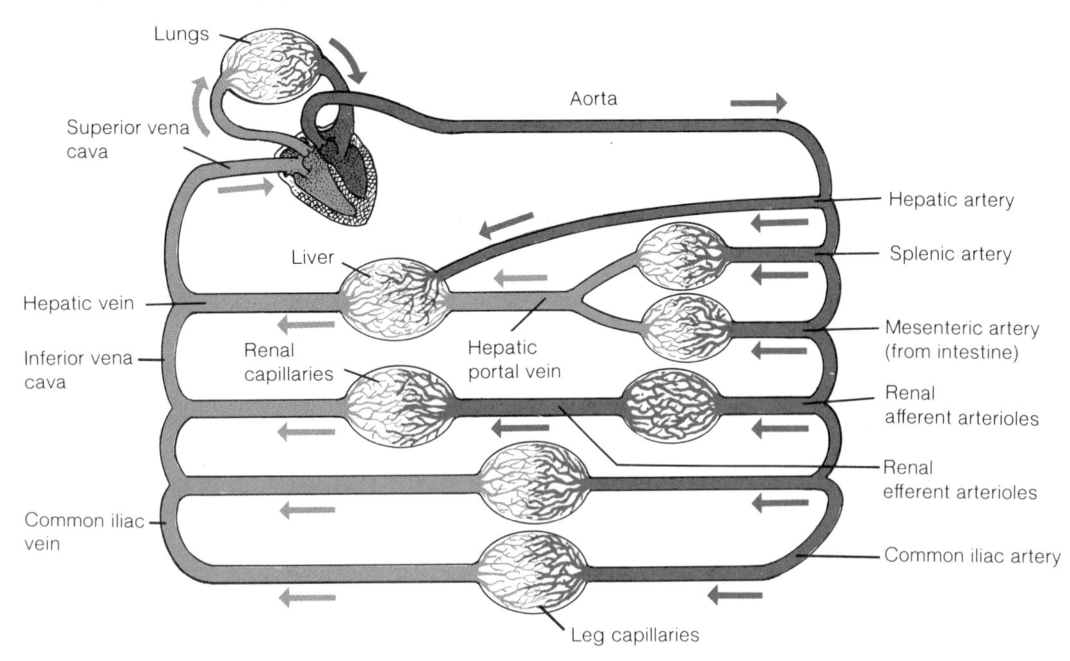

The deep veins of the lower leg, such as the *anterior* and *posterior tibial veins,* have names that correspond with the arteries they accompany. At the level of the knee, these vessels form a single trunk, the **popliteal vein.** This vein continues upward through the thigh as the **femoral vein,** which, in turn, becomes the **external iliac vein.**

The superficial veins of the foot and leg interconnect to form a complex network beneath the skin. These vessels drain into two major trunks—the small and great saphenous veins.

The **small saphenous** (sah-fe′nus) **vein** ascends along the back of the calf, enters the popliteal fossa, and joins the *popliteal vein.*

The **great saphenous vein,** which is the longest vein in the body, ascends in front of the medial malleolus and extends upward along the medial side of the leg. In the thigh, it penetrates deeply and joins the femoral vein. Near its termination, the great saphenous vein receives tributaries from a number of vessels that drain the upper thigh, groin, and lower abdominal wall.

In addition to communicating freely with each other, the *saphenous veins* communicate extensively with the deep veins of the leg. As a result, there are many pathways by which the blood can be returned to the heart from the lower extremities.

Whenever a person stands, the *saphenous veins* are subjected to increased blood pressure because of gravitational forces. When such pressure is prolonged, these veins are especially prone to developing abnormal dilations and becoming *varicose veins.*

In the pelvic region, the blood is carried away from the organs of the reproductive, urinary, and digestive systems by vessels leading to the **internal iliac vein.**

The internal iliac veins unite with the right and left *external iliac veins* to form the **common iliac veins.** These vessels, in turn, merge to produce the *inferior vena cava.*

Figure 15.31 provides a schematic illustration of the major blood vessels of the cardiovascular system.

1. Name the veins that return the blood to the right atrium.
2. What major veins drain the blood from the head? The abdominal viscera? The arm? The leg?

Clinical Terms Related to the Cardiovascular System

anastomosis (ah-nas″to-mo′sis)—an interconnection between two blood vessels, sometimes created by surgical means.

aneurysm (an′u-rizm)—saclike swelling in the wall of a blood vessel, usually an artery.

angiocardiography (an″je-o-kar″de-og′rah-fe)—the injection of radiopaque solution into the vascular system for an X-ray examination of the heart and pulmonary circuit.

angiospasm (an′je-o-spazm″)—muscular spasm in the wall of a blood vessel.

arteriography (ar″te-re-og′rah-fe)—the injection of radiopaque solution into the vascular system for an X-ray examination of arteries.

asystole (a-sis'to-le)—condition in which the myocardium fails to contract.

cardiac tamponade (kar-'de-ak tam"pon-ād)—compression of the heart by an accumulation of fluid within the pericardial cavity.

congestive heart failure (kon-jes'tiv hart fāl'yer)—condition in which the heart is unable to pump an adequate amount of blood to the body cells.

cor pulmonale (kor pul-mo-na'le)—heart-lung disorder characterized by pulmonary hypertension and hypertrophy of the right ventricle.

embolectomy (em"bo-lek'to-me)—the removal of an embolus through an incision in a blood vessel.

endarterectomy (en"dar-ter-ek'to-me)—the removal of the inner wall of an artery to reduce an arterial occlusion.

palpitation (pal"pĭ-ta'shun)—an awareness of a heartbeat that is unusually rapid, strong, or irregular.

pericardiectomy (per"ĭ-kar"de-ek'to-me)—an excision of the pericardium.

phlebitis (flĕ-bi'tis)—an inflammation of a vein, usually in the legs.

phlebosclerosis (fleb"o-sklĕ-ro'sis)—abnormal thickening or hardening of the walls of veins.

phlebotomy (flĕ-bot'o-me)—an incision of a vein for the purpose of withdrawing blood.

sinus rhythm (si'nus rithm)—the normal cardiac rhythm regulated by the S-A node.

thrombophlebitis (throm"bo-flĕ-bi'tis)—the formation of a blood clot in a vein in response to an inflammation of the venous wall.

valvotomy (val-vot'o-me)—an incision of a valve.

venography (ve-nog'rah-fe)—the injection of radiopaque solution into the vascular system for an X-ray examination of veins.

Chapter Summary

Introduction (page 390)

The cardiovascular system is vital for providing oxygen and nutrients to tissues and for removing wastes.

Structure of the Heart (page 390)

1. Size and location of the heart
   a. The heart is about 14 cm long and 9 cm wide.
   b. It is located within the mediastinum and rests on the diaphragm.
2. Coverings of the heart
   a. The heart is enclosed in a layered pericardium.
   b. The pericardial cavity is a potential space between the parietal and visceral layers of the pericardium.
3. Wall of the heart
   The wall of the heart is composed of an epicardium, a myocardium, and an endocardium.
4. Heart chambers and valves
   a. The heart is divided into two atria and two ventricles.
   b. Right chambers and valves
      (1) The right atrium receives blood from the venae cavae and coronary sinus.
      (2) The right chambers are separated by the tricuspid valve.
      (3) The base of the pulmonary trunk is guarded by a pulmonary valve.
   c. Left chambers and valves
      (1) The left atrium receives blood from the pulmonary veins.
      (2) The left chambers are separated by the bicuspid valve.
      (3) The base of the aorta is guarded by an aortic valve.
5. Skeleton of the heart
   The skeleton of the heart consists of fibrous rings that enclose the bases of the pulmonary artery and aorta.
6. Path of blood through the heart
   a. Blood that is relatively low in oxygen concentration and high in carbon dioxide concentration enters the right side of the heart and is pumped into the pulmonary circulation.
   b. After the blood is oxygenated in the lungs and some of its carbon dioxide is removed, it returns to the left side of the heart.
7. Blood supply to the heart
   a. Blood is supplied to the myocardium through the coronary arteries.
   b. It is returned to the right atrium through the cardiac veins and coronary sinus.

Actions of the Heart (page 397)

1. Cardiac cycle
   a. The atria contract while the ventricles relax; the ventricles contract while the atria relax.
   b. This series of events constitutes a cardiac cycle.
2. Heart sounds
   Heart sounds are due to the vibrations produced by the blood and valve movements.
3. Cardiac muscle fibers
   a. Cardiac muscle fibers are interconnected to form a functional syncytium.
   b. If any part of the syncytium is stimulated, the whole structure contracts as a unit.
4. Cardiac conduction system
   a. This system functions to initiate and conduct impulses through the myocardium.
   b. Impulses from the S-A node pass slowly to the A-V node; impulses travel rapidly along the A-V bundle and Purkinje fibers.
5. Electrocardiogram (ECG)
   a. An ECG is a recording of the electrical changes occurring in the myocardium during a cardiac cycle.
   b. The pattern contains several waves.
      (1) The P wave represents an atrial depolarization.
      (2) The QRS complex represents a ventricular depolarization.
      (3) The T wave represents a ventricular repolarization.

6. Regulation of the cardiac cycle
   a. The heartbeat is affected by physical exercise, body temperature, and the concentrations of various ions.
   b. The S-A and A-V nodes are innervated by branches of sympathetic and parasympathetic nerve fibers.
   c. Autonomic impulses are regulated by the cardiac center in the medulla oblongata.

Blood Vessels (page 402)

The blood vessels form a closed circuit of tubes.

1. Arteries and arterioles
   a. The arteries are adapted to carry high pressure blood away from the heart.
   b. The walls of arteries and arterioles consist of endothelium, smooth muscle, and connective tissue.
   c. The smooth muscles are innervated by autonomic fibers.
2. Capillaries
   a. The capillary wall consists of a single layer of epithelial cells.
   b. The openings in capillary walls, where adjacent endothelial cells overlap, vary in size and are responsible for the permeability of these walls.
   c. The blood flow into a capillary is controlled by a precapillary sphincter.
   d. Exchanges in the capillaries
      (1) Gases, nutrients, and metabolic by-products are exchanged between the capillary blood and tissue fluid.
      (2) Diffusion provides the most important means of transport.
      (3) Filtration causes a net outward movement of fluid at the arterial end of a capillary.
      (4) Osmosis causes a net inward movement of fluid at the venule end of a capillary.
3. Venules and veins
   a. Venules continue from capillaries and merge to form veins.
   b. Veins carry the blood to the heart.
   c. Venous walls are similar to arterial walls, but they are thinner and contain less muscle and elastic tissue.

Blood Pressure (page 402)

Blood pressure is the force exerted by the blood against the inside of the blood vessels.

1. Arterial blood pressure
   a. The arterial blood pressure is created primarily by heart action.
   b. The systolic pressure occurs when the ventricle contracts; the diastolic pressure occurs when the ventricle relaxes.
2. Factors that influence arterial blood pressure
   The arterial pressure increases as the cardiac output, blood volume, peripheral resistance, or blood viscosity increases.
3. Control of blood pressure
   a. Blood pressure is controlled in part by the mechanisms that regulate cardiac output and peripheral resistance.
   b. The more blood that enters the heart, the stronger the ventricular contraction, the greater the stroke volume, and the greater the cardiac output.
   c. Regulation of the heart rate and peripheral resistance involves the cardiac and vasomotor centers of the medulla oblongata.
4. Venous blood flow
   a. Venous blood flow depends on skeletal muscle contraction and breathing movements.
   b. Many veins contain flaplike valves that prevent the blood from backing up.
   c. Venoconstriction can increase the venous pressure and blood flow.

Paths of Circulation (page 411)

1. Pulmonary circuit
   The pulmonary circuit is composed of vessels that carry the blood from the right ventricle to the lungs and back to the left atrium.
2. Systemic circuit
   a. The systemic circuit is composed of vessels that lead from the heart to the body cells and back to the heart.
   b. It includes the aorta and its branches.

Arterial System (page 412)

1. Principal branches of the aorta
   a. The aorta is the largest artery.
   b. Its major branches include the coronary, brachiocephalic, left common carotid, and left subclavian arteries.
   c. The branches of the descending aorta include the thoracic and abdominal groups.
   d. The abdominal aorta terminates by dividing into the right and left common iliac arteries.
2. Arteries to the neck, head, and brain
   These include the branches of the subclavian and common carotid arteries.
3. Arteries to the thoracic and abdominal walls
   a. The thoracic wall is supplied by branches of the subclavian artery and thoracic aorta.
   b. The abdominal wall is supplied by branches of the abdominal aorta and other arteries.
4. Arteries to the shoulder and arm
   a. The subclavian artery passes into the upper arm and in various regions is called the axillary and the brachial artery.
   b. The branches of the brachial artery include the ulnar and radial arteries.
5. Arteries to the pelvis and leg
   The common iliac artery supplies the pelvic organs, gluteal region, and leg.

Venous System (page 417)

1. Characteristics of venous pathways
   a. Veins are responsible for returning the blood to the heart.
   b. The larger veins usually parallel the paths of major arteries.
2. Veins from the brain, head, and neck
   a. These regions are drained by the jugular veins.
   b. The jugular veins unite with subclavian veins to form the brachiocephalic veins.
3. Veins from the abdominal and thoracic walls
   These walls are drained by tributaries of the brachiocephalic and azygos veins.
4. Veins from the abdominal viscera
   a. The blood from the abdominal viscera generally enters the hepatic portal system and is carried to the liver.
   b. From the liver, the blood is carried by the hepatic veins to the inferior vena cava.
5. Veins from the arm and shoulder
   a. The arm is drained by sets of superficial and deep veins.
   b. The deep veins parallel arteries with similar names.
6. Veins from the leg and pelvis
   a. These regions are drained by sets of deep and superficial veins.
   b. The deep veins include the tibial veins, and the superficial veins include the saphenous veins.

Clinical Application of Knowledge

1. Based upon your understanding of the way capillary blood flow is regulated, do you think it is wiser to rest or to exercise following a heavy meal? Give a reason for your answer.
2. If a patient develops a blood clot in the femoral vein of the left leg and a portion of the clot breaks loose, where is the embolus likely to be carried? What symptoms is this condition likely to produce?
3. In the case of a patient with a heart weakened by damage to the myocardium, would it be better to keep the legs raised or lowered below the level of the heart? Why?
4. Cirrhosis of the liver is a disease commonly associated with alcoholism. In this condition, the blood flow through the hepatic blood vessels is often obstructed. As a result of such obstruction, the blood backs up, and the capillary pressure greatly increases in the organs drained by the hepatic portal system. What effects might this increasing capillary pressure produce, and which organs would be affected by it?

Review Activities

1. Describe the general structure, function, and location of the heart.
2. Describe the pericardium.
3. Compare the layers of the cardiac wall.
4. Identify and describe the locations of the chambers and valves of the heart.
5. Describe the skeleton of the heart, and explain its function.
6. Trace the path of the blood through the heart.
7. Trace the path of the blood through the coronary circulation.
8. Describe a cardiac cycle.
9. Explain the origin of heart sounds.
10. Distinguish between the S-A node and the A-V node.
11. Explain how the cardiac conduction system functions in controlling the cardiac cycle.
12. Describe a normal ECG pattern, and explain the significance of its waves.
13. Discuss how the nervous system functions in the regulation of the cardiac cycle.
14. Distinguish between an artery and an arteriole.
15. Explain how vasoconstriction and vasodilation are controlled.
16. Describe the structure and function of a capillary.
17. Explain how the blood flow through a capillary is controlled.
18. Explain how diffusion functions in the exchange of substances between blood plasma and tissue fluid.
19. Explain why water and dissolved substances leave the arteriole end of a capillary and enter the venule end.
20. Distinguish between a venule and a vein.
21. Explain how veins function as blood reservoirs.
22. Distinguish between systolic and diastolic blood pressures.
23. Name several factors that influence the blood pressure, and explain how each produces its effect.
24. Describe how the blood pressure is controlled.
25. List the major factors that promote the flow of venous blood.
26. Distinguish between the pulmonary and systemic circuits of the cardiovascular system.
27. Trace the path of the blood through the pulmonary circuit.
28. Describe the aorta, and name its principal branches.
29. On a diagram, locate and identify the major arteries that supply the abdominal visceral organs.
30. On a diagram, locate and identify the major arteries that supply parts of the head, neck, and brain.

31. On a diagram, locate and identify the major arteries that supply parts of the thoracic and abdominal walls.
32. On a diagram, locate and identify the major arteries that supply parts of the shoulder and arm.
33. On a diagram, locate and identify the major arteries that supply parts of the pelvis and leg.
34. Describe the relationship between the major venous pathways and the major arterial pathways.
35. On a diagram, locate and identify the major veins that drain parts of the brain, head, and neck.
36. On a diagram, locate and identify the major veins that drain parts of the abdominal and thoracic walls.
37. On a diagram, locate and identify the major veins that drain parts of the abdominal viscera.
38. On a diagram, locate and identify the major veins that drain parts of the arm and shoulder.
39. On a diagram, locate and identify the major veins that drain parts of the leg and pelvis.

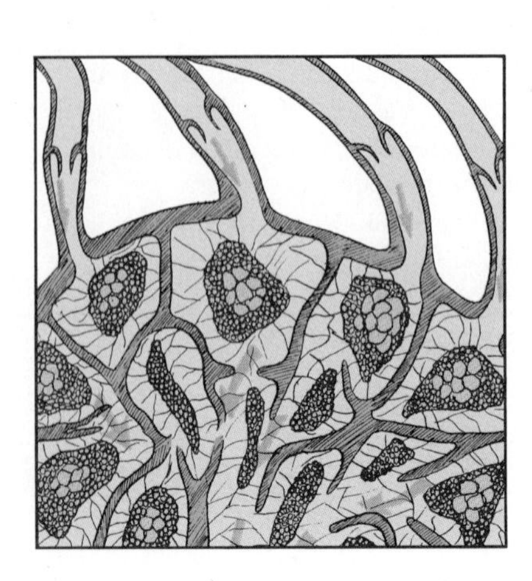

16 Lymphatic System and Immunity

When substances are exchanged between the blood and tissue fluid, more fluid normally leaves the blood capillaries than returns to them. If the fluid remaining in the tissue spaces were allowed to accumulate, the hydrostatic pressure in the tissues would increase. The *lymphatic system* helps prevent such an imbalance by providing pathways through which the tissue fluid can be transported as lymph from the tissue spaces to the veins, where it becomes part of the blood.

The lymphatic system also helps defend the tissues against infections by filtering particles from the lymph and by supporting the activities of the lymphocytes, which furnish immunity against specific disease-causing agents.

Chapter Objectives

After you have studied this chapter, you should be able to

1. Describe the general functions of the lymphatic system.
2. Describe the location of the major lymphatic pathways.
3. Describe how tissue fluid and lymph are formed, and explain the function of lymph.
4. Explain how lymphatic circulation is maintained.
5. Describe a lymph node and its major functions.
6. Discuss the functions of the thymus and spleen.
7. Distinguish between specific and nonspecific body defenses, and provide examples of each defense.
8. Explain how two major types of lymphocytes are formed, and explain how they function in immune mechanisms.
9. Name the major types of immunoglobulins, and discuss their origins and actions.
10. Distinguish between primary and secondary immune responses.
11. Distinguish between active and passive immunity.
12. Explain how allergic reactions and tissue rejection reactions are related to immune mechanisms.
13. Complete the review activities at the end of this chapter. Note that the items are worded in the form of specific learning objectives. You may want to refer to them before reading the chapter.

Key Terms

allergen (al′-er-jen)

antibody (an′tĭ-bod″e)

antigen (an′tĭ-jen)

clone (klōn)

immunity (ĭ-mu′nĭ-te)

immunoglobulin (im″u-no-glob′u-lin)

lymph (limf)

lymphatic pathway (lim-fat′ik path′wa)

lymph node (limf nōd)

lymphocyte (lim′fo-sīt)

pathogen (path′o-jen)

spleen (splēn)

thymus (thī′mus)

Aids to Understanding Words

gen-, to be produced: aller*gen*—a substance that stimulates an allergic response.

humor-, fluid: *humor*al immunity—immunity resulting from antibodies in body fluids.

immun-, free: *immun*ity—resistance to (freedom from) a specific disease.

inflamm-, setting on fire: *inflamm*ation—a condition characterized by localized redness, heat, swelling, and pain in the tissues.

nod-, knot: *nod*ule—a small mass of lymphocytes surrounded by connective tissue.

patho-, disease: *patho*gen—a disease-causing agent.

Introduction

THE LYMPHATIC SYSTEM is closely associated with the cardiovascular system because it includes a network of vessels that assist in circulating the body fluids. These vessels transport excess fluid away from the interstitial spaces that exist between the cells of most tissues and return it to the bloodstream. The organs of the lymphatic system also help defend the body against invasion by disease-causing agents (fig. 16.1).

Lymphatic Pathways

Lymphatic pathways begin as lymphatic capillaries. These tiny tubes merge to form larger lymphatic vessels, which, in turn, lead to the collecting ducts that unite with the veins in the thorax.

Lymphatic Capillaries

Lymphatic capillaries are microscopic, closed-ended tubes (fig. 16.2). They extend into the spaces within most tissues (interstitial spaces), forming complex networks that parallel the networks of the blood capillaries. The walls of the lymphatic capillaries, like those of the blood capillaries, consist of a single layer of squamous epithelial cells. This thin wall makes it possible for tissue fluid (intestitial fluid) from the interstitial spaces to enter the lymphatic capillary. Once the fluid is inside a lymphatic capillary, it is called **lymph.**

Lymphatic Vessels

Lymphatic vessels, which are formed by the merging of lymphatic capillaries, have walls similar to those of veins. Also like veins, lymphatic vessels have flaplike

Figure 16.1
Lymphatic vessels transport fluid from interstitial spaces to the bloodstream.

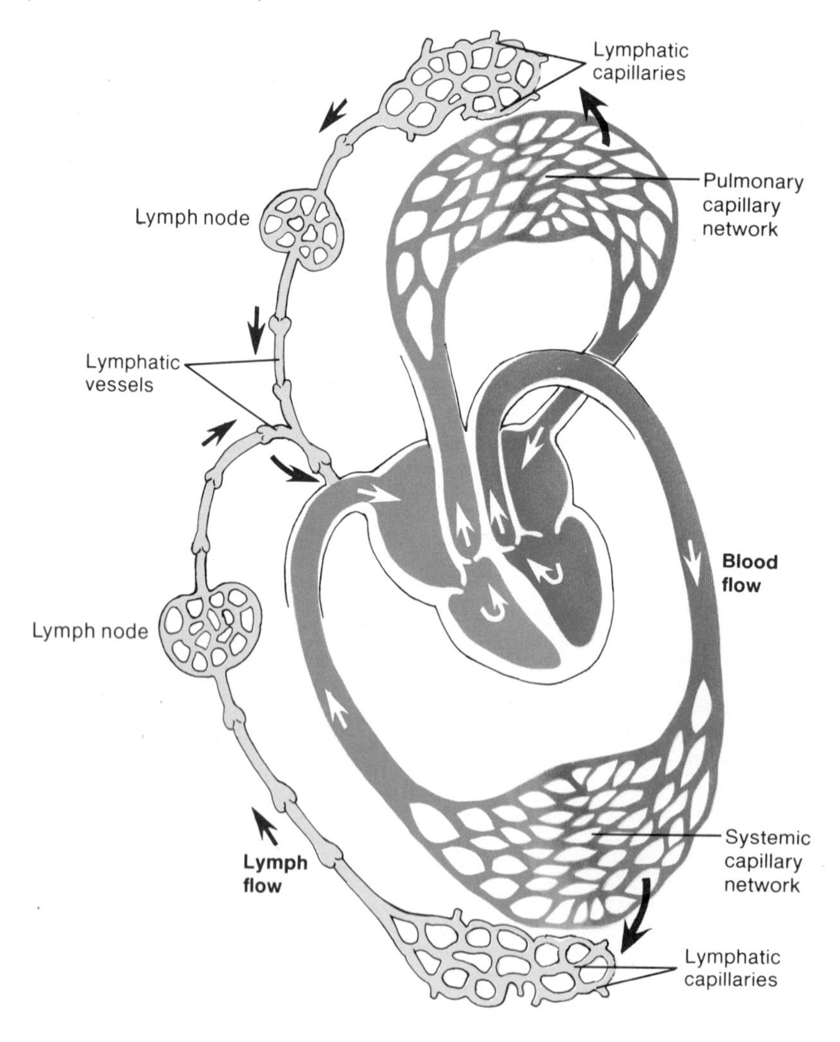

valves, which help to prevent the backflow of lymph. Figure 16.3 shows one of these valves.

Typically, the lymphatic vessels lead to specialized organs called **lymph nodes,** and after leaving these structures, the vessels merge to form still larger lymphatic trunks.

Lymphatic Trunks and Collecting Ducts

Lymphatic trunks, which drain lymph from relatively large portions of the body, are named for the regions they serve. These lymphatic trunks then join one of two **collecting ducts**—the thoracic duct or the right lymphatic duct (fig. 16.4).

The **thoracic duct** is the larger and longer of the two collecting ducts. It receives lymph from the lower body regions, left arm, and left side of the head and neck, and empties into the left subclavian vein near the junction of the left jugular vein.

The **right lymphatic duct** receives lymph from the right side of the head and neck, right arm, and right thorax. It empties into the right subclavian vein near the junction of the right jugular vein.

Figure 16.2
Lymphatic capillaries are microscopic closed-ended tubes that begin in the interstitial spaces of most tissues.

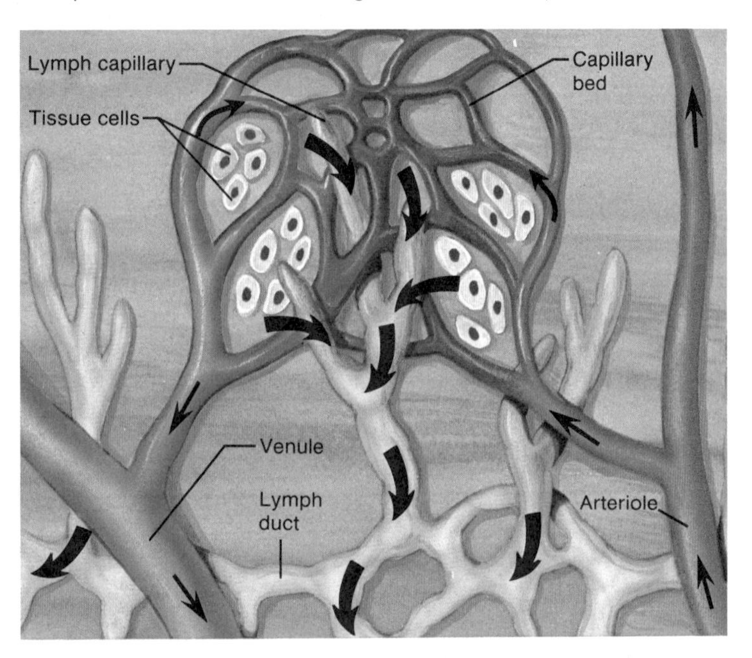

Figure 16.3
A light micrograph of the flaplike valve within a lymphatic vessel (×25). What is the function of this valve?

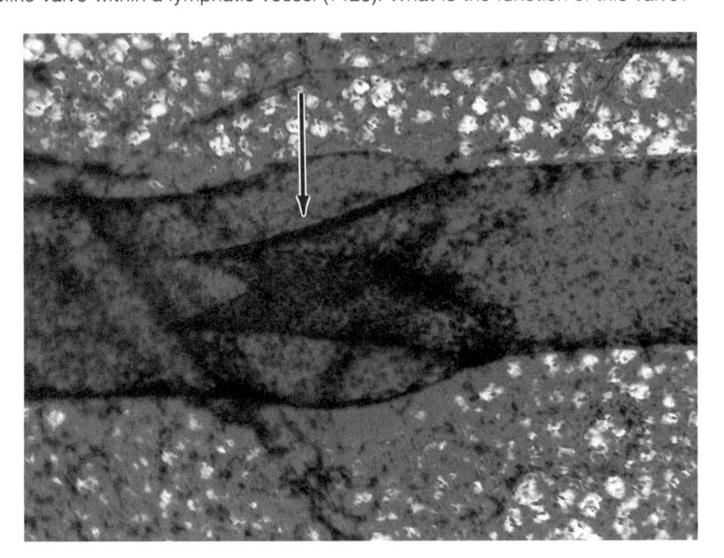

Figure 16.4
(*a*) The right lymphatic duct drains lymph from the upper right side of the body, and the thoracic duct drains lymph from the remainder; (*b*) lymph drainage of the right breast.

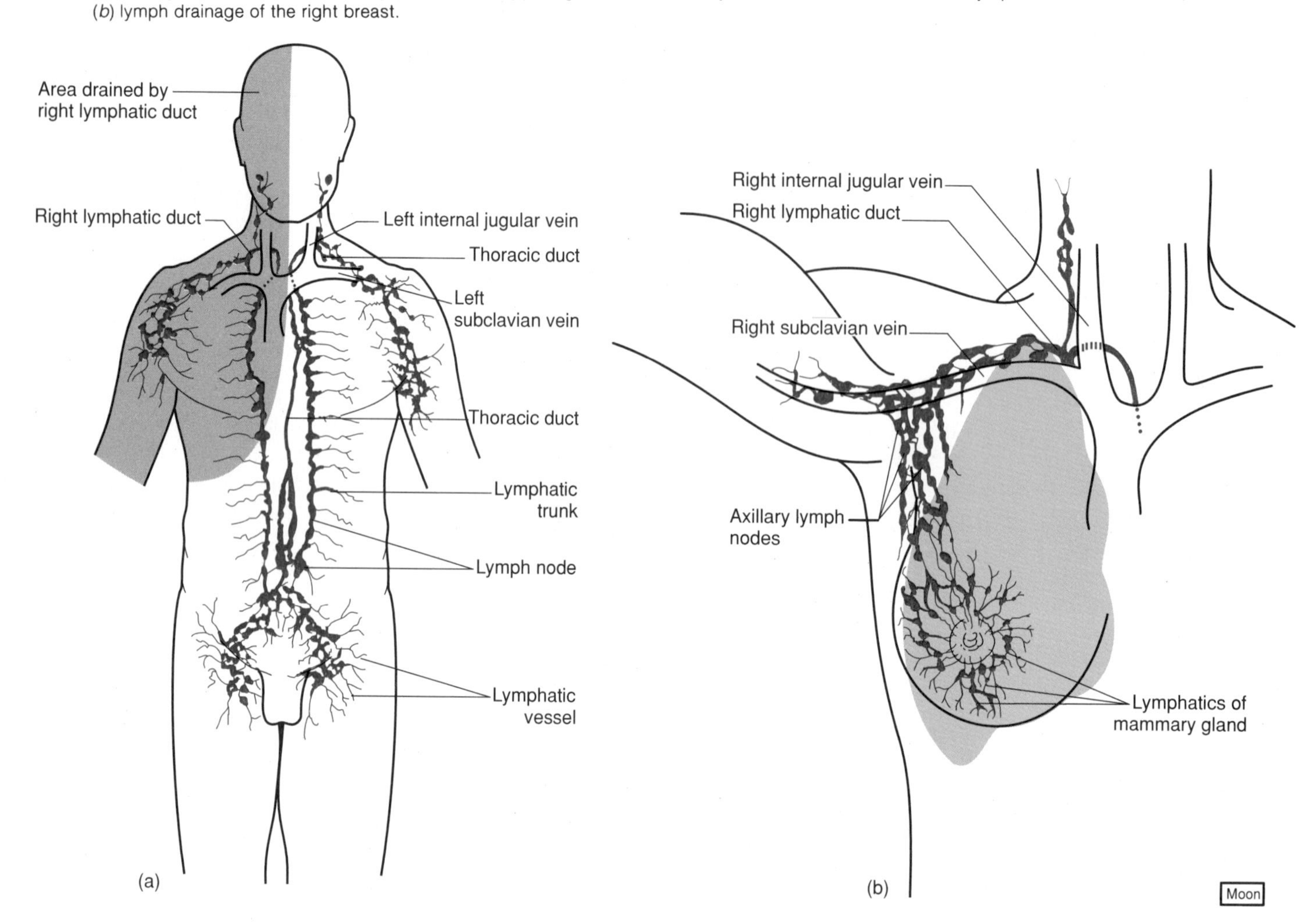

After leaving the collecting ducts, lymph enters the venous system and becomes part of the plasma just before the blood returns to the right atrium.

Chart 16.1 summarizes a typical lymphatic pathway.

The skin is richly supplied with lymphatic capillaries. Consequently, if the skin is broken or something is injected into it, such as venom from a stinging insect, foreign substances are likely to enter the lymphatic pathways relatively rapidly.

1. What is the general function of the lymphatic system?
2. Distinguish between the thoracic duct and the right lymphatic duct.

Tissue Fluid and Lymph

Lymph is essentially tissue fluid that has entered a lymphatic capillary. This formation of lymph is closely associated with the formation of tissue fluid.

Tissue Fluid Formation

As is explained in chapter 14, *tissue fluid* originates from blood plasma. Tissue fluid is composed of water and dissolved substances that leave the blood capillaries as a result of diffusion and filtration.

Although the tissue fluid contains various nutrients and gases found in the plasma, it generally lacks proteins of large molecular size. Some proteins with smaller molecules do leak out of the blood capillaries and enter the interstitial spaces. Usually, these smaller proteins are not reabsorbed when water and other dissolved substances move back into the venule ends of these capillaries by diffusion and osmosis. As a result, the protein concentration of the tissue fluid tends to rise, causing the *osmotic pressure* of the fluid to rise also.

Lymph Formation

As the osmotic pressure of the tissue fluid rises, this pressure interferes with the osmotic reabsorption of water by the blood capillaries. The volume of fluid in

Chart 16.1 Typical lymphatic pathway	
Tissue fluid	*Lymph node*
leaves the interstitial space and becomes	where lymph is filtered and leaves via a
Lymph	*Lymphatic vessel*
as it enters a	which merges with other vessels to form a
Lymphatic capillary	*Lymphatic trunk*
which merges with other capillaries to form a	which merges with other trunks and joins a
Lymphatic vessel	*Collecting duct*
which enters a	which empties into a
	Subclavian vein
	where lymph is added to the blood

the interstitial spaces then tends to increase, as does the pressure within the spaces.

This increasing interstitial pressure is responsible for forcing some of the tissue fluid into the lymphatic capillaries, where it becomes lymph (fig. 16.2).

Function of Lymph

Most of the protein molecules that leak out of the blood capillaries are carried away by lymph and are returned to the bloodstream. At the same time, lymph transports foreign particles, such as bacterial cells or viruses that may have entered the tissue fluids, to the lymph nodes.

1. Explain the relationship between the tissue fluid and lymph.
2. How does the presence of protein in the tissue fluid affect the formation of lymph?
3. What are the major functions of lymph?

Movement of Lymph

Although the entrance of lymph into the lymphatic capillaries is influenced by the *osmotic pressure* of the tissue fluid, the movement of lymph through the lymphatic vessels is controlled largely by *muscular activity*.

Flow of Lymph

Lymph, like venous blood, is under relatively low hydrostatic pressure and may not flow readily through the lymphatic vessels without the aid of outside forces. These forces include contraction of the skeletal muscles, the action of the breathing muscles, and contraction of the smooth muscles in the walls of the larger lymphatic vessels.

As the skeletal muscles contract, they compress the lymphatic vessels. This squeezing action causes the lymph inside a vessel to move, but since the lymphatic vessels contain valves that prevent backflow, the lymph only can move toward a collecting duct. Similarly, the smooth muscles in the walls of the larger lymphatic vessels may contract and compress the lymph inside. This action also helps force the fluid onward.

The breathing muscles aid in the circulation of lymph (as they do that of venous blood) by creating a relatively low pressure in the thorax during inhalation. At the same time, the pressure in the abdominal cavity is increased by the contracting diaphragm. Consequently, lymph (and venous blood as well) is squeezed out of the abdominal vessels and forced into the thoracic vessels. Once again, the backflow of lymph (and blood) is prevented by valves within the lymphatic (and blood) vessels.

Conditions sometimes occur that interfere with lymph movement, and tissue fluids accumulate in the interstitial spaces, causing a form of *edema.*

For example, lymphatic vessels may be obstructed as a result of surgical procedures in which portions of the lymphatic system are removed. Since the affected pathways can no longer drain lymph from the tissues, proteins tend to accumulate in the interstitial spaces. This causes an increase in the osmotic pressure of the tissue fluid, which, in turn, promotes the accumulation of water within the tissues.

1. How do the skeletal muscles promote the flow of lymph?
2. How does breathing aid in the movement of lymph?

Lymph Nodes

The **lymph nodes** (lymph glands) are structures located along the lymphatic pathways. They contain large numbers of *lymphocytes,* which are vital in the defense against invasion by microorganisms.

Figure 16.5
A section of a lymph node. What factors promote the flow of lymph through a node?

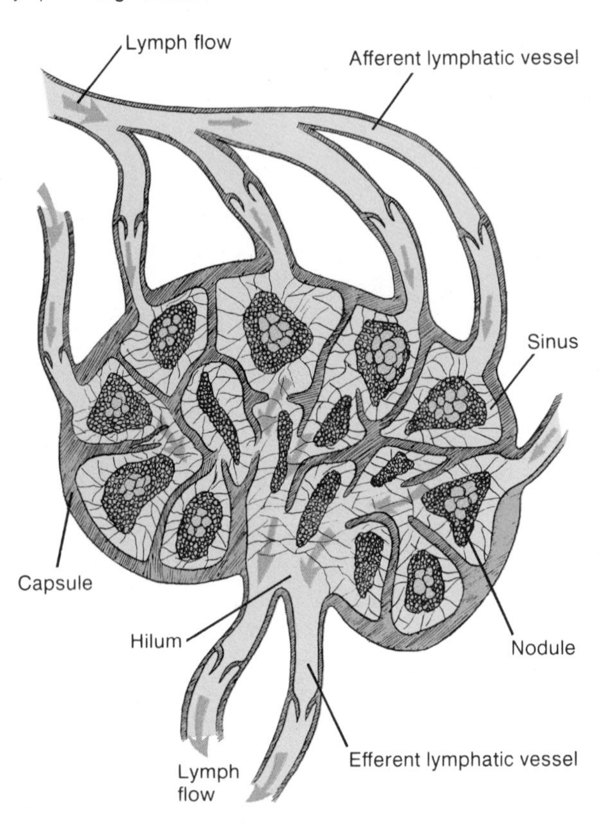

Figure 16.6
Lymph enters and leaves a lymph node through lymphatic vessels.

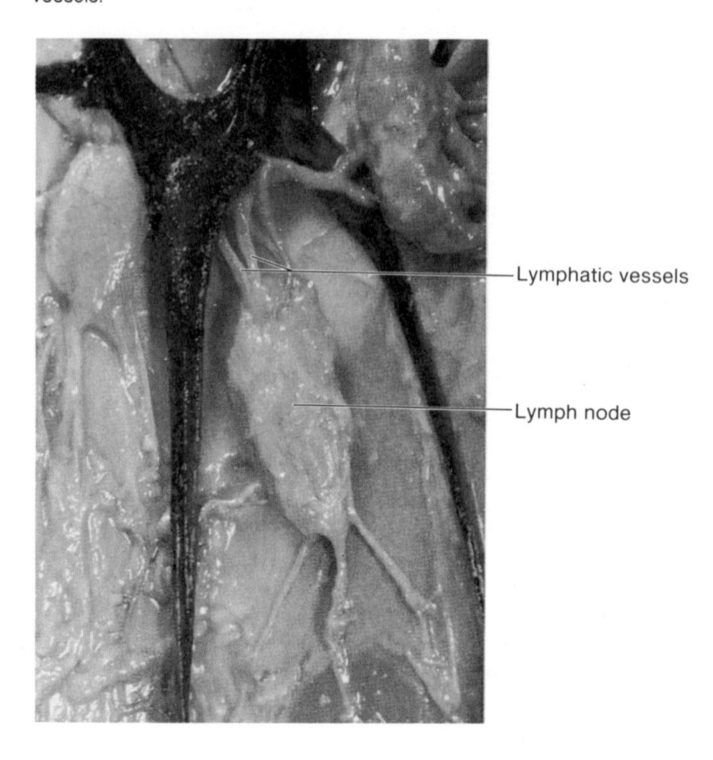

Structure of a Lymph Node

The lymph nodes vary in size and shape; however, they are usually less than 2.5 cm in length and are somewhat bean shaped. A section of a typical lymph node is illustrated in figures 16.5 and 16.6.

The indented region of a lymph node is called the **hilum,** and it is the portion through which the blood vessels and nerves connect with the structure. The lymphatic vessels leading to a node (afferent vessels) enter separately at various points on its convex surface, but the lymphatic vessels leaving the node (efferent vessels) exit from the hilum.

Each lymph node is subdivided into compartments, which contain dense masses of lymphocytes and macrophages. These masses, called **nodules,** represent the structural units of the node.

The spaces within the node, called **lymph sinuses,** provide a complex network of chambers and channels through which lymph circulates as it passes through the node.

Nodules also occur singly or in groups associated with the mucous membranes of the respiratory and digestive tracts. For example, the tonsils, described in chapter 12, are composed of partially encapsulated lymph nodules. Similarly, aggregations of nodules called *Peyer's patches* are scattered throughout the mucosal lining of the ileum of the small intestine.

Locations of Lymph Nodes

The lymph nodes generally occur in groups or chains along the paths of the larger lymphatic vessels. Although they are widely distributed throughout the body, lymph nodes are lacking in the tissues of the central nervous system.

The major lymph nodes are located and named in figure 16.7.

Sometimes the lymphatic vessels become inflamed due to a bacterial infection. When this happens in the superficial lymphatic vessels, painful reddish streaks may appear beneath the skin. This condition is called *lymphangitis,* and it is usually followed by *lymphadenitis,* an inflammation of the lymph nodes. The affected nodes may become greatly enlarged and quite painful.

Figure 16.7
Major locations of lymph nodes.

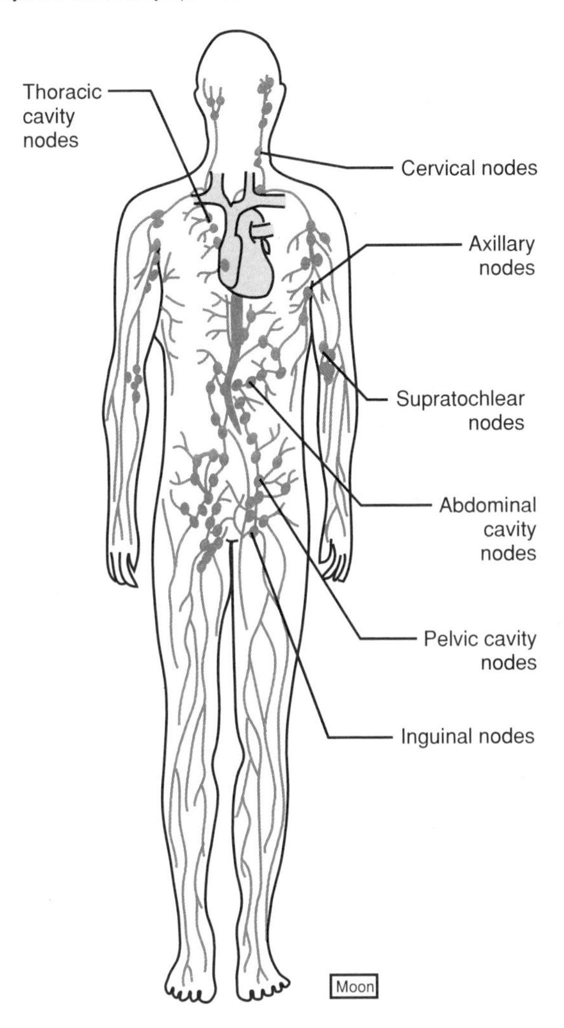

Functions of Lymph Nodes

As mentioned previously, the lymph nodes contain large numbers of lymphocytes. The lymph nodes, in fact, are centers for lymphocyte production, although such cells are produced in red bone marrow as well (see chapter 14). Lymphocytes act against foreign particles, such as bacterial cells and viruses, that are carried to the lymph nodes by the lymphatic vessels.

The lymph nodes also contain macrophages, which engulf and destroy foreign substances, damaged cells, and cellular debris. The functions of the cells associated with the lymph nodes are described in a subsequent section of this chapter.

1. How would you distinguish between a lymph node and a lymph nodule?
2. What are the major functions of the lymph nodes?

Thymus and Spleen

Two other lymphatic organs, whose functions are closely related to those of the lymph nodes, are the thymus and the spleen.

Thymus

The **thymus** (thymus gland) shown in figure 16.8, is a soft, bilobed structure whose lobes are surrounded by connective tissue. It is located in front of the aorta and behind the upper part of the sternum.

The thymus is composed of lymphatic tissue, which is subdivided into *lobules* by connective tissues. The lobules contain large numbers of lymphocytes. The majority of these cells (thymocytes) remain inactive; however, some of them develop into a group (T-lymphocytes) that leaves the thymus and functions in providing immunity.

The thymus may also secrete a hormone called *thymosin,* which is thought to stimulate the maturation of certain lymphocytes (T-lymphocytes) after they leave the thymus and migrate to other lymphatic tissues.

Spleen

The **spleen** is the largest of the lymphatic organs. It is located in the upper left portion of the abdominal cavity, just beneath the diaphragm and behind the stomach.

The spleen resembles a large lymph node and is subdivided into chambers or lobules. Unlike the sinuses in a lymph node, however, the spaces (venous sinuses) within the chambers of the spleen are filled with blood instead of lymph (fig. 16.9).

Within the lobules of the spleen, the tissues are called *pulp* and are of two types: white pulp and red pulp. The *white pulp* is distributed throughout the spleen in tiny islands. This tissue is composed of splenic nodules, which are similar to those found in lymph nodes, and they contain large numbers of lymphocytes. The *red pulp,* which fills the remaining spaces of the lobules, surrounds the venous sinuses. Red pulp contains many red blood cells, which are responsible for its color, along with numerous lymphocytes and macrophages.

The blood capillaries within the red pulp are quite permeable. Red blood cells can squeeze through the walls of these capillaries and enter the venous sinuses. The older, more fragile red blood cells may rupture as they make this passage, and the resulting cellular debris is removed by phagocytic macrophages located within the splenic sinuses.

Figure 16.8
(*a*) The thymus gland is a bilobed organ, located between the lungs and above the heart. (*b*) Note how the thymus is subdivided into lobules in this light micrograph (×10).

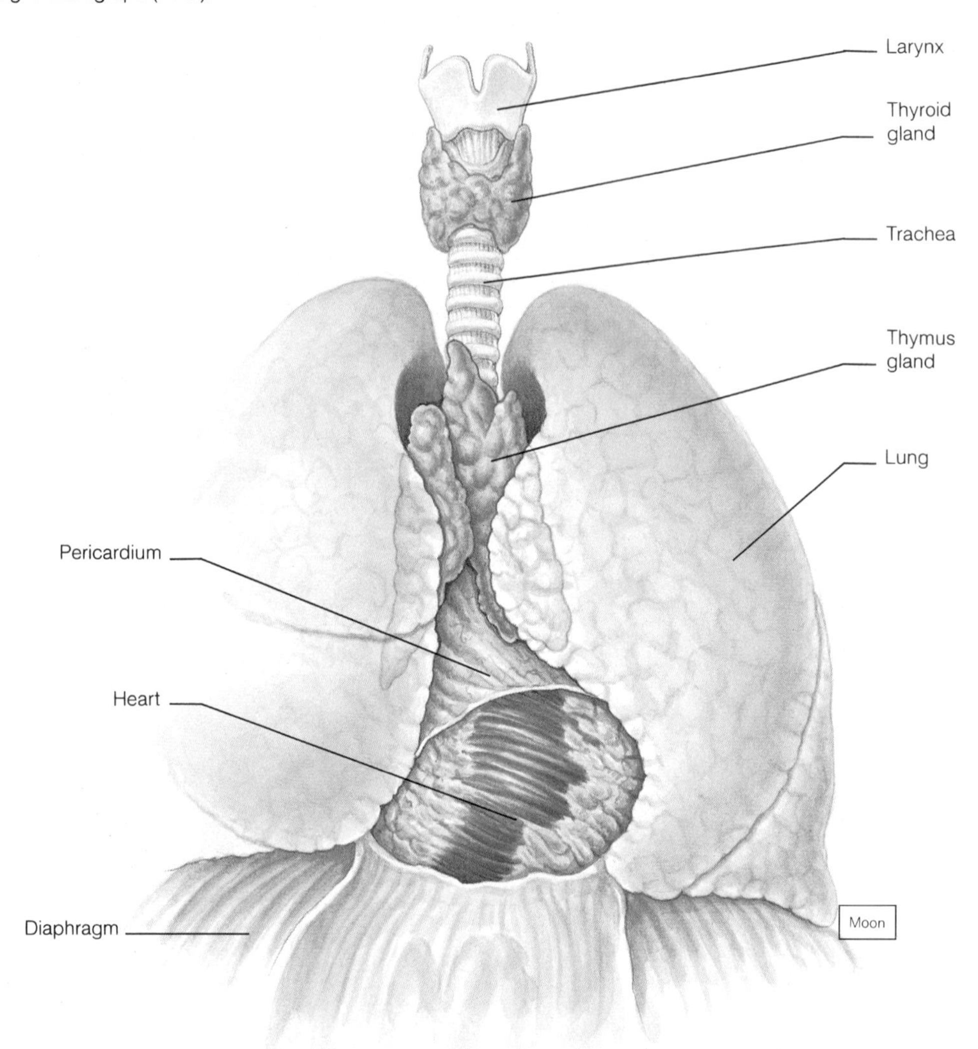

(a)

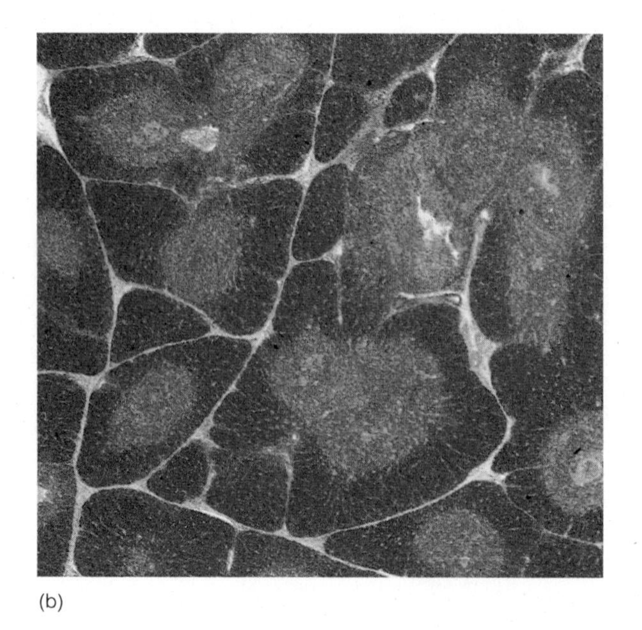

(b)

Figure 16.9
(*a*) The spleen resembles a large lymph node. (*b*) Light micrograph of the spleen (×40).

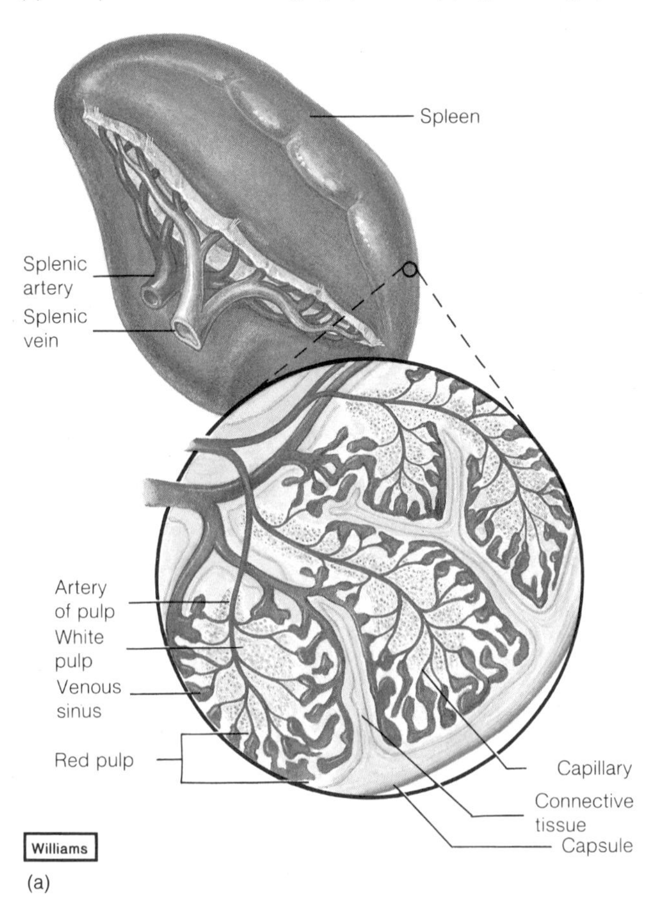

(a)

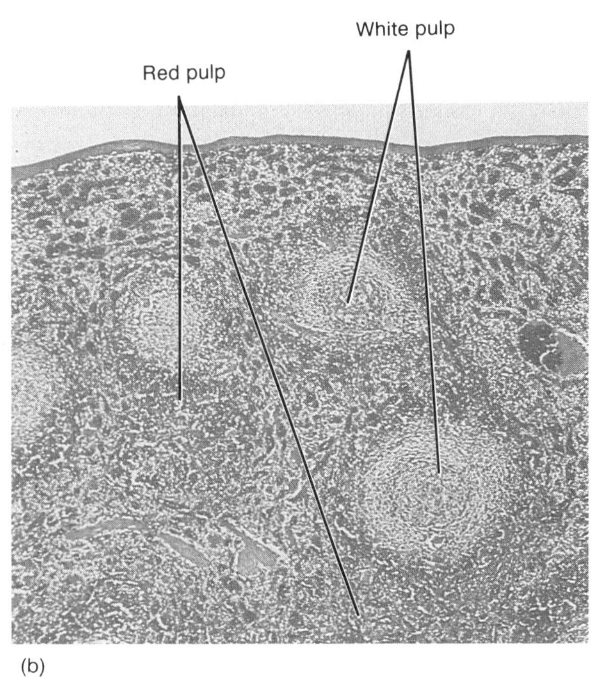

(b)

The macrophages also engulf and destroy foreign particles, such as bacteria, that may be carried in the blood as it flows through the splenic sinuses. Thus, the spleen filters the blood much as lymph nodes filter the lymph.

1. Why are the thymus and spleen considered to be organs of the lymphatic system?
2. What are the major functions of the thymus and the spleen?

Body Defenses against Infection

An infection is a condition caused by the presence of a disease-causing agent within the body. Such agents are termed **pathogens,** and they include viruses and microorganisms such as bacteria, fungi, and protozoans, as well as various other parasitic forms of life.

The human body is equipped with a variety of *defense mechanisms* that help prevent the entrance of pathogens or destroy them if they enter the tissues. Some defense mechanisms are quite general, or *nonspecific,* in that they protect against many types of pathogens. These mechanisms include species resistance, mechanical barriers, the actions of enzymes, interferon, inflammation, and phagocytosis.

Other defense mechanisms are very *specific* in their actions, providing protection against particular disease-causing agents. These mechanisms are responsible for the type of resistance called *immunity.*

Nonspecific Resistance

Species Resistance

Species resistance refers to the fact that a given kind of organism, or *species* (such as the human species, *Homo sapiens*), develops diseases that are unique to it. At the same time, a species may be resistant to diseases that affect other species, because its tissues somehow fail to provide the temperature or chemical environment needed by a particular pathogen. For example, humans are subject to infections by the microorganisms that cause measles, mumps, gonorrhea, and syphilis, but other animal species are generally resistant to these diseases.

Mechanical Barriers

The *skin* and *mucous membranes* lining the tubes of the respiratory, digestive, urinary, and reproductive systems create **mechanical barriers** against the entrance of infectious agents. As long as these barriers remain unbroken, many pathogens are unable to penetrate them.

Enzymatic Actions

Enzymatic actions against disease-causing agents are due to the presence of enzymes in various body fluids. Gastric juice, for example, contains the protein-splitting enzyme *pepsin* and has a low pH due to the presence of hydrochloric acid. The combined effect of these substances is lethal to many pathogens that enter the stomach. Similarly, tears contain the enzyme *lysozyme,* which has an antibacterial action against certain pathogens that get onto the surfaces of the eyes.

Interferon

Interferon is the name given to a group of proteins produced by the cells in response to the presence of *viruses* or the cells of certain *tumors.* Although the effect of interferon is nonspecific, it interferes with the proliferation of viruses and, thus, helps control diseases they cause. Interferon also stimulates the activity of certain cells that act against the growth of tumors.

Inflammation

Inflammation is a response that may accompany tissue invasion by pathogens.

The major symptoms of inflammation include localized redness, swelling, heat, and pain. The *redness* is a result of blood vessel dilation and the consequent increase in blood volume within the affected tissues. This effect, coupled with an increase in the permeability of the nearby capillaries, is responsible for tissue *swelling* (edema). The *heat* is due to the presence of blood from the deeper body parts, which is generally warmer than that near the surface, and the *pain* results from the stimulation of nearby pain receptors.

White blood cells tend to accumulate at the sites of inflammation, and some of these cells help to control pathogens by *phagocytosis.*

Body fluids also tend to collect in inflamed tissues. These fluids contain *fibrinogen* and other blood clotting factors. As a result of clotting, a network of fibrin threads may develop within the affected region. Later, *fibroblasts* may appear and form fibers around the area until it is enclosed in a sac of fibrous connective tissue. This action inhibits the spread of pathogens and toxic substances they may release to the adjacent tissues.

Phagocytosis

As mentioned in chapter 14, the most active phagocytic cells of the blood are *neutrophils* and *monocytes.* These wandering cells can leave the bloodstream by squeezing between the cells of the blood vessel walls. They are attracted toward sites of inflammation by the chemicals released from injured tissues (chemotaxis).

Monocytes give rise to macrophages (histiocytes), which become fixed in various tissues and attached to the inner walls of blood and lymphatic vessels. These relatively nonmotile, phagocytic cells can divide and produce new macrophages and are found in such organs as the lymph nodes, spleen, liver, and lungs. This diffuse group of macrophages constitutes the **reticuloendothelial tissue** (reticuloendothelial system, or RES).

As a result of reticuloendothelial activities, foreign particles are removed from the lymph as it moves from the interstitial spaces to the bloodstream. Any such foreign particles that reach the blood are likely to be removed by the phagocytes located in the vessels and tissues of the spleen, liver, or bone marrow.

---

1. What is meant by an infection?
2. Explain six nonspecific defense mechanisms.
3. Define *reticuloendothelial tissue.*

Immunity

Immunity is resistance to specific foreign agents, such as pathogens, or to the toxins they release. It involves a number of *immune mechanisms,* whereby certain cells recognize the presence of particular foreign substances and act against them. The cells that function in immune mechanisms include lymphocytes and macrophages.

Origin of Lymphocytes

During fetal development (before birth), undifferentiated lymphocytes (stem cells) are released from red bone marrow and carried away by the blood (figs. 16.10 and 16.11). About half of them reach the thymus gland, where they remain for a time. Within the thymus, these cells undergo special processing and become differentiated. Thereafter, they are called *T-lymphocytes* (thymus-derived lymphocytes), or *T-cells.* Later, these T-cells are transported away from the thymus by the blood, wherein they comprise 70% to 80% of the cir-

Figure 16.10
Bone marrow releases undifferentiated lymphocytes, which after processing become T-cells and B-cells.

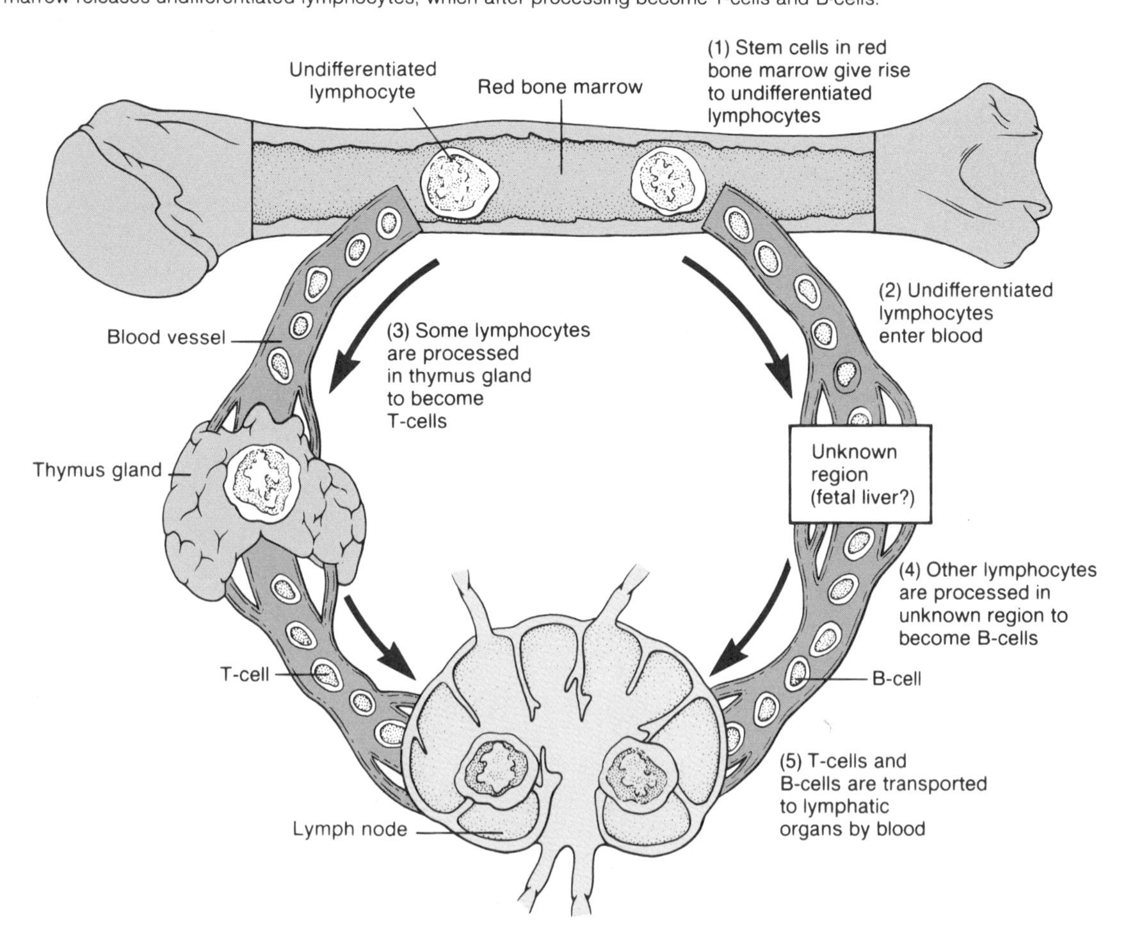

culating lymphocytes. They tend to reside in various organs of the lymphatic system and are particularly abundant in the lymph nodes, thoracic duct, and spleen.

Lymphocytes that have been released from the bone marrow and do not reach the thymus gland are processed in another part of the body (probably the fetal liver), and they differentiate into *B-lymphocytes,* or B-cells.

B-cells are distributed by the blood and constitute 20% to 30% of the circulating lymphocytes. They settle in the lymphatic organs with the T-cells and are abundant in the lymph nodes, spleen, bone marrow, secretory glands, intestinal lining, and reticuloendothelial tissue.

Antigens

Prior to birth, the body cells somehow make an inventory of the proteins and various other large molecules that are present in the body. Subsequently, the substances that are present (self-substances) can be

Figure 16.11
A scanning electron micrograph of a human circulating lymphocyte (×36,000).

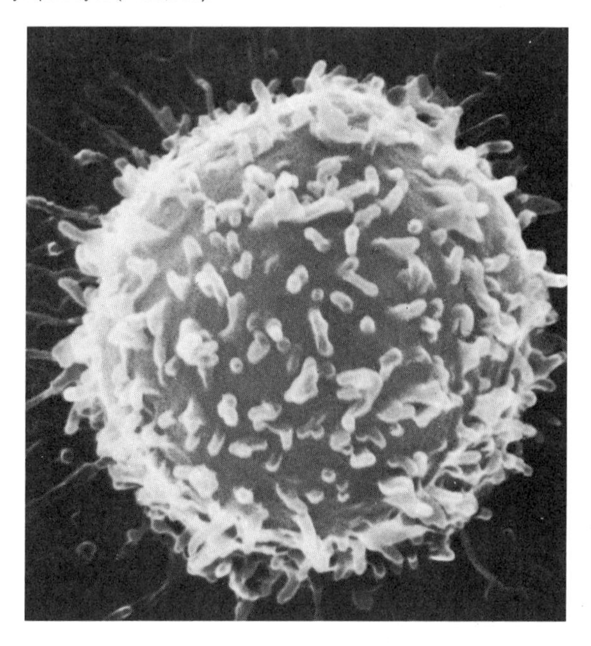

Chart 16.2 A comparison of T-cells and B-cells		
Characteristic	**T-cells**	**B-cells**
Origin of undifferentiated cells	Red bone marrow	Red bone marrow
Site of differentiation	Thymus gland	Region outside the thymus gland (probably the fetal liver)
Primary locations	Lymphatic tissues, 70%-80% of the circulating lymphocytes	Lymphatic tissues, 20%-30% of the circulating lymphocytes
Primary functions	Responsible for cell-mediated immunity in which T-cells interact with antigen-bearing agents directly	Responsible for antibody-mediated immunity in which B-cells interact with antigen-bearing agents indirectly by producing antibodies

distinguished from foreign substances (nonself-substances). More specifically, T-cells and B-cells develop receptors on their surfaces that allow them to recognize foreign substances. When such substances are recognized, these lymphocytes can produce immune reactions against them.

Foreign substances, such as proteins, polysaccharides, and some glycolipids, to which lymphocytes respond are called **antigens.**

1. Define *immunity.*
2. Explain how T-cells and B-cells originate.
3. What is an antigen?

Functions of Lymphocytes

T-cells and B-cells respond to antigens they recognize in different ways. For example, some T-cells attach themselves to antigen-bearing cells, such as bacterial cells, and interact with these foreign cells directly—that is, with cell-to-cell contact. This type of response is called **cell-mediated immunity** (CMI). B-cells, on the other hand, act indirectly against antigen-bearing agents by producing and secreting globular proteins (immunoglobulins) called *antibodies.* The antibodies, in turn, are carried by the body fluids and react in various ways to destroy specific antigens or antigen-bearing agents. This type of response is called **antibody-mediated immunity** (AMI) or humoral immunity. Chart 16.2 compares the characteristics of T-cells and B-cells.

Within the populations of T-cells and B-cells of each person are millions of varieties. The members of each variety originate from a single early cell, so that the members are all alike. Also, the members of each variety have a particular type of antigen-receptor on their surface membrane that is capable of responding only to a specific antigen. The term *clone* is used to refer to the members of each variety as a group.

Activation of B-cells.

Before a lymphocyte can respond to the presence of an antigen, the lymphocyte must be activated. A B-cell, for example, becomes activated when it encounters an antigen whose molecular shape fits the shape of the B-cell's antigen receptors. In response to the receptor-antigen combination, the B-cell proliferates by mitosis, and its clone is enlarged (fig. 16.12).

Some of the newly formed members of the activated B-cell's clone become *plasma cells,* and they make use of their DNA information and protein-synthesizing mechanism (see chapter 4) to produce antibody molecules. These antibody molecules are similar in structure to the antigen-receptor molecules that were present on the original B-cell's surface membrane. Thus, the antibodies are able to combine with the antigen-bearing agent that has invaded the body and react against it.

It is estimated that an individual's B-cells are able to produce as many as ten million to one billion different varieties of antibodies, each of which reacts against a specific antigen. This ability provides defense against a very large number of different disease-causing agents.

Types of Antibodies

The antibodies produced and secreted by B-cells are soluble globular proteins. These proteins are called **immunoglobulins,** and they constitute a portion of the *gamma globulin* fraction of the plasma proteins (see chapter 14).

The three most abundant types of immunoglobulins are immunoglobulin G, immunoglobulin A, and immunoglobulin M.

Immunoglobulin G (IgG) occurs in the plasma and tissue fluids and is particularly effective against bacterial cells, viruses, and various toxins. It functions in a variety of ways. For example, it may cause antigen-bearing particles to form insoluble precipitates, decompose the membranes of foreign cells, or alter cell membranes so the cells become more susceptible to phagocytosis.

Immunoglobulin A (IgA) is commonly found in the secretions of various *exocrine glands.* It occurs in milk, tears, nasal fluid, gastric juice, intestinal juice,

Figure 16.12
When a B-cell encounters an antigen that fits its antigen receptor, it becomes activated and proliferates, thus enlarging its clone.

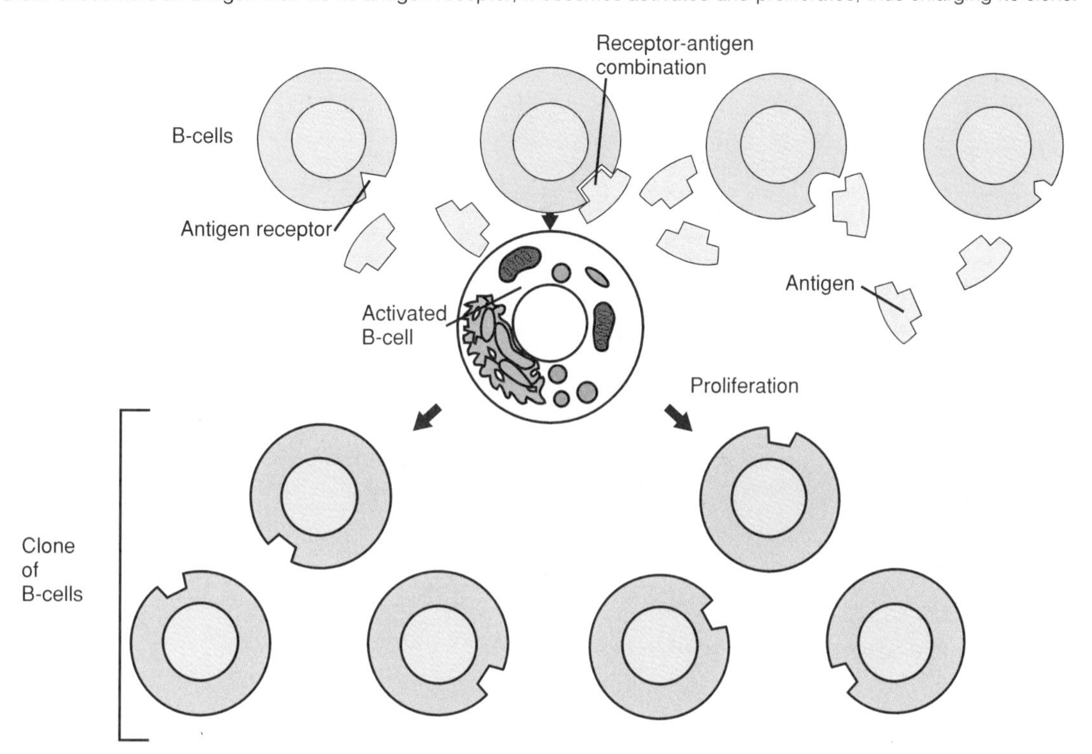

bile, and urine. IgA helps control certain respiratory viruses and various pathogens responsible for digestive disturbances. It causes antigens to form insoluble precipitates and causes antigen-bearing cells to clump together.

> IgA can pass from a nursing mother to her baby via milk and other breast secretions. These antibodies provide the infant with some protection against digestive disturbances that might otherwise cause serious problems.

Immunoglobulin M (IgM) is a type of antibody that develops in the blood plasma, apparently in response to contact with certain antigens in foods or bacteria. The agglutinins anti-A and anti-B described in chapter 14 are examples of IgM.

1. How do the functions of T-cells and B-cells differ?
2. How do B-cells become activated?
3. What is an immunoglobulin?
4. Which immunoglobulins are most abundant, and how do they differ?

Activation of T-cells.

Although a B-cell can by itself recognize the antigen for which it is specialized to react, a T-cell requires the presence of another kind of cell, called an *accessory cell* (antigen-presenting cell), before the T-cell can become activated. Macrophages, B-cells, and several other types of body cells can serve as accessory cells.

For example, when an antigen-bearing agent, such as an invading bacterium, is phagocytized by a macrophage, the agent is digested by the macrophage's lysosomes. Antigens that are released by the lysosomal digestive process are then moved to the macrophage's surface membrane and displayed on the membrane in association with certain protein molecules, called the *major histocompatibility complex* (MHC). A specialized type of T-cell, called a *T-helper cell,* may contact such a displayed antigen. If the antigen fits and combines with the T-helper cell's antigen receptors, the T-helper cell becomes activated.

When an activated T-helper cell encounters a B-cell that has already combined with an identical antigen, the T-helper cell interacts with the B-cell. As a result of this interaction, the T-helper cell releases some hormonelike substances called *lymphokines.* The lymphokines, in turn, enhance the B-cell reaction to the antigen by stimulating the B-cell to proliferate, thus enlarging its clone of antibody-producing cells (fig. 16.13). Lymphokines may also attract macrophages and leukocytes into inflammed tissues and help retain them there. Chart 16.3 summarizes the steps leading to antibody production as a result of B-cell activity or T-cell activity.

Figure 16.13
(*a*) After digesting antigen-bearing agents, a macrophage displays antigens on its surface; (*b*) T-helper cells become activated when they contact and remove displayed antigens that fit their antigen receptors; (*c*) an activated T-helper cell interacts with a B-cell that has combined with an identical antigen and causes the B-cell to proliferate.

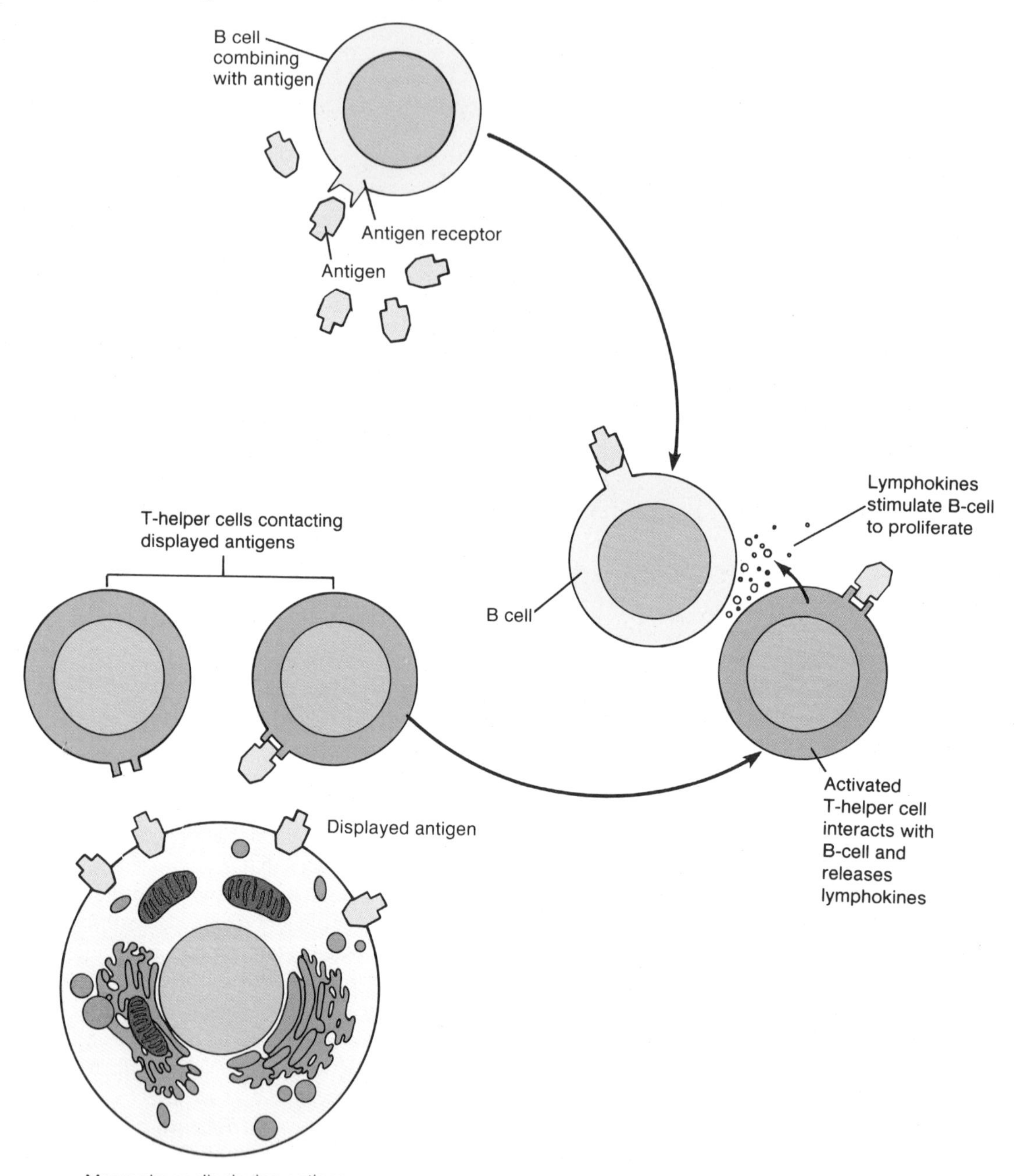

A second type of T-cell, called a *cytotoxic T-cell*, is specialized to recognize antigens that are produced in the cytoplasm of various body cells. For example, cells of tumors and cells that contain viruses usually produce specific kinds of antigens that are displayed on their surface membranes in association with MHC molecules. If a cytotoxic T-cell encounters an antigen that fits and combines with its antigen receptors, the cytotoxic T-cell becomes activated.

Once activated, a cytotoxic T-cell (killer T-cell) proliferates, enlarging its clone. Cytotoxic T-cells can bind to the surfaces of antigen-bearing cells, and the cytotoxic T-cells then release a protein that causes porelike openings to appear in the antigen-bearing cells' membranes, thus destroying such cells. In this manner, cytotoxic T-cells, which continually monitor the body cells, can recognize and eliminate abnormal tumor cells and cells infected with viruses.

Chart 16.3 Steps in the production of antibodies

B-cell activity

1. Antigen-bearing agents enter body tissues.
2. B-cell becomes activated when it encounters an antigen that fits its antigen receptors.
3. Activated B-cell proliferates, enlarging its clone.
4. Some of the newly formed B-cells become plasma cells.
5. Plasma cells synthesize and secrete antibodies whose molecular structure is similar to the activated B-cell's antigen receptors.
6. Antibodies combine with antigen-bearing agents, thus helping to destroy them.

T-cell activity

1. Antigen-bearing agents enter body tissues.
2. Accessory cell, such as a macrophage, phagocytizes antigen-bearing agent, and the macrophage's lysosomes digest the agent.
3. Antigens from the digested antigen-bearing agents are displayed on the surface membrane of the accessory cell.
4. T-helper cell becomes activated when it encounters a displayed antigen that fits its antigen receptors.
5. Activated T-helper cell releases lymphokines when it encounters a B-cell that has previously combined with an identical antigen-bearing agent.
6. Lymphokines stimulate the B-cell to proliferate.
7. Some of the newly formed B-cells become antibody-secreting plasma cells.
8. Antibodies combine with antigen-bearing agents, thus helping to destroy them.

Immune Responses

When B-cells or T-cells become activated after first encountering the antigens for which they are specialized to react, their actions constitute a **primary immune response.** During such a response, the plasma cells that are formed release antibodies into the lymph. The antibodies are transported to the blood and then to all body parts, where they help to destroy antigen-bearing agents. The production and release of the antibodies continues for several weeks.

Following a primary immune response, some of the B-cells that were produced during proliferation of the clone remain dormant and serve as *memory cells* (fig. 16.14). In this way, if the identical antigen is encountered in the future, the clones of these memory cells increase in size, and they can respond rapidly to the antigen to which they were previously sensitized. Such a subsequent reaction is called a **secondary immune response.**

As a result of a primary immune response, detectable concentrations of antibodies usually appear in the body fluids within five to ten days following an exposure to antigens. If the identical antigen is encountered some time later, a secondary immune response may produce additional antibodies within a day or two. Although such newly formed antibodies may persist in the body for only a few months or perhaps a few years, the memory cells live much longer. Consequently, the ability to produce a secondary immune response may be long-lasting.

1. How do T-cells become activated?
2. What is the function of lymphokines?
3. How do cytotoxic T-cells destroy antigen-bearing cells?
4. What is the difference between a primary and a secondary immune response?

Types of Immunity

One type of immunity is called *naturally acquired active immunity.* It occurs when a person who has been exposed to a live pathogen develops a disease and becomes resistant to that pathogen as a result of a primary immune response.

Another type of active immunity can be produced in response to a **vaccine.** Such a substance contains an antigen that can stimulate a primary immune response against a particular disease-causing agent, but does not cause severe disease symptoms.

A vaccine, for example, might contain bacteria or viruses that have been killed or weakened so that they cannot cause a serious infection; or it may contain a toxin of an infectious organism that has been chemically altered to destroy its toxic effects. In any case, the antigens present still retain the characteristics needed to stimulate a primary immune response. Thus, a person who has been vaccinated is said to develop *artificially acquired active immunity.*

Figure 16.14
An activated B-cell proliferates and gives rise to antibody-secreting plasma cells and dormant memory cells.

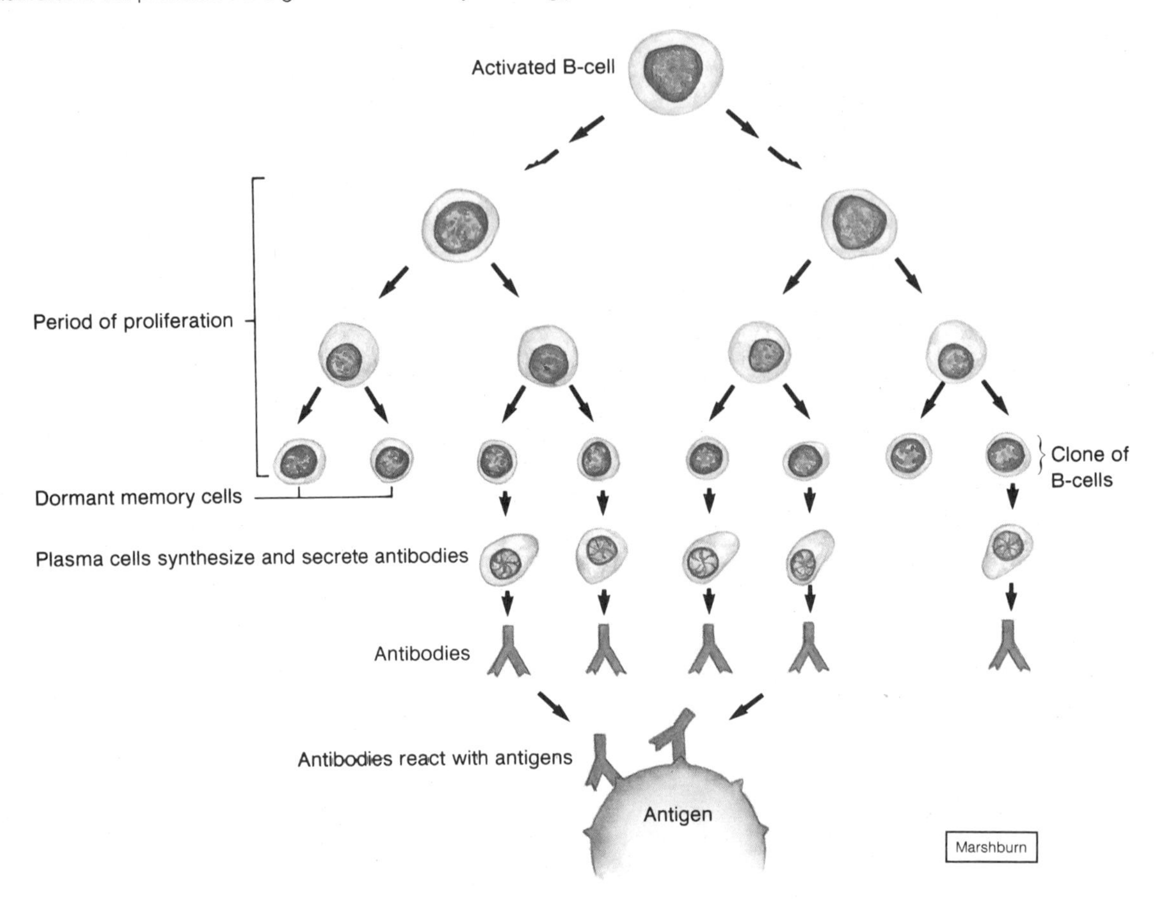

Sometimes a person needs protection against a disease-causing microorganism, but lacks the time needed to develop active immunity. In such a case, it may be possible to inject the person with ready-made antibodies. These antibodies may be obtained from *gamma globulin* (chapter 14) separated from the blood plasma of persons who have already developed immunity against the particular disease.

A person who receives an injection of gamma globulin is said to have *artificially acquired passive immunity*. This type of immunity is called passive because the antibodies involved are not produced by the recipient's cells. Such immunity is relatively short-term, seldom lasting more than a few weeks. Furthermore, because the recipient's lymphocytes have not been activated to respond against the pathogens for which the protection was needed, the person continues to be susceptible to those pathogens in the future.

During pregnancy, certain antibodies (IgG) are able to pass from the maternal blood through the placental membrane and into the fetal bloodstream. As a result, the fetus acquires a limited immunity against the pathogens for which the mother has developed active immunities. In this case, the fetus is said to have *naturally acquired passive immunity*, which may remain effective for six months to a year after birth.

Competence of the immune system tends to decline with advancing age. Consequently, elderly persons have a higher incidence of tumors and an increased susceptibility to infections, such as influenza, pneumonia, and tuberculosis. This decline in effectiveness is primarily due to a loss of T-cells. B-cell activity, on the other hand, is relatively unchanged with age.

The types of immunity are summarized in chart 16.4.

Allergic Reactions

Allergic reactions are closely related to immune responses in that both may involve the sensitizing of lymphocytes or the combining of antigens with antibodies. Allergic reactions, however, are likely to be excessive and to cause tissue damage.

Acquired Immune Deficiency Syndrome

A Current Topic

Acquired immune deficiency syndrome (AIDS) is a disease that was first recognized in 1981 and whose incidence is roughly doubling each year. It is caused by a virus called human immunodeficiency virus (HIV-1) that is widespread throughout the world. (A related virus, called HIV-2, which can also cause AIDS, occurs mainly in West Africa.) This virus infects T-helper cells—the cells required to activate B-lymphocytes and induce the production of antibodies—as well as macrophages, B-cells, endothelial cells, and certain neuroglial cells.

When the virus infects cells, its genes are incorporated into the host cell's DNA. Then, the virus may become dormant, failing to produce any noticeable effects for some time. During this latency period, which may last for several years, the person may show no symptoms of AIDS, but can probably transmit the virus to uninfected persons.

If the viral genes in the T-helper cells are activated by some unknown factor, they induce the production of new viral particles that can leave the host cell and infect other T-helper cells and kill them. As a consequence of widespread T-helper cell destruction, the person's immune system is greatly impaired, and the group of symptoms (syndrome) that characterize AIDS develops. These symptoms include enlargement of lymph nodes, weight loss, and fever, as well as severe infections caused by a variety of bacteria, protozoans, fungi, and viruses, (following depression of the immune system) and the appearance of certain forms of cancer (Kaposi's sarcoma, carcinoma of the skin or rectum, and B-cell lymphoma). Some patients develop serious neurological dysfunctions, including varying degrees of paralysis in the extremities, various sensory deficiencies, loss of deep tendon reflexes, exaggerated sensitivity to touch, and weakness in the urethral and anal sphincter muscles.

The transmission of this disease seems to require the direct introduction of the AIDS virus into the blood, as may result during certain sexual activities (such as anal intercourse in which the skin or mucous membranes are damaged), from the use of contaminated hypodermic needles, or from the transfusion of virus-containing blood or blood products. AIDS apparently is not spread by casual contact with AIDS patients. In fact, a study involving several hundred family members with AIDS patients living at home revealed that the AIDS virus had not been transmitted to any of the household members, except by sexual contact.

The persons who are at greatest risk for contracting AIDS are homosexual and bisexual males (who have comprised about 70% of the AIDS patients in recent years), heterosexual drug abusers (19%), and heterosexuals (male or female) whose sexual partners are infected (4%). The remaining patients (7%) do not belong to one of these high-risk groups and acquired the disease from some other source, such as a transfusion of or skin exposure to contaminated blood or blood products. The unborn fetuses of infected mothers, for example, acquire the disease in about 50% of the cases. As a result, one of the more rapidly growing groups of AIDS patients is comprised of children born to infected mothers who are intravenous drug users or whose sexual partners are intravenous drug users.

Antibodies appear in the blood of persons infected by the AIDS virus, and these antibodies can usually be detected with a simple blood test performed between two weeks and three months following exposure to the virus. If antibodies do not develop within several months, it is assumed that the person has not become infected.

At present no vaccine is available to prevent AIDS, and there is no cure for the disease. To prevent contact with the AIDS virus (as well as other blood-borne viruses, such as the hepatitis B virus), healthcare workers are advised to avoid puncturing their skin with needles contaminated with blood or blood products. They should also take special precautions to prevent skin and mucous membrane contact with body fluids of all patients, including vaginal secretions, semen, cerebrospinal fluid, synovial fluid, serous fluids from the pleural, peritoneal, or pericardial cavities, and any other body fluids that contain visible blood.

According to the Surgeon General of the U.S. Public Health Service, "The most certain way to avoid getting the AIDS virus and to control the AIDS epidemic in the United States is for individuals to avoid promiscuous sexual practices, to maintain mutually faithful monogamous sexual relationships, and to avoid injecting illicit drugs."

In the absence of monogamy, the best way for a sexually active person to prevent getting AIDS is by using a condom during sexual intercourse, although this method is not 100% effective in preventing infections.

Chart 16.4 Types of immunity

Type	Stimulus	Result
Naturally acquired active immunity	Exposure to live pathogens	Symptoms of a disease and stimulation of an immune response
Artificially acquired active immunity	Exposure to a vaccine containing weakened or dead pathogens	Stimulation of an immune response without the severe symptoms of a disease
Artificially acquired passive immunity	Injection of gamma globulin, containing antibodies	Immunity for a short time without stimulating an immune response
Naturally acquired passive immunity	Antibodies passed to a fetus from a mother with active immunity	Short-term immunity for the infant, without stimulating an immune response

Autoimmune Diseases

A Current Topic

Immune responses are usually directed toward foreign molecules (nonself substances), while molecules of the body (self-substances) are tolerated by the immune mechanism. Occasionally, something happens to change this, and the tolerance to self-substances is lost. As a result, cell-mediated and antibody-mediated immune responses may be directed toward a person's own tissues. Such a condition is called an *autoimmune disease* (autoallergy).

Although the cause of autoimmune diseases is not well understood, they are more common in older persons, and they may develop following viral or bacterial infections. Some investigators believe that the release of abnormally large quantities of antigens may occur when infectious agents damage tissues. These agents may also cause body proteins to change into forms that stimulate antibody production. At the same time, the activity of suppressor cells, which normally limits this type of reaction, seems to be repressed.

Some autoimmune diseases affect specific organs, while others produce more general effects. For example, in the disease *autoimmune thyroiditis*, the effects are directed toward the thyroid gland. In this condition, antibodies are produced by B-cells that respond abnormally to self-proteins associated with thyroid hormones and to substances in thyroid epithelial cell membranes. As a result, the thyroid gland may become inflamed, and much of its tissue may be destroyed.

Systemic lupus erythematosus (SLE), which occurs most commonly in young women, is an example of an autoimmune disease that affects many tissues and organs. In this case, the B-cells produce antibodies that react with cell nuclei and various cytoplasmic substances. The resulting antigen-antibody combinations accumulate in membranes of the kidneys, heart, lungs, blood vessels, and joints, creating functional problems in these organs. One of the early symptoms of SLE is an inflammation of the skin over the nose and cheeks.

Other autoimmune diseases include myasthenia gravis, juvenile rheumatoid arthritis, insulin-dependent diabetes mellitus, and multiple sclerosis.

One form of allergic reaction can occur in almost anyone, but another form affects only those people who have inherited from their parents an ability to produce exaggerated immune responses.

A *delayed-reaction allergy* is an allergic response that may occur in anyone. It results from repeated exposure of the skin to certain chemical substances—commonly, household or industrial chemicals or some cosmetics. As a consequence of these repeated contacts, the T-cells eventually become activated by the foreign substance, and a large number of T-cells collect in the skin. Their actions and the actions of the macrophages they attract cause the release of various chemical factors, which, in turn, cause eruptions and inflammation of the skin (dermatitis). This reaction is called *delayed* because it usually takes about forty-eight hours to occur.

In other cases, an allergic reaction may occur within minutes after contact with a nonself substance. Persons with this *immediate-reaction allergy* have an inherited ability to synthesize abnormally large quantities of antibodies (immunoglobulin E) in response to certain antigens. In this instance, the allergic reaction involves the activation of B-cells, and the antigen that triggers the reaction is called an **allergen.**

In an immediate-reaction allergy, the B-cells become activated when the allergen is first encountered, and subsequent exposures trigger reactions. The immunoglobulin involved (IgE), is attached to the membranes of certain widely distributed cells (mast cells), and when an allergen-antibody reaction occurs, the cells release substances such as *histamine* and *serotonin.* These substances, in turn, cause a variety of physiological effects, including the dilation of blood vessels, swelling of tissues, and contraction of smooth muscles. The result is a severe inflammation reaction that is responsible for the symptoms of the allergy—hives, hay fever, asthma, eczema, or gastric disturbances.

In most individuals, such severe responses to nonself substances are inhibited by special T-cells, called *suppressor cells.* In a person with an immediate-reaction allergy, however, this control function of T-cells seems to be less effective. Even so, the suppressor cells eventually interfere with the production of antibodies, and the allergic reaction is terminated.

Transplantation and Tissue Rejection

It is occasionally desirable to transplant some tissue or an organ, such as the skin, kidney, heart, or liver, from one person to another to replace a nonfunctional, damaged, or lost body part. In such cases, there is a danger that the recipient's cells may recognize the donor's tissues as being foreign. This triggers the recipient's immune mechanisms, which may act to destroy the donor tissue. Such a response is called a **tissue rejection reaction.**

Tissue rejection involves the activities of lymphocytes and both cell- and antibody-mediated responses—responses similar to those that occur when other nonself substances are encountered. The greater the antigenic difference between the proteins of the recipient tissues and the donor tissues, the more rapid and severe the rejection reaction will be. Thus, the reaction can sometimes be minimized by matching the recipient and donor tissues. This means locating a donor whose tissues are antigenically similar to those of the person needing a transplant—a procedure much like matching the blood of a donor with that of a recipient before giving a blood transfusion.

Another approach to reducing the rejection of transplanted tissue involves the use of *immunosuppressive drugs*. These substances act by interfering with the recipient's immune mechanisms. A drug may, for example, suppress the formation of antibodies, or it may cause the destruction of lymphocytes and so prevent them from producing antibodies.

Unfortunately, the use of immunosuppressive drugs leaves the recipient relatively unprotected against infections. Although the drug may prevent a tissue rejection reaction, the recipient may develop a serious infectious disease that is difficult to control.

1. Explain the difference between active and passive immunities.
2. In what ways is an allergic reaction related to an immune reaction?
3. In what ways is a tissue rejection reaction related to an immune response?

Clinical Terms Related to the Lymphatic System and Immunity

anaphylaxis (an″ah-fĭ-lak′sis)—immediate hypersensitivity to the presence of a foreign substance.
asplenia (ah-sple′ne-ah)—the absence of a spleen.
autograft (aw′to-graft)—the transplantation of tissue from one part of the body to another part of the same body.
histocompatibility (his″to-kom-pat*ĭ*-bil′ ĭ-te)—compatibility between the tissues of a donor and the tissues of a recipient based on antigenic similarities.
homograft (ho′mo-graft)—the transplantation of tissue from one person to another person.
immunocompetence (im″u-no-kom′pe-tens)—the ability to produce an immune response to the presence of antigens.
immunodeficiency (im″u-no-de-fish′en-se)—a lack of the ability to produce an immune response.
lymphadenectomy (lim-fad″ĕ-nek′to-me)—the surgical removal of lymph nodes.
lymphadenopathy (lim-fad″ĕ-nop′ah-the)—enlargement of the lymph nodes.
lymphadenotomy (lim-fad″ĕ-not′o-me)—an incision of a lymph node.
lymphocytopenia (lim″fo-si″to-pe′ne-ah)—an abnormally low concentration of lymphocytes in the blood.
lymphocytosis (lim″fo-si″to′sis)—an abnormally high concentration of lymphocytes in the blood.
lymphoma (lim-fo′mah)—a tumor composed of lymphatic tissue.
lymphosarcoma (lim″fo-sar-ko′mah)—a cancer within the lymphatic tissue.
splenectomy (sple-nek′to-me)—the surgical removal of the spleen.
splenitis (sple-ni′tis)—an inflammation of the spleen.
splenomegaly (sple″no-meg′ah-le)—an abnormal enlargement of the spleen.
splenotomy (sple-not′o-me)—an incision of the spleen.
thymectomy (thi-mek′to-me)—the surgical removal of the thymus gland.
thymitis (thi-mi′tis)—an inflammation of the thymus gland.

Chapter Summary

Introduction (page 428)

The lymphatic system is closely associated with the cardiovascular system, transports excess tissue fluid to the bloodstream, and helps defend the body against disease-causing agents.

Lymphatic Pathways (page 428)

1. Lymphatic capillaries
   a. Lymphatic capillaries are microscopic closed-ended tubes.
   b. They receive lymph through their thin walls.
2. Lymphatic vessels (page 428)
   a. Lymphatic vessels have walls similar to veins and possess valves.
   b. They lead to the lymph nodes and then merge into the lymphatic trunks.
3. Lymphatic trunks and collecting ducts
   a. The lymphatic trunks lead to the collecting ducts.
   b. The collecting ducts join the subclavian veins.

Tissue Fluid and Lymph (page 430)

1. Tissue fluid formation
   a. Tissue fluid originates from blood plasma.
   b. It generally lacks proteins, but some smaller protein molecules leak into the interstitial spaces.
   c. As the protein concentration of the tissue fluid increases, the osmotic pressure increases also.
2. Lymph formation
   a. Rising osmotic pressure in the tissue fluid interferes with the return of water to the blood.
   b. Increasing pressure within the interstitial spaces forces some tissue fluid into the lymphatic capillaries.
3. Function of lymph
   a. Lymph returns protein molecules to the bloodstream.
   b. It transports foreign particles to the lymph nodes.

Movement of Lymph (page 431)

Forces that aid in the movement of lymph include the squeezing action of skeletal muscles and breathing movements.

Lymph Nodes (page 431)

1. Structure of a lymph node
   a. The lymph nodes are subdivided into nodules.
   b. The nodules contain masses of lymphocytes and macrophages.
2. Locations of lymph nodes
   The lymph nodes generally occur in groups along the paths of the larger lymphatic vessels.
3. Functions of lymph nodes
   a. The lymph nodes are centers for the production of lymphocytes.
   b. They also contain phagocytic cells.

Thymus and Spleen (page 433)

1. Thymus
   a. The thymus is composed of lymphatic tissue, which is subdivided into lobules.
   b. Some lymphocytes leave the thymus and function in providing immunity.
2. Spleen
   a. The spleen resembles a large lymph node that is subdivided into lobules.
   b. Spaces within splenic lobules are filled with blood.
   c. The spleen contains numerous macrophages, which filter foreign particles and damaged red blood cells from the blood.

Body Defenses against Infection (page 435)

The body is equipped with specific and nonspecific defenses against infection.

Nonspecific Resistance (page 435)

1. Species resistance
   Each species of organism is resistant to certain diseases that may affect other species.
2. Mechanical barriers
   Mechanical barriers include the skin and mucous membranes, which prevent the entrance of some pathogens.
3. Enzymatic actions
   The enzymes of gastric juice and tears are lethal to some pathogens.
4. Interferon
   Interferon is a group of proteins produced by the cells in response to the presence of viruses and cells of tumors; it can interfere with the proliferation of viruses and stimulates cells that act against the growth of tumors.
5. Inflammation
   a. The inflammation response includes localized redness, swelling, heat, and pain.
   b. The chemicals released by damaged tissues attract various white blood cells to the site of inflammation.
   c. Fibrous connective tissue may form a sac around the injured tissue and prevent the spread of pathogens.
6. Phagocytosis
   a. The most active phagocytes of the blood are neutrophils and monocytes; monocytes give rise to macrophages, which remain fixed in the tissues.
   b. The macrophages associated with the linings of the blood vessels in the bone marrow, liver, spleen, and lymph nodes constitute the reticuloendothelial tissue.

Immunity (page 436)

1. Origin of lymphocytes
   a. Lymphocytes originate in red bone marrow and are released into the blood before they become differentiated.
   b. Some reach the thymus where they become T-cells.
   c. Those failing to reach the thymus become B-cells.
   d. Both T- and B-cells tend to reside in organs of the lymphatic system.
2. Antigens
   a. Before birth, the cells make an inventory of the proteins and other large molecules present in the body.
   b. After the inventory, the lymphocytes develop receptors that allow them to recognize foreign substances.
   c. Antigens are foreign substances that stimulate the lymphocytes to cause an immune reaction.
3. Functions of lymphocytes
   a. T-cells interact with antigen-bearing agents directly, producing cell-mediated immunity.
   b. B-cells interact with antigen-bearing agents indirectly, producing antibody-mediated immunity.
   c. A B-cell is activated when it encounters an antigen that fits the B-cell's antigen receptors.
   d. Some activated B-cells become antibody-secreting plasma cells.
   e. A T-helper cell is activated when it encounters an antigen displayed on the membrane of an accessory cell that fits the T-cell's antigen receptors.
   f. An activated T-helper cell releases lymphokines that stimulate some B-cells to proliferate and become antibody-secreting plasma cells.
4. Types of antibodies
   a. Antibodies are composed of soluble proteins called immunoglobulins.
   b. The three most abundant types of immunoglobulins are IgG, IgA, and IgM.
   c. Immunoglobulins act in various ways, causing antigens or antigen-bearing particles to be destroyed.
5. Immune responses
   a. The first response to an antigen is called a primary immune response.

(1) During this response, antibodies are produced for several weeks.
(2) Some B-cells remain dormant as memory cells.

b. A secondary immune response occurs rapidly if memory cells subsequently encounter the identical antigen.

6. Types of immunity
   a. A person who encounters a pathogen and has a primary immune response develops naturally acquired active immunity.
   b. A person who receives vaccine containing a dead or weakened pathogen develops artificially acquired active immunity.
   c. A person who receives an injection of gamma globulin that contains ready-made antibodies has artificially acquired passive immunity.
   d. When antibodies pass through a placental membrane from a pregnant woman to her fetus, the fetus develops naturally acquired passive immunity.
   e. Active immunity lasts much longer than passive immunity.
7. Allergic reactions
   a. Allergic reactions involve antigens combining with antibodies; such reactions are likely to be excessive or violent and may cause tissue damage.
   b. A delayed-reaction allergy results from repeated exposure to antigenic substances.
   c. A person with an inherited immediate-reaction allergy has an ability to produce an abnormally large amount of immunoglobulin.
   d. Allergic reactions may damage certain cells, which, in turn, release various chemicals.
   e. The released chemicals are responsible for the symptoms of the allergic reaction: hives, hay fever, asthma, eczema, or gastric disturbances.
8. Transplantation and tissue rejection
   a. If tissue is transplanted from one person to another, the recipient's cells may recognize the donor's tissue as foreign and act against it.
   b. The tissue rejection reaction may be reduced by matching the donor and recipient tissues.

Clinical Application of Knowledge

1. Based on your understanding of lymph nodes, how would you explain the fact that enlarged nodes are often removed for microscopic examination as an aid to diagnosing certain disease conditions?
2. Why is it that an injection into the skin is, to a large extent, an injection into the lymphatic system?
3. Explain why vaccination provides long-lasting protection against a disease, but gamma globulin provides only short-term protection.
4. If an infant was found to be lacking a thymus gland because of a developmental disorder, what could be predicted about the infant's susceptibility to infections? Why?

Review Activities

1. Explain how the lymphatic system is related to the cardiovascular system.
2. Trace the general pathway of lymph from the interstitial spaces to the bloodstream.
3. Describe the functions of lymph.
4. Explain why exercise promotes lymphatic circulation.
5. Describe the structure of a lymph node, and list its major functions.
6. Describe the structure and functions of the thymus gland.
7. Describe the structure and functions of the spleen.
8. Distinguish between specific and nonspecific body defenses against infection.
9. Define *species resistance.*
10. Explain how enzymes may provide resistance to invasion by pathogens.
11. Define *interferon.*
12. List the major symptoms of inflammation, and explain why each occurs.
13. Identify the major phagocytic cells in the blood and other tissues.
14. Explain where *reticuloendothelial tissue* is located.
15. Review the origins of T-cells and B-cells.
16. Distinguish between an antigen and an antibody.
17. Explain what is meant by cell-mediated immunity.
18. Explain what is meant by antibody-mediated immunity.
19. Define a *clone.*
20. Explain how B-cells are activated.
21. Define *plasma cell.*
22. Explain how a T-helper cell becomes activated.
23. Describe the source and action of lymphokines.
24. Review the function of cytotoxic T-cells.
25. List three types of immunoglobulin, and describe where each occurs.
26. Distinguish between a primary and a secondary immune response.
27. Distinguish between active and passive immunity.
28. Define *vaccine.*
29. Explain how a vaccine causes its effect.
30. Explain the relationship between an allergic reaction and an immune response.
31. Distinguish between an antigen and an allergen.
32. List the major events leading to a delayed-reaction allergic response.
33. Describe how an immediate-reaction allergic response may occur.
34. Explain the relationship between a tissue rejection and an immune response.
35. Describe a method used to reduce the severity of a tissue rejection reaction.

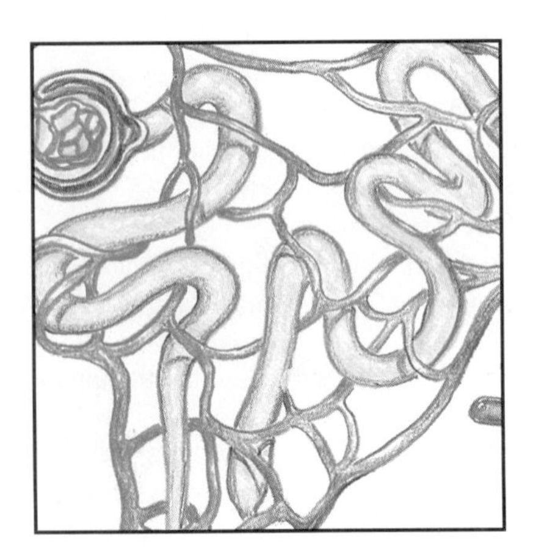

17
Urinary System

The body cells form a variety of wastes, and if these substances are allowed to accumulate, their effects are likely to be toxic. The body fluids, such as the blood and lymph, carry wastes away from the tissues that produce them. Other parts remove these wastes from the blood and transport them to the outside. The respiratory system, for example, removes carbon dioxide from the blood, and the *urinary system* removes various salts and nitrogenous wastes.

The urinary system also helps to maintain the normal concentrations of water and electrolytes within the body fluids, to regulate the pH and volume of body fluids, and to control red blood cell production and blood pressure.

Chapter Objectives

After you have studied this chapter, you should be able to

1. Name the organs of the urinary system, and list their general functions.
2. Describe the location of the kidneys and the structure of a kidney.
3. List the functions of the kidneys.
4. Trace the pathway of the blood through the major vessels within a kidney.
5. Describe a nephron, and explain the functions of its major parts.
6. Explain how the glomerular filtrate is produced, and describe its composition.
7. Explain how various factors affect the rate of glomerular filtration, and explain how this rate is regulated.
8. Discuss the role of tubular reabsorption in urine formation.
9. Define *tubular secretion,* and explain its role in urine formation.
10. Describe the structure of the ureters, urinary bladder, and urethra.
11. Discuss the process of micturition, and explain how it is controlled.
12. Complete the review activities at the end of this chapter. Note that the items are worded in the form of specific learning objectives. You may want to refer to them before reading the chapter.

Key Terms

afferent arteriole (af′er-ent ar-te′re-ōl)

Bowman's capsule (bo′manz kap′sūl)

detrusor muscle (de-truz′or mus′l)

efferent arteriole (ef′er-ent ar-te′re-ōl)

glomerulus (glo-mer′u-lus)

juxtaglomerular apparatus (juks″tah-glo-mer′u-lar ap″ah-ra′tus)

micturition (mik″tu-rish′un)

nephron (nef′ron)

peritubular capillary (per″ĭ-tu′bu-lar kap′ĭ-ler″e)

renal corpuscle (re′nal kor′pusl)

renal cortex (re′nal kor′teks)

renal medulla (re′nal mĕ-dul′ah)

renal tubule (re′nal tu′būl)

retroperitoneal (re″tro-per″ĭ-to-ne′al)

Aids to Understanding Words

calyc-, small cup: major *calyc*es—cuplike divisions of the renal pelvis.

cort-, covering: renal *cort*ex—the shell of tissues surrounding the inner region of a kidney.

detrus-, to force away: *detrus*or muscle—a muscle within the bladder wall that causes urine to be expelled.

glom-, little ball: *glom*erulus—a cluster of capillaries within a renal corpuscle.

nephr-, pertaining to the kidney: *nephr*on—the functional unit of a kidney.

mict-, to pass urine: *mict*urition—the process of expelling urine from the bladder.

papill-, nipple: renal *papill*ae—small elevations that project into a renal calyx.

trigon-, triangular shape: *trigone*—a triangular area on the internal floor of the bladder.

Introduction

THE URINARY SYSTEM consists of the following parts: a pair of *kidneys,* which remove substances from the blood, form urine, and help regulate various metabolic processes; a pair of tubular *ureters,* which transport urine away from the kidneys; a saclike *urinary bladder,* which serves as a urine reservoir; and a tubular *urethra,* which conveys urine to the outside of the body. These organs are shown in figure 17.1.

Kidneys

A **kidney** is a reddish brown, bean-shaped organ with a smooth surface. It is about 12 cm long, 6 cm wide, and 3 cm thick in an adult, and it is enclosed in a tough, fibrous capsule (fig. 17.2).

Location of the Kidneys

The kidneys lie on either side of the vertebral column in a depression high on the posterior wall of the abdominal cavity.

The upper and lower borders of the kidneys are generally at the levels of the twelfth thoracic and third lumbar vertebrae, respectively. The left kidney is usually about 1.5–2.0 cm higher than the right one.

The kidneys are positioned *retroperitoneally,* which means they are behind the parietal peritoneum and against the deep muscles of the back. They are held in position by the connective tissue and masses of adipose tissue that surround them (see figure 1.9).

Structure of a Kidney

The lateral surface of each kidney is convex, but its medial side is deeply concave. The resulting medial depression leads into a hollow chamber called the **renal sinus.** The entrance to this sinus is termed the *hilum,* and through it pass various blood vessels, nerves, lymphatic vessels, and the ureter (figs. 17.1 and 17.2).

The superior end of the ureter is expanded to form a funnel-shaped sac called the **renal pelvis,** which is located inside the renal sinus. The pelvis is subdivided into two or three tubes, called *major calyces* (sing., *calyx*), and they, in turn, are subdivided into several *minor calyces.*

Figure 17.1
The urinary system includes the kidneys, ureters, urinary bladder, and urethra. What are the general functions of this system?

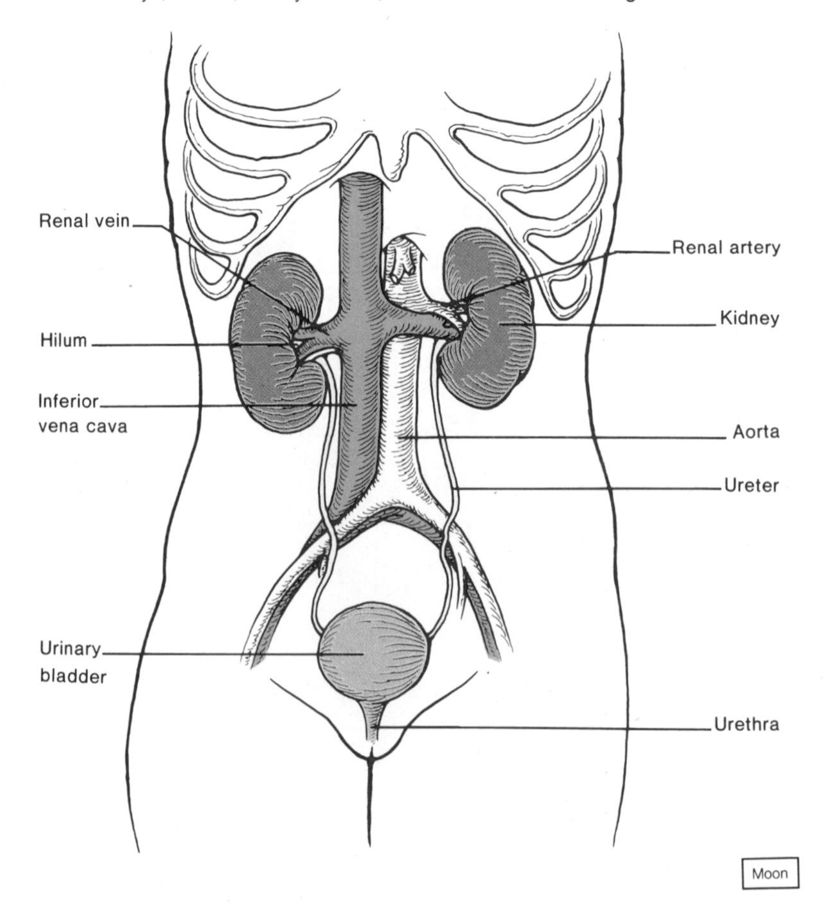

Figure 17.2
(*a*) A longitudinal section of a kidney; (*b*) a renal pyramid containing nephrons; (*c*) a single nephron.

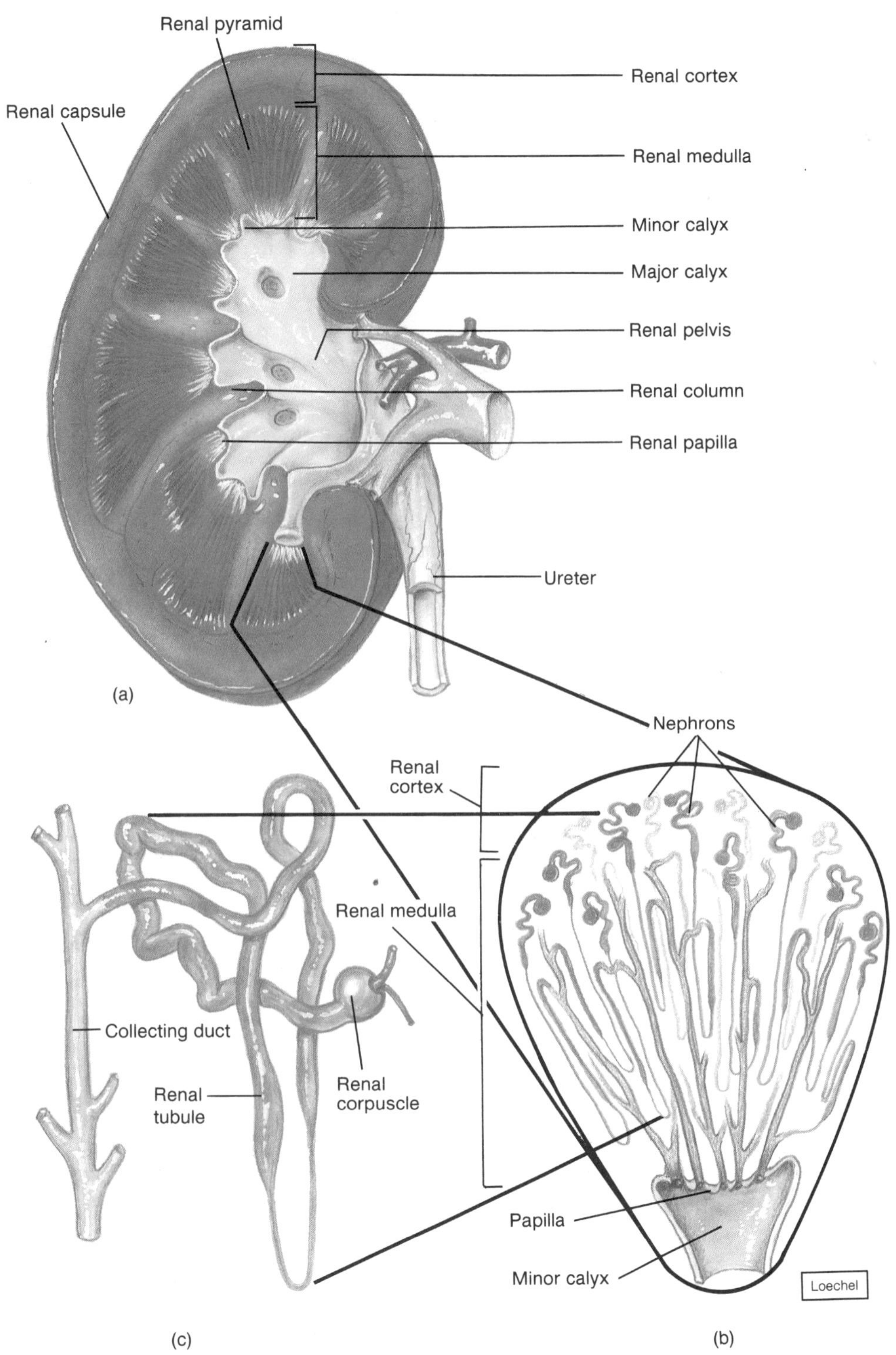

Figure 17.3
Main branches of the renal artery and renal vein.

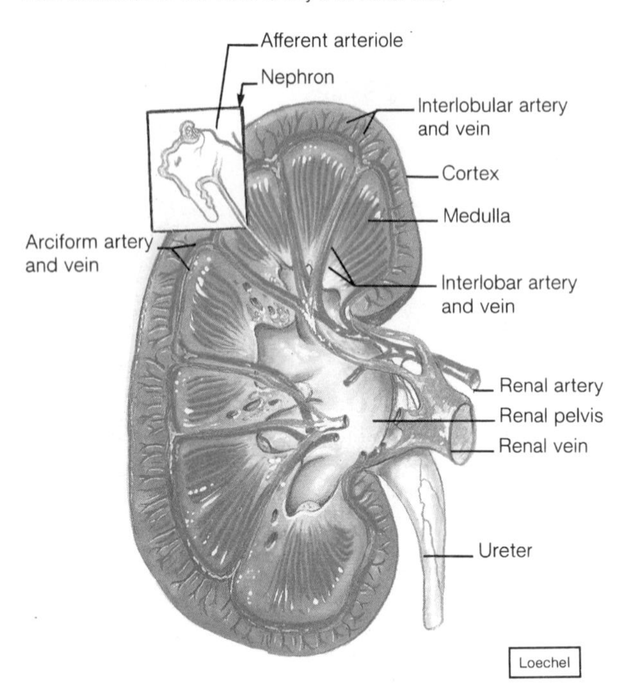

Figure 17.4
A scanning electron micrograph of a cast of the renal blood vessels associated with glomeruli (×260).
From *Tissues and Organs: A Text-Atlas of Scanning Electron Microscopy* by Richard G. Kessel and Randy H. Kardon. W. H. Freeman and Company. Copyright © 1979.

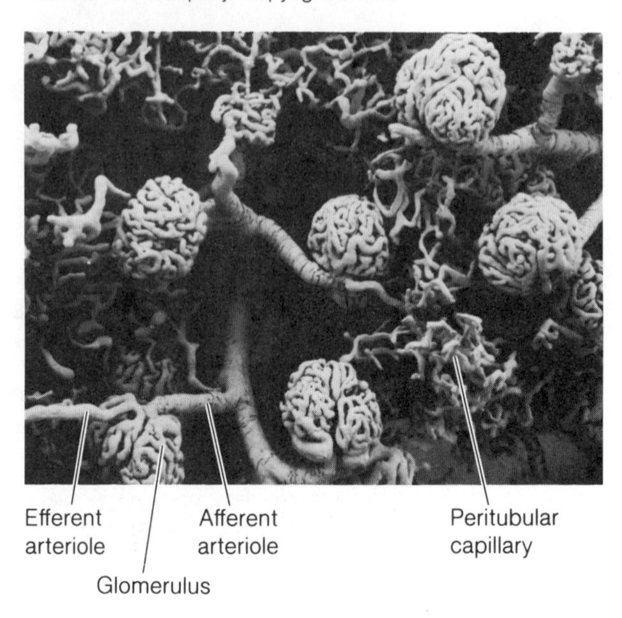

A series of small elevations project into the renal sinus from its wall. These projections are called *renal papillae,* and each is pierced by tiny openings that lead into a minor calyx.

The substance of the kidney includes two distinct regions—an inner medulla and an outer cortex. The **renal medulla** is composed of conical masses of tissue called *renal pyramids.*

The **renal cortex,** which appears somewhat granular in a sectioned kidney, forms a shell around the medulla. Its tissue dips into the medulla between adjacent renal pyramids, forming *renal columns.* The granular appearance of the cortex is due to the random arrangement of tiny tubules associated with the **nephrons,** the functional units of the kidney.

1. Where are the kidneys located?
2. Describe the structure of a kidney.
3. Name the functional unit of the kidney.

Functions of the Kidneys

The kidneys remove metabolic wastes from the blood and excrete them to the outside. They also carry on a variety of equally important regulatory activities, including helping control the rate of red blood cell formation by secreting the hormone *erythropoietin* (see chapter 14) and helping regulate the blood pressure by secreting the enzyme *renin.*

The kidneys also help regulate the volume, composition, and pH of the body fluids. These functions involve complex mechanisms that lead to the formation of urine. They are discussed in a subsequent section of this chapter and are explored in still more detail in chapter 18.

Renal Blood Vessels

Blood is supplied to the kidneys by means of the **renal arteries,** which arise from the abdominal aorta. These arteries transport a relatively large volume of blood; in fact, when a person is at rest the renal arteries usually carry from 15% to 30% of the total cardiac output into the kidneys.

A renal artery enters a kidney through the hilum and gives off several branches, called the *interlobar arteries,* which pass between the renal pyramids (fig. 17.3). At the junction between the medulla and cortex, the interlobar arteries branch to form a series of incomplete arches, the *arciform arteries* (arcuate arteries), which, in turn, give rise to *interlobular arteries.* The lateral branches of the interlobular arteries, called **afferent arterioles,** lead to the nephrons (figs. 17.3 and 17.4).

Venous blood is returned through a series of vessels that correspond generally to the arterial pathways. The **renal vein** then joins the inferior vena cava as it courses through the abdominal cavity (fig. 17.1).

> Patients with end-stage renal disease are sometimes treated with a *kidney transplant*. In this procedure, a kidney from a living donor or a cadaver, whose tissues are antigenically similar (histocompatible) to those of the recipient, is placed in the depression on the medial surface of the right or left ilium (iliac fossa). The renal artery and vein of the donor kidney are connected to the recipient's iliac artery and vein respectively, and the ureter of the kidney is attached to the dome of the recipient's urinary bladder.

Nephrons

Structure of a Nephron. A kidney contains about one million nephrons, each consisting of a **renal corpuscle** and a **renal tubule** (fig. 17.2).

A renal corpuscle is composed of a tangled cluster of blood capillaries, called a **glomerulus,** which is surrounded by a thin-walled, saclike structure, called a **Bowman's capsule** (figs. 17.5 and 17.6).

The Bowman's capsule is an expansion at the closed end of a renal tubule. The renal tubule leads away from the Bowman's capsule and becomes highly coiled; this coiled portion of the tubule is named the *proximal convoluted tubule.*

Figure 17.5
Light micrograph of a section of the human renal cortex (×240). What features can you identify?

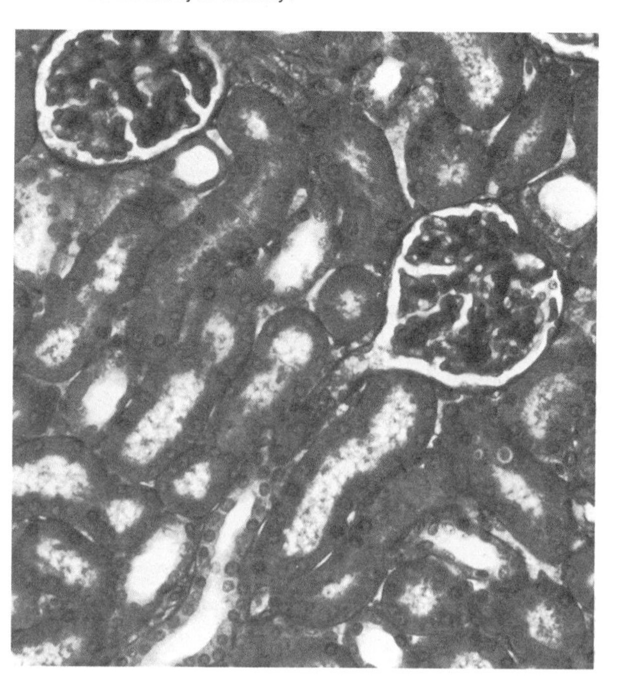

Figure 17.6
The structure of a nephron and the blood vessels associated with it.

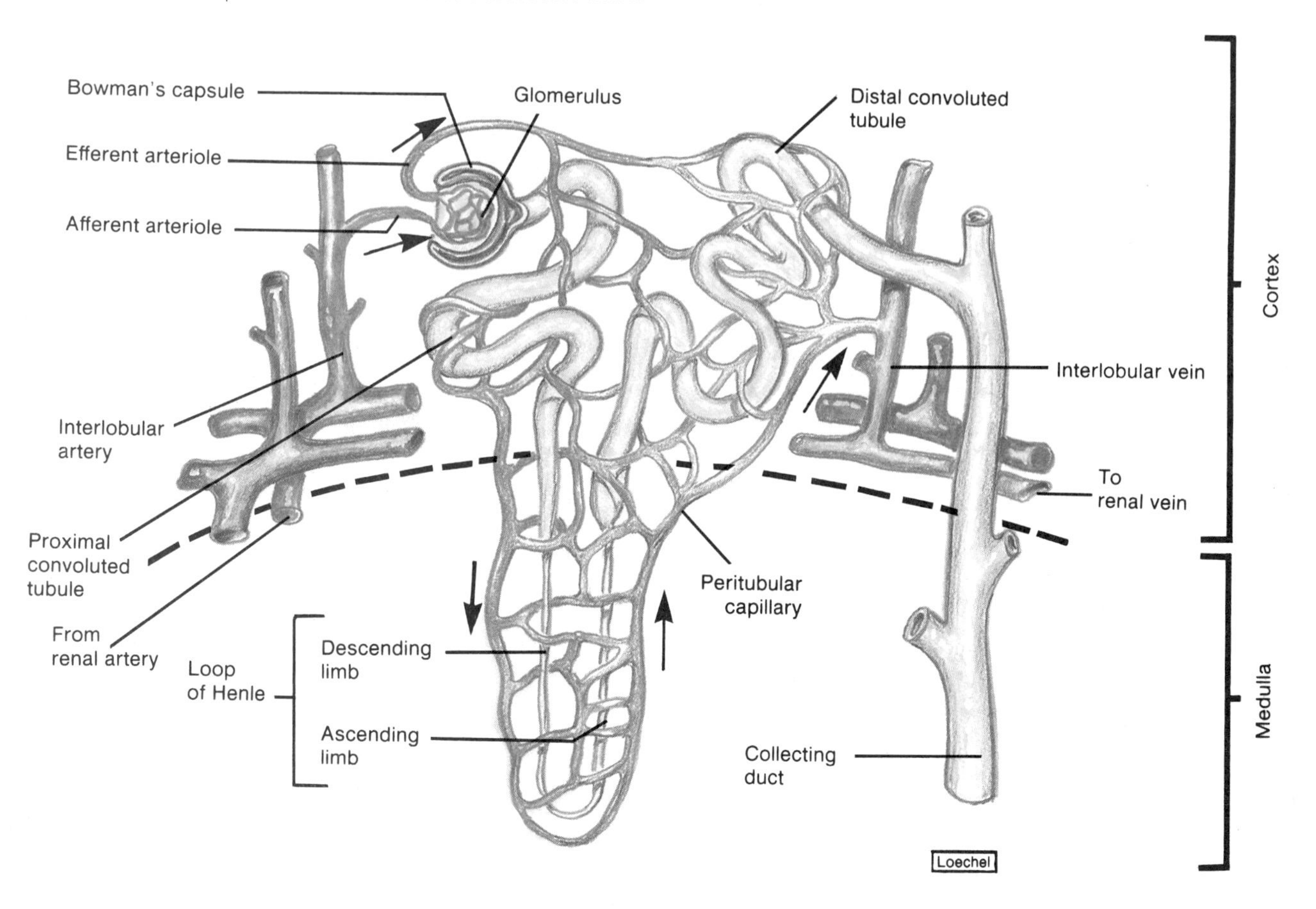

Figure 17.7
(*a*) Location of the juxtaglomerular apparatus; (*b*) enlargement of the juxtaglomerular apparatus, which consists of the macula densa and juxtaglomerular cells.

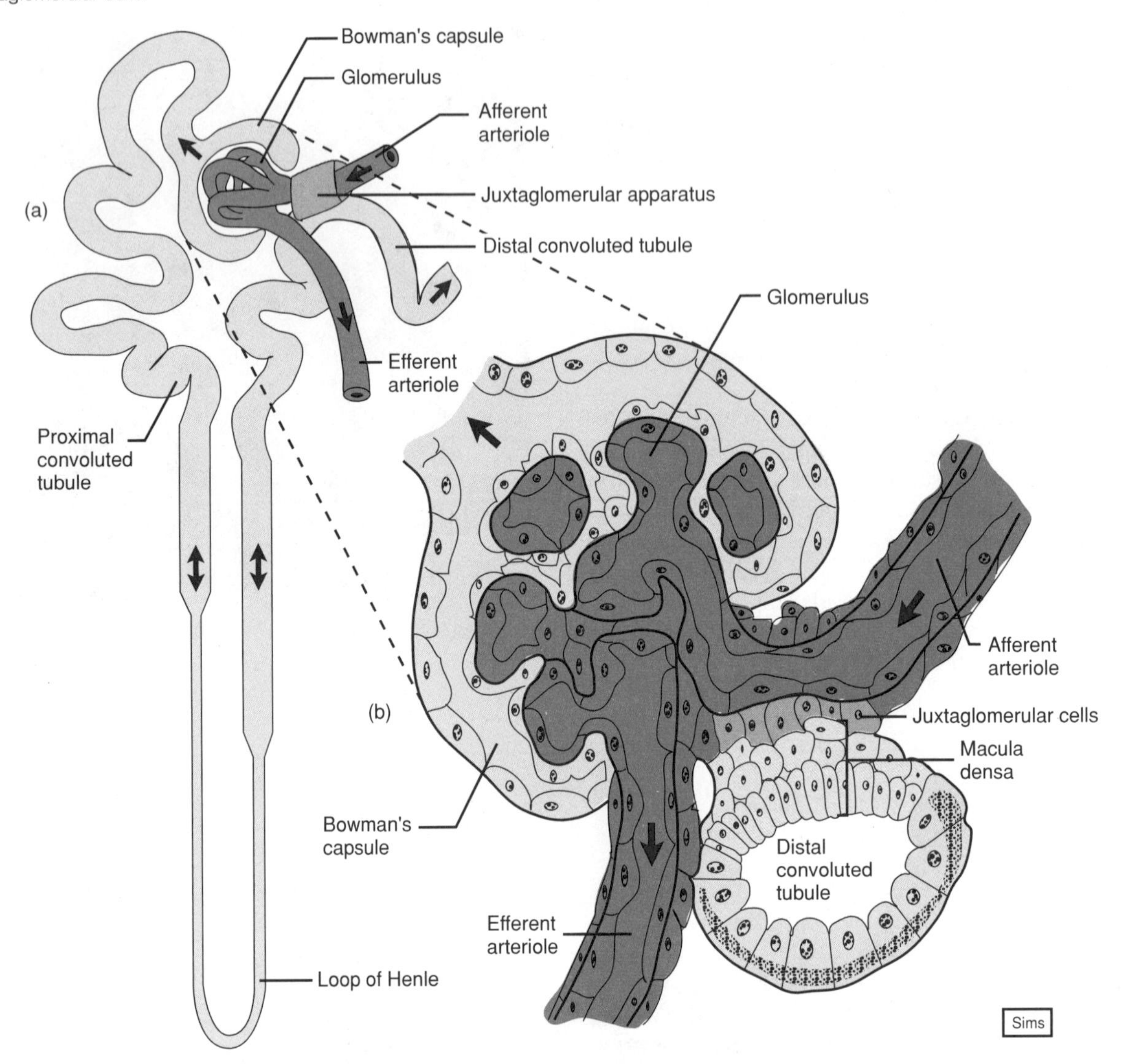

The proximal convoluted tubule dips toward the renal pelvis to become the *descending limb of the loop of Henle.* The tubule then curves back toward its renal corpuscle and forms the *ascending limb of the loop of Henle.* The ascending limb returns to the region of the renal corpuscle, where it becomes highly coiled again, and is called the *distal convoluted tubule.*

Several distal convoluted tubules merge in the renal cortex to form a *collecting duct,* which, in turn, passes into the renal medulla, becoming larger and larger as it is joined by other collecting ducts. The resulting tube empties into a minor calyx through an opening in a renal papilla. Figure 17.6 summarizes the structure of a nephron and the blood vessels associated with it.

1. List the general functions of the kidneys.
2. Trace the blood supply to the kidney.
3. Name the parts of a nephron.

Blood Supply of a Nephron

The cluster of capillaries that forms a glomerulus arises from an afferent arteriole. After passing through the capillary of the glomerulus, the blood enters an **efferent arteriole** (rather than a venule), whose diameter is somewhat less than that of the afferent vessel.

Because of its small diameter, the efferent arteriole creates some resistance to blood flow. This causes blood to back up into the glomerulus, producing a relatively high pressure in the glomerular capillary.

Figure 17.8

(*a*) The first step in urine formation is the filtration of substances through the glomerular membrane into Bowman's capsule. (*b*) The glomerular filtrate passes through fenestrae of the capillary endothelium.

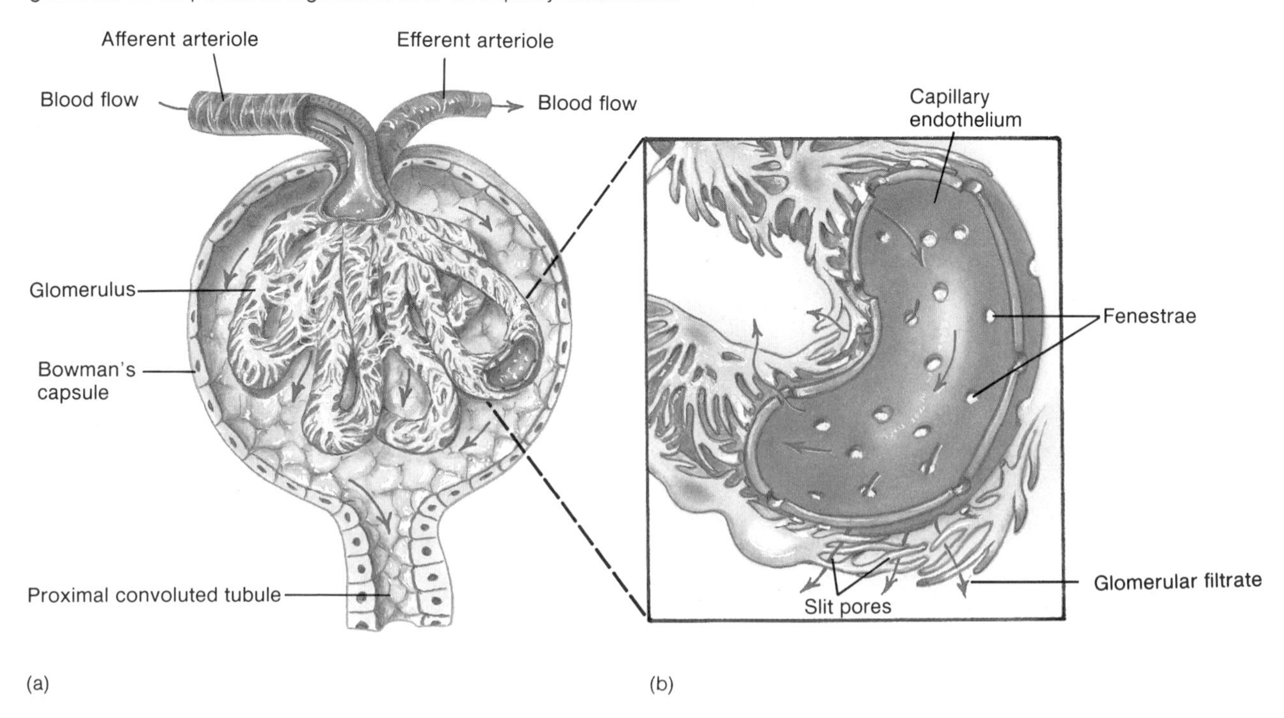

The efferent arteriole branches into a complex, freely interconnecting network of capillaries, which surrounds the various portions of the renal tubule. This network is called the **peritubular capillary system,** and the blood it contains is under relatively low pressure (fig. 17.6).

After flowing through the capillary network, the blood is returned to the renal cortex, where it joins the blood from other branches of the peritubular capillary system and enters the venous system of the kidney.

Juxtaglomerular Apparatus

Near its beginning, the distal convoluted tubule passes between the afferent and efferent arterioles and contacts them. At the point of contact, the epithelial cells of the distal tubule are quite narrow and densely packed. These cells comprise a structure called the *macula densa* (fig. 17.7).

Close by, in the walls of the arterioles near their attachments to the glomerulus, are some enlarged smooth muscle cells. They are called *juxtaglomerular cells,* and together with the cells of the macula densa, they constitute the **juxtaglomerular apparatus** (complex). The function of this apparatus is described in a subsequent section of this chapter.

1. Describe the system of blood vessels that supplies blood to a nephron.
2. What parts comprise the juxtaglomerular apparatus?

Urine Formation

The primary functions of the nephrons include removing waste substances from the blood and regulating water and electrolyte concentrations within the body fluids. The end product of these functions is **urine,** which is excreted to the outside of the body, carrying with it wastes, excess water, and excess electrolytes.

Urine formation involves glomerular filtration, tubular reabsorption, and tubular secretion.

Glomerular Filtration

Urine formation begins when water and various dissolved substances are filtered out of the glomerular capillaries and into the Bowman's capsules. The filtration of these materials through the capillary walls is much like the filtration that occurs at the arteriole ends of other capillaries throughout the body. The glomerular capillaries, however, are many times more permeable than the capillaries in other tissues due to the presence of numerous tiny openings (fenestrae) in their walls (fig. 17.8).

Filtration Pressure

As in the case of other capillaries, the main force responsible for moving substances through the glomerular capillary wall is the pressure of the blood inside (glomerular hydrostatic pressure). This movement is also influenced by the osmotic pressure of the plasma

Chart 17.1 Relative concentrations of substances in the plasma, glomerular filtrate, and urine			
	Concentrations (mEq/l)		
Substance	*Plasma*	*Glomerular filtrate*	*Urine*
Sodium (Na^+)	142	142	128
Potassium (K^+)	5	5	60
Calcium (Ca^{+2})	4	4	5
Magnesium (Mg^{+2})	3	3	15
Chlorine (Cl^-)	103	103	134
Bicarbonate (HCO_3^-)	27	27	14
Sulfate (SO_4^{-2})	1	1	33
Phosphate (PO_4^{-3})	2	2	40
	Concentrations (mg/100 ml)		
Substance	*Plasma*	*Glomerular filtrate*	*Urine*
Glucose	100	100	0
Urea	26	26	1820
Uric acid	4	4	53

in the glomerulus and by the hydrostatic pressure inside the Bowman's capsule. An increase in either of these pressures will oppose movement out of the capillary and, thus, reduce filtration. The net pressure acting to force substances out of the glomerulus is called the **filtration pressure.**

The **glomerular filtrate** consists of substances that enter the space within the Bowman's capsule, and it has about the same composition as the filtrate that becomes tissue fluid elsewhere in the body. That is, glomerular filtrate is largely water and contains essentially the same substances as the blood plasma, except for the larger protein molecules, which the filtrate lacks. The relative concentrations of some of the substances in the plasma, glomerular filtrate, and urine are shown in chart 17.1.

Filtration Rate

The rate of glomerular filtration is directly proportional to the filtration pressure. Consequently, the factors that affect the glomerular hydrostatic pressure, glomerular plasma osmotic pressure, or hydrostatic pressure in the Bowman's capsule will also affect the rate of filtration.

For example, since the glomerular capillary is located between two arterioles—the *afferent* and *efferent arterioles*—any change in the diameters of these vessels is likely to cause a change in the glomerular hydrostatic pressure and will be accompanied by a change in the glomerular filtration rate. The afferent arteriole, through which the blood enters the glomerulus, may constrict as a result of mild stimulation by sympathetic nerve impulses. If this occurs, the blood flow diminishes, the glomerular hydrostatic pressure decreases, and the filtration rate drops. If, on the other hand, the efferent arteriole (through which the blood leaves the glomerulus) constricts, the blood backs up into the glomerulus, the glomerular hydrostatic pressure increases, and the filtration rate rises. Converse effects are produced by vasodilation of these vessels.

In the capillaries, the blood pressure, acting to force water and dissolved substances outward, is opposed by the effect of the plasma osmotic pressure that attracts water inward (chapter 14). As filtration occurs through the capillary wall, the proteins remaining in the plasma cause the osmotic pressure within the glomerular capillary to rise. When this pressure reaches a certain high level, filtration ceases. Conversely, conditions that tend to decrease plasma osmotic pressure, such as a decrease in plasma protein concentration, cause an increase in the filtration rate.

In the condition called *glomerulonephritis,* the glomerular capillaries are inflamed and become more permeable to proteins. Consequently, proteins appear in the glomerular filtrate and are excreted in the urine (proteinuria). At the same time, the protein concentration in the blood plasma decreases (hypoproteinemia), and this causes a drop in the osmotic pressure of the blood. As a result, the movement of tissue fluid into the capillaries is decreased, and edema develops.

The hydrostatic pressure in Bowman's capsule sometimes changes as a result of an obstruction, such as may be caused by a stone in a ureter or an enlarged prostate gland pressing on the urethra. If this occurs, fluids tend to back up into the renal tubules and cause the hydrostatic pressure in Bowman's capsules to rise. Since any increase in capsular pressure opposes glomerular filtration, the rate of filtration may decrease significantly.

In an average adult, the glomerular filtration rate for the nephrons of both kidneys is about 125 milliliters per minute, or 180,000 milliliters (180 liters) in 24 hours. Since this 24-hour volume is nearly 45 gallons, it is obvious that not all of it is excreted as urine. Instead, most of the fluid that passes through the renal tubules is reabsorbed and reenters the plasma.

1. What general processes are involved with urine formation?
2. What forces affect the filtration pressure?
3. What factors influence the rate of glomerular filtration?

Figure 17.9
Two mechanisms involving the macula densa ensure a constant blood flow through the glomerulus.

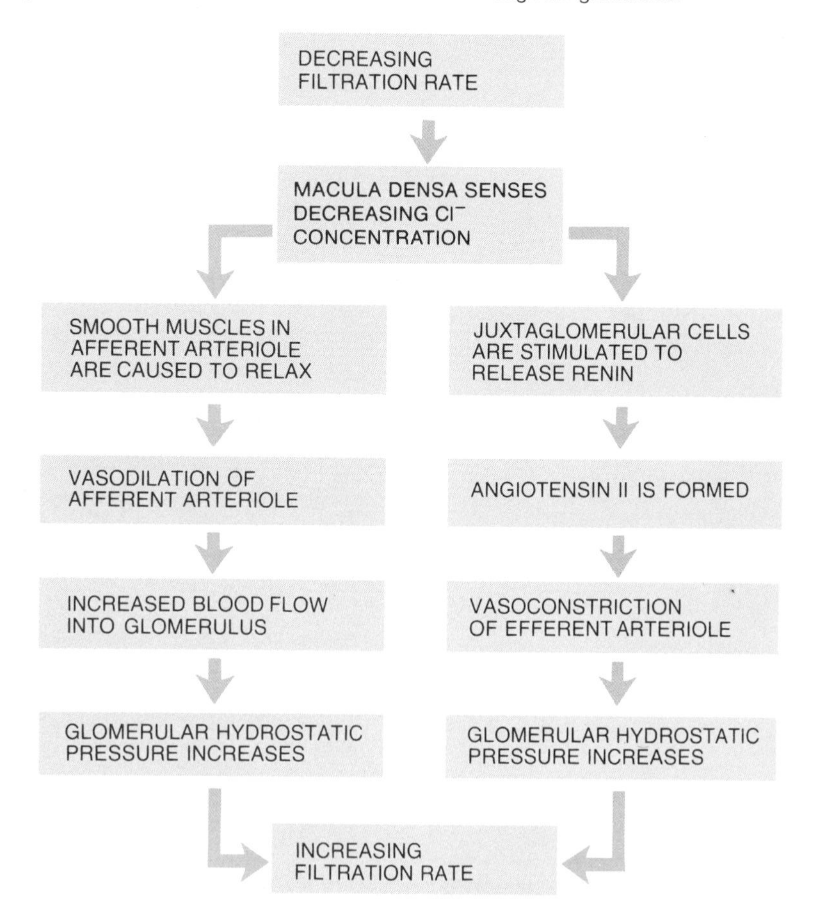

Regulation of Filtration Rate. The regulation of glomerular filtration rate involves the *juxtaglomerular apparatus,* which was described previously (fig. 17.7), and two negative feedback mechanisms (fig. 17.9). These mechanisms are triggered whenever the filtration rate is decreasing. For example, as the rate decreases, the concentration of chloride ions reaching the *macula densa* in the distal convoluted tubule also decreases. In response, the macula densa signals the smooth muscles in the wall of the afferent arteriole to relax, and the vessel becomes dilated. This action allows more blood to flow into the glomerulus, increasing the glomerular pressure, and as a consequence, the filtration rate rises toward its previous level.

At the same time that the macula densa signals the afferent arteriole to dilate, it stimulates the *juxtaglomerular cells* to release *renin.* This enzyme causes a plasma globulin (angiotensinogen) to form a substance called *angiotensin I,* which is converted quickly to *angiotensin II* by an enzyme (converting enzyme) present in the lungs and plasma.

Angiotensin II is a vasoconstrictor, and it stimulates the smooth muscle cells in the wall of the efferent arteriole to contract, constricting the vessel. As a result of this action, blood tends to back up into the glomerulus, and as the glomerular hydrostatic pressure increases, the filtration rate also increases.

These two mechanisms operate together to ensure a constant blood flow through the glomerulus and a relatively stable glomerular filtration rate, in spite of marked changes occurring in the arterial blood pressure.

1. What is the function of the macula densa?
2. How does renin help to regulate the filtration rate?

Tubular Reabsorption

If the composition of the glomerular filtrate entering the renal tubule is compared with that of the urine leaving the tubule, it is clear that changes occur as the fluid passes through the tubule (chart 17.1). For example, glucose is present in the filtrate, but absent in the urine. Also, urea and uric acid are considerably

Figure 17.10
Reabsorption is the process by which substances are transported from the glomerular filtrate into the blood of the peritubular capillary. What substances are reabsorbed in this manner?

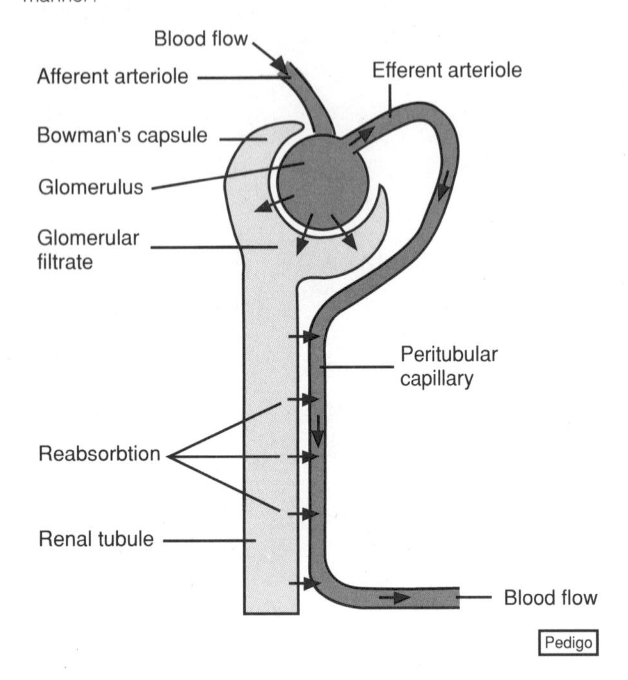

more concentrated in the urine than they are in the glomerular filtrate. Such changes in fluid composition are largely the result of **tubular reabsorption,** a process by which substances are transported out of the glomerular filtrate, through the epithelium of the renal tubule, and into the blood of the peritubular capillary (fig. 17.10).

Since the efferent arteriole is narrower than the peritubular capillary, the blood flowing from the former into the latter is under relatively low pressure. Also, the wall of this capillary is more permeable than that of other capillaries. Both of these factors enhance the rate of fluid reabsorption from the renal tubule.

Although tubular reabsorption occurs throughout the renal tubule, most of it occurs in the proximal convoluted portion. The epithelial cells in this portion have numerous microscopic projections, called *microvilli,* that form a "brush border" on their free surfaces. These tiny extensions greatly increase the surface area exposed to the glomerular filtrate and enhance the reabsorption process.

Various segments of the renal tubule are adapted to reabsorb specific substances, using particular modes of transport. Glucose reabsorption, for example, occurs primarily through the walls of the proximal tubule by active transport. Water also is reabsorbed rapidly through the epithelium of the proximal tubule by osmosis; however, portions of the distal tubule are almost impermeable to water. This characteristic of the distal tubule is important in the regulation of urine concentration and volume, as is described in a subsequent section of this chapter.

Active transport depends on the presence of carrier molecules in a cell membrane (see chapter 3). These carriers transport passenger molecules through the membrane, release them, and return to the other side to transport more passenger molecules. Such a mechanism has a *limited transport capacity;* that is, it can only transport a certain number of molecules in a given amount of time, because the number of carriers is limited.

Usually all of the glucose in the glomerular filtrate is reabsorbed, because there are enough carrier molecules to transport it. Sometimes, however, the plasma glucose concentration increases, and if it reaches a critical level, called the *renal plasma threshold,* there will be more glucose molecules in the filtrate than the active transport mechanism can handle. As a result, some glucose will remain in the filtrate and be excreted in the urine.

The appearance of glucose in the urine is called *glucosuria* (or *glycosuria*). This condition may occur following the administration of glucose intravenously or in a patient with insulin-dependent (type I) diabetes mellitus. If the cause is diabetes mellitus, the blood glucose concentration rises because of insufficient insulin from the pancreas (chapter 11).

Amino acids also enter the glomerular filtrate and are reabsorbed in the proximal convoluted tubule, apparently by three different active transport mechanisms. Each mechanism is thought to reabsorb a different group of amino acids, whose members have molecular similarities. As a result of their actions, only a trace of amino acids usually remains in the urine.

Although the glomerular filtrate is nearly free of protein, some *albumin* may be present. These proteins have relatively small molecules, and they are reabsorbed by *pinocytosis* through the brush border of the epithelial cells lining the proximal convoluted tubule. Once they are inside an epithelial cell, the proteins are converted to amino acids and moved into the blood of the peritubular capillary.

Other substances reabsorbed by the epithelium of the proximal convoluted tubule include creatine, lactic acid, citric acid, uric acid, ascorbic acid (vitamin C), phosphate ions, sulfate ions, calcium ions, potassium ions, and sodium ions. As a group, these substances are reabsorbed by active transport mechanisms with limited transport capacities. Such a substance usually does not appear in the urine until its concentration in the glomerular filtrate exceeds its particular threshold.

Figure 17.11
Water reabsorption by osmosis occurs in response to the reabsorption of sodium by active transport.

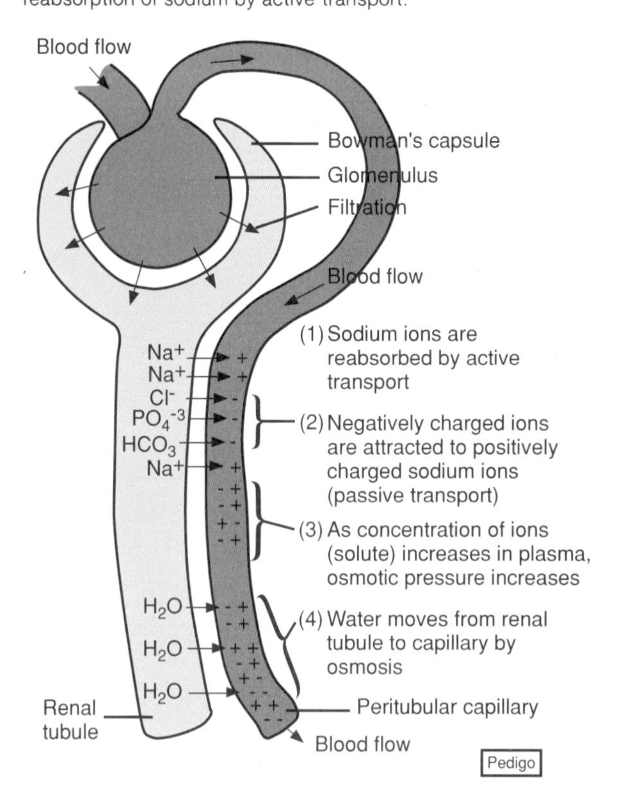

Sodium and Water Reabsorption

Substances that remain in the renal tubule tend to become more and more concentrated as water is reabsorbed from the filtrate. Most water reabsorption occurs *passively* by *osmosis* in the proximal convoluted tubule and is closely associated with the active reabsorption of sodium ions (fig 17.11). In fact, if sodium reabsorption increases, water reabsorption increases; if sodium reabsorption decreases, water reabsorption decreases also.

About 70% of *sodium ion reabsorption* occurs in the proximal segment of the renal tubule by active transport (the sodium pump mechanism). As these positively charged ions (Na^+) are moved through the tubular wall, negatively charged ions including chloride ions (Cl^-), phosphate ions (PO_4^{-3}), and bicarbonate ions (HCO_3^-) accompany them. This movement of negatively charged ions is due to the electrochemical attraction between particles of opposite charge. It is termed **passive transport** because it does not require a direct expenditure of cellular energy.

As more and more sodium ions are actively transported into the peritubular capillary, along with various negatively charged ions, the concentration of solutes within the peritubular blood is increased. Furthermore, since water moves through cell membranes from regions of lesser solute concentration (hypotonic) toward regions of greater solute concentration (hypertonic), water is transported by osmosis from the renal tubule into the peritubular capillary. Because of the movement of solutes and water into the peritubular capillary, the volume of fluid within the renal tubule is greatly reduced.

1. What substances present in the glomerular filtrate are not normally present in the urine?
2. What mechanisms are responsible for the reabsorption of solutes from the glomerular filtrate?
3. Describe the role of passive transport in urine formation.

Regulation of Urine Concentration and Volume

Sodium ions continue to be reabsorbed by active transport as the tubular fluid moves through the loop of Henle, the distal convoluted segment, and the collecting duct. As a result, almost all the sodium that enters the renal tubule as part of the glomerular filtrate may be reabsorbed before the urine is excreted, and consequently, water also continues to be reabsorbed passively by osmosis in various segments of the renal tubule.

Additional water may be reabsorbed due to the action of aldosterone and antidiuretic hormone (ADH). As discussed in chapter 11, aldosterone is secreted by cells of the adrenal cortex in response to changes in the blood concentrations of sodium ions and potassium ions. The effect of aldosterone is to cause sodium ions and water molecules to be conserved, thus reducing urine output.

ADH is produced by neurons in the *hypothalamus*. It is released from the posterior lobe of the pituitary gland in response to a decreasing concentration of water in the blood. When the ADH reaches the kidney, it causes an increase in the permeability of the epithelial linings of the distal convoluted tubule and collecting duct, and water moves rapidly out of these segments by osmosis. Consequently, the urine volume is reduced, and the urine becomes more concentrated.

Thus ADH stimulates the production of concentrated urine, which contains soluble wastes and other substances in a minimum of water; it also inhibits the loss of body fluids whenever there is a danger of dehydration (fig. 17.12). If the water concentration of the body fluids is excessive, ADH secretion is decreased. In the absence of ADH, the epithelial linings of the distal segment and collecting duct become less permeable to water, less water is reabsorbed, and the urine tends to be more dilute. Chart 17.2 summarizes the role of ADH on urine production.

Figure 17.12
(*a*) The distal convoluted tubule and collecting duct are impermeable to water, so water may be excreted as dilute urine; (*b*) if ADH is present, these segments become permeable, and water is reabsorbed by osmosis.

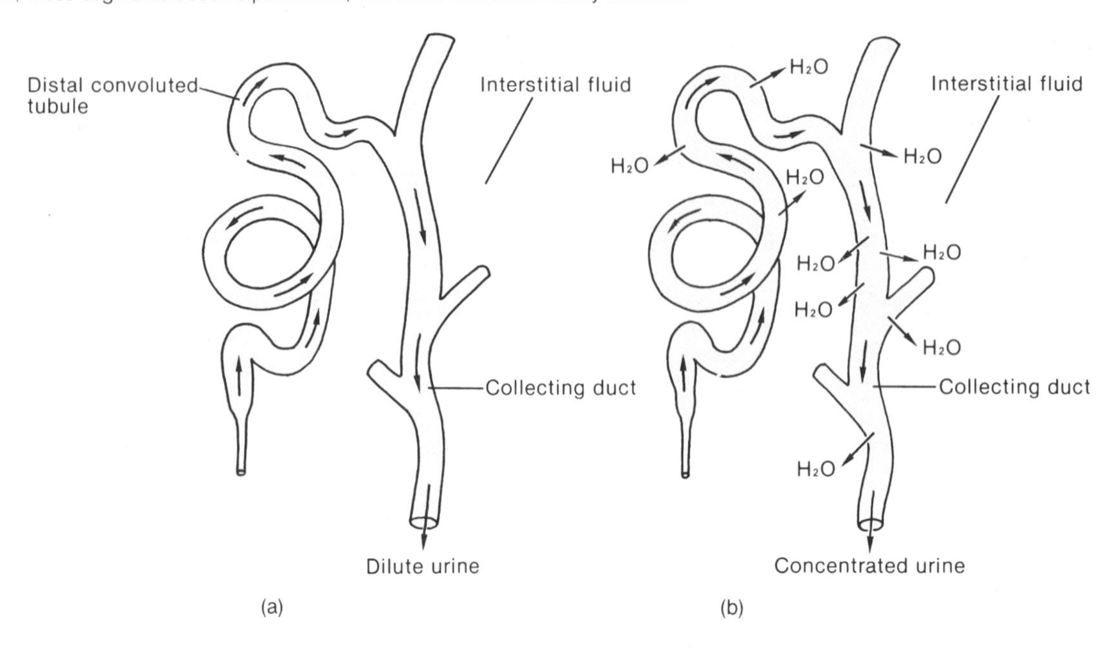

Chart 17.2 Role of ADH in the regulation of urine concentration and volume

1. Concentration of water in the blood decreases
2. Osmoreceptors in the hypothalamus of the brain are stimulated by an increase in the osmotic pressure of the body fluid
3. Hypothalamus signals the posterior pituitary gland to release ADH
4. Blood carries ADH to the kidneys
5. ADH causes the distal convoluted tubules and collecting ducts to increase water reabsorption by osmosis
6. Urine becomes more concentrated and the urine volume decreases

Urea and Uric Acid Excretion

Urea is a by-product of amino acid metabolism. Consequently, its plasma concentration is directly related to the amount of protein in the diet. Urea enters the renal tubule by filtration, and about 50% of it is reabsorbed (passively) by diffusion, but the remainder is excreted in the urine.

Uric acid, which results from the metabolism of certain organic bases in nucleic acids, is reabsorbed by active transport. Although this mechanism seems able to reabsorb all the uric acid normally present in glomerular filtrate, about 10% of the amount filtered is excreted in the urine. This amount is apparently *secreted* into the renal tubule.

Gout is a disorder in which the plasma concentration of uric acid becomes abnormally high. Since uric acid is a relatively insoluble substance, it tends to precipitate when it is present in excess. As a result, crystals of uric acid may be deposited in the joints and other tissues, where they produce inflammation and extreme pain. The joints of the great toes are affected most commonly, but other joints in the hands and feet may also be involved.

This condition, which often seems to be inherited, is sometimes treated by administering drugs that inhibit the reabsorption of uric acid and, thus, cause its excretion to increase.

1. What role does the hypothalamus play in regulating urine concentration and volume?
2. Explain how urea and uric acid are excreted.

Tubular Secretion

Tubular secretion is the process by which certain substances are transported from the plasma of the peritubular capillary into the fluid of the renal tubule (fig. 17.13). As a result, the amount of a particular substance excreted into the urine may be greater than the amount filtered from the plasma in the glomerulus.

Some substances are secreted by active transport mechanisms similar to those that function in reabsorption. *Secretory mechanisms,* however, transport sub-

Figure 17.13
Secretory mechanisms act to move substances from the plasma of the peritubular capillary into the fluid of the renal tubule.

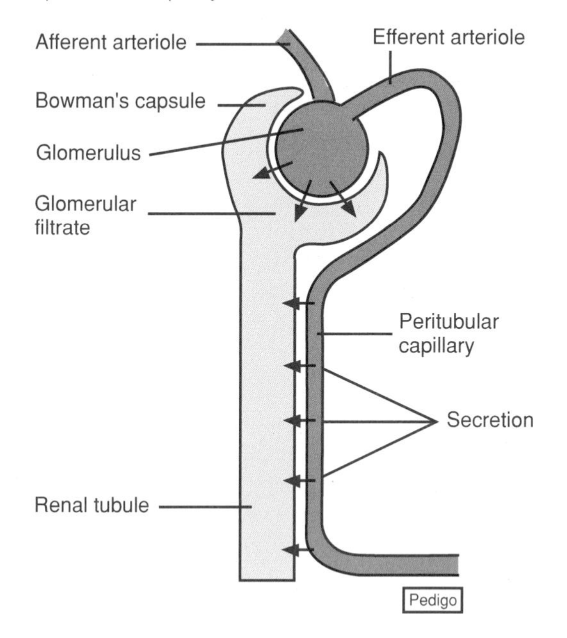

stances in the opposite direction. For example, certain organic compounds, including penicillin, creatinine, and histamine, are actively secreted into the tubular fluid by the epithelium of the proximal convoluted segment.

Hydrogen ions are also actively secreted. In this case, the proximal segment of the renal tubule is specialized to secrete large quantities of hydrogen ions between the plasma and the tubular fluid, where the hydrogen ion concentrations are similar. This secretion of hydrogen ions plays an important role in the regulation of the pH of the body fluids, as is explained in chapter 18.

Although most of the *potassium ions* in the glomerular filtrate are actively reabsorbed in the proximal convoluted tubule, some may be secreted passively in the distal segment and collecting duct. During this process, the active reabsorption of sodium ions out of the tubular fluid creates a negative electrical charge within the tube. Because the positively charged potassium ions (K^+) and hydrogen ions (H^+) are attracted to regions that are negatively charged, these ions move through the tubular epithelium and enter the tubular fluid (fig. 17.14).

Chart 17.3 summarizes the functions of various parts of the nephron.

Figure 17.14
Passive secretion of potassium ions (or hydrogen ions) may occur in response to the active reabsorption of sodium ions.

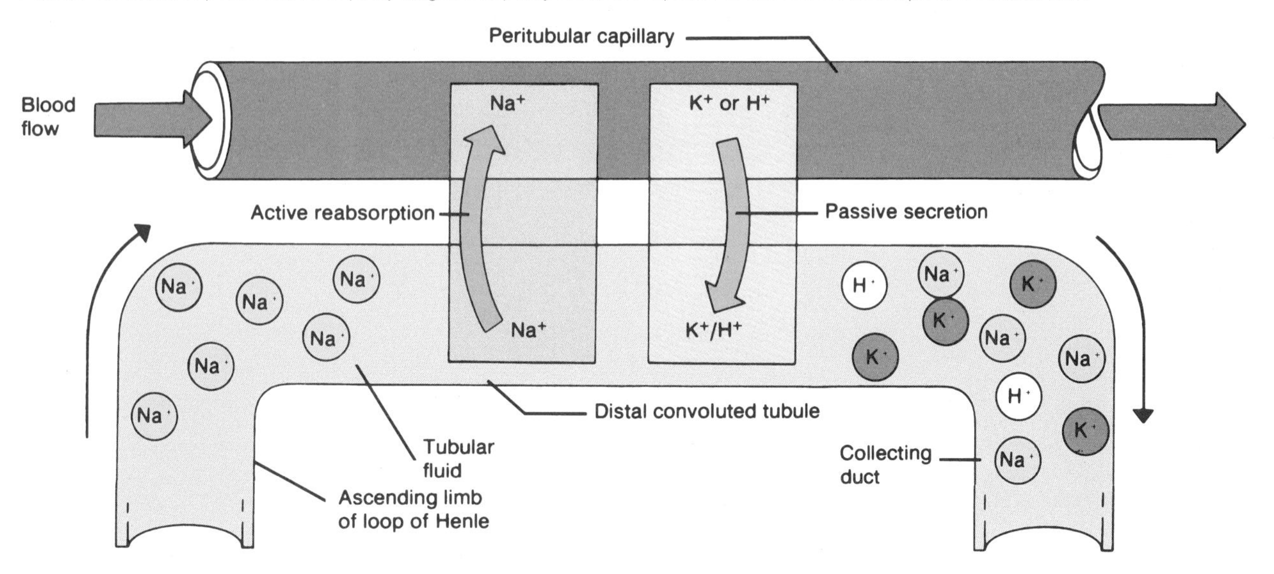

Chart 17.3 Functions of the nephron parts

Part	Function
Renal Corpuscle	
Glomerulus	Filtration of water and dissolved substances from the plasma
Bowman's capsule	Receives the glomerular filtrate
Renal Tubule	
Proximal convoluted tubule	Reabsorption of glucose, amino acids, lactic acid, uric acid, creatine, citric acid, ascorbic acid, phosphate ions, sulfate ions, calcium ions, potassium ions, and sodium ions by active transport
	Reabsorption of water by osmosis
	Reabsorption of chloride ions and other negatively charged ions by electrochemical attraction
	Active secretion of substances such as penicillin, histamine, creatinine, and hydrogen ions
Descending limb of the loop of Henle	Reabsorption of water by osmosis
Ascending limb of the loop of Henle	Reabsorption of chloride ions by active transport and passive reabsorption of sodium ions
Distal convoluted tubule	Reabsorption of sodium ions by active transport
	Reabsorption of water by osmosis
	Active secretion of hydrogen ions
	Passive secretion of potassium ions by electrochemical attraction

1. Define *tubular secretion.*
2. What substances are actively secreted?
3. How does the reabsorption of sodium affect the secretion of potassium?

Composition of Urine

The composition of urine varies considerably from time to time because of variations in dietary intake and physical activity. In addition to containing about 95% water, urine usually contains *urea* and *uric acid.* It may also contain a trace of *amino acids,* as well as a variety of *electrolytes,* whose concentrations tend to vary directly with the amounts included in the diet (chart 17.1). Normal concentrations of urine components are listed in the Appendix, page 531.

The volume of urine produced usually varies between 0.6 and 2.5 liters per day. The exact volume is influenced by such factors as fluid intake, environmental temperature, relative humidity of the surrounding air, and the person's emotional condition, respiratory rate, and body temperature. An output of 50–60 cc of urine per hour is considered normal, and an output of less than 30 cc per hour may be an indication of kidney failure.

1. List the normal constituents of urine.
2. What factors affect the volume of urine produced?

Elimination of Urine

After being formed by the nephrons, urine passes from the collecting ducts through openings in the renal papillae and enters the calyces of the kidney (fig. 17.2). From there, it passes through the renal pelvis and is conveyed by a ureter to the urinary bladder (fig. 17.1). Urine is excreted to the outside by means of the urethra.

Ureters

Each **ureter** is a tubular organ about 25 cm long, which begins as the funnel-shaped renal pelvis. It extends downward behind the parietal peritoneum and parallel to the vertebral column. Within the pelvic cavity, it courses forward and medially to join the urinary bladder from underneath.

The wall of a ureter is composed of three layers. The inner layer (*mucous coat*) is continuous with the linings of the renal tubules and the urinary bladder. The middle layer (*muscular coat*) consists largely of smooth muscle fibers. The outer layer (*fibrous coat*) is composed of connective tissue (fig. 17.15).

Kidney stones, which are usually composed of uric acid, calcium oxalate, calcium phosphate, or magnesium phosphate, sometimes form in the collecting ducts and renal pelvis. If such a stone passes into a ureter, it may stimulate severe pain. This pain commonly begins in the region of the kidney and tends to radiate into the abdomen, pelvis, and legs. It may also be accompanied by nausea and vomiting.

Although about 60% of kidney stone patients pass their stones spontaneously, the others must have the stones removed. In the past, such removal required surgery or tubular instruments that could be passed through the tubes of the urinary tract and used to capture or crush the stones. More recently, kidney stones have been fragmented by shock waves generated outside the body. The resulting sandlike fragments are then eliminated with the urine. This procedure, called *extracorporeal shock-wave lithotripsy* (ESWL), involves placing the patient in a stainless steel tub filled with water. The shock waves are created underwater by a spark-gap electrode, and the waves are focused on the kidney stones by means of a reflector that concentrates the shock-wave energy.

Figure 17.15
Cross section of a ureter (×160).

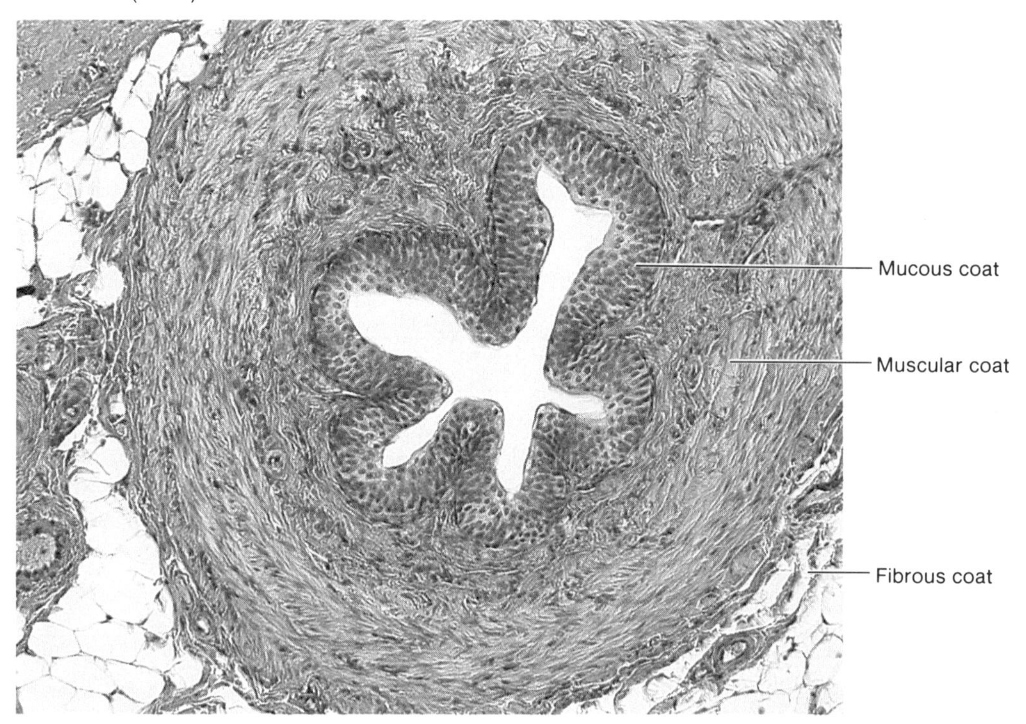

Although the ureter is simply a tube leading from the kidney to the urinary bladder, its muscular wall helps move the urine. Muscular peristaltic waves, originating in the renal pelvis, force the urine along the length of the ureter.

When such a peristaltic wave reaches the urinary bladder, it causes a jet of urine to spurt into the bladder. The opening through which the urine enters is covered by a flaplike fold of mucous membrane. This fold acts as a valve, allowing urine to enter the bladder from the ureter but preventing it from backing up.

1. Describe the structure of a ureter.
2. How is urine moved from the renal pelvis to the urinary bladder?
3. What prevents urine from backing up from the urinary bladder into the ureters?

Urinary Bladder

The **urinary bladder** is a hollow, distensible, muscular organ (fig. 17.1). It is located within the pelvic cavity, behind the symphysis pubis, and beneath the parietal peritoneum.

Although the bladder is somewhat spherical, its shape is altered by the pressures of surrounding organs. When it is empty, the inner wall of the bladder is thrown into many folds, but as it fills with urine, the wall becomes smoother. At the same time, the superior surface of the bladder expands upward into a dome.

The internal floor of the bladder includes a triangular area called the *trigone,* which has an opening at each of its three angles (fig. 17.16). Posteriorly, at the base of the trigone, the openings are those of the ureters. Anteriorly, at the apex of the trigone, there is a short, funnel-shaped extension, called the *neck* of the bladder. This part contains the opening into the urethra.

The wall of the urinary bladder consists of four layers. The inner layer, or *mucous coat,* includes several thicknesses of transitional epithelial cells. The thickness of this tissue changes as the bladder expands and contracts. Thus, during distension the tissue may be only two or three cells thick; but during contraction, it may be five or six cells thick (see chapter 5).

The second layer of the wall is the *submucous coat.* It consists of connective tissue and contains many elastic fibers.

The third layer of the bladder wall, or *muscular coat,* is composed primarily of coarse bundles of smooth muscle fibers. These bundles are interlaced in all directions and at all depths, and together they comprise the **detrusor muscle.** The portion of the detrusor muscle that surrounds the neck of the bladder forms an *internal urethral sphincter.* Sustained contraction of this muscle prevents the bladder from emptying until the pressure within the bladder increases to a certain level.

Figure 17.16
A frontal section of the male urinary bladder.

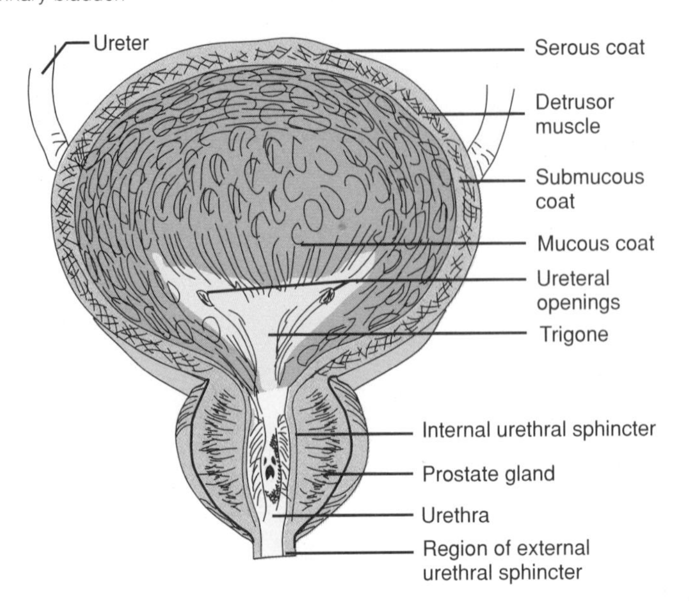

The detrusor muscle is supplied with parasympathetic nerve fibers that function in the micturition reflex, which is discussed in the following section.

The outer layer of the bladder wall, or *serous coat,* consists of the parietal peritoneum. This layer occurs only on the upper surface of the bladder. Elsewhere, the outer coat is composed of fibrous connective tissue.

Because the linings of the ureters and the urinary bladder are continuous, infectious agents, such as bacteria, may ascend from the bladder into the ureters. An inflammation of the bladder, which is called *cystitis,* occurs more commonly in women than men because the female urethral pathway is shorter. An inflammation of the ureter is called *ureteritis.*

1. Describe the trigone of the urinary bladder.
2. Describe the structure of the bladder wall.
3. What kind of nerve fibers supply the detrusor muscle?

Micturition

Micturition (urination) is the process by which urine is expelled from the urinary bladder. It involves the contraction of the detrusor muscle and may be aided by contractions of muscles in the abdominal wall and pelvic floor and by fixation of the thoracic wall and diaphragm. Micturition also involves the relaxation of the *external urethral sphincter.* This muscle, which is part of the urogenital diaphragm described in chapter 8, surrounds the urethra about 3 cm from the bladder and is composed of voluntary skeletal muscle tissue.

The need to urinate is usually stimulated by distension of the bladder wall as it fills with urine. When the wall expands, stretch receptors are stimulated, and the micturition reflex is triggered.

The *micturition reflex center* is located in the sacral segments of the spinal cord. When it is signaled by sensory impulses from the stretch receptors, *parasympathetic motor impulses* travel out to the detrusor muscle, which undergoes rhythmic contractions in response. This action is accompanied by a sensation of urgency.

Although the urinary bladder may hold as much as 500 ml of urine before pain receptors are stimulated, the desire to urinate is usually experienced when it contains about 150 ml. Then, as the volume of urine increases to 300 ml or more, the sensation of fullness becomes increasingly uncomfortable.

As the bladder continues to fill with urine, and its internal pressure increases, contractions of its wall become more and more powerful. When these contractions become strong enough to force the internal urethral sphincter to open, another reflex begins to operate. This second reflex signals the external urethral sphincter to relax and the bladder may empty.

However, because the external urethral sphincter is composed of skeletal muscle, it can be consciously controlled. Thus, the sphincter muscle ordinarily remains contracted until a decision is made to urinate. This control is aided by nerve centers in the *brain stem* and *cerebral cortex* that are able to partially inhibit the micturition reflex. When a person decides to urinate, the external urethral sphincter is allowed to relax, and the micturition reflex is no longer inhibited. Nerve centers within the *pons* and the *hypothalamus* of the brain function to make the micturition reflex more effective.

Figure 17.17
Cross section through the urethra (×10).

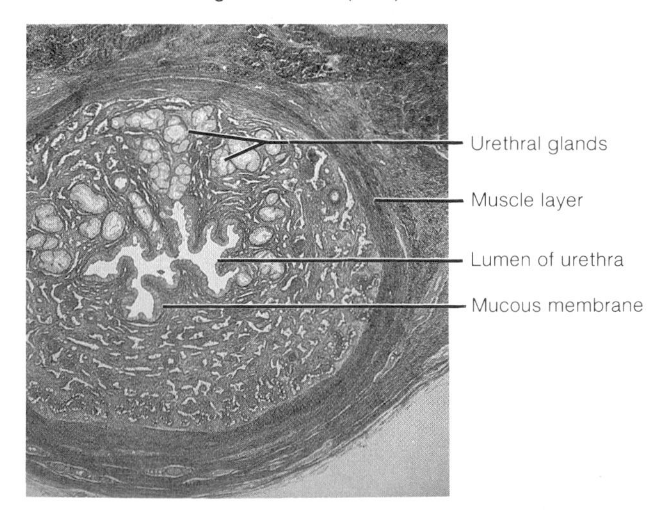

Consequently, the detrusor muscle contracts, and urine is excreted to the outside through the urethra. Within a few moments, the neurons of the micturition reflex fatigue, the detrusor muscle relaxes, and the bladder begins to fill with urine again.

Damage to the spinal cord above the sacral region may result in the loss of voluntary control of urination. If the micturition reflex center and its sensory and motor fibers are uninjured, however, micturition may continue to occur reflexly. In this case, the bladder collects urine until its walls are stretched enough to trigger a micturition reflex, and the detrusor muscle contracts in response. This condition is called an *automatic bladder.*

Urethra

The **urethra** is a tube that conveys urine from the urinary bladder to the outside (fig. 17.1). Its wall is lined with mucous membrane and contains a relatively thick layer of smooth muscle tissue, whose fibers are generally directed longitudinally. It also contains numerous mucous glands, called *urethral glands,* which secrete mucus into the urethral canal (fig. 17.17).

1. Describe process of micturition.
2. How is it possible to consciously inhibit the micturition reflex?
3. Describe the structure of the urethra.

Clinical Terms Related to the Urinary System

anuria (ah-nu′re-ah)—an absence of urine due to failure of kidney function or to an obstruction in a urinary pathway.

bacteriuria (bak-te″re-u′re-ah)—bacteria in the urine.

cystectomy (sis-tek′to-me)—the surgical removal of the urinary bladder.

cystitis (sis-ti′tis)—an inflammation of the urinary bladder.

cystoscope (sis′to-skōp)—an instrument used for visual examination of the interior of the urinary bladder.

cystotomy (sis-tot′o-me)—an incision of the wall of the urinary bladder.

diuresis (di″u-re′sis)—an increased excretion of urine.

diuretic (di″u-ret′ik)—a substance that causes an increased production of urine.

dysuria (dis-u′re-ah)—painful or difficult urination.

enuresis (en″u-re′sis)—uncontrolled urination.

hematuria (hem″ah-tu′re-ah)—blood in the urine.

incontinence (in-kon′ti-nens)—an inability to control urination and/or defecation reflexes.

nephrectomy (nĕ-frek′to-me)—the surgical removal of a kidney.

nephrolithiasis (nef″ro-lĭ-thi′ah-sis)—the presence of a stone or stones in the kidney.

nephroptosis (nef″rop-to′sis)—a movable or displaced kidney.

oliguria (ol″ĭ-gu′re-ah)—a scanty output of urine.

polyuria (pol″e-u′re-ah)—an excessive output of urine.

pyelolithotomy (pi″ĕ-lo-lĭ-thot′o-me)—the removal of a stone from the renal pelvis.

pyelonephritis (pi″e-lon-ne-fri′tis)—an inflammation of the renal pelvis.

pyelotomy (pi″ĕ-lot′o-me)—an incision into the renal pelvis.

pyuria (pi-u′re-ah)—pus (white blood cells) in the urine.

uremia (u-re′me-ah)—a condition in which substances ordinarily excreted in the urine accumulate in the blood.

ureteritis (u-re″ter-i′tis)—an inflammation of the ureter.

urethritis (u″re-thri′tis)—an inflammation of the urethra.

Chapter Summary

Introduction (page 450)
The urinary system consists of the kidneys, ureters, urinary bladder, and urethra.

Kidneys (page 450)
1. Location of the kidneys
   a. The kidneys are high on the posterior wall of the abdominal cavity.
   b. They are positioned behind the parietal peritoneum.
2. Structure of a kidney
   a. A kidney contains a hollow renal sinus.
   b. The ureter expands into the renal pelvis.
   c. Renal papillae project into the renal sinus.
   d. Kidney tissue is divided into a medulla and a cortex.
3. Functions of the kidneys
   a. The kidneys remove metabolic wastes from the blood and excrete them to the outside.
   b. They also help to regulate red blood cell production; blood pressure; and the volume, composition, and pH of the blood.

4. Renal blood vessels
   a. Arterial blood flows through the renal artery, interlobar arteries, arciform arteries, interlobular arteries, and afferent arterioles.
   b. Venous blood returns through a series of vessels that correspond to those of the arterial pathways.
5. Nephrons
   a. Structure of a nephron
      (1) A nephron is the functional unit of the kidney.
      (2) It consists of a renal corpuscle and a renal tubule.
         (a) The corpuscle consists of a glomerulus and Bowman's capsule.
         (b) Portions of the renal tubule include the proximal convoluted tubule, loop of Henle (ascending and descending limbs), distal convoluted tubule, and collecting duct.
      (3) The collecting duct empties into the minor calyx of the renal pelvis.
   b. Blood supply of a nephron
      (1) The glomerular capillary receives blood from the afferent arteriole and passes it to the efferent arteriole.
      (2) The efferent arteriole gives rise to the peritubular capillary system, which surrounds the renal tubule.
   c. Juxtaglomerular apparatus
      (1) The juxtaglomerular apparatus is located at the point of contact between the distal convoluted tubule and the afferent and efferent arterioles.
      (2) It consists of the macula densa and juxtaglomerular cells.

Urine Formation (page 455)

The nephrons function to remove wastes from the blood and to regulate water and electrolyte concentrations. Urine is the end product of these functions.

1. Glomerular filtration
   a. Urine formation begins when water and dissolved materials are filtered out of the glomerular capillary.
   b. The glomerular capillaries are much more permeable than the capillaries in other tissues.
   c. Filtration pressure
      (1) Filtration is due mainly to hydrostatic pressure inside the glomerular capillaries.
      (2) The osmotic pressure of the plasma and hydrostatic pressure in the Bowman's capsule also affect filtration.
      (3) Filtration pressure is the net force acting to move material out of the glomerulus and into the Bowman's capsule.
      (4) The composition of the filtrate is similar to that of tissue fluid.
   d. Filtration rate
      (1) The rate of filtration varies with the filtration pressure.
      (2) The filtration pressure changes with the diameters of the afferent and efferent arterioles.
      (3) As the osmotic pressure in the glomerulus increases, filtration decreases.
      (4) As the hydrostatic pressure in a Bowman's capsule increases, the filtration rate decreases.
      (5) The kidneys produce about 125 ml of glomerular fluid per minute, most of which is reabsorbed.
      (6) Regulation of filtration rate
         (a) When the filtration rate decreases, the macula densa causes the afferent arteriole to dilate, increasing blood flow through the glomerulus and increasing the filtration rate.
         (b) The macula densa also causes the juxtaglomerular cells to release renin, which triggers a series of changes leading to constriction of the efferent arteriole, increasing the glomerular hydrostatic pressure and increasing the filtration rate.
2. Tubular reabsorption
   a. Substances are selectively reabsorbed.
   b. The peritubular capillary is adapted for reabsorption by being very permeable.
   c. Most reabsorption occurs in the proximal tubule, where the epithelial cells possess microvilli.
   d. Various substances are reabsorbed in particular segments of the renal tubule by different modes of transport.
      (1) Glucose and amino acids are reabsorbed by active transport.
      (2) Water is reabsorbed by osmosis.
   e. Active transport mechanisms have limited transport capacities.
   f. Substances that remain in the filtrate are concentrated as water is reabsorbed.
   g. Sodium ions are reabsorbed by active transport.
      (1) As positively charged sodium ions are transported out of the filtrate, negatively charged ions accompany them.
      (2) Water is passively reabsorbed by osmosis.
3. Regulation of urine concentration and volume
   a. Most of the sodium is reabsorbed before the urine is excreted.
   b. ADH causes the permeability of the distal tubule and collecting duct to increase, and, thus, promotes the reabsorption of water.
4. Urea and uric acid excretion
   a. Urea is reabsorbed passively by diffusion.
   b. Uric acid is reabsorbed by active transport and secreted into the renal tubule.

5. Tubular secretion
    a. Secretion is the process by which certain substances are transported from the plasma to the tubular fluid.
    b. Various organic compounds and hydrogen ions are secreted actively.
    c. Potassium ions may be secreted passively.
6. Composition of urine
    a. Urine is about 95% water, and it usually contains urea and uric acid.
    b. It may contain a trace of amino acids and varying amounts of electrolytes.
    c. The volume of urine varies with the fluid intake and with certain environmental factors.

Elimination of Urine

1. Ureters (page 462)
    a. The ureter extends from the kidney to the urinary bladder.
    b. Peristaltic waves in the ureter force urine to the bladder.
2. Urinary bladder
    a. The urinary bladder stores urine and forces it into the urethra.
    b. The openings for the ureters and urethra are located at the three angles of the trigone.
    c. A portion of the detrusor muscle forms an internal urethral sphincter.
3. Micturition
    a. Micturition is the process by which urine is expelled.
    b. It involves contraction of the detrusor muscle and relaxation of the external urethral sphincter.
    c. Micturition reflex
        (1) Stretch receptors in the bladder wall are stimulated by distension.
        (2) The micturition reflex center in the sacral spinal cord sends parasympathetic motor impulses to the detrusor muscle.
        (3) As the bladder fills, its internal pressure increases, and the internal urethral sphincter is forced open.
        (4) A second reflex causes the external urethral sphincter to relax, unless its contraction is maintained by voluntary control.
        (5) The control of urination is aided by nerve centers in the cerebral cortex and brain stem.
4. Urethra
    The urethra conveys urine from the bladder to the outside.

Clinical Application of Knowledge

1. If an infant is born with narrowed renal arteries, what effect would this condition have on the volume of urine produced?
2. If a patient who has had major abdominal surgery receives intravenous fluids equal to the volume of the blood lost during surgery, would you expect the volume of urine produced to be greater or less than normal? Why?
3. If a physician prescribed oral penicillin therapy for a patient with an infection of the urinary bladder, how would you describe to the patient the route by which the drug would reach the bladder?
4. If the blood pressure decreases excessively in a patient who is in shock as a result of a severe injury, how would you expect the volume of urine produced by the patient's kidneys to change? Why?

Review Activities

1. Name the organs of the urinary system, and list their general functions.
2. Describe the external and internal structure of a kidney.
3. List the functions of the kidneys.
4. Name the vessels through which the blood passes as it travels from the renal artery to the renal vein.
5. Distinguish between a renal corpuscle and a renal tubule.
6. Name the parts through which fluid passes as it travels from the glomerulus to the collecting duct.
7. Describe the location and structure of the juxtaglomerular apparatus.
8. Define *filtration pressure.*
9. Compare the composition of the glomerular filtrate with that of the blood plasma.
10. Explain how the diameters of the afferent and efferent arterioles affect the rate of glomerular filtration.
11. Explain how changes in the osmotic pressure of the blood plasma may affect the rate of glomerular filtration.
12. Explain how the hydrostatic pressure of a Bowman's capsule affects the rate of glomerular filtration.
13. Describe two mechanisms by which the juxtaglomerular apparatus helps regulate the filtration rate.
14. Discuss how tubular reabsorption is a selective process.
15. Explain how the peritubular capillary is adapted for reabsorption.
16. Explain how epithelial cells of the proximal convoluted tubule are adapted for reabsorption.

17. Explain why active transport mechanisms have limited transport capacities.
18. Define *renal plasma threshold.*
19. Explain how amino acids and proteins are reabsorbed.
20. Describe the effect of sodium reabsorption on the reabsorption of negatively charged ions.
21. Explain how sodium reabsorption affects water reabsorption.
22. Describe the function of ADH.
23. Compare the processes by which urea and uric acid are reabsorbed.
24. Explain how potassium ions may be secreted passively.
25. List the more common substances found in urine and their sources.
26. List some of the factors that affect the volume of urine produced each day.
27. Describe the structure and function of a ureter.
28. Explain how the muscular wall of the ureter aids in moving urine.
29. Describe the structure and location of the urinary bladder.
30. Define *detrusor muscle.*
31. Distinguish between the internal and external urethral sphincters.
32. Describe the micturition reflex.
33. Explain how the micturition reflex can be voluntarily controlled.

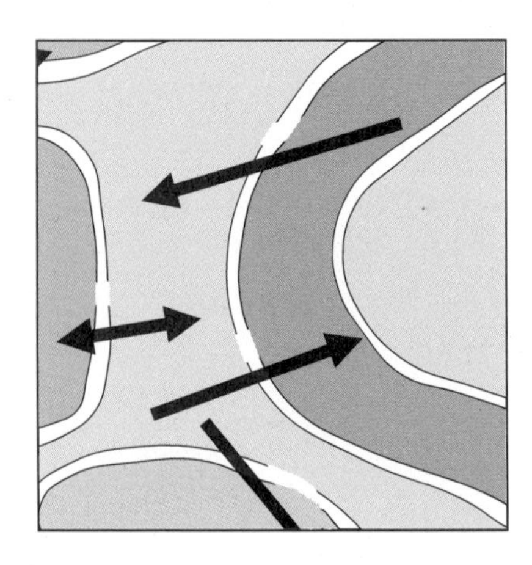

18 Water, Electrolyte, and Acid-Base Balance

Cell functions and, indeed, cell survival depend upon homeostasis—the existence of a stable cellular environment. In such an environment, the body cells are continually supplied with oxygen and nutrients, and the waste products resulting from their metabolic activities are continually carried away. At the same time, the concentrations of water and dissolved electrolytes and the pH in the cellular fluids remain constant. This condition requires the maintenance of a *water, electrolyte,* and *acid-base balance.*

Chapter Objectives

After you have studied this chapter, you should be able to

1. Explain what is meant by water and electrolyte balance, and discuss the importance of this balance.
2. Describe how the body fluids are distributed within compartments, how the fluid composition differs between compartments, and how the fluids move from one compartment to another.
3. List the routes by which water enters and leaves the body, and explain how water input and output are regulated.
4. Explain how electrolytes enter and leave the body, and explain how the input and output of electrolytes are regulated.
5. List the major sources of hydrogen ions in the body.
6. Distinguish between strong and weak acids and bases.
7. Explain how changing pH values of the body fluids are minimized by chemical buffer systems, the respiratory center, and the kidneys.
8. Complete the review activities at the end of this chapter. Note that the items are worded in the form of specific learning objectives. You may want to refer to them before reading the chapter.

Key Terms

acid (as′id)

base (bās)

buffer system (buf′er sis′tem)

electrolyte balance (e-lek′tro-līt bal′ans)

extracellular (ek″strah-sel′u-lar)

intracellular (in″trah-sel′u-lar)

transcellular (trans-sel′u-lar)

water balance (wot′er bal′ans)

Aids to Understanding Words

de-, separation from: *de*hydration—the removal of water from the cells or body fluids.

extra-, outside: *extra*cellular fluid—the fluid outside of the body cells.

im- (or in-), not: *im*balance—a condition in which factors are not in equilibrium.

intra-, within: *intra*cellular fluid—the fluid within the body cells.

neutr-, neither one nor the other: *neutr*al—a solution that is neither acidic nor basic.

Introduction

THE TERM *BALANCE* suggests a state of equilibrium, and in the case of water and electrolytes, it means that the quantities entering the body are equal to the quantities leaving it. Maintaining such a balance requires mechanisms to ensure that lost water and electrolytes will be replaced and that any excesses will be expelled. As a result, the quantities within the body are relatively stable at all times.

It is important to remember that water balance and electrolyte balance are interdependent, because the electrolytes are dissolved in the water of the body fluids. Consequently, anything that alters the concentrations of the electrolytes will necessarily alter the concentration of the water by adding solutes to it or by removing solutes from it. Likewise, anything that changes the concentration of the water will change the concentrations of the electrolytes by making them either more concentrated or more diluted.

Distribution of Body Fluids

Water and electrolytes are not uniformly distributed throughout the tissues. Instead, they occur in regions, or *compartments,* that contain fluids of varying compositions. The movement of water and electrolytes between these compartments is regulated, so that their distribution remains stable.

Fluid Compartments

The body of an average adult male is about 63% water by weight. This water (about 40 liters), together with its dissolved electrolytes, is distributed into two major compartments—an intracellular fluid compartment and an extracellular fluid compartment (fig. 18.1).

The **intracellular fluid compartment** includes all the water and electrolytes that are enclosed by cell membranes. In other words, intracellular fluid is the fluid within the cells, and in an adult, it represents about 63% by volume of the total body water.

The **extracellular fluid compartment** includes all the fluid outside the cells—within the tissue spaces (interstitial fluid), blood vessels (plasma), and lymphatic vessels (lymph). A specialized fraction of the extracellular fluid is separated from other extracellular fluid by a layer of epithelium. It is called **transcellular fluid** and includes the *cerebrospinal fluid* of the central nervous system, *aqueous and vitreous humors* of the eyes, *synovial fluid* of the joints, *serous fluid* within the various body cavities, and fluid *secretions* of the glands. All together, the fluids of the extracellular compartment constitute about 37% by volume of the total body water.

Composition of the Body Fluids

Generalizations about the concentrations of the components of extracellular fluids relative to those of intracellular fluid can be made (fig. 18.2). *Extracellular fluids* generally have similar compositions. They are characterized by relatively high concentrations of sodium, chloride, and bicarbonate ions, and lesser concentrations of potassium, calcium, magnesium, phosphate, and sulfate ions. The plasma fraction of extracellular fluid contains considerably more protein than do either interstitial fluid or lymph.

Intracellular fluid contains relatively high concentrations of potassium, phosphate, and magnesium ions. It includes a somewhat greater concentration of sulfate ions, and lesser concentrations of sodium, chloride, and bicarbonate ions than does extracellular fluids. Intracellular fluid also has a greater concentration of protein than plasma.

Figure 18.1
Fluid within the intracellular compartment is separated from fluid in the extracellular compartment by cell membranes. What membranes separate the various components of the extracellular fluid compartment?

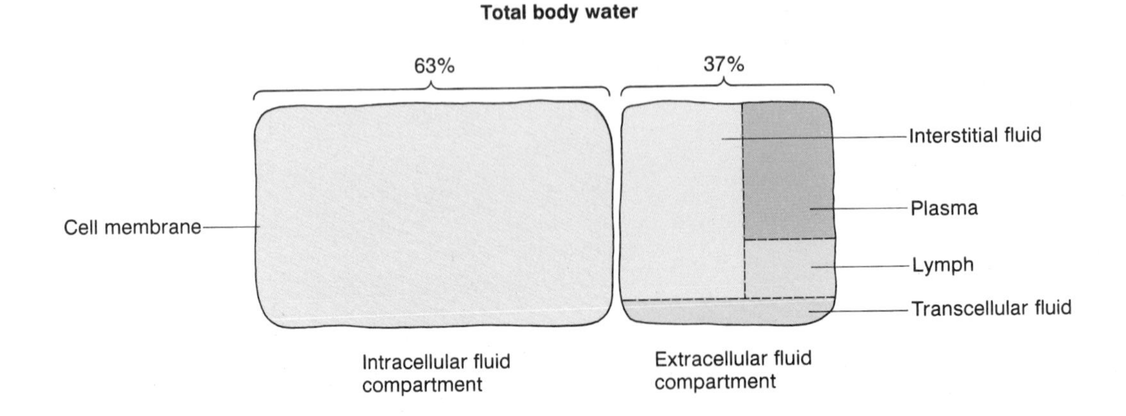

1. Describe the normal distribution of water within the body.
2. Which electrolytes are in higher concentrations in extracellular fluids? In intracellular fluid?
3. How does the concentration of protein vary in the various body fluids?

Movement of Fluid between the Compartments

The movement of water and electrolytes from one compartment to another is largely regulated by two factors—hydrostatic pressure and osmotic pressure (fig. 18.3). For example, as explained in chapter 15, fluid leaves the plasma at the arteriole ends of capillaries and enters the interstitial spaces because of the net outward force of *hydrostatic pressure* (blood pressure). Fluid returns to the plasma from the interstitial spaces at the venule ends of capillaries because of the net inward force of *osmotic pressure.* Likewise, as is mentioned in chapter 16, fluid leaves the interstitial spaces and enters the lymph capillaries because of the osmotic pressure that develops within these interstitial spaces. As a result of the circulation of lymph, interstitial fluid is returned to the plasma.

The movement of fluid between the intracellular and extracellular compartments is similarly controlled by such pressures. Because the hydrostatic pressure within the cells and surrounding interstitial fluid is ordinarily equal and remains stable, however, any fluid movement that occurs is likely to be the result of changes in osmotic pressure.

Figure 18.2
Extracellular fluid has relatively high concentrations of sodium, chloride, and bicarbonate ions; intracellular fluid has relatively high concentrations of potassium, magnesium, and phosphate ions.

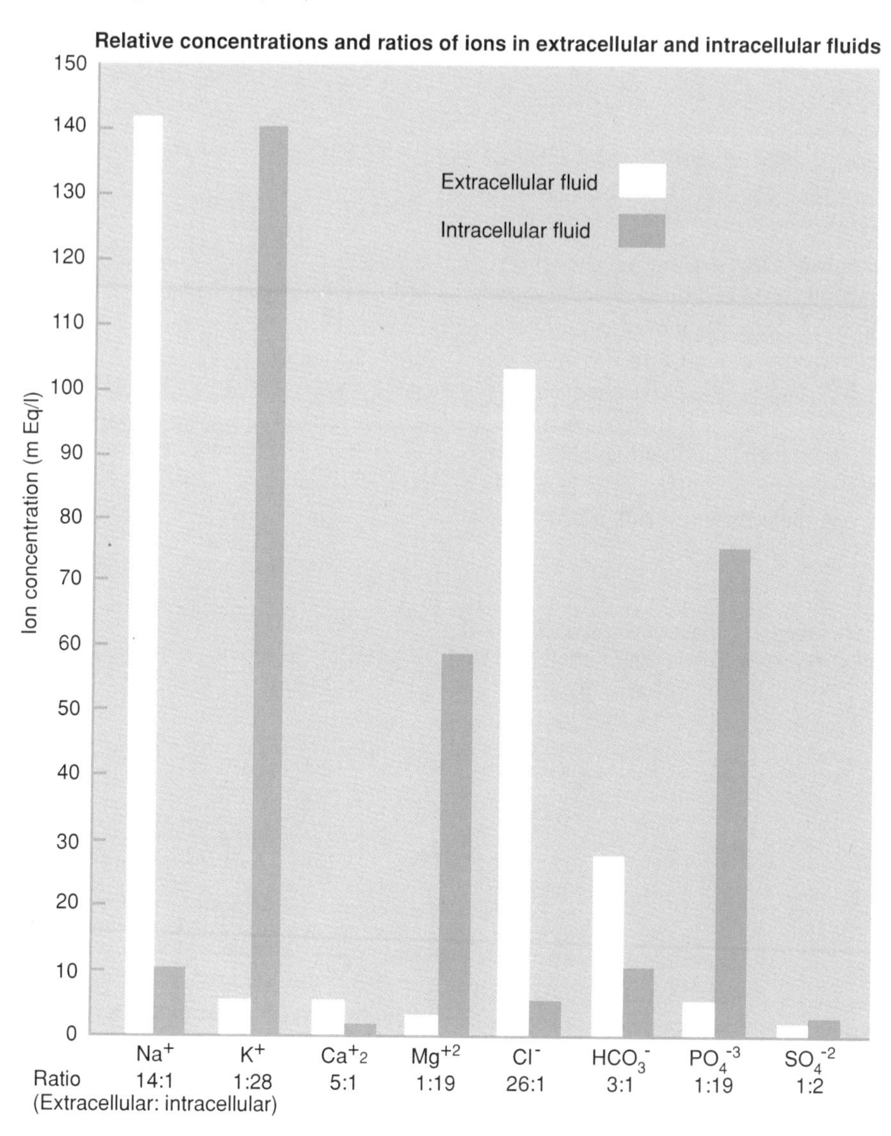

Figure 18.3
The movement of fluids between compartments results from differences in the hydrostatic and osmotic pressures.

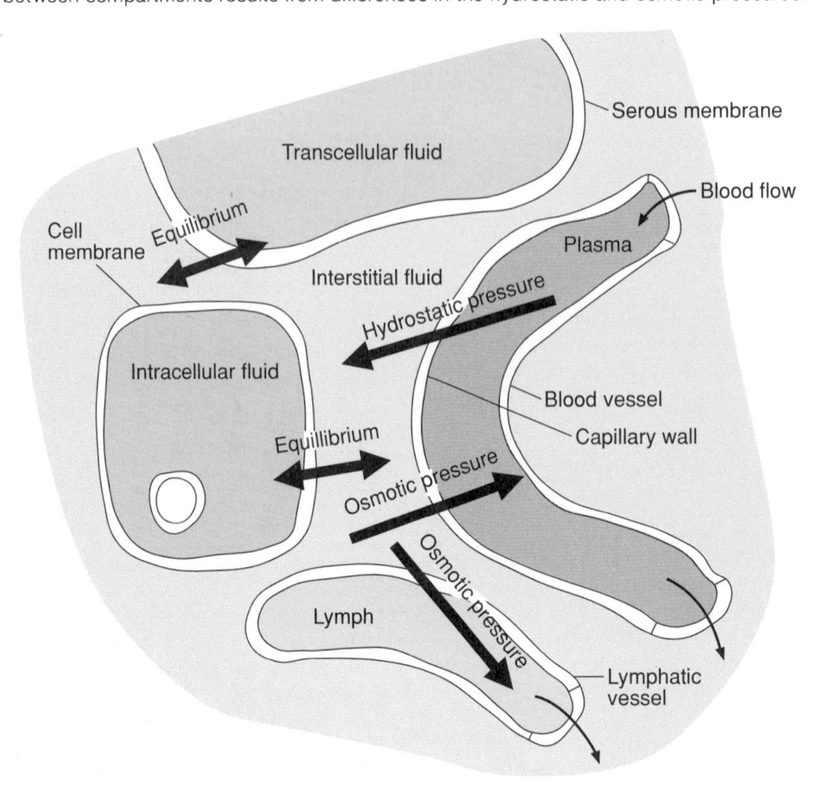

For example, because the sodium ion concentration in extracellular fluids is especially high, a decrease in this concentration will cause a net movement of water from the extracellular compartment into the intracellular compartment by osmosis. As a consequence, the cells will tend to swell. Conversely, if the concentration of sodium ions in the interstitial fluid increases, the net movement of water will be outward from the intracellular compartment, and the cells will shrink as they lose water.

1. What factors control the movement of water and electrolytes from one fluid compartment to another?
2. How does the sodium ion concentration within body fluids affect the net movement of water between the compartments?

Water Balance

Water balance exists when the total intake of water is equal to the total loss of water.

Water Intake

Although the volume of water gained each day varies from individual to individual, an average adult living in a moderate environment takes in about 2,500 ml (fig. 18.4).

Figure 18.4
Major sources of body water.

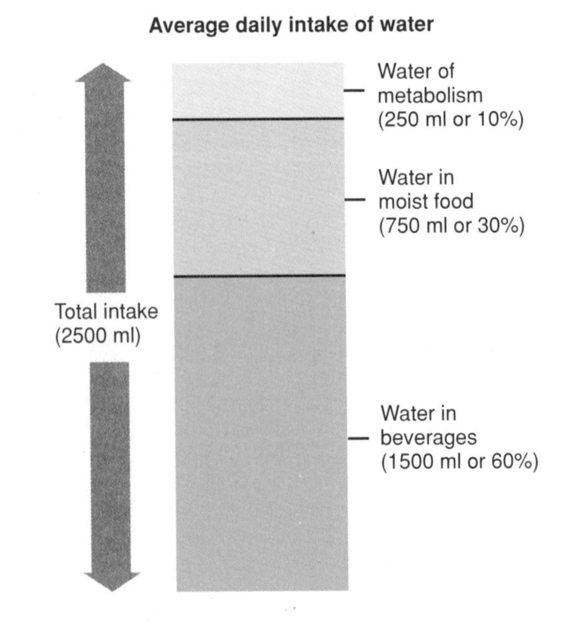

Of this amount, probably 60% will be obtained in drinking water or beverages, and another 30% will be gained from moist foods. The remaining 10% will be a by-product of the oxidative metabolism of various nutrients, which is called the **water of metabolism.**

Water Balance Disorders
A Current Topic

Among the more common water balance disorders are dehydration and water intoxication.

Dehydration occurs when the body's output of water exceeds its intake. This condition may develop following excessive sweating, or as a result of prolonged water deprivation accompanied by continued water output. In either case, as water is lost, the extracellular fluid becomes increasingly concentrated (hypertonic), and water leaves body cells by osmosis (chapter 3). Dehydration may also accompany illnesses in which excessive fluids are lost as a result of prolonged vomiting or diarrhea.

During dehydration, the skin and mucous membranes of the mouth feel dry, and body weight is lost. Also, a high fever may develop as the body temperature-regulating mechanism becomes less effective due to a deficiency of water needed for sweating. In severe cases, as waste products accumulate in the concentrated extracellular fluid, symptoms of cerebral disturbances, including mental confusion, delirium, and coma, may develop.

The treatment of dehydration involves replacing lost water and electrolytes. It is important to note that if the water alone is replaced, the extracellular fluid may be more dilute than normal, and this may produce the condition called water intoxication.

Water intoxication may occur as a result of administering pure water to a severely dehydrated person, or it can develop in a person who drinks water for a prolonged period faster than the kidneys can excrete the excess. In either instance, as the extra water is absorbed, the extracellular fluid becomes more diluted (hypotonic). Consequently, the body cells swell as water enters them by osmosis.

The symptoms of water intoxication are mainly related to a decrease in the extracellular sodium ion concentration, and they include painful muscular contractions (heat cramps), convulsions, confusion, and coma. Treatment for this condition involves restricting water intake and administering hypertonic electrolyte solutions.

Regulation of Water Intake

The primary regulator of water intake is thirst. Although the thirst mechanism is poorly understood, it seems to involve the osmotic pressure of extracellular fluids and a *thirst center* in the hypothalamus of the brain.

As water is lost from the body, the osmotic pressure of the extracellular fluids increases. *Osmoreceptors* in the thirst center are stimulated by such a change, and as a result, the hypothalamus causes the person to feel thirsty and to seek water.

The thirst mechanism is normally triggered whenever the total body water is decreased by 1% or 2%. As a person drinks water in response to thirst, the act of drinking and the resulting distension of the stomach wall seem to trigger nerve impulses that inhibit the thirst mechanism. Thus, the person usually stops drinking long before the swallowed water has been absorbed.

1. What is meant by water balance?
2. Where is the thirst center located?
3. What mechanism stimulates fluid intake? What inhibits it?

Water Output

Water normally enters the body through the mouth, but it can be lost by a variety of routes. These include losses in urine, feces, and sweat (sensible perspiration), as well as less obvious losses, which occur by the diffusion of water through the skin (insensible perspiration) and the evaporation of water from the lungs during breathing (fig. 18.5).

If an average adult takes in 2,500 ml of water each day, then 2,500 ml must be eliminated if water balance is to be maintained. Of this volume, perhaps 60% will be lost in urine, 6% in feces, and 6% in sweat. About 28% will be lost by diffusion through the skin

Figure 18.5
Routes by which body water is lost. What factors influence water loss by each of these routes?

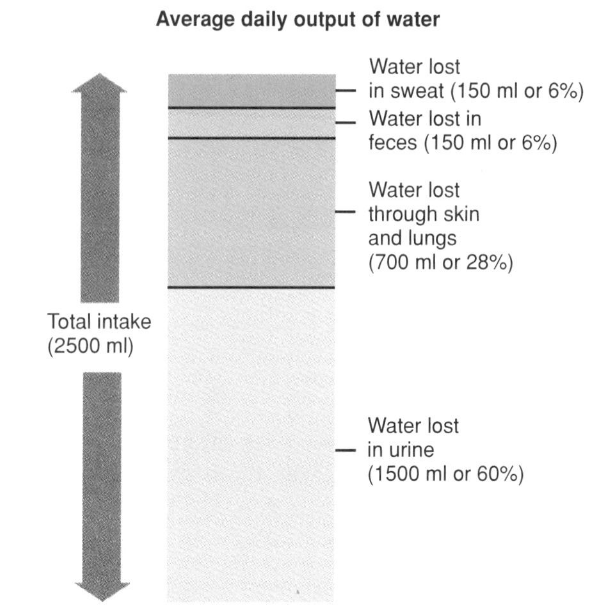

and by evaporation from the lungs. These percentages will, of course, vary with such environmental factors as temperature and relative humidity and with physical exercise.

Regulation of Water Balance

The primary regulator of water output is urine production. As discussed in chapter 17, the volume of water excreted in the urine is regulated mainly by activity in the distal convoluted tubules and collecting ducts of the nephrons. The epithelial linings of these segments of the renal tubule remain relatively impermeable to water unless antidiuretic hormone (ADH) is present. The action of ADH causes water to be reabsorbed in these segments and, thus, to be conserved. In the absence of ADH, less water is reabsorbed and the urine volume increases (see chapter 17).

Diuretics are substances that promote the production of urine. A number of common substances, such as caffeine in coffee and tea, have diuretic effects, as do a variety of drugs used to reduce the volume of the body fluids.

Various diuretics produce their effects in different ways. For example, some, such as alcohol and various narcotic drugs, promote urine formation by inhibiting the release of ADH. Others inhibit the reabsorption of sodium ions in the renal tubules. As a consequence, the osmotic pressure of the tubular fluid increases, and the reabsorption of water by osmosis and its return to the blood are reduced.

1. By what routes is water lost from the body?
2. What role do the renal tubules play in the regulation of water balance?

Electrolyte Balance

An **electrolyte balance** exists when the quantities of the various electrolytes gained by the body are equal to those lost.

Electrolyte Intake

The electrolytes (substances that release ions in water) of greatest importance to cellular functions are those that release the ions of sodium, potassium, calcium, magnesium, chloride, sulfate, phosphate, and bicarbonate. These substances are obtained primarily from foods, but they may also occur in drinking water and other beverages. In addition, some electrolytes occur as by-products of various metabolic reactions.

Regulation of Electrolyte Intake

Ordinarily, a person obtains sufficient electrolytes by responding to hunger and thirst. When there is a severe electrolyte deficiency, however, a person may experience a *salt craving,* which is a strong desire to eat salty foods.

Electrolyte Output

Some electrolytes are lost by perspiration. The quantities of electrolytes leaving will vary with the amount of perspiration produced. Greater quantities are lost on warmer days and during times of strenuous exercise. Also, varying amounts of electrolytes are lost in feces. The greatest electrolyte output, however, occurs as a result of kidney function and urine production.

1. What electrolytes are most important to cellular functions?
2. What mechanisms ordinarily regulate electrolyte intake?
3. By what routes are electrolytes lost from the body?

Regulation of Electrolyte Balance

The concentrations of positively charged ions, such as sodium (Na^+), potassium (K^+), and calcium (Ca^{+2}), are particularly important. Certain concentrations of these ions, for example, are necessary for the conduction of nerve impulses, contraction of muscle fibers, and maintenance of cell membrane permeability.

Sodium ions account for nearly 90% of the positively charged ions in extracellular fluids. The primary mechanism regulating these ions involves the kidneys and the hormone *aldosterone*. As is explained in chapter 11, this hormone is secreted by the adrenal cortex. Its presence causes an increase in sodium reabsorption in the collecting ducts of the renal tubules.

Aldosterone also functions in regulating potassium (fig. 18.6). In fact, the most important stimulus for aldosterone secretion is a rising potassium ion concentration, which seems to stimulate the cells of the adrenal cortex directly. This hormone enhances the reabsorption of sodium ions and at the same time causes the secretion of potassium ions into the renal filtrate.

As mentioned in chapter 11, the concentration of calcium ions in extracellular fluids is regulated mainly by the parathyroid glands. Whenever the calcium ion concentration drops below normal, these glands are stimulated directly, and they secrete *parathyroid hor-*

Sodium and Potassium Imbalances

A Current Topic

Extracellular fluids usually have relatively high sodium ion concentrations, and intracellular fluids have relatively high potassium ion concentrations. The renal regulation of the sodium concentration is closely related to that of potassium, because active reabsorption of sodium ions under the influence of aldosterone is accompanied by passive secretion (and excretion) of potassium ions. Consequently, conditions that cause imbalance of sodium are closely related to those that cause imbalance of potassium.

Sodium and potassium imbalances include the following:

1. *Low sodium concentration* (hyponatremia) may result from prolonged sweating, vomiting, or diarrhea; renal disease in which sodium is inadequately reabsorbed; adrenal cortex disorders in which the secretion of aldosterone is reduced; or an excessive intake of water. As a consequence of low sodium, extracellular fluid becomes hypotonic, and water enters body cells excessively by osmosis, producing the symptoms of water intoxication.
2. *High sodium concentration* (hypernatremia) may occur when water is lost excessively by diffusion and evaporation during high fever or when water loss is increased as a result of a reduced secretion of ADH. Possible effects of hypernatremia include disturbances within the central nervous system, such as confusion, stupor, and coma.
3. *Low potassium concentration* (hypokalemia) may be caused by an excessive release of aldosterone (Cushing's syndrome) that stimulates increased renal secretion of potassium; use of diuretic drugs that promote potassium excretion; renal disorders in which potassium excretion is increased; or prolonged vomiting and diarrhea. Effects of hypokalemia may include muscular weakness or paralysis, difficulty in breathing, and severe cardiac disturbances, such as arrhythmias.
4. *High potassium concentration* (hyperkalemia) may result from renal disease in which potassium excretion is decreased; use of drugs that promote renal conservation of potassium; decreased release of aldosterone (Addison's disease); increased hydrogen ion concentration (acidosis) which causes potassium ions to move from intracellular fluid to extracellular fluid. Possible effects of hyperkalemia include paralysis of skeletal muscles and severe cardiac disorders, such as cardiac arrest.

Figure 18.6
If the concentration of potassium ions increases, the kidneys act to conserve sodium ions and to excrete potassium ions.

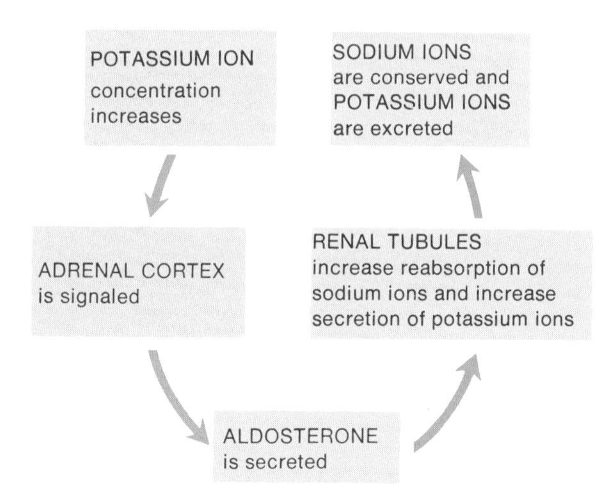

mone in response. Parathyroid hormone causes the concentrations of calcium and phosphate ions in the extracellular fluids to increase.

Generally, the concentrations of negatively charged ions are controlled secondarily by the regulatory mechanisms that control positively charged ions. For example, chloride ions (Cl^-), the most abundant negatively charged ions in the extracellular fluids, are passively reabsorbed from the renal tubules as a result of the active reabsorption of sodium ions. That is, the negatively charged chloride ions are electrically attracted to the positively charged sodium ions and accompany them as they are reabsorbed (see chapter 17).

Some negatively charged ions, such as those of phosphate (PO_4^{-3}) and sulfate (SO_4^{-2}), are also partially regulated by active transport mechanisms that have limited transport capacities. Thus, if the extracellular phosphate ion concentration is low, the phosphate ions in the renal tubules are conserved. On the other hand, if the renal plasma threshold is exceeded, the excess phosphate will be excreted in the urine.

1. How are the concentrations of sodium and potassium ions controlled?
2. How is calcium regulated?
3. What mechanism functions to regulate the concentrations of most negatively charged ions?

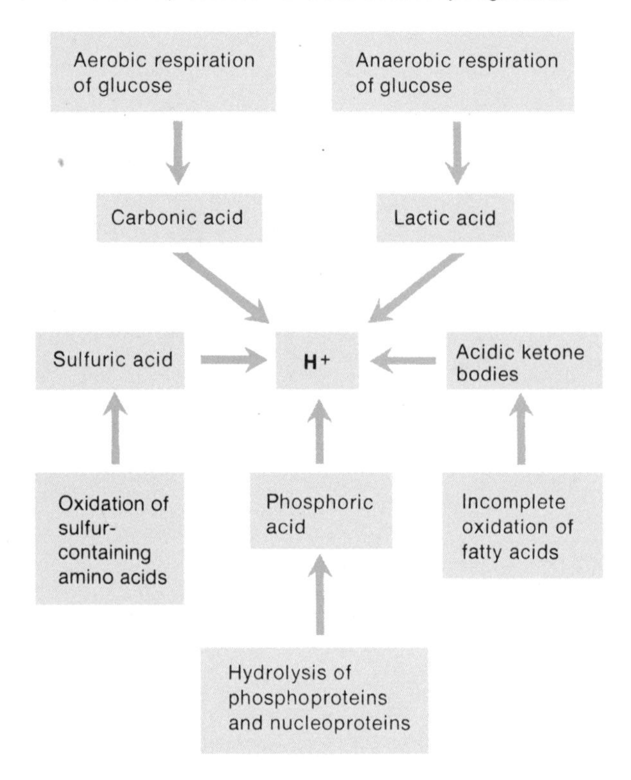

Figure 18.7
Some metabolic processes that are sources of hydrogen ions.

Acid-Base Balance

As discussed in chapter 2, electrolytes that ionize in water and release hydrogen ions are called *acids*. Substances that combine with hydrogen ions are called *bases*. The concentrations of acids and bases within the body fluids must be controlled if homeostasis is to be maintained.

Sources of Hydrogen Ions

Most of the hydrogen ions in the body fluids originate as by-products of metabolic processes, although small quantities may enter in foods (fig. 18.7).

The major metabolic sources of hydrogen ions include the following:

1. **Aerobic respiration of glucose.** This process results in the production of carbon dioxide and water. Carbon dioxide diffuses out of the cells and reacts with water in the extracellular fluids to form *carbonic acid*. The resulting carbonic acid then ionizes to release hydrogen ions and bicarbonate ions.

$$H_2CO_3 \rightarrow H^+ + HCO_3^-$$

2. **Anaerobic respiration of glucose.** When glucose is utilized anaerobically, *lactic acid* is produced and hydrogen ions are added to the body fluids.
3. **Incomplete oxidation of fatty acids.** The incomplete oxidation of fatty acids results in the production of *acidic ketone bodies,* which cause the hydrogen ion concentration to increase.
4. **Oxidation of amino acids containing sulfur.** The oxidation of sulfur-containing amino acids results in the production of *sulfuric acid* (H_2SO_4), which ionizes to release hydrogen ions and sulfate ions.
5. **Breakdown of phosphoproteins and nucleoproteins.** Phosphoproteins and nucleoproteins contain phosphorus. Their oxidation results in the production of *phosphoric acid* (H_3PO_4), which ionizes to release hydrogen ions.

The acids produced by various metabolic processes vary in strength, and, thus, their effects on the hydrogen ion concentration of the body fluids vary also.

1. Distinguish between an acid and a base.
2. What are the major sources of hydrogen ions in the body?

Strengths of Acids and Bases

Acids that ionize more completely are *strong acids,* and those that ionize less completely are *weak acids.* For example, the hydrochloric acid (HCl) of gastric juice is a strong acid, but the carbonic acid (H_2CO_3) produced when carbon dioxide reacts with water is weak.

Bases are substances, like hydroxyl ions (OH^-), that will combine with hydrogen ions. *Chloride ions* (Cl^-) and *bicarbonate ions* (HCO_3^-) are also bases. Furthermore, since chloride ions combine less readily with hydrogen ions, they are *weak* bases, but the bicarbonate ions, which combine more readily with hydrogen ions, are *strong bases.*

Regulation of Hydrogen Ion Concentration

The concentration of hydrogen ions in the body fluids, as measured by the pH, is regulated primarily by acid-base buffer systems, the activity of the respiratory center in the brain stem, and the functions of the nephrons in the kidneys.

Acid-Base Buffer Systems

Acid-base buffer systems occur in all the body fluids and are usually composed of sets of two or more chemical substances. Such chemicals can combine with acids or bases when they occur in excess. More specifically, the substances of a buffer system can convert strong acids, which tend to release large quantities of hydrogen ions, into weak acids, which release fewer hydrogen ions. Likewise, these buffering substances can combine with strong bases and change them into weak bases. Such actions help to minimize the pH changes in body fluids.

The three most important acid-base buffer systems in body fluids are these:

1. **Bicarbonate buffer system.** The bicarbonate buffer system, which is present in the intracellular and extracellular body fluids, consists of carbonic acid (H_2CO_3) and sodium bicarbonate ($NaHCO_3$). If a strong acid, like hydrochloric acid, is present, it reacts with the sodium bicarbonate. The products of the reaction are carbonic acid, which is a weaker acid, and sodium chloride. Consequently, an increase in the hydrogen ion concentration in the body fluid is minimized.

$$\underset{\text{(strong acid)}}{HCl} + NaHCO_3 \rightarrow \underset{\text{(weaker acid)}}{H_2CO_3} + NaCl$$

If, on the other hand, a strong base like sodium hydroxide (NaOH) is present, it reacts with the carbonic acid. The products are sodium bicarbonate ($NaHCO_3$), which is a weaker base, and water. Thus, a shift toward a more basic (alkaline) state is minimized.

$$\underset{\text{(strong base)}}{NaOH} + H_2CO_3 \rightarrow \underset{\text{(weaker base)}}{NaHCO_3} + H_2O$$

2. **Phosphate buffer system.** The phosphate acid-base buffer system is also present in the intracellular and extracellular body fluids. It is particularly important as a regulator of the hydrogen ion concentration in the tubular fluid of the nephrons and in urine. This buffer system consists of two phosphate compounds—sodium monohydrogen phosphate (Na_2HPO_4) and sodium dihydrogen phosphate (NaH_2PO_4).

   Sodium monohydrogen phosphate can react with strong acids to produce weaker ones, and sodium dihydrogen phosphate can react with strong bases to produce weaker ones.
3. **Protein buffer system.** The protein acid-base buffer system consists of the plasma proteins, such as albumin, and various proteins within the cells, including the hemoglobin of red blood cells.

   As is discussed in chapter 2, proteins are composed of amino acids bound together in complex chains. Some of these amino acids have freely exposed groups of atoms, called *carboxyl groups,* (—COOH). Under certain conditions, a carboxyl group can become ionized, and a hydrogen ion is released.

$$—COOH \rightarrow —COO^- + H^+$$

Some of the amino acids within a protein molecule also contain freely exposed *amino groups* (—NH_2). Under certain conditions, these amino groups can accept hydrogen ions.

$$—NH_2 + H^+ \rightarrow —NH_3^+$$

Thus, protein molecules can function as acids by releasing hydrogen ions from their carboxyl groups or as bases by accepting hydrogen ions into their amino groups. This special property allows protein molecules to operate as an acid-base buffer system. In the presence of excess hydrogen ions, the —COO^- portions of the protein molecules accept hydrogen ions and become —COOH groups again. This action decreases the number of free hydrogen ions in body fluid and minimizes the pH change.

In the presence of excess hydroxyl ions (OH^+), the —NH_3^+ groups of protein molecules give up hydrogen ions and become —NH_2 groups again. These hydrogen ions then combine with the hydroxyl ions to form water molecules. Once again, the pH change is reduced, because water is a neutral substance.

$$H^+ + OH^- \rightarrow H_2O$$

Chart 18.1 summarizes the actions of the three major buffer systems.

Chart 18.1 Chemical acid-base buffer systems

Buffer system	Constituents	Actions
Bicarbonate system	Sodium bicarbonate ($NaHCO_3$)	Converts a strong acid into a weak acid
	Carbonic acid (H_2CO_3)	Converts a strong base into a weak base
Phosphate system	Sodium monohydrogen phosphate (Na_2HPO_4)	Converts a strong acid into a weak acid
	Sodium dihydrogen phosphate (NaH_2PO_4)	Converts a strong base into a weak base
Protein system (and amino acids)	—COO^- group of a molecule	Accepts hydrogen ions in the presence of excess acid
	—NH_3^+ group of a molecule	Releases hydrogen ions in the presence of excess base

Acid-Base Imbalances

A Current Topic

Ordinarily, the hydrogen ion concentration of body fluids is maintained within a very narrow pH range by the actions of chemical and physiological buffer systems. However, disease conditions may disturb the normal acid-base balance and produce serious consequences. For example, the pH of arterial blood is normally about 7.4, and if this value drops below 7.4, the person is said to have *acidosis*. If the pH rises above 7.4, the condition is called *alkalosis*. Such shifts in the pH of body fluids may be life threatening, and in fact, a person usually does not survive if the pH drops to 6.8 or rises to 8.0 for more than a few hours.

Acidosis is caused by an accumulation of acid or a loss of base, resulting in an increase in the hydrogen ion concentration of body fluids. Conversely, alkalosis is caused by a loss of acid or an accumulation of base, accompanied by a decrease in hydrogen ion concentration.

Two types of acidosis are *respiratory acidosis* and *metabolic acidosis*. Respiratory acidosis is caused by conditions that hinder pulmonary ventilation, such as injuries to the respiratory center in the brain stem, obstructions in the air passages, or diseases that reduce the surface area of the respiratory membrane and decrease the volume of gas exchanged in the lungs.

The symptoms of respiratory acidosis include labored breathing, cyanosis, and depression of the central nervous system, characterized by drowsiness, disorientation, and stupor.

Metabolic acidosis occurs when nonrespiratory acids accumulate or bases are lost. Conditions that produce metabolic acidosis include renal disease in which the excretion of acids produced by metabolic processes is decreased; vomiting in which the alkaline contents of the upper intestine are lost; prolonged diarrhea in which the alkaline intestinal secretions are lost; diabetes mellitus in which fatty acids are converted into ketone bodies.

The symptoms of metabolic acidosis are nausea, vomiting, and rapid and deep breathing.

Two types of alkalosis are *respiratory alkalosis* and *metabolic alkalosis*. Respiratory alkalosis develops as a result of hyperventilation (chapter 13), which is accompanied by an excessive loss of carbon dioxide and a consequent decrease in carbonic acid and hydrogen ion concentration of body fluids. Hyperventilation most commonly occurs during periods of anxiety, although it may also accompany fever or the toxic effects of certain drugs, such as aspirin.

The symptoms of respiratory alkalosis include light headedness, agitation, dizziness, and tingling sensations. In severe cases, impulses may be spontaneously triggered on motor neurons, and skeletal muscles may undergo tetanic contractions in response.

Metabolic alkalosis results from excessive loss of hydrogen ions or from a gain in bases. This condition may occur following the removal of gastric juice (lavage), prolonged vomiting in which only the stomach contents are lost, or the ingestion of excessive amounts of antacids, such as sodium bicarbonate.

The symptoms of metabolic alkalosis include a decreased rate and depth of breathing, irritability, muscular weakness, and decreased intestinal motility.

---

1. What is the difference between a strong acid or base and a weak acid or base?
2. How does a chemical buffer system help to regulate the pH?
3. List the major buffer systems of the body.

The Respiratory Center

The **respiratory center** in the brain stem helps regulate hydrogen ion concentrations in the body fluids by controlling the rate and depth of breathing (see chapter 13). Specifically, if the cells increase their production of carbon dioxide, as occurs during periods of physical exercise, the production of carbonic acid increases (fig. 18.8). As the carbonic acid dissociates, the concentration of hydrogen ions increases, and the pH of the fluids tends to drop.

Such an increasing concentration of carbon dioxide and the consequent increase in hydrogen ion concentration in the plasma stimulates chemosensitive areas within the respiratory center. In response, the respiratory center causes the depth and rate of breathing to increase, so that a greater amount of carbon dioxide is excreted through the lungs. This loss of carbon dioxide is accompanied by a drop in the hydrogen ion concentration in the body fluids, because the released carbon dioxide comes from carbonic acid (fig. 18.8).

$$H_2CO_3 \rightarrow CO_2 + H_2O$$

Conversely, if the body cells are less active, the concentrations of carbon dioxide and hydrogen ions in the plasma remain relatively low. As a consequence, the breathing rate and depth are decreased.

The Kidneys

The nephrons of the kidneys help regulate the hydrogen ion concentrations of the body fluids by secreting hydrogen ions. As discussed in chapter 17, these ions are secreted into the urine by the epithelial cells that line certain segments of the renal tubules.

The various regulators of hydrogen ion concentration operate at different rates (fig. 18.9). Acid-base

Figure 18.8
An increase in carbon dioxide production is followed by an increase in carbon dioxide elimination.

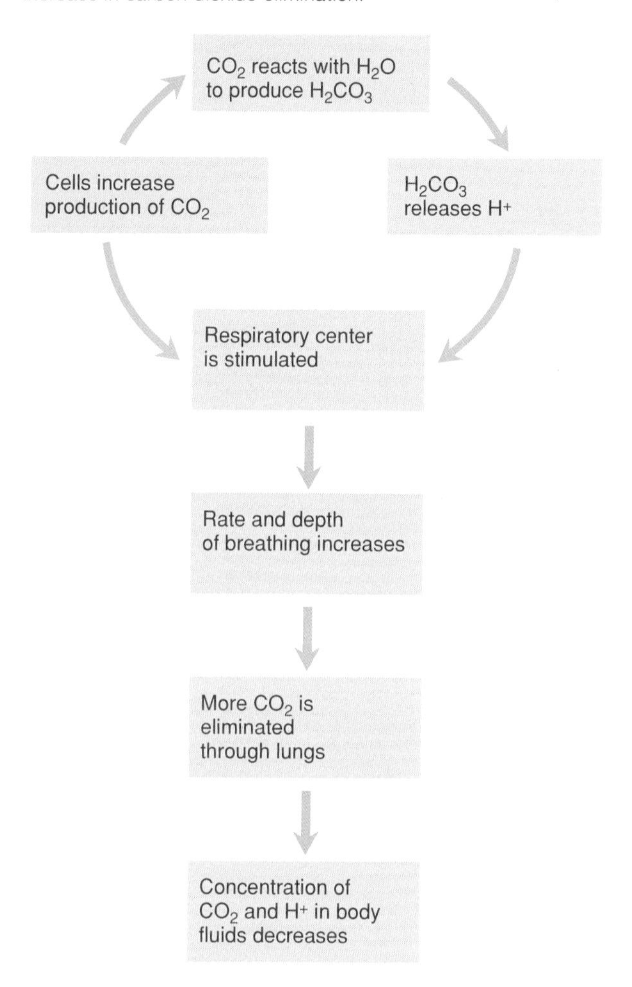

buffers, for example, function rapidly and can convert strong acids or bases into weak acids or bases almost immediately. For this reason, chemical buffer systems sometimes are called the body's *first line of defense* against shifts in the pH.

Physiological buffer systems, such as the respiratory and renal mechanisms, function more slowly, and constitute *secondary defenses*. The respiratory mechanism may require several minutes to begin resisting a change in the pH, and the renal mechanism may require one to three days to regulate a changing hydrogen ion concentration.

1. How does the respiratory system help to regulate acid-base balance?
2. How do the kidneys respond to excessive concentrations of hydrogen ions?
3. How do the rates at which chemical and physiological buffer systems act differ?

Figure 18.9
Chemical buffers act rapidly, but physiological buffers may require several minutes to begin resisting a change in pH.

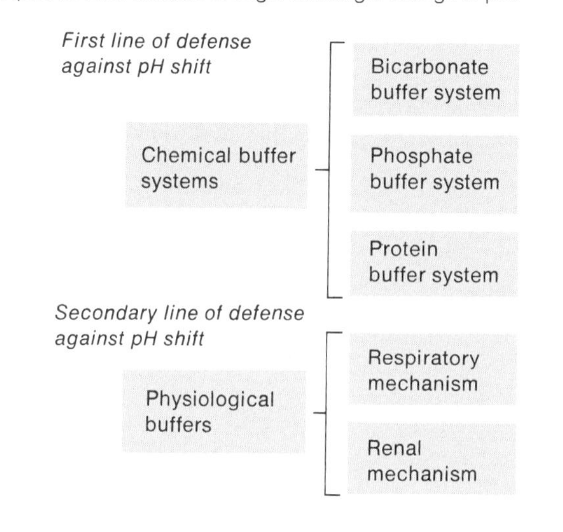

Clinical Terms Related to Water and Electrolyte Balance

acetonemia (as″ĕ-to-ne′me-ah)—the presence of abnormal amounts of acetone in the blood.

acetonuria (as″ĕ-to-nu′re-ah)—the presence of abnormal amounts of acetone in the urine.

albuminuria (al-bu″mĭ-nu′re-ah)—the presence of albumin in the urine.

anasarca (an″ah-sar′kah)—a widespread accumulation of tissue fluid.

antacid (ant-as′id)—a substance that neutralizes an acid.

anuria (ah-nu′re-ah)—the absence of urine excretion.

azotemia (az″o-te′me-ah)—an accumulation of nitrogenous wastes in the blood.

diuresis (di″u-re′sis)—the increased production of urine.

glycosuria (gli″ko-su′re-ah)—the presence of excessive sugar in the urine.

hyperglycemia (hi″per-gli-se′me-ah)—an abnormally high concentration of blood sugar.

hyperkalemia (hi″per-kah-le′me-ah)—the presence of excessive potassium in the blood.

hypernatremia (hi″per-na-tre′me-ah)—the presence of excessive sodium in the blood.

hyperuricemia (hi″per-u″rĭ-se′me-ah)—the presence of excessive uric acid in the blood.

hypoglycemia (hi″po-gli-se′me-ah)—an abnormally low level of blood sugar.

ketonuria (ke″to-nu′re-ah)—the presence of ketone bodies in the urine.

ketosis (ke″to′sis)—acidosis due to the presence of excessive ketone bodies in the body fluids.

proteinuria (pro″te-ĭ-nu′re-ah)—the presence of protein in the urine.

uremia (u-re′me-ah)—a toxic condition resulting from the presence of excessive amounts of nitrogenous wastes in the blood.

Chapter Summary

Introduction (page 472)

The maintenance of water and electrolyte balance requires that the quantities of these substances entering the body equal the quantities leaving it. Altering the water balance necessarily affects the electrolyte balance.

Distribution of Body Fluids (page 472)

1. Fluid compartments
   a. The intracellular fluid compartment includes the fluids and electrolytes enclosed by cell membranes.
   b. The extracellular fluid compartment includes all the fluids and electrolytes outside the cell membranes.
2. Composition of the body fluids
   a. Extracellular fluids
      (1) Extracellular fluids are characterized by high concentrations of sodium, chloride, and bicarbonate ions with lesser amounts of potassium, calcium, magnesium, phosphate, and sulfate ions.
      (2) Plasma contains more protein than either interstitial fluid or lymph.
   b. Intracellular fluid contains relatively high concentrations of potassium, phosphate, and magnesium ions; it also contains a greater concentration of sulfate ions and lesser concentrations of sodium, chloride, and bicarbonate ions than extracellular fluids contain.
3. Movement of fluid between compartments
   a. Fluid movements are regulated by hydrostatic pressure and osmotic pressure.
      (1) Fluid leaves plasma because of hydrostatic pressure and returns to plasma because of osmotic pressure.
      (2) Fluid enters the lymph vessels because of osmotic pressure.
      (3) Fluid movement in and out of the cells is regulated primarily by osmotic pressure.
   b. Sodium ion concentrations are especially important in the regulation of fluid movements.

Water Balance (page 474)

1. Water intake
   a. Most water is gained in liquid or in moist foods.
   b. Some water is produced by oxidative metabolism.
2. Regulation of water intake
   a. The thirst mechanism is the primary regulator of water intake.
   b. The act of drinking and the resulting distension of the stomach inhibit the thirst mechanism.
3. Water output
   Water is lost in urine, feces, and sweat, and by evaporation from the skin and lungs.
4. Regulation of water balance
   Water balance is regulated mainly by the distal convoluted tubules and collecting ducts of the nephrons.

Electrolyte Balance (page 476)

1. Electrolyte intake
   a. The most important electrolytes in the body fluids are those that release ions of sodium, potassium, calcium, magnesium, chloride, sulfate, phosphate, and bicarbonate.
   b. These ions are obtained in foods and beverages or as by-products of metabolic processes.
2. Regulation of electrolyte intake
   a. Electrolytes are usually obtained in sufficient quantities in response to hunger and thirst mechanisms.
   b. In a severe electrolyte deficiency, a person may experience a salt craving.
3. Electrolyte output
   a. Electrolytes are lost through perspiration, feces, and urine.
   b. Quantities lost vary with temperature and physical exercise.
   c. The greatest electrolyte loss occurs as a result of kidney functions.
4. Regulation of electrolyte balance
   a. Concentrations of sodium, potassium, and calcium ions in the body fluids are particularly important.
   b. The regulation of sodium and potassium ions involves the secretion of aldosterone from the adrenal glands.
   c. The regulation of calcium ions involves parathyroid hormone.
   d. In general, negatively charged ions are regulated secondarily by the mechanisms that control positively charged ions.

Acid-Base Balance (page 478)

Acids are electrolytes that release hydrogen ions. Bases are substances that combine with hydrogen ions.

1. Sources of hydrogen ions
   a. The aerobic respiration of glucose results in carbonic acid.
   b. The anaerobic respiration of glucose gives rise to lactic acid.
   c. The incomplete oxidation of fatty acids gives rise to acidic ketone bodies.
   d. The oxidation of sulfur-containing amino acids gives rise to sulfuric acid.
   e. The oxidation of phosphoproteins and nucleoproteins gives rise to phosphoric acid.
2. Strengths of acids and bases
   a. Acids vary in the extent to which they ionize.
      (1) Strong acids, such as hydrochloric acid, ionize more completely.
      (2) Weak acids, such as carbonic acid, ionize less completely.
   b. Bases also vary in strength.
3. Regulation of hydrogen ion concentration
   a. Acid-base buffer systems
      (1) Buffer systems function to convert strong acids into weaker acids or strong bases into weaker bases.

(2) They include the bicarbonate buffer system, phosphate buffer system, and protein buffer system.
(3) Buffer systems minimize pH changes.

b. The respiratory center helps regulate the pH by controlling the rate and depth of breathing.
c. Kidneys
The nephrons help regulate pH by secreting hydrogen ions.
d. Chemical buffers act rapidly; physiological buffers act more slowly.

Clinical Application of Knowledge

1. Some time ago, a news story reported the death of several newborn infants due to an error in which sodium chloride was substituted for sugar in their formula. What symptoms would this produce? Why do you think infants are more prone to the hazard of excess salt intake than adults?
2. An elderly, semiconscious patient is tentatively diagnosed as having acidosis. What component of arterial blood will be most valuable in determining if the acidosis is of respiratory origin?
3. During radiation therapy, the mucosa of the stomach and intestines may be damaged. Explain the effect this might have on the patient's electrolyte balance?
4. If the right ventricle of a patient's heart is failing, and the venous blood pressure increases as a result, what changes might occur in the patient's extracellular fluid compartments?

Review Activities

1. Explain how water balance and electrolyte balance are interdependent.
2. Name the body fluid compartments, and describe their locations.
3. Explain how the fluids within these compartments differ in composition.
4. Describe how fluid movements between the compartments are controlled.
5. List the sources of normal water gain and loss to illustrate how the input of water equals the output of water.
6. Define *water of metabolism*.
7. Explain how water intake is regulated.
8. Explain how the nephrons function in the regulation of water output.
9. List the most important electrolytes in the body fluids.
10. Explain how electrolyte intake is regulated.
11. List the routes by which electrolytes leave the body.
12. Explain how the adrenal cortex functions in the regulation of electrolyte output.
13. Describe the role of the parathyroid glands in regulating electrolyte balance.
14. Distinguish between an acid and a base.
15. List five sources of hydrogen ions in the body fluids, and name an acid that originates from each source.
16. Distinguish between a strong acid and a weak acid, and name an example of each.
17. Distinguish between a strong base and a weak base, and name an example of each.
18. Explain how an acid-base buffer system functions.
19. Describe how the bicarbonate buffer system resists changes in pH.
20. Explain why a protein has acidic as well as basic properties.
21. Explain how the respiratory center functions in the regulation of the acid-base balance.
22. Explain how the kidneys function in the regulation of the acid-base balance.
23. Distinguish between a chemical buffer system and a physiological buffer system.

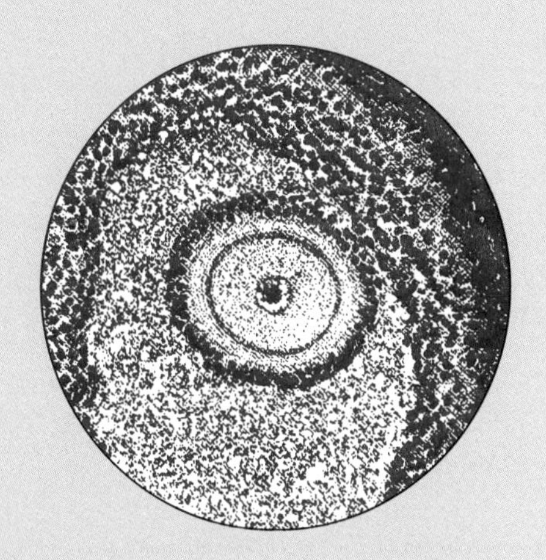

Unit 5
Reproduction

The chapters of unit 5 are concerned with the reproduction, growth, and development of the human organism. They describe how the organs of the male and female reproductive systems function to produce an embryo and how this offspring grows and develops as it passes through the phases of its life cycle, before and after birth. This unit includes:

These circular images are a computer-manipulated version of figure 19.10.

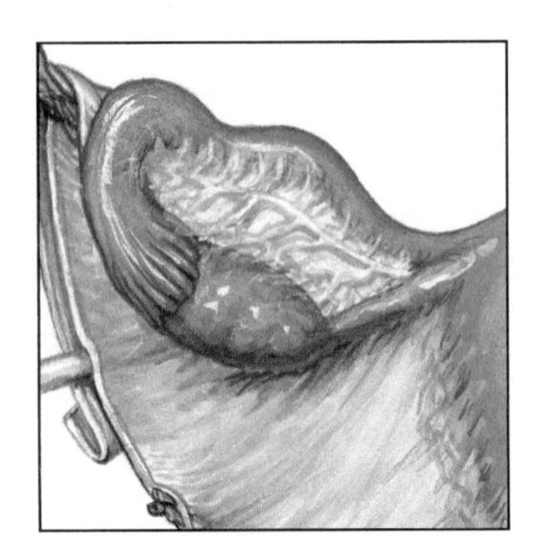

19 Reproductive Systems

The male and female *reproductive systems* are specialized to produce offspring. These systems are unique in that their functions are not necessary for the survival of each individual. Instead, their functions are vital to the continuation of the human species. Without reproduction by at least some of its members, the species would soon become extinct.

The organs of the male and female reproductive systems are adapted to produce sex cells, to sustain these cells, and to transport them to a location where fertilization may occur. Some of the reproductive organs also secrete hormones that play vital roles in the development and maintenance of sexual characteristics and in the regulation of reproductive physiology.

Chapter Objectives

After you have studied this chapter, you should be able to

1. Name the parts of the male reproductive system, and describe the general functions of each part.
2. Describe the structure of a testis, and explain how sperm cells are formed.
3. Trace the path followed by sperm cells from the site of their formation to the outside.
4. Explain how hormones control the activities of the male reproductive organs.
5. Name the parts of the female reproductive system, and describe the general functions of each part.
6. Describe the structure of an ovary, and explain how egg cells and follicles are formed.
7. Trace the path followed by an egg cell after ovulation.
8. Describe how hormones control the activities of the female reproductive organs.
9. Describe the major events that occur during a menstrual cycle.
10. Review the structure of the mammary glands.
11. Complete the review activities at the end of this chapter. Note that the items are worded in the form of specific learning objectives. You may want to refer to them before reading the chapter.

Key Terms

androgen (an′dro-jen)

ejaculation (e-jak″u-la′shun)

emission (e-mish′un)

estrogen (es′tro-jen)

follicle (fol′i-kl)

gonadotropin (gon″ah-do-tro′pin)

meiosis (mi-o′sis)

menstrual cycle (men′stroo-al si′kl)

oogenesis (ō″o-jen′ĕ-sis)

orgasm (or′gazm)

ovulation (o″vu-la′shun)

progesterone (pro-jes′tĕ-rōn)

puberty (pu′ber-te)

testosterone (tes-tos′te-rōn)

spermatogenesis (sper″mah-to-jen′ĕ-sis)

zygote (zi′gōt)

Aids to Understanding Words

andr-, man: *andr*ogens—male sex hormones.

ejacul-, to shoot forth: *ejacul*ation—the process by which semen is expelled from the male reproductive tract.

fimb-, fringe: *fimb*riae—irregular extensions on the margin of the infundibulum of the uterine tube.

follic-, small bag: *follic*le—the ovarian structure that contains an egg.

genesis-, origin: spermato*genesis*—the process by which sperm cells are formed.

germ-, to bud or sprout: *germ*inal epithelium—the tissue that gives rise to sex cells by cellular division.

labi-, lip: *labi*a minora—flattened, longitudinal folds that extend along the margins of the vestibule.

mens-, month: *mens*trual cycle—the monthly female reproductive cycle.

mons-, mountain: *mons* pubis—the rounded elevation overlying the pubic symphysis in females.

puber-, adult: *puber*ty—the time when a person becomes able to function reproductively.

Introduction

THE ORGANS OF the male and female reproductive systems are concerned with the process of creating offspring, and each organ is adapted for specialized tasks. Some, for example, are concerned with the production of sex cells, and others sustain the lives of these cells or transport them from one location to another. Still other parts produce and secrete hormones, which regulate the formation of sex cells and play roles in the development and maintenance of male and female sexual characteristics.

Organs of the Male Reproductive System

The main function of the male reproductive system is to produce and maintain male sex cells, called **sperm cells** (spermatozoa). Parts of this system are specialized to transport sperm cells and various fluids to the outside and to produce and secrete sex hormones. The *primary sex organs* (gonads), of the male system are the *testes* (singular, *testis*), in which the sperm cells and male sex hormones are formed (fig. 19.1). The other structures of the male reproductive system are termed *accessory organs,* and they include two groups—the internal and external reproductive organs.

Testes

The **testes** are ovoid structures, about 5 cm in length and 3 cm in diameter. Each testis is contained within the cavity of the saclike *scrotum.*

Structure of the Testes

A testis is enclosed by a tough, white fibrous capsule (fig. 19.2). Along its posterior border, the connective tissue thickens and extends into the organ forming thin septa, which subdivide it into about 250 *lobules.*

Each lobule contains one to four highly coiled, convoluted **seminiferous tubules,** each of which is approximately 70 cm long when uncoiled. These tubules course posteriorly and unite to form a complex network of channels. These channels, in turn, give rise to several ducts that join a tube called the *epididymis.* The epididymis is coiled on the outer surface of the testis.

The seminiferous tubules are lined with a specialized stratified epithelium, which includes spermatogenic cells that give rise to sperm cells. Other specialized cells, called **interstitial cells,** are located in the spaces between the seminiferous tubules. They produce and secrete male sex hormones.

Figure 19.1
A sagittal view of the male reproductive organs.

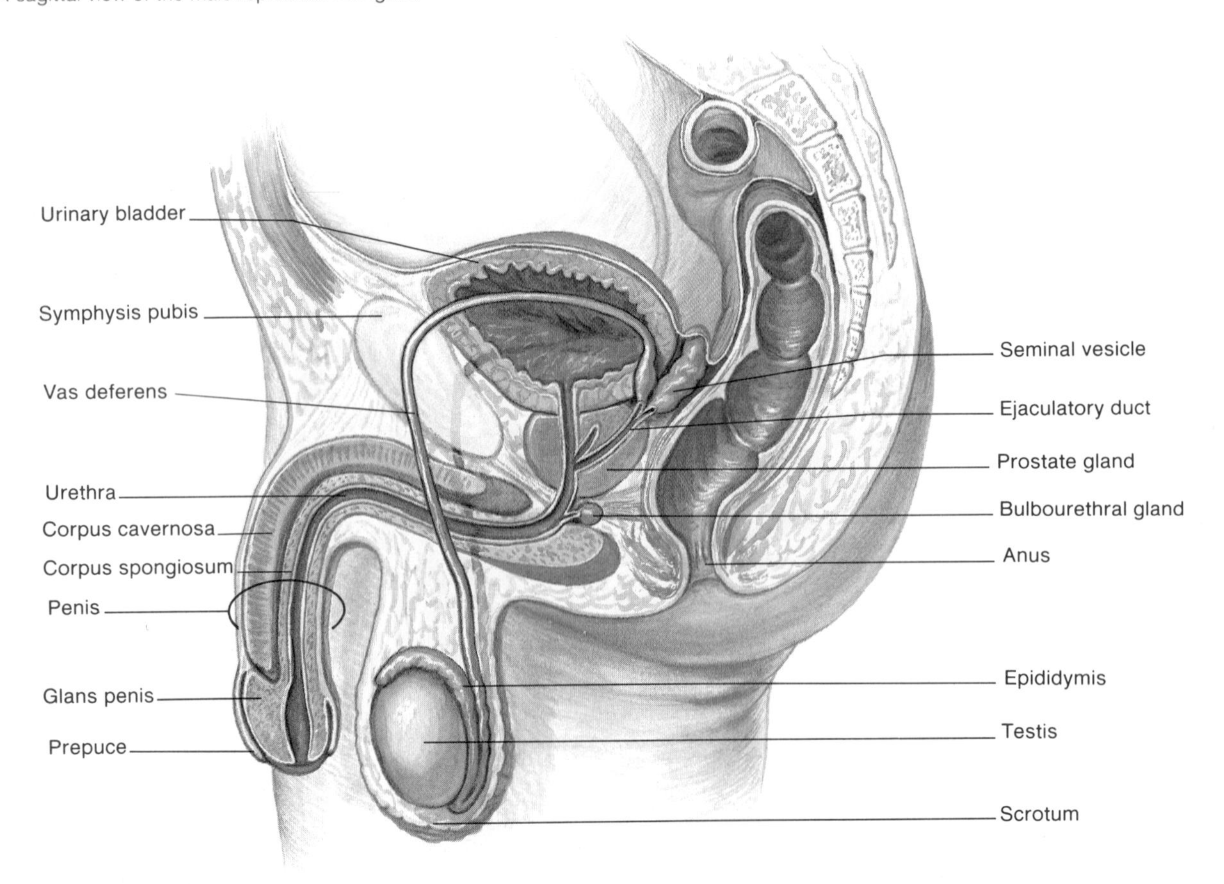

The epithelial cells of the seminiferous tubules sometimes give rise to *testicular cancer*, one of the more common types of cancer that occurs in young men. In most cases, the first sign of this condition is a painless enlargement of a testis or a scrotal mass that seems to be attached to a testis. When testicular cancer is suspected, a tissue sample is usually removed (biopsied) and examined microscopically. If cancer cells are present, the affected testis is surgically removed (orchiectomy). Depending upon the type of cancerous tissue present and the extent of the disease, a patient may be treated with radiation or with chemotherapy employing one or more drugs. As a result of such treatment, the cure rate for testicular cancer is high.

1. Describe the structure of a testis.
2. Where are the sperm cells produced?
3. What cells produce male sex hormones?

Sperm Cell Formation

The epithelium of the seminiferous tubules includes two types of cells—supporting cells and spermatogenic cells. The *supporting cells* support, nourish, and regulate the *spermatogenic cells,* which, in turn, give rise to the sperm cells.

Sperm cell production occurs continually throughout the reproductive life of a male. The resulting sperm cells collect in the lumen of each seminiferous tubule. Then they pass to the epididymis, where they remain for a time and mature.

A mature sperm cell is a tiny, tadpole-shaped structure about 0.06 mm long. It consists of a flattened head, a cylindrical body, and an elongated tail (fig. 19.3).

The *head* of a sperm cell is composed primarily of a nucleus, which contains 23 chromosomes in highly compact form. A small part of the head at its anterior end, called the *acrosome,* contains enzymes that aid the sperm cell in penetrating an egg cell at the time of fertilization. (This process is described in chapter 20.)

Spermatogenesis

In a young male, the spermatogenic cells are undifferentiated and are called *spermatogonia* (fig. 19.4). Each of these cells contains 46 chromosomes in its nucleus, which is the usual number for human body cells. During early adolescence, hormones stimulate the spermatogonia to undergo mitosis (see chapter 3), and some of them enlarge to become *primary spermatocytes.* When these primary spermatocytes divide, however, they do so by a special type of cell division called **meiosis.**

In the course of meiosis, the primary spermatocytes each divide to form two *secondary spermatocytes.* Each of these cells, in turn, divides to form two *spermatids,* which mature into sperm cells. Also during meiosis, the number of chromosomes in each cell is reduced by one-half. Consequently, for each primary spermatocyte that undergoes meiosis, four sperm cells with 23 chromosomes in each of their nuclei are formed (fig. 19.5). This process by which sperm cells are produced is called **spermatogenesis.**

1. Explain the function of the supporting cells in the seminiferous tubules.
2. Describe the structure of a sperm cell.
3. Describe the process of spermatogenesis.

Figure 19.2
(*a*) A sagittal section of a testis; (*b*) a cross section of a seminiferous tubule.

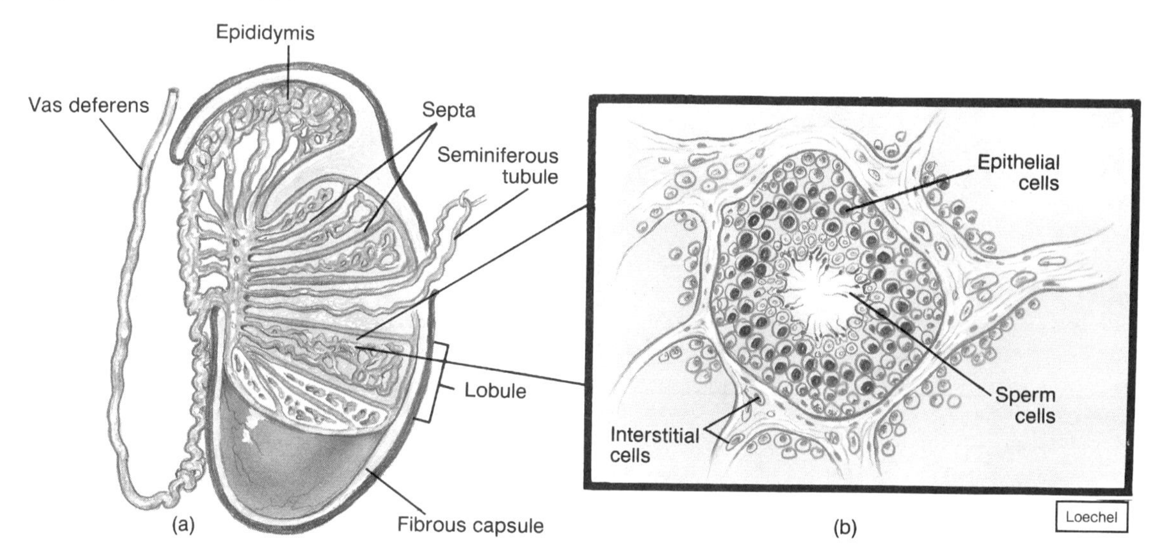

Figure 19.3
(*a*) The structure of a human sperm cell; (*b*) a scanning electron micrograph of human sperm cells (×4,000).

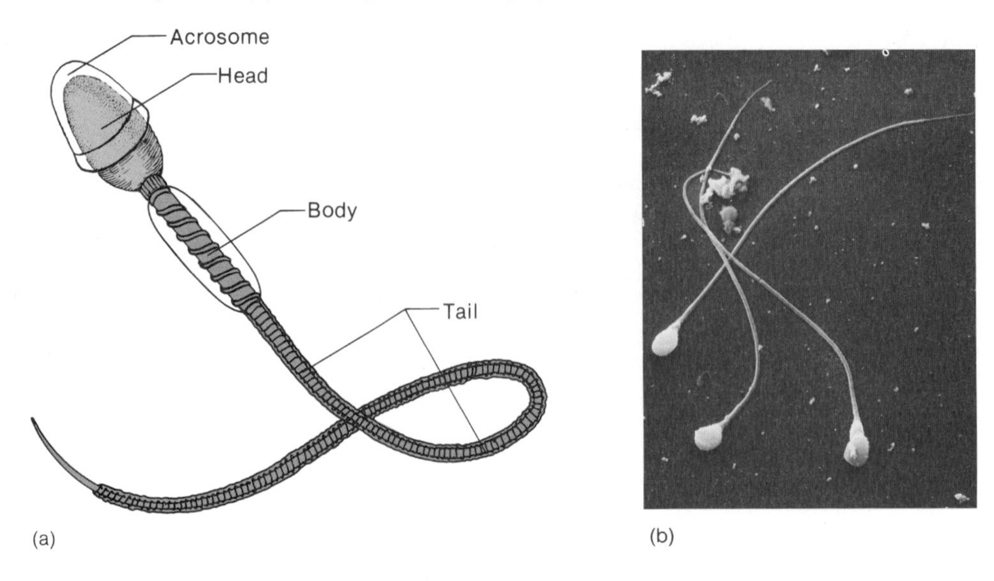

Figure 19.4
(*a*) What features can you identify in this light micrograph of seminiferous tubules (×200)? (*b*) Spermatogonia give rise to primary spermatocytes by mitosis; the spermatocytes, in turn, give rise to sperm cells by meiosis.

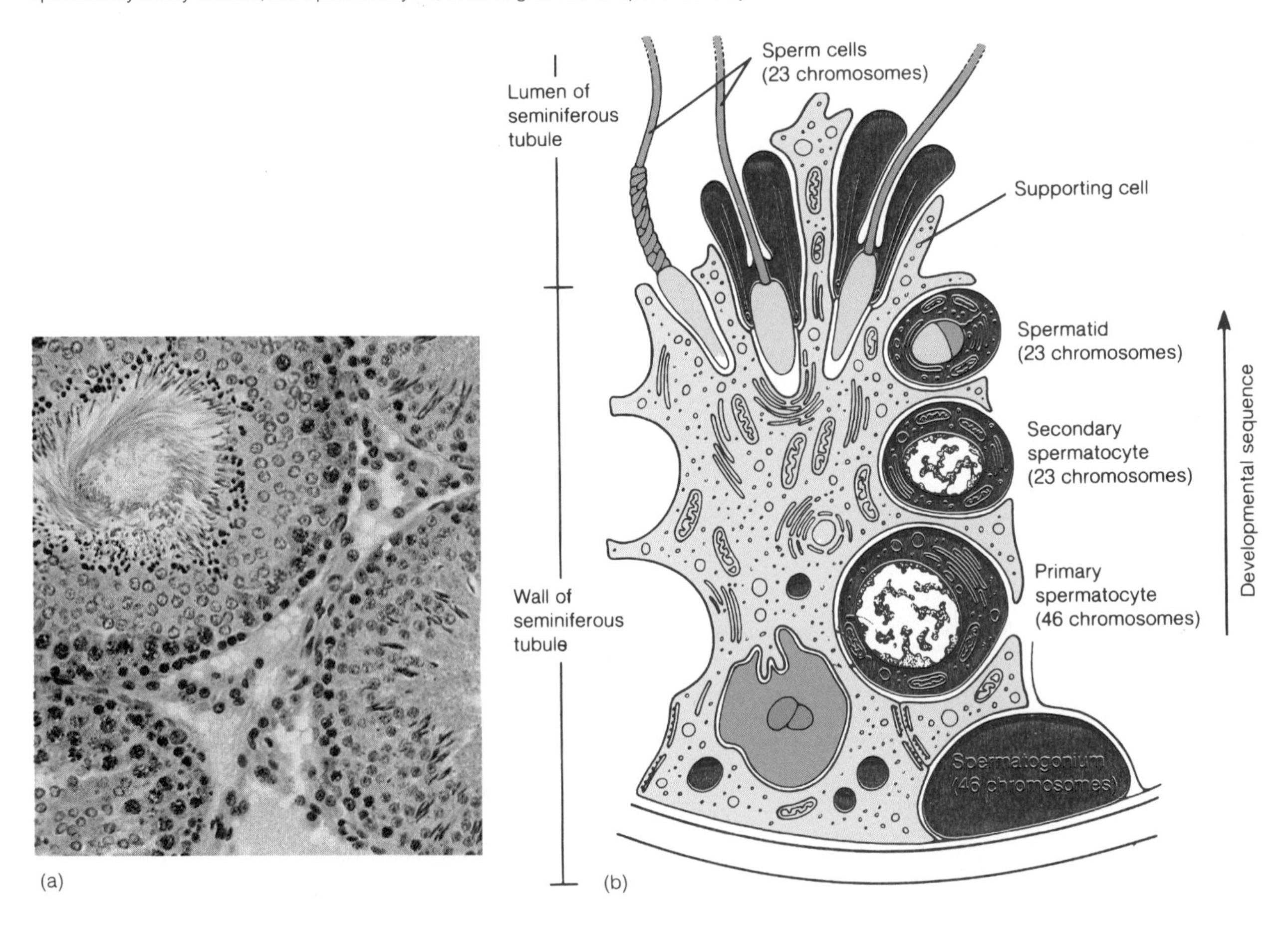

Figure 19.5
Spermatogenesis is a meiotic process that involves two successive divisions.

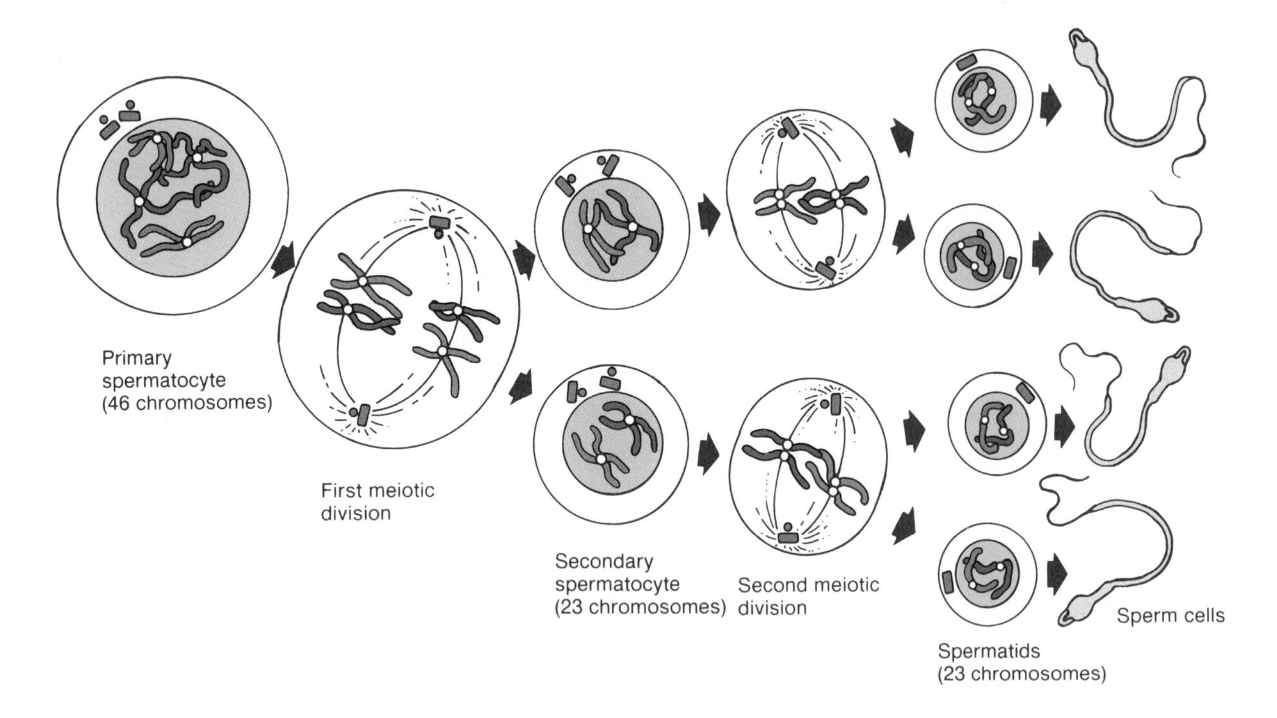

Male Internal Accessory Organs

The *internal accessory organs* of the male reproductive system include the epididymides, vasa deferentia, ejaculatory ducts, and urethra, as well as the seminal vesicles, prostate gland, and bulbourethral glands.

Epididymis

Each **epididymis** (pl. *epididymides*) is a tightly coiled, threadlike tube that is about 6 meters long (figs. 19.1 and 19.2). This tube is connected to ducts within a testis. It emerges from the top of the testis, descends along its posterior surface, and then courses upward to become the *vas deferens.*

When immature sperm cells reach the epididymis, they are nonmobile. However, as they travel through the epididymis, as a result of rhythmic peristaltic contractions, they undergo *maturation.* Following this aging process, the sperm cells are capable of moving independently and of fertilizing egg cells.

Vas Deferens

Each **vas deferens** (pl. *vasa deferentia*) is a muscular tube about 45 cm long (fig. 19.1). It passes upward along the medial side of a testis, through a passage in the lower abdominal wall (inguinal canal), enters the abdominal cavity, and ends behind the urinary bladder.

Just outside the prostate gland, the vas deferens unites with the duct of a seminal vesicle. The fusion of these two ducts forms an **ejaculatory duct,** which passes through the substance of the prostate gland and empties into the urethra.

Seminal Vesicle

A **seminal vesicle** (fig. 19.1) is a convoluted, saclike structure about 5 cm long that is attached to the vas deferens near the base of the urinary bladder.

The glandular tissue lining the inner wall of a seminal vesicle secretes a slightly alkaline fluid. This fluid is thought to help regulate the pH of the tubular contents as the sperm cells are conveyed to the outside. The secretion of the seminal vesicle also contains a variety of nutrients. For instance, it is rich in *fructose,* a monosaccharide that provides the sperm cells with an energy source.

1. Describe the structure of the epididymis.
2. Trace the path of the vas deferens.
3. What is the function of the seminal vesicle?

Prostate Gland

The **prostate gland** (fig. 19.1) is a chestnut-shaped structure about 4 cm across and 3 cm thick that surrounds the beginning of the urethra, just below the urinary bladder. It is enclosed by connective tissue and composed of many branched tubular glands, whose ducts open into the urethra.

Male Infertility
A Current Topic

In a male, *infertility* (sterility) is a lack of ability to induce the fertilization of an egg cell. This condition can result from a variety of disorders. For example, if during development the testes fail to descend into the scrotum from the abdominal cavity where they are formed, the higher temperature of the abdominal cavity prevents the formation of sperm cells by causing the cells in the seminiferous tubules to degenerate. Similarly, certain diseases, such as mumps, may cause an inflammation of the testes (orchitis) and produce infertility by destroying the cells of the seminiferous tubules.

Other males are infertile because of a deficiency of sperm cells in their semen. Normally, a milliliter of this fluid contains about 120 million sperm cells, and although estimates of the number of cells necessary for fertility vary, 20 million sperm cells per milliliter in a release of 3 to 5 milliliters often is cited as the minimum for fertility.

Even though a single sperm cell is needed to fertilize an egg cell, many sperm must be present at the time. This is because each sperm cell can release some enzymes that are stored in its acrosome. A certain concentration of these enzymes is apparently necessary to remove the layers of cells that normally surround an egg cell, thus exposing the egg cell for fertilization.

Still other males seem to be infertile because of the quality of the sperm cells they produce. In such cases, the sperm cells may have poor motility, or they may have abnormal shapes related to the presence of defective genetic material.

The prostate gland secretes a thin, milky fluid with an alkaline pH. This secretion functions to neutralize the sperm cell-containing fluid (semen), which is acidic due to an accumulation of metabolic wastes produced by the stored sperm cells. It also enhances sperm cell motility.

Although the prostate gland is relatively small in male children, it begins to grow in early adolescence and reaches its adult size a few years later. As a rule, its size remains unchanged between the ages of twenty and fifty years. In older males, the prostate gland commonly enlarges. As this happens, it may squeeze the urethra and interfere with urine excretion.

The treatment of an abnormally enlarged prostate gland is usually surgical. If the obstruction created by the gland is slight, the procedure may be performed through the urethral canal (*transurethral prostatic resection*).

Bulbourethral Glands

The **bulbourethral glands** (Cowper's glands) (fig. 19.1) are two small structures about the size of peas, which are located below the prostate gland and enclosed by muscle fibers of the external urethral sphincter.

These glands are composed of numerous tubes, whose epithelial linings secrete a mucuslike fluid. This fluid is released in response to sexual stimulation and provides some lubrication to the end of the penis in preparation for sexual intercourse. Most of the lubricating fluid for intercourse, however, is secreted by the female reproductive organs.

Semen

The fluid conveyed by the urethra to the outside as a result of sexual stimulation is called **semen**. It consists of sperm cells from the testes and secretions of the seminal vesicles, prostate gland, and bulbourethral glands. Semen has a slightly alkaline pH (about 7.5) and contains a variety of nutrients. It also contains prostaglandins (see chapter 11), which enhance sperm cell survival and movement through the female reproductive tract.

The volume of semen released at one time varies from 2 to 6 milliliters, and the average number of sperm cells present in the fluid is about 120 million per milliliter.

Sperm cells remain immobile while they are in the ducts of the testis and epididymis, but become activated as they are mixed with the secretions of the accessory glands. However, the sperm cells remain unable to fertilize an egg cell until they enter the female reproductive tract. The development of this ability is called *capacitation*, and it involves changes that weaken the acrosomal membranes of the sperm cells.

1. Where is the prostate gland located?
2. What is the function of the prostatic secretion?
3. What are the characteristics of semen?

Male External Reproductive Organs

The male external reproductive organs are the scrotum, which encloses the testis, and the penis, through which the urethra passes.

Scrotum

The **scrotum** (fig. 19.1) is a pouch of skin and subcutaneous tissue that hangs from the lower abdominal region behind the penis. It is subdivided into chambers by a medial septum, and each chamber is occupied by a testis. Each chamber also contains a serous membrane, which provides a covering and helps ensure that the testis will move smoothly within the scrotum.

Penis

The **penis** is a cylindrical organ that conveys urine and semen through the urethra (fig. 19.1). It is also specialized to become enlarged and stiffened by a process called *erection,* so that it can be inserted into the female vagina during sexual intercourse.

The *body,* or shaft, of the penis is composed of three columns of erectile tissue, which include a pair of dorsally located *corpora cavernosa* and a single *corpus spongiosum* below. The penis is enclosed by skin, a thin layer of subcutaneous tissue, and a layer of elastic tissue. In addition, each column is surrounded by a tough capsule of white fibrous connective tissue.

The corpus spongiosum, through which the urethra extends, is enlarged at its distal end to form a sensitive, cone-shaped **glans penis.** The glans covers the ends of the corpora cavernosa and bears the urethral opening (external urethral meatus). The skin of the glans is very thin and hairless. A loose fold of skin, called the *prepuce* (foreskin), begins just behind the glans and extends forward to cover it as a sheath. The prepuce is sometimes removed by a surgical procedure called *circumcision.*

1. Describe the structure of the penis.
2. What is circumcision?

Erection, Orgasm, and Ejaculation

The erectile tissue within the body of the penis contains a network of vascular spaces (venous sinusoids). These spaces are lined with endothelium and are separated from each other by cross bars of smooth muscle and connective tissue.

Ordinarily, the vascular spaces remain small as a result of partial contractions in the smooth muscle fibers that surround them. During sexual stimulation, however, the smooth muscles become relaxed. At the same time, *parasympathetic* nerve impulses pass from the sacral portion of the spinal cord to the arteries leading into the penis, causing them to dilate. These impulses also stimulate the veins leading away from the penis to constrict. As a result, arterial blood under relatively high pressure enters the vascular spaces, and the flow of venous blood away from the penis is reduced. Consequently, blood accumulates in the erectile tissues, and the penis swells, elongates, and becomes erect.

The culmination of sexual stimulation is called **orgasm** and involves a pleasurable feeling of physiological and psychological release. Also, orgasm in the male is accompanied by emission and ejaculation.

Emission is the movement of sperm cells from the testes and secretions from the prostate gland and seminal vesicles into the urethra, where they are mixed to form semen. Emission occurs in response to *sympathetic nerve* impulses traveling from the spinal cord, which cause peristaltic contractions in the walls of the testicular ducts, epididymides, vasa deferentia, and ejaculatory ducts. At the same time, other sympathetic impulses stimulate rhythmic contractions of the seminal vesicles and prostate gland.

As the urethra fills with semen, sensory impulses are stimulated and pass into the sacral portion of the spinal cord. In response, motor impulses are transmitted from the cord to certain skeletal muscles at the base of the erectile columns of the penis, causing them to contract rhythmically. This increases the pressure within the erectile tissues and aids in forcing the semen through the urethra to the outside—a process called **ejaculation.**

The sequence of events during emission and ejaculation is regulated so that the fluid from the bulbourethral glands is expelled first. This is followed by the release of fluid from the prostate gland, the passage of the sperm cells, and finally the ejection of fluid from the seminal vesicles.

Immediately after ejaculation, sympathetic impulses cause vasoconstriction of the arteries that supply the erectile tissue with blood, reducing the inflow of blood. The smooth muscles within the walls of the vascular spaces partially contract again, and the veins of the penis carry the excess blood out of these spaces. Thus, the penis gradually returns to its former flaccid condition.

The functions of the male reproductive organs are summarized in chart 19.1.

1. How is the blood flow into the erectile tissues of the penis controlled?
2. Distinguish between orgasm, emission, and ejaculation.
3. Review the events associated with emission and ejaculation.

Hormonal Control of Male Reproductive Functions

Male reproductive functions are largely controlled by hormones secreted by the *hypothalamus, anterior pituitary gland,* and *testes.*

Hypothalamic and Pituitary Hormones

For about ten years, the young male is reproductively immature. During this time, the body remains childlike, and the spermatogenic cells of the testes remain undifferentiated. Then a series of changes occur leading to the development of a reproductively functional adult. Although the mechanism that initiates such changes is not well understood, it involves the hypothalamus.

Chart 19.1 Functions of the male reproductive organs	
Organ	**Function**
Testis	
Seminiferous tubules	Production of sperm cells
Interstitial cells	Production and secretion of male sex hormones
Epididymis	Storage and maturation of sperm cells; conveys sperm cells to the vas deferens
Vas deferens	Conveys sperm cells to the ejaculatory duct
Seminal vesicle	Secretes an alkaline fluid containing nutrients and prostaglandins; this fluid helps neutralize the acidic semen
Prostate gland	Secretes an alkaline fluid, which helps neutralize the acidic semen and enhances the motility of sperm cells
Bulbourethral gland	Secretes a fluid that lubricates end of the penis
Scrotum	Encloses and protects the testes
Penis	Conveys urine and semen to outside of the body; inserted into the vagina during sexual intercourse; the glans penis is richly supplied with sensory nerve endings, associated with feelings of pleasure during sexual stimulation

As explained in chapter 11, the hypothalamus secretes gonadotropin-releasing hormone (GnRH), which enters the blood vessels leading to the anterior pituitary gland. In response, the anterior pituitary gland secretes **gonadotropins** called *luteinizing hormone* (LH) and *follicle-stimulating hormone* (FSH). LH, which in males is also called interstitial cell-stimulating hormone (ICSH), promotes the development of the interstitial cells of the testes, and they, in turn, secrete male sex hormones. FSH causes the supporting cells of the germinal epithelium to become responsive to the effects of the male sex hormone *testosterone.* Then, in the presence of FSH and testosterone, these supporting cells stimulate the spermatogenic cells to undergo spermatogenesis, giving rise to sperm cells.

Male Sex Hormones

As a group, the male sex hormones are termed **androgens,** and although most of them are produced by the interstitial cells of the testes, small amounts are synthesized in the adrenal cortex (see chapter 11).

The hormone called **testosterone** is the most abundant of the androgens, and when it is secreted, it is transported in the blood loosely attached to certain plasma proteins.

Although the secretion of testosterone begins during fetal development and continues for a few weeks following birth, its secretion nearly ceases during childhood. Sometime between the ages of thirteen and fifteen, however, androgen production usually increases rapidly. This phase in development, during which an individual becomes reproductively functional, is called **puberty.** After puberty, testosterone secretion continues throughout the life of a male.

Actions of Testosterone

During puberty, testosterone stimulates an enlargement of the testes and various accessory organs of the reproductive system, and it causes the development of the male *secondary sexual characteristics.* (Note: the primary male sexual characteristic is the presence of the testes.) Secondary sexual characteristics are special features associated with the adult male body, and they include the following:

1. Increased growth of body hair, particularly on the face, chest, axillary region, and pubic region, but sometimes accompanied by decreased growth of hair on the scalp.
2. Enlargement of the larynx and thickening of the vocal folds, accompanied by the development of a lower-pitched voice.
3. Thickening of the skin.
4. Increased muscular growth accompanied by the development of broader shoulders and a relatively narrow waist.
5. Thickening and strengthening of bones.

The other actions of testosterone include increasing the rate of cellular metabolism and the production of red blood cells, so that the average number of red blood cells in a cubic millimeter of blood usually is greater in males than in females. Testosterone also stimulates sexual activity by influencing certain portions of the brain.

Regulation of Male Sex Hormones

The extent to which the male secondary sexual characteristics develop is directly related to the amount of testosterone secreted by the interstitial cells. This quantity is regulated by a negative feedback system, involving the hypothalamus (fig. 19.6).

As the concentration of testosterone in the blood increases, the hypothalamus becomes inhibited, and its stimulation of the anterior pituitary gland by GnRH is decreased. As the pituitary's secretion of LH (ICSH) is reduced in response, the amount of testosterone released by the interstitial cells is also reduced.

As the blood concentration of testosterone drops, the hypothalamus becomes less inhibited, and it once again stimulates the pituitary gland to release LH. The

Figure 19.6
A negative feedback mechanism operating between the anterior lobe of the pituitary gland and the testes controls the concentration of testosterone.

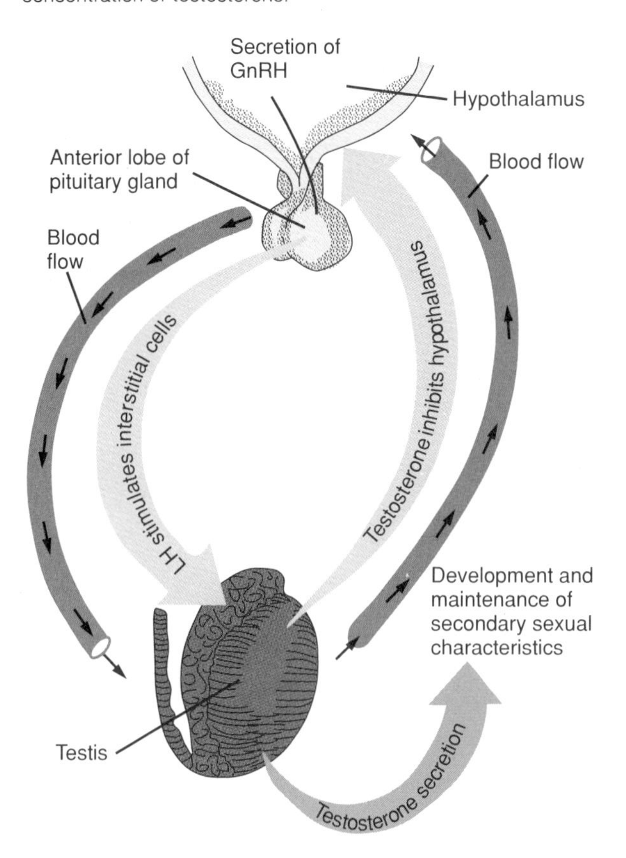

increasing secretion of LH causes the interstitial cells to release more testosterone, and the blood concentration of testosterone rises.

Thus, the concentration of testosterone in the male body is regulated so that it remains relatively constant.

The amount of testosterone secreted by the testes usually declines gradually after about forty years of age. Consequently, even though sexual activity may be continued into old age, males typically experience a decrease in sexual functions as they grow older. This decrease is sometimes called the *male climacteric.*

1. What hormone initiates the changes associated with male sexual maturity?
2. Describe several of the male secondary sexual characteristics.
3. List the functions of testosterone.
4. Explain how the secretion of male sex hormones is regulated.

Organs of the Female Reproductive System

The organs of the female reproductive system are specialized to produce and maintain the female sex cells, or **egg cells,** to transport these cells to the site of fertilization, to provide a favorable environment for a developing offspring, to move the offspring to the outside, and to produce female sex hormones.

The *primary sex organs* (gonads) of this system are the *ovaries,* which produce the female sex cells and sex hormones. The other parts of the reproductive system comprise the internal and external *accessory organs.*

Ovaries

The **ovaries** are solid, ovoid structures measuring about 3.5 cm in length, 2 cm in width, and 1 cm in thickness. They are located, one on each side, in a shallow depression in the lateral wall of the pelvic cavity (fig. 19.7).

Structure of the Ovaries

The tissues of an ovary can be divided into two rather indistinct regions, an inner *medulla* and an outer *cortex.*

The ovarian medulla is largely composed of loose connective tissue and contains numerous blood vessels, lymphatic vessels, and nerve fibers. The ovarian cortex is composed of more compact tissue and has a somewhat granular appearance due to the presence of tiny masses of cells called *ovarian follicles.*

The free surface of the ovary is covered by a layer of cuboidal epithelial cells. Just beneath this epithelium is a dense layer of connective tissue.

1. What are the primary sex organs of the female?
2. Describe the structure of an ovary.

Primordial Follicles

During prenatal development (before birth), small groups of cells in the outer region of the ovarian cortex form several million **primordial follicles.** Each of these structures consists of a single, large cell, called a *primary oocyte,* and several epithelial cells, called *follicular cells,* that closely surround the oocyte.

Also early in development, the primary oocytes begin to undergo *meiosis,* but the process soon halts and is not continued until puberty. Once the primordial follicles have appeared, no new ones are formed. Instead, the number of oocytes in the ovary steadily declines, as many of the oocytes degenerate. Of the several million oocytes formed originally, only a million or so remain at the time of birth, and perhaps 400,000 are

Figure 19.7
Organs of the female reproductive system.

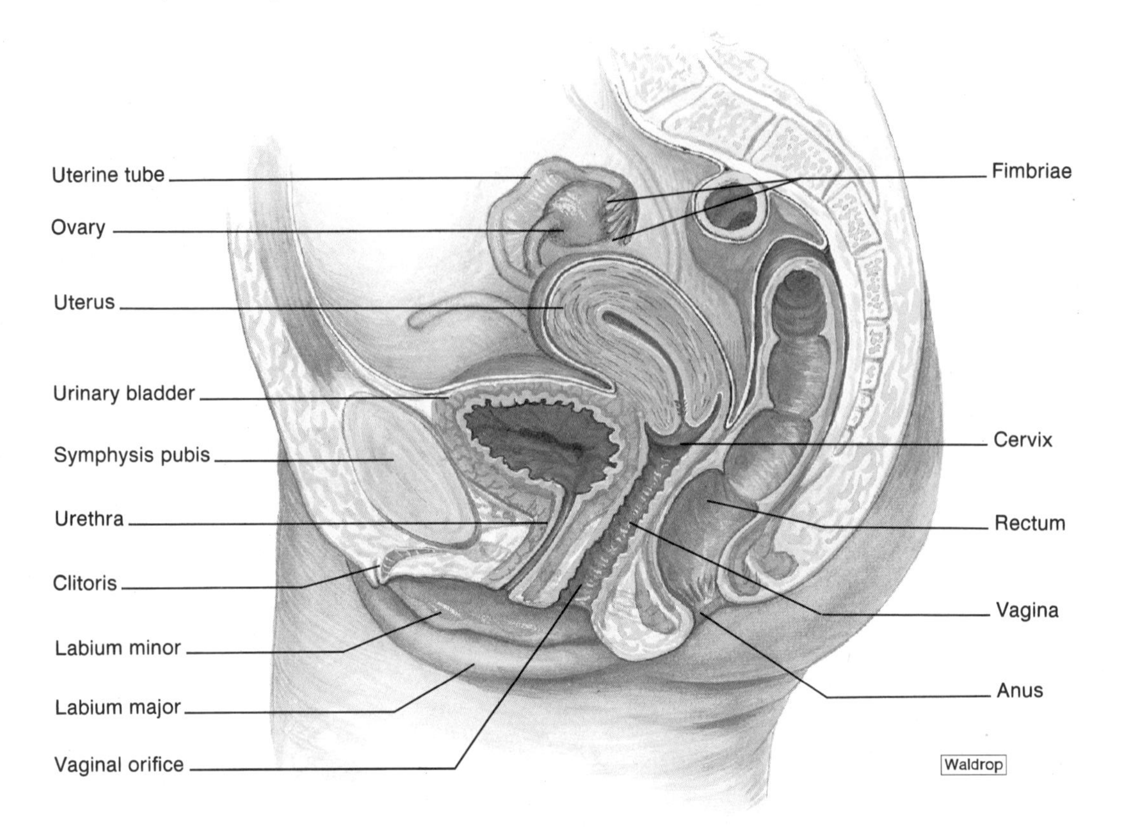

present at puberty. Of these, probably fewer than 1,000 will be released from the ovary during the reproductive life of a female.

Oogenesis

Beginning at puberty, some of the primary oocytes are stimulated to continue meiosis. As in the case of sperm cells, the resulting cells have one-half as many chromosomes (23) in their nuclei as their parent cells.

When a primary oocyte divides, the division of the cellular cytoplasm is very unequal. One of the resulting cells, called a *secondary oocyte,* is quite large, and the other, called the *first polar body,* is very small (fig. 19.8).

The large secondary oocyte represents a future *egg cell* (ovum) in that it can be fertilized by uniting with a sperm cell. If this happens, the oocyte divides unequally to produce a tiny *second polar body* and a relatively large fertilized egg cell, or **zygote.**

Thus, the result of this process, which is called **oogenesis,** is one secondary oocyte (egg cell) and one polar body. After being fertilized, the secondary oocyte divides to produce a second polar body and a zygote, which can give rise to an embryo. The polar bodies have no further function, and they soon degenerate.

Figure 19.8
During oogenesis, a single egg cell (secondary oocyte) results from the meiosis of a primary oocyte. If the egg cell is fertilized, it forms a second polar body and becomes a zygote.

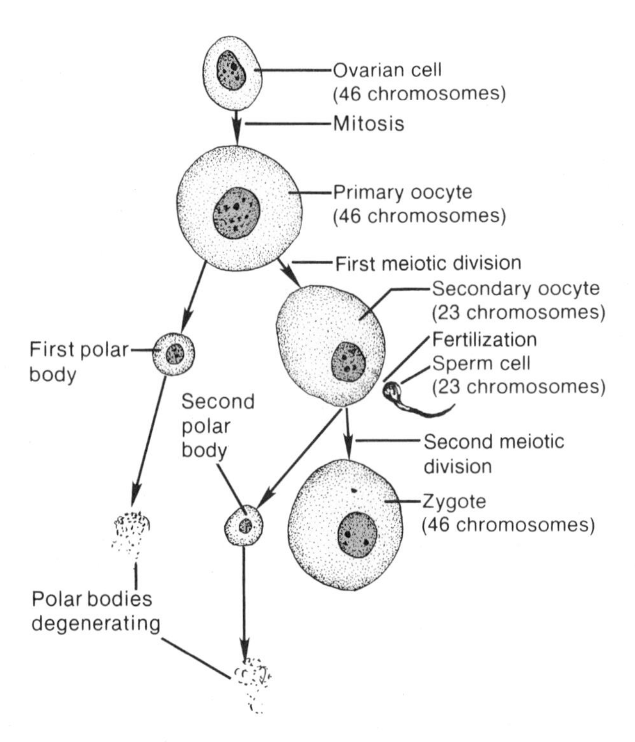

Figure 19.9
As a follicle matures, the egg cell enlarges and becomes surrounded by a mantle of follicular cells and fluid. Eventually, the mature follicle ruptures and the egg cell is released.

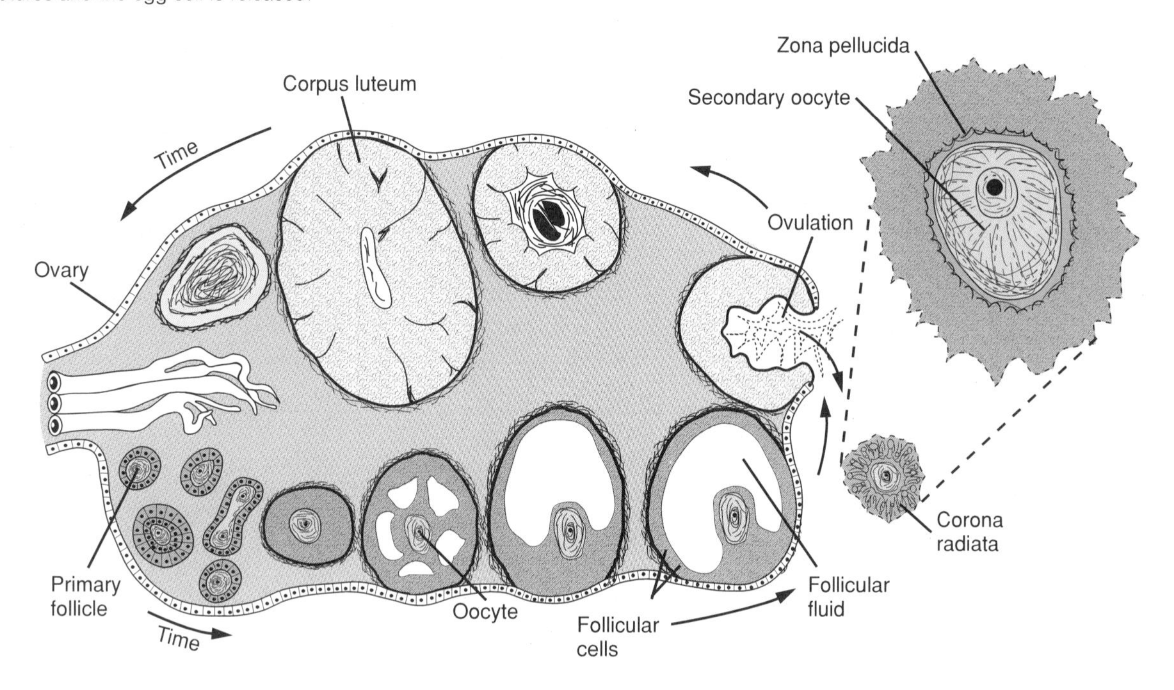

1. How does the timing of egg cell production differ from that of the sperm cells?
2. Describe the major events of oogenesis.

Maturation of a Follicle

The primordial follicles remain relatively unchanged throughout childhood. At puberty, however, the anterior pituitary gland secretes a greatly increased amount of FSH, and the ovaries begin to enlarge in response. At the same time, some of the primordial follicles begin to undergo maturation, becoming *primary follicles* (fig. 19.9).

During maturation, the oocyte of a primary follicle grows larger, and the follicular cells surrounding it divide actively by mitosis. These follicular cells become organized into layers, and soon a cavity appears in the cellular mass. As the cavity forms, it becomes filled with a clear *follicular fluid,* which bathes the oocyte.

The fluid-filled follicular cavity continues to enlarge, and the oocyte is pressed to one side. In time, the follicle reaches a diameter of 10 mm or more and bulges outward on the surface of the ovary like a blister.

The oocyte within such a mature follicle is a large, spherical cell, surrounded by a membrane (zona pellucida) and enclosed by a mantle of follicular cells (corona radiata). Processes from these follicular cells extend through the zona pellucida and are thought to supply the oocyte with nutrients.

Figure 19.10
What features can you identify in this light micrograph of a maturing follicle (×250)?

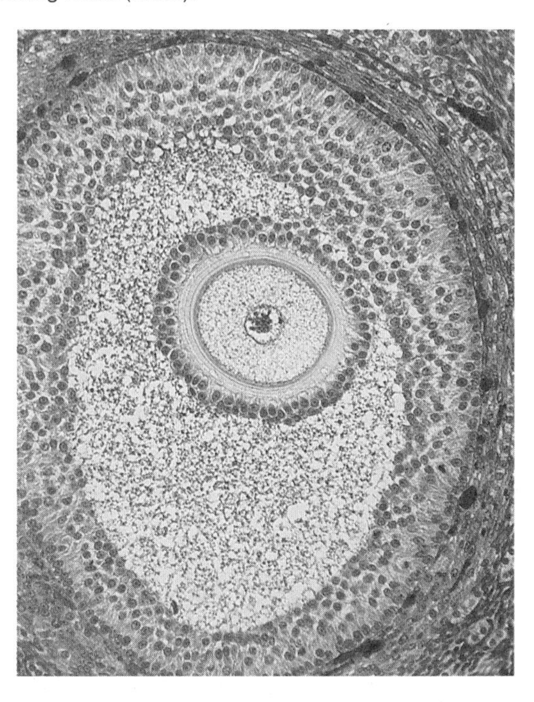

Although as many as twenty primary follicles may begin the process of maturation at any one time, one follicle usually outgrows the others. This single large follicle typically reaches full development, and the others degenerate (fig. 19.10).

Figure 19.11
The funnel-shaped infundibulum of the uterine tube partially encircles the ovary. What factors aid the movement of an egg cell into the infundibulum following ovulation?

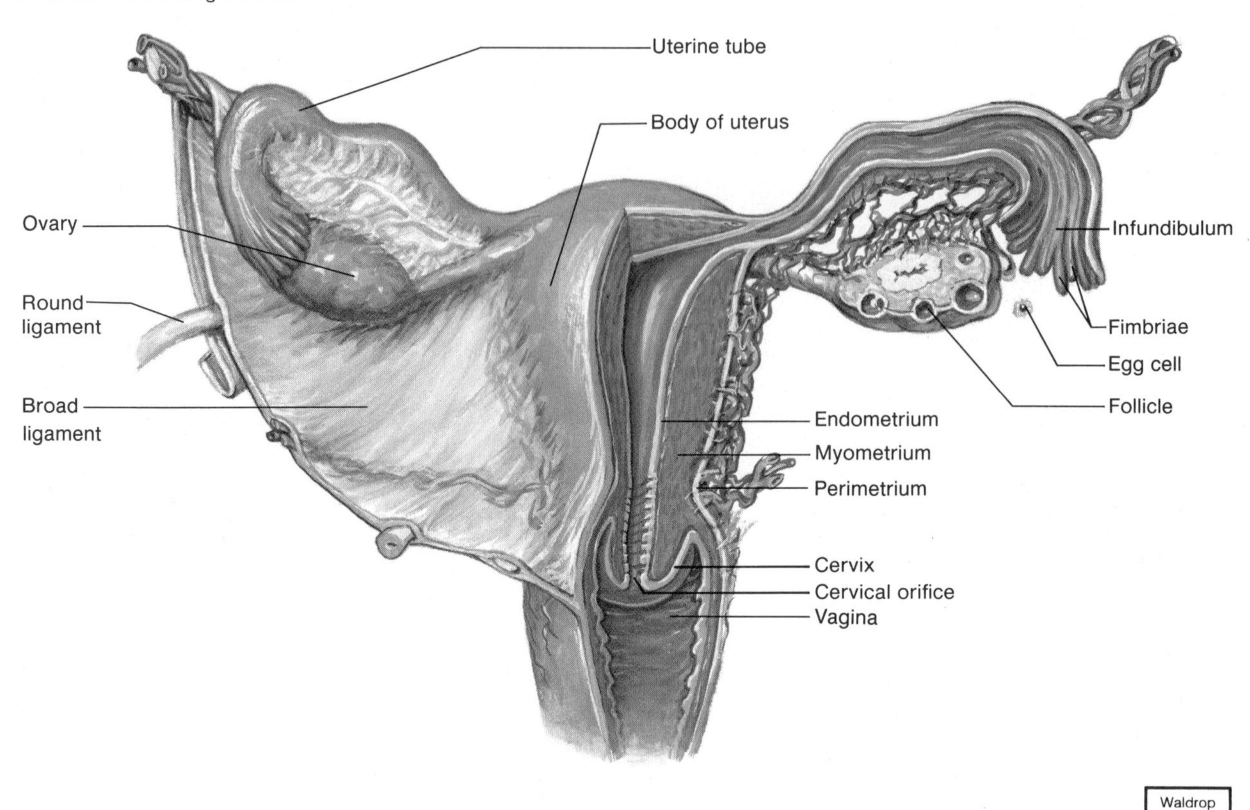

Ovulation

As a follicle matures, its primary oocyte undergoes oogenesis, giving rise to a secondary oocyte and a first polar body. These cells are released from the follicle by the process called ovulation.

Ovulation is stimulated by hormones from the anterior pituitary gland, which are described in a subsequent section of this chapter. These hormones cause the mature follicle to swell rapidly and its wall to weaken. Eventually, the wall ruptures and follicular fluid, accompanied by the oocyte, oozes outward from the surface of the ovary and enters the peritoneal cavity (fig. 19.9).

After ovulation, the oocyte and one or two layers of follicular cells surrounding it are usually propelled to the opening of the nearby *uterine tube* (fig. 19.11). If the oocyte is not fertilized by union with a sperm cell within a relatively short time, it will degenerate.

1. What changes occur in a follicle and its oocyte during maturation?
2. What causes ovulation?
3. What happens to an egg cell following ovulation?

Female Internal Accessory Organs

The *internal accessory organs* of the female reproductive system include a pair of uterine tubes, a uterus, and a vagina.

Uterine Tubes

The **uterine tubes** (fallopian tubes or oviducts) have openings near the ovaries (fig. 19.11). Each tube, which is about 10 cm long and 0.7 cm in diameter, passes medially to the uterus, penetrates its wall, and opens into the uterine cavity.

Near each ovary, a uterine tube expands to form a funnel-shaped *infundibulum,* which partially encircles the ovary medially. On its margin, the infundibulum bears a number of extensions called *fimbriae.* Although the infundibulum generally does not touch the ovary, one of its larger extensions is connected directly to the ovary.

The wall of a uterine tube is lined with simple columnar epithelial cells, some of which are *ciliated.* The epithelium secretes mucus, and the cilia beat toward the uterus. These actions help draw the egg cell and expelled follicular fluid into the infundibulum following ovulation.

Ciliary action also aids the transport of the egg cell down the uterine tube, and peristaltic contractions of the tube's muscular layer help force the egg along.

Uterus

The **uterus** receives the embryo that results from fertilization of an egg cell, and sustains its life during development. It is a hollow, muscular organ, shaped somewhat like an inverted pear (fig. 19.11).

Although the size of the uterus changes greatly during pregnancy, in its nonpregnant state it is about 7 cm long, 5 cm wide (at its broadest point), and 2.5 cm in diameter. The uterus is located medially within the anterior portion of the pelvic cavity, above the vagina, and is usually bent forward over the urinary bladder.

The upper two-thirds or *body* of the uterus has a dome-shaped top and is joined by the uterine tubes, which enter its wall at its broadest part.

The lower one-third of the uterus is called the **cervix.** This tubular part extends downward into the upper portion of the vagina.

Cancer developing within the tissues of the cervix can usually be detected by means of a relatively simple and painless procedure called the *Pap* (Papanicolaou) *smear test.* This technique involves scraping a tiny sample from the cervical tissue, smearing the sample on a glass slide, staining it, and examining it for the presence of abnormal cells.

Because this test can reveal certain types of cervical cancer in the early stages of development, when it may be cured completely, the American Cancer Society recommends that women between the ages of twenty and sixty-five have a Pap test every three years.

The uterine wall is relatively thick and is composed of three layers—the endometrium, myometrium, and perimetrium. The **endometrium** forms the inner mucosal layer, lining the uterine cavity. This lining is covered with columnar epithelium and contains numerous tubular glands. The **myometrium,** a very thick, muscular layer, consists largely of bundles of smooth muscle fibers. During the monthly female reproductive cycles and during pregnancy, the endometrium and myometrium undergo extensive changes. The **perimetrium** consists of an outer serosal layer, which covers the body of the uterus and part of the cervix (fig. 19.11).

Vagina

The **vagina** is a fibromuscular tube, about 9 cm in length, extending from the uterus to the outside (fig. 19.7). It conveys uterine secretions, receives the erect penis during sexual intercourse, and transports the offspring during the birth process.

The vagina extends upward and back into the pelvic cavity. It is posterior to the urinary bladder and urethra and anterior to the rectum and attached to these parts by connective tissues.

The *vaginal orifice* is partially closed by a thin membrane of connective tissue and stratified squamous epithelium, called the **hymen.** A central opening of varying size allows uterine and vaginal secretions to pass to the outside.

The vaginal wall consists of three layers. The inner *mucosal layer* consists of stratified squamous epithelium and underlying connective tissue. This layer is devoid of mucous glands; the mucus found in the lumen of the vagina comes from the glands of the uterus.

The middle *muscular layer* consists mainly of smooth muscle fibers. At the lower end of the vagina is a thin band of striated muscle. This band helps close the vaginal opening; however, a voluntary muscle (bulbospongiosus) is primarily responsible for closing this orifice.

The outer *fibrous layer* consists of dense fibrous connective tissue interlaced with elastic fibers, and it attaches the vagina to the surrounding organs.

1. How is an egg cell moved along a uterine tube?
2. Describe the structure of the uterus.
3. Describe the structure of the vagina.

Female External Reproductive Organs

The *external accessory organs* of the female reproductive system include the labia majora, labia minora, clitoris, and vestibular glands. As a group, these structures that surround the openings of the urethra and vagina compose the **vulva** (fig. 19.7).

Labia Majora

The **labia majora** (sing. *labium majus*) enclose and protect the other external reproductive organs. They correspond to the scrotum of the male and are composed primarily of rounded folds of adipose tissue and a thin layer of smooth muscle, covered by skin.

The labia majora lie closely together and are separated longitudinally by a cleft, which includes the urethral and vaginal openings. At their anterior ends, the labia merge to form a medial, rounded elevation of fatty tissue, called the *mons pubis,* which overlies the symphysis pubis (fig. 19.7).

Labia Minora

The **labia minora** (sing. *labium minus*) are flattened longitudinal folds located within the cleft between the labia majora (fig. 19.7). They are composed largely of connective tissue, which is richly supplied with blood vessels, causing a pinkish appearance.

Posteriorly, the labia minora merge with the labia majora, and anteriorly, they converge to form a hood-like covering around the clitoris.

Clitoris

The **clitoris** is a small projection at the anterior end of the vulva between the labia minora (fig. 19.7). Although most of it is embedded in the surrounding tissues, it is usually about 2 cm long and 0.5 cm in diameter. The clitoris corresponds to the male penis and has a similar structure. More specifically, it is composed of two columns of erectile tissue, called *corpora cavernosa.* At its anterior end, a small mass of erectile tissue forms a **glans,** which is richly supplied with sensory nerve fibers.

Vestibule

The **vestibule** of the vulva is the space enclosed by the labia minora. The vagina opens into the posterior portion of the vestibule, and the urethra opens in the midline, just in front of the vagina and about 2.5 cm behind the glans of the clitoris.

A pair of **vestibular glands,** which correspond to the male bulbourethral glands, lie one on either side of the vaginal opening.

Beneath the mucosa of the vestibule on either side is a mass of vascular erectile tissue called the *vestibular bulb.*

1. What is the male counterpart of the labia majora? Of the clitoris?
2. What structures are located within the vestibule?

Erection, Lubrication, and Orgasm

Erectile tissues located in the clitoris and around the vaginal entrance respond to sexual stimulation. Following such stimulation, *parasympathetic nerve impulses* pass out from the sacral portion of the spinal cord, causing the arteries associated with the erectile tissues to dilate and the veins to constrict. As a result, the inflow of blood increases, the outflow of blood decreases, and the erectile tissues swell.

At the same time, the vagina expands and elongates. If sexual stimulation is sufficiently intense, parasympathetic impulses cause the vestibular glands to secrete mucus into the vestibule. This secretion moistens and lubricates the tissues surrounding the vestibule and the lower end of the vagina, facilitating the insertion of the penis into the vagina.

The clitoris is abundantly supplied with sensory nerve fibers, which are especially sensitive to local stimulation. The culmination of such stimulation is the pleasurable sense of physiological and psychological release called **orgasm.**

Just prior to orgasm, the tissues of the outer third of the vagina become engorged with blood and swell. This action increases the friction on the penis during intercourse. As orgasm occurs, a series of reflexes involving the sacral and lumbar portions of the spinal cord are initiated.

In response to these reflexes, muscles of the perineum contract rhythmically, and the muscular walls of the uterus and uterine tubes become active. These muscular contractions are thought to aid the transport of sperm cells through the female reproductive tract toward the upper ends of the uterine tubes.

The various functions of the female reproductive organs are summarized in chart 19.2.

Chart 19.2 Functions of the female reproductive organs

Organ	Function
Ovary	Production of egg cells and the female sex hormones
Uterine tube	Conveys the egg cell toward the uterus; the site of fertilization; conveys the developing embryo to the uterus
Uterus	Protects and sustains the life of the embryo during pregnancy
Vagina	Conveys uterine secretions; receives the erect penis during sexual intercourse; transports the fetus during the birth process
Labia majora	Enclose and protect other external reproductive organs
Labia minora	Form the margins of the vestibule; protect openings of the vagina and urethra
Clitoris	Glans is richly supplied with sensory nerve endings, associated with the feeling of pleasure during sexual stimulation
Vestibule	Space between the labia minora that includes the vaginal and urethral openings
Vestibular glands	Secrete a fluid that moistens and lubricates the vestibule

1. What events result from parasympathetic stimulation of the female reproductive organs?
2. What changes take place in the vagina just prior to and during female orgasm?

Hormonal Control of Female Reproductive Functions

Female reproductive functions are largely controlled by hormones secreted by the *hypothalamus, anterior pituitary gland,* and *ovaries.* These hormones are responsible for the development and maintenance of female secondary sexual characteristics, the maturation of female sex cells, and the changes that occur during the monthly reproductive cycles.

Female Sex Hormones

A female child's body remains reproductively immature until about eight years of age. At that time, the hypothalamus begins to secrete increasing amounts of gonadotropin-releasing hormone (GnRH), which, in turn, stimulates the anterior pituitary gland to release the gonadotropins FSH and LH. These hormones play primary roles in the control of female sex cell maturation and the production of sex hormones.

Several different female sex hormones are secreted by various tissues, including the ovaries, adrenal cortices, and placenta (during pregnancy). These hormones belong to two major groups called **estrogen** and **progesterone.**

The primary source of *estrogen* (in a nonpregnant female) is the ovaries. At puberty, under the influence of the anterior pituitary gland, these organs secrete increasing amounts of the hormone. Estrogen stimulates enlargement of various accessory organs, including the vagina, uterus, uterine tubes, ovaries, and the external reproductive structures. Estrogen is also responsible for the development and maintenance of the female *secondary sexual characteristics,* which include:

1. Development of the breasts and the ductile system of the mammary glands within the breasts.
2. Increased deposition of adipose tissue in the subcutaneous layer generally and particularly in the breasts, thighs, and buttocks.
3. Increased vascularization of the skin.

The ovaries are also the primary source of *progesterone* (in a nonpregnant female). This hormone promotes changes that occur in the uterus during the female reproductive cycles. In addition, it affects the mammary glands and helps to regulate the secretion of gonadotropins from the anterior pituitary gland.

Certain other changes that occur in females at puberty seem to be related to *androgen* concentrations. For example, increased growth of hair in the pubic and axillary regions seems to be due to the presence of androgen, secreted by the adrenal cortices. Conversely, the development of the female skeletal configuration, which includes narrow shoulders and broad hips, seems to be related to a low concentration of androgen.

1. What factors initiate sexual maturation in a female?
2. What are the functions of estrogen?
3. What is the function of androgen in a female?

Female Reproductive Cycle

The female reproductive cycle, or **menstrual cycle,** is characterized by regular, recurring changes in the uterine lining, which culminate in menstrual bleeding. Such cycles usually begin near the thirteenth year of life and continue into middle age.

Women athletes sometimes experience disturbances in their menstrual cycles, ranging from diminished menstrual flow (oligomenorrhea) to the complete stoppage of menses (amenorrhea). The incidence of menstrual disorders generally increases with the intensity and duration of exercise periods, occurring most commonly in athletes who perform the most strenuous activities and who follow the most intense training schedules. This effect seems to be related to a loss of adipose tissue and a consequent decline in estrogen, which is synthesized in adipose tissue from adrenal androgens.

A female's first menstrual cycle is initiated after the ovaries and other organs of the female reproductive control system have become mature and responsive to certain hormones. Then, the hypothalamus secretion of gonadotropin-releasing hormone (GnRH) stimulates the anterior pituitary gland to release threshold levels of follicle-stimulating hormone (FSH) and luteinizing hormone (LH). The FSH acts upon an ovary to stimulate the maturation of a follicle, and during its development, the follicular cells produce increasing amounts of estrogen and some progesterone. LH also plays a role in the production of estrogen by stimulating certain ovarian cells to secrete the necessary precursor molecules (testosterone) from which estrogen is synthesized.

In a young female, estrogen stimulates the development of various secondary sexual characteristics. The estrogen secreted during subsequent menstrual cycles is responsible for continuing the development of these traits and maintaining them.

An increasing concentration of estrogen during the first week or so of a menstrual cycle causes changes in the uterine lining, including thickening of the glandular endometrium (fig. 19.12). Meanwhile, the developing follicle has completed its maturation, and by the fourteenth day of the cycle, the follicle appears on the surface of the ovary as a blisterlike bulge.

Within the follicle, the follicular cells, which surround the oocyte and connect it to the inner wall of the follicle, have become loosened. Also, the follicular fluid has increased rapidly.

While the follicle is maturing, LH is produced and stored in the anterior pituitary gland; however, the estrogen secreted by the follicular cells inhibits the release of the LH from the anterior pituitary gland. The estrogen also causes the anterior pituitary cells to

Figure 19.12
Major events in the female reproductive cycle.

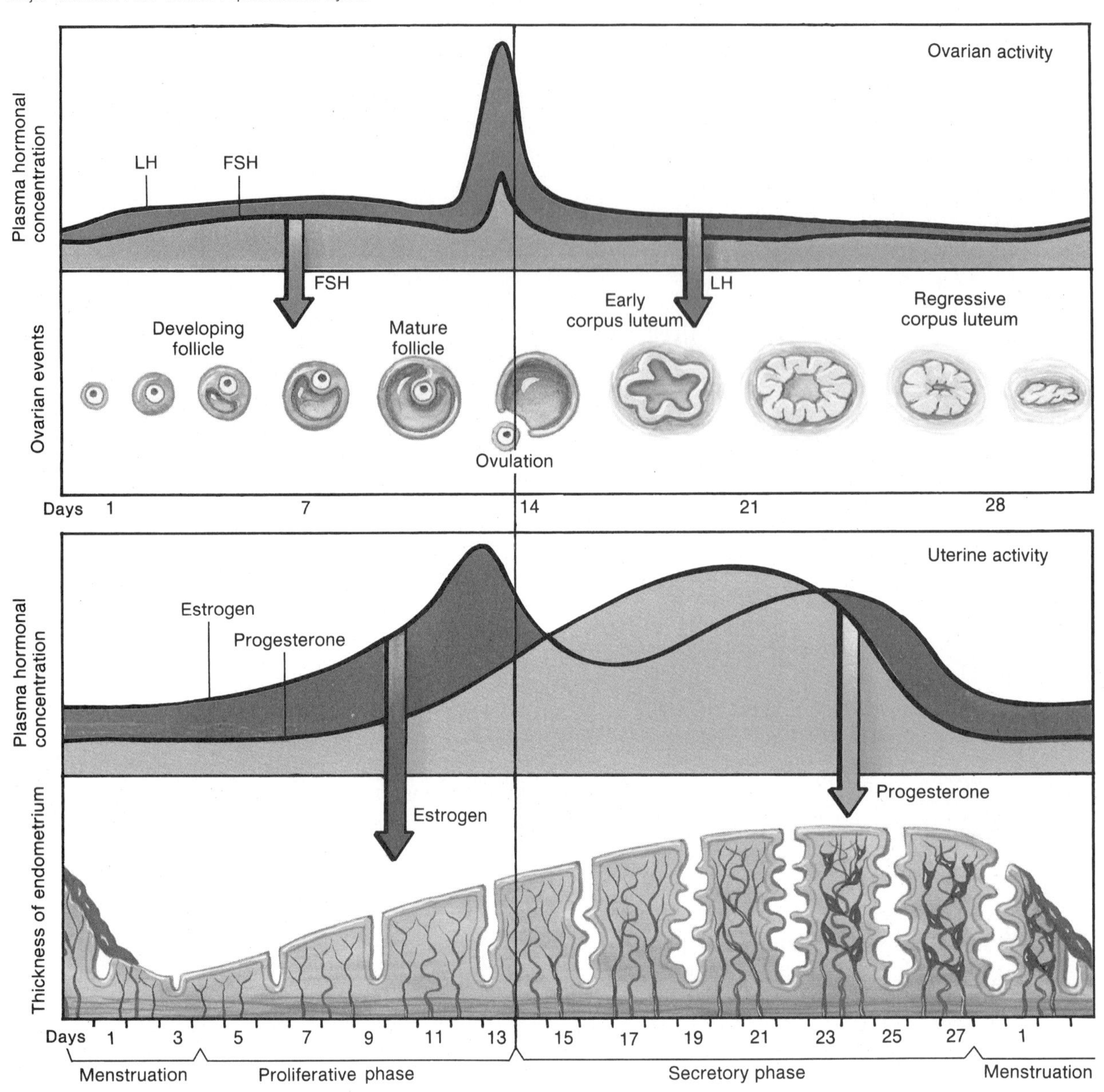

become more sensitive to the action of GnRH, which is released from the hypothalamus in rhythmic pulses about ninety minutes apart.

Near the fourteenth day of follicular development, the anterior pituitary cells finally respond to the pulses of GnRH and to the increasing quantity of progesterone released by the follicular cells. In response, the pituitary cells release the LH that they have stored. The resulting surge in LH concentration, which lasts about thirty-six hours, causes the bulging follicular wall to weaken and rupture. At the same time, the oocyte and follicular fluid escape from the ovary in the process of *ovulation.*

Following ovulation, the remnants of the follicle within the ovary undergo rapid changes. The space occupied by the follicular fluid fills with blood, which soon clots. Under the influence of LH, the follicular cells enlarge to form a temporary glandular structure, called a **corpus luteum.**

Although the follicular cells secrete minute quantities of progesterone during the first part of the menstrual cycle, corpus luteum cells secrete very large quantities of progesterone and estrogen during the last half of the cycle. Consequently, as a corpus luteum becomes established, the blood concentration of progesterone increases sharply.

Progesterone acts on the endometrium of the uterus, causing it to become more vascular and glandular. It also stimulates the uterine glands to secrete increasing quantities of glycogen and lipids. As a result, the endometrial tissues of the uterus become filled with fluids containing nutrients and electrolytes, which can provide a favorable environment for the development of an embryo.

Estrogen and progesterone inhibit the release of LH from the anterior pituitary gland. The corpus luteum also secretes a hormone called *inhibin*, which inhibits the secretion of FSH. Consequently, no other follicles are stimulated to develop during the time that corpus luteum is active. If the oocyte that was released at ovulation is not fertilized by a sperm cell, however, the corpus luteum begins to degenerate about the twenty-fourth day of the cycle.

When the corpus luteum ceases to function, the concentrations of estrogen and progesterone decline rapidly, and in response, the blood vessels in the endometrium become constricted. This action reduces the supply of oxygen and nutrients to the thickened uterine lining, and the lining tissues soon disintegrate and slough away. At the same time, blood escapes from damaged capillaries, creating a flow of blood and cellular debris, which passes through the vagina as the *menstrual flow* (menses). This flow usually begins about the twenty-eighth day of the cycle and continues for three to five days while the estrogen concentration is relatively low.

The beginning of the menstrual flow marks the end of a menstrual cycle and the beginning of a new cycle. This cycle is summarized in chart 19.3.

Since the blood concentrations of estrogen and progesterone are low at the beginning of the menstrual cycle, the hypothalamus and pituitary gland are no longer inhibited. Consequently, the concentrations of FSH and LH soon increase, and a new follicle is stimulated to mature. As this follicle secretes estrogen, the uterine lining undergoes repair and the endometrium begins to thicken again.

Chart 19.3 Major events in a menstrual cycle

1. The anterior pituitary gland secretes FSH and LH.
2. FSH stimulates maturation of a follicle.
3. Follicular cells produce and secrete estrogen.
   a. Estrogen maintains secondary sexual traits.
   b. Estrogen causes the uterine lining to thicken.
4. The anterior pituitary gland secretes a surge of LH, which stimulates ovulation.
5. Follicular cells become corpus luteum cells, which secrete estrogen and progesterone.
   a. Estrogen continues to stimulate uterine wall development.
   b. Progesterone stimulates the uterine lining to become more glandular and vascular.
   c. Estrogen and progesterone inhibit the secretion of LH and inhibin inhibits the secretion of FSH from the anterior pituitary gland.
6. If the egg cell is not fertilized, the corpus luteum degenerates, and no longer secretes estrogen and progesterone.
7. As the concentrations of luteal hormones decline, the blood vessels in the uterine lining constrict.
8. The uterine lining disintegrates and sloughs away, producing a menstrual flow.
9. The anterior pituitary gland, which is no longer inhibited, again secretes FSH and LH.
10. The menstrual cycle is repeated.

After puberty, menstrual cycles normally continue to occur at more or less regular intervals into the late forties or early fifties, at which time they usually become increasingly irregular. After a few months or years, the cycles cease altogether. This period of life is called *menopause*.

The cause of menopause seems to be an aging of the ovaries. After about thirty-five years of cycling, apparently few primary follicles remain to be stimulated by pituitary gonadotropins. Consequently, the follicles no longer mature, ovulation does not occur, and the blood concentration of estrogen decreases greatly.

As a result of low estrogen concentration and lack of progesterone, the female secondary sexual characteristics undergo varying degrees of change. For example, the vagina, uterus, and uterine tubes may decrease in size, as may the external reproductive organs. The pubic and axillary hair may become thinner, and the breasts may regress.

1. Trace the events of the female menstrual cycle.
2. What causes the menstrual flow?

Birth Control
A Current Topic

Birth control is the voluntary regulation of the number of offspring produced and the time they will be conceived. Methods of birth control (contraception) are designed to prevent the fertilization of an egg cell following sexual intercourse or to prevent the implantation of an embryo if fertilization occurs. Commonly used methods of contraception include the following:

1. *Coitus interuptus* involves withdrawing the penis from the vagina before ejaculation. This method often results in pregnancy, because some males find it emotionally difficult to withdraw just prior to ejaculation. Also, small quantities of semen may be expelled from the penis before ejaculation occurs.
2. *Rhythm method* requires abstinence from sexual intercourse a few days before and a few days after ovulation. Theoretically, ovulation occurs on the fourteenth day of a twenty-eight day cycle. However, few women have absolutely regular menstrual cycles, and the lengths of their cycles vary from time to time. Furthermore, the variable (unpredictable) part of a menstrual cycle occurs before ovulation, because, regardless of the length of the cycle, the menstrual flow almost always begins thirteen to fifteen days following ovulation. Consequently, since it is almost impossible to accurately predict the time of ovulation, the rhythm method results in a relatively high rate of pregnancy.

   The effectiveness of the rhythm method can sometimes be increased by measuring and recording a woman's body temperature when she awakes each morning for several months. Because the body temperature typically rises about 0.6 degrees Fahrenheit immediately following ovulation, this procedure may allow a woman to more accurately predict the "unsafe times" in her reproductive cycle. Many women, however, apparently do not show a change in body temperature at ovulation, and furthermore, body temperature may vary in response to other factors, such as illness or emotional upsets.
3. *Mechanical barriers* are used to prevent sperm cells from entering the female reproductive tract during sexual intercourse. One such device is a *condom.* It consists of a thin rubber sheath that is placed over the erect penis before intercourse to prevent semen from entering the vagina. A rubber barrier, used by women, is a *diaphragm.* It is a cup-shaped device with a flexible rim that is inserted into the vagina so that it covers the cervix, thus preventing the entrance of sperm cells into the uterus. To be effective, a diaphragm must be fitted for size by a physician. It must also be inserted properly before sexual contact, be used in conjunction with a chemical spermicide, and be left in position for several hours following sexual intercourse.
4. *Chemical barriers* include a variety of creams, foams, and jellies with spermicidal properties. Within the vagina, such chemicals create an environment that is unfavorable for sperm cells. However, when used alone, chemical barriers

Mammary Glands

The **mammary glands** are accessory organs of the female reproductive system that are specialized to secrete milk following pregnancy (fig. 19.13). They are located in the subcutaneous tissue of the anterior thorax within elevations called *breasts.* The breasts overlie the *pectoralis major* muscles and extend from the second to the sixth ribs and from the sternum to the axillae.

A *nipple* is located near the tip of each breast at about the level of the fourth intercostal space, and it is surrounded by a circular area of pigmented skin, called the *areola.*

A mammary gland is composed of fifteen to twenty lobes. Each lobe contains tubular glands, called *alveolar glands,* and a duct (lactiferous duct) that leads to the nipple and opens to the outside. The lobes are result in a relatively high pregnancy rate. They are most effective when used together with a rubber diaphragm.

5. *Oral contraceptives*, commonly called "the pill," contain synthetic estrogenlike and progesteronelike substances that must be prescribed by a physician. When taken daily, these drugs disrupt the normal pattern of reproductive hormonal secretions and prevent ovulation. Successful use of oral contraceptives requires careful adherence to instructions, but if they are used correctly, they prevent pregnancy nearly 100% of the time. However, in some women oral contraceptives may cause undesirable side effects, such as nausea, retention of body fluids, and breast tenderness.
6. *Intrauterine devices* (IUD) are small solid objects, often with exposed copper parts. Such a device can be placed within the uterine cavity by a physician and is relatively effective in preventing pregnancy. Intrauterine devices seem to interfere with the implantation of embryos within the uterine wall by causing inflammatory reactions in the uterine tissues. Because an IUD may cause unpleasant and potentially serious health problems, it should be checked at regular intervals by a physician.
7. *Surgical methods* are performed to make individuals incapable of reproducing (sterile). In the male, a small section of each vas deferens is removed, and the cut ends of the ducts are tied. This procedure is called *vasectomy*, and it prevents sperm cells from leaving the epididymis. The corresponding procedure in the woman is called *tubal ligation*, and it involves cutting and tying the uterine tubes so that sperm cells cannot reach the egg cells. Neither procedure produces changes in the hormonal concentrations or the sexual drives of the individuals involved. These surgical methods provide the most reliable forms of contraception.

The relative effectiveness of these common contraceptive methods is summarized in chart 19.4.

Chart 19.4 Effectiveness of contraceptive methods (from various studies)

Method	Failure rate* (%)
Abstinence	0
Rhythm method	15.5–24.0
Condom	9.5–12.0
Diaphragm (with spermicide)	14.5–19.0
Creams, foams, and jellies	12.0–21.0
Oral contraceptive	2.5–3.0
IUD	4.5–6.0
Vasectomy	less than 0.5
Tubal ligation	less than 0.5
Chance (no protection)	90

*Failure rate equals the number of women per hundred who use a contraceptive method and become pregnant within one year.

separated from each other by dense connective and adipose tissues. These tissues also support the glands and attach them to the fascia of the underlying pectoral muscles. Other connective tissue, which forms dense strands called *suspensory ligaments*, extends inward from the dermis of the breast to the fascia, helping to support the weight of the breast.

The mammary glands of male and female children are similar. As children reach *puberty*, however, the male glands fail to develop, and the female glands are stimulated to develop by the ovarian hormones. As a result, the aveolar glands and ducts enlarge, and fat is deposited so that the glands become surrounded by adipose tissue.

Figure 19.13
The structure of the breast.

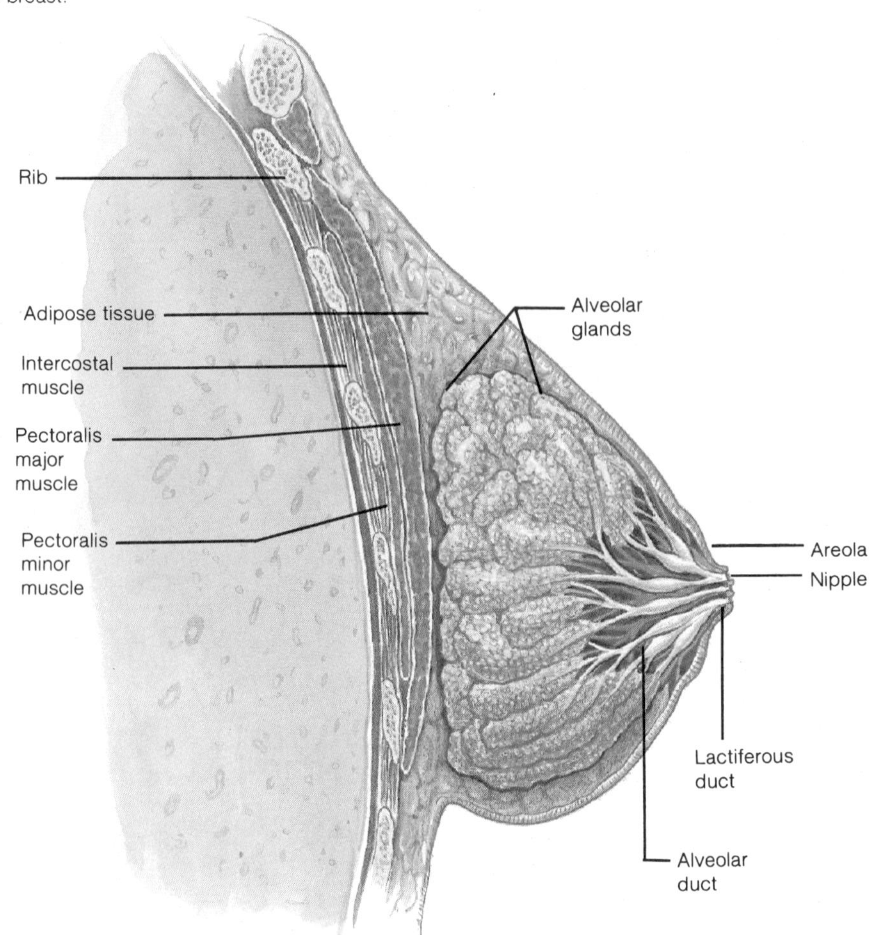

Breast cancer, which is one of the more common types of cancer among women, usually begins as a small, painless lump. Because early diagnosis is of prime importance in successful treatment, the American Cancer Society recommends that after age twenty women examine their breasts each month, paying particular attention to the upper, outer portions. The examination should be made just after menstruation, when the breasts are usually soft, and any lump that is discovered should be checked immediately by a physician.

It also is recommended that after age thirty-five women have their breasts examined at regular intervals with *mammography*—a breast cancer detection technique, which makes use of relatively low dosage X rays. More specifically, the American Cancer Society recommends that women between the ages of thirty-five and forty have a baseline mammogram, that those between forty and forty-nine have a mammogram every one or two years, and that those over the age of fifty have a mammogram yearly. A breast cancer can be detected on a mammogram when the growing tumor is relatively small, perhaps two years before a lump can be felt (fig. 19.14).

Whenever a questionable lump is detected by physical examination or mammography, a biopsy should be performed so that the tissue can be observed microscopically. As a rule, such a microscopic examination is necessary before a breast cancer can be diagnosed with certainty.

Figure 19.14
Mammogram of a breast with a tumor (*arrow*).

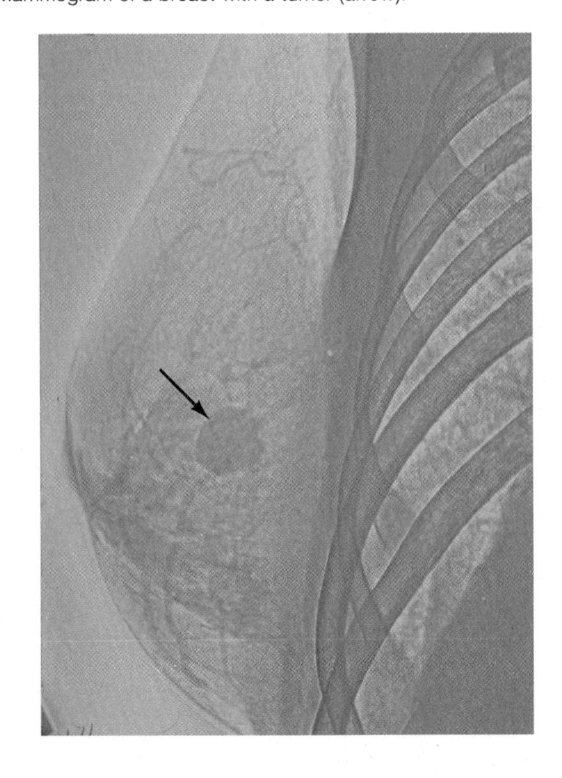

1. Describe the structure of a mammary gland.
2. What changes occur in these glands as a result of the ovarian hormones?

Clinical Terms Related to the Reproductive Systems

amenorrhea (a-men″o-re′ah)—the absence of menstrual flow, usually due to a disturbance in hormonal levels.
conization (ko″nĭ-za′shun)—the surgical removal of a cone of tissue from the cervix for examination.
curettage (ku″rĕ-tahzh′)—a surgical procedure in which the cervix is dilated and the endometrium of the uterus is scraped (commonly called D and C, for dilation and curettage).
dysmenorrhea (dis″men-ŏ-re′ah)—painful menstruation.
endometriosis (en″do-me″tre-o′sis)—a condition in which tissue similar to the inner lining of the uterus occurs within the pelvic cavity.
endometritis (en″do-mĕ-tri′tis)—an inflammation of the uterine lining.
epididymitis (ep″ĭ-did″ĭ-mi′tis)—an inflammation of the epididymis.
hematometra (hem″ah-to-me′trah)—an accumulation of menstrual blood within the uterine cavity.
hysterectomy (his″tĕ-rek′to-me)—the surgical removal of the uterus.
mastitis (mas″ti′tis)—an inflammation of a mammary gland.
oophorectomy (o″of-o-rek′to-me)—the surgical removal of an ovary.
oophoritis (o″of-o-ri′tis)—an inflammation of an ovary.
orchiectomy (or″ke-ĕk′to-me)—the surgical removal of a testis.
orchitis (or-ki′tis)—an inflammation of a testis.
prostatectomy (pros″tah-tek′to-me)—the surgical removal of a portion or all of the prostate gland.
prostatitis (pros″tah-ti′tis)—an inflammation of the prostate gland.
salpingectomy (sal″pin-jek′to-me)—the surgical removal of a uterine tube.
vaginitis (vaj″ĭ-ni′tis)—an inflammation of the vaginal lining.
varicocele (var′ĭ-ko-sĕl″)—a distension of the veins within the spermatic cord.

Chapter Summary

Introduction (page 488)

Various reproductive organs produce sex cells and sex hormones, help sustain the lives of these cells, or transport them from place to place.

Organs of the Male Reproductive System (page 488)

The primary organs are the testes, which produce sperm cells and male sex hormones.

1. Testes
   a. Structure of the testes
      (1) The testes are composed of lobules, separated by connective tissue and filled with seminiferous tubules.
      (2) The seminiferous tubules are lined with epithelium, which produces sperm cells.
      (3) The interstitial cells produce male sex hormones.
   b. Formation of sperm cells
      (1) The epithelium lining the seminiferous tubules consists of the supporting cells and spermatogenic cells.
         (a) The supporting cells support and nourish the spermatogenic cells.
         (b) The spermatogenic cells give rise to the sperm cells.
      (2) A sperm cell consists of a head, body, and tail.
   c. Spermatogenesis
      (1) Sperm cells are produced from spermatogonia.
      (2) The number of chromosomes in sperm cells is reduced by one-half (46 to 23) by meiosis.
      (3) Spermatogenesis produces four sperm cells from each primary spermatocyte.
2. Male internal accessory organs
   a. Epididymis
      (1) The epididymis is a tightly coiled tube that leads into the vas deferens.
      (2) It stores immature sperm cells.
   b. Vas deferens
      (1) The vas deferens is a muscular tube that passes along the medial side of the testis.
      (2) It passes through the inguinal canal.
      (3) It fuses with the duct from the seminal vesicle to form the ejaculatory duct.
   c. Seminal vesicle
      (1) The seminal vesicle is a saclike structure attached to the vas deferens.
      (2) It secretes an alkaline fluid, which contains nutrients, such as fructose.
   d. Prostate gland
      (1) This gland surrounds the urethra just below the urinary bladder.
      (2) It secretes a thin, milky fluid, which neutralizes semen.
   e. Bulbourethral glands
      (1) These glands are two small structures beneath the prostate gland.
      (2) They secrete a fluid that serves as a lubricant for the penis.

f. Semen
(1) Semen is composed of sperm cells and secretions of the seminal vesicles, prostate gland, and bulbourethral glands.
(2) This fluid is slightly alkaline and contains nutrients and prostaglandins.
(3) The contents of semen activates sperm cells, but these cells are unable to fertilize egg cells until they enter the female reproductive tract.

3. Male external reproductive organs
a. Scrotum
The scrotum is a pouch of skin and subcutaneous tissue that encloses the testes.
b. Penis
(1) The penis is specialized to become erect for insertion into the vagina during sexual intercourse.
(2) Its body is composed of three columns of erectile tissue.

4. Erection, orgasm, and ejaculation
a. During erection, the vascular spaces within the erectile tissue become engorged with blood.
b. Orgasm is the culmination of sexual stimulation and is accompanied by emission and ejaculation.
c. The movement of semen occurs as a result of sympathetic reflexes.

Hormonal Control of Male Reproductive Functions (page 493)

1. Hypothalamic and pituitary hormones
The male body remains reproductively immature until the hypothalamus releases GnRH, which stimulates the anterior pituitary gland to release gonadotropins.
a. FSH stimulates spermatogenesis.
b. LH (ICSH) stimulates the interstitial cells to produce male sex hormones.

2. Male sex hormones
Male sex hormones are called androgens. Testosterone is the most important androgen. Androgen production increases rapidly at puberty.
a. Actions of testosterone
(1) Testosterone stimulates the development of the male reproductive organs.
(2) It is responsible for the development and maintenance of male secondary sexual characteristics.
b. Regulation of male sex hormones
(1) Testosterone concentration is regulated by a negative feedback mechanism.
(a) As its concentration rises, the hypothalamus is inhibited and the pituitary secretion of gonadotropins is reduced.
(b) As the concentration falls, the hypothalamus signals the pituitary to secrete gonadotropins.
(2) The concentration of testosterone remains relatively stable from day to day.

Organs of the Female Reproductive System (page 495)

The primary organs are the ovaries, which produce female sex cells and sex hormones.

1. Ovaries
a. Structure of the ovaries
(1) The ovaries are subdivided into a medulla and a cortex.
(2) The medulla is composed of connective tissue, blood vessels, lymphatic vessels, and nerves.
(3) The cortex contains ovarian follicles and is covered by cuboidal epithelium.
b. Primordial follicles
(1) During development, groups of cells in the ovarian cortex form millions of primordial follicles.
(2) Each primordial follicle contains a primary oocyte and several follicular cells.
(3) The primary oocyte begins to undergo meiosis, but the process is halted and is not continued until puberty.
(4) The number of oocytes steadily declines throughout the life of a female.
c. Oogenesis
(1) Beginning at puberty, some oocytes are stimulated to continue meiosis.
(2) When a primary oocyte undergoes oogenesis, it gives rise to a secondary oocyte, in which the original chromosome number is reduced by one-half.
(3) A secondary oocyte is an egg cell that can be fertilized to produce a zygote.
d. Maturation of a follicle
(1) At puberty, FSH stimulates the primordial follicles to become primary follicles.
(2) During maturation, the oocyte enlarges, the follicular cells multiply, and a fluid-filled cavity appears.
(3) Usually only one follicle reaches full development.
e. Ovulation
(1) Oogenesis is completed as the follicle matures.
(2) The resulting oocyte is released when the follicle ruptures.
(3) After ovulation, the oocyte is drawn into the opening of the uterine tube.

2. Female internal accessory organs
a. Uterine tubes
(1) The end of each uterine tube is expanded, and its margin bears irregular extensions.
(2) The movement of an egg cell into the uterine tube is aided by ciliated cells that line the tube and by peristaltic contractions in the wall of the tube.
b. Uterus
(1) The uterus receives the embryo and sustains its life during development.
(2) The uterine wall includes an endometrium, myometrium, and perimetrium.

c. Vagina
    (1) The vagina receives the erect penis, conveys uterine secretions, and transports the offspring during birth.
    (2) Its wall consists of mucosal, muscular, and fibrous layers.
3. Female external reproductive organs
    a. Labia majora
        (1) The labia majora are rounded folds of fatty tissue and skin.
        (2) The upper ends form a rounded, fatty elevation over the symphysis pubis.
    b. Labia minora
        (1) The labia minora are flattened, longitudinal folds between the labia majora.
        (2) They are well supplied with blood vessels.
    c. Clitoris
        (1) The clitoris is a small projection at the anterior end of the vulva, which corresponds to the male penis.
        (2) It is composed of two columns of erectile tissue.
    d. Vestibule
        (1) The vestibule is the space between the labia majora.
        (2) The vestibular glands secrete mucus into the vestibule during sexual stimulation.
4. Erection, lubrication, and orgasm
    a. During periods of sexual stimulation, the erectile tissues of the clitoris and vestibular bulbs become engorged with blood and swollen.
    b. The vestibular glands secrete mucus into the vestibule.
    c. During orgasm, the muscles of the perineum, uterine wall, and uterine tubes contract rhythmically.

Hormonal Control of Female Reproductive Functions (page 500)

Hormones from the hypothalamus, anterior pituitary gland, and ovaries play important roles in the control of sex cell maturation and the development and maintenance of female secondary sexual characteristics.

1. Female sex hormones
    a. A female body remains reproductively immature until about eight years of age when gonadotropin secretion increases.
    b. The most important female sex hormones are estrogen and progesterone.
        (1) Estrogen from the ovaries is responsible for the development and maintenance of most female secondary sexual characteristics.
        (2) Progesterone functions to cause changes in the uterus.
2. Female reproductive cycle
    a. A menstrual cycle is initiated by FSH, which stimulates the maturation of a follicle.
    b. The maturing follicular cells secrete estrogen, which is responsible for maintaining the secondary sexual traits and causing the uterine lining to thicken.
    c. Ovulation is triggered when the anterior pituitary gland secretes a relatively large amount of LH.
    d. Following ovulation, the follicular cells give rise to the corpus luteum.
        (1) The corpus luteum secretes progesterone, which causes the uterine lining to become more vascular and glandular.
        (2) If an egg cell is not fertilized, the corpus luteum begins to degenerate.
        (3) As the concentrations of estrogen and progesterone decline, the uterine lining disintegrates, causing menstrual flow.
    e. During this cycle, estrogen and progesterone inhibit the release of LH, and inhibin inhibits the release of FSH; as the concentrations of these hormones decline, the pituitary secretes FSH and LH again, stimulating a new menstrual cycle.

Mammary Glands (page 504)

1. The mammary glands are located in the subcutaneous tissue of the anterior thorax.
2. They are composed of lobes that contain tubular glands.
3. Lobes are separated by dense connective and adipose tissues.
4. Estrogen stimulates female breast development.
    a. Alveolar glands and ducts enlarge.
    b. Fat is deposited within the breasts.

Clinical Application of Knowledge

1. What changes, if any, might occur in the secondary sexual characteristics of an adult male following the removal of one testis? Following the removal of both testes? Following the removal of the prostate gland?
2. What affect would it have on a female's menstrual cycle if a single ovary was removed surgically? What affect if both ovaries were removed?
3. Which methods of contraception are theoretically most effective in preventing unwanted pregnancies? Which methods are least effective?
4. A woman who is considering having a tubal ligation asks, "Will the operation cause me to go through my change of life early?" How would you answer?

Review Activities

1. List the general functions of the reproductive systems.
2. Distinguish between the primary and accessory male reproductive organs.
3. Describe the structure of a testis.

4. Review the process of meiosis.
5. Describe the epididymis, and explain its function.
6. Trace the path of the vas deferens from the epididymis to the ejaculatory duct.
7. On a diagram, locate the seminal vesicles, prostate gland, and bulbourethral glands, and describe the composition of their secretions.
8. Define *semen.*
9. Describe the structure of the penis.
10. Explain the mechanism that produces an erection of the penis.
11. Distinguish between emission and ejaculation.
12. Describe the mechanism of ejaculation.
13. Explain the role of GnRH in the control of male reproductive functions.
14. List several male secondary sexual characteristics.
15. Explain how the concentration of testosterone is regulated.
16. Describe the structure of an ovary.
17. Describe how a follicle matures.
18. On a diagram, locate the uterine tubes, and explain their function.
19. Describe the structure of the uterus.
20. On a diagram, locate the clitoris, and describe its structure.
21. Explain the role of GnRH in regulating female reproductive functions.
22. List several female secondary sexual characteristics.
23. Define *menstrual cycle.*
24. Explain the roles of estrogen and progesterone in the menstrual cycle.
25. Summarize the major events in a menstrual cycle.
26. Describe the structure of a mammary gland.
27. Describe the changes that occur in the female mammary glands at puberty.

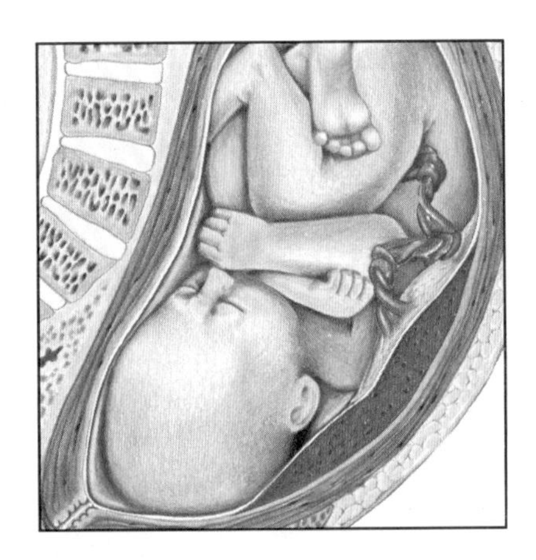

20 Pregnancy, Growth, and Development

The products of the female and male reproductive systems are egg cells and sperm cells, respectively. When an egg cell and a sperm cell unite, a zygote is formed by the process of fertilization. Such a single-celled zygote is the first cell of an offspring, and it is capable of giving rise to an adult of the subsequent generation. The processes by which this is accomplished are called *growth* and *development*.

Chapter Objectives

After you have studied this chapter, you should be able to

1. Distinguish between growth and development.
2. Define *pregnancy,* and describe the process of fertilization.
3. Describe the major events that occur during the period of cleavage.
4. Describe the hormonal changes that occur in the maternal body during pregnancy.
5. Explain how the primary germ layers originate, and list the major structures produced by each layer.
6. Describe the formation and function of the placenta.
7. Define *fetus,* and describe the major events that occur during the fetal stage of development.
8. Trace the general path of blood through the fetal circulatory system.
9. Describe the birth process, and explain the role of hormones in the process.
10. Describe the major circulatory and physiological adjustments that occur in the newborn.
11. Complete the review activities at the end of this chapter. Note that the items are worded in the form of specific learning objectives. You may want to refer to them before reading the chapter.

Key Terms

amnion (am′ne-on)

chorion (ko′re-on)

cleavage (klēv′ij)

embryo (em′bre-o)

fertilization (fer″tĭ-lĭ-za′shun)

fetus (fe′tus)

germ layer (jerm la′er)

neonatal (ne″o-na′tal)

placenta (plah-sen′tah)

postnatal (pōst-na′tal)

prenatal (pre-na′tal)

umbilical cord (um-bil′ĭ-kal kord)

zygote (zi′gōt)

Aids to Understanding Words

allant-, sausage-shaped: *allant*ois—a tubelike structure that extends from the yolk sac into the connecting stalk of the embryo.

chorio-, skin: *chorion*—the outermost membrane that surrounds the fetus and other fetal membranes.

cleav-, to divide: *cleav*age—the period of development characterized by a division of the zygote into smaller and smaller cells.

lacun-, pool: *lacun*a—the space between the chorionic villi that is filled with maternal blood.

morul-, mulberry: *morul*a—an embryonic structure consisting of a solid ball of about 16 cells; thus, looking somewhat like a mulberry.

nat-, to be born: pre*nat*al—the period of development before birth.

troph-, nourishment: *troph*oblast—the cellular layer that surrounds the inner cell mass and helps nourish it.

umbil-, navel: *umbil*ical cord—the structure attached to the fetal navel (umbilicus) that connects the fetus to the placenta.

Introduction

GROWTH REFERS TO an increase in size, and in a human, it usually reflects an increase in cell number, followed by an enlargement of the newly formed cells. Development, on the other hand, is a continuous process by which an individual changes from one life phase to another. These phases include a *prenatal period,* which begins with fertilization and ends at birth, and a *postnatal period,* which begins at birth and ends at death.

Pregnancy

Pregnancy is the condition characterized by the presence of a developing offspring within the uterus. It results from the union of an egg cell with a sperm cell—an event called *fertilization.*

Transport of Sex Cells

Ordinarily, before fertilization can occur, an egg cell (secondary oocyte) must be released by ovulation and enter a uterine tube.

During sexual intercourse, semen containing sperm cells is deposited in the vagina near the cervix. To reach the egg cell, the sperm cells must move upward through the uterus and uterine tube. This movement is aided by the lashing of the sperm tails and by muscular contractions within the walls of the uterus and uterine tube, which are stimulated by prostaglandins in the semen. Also, under the influence of the high estrogen concentrations during the first part of the menstrual cycle, the uterus and cervix contain a thin, watery secretion that promotes sperm transport and survival. Conversely, during the latter part of the cycle, when progesterone concentration is relatively high, these parts secrete a viscous fluid that is unfavorable for sperm transport and survival.

Sperm cells are thought to reach the upper portions of the uterine tube within an hour following sexual intercourse. Although many sperm cells may reach an egg cell, only one actually fertilizes the egg (fig. 20.1).

Figure 20.1
Scanning electron micrograph of sperm cells on the surface of an egg cell (×1200).

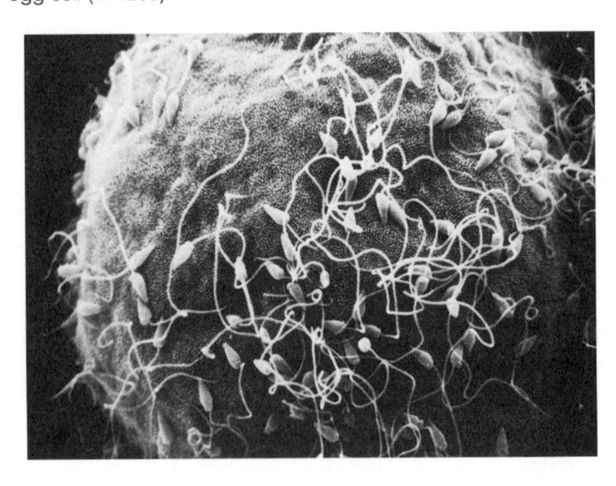

Fertilization

When a sperm cell reaches an egg cell, it invades the follicular cells that adhere to the egg's surface (corona radiata) and penetrates the *zona pellucida,* which surrounds the egg cell membrane. This penetration is aided by an enzyme (hyaluronidase), released by the acrosome of the sperm.

As the sperm cell penetrates the zona pellucida, this covering becomes impenetrable to any other sperm cells (fig. 20.2). The mechanism responsible for this seems to involve structural changes in the membrane, which occur as the first sperm cell penetrates it, and the release of enzymes that prevent other sperm cells from attaching to the membrane.

Once a sperm cell reaches the egg cell membrane, it passes through the membrane and enters the cytoplasm. During this process, the sperm cell loses its tail, and the nucleus in its head swells. As explained previously, the egg cell (secondary oocyte) then divides unequally to form a relatively large cell and a tiny second polar body, which is expelled later. The nuclei of the egg cell and sperm cell come together in the center of the larger cell. Their nuclear membranes disappear, and their chromosomes combine, thus completing the process of **fertilization.**

Because the sperm cell and the egg cell each provide 23 chromosomes, the product of fertilization is a cell with 46 chromosomes—the usual number in a human cell. This cell, called a **zygote,** is the first cell of the future offspring.

1. What factors enhance motility of sperm cells within the female reproductive tract?
2. Where in the female reproductive tract does fertilization normally take place?
3. List the events that occur during fertilization.

Female Infertility

A Current Topic

It is estimated that about 60% of infertile marriages are the result of female disorders. One of the more common of these disorders is hyposecretion of gonadotropic hormones from the anterior pituitary gland, followed by failure of the female to ovulate (anovulation).

This type of anovulatory cycle can sometimes be detected by testing the female's urine for the presence of *pregnanediol,* a product of progesterone metabolism. Since the concentration of progesterone normally rises following ovulation, no increase in pregnanediol in the urine during the latter part of the menstrual cycle suggests a lack of ovulation.

The treatment of such a disorder may include the administration of the hormone HCG (human chorionic gonadotropin), which is obtained from human placentas. This substance has effects similar to those of LH and can stimulate ovulation. Another substance, HMG (human menopausal gonadotropin), which can be obtained from the urine of postmenopausal women, is rich in LH and FSH, and may also be used to treat females with gonadotropin deficiencies. Either treatment, however, may overstimulate the ovaries and cause many follicles to release egg cells simultaneously, resulting in multiple births later.

Another cause of female infertility is *endometriosis,* in which tissue resembling the inner lining of the uterus (endometrium) is present abnormally in the abdominal cavity. Some investigators believe that small pieces of the endometrium may move up through the uterine tubes during menses and become implanted in the abdominal cavity. In any case, once this tissue is present in the cavity, it undergoes changes similar to those that take place in the uterine lining during the menstrual cycle. However, when the tissue begins to break down at the end of the cycle, it cannot be expelled to the outside. Instead, its products remain in the abdominal cavity where they may irritate its lining (peritoneum) and cause considerable abdominal pain. These products also tend to stimulate the formation of fibrous tissue (fibrosis), which, in turn, may encase the ovary, preventing ovulation mechanically, or may obstruct the uterine tubes.

Still other women become infertile as a result of infections, such as gonorrhea, which may cause the uterine tubes to become inflamed and obstructed, or may stimulate the production of viscous mucus that can plug the cervix and prevent the entrance of sperm cells.

Figure 20.2
Steps in the fertilization process: (*1*) Sperm cell reaches corona radiata surrounding the egg cell. (*2*) Acrosome of sperm cell releases protein-digesting enzyme. (*3* and *4*) Sperm cell penetrates zona pellucida surrounding egg cell. (*5*) Sperm cell's plasma membrane fuses with egg cell membrane.

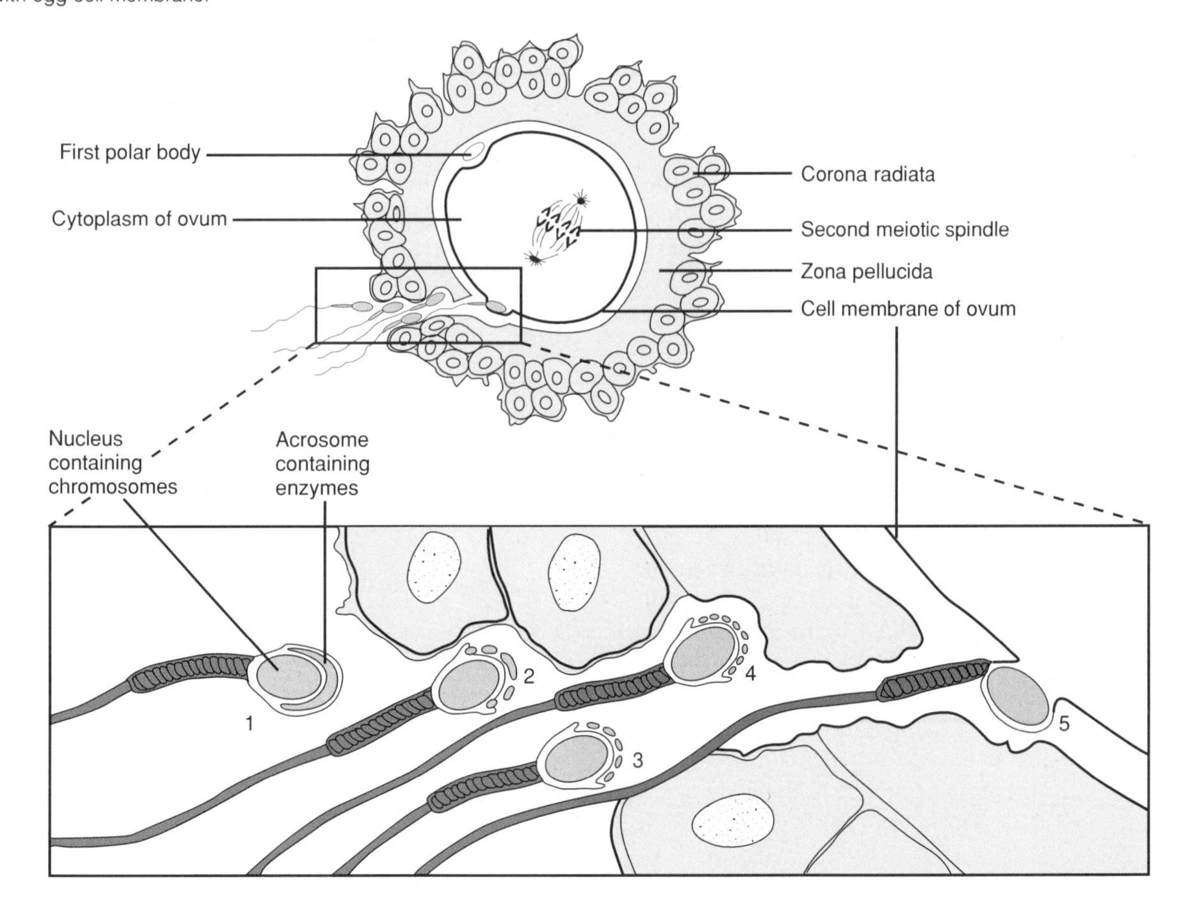

Figure 20.3
Stages in early human development.

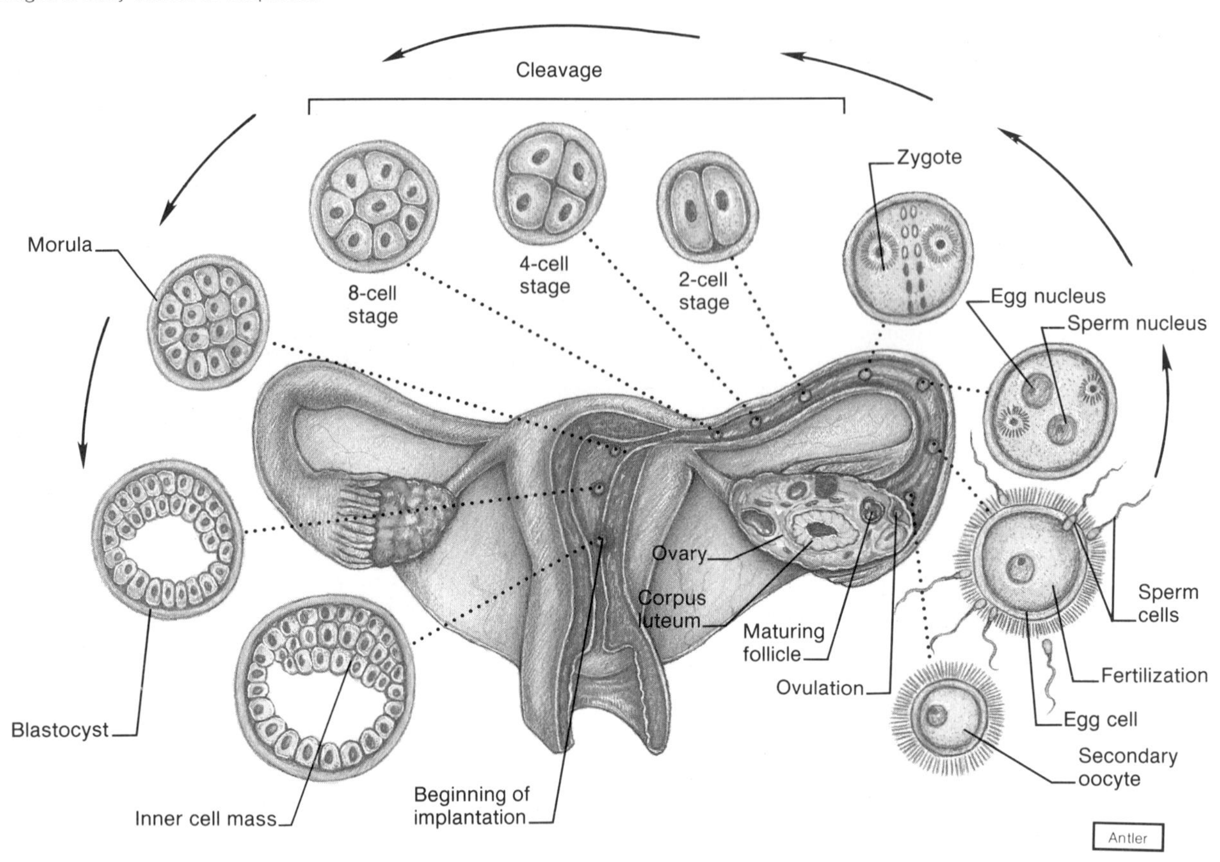

Prenatal Period

Early Embryonic Development

About thirty hours after forming, a zygote undergoes *mitosis,* giving rise to two new cells (blastomeres) (fig. 20.3). These cells, in turn, divide into four cells, which divide into eight cells, and so forth. With each subsequent division, the resulting cells are smaller and smaller. This phase of development is termed **cleavage.**

Meanwhile, the tiny mass of cells is moved through the uterine tube to the uterine cavity. The trip to the uterus takes about three days, and by then the structure consists of a solid ball (morula) of about sixteen cells.

The structure remains free within the uterine cavity for about three days. During this stage, the zona pellucida of the original egg cell degenerates, and the structure, which now consists of a hollow ball of cells (blastocyst), begins to attach itself to the uterine lining. By the end of the first week of development, it is superficially *implanted* in the endometrium.

About the time of implantation, certain cells within the blastocyst become organized into a group (inner cell mass) that will give rise to the body of the offspring (fig. 20.3). This marks the beginning of the *embryonic period* of development. The offspring is termed an **embryo** until the end of the eighth week, after which and until birth, it is called a **fetus.**

Eventually, the outer cells of the embryo together with cells of the maternal endometrium form a complex vascular structure called the **placenta.** This organ attaches the embryo to the uterine wall and exchanges nutrients, gases, and wastes between the maternal blood and the embryonic blood. The placenta also secretes hormones.

1. What is meant by cleavage?
2. What is meant by implantation?
3. What is the difference between an embryo and a fetus?

Hormonal Changes during Pregnancy

During a typical menstrual cycle, the corpus luteum degenerates about two weeks after ovulation (chapter 19). Consequently, the estrogen and progesterone concentrations decline rapidly, the uterine lining is no longer maintained, and the endometrium sloughs away as menstrual flow. If this occurs following implantation, the embryo will be lost.

In Vitro Fertilization

A Current Topic

A woman who is infertile because her uterine tubes are blocked may become pregnant by means of *in vitro fertilization.* In this technique, oocytes are removed from the woman's ovary and mixed with sperm cells in a laboratory dish to achieve fertilization. Later, the resulting embryos are transferred into the woman's uterus for development.

The in vitro fertilization procedure usually begins with the administration of a substance that will induce the development of ovarian follicles, such as the hormone HMG (human menopausal gonadotropin). The growth of follicles can be monitored using ultrasonography, a noninvasive technique that makes use of ultrasonic sound waves, and when the follicles have reached a certain size, the patient is given the hormone HCG (human chorionic gonadotropin) to induce ovulation.

Oocytes released from the ovary are collected with the aid of a *laparoscope*—an optical instrument used to examine the abdominal interior. The oocytes are incubated at 37° C in a buffered medium with a pH of 7.4, and when they are mature, they are mixed in a laboratory dish with sperm that have been washed to remove various inhibitory factors.

Fertilized eggs are then incubated in a special medium. After fifty to sixty hours, when the developing embryos have reached the 8- or 16-cell stage, normal ones are transferred through the woman's cervix and into her uterus with the aid of a specially designed catheter. She is subsequently treated with progesterone to promote a favorable uterine environment for implantation of the embryos. As a result of this procedure, successful implantation occurs in about 20% to 30% of the cases.

Figure 20.4
(*a*) About the sixth day of development, the blastocyst contacts the uterine wall and (*b*) begins to become implanted within the wall.

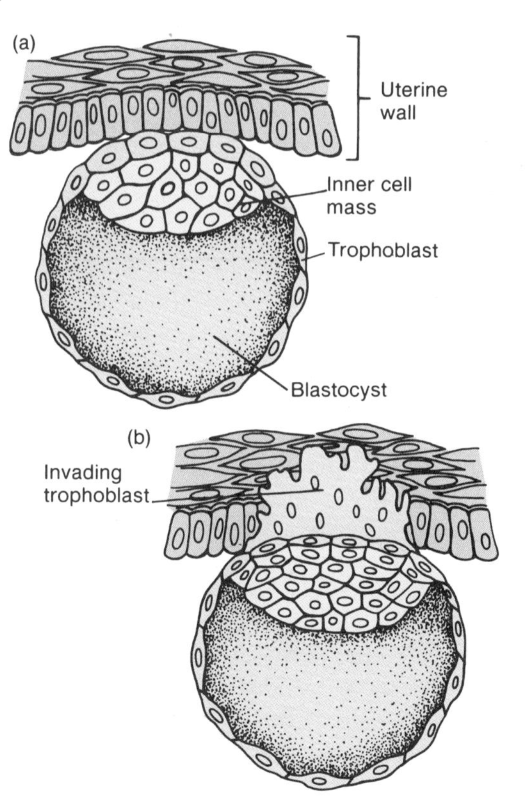

Figure 20.5
Relative concentrations of 3 hormones in maternal blood during pregnancy.

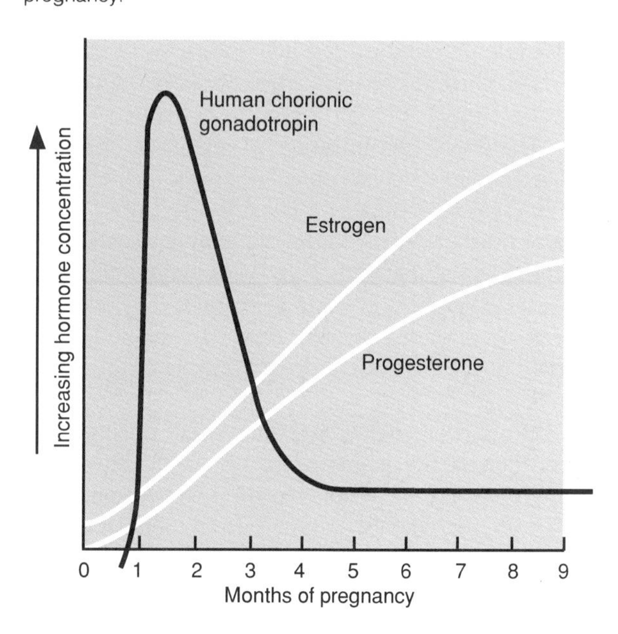

The mechanism that normally prevents such a termination of pregnancy involves a hormone, called HCG (human chorionic gonadotropin). This hormone is secreted by a layer of embryonic cells (trophoblast) that surrounds the developing embryo and, later, is involved in the formation of the placenta (fig. 20.4). HCG has properties similar to those of LH, and it causes the corpus luteum to be maintained and to continue secreting relatively large amounts of estrogen and progesterone. Thus, the uterine wall continues to grow and develop. At the same time, the estrogen and progesterone suppress the release of FSH and LH from the anterior pituitary gland, so that normal menstrual cycles are inhibited.

The secretion of HCG continues at a high level for about two months, then declines to a relatively low level by the end of four months (fig. 20.5). Although the corpus luteum is maintained throughout pregnancy, its function as a source of hormones becomes less important after the first three months (first trimester). This is due to the fact that the placenta is usually well developed by then, and the placental tissues secrete sufficient estrogen and progesterone.

HCG secretion by the embryonic tissues begins shortly after fertilization and increases to a peak in about 50 to 60 days. Thereafter, the concentration of HCG drops to a much lower level, and its concentration remains relatively stable throughout the pregnancy.

Since HCG is excreted in urine, its presence in urine can be used to detect the presence of an embryo. Such a pregnancy test may indicate positive results within about 8 to 10 days of fertilization.

For the remainder of the pregnancy, *placental estrogen* and *progesterone* maintain the uterine wall. The placenta also secretes a hormone called **placental lactogen.** This hormone stimulates breast development and preparation for milk secretion, a function that is aided by placental estrogen and progesterone. Placental progesterone and *relaxin,* a polypeptide hormone from the corpus luteum, inhibit the smooth muscles in the myometrium so that uterine contractions are suppressed until it is time for the birth process to begin.

The high concentration of placental estrogen during pregnancy causes an enlargement of the vagina and external reproductive organs. It also causes relaxation of the ligaments holding the symphysis pubis and sacroiliac joints together. This latter action allows for a greater movement at these joints and, thus, aids the passage of the fetus through the birth canal. The relaxation of these ligaments and softening of the cervix near the time of birth may be aided by relaxin.

Other hormonal changes that occur during pregnancy include the increased secretion of aldosterone from the adrenal cortex and parathyroid hormone from the parathyroid glands. Aldosterone promotes the renal reabsorption of sodium, leading to fluid retention, and parathyroid hormone helps to maintain a high concentration of maternal blood calcium (see chapter 11).

Chart 20.1 summarizes the hormonal changes of pregnancy.

1. What mechanism is responsible for maintaining the uterine wall during pregnancy?
2. What is the source of the hormones that sustain the uterine wall during pregnancy?
3. What other hormonal changes occur during pregnancy?

Chart 20.1 Hormonal changes during pregnancy

1. Following implantation, the embryonic cells begin to secrete HCG.
2. HCG causes the corpus luteum to be maintained and to continue secreting estrogen and progesterone.
3. As the placenta develops, it secretes large quantities of estrogen and progesterone.
4. Placental estrogen and progesterone act to
   a. Stimulate the uterine lining to continue development.
   b. Maintain the uterine lining.
   c. Inhibit the secretion of FSH and LH from the anterior pituitary gland.
   d. Stimulate the development of the mammary glands.
   e. Progesterone inhibits uterine contractions.
   f. Estrogen causes an enlargement of the reproductive organs and relaxation of the ligaments of the pelvic joints.
5. Relaxin from the corpus luteum also inhibits uterine contractions and causes the pelvic ligaments to relax.
6. The placenta secretes placental lactogen that stimulates breast development.
7. Aldosterone from the adrenal cortex promotes reabsorption of sodium.
8. Parathyroid hormone from the parathyroid glands helps maintain a high concentration of maternal blood calcium.

Embryonic Stage of Development

The **embryonic stage** extends from the second week through the eighth week of development and is characterized by the formation of the placenta, the development of the main internal organs, and the appearance of the major external body structures.

Early in this stage, the cells of the inner cell mass become organized into a flattened **embryonic disk,** which consists of two distinct layers—an outer *ectoderm* and an inner *endoderm* (fig. 20.6). A short time later, a third layer of cells, the *mesoderm,* forms between the ectoderm and endoderm. These three layers of cells are called the **primary germ layers,** and they are responsible for forming all the body organs.

More specifically, *ectodermal cells* give rise to the nervous system and portions of special sensory organs, as well as the epidermis, hair, nails, glands of the skin, and linings of the mouth and anal canal. *Mesodermal cells* form all types of muscle tissue, bone tissue, and bone marrow, as well as the blood, blood vessels, lymphatic vessels, various connective tissues, internal reproductive organs, kidneys, and epithelial linings of the body cavities. *Endodermal cells* produce the epithelial linings of the digestive tract, respiratory tract, urinary bladder, and urethra.

During the fourth week of development, the flat embryonic disk is transformed into a cylindrical struc-

Figure 20.6
Early in the embryonic stage of development, 3 primary germ layers are formed.

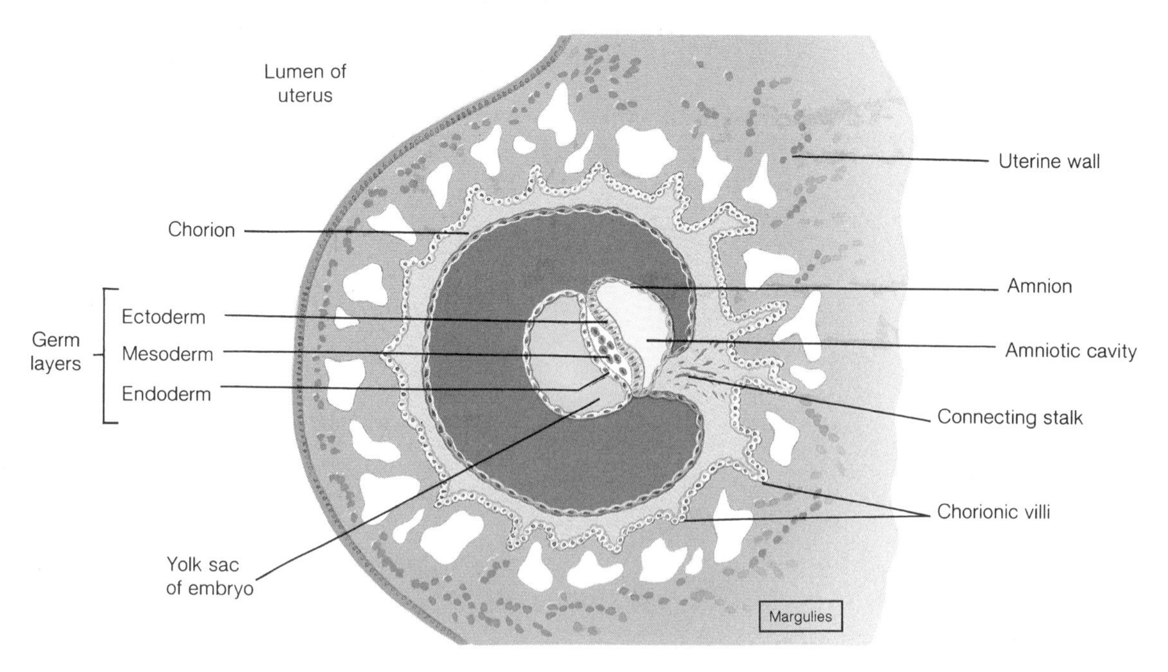

ture, which is attached to the developing placenta by a *connecting stalk* (fig. 20.7). By this time, the head and jaws are appearing, the heart is beating and forcing blood through the blood vessels, and tiny buds are forming, which will give rise to the arms and legs.

During the fifth through the seventh weeks, as shown in figure 20.8, the head grows rapidly and becomes rounded and erect. The face, which is developing the eyes, nose, and mouth, becomes more humanlike. The arms and legs elongate, and fingers and toes appear.

By the end of the seventh week, all the main internal organs have become established, and as these structures enlarge, they affect the shape of the body. Consequently, the body takes on a humanlike appearance.

Meanwhile, the embryo continues to become implanted within the uterus. Early in this process slender projections grow out from the wall of the blastocyst into the surrounding endometrium of the uterine wall. These extensions, which are called **chorionic villi** (fig. 20.7), become branched; and by the end of the fourth week, they are well formed.

While the chorionic villi are developing, embryonic blood vessels appear within them, and these vessels are continuous with those passing through the connecting stalk to the body of the embryo. At the same time, irregular spaces, called **lacunae,** are eroded around

Figure 20.7
(*a*) During the third week, the embryonic disk is flat; (*b*) during the fourth week, the disk becomes a cylindrical structure.

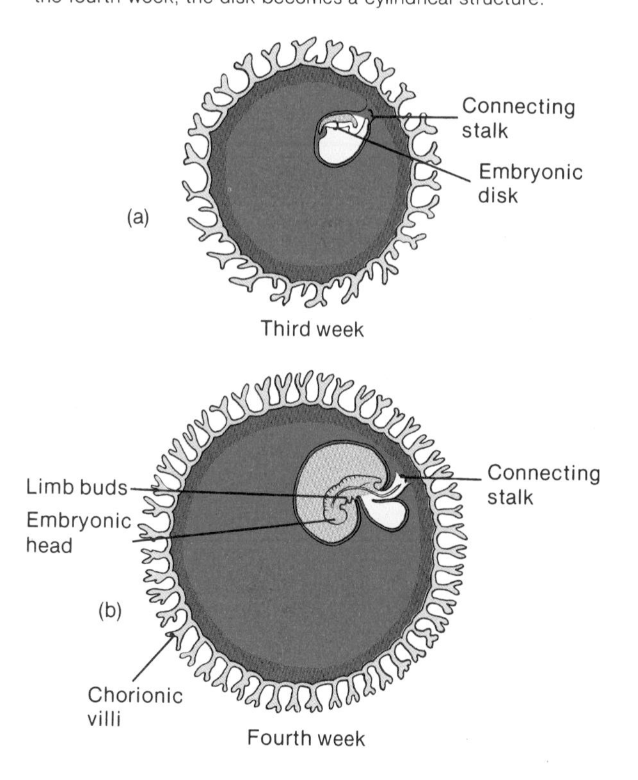

Figure 20.8
In the fifth through the seventh weeks of development, the embryonic body and face develop a humanlike appearance.

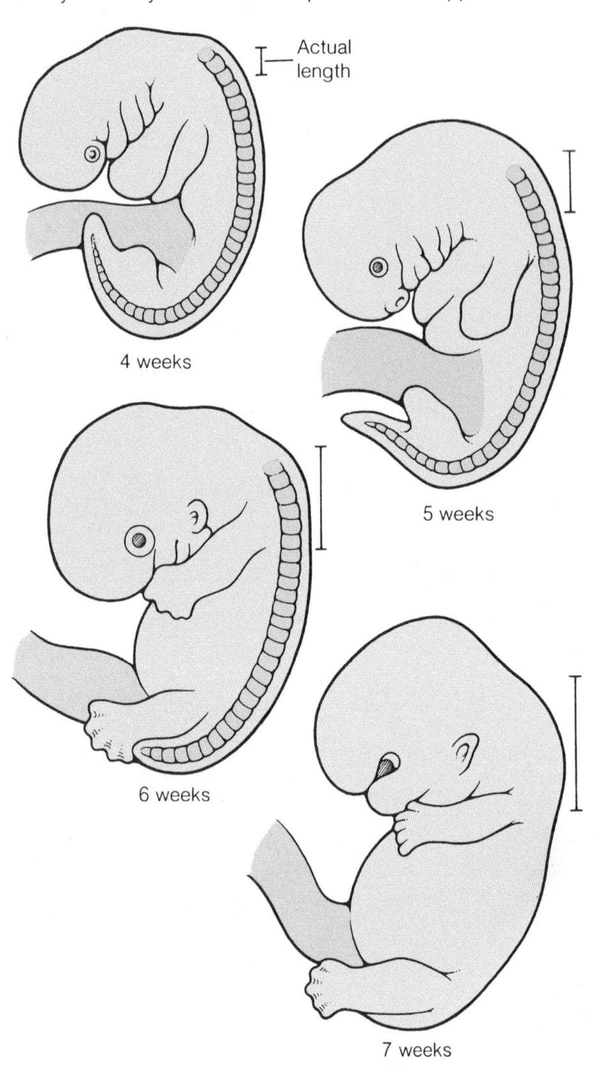

Figure 20.9
As illustrated in the section of a villus, the placental membrane consists of the epithelial wall of an embryonic capillary and the epithelial wall of a chorionic villus. What is the significance of this membrane?

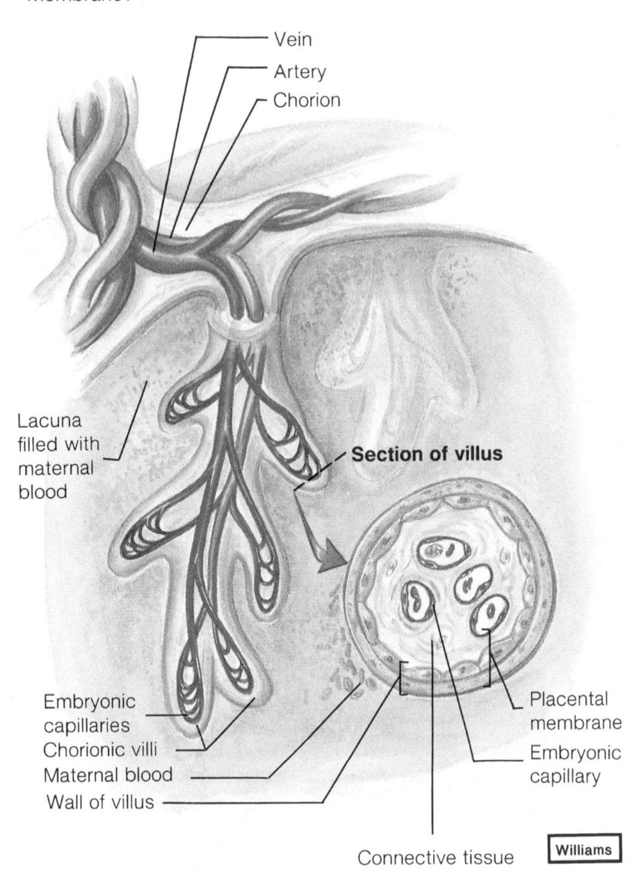

and between the villi (fig. 20.9). These spaces become filled with the maternal blood that escapes from eroded endometrial blood vessels.

A thin membrane separates the embryonic blood within the capillary of a chorionic villus from the maternal blood in a lacuna. This membrane, called the **placental membrane,** is composed of the epithelium of the villus and the epithelium of the capillary. Through this membrane, exchanges take place between the maternal blood and the embryonic blood. Oxygen and nutrients diffuse from the maternal blood into the embryonic blood, and carbon dioxide and other wastes diffuse from the embryonic blood into the maternal blood. Various substances also move through the placental membrane by active transport and pinocytosis.

Because many drugs are able to pass freely through the placental membrane, substances ingested by the mother may affect the fetus. Thus, fetal drug addiction may occur following the mother's use of various addicting drugs, such as heroin.

Similarly, depressant drugs administered to the mother during labor can produce effects within the fetus and may, for example, depress the activity of its respiratory system.

1. What major events occur during the embryonic stage of development?
2. What tissues and structures develop from ectoderm? From mesoderm? From endoderm?
3. How are substances exchanged between the embryonic blood and the maternal blood?

Until about the end of the eighth week, the chorionic villi cover the entire surface of the former blastocyst. The membrane that contains these villi and

Figure 20.10
(*a, b, c*) As the amnion develops, it surrounds the embryo and (*d*) the umbilical cord is formed. What structures comprise this cord?

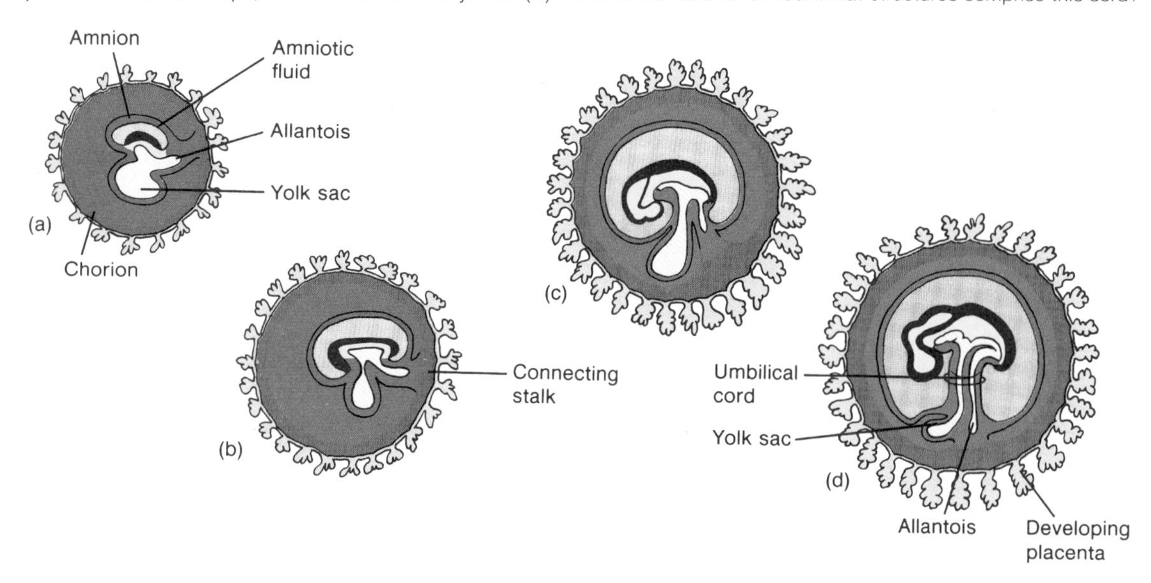

surrounds the developing embryo is called the **chorion.** As the embryo and the chorion continue to enlarge, only those villi that remain in contact with the endometrium endure. The others degenerate, and the portions of the chorion to which they were attached become smooth. Thus, the region of the chorion still in contact with the uterine wall is restricted to a disk-shaped area that becomes the **placenta.**

The embryonic portion of the placenta is composed of the chorion and its villi; the maternal portion is composed of the area of the uterine wall to which the villi are attached. When it is fully formed, the placenta appears as a reddish brown disk about 20 cm long and 2.5 cm thick. It usually weighs about 0.5 kg.

While the placenta is forming, another membrane, called the **amnion,** develops around the embryo (fig. 20.10). This second membrane begins to appear during the second week. Its margin is attached around the edge of the embryonic disk, and fluid, called **amniotic fluid,** fills the space between the amnion and embryonic disk.

As the embryo is transformed into a cylindrical structure, the margins of the amnion are folded around it so that the embryo becomes enclosed by the amnion and surrounded by amniotic fluid. As this process continues, the amnion envelops the tissues on the underside of the embryo, by which it is attached to the chorion and developing placenta. In this manner, the **umbilical cord** is formed.

The umbilical cord contains three blood vessels—two *umbilical arteries* and one *umbilical vein*—through which blood passes between the embryo and placenta (fig. 20.11).

Figure 20.11
The placenta consists of an embryonic portion and a maternal portion.

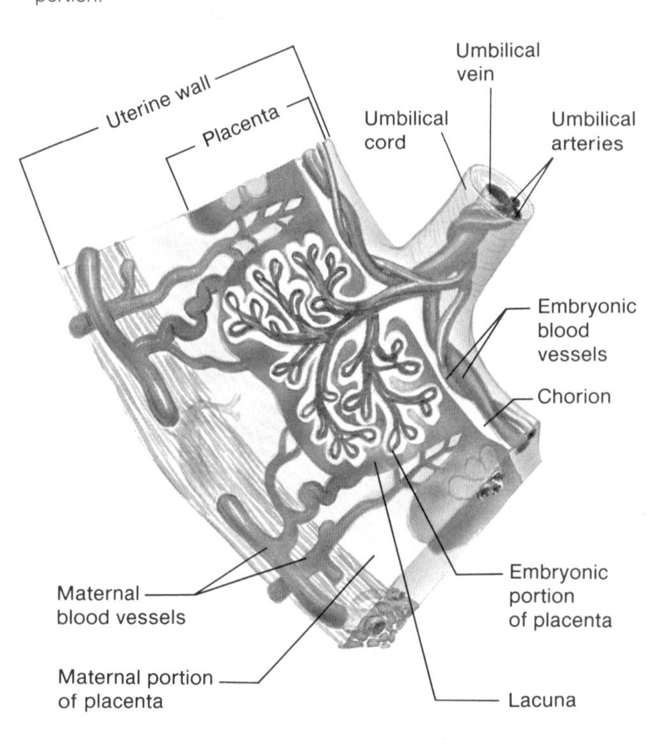

The umbilical cord also suspends the embryo in the *amniotic cavity,* and the amniotic fluid provides a watery environment in which the embryo can grow freely without being compressed by surrounding tissues. The amniotic fluid also protects the embryo from being jarred by the movements of the mother's body.

Teratogens

A Current Topic

Factors that can cause congenital malformations by affecting an embryo during its period of rapid growth and development are called *teratogens*. Such agents include various drugs and certain microorganisms.

The molecules of many drugs are able to pass freely from the maternal blood to the embryonic blood through the placental membrane. Consequently, substances ingested by the mother-to-be may affect her embryo. For example, when a pregnant woman drinks alcohol, her embryo is exposed to an alcohol concentration equal to that of her blood, and the alcohol can cause a wide range of effects on the developing offspring.

More specifically, alcohol is able to produce a set of physical and mental abnormalities called the *fetal alcohol syndrome* (FAS). The symptoms of this syndrome include prenatal and postnatal growth retardation, abnormal facial features, reduced head size, organ malformations, and mental retardation. FAS is thought to result when a pregnant woman drinks as little as three ounces of alcohol per day or engages in a single drinking binge. The Surgeon General of the United States has recommended that women who are pregnant and those attempting to become pregnant should abstain from drinking alcohol.

The rubella virus that causes German measles can also cause congenital malformations. If a pregnant woman develops a rubella infection during the first four or five weeks of embryonic development, the embryonic eyes, ears, and heart may be malformed. Consequently, the offspring may be blind, be deaf, and have various heart disorders. Exposure to the rubella virus during the later stages of development may result in functional defects in the central nervous system.

If there is any chance that a woman may be pregnant, it is prudent for her to avoid taking any drugs—even seemingly harmless ones—without the advice of her physician.

In addition to the amnion and chorion, two other embryonic membranes appear during development. They are the yolk sac and the allantois (fig. 20.10).

The **yolk sac** appears during the second week, and it is attached to the underside of the embryonic disk. It forms blood cells in the early stages of development and gives rise to the cells that later become sex cells.

The **allantois** forms during the third week as a tube extending from the early yolk sac into the connecting stalk of the embryo. It, too, forms blood cells and gives rise to the umbilical arteries and vein.

By the beginning of the eighth week, the embryo is usually 30 mm in length and weighs less than 5 gm. Although its body is quite unfinished, it is clearly recognizable as a human being (fig. 20.12).

1. Describe how the placenta forms.
2. What is the function of amniotic fluid?
3. What is the significance of the yolk sac?

Figure 20.12
What structures can you identify in this photograph of a 7-week-old human embryo?

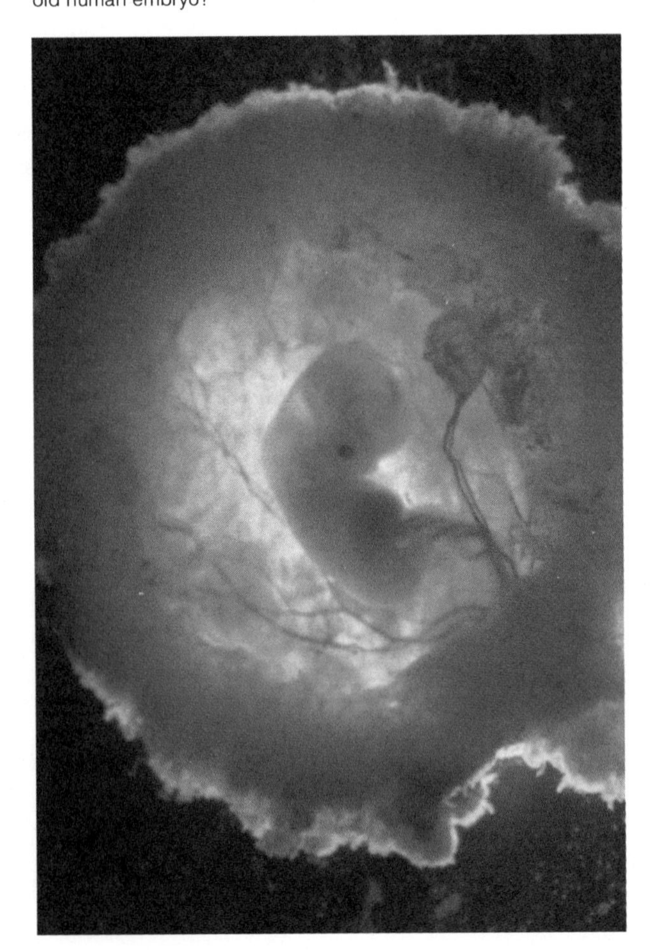

Fetal Stage of Development

The **fetal stage** begins at the end of the eighth week of development and lasts until birth. During this period, the existing body structures continue to grow and mature, and only a few new parts appear. The rate of growth is great, however, and the body proportions change considerably. For example, at the beginning of the fetal stage, the head is disproportionately large and the legs are relatively short (fig. 20.13).

During the third lunar month, the growth in body length is accelerated, but the growth of the head slows.

Figure 20.13
A human fetus after about 10 weeks of development.

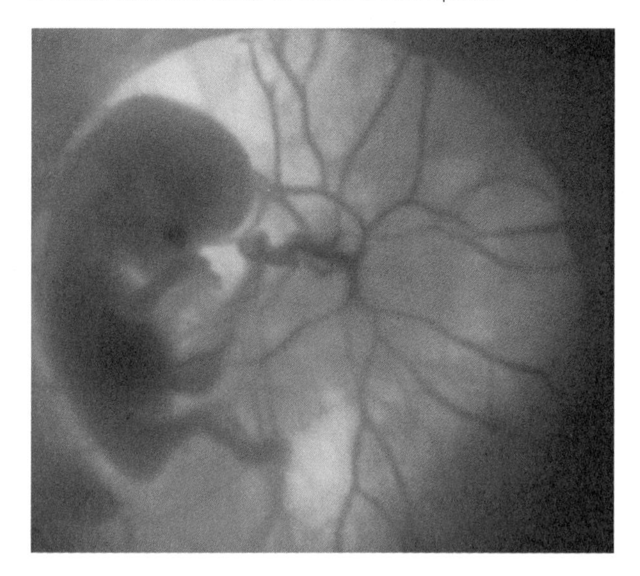

(Note: A lunar month equals 28 days and is commonly used in studies of human development.) The arms achieve the relative length they will maintain throughout development, and ossification centers appear in most of the bones. By the twelfth week, the external reproductive organs are distinguishable as male or female.

In the fourth lunar month, the body grows very rapidly and reaches a length of 13–17 cm. The legs lengthen considerably, and the skeleton continues to ossify.

In the fifth lunar month, the rate of growth decreases somewhat. The legs achieve their final relative proportions, and the skeletal muscles become active, so that the mother may feel fetal movements. Some hair appears on the head, and the skin becomes covered with fine, downy hair. The skin is also coated with a cheesy mixture of sebum from the sebaceous glands and dead epidermal cells.

During the sixth lunar month, the body gains a substantial amount of weight. The eyebrows and eyelashes appear. The skin is quite wrinkled and translucent. It is also reddish, due to the presence of dermal blood vessels.

In the seventh lunar month, the skin becomes smoother as fat is deposited in the subcutaneous tissues. The eyelids, which fused together during the third month, reopen. At the end of this month, a fetus is about 37 cm in length.

In the eighth lunar month, the fetal skin is still reddish and somewhat wrinkled. The testes of males descend from regions near the developing kidneys, through the inguinal canals, and into the scrotum.

If a fetus is born prematurely, its chance of surviving increases directly with its age. One of the factors involved is the development of the lungs. Thus, fetuses have increased chances of surviving if their lungs are sufficiently developed so that they have the thin respiratory membranes necessary for a rapid exchange of oxygen and carbon dioxide and if the lungs produce enough surfactant to reduce the alveolar surface tension (see chapter 13). A fetus of less than 24 weeks of development or weighing less than 600 grams at birth seldom survives, even when given intensive medical care.

During the ninth lunar month, the fetus reaches a length of about 47 cm. The skin is smooth, and the body appears chubby due to an accumulation of subcutaneous fat. The reddishness of the skin fades to pinkish or bluish pink, even in fetuses of dark-skinned parents, because melanin is not produced until the skin is exposed to light.

At the end of the tenth lunar month, the fetus is said to be *full term.* It is about 50 cm long and weighs 2.7 to 3.6 kg. The skin has lost its downy hair but is still coated with sebum and dead epidermal cells. The scalp is usually covered with hair, the fingers and toes have well-developed nails, and the skull bones are largely ossified. As figure 20.14 shows, the full-term fetus is usually positioned with its head toward the cervix.

1. What major changes characterize the fetal stage of development?
2. Describe a full-term fetus.

Fetal Blood and Circulation

Throughout the fetal stage of development, maternal blood supplies the fetus with oxygen and nutrients and carries away its wastes. These substances diffuse between the maternal and fetal blood through the placental membrane, and they are carried to and from the fetal body by the umbilical blood vessels. Consequently, the fetal blood and vascular system must be adapted to intrauterine life in special ways.

For example, the concentration of oxygen-carrying hemoglobin in fetal blood is about 50% greater than in maternal blood. Also, fetal hemoglobin is chemically slightly different, so that it has a greater attraction for oxygen than maternal hemoglobin. Thus, at a particular oxygen pressure, fetal hemoglobin can carry 20% to 30% more oxygen than maternal hemoglobin.

In the fetal circulatory system, the *umbilical vein* transports blood rich in oxygen and nutrients from the placenta to the fetal body (fig. 20.15). This vein enters the body and travels along the anterior abdominal wall

Figure 20.14
A full-term fetus usually becomes positioned in the uterus with its head near the cervix.

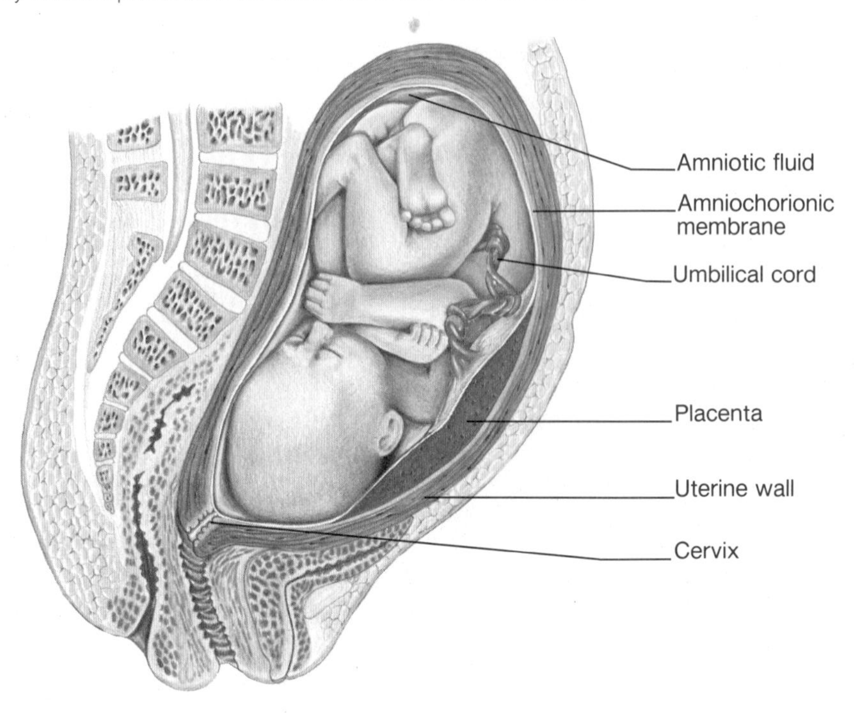

Figure 20.15
The general pattern of fetal circulation. How does this pattern of circulation differ from that of an adult?

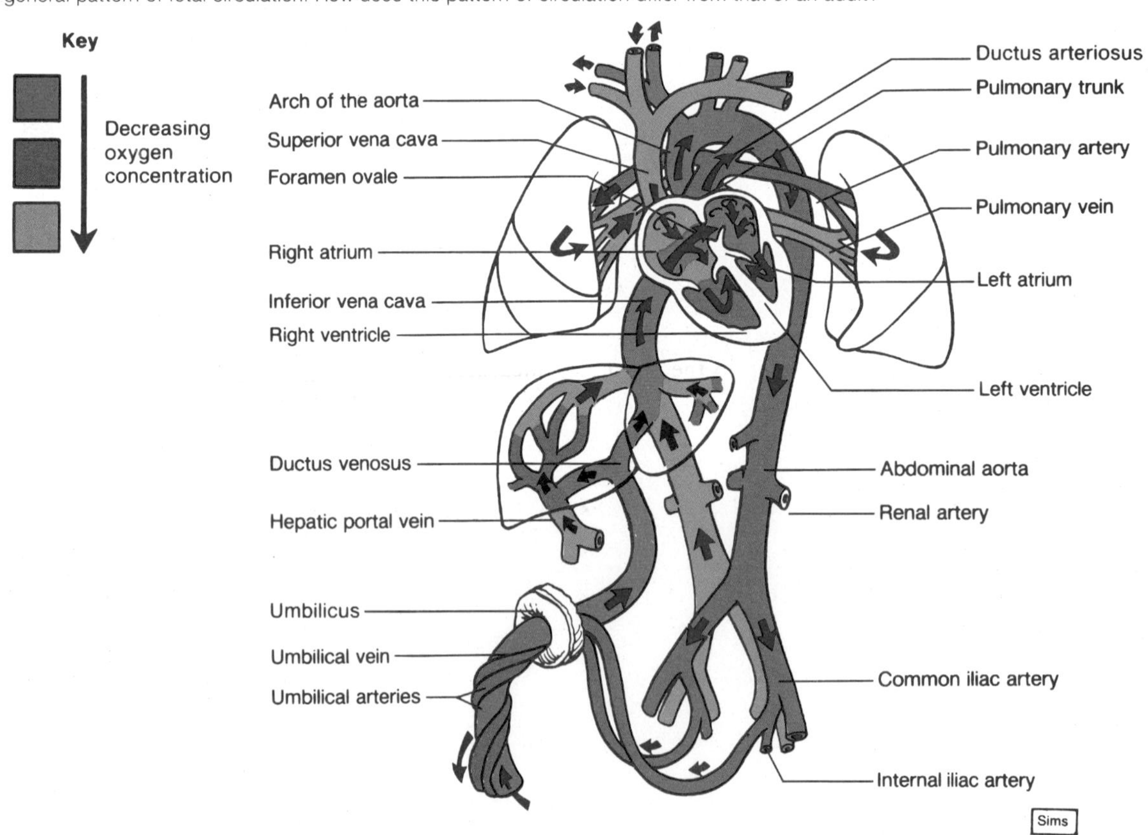

to the liver. About half the blood it carries passes into the liver, and the rest enters a vessel, called the **ductus venosus,** that bypasses the liver.

The ductus venosus travels a short distance and joins the inferior vena cava. There the oxygenated blood from the placenta is mixed with the deoxygenated blood from the lower parts of the fetal body. This mixture continues through the vena cava to the right atrium.

In an adult heart, the blood from the right atrium enters the right ventricle and is pumped through the pulmonary trunk and arteries to the lungs (see chapter 15). In the fetus, however, the lungs are nonfunctional, and the blood largely bypasses them. More specifically, as the blood from the inferior vena cava enters the fetal right atrium, a large proportion of it is shunted directly into the left atrium through an opening in the atrial septum. This opening is called the **foramen ovale,** and the blood passes through it because the blood pressure in the right atrium is somewhat greater than that in the left atrium. Furthermore, a small valve located on the left side of the atrial septum overlies the foramen ovale and helps to prevent blood from moving in the reverse direction.

The rest of the fetal blood entering the right atrium from the inferior vena cava, as well as a large proportion of the deoxygenated blood entering from the superior vena cava, passes into the right ventricle and out through the pulmonary trunk.

Only a small volume of the blood enters the pulmonary circuit, however, because the lungs are collapsed, and their blood vessels have a high resistance to blood flow. But enough blood does reach the lung tissues to sustain them.

Most of the blood in the pulmonary trunk bypasses the lungs by entering a fetal vessel called the **ductus arteriosus,** which connects the left pulmonary artery to the descending portion of the aortic arch. As a result of this connection, blood with a relatively low oxygen concentration, which is returning to the heart through the superior vena cava, bypasses the lungs. At the same time, it is prevented from entering the portion of the aorta that provides branches leading to the heart and brain.

The more highly oxygenated blood that enters the left atrium through the foramen ovale is mixed with a small amount of deoxygenated blood returning from the pulmonary veins. This mixture moves into the left ventricle and is pumped into the aorta. Some of it reaches the myocardium through the coronary arteries, and some reaches the brain tissues through the carotid arteries.

The blood carried by the descending aorta is partially oxygenated and partially deoxygenated. Some of it is carried into the branches of the aorta that lead to various parts of the lower regions of the body. The rest passes into the *umbilical arteries,* which branch from the internal iliac arteries and lead to the placenta. There the blood is reoxygenated.

Chart 20.2 Fetal circulatory adaptations

Adaptation	Function
Umbilical vein	Carries oxygenated blood from the placenta to the fetus
Ductus venosus	Conducts about half the blood from the umbilical vein directly to the inferior vena cava, thus bypassing the liver
Foramen ovale	Conveys a large proportion of the blood entering the right atrium from the inferior vena cava, through the atrial septum, and into the left atrium, thus bypassing the lungs
Ductus arteriosus	Conducts some blood from the pulmonary artery to the aorta, thus bypassing the lungs
Umbilical arteries	Carry the blood from the internal iliac arteries to the placenta

Chart 20.2 summarizes the major features of fetal circulation. At birth, important adjustments must occur in this circulatory system when the placenta ceases to function and the newborn begins to breathe.

1. Which umbilical vessel carries oxygen-rich blood to the fetus?
2. What is the function of the ductus venosus?
3. How does the fetal circulation allow blood to bypass the lungs?

Birth Process

Pregnancy usually continues for 40 weeks (280 days) or about 9 calendar months (10 lunar months), if it is measured from the beginning of the mother's last menstrual cycle. Pregnancy terminates with the *birth process.*

Although the mechanism that initiates birth is not well understood, a variety of factors seem to be involved. For example, progesterone suppresses uterine contractions during pregnancy. Estrogen, however, tends to excite such contractions. After the seventh month, the placental secretion of estrogen increases to a greater degree than the secretion of progesterone. As a result of the rising estrogen concentration, the contractility of the uterine wall is enhanced.

The stretching of the uterine and vaginal tissues late in pregnancy is thought to initiate nerve impulses to the hypothalamus. The hypothalamus, in turn, signals the posterior pituitary gland, which responds by releasing the hormone **oxytocin** (see chapter 11).

Figure 20.16
Stages in the birth process. (*a*) Fetal position before labor; (*b*) dilation of cervix; (*c*) expulsion of fetus; and (*d*) expulsion of placenta.

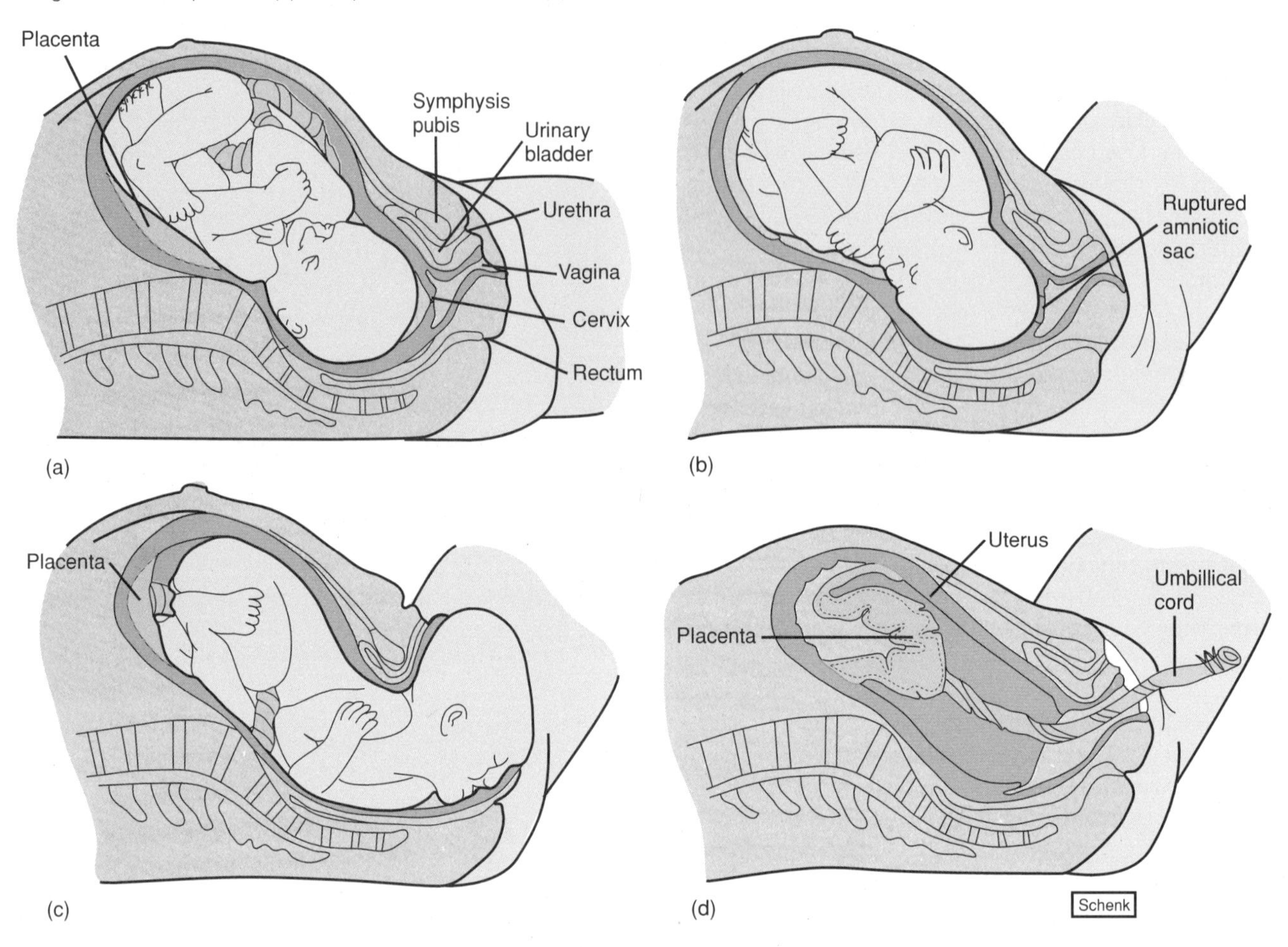

Oxytocin is a powerful stimulator of uterine contractions, and its effect, combined with the greater excitability of the myometrium of the uterus, may be involved in initiating labor.

Labor refers to the process whereby muscular contractions force the fetus through the birth canal. Once labor starts, rhythmic contractions that begin at the top of the uterus and travel down its length force the uterine contents toward the cervix.

Since the fetus is usually positioned with its head downward, labor contractions force the head against the cervix (fig. 20.16). This action causes stretching of the cervix, which is thought to elicit a reflex that stimulates still stronger labor contractions. Thus, a positive feedback system operates (chapter 14), in which uterine contractions produce more intense uterine contractions until a maximum effort is achieved. At the same time, dilation of the cervix reflexly stimulates an increased release of oxytocin from the posterior pituitary gland.

As labor continues, the abdominal wall muscles are stimulated to contract by a positive feedback mechanism, and they also aid in forcing the fetus to the outside.

Following the birth of the fetus (usually within ten to fifteen minutes), the placenta separates from the uterine wall and is expelled by uterine contractions through the birth canal. This expulsion, termed the *afterbirth,* is accompanied by bleeding, because vascular tissues are damaged in the process. This loss of blood is usually minimized, however, by the continued contraction of the uterus, which constricts the bleeding vessels.

1. Describe the events thought to initiate labor.
2. Explain how dilation of the cervix affects labor.

Postnatal Period

Following birth, a variety of physiological and structural changes occur in the mother and the newborn.

Production and Secretion of Milk

During pregnancy, placental estrogen and progesterone stimulate changes in the mammary glands. Estrogen causes the ductile systems to grow and become branched and to have large quantities of fat deposited around them. Progesterone, on the other hand, stimulates the development of the alveolar glands (fig. 19.13). These changes are also promoted by placental lactogen, which was mentioned previously.

As a consequence of hormonal activity, the breasts typically double in size during pregnancy, and the mammary glands become capable of secreting milk. No milk is produced, however, because the high concentrations of estrogen and progesterone that occur during pregnancy inhibit the hypothalamus. The hypothalamus, in turn, suppresses the secretion of the hormone **prolactin** by the anterior pituitary gland (see chapter 11).

Following childbirth and the expulsion of the placenta, the maternal blood concentrations of estrogen and progesterone decline rapidly. Consequently, the hypothalamus is no longer inhibited, and it signals the anterior pituitary gland to release prolactin.

Prolactin stimulates the mammary glands to secrete large quantities of milk. This hormonal effect does not occur for two or three days following birth, and meanwhile, the glands secrete a thin, watery fluid called *colostrum.* Although colostrum is rich in proteins, it contains fewer carbohydrates and fats than milk.

The milk produced under the influence of prolactin does not flow readily through the ductile system of the mammary gland, but must be actively ejected by the contraction of specialized *myoepithelial cells* surrounding the ducts. The contraction of these cells and the consequent ejection of milk through the nipple result from a reflex action.

This reflex is elicited when the breast is suckled or the nipple or areola is otherwise stimulated. When this occurs, sensory impulses travel to the hypothalamus, which signals the posterior pituitary gland to release oxytocin. The oxytocin reaches the breast by means of the blood, and the hormone stimulates the myoepithelial cells of the ductile system to contract. Consequently, milk is ejected into a suckling infant's mouth in about thirty seconds.

As long as milk is removed from the breasts, prolactin and oxytocin continue to be released, and milk continues to be produced. If milk is not removed regularly, the hypothalamus causes the secretion of prolactin to be inhibited, and within about one week, the mammary glands lose their capacity to produce milk.

1. How does pregnancy affect the mammary glands?
2. What stimulates the mammary glands to produce milk?
3. How is milk stimulated to flow into the ductile system of a mammary gland?

Neonatal Period of Development

The **neonatal period,** which extends from birth to the end of the first four weeks, begins abruptly at birth. At that moment, physiological adjustments must be made quickly, because the newborn must suddenly do for itself those things that the mother's body had been doing for it. The newborn must carry on respiration, obtain nutrients, digest nutrients, excrete wastes, regulate body temperature, and so forth. Its most immediate need, however, is to obtain oxygen and excrete carbon dioxide, so its first breath is critical.

The first breath must be particularly forceful because the newborn's lungs are collapsed and the airways are small, offering considerable resistance to air movement. Also, surface tension tends to hold the moist membranes of the lungs together. However, the lungs of a full-term fetus continuously secrete *surfactant* (see chapter 13), which reduces surface tension. Thus, after the first powerful breath, which begins to expand the lungs, breathing becomes easier.

The newborn has a relatively high rate of metabolism, and its liver, which is not fully mature, may be unable to supply enough glucose to support its metabolic needs. Consequently, the newborn typically utilizes stored fat as an energy source.

1. Why must the first breath of an infant be particularly forceful?
2. What does a newborn use for an energy supply during the first few days after birth?

As a rule, the newborn's kidneys are unable to produce concentrated urine, so they excrete a relatively dilute fluid. For this reason, the newborn may become dehydrated and develop a water and electrolyte imbalance. Also, some of the newborn's homeostatic control mechanisms may function imperfectly. The temperature-regulating system, for example, may be unable to maintain a constant body temperature.

When the placenta ceases to function and breathing begins, changes also occur in the circulatory system.

For example, following birth, the umbilical vessels constrict. The arteries close first, and if the umbilical cord is not clamped or severed for a minute or so, blood continues to flow from the placenta to the newborn through the umbilical vein, adding to the newborn's blood volume.

Similarly, the ductus venosus constricts shortly after birth and is represented in the adult as a fibrous cord (ligamentum venosum), which is superficially embedded in the wall of the liver (fig. 20.17).

The foramen ovale closes as a result of the blood pressure changes occurring in the right and left atria as the fetal vessels constrict. More precisely, as blood

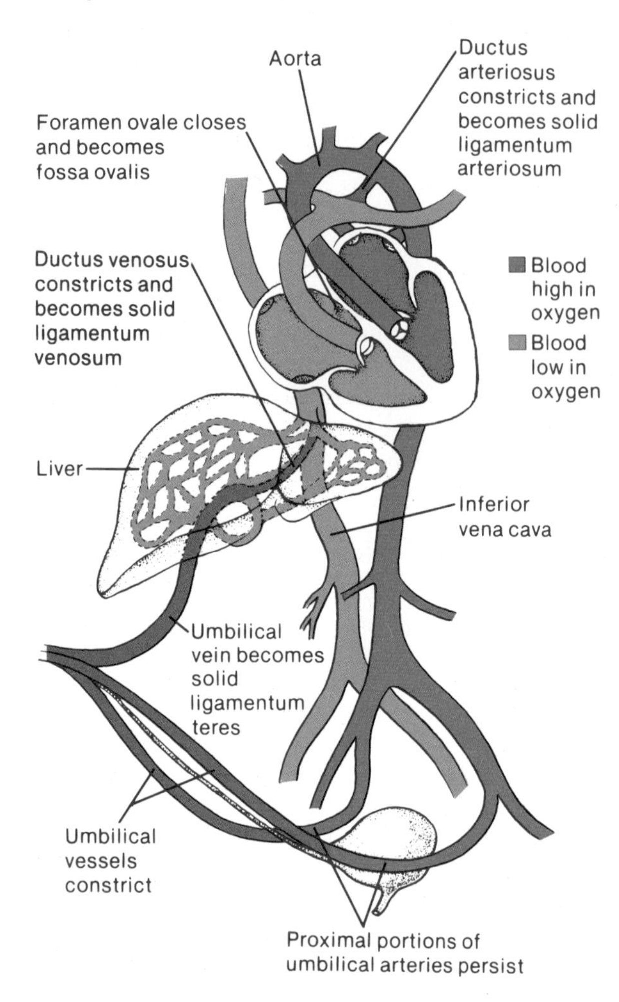

Figure 20.17
Changes that occur in the newborn's circulatory system.

ceases to flow from the umbilical vein into the inferior vena cava, the blood pressure in the right atrium falls. Also, as the lungs expand with the first breathing movements, the resistance to blood flow through the pulmonary circuit decreases, more blood enters the left atrium through the pulmonary veins, and the blood pressure in the left atrium increases.

As the pressure in the left atrium rises and that in the right atrium falls, the valve on the left side of the atrial septum closes the foramen ovale. In most individuals, this valve gradually fuses with the tissues along the margin of the foramen. In an adult, the site of the previous opening is marked by a depression called the *fossa ovalis*.

In some newborns, the foramen ovale remains open, and because the blood pressure in the left atrium is greater than that in the right atrium, some blood may flow from the left chamber to the right chamber. Although this condition usually produces no symptoms, it may cause a heart murmur.

The ductus arteriosus, like the other fetal vessels, constricts after birth. This constriction seems to be stimulated by a substance called *bradykinin*, which is released from the lungs during their initial expansions. Bradykinin is thought to act when the oxygen concentration of the aortic blood rises as a result of breathing. After the ductus arteriosus has closed it also becomes a solid cord (ligamentum arteriosum), and blood can no longer bypass the lungs by moving from the pulmonary artery directly into the aorta.

The changes in the newborn's circulatory system do not occur very rapidly. Although the constriction of the ductus arteriosus may be functionally complete within fifteen minutes, the permanent closure of the foramen ovale may take up to a year.

As mentioned previously, fetal hemoglobin is slightly different and has a greater affinity for oxygen than the adult type. As a rule, however, fetal hemoglobin is not produced after birth, and by the age of four months, most of the circulating hemoglobin is the adult type.

1. How do the kidneys of a newborn differ from those of an adult?
2. What changes occur in the newborn's circulatory system?

Clinical Terms Related to Growth and Development

ablatio placentae (ab-la'she-o plah-cen'tā)—a premature separation of the placenta from the uterine wall.

amniocentesis (am″ne-o-sen-te'sis)—a technique in which a sample of amniotic fluid is withdrawn from the amniotic cavity by inserting a hollow needle through the mother's abdominal wall.

dizygotic twins (di″zi-got'ik twinz)—twins resulting from the fertilization of two ova by two sperm cells.

hydatid mole (hi'dah-tid mōl)—a type of uterine tumor that originates from placental tissue.

hydramnios (hi-dram'ne-os)—the presence of excessive amniotic fluid.

intrauterine transfusion (in″trah-u'ter-in trans-fu'zhun)—a transfusion administered by injecting blood into the fetal peritoneal cavity before birth.

lochia (lo'ke-ah)—vaginal discharge following childbirth.

meconium (mĕ-ko'ne-um)—the first fecal discharge of a newborn.

monozygotic twins (mon″o-zi-got'ik twinz)—twins resulting from the fertilization of one ovum by one sperm cell.

perinatology (per″i-na-tol'o-je)—the branch of medicine concerned with the fetus after twenty-five weeks of development and with the newborn for the first four weeks after birth.

postpartum (pōst-par'tum)—occurring after birth.

teratology (ter″ah-tol′o-je)—the study of abnormal development and congenital malformations.

trimester (tri-mes′ter)—each third of the total period of pregnancy.

ultrasonography (ul″trah-son-og′rah-fe)—a technique used to visualize the size and position of fetal structures by means of ultrasonic sound waves.

Chapter Summary

Introduction (page 514)

Growth refers to an increase in size; development is the process of changing from one life phase to another.

Pregnancy (page 514)

Pregnancy is characterized by the presence of a developing offspring in the uterus.

1. Transport of sex cells
   a. Semen is deposited into the vagina during sexual intercourse.
   b. Sperm movement is aided by the lashing of sperm tails and by the muscular contractions in the uterus and uterine tube.
2. Fertilization
   a. A sperm cell penetrates an egg cell with the aid of an enzyme.
   b. When a sperm cell penetrates an egg cell, the entrance of any other sperm cells is prevented by structural changes that occur in the egg cell membrane and by enzyme actions.
   c. When the nuclei of a sperm and an egg cell fuse, the process of fertilization is complete.
   d. The product of fertilization is a zygote with 46 chromosomes.

Prenatal Period (page 516)

1. Early embryonic development
   a. Cells undergo mitosis, giving rise to smaller and smaller cells during cleavage.
   b. The developing offspring is moved down the uterine tube to the uterus, where it becomes implanted in the endometrium.
   c. The offspring is called an embryo from the second through the eighth week of development; thereafter, it is a fetus.
   d. Eventually, the embryonic and maternal cells together form a placenta.
2. Hormonal changes during pregnancy
   a. Embryonic cells produce HCG, which causes the corpus luteum to be maintained.
   b. Placental tissue produces high concentrations of estrogen and progesterone.
      (1) Estrogen and progesterone maintain the uterine wall and inhibit the secretion of FSH and LH.
      (2) Progesterone and relaxin cause uterine contractions to be suppressed.
      (3) Estrogen causes enlargement of the vagina and relaxation of the ligaments that hold the pelvic joints together.
      (4) Relaxin may also help to soften the cervix and relax the pelvic ligaments.
   c. The placenta secretes placental lactogen, which stimulates the development of the breasts.
   d. During pregnancy, the increasing secretion of aldosterone promotes the retention of sodium and body fluid, and the increasing secretion of parathyroid hormone helps to maintain a high concentration of maternal blood calcium.
3. Embryonic stage of development
   a. The embryonic stage extends from the second through the eighth week.
   b. It is characterized by the development of the placenta and main body structures.
   c. The cells of the inner cell mass become arranged into primary germ layers.
   d. The embryonic disk becomes cylindrical and is attached to the developing placenta.
   e. The placental membrane consists of the epithelium of the chorionic villi and the epithelium of the capillaries inside the villi.
      (1) Oxygen and nutrients diffuse from the maternal blood through the placental membrane and into the fetal blood.
      (2) Carbon dioxide and other wastes diffuse from the fetal blood through the placental membrane and into the maternal blood.
   f. A fluid-filled amnion develops around the embryo.
   g. The umbilical cord is formed as the amnion envelops the tissues attached to the underside of the embryo.
   h. The yolk sac forms on the underside of the embryonic disk.
   i. The allantois extends from the yolk sac into the connecting stalk.
   j. By the beginning of the eighth week, the embryo is recognizable as a human.
4. Fetal stage of development
   a. This stage extends from the end of the eighth week and continues until birth.
   b. Existing structures grow and mature; only a few new parts appear.
   c. A fetus is full term at the end of the tenth lunar month.
   d. Fetal blood and circulation
      (1) Blood is carried between the placenta and fetus by the umbilical vessels.
      (2) Fetal blood carries a greater concentration of oxygen than maternal blood.
      (3) Blood enters the fetus through the umbilical vein and partially bypasses the liver.
      (4) Blood enters the right atrium and partially bypasses the lungs by means of the foramen ovale.

(5) Blood entering the pulmonary artery partially bypasses the lungs by means of the ductus arteriosus.
(6) Blood enters the umbilical arteries by means of the internal iliac arteries.

5. Birth process
   a. During pregnancy, estrogen excites uterine contractions and progesterone inhibits uterine contractions.
   b. A variety of factors are involved in the birth process.
      (1) The secretion of progesterone decreases.
      (2) The posterior pituitary gland releases oxytocin.
      (3) The uterine muscles are stimulated to contract, and labor begins.
   c. Following birth, the placental tissues are expelled.

Postnatal Period (page 526)

1. Production and secretion of milk
   a. Following birth, the maternal anterior pituitary secretes prolactin, and the mammary glands begin to secrete milk.
   b. A reflex response to mechanical stimulation of the nipple signals the release of oxytocin, which causes milk to be ejected from ducts.
2. Neonatal period of development
   a. This period extends from birth to the end of the fourth week.
   b. The newborn must begin to carry on respiration, obtain nutrients, excrete wastes, and regulate its body temperature.
   c. The first breath must be powerful in order to expand the lungs.
   d. The liver is immature and unable to supply sufficient glucose, so the newborn depends primarily on stored fat.
   e. The immature kidneys cannot concentrate urine very well.
   f. Homeostatic mechanisms may function imperfectly.
   g. The circulatory system undergoes changes when placental circulation ceases.
      (1) The umbilical vessels constrict.
      (2) The ductus venosus constricts.
      (3) The foramen ovale is closed by a valve.
      (4) Bradykinin released from the lungs stimulates the constriction of the ductus arteriosus.
      (5) Fetal hemoglobin is replaced by the adult-type of hemoglobin within a few months following birth.

Clinical Application of Knowledge

1. How would you explain the observation that it is sometimes difficult to determine the racial origin of a newborn by its skin color during the first few days following birth?
2. What symptoms might occur in a newborn if its ductus arteriosus fails to close? Explain the reason for your answer.
3. One of the more common congenital cardiac disorders is a ventricular septum defect in which an opening remains between the right and left ventricles. What problem would such a defect create as blood moves through the heart?

Review Activities

1. Distinguish between growth and development.
2. Define *pregnancy*.
3. Describe how male sex cells are transported within the female reproductive tract.
4. Describe the process of fertilization.
5. Define *cleavage*.
6. Define *implantation*.
7. Define *embryo*.
8. Explain the major hormonal changes that occur in the maternal body during pregnancy.
9. Explain how the primary germ layers form.
10. List the major body parts derived from ectoderm, mesoderm, and endoderm.
11. Describe the formation of the placenta, and explain its functions.
12. Define *placental membrane*.
13. Distinguish between the chorion and the amnion.
14. Explain the function of amniotic fluid.
15. Describe the formation of the umbilical cord.
16. Explain how the yolk sac and allantois form.
17. Define *fetus*.
18. List the major changes that occur during the fetal stage of development.
19. Describe a full-term fetus.
20. Compare fetal hemoglobin with maternal hemoglobin.
21. Trace the pathway of blood from the placenta to the fetus and back to the placenta.
22. Discuss the events that occur during the birth process.
23. Explain the roles of prolactin and oxytocin in milk production and secretion.
24. Explain why a newborn's first breath must be particularly forceful.
25. Explain why newborns tend to develop water and electrolyte imbalances.
26. Describe the circulatory changes that occur in the newborn.

Appendix
Some Laboratory Tests of Clinical Importance

Common tests performed on blood		
Test	**Normal values* (adult)**	**Clinical significance**
Albumin (serum)	3.2–5.5 gm/100 ml	Values increase in multiple myeloma and decrease with proteinuria and as a result of severe burns.
Albumin-globulin ratio or A/G ratio (serum)	1.5:1 to 2.5:1	Ratio of albumin to globulin is lowered in kidney diseases and malnutrition.
Ammonia	12–55 μ mol/liter	Values increase in severe liver disease, pneumonia, shock, and congestive heart failure.
Amylase (serum)	4–25 units/ml	Values increase in acute pancreatitis, intestinal obstructions, and mumps. They decrease in chronic pancreatitis, cirrhosis of the liver, and toxemia of pregnancy.
Bilirubin, total (serum)	0–1.0 mg/100 ml	Values increase in conditions causing red blood cell destruction or biliary obstruction.
Blood urea nitrogen or BUN (plasma or serum)	8–25 mg/100 ml	Values increase in various kidney disorders and decrease in liver failure and during pregnancy.
Calcium (serum)	8.5–10.5 mg/100 ml	Values increase in hyperparathyroidism, hypervitaminosis D, and respiratory conditions that cause a rise in CO_2 concentration. They decrease in hypoparathyroidism, malnutrition, and severe diarrhea.
Carbon dioxide (serum)	24–30 mEq/l	Values increase in respiratory diseases, intestinal obstruction, and vomiting. They decrease in acidosis, nephritis, and diarrhea.
Chloride (serum)	100–106 mEq/l	Values increase in nephritis, Cushing's syndrome, dehydration, and hyperventilation. They decrease in metabolic acidosis, Addison's disease, diarrhea, and following severe burns.
Cholesterol, total (serum)	120–220 mg/100 ml (below 200 mg/100 ml recommended by the American Heart Association)	Values increase in diabetes mellitus and hypothyroidism. They decrease in pernicious anemia, hyperthyroidism, and acute infections.
High-density lipoprotein cholesterol (HDL)	Women: 30–80 mg/100 ml Men: 30–70 mg/100 ml	Values increase in liver disease. Decreased values are associated with an increased risk of atherosclerosis.
Low-density lipoprotein cholesterol (LDL)	62–185 mg/100 ml	Increased values are associated with an increased risk of atherosclerosis.
Creatine (serum)	0.2–0.8 mg/100 ml	Values increase in muscular dystrophy, nephritis, severe damage to muscle tissue, and during pregnancy.
Creatinine (serum)	0.6–1.5 mg/100 ml	Values increase in various kidney diseases.

*These values may vary with hospital, physician, and type of equipment used to make measurements.

Test	Normal values* (adult)	Clinical significance
Erythrocyte count or red cell count (whole blood)	Men: 4,600,000–6,200,000/cu mm Women: 4,200,000–5,400,000/cu mm Children: 4,500,000–5,100,000/cu mm (varies with age)	Values increase as a result of severe dehydration or diarrhea and decrease in anemia, leukemia, and following severe hemorrhage.
Ferritin (serum)	Men: 10–270 μg/100 ml Women: 5–280 μg/100 ml	Values correlate with total body iron store. They decrease with iron deficiency.
Globulin (serum)	2.3–3.5 gm/100 ml	Values increase as a result of chronic infections.
Glucose (plasma)	70–110 mg/100 ml	Values increase in diabetes mellitus, liver diseases, nephritis, hyperthyroidism, and pregnancy. They decrease in hyperinsulinism, hypothyroidism, and Addison's disease.
Hematocrit (whole blood)	Men: 40–54 ml/100 ml Women: 37–47 ml/100 ml Children: 35–49 ml/100 ml (varies with age)	Values increase in polycythemia due to dehydration or shock. They decrease in anemia and following severe hemorrhage.
Hemoglobin (whole blood)	Men: 14–18 gm/100 ml Women: 12–16 gm/100 ml Children: 11.2–16.5 gm/100 ml (varies with age)	Values increase in polycythemia, obstructive pulmonary diseases, congestive heart failure, and at high altitudes. They decrease in anemia, pregnancy, and as a result of severe hemorrhage or excessive fluid intake.
Iron (serum)	50–150 μg/100 ml	Values increase in various anemias and liver disease. They decrease in iron deficiency anemia.
Iron-binding capacity (serum)	250–410 μg/100 ml	Values increase in iron deficiency anemia and pregnancy. They decrease in pernicious anemia, liver disease, and chronic infections.
Lactic acid (whole blood)	0.6–1.8 mEq/liter	Values increase with muscular activity and in congestive heart failure, severe hemorrhage, and shock.
Lactic dehydrogenase, or LDH (serum)	45–90 U/liter	Values increase in pernicious anemia, myocardial infarction, liver diseases, acute leukemia, and widespread carcinoma.
Lipids, total (serum)	450–850 mg/100 ml	Values increase in hypothyroidism, diabetes mellitus, and nephritis. They decrease in hyperthyroidism.
Magnesium	1.3–2.1 mEq/liter	Values increase in renal failure, hypothyroidism, and Addison's disease. They decrease in renal disease, liver disease, and pancreatitis.
Mean corpuscular hemoglobin (MCH)	26–32 picograms/RBC	Values increase in macrocytic anemia. They decrease in microcytic anemia.
Mean corpuscular volume (MCV)	86–98 μm^3/RBC	Values increase in liver disease and pernicious anemia. They decrease in iron-deficiency anemia.
Osmolality	275–295 mosm/kg	Values increase in dehydration, hypercalcemia, and diabetes mellitus. They decrease in hyponatremia, Addison's disease, and water intoxication.
Oxygen saturation (whole blood)	Arterial: 94%-100% Venous: 60%-85%	Values increase in polycythemia and decrease in anemia and obstructive pulmonary diseases.
pH (whole blood)	7.35–7.45	Values increase due to vomiting, Cushing's syndrome, and hyperventilation. They decrease as a result of hypoventilation, severe diarrhea, Addison's disease, and diabetic acidosis.
Phosphatase, acid (serum)	Women: 0.01–0.56 Sigma U/ml Men: 0.13–0.63 Sigma U/ml	Values increase in cancer of the prostate gland, hyperparathyroidism, certain liver diseases, myocardial infarction, and pulmonary embolism.

Test	Normal values* (adult)	Clinical significance
Phosphatase, alkaline (serum)	13–39 U/liter	Values increase in hyperparathyroidism (and in other conditions that promote resorption of bone), liver diseases, and pregnancy.
Phosphorus (serum)	3.0–4.5 mg/100 ml	Values increase in kidney diseases, hypoparathyroidism, acromegaly, and hypervitaminosis D. They decrease in hyperparathyroidism.
Platelet count (whole blood)	150,000–350,000/cu mm	Values increase in polycythemia and certain anemias. They decrease in acute leukemia and aplastic anemia.
Potassium (serum)	3.5–5.0 mEq/l	Values increase in Addison's disease, hypoventilation, and conditions that cause severe cellular destruction. They decrease in diarrhea, vomiting, diabetic acidosis, and chronic kidney disease.
Protein, total (serum)	6.0–8.4 gm/100 ml	Values increase in severe dehydration and shock. They decrease in severe malnutrition and hemorrhage.
Prothrombin time (serum)	12–14 sec (one stage)	Values increase in certain hemorrhagic diseases, liver disease, vitamin K deficiency, and following the use of various drugs.
Red cell distribution width (RDW)	8.5–11.5 microns	Variation in cell width changes with pernicious anemia.
Sedimentation rate, Erythrocyte (whole blood)	Men: 1–13 mm/hr Women: 1–20 mm/hr	Values increase in infectious diseases, menstruation, pregnancy, and as a result of severe tissue damage.
Serum glutamic pyruvic transaminase (SGPT)	Women: 4–17 U/liter Men: 6–24 U/liter	Values increase in liver disease, pancreatitis, and acute myocardial infarction.
Sodium (serum)	135–145 mEq/l	Values increase in nephritis and severe dehydration. They decrease in Addison's disease, myxedema, kidney disease, and diarrhea.
Thromboplastin time, partial (plasma)	35–45 sec	Values increase in deficiencies of blood factors VIII, IX, and X.
Thyroid-stimulating hormone (TSH)	0.5–5.0 μgU/ml	Values increase in hypothyroidism and decrease in hyperthyroidism.
Thyroxine or T_4 (serum)	4–12 μg/100 ml	Values increase in hyperthyroidism and pregnancy. They decrease in hypothyroidism.
Transaminases or SGOT (serum)	7–27 units/ml	Values increase in myocardial infarction, liver disease, and diseases of the skeletal muscles.
Triglycerides	40–150 mg/100 ml	Values increase in liver disease, nephrotic syndrome, hypothyroidism, and pancreatitis. They decrease in malnutrition and hyperthyroidism.
Triiodothyronine or T3 (serum)	75–195 ng/100 ml	Values increase in hyperthyroidism and decrease in hypothyroidism.
Uric acid (serum)	Men: 2.5–8.0 mg/100 ml Women: 1.5–6.0 mg/100 ml	Values increase in gout, leukemia, pneumonia, toxemia of pregnancy, and as a result of severe tissue damage.
White blood cell count, differential (whole blood)	Neutrophils 54%-62% Eosinophils 1%-3% Basophils 0–1% Lymphocytes 25%-33% Monocytes 3%-7%	Neutrophils increase in bacterial diseases; lymphocytes and monocytes increase in viral diseases; eosinophils increase in collagen diseases, in allergies, and in the presence of intestinal parasites.
White blood cell count, total (whole blood)	5,000–10,000/cu mm	Values increase in acute infections, acute leukemia, and following menstruation. They decrease in aplastic anemia and as a result of drug toxicity.

Common tests performed on urine		
Test	**Normal values (adult)**	**Clinical significance**
Acetone and acetoacetate	0	Values increase in diabetic acidosis.
Albumin, qualitative	0 to trace	Values increase in kidney disease, hypertension, and heart failure.
Ammonia	20–70 mEq/l	Values increase in diabetes mellitus and liver diseases.
Bacterial count	Under 10,000/ml	Values increase in urinary tract infection.
Bile and bilirubin	0	Values increase in melanoma and biliary tract obstruction.
Calcium	Under 300 mg/24 hr	Values increase in hyperparathyroidism and decrease in hypoparathyroidism.
Creatinine	15–25 mg/kg body weight/day	Values increase in infections and decrease in muscular atrophy, anemia, leukemia, and kidney diseases.
Creatinine clearance (24 hour)	100–140 ml/min	Values increase in renal diseases.
Glucose	0	Values increase in diabetes mellitus and various pituitary gland disorders.
Hemoglobin	0	Blood may occur in urine as a result of extensive burns, crushing injuries, hemolytic anemia, or blood transfusion reactions.
17-hydroxycorticosteroids	3–8 mg/24 hr	Values increase in Cushing's syndrome and decrease in Addison's disease.
Osmolality	850 osmol/kg	Values increase in hepatic cirrhosis, congestive heart failure, and Addison's disease. They decrease in hypokalemia, hypercalcemia, and diabetes insipidus.
pH	4.6–8.0	Values increase in urinary tract infections and chronic renal failure. They decrease in diabetes mellitus, emphysema, and starvation.
Phenylpyruvic acid	0	Values increase in phenylketonuria.
Specific gravity (SG)	1.003–1.035	Values increase in diabetes mellitus, nephrosis, and dehydration. They decrease in diabetes insipidus, glomerulonephritis, and severe renal injury.
Urea	25–35 gm/24 hr	Values increase as a result of excessive protein breakdown. They decrease as a result of impaired renal function.
Urea clearance	Over 40 ml blood cleared of urea/min	Values increase in renal diseases.
Uric acid	0.6–1.0 gm/24 hr as urate	Values increase in gout and decrease in various kidney diseases.
Urobilinogen	0–4 mg/24 hr	Values increase in liver diseases and hemolytic anemia. They decrease in complete biliary obstruction and severe diarrhea.

Glossary

The words in this glossary are followed by a phonetic guide to pronunciation.

In this guide, any unmarked vowel that ends a syllable or stands alone as a syllable is long. Thus, the word *play* would be spelled *pla.*

Any unmarked vowel that is followed by a consonant has the short sound. The word *tough,* for instance, would be spelled *tuf.*

If a long vowel appears in the middle of a syllable (followed by a consonant), then it is marked with the macron (-), the sign for a long vowel. For instance, the word *plate* would be phonetically spelled *plāt.*

Similarly, if a vowel stands alone or ends a syllable, but should have the short sound, it is mark with a breve (˘).

a

abdomen (ab-do′men) That portion of the body between the diaphragm and the pelvis.
abduction (ab-duk′shun) The movement of a body part away from the midline.
absorption (ab-sorp′shun) The taking in of substances by cells or membranes.
accessory organs (ak-ses′o-re or′ganz) Organs that supplement the functions of other organs.
accommodation (ah-kom″o-da′shun) Adjustment of the lens for close vision.
acetylcholine (as″ĕ-til-ko′lēn) A substance secreted at the axon ends of many neurons, which transmits a nerve impulse across a synapse.
acetyl coenzyme A (as′ĕ-til ko-en′zīm) An intermediate compound produced during the oxidation of carbohydrates and fats.
acid (as′id) A substance that ionizes in water to release hydrogen ions.
acidosis (as″ĭ-do′sis) A condition resulting from an excessive accumulation of acid in the body fluids.
ACTH Adrenocorticotropic hormone.
actin (ak′tin) A protein in a muscle fiber that, together with myosin, is responsible for contraction and relaxation.
action potential (ak′shun po-ten′shal) The sequence of electrical changes occurring when a nerve cell membrane is exposed to a stimulus that exceeds its threshold.
active transport (ak′tiv trans′port) A process that requires an expenditure of energy to move a substance across a cell membrane.
adaptation (ad″ap-ta′shun) An adjustment to environmental conditions.
adduction (ah-duk′shun) The movement of a body part toward the midline.
adenosine diphosphate (ah-den′o-sēn di-fos′fāt) ADP; a molecule created when the terminal phosphate is lost from a molecule of adenosine triphosphate.
adenosine triphosphate (ah-den′o-sēn tri-fos′fāt) An organic molecule that stores energy and releases energy for use in cellular processes.
adenylate cyclase (ah-den′i-lāt si′klās) An enzyme that is activated when certain hormones combine with receptors on cell membranes, causing ATP to become cyclic AMP.
ADH Antidiuretic hormone.
adipose tissue (ad′ĭ-pōs tish′u) Fat-storing tissue.
ADP Adenosine diphosphate.
adrenal cortex (ah-dre′nal kor′teks) The outer portion of the adrenal gland.
adrenal glands (ah-dre′nal glandz) The endocrine glands located on the tops of the kidneys.
adrenal medulla (ah-dre′nal me-dul′ah) The inner portion of the adrenal gland.
adrenergic fiber (ad″ren-er′jik fi′ber) A nerve fiber that secretes norepinephrine at the terminal end of its axon.
adrenocorticotropic hormone (ad-re″no-kor″te-ko-trōp′ik hor′mōn) ACTH; a hormone secreted by the anterior lobe of the pituitary gland that stimulates activity in the adrenal cortex.
aerobic respiration (a″er-o′bik res″pi-ra′shun) A phase of cellular respiration that requires the presence of oxygen.
afferent arteriole (af′er-ent ar-te′re-ōl) A vessel that supplies blood to the glomerulus of a nephron.
agglutination (ah-gloo″tĭ-na′shun) The clumping together of blood cells in response to a reaction between an agglutinin and an agglutinogen.
agglutinin (ah-gloo′tĭ-nin) A substance that reacts with an agglutinogen; an antibody.
agglutinogen (ag″loo-tin′o-jen) A substance that stimulates the formation of agglutinins.
agranulocyte (a-gran′u-lo-sīt) A nongranular leukocyte.
albumin (al-bu′min) A plasma protein that helps to regulate the osmotic concentration of the blood.
aldosterone (al′do-ster-ōn″) A hormone, secreted by the adrenal cortex, that functions in regulating sodium and potassium concentrations.
alimentary canal (al″i-men′tar-e kah-nal′) The tubular portion of the digestive tract that leads from the mouth to the anus.
alkaline (al′kah-līn) Pertaining to or having the properties of a base.
alkalosis (al″kah-lo′sis) A condition resulting from an excessive accumulation of base in the body fluids.
allantois (ah-lan′to-is) A structure that appears during embryonic development and functions in the formation of umbilical blood vessels.
allergen (al′er-jen) A foreign substance capable of stimulating an allergic reaction.

all-or-none response (al'or-nun' re-spons') A phenomenon in which a muscle fiber contracts completely when it is exposed to a stimulus of threshold strength.

alveolar ducts (al-ve'o-lar dukts') Fine tubes that carry air to the air sacs of the lungs.

alveolus (al-ve'o-lus) An air sac of a lung; a saclike structure.

amino acid (ah-me'no as'id) An organic compound of a relatively small molecular size that contains an amino group ($-NH_2$) and a carboxyl group ($-COOH$); the structural unit of a protein molecule.

amnion (am'ne-on) An embryonic membrane that encircles a developing fetus and contains amniotic fluid.

amphiarthrosis (am"fe-ar-thro'sis) A joint in which bones are connected by fibrocartilage or ligaments; a slightly movable joint.

ampulla (am-pul'ah) An expansion at the end of each semicircular canal that contains a crista ampullaris.

amylase (am'i-lās) An enzyme that functions to hydrolyze starch.

anabolic metabolism (an"ah-bol'ik mĕ-tab'o-lizm) A metabolic process by which larger molecules are formed from smaller ones; anabolism.

anaerobic respiration (an-a"er-o'bik res"pi-ra'shun) A phase of cellular respiration that occurs in the absence of oxygen.

anaphase (an'ah-fāz) The stage in mitosis during which duplicate chromosomes move to opposite poles of the cell.

anaplasia (an"ah-pla'ze-ah) A change in which mature cells become more primitive; failure to differentiate.

anatomy (ah-nat'o-me) The branch of science dealing with the form and structure of body parts.

androgen (an'dro-jen) A male sex hormone, such as testosterone.

antagonist (an-tag'o-nist) A muscle that acts in opposition to a prime mover.

antebrachium (an"te-bra'ke-um) The forearm.

antecubital (an"te-ku'bĭ-tal) The region in front of the elbow joint.

anterior (an-te're-or) Pertaining to the front; the opposite of posterior.

anterior pituitary (an-te're-or pĭ-tu"ĭ-tār"e) The front lobe of the pituitary gland.

antibody (an'tĭ-bod"e) A specific substance produced by cells in response to the presence of an antigen.

antibody-mediated immunity (an'tĭ-bod"e me'de-āt id ĭmu'nĭ-te) Resistance to disease-causing agents resulting from the production of specific antibodies by B-cells; humoral immunity.

anticodon (an"tĭ-ko'don) A set of three nucleotides of a transfer RNA molecule that bonds with a complementary set of three nucleotides of a messenger RNA molecule.

antidiuretic hormone (an"tĭ-di"u-ret'ik hor'mōn) A hormone released from the posterior lobe of the pituitary gland that enhances the conservation of water.

antigen (an'tĭ-jen) A substance that stimulates cells to produce antibodies.

aorta (a-or'tah) The major systemic artery that receives blood from the left ventricle.

aortic valve (a-or'tik valv) Structure in the wall of the aorta near its origin that prevents blood from returning to the left ventricle of the heart.

aortic sinus (a-or'tik si'nus) A swelling in the wall of the aorta that contains pressoreceptors.

apocrine gland (ap'o-krin gland) A type of sweat gland that responds during periods of emotional stress.

aponeurosis (ap"o-nu-ro'sis) A sheetlike tendon by which certain muscles are attached to other parts.

appendicular (ap"en-dik'u-lar) Pertaining to the arms or legs.

aqueous humor (a'kwe-us hu'mor) The watery fluid that fills the anterior and posterior chambers of the eye.

arachnoid mater (ah-rak'noid ma'ter) The delicate, weblike middle layer of the meninges.

arrector pili muscle (ah-rek'tor pil'i mus'l) A smooth muscle in the skin associated with a hair follicle.

arrhythmia (ah-rith'me-ah) Abnormal heart action characterized by a loss of rhythm.

arteriole (ar-te're-ōl) A small branch of an artery that communicates with a capillary network.

artery (ar'ter-e) A vessel that transports blood away from the heart.

articular cartilage (ar-tik'u-lar kar'tĭ-lij) Hyaline cartilage that covers the ends of bones in synovial joints.

articulation (ar-tik"u-la'shun) The joining together of parts at a joint.

ascending tracts (ah-send'ing trakts) Groups of nerve fibers in the spinal cord that transmit sensory impulses upward to the brain.

ascorbic acid (as-kor'bik as'id) One of the water-soluble vitamins; vitamin C.

assimilation (ah-sim"ĭ-la'shun) The action of changing absorbed substances into forms that differ chemically from those entering.

association area (ah-so"se-a'shun a're-ah) A region of the cerebral cortex related to memory, reasoning, judgment, and emotional feelings.

astrocyte (as'tro-sīt) A type of neuroglial cell that functions to connect neurons to blood vessels.

atmospheric pressure (at"mos-fer'ik presh'ur) The pressure exerted by the weight of the air.

atom (at'om) The smallest particle of an element that has the properties of that element.

atomic number (ah-tom'ik num'ber) A number equal to the number of protons in an atom of an element.

atomic weight (ah-tom'ik wāt) Number approximately equal to the number of protons plus the number of neutrons in an atom of an element.

ATP Adenosine triphosphate.

atrioventricular bundle (a"tre-o-ven-trik'u-lar bun'dl) A group of specialized fibers that conduct impulses from the atrioventricular node to the ventricular muscle of the heart; A-V bundle.

atrioventricular node (a"tre-o-ven-trik'u-lar nōd) A specialized mass of muscle fibers located in the interatrial septum of the heart; A-V node.

atrium (a'tre-um) A chamber of the heart that receives blood from the veins.

atrophy (at'ro-fe) A wasting away or decrease in size of an organ or a tissue.

auditory (aw'di-to"re) Pertaining to the ear or to the sense of hearing.

auditory ossicle (aw'di-to"re os'i-kl) A bone of the middle ear.

auditory tube (aw'di-to"re tūb) The channel that connects the middle ear with the pharynx; the eustachian tube.

auricle (aw'ri-kl) An earlike structure; the portion of the heart that forms the wall of an atrium.

autonomic nervous system (aw"to-nom'ik ner'vus sis'tem) The portion of the nervous system that functions to control the actions of the visceral organs and skin.

A-V bundle (bun'dl) A group of fibers that conducts cardiac impulses from the A-V node to the Purkinje fibers; the bundle of His.

A-V node (nōd) The atrioventricular node.

axial skeleton (ak'se-al skel'ĕ-ton) That portion of the skeleton that supports and protects the organs of the head, neck, and trunk.

axillary (ak'sĭ-ler"e) Pertaining to the armpit.

axon (ak'son) A nerve fiber that conducts a nerve impulse away from a neuron cell body.

b

basal ganglion (ba'sal gang'gle-on) A mass of gray matter located deep within a cerebral hemisphere of the brain.

base (bās) A substance that ionizes in water to release hydroxyl ions (OH^-) or other ions that combine with hydrogen ions.

basement membrane (bās'ment mem'brān) layer of nonliving material that anchors epithelial tissue to underlying connective tissue.

basophil (ba'so-fil) A white blood cell characterized by the presence of cytoplasmic granules that become stained by basophilic dye.

B-cell (sel) A lymphocyte that reacts against foreign substances in the body by producing and secreting antibodies.

beta oxidation (ba'tah ok"sĭ-da'shun) The chemical process by which fatty acids are converted to molecules of acetyl coenzyme A.

bicuspid valve (bi-kus'pid valv) The heart valve located between the left atrium and the left ventricle; the mitral valve.

bile (bīl) A fluid secreted by the liver and stored in the gallbladder.

bilirubin (bil"ĭ-roo'bin) A bile pigment produced as a result of hemoglobin breakdown.

biliverdin (bil"ĭ-ver'din) A bile pigment produced as a result of hemoglobin breakdown.

biotin (bi'o-tin) A water-soluble vitamin; a member of the vitamin B complex.

blastocyst (blas'to-sist) An early stage of embryonic development that consists of a hollow ball of cells.

Bowman's capsule (bo'manz kap'sūl) The proximal portion of a renal tubule that encloses the glomerulus of a nephron.

brachial (bra'ke-al) Pertaining to the arm.

brain stem (brān stem) That portion of the brain that includes the midbrain, pons, and medulla oblongata.

Broca's area (bro'kahz a're-ah) The region of the frontal lobe that coordinates complex muscular actions of the mouth, tongue, and larynx, making speech possible.

bronchial tree (brong'ke-al tre) The bronchi and their branches that function to carry air from the trachea to the alveoli of the lungs.

bronchiole (brong'ke-ōl) A small branch of a bronchus within the lung.

bronchus (brong'kus) A branch of the trachea that leads to a lung.

buccal (buk'al) Pertaining to the mouth and the inner lining of the cheeks.

buffer (buf'er) A substance that can react with a strong acid or base to form a weaker acid or base and, thus, resist a change in pH.

bulbourethral glands (bul"bo-u-re'thral glandz) Glands that secrete a viscous fluid into the male urethra at times of sexual excitement; Cowper's glands.

bursa (bur'sah) A saclike, fluid-filled structure, lined with synovial membrane, that occurs near a joint.

C

calcitonin (kal"sĭ-to'nin) A hormone secreted by the thyroid gland that helps to regulate the concentration of blood calcium.

calorie (kal'o-re) A unit used in the measurement of heat energy and the energy values of foods.

canaliculus (kan"ah-lik'u-lus) Microscopic canal that interconnects the lacunae of bone tissue.

cancellous bone (kan'sĕ-lus bōn) Bone tissue with a lattice-work structure; spongy bone.

capacitation (kah-pas"i-ta'shun) The development by a sperm cell of the ability to fertilize an egg cell.

capillary (kap'ĭ-ler"e) A small blood vessel that connects an arteriole and a venule.

carbaminohemoglobin (kar"bah-me'no-he"mo-glo'bin) The compound formed by the union of carbon dioxide and hemoglobin.

carbohydrate (kar"bo-hi'drāt) An organic compound that contains carbon, hydrogen, and oxygen, with a 2:1 ratio of hydrogen to oxygen atoms.

carbonic anhydrase (kar-bon'ik an-hi'drās) An enzyme that promotes the reaction between carbon dioxide and water to form carbonic acid.

carboxypeptidase (kar-bok"se-pep'ti-dās) A protein-splitting enzyme found in pancreatic juice.

cardiac conduction system (kar'de-ak kon-duk'shun sis'tem) The system of specialized muscle fibers that conducts cardiac impulses from the S-A node into the myocardium.

cardiac cycle (kar'de-ak si'kl) A series of myocardial contractions that constitutes a complete heartbeat.

cardiac muscle (kar'de-ak mus'l) A specialized type of muscle tissue found only in the heart.

cardiac output (kar'de-ak owt'poot) A quantity calculated by multiplying the stroke volume in milliliters by the heart rate in beats per minute.

carpals (kar'pals) Bones of the wrist.

carpus (kar'pus) The wrist; the wrist bones as a group.

cartilage (kar'tĭ-lij) The type of connective tissue in which cells are located within lacunae and are separated by a semisolid matrix.

catabolic metabolism (kat"ah-bol'ik mĕ-tab'o-lism) The metabolic process by which large molecules are broken down into smaller ones; catabolism.

celiac (se'le-ak) Pertaining to the abdomen.

cell (sel) The structural and functional unit of an organism.

cell body (sel bod'e) The portion of a nerve cell that includes a cytoplasmic mass and a nucleus and from which the nerve fibers extend.

cell-mediated immunity (sel me'de-ā"tid ĭ-mu'nĭ-te) The resistance to invasion by foreign cells that is characterized by the direct attack of T-cells.

cellular respiration (sel'u-lar res"pĭ-ra'shun) The process by which energy is released from organic compounds within cells.

cellulose (sel'u-lōs) A polysaccharide that is very abundant in plant tissues, but cannot be digested by human enzymes.

cementum (se-men'tum) A bonelike material that surrounds the root of a tooth.

central canal (sen'tral kah-nal') A tube within the spinal cord that is continuous with the ventricles of the brain and contains cerebrospinal fluid.

central nervous system (sen'tral ner'vus sis'tem) That portion of the nervous system that consists of the brain and spinal cord; CNS.

centriole (sen'tre-ōl) A cellular organelle that functions in the organization of the spindle during mitosis.

centromere (sen'tro-mēr) The portion of a chromosome to which the spindle fiber attaches during mitosis.

centrosome (sen'tro-sōm) A cellular organelle consisting of two centrioles.

cephalic (sĕ-fal'ik) Pertaining to the head.

cerebellar cortex (ser"ĕ-bel'ar kor'teks) The outer layer of the cerebellum.

cerebellum (ser"ĕ-bel'um) The portion of the brain that coordinates skeletal muscle movement.

cerebral cortex (ser'ĕ-bral kor'teks) The outer layer of the cerebrum.

cerebral hemisphere (ser'ĕ-bral hem'ĭ-sfēr) One of the large, paired structures that together constitute the cerebrum of the brain.

cerebrospinal fluid (ser"ĕ-bro-spi'nal floo'id) The fluid that occupies the ventricles of the brain, the subarachnoid space of the meninges, and the central canal of the spinal cord.

cerebrum (ser'ĕ-brum) The portion of the brain that occupies the upper part of the cranial cavity and is concerned with higher mental functions.

cervical (ser'vĭ-kal) Pertaining to the neck or to the cervix of the uterus.

cervix (ser'viks) The narrow, inferior end of the uterus that leads into the vagina.

chemoreceptor (ke"mo-re-sep'tor) A receptor that is stimulated by the presence of certain chemical substances.

chief cell (chēf sel) A cell of a gastric gland that secretes various digestive enzymes, including pepsinogen.

cholecystokinin (ko"le-sis"to-ki'nin) A hormone secreted by the small intestine that stimulates the release of pancreatic juice from the pancreas and bile from the gallbladder.

cholesterol (ko-les'ter-ol) A lipid, produced by the body cells, that is used in the synthesis of steroid hormones.

cholinergic fiber (ko"lin-er'jik fi'ber) A nerve fiber that secretes acetylcholine at the terminal end of its axon.

cholinesterase (ko"lin-es'ter-ās) An enzyme that causes the decomposition of acetylcholine.

chondrocyte (kon'dro-sīt) A cartilage cell.

chorion (ko're-on) The embryonic membrane that forms the outermost covering around a developing fetus.

chorionic villi (ko"re-on'ik vil'i) The projections that extend from the outer surface of the chorion.

choroid coat (ko'roid kōt) The vascular, pigmented middle layer of the wall of the eye.

choroid plexus (ko'roid plek'sus) A mass of specialized capillaries from which cerebrospinal fluid is secreted.

chromatid (kro'mah-tid) A member of a duplicate pair of chromosomes.

chromatin (kro′mah-tin) The nuclear material that gives rise to chromosomes during mitosis.
chromosome (kro′mo-sōm) A rodlike structure that appears in the nucleus of a cell during mitosis.
chylomicron (ki″lo-mi′kron) A microscopic droplet of fat, found in the blood following the digestion of fats.
chyme (kīm) A semifluid mass of food material that passes from the stomach to the small intestine.
chymotrypsin (ki″mo-trip′sin) A protein-splitting enzyme found in pancreatic juice.
cilia (sil′e-ah) Microscopic, hairlike processes on the exposed surfaces of certain epithelial cells.
ciliary body (sil′e-er″e bod′e) A structure, associated with the choroid layer of the eye, that secretes aqueous humor and contains the ciliary muscle.
circadian rhythm (ser″kah-de′an rithm) A pattern of repeated behavior associated with the cycles of night and day.
circle of Willis (ser′kl uv wil′is) An arterial ring located on the ventral surface of the brain.
circular muscles (ser′ku-lar mus′lz) Muscles whose fibers are arranged in circular patterns, usually around an opening or in the wall of a tube.
circumduction (ser″kum-duk′shun) The movement of a body part, such as a limb, so that the end follows a circular path.
cisternae (sis-ter′ne) The enlarged portions of the sarcoplasmic reticulum near the actin and myosin filaments of a muscle fiber.
citric acid cycle (sit′rik as′id si′kl) A series of chemical reactions by which various molecules are oxidized and energy is released from them; Krebs cycle.
cleavage (klēv′ij) The early successive divisions of embryonic cells into smaller and smaller cells.
clitoris (kli′to-ris) A small erectile organ located in the anterior portion of the female vulva, corresponding to the penis of the male.
clone (klōn) A group of identical cells that originated from a single early cell.
CNS Central nervous system.
coagulation (ko-ag″u-la′shun) The clotting of blood.
cochlea (kok′le-ah) The portion of the inner ear that contains the receptors of hearing.
codon (ko′don) A set of three nucleotides of a messenger RNA molecule that is complementary to a set of three nucleotides of a DNA molecule and represents a single amino acid.
coenzyme (ko-en′zīm) A nonprotein substance that is necessary to complete the structure of an enzyme molecule.
cofactor (ko′fak-tor) A nonprotein substance that must be combined with the protein portion of an enzyme before the enzyme can act.
collagen (kol′ah-jen) Protein that occurs in the white fibers of connective tissues.
common bile duct (kom′mon bīl dukt) The tube that transports bile from the cystic duct to the duodenum.
complete protein (kom-plēt′ pro′te-in) A protein that contains adequate amounts of the essential amino acids.
compound (kom′pownd) A substance composed of two or more elements joined by chemical bonds.
condyle (kon′dīl) A rounded process of a bone, usually at the articular end.
cones (kōns) Color receptors located in the retina of the eye.
conjunctiva (kon″junk-ti′vah) A membranous covering on the anterior surface of the eye.
connective tissue (kō-nek′tiv tish′u) One of the basic types of tissue, which includes bone, cartilage, and various fibrous tissues.
convergence (kon-ver′jens) The coming together of nerve impulses from different parts of the nervous system so that they reach the same neuron.
convolution (kon″vo-lu′shun) An elevation on the surface of a structure.
cornea (kor′ne-ah) The transparent anterior portion of the outer layer of the eye wall.
coronary artery (kor′o-na″re ar′ter-e) An artery that supplies blood to the wall of the heart.
coronary sinus (kor′o-na″re si′nus) A large vessel on the posterior surface of the heart into which the cardiac veins drain.
corpus callosum (kor′pus kah-lo′sum) A mass of white matter within the brain, composed of nerve fibers connecting the right and left cerebral hemispheres.
corpus luteum (kor′pus loot′e-um) A structure that forms from the tissues of a ruptured ovarian follicle and functions to secrete female hormones.
cortex (kor′teks) The outer layer of an organ.
cortisol (kor′tī-sol) A glucocorticoid secreted by the adrenal cortex.
costal (kos′tal) Pertaining to the ribs.
covalent bond (ko′va-lent bond) A chemical bond created by the sharing of electrons between atoms.
cranial (kra′ne-al) Pertaining to the cranium.
cranial nerve (kra′ne-al nerv) A nerve that arises from the brain.
crenation (kre-na′shun) Shrinkage of a cell caused by contact with a hypertonic solution.
crest (krest) A ridgelike projection of a bone.
cricoid cartilage (kri′koid kar′tī-lij) A ringlike cartilage that forms the lower end of the larynx.
crista ampullaris (kris′tah am-pul′ar-is) A sensory organ located within a semicircular canal.
cubital (ku′bi-tal) Pertaining to the forearm.
cutaneous (ku-ta′ne-us) Pertaining to the skin.
cystic duct (sis′tik dukt) The tube that connects the gallbladder to the common bile duct.
cytoplasm (si′to-plazm) The contents of a cell surrounding its nucleus.

d

deamination (de-am″ī-na′shun) The chemical process by which amino groups ($-NH_2$) are removed from amino acid molecules.
deciduous teeth (de-sid′u-us tēth) Teeth that are shed and replaced by permanent teeth; primary teeth.
decomposition (de-kom″po-zish′un) The breakdown of molecules into simpler compounds.
defecation (def″ĕ-ka′shun) The discharge of feces from the rectum through the anus.
dehydration synthesis (de″hi-dra′shun sin′thĕ-sis) An anabolic process by which molecules are joined together to form larger molecules.
dendrite (den′drīt) A nerve fiber that transmits impulses toward a neuron cell body.
dentin (den′tin) The bonelike substance that forms the bulk of a tooth.
deoxyhemoglobin (de-ok″se-he″mo-glo′bin) Hemoglobin that lacks oxygen.
depolarization (de-po″lar-ī-za′shun) The loss of an electrical charge on the surface of a membrane.
dermis (der′mis) The thick layer of the skin beneath the epidermis.
descending tracts (de-send′ing trakts) Groups of nerve fibers that carry nerve impulses downward from the brain through the spinal cord.
detrusor muscle (de-trūz′or mus′l) The muscular wall of the urinary bladder.
diapedesis (di″ah pē de′sis) The process by which leukocytes squeeze between the cells of blood vessel walls.
diaphragm (di′ah-fram) A sheetlike structure that separates the thoracic and abdominal cavities.
diaphysis (di-af′ĭ-sis) The shaft of a long bone.
diarthrosis (di″ar-thro′sis) A freely movable joint.
diastole (di-as′tol-le) The phase of the cardiac cycle during which a heart chamber wall is relaxed.
diastolic pressure (di-a-stol′ik presh′ur) The arterial blood pressure during the diastolic phase of the cardiac cycle.
diencephalon (di″en-sef′ah-lon) A portion of the brain in the region of the third ventricle that includes the thalamus and hypothalamus.
differentiation (dif″er-en″she-a′shun) The process by which cells become structurally and functionally specialized.

diffusion (dī-fu'zhun) The random movement of molecules from a region of higher concentration toward one of lower concentration.
digestion (di-jes'chun) The process by which larger molecules of food substances are broken down into smaller molecules that can be absorbed; hydrolysis.
dipeptide (di-pep'tīd) A molecule composed of two amino acids joined together.
disaccharide (di-sak'ah-rīd) The sugar produced by the union of two monosaccharide molecules.
distal (dis'tal) Farther from the midline or origin; the opposite of proximal.
divergence (di-ver'jens) The arrangement of nerve fibers by which a nerve impulse can spread to increasing numbers of neurons within a neuronal pool.
DNA Deoxyribonucleic acid.
dopa (do'pah) A substance formed from tyrosine in the synthesis of epinephrine.
dopamine (do'pah-mēn) A substance formed from dopa in the synthesis of epinephrine.
dorsal root (dor'sal root) The sensory branch of a spinal nerve by which it joins the spinal cord.
dorsal root ganglion (dor'sal root gang'gle-on) A mass of sensory neuron cell bodies located in the dorsal root of a spinal nerve.
dorsum (dors'um) Pertaining to the back surface of a body part.
ductus arteriosus (duk'tus ar-te"re-o'sus) The blood vessel that connects the pulmonary artery and the aorta in a fetus.
ductus venosus (duk'tus ven-o'sus) The blood vessel that connects the umbilical vein and the inferior vena cava in a fetus.
dura mater (du'rah ma'ter) The tough outer layer of the meninges.
dynamic equilibrium (di-nam'ik e"kwi-lib're-um) The maintenance of balance when the head and body are suddenly moved or rotated.

e

eccrine gland (ek'rin gland) Sweat gland that functions in the maintenance of body temperature.
ECG An electrocardiogram; EKG.
ectoderm (ek'to-derm) The outermost layer of the primary germ layers, responsible for forming certain embryonic body parts.
edema (ē-de'mah) An excessive accumulation of fluid within the tissue spaces.
effector (ē-fek'tor) An organ, such as a muscle or gland, that responds to stimulation.
efferent arteriole (ef'er-ent ar-te're-ol) An arteriole that conducts blood away from the glomerulus of a nephron.
ejaculation (e-jak"u-la'shun) The discharge of sperm-containing semen from the male urethra.
elastin (e-las'tin) A protein that comprises the yellow, elastic fibers of connective tissue.
electrocardiogram (el-lek"tro-kar'de-o-gram") A recording of the electrical activity associated with the heartbeat; an ECG or EKG.
electrolyte (e-lek'tro-līt) A substance that ionizes in a water solution.
electrolyte balance (e-lek'tro-līt bal'ans) The condition that exists when the quantities of electrolytes entering the body equal those leaving it.
electron (e-lek'tron) A small, negatively charged particle that revolves around the nucleus of an atom.
electrovalent bond (e-lek"tro-va'lent bond) The chemical bond formed between two ions as a result of the transfer of electrons.
element (el'ē-ment) A basic chemical substance.
embolus (em'bo-lus) A substance, such as a blood clot or bubble of gas, that is carried by the blood and obstructs a blood vessel.
embryo (em'bre-o) An organism in its earliest stages of development.
emission (e-mish'un) The movement of sperm cells from the vas deferens into the ejaculatory duct and urethra.
emulsification (e-mul"sī-fi-ka'shun) The process by which fat globules are caused to break up into smaller droplets.
enamel (e-nam'el) The hard covering on the exposed surface of a tooth.
endocardium (en"do-kar'de-um) The inner lining of the heart chambers.
endochondral bone (en"do-kon'dral bōn) Bone that begins as hyaline cartilage that is subsequently replaced by bone tissue.
endocrine gland (en'do-krin gland) A gland that secretes hormones directly into the blood or body fluids.
endocytosis (en"do-si-to'sis) The process by which substances move through a cell membrane and involves the formation of tiny cytoplasmic vacuoles.
endoderm (en'do-derm) The innermost layer of the primary germ layers responsible for forming certain embryonic body parts.
endometrium (en"do-me'tre-um) The inner lining of the uterus.
endoplasmic reticulum (en-do-plaz'mic rē-tik'-u-lum) A cytoplasmic organelle composed of a system of interconnected membranous tubules and vesicles.
endothelium (en"do-the'le-um) The layer of epithelial cells that forms the inner lining of the blood vessels and heart chambers.
energy (en'er-je) An ability to cause something to move and, thus, to do work.
enterogastrone (en"ter-o-gas'trōn) A hormone secreted from the intestinal wall that inhibits gastric secretion and motility.
enzyme (en'zīm) A protein that is synthesized by a cell and acts as a catalyst in a specific cellular reaction.
eosinophil (e"o-sin'o-fil) A white blood cell characterized by the presence of cytoplasmic granules that become stained by acidic dye.
ependyma (ē-pen'dī-mah) Neuroglial cells that line the ventricles of the brain.
epicardium (ep"ī-kar'de-um) The visceral portion of the pericardium.
epicondyle (ep"ī-kon'dīl) A projection of a bone located above a condyle.
epidermis (ep"ī-der'mis) The outer epithelial layer of the skin.
epididymis (ep"ī-did'ī-mis) The highly coiled tubule that leads from the seminiferous tubules of the testis to the vas deferens.
epidural space (ep"ī-du'ral spās) The space between the dural sheath of the spinal cord and the bone of the vertebral canal.
epigastric region (ep"ī-gas'trik re'jun) The upper middle portion of the abdomen.
epiglottis (ep"ī-glot'is) A flaplike cartilaginous structure located at the back of the tongue near the entrance to the trachea.
epinephrine (ep"ī-nef'rin) A hormone secreted by the adrenal medulla during times of stress.
epiphyseal disk (ep"ī-fiz'e-al disk) A cartilaginous layer within the epiphysis of a long bone that functions as a growing region.
epiphysis (ē-pif'ī-sis) The end of a long bone.
epithelium (ep"ī-the'le-um) The type of tissue that covers all free body surfaces.
equilibrium (e"kwī-lib're-um) A state of balance between two opposing forces.
erythroblast (ē-rith'ro-blast) An immature red blood cell.
erythrocyte (ē-rith'ro-sīt) A red blood cell.
erythropoiesis (ē-rith"ro-poi-e'sis) Red blood cell formation.
erythropoietin (ē-rith"ro-poi'ē-tin) A substance released by the kidneys and liver that promotes red blood cell formation.
esophagus (ē-sof'ah-gus) The tubular portion of the digestive tract that leads from the pharynx to the stomach.
essential amino acid (ē-sen'shal ah-me'no as'id) An amino acid required for health that cannot be synthesized in adequate amounts by the body cells.
essential fatty acid (ē-sen'shal fat'e as'id) A fatty acid required for health that cannot be synthesized in adequate amounts by the body cells.
estrogen (es'tro-jen) A hormone that stimulates the development of female secondary sexual characteristics.
evaporation (e"vap'o-ra-shun) The process by which a liquid changes into a gas.
eversion (e-ver'zhun) The movement in which the sole of the foot is turned outward.

excretion (ek-skre′shun) The process by which metabolic wastes are eliminated.

exocrine gland (ek′so-krin gland) A gland that secretes its products into a duct or onto a body surface.

expiration (ek″spi̅-ra′shun) The process of expelling air from the lungs.

extension (ek-sten′shun) The movement by which the angle between the parts at a joint is increased.

extracellular (ek″strah-sel′u-lar) Outside of cells.

extrapyramidal tract (ek″strah-pi̅-ram′i-dal trakt) The nerve tracts, other than the corticospinal tracts, that transmit impulses from the cerebral cortex into the spinal cord.

extremity (ek-strem′i-te) A limb; an arm or a leg.

f

facet (fas′et) A small, flattened surface of a bone.

facilitated diffusion (fah-sil″i-tat′ed di̅-fu′zhun) The movement of molecules from a region of higher concentration toward one of lower concentration that involves special carrier molecules within a cell membrane.

facilitation (fah-sil″i-ta′shun) The process by which a neuron becomes more excitable as a result of incoming stimulation.

fallopian tube (fah-lo′pe-an tu̅b) The tube that transports an egg cell from the region of the ovary to the uterus; an oviduct or a uterine tube.

fascia (fash′e-ah) A sheet of fibrous connective tissue that encloses a muscle.

fat (fat) Adipose tissue; or an organic substance whose molecules contain glycerol and fatty acids.

fatty acid (fat′e as′id) An organic substance that serves as a building block for a fat molecule.

feces (fe′se̅z) Material expelled from the digestive tract during defecation.

fertilization (fer″ti-li-za′shun) The union of an egg cell and a sperm cell.

fetus (fe′tus) A human embryo after eight weeks of development.

fibril (fi′bril) A tiny fiber or filament.

fibrin (fi′brin) An insoluble, fibrous protein formed from fibrinogen during blood coagulation.

fibrinogen (fi-brin′o-jen) The plasma protein that is converted into fibrin during blood coagulation.

fibroblast (fi′bro-blast) A cell that functions to produce fibers and other intercellular materials in connective tissues.

filtration (fil-tra′shun) The movement of a material through a membrane as a result of hydrostatic pressure.

fissure (fish′ur) A narrow cleft separating parts, such as the lobes of the cerebrum.

flexion (flek′shun) Bending at a joint so that the angle between bones is decreased.

follicle (fol′i-kl) A pouchlike depression or cavity.

follicle-stimulating hormone (fol′i-kl stim′u-la″ting hor′mo̅n) A substance secreted by the anterior pituitary gland that stimulates the development of an ovarian follicle in a female or the production of sperm cells in a male; FSH.

follicular cells (fo̅-lik′u-lar selz) Ovarian cells that surround a developing egg cell and secrete female sex hormones.

fontanel (fon″tah-nel′) A membranous region located between certain cranial bones in the skull of a fetus.

foramen (fo-ra′men) An opening, usually in a bone or membrane (plural, *foramina*).

foramen magnum (fo-ra′men mag′num) The opening in the occipital bone of the skull through which the spinal cord passes.

foramen ovale (fo-ra′men o-val′e) The opening in the interatrial septum of the fetal heart.

formula (fo̅r′mu-lah) A group of symbols and numbers used to express the composition of a compound.

fossa (fos′ah) A depression in a bone or other part.

fovea (fo′ve-ah) A tiny pit or depression.

fovea centralis (fo′ve-ah sen-tral′is) The region of the retina consisting of densely packed cones.

frontal (frun′tal) Pertaining to the forehead.

FSH Follicle-stimulating hormone.

g

gallbladder (gawl′blad-er) A saclike organ, associated with the liver, that stores and concentrates bile.

ganglion (gang′gle-on) A mass of neuron cell bodies, usually outside the central nervous system.

gastric gland (gas′trik gland) A gland within the stomach wall that secretes gastric juice.

gastric juice (gas′trik jo̅o̅s) The secretion of the gastric glands within the stomach.

gastrin (gas′trin) A hormone secreted by the stomach lining that stimulates the secretion of gastric juice.

gene (je̅n) The portion of a DNA molecule that contains the information needed to synthesize an enzyme.

genetic code (je̅-net′ik ko̅d) The system by which information for synthesizing proteins is built into the structure of DNA molecules.

germinal epithelium (jer′mi̅-nal ep″i-the′le-um) A layer of cuboidal epithelial cells on the free surface of an ovary.

germ layers (jerm la′ers) Layers of cells within an embryo that form the body organs during development; ectoderm, mesoderm, and endoderm.

gland (gland) A cell or group of cells that secrete a product.

globin (glo′bin) The protein portion of a hemoglobin molecule.

globulin (glob′u-lin) A type of protein that occurs in blood plasma.

glomerulus (glo-mer′u-lus) A capillary tuft located within the Bowman's capsule of a nephron.

glottis (glot′is) A slitlike opening between the true vocal folds.

glucagon (gloo′kah-gon) The hormone secreted by the pancreatic islets of Langerhans that causes the release of glucose from glycogen.

glucocorticoid (gloo″ko-kor′ti̅-koid) Any one of a group of hormones secreted by the adrenal cortex that influences carbohydrate, fat, and protein metabolism.

glucose (gloo′ko̅s) A monosaccharide, found in the blood, which serves as the primary source of cellular energy.

gluteal (gloo′te-al) Pertaining to the buttocks.

glycerol (glis′er-ol) An organic compound that serves as a building block for fat molecules.

glycogen (gli′ko-jen) A polysaccharide that functions to store glucose in the liver and muscles.

glycoprotein (gli″ko-pro′te-in) A substance composed of a carbohydrate combined with a protein.

goblet cell (gob′let sel) An epithelial cell that is specialized to secrete mucus.

Golgi apparatus (gol′je ap″ah-ra′tus) A cytoplasmic organelle that functions in preparing cellular products for secretion.

gonadotropin (go-nad″o-tro̅p′in) A hormone that stimulates activity in the gonads.

granulocyte (gran′u-lo-si̅t) A leukocyte that contains granules in its cytoplasm.

gray matter (gra mat′er) A region of the central nervous system that generally lacks myelin.

groin (groin) The region of the body between the abdomen and thighs.

growth (gro̅th) The process by which a structure enlarges.

growth hormone (gro̅th hor′mo̅n) A hormone released by the anterior lobe of the pituitary gland that promotes growth.

h

hair follicle (ha̅r fol′i-kl) A tubelike depression in the skin in which a hair develops.

head (hed) An enlargement on the end of a bone.

hematocrit (he-mat′o-krit) The percentage by volume of red blood cells in a blood sample.

hematoma (he″mah-to′mah) A mass of coagulated blood within the tissues or a body cavity.

hematopoiesis (hem″ah-to-poi-e′sis) The production of blood and blood cells; hemopoiesis.

heme (hem) The iron-containing portion of a hemoglobin molecule.

hemoglobin (he″mo-glo′bin) A pigment of red blood cells responsible for the transport of oxygen.

hemorrhage (hem′o-rij) The loss of blood from the circulatory system; bleeding.
hemostasis (he″mo-sta′sis) The stoppage of bleeding.
hepatic (hē-pat′ik) Pertaining to the liver.
hepatic lobule (hē-pat′ik lob′ul) A functional unit of the liver.
homeostasis (ho″me-o-sta′sis) A state of equilibrium in which the internal environment of the body remains relatively constant.
hormone (hor′mōn) A substance secreted by an endocrine gland, which is transmitted in the blood or body fluids.
humoral immunity (hu′mor-al ĭ-mu′nĭ-te) Resistance to the effects of specific disease-causing agents due to the presence of circulating antibodies.
hydrolysis (hi-drol′ĭ-sis) The splitting of a molecule into smaller portions by the addition of a water molecule.
hydrostatic pressure (hi″dro-stat′ik presh′ur) The pressure exerted by a fluid.
hydroxyl ion (hi-drok′sil i′on) OH^-.
hymen (hi′men) A membranous fold of tissue that partially covers the vaginal opening.
hyperplasia (hi″per-pla′ze-ah) An increased production and growth of new cells.
hypertonic (hi″per-ton′ik) A condition in which a solution contains a greater concentration of dissolved particles than the solution with which it is compared.
hypertrophy (hi″per′tro-fe) The enlargement of an organ or tissue.
hyperventilation (hi″per-ven″tĭ-la′shun) Breathing that is abnormally deep and prolonged.
hypochondriac region (hi″po-kon′dre-ak re′jun) The portion of the abdomen on either side of the middle, or epigastric, region.
hypogastric region (hi″po-gas′trik re′jun) The lower middle portion of the abdomen.
hypothalamus (hi″po-thal′ah-mus) A portion of the brain located below the thalamus.
hypotonic (hi″po-ton′ik) A condition in which a solution contains a lesser concentration of dissolved particles than the solution to which it is compared.

i

iliac region (il′e-ak re′jun) The portion of the abdomen on either side of the lower middle, or hypogastric region.
ilium (il′e-um) One of the bones of a coxal bone or hipbone.
immunity (i-mu′nĭ-te) Resistance to the effects of specific disease-causing agents.
immunoglobulin (im″u-no-glob′u-lin) Globular plasma proteins that function as antibodies of immunity.
implantation (im″plan-ta′shun) The embedding of an embryo in the lining of the uterus.
impulse (im′puls) A wave of depolarization conducted along a nerve fiber or a muscle fiber.
incomplete protein (in″kom-plēt′ pro′te-in) A protein that lacks essential amino acids.
inferior (in-fer′e-or) Situated below something else; pertaining to the lower surface of a part.
inflammation (in″flah-ma′shun) A tissue response to stress that is characterized by the dilation of blood vessels and an accumulation of fluid in the affected region.
inguinal (ing′gwĭ-nal) Pertaining to the groin region.
inorganic (in″or-gan′ik) Pertaining to chemical substances that lack carbon.
insertion (in-ser′shun) The end of a muscle that is attached to a movable part.
inspiration (in″spĭ-ra′shun) The act of breathing in; inhalation.
insula (in′su-lah) A cerebral lobe located deep within the lateral sulcus.
insulin (in′su-lin) A hormone secreted by the pancreatic islets of Langerhans, which functions in the control of carbohydrate metabolism.
integumentary (in-teg-u-men′tar-e) Pertaining to the skin and its accessory organs.
intercalated disk (in-ter″kah-lāt′ed disk) The membranous boundary between adjacent cardiac muscle cells.
intercellular (in″ter-sel′u-lar) Between cells.
intercellular fluid (in″ter-sel′u-lar floo′id) Tissue fluid located between cells other than blood cells.
interneuron (in″ter-nu′ron) A neuron located between a sensory neuron and a motor neuron; also called intercalated, internuncial, or association neuron.
interphase (in′ter-fāz) The period between two cell divisions.
interstitial cell (in″ter-stish′al sel) A hormone-secreting cell located between the seminiferous tubules of the testis.
interstitial fluid (in″ter-stish′al floo′id) The same as intercellular fluid.
intervertebral disk (in″ter-ver′tē-bral disk) A layer of fibrocartilage located between the bodies of adjacent vertebrae.
intestinal gland (in-tes′tĭ-nal gland) A tubular gland located at the base of a villus within the intestinal wall.
intestinal juice (in-tes′tĭ-nal jo̅o̅s) The secretion of the intestinal glands.
intracellular (in″trah-sel′u-lar) Within cells.
intracellular fluid (in″trah-sel′u-lar floo′id) Fluid within cells.
intramembranous bone (in″trah-mem′brah-nus bōn) Bone that develops from layers of membranous connective tissue.
intrinsic factor (in-trin′sik fak′tor) A substance produced by the gastric glands that promotes the absorption of vitamin B_{12}.
inversion (in-ver′zhun) A movement in which the sole of the foot is turned inward.
involuntary (in-vol′un-tar″e) Not consciously controlled; functions automatically.
ion (i′on) An atom or a group of atoms with an electrical charge.
ionization (i″on-ĭ-za′shun) The chemical process by which substances dissociate into ions.
iris (i′ris) The colored muscular portion of the eye, which surrounds the pupil.
irritability (ir″ĭ-tah-bil′ ĭ-te) The ability of an organism to react to changes taking place in its environment.
ischemia (is-ke′me-ah) A deficiency of blood in a body part.
isotonic solution (i″so-ton′ik so-lu′shun) A solution that has the same concentration of dissolved particles as the solution with which it is compared.
isotope (i′so-tōp) An atom that has the same number of protons as other atoms of an element but has a different number of neutrons in its nucleus.

j

joint (joint) The union of two or more bones; an articulation.
juxtaglomerular apparatus (juks″tah-glo-mer′u-lar ap″ah-ra′tus) A structure located in the walls of arterioles near the glomerulus that plays an important role in regulating renal blood flow.

k

keratin (ker′ah-tin) A protein present in the epidermis, hair, and nails.
keratinization (ker″ah-tin″ĭ-za′shun) The process by which cells form fibrils of keratin.
ketone body (ke′tōn bod′e) A type of compound produced during fat catabolism.
kinase (ki′nās) An enzyme that transfers phosphate groups from ATP molecules to various proteins.
Kupffer cell (koop′fer sel) A large, fixed phagocyte in the liver, which removes bacterial cells from the blood.

l

labor (la′bor) The process of childbirth.
labyrinth (lab′ĭ-rinth) The system of interconnecting tubes within the inner ear.
lacrimal gland (lak′rĭ-mal gland) A tear-secreting gland.
lactation (lak-ta′shun) The production of milk by the mammary glands.
lacteal (lak′te-al) A lymphatic vessel associated with a villus of the small intestine.

lactic acid (lak′tik as′id) An organic substance formed from pyruvic acid during anaerobic respiration.

lacuna (lah-ku′nah) A hollow cavity.

laryngopharynx (lah-ring″go-far′ingks) The lower portion of the pharynx near the opening to the larynx.

larynx (lar′ingks) A structure, located between the pharynx and trachea, that houses the vocal cords.

lateral (lat′er-al) Pertaining to the side.

leukocyte (lu′ko-sīt) A white blood cell.

lever (lev′er) A simple mechanical device consisting of a rod, fulcrum, weight, and a source of energy.

ligament (lig′ah-ment) A cord or sheet of connective tissue by which two or more bones are bound together at a joint.

limbic system (lim′bik sis′tem) A group of interconnected structures within the brain that functions to produce various emotional feelings.

lingual (ling′gwal) Pertaining to the tongue.

lipase (li′pās) A fat-digesting enzyme.

lipid (lip′id) A fat, oil, or fatlike compound.

lipoprotein (lip″o-pro′te-in) A complex of lipid and protein

lumbar (lum′bar) Pertaining to the loins.

lumen (lu′men) The space within a tubular structure.

luteinizing hormone (lu′te-in-īz″ing hor′mōn) Hormone secreted by the anterior pituitary gland that controls the formation of the corpus luteum in females and the secretion of testosterone in males; LH.

lymph (limf) The fluid transported by the lymphatic vessels.

lymph node (limf nōd) A mass of lymphoid tissue located along the course of a lymphatic vessel.

lymphocyte (lim′fo-sīt) A type of white blood cell that plays important roles in immunity.

lymphokine (lim′fo-kīn) A hormonelike substance secreted by T-helper cells that stimulates B-cells to proliferate.

lysosome (li′so-sōm) A cytoplasmic organelle that contains digestive enzymes.

m

macrophage (mak′ro-fāj) A large phagocytic cell.

macroscopic (mak″ro-skop′ik) Large enough to be seen with the unaided eye.

macula lutea (mak′u-lah lu′te-ah) A yellowish depression in the retina of the eye.

malignant (mah-lig′nant) The power to threaten life; cancerous.

mammary (mam′ar-e) Pertaining to the breast.

marrow (mar′o) Connective tissue that occupies the spaces within bones.

mast cell (mast sel) A cell to which antibodies, formed in response to allergens, become attached.

mastication (mas″ti-ka′shun) Chewing movements.

matrix (ma′triks) The intercellular substance of connective tissue.

matter (mat′er) Anything that has weight and occupies space.

meatus (me-a′tus) A passageway or channel, or the external opening of such a passageway.

mechanoreceptor (mek″ah-no-re-sep′tor) A sensory receptor that is sensitive to mechanical stimulation.

medial (me′de-al) Toward or near the midline.

mediastinum (me″de-ah-sti′num) The tissues and organs of the thoracic cavity that form a septum between the lungs.

medulla (mē-dul′ah) The inner portion of an organ.

medulla oblongata (mē-dul′ah ob″long-gah′tah) The portion of the brain stem located between the pons and the spinal cord.

medullary cavity (med′u-lār″e kav′ĭ-te) A cavity within the diaphysis of a long bone occupied by marrow.

meiosis (mi-o′sis) The process of cell division by which egg and sperm cells are formed.

melanin (mel′ah-nin) A dark pigment normally found in the skin and hair.

melanocyte (mel′ah-no-sīt″) Melanin-producing cell.

melatonin (mel″ah-to′in) A hormone secreted by the pineal gland.

memory cell (mem′o-re sel) A B- or T-cell, produced in response to a primary immune response, that remains dormant and can respond rapidly if the same antigen is encountered in the future.

meninges (mē-nin′jēz) A group of three membranes that covers the brain and spinal cord (singular, *meninx*).

menisci (men-is′si) Pieces of fibrocartilage that separate the articulating surfaces of bones in the knee.

menopause (men′o-pawz) The termination of menstrual cycles.

menstrual cycle (men′stroo-al si′kl) The female reproductive cycle that is characterized by regularly reoccurring changes in the uterine lining.

menstruation (men″stroo-a′shun) The loss of blood and tissue from the uterine lining at the end of a female reproductive cycle.

mesentery (mes′en-ter″e) A fold of peritoneal membrane that attaches an abdominal organ to the abdominal wall.

mesoderm (mez′o-derm) The middle layer of the primary germ layers, responsible for forming certain embryonic body parts.

messenger RNA (mes′in-jer) A molecule of RNA that transmits information for protein synthesis from the nucleus of a cell to the cytoplasm.

metabolic rate (met″ah-bol′ic rāt) The rate at which chemical changes occur within the body.

metabolism (mē-tab′o-lizm) The totality of chemical changes that occur within cells.

metacarpals (met″ah-kar′pals) The bones of the hand between the wrist and finger bones.

metaphase (met′ah-fāz) The stage in mitosis when the chromosomes become aligned in the middle of the spindle.

metatarsals (met″ah-tar′sals) The bones of the foot between the ankle and toe bones.

microfilament (mi″kro-fil′ah-ment) A tiny rod of protein that occurs in cytoplasm and functions in causing various cellular movements.

microglia (mi-krog′le-ah) A type of neuroglial cell that helps support neurons and acts to carry on phagocytosis.

microscopic (mi″kro-skop′ik) Too small to be seen by the unaided eye.

microtubule (mi″kro-tu′būl) A minute, hollow rod found in the cytoplasm of cells.

microvilli (mi″kro-vil′i) Tiny, cylindrical processes that extend outward from some epithelial cell membranes and increase the membrane surface area.

micturition (mik″tu-rish′un) Urination.

midbrain (mid′brān) A small region of the brain stem located between the diencephalon and pons.

mineralocorticoid (min″er-al-o-kor′tĭ-koid) Any one of a group of hormones secreted by the adrenal cortex that influences the concentrations of electrolytes in the body fluids.

mitochondrion (mi″to-kon′dre-on) A cytoplasmic organelle that contains enzymes responsible for aerobic respiration (plural, *mitochondria*).

mitosis (mi-to′sis) The process by which body cells divide to form two identical cells.

mixed nerve (mikst nerv) A nerve that includes both sensory and motor nerve fibers.

molecular formula (mo-lek′u-lar fōr′mu-lah) A representation of a molecule using symbols for elements and numbers to indicate how many atoms of each element are present.

molecule (mol′ē-kūl) A particle composed of two or more atoms bound together.

monocyte (mon′o-sīt) A type of white blood cell that functions as a phagocyte.

monosaccharide (mon″o-sak′ah-rīd) A simple sugar, such as glucose or fructose, that represents the structural unit of a carbohydrate.

motor area (mo′tor a′re-ah) A region of the brain from which impulses to muscles or glands originate.

motor end plate (mo′tor end plāt) A specialized region of a muscle fiber where it is joined by the end of a motor nerve fiber.

motor nerve (mo′tor nerv) A nerve that consists of motor nerve fibers.

motor neuron (mo′tor nu′ron) A neuron that transmits impulses from the central nervous system to an effector.
motor unit (mo′tor u′nit) A motor neuron and the muscle fibers associated with it.
mucosa (mu-ko′sah) The membrane that lines the tubes and body cavities that open to the outside of the body; the mucous membrane.
mucous cell (mu′kus sel) A glandular cell that secretes mucus.
mucous membrane (mu′kus mem′brān) The mucosa.
mucus (mu′kus) A fluid secretion of the mucous cells.
mutagenic (mu′tah-jen′ik) Pertaining to a factor that can cause mutations.
mutation (mu-ta′shun) A change in the genetic information of a chromosome.
myelin (mi′ĕ-lin) The fatty material that forms a sheathlike covering around some nerve fibers.
myocardium (mi″o-kar′de-um) The muscle tissue of the heart.
myofibril (mi″o-fi′bril) One of the contractile fibers found within muscle cells.
myoglobin (mi″o-glo′bin) A pigmented compound found in muscle tissue that acts to store oxygen.
myogram (mi′o-gram) A recording of a muscular contraction.
myometrium (mi″o-me′tre-um) The layer of smooth muscle tissue within the uterine wall.
myosin (mi′o-sin) A protein that, together with actin, is responsible for muscular contraction and relaxation.

n

nasal cavity (na′zal kav′ĭ-te) The space within the nose.
nasal concha (na′zal kong′kah) A shelllike bone extending outward from the wall of the nasal cavity.
nasal septum (na′zal sep′tum) A wall of bone and cartilage that separates the nasal cavity into two portions.
nasopharynx (na″zo-far′ingks) The portion of the pharynx associated with the nasal cavity.
negative feedback (neg′ah-tiv fēd′bak) A mechanism that is activated by an imbalance and acts to correct it.
neonatal (ne″o-na′tal) Pertaining to the period of life from birth to the end of the fourth week.
nephron (nef′ron) The functional unit of a kidney, consisting of a renal corpuscle and a renal tubule.
nerve (nerv) A bundle of nerve fibers.
nerve impulse (nerv im′puls) A wave of depolarization conducted along a nerve fiber.
neurilemma (nu″rĭ-lem′ah) Sheath on the outside of some nerve fibers due to the presence of Schwann cells.
neuroglia (nu-rog′le-ah) The supporting tissue within the brain and spinal cord, composed of neuroglial cells.
neuromuscular junction (nu″ro-mus′ku-lar jungk′shun) The union between a nerve fiber and a muscle fiber.
neuron (nu′ron) A nerve cell that consists of a cell body and its processes.
neuronal pool (nu′ro-nal pōōl) A group of neurons within the central nervous system that receives impulses from input nerve fibers, processes the impulses, and conducts the impulses away on output fibers.
neurotransmitter (nu″ro-trans′mit-er) A chemical substance secreted by the terminal end of an axon that stimulates a muscle fiber contraction or an impulse in another neuron.
neutral (nu′tral) Neither acid nor alkaline.
neutron (nu′tron) An electrically neutral particle found in an atomic nucleus.
neutrophil (nu′tro-fil) A type of phagocytic leukocyte.
niacin (ni′ah-sin) A vitamin of the B-complex group; nicotinic acid.
Nissl bodies (nis′l bod′ēz) Membranous sacs that occur within the cytoplasm of nerve cells and have ribosomes attached to their surfaces.
nonelectrolyte (non″e-lek′tro-līt) A substance that does not dissociate into ions when it is dissolved.
nonprotein nitrogenous substance (non-pro′te-in ni-troj′ĕ-nus sub′stans) Substance, such as urea or uric acid, that contains nitrogen but is not a protein.
norepinephrine (nor″ep-ĭ-nef′rin) A neurotransmitter substance released from the axon ends of some nerve fibers.
nuclease (nu′kle-ās) An enzyme that causes nucleic acids to decompose.
nucleic acid (nu-kle′ik as′id) A substance composed of nucleotides joined together; RNA or DNA.
nucleolus (nu-kle′o-lus) A small structure that occurs within the nucleus of a cell and contains RNA and proteins.
nucleotide (nu′kle-o-tīd″) A component of a nucleic acid molecule, consisting of a sugar, a nitrogenous base, and a phosphate group.
nucleus (nu′kle-us) A cellular organelle that is enclosed by a double-layered, porous membrane and contains DNA; the dense core of an atom that is composed of protons and neutrons.
nutrient (nu′tre-ent) A chemical substance that must be supplied to the body from its environment.

o

occipital (ok-sip′ĭ-tal) Pertaining to the lower, back portion of the head.
olfactory (ol-fak′to-re) Pertaining to the sense of smell.
olfactory nerves (ol-fak′to-re nervz) The first pair of cranial nerves that conduct impulses associated with the sense of smell.
oligodendrocyte (ol″ĭ-go-den′dro-sīt) A type of neuroglial cell that functions to connect neurons to blood vessels and to form myelin.
oocyte (o′o-sīt) An immature egg cell.
oogenesis (o″o-jen′ĕ-sis) The process by which an egg cell forms from an oocyte.
ophthalmic (of-thal′mik) Pertaining to the eye.
optic (op′tik) Pertaining to the eye.
optic chiasma (op′tik ki-az′mah) An X-shaped structure on the underside of the brain, created by a partial crossing over of fibers in the optic nerves.
optic disk (op′tik disk) The region in the retina of the eye where nerve fibers leave to become part of the optic nerve.
oral (o′ral) Pertaining to the mouth.
organ (or′gan) A structure consisting of a group of tissues that performs a specialized function.
organelle (or″gah-nel′) A living part of a cell that performs a specialized function.
organic (or-gan′ik) Pertaining to carbon-containing substances.
organism (or′gah-nizm) An individual living thing.
orifice (or′ĭ-fis) An opening.
origin (or′ĭ-jin) The end of a muscle that is attached to a relatively immovable part.
oropharynx (o″ro-far′ingks) The portion of the pharynx in the posterior part of the oral cavity.
osmoreceptor (oz″mo-re-sep′tor) A receptor that is sensitive to changes in the osmotic pressure of the body fluids.
osmosis (oz-mo′sis) The diffusion of water through a selectively permeable membrane.
osmotic (oz-mot′ik) Pertaining to osmosis.
osmotic pressure (oz-mot′ik presh′ur) The amount of pressure needed to stop osmosis; the potential pressure of a solution due to the presence of nondiffusible solute particles in the solution.
ossification (os″ĭ-fĭ-ka′shun) The formation of bone tissue.
osteoblast (os′te-o-blast″) A bone-forming cell.
osteoclast (os′te-o-klast″) A cell that causes the erosion of bone.
osteocyte (os′te-o-sīt) A bone cell.
osteon (os′te-on) A group of bone cells and the osteonic canal that they surround; the basic unit of structure in bone tissue; haversian system.
osteonic canal (os′te-on-ik kah-nal′) A tiny channel in bone tissue that contains a blood vessel; haversian canal.
otolith (o′to-lith) A small particle of calcium carbonate associated with the receptors of equilibrium.
oval window (o′val win′do) The opening between the stapes and the inner ear
ovarian (o-va′re-an) Pertaining to the ovary.
ovary (o′vah-re) The primary reproductive organ of a female; an egg-cell-producing organ.

oviduct (o'vĭ-dukt) A tube that leads from the ovary to the uterus; the uterine tube or fallopian tube.
ovulation (o"vu-la'shun) The release of an egg cell from a mature ovarian follicle.
oxidation (ok"sĭ-da'shun) The process by which oxygen is combined with a chemical substance.
oxygen debt (ok'sĭ-jen det) The amount of oxygen that must be supplied following physical exercise to convert the accumulated lactic acid to glucose.
oxyhemoglobin (ok"si-he"mo-glo'bin) The compound formed when oxygen combines with hemoglobin.
oxytocin (ok"sĭ-to'sin) A hormone released by the posterior lobe of the pituitary gland that causes contraction of the smooth muscles in the uterus and mammary glands.

p

pacemaker (pās'māk-er) The mass of specialized muscle tissue that controls the rhythm of the heartbeat; the sinoatrial node.
pain receptor (pān re"sep'tor) A sensory nerve ending associated with the feeling of pain.
palate (pal'at) The roof of the mouth.
palatine (pal'ah-tīn) Pertaining to the palate.
palmar (pahl'mar) Pertaining to the palm of the hand.
pancreas (pan'kre-as) A glandular organ in the abdominal cavity that secretes hormones and digestive enzymes.
pancreatic (pan"kre-at'ik) Pertaining to the pancreas.
pantothenic acid (pan"to-then'ik as'id) A vitamin of the B-complex group.
papilla (pah-pil'ah) A tiny nipplelike projection.
papillary muscle (pap'ĭ-ler"e mus'l) The muscle that extends inward from the ventricular wall of the heart.
parasympathetic division (par"ah-sim"pah-thet'ik dĭ-vizh'un) The portion of the autonomic nervous system that arises from the brain and sacral region of the spinal cord.
parathyroid glands (par"ah-thi'roid glandz) The small endocrine glands that are embedded in the posterior portion of the thyroid gland.
parathyroid hormone (par"ah-thi'roid hor'mōn) The hormone secreted by the parathyroid glands, which helps to regulate the concentration of blood calcium and phosphate ions.
parietal (pah-ri'ĕ-tal) Pertaining to the wall of an organ or a cavity.
parietal cell (pah-ri'ĕ-tal sel) The cell of a gastric gland that secretes hydrochloric acid and intrinsic factor.
parietal pleura (pah-ri'ĕ-tal ploo'rah) The membrane that lines the inner wall of the thoracic cavity.
parotid glands (pah-rot'id glandz) The large salivary glands located on the sides of the face just in front and below the ears.
partial pressure (par'shal presh'ur) The pressure produced by one gas in a mixture of gases.
pectoral (pek'tor-al) Pertaining to the chest.
pectoral girdle (pek'tor-al ger'dl) The portion of the skeleton that provides support and attachment for the arms.
pelvic (pel'vik) Pertaining to the pelvis.
pelvic girdle (pel'vik ger'dl) The portion of the skeleton to which the legs are attached.
pelvis (pel'vis) The bony ring formed by the sacrum and coxal bones.
penis (pe'nis) The external reproductive organ of the male through which the urethra passes.
pepsin (pep'sin) A protein-splitting enzyme secreted by the gastric glands of the stomach.
pepsinogen (pep-sin'o-jen) An inactive form of pepsin.
peptide (pep'tīd) A compound composed of two or more amino acid molecules joined together.
peptide bond (pep'tīd bond) The bond that forms between the carboxyl group of one amino acid and the amino group of another.
pericardial (per"ĭ-kar'de-al) Pertaining to the pericardium.
pericardium (per"ĭ-kar'de-um) The serous membrane that surrounds the heart.
perichondrium (per"ĭ-kon'dre-um) The layer of fibrous connective tissue that encloses cartilaginous structures.
perilymph (per'ĭ-limf) The fluid contained in the space between the membranous and osseous labyrinths of the inner ear.
perineal (per"ĭ-ne'al) Pertaining to the perineum.
perineum (per"ĭ-ne'um) The body region between the scrotum or urethral opening and the anus.
periodontal ligament (per"e-o-don'tal lig'ah-ment) The fibrous membrane that surrounds a tooth and attaches it to the bone of the jaw.
periosteum (per"e-os'te-um) The covering of fibrous connective tissue on the surface of a bone.
peripheral (pĕ-rif'er-al) Pertaining to parts located near the surface or toward the outside.
peripheral nervous system (pĕ-rif'er-al ner'vus sis'tem) The portions of the nervous system outside the central nervous system.
peripheral resistance (pĕ-rif'er-al re-zis'tans) Resistance to blood flow due to friction between the blood and the walls of the blood vessels.
peristalsis (per"ĭ-stal'sis) Rhythmic waves of muscular contraction that occur in the walls of various tubular organs.
peritoneal (per"ĭ-to-ne'al) Pertaining to the peritoneum.
peritoneal cavity (per"ĭ-to-ne'al kav'ĭ-te) The potential space between the parietal and visceral peritoneal membranes.
peritoneum (per"ĭ-to-ne'um) A serous membrane that lines the abdominal cavity and encloses the abdominal viscera.
peritubular capillary (per"ĭ-tu'bu-lar kap'ĭ-ler"e) A capillary that surrounds a renal tubule of a nephron.
permeable (per'me-ah-bl) Open to passage or penetration.
pH The measurement unit of the hydrogen ion concentration used to indicate the acid or alkaline condition of a solution.
phagocytosis (fag"o-si-to'sis) The process by which a cell engulfs and digests solid substances.
phalanx (fa'langks) A bone of a finger or toe.
pharynx (far'ingks) The portion of the digestive tube between the mouth and esophagus.
phospholipid (fos"fo-lip'id) A lipid that contains two fatty acid molecules and a phosphate group combined with a glycerol molecule.
photoreceptor (fo"to-re-sep'tor) A nerve ending that is sensitive to light energy.
physiology (fiz"e-ol'o-je) The branch of science dealing with the study of body functions.
pia mater (pi'ah ma'ter) The inner layer of the meninges that encloses the brain and spinal cord.
pineal gland (pin'e-al gland) A small structure located in the central part of the brain.
pinocytosis (pin"o-si-to'sis) The process by which a cell engulfs droplets of fluid from its surroundings.
pituitary gland (pĭ-tu'ĭ-tār"e gland) The endocrine gland that is attached to the base of the brain and consists of anterior and posterior lobes.
placenta (plah-sen'tah) The structure by which an unborn child is attached to its mother's uterine wall and through which it is nourished.
plantar (plan'tar) Pertaining to the sole of the foot.
plasma (plaz'mah) The fluid portion of circulating blood.
plasma cell (plaz'mah sel) An antibody-producing cell that is formed as a result of the proliferation of sensitized B-cells.
plasma protein (plaz'mah pro'te-in) Any of several proteins normally found dissolved in the blood plasma.
platelet (plāt'let) A cytoplasmic fragment formed in the bone marrow, which functions in blood coagulation.
pleural (ploo'ral) Pertaining to the pleura or membranes surrounding the lungs.
pleural cavity (ploo'ral kav'ĭ-te) The potential space between the pleural membranes.
pleural membranes (ploo'ral mem'brānz) The serous membranes that enclose the lungs.
plexus (plek'sus) A network of interlaced nerves or blood vessels.
polar body (po'lar bod'e) The small, nonfunctional cell produced as a result of meiosis during egg cell formation.

polarization (po″lar-i-za′shun) The development of an electrical charge on the surface of a cell membrane.
polypeptide (pol″e-pep′tīd) A compound formed by the union of many amino acid molecules.
polysaccharide (pol″e-sak′ah-rīd) A carbohydrate composed of many monosaccharide molecules joined together.
polyunsaturated fatty acid (pol″e-un-sach′ĕ-ra-ted fat′e as′id) A fatty acid that contains one or more double bonds in its carbon atom chain.
pons (ponz) A portion of the brain stem above the medulla oblongata and below the midbrain.
popliteal (pop″li-te′al) Pertaining to the region behind the knee.
positive feedback (poz′i-tiv fēd′bak) The process by which changes cause more changes of a similar type, producing unstable conditions.
posterior (pos-tēr′e-or) Toward the back; the opposite of anterior.
postganglionic fiber (pōst″gang-gle-on′ik fi′ber) An autonomic nerve fiber located on the distal side of a ganglion.
postnatal (pōst-na′tal) After birth.
preganglionic fiber (pre″gang-gle-on′ik fi′ber) An autonomic nerve fiber located on the proximal side of a ganglion.
pregnancy (preg′nan-se) The condition in which a female has a developing offspring in her uterus.
prenatal (pre-na′tal) Before birth.
pressoreceptor (pres″o-re-sep′tor) A receptor that is sensitive to changes in pressure.
primary reproductive organs (pri′ma-re re″pro-duk′tiv or′ganz) The sex-cell-producing parts; the testes in males and the ovaries in females.
prime mover (prīm moov′er) A muscle that is mainly responsible for a particular body movement.
process (pros′es) A prominent projection on a bone.
progesterone (pro-jes′tĕ-rōn) A female hormone secreted by the corpus luteum of the ovary and by the placenta.
projection (pro-jek′shun) The process by which the brain causes a sensation to seem to come from the region of the body being stimulated.
prolactin (pro-lak′tin) The hormone secreted by the anterior pituitary gland that stimulates the production of milk from the mammary glands; PRL.
pronation (pro-na′shun) A movement in which the palm of the hand is moved downward or backward.
prophase (pro′fāz) The stage of mitosis during which the chromosomes become visible.
proprioceptor (pro″pre-o-sep′tor) A sensory nerve ending that is sensitive to changes in the tension of a muscle or tendon.
prostaglandins (pros″tah-glan′dins) A group of compounds that have powerful, hormonelike effects.
prostate gland (pros′tāt gland) The gland that surrounds the male urethra below the urinary bladder.
protein (pro′te-in) A nitrogen-containing organic compound composed of amino acid molecules joined together.
prothrombin (pro-throm′bin) A plasma protein that functions in the formation of blood clots.
proton (pro′ton) A positively charged particle found in an atomic nucleus.
protraction (pro-trak′shun) A forward movement of a body part.
proximal (prok′si-mal) Closer to the midline or origin; the opposite of distal.
puberty (pu′ber-te) The stage of development in which the reproductive organs become functional.
pulmonary (pul′mo-ner″e) Pertaining to the lungs.
pulmonary circuit (pul′mo-ner″e ser′kit) The system of blood vessels that carries the blood between the heart and lungs.
pulse (puls) The surge of blood felt through the walls of arteries due to the contraction of the ventricles of the heart.
pupil (pu′pil) The opening in the iris through which light enters the eye.
Purkinje fibers (pur-kin′je fi′berz) Specialized muscle fibers that conduct the cardiac impulse from the A-V bundle into the ventricular walls.
pyramidal cell (pi-ram′i-dal sel) A large, pyramid-shaped neuron found within the cerebral cortex.
pyruvic acid (pi-roo′vik as′id) An intermediate product of carbohydrate oxidation.

r

radioactive (ra″de-o-ak′tiv) property of an unstable isotope by which it releases energy or atomic particles.
receptor (re″sep′tor) A part located at the distal end of a sensory dendrite that is sensitive to stimulation.
red marrow (red mar′o) Blood-cell-forming tissue located in spaces within bones.
referred pain (re-ferd′ pān) Pain that feels as if it is originating from a part other than the site being stimulated.
reflex (re′fleks) A rapid, automatic response to a stimulus.
reflex arc (re′fleks ark) A nerve pathway, consisting of a sensory neuron, interneuron, and motor neuron, which forms the structural and functional bases for a reflex.
refraction (re-frak′shun) A bending of light as it passes from one medium into another medium with a different density.
relaxin (re-lak′sin) A hormone from the corpus luteum that inhibits uterine contractions during pregnancy.
renal (re′nal) Pertaining to the kidney.
renal corpuscle (re′nal kor′pusl) The part of a nephron that consists of a glomerulus and a Bowman's capsule.
renal cortex (re′nal kor′teks) The outer portion of a kidney.
renal medulla (re′nal mĕ-dul′ah) The inner portion of a kidney.
renal pelvis (re′nal pel′vis) The hollow cavity within a kidney.
renal tubule (re′nal tu′būl) The portion of a nephron that extends from the renal corpuscle to the collecting duct.
renin (re′nin) The enzyme released from the kidneys that triggers a mechanism leading to a rise in blood pressure.
reproduction (re″pro-duk′shun) The process by which an offspring is formed.
resorption (re-sorp′shun) The process by which something that has been secreted is absorbed.
respiration (res″pi-ra′shun) The cellular process by which energy is released from nutrients.
respiratory center (re-spi′rah-to″re sen′ter) The portion of the brain stem that controls the depth and rate of breathing.
respiratory membrane (re-spi′rah-to″re mem′brān) The membrane composed of a capillary wall and an alveolar wall through which gases are exchanged between the blood and air.
response (re-spons′) The action resulting from a stimulus.
resting potential (res′ting po-ten′shal) The difference in electrical charge between the inside and outside of an undisturbed nerve cell membrane.
reticular formation (rĕ-tik′u-lar for-ma′shun) A complex network of nerve fibers within the brain stem that functions in arousing the cerebrum.
reticuloendothelial tissue (rĕ-tik″u-lo-en″do-the′le-al tish′u) Tissue composed of widely scattered phagocytic macrophages.
retina (ret′i-nah) The inner layer of the eye wall that contains the visual receptors.
retinal (ret′i-nal) A substance used in the production of rhodopsin by the rods of the retina.
retraction (rĕ-trak′shun) The movement of a part toward the back.
rhodopsin (ro-dop′sin) A light-sensitive substance that occurs in the rods of the retina.
riboflavin (ri″bo-fla′vin) A vitamin of the B complex group; vitamin B_2.
ribonucleic acid (ri″bo-nu-kle′ik as′id) A nucleic acid that contains ribose sugar; RNA.
ribose (ri′bos) A five-carbon sugar found in RNA molecules.
ribosome (ri′bo-sōm) A cytoplasmic organelle that functions in the synthesis of proteins.
RNA Ribonucleic acid.
rod (rod) A type of light receptor that is responsible for colorless vision.
rotation (ro-ta′shun) The movement by which a body part is turned on its longitudinal axis.
round window (rownd win′do) A membrane-covered opening between the inner ear and the middle ear.

S

sagittal (saj'ĭ-tal) A plane or section that divides a structure into right and left portions.

S-A node (nōd) The sinoatrial node.

sarcomere (sar'ko-mēr) The structural and functional unit of a myofibril.

sarcoplasmic reticulum (sar"ko-plaz'mik rē-tik'u-lum) A membranous network of channels and tubules within a muscle fiber.

saturated fatty acid (sat'u-rāt"ed fat'e as'id) A fatty acid molecule that lacks double bonds between the atoms of its carbon chain.

Schwann cell (shwahn sel) A type of neuroglial cell that surrounds a fiber of a peripheral nerve.

sclera (skle'rah) The white fibrous outer layer of the eyeball.

scrotum (skro'tum) A pouch of skin that encloses the testes.

sebaceous gland (se-ba'shus gland) A gland of the skin that secretes sebum.

sebum (se'bum) An oily secretion of the sebaceous glands.

secretin (se-kre'tin) A hormone secreted from the small intestine that stimulates the release of pancreatic juice.

semen (se'men) The fluid discharged from the male reproductive tract at ejaculation.

semicircular canal (sem"ĭ-ser'ku-lar kah-nal') The tubular structure within the inner ear that contains the receptors responsible for the sense of dynamic equilibrium.

seminiferous tubule (sem"ĭ-nif'er-us tu'būl) Tubule within the testes in which sperm cells are formed.

semipermeable (sem"ĭ-per'me-ah-bl) A condition in which a membrane is permeable to some molecules and not to others; selectively permeable.

sensation (sen-sa'shun) A feeling resulting from the interpretation of sensory nerve impulses by the brain.

sensory area (sen'so-re a're-ah) A portion of the cerebral cortex that receives and interprets sensory nerve impulses.

sensory nerve (sen'so-re nerv) A nerve composed of sensory nerve fibers.

sensory neuron (sen'so-re nu'ron) A neuron that transmits an impulse from a receptor to the central nervous system.

serotonin (se"ro-to'nin) A vasoconstricting substance that is released by blood platelets when blood vessels are broken.

serous cell (se'rus sel) A glandular cell that secretes a watery fluid with a high enzyme content.

serous fluid (ser'us floo'id) The secretion of a serous cell.

serous membrane (ser'us mem'brān) A membrane that lines a cavity without an opening to the outside of the body.

serum (se'rum) The fluid portion of coagulated blood.

sesamoid bone (ses'ah-moid bōn) A round bone that may occur in tendons adjacent to joints.

simple sugar (sim'pl shoog'ar) A monosaccharide.

sinoatrial node (si"no-a'tre-al nōd) The group of specialized tissue in the wall of the right atrium that initiates cardiac cycles; the pacemaker; the S-A node.

sinus (si'nus) A cavity or hollow space in a bone or other body part.

skeletal muscle (skel'ĭ-tal mus'l) The type of muscle tissue found in muscles attached to skeletal parts.

smooth muscle (smooth mus'l) The type of muscle tissue found in the walls of hollow visceral organs; visceral muscle.

solute (sol'ūt) A substance that is dissolved in a solution.

solution (so-lu'shun) A homogenous mixture of molecular or ionic substances (solutes) within a dissolving medium (solvent).

solvent (sol'vent) The liquid portion of a solution in which a solute is dissolved.

somatic cell (so-mat'ik sel) Any cell of the body other than the sex cells.

special sense (spesh'al sens) A sense that involves receptors associated with specialized sensory organs, such as the eyes and ears.

spermatid (sper'mah-tid) An intermediate stage in the formation of sperm cells.

spermatocyte (sper-mat'o-sīt) An early stage in the formation of sperm cells.

spermatogenesis (sper"mah-to-jen'ĕ-sis) The production of sperm cells.

spermatogonium (sper'mah-to-go'ne-um) An undifferentiated spermatogenic cell.

sphincter (sfingk'ter) A circular muscle that functions to close an opening or the lumen of a tubular structure.

spinal (spi'nal) Pertaining to the spinal cord or to the vertebral canal.

spinal cord (spi'nal kord) The portion of the central nervous system extending downward from the brain stem through the vertebral canal.

spinal nerve (spi'nal nerv) A nerve that arises from the spinal cord.

spleen (splēn) A large, glandular organ located in the upper left region of the abdomen.

spongy bone (spunj'e bōn) Bone that consists of bars and plates separated by irregular spaces; cancellous bone.

squamous (skwa'mus) Flat or platelike.

starch (starch) A polysaccharide that is common in foods of plant origin.

static equilibrium (stat'ik e"kwi-lib're-um) The maintenance of balance when the head and body are motionless.

stimulus (stim'u-lus) A change in the environmental conditions that is followed by a response by an organism or a cell.

stomach (stum'ak) The digestive organ located between the esophagus and the small intestine.

stratified (strat'ĭ-fīd) Arranged in layers.

stratum basale (stra'tum ba'sal-e) The deepest layer of the epidermis in which the cells undergo mitosis.

stratum corneum (stra'tum kor'ne-um) The outer horny layer of the epidermis.

stressor (stres'or) A factor capable of stimulating a stress response.

stroke volume (strōk vol'ūm) The amount of blood discharged from the ventricle with each heartbeat.

structural formula (struk'cher-al for'mu-lah) A representation of the way atoms are bound together within a molecule, using the symbols for each element present and lines to indicate chemical bonds.

subarachnoid space (sub"ah-rak'noid spās) The space within the meninges between the arachnoid mater and the pia mater.

subcutaneous (sub"ku-ta'ne-us) Beneath the skin.

sublingual (sub-ling'gwal) Beneath the tongue.

submucosa (sub"mu-ko'sah) A layer of connective tissue that underlies a mucous membrane.

substrate (sub'strāt) The substance upon which an enzyme acts.

sucrose (soo'krōs) A disaccharide; table sugar.

sulcus (sul'kus) A shallow groove, such as that between adjacent convolutions on the surface of the brain.

superficial (soo"per-fish'al) Near the surface.

superior (soo-pe're-or) Pertaining to a structure that is higher than another structure.

supination (soo"pĭ-na'shun) Rotation of the forearm so that the palm faces upward when the arm is outstretched.

suppressor cell (sŭ-pres'or sel) A special type of T-cell that interferes with the production of antibodies responsible for allergic reactions.

surface tension (ser'fas ten'shun) The force that tends to hold moist membranes together.

surfactant (ser-fak'tant) A substance, produced by the lungs, that reduces the surface tension within the alveoli.

suture (soo'cher) An immovable joint, such as that between adjacent flat bones of the skull.

sympathetic nervous system (sim"pah-thet'ik ner'vus sis'tem) The portion of the autonomic nervous system that arises from the thoracic and lumbar regions of the spinal cord.

symphysis (sim'fĭ-sis) A slightly movable joint between bones separated by a pad of fibrocartilage.

synapse (sin'aps) The junction between the axon end of one neuron and the dendrite or cell body of another neuron.

synaptic knob (sĭ-nap'tik nob) The tiny enlargement at the end of an axon that secretes a neurotransmitter substance.

synarthrosis (sin″ar-thro′sis) A joint in which the bones are fastened tightly together by a thin layer of fibrous connective tissue; an immovable joint.
syncytium (sin-sish′e-um) A mass of merging cells.
synovial fluid (sĭ-no′ve-al floo′id) The fluid secreted by the synovial membrane.
synovial joint (sĭ-no′ve-al joint) A freely movable joint.
synovial membrane (sĭ-no′ve-al mem′brān) The membrane that forms the inner lining of the capsule of a freely movable joint.
synthesis (sin′thĕ-sis) The process by which substances are united to form more complex substances.
system (sis′tem) A group of organs that act together to carry on a specialized function.
systemic circuit (sis-tem′ik ser′kit) The vessels that conduct blood between the heart and all the body tissues except the lungs.
systole (sis′to-le) The phase of the cardiac cycle during which a heart chamber wall is contracted.
systolic pressure (sis-tol′ik presh′ur) The arterial blood pressure during the systolic phase of the cardiac cycle.

t

target tissue (tar′get tish′u) The specific tissue on which a hormone acts.
tarsus (tar′sus) The bones that form the ankle.
taste bud (tāst bud) An organ containing the receptors associated with the sense of taste.
T-cell (sel) A type of lymphocyte that interacts directly with antigen-bearing particles and is responsible for cell-mediated immunity.
telophase (tel′o-fāz) The stage in mitosis during which newly formed cells become separate structures.
tendon (ten′don) A cordlike or bandlike mass of white fibrous connective tissue that connects a muscle to a bone.
testis (tes′tis) The primary reproductive organ of a male; a sperm-cell-producing organ.
testosterone (tes-tos′tĕ-rōn) A male sex hormone.
tetany (tet′ah-ne) A continuous, forceful muscular contraction.
thalamus (thal′ah-mus) A mass of gray matter located at the base of the cerebrum in the wall of the third ventricle.
thermoreceptor (ther″mo-re-sep′tor) A sensory receptor that is sensitive to changes in temperature; a heat receptor.
thiamine (thi′ah-min) Vitamin B_1.
thoracic (tho-ras′ik) Pertaining to the chest.
threshold stimulus (thresh′old stim′u-lus) The level of stimulation that must be exceeded to elicit a nerve impulse or a muscle contraction.
thrombus (throm′bus) A blood clot in a blood vessel that remains at its site of formation.
thymus (thi′mus) A glandular organ located in the mediastinum behind the sternum and between the lungs.
thyroid gland (thi′roid gland) An endocrine gland located just below the larynx and in front of the trachea.
thyroxine (thī-rok′sin) A hormone secreted by the thyroid gland.
tissue (tish′u) A group of similar cells that performs a specialized function.
trachea (tra′ke-ah) The tubular organ that leads from the larynx to the bronchi.
transcellular fluid (trans″sel′u-lar floo′id) A portion of the extracellular fluid, including the fluid within special body cavities.
transfer RNA (trans′fer) A molecule of RNA that carries an amino acid to a ribosome in the process of protein synthesis.
transverse tubule (trans-vers′ tu′būl) A membranous channel that extends inward from a muscle fiber membrane and passes through the fiber.
tricuspid valve (tri-kus′pid valv) The heart valve located between the right atrium and the right ventricle.
triglyceride (tri-glis′er-īd) A lipid composed of three fatty acids combined with a glycerol molecule.
triiodothyronine (tri″i-o″do-thi′ro-nēn) One of the thyroid hormones.
trochanter (tro-kan′ter) A broad process on a bone.
trochlea (trok′le-ah) A pulley-shaped structure.
tropic hormone (tro′pik hor′mōn) A hormone that has an endocrine gland as its target tissue.
trypsin (trip′sin) An enzyme in pancreatic juice that acts to break down protein molecules.
tubercle (tu′ber-kl) A small, rounded process on a bone.
tuberosity (tu″bĕ-ros′ĭ-te) An elevation or a protuberance on a bone.
twitch (twich) A brief muscular contraction followed by relaxation.
tympanic membrane (tim-pan′ik mem′brān) A thin membrane that covers the auditory canal and separates the external ear from the middle ear; the eardrum.

u

umbilical cord (um-bil′ĭ-kal kord) The cordlike structure that connects the fetus to the placenta.
umbilical region (um-bil′ĭ-kal re′jun) The central portion of the abdomen.
unsaturated fatty acid (un-sat′u-rāt″ed fat′e as′id) A fatty acid molecule that has one or more double bonds between the atoms of its carbon chain.
urea (u-re′ah) A nonprotein nitrogenous substance produced as a result of protein metabolism.
ureter (u-re′ter) A muscular tube that carries urine from the kidney to the urinary bladder.
urethra (u-re′thrah) A tube leading from the urinary bladder to the outside of the body.
uterine (u′ter-in) Pertaining to the uterus.
uterine tube (u′ter-in tūb) A tube that extends from the uterus on each side toward an ovary and functions to transport sex cells; a fallopian tube or an oviduct.
uterus (u′ter-us) A hollow muscular organ located within the female pelvis in which a fetus develops.
utricle (u′trĭ-kl) An enlarged portion of the membranous labyrinth of the inner ear.
uvula (u′vu-lah) A fleshy portion of the soft palate that hangs down above the root of the tongue.

v

vaccine (vak′sēn) A substance that contains antigens and is used to stimulate the production of antibodies.
vagina (vah-ji′nah) A tubular organ that leads from the uterus to the vestibule of the female reproductive tract.
Valsalva's maneuver (Val-sal′vahz mah′noo′ver) Increasing the intrathoracic pressure by forcing air from the lungs against a closed glottis.
vascular (vas′ku-lar) Pertaining to the blood vessels.
vas deferens (vas def′er-ens) A tube that leads from the epididymis to the urethra of the male reproductive tract (plural, *vasa deferentia*).
vasoconstriction (vas″o-kon-strik′shun) A decrease in the diameter of a blood vessel.
vasodilation (vas″o-di-la′shun) An increase in the diameter of a blood vessel.
vein (vān) A vessel that carries blood toward the heart.
vena cava (ve′nah kav′ah) One of two large veins that convey deoxygenated blood to the right atrium of the heart.
ventral root (ven′tral root) The motor branch of a spinal nerve by which it is attached to the spinal cord.
ventricle (ven′trĭ-kl) A cavity, such as those of the brain that are filled with cerebrospinal fluid or those of the heart that contain blood.
venule (ven′ūl) A vessel that carries blood from the capillaries to a vein.
vesicle (ves′ĭ-kl) A membranous sac within the cytoplasm that forms by an infolding of the cell membrane.
villus (vil′us) A tiny, fingerlike projection.
visceral (vis′er-al) Pertaining to the contents of a body cavity.
visceral peritoneum (vis′er-al per″ĭ-to-ne′um) A membrane that covers the surfaces of organs within the abdominal cavity.

visceral pleura (vis′er-al ploo′rah) A membrane that covers the surfaces of the lungs.

viscosity (vis-kos′ĭ-te) The tendency for a fluid to resist flowing due to the friction of its molecules on one another.

vitamin (vi′tah-min) An organic substance other than a carbohydrate, lipid, or protein that is needed for normal metabolism but cannot be synthesized in adequate amounts by the body.

vitreous humor (vit′re-us hu′mor) The substance that occupies the space between the lens and retina of the eye.

vocal cords (vo′kal kordz) Folds of tissue within the larynx that create vocal sounds when they vibrate.

voluntary (vol′un-tār″e) Capable of being consciously controlled.

vulva (vul′vah) The external reproductive parts of the female that surround the opening of the vagina.

w

water balance (wot′er bal′ans) A condition in which the quantity of water entering the body is equal to the quantity leaving it.

water of metabolism (wot′er uv mĕ-tab′o-lizm) Water produced as a by-product of metabolic processes.

y

yellow marrow (yel′o mar′o) Fat storage tissue found in the cavities within bones.

z

zygote (zi′gōt) A cell produced by the fusion of an egg and a sperm; a fertilized egg cell.

zymogen granule (zi-mo′jen gran′ūl) A cellular structure that stores inactive forms of protein-splitting enzymes in a pancreatic cell.

Credits

PHOTOGRAPHS

Table of Contents

Page vii: © Stephen Marks/Stockphotos, Inc.; **p. ix:** © Manfred Kage/Peter Arnold, Inc.; **p. xi:** © Edwin A. Reschke; **p. xii:** © Fred Hossler/Visuals Unlimited; **p. xiii, xvi:** © Edwin A. Reschke

Unit Openers

Man: © Stephen Marks/Stockphotos, Inc. **Micrograph: 1:** © Manfred Kage/Peter Arnold, Inc.; **2:** © Edwin A. Reschke; **3:** © Fred Hossler/Visuals Unlimited; **4, 5:** © Edwin A. Reschke

Chapter 1

1.12A: © Science Photo Library/Photo Researchers; **1.12B:** © CNRI/Science Photo Library/Photo Researchers

Chapter 2

2.8 (both), 2.16: Courtesy of Ealing Corporation, South Natick

Chapter 3

3.3B: © Gordon Leedale/BioPhoto Associates; **3.4:** © David Zieleniec; **3.6:** © Keith Porter/Science Source/Photo Researchers, Inc.; **3.7A:** © Gordon Leedale/BioPhoto Associates; **3.8:** © Reproduced with permission of the American Lung Association; **3.9A:** © Stephen L. Wolfe; **3.18:** © Edwin A. Reschke

Chapter 4

4.13: Courtesy of Ealing Corporation, South Natick

Chapter 5

5.1B&C, 5.2B: © Edwin A. Reschke; **5.3B:** © Manfred Kage/Peter Arnold, Inc.; **5.4B:** © Edwin A. Reschke; **5.5B:** © Fred Hossler/Visuals Unlimited; **5.6A:** © Edwin A. Reschke; **5.8:** © David M. Phillips/Visuals Unlimited; **5.9:** © Jean-Paul Revel; **5.10:** © Veronica Burmeister/Visuals Unlimited; **5.11B, 5.12B, 5.13B, 5.14B, 5.15B:** © Edwin A. Reschke; **5.16B:** © John D. Cunningham; **5.17B:** © Victor B. Eichler; **5.17C:** © BioPhoto Associates/Photo Researchers, Inc.; **5.18B, 5.19B, 5.20B:** © Edwin A. Reschke; **5.21B, 5.22B:** © Manfred Kage/Peter Arnold, Inc.

Chapter 6

6.2A: © Victor B. Eichler; **6.2B:** © M. Schlwia/Visuals Unlimited; **6.3B:** © Victor B. Eichler; **6.4:** © Per H. Kjeldson, University of Michigan at Ann Arbor

Chapter 7

7.3: © BioPhoto Associates/Photo Researchers, Inc.; **7.21:** Courtesy of Kodak; **7.28:** © Martin M. Rotker/Science Photo Library/Taurus Photos, Inc.; **7.D (both):** Courtesy of Jack Hole

Chapter 8

8.3: Courtesy of H. E. Huxley, Cambridge University; **8.C (both):** © John Carl Mese

Chapter 9

9.1: © Edwin A. Reschke; **9.4:** "TEM of myelinated and unmyelinated axons in cross section," by H. Webster and Hubbard John from THE VERTEBRATE PERIPHERAL NERVOUS SYSTEM, 1974. **9.14:** © Don Fawcett/Photo Researchers, Inc.; **9.21B:** © Per H. Kjeldson, University of Michigan at Ann Arbor

Chapter 10

10.11: © Fred Hossler/Visuals Unlimited; **10.21:** © Per H. Kjeldson, University of Michigan at Ann Arbor; **10.22:** © Carroll W. Weiss/Camera M.D. Studios; **10.24C:** © Frank S. Werblin

Chapter 11

11.15: © Edwin A. Reschke

Chapter 12

12.5: Courtesy of Jack Hole; **12.11:** © Edwin A. Reschke; **12.17:** © Victor B. Eichler; **12.18:** © Carroll H. Weiss, RBP, 1973/Camera M.D.; **12.21:** © Armed Forces Institute of Pathology; **12.23:** © Manfred Kage/Peter Arnold, Inc.; **12.26:** © James L. Shaffer

Chapter 13

13.4C: © Phototake, Inc.; **13.8, 13.10 A&B:** © Victor B. Eichler

Chapter 14

14.2B: © Bill Longcore/Science Source/Photo Researchers, Inc.; **14.6:** © Warren Rosenberg, Iona College/Biological Photo Service; **14.7:** © Edwin A. Reschke/Peter Arnold, Inc.; **14.8:** © Alfred Owzarzak/Taurus Photos, Inc.; **14.9, 14.10:** © Edwin A. Reschke; **14.12A&B:** © Victor B. Eichler; **14.15:** © David M. Phillips/Visuals Unlimited; **14.16A&B:** © Camera M.D. Studios, Inc.

Chapter 15

15.5: Courtesy of Igaku Shoin LTD.; **15.17:** T. Kuwabara, from Bloom, W. and Fawcett, D. W., A Textbook of Histology, 10/E, © 1975 W.B. Saunders, Co.; **15.19A:** © Edwin A. Reschke; **15.19B:** © Victor B. Eichler; **15.22:** Courtesy of Dr. Patricia Phelps

Chapter 16

16.3: © Edwin A. Reschke; **16.6:** Courtesy of Igaku Shoin LTD.; **16.8B:** © John D. Cunningham/Visuals Unlimited; **16.9B:** © BioPhoto Associates/Photo Researchers, Inc.; **16.11:** Courtesy of Nina Lampen/Memorial Sloan-Kettering Cancer Center

Chapter 17

17.4: © *From Tissues and Organs: A Text Atlas of Scanning Electron Microscopy*, by Richard G. Kessel and Randy H. Kardon, W. H. Freeman and Company, copyright 1979; **17.5:** © BioPhoto Associates/Photo Researchers, Inc.; **17.15:** © Per H. Kjeldson, University of Michigan at Ann Arbor; **17.17:** © Edwin A. Reschke

Chapter 19

19.3B: © David M. Phillips/The Population Council 1986/Taurus Photos, Inc.; **19.4A:** © BioPhoto Associates/Photo Researchers, Inc.; **19.10:** © Edwin A. Reschke/Peter Arnold, Inc.; **19.14:** SIU School of Medicine

Chapter 20

20.1: © "Sea Urchin Sperm-Egg Interactions Studied with the Scanning Electron Microscope," Tegner, M. J. and Epel, D. Science Vol. 179, pp. 685–688, fig. 2, 16 February 1973, copyright 1973 by the AAAS; **20.12, 20.13:** © Petit Format/Nestle/Photo Researchers, Inc.

ILLUSTRATORS

Antler, Laurel: Figures 3.7B, 3.19, 7.5, 20.3
Ayres Associates: Figure 3.15
Carolina Biological Supply: Figure 16.8A
Collins, Samuel: Figure 9.31
Cousins, Barbara: Figure 1.2
Creek, Chris: Figure 1.5
Fineline Illustrations: Figures 1.3, 1.11, 2.1, 2.2, 2.3, 2.4, 2.5, 2.6, 2.7, 2.10, 2.11, 2.12, 2.13, 2.14, 2.15, 2.17, 2.18, 3.10, 3.11, 3.13, 3.14, 3.16, 4.1, 4.2, 4.3, 4.5, 4.6, 4.7, 4.8, 4.9, 4.10, 4.11, 4.14, 4.15, 4.16, 8.9, 8.10, 8.11, 8.12, 9.6, 9.7, 9.8, 9.9, 9.10, 9.11, 10.23, 11.5, 11.13, 12.29, 12.30, 13.15, 13.19, 13.20, 13.21, 13.22, 14.17, 14.18, 14.19, 14.20, 14.21, 15.13B, 15.21, 17.9, 18.1, 18.2, 18.3, 18.4, 18.5, 18.6, 18.7, 18.8, 18.9, 20.5
J & R Art Service: Figure 4.18
Kasnot, Keith: Figures 9.32, 9.33, 10.12, 10.13, 10.17, 10.18
Khoury, Lila: Figure 5.17A
Krabach, Ruth: Figures 7.1, 7.2, 7.4, 7.15, 7.18, 7.19, 7.26, 7.34, 7.35, 7.36, 7.37, 7.38, 20.14
Loechel, William: Figures 6.3A, 10.1, 11.12, 11.16, 12.2, 12.9, 12.10, 12.14, 12.15, 12.16, 12.19, 12.20, 12.22, 12.25, 12.27, 13.2, 13.11, 13.18, 17.2, 17.3, 17.6, 17.8, 19.2
Lynch, Patrick: Figures 1.14A&B, 5.13A, 15.26, 15.30
Margulies, Robert: Figures 14.11, 20.6
Margulies, Robert/Waldrop, Tom: Figures 8.14, 8.15
Marshburn, Nancy: Figures 3.1, 3.3A, 3.5, 4.12, 4.17, 4.19, 9.2, 9.20, 9.22, 9.29, 10.20, 10.24A,B, 11.4, 12.3, 14.1, 14.2A, 15.4B, 15.18, 16.14
Mclean, Ron: Figures 8.7, 19.9
Moon, Steve: Plates 1–7 and figures 3.2, 6.6, 9.5, 9.12, 9.13, 9.15, 10.4, 10.5, 10.7, 10.10A, 10.14, 10.16, 11.9, 11.10, 12.4, 12.8, 13.3, 13.4A,B, 13.5, 13.16, 13.17, 14.5, 14.13, 15.2, 15.10, 15.23, 16.4, 16.7, 17.1, 19.13
Nelson, Diane: Figures 15.28, 15.29
Diane Nelson & Associates: Figures 1.10, 5.1A, 5.2A, 5.3A, 5.4A, 5.5A, 5.12A, 5.14A, 5.15A, 5.16A, 5.19, 5.20A, 5.21A, 5.22A, 7.6, 7.8, 7.9, 7.10, 7.11, 7.12, 7.13, 7.14, 7.16, 7.17, 7.20, 7.22, 7.23, 7.24, 7.25, 7.27, 7.29, 7.30, 7.31, 7.32, 7.33, 9.23, 10.2, 10.6, 10.8, 10.9, 10.25, 11.7, 11.8, 12.12, 12.24, 13.7, 13.9, 14.3, 14.14, 15.1, 15.6, 15.9A, 15.9B, 15.12, 15.20, 15.24, 15.25, 15.27, 16.1, 16.5, 17.10, 17.11,

17.12, 17.13, 19.3A, 19.6, 19.8, 20.7, 20.10, 20.11, 20.17
Nyquist, John: Figure 15.8
Reschke, Edwin A.: Figure 5.6B,C
Rinehart, Mildred: Figure 6.5
Rolin Graphics: Figure 12.28
Schenk, Michael: Figures 1.6A, 1.6B, 1.13, 5.7, 7.7, 11.14, 12.1, 12.7, 13.1, 13.6, 13.12, 19.4B, 20.2, 20.8, 20.16
Sims, Tom: Figures 10.3, 15.14, 17.7, 20.15
Sims, Tom/Schenk, Mike: Figures 1.7, 9.24, 10.19
Twomey, Catherine: Figures 3.17, 16.12, 16.13, 19.5, 20.4
Waldrop, Tom: Figures 1.8, 1.9, 8.1, 8.2, 8.4, 8.5, 8.6, 8.13, 8.16A, 8.16B, 8.17, 8.18, 8.19, 8.20, 8.21, 8.22, 8.23, 8.24, 8.25, 8.26, 8.27, 8.28, 8.29, 9.18, 9.28, 11.6, 19.1, 19.7, 19.11, 19.12
John Walters & Associates: Figure 15.31
Williams, Marcia: Figures 6.1, 7.39, 7.40, 7.41, 9.3, 9.16, 9.17, 13.13, 13.14, 14.4, 15.3, 15.4A, 15.7, 15.11, 16.9A, 20.9
Wood, Bill: Figures 9.25, 9.26
Wood, Charles: Figure 10.15

LINE ART

Chapter 1

Figure 1.1: From C. M. Saunders and Charles P. O'Malley, *Works of Andreas Vesalius of Brussels.* Copyright © 1973 Dover Publications, Inc., Mineola, NY. Used with permission.
Figures 1.6a, b, 1.13: From Kent M. Van De Graaff, *Human Anatomy,* 2d ed. Copyright © 1988 Wm. C. Brown Publishers, Dubuque, Iowa. All Rights Reserved. Reprinted by permission.
Figure 1.7: From Kent M. Van De Graaff and Stuart Ira Fox, *Concepts of Human Anatomy and Physiology.* Copyright © 1986 Wm. C. Brown Publishers, Dubuque, Iowa. All Rights Reserved. Reprinted by permission.

Chapter 3

Figure 3.12: From Kent M. Van De Graaff and Stuart Ira Fox, *Concepts of Human Anatomy and Physiology,* 2d ed. Copyright © 1989 Wm. C. Brown Publishers, Dubuque, Iowa. All Rights Reserved. Reprinted by permission.
Figure 3.15: From Stuart Ira Fox, *Human Physiology,* 3d ed. Copyright © 1990 Wm. C. Brown Publishers, Dubuque, Iowa. All Rights Reserved. Reprinted by permission.

Chapter 4

Figure 4.18: From Stuart Ira Fox, *Human Physiology,* 3d ed. Copyright © 1990 Wm. C. Brown Publishers, Dubuque, Iowa. All Rights Reserved. Reprinted by permission.

Chapter 5

Figure 5.7: From Kent M. Van De Graaff, *Human Anatomy,* 2d ed. Copyright © 1988 Wm. C. Brown Publishers, Dubuque, Iowa. All Rights Reserved. Reprinted by permission.

Chapter 7

Figures 7.7, 7.8, 7.10, 7.11, 7.12, 7.13: From Kent M. Van De Graaff, *Human Anatomy,* 2d ed. Copyright © 1988 Wm. C. Brown Publishers, Dubuque, Iowa. All Rights Reserved. Reprinted by permission.
Figure 7.15: From Kent M. Van De Graaff and Stuart Ira Fox, *Concepts of Human Anatomy and Physiology,* 2d ed. Copyright © 1989 Wm. C. Brown Publishers, Dubuque, Iowa. All Rights Reserved. Reprinted by permission.

Chapter 8

Figures 8.14, 8.15: From Kent M. Van De Graaff and Stuart Ira Fox, *Concepts of Human Anatomy and Physiology,* 2d ed. Copyright © 1989 Wm. C. Brown Publishers, Dubuque, Iowa. All Rights Reserved. Reprinted by permission.
Figures 8.27, 8.28: From Kent M. Van De Graaff, *Human Anatomy,* 2d ed. Copyright © 1988 Wm. C. Brown Publishers, Dubuque, Iowa. All Rights Reserved. Reprinted by permission.

Chapter 9

Figure 9.19: From Harold J. Benson, et al., *Anatomy and Physiology Laboratory Textbook, Complete Version,* 4th ed. Copyright © 1988 Wm. C. Brown Publishers, Dubuque, Iowa. All Rights Reserved. Reprinted by permission.
Figure 9.21a: From Kent M. Van De Graaff, *Human Anatomy,* 2d ed. Copyright © 1988 Wm. C. Brown Publishers, Dubuque, Iowa. All Rights Reserved. Reprinted by permission.
Figures 9.24, 9.31: From Kent M. Van De Graaff and Stuart Ira Fox, *Concepts of Human Anatomy and Physiology,* 2d ed. Copyright © 1989 Wm. C. Brown Publishers, Dubuque, Iowa. All Rights Reserved. Reprinted by permission.

Chapter 10

Figures 10.3, 10.10a, 10.19: From Kent M. Van De Graaff and Stuart Ira Fox, *Concepts of Human Anatomy and Physiology,* 2d ed. Copyright © 1989 Wm. C. Brown Publishers, Dubuque, Iowa. All Rights Reserved. Reprinted by permission.
Figure 10.10b: From J. R. McClintic, *Physiology of the Human Body,* 2d ed. Copyright © 1978 John Wiley & Sons, Inc., New York, NY. Reprinted by permission of John Wiley & Sons, Inc.

Chapter 11

Figure 11.14: From Kent M. Van De Graaff and Stuart Ira Fox, *Concepts of Human Anatomy and Physiology,* 2d ed. Copyright © 1989 Wm. C. Brown Publishers, Dubuque, Iowa. All Rights Reserved. Reprinted by permission.

Chapter 12

Opener, figures 12.1, 12.7, 12.13: From Kent M. Van De Graaff, *Human Anatomy,* 2d ed. Copyright © 1988 Wm. C. Brown Publishers, Dubuque, Iowa. All Rights Reserved. Reprinted by permission.
Figures 12.2, 12.16, 12.25, 12.27: From Kent M. Van De Graaff and Stuart Ira Fox, *Concepts of Human Anatomy and Physiology,* 2d ed. Copyright © 1989 Wm. C. Brown Publishers, Dubuque, Iowa. All Rights Reserved. Reprinted by permission.

Chapter 13

Opener, figures 13.1, 13.6, 13.12, 13.16: From Kent M. Van De Graaff, *Human Anatomy,* 2d ed. Copyright © 1988 Wm. C. Brown Publishers, Dubuque, Iowa. All Rights Reserved. Reprinted by permission.

Chapter 15

Opener, figure 15.2: From Kent M. Van De Graaff and Stuart Ira Fox, *Concepts of Human Anatomy and Physiology,* 2d ed. Copyright © 1989 Wm. C. Brown Publishers, Dubuque, Iowa. All Rights Reserved. Reprinted by permission.
Figures 15.9a, b, 15.15: From Kent M. Van De Graaff, *Human Anatomy,* 2d ed. Copyright © 1988 Wm. C. Brown Publishers, Dubuque, Iowa. All Rights Reserved. Reprinted by permission.
Figure 15.10: From James E. Crouch, *Functional Human Anatomy,* 2d ed. Copyright © 1972 Lea & Febiger, Philadelphia, PA. Reprinted by permission.
Figure 15.31: From Stuart Ira Fox, *Human Physiology,* 3d ed. Copyright © 1990 Wm. C. Brown Publishers, Dubuque, Iowa. All Rights Reserved. Reprinted by permission.

Chapter 16

Figure 16.2: From Kent M. Van De Graaff, *Human Anatomy,* 2d ed. Copyright © 1988 Wm. C. Brown Publishers, Dubuque, Iowa. All Rights Reserved. Reprinted by permission.

Chapter 17

Figure 17.7: From Kent M. Van De Graaff and Stuart Ira Fox, *Concepts of Human Anatomy and Physiology,* 2d ed. Copyright © 1989 Wm. C. Brown Publishers, Dubuque, Iowa. All Rights Reserved. Reprinted by permission.
Figure 17.14: From Stuart Ira Fox, *Human Physiology,* 3d ed. Copyright © 1990 Wm. C. Brown Publishers, Dubuque, Iowa. All Rights Reserved. Reprinted by permission.
Figure 17.16: From Kent M. Van De Graaff, *Human Anatomy,* 2d ed. Copyright © 1988 Wm. C. Brown Publishers, Dubuque, Iowa. All Rights Reserved. Reprinted by permission.

Chapter 19

Opener, figures 19.4b, 19.11, 19.13: From Kent M. Van De Graaff and Stuart Ira Fox, *Concepts of Human Anatomy and Physiology,* 2d ed. Copyright © 1989 Wm. C. Brown Publishers, Dubuque, Iowa. All Rights Reserved. Reprinted by permission.

Chapter 20

Figure 20.2: From Kent M. Van De Graaff, *Human Anatomy,* 2d ed. Copyright © 1988 Wm. C. Brown Publishers, Dubuque, Iowa. All Rights Reserved. Reprinted by permission.
Figure 20.15: From Kent M. Van De Graaff and Stuart Ira Fox, *Concepts of Human Anatomy and Physiology.* Copyright © 1986 Wm. C. Brown Publishers, Dubuque, Iowa. All Rights Reserved. Reprinted by permission.
Figure 20.16: From Kent M. Van De Graaff and Stuart Ira Fox, *Concepts of Human Anatomy and Physiology,* 2d ed. Copyright © 1989 Wm. C. Brown Publishers, Dubuque, Iowa. All Rights Reserved. Reprinted by permission.

Index

In this index, page numbers set in *italics* designate figures; page numbers followed by *t* designate tables (charts). *See also* cross-references direct the reader to related topics or to more detailed topic breakdowns. *See* cross-references direct the reader to synonymous terms, which are listed in the form used predominantly in this text.

a

d

W

Z